PRINCIPLES OF Virology

4TH EDITION

VOLUME I *Molecular Biology*

PRINCIPLES OF Virology

4TH EDITION

Jane Flint
Department of Molecular Biology
Princeton University
Princeton, New Jersey

Vincent R. Racaniello
Department of Microbiology & Immunology
College of Physicians and Surgeons
Columbia University
New York, New York

Glenn F. Rall
Fox Chase Cancer Center
Philadelphia, Pennsylvania

Anna Marie Skalka
Fox Chase Cancer Center
Philadelphia, Pennsylvania

with

Lynn W. Enquist
Department of Molecular Biology
Princeton University
Princeton, New Jersey

WASHINGTON, DC

Library of Congress Cataloging-in-Publication Data

Flint, S. Jane, author.
Principles of virology / Jane Flint, Department of Molecular Biology, Princeton University, Princeton, New Jersey; Vincent R. Racaniello, Department of Microbiology, College of Physicians and Surgeons, Columbia University, New York, New York; Glenn F. Rall, Fox Chase Cancer Center, Philadelphia, Pennsylvania; Anna Marie Skalka, Fox Chase Cancer Center, Philadelphia, Pennsylvania; with Lynn W. Enquist, Department of Molecular Biology, Princeton University, Princeton, New Jersey.—4th edition.
pages cm
Revision of: Principles of virology / S.J. Flint ... [et al.]. 3rd ed.
Includes bibliographical references and index.
ISBN 978-1-55581-933-0 (v. 1 pbk.)—ISBN 978-1-55581-934-7 (v. 2 pbk.)—ISBN 978-1-55581-951-4 (set pbk.)—ISBN 978-1-55581-952-1 (set ebook) 1. Virology. I. Racaniello, V. R. (Vincent R.), author. II. Rall, Glenn F., author. III. Skalka, Anna M., author. IV. Enquist, L. W. (Lynn W.), author. V. Title.
QR360.P697 2015
616.9'101--dc23
2015026213
doi:10.1128/9781555818951 (Volume I)
doi:10.1128/9781555818968 (Volume II)
doi:10.1128/9781555819521 (e-bundle)

10 9 8 7 6 5 4 3 2 1

Printed in the United States of America

Address editorial correspondence to ASM Press, 1752 N St., N.W., Washington, DC 20036-2904, USA

Send orders to ASM Press, P.O. Box 605, Herndon, VA 20172, USA
Phone: 800-546-2416; 703-661-1593
Fax: 703-661-1501
E-mail: books@asmusa.org
Online: http://www.asmscience.org

Illustrations and illustration concepting: Patrick Lane, ScEYEnce Studios
Cover and interior design: Susan Brown Schmidler
Cover image: Courtesy of Jason A. Roberts (Victorian Infectious Diseases Reference Laboratory, Doherty Institute, Melbourne, Australia)
Back cover photos: Peter Kurilla Photography

We dedicate this book to the students, current and future scientists, physicians, and all those with an interest in the field of virology, for whom it was written.
We kept them ever in mind.

We also dedicate it to our families:
Jonn, Gethyn, and Amy Leedham
Doris, Aidan, Devin, and Nadia
Eileen, Kelsey, and Abigail
Rudy, Jeanne, and Chris
And
Kathy and Brian

Oh, be wiser thou!
Instructed that true knowledge leads to love.

William Wordsworth
Lines left upon a Seat in a Yew-tree
1888

Contents

Preface *xvii*
Acknowledgments *xxi*
About the Authors *xxiii*

PART I
The Science of Virology 1

1 Foundations 2

Luria's Credo 3
Why We Study Viruses 3
Viruses Are Everywhere 3
Viruses Can Cause Human Disease 3
Viruses Infect All Living Things 3
Viruses Can Be Beneficial 4
Viruses Can Cross Species Boundaries 4
Viruses "R" Us 4
Viruses Are Unique Tools To Study Biology 5
Virus Prehistory 6
Viral Infections in Antiquity 6
The First Vaccines 7
Microorganisms as Pathogenic Agents 9
Discovery of Viruses 10
The Definitive Properties of Viruses 12
The Structural Simplicity of Virus Particles 12
The Intracellular Parasitism of Viruses 14
Viruses Defined 17
Cataloging Animal Viruses 17
The Classical System 17
Classification by Genome Type: the Baltimore System 20

A Common Strategy for Viral Propagation 21
Perspectives 21
References 23

2 The Infectious Cycle 24

Introduction 25
The Infectious Cycle 25
The Cell 25
The Architecture of Cell Surfaces 27
The Extracellular Matrix: Components and Biological Importance 27
Properties of the Plasma Membrane 29
Cell Membrane Proteins 30
Entering Cells 31
Making Viral RNA 31
Making Viral Proteins 31
Making Viral Genomes 31
Forming Progeny Virus Particles 31
Viral Pathogenesis 32
Overcoming Host Defenses 32
Cultivation of Viruses 32
Cell Culture 32
Embryonated Eggs 35
Laboratory Animals 35
Assay of Viruses 36
Measurement of Infectious Units 36
Efficiency of Plating 39
Measurement of Virus Particles and Their Components 39
Viral Reproduction: the Burst Concept 46
The One-Step Growth Cycle 46
Initial Concept 46
One-Step Growth Analysis: a Valuable Tool for Studying Animal Viruses 49
Systems Biology 50
Perspectives 51
References 52

PART II
Molecular Biology 53

3 Genomes and Genetics 54

Introduction 55
Genome Principles and the Baltimore System 55

Structure and Complexity of Viral Genomes 55
DNA Genomes 56
RNA Genomes 58
What Do Viral Genomes Look Like? 59
Coding Strategies 60
What Can Viral Sequences Tell Us? 60
The Origin of Viral Genomes 61
The "Big and Small" of Viral Genomes: Does Size Matter? 65
Genetic Analysis of Viruses 65
Classical Genetic Methods 66
Engineering Mutations into Viral Genomes 67
Engineering Viral Genomes: Viral Vectors 73
Perspectives 78
References 79

4 Structure 80

Introduction 81
Functions of the Virion 81
Nomenclature 82
Methods for Studying Virus Structure 83
Building a Protective Coat 86
Helical Structures 86
Capsids with Icosahedral Symmetry 89
Other Capsid Architectures 102
Packaging the Nucleic Acid Genome 104
Direct Contact of the Genome with a Protein Shell 104
Packaging by Specialized Viral Proteins 105
Packaging by Cellular Proteins 105
Viruses with Envelopes 106
Viral Envelope Components 106
Simple Enveloped Viruses: Direct Contact of External Proteins with the Capsid or Nucleocapsid 109
Enveloped Viruses with an Additional Protein Layer 109
Large Viruses with Multiple Structural Elements 111
Bacteriophage T4 111
Herpesviruses 112
Poxviruses 113
Giant Viruses 114
Other Components of Virions 116
Enzymes 116
Other Viral Proteins 116
Nongenomic Viral Nucleic Acid 117
Cellular Macromolecules 117
Perspectives 119
References 119

5 Attachment and Entry 122

Introduction 123

Attachment of Virus Particles to Cells 123

General Principles 123

Identification of Receptors for Virus Particles 124

Virus-Receptor Interactions 126

Entry into Cells 132

Uncoating at the Plasma Membrane 132

Uncoating during Endocytosis 135

Membrane Fusion 137

Movement of Viral and Subviral Particles within Cells 147

Virus-Induced Signaling via Cell Receptors 148

Import of Viral Genomes into the Nucleus 148

Nuclear Localization Signals 149

The Nuclear Pore Complex 149

The Nuclear Import Pathway 150

Import of Influenza Virus Ribonucleoprotein 151

Import of DNA Genomes 151

Import of Retroviral Genomes 151

Perspectives 153

References 154

6 Synthesis of RNA from RNA Templates 156

Introduction 157

The Nature of the RNA Template 157

Secondary Structures in Viral RNA 157

Naked or Nucleocapsid RNA 158

The RNA Synthesis Machinery 159

Identification of RNA-Dependent RNA Polymerases 159

Sequence Relationships among RNA Polymerases 161

Three-Dimensional Structure of RNA-Dependent RNA Polymerases 161

Mechanisms of RNA Synthesis 164

Initiation 164

Capping 168

Elongation 168

Template Specificity 169

Unwinding the RNA Template 169

Role of Cellular Proteins 170

Paradigms for Viral RNA Synthesis 170

(+) Strand RNA 171

Synthesis of Nested Subgenomic mRNAs 172

(−) Strand RNA 173

Ambisense RNA 174

Double-Stranded RNA 175
Unique Mechanisms of mRNA and Genome Synthesis of Hepatitis Delta Satellite Virus 176
Why Are (−) and (+) Strands Made in Unequal Quantities? 177
Do Ribosomes and RNA Polymerases Collide? 179
Cellular Sites of Viral RNA Synthesis 179
Origins of Diversity in RNA Virus Genomes 182
Misincorporation of Nucleotides 182
Segment Reassortment and RNA Recombination 183
RNA Editing 185
Perspectives 185
References 185

7 Reverse Transcription and Integration 188

Retroviral Reverse Transcription 189
Discovery 189
Impact 189
The Process of Reverse Transcription 189
General Properties and Structure of Retroviral Reverse Transcriptases 198
Other Examples of Reverse Transcription 202
Retroviral DNA Integration Is a Unique Process 204
The Pathway of Integration: Integrase-Catalyzed Steps 205
Integrase Structure and Mechanism 210
Hepadnaviral Reverse Transcription 214
A DNA Virus with Reverse Transcriptase 214
The Process of Reverse Transcription 216
Perspectives 221
References 222

8 Synthesis of RNA from DNA Templates 224

Introduction 225
Properties of Cellular RNA Polymerases That Transcribe Viral DNA 225
Some Viral Genomes Must Be Converted to Templates Suitable for Transcription 226
Transcription by RNA Polymerase II 228
Regulation of RNA Polymerase II Transcription 228
Common Properties of Proteins That Regulate Transcription 234
The Cellular Machinery Alone Can Transcribe Viral DNA Templates 235
Viral Proteins That Govern Transcription of Viral DNA Templates 237
Patterns of Regulation 237
The Human Immunodeficiency Virus Type 1 Tat Protein Autoregulates Transcription 237

The Transcriptional Cascades of DNA Viruses 245
Entry into One of Two Alternative Transcriptional Programs 254
Transcription of Viral Genes by RNA Polymerase III 257
The VA-RNA I Promoter 257
Regulation of VA-RNA Gene Transcription 259
Inhibition of the Cellular Transcriptional Machinery 259
Unusual Functions of Cellular Transcription Components 260
A Viral DNA-Dependent RNA Polymerase 260
Perspectives 262
References 263

9 Replication of DNA Genomes 266

Introduction 267
DNA Synthesis by the Cellular Replication Machinery 269
Eukaryotic Replicons 269
Cellular Replication Proteins 270
Mechanisms of Viral DNA Synthesis 271
Lessons from Simian Virus 40 271
Replication of Other Viral DNA Genomes 275
Properties of Viral Replication Origins 278
Recognition of Viral Replication Origins 280
Viral DNA Synthesis Machines 286
Resolution and Processing of Viral Replication Products 287
Exponential Accumulation of Viral Genomes 288
Viral Proteins Can Induce Synthesis of Cellular Replication Proteins 288
Synthesis of Viral Replication Machines and Accessory Enzymes 290
Viral DNA Replication Independent of Cellular Proteins 291
Delayed Synthesis of Structural Proteins Prevents Premature Packaging of DNA Templates 291
Inhibition of Cellular DNA Synthesis 291
Viral DNAs Are Synthesized in Specialized Intracellular Compartments 292
Limited Replication of Viral DNA Genomes 296
Integrated Parvoviral DNA Can Replicate as Part of the Cellular Genome 296
Different Viral Origins Regulate Replication of Epstein-Barr Virus 297
Limited and Amplifying Replication from a Single Origin: the Papillomaviruses 299
Origins of Genetic Diversity in DNA Viruses 301
Fidelity of Replication by Viral DNA Polymerases 301
Inhibition of Repair of Double-Strand Breaks in DNA 303
Recombination of Viral Genomes 304
Perspectives 307
References 307

10 Processing of Viral Pre-mRNA 310

Introduction 311

Covalent Modification during Viral Pre-mRNA Processing 312

Capping the 5′ Ends of Viral mRNA 312
Synthesis of 3′ Poly(A) Segments of Viral mRNA 315
Splicing of Viral Pre-mRNA 317
Alternative Processing of Viral Pre-mRNA 322
Editing of Viral mRNAs 325

Export of RNAs from the Nucleus 327

The Cellular Export Machinery 327
Export of Viral mRNA 327

Posttranscriptional Regulation of Viral or Cellular Gene Expression by Viral Proteins 330

Temporal Control of Viral Gene Expression 330
Viral Proteins Can Inhibit Cellular mRNA Production 333

Regulation of Turnover of Viral and Cellular mRNAs in the Cytoplasm 335

Regulation of mRNA Stability by Viral Proteins 336
mRNA Stabilization Can Facilitate Transformation 338

Production and Function of Small RNAs That Inhibit Gene Expression 338

Small Interfering RNAs, Micro-RNAs, and Their Synthesis 338
Viral Micro-RNAs 342
Viral Gene Products That Block RNA Interference 345

Perspectives 345

References 346

11 Protein Synthesis 348

Introduction 349

Mechanisms of Eukaryotic Protein Synthesis 349

General Structure of Eukaryotic mRNA 349
The Translation Machinery 350
Initiation 351
Elongation and Termination 360

The Diversity of Viral Translation Strategies 362

Polyprotein Synthesis 363
Leaky Scanning 365
Reinitiation 366
Suppression of Termination 366
Ribosomal Frameshifting 368
Bicistronic mRNAs 368

Regulation of Translation during Viral Infection 368

Inhibition of Translation Initiation after Viral Infection 369
Regulation of eIF4F 372

Regulation of Poly (A)-Binding Protein Activity 376
Regulation of eIF3 376
Interfering with RNA 376
Stress-Associated RNA Granules 377

Perspectives 377

References 379

12 Intracellular Trafficking 380

Introduction 381

Assembly within the Nucleus 382

Import of Viral Proteins for Assembly 383

Assembly at the Plasma Membrane 384

Transport of Viral Membrane Proteins to the Plasma Membrane 386
Sorting of Viral Proteins in Polarized Cells 401
Disruption of the Secretory Pathway in Virus-Infected Cells 404
Signal Sequence-Independent Transport of Viral Proteins to the Plasma Membrane 406

Interactions with Internal Cellular Membranes 409

Localization of Viral Proteins to Compartments of the Secretory Pathway 410
Localization of Viral Proteins to the Nuclear Membrane 411

Transport of Viral Genomes to Assembly Sites 411

Transport of Genomic and Pregenomic RNA from the Nucleus to the Cytoplasm 411
Transport of Genomes from the Cytoplasm to the Plasma Membrane 411

Perspectives 413

References 414

13 Assembly, Exit, and Maturation 416

Introduction 417

Methods of Studying Virus Assembly and Egress 418

Structural Studies of Virus Particles 418
Visualization of Assembly and Exit by Microscopy 418
Biochemical and Genetic Analyses of Assembly Intermediates 418
Methods Based on Recombinant DNA Technology 421

Assembly of Protein Shells 421

Formation of Structural Units 421
Capsid and Nucleocapsid Assembly 423
Self-Assembly and Assisted Assembly Reactions 425

Selective Packaging of the Viral Genome and Other Components of Virus Particles 430

Concerted or Sequential Assembly 430
Recognition and Packaging of the Nucleic Acid Genome 431
Incorporation of Enzymes and Other Nonstructural Proteins 438

Acquisition of an Envelope 439
Sequential Assembly of Internal Components and Budding from a Cellular Membrane 439
Coordination of the Assembly of Internal Structures with Acquisition of the Envelope 440
Release of Virus Particles 441
Assembly and Budding at the Plasma Membrane 441
Assembly at Internal Membranes: the Problem of Exocytosis 444
Release of Nonenveloped Viruses 450
Maturation of Progeny Virus Particles 450
Proteolytic Processing of Structural Proteins 450
Other Maturation Reactions 456
Cell-to-Cell Spread 457
Perspectives 460
References 460

14 The Infected Cell 464

Introduction 465
Signal Transduction 465
Signaling Pathways 465
Signaling in Virus-Infected Cells 466
Gene Expression 470
Inhibition of Cellular Gene Expression 470
Differential Regulation of Cellular Gene Expression 474
Metabolism 477
Methods To Study Metabolism 477
Glucose Metabolism 479
The Citric Acid Cycle 483
Electron Transport and Oxidative Phosphorylation 484
Lipid Metabolism 486
Remodeling of Cellular Organelles 491
The Nucleus 491
The Cytoplasm 495
Perspectives 498
References 500

APPENDIX **Structure, Genome Organization, and Infectious Cycles 501**

Glossary 537

Index 543

Preface

The enduring goal of scientific endeavor, as of all human enterprise, I imagine, is to achieve an intelligible view of the universe. One of the great discoveries of modern science is that its goal cannot be achieved piecemeal, certainly not by the accumulation of facts. To understand a phenomenon is to understand a category of phenomena or it is nothing. Understanding is reached through creative acts.

A. D. HERSHEY
Carnegie Institution Yearbook 65

All four editions of this textbook have been written according to the authors' philosophy that the best approach to teaching introductory virology is by emphasizing shared principles. Studying the phases of the viral reproductive cycle, illustrated with a set of representative viruses, provides an overview of the steps required to maintain these infectious agents in nature. Such knowledge cannot be acquired by learning a collection of facts about individual viruses. Consequently, the major goal of this book is to define and illustrate the basic principles of animal virus biology.

In this information-rich age, the quantity of data describing any given virus can be overwhelming, if not indigestible, for student and expert alike. The urge to write more and more about less and less is the curse of reductionist science and the bane of those who write textbooks meant to be used by students. In the fourth edition, we continue to distill information with the intent of extracting essential principles, while providing descriptions of how the information was acquired. Boxes are used to emphasize major principles and to provide supplementary material of relevance, from explanations of terminology to descriptions of trail-blazing experiments. Our goal is to illuminate process and strategy as opposed to listing facts and figures. In an effort to make the book readable, rather than comprehensive, we are selective in our choice of viruses and examples. The encyclopedic *Fields Virology* (2013) is recommended as a resource for detailed reviews of specific virus families.

What's New

This edition is marked by a change in the author team. Our new member, Glenn Rall, has brought expertise in viral immunology and pathogenesis, pedagogical clarity, and down-to-earth humor to our work. Although no longer a coauthor, our colleague Lynn Enquist has continued to provide insight, advice, and comments on the chapters.

Each of the two volumes of the fourth edition has a unique appendix and a general glossary. Links to Internet resources such as websites, podcasts, blog posts, and movies are provided; the digital edition provides one-click access to these materials.

A major new feature of the fourth edition is the incorporation of in-depth video interviews with scientists who have made a major contribution to the subject of each chapter. Students will be interested in these conversations, which also explore the factors that motivated the scientists' interest in the field and the personal stories associated with their contributions.

Volume I covers the molecular biology of viral reproduction, and Volume II focuses on viral pathogenesis, control of virus infections, and virus evolution. The organization into two volumes follows a natural break in pedagogy and provides considerable flexibility and utility for students and teachers alike. The volumes can be used for two courses, or as two parts of a one-semester course. The two volumes differ in content but are integrated in style and presentation. In addition to updating the chapters and Appendices for both volumes, we have organized the material more efficiently and new chapters have been added.

As in our previous editions, we have tested ideas for inclusion in the text in our own classes. We have also received constructive comments and suggestions from other virology instructors and their students. Feedback from students was particularly useful in finding typographical errors, clarifying confusing or complicated illustrations, and pointing out inconsistencies in content.

For purposes of readability, references are generally omitted from the text, but each chapter ends with an updated list of relevant books, review articles, and selected research papers for readers who wish to pursue specific topics. In general, if an experiment is featured in a chapter, one or more references are listed to provide more detailed information.

Principles Taught in Two Distinct, but Integrated Volumes

These two volumes outline and illustrate the strategies by which all viruses reproduce, how infections spread within a host, and how they are maintained in populations. The principles of viral reproduction established in Volume I are essential for understanding the topics of viral disease, its control, and the evolution of viruses that are covered in Volume II.

Volume I The Science of Virology and the Molecular Biology of Viruses

This volume examines the molecular processes that take place in an infected host cell. It begins with a general introduction and historical perspectives, and includes descriptions of the unique properties of viruses (Chapter 1). The unifying principles that are the foundations of virology, including the concept of a common strategy for viral propagation, are then described. An introduction to cell biology, the principles of the infectious cycle, descriptions of the basic techniques for cultivating and assaying viruses, and the concept of the single-step growth cycle are presented in Chapter 2.

The fundamentals of viral genomes and genetics, and an overview of the surprisingly limited repertoire of viral strategies for genome replication and mRNA synthesis, are topics of Chapter 3. The architecture of extracellular virus particles in the context of providing both protection and delivery of the viral genome in a single vehicle are considered in Chapter 4. Chapters 5 through 13 address the broad spectrum of molecular processes that characterize the common steps of the reproductive cycle of viruses in a single cell, from decoding genetic information to genome replication and production of progeny virions. We describe how these common steps are accomplished in cells infected by diverse but representative viruses, while emphasizing common principles. Volume I concludes with a new chapter, "The Infected Cell," which presents an integrated description of cellular responses to illustrate the marked, and generally, irreversible, impact of virus infection on the host cell.

The appendix in Volume I provides concise illustrations of viral life cycles for members of the main virus families discussed in the text; five new families have been added in the fourth edition. It is intended to be a reference resource when reading individual chapters and a convenient visual means by which specific topics may be related to the overall infectious cycles of the selected viruses.

Volume II Pathogenesis, Control, and Evolution

This volume addresses the interplay between viruses and their host organisms. The first five chapters have been reorganized and rewritten to reflect our growing appreciation of the host immune response and how viruses cause disease. In Chapter 1 we introduce the discipline of epidemiology, provide historical examples of epidemics in history, and consider basic aspects that govern how the susceptibility of a population is controlled and measured. With an understanding of how viruses affect human populations, subsequent chapters focus on the impact of viral infections on hosts, tissues and individual cells. Physiological barriers to virus infections, and how viruses spread in a host, invade organs, and spread to other hosts are the topics of Chapter 2. The early host response to infection, comprising cell autonomous (intrinsic) and innate immune responses, are the topics of Chapter 3, while the next chapter considers adaptive immune defenses, that are tailored to the pathogen, and immune memory. Chapter 5 focuses on the classic patterns of virus infection within cells and hosts, the myriad ways that viruses cause illness, and the value of animal models in uncovering new principles of viral pathogenesis. In Chapter 6, we discuss virus infections that transform cells in culture and promote oncogenesis (the formation of tumors) in animals. Chapter 7 is devoted entirely to the AIDS virus, not only because it is the causative agent of the most serious current world-wide epidemic, but also because of its unique and informative interactions with the human immune defenses.

Next, we consider the principles involved in treatment and control of infection. Chapter 8 focuses on vaccines, and Chapter 9 discusses the approaches and challenges of antiviral drug discovery. The topics of viral evolution and emergence have now been divided into two chapters. The origin of viruses, the drivers of viral evolution, and host-virus conflicts are the subjects of Chapter 10. The principles of emerging virus infections, and humankind's experiences with epidemic and pandemic viral infections, are considered in Chapter 11. Volume II ends with a new chapter on unusual infectious agents, viroids, satellites, and prions.

The Appendix of Volume II provides snapshots of the pathogenesis of common human viruses. This information is presented in four illustrated panels that summarize the viruses and diseases, epidemiology, disease mechanisms, and human infections.

Reference

Knipe DM, Howley PM (ed). 2013. *Fields Virology*, 6th ed. Lippincott Williams & Wilkins, Philadelphia, PA.

For some behind-the-scenes information about how the authors created the fourth edition of *Principles of Virology*, see: http://bit.ly/Virology_MakingOf

Acknowledgments

These two volumes of *Principles* could not have been composed and revised without help and contributions from many individuals. We are most grateful for the continuing encouragement from our colleagues in virology and the students who use the text. Our sincere thanks also go to colleagues (listed in the Acknowledgments for the third edition) who have taken considerable time and effort to review the text in its evolving manifestations. Their expert knowledge and advice on issues ranging from teaching virology to organization of individual chapters and style were invaluable, and are inextricably woven into the final form of the book.

We also are grateful to those who gave so generously of their time to serve as expert reviewers of individual chapters or specific topics in these two volumes: Siddharth Balachandran (Fox Chase Cancer Center), Patrick Moore (University of Pittsburgh), Duane Grandgenett (St. Louis University), Frederick Hughson (Princeton University), Bernard Moss (Laboratory of Viral Diseases, National Institutes of Health), Christoph Seeger (Fox Chase Cancer Center), and Thomas Shenk (Princeton University). Their rapid responses to our requests for details and checks on accuracy, as well as their assistance in simplifying complex concepts, were invaluable. All remaining errors or inconsistencies are entirely ours.

Since the inception of this work, our belief has been that the illustrations must complement and enrich the text. Execution of this plan would not have been possible without the support of Christine Charlip (Director, ASM Press), and the technical expertise and craft of our illustrator. The illustrations are an integral part of the text, and credit for their execution goes to the knowledge, insight, and artistic talent of Patrick Lane of ScEYEnce Studios. We also are indebted to Jason Roberts (Victorian Infectious Diseases Reference Laboratory, Doherty Institute, Melbourne, Australia) for the computational expertise and time he devoted to producing the beautiful renditions of poliovirus particles on our new covers. As noted in the figure legends, many could not have been completed without the help and generosity of numerous colleagues who provided original images. Special thanks go to those who crafted figures or videos tailored specifically to our needs, or provided multiple pieces: Chantal Abergel (CNRS, Aix-Marseille Université, France), Mark Andrake (Fox Chase Cancer Center), Timothy Baker (University of California), Bruce Banfield (The University of Colorado), Christopher Basler and Peter Palese (Mount Sinai School of Medicine), Ralf Bartenschlager (University of Heidelberg, Germany), Eileen Bridge (Miami University, Ohio), Richard Compans (Emory University), Kartik Chandran (Albert Einstein College of Medicine), Paul Duprex (Boston University School of Medicine), Ramón González (Universidad Autónoma del Estado de Morelos), Urs Greber (University of Zurich), Reuben Harris (University of Minnesota), Hidesaburo Hanafusa (deceased), Ari Helenius (University of Zurich), David Knipe (Harvard Medical School), J. Krijnse-Locker (University of Heidelberg, Germany), Petr G. Leiman (École Polytechnique Fédérale de Lausanne), Stuart Le Grice (National Cancer Institute, Frederick MD), Hongrong Liu (Hunan Normal University), David McDonald (Ohio State University), Thomas Mettenleiter (Federal Institute for Animal Diseases, Insel Reims, Germany), Bernard Moss (Laboratory of Viral Diseases, National Institutes of Health), Norm Olson (University of California), B. V. Venkataram Prasad (Baylor College of Medicine), Andrew Rambaut (University of Edinburgh), Jason Roberts (Victorian Infectious Diseases Reference Laboratory, Doherty Institute, Melbourne, Australia), Felix Rey (Institut Pasteur, Paris, France), Michael Rossmann (Purdue University), Anne Simon (University of Maryland), Erik Snijder (Leiden University Medical Center), Alasdair Steven (National Institutes of Health), Paul Spearman (Emory University), Wesley Sundquist (University of Utah), Livia Varstag (Castleton State College, Vermont), Jiri Vondrasek (Institute of organic Chemistry and Biochemistry, Czech Republic), Matthew Weitzman (University of Pennsylvania), Sandra Weller (University of Connecticut Health Sciences Center, Connecticut), Tim Yen (Fox Chase Cancer Center), and Z. Hong Zhou (University of California, Los Angeles).

The collaborative work undertaken to prepare the fourth edition was facilitated greatly by several authors' retreats. ASM Press generously provided financial support for these retreats as well as for our many other meetings.

We thank all those who guided and assisted in the preparation and production of the book: Christine Charlip (Director, ASM Press) for steering us through the complexities inherent in a team effort, Megan Angelini and John Bell (Production Managers, ASM Press) for keeping us on track during production, and Susan Schmidler for her elegant and creative designs for the layout and cover. We are also grateful for the expert secretarial and administrative support from Ellen Brindle-Clark (Princeton University) that facilitated preparation of this text. Special thanks go to Ellen for obtaining many of the permissions required for the figures.

There is little doubt in undertaking such a massive effort that inaccuracies still remain, despite our best efforts to resolve or prevent them. We hope that the readership of this edition will draw our attention to them, so that these errors can be eliminated from future editions of this text.

This often-consuming enterprise was made possible by the emotional, intellectual, and logistical support of our families, to whom the two volumes are dedicated.

About the Authors

Jane Flint is a Professor of Molecular Biology at Princeton University. Dr. Flint's research focuses on investigation of the molecular mechanisms by which viral gene products modulate host cell pathways and antiviral defenses to allow efficient reproduction in normal human cells of adenoviruses, viruses that are widely used in such therapeutic applications as gene transfer and cancer treatment. Her service to the scientific community includes membership of various editorial boards and several NIH study sections and other review panels. Dr. Flint is currently a member of the Biosafety Working Group of the NIH Recombinant DNA Advisory Committee.

Vincent Racaniello is Higgins Professor of Microbiology & Immunology at Columbia University Medical Center. Dr. Racaniello has been studying viruses for over 35 years, including poliovirus, rhinovirus, enteroviruses, and hepatitis C virus. He teaches virology to graduate, medical, dental, and nursing students and uses social media to communicate the subject outside of the classroom. His Columbia University undergraduate virology lectures have been viewed by thousands at iTunes University, Coursera, and on YouTube. Vincent blogs about viruses at virology.ws and is host of the popular science program This Week in Virology.

Glenn Rall is a Professor and the Co-Program Leader of the Blood Cell Development and Function Program at the Fox Chase Cancer Center in Philadelphia. At Fox Chase, Dr. Rall is also the Associate Chief Academic Officer and Director of the Postdoctoral Program. He is an Adjunct Professor in the Microbiology and Immunology departments at the University of Pennsylvania, Thomas Jefferson, Drexel, and Temple Universities. Dr. Rall's laboratory studies viral infections of the brain and the immune responses to those infections, with the goal of defining how viruses contribute to disease in humans. His service to the scientific community includes membership on the Autism Speaks Scientific Advisory Board, Opinions Editor of *PLoS Pathogens*, chairing the Education and Career Development Committee of the American Society for Virology, and membership on multiple NIH grant review panels.

Anna Marie Skalka is a Professor and the W.W. Smith Chair in Cancer Research at Fox Chase Cancer Center in Philadelphia and an Adjunct Professor at the University of Pennsylvania. Dr. Skalka's major research interests are the molecular aspects of the replication of retroviruses. Dr. Skalka is internationally recognized for her contributions to the understanding of the biochemical mechanisms by which such viruses (including the AIDS virus) replicate and insert their genetic material into the host genome. Both an administrator and researcher, she has been deeply involved in state, national, and international advisory groups concerned with the broader, societal implications of scientific research, including the NJ Commission on Cancer Research and the U.S. Defense Science Board. Dr. Skalka has served on the editorial boards of peer-reviewed scientific journals and has been a member of scientific advisory boards including the National Cancer Institute Board of Scientific Counselors, the General Motors Cancer Research Foundation Awards Assembly, the Board of Governors of the American Academy of Microbiology, and the National Advisory Committee for the Pew Biomedical Scholars Program.

PART I
The Science of Virology

1 Foundations
2 The Infectious Cycle

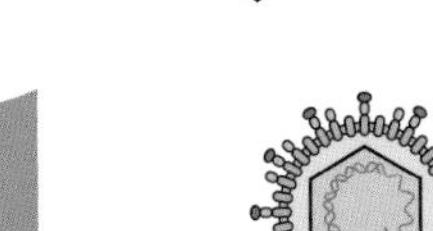

1 Foundations

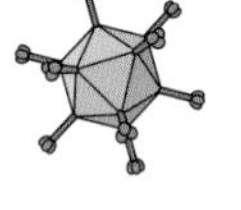

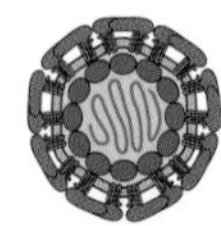

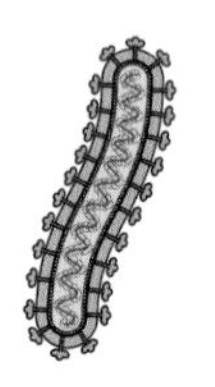

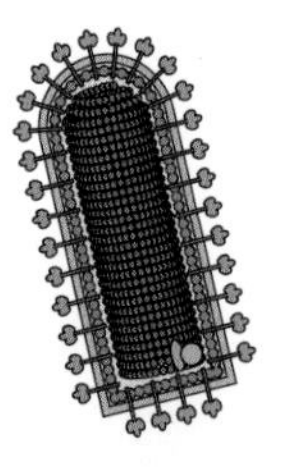

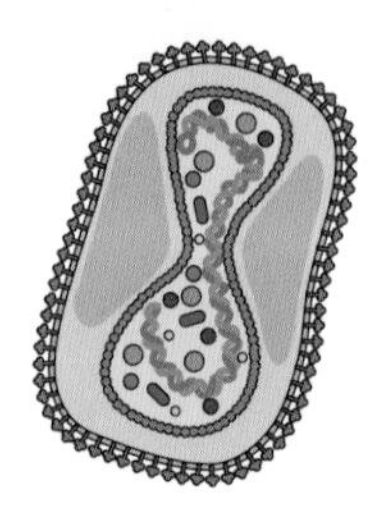

Luria's Credo

Why We Study Viruses

Viruses Are Everywhere

Viruses Can Cause Human Disease

Viruses Infect All Living Things

Viruses Can Be Beneficial

Viruses Can Cross Species Boundaries

Viruses "R" Us

Viruses Are Unique Tools To Study Biology

Virus Prehistory

Viral Infections in Antiquity

The First Vaccines

Microorganisms as Pathogenic Agents

Discovery of Viruses

The Definitive Properties of Viruses

The Structural Simplicity of Virus Particles

The Intracellular Parasitism of Viruses

Viruses Defined

Cataloging Animal Viruses

The Classical System

Classification by Genome Type: the Baltimore System

A Common Strategy for Viral Propagation

Perspectives

References

LINKS FOR CHAPTER 1

- ***Video: Interview with Dr. Donald Henderson***
  http://bit.ly/Virology_Henderson
- ***This Week in Virology (TWIV): A weekly podcast about viruses featuring informal yet informative discussions and interviews with guests about the latest topics in the field.***
  http://www.twiv.tv
- ***Marine viruses and insect defense***
  http://bit.ly/Virology_Twiv301
- ***Giants among viruses***
  http://bit.ly/Virology_Twiv261
- ***Latest update of virus classification from the ICTV.***
  http://www.ictvonline.org/virusTaxonomy.asp?bhcp=1
- ***The abundant and diverse viruses of the seas.***
  http://bit.ly/Virology_3-20-09
- ***How many viruses on Earth?***
  http://bit.ly/Virology_9-6-13

> *Thus, we cannot reject the assumption that the effect of the filtered lymph is not due to toxicity, but rather to the ability of the agent to replicate.*
>
> F. LOEFFLER *1898*

Luria's Credo

"There is an intrinsic simplicity of nature and the ultimate contribution of science resides in the discovery of unifying and simplifying generalizations, rather than in the description of isolated situations—in the visualization of simple, overall patterns rather than in the analysis of patchworks." More than half a century has passed since Salvador Luria wrote this credo in the introduction to the classic textbook *General Virology*.

Despite an explosion of information in biology since Luria wrote these words, his vision of unity in diversity is as relevant now as it was then. That such unifying principles exist may not be obvious considering the bewildering array of viruses, genes, and proteins recognized in modern virology. Indeed, new viruses are being described regularly, and viral diseases such as acquired immunodeficiency syndrome (AIDS), hepatitis, and influenza continue to challenge our efforts to control them. Yet Luria's credo still stands: even as our knowledge continues to increase, it is clear that all viruses follow the same simple strategy to ensure their survival. This insight has been hard-won over many years of observation, research, and debate; the history of virology is rich and instructive.

Why We Study Viruses

Viruses Are Everywhere

Viruses are all around us, comprising an enormous proportion of our environment, in both number and total mass (Box 1.1). All living things encounter billions of virus particles every day. For example, they enter our lungs in the 6 liters of air each of us inhales every minute; they enter our digestive systems with the food we eat; and they are transferred to our eyes, mouths, and other points of entry from the surfaces we touch and the people with whom we interact. Our bodies are reservoirs for viruses that reside in our respiratory, gastrointestinal, and urogenital tracts. In addition to viruses that can infect us, our intestinal tracts are loaded with myriad plant and insect viruses, as well as hundreds of bacterial species that harbor their own constellations of viruses.

Viruses Can Cause Human Disease

With such constant exposure, it is nothing short of amazing that the vast majority of viruses that infect us have little or no impact on our health or well-being. As described in Volume II, we owe such relative safety to our elaborate immune defense systems, which have evolved to fight microbial infection. When these defenses are compromised, even the most common infection can be lethal. Despite such defenses, some of the most devastating human diseases have been or still are caused by viruses; these diseases include smallpox, yellow fever, poliomyelitis, influenza, measles, and AIDS. Viral infections can lead to life-threatening diseases that impact virtually all organs, including the lungs, liver, central nervous system, and intestines. Viruses are responsible for approximately 20% of the human cancer burden, and viral infections of the respiratory and gastrointestinal tracts kill millions of children in the developing world each year. As summarized in Volume II, Appendix, there is no question about the biomedical importance of these agents.

Viruses Infect All Living Things

While most of this textbook focuses on viral infections of humans, it is important to bear in mind that viruses also infect pets, food animals, plants, insects, and wildlife throughout

PRINCIPLES *Foundations*

- The field of virology encompasses viral discovery, the study of virus structure and reproduction, and the importance of viruses in biology and disease.
- While this text focuses primarily on viruses that infect vertebrates, especially humans, it is important to keep in mind that viruses infect **all** living things including insects, plants, bacteria, and even other viruses.
- Viruses are not solely pathogenic nuisances; they can be beneficial. Viruses contribute to ecological homeostasis, keep our immune responses activated and alert, and can be used as molecular flashlights to illuminate cellular processes.
- Viruses have been part of all of human history: they were present long before *Homo sapiens* evolved, and the majority of human infections were likely acquired from other animals (zoonoses). As viruses continue to be discovered, our understanding of how human health and well-being are affected by these agents remains incomplete.
- Viruses are obligate intracellular parasites and depend on their host cell for all aspects of the viral life cycle.
- While Koch's postulates were essential for defining many agents of disease, not all pathogenic viruses fulfilled these criteria.
- Viruses can be cataloged based on their appearance, the hosts they infect, or the nature of their nucleic acid genome.
- The Baltimore classification allows relationships among various viral genomes and the pathway to mRNA to be determined.
- A common strategy underlies the propagation of all viruses. This textbook describes that strategy and the similarities and differences in the manner in which it is accomplished by different viruses.

BOX 1.1

BACKGROUND

Some astounding numbers

- Viruses are the most abundant entities in the biosphere. The biomass on our planet of bacterial viruses *alone* exceeds that of all of Earth's elephants by more than 1,000-fold. There are more than 10^{30} bacteriophage particles in the world's oceans, enough to extend out into space for 200 million light-years if arranged head to tail (http://www.virology.ws/2009/03/20/the-abundant-and-diverse-viruses-of-the-seas/).
- Whales are commonly infected with a member of the virus family *Caliciviridae* that causes rashes, blisters, intestinal problems, and diarrhea and can also infect humans. Infected whales excrete more than 10^{13} calicivirus particles daily.
- The average human body contains approximately 10^{13} cells, but these are outnumbered 10-fold by bacteria and as much as 100-fold by virus particles.
- With about 10^{16} human immunodeficiency virus (HIV) genomes on the planet today, it is highly probable that somewhere there exist HIV genomes that are resistant to every one of the antiviral drugs that we have now or are likely to have in the future.

Earth and its oceans. Courtesy: NASA/Goddard Space Flight Center.

the world. They infect microbes such as algae, fungi, and bacteria, and some even interfere with the reproduction of other viruses. Viral infection of agricultural plants and animals can have enormous economic and societal impact. Outbreaks of infection by foot-and-mouth disease and avian influenza viruses have led to the destruction (**culling**) of millions of cattle, sheep, and poultry to prevent further spread. Losses in the United Kingdom during the 2001 outbreak of foot-and-mouth disease ran into billions of dollars and caused havoc for both farmers and the government (Box 1.2). More recent outbreaks of the avian influenza virus H5N1 in Asia have resulted in similar disruption and economic loss. Viruses that infect crops such as potatoes and fruit trees are common and can lead to serious food shortages as well as financial devastation.

Viruses Can Be Beneficial

Despite the appalling statistics from human and agricultural epidemics, it is important to realize that viruses can also be beneficial. Such benefit can be seen most clearly in marine ecology, where virus particles are the most abundant biological entities (Box 1.1). Indeed, they comprise 94% of all nucleic acid-containing particles in the oceans and are 15 times more abundant than the *Bacteria* and *Archaea*. Viral infections in the ocean kill 20 to 40% of marine microbes daily, converting these living organisms into particulate matter, and in so doing release essential nutrients that supply phytoplankton at the bottom of the ocean's food chain, as well as carbon dioxide and other gases that affect the climate of the earth. Pathogens can also influence one another: infection by one virus can have an ameliorating effect on the pathogenesis of a second virus or even bacteria. For example, human immunodeficiency virus-infected AIDS patients show a substantial decrease in their disease progression if they are persistently infected with hepatitis G virus, and mice latently infected with some murine herpesviruses are resistant to infection with the bacterial pathogens *Listeria monocytogenes* and *Yersinia pestis*. The idea that viruses are solely agents of disease is giving way to the notion that they can exert positive, even necessary, effects.

Viruses Can Cross Species Boundaries

Although viruses generally have a limited host range, they can and do spread across species barriers. As the world's human population continues to expand and impinge on the wilderness, cross-species (**zoonotic**) infections of humans are occurring with increasing frequency. In addition to the AIDS pandemic, the highly fatal Ebola hemorrhagic fever and the severe acute respiratory syndrome (SARS) are recent examples of viral diseases to emerge from zoonotic infections. The current pandemic of influenza virus H5N1 in avian species has much of the world riveted by the frightening possibility that a new, highly pathogenic strain might emerge following transmission from birds to human hosts. Indeed, given the eons over which viruses have had the opportunity to interact with various species, today's "natural" host may simply be a way station in viral evolution.

Viruses "R" Us

Every cell in our body contains viral DNA. Human endogenous retroviruses, and elements thereof, make up about 5 to 8% of our DNA. Most are inactive, fossil remnants from infections of germ cells that have occurred over millions of years during our evolution. Some of them are suspected to be associated with specific diseases, but the protein products of other endogenous retroviruses are essential for placental development.

Recent genomic studies have revealed that our viral "heritage" is not limited to retroviruses. Human and other

BOX 1.2

DISCUSSION

The first animal virus discovered remains a scourge today

Foot-and-mouth disease virus infects domestic cattle, pigs, and sheep, as well as many species of wild animals. Although mortality is low, morbidity is high and infected farm animals lose their commercial value. The virus is highly contagious, and the most common and effective method of control is by the slaughter of entire herds in affected areas.

Outbreaks of foot-and-mouth disease were widely reported in Europe, Asia, Africa, and South and North America in the 1800s. The largest epidemic ever recorded in the United States occurred in 1914. After gaining entry into the Chicago stockyards, the virus spread to more than 3,500 herds in 22 states. This calamity accelerated epidemiological and disease control programs, eventually leading to the field- and laboratory-based systems maintained by the U.S. Department of Agriculture to protect domestic livestock from foreign animal and plant diseases. Similar control systems have been established in other Western countries, but this virus still presents a formidable challenge throughout the world. A 1997 outbreak of foot-and-mouth disease among pigs in Taiwan resulted in economic losses of greater than $10 billion.

In 2001, an epidemic outbreak in the United Kingdom spread to other countries in Europe and led to the slaughter of more than 3 million infected and uninfected farm animals. The associated economic, societal, and political costs threatened to bring down the British government. Images of mass graves and horrific pyres consuming the corpses of dead animals (see figure) sensitized the public as never before. Recent outbreaks and societal unrest in Turkey and regions of North Africa, including Libya and Egypt, make the threat of further spread a serious concern for other countries.

Hunt J. 3 January 2013. Foot-and-mouth is knocking on Europe's door. *Farmers Weekly*. http://www.fwi.co.uk/articles/03/01/2013/136943/foot-and-mouth-is-knocking-on-europe39s-door.htm.

Murphy FA, Gibbs EPJ, Horzinek MC, Studdert MJ. 1999. *Veterinary Virology*, 3rd ed. Academic Press, Inc, San Diego, CA.

Mass burning of cattle carcasses during the 2001 foot-and-mouth disease outbreak in the United Kingdom.

vertebrate genomes harbor sequences derived from several DNA and RNA viruses that, in contrast to the retroviruses, lack mechanisms to invade host DNA. As many of these insertions are estimated to have occurred some 40 million to 90 million years ago, this knowledge has provided unique insight into the ages and evolution of some currently circulating viruses. Furthermore, the conservation of some of the viral sequences in vertebrate genomes suggests that they may have been selected for beneficial properties over evolutionary time.

Viruses Are Unique Tools To Study Biology

Because viruses are dependent on their hosts for propagation, studies that focus on viral reprogramming of cellular mechanisms have provided unique insights into cellular biology and functioning of host defenses. Groundbreaking studies of viruses that infect bacteria, the bacteriophages, laid the foundations of modern molecular biology (Table 1.1), and crystallization of the plant virus tobacco mosaic virus was a landmark in structural biology. Studies of animal viruses established many fundamental principles of cellular function, including the presence of intervening sequences in eukaryotic genes. The study of cancer (transforming) viruses revealed the genetic basis of this disease. It seems clear that studies of viruses will continue to open up such paths of discovery in the future.

With the development of recombinant DNA technology and our increased understanding of some viral systems, it has become possible to use viral genomes as vehicles for the delivery of genes to cells and organisms for both scientific and therapeutic purposes. The use of viral vectors to introduce genes into various cells and organisms to study their function has become a standard method in biology. Viral vectors are

Table 1.1 Bacteriophages: landmarks in molecular biology[a]

Year	Discovery (discoverer[s])
1939	One-step growth of viruses (Ellis and Delbrück)
1946	Mixed phage infection leads to genetic recombination (Delbrück)
1947	Mutation and DNA repair (multiplicity reactivation) (Luria)
1952	Transduction of genetic information (Zinder and Lederberg)
1952	DNA, not protein, found to be the genetic material (Hershey and Chase)
1952	Restriction and modification of DNA (Luria)
1955	Definition of a gene (*cis-trans* test) (Benzer)
1958	Mechanisms of control of gene expression by repressors and activators established (Pardee, Jacob, and Monod)
1958	Definition of the episome (Jacob and Wollman)
1961	Discovery of mRNA (Brenner, Jacob, and Meselson)
1961	Elucidation of the triplet code by genetic analysis (Crick, Barnett, Brenner, and Watts-Tobin)
1961	Genetic definition of nonsense codons as stop signals for translation (Campbell, Epstein, and Bernstein)
1964	Colinearity of the gene with the polypeptide chain (Sarabhai, Stretton, and Brenner)
1966	Pathways of macromolecular assembly (Edgar and Wood)
1974	Vectors for recombinant DNA technology (Murray and Murray, Thomas, Cameron, and Davis)

[a]Sources: T. D. Brock, *The Emergence of Bacterial Genetics* (Cold Spring Harbor Laboratory Press, Cold Spring Harbor, NY, 1990); K. Denniston and L. Enquist, *Recombinant DNA. Benchmark Papers in Microbiology*, vol. 15 (Dowden, Hutchinson and Ross, Inc., Stroudsburg, PA, 1981); and C. K. Mathews, E. Kutter, G. Mosig, and P. Berget, *Bacteriophage T4* (American Society for Microbiology, Washington, DC, 1983).

also being used to treat human disease via "gene therapy," in which functional genes delivered by viral vectors compensate for faulty genes in the host cells.

Virus Prehistory

Although viruses have been known as distinct biological entities for little more than 100 years, evidence of viral infection can be found among the earliest recordings of human activity, and methods for combating viral disease were practiced long before the first virus was recognized. Consequently, efforts to understand and control these important agents of disease are phenomena of the 20th century.

Viral Infections in Antiquity

Reconstruction of the prehistoric past to provide a plausible account of when or how viruses established themselves in human populations is a challenging task. However, extrapolating from current knowledge, we can deduce that some modern viruses were undoubtedly associated with the earliest precursors of mammals and coevolved with humans. Other viruses entered human populations only recently. The last 10,000 years of history was a time of radical change for humans and our viruses: animals were domesticated, the human population increased dramatically, large population centers appeared, and commerce drove worldwide travel and interactions among unprecedented numbers of people.

Viruses that established themselves in human populations were undoubtedly transmitted from animals, much as still happens today. Early human groups that domesticated and lived with their animals were almost certainly exposed to different viruses than were nomadic hunter societies. Similarly, as many different viruses are **endemic** in the tropics, human societies in that environment must have been exposed to a greater variety of viruses than societies established in temperate climates. When nomadic groups met others with domesticated animals, human-to-human contact could have provided new avenues for virus spread. Even so, it seems unlikely that viruses such as those that cause measles or smallpox could have entered a permanent relationship with small groups of early humans. Such highly virulent viruses, as we now know them to be, either kill their hosts or induce lifelong immunity. Consequently, they can survive only when large, interacting host populations offer a sufficient number of naive and permissive hosts for their continued propagation. Such viruses could not have been established in human populations until large, settled communities appeared. Less virulent viruses that enter into a long-term relationship with their hosts were therefore more likely to be the first to become adapted to reproduction in the earliest human populations. These viruses include the modern retroviruses, herpesviruses, and papillomaviruses.

Evidence of several viral diseases can be found in ancient records. The Greek poet Homer characterizes Hector as

A

Here this firebrand, rabid Hector, leads the charge.
HOMER, *The Iliad*,
translated by Robert Fagels
(Viking Penguin)

B

Figure 1.1 References to viral diseases abound in the ancient literature. (A) An image of Hector from an ancient Greek vase. Courtesy of the University of Pennsylvania Museum (object 30-44-4). **(B)** An Egyptian stele, or stone tablet, from the 18th dynasty (1580–1350 B.C.) depicting a man with a withered leg and the "drop foot" syndrome characteristic of polio. Panel B is reprinted from W. Biddle, *A Field Guide to Germs* (Henry Holt and Co., LLC, New York, NY, 1995; © 1995 by Wayne Biddle), with permission from the publisher.

"rabid" in *The Iliad* (Fig. 1.1A), and Mesopotamian laws that outline the responsibilities of the owners of rabid dogs date from before 1000 B.C. Their existence indicates that the communicable nature of this viral disease was already well-known by that time. Egyptian hieroglyphs that illustrate what appear to be the consequences of poliovirus infection (a withered leg typical of poliomyelitis [Fig. 1.1B]) or pustular lesions characteristic of smallpox also date from that period. The smallpox virus, which was probably endemic in the Ganges River basin by the fifth century B.C. and subsequently spread to other parts of Asia and Europe, has played an important part in human history. Its introduction into the previously unexposed native populations of Central and South America by colonists in the 16th century led to lethal epidemics, which are considered an important factor in the conquests achieved by a small number of European soldiers. Other viral diseases known in ancient times include mumps and, perhaps, influenza. Yellow fever has been described since the discovery of Africa by Europeans, and it has been suggested that this scourge of the tropical trade was the basis for legends about ghost ships, such as the *Flying Dutchman*, in which an entire ship's crew perished mysteriously.

Humans have not only been subject to viral disease throughout much of their history but have also manipulated these agents, albeit unknowingly, for much longer than might be imagined. One classic example is the cultivation of marvelously patterned tulips, which were of enormous value in 17th-century Holland. Such efforts included deliberate spread of a virus (tulip breaking virus or tulip mosaic virus) that we now know causes the striping of tulip petals so highly prized at that time (Fig. 1.2). Attempts to control viral disease have an even more venerable history.

The First Vaccines

Measures to control one viral disease have been used with some success for the last millennium. The disease is smallpox

Figure 1.2 ***Three Broken Tulips.*** A painting by Nicolas Robert (1624–1685), now in the collection of the Fitzwilliam Museum, Cambridge, United Kingdom. Striping patterns (color breaking) in tulips were described in 1576 in western Europe and were caused by a viral infection. This beautiful image depicts the remarkable consequences of infection with the tulip mosaic virus. Courtesy of the Fitzwilliam Museum, University of Cambridge.

(Fig. 1.3), and the practice is called **variolation**, inoculation of healthy individuals with material from a smallpox pustule into a scratch made on the arm. Variolation, widespread in China and India by the 11th century, was based on the recognition that smallpox survivors were protected against subsequent bouts of the disease. Variolation later spread to Asia Minor, where its value was recognized by Lady Mary Wortley Montagu, wife of the British ambassador to the Ottoman Empire. She introduced this practice into England in 1721, where it became quite widespread following the successful inoculation of children of the royal family. George Washington is said to have introduced variolation among Continental Army soldiers in 1776. However, the consequences of variolation were unpredictable and never pleasant: serious skin lesions invariably developed at the site of inoculation and were often accompanied by more generalized rash and disease, with a fatality rate of 1 to 2%. From the comfortable viewpoint of an affluent country in the 21st century, such a death rate seems unacceptably high. However, in the 18th century, variolation

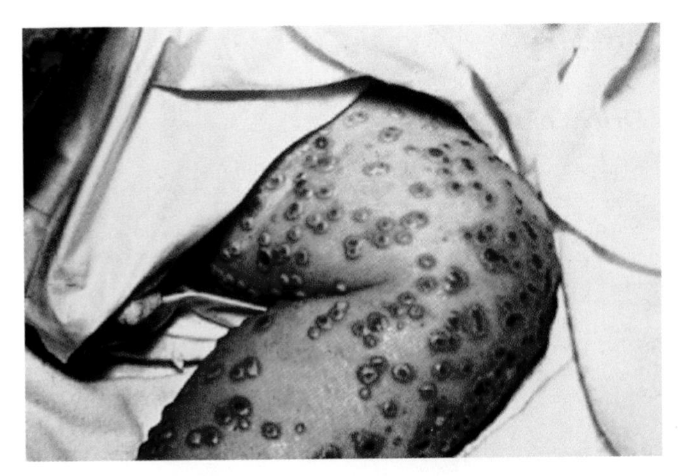

Figure 1.3 Characteristic smallpox lesions in a young victim. Illustrations like these were used as examples to track down individuals infected with the smallpox virus (variola virus) during the World Health Organization campaign to eradicate the disease. Photo courtesy of the Immunization Action Coalition (original source: Centers for Disease Control and Prevention). (See also the interview with Dr. Donald Henderson: http://bit.ly/Virology_Henderson)

was perceived as a much better alternative than contracting natural smallpox, a disease with a fatality rate of 25% in the whole population and 40% in babies and young children.

In the 1790s, Edward Jenner, an English country physician, recognized the principle on which modern methods of viral immunization are based, even though viruses themselves were not to be identified for another 100 years. Jenner himself was variolated as a boy and also practiced this procedure. He was undoubtedly familiar with its effects and risks. Perhaps this experience spurred his great insight upon observing that milkmaids were protected against smallpox if they previously contracted cowpox (a mild disease in humans). Jenner followed up this astute observation with direct experiments. In 1794 to 1796, he demonstrated that inoculation with extracts from cowpox lesions induced only mild symptoms but protected against the far more dangerous disease. It is from these experiments with cowpox that we derive the term **vaccination** (*vacca* = "cow" in Latin); Louis Pasteur coined this term in 1881 to honor Jenner's accomplishments.

Initially, the only way to propagate and maintain the cowpox vaccine was by serial infection of human subjects. This method was eventually banned, as it was often associated with transmission of other diseases such as syphilis and hepatitis. By 1860, the vaccine had been passaged in cows; later, sheep and water buffaloes were also used. While Jenner's original vaccine was based on the virus that causes cowpox, sometime during the human-to-human or cow-to-cow transfers, the poxvirus now called vaccinia virus replaced the cowpox virus. Vaccinia virus is the basis for the modern smallpox vaccine, but its origins remain a mystery: it exhibits limited genetic similarity to the viruses that cause cowpox or smallpox, or to many of the

BOX 1.3

DISCUSSION

Origin of vaccinia virus

Over the years, at least three hypotheses have been advanced to explain the curious substitution of cowpox virus by vaccinia virus:

1. Recombination of cowpox virus with smallpox virus after variolation of humans
2. Recombination between cowpox virus and animal poxviruses during passage in various animals
3. Genetic drift of cowpox virus after repeated passage in humans and animals

None of these hypotheses has been proven conclusively, and all fail to account fully for the origins of the sequences in the vaccinia virus genome.

Evans DH. 2 June 2013. Episode 235, *This Week in Virology*. http://www.twiv.tv/2013/06/02/twiv-235-live-in-edmonton-eh/

Qin L, Upton C, Hazes B, Evans DH. 2011. Genomic analysis of the vaccinia virus strain variants found in Dryvax vaccine. *J Virol* **24:**13049–13060.

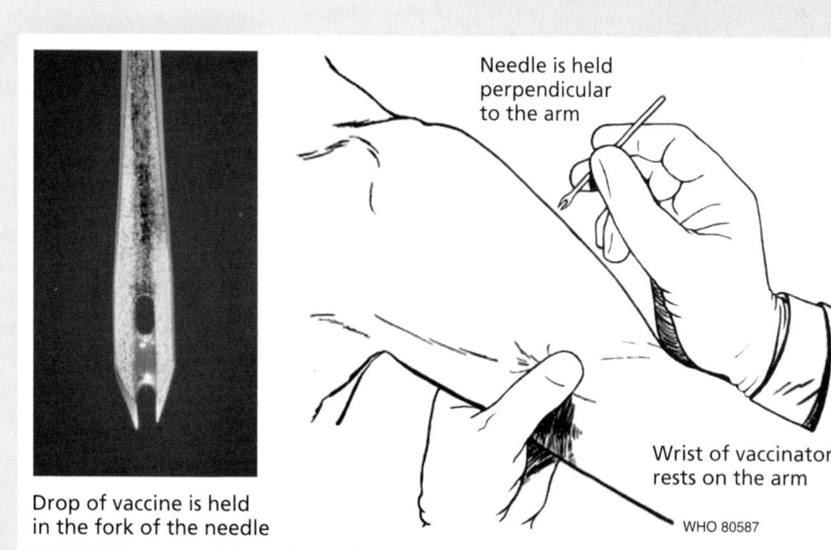

Smallpox vaccine is delivered via multiple punctures with a special two-pronged needle (inset) that has been dipped in the vaccine (Adapted from WHO, with permission).

other known members of the poxvirus family. Scientists have recovered the smallpox vaccine used in New York in 1876 and have verified that it contains vaccinia virus and not cowpox virus. Speculation about when and how the switch occurred has produced some possible scenarios (Box 1.3).

The first deliberately attenuated viral vaccine was made by Louis Pasteur, although he had no idea at the time that the relevant agent was a virus. In 1885, he inoculated rabbits with material from the brain of a cow suffering from rabies and then used aqueous suspensions of dried spinal cords from these animals to infect other rabbits. After several such passages, the resulting preparations caused mild disease (i.e., were **attenuated**) yet produced effective immunity against rabies. Safer and more efficient methods for the production of larger quantities of these first vaccines awaited the recognition of viruses as distinctive biological entities and parasites of cells in their hosts. Indeed, it took almost 50 years to discover the next antiviral vaccines: a vaccine for yellow fever virus was developed in 1935, and an influenza vaccine was available in 1936. These advances became possible only with radical changes in our knowledge of living organisms and of the causes of disease.

Microorganisms as Pathogenic Agents

The 19th century was a period of revolution in scientific thought, particularly in ideas about the origins of living things. The publication of Charles Darwin's *The Origin of Species* in 1859 crystallized startling (and, to many people, shocking) new ideas about the origin of diversity in plants and animals, until then generally attributed directly to the hand of God. These insights permanently undermined the perception that humans were somehow set apart from all other members of the animal kingdom. From the point of view of the science of virology, the most important changes were in ideas about the causes of disease.

The diversity of macroscopic organisms has been appreciated and cataloged since the dawn of recorded human history. A vast new world of organisms too small to be visible to the naked eye was revealed through the microscopes of Antony van Leeuwenhoek (1632–1723). Van Leeuwenhoek's vivid and exciting descriptions of living microorganisms, the "wee animalcules" present in such ordinary materials as rain or seawater, included examples of protozoa, algae, and bacteria. By the early 19th century, the scientific community had accepted the existence of microorganisms and turned to the question of their origin, a topic of fierce debate. Some believed that microorganisms arose spontaneously, for example in decomposing matter, where they were especially abundant. Others held the view that all were generated by the reproduction of like microorganisms, as were macroscopic organisms. The death knell of the spontaneous-generation hypothesis was sounded with the famous experiments of Pasteur. He demonstrated

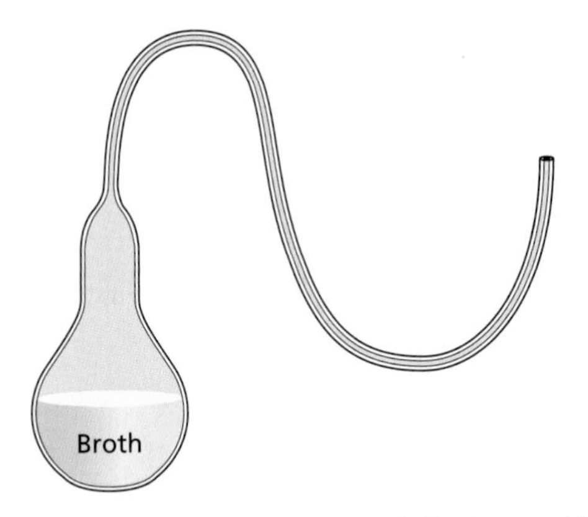

Figure 1.4 Pasteur's famous swan-neck flasks provided passive exclusion of microbes from the sterilized broth. Although the flask was freely open to the air at the end of the long curved stem, the broth remained sterile as long as the microbe-bearing dust that collected in the neck of the stem did not reach the liquid.

that boiled (i.e., sterilized) medium remained free of microorganisms as long as it was maintained in special flasks with curved, narrow necks designed to prevent entry of airborne microbes (Fig. 1.4). Pasteur also established that particular microorganisms were associated with specific processes, such as fermentation, an idea that was crucial in the development of modern explanations for the causes of disease.

From the earliest times, poisonous air (miasma) was generally invoked to account for **epidemics** of contagious diseases, and there was little recognition of the differences among causative agents. The association of particular microorganisms, initially bacteria, with specific diseases can be attributed to the ideas of the German physician Robert Koch. He developed and applied a set of criteria for identification of the agent responsible for a specific disease (a **pathogen**), articulated in an 1890 presentation in Berlin. These criteria, **Koch's postulates**, can be summarized as follows.

- The organism must be regularly associated with the disease and its characteristic lesions.
- The organism must be isolated from the diseased host and grown in culture.
- The disease must be reproduced when a pure culture of the organism is introduced into a healthy, susceptible host.
- The same organism must be reisolated from the experimentally infected host (Box 1.4).

By applying his criteria, Koch demonstrated that anthrax, a common disease of cattle, was caused by a specific bacterium (designated *Bacillus anthracis*) and that a second, distinct bacterial species caused tuberculosis in humans. Guided by these postulates and the methods for the sterile culture and isolation of pure preparations of bacteria developed by Pasteur, Joseph Lister, and Koch, many pathogenic bacteria (as well as yeasts and fungi) were identified and classified during the last part of the 19th century (Fig. 1.5). From these beginnings, investigation into the causes of infectious disease was placed on a secure scientific foundation, the first step toward rational treatment and ultimately control. Furthermore, during the last decade of the 19th century, failures of the paradigm that bacterial or fungal agents are responsible for **all** diseases led to the identification of a new class of infectious agents—submicroscopic pathogens that came to be called **viruses**.

Discovery of Viruses

The first report of a pathogenic agent smaller than any known bacterium appeared in 1892. The Russian scientist Dimitrii Ivanovsky observed that the causative agent of tobacco mosaic disease was not retained by the unglazed filters used at

BOX 1.4

DISCUSSION

New methods extend Koch's principles

While it is clear that a microbe that fulfills Koch's postulates is almost certainly the cause of the disease in question, we now know that microbes that do not fulfill such criteria may still represent the etiological agents of disease. In the latter part of the 20th century, new methods were developed to associate particular viruses with disease based on immunological evidence of infection, for example, the presence of antibodies in blood. The availability of these methods led to the proposal of modified "molecular Koch's postulates" based on the application of molecular techniques to monitor the role played by virulence genes in bacteria.

The most revolutionary advances in our ability to link particular viruses with disease (or benefit) come from the more recent development of high-throughput nucleic acid sequencing methods and bioinformatics tools that allow detection of viral genetic material directly in environmental or biological samples, an approach called viral metagenomics. Based on these developments, alternative "metagenomic Koch's postulates" have been proposed in which (i) the definitive traits are molecular markers such as genes or full genomes that can uniquely distinguish samples obtained from diseased subjects from those obtained from matched, healthy control subjects and (ii) inoculating a healthy individual with a sample from a diseased subject results in transmission of the disease as well as the molecular markers.

Falkow S. 1988. Molecular Koch's postulates applied to microbial pathogenicity. *Rev Infect Dis* **10**(Suppl 2):S274–S276.

Fredericks DN, Relman DA. 1996. Sequence-based identification of microbial pathogens: a reconsideration of Koch's postulates. *Clin Microbiol Rev* **9:**18–33.

Mokili JL, Rohwer F, Dutilh BE. 2012. Metagenomics and future perspectives in virus discovery. *Curr Opin Virol* **2:**63–77.

Racaniello V. 22 January 2010. Koch's postulates in the 21st century. Virology Blog. http://www.virology.ws/2010/01/22/kochs-postulates-in-the-21st-century/

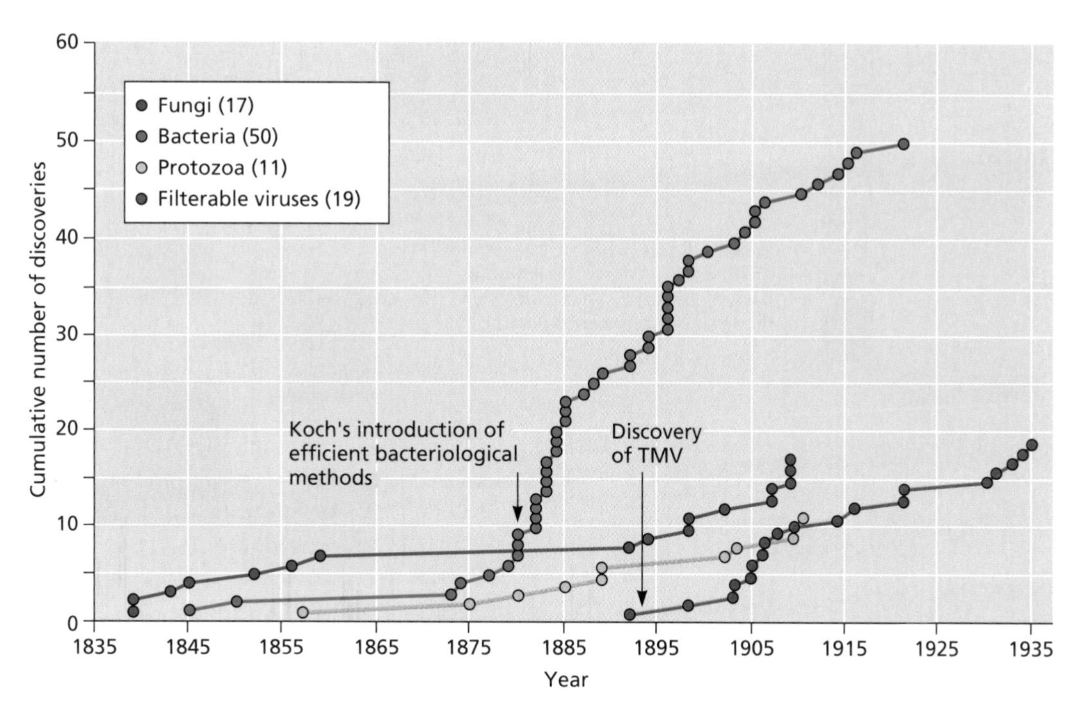

Figure 1.5 The pace of discovery of new infectious agents in the 19th and 20th centuries. Koch's introduction of efficient bacteriological techniques spawned an explosion of new discoveries of bacterial agents in the early 1880s. Similarly, the discovery of filterable agents launched the field of virology in the early 1900s. Despite an early surge of virus discovery, only 19 distinct human viruses had been reported by 1935. TMV, tobacco mosaic virus. Adapted from K. L. Burdon, *Medical Microbiology* (Macmillan Co., New York, NY, 1939), with permission.

that time to remove bacteria from extracts and culture media (Fig. 1.6A). Six years later in Holland, Martinus Beijerinck independently made the same observation. More importantly, Beijerinck made the conceptual leap that this must be a distinctive agent, because it was so small that it could pass through filters that trapped all known bacteria. However, Beijerinck thought that the agent was an infectious liquid. It was two former students and assistants of Koch, Friedrich Loeffler and Paul Frosch, who in the same year (1898) deduced that such infectious filterable agents comprised small particles: they observed that while the causative agent of foot-and-mouth disease (Box 1.2) passed through filters that held back bacteria, it could be retained by a finer filter.

Not only were the tobacco mosaic and foot-and-mouth disease pathogens much smaller than any previously recognized microorganism, but also they were replicated **only** in their host organisms. For example, extracts of an infected tobacco plant diluted into sterile solution produced no additional infectious agents until introduced into leaves of healthy plants, which subsequently developed tobacco mosaic disease. The serial transmission of infection by diluted extracts established that these diseases were not caused by a bacterial toxin present in the original preparations derived from infected tobacco plants or cattle. The failure of both pathogens to multiply in solutions that readily supported the growth of bacteria, as well as their dependence on host organisms for reproduction, further distinguished these new agents from pathogenic bacteria. Beijerinck termed the submicroscopic agent responsible for tobacco mosaic disease *contagium vivum fluidum* to emphasize its infectious nature and distinctive reproductive and physical properties. Agents passing through filters that retain bacteria came to be called ultrafilterable viruses, appropriating the term "*virus*" from the Latin for "poison." This term eventually was simplified to "virus."

The discovery of the first virus, tobacco mosaic virus, is often attributed to the work of Ivanovsky in 1892. However, he did not identify the tobacco mosaic disease pathogen as a distinctive agent, nor was he convinced that its passage through bacterial filters was not the result of some technical failure. It may be more appropriate to attribute the founding of the field of virology to the astute insights of Beijerinck, Loeffler, and Frosch, who recognized the distinctive nature of the plant and animal pathogens they were studying more than 100 years ago.

The pioneering work on tobacco mosaic and foot-and-mouth disease viruses was followed by the identification of viruses associated with specific diseases in many other organisms. Important landmarks from this early period include the identification of viruses that cause leukemias or solid tumors in chickens by Vilhelm Ellerman and Olaf Bang in 1908 and Peyton Rous in 1911, respectively. The study of viruses associated with cancers in chickens, particularly Rous

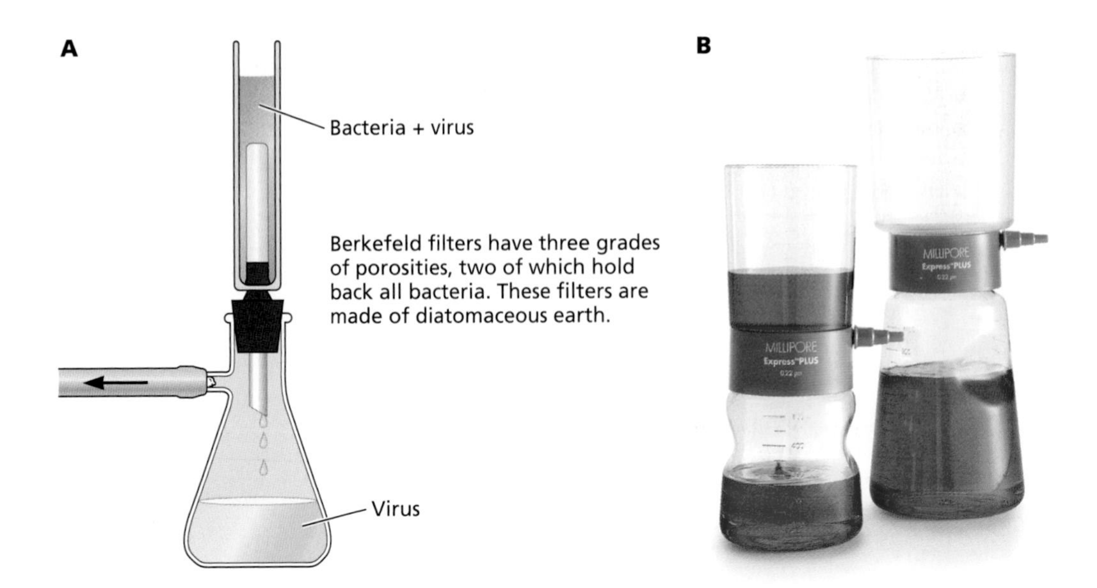

Figure 1.6 Filter systems used to characterize/purify virus particles. (A) The earliest, the Berkefeld filter, was invented in Germany in 1891. It was a "candle"-style filter comprising diatomaceous earth, or Kieselguhr, pressed into the shape of a hollow candle. The white candle is in the upper chamber of the apparatus, which is open at the top to receive the liquid to be filtered into the suction flask. The smallest pore size retained bacteria and allowed virus particles to pass through. Such filters were probably used by Ivanovsky, Loeffler, and Frosch to isolate the first plant and animal viruses. **(B)** A typical Millipore membrane filter apparatus. Such modern-day filter systems are disposable plastic laboratory items in which the upper and lower chambers are separated by a biologically inert membrane, available in a variety of pore sizes. Such filtration approaches may have limited our detection of giant viruses. Image provided courtesy of EMD Millipore Corporation.

sarcoma virus, eventually led to an understanding of the molecular basis of cancer (Volume II, Chapter 7).

The fact that bacteria could also be hosts to viruses was first recognized by Frederick Twort in 1915 and Félix d'Hérelle in 1917. d'Hérelle named such viruses **bacteriophages** because of their ability to lyse bacteria on the surface of agar plates ("phage" is derived from the Greek for "eating"). In an interesting twist of serendipity, Twort made his discovery of bacterial viruses while testing the smallpox vaccine virus to see if it would grow on simple media. He found bacterial contaminants, some of them appearing more transparent, which proved to be the result of lysis by a bacteriophage. Investigation of bacteriophages established the foundations for the field of molecular biology, as well as fundamental insights into how viruses interact with their host cells.

The Definitive Properties of Viruses

Throughout the early period of virology when many viruses of plants, animals, and bacteria were cataloged, ideas about the origin and nature of these distinctive infectious agents were quite controversial. Arguments centered on whether viruses originated from parts of a cell or were built from unique components. Little progress was made toward resolving these issues and establishing the definitive properties of viruses until the development of new techniques that allowed their visualization or propagation in cultured cells.

The Structural Simplicity of Virus Particles

Dramatic confirmation of the structural simplicity of virus particles came in 1935, when Wendell Stanley obtained crystals of tobacco mosaic virus. At that time, nothing was known of the structural organization of any biologically important macromolecules, such as proteins and DNA. Indeed, the crucial role of DNA as genetic material had not even been recognized. The ability to obtain an infectious agent in crystalline form, a state that is more generally associated with inorganic material, created much wonder and speculation about whether a virus is truly a life form. In retrospect, it is obvious that the relative ease with which tobacco mosaic virus could be crystallized was a direct result of both its structural simplicity and the ability of many particles to associate in regular arrays.

The 1930s saw the introduction of the instrument that rapidly revolutionized virology: the electron microscope. The great magnifying power of this instrument (eventually more than 100,000-fold) allowed direct visualization of virus particles for the first time. It has always been an exciting experience for investigators to obtain images of viruses, especially as they appear to be remarkably elegant (Fig. 1.7). Images of many different virus particles confirmed that these agents are very small (Fig. 1.8) and that most are far simpler in structure than any cellular organism. Many appeared

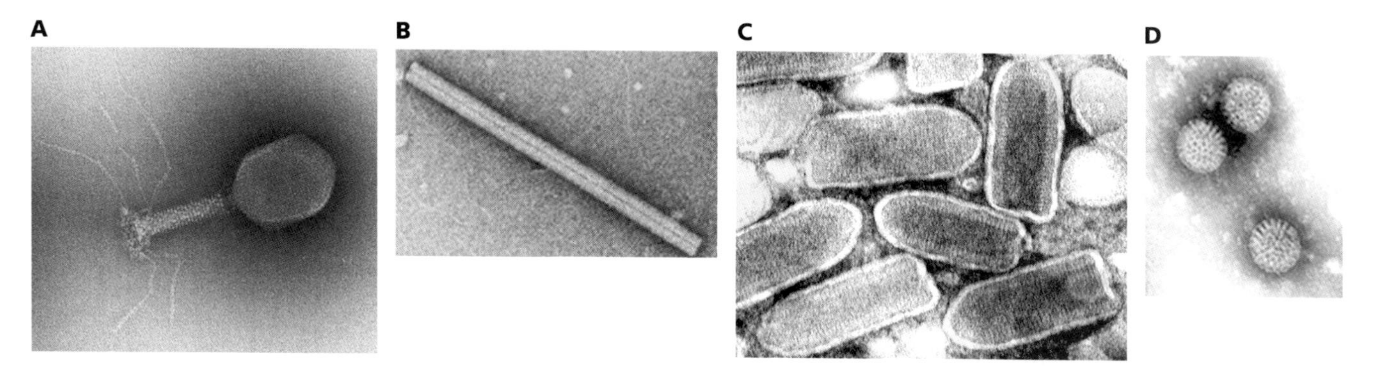

Figure 1.7 Electron micrographs of virus particles following negative staining. (A) The complex, nonenveloped virus bacteriophage T4. Note the intricate tail and tail fibers. Courtesy of R. L. Duda, University of Pittsburgh, Pittsburgh, PA. **(B)** The helical, nonenveloped particle of tobacco mosaic virus. Reprinted from the Universal Virus Database of the International Committee on Taxonomy of Viruses (http://ictvonline.org/), with permission. **(C)** Enveloped particles of the rhabdovirus vesicular stomatitis virus. Courtesy of F. P. Williams, University of California, Davis. **(D)** Nonenveloped, icosahedral human rotavirus particles. Courtesy of F. P. Williams, U.S. Environmental Protection Agency, Washington, DC.

Figure 1.8 Size matters. (A) Sizes of animal and plant cells, bacteria, viruses, proteins, molecules, and atoms are indicated. The resolving powers of various techniques used in virology, including light microscopy, electron microscopy, X-ray crystallography, and nuclear magnetic resonance (NMR) spectroscopy, are indicated. Viruses span a broad range from that equal to some small bacteria to just under ribosome size. The units commonly used in descriptions of virus particles or their components are the nanometer (nm [10^{-9} m]) and the angstrom (Å [10^{-10} m]). Adapted from A. J. Levine, *Viruses* (Scientific American Library, New York, NY, 1991); used with permission of Henry Holt and Company, LLC. **(B)** Illustration of the size differences among two viruses and a typical host cell.

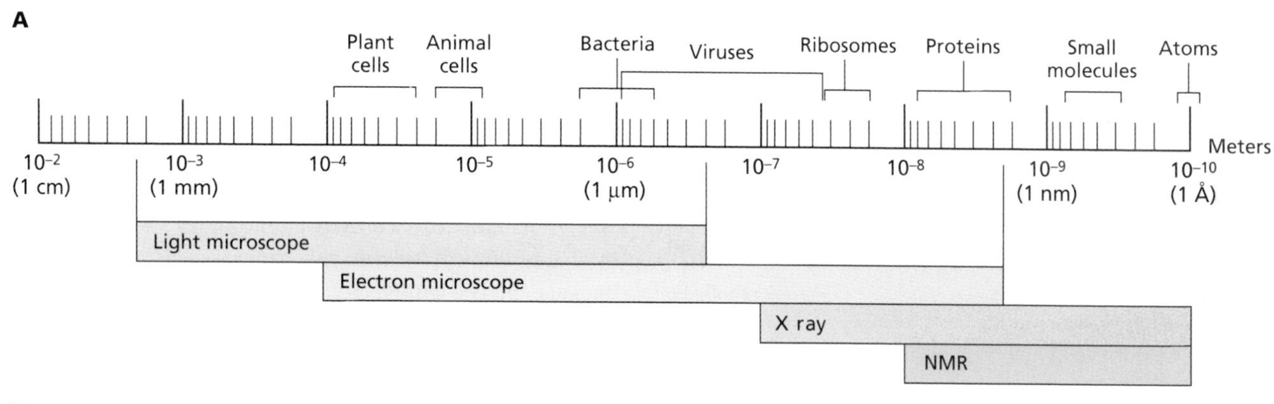

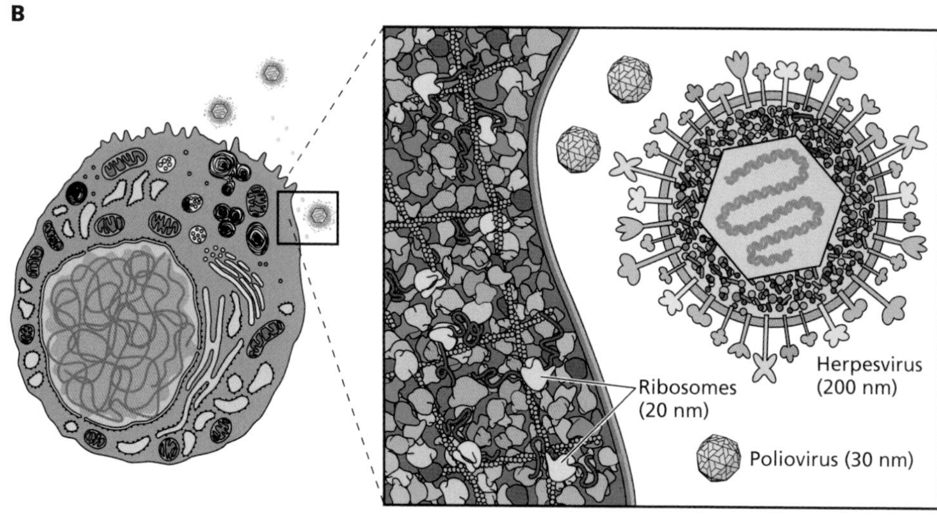

as regular helical or spherical particles. The description of the morphology of virus particles made possible by electron microscopy also opened the way for the first rational classification of viruses.

The Intracellular Parasitism of Viruses

Organisms as Hosts

The defining characteristic of viruses is their absolute dependence on a living host for reproduction: they are **obligate parasites**. Transmission of plant viruses such as tobacco mosaic virus can be achieved readily, for example, by applying extracts of an infected plant to a scratch made on the leaf of a healthy plant. Furthermore, as a single infectious particle of many plant viruses is sufficient to induce the characteristic lesion (Fig. 1.9), the concentration of the infectious agent could be measured. Plant viruses were therefore the first to be studied in detail. Some viruses of humans and other species could also be propagated in laboratory animals, and methods were developed to quantify them by determining the lethal dose. The transmission of yellow fever virus to mice by Max Theiler in 1930 was an achievement that led to the isolation of an attenuated strain, still considered one of the safest and most effective ever produced for the vaccination of humans.

After specific viruses and host organisms were identified, it became possible to produce sufficient quantities of virus particles for study of their physical and chemical properties and the consequences of infection for the host. Features such as the incubation period, symptoms of infection, and effects on specific tissues and organs were investigated. Laboratory animals remain an essential tool in investigations of the pathogenesis of viruses that cause disease. However, real progress toward understanding the mechanisms of virus reproduction was made only with the development of cell culture systems. Among the simplest, but crucial to both virology and molecular biology, were cultures of bacterial cells.

Figure 1.9 Lesions induced by tobacco mosaic virus on an infected tobacco leaf. In 1886, Adolph Mayer first described the characteristic patterns of light and dark green areas on the leaves of tobacco plants infected with tobacco mosaic virus. He demonstrated that the mosaic lesions could be transmitted from an infected plant to a healthy plant by aqueous extracts derived from infected plants. The number of local necrotic lesions that result is directly proportional to the number of infectious particles in the preparation. Courtesy J. P. Krausz; Reproduced, by permission of APS, from Scholthof, K.-B. G. 2000. Tobacco mosaic virus. The Plant Health Instructor. doi:10.1094/PHI-I-2000-1010-01.

Lessons from Bacteriophages

In the late 1930s and early 1940s, bacteriophages, or "phages," received increased attention as a result of controversy centering on how they were formed. John Northrup, a biochemist at the Rockefeller Institute in Princeton, NJ, championed the theory that a phage was a metabolic product of a bacterium. On the other hand, Max Delbrück, in his work with Emory Ellis and later with Luria, regarded phages as autonomous, stable, self-replicating entities characterized by heritable traits. According to this paradigm, phages were seen as ideal tools with which to investigate the nature of genes and heredity. Probably the most critical early contribution of Delbrück and Ellis was the perfection of the one-step growth method for synchronization of the reproduction of phages, an achievement that allowed analysis of a single cycle of phage reproduction in a population of bacteria. This approach introduced highly quantitative methods to virology, as well as an unprecedented rigor of analysis. The first experiments showed that phages indeed multiplied in the bacterial host and were liberated in a "burst" by lysis of the cell.

Delbrück was a zealot for phage research and recruited talented scientists to pursue the fundamental issues of what is now known as the field of molecular biology. This group of scientists, working together in what came to be called the "phage school," focused their attention on specific phages of the bacterium *Escherichia coli*. Progress was rapid, primarily because of the simplicity of the phage infectious cycle. Phages reproduce in bacterial hosts, which can be obtained in large numbers by overnight culture. By the mid-1950s, it was evident that viruses from bacteria, animals, and plants share many fundamental properties. However, the phages provided a far more tractable experimental system. Consequently, their study had a profound impact on the field of virology.

One critical lesson came from a definitive experiment that established that viral nucleic acid carries genetic information. It was known from studies of the "transforming principle" of pneumococcus by Oswald Avery, Colin MacLeod, and Maclyn McCarty (1944) that nucleic acid was both necessary and sufficient for the transfer of genetic traits of bacteria. However, in the early 1950s, protein was still suspected to be an important component of viral heredity. In a brilliantly simple experiment that included the use of a common kitchen food blender, Alfred Hershey and Martha Chase showed that this hypothesis was incorrect (Box 1.5).

BOX 1.5

EXPERIMENTS

The Hershey-Chase experiment

By differentially labeling the nucleic acid and protein components of virus particles with radioactive phosphorus (^{32}P) and radioactive sulfur (^{35}S), respectively, Alfred Hershey and Martha Chase showed that the protein coat of the infecting virus could be removed soon after infection by agitating the bacteria for a few minutes in a blender. In contrast, ^{32}P-labeled phage DNA entered and remained associated with the bacterial cells under these conditions. Because such blended cells produced a normal burst of new virus particles, it was clear that the DNA contained all of the information necessary to produce progeny phages.

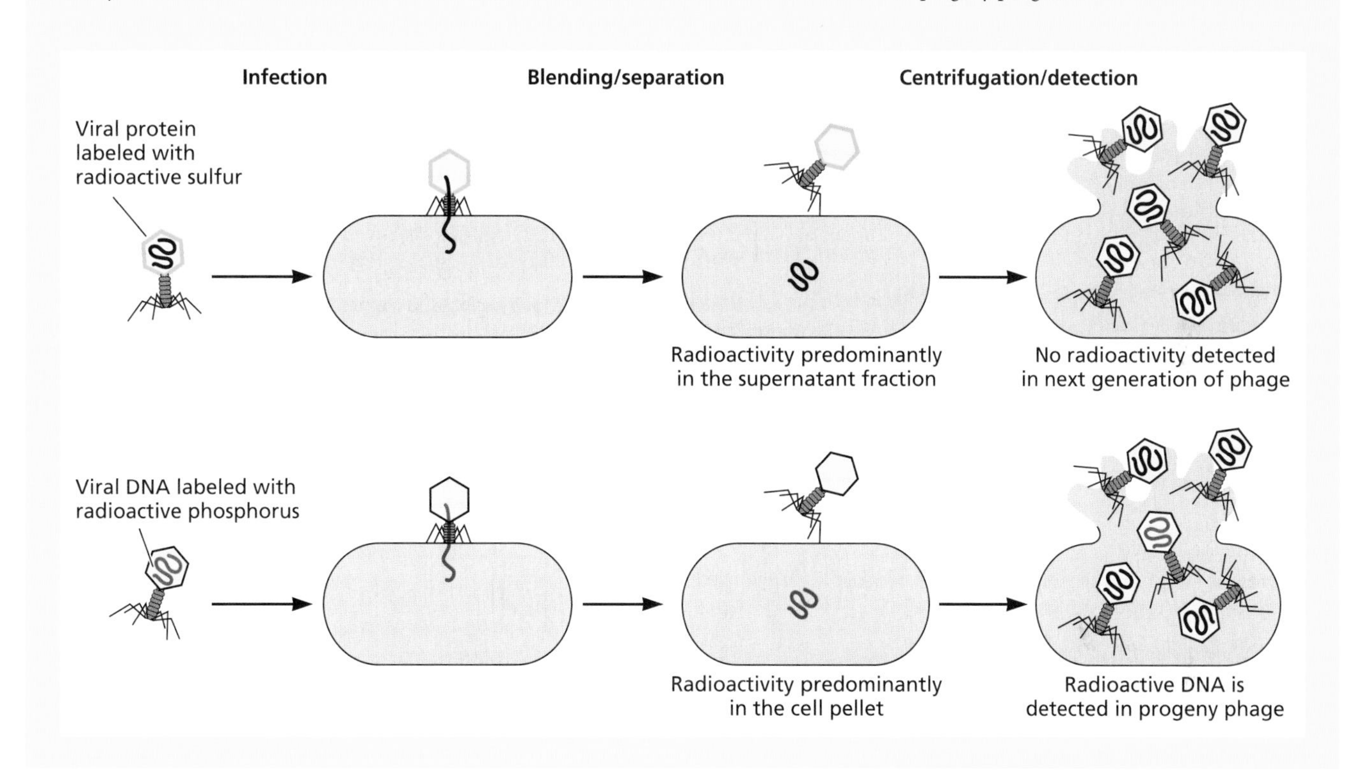

Bacteriophages were originally thought to be lethal agents, killing their host cells after infection. In the early 1920s, a previously unknown interaction was discovered, in which the host cell not only survived the infection but also stably inherited the genetic information of the virus. It was also observed that certain bacterial strains could lyse spontaneously and produce bacteriophages after a period of growth in culture. Such strains were called **lysogenic**, and the phenomenon, **lysogeny**. Studies of lysogeny uncovered many previously unrecognized features of virus-host cell interactions (Box 1.6). Recognition of this phenomenon came from the work of many scientists, but it began with the elegant experiments of André Lwoff and colleagues at the Institut Pasteur in Paris. Lwoff showed that a viral genome exists in lysogenic cells in the form of a silent genetic element called the **prophage**. This element determined the ability of lysogenic bacteria to produce infectious bacteriophages. Subsequent studies of the *E. coli* phage lambda established a paradigm for one mechanism of lysogeny, the integration of a phage genome into a specific site on the bacterial chromosome.

Bacteriophages became inextricably associated with the new field of molecular biology (Table 1.1). Their study established many fundamental principles: for example, control of the decision to enter a lysogenic or a lytic pathway is encoded in the genome of the virus. The first mechanisms discovered for the control of gene expression, exemplified by the elegant operon theory of Nobel laureates François Jacob and Jacques Monod, were deduced in part from studies of lysogeny by phage lambda. The biology of phage lambda provided a fertile ground for work on gene regulation, but study of virulent T phages (T1 to T7, where T stands for "type") of *E. coli* paved the way for many other important advances (Table 1.1). As we shall see, these systems also

BOX 1.6

BACKGROUND

Properties of lysogeny shared with animal viruses

Lytic versus Lysogenic Response to Infection

Some bacterial viruses can enter into either destructive (lytic) or relatively benign (lysogenic) relationships with their host cells. Such bacteriophages were called temperate. In a lysogenic bacterial cell, viral genetic information persists but viral gene expression is repressed. Such cells are called lysogens, and the quiescent viral genome, a prophage. By analogy with the prophage, an integrated DNA copy of a retroviral genome in an animal genome is termed a provirus.

Propagation as a Prophage

For some bacteriophages like lambda and Mu (Mu stands for "mutator"), prophage DNA is integrated into the host genome of lysogens and passively replicated by the host. Virally encoded enzymes, known as integrase (lambda) and transposase (Mu), mediate the covalent insertion of viral DNA into the chromosome of the host bacterium, establishing it as a prophage. The prophage DNA of other bacteriophages, such as P1, exists as a plasmid, a self-replicating, autonomous chromosome in a lysogen. Both forms of propagation have been identified in certain animal viruses.

Insertional Mutagenesis

Bacteriophage Mu inserts its genome into many random locations on the host chromosome, causing numerous mutations. This process is called insertional mutagenesis and is a phenomenon observed with retroviruses.

Gene Repression and Induction

Prophage gene expression in lysogens is turned off by the action of viral proteins called repressors. Expression can be turned on when repressors are inactivated (a process called induction). Elucidation of the mechanisms of these processes set the stage for later investigation of the control of gene expression in experiments with other viruses and their host cells.

Transduction of Host Genes

Bacteriophage genomes can pick up cellular genes and deliver them to new cells (a process known as transduction). The process can be generalized, with the acquisition by the virus of any segment from the host chromosome, or specialized, as is the case for viruses that integrate into specific sites in the host chromosome. For example, occasional mistakes in excision of the lambda prophage after induction result in production of unusual progeny phage that have lost some of their own DNA but have acquired the bacterial DNA adjacent to the prophage. As described in Volume II, Chapter 7, the acute transforming retroviruses also arise via capture of genes in the vicinity of their integration as proviruses. These cancer-inducing cellular genes are then transduced along with viral genes during subsequent infection.

Pioneers in the study of lysogeny: Nobel laureates François Jacob, Jacques Monod, and André Lwoff.

provided an extensive preview of mechanisms of animal virus reproduction (Box 1.7).

Animal Cells as Hosts

The culture of animal cells in the laboratory was initially more of an art than a science, restricted to cells that grew out of organs or tissues maintained in nutrient solutions under sterile conditions. The finite life span of such **primary cells**; their dependence for growth on natural components in their media such as lymph, plasma, or chicken embryo extracts; and the technical demands of sterile culture prior to the discovery of antibiotics made reproducible experimentation very difficult. However, by 1955, the work of many investigators had led to a series of important methodological advances. These included the development of defined media optimal for growth of mammalian cells, incorporation of antibiotics into cell culture media, and development of immortal cell lines such as the mouse L and human HeLa cells that are still in widespread use. These advances allowed growth of animal cells in culture to become a routine, reproducible exercise.

The availability of well-characterized cell cultures had several important consequences for virology. It allowed the discovery of new human viruses, such as adenovirus, measles virus, and rubella virus, for which animal hosts were not available. In 1949, John Enders and colleagues used cell cultures to propagate poliovirus, a feat that led to the development of polio vaccines a few years later. Cell culture technology revolutionized the ability to investigate the reproduction of viruses. Viral infectious cycles could be studied

BOX 1.7

TERMINOLOGY
The episome

In 1958, François Jacob and Elie Wollman realized that lambda prophage and the *E. coli* F sex factor had many common properties. This remarkable insight led to the definition of the episome.

An episome is an exogenous genetic element that is not necessary for cell survival. Its defining characteristic is the ability to reproduce in two alternative states: while integrated in the host chromosome or autonomously. However, this term is often applied to genomes that can be maintained in cells by autonomous replication and never integrate, for example, the DNA genomes of certain animal viruses.

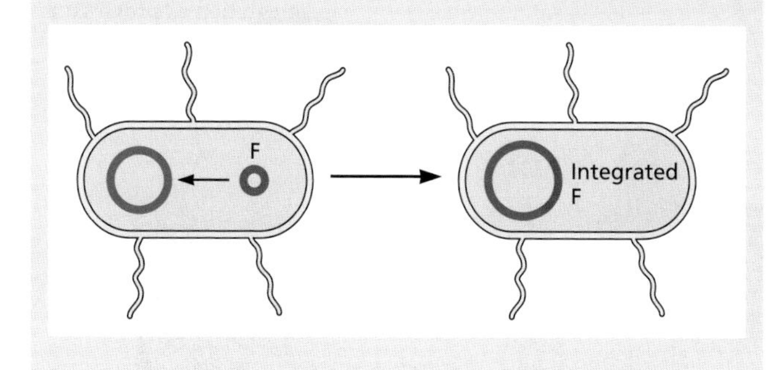

under precisely controlled conditions by employing the analog of the one-step growth cycle of bacteriophages and simple methods for quantification of infectious particles described in Chapter 2. Our current understanding of the molecular basis of viral parasitism, the focus of this volume, is based almost entirely on analyses of one-step growth cycles in cultured cells. Such studies established that viruses are **molecular** parasites: for example, their reproduction depends absolutely on their host cell's biosynthetic machinery for synthesis of the components from which they are built. In contrast to cells, viruses are not reproduced by growth and division. Rather, the infecting genome contains the information necessary to redirect cellular systems to the production of many copies of all the components needed for the *de novo* assembly of new virus particles.

Viruses Defined

Advances in knowledge of the structure of virus particles and the mechanisms by which they are produced in their host cells have been accompanied by increasingly accurate definitions of these unique agents. The earliest pathogenic agents, distinguished by their small size and dependence on a host organism for reproduction, emphasized the importance of viruses as agents of disease. We can now provide a much more precise definition, elaborating their relationship with the host cell and the important features of virus particles. The definitive properties of viruses are summarized as follows:

- A virus is an infectious, obligate intracellular parasite.
- The viral genome comprises DNA or RNA.
- The viral genome directs the synthesis of viral components by cellular systems within an appropriate host cell.
- Infectious progeny virus particles, called **virions**, are formed by *de novo* self-assembly from newly synthesized components.
- A progeny virion assembled during the infectious cycle is the vehicle for transmission of the viral genome to the next host cell or organism, where its disassembly initiates the next infectious cycle.

While viruses lack the complex energy-generating and biosynthetic systems necessary for independent existence (Box 1.8), they are **not** the simplest biologically active agents: **viroids**, which are infectious agents of a variety of economically important plants, comprise a single small molecule of noncoding RNA, whereas other agents, termed **prions**, are thought to be single protein molecules (Volume II, Chapter 12).

Cataloging Animal Viruses

Virus classification was at one time a subject of colorful and quite heated controversy (Box 1.9). As new viruses were being discovered and studied by electron microscopy, the virus world was seen to be a veritable zoo of particles with different sizes, shapes, and compositions (see, for example, Fig. 1.10). Very strong opinions were advanced concerning classification and nomenclature. One camp pointed to the inability to infer, from the known properties of viruses, anything about their evolutionary origin or their relationships to one another—the major goal of classical taxonomy. The other camp maintained that despite such limitations, there were significant practical advantages in grouping isolates with similar properties. A major sticking point, however, was finding agreement on **which** properties should be considered most important in constructing a scheme for virus classification.

The Classical System

Lwoff, Robert Horne, and Paul Tournier, in 1962, advanced a comprehensive scheme for the classification of all viruses (bacterial, plant, and animal) under the classical Linnaean hierarchical system consisting of phylum, class, order, family, genus, and species. Although a subsequently formed international committee on the nomenclature of viruses did not adopt this system *in toto*, its designation of families, genera, and species was used for the classification of animal viruses.

One of the most important principles embodied in the system advanced by Lwoff and his colleagues was that viruses

BOX 1.8

DISCUSSION

Are viruses living entities? What can/can't they do?

Viruses can be viewed as microbes that exist in two phases: an inanimate phase, the virion; and a multiplying phase in an infected cell. Some researchers have promoted the idea that viruses are organisms and that the inanimate virions may be viewed as "spores" that come "alive" in cells, or in factories within cells. This has long been a topic of intense discussion, stimulated most recently by the discovery of giant viruses such as the mimiviruses and pandoraviruses. Check out what the contemporary general public feels about this topic (http://www.virology.ws/are-viruses-alive/).

Apart from attributing "life" to viruses, many scientists have succumbed to the temptation of ascribing various **actions** and **motives** when discussing them. While remarkably effective in enlivening a lecture or an article, anthropomorphic characterizations are inaccurate and also quite misleading. Infected cells and hosts respond in many ways after infection, but viruses are **passive** agents, totally at the mercy of their environments. Therefore viruses cannot employ, ensure, synthesize, exhibit, display, destroy, deploy, depend, reprogram, avoid, retain, evade, exploit, generate, etc.

As virologists can be very passionate about their subject, it is exceedingly difficult to purge such anthropomorphic terms from virology communications. Indeed, hours were spent doing so in the preparation of this textbook, though undoubtedly there remain examples in which actions are attributed to viruses. Should you find them, let us know!

Bândea CI. 1983. A new theory on the origin and the nature of viruses. *J Theor Biol* **105:**591–602.

Claverie JM, Abergel C. 2013. Open questions about giant viruses. *Adv Virus Res* **85:**25-56.

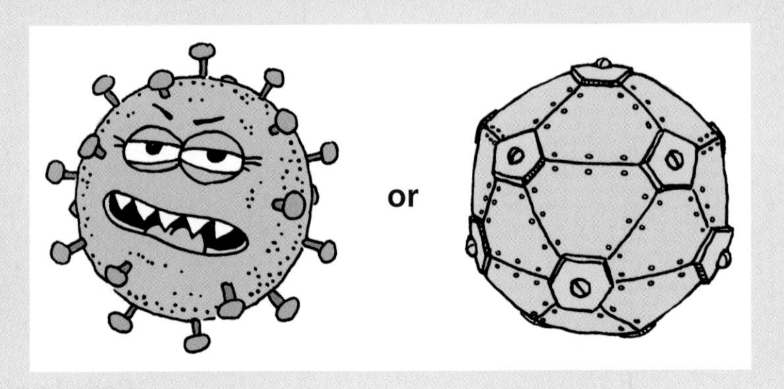

should be grouped according to **their** shared properties rather than the properties of the cells or organisms they infect. A second principle was a focus on the nucleic acid genome as the primary criterion for classification. The importance of the genome had become clear when it was inferred from the Hershey-Chase experiment that viral nucleic acid alone can be infectious (Box 1.5). Four characteristics were to be used in the classification of all viruses:

1. Nature of the nucleic acid in the virion (**DNA** or **RNA**)
2. Symmetry of the protein shell (**capsid**)
3. Presence or absence of a lipid membrane (**envelope**)
4. Dimensions of the virion and capsid

The elucidation of evolutionary relationships by analyses of nucleic acid and protein sequence similarities is now a standard method for assigning viruses to a particular family and to order members within a family. For example, hepatitis C virus was classified as a member of the family *Flaviviridae* and the Middle East respiratory SARS-like virus (MERS) was assigned to the *Coronaviridae* based on their genome sequences. However, as our knowledge of molecular

BOX 1.9

TERMINOLOGY

Complexities of viral nomenclature

No consistent system for naming viral isolates has been established by their discoverers. For example, among the vertebrate viruses, some are named for the associated diseases (e.g., poliovirus, rabies virus), for the specific type of disease they cause (e.g., murine leukemia virus), or for the sites in the body that are affected or from which they were first isolated (e.g., rhinovirus and adenovirus). Others are named for the geographic locations in which they were first isolated (e.g., Sendai virus [Sendai, Japan] and Coxsackievirus [Coxsackie, NY]) or for the scientists who first discovered them (e.g., Epstein-Barr virus). In these cases the virus names are often capitalized. Some viruses are even named for the way in which people imagined they were contracted (e.g., influenza, for the "influence" of bad air), how they were first perceived (e.g., the giant mimiviruses [Box 1.10], for the fact that they "mimic" bacteria), or totally by whimsy (e.g., pandoravirus). Finally, combinations of the above designations are also used (e.g., Rous sarcoma virus).

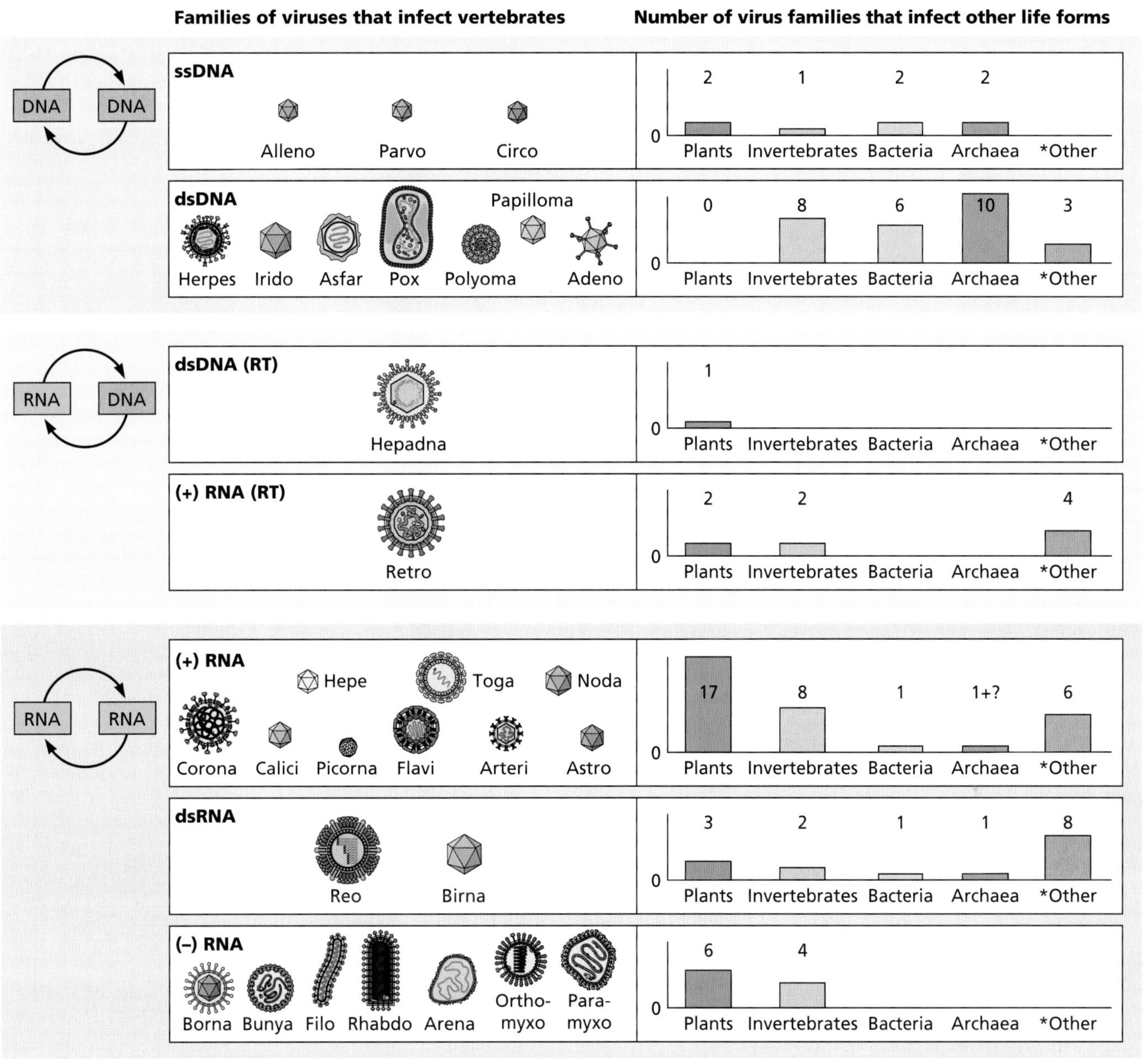

Figure 1.10 Viral families sorted according to the nature of the viral genomes. A wide variety of sizes and shapes are illustrated for the families of viruses that infect vertebrates. Similar diversity exists for the families of viruses that infect other life forms, but the chart illustrates only the number found to date in each category. As noted, in some categories there are as yet no examples. The notation "1+?" for RNA archaeal viruses indicates that additional viral genomes have been predicted from metagenomic analyses but are not yet confirmed. Abbreviations: ds, double-stranded; ss, single-stranded. Adapted from A. M. Q. King et al. (ed.), *Virus Taxonomy: Classification and Nomenclature of Viruses*, Ninth Report of the International Committee on Taxonomy of Viruses (Academic Press, Inc., San Diego, CA, 2012).

properties of viruses and their reproduction has increased, other relationships have become apparent. *Hepadnaviridae*, *Retroviridae*, and some plant viruses are classified as different families on the basis of the nature of their genomes. Nevertheless, they are all related by the fact that reverse transcription is an essential step in their reproductive cycles, and the viral polymerases that perform this task exhibit important similarities in amino acid sequence. Another example is the classification of the giant protozoan *Mimiviridae* and pandoraviruses as members of a related group called nucleocytoplasmic large

BOX 1.10

DISCUSSION

Viral giants and a new satellite

The mimivirus virion, the prototype member of the *Mimiviridae*, is large enough to be visible in a light microscope, and it was the first of the giant viruses to be discovered. The mimivirus was isolated from water in a cooling tower in England in 1992 and was initially thought to be an intracellular bacterium in its amoeba host. Not until a brief note in 2003 was it made apparent that this giant is a member of a group of nucleocytoplasmic large DNA viruses (NCLDVs) that include the poxviruses and several aquatic viruses. The mimiviruses' fiber-coated icosahedral capsid (0.75 μm in diameter) is just the right size for phagocytosis by its amoeba host. The mimivirus genome of 1.2 kbp encodes more than 900 proteins and is larger than that of some bacteria. Many of these proteins are components of the protein translational apparatus, a function for which other viruses rely entirely on the host. These unusual properties have prompted speculation that the giant viruses might represent a separate branch in the tree of life, or that they arose by reductive evolution from the nucleus of a primitive cellular life form.

The discovery of a second isolate in the family *Mimiviridae*, called mamavirus, produced yet another surprise. Mamavirus was associated with a 50-nm icosahedral satellite virus particle, called Sputnik, which differs from known satellites in that it contains a double-stranded DNA genome (<18 kbp). Nevertheless, like other satellites, Sputnik depends on proteins of its helper, mamavirus, for propagation. Because it replicates in the mamavirus "factories" within the coinfected host cell and reduces virus titer, Sputnik's discoverers consider it to be a parasite of mamavirus. For this and other reasons, they argue that Sputnik represents a new, and as yet uncharacterized, family of viruses and placed it in a new a classification, which they called virophages in analogy to the functional relationship between bacteria and bacteriophages. Other investigators have argued that the general biological behavior and genetic properties of Sputnik do not differ substantially from other known satellites, and that the virophage classification can be misleading. A metagenomic analysis of Sputnik and Sputnik-like genomes has shown that they are abundant in almost all geographical zones. The overall low sequence similarity between the shared homologous genes in the three distinguished lineages, and their distant phylogenetic relationships, suggest that the genetic diversity of these satellite viruses is much beyond what we know thus far.

Another giant virus that infects amoebae was discovered in 2013. Called pandoravirus, the largest isolate has a genome that is twice the size of that of the *Mimiviridae* and encodes 2,556 putative protein-coding sequences, most of them never seen before. All of this serves to remind us that life is a continuum, and the delineation of distinct categories can sometimes be quite difficult.

La Scola B, Audic S, Robert C, Jungang L, de Lamballerie X, Drancourt M, Birtles R, Claverie JM, Raoult D. 2003. A giant virus in amoebae. *Science* **299:**2033.

Claverie JM, Abergel C. 2009. Mimivirus and its virophage. *Annu Rev Genet* **43:**49–66.

La Scola B, Desnues C, Pagnier I, Robert C, Barrassi L, Fournous G, Merchat M, Suzan-Monti M, Forterre P, Koonin E, Raoult D. 2008. The virophage as a unique parasite of the giant mimivirus. *Nature* **455:**100-104.

Kuprovic M, Cvirkaite-Kuprovic V. 2011. Virophages or satellite viruses? *Nat Rev Microbiol* **9:**762–763.

Zhou J, Zhang W, Yan S, Xiao J, Zhang Y, Li B, Pan Y, Wang Y. 2013. Diversity of virophages in metagenomic data sets. *J Virol* **87:**4225–4236.

Philippe N, Legendre M, Doutre G, Couté Y, Poirot O, Lescot M, Arslan D, Seltzer V, Bertaux L, Bruley C, Garin J, Claverie JM, Abergel C. 2013. Pandoraviruses: amoeba viruses with genomes up to 2.5 Mb reaching that of parasitic eukaryotes. *Science* **341:**281–286.

For illustrations of sputnik and mimivirus structure see: http://viralzone.expasy.org/all_by_species/670.html

See also TWiV 261: Giants among viruses. Interview with Drs. Chantal Abergel and Jean-Michel Claverie at http://www.twiv.tv/2013/12/01/twiv-261-giants-among-viruses/

DNA viruses (NCLDVs), which includes the *Poxviridae* that infect vertebrates (Box 1.10).

The 2012 report of the International Committee on Taxonomy of Viruses (ICTV) lists 2,618 virus and viroid species distributed amongst 420 genera, 22 subfamilies, 96 families, and 7 orders, as well as numerous viruses that have not yet been classified and are probably representatives of new genera and/or families. The ICTV report also includes descriptions of subviral agents (**satellites**, viroids, and prions) and a list of viruses for which information is still insufficient to make assignments. Satellites are composed of nucleic acid molecules that depend for their multiplication on coinfection of a host cell with a helper virus. However, they are not related to this helper. When a satellite encodes the coat protein in which its nucleic acid is encapsidated, it is referred to as a **satellite virus** (e.g., hepatitis delta virus is a satellite virus). The pace of discovery of new viruses has been accelerated greatly with the application of **metagenomic analyses**, direct sequencing of genomes from environmental samples, suggesting that we have barely begun to chart the viral universe.

The ICTV nomenclature has been applied widely in both the scientific and medical literature, and therefore we adopt it in this text. In this nomenclature, the Latinized virus family names are recognized as starting with capital letters and ending with *-viridae*, as, for example, in the family name *Parvoviridae*. These names are used interchangeably with their common derivatives, as, for example, parvoviruses.

Classification by Genome Type: the Baltimore System

Because the viral genome carries the entire blueprint for virus propagation, molecular virologists have long considered it the most important characteristic for classification purposes. Therefore, although individual virus families are known by their classical designations, they are more commonly placed in groups according to their genome types, as illustrated

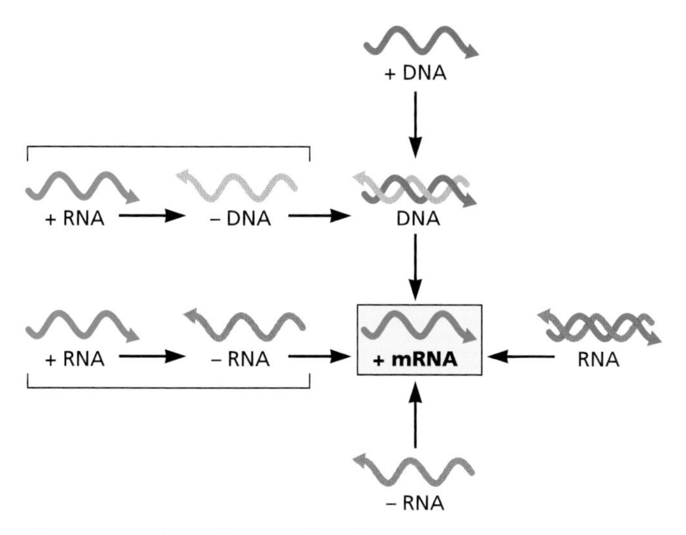

Figure 1.11 The Baltimore classification. All viruses must produce mRNA that can be translated by cellular ribosomes. In this classification system, the unique pathways from various viral genomes to mRNA define specific virus classes on the basis of the nature and polarity of their genomes.

in Fig. 1.10. There are seven genome types for all families of viruses, and all seven are represented in viruses that infect vertebrates.

Francis Crick conceptualized the central dogma for flow of genetic information:

DNA → RNA → protein

All viruses must direct the synthesis of mRNA that is decoded by the host's translational machinery. Appreciation of the essential role of the translational machinery inspired David Baltimore to devise a complementary classification scheme (Fig. 1.11). This scheme describes the pathways for formation of mRNA for viruses with either RNA or DNA genomes.

By convention, mRNA is defined as a **positive [(+)] strand** because it contains immediately translatable information. In the Baltimore classification, a strand of DNA that is of equivalent sequence is also designated a (+) strand. The RNA and DNA complements of (+) strands are designated **negative [(−)] strands**. The information embodied in this classification provides virologists with immediate insight into the steps that must take place to initiate replication and expression of the viral genome.

A Common Strategy for Viral Propagation

The basic thesis of this textbook is that **all** viral propagation can be described in the context of three fundamental properties.

- All viral genomes are packaged inside particles that mediate their transmission from host to host.
- The viral genome contains the information for initiating and completing an infectious cycle within a susceptible, permissive cell. An infectious cycle includes attachment and entry, decoding of genome information, genome replication, and assembly and release of particles containing the genome.
- All successful viruses are able to establish themselves in a host population so that virus propagation is ensured.

Perspectives

The study of viruses has increased our understanding of the importance and ubiquitous existence of these diverse agents and, in many cases, yielded new and unexpected insight into the molecular biology of host cells and organisms. Indeed, as viruses are obligate molecular parasites, every tactical solution encountered in their reproduction and propagation must of necessity tell us something about the host as well as the virus. Some of the important landmarks and achievements in the field of animal virology are summarized in Fig. 1.12. It is apparent that much has been discovered about the biology of viruses and about host defenses against them. Yet the more we learn, the more we realize that much is still unknown.

In the first edition of this textbook (published in 2000), we noted that the most recent (1995) report of the ICTV listed 71 different virus families, which covered most new isolates. We speculated therefore that: "As few new virus families had been identified in recent years, it seems likely that a significant fraction of all existing virus families are now known." In the intervening years, this prediction has been shattered, not only by the discovery of new families of viruses, including giant viruses with genome sizes that surpass those of some bacteria, but also by results from metagenomic analyses. For example, the fact that a high percentage (93%) of protein-coding sequences in the genomes of the giant pandoraviruses have **no** homologs in the current databases and the unusual morphological features and atypical reproduction process of these viruses were totally unexpected. It is also mind-boggling to contemplate that of almost 900,000 viral sequences identified in samples of only one type of ecosystem (raw sewage), more than 66% bore **no** relationship to any viral family in the current database. From these analyses and similar studies of other ecosystems (i.e., oceans and soil), it has been estimated that less than 1% of the extant viral diversity has been explored to date. Clearly, the viral universe is far more vast and diverse than suspected only a decade ago, and there is much fertile ground for gaining a deeper understanding of the biology of viruses, and their host cells and organisms, in the future.

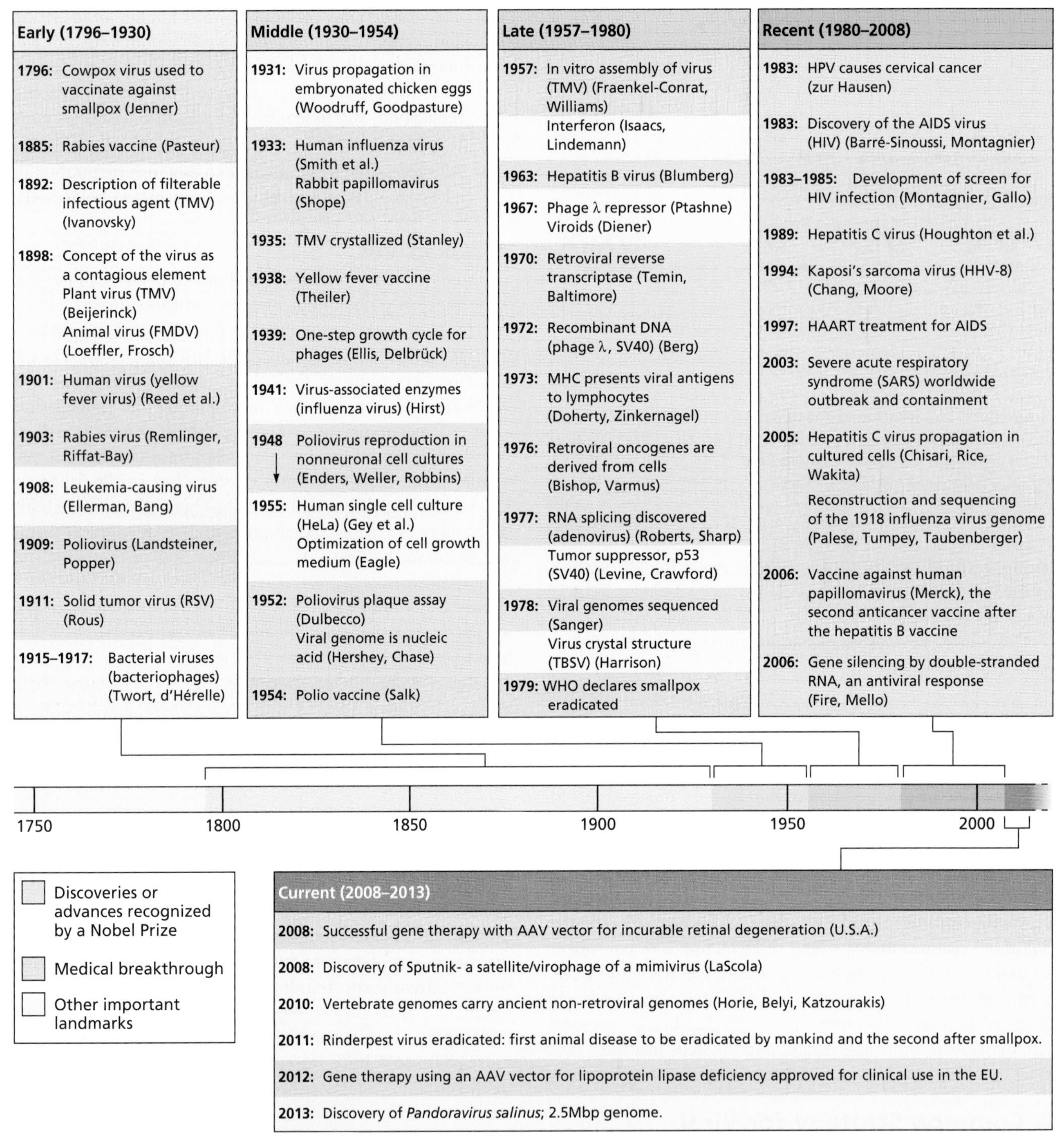

Figure 1.12 Landmarks in the study of animal viruses. Key discoveries and technical advances are listed for each time interval. The pace of discovery has increased exponentially over time. Abbreviations: AAV, adenovirus associated virus; EU, European Union; HAART, highly active antiretroviral therapy; HIV, human immunodeficiency virus; HPV, human papillomavirus; TBSV, tomato bushy stunt virus; TMV, tobacco mosaic virus; SV40, simian virus 40; FMDV, foot-and-mouth disease virus; WHO, World Health Organization; MHC, major histocompatibility complex; HHV-8, human herpesvirus 8; RSV, Rous sarcoma virus.

References

Books

Barry JM. 2005. *The Great Influenza*. Penguin Books, New York, NY.

Brock TD. 1990. *The Emergence of Bacterial Genetics*. Cold Spring Harbor Laboratory Press, Cold Spring Harbor, NY.

Brothwell D, Sandison AT (ed). 1967. *Diseases in Antiquity*. Charles C Thomas, Publisher, Springfield, IL.

Cairns J, Stent GS, Watson JD (ed). 1966. *Phage and the Origins of Molecular Biology*. Cold Spring Harbor Laboratory for Quantitative Biology, Cold Spring Harbor, NY.

Creager ANH. 2002. *The Life of a Virus: Tobacco Mosaic Virus as an Experimental Model, 1930–1965*. University of Chicago Press, Chicago, IL.

Denniston K, Enquist L. 1981. *Recombinant DNA. Benchmark Papers in Microbiology*, vol. 15. Dowden, Hutchinson and Ross, Inc, Stroudsburg, PA.

Fiennes R. 1978. *Zoonoses and the Origins and Ecology of Human Disease*. Academic Press, Inc, New York, NY.

Hughes SS. 1977. *The Virus: a History of the Concept*. Heinemann Educational Books, London, United Kingdom.

Karlen A. 1996. *Plague's Progress, a Social History of Man and Disease*. Indigo, Guernsey Press Ltd, Guernsey, Channel Islands.

King AMQ, Adams MJ, Carstens EB, Lefkowitz EJ (ed). 2011. *Virus Taxonomy: Classification and Nomenclature of Viruses*. Ninth Report of the International Committee on Taxonomy of Viruses. Academic Press, Inc, San Diego, CA.

Knipe DM, Howley PM (ed). 2013. *Fields Virology*, 6th ed. Lippincott Williams & Wilkins, Philadelphia, PA.

Luria SE. 1953. *General Virology*. John Wiley & Sons, Inc, New York, NY.

Murphy FA, Fauquet CM, Bishop DHL, Ghabrial SA, Jarvis AW, Rasmussen N. 1997. *Picture Control: the Electron Microscope and the Transformation of Biology in America 1940–1960*. Stanford University Press, Stanford, CA.

Oldstone MBA. 1998. *Viruses, Plagues and History*. Oxford University Press, New York, NY.

Stent GS. 1960. *Papers on Bacterial Viruses*. Little, Brown & Co, Boston, MA.

Waterson AP, Wilkinson L. 1978. *An Introduction to the History of Virology*. Cambridge University Press, Cambridge, United Kingdom.

Papers of Special Interest

Baltimore D. 1971. Expression of animal virus genomes. *Bacteriol Rev* **35:**235–241.

Brown F, Atherton J, Knudsen D. 1989. The classification and nomenclature of viruses: summary of results of meetings of the International Committee on Taxonomy of Viruses. *Intervirology* **30:**181–186.

Burnet FM. 1953. Virology as an independent science. *Med J Aust* **40:**842.

Crick FHC, Watson JD. 1956. Structure of small viruses. *Nature* **177:**473–475.

Culley AI, Lang AS, Suttle CA. 2006. Metagenomic analysis of coastal RNA virus communities. *Science* **312:**1795–1798.

DiMaio D. 2012. Viruses, masters at downsizing. *Cell Host Microbe* **11:**560–561.

Lustig A, Levine AJ. 1992. One hundred years of virology. *J Virol* **66:**4629–4631.

Murray NE, Gann A. 2007.What has phage lambda ever done for us? *Curr Biol* **17:**R305–R312.

Suttle CA. 2007. Marine viruses—major players in the global ecosystem. *Nat Rev Microbiol* **5:**801–812.

Van Helvoort T. 1993. Research styles in virus studies in the twentieth century: controversies and the formation of consensus. Doctoral dissertation. University of Limburg, Maastricht, The Netherlands.

Websites

http://www.ictvonline.org/virusTaxonomy.asp?bhcp=1 *Latest update of virus classification from the ICTV.*

http://ictvonline.org/ *ICTV-approved virus names and other information as well as links to virus databases.*

http://www.twiv.tv *A weekly netcast about viruses featuring informal yet informative interviews with guest virologists who discuss their recent findings and other topics of general interest.*

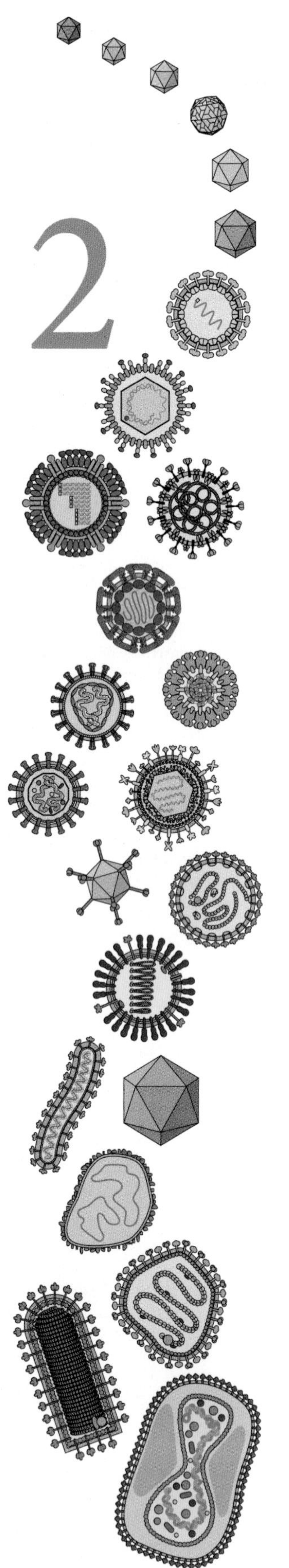

2 The Infectious Cycle

Introduction

The Infectious Cycle

The Cell

The Architecture of Cell Surfaces

The Extracellular Matrix: Components and Biological Importance

Properties of the Plasma Membrane

Cell Membrane Proteins

Entering Cells

Making Viral RNA

Making Viral Proteins

Making Viral Genomes

Forming Progeny Virus Particles

Viral Pathogenesis

Overcoming Host Defenses

Cultivation of Viruses

Cell Culture

Embryonated Eggs

Laboratory Animals

Assay of Viruses

Measurement of Infectious Units

Efficiency of Plating

Measurement of Virus Particles and Their Components

Viral Reproduction: the Burst Concept

The One-Step Growth Cycle

Initial Concept

One-Step Growth Analysis: a Valuable Tool for Studying Animal Viruses

Systems Biology

Perspectives

References

LINKS FOR CHAPTER 2

- *Video: Interview with Dr. Thomas Hope*
  http://bit.ly/Virology_Hope
- *Cloning HeLa cells with Philip I. Marcus*
  http://bit.ly/Virology_Twiv197
- *Ode to a plaque*
  http://bit.ly/Virology_Twiv68
- *Movie of vaccinia virus plaque formation*
  http://www.sciencemag.org/content/suppl/2010/01/19/science.1183173.DC1/1183173s1.mov
- *Think globally, act locally*
  http://bit.ly/Virology_Twim90

> *You know my methods, Watson.*
>
> SIR ARTHUR CONAN DOYLE

Introduction

Viruses are unique: they are exceedingly small, often made up of nothing more than a nucleic acid molecule within a protein shell, yet when they enter cells, they parasitize the cellular machinery to produce thousands of progeny. This simplicity is misleading: viruses can infect all known life forms, and they comprise a variety of structures and genomes. Despite such variety, viruses are amenable to study because all viral propagation can be described in the context of three fundamental properties, as described in Chapter 1: all viral genomes are packaged inside particles that mediate their transmission from cell to cell; the viral genome contains the information for initiating and completing an infectious cycle; and all viruses can establish themselves in a host population to ensure virus survival.

The objective of research in virology is to understand how viruses enter individual cells, replicate, and assemble new infectious particles. These studies are usually carried out with cell cultures, because they are a much simpler and more homogeneous experimental system than animals. Cells can be infected in such a way as to ensure that a single replication cycle occurs synchronously in every infected cell, the **one-step growth curve**. Because all viral infections take place within a cell, a full understanding of viral life cycles also requires knowledge of cell biology and cellular architecture. These are the topics of this chapter: the cell surface (the site at which viruses enter and exit cells), methods for detecting viruses and viral growth, and one-step growth analysis.

The Infectious Cycle

The production of new infectious viruses can take place only within a cell (Fig. 2.1). Virologists divide the viral infectious cycle into discrete steps to facilitate their study, although in virus-infected cells no such artificial boundaries occur. The infectious cycle comprises attachment and entry of the particle, production of viral mRNA and its translation by host ribosomes, genome replication, and assembly and release of particles containing the genome. New virus particles produced during the infectious cycle may then infect other cells. The term **virus reproduction** is another name for the sum total of all events that occur during the infectious cycle.

There are events common to virus replication in animals and in cells in culture, but there are also many important differences. While viruses readily attach to cells in culture, in nature a virus particle must encounter a host, no mean feat for nanoparticles without any means of locomotion. After encountering a host, the virus particle must pass through physical host defenses, such as dead skin, mucous layers, and the extracellular matrix. Host defenses such as antibodies and immune cells, which exist to combat virus infections, are not found in cell cultures. Virus infection of cells in culture has been a valuable tool for understanding viral infectious cycles, but the differences compared with infection of a living animal must always be considered.

The Cell

Viruses require many different functions of the host cell (Fig. 2.2) for propagation. Examples include the machinery for translation of viral mRNAs, sources of energy, and enzymes for genome replication. The cellular transport apparatus brings

PRINCIPLES *The infectious cycle*

- Many distinct functions of the host cell are required to complete a viral life cycle.
- A productive infection requires target cells that are both susceptible (i.e., allow virus entry) and permissive (i.e., support virus reproduction).
- Viral nucleic acids must be shielded from harsh environmental conditions as extracellular particles, but be readily accessible for replication once inside the cell.
- To advance their study, viruses may be propagated in cells within a laboratory animal or in cell cultures, which include immortalized cells or primary cultures derived from the natural host or other animals.
- Plaque assays are the major way to determine the concentration of infectious virus particles in a sample, though alternative strategies exist for viruses that do not form plaques.
- While the goals of quantifying and characterizing virus particles remain fundamental for research in virology, the specific techniques used evolve rapidly, based on developments in detection, ease, cost, safety, utility in the field, and amenability to large-scale implementation.
- Viral nucleic acids can be detected and characterized by multiple methods, including direct sequencing of genomes and mRNAs, PCR, and microarrays.
- Relationships among viruses can be deduced from phylogenetic trees generated from protein or nucleic acid sequences.
- Viral reproduction is distinct from cellular or bacterial replication: rather than doubling with each cycle, each single cell cycle of viral reproduction is typically characterized by the release of many (often thousands) of progeny virions.
- The multiplicity of infection (MOI) is the number of infectious units added per cell in an experimental setting; the probability that any one target cell will become infected based on the MOI can be calculated from the Poisson distribution.
- Application of systems biology approaches to virology can implicate particular cellular pathways in viral reproduction and can reveal signatures of virus-induced lethality or immune protection.

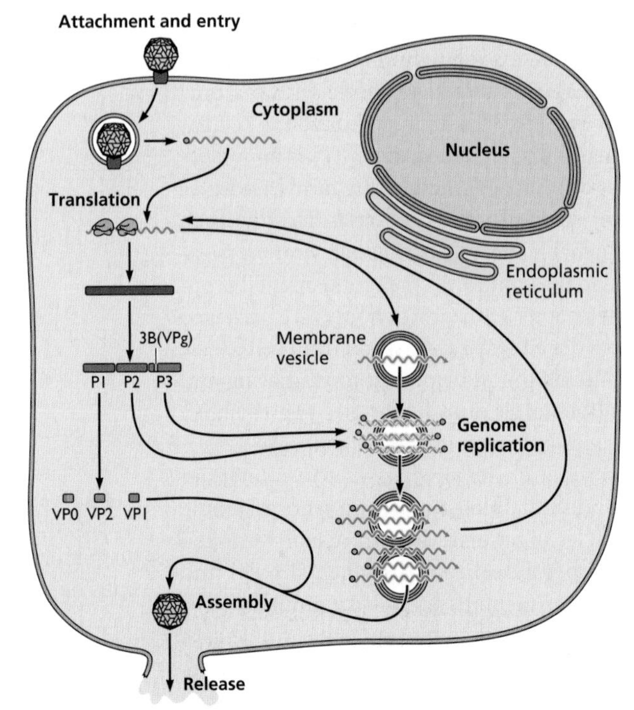

Figure 2.1 The viral infectious cycle. The infectious cycle of poliovirus is shown as an example, illustrating the steps common to all viral life cycles: attachment and entry, translation, genome replication, assembly, and release. See Appendix, Figures 21 and 22 for explanation of abbreviations.

Figure 2.2 The mammalian cell. Illustrated schematically are the nucleus, the major membrane-bound compartments of the cytoplasm, and components of the cytoskeleton that play important roles in virus reproduction. A small part of the cytoplasm is magnified, showing the crowded contents. The figure is not drawn to scale.

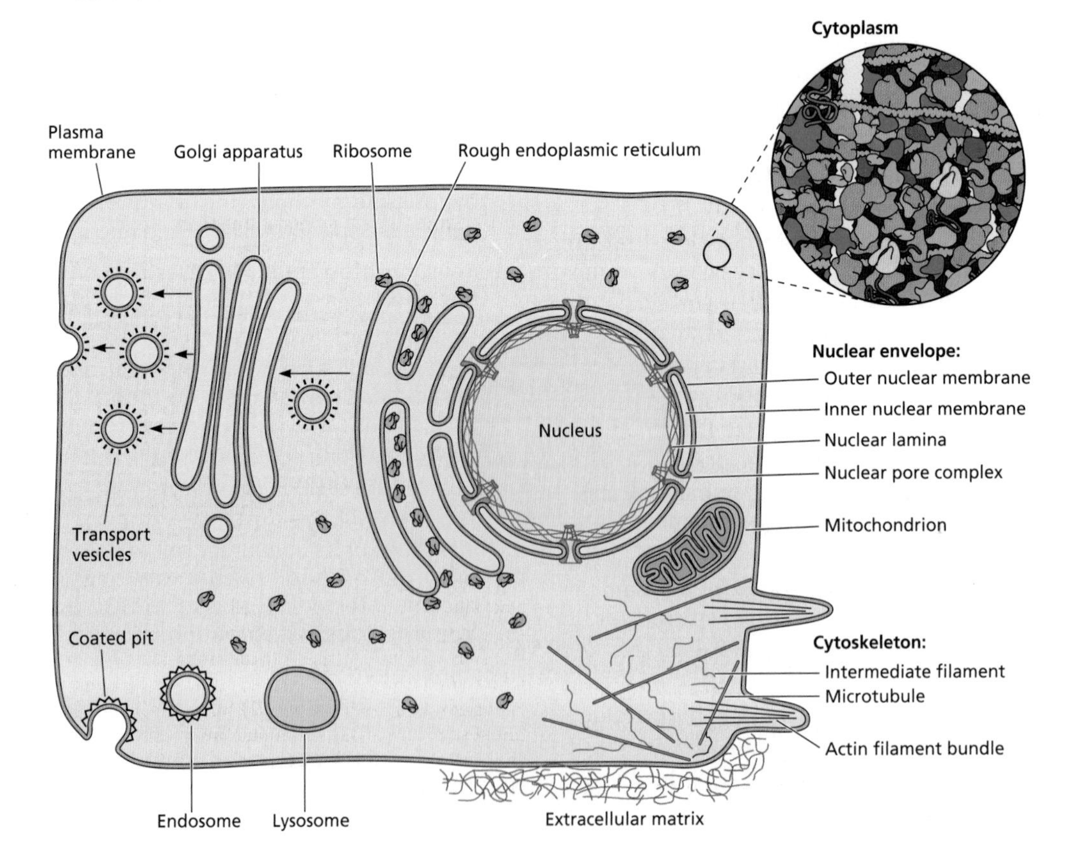

viral genomes to the correct cellular compartment and ensures that viral subunits reach locations where they may be assembled into virus particles. Subsequent chapters include a discussion of cellular functions that are important for individual steps in the viral replication cycle. In the following section, we consider the architecture of cell surfaces. The cell membrane merits this special focus because it is not only the portal of entry for all animal viruses, but also the site from which many viruses leave the cell.

The Architecture of Cell Surfaces

In animals, viral infections usually begin at the surfaces of the body that are exposed to the environment (Fig. 2.3). Epithelial cells cover these surfaces, and the region of these cells exposed to the environment is called the **apical surface**. Conversely, the **basolateral surfaces** of such cells are in contact with adjacent or underlying cells or tissues. These cells exhibit a differential (polar) distribution of proteins and lipids in the plasma membranes that creates the two distinct surface domains. As illustrated in Fig. 2.3, these cell layers differ in thickness and organization. Movement of macromolecules between the cells in the epithelium is prevented by **tight junctions**, which circumscribe the cells at the apical edges of their lateral membranes. Many viral infections are initiated upon entry into epithelial or endothelial cells (which line the interior surface of blood and lymphatic vessels) at their exposed apical surfaces, often by attaching to cell surface molecules specific for these domains. Viruses that both enter and are released at apical membranes can be transmitted laterally from cell to cell without ever traversing the epithelial or endothelial layers; they generally cause localized infections. In other cases, progeny virus particles are transported to the basolateral surface and released into the underlying cells and tissues, a process that facilitates viral spread to other sites of replication.

There are also more-specific pathways by which viruses reach susceptible cells. For example, some epithelial tissues contain M cells, specialized cells that overlie the collections of lymphoid cells in the gut known as Peyer's patches. M cells transport intestinal contents to Peyer's patches by a mechanism called **transcytosis**. Certain viruses, such as poliovirus and human immunodeficiency virus type 1, can be transported through them to gain access to underlying tissues. Such specialized pathways of invasion are considered in Volume II, Chapter 2. Below, we describe briefly the structures that surround cells and tissues, as well as the membrane components that are relevant to virus replication.

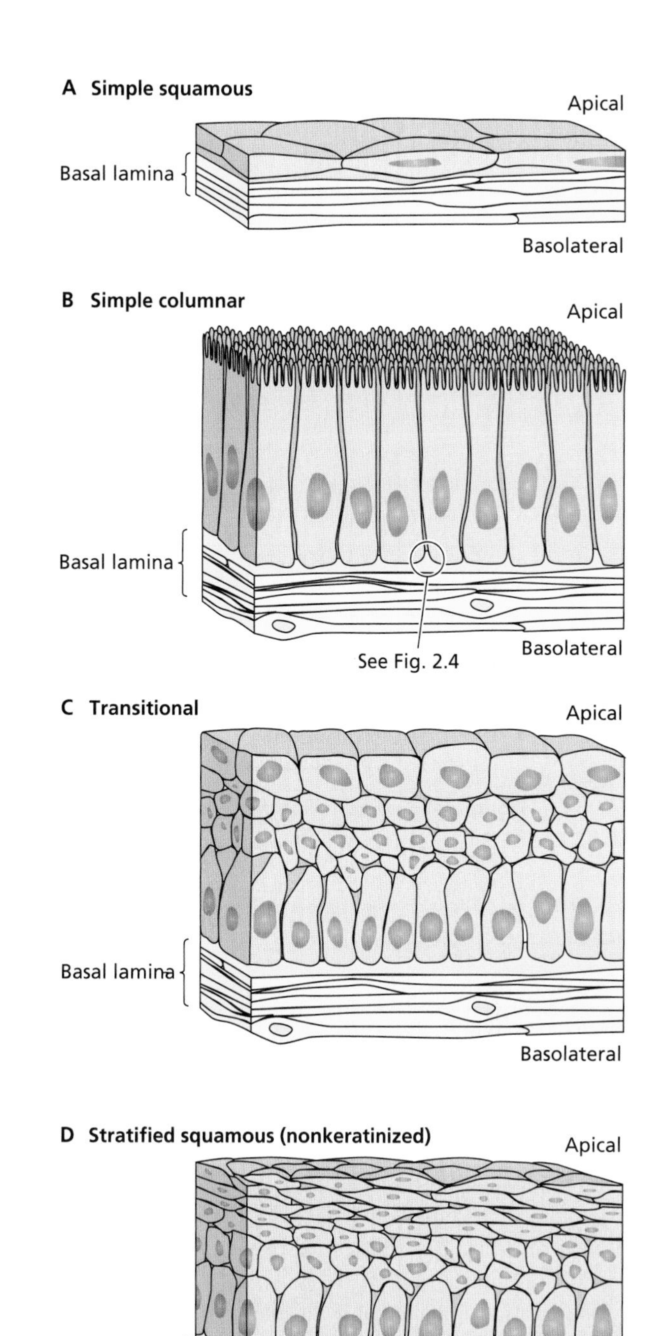

Figure 2.3 Major types of epithelia. (A) Simple squamous epithelium made up of thin cells such as those lining blood vessels and many body cavities. **(B)** Simple columnar epithelium found in the stomach, cervical tract, and small intestine. **(C)** Transitional epithelium, which lines cavities, such as the urinary bladder, that are subject to expansion and contraction. **(D)** Stratified, nonkeratinized epithelium lining surfaces such as the mouth and vagina. Adapted from H. Lodish et al., *Molecular Cell Biology*, 3rd ed. (W. H. Freeman & Co., New York, NY, 1995), with permission.

The Extracellular Matrix: Components and Biological Importance

Extracellular matrices, which hold the cells and tissues of the body together, are made up of two main classes of macromolecules (Fig. 2.4). The first comprises glycosaminoglycans (such as heparan sulfate and chondroitin sulfate), which are unbranched polysaccharides made of repeating disaccharides. Glycosaminoglycans are usually linked to proteins to

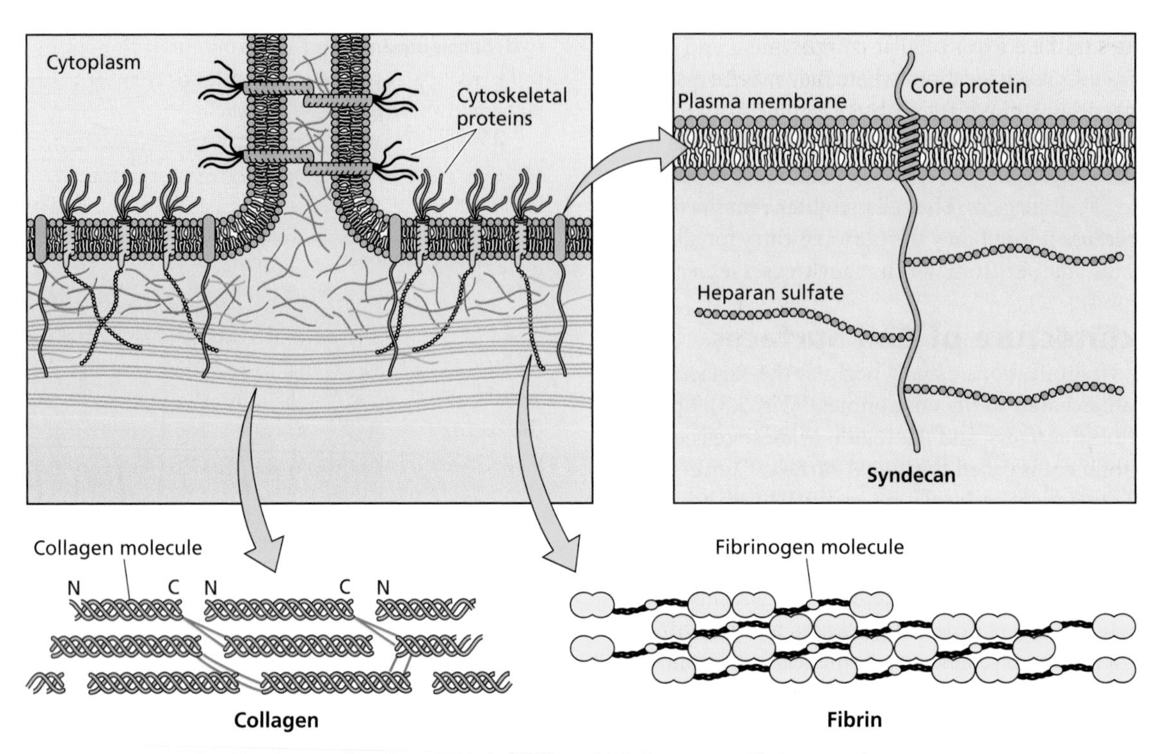

Figure 2.4 Cell adhesion molecules and components of the extracellular matrix. The diagram (expanded from Fig. 2.3B) illustrates some of the many cell surface components that contribute to cell-cell adhesion and attachment to the extracellular matrix. Adapted from G. M. Cooper, *The Cell: a Molecular Approach* (ASM Press, Washington, DC, and Sinauer Associates, Sunderland, MA, 1997).

form **proteoglycans**. The second class of macromolecules in the extracellular matrix consists of fibrous proteins with structural (**collagen** and **elastin**) or adhesive (**fibronectin** and **laminin**) functions. The proteoglycan molecules in the matrix form hydrated gels in which the fibrous proteins are embedded, providing strength and resilience. The gel provides resistance to compression and allows the diffusion of nutrients between blood and tissue cells. The extracellular matrix of each cell type is specialized for the particular function required, varying in thickness, strength, elasticity, and degree of adhesion.

Most organized groups of cells, like epithelial cells of the skin (Fig. 2.3 and 2.5), are bound tightly on their basal surface to a thin layer of extracellular matrix called the **basal lamina**. This matrix is linked to the basolateral membrane by specific receptor proteins called **integrins** (which are discussed in "Cell Membrane Proteins" below). Integrins are anchored to the intracellular structural network (the **cytoskeleton**) at the inner surface of the cell membrane. The basal lamina is attached to collagen and other material in the underlying loose connective tissue found in many organs of the body (Fig. 2.5). Capillaries, glands, and specialized cells are embedded in the connective tissue. Some viruses gain access to susceptible cells by attaching specifically to components of the extracellular matrix, including cell adhesion proteins or proteoglycans.

Figure 2.5 Cross section through skin. In this diagram of skin from a pig, the precursor epidermal cells rest on a thin layer of extracellular matrix called the basal lamina. Underneath is loose connective tissue consisting mostly of extracellular matrix. Fibroblasts in the connective tissue synthesize the connective tissue proteins, hyaluronan, and proteoglycans. Blood and lymph capillaries are also located in the loose connective tissue layer. Adapted from H. Lodish et al., *Molecular Cell Biology*, 3rd ed. (W. H. Freeman & Co., New York, NY, 1995), with permission.

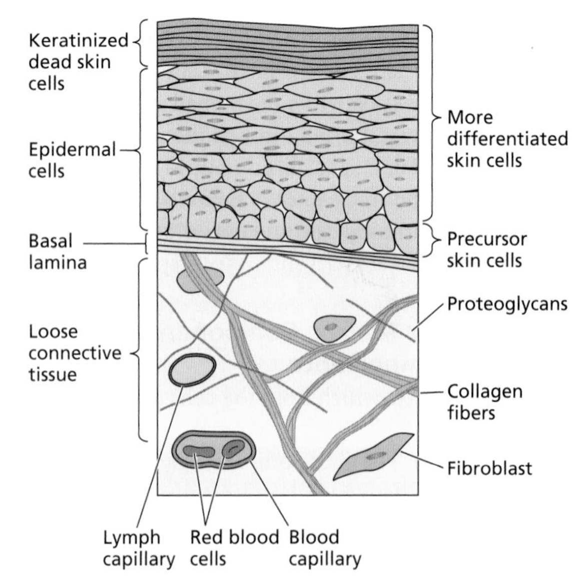

Properties of the Plasma Membrane

The plasma membrane of every mammalian cell type is composed of a similar phospholipid/glycolipid bilayer, but different sets of membrane proteins and lipids allow the cells of different tissues to carry out their specialized functions. The lipid bilayer is constructed from molecules that possess both hydrophilic and hydrophobic portions; they are known as **amphipathic** molecules, from the Greek word *amphi* (meaning "on both sides") (Fig. 2.6). They form a sheet-like structure in which polar head groups face the aqueous environment of the cell's cytoplasm (inner surface) or the surrounding environment (outer surface). The polar head groups of the inner and outer leaflets bear side chains with different lipid compositions. The fatty acyl side chains form a continuous hydrophobic interior about 3 nm thick. Hydrophobic interactions are the driving force for formation of the bilayer. However, hydrogen bonding and electrostatic interactions among the polar groups and water molecules or membrane proteins also stabilize the structure.

Thermal energy permits the phospholipid and glycolipid molecules comprising natural cell membranes to rotate freely around their long axes and diffuse laterally. If unencumbered, a lipid molecule can diffuse the length of an animal cell in only 20 s at 37°C. In most cases, phospholipids and glycolipids do not flip-flop from one side of a bilayer to the other, and the outer and inner leaflets of the bilayer remain separate. Similarly, membrane proteins not anchored to the extracellular matrix and/or the underlying structural network of the cell can diffuse rapidly, moving laterally like icebergs in this fluid bilayer. In this way, certain membrane proteins can form functional aggregates. Intracellular organelles such as the nucleus, endoplasmic reticulum, and lysosomes are also

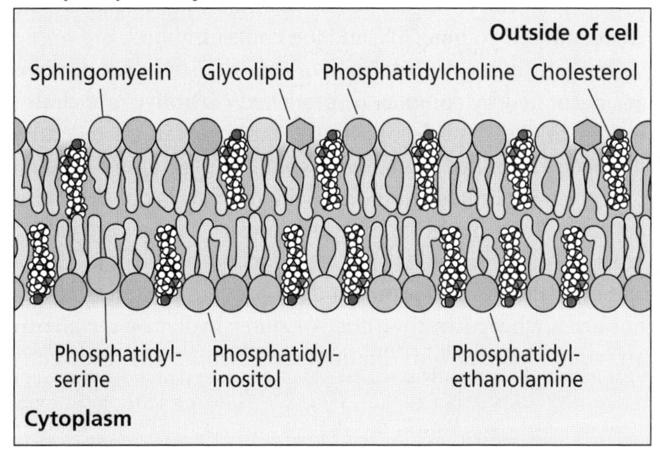

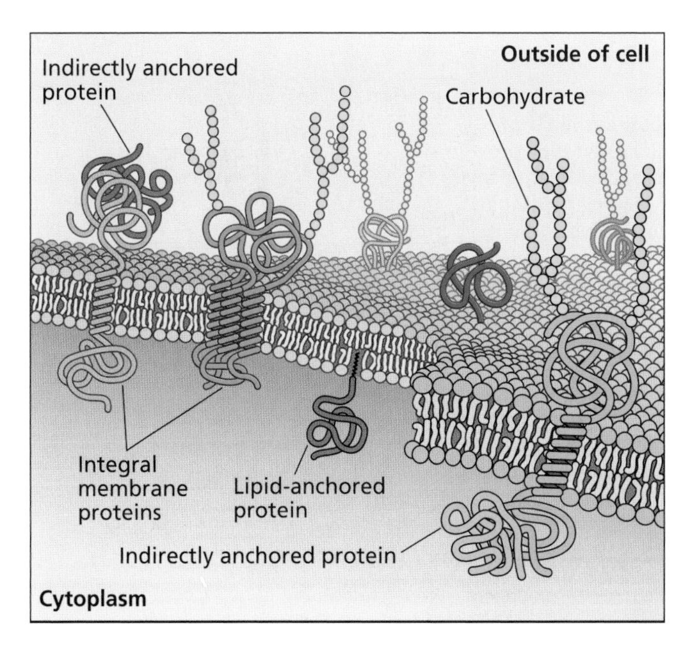

Figure 2.6 The plasma membrane. (Top) Lipid components of the plasma membrane. The membrane consists of two layers (leaflets) of phospholipid and glycolipid molecules. Their fatty acid tails converge to form the hydrophobic interior of the bilayer; the polar hydrophilic head groups (shown as balls) line both surfaces. (Bottom) Different types of membrane proteins are illustrated. Some integral membrane proteins are transmembrane proteins and are exposed on both sides of the bilayer. Adapted from G. M. Cooper, *The Cell: a Molecular Approach* (ASM Press, Washington, DC, and Sinauer Associates, Sunderland, MA, 1997), with permission.

enclosed in lipid bilayers, although their composition and physical properties differ.

The plasma membrane was once viewed as a uniform and fluid sea, in which lipid and protein components diffused randomly in the plane of the membrane. This simplistic model has been dispelled by experimental findings, which demonstrate that plasma membranes comprise **microdomains**, regions with distinct lipid and protein composition (Box 2.1). The **lipid raft** is one type of microdomain that is important for virus replication. Lipid rafts are enriched in cholesterol and saturated fatty acids, and are more densely packed and less fluid than other regions of the membrane. The assembly of a variety of viruses takes place at lipid rafts (see Chapter 13). Furthermore, the entry of some viruses requires lipid rafts. For example, particles of human immunodeficiency virus type 1 and Ebola virus enter cells at lipid rafts. Treatment of cells with compounds that disrupt these microdomains blocks entry. One explanation for this requirement might be that cell membrane proteins required for entry are concentrated in lipid rafts: receptors and coreceptors for human immunodeficiency virus are preferentially located in these domains.

Cell Membrane Proteins

Membrane proteins are classified into two broad categories, **integral membrane proteins** and **indirectly anchored proteins**, names that describe the nature of their interactions with the plasma membrane (Fig. 2.6).

Integral membrane proteins are embedded in the lipid bilayer, via one or more **membrane-spanning domains**, and contain portions that protrude out into the exterior and interior of the cell (Fig. 2.6). Many membrane-spanning domains consist of an α-helix typically 3.7 nm long. Such a domain includes 20 to 25 generally hydrophobic or uncharged residues embedded in the membrane, with the hydrophobic side chains protruding outward to interact with the fatty acyl chains of the lipid bilayer. The first and last residues are often positively charged amino acids (lysine or arginine) that can interact with the negatively charged polar head groups of the phospho- or glycolipids to stabilize the membrane-spanning domain. Some proteins with membrane-spanning domains enable the cell to respond to external signals. Such membrane proteins are designed to bind external ligands (e.g., hormones, cytokines, or membrane proteins on the same cell or on other cells) and to signal the occurrence of such interactions to molecules in the interior of the cell. Other proteins with multiple membrane-spanning domains (Fig. 2.6) form critical components of molecular pores or pumps, which mediate the internalization of required nutrients or the expulsion of undesirable material from the cell, or maintain homeostasis with respect to cell volume, pH, and ion concentration.

In many cases, the external portions of membrane proteins are decorated by complex or branched **carbohydrate chains** linked to the peptide backbone. Linkage can be to either nitrogen (**N linked**) in the side chain of asparagine residues or oxygen (**O linked**) in the side chains of serine or threonine residues. Such membrane **glycoproteins**, as they are called, quite frequently serve as viral receptors.

Some membrane proteins do not span the lipid bilayer, but are anchored in the inner or outer leaflet by covalently attached hydrocarbon chains (see Chapter 12). Indirectly

BOX 2.1

BACKGROUND

Plasma membrane microdomains

According to the Singer-Nicholson fluid mosaic model of membrane structure proposed in 1972, membranes are two-dimensional fluids with proteins inserted into the lipid bilayers (Fig. 2.6). Although the model accurately predicts the general organization of membranes, one of its conclusions has proven incorrect: that proteins and lipids are randomly distributed because they can freely rotate and laterally diffuse within the plane of the membrane. Beginning in the 1990s, the results of a series of experiments indicated that the movement of most proteins in the plasma membrane is partially restricted. In particular, these studies provided evidence for the existence of plasma membrane microdomains that are enriched in glycosphingolipids, cholesterol, glycosylphosphatidylinositol-anchored proteins, and certain intracellular signaling proteins. These microdomains, called lipid rafts, are defined experimentally as being resistant to extraction in cold 1% Triton X-100 and floating in the top half of a 5 to 30% sucrose density gradient. A major component of lipid rafts was found to be caveolin-1, a major coat protein of caveolae. These flask-shaped invaginations of the plasma membrane take up sphingolipids and integrins, as well as viruses, bacteria, and toxins. There are many noncaveolar lipid raft domains in the plasma membrane: detergent-insoluble microdomains are also present in cells that lack caveolin-1.

Sonnino S, Prinetti A. 2013. Membrane domains and the "lipid raft" concept. *Curr Med Chem* **20:**4–21.

anchored proteins are bound to the plasma membrane lipid bilayer by interacting either with integral membrane proteins or with the charged sugars of the glycolipids. Fibronectin, a protein in the extracellular matrix that binds to integrins (Fig. 2.4), is an example.

Entering Cells

Viral infection is initiated by a collision between the virus particle and the cell, a process that is governed by chance. Consequently, a higher concentration of virus particles increases the probability of infection. However, a virion may not infect every cell it encounters; it must first come in contact with the cells and tissues to which it can bind. Such cells are normally recognized by means of a specific interaction of a virus particle with a cell surface receptor. These molecules do not exist for the benefit of viruses: they all have cellular functions, and viruses have evolved to bind them for cell entry. Virus-receptor interactions can be either promiscuous or highly selective, depending on the virus and the distribution of the cell receptor. The presence of such receptors determines whether the cell will be **susceptible** to the virus. However, whether a cell is **permissive** for the reproduction of a particular virus depends on other, intracellular components found only in certain cell types. Cells must be both susceptible **and** permissive if an infection is to be successful.

Viruses have no intrinsic means of locomotion, but their small size facilitates diffusion driven by Brownian motion. Propagation of viruses is dependent on essentially random encounters with potential hosts and host cells. Features that increase the probability of favorable encounters are very important. In particular, viral propagation is critically dependent on the production of large numbers of progeny virus particles with surfaces composed of many copies of structures that enable the attachment of virus particles to susceptible cells.

Successful entry of a virus into a host cell requires traversal of the plasma membrane and in some cases the nuclear membrane. The virus particle must be partially or completely disassembled, and the nucleic acid must be targeted to the correct cellular compartment. These are not simple processes. Furthermore, virus particles or critical subassemblies are brought across such barriers by specific transport pathways. To survive in the extracellular environment, the viral genome must be encapsidated in a protective coat that shields viral nucleic acid from the variety of potentially harsh conditions that may be met during transit from one host cell or organism to another. For example, UV irradiation (from sunlight), extremes of pH (in the gastrointestinal tract), dehydration (in the air), and enzymatic attack (in body fluids) are all capable of damaging viral nucleic acids. However, once in the host cell, the protective structures must become sufficiently unstable to release the genome. Virus particles cannot be viewed only as passive vehicles: they must be able to undergo structural transformations that are important for attachment and entry into a new host cell and for the subsequent disassembly required for viral replication.

Making Viral RNA

Although the genomes of viruses come in a number of configurations, they share a common requirement: they must be efficiently copied into mRNAs for the synthesis of viral proteins and progeny genomes for assembly. The synthesis of RNA molecules in cells infected with RNA viruses is a unique process that has no counterpart in the cell. With the exception of retroviruses, all RNA viruses encode an RNA-dependent RNA polymerase to catalyze the synthesis of mRNAs and genomes. For the majority of DNA viruses and retroviruses, synthesis of viral mRNA is accomplished by RNA polymerase II, the enzyme that produces cellular mRNA. Much of our current understanding of the mechanisms of cellular transcription comes from study of the transcription of viral templates.

Making Viral Proteins

Because viruses are parasites of translation, all viral mRNAs must be translated by the host's cytoplasmic protein-synthesizing machinery (see Chapter 11). However, viral infection often results in modification of the host's translational apparatus so that viral mRNAs are translated selectively. The study of such modifications has revealed a great deal about mechanisms of protein synthesis. Analysis of viral translation has also revealed new strategies, such as internal ribosome binding and leaky scanning, that have been subsequently found to occur in uninfected cells.

Making Viral Genomes

Many viral genomes are copied by the cell's synthetic machinery in cooperation with viral proteins (see Chapters 6 through 9). The cell provides nucleotide substrates, energy, enzymes, and other proteins. Transport systems are required because the cell is compartmentalized: essential components might be found only in the nucleus, the cytoplasm, or cellular membranes. Study of the mechanisms of viral genome replication has established fundamental principles of cell biology and nucleic acid synthesis.

Forming Progeny Virus Particles

The various components of a virus particle—the nucleic acid genome, capsid protein(s), and in some cases envelope proteins—are often synthesized in different cellular compartments. Their trafficking through and among the

BOX 2.2

EXPERIMENTS

In vitro *assembly of tobacco mosaic virus*

The ability of the primary sequence of viral structural proteins to specify assembly is exemplified by the coat protein of tobacco mosaic virus. Heinz Fraenkel-Conrat and Robley Williams showed in 1955 that purified tobacco mosaic virus RNA and capsid protein assemble into infectious particles when mixed and incubated for 24 h. When examined by electron microscopy, the particles produced *in vitro* were identical to the rod-shaped virions produced from infected tobacco plants (Fig. 1.7B). Neither the purified viral RNA nor the capsid protein was infectious. These results indicate that the viral coat protein contains all the information needed for assembly of a virion. The spontaneous formation of tobacco mosaic virions *in vitro* from protein and RNA components is the paradigm for self-assembly in biology.

Fraenkel-Conrat H, Williams RC. 1955. Reconstitution of active tobacco mosaic virus from its inactive protein and nucleic acid components. *Proc Natl Acad Sci U S A* **40:**690–698.

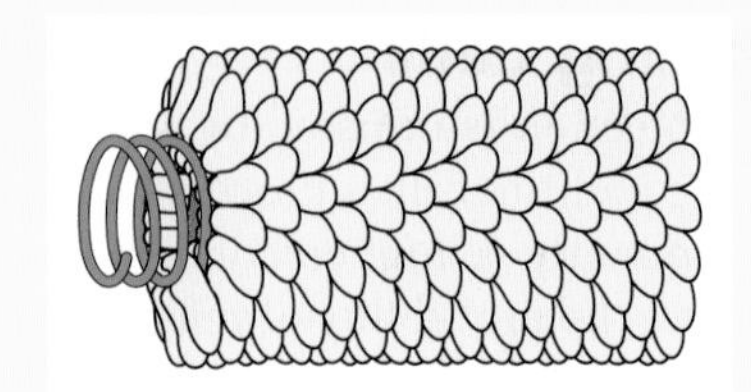

cell's compartments and organelles requires that they be equipped with the proper homing signals (see Chapter 12). Components of virus particles must be assembled at some central location, and the information for assembly must be preprogrammed in the component molecules (see Chapter 13). The primary sequences of viral structural proteins contain sufficient information to specify assembly; this property is exemplified by the remarkable *in vitro* assembly of tobacco mosaic virus from coat protein and RNA (Box 2.2). Successful virus reproduction depends on redirection of the host cell's metabolic and biosynthetic capabilities, signal transduction pathways, and trafficking systems (see Chapter 14).

Viral Pathogenesis

Viruses command our attention because of their association with animal and plant diseases. The process by which viruses cause disease is called **viral pathogenesis**. To study this process, we must investigate not only the relationships of viruses with the specific cells that they infect but also the consequences of infection for the host organism. The nature of viral disease depends on the effects of viral reproduction on host cells, the responses of the host's defense systems, and the ability of the virus to spread in and among hosts (Volume II, Chapters 1 to 5).

Overcoming Host Defenses

Organisms have many physical barriers to protect themselves from dangers in their environment such as invading parasites. In addition, vertebrates possess an effective immune system to defend against anything recognized as nonself or dangerous. Studies of the interactions between viruses and the immune system are particularly instructive, because of the many viral countermeasures that can frustrate this system. Elucidation of these measures continues to teach us much about the basis of immunity (Volume II, Chapters 2 to 4).

Cultivation of Viruses

Cell Culture

Types of Cell Culture

Although human and other animal cells were first cultured in the early 1900s, contamination with bacteria, mycoplasmas, and fungi initially made routine work with such cultures extremely difficult. For this reason, most viruses were grown in laboratory animals. In 1949, John Enders, Thomas Weller, and Frederick Robbins made the discovery that poliovirus could multiply in cultured cells. As noted in Chapter 1, this revolutionary finding, for which these three investigators were awarded the Nobel Prize in Physiology or Medicine in 1954, led the way to the propagation of many other viruses in cells in culture, the discovery of new viruses, and the development of viral vaccines such as those against poliomyelitis, measles, and rubella. The ability to infect cultured cells synchronously permitted studies of the biochemistry and molecular biology of viral replication. Large-scale growth and purification allowed studies of the composition of virus particles, leading to the solution of high-resolution, three-dimensional structures, as discussed in Chapter 4.

Cells in culture are still the most commonly used hosts for the propagation of animal viruses. To prepare a cell culture, tissues are dissociated into a single-cell suspension by mechanical disruption followed by treatment with proteolytic enzymes. The cells are then suspended in culture medium and placed in plastic flasks or covered plates. As the cells divide, they cover the plastic surface. Epithelial and fibroblastic cells attach to the plastic and form a **monolayer**, whereas blood cells such as lymphocytes settle, but do not adhere. The cells are grown in a chemically defined and buffered medium optimal for their growth. Commonly used cell lines double in number in 24 to 48 h in such media. Most cells retain viability after being frozen at low temperatures (−70 to −196°C).

A 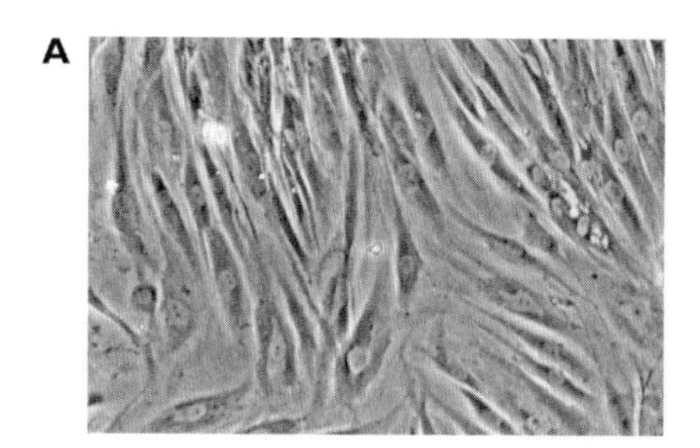

B 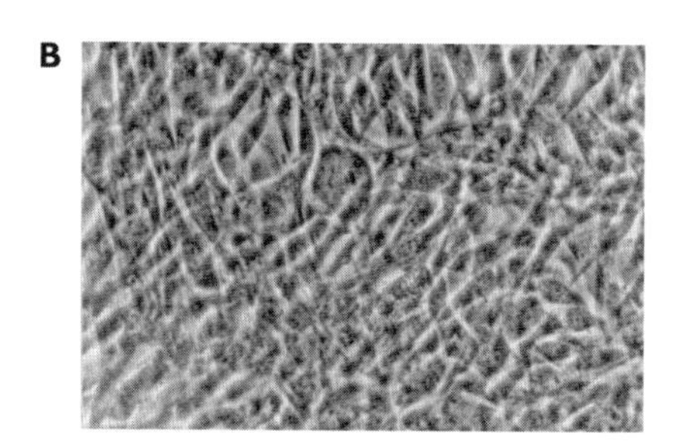

C 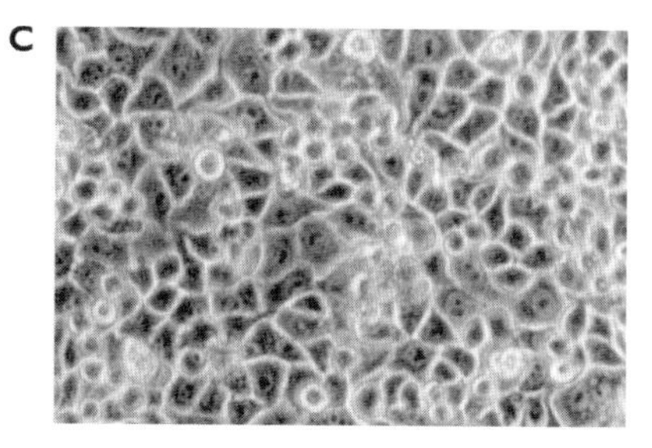

Figure 2.7 Different types of cell culture used in virology. Confluent cell monolayers photographed by low-power light microscopy. **(A)** Primary human foreskin fibroblasts; **(B)** established line of mouse fibroblasts (3T3); **(C)** continuous line of human epithelial cells (HeLa [Box 2.3]). The ability of transformed HeLa cells to overgrow one another is the result of a loss of contact inhibition. Courtesy of R. Gonzalez, Princeton University.

There are three main kinds of cell cultures (Fig. 2.7), each with advantages and disadvantages. **Primary cell cultures** are prepared from animal tissues as described above. They have a limited life span, usually no more than 5 to 20 cell divisions. Commonly used primary cell cultures are derived from monkey kidneys, human embryonic amnion and kidneys, human foreskins and respiratory epithelium, and chicken or mouse embryos. Such cells are used for experimental virology when the state of cell differentiation is important or when appropriate cell lines are not available. They are also used in vaccine production: for example, live attenuated poliovirus vaccine strains may be propagated in primary monkey kidney cells. Primary cell cultures were mandated for the growth of viruses to be used as human vaccines to avoid contamination of the product with potentially oncogenic DNA from continuous cell lines (see below). Some viral vaccines are now prepared in **diploid cell strains**, which consist of a homogeneous population of a single type and can divide up to 100 times before dying. Despite the numerous divisions, these cell strains retain the diploid chromosome number. The most widely used diploid cells are those established from human embryos, such as the WI-38 strain derived from human embryonic lung.

Continuous cell lines consist of a single cell type that can be propagated indefinitely in culture. These immortal lines are usually derived from tumor tissue or by treating a primary cell culture or a diploid strain with a mutagenic chemical or a tumor virus. Such cell lines often do not resemble the cell of origin; they are less differentiated (having lost the morphology and biochemical features that they possessed in the organ), are often abnormal in chromosome morphology and number (**aneuploid**), and can be tumorigenic (i.e., they produce tumors when inoculated into immunodeficient mice). Examples of commonly used continuous cell lines include those derived from human carcinomas (e.g., HeLa [Henrietta Lacks] cells; Box 2.3) and from mice (e.g., L and 3T3 cells). Continuous cell lines provide a uniform population of cells that can be infected synchronously for growth curve analysis (see "The One-Step Growth Cycle" below) or biochemical studies of virus replication.

In contrast to cells that grow in monolayers on plastic dishes, others can be maintained in **suspension cultures**, in which a spinning magnet continuously stirs the cells. The advantage of suspension culture is that a large number of cells can be grown in a relatively small volume. This culture method is well suited for applications that require large quantities of virus particles, such as X-ray crystallography or production of vectors.

Because viruses are obligatory intracellular parasites, they cannot reproduce outside a living cell. An exception comes from the demonstration in 1991 that infectious poliovirus could be produced in an extract of human cells incubated with viral RNA. Similar extracellular replication of the complete viral infectious cycle has not been achieved for any other virus. Consequently, most analysis of viral replication is done using cultured cells, embryonated eggs, or laboratory animals (Box 2.4).

Evidence of Viral Growth in Cultured Cells

Some viruses kill the cells in which they reproduce, and the infected cells may eventually detach from the cell culture plate. As more cells are infected, the changes become visible and are called **cytopathic effects** (Table 2.1). Many types of cytopathic effect can be seen with a simple light or phase-contrast microscope at low power, without fixing or staining the cells. These changes include the rounding up and detachment of cells from the culture dish, cell lysis, swelling of nuclei, and sometimes the formation of a group of fused cells called a syncytium (Fig. 2.8). Observation of other cytopathic effects requires high-power microscopy. These cytopathic effects include the development of intracellular masses of virus particles or unassembled viral components in the nucleus and/or cytoplasm (inclusion bodies), formation of crystalline arrays of viral proteins, membrane blebbing, duplication of membranes, and fragmentation of organelles. The time required for the development of cytopathology varies

BOX 2.3

BACKGROUND

The cells of Henrietta Lacks

The most widely used continuous cell line in virology, the HeLa cell line, was derived from Henrietta Lacks. In 1951, the 31-year-old mother of five visited a physician at Johns Hopkins Hospital in Baltimore and found that she had a malignant tumor of the cervix. A sample of the tumor was taken and given to George Gey, head of tissue culture research at Hopkins. Gey had been attempting for years, without success, to produce a line of human cells that would live indefinitely. When placed in culture, Henrietta Lacks' cells propagated as no other cells had before.

On the day in October that Henrietta Lacks died, Gey appeared on national television with a vial of her cells, which he called HeLa cells. He said, "It is possible that, from a fundamental study such as this, we will be able to learn a way by which cancer can be completely wiped out." Soon after, HeLa cells were used to propagate poliovirus, which was causing poliomyelitis throughout the world, and they played an important role in the development of poliovirus vaccines. Henrietta Lacks' HeLa cells started a medical revolution: not only was it possible to propagate many different viruses in these cells, but the work set a precedent for producing continuous cell lines from many human tissues. Sadly, the family of Henrietta Lacks did not learn about HeLa cells, or the revolution they started, until 24 years after her death. Her family members were shocked that cells from Henrietta lived in so many laboratories, and hurt that they had not been told that any cells had been taken from her.

The story of HeLa cells is an indictment of the lack of informed consent that pervaded medical research in the 1950s. Since then, biomedical ethics have changed greatly, and now there are strict regulations about clinical research: physicians may not take samples from patients without permission. Nevertheless, in early 2013, HeLa cells generated more controversy when a research group published the cells' genome sequence. The Lacks family objected to the publication, claiming that the information could reveal private medical information about surviving family members. As a result, the sequence was withdrawn from public databases. Months later, a second HeLa cell genome sequence was published, but this time the authors were bound by an agreement brokered by the National Institutes of Health, which required an application process for any individual wishing to view the sequence.

Adey A, Burton JN, Kitzman JO, Hiatt JB, Lewis AP, Martin BK, Qiu R, Lee C, Shendure J. 2013. The haplotype-resolved genome and epigenome of the aneuploid HeLa cancer cell line. *Nature* **500:**207–211.

Callaway E. 2013. Deal done over HeLa cell line. *Nature* **500:**132–133.

Callaway E. 2013. HeLa publication brews bioethical storm. *Nature* doi:10.1038/nature.2013.12689.

Skloot R. April 2000. Henrietta's dance. *Johns Hopkins Magazine.* http://pages.jh.edu/~jhumag/0400web/01.html.

Skloot R. 2011. *The Immortal Life of Henrietta Lacks.* Broadway Books, New York, NY.

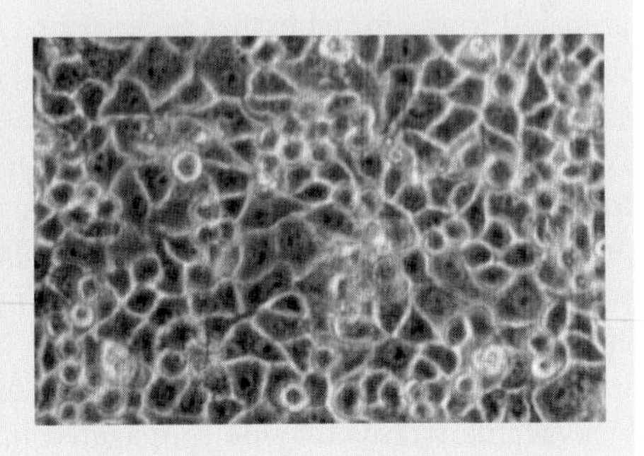

BOX 2.4

TERMINOLOGY

In vitro *and* in vivo

The terms "*in vitro*" and "*in vivo*" are common in the virology literature. *In vitro* means "in glass" and refers to experiments carried out in an artificial environment, such as a glass test tube. Unfortunately, the phrase "experiments performed *in vitro*" is used to designate not only work done in the cell-free environment of a test tube but also work done within cultured cells. The use of the phrase *in vitro* to describe living cultured cells leads to confusion and is inappropriate.

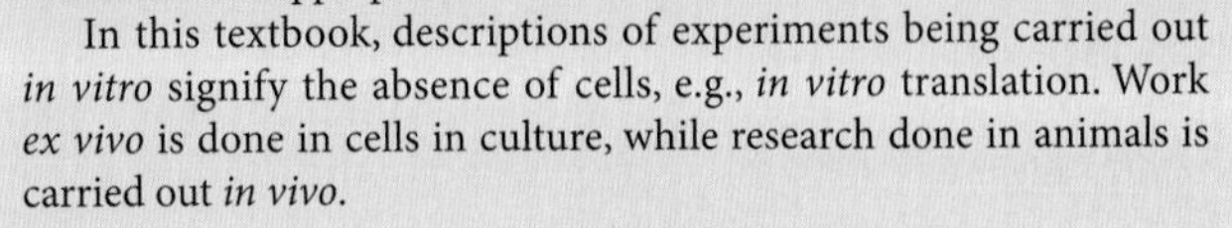

In this textbook, descriptions of experiments being carried out *in vitro* signify the absence of cells, e.g., *in vitro* translation. Work *ex vivo* is done in cells in culture, while research done in animals is carried out *in vivo.*

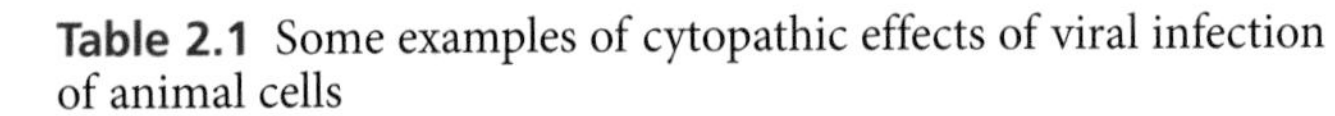

Table 2.1 Some examples of cytopathic effects of viral infection of animal cells

Cytopathic effect(s)	Virus(es)
Morphological alterations	
Nuclear shrinking (pyknosis), proliferation of membrane	Picornaviruses
Proliferation of nuclear membrane	Alphaviruses, herpesviruses
Vacuoles in cytoplasm	Polyomaviruses, papillomaviruses
Syncytium formation (cell fusion)	Paramyxoviruses, coronaviruses
Margination and breaking of chromosomes	Herpesviruses
Rounding up and detachment of cultured cells	Herpesviruses, rhabdoviruses, adenoviruses, picornaviruses
Inclusion bodies	
Virions in nucleus	Adenoviruses
Virions in cytoplasm (Negri bodies)	Rabies virus
"Factories" in cytoplasm (Guarnieri bodies)	Poxviruses
Clumps of ribosomes in virions	Arenaviruses
Clumps of chromatin in nucleus	Herpesviruses

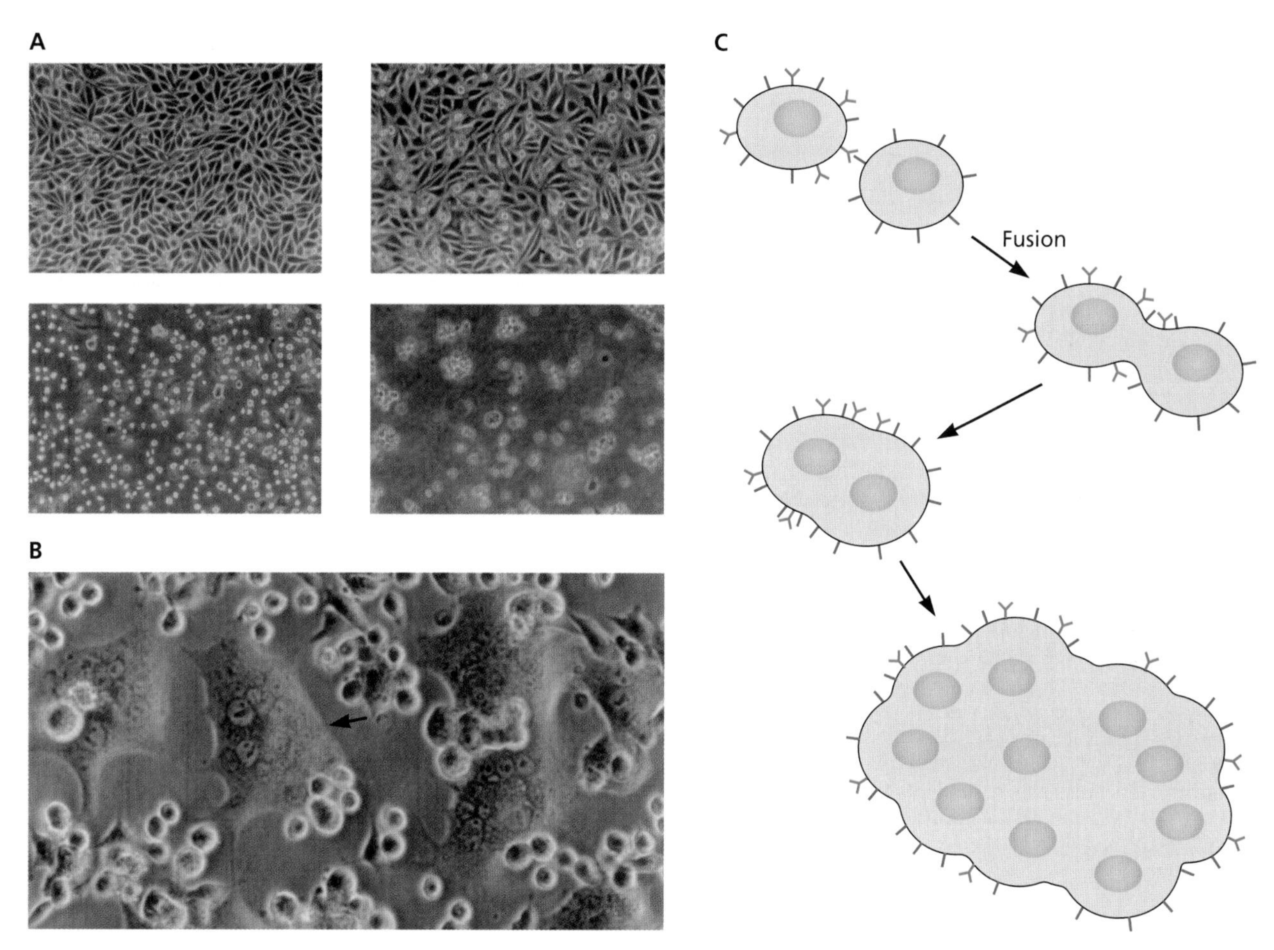

Figure 2.8 Development of cytopathic effect. (A) Cell rounding and lysis during poliovirus infection. (Upper left) Uninfected cells; (upper right) 5.5 h after infection; (lower left) 8 h after infection; (lower right) 24 h after infection. **(B)** Syncytium formation induced by murine leukemia virus. The field shows a mixture of individual small cells and syncytia, indicated by the arrow, which are large, multinucleate cells. Courtesy of R. Compans, Emory University School of Medicine. **(C)** Schematic illustration of syncytium formation. Viral glycoproteins on the surface of an infected cell bind receptors on a neighboring cell, causing fusion.

greatly among animal viruses. For example, depending on the size of the inoculum, enteroviruses and herpes simplex virus can cause cytopathic effects in 1 to 2 days and destroy the cell monolayer in 3 days. In contrast, cytomegalovirus, rubella virus, and some adenoviruses may not produce such effects for several weeks.

The development of characteristic cytopathic effects in infected cell cultures is frequently monitored in diagnostic virology during isolation of viruses from specimens obtained from infected patients or animals. However, cytopathic effect is also of value in the research laboratory: it can be used to monitor the progress of an infection, and is often one of the phenotypic traits by which mutant viruses are characterized.

Some viruses multiply in cells without causing obvious cytopathic effects. For example, many members of the families *Arenaviridae*, *Paramyxoviridae*, and *Retroviridae* do not cause obvious damage to cultured cells. Infection by such viruses must therefore be assayed using alternative methods, as described in "Assay of Viruses" below.

Embryonated Eggs

Before the advent of cell culture, many viruses were propagated in embryonated chicken eggs (Fig. 2.9). At 5 to 14 days after fertilization, a hole is drilled in the shell and virus is injected into the site appropriate for its replication. This method of virus propagation is now routine only for influenza virus. The robust yield of this virus from chicken eggs has led to their widespread use in research laboratories and for vaccine production.

Laboratory Animals

In the early 1900s, when viruses were first isolated, freezers and cell cultures were not available and it was necessary to maintain virus stocks by continuous passage from animal to

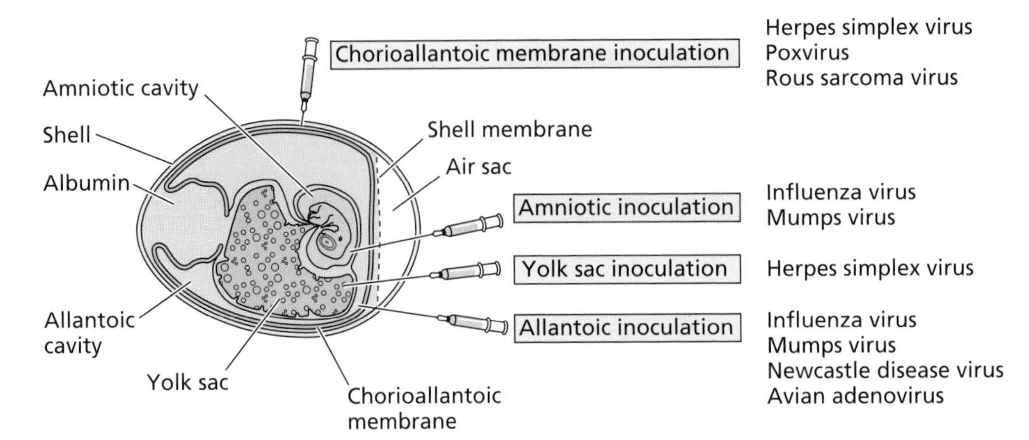

Figure 2.9 Growth of viruses in embryonated eggs. The cutaway view of an embryonated chicken egg shows the different routes by which viruses are inoculated into eggs and the different compartments in which viruses may grow. Adapted from F. Fenner et al., *The Biology of Animal Viruses* (Academic Press, New York, NY, 1974), with permission.

animal. This practice not only was inconvenient but also, as we shall see in Volume II, Chapter 8, led to the selection of viral mutants. For example, monkey-to-monkey intracerebral passage of poliovirus selected a mutant that could no longer infect chimpanzees by the oral route, the natural means of infection. Cell culture has largely supplanted the use of animals for propagating viruses, but some viruses cannot yet be grown in this way.

Experimental infection of laboratory animals has always been, and will continue to be, obligatory for studying the processes by which viruses cause disease. The use of monkeys in the study of poliomyelitis, the paralytic disease caused by poliovirus, led to an understanding of the basis of this disease and was instrumental in the development of a successful vaccine. Similarly, the development of vaccines against hepatitis B virus would not have been possible without experimental studies with chimpanzees. Understanding how the immune system or any complex organ reacts to a virus cannot be achieved without research on living animals. The development of viral vaccines, antiviral drugs, and diagnostic tests for veterinary medicine has also benefited from research on diseases in laboratory animals.

Assay of Viruses

There are two main types of assay for detecting viruses: biological and physical. Because viruses were first recognized by their infectivity, the earliest assays focused on this most sensitive and informative property. However, biological assays such as the plaque assay and end-point titration methods do not measure noninfectious particles. All such particles are included when measured by physical assays such as electron microscopy or by immunological methods. Knowledge of the number of noninfectious particles is useful for assessing the quality of a virus preparation.

Measurement of Infectious Units

One of the most important procedures in virology is measuring the **virus titer**, the concentration of a virus in a sample. This parameter is determined by inoculating serial dilutions of virus into host cell cultures, chicken embryos, or laboratory animals and monitoring for evidence of virus multiplication. The response may be quantitative (as in assays for plaques, fluorescent foci, infectious centers, or transformation) or all-or-none, in which the presence or absence of infection is measured (as in an end-point dilution assay).

Plaque Assay

In 1952, Renato Dulbecco modified the plaque assay developed to determine the titers of bacteriophage stocks for use in animal virology. The plaque assay was adopted rapidly for reliable determination of the titers of a wide variety of viruses. In this procedure, monolayers of cultured cells are incubated with a preparation of virus to allow adsorption to cells. After removal of the inoculum, the cells are covered with nutrient medium containing a supplement, most commonly agar, which forms a gel. When the original infected cells release new progeny particles, the gel restricts the spread of viruses to neighboring uninfected cells. As a result, each infectious particle produces a circular zone of infected cells, a **plaque**. If the infected cells are damaged, the plaque can be distinguished from the surrounding monolayer. In time, the plaque becomes large enough to be seen with the naked eye (Fig. 2.10). Only viruses that cause visible damage of cultured cells can be assayed in this way.

For the majority of animal viruses, there is a linear relationship between the number of infectious particles and the plaque count (Fig. 2.11). One infectious particle is therefore sufficient to initiate infection, and the virus is said to infect cells with **one-hit kinetics**. Some examples of **two-hit kinetics**,

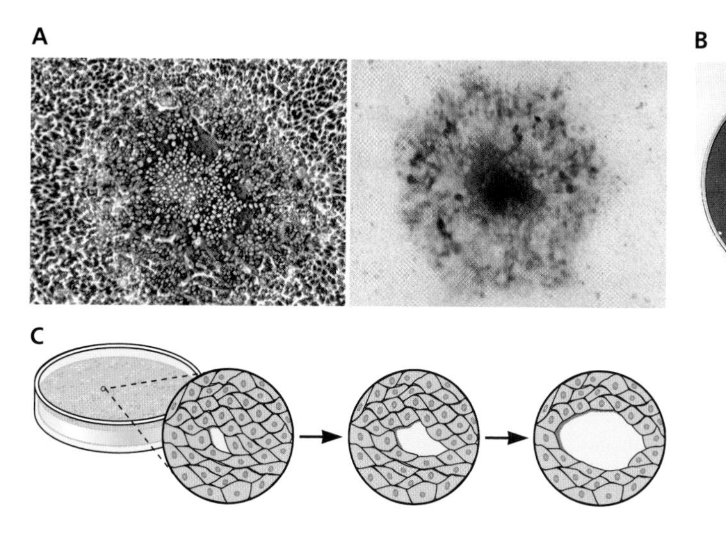

Figure 2.10 Plaques formed by different animal viruses. Plaque sizes reflect the reproductive cycle of a virus in a particular cell type. **(A)** Photomicrograph of a single plaque formed by pseudorabies virus in Georgia bovine kidney cells. (Left) Unstained cells. (Right) Cells stained with the chromogenic substrate X-Gal (5-bromo-4-chloro-3-indolyl-β-D-galactopyranoside), which is converted to a blue compound by the product of the *lacZ* gene carried by the virus. Courtesy of B. Banfield, Princeton University. **(B)** Plaques formed by poliovirus on human HeLa cells stained with crystal violet. **(C)** Illustration of the spread of virus from an initial infected cell to neighboring cells, resulting in a plaque.

in which two different types of virus particle must infect a cell to ensure replication, have been recognized. For example, the genomes of some (+) strand RNA viruses of plants consist of two RNA molecules that are encapsidated separately. Both RNAs are required for infectivity. The dose-response curve in plaque assays for these viruses is parabolic rather than linear (Fig. 2.11).

Figure 2.11 The dose-response curve of the plaque assay. The number of plaques produced by a virus with one-hit kinetics (red) or two-hit kinetics (blue) is plotted against the relative concentration of the virus. In two-hit kinetics, there are two classes of uninfected cells, those receiving one particle and those receiving none. The Poisson distribution can be used to determine the proportion of cells in each class: they are e^{-m} and me^{-m} (Box 2.10). Because one particle is not sufficient for infection, $P(0) = e^{-m}(1 + m)$. At a very low multiplicity of infection, this equation becomes $P(\mathrm{i}) = (1/2)m^2$ (where i = infection), which gives a parabolic curve. Adapted from B. D. Davis et al., *Microbiology* (J. B. Lippincott Co., Philadelphia, PA, 1980), with permission.

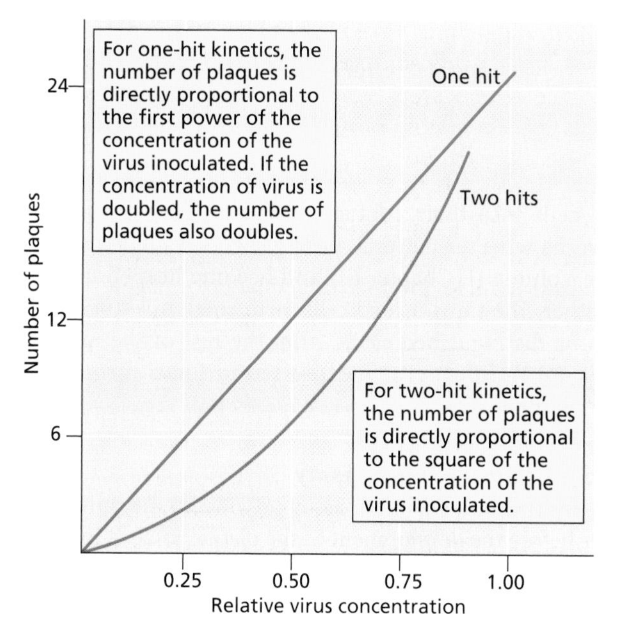

The titer of a virus stock can be calculated in **plaque-forming units (PFU) per milliliter** (Box 2.5). When one infectious virus particle initiates a plaque, the viral progeny within the plaque are clones, and virus stocks prepared from a single plaque are known as **plaque purified.** Plaque purification is employed widely in virology to establish clonal virus stocks. The tip of a small pipette is plunged into the overlay above the plaque, and the plug of agar containing the virus is recovered. The virus within the agar plug is eluted into buffer and used to prepare virus stocks. To ensure purity, this process is usually repeated at least one more time.

Fluorescent-Focus Assay

The fluorescent-focus assay, a modification of the plaque assay, can be done more rapidly and is useful in determining the titers of viruses that do not form plaques. The initial procedure is the same as in the plaque assay. However, after a period sufficient for adsorption and gene expression, cells are made permeable and incubated with an antibody raised against a viral protein. A second antibody, which recognizes the first, is then added. This second antibody is usually conjugated to a fluorescent molecule. The cells are then examined under a microscope at an appropriate wavelength. The titer of the virus stock is expressed in fluorescent-focus-forming units per milliliter. When the gene encoding a fluorescent protein is incorporated into the viral genome, foci may be detected without the use of antiviral antibodies.

Infectious-Centers Assay

Another modification of the plaque assay, the infectious-centers assay, is used to determine the fraction of cells in a culture that are infected with a virus. Monolayers of infected cells are suspended before progeny viruses are produced.

BOX 2.5

METHODS

Calculating virus titer from the plaque assay

To calculate the titer of a virus in plaque-forming units (PFU) per milliliter, 10-fold serial dilutions of a virus stock are prepared, and 0.1-ml aliquots are inoculated onto susceptible cell monolayers (see figure). After a suitable incubation period, the monolayers are stained and the plaques are counted. To minimize error in calculating the virus titer, only plates containing between 10 and 100 plaques are counted, depending on the area of the cell culture vessel. Plates with >100 plaques are generally not counted because the plaques may overlap, causing inaccuracies. According to statistical principles, when 100 plaques are counted, the sample titer varies by ±10%. For accuracy, each dilution is plated in duplicate or triplicate (not shown in the figure). In the example shown in the figure, 17 plaques are observed on the plate produced from the 10^{-6} dilution. Therefore, the 10^{-6} dilution tube contains 17 PFU per 0.1 ml, or 170 PFU per ml, and the titer of the virus stock is 170×10^6 or 1.7×10^8 PFU/ml.

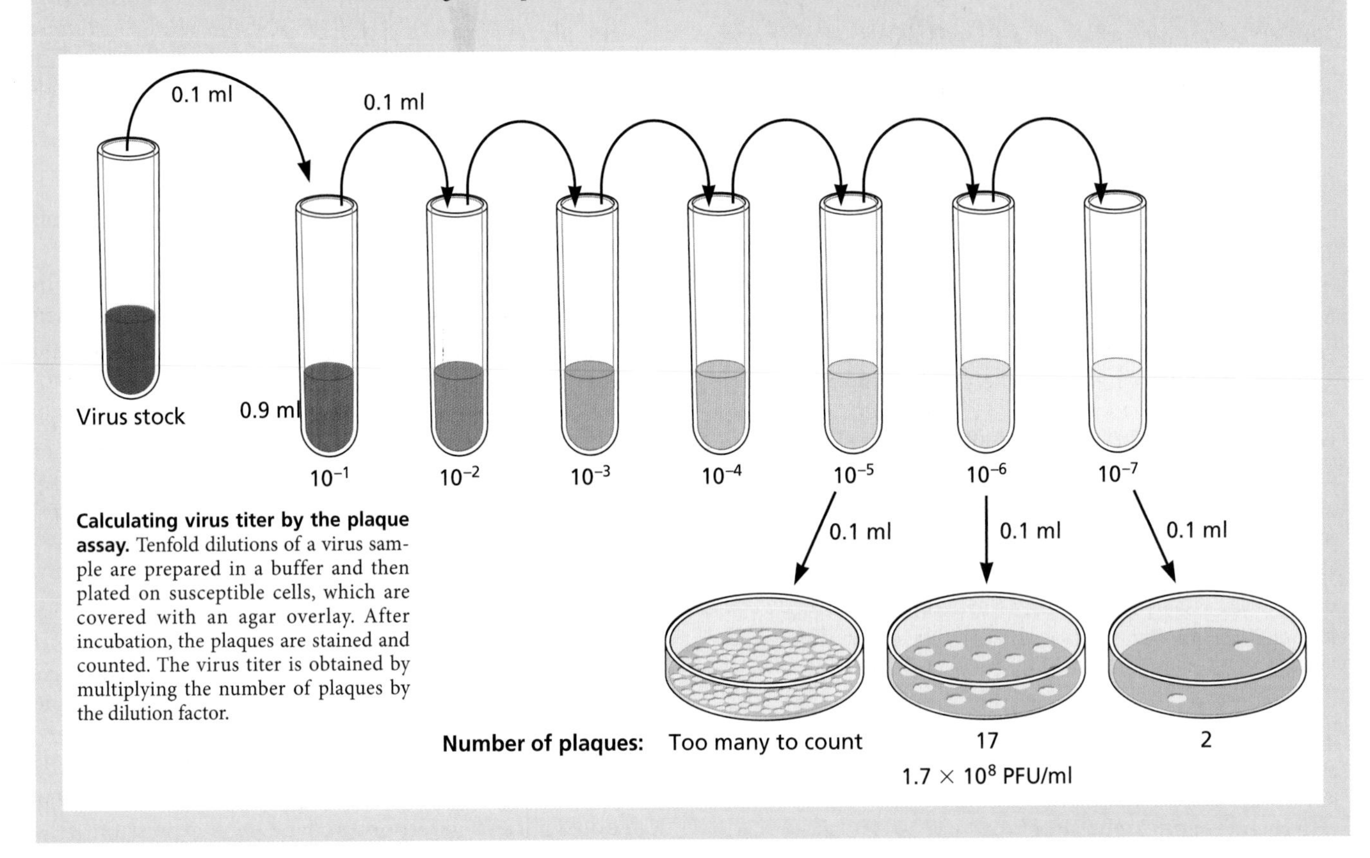

Calculating virus titer by the plaque assay. Tenfold dilutions of a virus sample are prepared in a buffer and then plated on susceptible cells, which are covered with an agar overlay. After incubation, the plaques are stained and counted. The virus titer is obtained by multiplying the number of plaques by the dilution factor.

Dilutions of a known number of infected cells are then plated on monolayers of susceptible cells, which are covered with an agar overlay. The number of plaques that form on the indicator cells is a measure of the number of cells infected in the original population. The fraction of infected cells can therefore be determined. A typical use of the infectious-centers assay is to measure the proportion of infected cells in persistently infected cultures.

Transformation Assay

The transformation assay is useful for determining the titers of some retroviruses that do not form plaques. For example, when Rous sarcoma virus transforms chicken embryo cells, the cells lose their contact inhibition (the property that governs whether cultured cells grow as a single monolayer [see Volume II, Chapter 6]) and become heaped up on one another. The transformed cells form small piles, or **foci**, that can be distinguished easily from the rest of the monolayer (Fig. 2.12). Infectivity is expressed in focus-forming units per milliliter.

End-Point Dilution Assay

The end-point dilution assay provided a measure of virus titer before the development of the plaque assay. It is still used for measuring the titers of certain viruses that do not form plaques or for determining the virulence of a virus in animals.

A

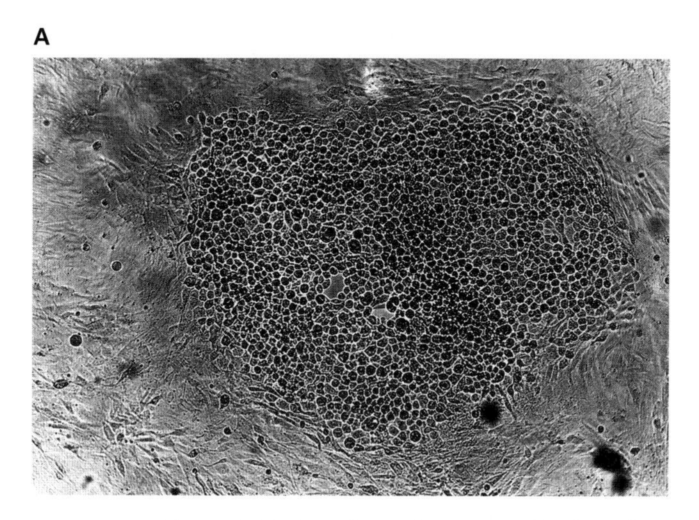

B

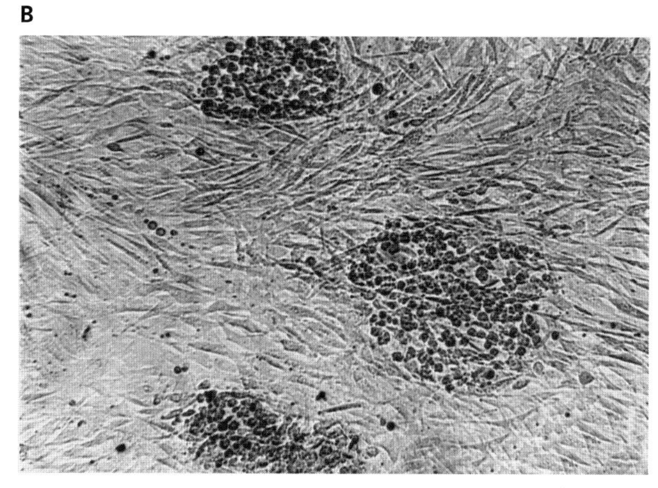

Figure 2.12 Transformation assay. Chicken cells transformed by two different strains of Rous sarcoma virus are shown. Loss of contact inhibition causes cells to pile up rather than grow as a monolayer. One focus is seen in panel **A**, and three foci are seen in panel **B** at the same magnification. Courtesy of H. Hanafusa, Osaka Bioscience Institute.

Serial dilutions of a virus stock are inoculated into replicate test units (typically 8 to 10), which can be cell cultures, eggs, or animals. The number of test units that have become infected is then determined for each virus dilution. When cell culture is used, infection may be determined by the development of cytopathic effect; in eggs or animals, infection is gauged by death or disease. An example of an end-point dilution assay using cell cultures is shown in Box 2.6. At high dilutions, none of the cell cultures are infected because no infectious particles are delivered to the cells; at low dilutions, every culture is infected. The **end point** is the dilution of virus that affects 50% of the test units. This number can be calculated from the data and expressed as 50% infectious dose (ID_{50}) per milliliter. The first preparation illustrated in Box 2.6 contains 10^5 ID_{50} per ml. This type of assay is suitable for high-throughput applications.

When the end-point dilution assay is used to assess the virulence of a virus or its capacity to cause disease (Volume II, Chapter 1), the result of the assay can be expressed in terms of 50% lethal dose (LD_{50}) per milliliter or 50% paralytic dose (PD_{50}) per milliliter, end points of death and paralysis, respectively. If the virus titer can be determined separately by plaque assay, the 50% end point determined in an animal host can be related to this parameter. In this way, the effects of the route of inoculation or specific mutations on viral virulence can be quantified.

Efficiency of Plating

Efficiency of plating is defined as the virus titer (in PFU/ml) divided by the number of virus particles in the sample. The **particle-to-plaque-forming-unit (PFU) ratio**, a term more commonly used today, is the inverse value (Table 2.2). For many bacteriophages, the particle-to-PFU ratio approaches 1, the lowest value that can be obtained. However, for animal viruses, this value can be much higher, ranging from 1 to 10,000. These high values have complicated the study of animal viruses. For example, when the particle-to-PFU ratio is high, it is never certain whether properties measured biochemically are in fact those of the infectious particle or those of the noninfectious component.

Although the linear nature of the dose-response curve indicates that a single particle is capable of initiating an infection (one-hit kinetics) (Fig. 2.11), the high particle-to-PFU ratio for many viruses demonstrates that not all virus particles are successful. High values are sometimes caused by the presence of noninfectious particles with genomes that harbor lethal mutations or that have been damaged during growth or purification. An alternative explanation is that although all viruses in a preparation are in fact capable of initiating infection, not all of them succeed because of the complexity of the infectious cycle. Failure at any one step in the cycle prevents completion. A high particle-to-PFU ratio indicates not that most particles are defective but, rather, that they failed to complete the infection.

Measurement of Virus Particles and Their Components

Although the numbers of virus particles and infectious units are often not equal, assays for particle number are frequently used to approximate the number of infectious particles present in a sample. For example, the concentration of viral DNA or protein can be used to estimate the particle number, assuming that the ratio of infectious units to physical particles is constant. Biochemical or physical assays are usually more

BOX 2.6

METHODS

End-point dilution assays

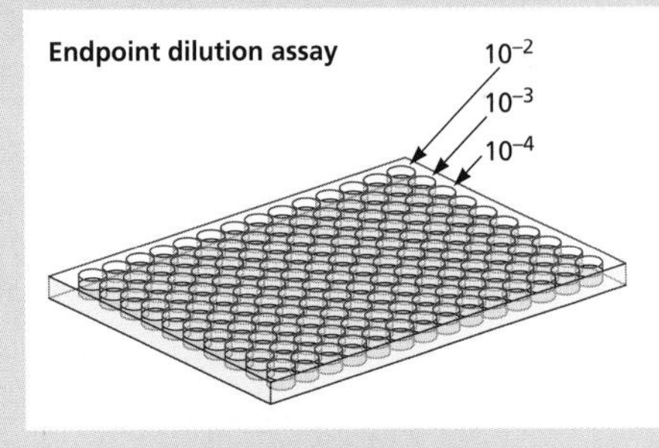

Virus dilution	Cytopathic effect									
10^{-2}	+	+	+	+	+	+	+	+	+	+
10^{-3}	+	+	+	+	+	+	+	+	+	+
10^{-4}	+	+	−	+	+	+	+	+	+	+
10^{-5}	−	+	+	−	+	−	−	+	−	+
10^{-6}	−	−	−	−	−	−	+	−	−	−
10^{-7}	−	−	−	−	−	−	−	−	−	−

End-point dilution assays are usually carried out in multiwell plastic plates (see the figure). In the example shown in the first table, 10 monolayer cell cultures were infected with each virus dilution. After the incubation period, plates that displayed cytopathic effect were scored +. Fifty percent of the cell cultures displayed cytopathic effect at the 10^{-5} dilution, and therefore the virus stock contains 10^5 $TCID_{50}$ units.

In most cases, the 50% end point does not fall on a dilution tested as shown in the example; for this reason, various statistical procedures have been developed to calculate the end point of the titration. In one popular method, the dilution containing the ID_{50} is identified by interpolation between the dilutions on either side of this value. The assumption is made that the location of the 50% end point varies linearly with the log of the dilution. Because the number of test units used at each dilution is usually small, the accuracy of this method is relatively low. For example, if six test units are used at each 10-fold dilution, differences in virus titer of only 50-fold or more can be detected reliably. The method is illustrated in the second example, in which the lethality of poliovirus in mice is the end point. Eight mice were inoculated per dilution. In the method of Reed and Muench, the results are pooled, as shown in the table, which equalizes chance variations (another way to achieve the same result would be to utilize greater numbers of animals at each dilution). The interpolated value of the 50% end point, which in this case falls between the 5th and 6th dilutions, is calculated to be $10^{-6.5}$. The virus sample therefore contains $10^{6.5}$ LD_{50}s. The LD_{50} may also be calculated as the concentration of the stock virus in PFU per milliliter (1×10^9) times the 50% end-point titer. In the example shown, the LD_{50} is 3×10^2 PFU.

Reed LJ, Muench H. 1938. A simple method of estimating fifty per cent endpoints. *Am J Hyg* **27:**493–497.

Dilution	Alive	Dead	Total alive	Total dead	Mortality ratio	Mortality (%)
10^{-2}	0	8	0	40	0/40	100
10^{-3}	0	8	0	32	0/32	100
10^{-4}	1	7	1	24	1/25	96
10^{-5}	0	8	1	17	1/18	94
10^{-6}	2	6	3	9	3/12	75
10^{-7}	5	3	8	3	8/11	27

Table 2.2 Particle-to-PFU ratios of some animal viruses

Virus	Particle/PFU ratio
Papillomaviridae	
Papillomavirus	10,000
Picornaviridae	
Poliovirus	30–1,000
Herpesviridae	
Herpes simplex virus	50–200
Polyomaviridae	
Polyomavirus	38–50
Simian virus 40	100–200
Adenoviridae	20–100
Poxviridae	1–100
Orthomyxoviridae	
Influenza virus	20–50
Reoviridae	
Reovirus	10
Alphaviridae	
Semliki Forest virus	1–2

rapid and easier to carry out than assays for infectivity, which may be slow, cumbersome, or not possible. Assays for subviral components also provide information on particle number if the stoichiometry of these components in the virus particle is known.

Imaging Particles

Electron microscopy. With few exceptions, virus particles are too small to be observed directly by light microscopy. However, they can be seen readily in the electron microscope. If a sample contains only one type of virus, the particle count can be determined. First, a virus preparation is mixed with a known concentration of latex beads. The numbers of virus particles and beads are then counted, allowing the concentration of the virus particles in the sample to be determined by comparison.

Live-cell imaging of single fluorescent virions. The discovery of green fluorescent protein revolutionized the study of the cell biology of virus infection. This protein, isolated from the jellyfish *Aequorea victoria*, is a convenient reporter for monitoring gene expression, because it is directly visible in living cells without the need for fixation, substrates, or coenzymes. Similar proteins isolated from different organisms, which emit light of different wavelengths, are also widely used in virology. The use of fluorescent proteins has allowed visualization of single virus particles in living cells. The coding sequence for the fluorescent protein is inserted into the viral genome, often fused to the coding region of a virion protein. The fusion protein is incorporated into the viral particle, which is visible in cells by fluorescence microscopy (Fig. 2.13). Using this approach, entry, uncoating, replication, assembly, and egress of single particles can all theoretically be observed in living cells.

Hemagglutination

Members of the *Adenoviridae*, *Orthomyxoviridae*, and *Paramyxoviridae*, among others, contain proteins that can bind to erythrocytes (red blood cells); these viruses can link multiple cells, resulting in formation of a lattice. This property is called **hemagglutination**. For example, influenza viruses contain an envelope glycoprotein called hemagglutinin, which binds to *N*-acetylneuraminic acid-containing glycoproteins on erythrocytes. In practice, 2-fold serial dilutions of the virus stock are prepared, mixed with a known quantity of red blood cells, and added to small wells in a plastic tray (Fig. 2.14). Unadsorbed red blood cells tumble to the bottom of the well and form a sharp dot or button. In contrast, agglutinated red blood cells form a diffuse lattice that coats the well. Because the assay is rapid (30 min), it is often used as a quick indicator of the relative quantities of virus particles. However, it is not sufficiently sensitive to detect small numbers of particles.

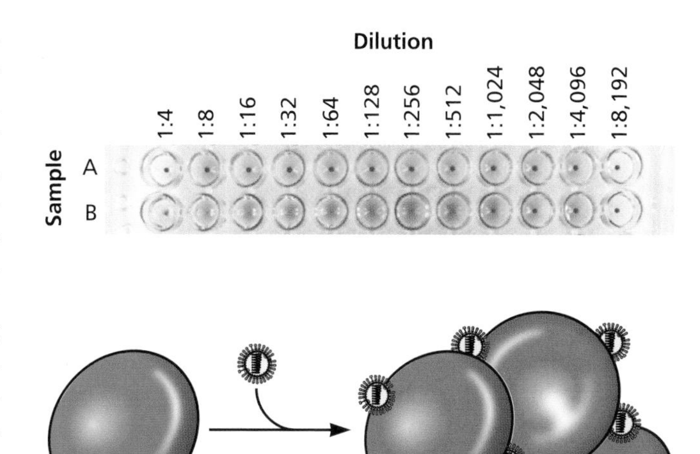

Figure 2.14 Hemagglutination assay. (Top) Samples of different influenza viruses were diluted, and a portion of each dilution was mixed with a suspension of chicken red blood cells and added to the wells. After 30 min at 4°C, the wells were photographed. Sample A does not contain virus. Sample B causes hemagglutination until a dilution of 1:512 and therefore has a hemagglutination titer of 512. Elution of the virus from red blood cells at the 1:4 dilution is caused by neuraminidase in the virus particle. This enzyme cleaves *N*-acetylneuraminic acid from glycoprotein receptors and elutes bound viruses from red blood cells. (Bottom) Schematic illustration of hemagglutination of red blood cells by influenza virus. Top, Courtesy of C. Basler and P. Palese, Mount Sinai School of Medicine of the City University of New York.

Figure 2.13 Live-cell imaging of single virus particles by fluorescence. Single-virus-particle imaging with green fluorescent protein illustrates microtubule-dependent movement of human immunodeficiency virus type 1 particles in cells. Rhodamine-tubulin was injected into cells to label microtubules (red). The cells were infected with virus particles that contain a fusion of green fluorescent protein with Vpr. Virus particles can be seen as green dots. Bar, 5 μm. Courtesy of David McDonald, University of Illinois.

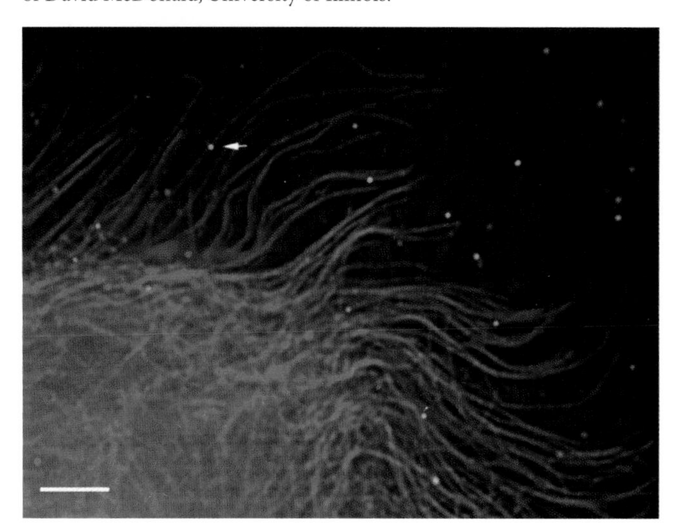

Measurement of Viral Enzyme Activity

Some animal virus particles contain nucleic acid polymerases, which can be assayed by mixing permeabilized particles with radioactively labeled precursors and measuring the incorporation of radioactivity into nucleic acid. This type of assay is used most frequently for retroviruses, many of which neither transform cells nor form plaques. The reverse transcriptase incorporated into the virus particle is assayed by mixing cell culture supernatants with a mild detergent (to permeabilize the viral envelope), an RNA template and primer, and a radioactive nucleoside triphosphate. If reverse transcriptase is present, a radioactive product will be produced by priming on the template. This product can be detected by precipitation or bound to a filter and quantified.

Because enzymatic activity is proportional to particle number, this assay allows rapid tracking of virus production in the course of an infection. Many of these assays have been modified to permit the use of safer, nonradioactive substrates. For example, when nucleoside triphosphates conjugated to biotin are used, the product can be detected with streptavidin (which binds biotin) conjugated to a fluorochrome. Alternatively, the reaction products may be quantified by quantitative real-time PCR (see "Detection of Viral Nucleic Acids" below).

Serological Methods

The specificity of the antibody-antigen reaction has been used to design a variety of assays for viral proteins and antiviral antibodies. These techniques, such as immunostaining, immunoprecipitation, immunoblotting, and the enzyme-linked immunosorbent assay, are by no means limited to virology: all these approaches have been used extensively to study the structures and functions of cellular proteins.

Virus neutralization. When a virus preparation is inoculated into an animal, an array of antibodies is produced. These antibodies can bind to virus particles, but not all of them can block infectivity (**neutralize**), as discussed in Volume II, Chapter 4. Virus neutralization assays are usually conducted by mixing dilutions of antibodies with virus; incubating them; and assaying for remaining infectivity in cultured cells, eggs, or animals. The end point is defined as the highest dilution of antibody that inhibits the development of cytopathic effect in cells or virus replication in eggs or animals.

Some neutralizing antibodies define **type-specific antigens** on the virus particle. For example, the three **serotypes** of poliovirus are distinguished on the basis of neutralization tests: type 1 poliovirus is neutralized by antibodies to type 1 virus but not by antibodies to type 2 or type 3 poliovirus, and so forth. The results of neutralization tests were once used for virus classification, a process now accomplished largely by comparing viral genome sequences. Nevertheless, the detection of antiviral antibodies in animal sera is still extremely important for identifying infected hosts. These antibodies may also be used to map the three-dimensional structure of neutralization antigenic sites on the virus particle (Box 2.7).

Hemagglutination inhibition. Antibodies against viral proteins with hemagglutination activity can block the ability of virus to bind red blood cells. In this assay, dilutions of antibodies are incubated with virus, and erythrocytes are added as outlined above. After incubation, the hemagglutination inhibition titer is read as the highest dilution of antibody that inhibits hemagglutination. This test is sensitive, simple, inexpensive, and rapid: it is the method of choice for assaying antibodies to any virus that causes hemagglutination. It can be used to detect antibodies to viral hemagglutinin in animal and human sera or to identify the origin of the hemagglutinin of influenza viruses produced in cells coinfected with two parent

BOX 2.7

DISCUSSION

Neutralization antigenic sites

Knowledge of the antigenic structure of a virus is useful in understanding the immune response to these agents and in designing new vaccination strategies. The use of **monoclonal antibodies** (antibodies of a single specificity made by a clone of antibody-producing cells) in neutralization assays permits mapping of antigenic sites on a virus particle, or of the amino acid sequences that are recognized by neutralizing antibodies. Each monoclonal antibody binds specifically to a short amino acid sequence (8 to 12 residues) that fits into the antibody-combining site. This amino acid sequence, which may be linear or nonlinear, is known as an **epitope**. In contrast, **polyclonal antibodies** comprise the repertoire produced in an animal against the many epitopes of an antigen. Antigenic sites may be identified by cross-linking the monoclonal antibody to the virus and determining which protein is the target of the antibody.

Epitope mapping may also be performed by assessing the abilities of monoclonal antibodies to bind synthetic peptides representing viral protein sequences. When the monoclonal antibody recognizes a linear epitope, it may react with the protein in Western blot analysis, facilitating direct identification of the viral protein harboring the antigenic site. The most elegant understanding of antigenic structures has come from the isolation and study of variant viruses that are resistant to neutralization with specific monoclonal antibodies (called **monoclonal antibody-resistant variants**). By identifying the amino acid change responsible for this phenotype, the antibody-binding site can be located and, together with three-dimensional structural information, can provide detailed information on the nature of antigenic sites that are recognized by neutralizing antibodies (see the figure).

Locations of neutralization antigenic sites on the capsid of poliovirus type 1. Amino acids that change in viral mutants selected for resistance to neutralization by monoclonal antibodies are shown in white on a model of the viral capsid. These amino acids are in VP1 (blue), VP2 (green), and VP3 (red) on the surface of the virus particle. Figure courtesy of Jason Roberts, Victorian Infectious Diseases Reference Laboratory, Doherty Institute, Melbourne, Australia.

viruses. For example, hemagglutination inhibition assays were used to identify individuals who had been infected with the newly discovered avian influenza A (H7N9) virus in China during the 2013 outbreak.

Immunostaining. Antibodies can be used to visualize viral proteins in infected cells or tissues. In direct immunostaining, an antibody that recognizes a viral protein is coupled directly to an indicator such as a fluorescent dye or an enzyme (Fig. 2.15). A more sensitive approach is indirect immunostaining, in which a second antibody is coupled to the indicator. The second antibody recognizes a common region on the virus-specific antibody. Multiple second-antibody molecules bind to the first antibody, resulting in an increased signal from the indicator compared with that obtained with direct immunostaining. Furthermore, a single indicator-coupled second antibody can be used in many assays, avoiding the need to purify and couple an indicator to multiple first antibodies.

In practice, virus-infected cells (unfixed or fixed with acetone, methanol, or paraformaldehyde) are incubated with **polyclonal** or **monoclonal antibodies** directed against viral antigen. Excess antibody is washed away, and in direct immunostaining, cells are examined by microscopy. For indirect immunostaining, the second antibody is added before examination of the cells by microscopy. Commonly used indicators include fluorescein and rhodamine, which fluoresce on exposure to UV light. Filters are placed between the specimen and the eyepiece to remove blue and UV light so that the field is dark, except for cells to which the antibody has bound, which emit green (fluorescein) or red (rhodamine) light (Fig. 2.15). Even though these colors are at the opposite ends of the visible light spectrum, bleeding of red into green and vice versa still occur. Today's optics are much better at keeping the wavelengths separate, and many more colors in between red and green are now available. Antibodies can also be coupled to molecules other than fluorescent indicators, including enzymes such as alkaline phosphatase, horseradish peroxidase, and β-galactosidase, a bacterial enzyme that in a test system converts the chromogenic substrate X-Gal (5-bromo-4-chloro-3-indolyl-β-d-galactopyranoside) to a blue product. After excess antibody is washed away, a suitable chromogenic substrate is added, and the presence of the indicator antibody is revealed by the development of a color that can be visualized.

Immunostaining has been applied widely in the research laboratory for determining the subcellular localization of proteins in cells (Fig. 2.15), monitoring the synthesis of viral proteins, determining the effects of mutation on protein production, and localizing the sites of virus replication in animal hosts. It is the basis of the fluorescent-focus assay.

Figure 2.15 Direct and indirect methods for antigen detection. (A) The sample (tissue section, smear, or bound to a solid phase) is incubated with a virus-specific antibody (Ab). In direct immunostaining, the antibody is linked to an indicator such as fluorescein. In indirect immunostaining, a second antibody, which recognizes a general epitope on the virus-specific antibody, is coupled to the indicator. Mab, monoclonal antibody. **(B)** Use of immunofluorescence to determine the intracellular location of mumps virus by direct and indirect immunofluorescence using confocal laser scanning microscopy. A mumps virus small hydrophobic protein-enhanced green fluorescent protein fusion protein was produced following transfection of Vero cells with a plasmid encoding the fusion protein. The small hydrophobic protein-enhanced green fluorescent protein was visualized by virtue of its autofluorescence by excitation at 488 nm (green). Protein disulfide isomerase, an enzyme in the endoplasmic reticulum, was detected using a monoclonal antibody and visualized indirectly by excitation at 647 nm of a fluorescent molecule, cyanine 5 (Cy5) conjugated to a secondary antibody (blue). Filamentous actin, a major component of the cytoskeleton, was detected using phalloidin conjugated to tetramethylrhodamine and visualized directly by excitation at 586 nm (red). Nuclei were counterstained using 4′,6-diamidino-2-phenylindole and visualized directly by excitation at 405 nm (yellow). Colocalization of the mumps virus small hydrophobic protein-enhanced green fluorescent protein and protein disulfide isomerase (cyan) demonstrated that the protein was present in the endoplasmic reticulum. Image provided by Paul Duprex, Boston University Medical School, with permission of the Wellcome Trust.

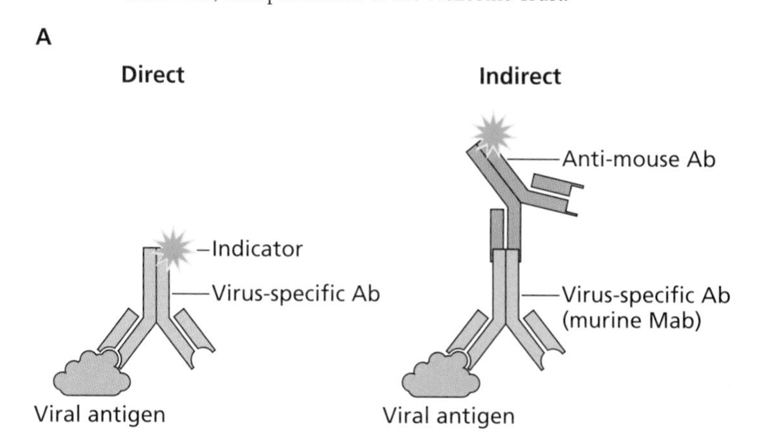

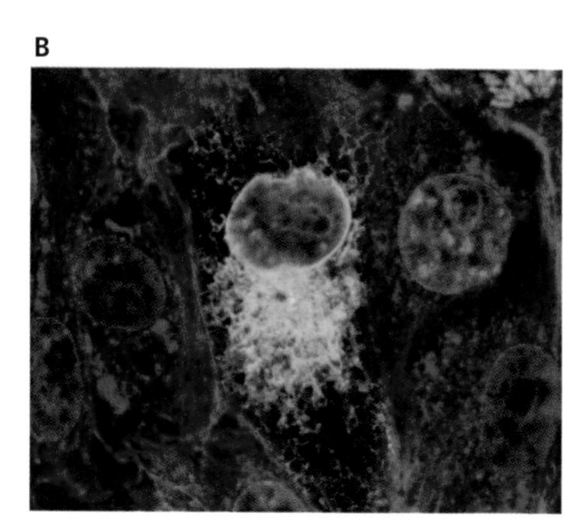

Immunostaining of viral antigens in smears of clinical specimens may be used to diagnose viral infections. For example, direct and indirect immunofluorescence assays with nasal swabs or washes are used to diagnose a variety of viruses, including influenza virus and measles virus. Viral proteins or nucleic acids may also be detected in infected animals by immunohistochemistry. In this procedure, tissues are embedded in a solid medium such as paraffin, and thin slices are produced using a microtome. Viral antigens can be detected within the cells of the sections by direct and indirect immunofluorescence assays.

Recent improvements in microscopy technology and computational image manipulation have led to unprecedented levels of resolution and contrast and the ability to reconstruct three-dimensional structures from captured images. An example is **confocal microscopy**, which utilizes a scanning point of light instead of full-sample illumination, providing improvements in optical sectioning. **Super-resolution microscopy** combines the advantages of fluorescent imaging (multicolor labeling and live-cell imaging) with the high resolution of electron microscopy. While conventional fluorescent microscopy has a resolution of 200 to 500 nm, single-molecule localization microscopy can achieve resolution below 1 nm. This resolution is achieved by combining sequential acquisition of images with random switching of fluorophores on and off. From several hundred to thousands of images are collected and processed to generate a super-resolution dataset that can resolve cellular ultrastructure.

Enzyme immunoassay. Detection of viral antigens or antiviral antibodies can be accomplished by solid-phase methods, in which antiviral antibody or protein is adsorbed to a plastic surface. To detect viral antigens in serum or clinical samples, a "capture" antibody, directed against the virus, is linked to a solid support, a plastic dish or bead (Fig. 2.16A). The specimen is added to the plastic support, and if viral antigens are present, they will be captured by the bound antibody. Bound viral antigen is detected by using a second antibody linked to an enzyme. A chromogenic molecule that is converted by the enzyme to an easily detectable product is then added. The enzyme amplifies the signal because a single catalytic enzyme molecule can generate many product molecules. To detect IgG antibodies to viruses, viral protein is first linked to the plastic support, and then the specimen is added (Fig. 2.16B). If antibodies against the virus are present in the specimen, they will bind to the immobilized antibody. The bound antibodies are then detected by using a second antibody directed against a common region on the first antibody. Like other detection methods, enzyme immunoassays are used in both experimental and diagnostic virology. In the clinical laboratory, enzyme immunoassays are used to detect a variety of viruses including rotavirus, herpes simplex virus, and human immunodeficiency virus.

A modification of the enzyme immunoassay is the lateral flow immunochromatographic assay, which has been used in rapid antigen detection test kits. In this assay, a sample is applied to a membrane and is drawn across it by capillary action. Antigens in the sample react with a specific

Figure 2.16 Detection of viral antigen or antibodies against viruses by enzyme-linked immunosorbent assay. (A) To detect viral proteins in a sample, antibodies specific for the virus are immobilized on a solid support such as a plastic well. The sample is placed in the well, and viral proteins are "captured" by the immobilized antibody. After washing to remove unbound proteins, a second antibody against the virus is added, which is linked to an indicator. Another wash is done to remove unbound second antibody. If viral antigen has been captured by the first antibody, the second antibody will bind and the complex will be detected by the indicator. **(B)** To detect antibodies to a virus in a sample, viral antigen is immobilized on a solid support. The sample is placed in the well, and viral antibodies bind the immobilized antigen. After washing to remove unbound antibodies, a second antibody, directed against a general epitope on the first antibody, is added. Another wash removes unbound second antibody. If viral antibodies are bound by the immobilized antigen, the second antibody will bind and the complex will be detected by the indicator.

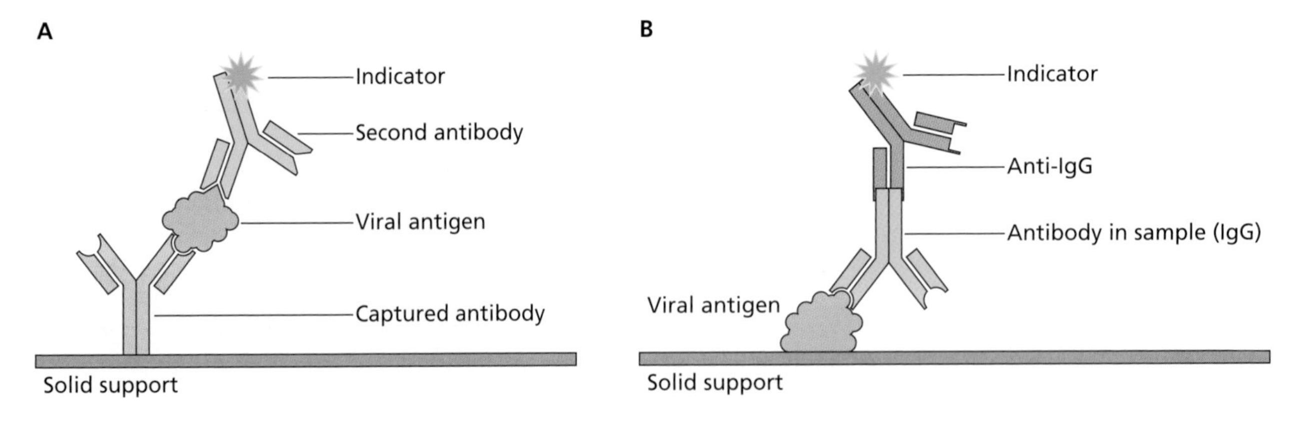

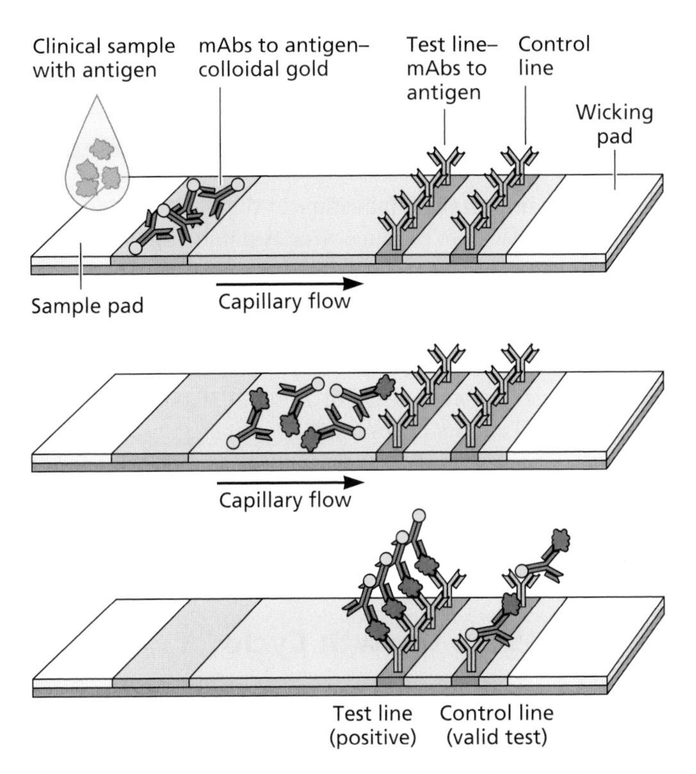

Figure 2.17 Lateral flow immunochromatographic assay. A slide or "dipstick" covered with a membrane is used to assay for the presence of viral antigens. The clinical specimen is placed on an absorbent pad at one end and is drawn across the slide by capillary action. Antigens in the sample react with a specific antibody, which is conjugated to a detector. The antigen-antibody complexes move across the membrane until they are captured by a second antibody. At this point a line becomes visible, indicating that viral antigen is present in the sample.

antibody, which is conjugated to a detector. The antigen-antibody complexes move across the membrane until they are captured by a second antibody. At this point a line becomes visible, indicating that viral antigen is present in the sample (Fig. 2.17). The lateral flow immunochromatographic assay does not require instrumentation and can be read in 5 to 20 min in a physician's office or in the field. Commercial rapid antigen detection assays are currently available for influenza virus, respiratory syncytial virus, and rotavirus.

Detection of Viral Nucleic Acids

The use of cell culture to detect viruses is being increasingly supplanted by molecular methods such as the polymerase chain reaction, DNA microarrays, and high-throughput sequencing, especially for discovery of new viruses associated with human diseases. These methods can be used to identify viruses that cannot be propagated in cell culture, necessitating new ways to fulfill Koch's postulates (Box 1.4).

Polymerase chain reaction. In this technique, specific oligonucleotides are used to amplify viral DNA sequences from infected cells or clinical specimens. Amplification is done in cycles, using a thermostable DNA polymerase. Each cycle consists of primer annealing, extension, and thermal denaturation carried out by automated cycler machines. The result is exponential amplification (a $2n$-fold increase after n cycles of amplification) of the target sequence that is located between the two DNA primers. Clinical laboratories use PCR assays to detect evidence for infection by a single type of virus (singleplex PCR), while screening for the presence of up to 30 different viruses can be done using multiplex PCR. In contrast to conventional PCR, real-time PCR can be used to quantitate the amount of DNA or RNA in a sample. In this procedure, also called quantitative PCR, the amplified DNA is detected as the reaction progresses, for example, by the use of fluorescent dyes that intercalate nonspecifically into DNA. The number of cycles needed to detect fluorescence above background can then be compared between standard and experimental samples.

DNA microarrays. This approach provides a method for studying the gene expression profile of a cell in response to virus infection (Chapter 8), and can also be used to discover new viruses. In this method, millions of unique viral DNA sequences fixed to glass or silicon wafers are incubated with complementary sequences amplified from clinical and environmental samples. Binding is usually detected by using fluorescent molecules incorporated into amplified nucleic acids.

High-throughput sequencing. Sequencing of thousands to millions of DNA molecules at the same time is a feature of this method, also known as next-generation sequencing to distinguish it from older methods. The newer approaches have not only made sequencing of DNA cheaper and faster but also helped create innovative experimental approaches to study genome organization, function, and evolution. Current high-throughput sequencing methods include 454, Illumina, and SOLiD (sequencing by oligonucleotide ligation and detection). Each has advantages and limitations in terms of read length, cost, and speed. Illumina technology provides extremely high throughput, enabling sequencing of 1 human genome per 24-h period. SOLiD is comparable in performance to Illumina technology but at a higher price.

The use of high-throughput sequencing has led to the discovery of new viruses and has given birth to the field of metagenomics, the analysis of sequences from clinical or environmental samples. These sequencing technologies can be used to study the **virome,** the collection of all viruses in a specific environment, such as sewage, the human body, or the intestinal tract.

The generation of nucleotide sequences at an unprecedented rate has spawned a new branch of bioinformatics to develop algorithms for assembling sequence reads into continuous strings and to determine whether they are from a virus, and if so, whether it is novel or previously discovered. Storing, analyzing, and sharing massive quantities of data, an estimated 15 quadrillion nucleotides per year, is an immense challenge. While these virus detection technologies are extremely powerful, the results obtained must be interpreted with caution. It is very easy to detect traces of a viral contaminant when searching for new agents of human disease (Box 2.8).

Genome sequences can provide considerable insight into the evolutionary relationships among viruses. Such information can be used to understand the origin of viruses and how selection pressures change viral genomes, and to assist in epidemiological investigations of viral outbreaks. When few viral genome sequences were available, pairwise homologies were often displayed in simple tables. As sequence databases increased in size, tables of multiple alignments were created, but these were still based only on pairwise comparisons. Today, phylogenetic trees are used to illustrate the relationships among numerous viruses or viral proteins (Box 2.9). Not only are such trees important tools for understanding evolutionary relationships, but they may allow conclusions to be drawn about biological functions: examination of a phylogenetic tree may allow determination of how closely or distantly a sequence relates to one of known function.

Viral Reproduction: the Burst Concept

A fundamental and important principle is that viruses replicate by the assembly of preformed components into particles. The parts are first made in cells and then assembled into the final product. The reproduction of viruses is very different from that of cells, which multiply by binary fission. This simple build-and-assemble strategy is unique to viruses, but the details for members of different virus families are astoundingly different. There are many ways to build a virus particle, and each one tells us something new about virus structure and assembly.

Modern studies of virus replication strategies have their origins in the work of Max Delbrück and colleagues, who studied the T-even bacteriophages starting in 1937. Delbrück believed that these bacteriophages were perfect models for understanding virus replication. He also thought that phages were excellent models for studying the gene: they were self-replicating (a hallmark of a gene); their mutations were inherited; and they were small, easily manipulated entities with short reproductive cycles.

Delbrück focused his attention on the fact that one bacterial cell usually makes hundreds of progeny virus particles. The yield from one cell is one viral generation; it was called the **burst** because viruses literally burst from the infected cell. Under carefully controlled laboratory conditions, most cells make, on average, about the same number of bacteriophages per cell. For example, in one of Delbrück's experiments, the average number of bacteriophage T4 particles produced from individual single-cell bursts from *Escherichia coli* cells was 150 particles per cell. If this experiment were done today, using comparable experimental conditions, the average burst would be similar.

Another important implication of the burst is that a cell has a finite capacity to produce virus. A number of parameters limit the number of particles produced per cell, such as metabolic resources, the number of sites for replication in the cell, the regulation of release of virus particles, and host defenses. In general, larger cells (e.g., eukaryotic cells) produce more virus particles per cell: yields of 1,000 to 10,000 virions per eukaryotic cell are not uncommon.

A burst occurs for viruses that kill the cell after infection, namely, the cytopathic viruses. However, some viruses do **not** kill their host cells, and virus particles are produced as long as the cell is alive. Examples include filamentous bacteriophages, some retroviruses, and hepatitis viruses.

The One-Step Growth Cycle

Initial Concept

The idea that one-step growth analysis can be used to study the single-cell life cycle of viruses originated from the work on bacteriophages by Emory Ellis and Delbrück. In their classic experiment, they added virus particles to a culture of rapidly growing *E. coli* cells. These particles adsorbed quickly to the cells. The infected culture was then diluted, preventing further adsorption of unbound particles. This simple dilution step is the key to the experiment: it reduces further binding of virus to cells and effectively synchronizes the infection. Samples of the diluted culture were then taken every few minutes and analyzed for the number of infectious bacteriophages. When the results were plotted, several key observations emerged. The results were surprising in that they did not resemble the growth curves of bacteria or cultured cells. After a short lag, bacterial cell growth becomes exponential (i.e., each progeny cell is capable of dividing) and follows a straight line (Fig. 2.18A). Exponential growth continues until the nutrients in the medium are exhausted. In contrast, numbers of new viruses do not increase in a linear fashion from the start of the infection (Fig. 2.18B, left). There is an initial lag, followed by a rapid increase in virus production, which then plateaus. This single cycle of virus reproduction produces the "burst" of virus progeny. If the experiment is repeated, so that only a few cells are initially infected, the graph looks different (Fig. 2.18B, right). Instead of a single cycle, there is a stepwise increase in numbers of new viruses with time. Each step represents one cycle of virus infection.

Once the nature of the viral growth cycle was explored using the one-step growth curve, questions emerged about what was happening in the cell before the burst. What was the fate of the incoming virus? Did it disappear? How were more

BOX 2.8

EXPERIMENTS

Pathogen de-discovery

Deep sequencing of nucleic acids has accelerated the pace of virus discovery, but at a cost: contaminants are much easier to detect.

During a search for the causative agent of seronegative hepatitis (disease not caused by hepatitis A, B, C, D, or E virus) in Chinese patients, a new virus was discovered in sera by next-generation sequencing. This virus, provisionally called NIH-Chongquing (NIH-CQV) has a single-stranded DNA genome that is a hybrid between that of parvoviruses and circoviruses. When human sera were screened by PCR, 63 of 90 patient samples (70%) were positive for the virus, while sera from 45 healthy controls were negative. Furthermore, 84% of patients were positive for IgG antibodies against the virus and 31% were positive for IgM antibodies (suggesting a recent infection). Among healthy controls, 78% were positive for IgG and all were negative for IgM. The authors concluded that this virus was highly prevalent in some patients with seronegative hepatitis.

A second independent laboratory also identified the same virus (which they called parvovirus-like hybrid virus, PHV-1) in sera from patients in the United States with non-A-to-E hepatitis, while a third group identified the virus in diarrheal stool samples from Nigeria.

The first clue that something was amiss was the observation that the new virus identified in all three laboratories shared 99% nucleotide and amino acid identity: this similarity would not be expected in virus samples from such geographically, temporally, and clinically diverse samples. Another problem was that in the U.S. non-A-to-E hepatitis study, all patient sample pools were positive for viral sequences. These observations suggested the possibility of viral contamination.

When nucleic acids were repurified from the U.S. non-A-to-E hepatitis samples using a different method, **none** of the samples were positive for the new virus. The presence of the virus was ultimately traced to the use of column-based purification kits manufactured by Qiagen, Inc. Nearly the entire viral genome could be detected by deep sequencing of water that was passed through these columns.

The nucleic acid purification columns contaminated with the new virus were used to purify nucleic acid from patient samples. These columns, produced by a number of manufacturers, are typically an inch in length and contain a silica gel membrane that binds nucleic acids. The clinical samples are added to the column, which is then centrifuged briefly to remove liquids (hence the name "spin" columns). The nucleic acid adheres to the silica gel membrane. Contaminants are washed away, and then the nucleic acids are released from the silica by the addition of a buffer.

Why were the Qiagen spin columns contaminated with the parvovirus-circovirus hybrid? A search of the publicly available environmental metagenomic datasets revealed the presence of sequences highly related to PHV-1 (87 to 99% nucleotide identity). The datasets containing PHV-1 sequences were obtained from sampled seawater off the Pacific coast of North America and coastal regions of Oregon and Chile. Silica, a component of spin columns, may be produced from diatoms. The source of contamination could be explained if the silica in the Qiagen spin columns was produced from diatoms and if PHV-1 is a virus of ocean-dwelling diatoms.

In retrospect, it was easy to be fooled into believing that NIH-CQV might be a human pathogen because it was detected only in sick and not healthy patients. Why antibodies to the virus were detected in samples from both sick and healthy patients remains to be explained. However, NIH-CQV/PHV-1 is likely not associated with any human illness: when non-Qiagen spin columns were used, PHV-1 was not found in any patient sample.

The lesson to be learned from this story is clear: deep sequencing is a very powerful and sensitive method, but must be applied with great care. Every step of the virus discovery process must be carefully controlled, from the water used to the plastic reagents. Most importantly, laboratories involved in pathogen discovery must share their sequence data, something that took place during this study.

Naccache SN, Greninger AL, Lee D, Coffey LL, Phan T, Rein-Weston A, Aronsohn A, Hackett J, Jr, Delwart EL, Chiu CY. 2013. The perils of pathogen discovery: origin of a novel parvovirus-like hybrid genome traced to nucleic acid extraction spin columns. *J Virol* **87:**11966–11977.

Xu B Zhi N, Hu G, Wan Z, Zheng X, Liu X, Wong S, Kajigaya S, Zhao K, Mao Q, Young NS. 2013. Hybrid DNA virus in Chinese patients with seronegative hepatitis discovered by deep sequencing. *Proc Natl Acad Sci U S A* **110:**10264–10269.

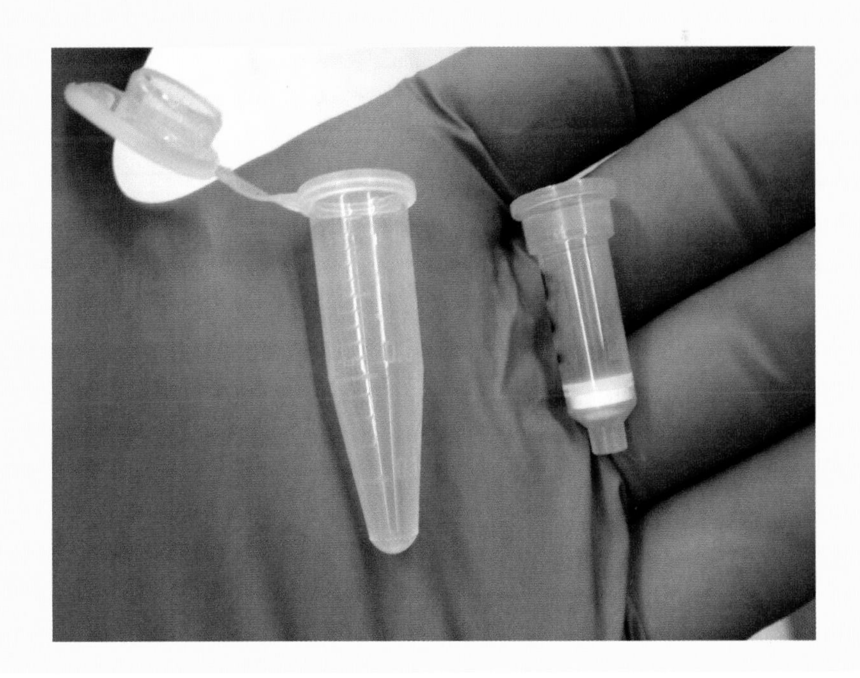

BOX 2.9

METHODS

How to read a phylogenetic tree

Phylogenetic dendrograms, or trees, provide information about the inferred evolutionary relationships between viruses. The example shown in the figure is a phylogenetic tree for 10 viral isolates from different individuals whose genome sequences have been determined. The horizontal dimension of the tree represents the amount of genetic change, and the scale (0.07) is the number of changes divided by the length of the sequence (in some trees this may be expressed as % change). The blue circles, called nodes, represent putative ancestors of the sampled viruses. Therefore, the branches represent chains of infections that have led to sampled viruses. The vertical distances have no significance.

The tree in the figure is *rooted*, which means that we know the common ancestor of all the sampled viruses. A rooted tree gives the order of branching from left to right: virus A existed before B, although the unit of time might not be known. The numbers next to each node represent the measure of support; these are computed by a variety of statistical approaches including "bootstrapping" and "Bayesian posterior probabilities." A value close to 1 indicates strong evidence that sequences to the right of the node cluster together better than any other sequences. Often there is no known isolate corresponding to the root of the tree; in this case, an arbitrary root may be estimated, or the tree will be unrooted. In these cases, it can no longer be assumed that the order of ancestors proceeds from left to right.

Phylogenetic trees can also be constructed by grouping sampled viruses by host of isolation. Such an arrangement sometimes makes it possible to identify the animal source of a human virus. Circular forms, such as a radial format tree, are often displayed when the root is unknown.

Trees relating nucleic acid sequences depict the relationships as if sampled and intermediary sequences were on a trajectory to the present sequences. This deduction is an oversimplification, because any intermediate that was lost during evolution will not be represented in the tree. In addition, any recombination or gene exchange by coinfection with similar viral genomes will scramble ordered lineages.

A fair question is whether we can predict the future trajectory or branches of the tree. We can never answer this question for two reasons: any given sample may not represent the diversity of any given virus population in an ecosystem, and we cannot predict the selective pressures that will be imposed.

Hall BG. 2011. *Phylogenetic Trees Made Easy: a How-to Manual*, 4th ed. Sinauer Associates, Sunderland, MA.

ViralZone. Phylogenetics of animal pathogens: basic principles and applications (a tutorial). http://viralzone.expasy.org/e_learning/phylogenetics/content.html

Rooted phylogenetic tree of 10 viral genome sequences. Adapted from http://epidemic.bio.ed.ac.uk/how_to_read_a_phylogeny, with permission.

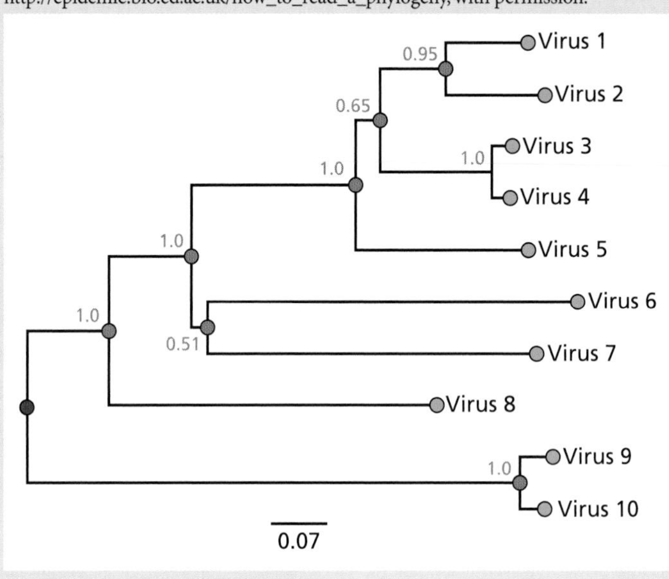

virus particles produced? These questions were answered by looking inside the infected cell. Instead of sampling the diluted culture for virus after various periods of infection, researchers prematurely lysed the infected cells as the infection proceeded and then assayed for infectious virus. The results were extremely informative. Immediately after dilution, there was a complete loss, or eclipse, of infectious virus for 10 to 15 min (Fig. 2.18B). In other words, input virions disappeared and no new phage particles were produced during this period. The loss of infectivity is a consequence of the release of the genome from the virion, to allow for subsequent transcription of viral genes. Particle infectivity is lost during this phase because the released genome is not infectious under the conditions of the plaque assay. Next, new infectious particles were detected inside the cell, before they were released into the medium. These were newly assembled virus particles that had not yet been released by cell lysis. The results of these experiments defined two new terms in virology: the **eclipse period**, the phase in which infectivity is lost when virions are disassembled after penetrating cells; and the **latent period**, the time it takes to replicate, assemble, and release new virus particles before lysis, ~20 to 25 min for *E. coli* bacteriophages.

Synchronous infection, the key to the one-step growth cycle, is usually accomplished by infecting cells with a sufficient number of virus particles to ensure that most of the cells are infected rapidly.

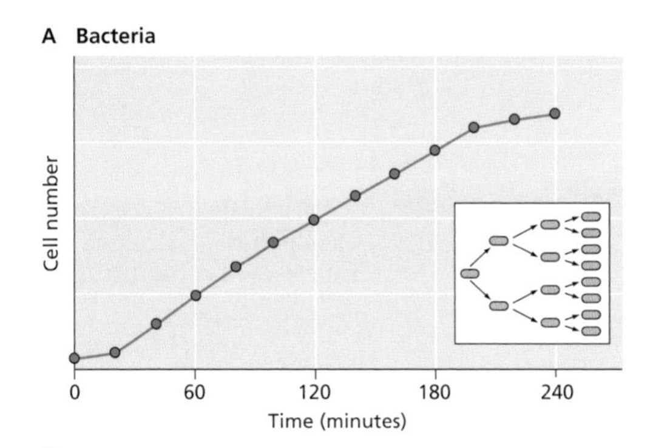

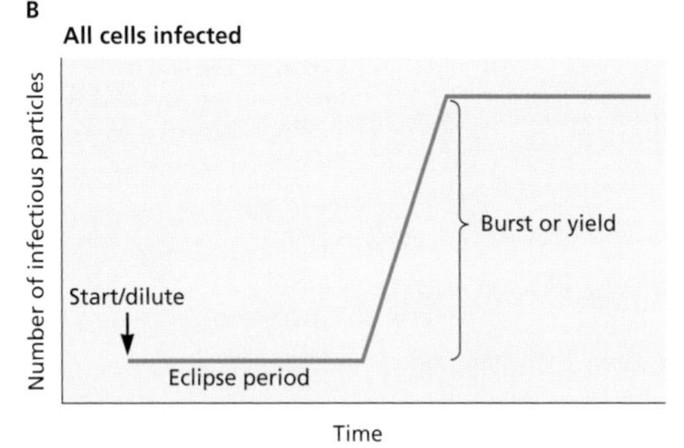

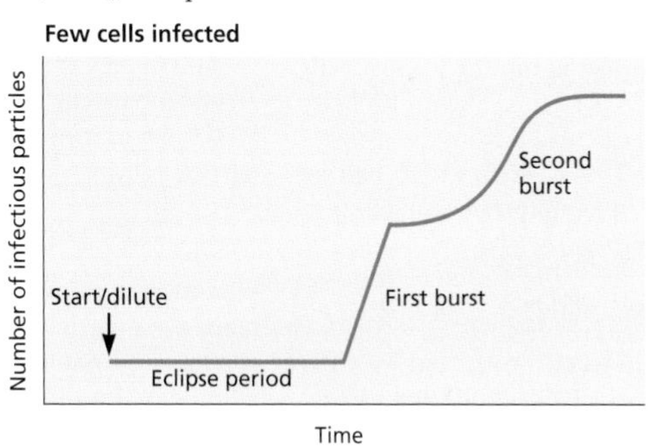

Figure 2.18 Comparison of bacterial and viral reproduction. (A) Growth curve for a bacterium. The number of bacteria is plotted as a function of time. One bacterium is added to the culture at time zero; after a brief lag, the bacterium begins to divide. The number of bacteria doubles every 20 min until nutrients in the medium are depleted and the growth rate decreases. The inset illustrates the growth of bacteria by binary fission. One- and two-step growth curves of bacteriophages **(B)** Growth of a bacteriophage in *E. coli* under conditions when all cells are infected (left) and when only a few cells are infected (right). Panel A adapted from B. Voyles, *The Biology of Viruses* (McGraw-Hill, New York, NY, 1993), with permission.

One-Step Growth Analysis: a Valuable Tool for Studying Animal Viruses

One-step growth analysis soon became adapted for studying the replication of animal viruses. The experiment begins with removal of the medium from the cell monolayer and addition of virus in a small volume to promote rapid adsorption. After ~1 h, unadsorbed inoculum containing virus particles is removed, the cells are washed, and fresh medium is added. At different times after infection, samples of the cell culture supernatant are collected and the virus titer is determined. The kinetics of intracellular virus production can be monitored by removing the medium containing extracellular particles, scraping the cells into fresh medium, and lysing them. A cell extract is prepared after removal of cellular debris by centrifugation, and the virus titer in the extract is measured.

The results of a one-step growth experiment establish a number of important features about viral replication. In the example shown in Fig. 2.19A, the first 11 h after infection constitutes the eclipse period, during which the viral nucleic acid is uncoated from its protective shell and no infectious virus can be detected inside cells. The small number of infectious particles detected during this period probably results from adsorbed virus that was not uncoated. Beginning at 12 h after adsorption, the quantity of intracellular infectious virus begins to increase, marking the onset of the synthetic phase, during which new virus particles are assembled. During the latent period, no extracellular virus can be detected. At 18 h after adsorption, virions are released from cells into the extracellular medium. Ultimately, virus production plateaus as the

Figure 2.19 One-step growth curves of animal viruses. (A) Growth of a nonenveloped virus, adenovirus type 5. The inset illustrates the concept that viruses multiply by assembly of preformed components into particles. **(B)** Growth of an enveloped virus, Western equine encephalitis virus, a member of the *Togaviridae*. This virus acquires infectivity after maturation at the plasma membrane, and therefore little intracellular virus can be detected. The small quantities observed at each time point probably represent released virus contaminating the cell extract. Adapted from B. D. Davis et al., *Microbiology* (J. B. Lippincott Co., Philadelphia, PA, 1980), with permission.

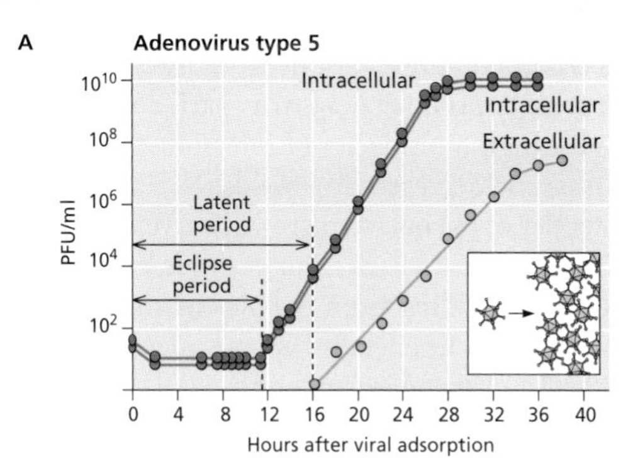

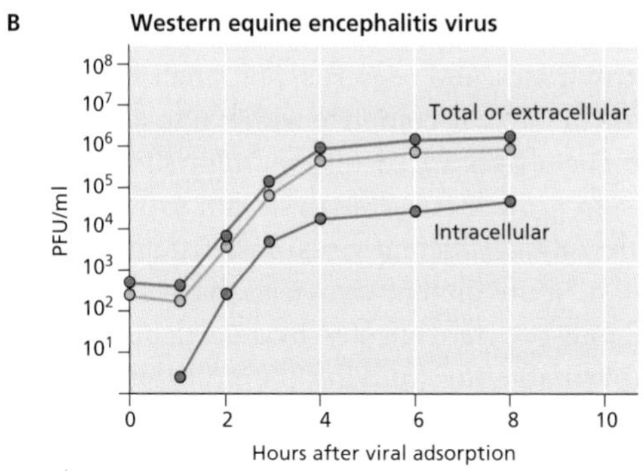

BOX 2.10

DISCUSSION

Multiplicity of infection (MOI)

Infection depends on the random collision of cells and virus particles. When susceptible cells are mixed with a suspension of virus particles, some cells are uninfected and other cells receive one, two, three, etc., particles. The distribution of virus particles per cell is best described by the Poisson distribution:

$$P(k) = e^{-m}m^k/k!$$

In this equation, $P(k)$ is the fraction of cells infected by k virus particles. The multiplicity of infection, m, is calculated from the proportion of uninfected cells, $P(0)$, which can be determined experimentally. If k is made 0 in the above equation, then

$$P(0) = e^{-m} \text{ and } m = -\ln P(0)$$

The fraction of cells receiving 0, 1, and >1 virus particle in a culture of 10^6 cells infected with an MOI of 10 can be determined as follows.

The fraction of cells that receive 0 particles is

$$P(0) = e^{-10} = 4.5 \times 10^{-5}$$

and in a culture of 10^6 cells, this equals 45 uninfected cells.

The fraction of cells that receive 1 particle is

$$P(1) = 10 \times 4.5 \times 10^{-5} = 4.5 \times 10^{-4}$$

and in a culture of 10^6 cells, 450 cells receive 1 particle.

The fraction of cells that receive >1 particle is

$$P(>1) = 1 - e^{-m}(m + 1) = 0.9995$$

and in a culture of 10^6 cells, 999,500 cells receive >1 particle. [The value in this equation is obtained by subtracting from 1 (the sum of all probabilities for any value of k) the probabilities $P(0)$ and $P(1)$.]

The fraction of cells receiving 0, 1, and >1 virus particle in a culture of 10^6 cells infected with an MOI of 0.001 is

$P(0) = 99.99\%$
$P(1) = 0.0999\%$ (for 10^6 cells, 10^4 are infected)
$P(>1) = 10^{-6}$

The MOI required to infect 99% of the cells in a cell culture dish is

$P(0) = 1\% = 0.01$
$m = -\ln(0.01) = 4.6$ PFU per cell

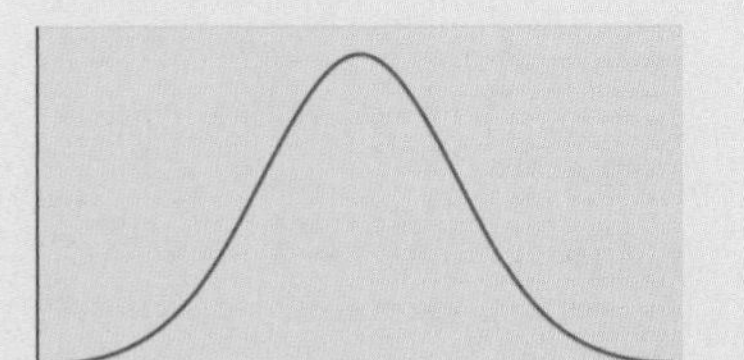

cells become metabolically and structurally incapable of supporting additional replication.

The yield of infectious virus per cell can be calculated from the data collected during a one-step growth experiment (Fig. 2.19). This value varies widely among different viruses and with different virus-host cell combinations. For many viruses, increasing the **multiplicity of infection** (Box 2.10) above a certain point does not increase the yield: cells have a finite capacity to produce new virus particles. In fact, infecting at a very high multiplicity of infection can cause premature cell lysis and decrease virus yields.

The kinetics of the one-step growth curve can vary dramatically among different viruses. For example, enveloped viruses that mature by budding from the plasma membrane, as discussed in Chapter 13, generally become infectious only as they leave the cell, and therefore little intracellular infectious virus can be detected (Fig. 2.19B). The curve shown in Fig. 2.19A illustrates the pattern observed for a DNA virus with the long latent and synthetic phases typical of many DNA viruses, some retroviruses, and reovirus. For small RNA viruses, the entire growth curve is complete within 6 to 8 h, and the latent and synthetic phases are correspondingly shorter.

One-step growth curve analysis can provide quantitative information about different virus-host systems. It is frequently employed to study mutant viruses to determine what parts of the infectious cycle are affected by a particular genetic lesion. It is also valuable for studying the multiplication of a new virus or viral replication in a new virus-host cell combination.

When cells are infected at a low multiplicity of infection, several cycles of viral replication may occur (Fig. 2.18). Growth curves established under these conditions can also provide useful information. When infection is done at a high multiplicity of infection, a mutation may fail to have an obvious effect on viral replication. The defect may become obvious following a low-multiplicity infection. Because the effect of a mutation in each cycle is multiplied over several cycles, a small effect can be amplified. Defects in the ability of viruses to spread from cell to cell may also be revealed when multiple cycles of replication occur.

Systems Biology

The use of one-step growth analysis to study the replication cycles of many viruses has allowed a reductionist approach to understanding and defining the steps of virus attachment, entry, replication, and assembly. New experimental and computational tools that permit global analysis of viral, cellular, and host responses to infection have been developed. Systems biology analysis uses high-throughput technologies (such as next-generation sequencing of DNA and RNA, and mass spectrometry) to measure system-wide changes in DNA, RNA, proteins, and metabolites during virus infection of cells, tissues, or entire organisms. Data obtained from high-throughput measurements are integrated and analyzed using mathematical algorithms to generate models that are predictive of the system. For example, virus infections of different animals are characterized by the induction of distinct sets of cytokine genes, which can be correlated with different pathogenic outcomes. When a model has been developed, it can be further refined by the use of viral mutants or targeted inhibition of host genes or pathways. Systems virology is therefore a holistic, host-directed approach that complements traditional approaches to studying viruses.

Examples of systems virology approaches include the use of genome-wide transcriptional profiling to study the host response to infection. Infection of mice with the 1918 strain of influenza virus leads to a rapidly fatal infection characterized by sustained induction of proinflammatory cytokine and chemokine genes. Understanding the gene expression signature that correlates with lethality is one goal of these studies. Systems virology approaches can also be used to predict signatures of vaccine efficacy. In one study, transcriptional profiling of peripheral blood mononuclear cells from vaccinated subjects revealed that the yellow fever virus vaccine induces the expression of genes encoding members of the complement system and stress response proteins. This pattern accurately predicts subsequent $CD8^+$ T cell activation ($CD8^+$ T cell and antibody responses that are thought to mediate protection from infection with yellow fever virus). A separate signature was also identified that accurately predicts neutralizing antibody synthesis during infection. Systems virology approaches also can be used to identify and analyze all interactions between cellular and viral proteins and the roles of such interactions in replication (Box 2.11).

Perspectives

The one-step growth analysis is used nearly universally to study virus replication. When cells are infected at a high multiplicity of infection, sufficient viral nucleic acid or protein can be isolated to allow a study of events during the replication cycle. Synchronous infection is the key to this approach, because under this condition, the same steps of the reproduction cycle typically occur in all cells at the same time. Many of the experimental results discussed in subsequent chapters of this book were obtained using one-step growth analysis. The power of this approach is such that it reports on all stages of the reproduction cycle in a simple and quantitative fashion. With modest expenditure of time and reagents, virologists can deduce a great deal about viral translation, replication, or assembly.

From the humble beginnings of the one-step growth curve, many new methods have been developed that have propelled

BOX 2.11

WARNING

Determining a role for cellular proteins in viral replication can be quite difficult

Understanding the roles of both viral and cellular proteins at various stages of viral reproduction is essential for elucidating molecular mechanisms and for developing strategies for blocking pathogenic infections. As viral genomes have a limited set of genes, the viral proteins or genetic elements that are essential at each step can be deduced by introducing mutations and observing phenotypes. Identifying critical cellular genes and proteins is much more difficult. A general approach to select likely candidates has been to identify cellular proteins that are included in virus particles and/or bind to viral proteins (*in vitro* or in cells).

Once candidates are identified, the contribution of the cellular protein to viral reproduction may be evaluated by observing the effects of

- specific small-molecule inhibitors of the protein's function (inhibitory drugs)
- synthesis of an altered protein, known to have a dominant-negative effect on its normal function
- treatment with small RNAs that induce mRNA degradation (see Chapter 10) and reduce the concentration of the cellular protein
- reproduction in cells in which the candidate gene has been mutated or deleted

Even after applying the multiple approaches and methods described above, identifying relevant cellular proteins and evaluating their roles in viral reproduction is seldom easy. The problems encountered include the following.

- More than one protein may provide the required function (redundancy).
- The function of the protein might be essential to the cell, and mutation of the gene that encodes it (or inhibition of protein production) could be lethal.
- Only small quantities of the protein might be required, and reducing its activity with an inhibitor, or its concentration may be insufficient to observe a defect in viral reproduction.
- The cellular protein might provide a slight enhancement to viral reproduction that could be difficult to detect, but may be physiologically significant.
- Synthesis of an altered cellular gene or overexpression of a normal cellular gene may produce changes that affect virus reproduction for reasons that are irrelevant to the natural infection (artifacts).

Given these difficulties, it is not surprising that the literature in this area is sometimes contradictory and the results can be controversial.

our understanding of viruses and infected cells to greater depths and at unprecedented speed. These abilities are illustrated by the new human coronavirus, Middle East respiratory syndrome coronavirus, isolated in early 2013. Within 6 months, not only had the virus been isolated and studied by one-step growth analysis, but the genome was molecularly cloned and its sequence was determined; an infectious DNA clone was made and used to produce viral mutants; the cellular receptor was identified; immunofluorescence was used to study infection of various cell types; and serological assays, including neutralization and enzyme immunoassays, were used to screen animal sera to determine the origin of the virus. We are truly in a remarkable era, when few experimental questions are beyond the reach of the techniques that are currently available.

References

Books

Ausubel FM, Brent R, Kingston RE, Moore DD, Seidman JG, Smith JA, Struhl K (ed). 1998. *Current Protocols in Molecular Biology.* John Wiley & Sons, Inc, New York, NY. Updated frequently: http://www.currentprotocols.com.

Cann AJ. 2000. *Virus Culture: a Practical Approach.* Oxford University Press, Oxford, United Kingdom.

Freshney IA. 2005. *Culture of Animal Cells: a Manual of Basic Techniques.* Wiley-Liss, Hoboken, NJ.

Harlow E, Lane D. 1998. *Using Antibodies: a Laboratory Manual.* Cold Spring Harbor Laboratory Press, Cold Spring Harbor, NY.

Maramorosch K, Koprowski H (ed). *Methods in Virology.* Academic Press, New York, NY. A series of volumes begun in the 1960s that contain review articles on classic virological methods.

Sambrook J, Russell D. 2001. *Molecular Cloning: a Laboratory Manual,* 3rd ed. Cold Spring Harbor Laboratory Press, Cold Spring Harbor, NY.

Review Articles

Dulbecco R, Vogt M. 1953. Some problems of animal virology as studied by the plaque technique. *Cold Spring Harb Symp Quant Biol* **18:**273–279.

Ekblom R, Wolf JB. 2014. A field guide to whole-genome sequencing, assembly and annotation. *Evol Appl* **7:**1026–1042.

Laude AJ, Prior IA. 2004. Plasma membrane microdomains: organization, function and trafficking. *Mol Membr Biol* **21:**193–205.

Law GL, Korth MJ, Benecke AG, Katze MG. 2013. Systems virology: host-directed approaches to viral pathogenesis and drug targeting. *Nat Rev Microbiol* **11:**455–466.

Papers of Special Interest

Elliott G, O'Hare P. 1999. Live-cell analysis of a green fluorescent protein-tagged herpes simplex virus infection. *J Virol* **73:**4110–4119.

Ellis EL, Delbrück M. 1939. The growth of bacteriophage. *J Gen Physiol* **22:** 365–384.

Raj VS, Mou H, Smits SL, Dekkers DH, Müller MA, Dijkman R, Muth D, Demmers JA, Zaki A, Fouchier RA, Thiel V, Drosten C, Rottier PJ, Osterhaus AD, Bosch BJ, Haagmans BL. 2013. Dipeptidyl peptidase 4 is a functional receptor for the emerging human coronavirus-EMC. *Nature* **495:**251–254.

Reed LJ, Muench H. 1932. A simple method for estimating fifty per cent endpoints. *Am J Hyg* **27:**493–497.

PART II
Molecular Biology

3 Genomes and Genetics
4 Structure
5 Attachment and Entry
6 Synthesis of RNA from RNA Templates
7 Reverse Transcription and Integration
8 Synthesis of RNA from DNA Templates
9 Replication of DNA Genomes
10 Processing of Viral Pre-mRNA
11 Protein Synthesis
12 Intracellular Trafficking
13 Assembly, Exit, and Maturation
14 The Infected Cell

3 Genomes and Genetics

Introduction

Genome Principles and the Baltimore System

Structure and Complexity of Viral Genomes

DNA Genomes

RNA Genomes

What Do Viral Genomes Look Like?

Coding Strategies

What Can Viral Sequences Tell Us?

The Origin of Viral Genomes

The "Big and Small" of Viral Genomes: Does Size Matter?

Genetic Analysis of Viruses

Classical Genetic Methods

Engineering Mutations into Viral Genomes

Engineering Viral Genomes: Viral Vectors

Perspectives

References

LINKS FOR CHAPTER 3

- *Video: Interview with Dr. Katherine High*
  http://bit.ly/Virology_High
- *Virocentricity with Eugene Koonin*
  http://bit.ly/Virology_Twiv275
- *What if influenza virus did not reassort?*
  http://bit.ly/Virology_9-15-09

> *... everywhere an interplay between nucleic acids and proteins; a spinning wheel in which the thread makes the spindle and the spindle the thread.*
>
> Erwin Chargaff
> as quoted in A. N. H. Creager, *The Life of a Virus*, *University of Chicago Press*, 2002

Introduction

Earth abounds with uncountable numbers of viruses of great diversity. However, because taxonomists have devised methods of classifying viruses, the number of identifiable groups is manageable (Chapter 1). One of the contributions of molecular biology has been a detailed analysis of the genetic material of representatives of these major virus families. From these studies emerged the principle that the **viral genome** is the nucleic acid-based repository of the information needed to build, reproduce, and transmit a virus (Box 3.1). These analyses also revealed that the thousands of distinct viruses defined by classical taxonomic methods can be organized into seven groups, based on the structures of their genomes.

Genome Principles and the Baltimore System

A universal function of viral genomes is to specify proteins. However, these genomes do not encode the complete machinery needed to carry out protein synthesis. Consequently, one important principle is that all viral genomes must be copied to produce messenger RNAs (mRNAs) that can be read by host ribosomes. Literally, all viruses are parasites of their host cells' translation system.

A second principle is that there is unity in diversity: evolution has led to the formation of only seven major types of viral genome. The Baltimore classification system integrates these two principles to construct an elegant molecular algorithm for virologists (see Fig. 1.11). When the bewildering array of viruses is classified by this system, we find fewer than 10 pathways to mRNA. The value of the Baltimore system is that by knowing only the nature of the viral genome, one can deduce the basic steps that must take place to produce mRNA. Perhaps more pragmatically, the system simplifies comprehension of the extraordinary life cycles of viruses.

The Baltimore system omits the second universal function of viral genomes, to serve as a template for synthesis of progeny genomes. Nevertheless, there is also a finite number of nucleic acid-copying strategies, each with unique primer, template, and termination requirements. We shall combine this principle with that embodied in the Baltimore system to define seven strategies based on mRNA synthesis **and** genome replication.

Replication and mRNA synthesis present no obvious challenges for most viruses with DNA genomes, as all cells use DNA-based mechanisms. In contrast, animal cells possess no known mechanisms to copy viral RNA templates and to produce mRNA from them. For RNA viruses to survive, their RNA genomes must, by definition, encode a nucleic acid polymerase.

Structure and Complexity of Viral Genomes

Despite the simplicity of expression strategies, the composition and structures of viral genomes are more varied than those seen in the entire archaeal, bacterial, or eukaryotic kingdoms. Nearly every possible method for encoding information in nucleic acid can be found in viruses. Viral genomes can be

- DNA or RNA
- DNA with short segments of RNA
- DNA or RNA with covalently attached protein
- single stranded (+) strand, (−) strand, or ambisense (Box 3.2)
- double stranded
- linear
- circular
- segmented
- gapped

PRINCIPLES *Genomes and genetics*

- Viral genomes specify some, but never all, of the proteins needed to complete the viral life cycle.
- That only seven viral genome replication strategies exist for thousands of known viruses implies unity in viral diversity.
- Some genomes can enter the reproduction cycle upon entry into a target cell, whereas others require prior modification or other viral nucleic acids before replication can proceed.
- Although the details of replication differ, all viruses with RNA genomes must encode either an RNA-dependent RNA polymerase to synthesize RNA from an RNA template, or a reverse transcriptase to convert viral RNA to DNA.
- The information encoded in viral genomes is optimized by a variety of mechanisms; the smaller the genome, the greater the compression of genetic information.
- The genome sequence of a virus is at best a biological "parts list" and tells us little about how the virus interacts with its host.
- The genomes of viruses range from those that are extraordinarily small (<2 kb) to those that are extraordinarily large (>2,500 kbp); the diversity in size likely provides advantages in the niche in which particular viruses exist.
- Technical advances allowing the introduction of mutations into any viral gene or genome sequence were responsible for much of what we know about viruses and their lifestyles.

BOX 3.1

BACKGROUND

What information is encoded in a viral genome?

Gene products and regulatory signals required for

- replication of the genome
- efficient expression of the genome
- assembly and packaging of the genome
- regulation and timing of the reproduction cycle
- modulation of host defenses
- spread to other cells and hosts

Information **not** contained in viral genomes:

- genes encoding a complete protein synthesis machinery (e.g., no ribosomal RNA and no ribosomal or translation proteins); note: the genomes of some large DNA viruses contain genes for transfer RNAs (tRNAs), aminoacyl-tRNA synthetases, and enzymes that participate in sugar and lipid metabolism
- genes encoding proteins of energy metabolism or membrane biosynthesis
- telomeres (to maintain genomes) or centromeres (to ensure segregation of genomes)

BOX 3.2

TERMINOLOGY

Important conventions: plus (+) and minus (−) strands

mRNA is defined as the positive (+) strand, because it can be translated. A strand of DNA of the equivalent polarity is also designated as a (+) strand; i.e., if it were mRNA, it would be translated into protein.

The RNA or DNA complement of the (+) strand is called the (−) strand. The (−) strand cannot be translated; it must first be copied to make the (+) strand. Ambisense RNA contains both (+) and (−) sequences.

The seven strategies for expression and replication of viral genomes are illustrated in Fig. 3.1 through 3.7. In some cases, genomes can enter the replication cycle directly, but in others, genomes must first be modified and additional viral nucleic acids participate in the replication cycle. Examples of specific viruses in each class are provided.

DNA Genomes

The strategy of having DNA as a viral genome appears at first glance to be simplicity itself: the host genetic system is based on DNA, so viral genome replication and expression could simply emulate the host system. Many surprises await those who believe that this is all such a strategy entails.

Double-Stranded DNA (dsDNA) (Fig. 3.1)

There are 32 families of viruses with dsDNA genomes. Those that include vertebrate viruses are the *Adenoviridae*, *Asfarviridae*, *Herpesviridae*, *Papillomaviridae*, *Polyomaviridae*,

Figure 3.1 Structure and expression of viral double-stranded DNA genomes. (A) Synthesis of genomes, mRNA, and protein. The icon represents a polyomavirus particle. **(B to E)** Genome configurations. Ori, origin of replication; ITR, inverted terminal repeat; TP, terminal protein; L, long region; S, short region; UL, US, long and short unique regions; IRL, internal repeat sequence, long region; IRS, internal repeat sequence, short region; TRL, terminal repeat sequence, long region; TRS, terminal repeat sequence, short region; OriL, origin of replication of the long region; OriS, origin of replication of the short region.

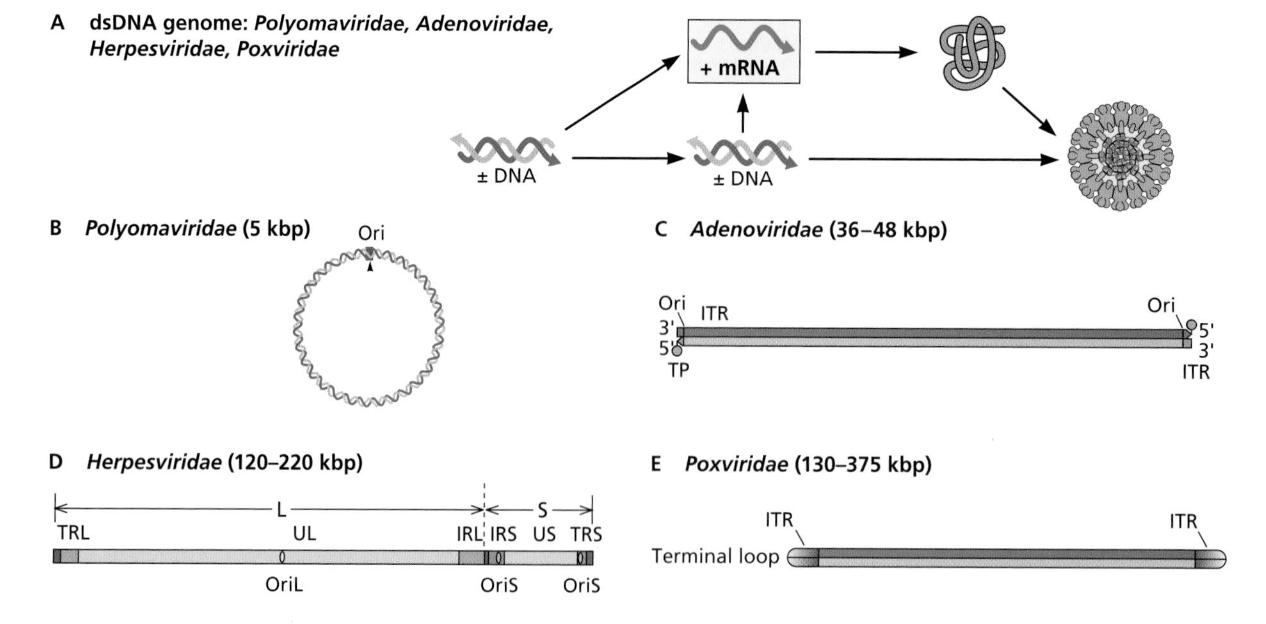

A Gapped, circular, dsDNA genome: *Hepadnaviridae*

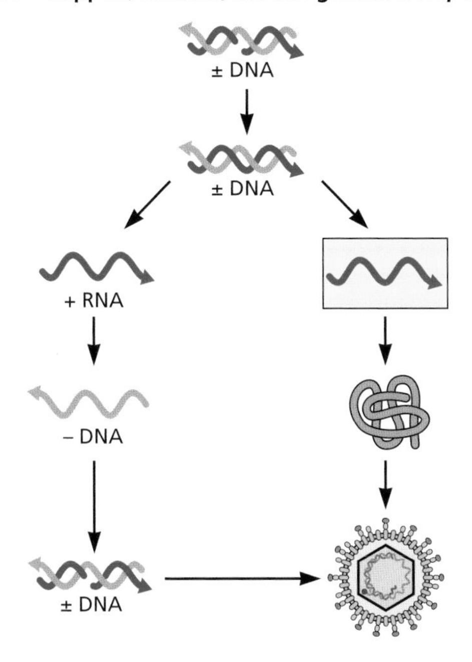

B *Hepadnaviridae* (3.4 kbp)

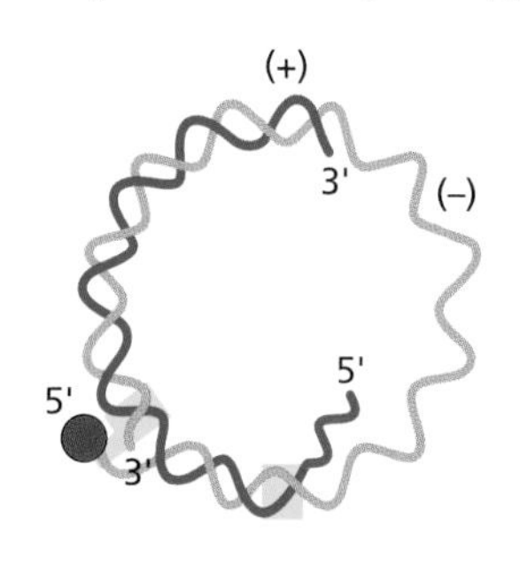

Figure 3.2 Structure and expression of viral gapped, circular, double-stranded DNA genomes. (A) Synthesis of genome, mRNA, and protein. **(B)** Configuration of the hepadnavirus genome.

Iridoviridae, and *Poxviridae* (Fig. 1.10). These genomes may be linear or circular. Genome replication and mRNA synthesis are accomplished by host or viral DNA-dependent DNA and RNA polymerases.

Gapped DNA (Fig. 3.2)

As the gapped DNA genome is partially double stranded, the gaps must be filled to produce perfect duplexes. This repair process must precede mRNA synthesis because the host RNA polymerase can transcribe only fully dsDNA. The unusual gapped DNA genome is produced from an RNA template by a virus-encoded enzyme, reverse transcriptase. Members of one virus family that infect vertebrates, the *Hepadnaviridae*, have a gapped DNA genome.

Single-Stranded DNA (ssDNA) (Fig. 3.3)

Seven families of viruses containing ssDNA genomes have been recognized; the families *Anelloviridae*, *Circoviridae*, and *Parvoviridae* include viruses that infect vertebrates. ssDNA must be copied into mRNA before proteins can be produced.

Figure 3.3 Structure and expression of viral single-stranded DNA genomes. (A) Synthesis of genomes, mRNA, and protein. **(B, C)** Genome configurations.

A ssDNA genome: *Circoviridae, Parvoviridae*

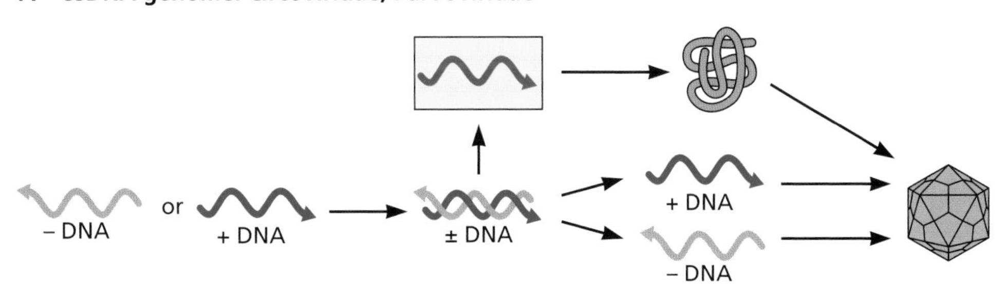

B *Circoviridae* (1.7–2.2 kb)

C *Parvoviridae* (4–6 kb)

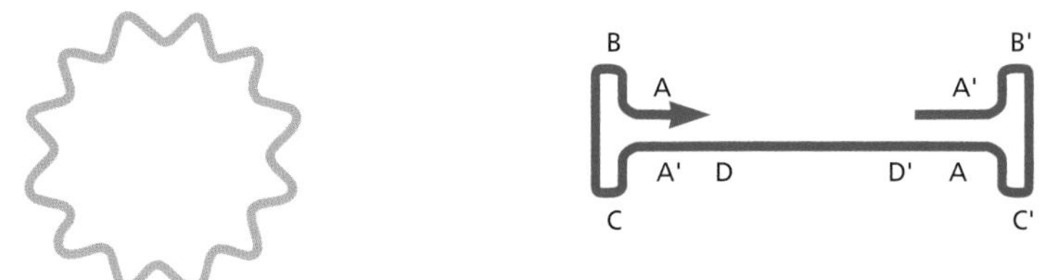

BOX 3.3

BACKGROUND

RNA synthesis in cells

There are no known host cell enzymes that can copy the genomes of RNA viruses. However, at least one enzyme, RNA polymerase II, can copy an RNA template. The 1.7-kb circular, ssRNA genome of hepatitis delta satellite virus is copied by RNA polymerase II to form multimeric RNAs (see the figure). How RNA polymerase II, an enzyme that produces pre-mRNAs from DNA templates, is reprogrammed to copy a circular RNA template is not known.

Hepatitis delta satellite (−) strand genome RNA is copied by RNA polymerase II at the indicated position. The polymerase passes the poly(A) signal (purple box) and the self-cleavage domain (red circle). For more information, see Fig. 6.14. Redrawn from J. M. Taylor, *Curr Top Microbiol Immunol* **239:**107–122, 1999, with permission.

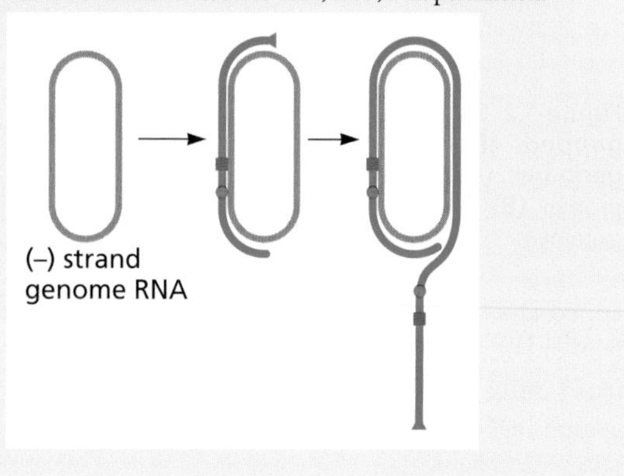

However, RNA can be made only from a dsDNA template, whatever the sense of the ssDNA. Consequently, some DNA synthesis **must** precede mRNA production in the replication cycles of these viruses. The single-stranded viral genome is produced by cellular DNA polymerases.

RNA Genomes

Cells have no RNA-dependent RNA polymerases that can replicate the genomes of RNA viruses or make mRNA from RNA templates (Box 3.3). One solution to this problem is that RNA virus genomes encode RNA-dependent RNA polymerases that produce RNA from RNA templates. The other solution, exemplified by retrovirus genomes, is reverse transcription of the genome to dsDNA, which can be transcribed by host RNA polymerase.

dsRNA (Fig. 3.4)

There are eight families of viruses with dsRNA genomes. The number of dsRNA segments ranges from 1 (*Totiviridae* and *Endornaviridae*, viruses of fungi, protozoa, and plants) to 9 to 12 (*Reoviridae*, viruses of fungi, invertebrates, plants, protozoa, and vertebrates). While dsRNA contains a (+) strand, it cannot be translated as part of a duplex to synthesize viral proteins. The (−) strand of the genomic dsRNA is first copied into mRNAs by a viral RNA-dependent RNA polymerase. Newly synthesized mRNAs are encapsidated and then copied to produce dsRNAs.

(+) Strand RNA (Fig. 3.5)

The (+) strand RNA viruses are the most plentiful on this planet; 29 families have been recognized [not counting (+) strand RNA viruses with DNA intermediates]. The families *Arteriviridae*, *Astroviridae*, *Caliciviridae*, *Coronaviridae*, *Flaviviridae*, *Hepeviridae*, *Nodaviridae*, *Picornaviridae*, and *Togaviridae* include viruses that infect vertebrates. (+) strand RNA genomes usually can be translated directly into protein by host ribosomes. The genome is replicated in two steps. The (+) strand genome is first copied into a full-length (−) strand, and the (−) strand is then copied into full-length (+) strand genomes. In some cases, a subgenomic mRNA is produced.

A dsRNA genome: *Reoviridae*

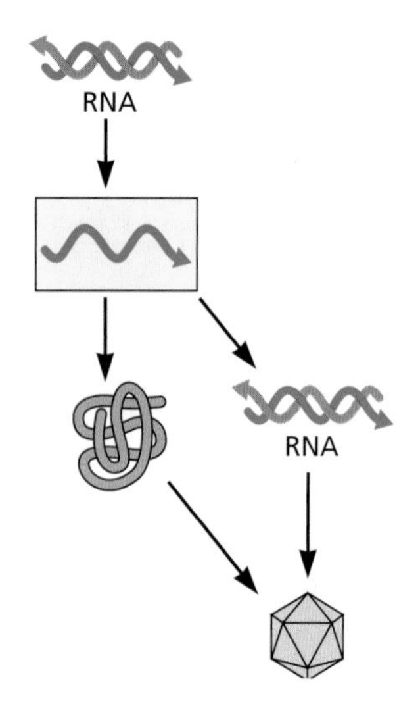

B *Reoviridae* (19–32 kbp in 10 dsRNA segments)

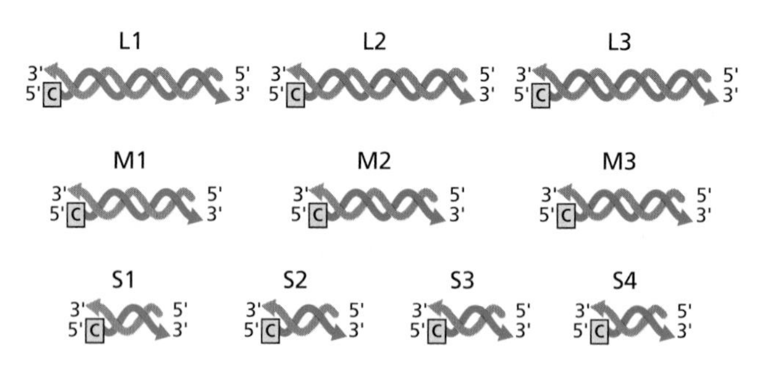

Figure 3.4 Structure and expression of viral double-stranded RNA genomes. (A) Synthesis of genomes, mRNA, and protein. **(B)** Genome configuration.

A ss (+) RNA: *Coronaviridae, Flaviviridae, Picornaviridae, Togaviridae*

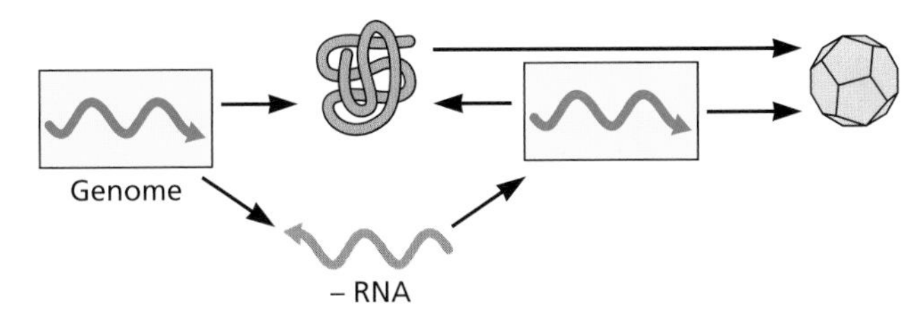

B *Coronaviridae* (28–33 kb)

B *Flaviviridae* (10–12 kb)

B *Picornaviridae* (7–8.5 kb)

B *Togaviridae* (10–13 kb)

Figure 3.5 Structure and expression of viral single-stranded (+) RNA genomes. (A) Synthesis of genomes, mRNA, and protein. **(B)** Genome configurations. UTR, untranslated region.

A ss (+) RNA with DNA intermediate: *Retroviridae*

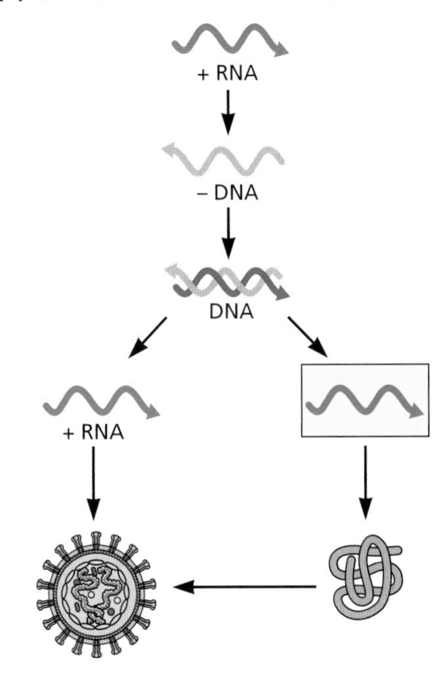

B *Retroviridae* (7–10 kb)

Figure 3.6 Structure and expression of viral single-stranded (+) RNA genomes with a DNA intermediate. (A) Synthesis of genomes, mRNA, and protein. **(B)** Genome configuration.

(+) Strand RNA with DNA Intermediate (Fig. 3.6)

In contrast to other (+) strand RNA viruses, the (+) strand RNA genome of retroviruses is converted to a dsDNA intermediate by viral RNA-dependent DNA polymerase (reverse transcriptase). This DNA then serves as the template for viral mRNA and genome RNA synthesis by cellular enzymes. There are three families of (+) strand RNA viruses with a DNA intermediate; members of the *Retroviridae* infect vertebrates.

(−) Strand RNA (Fig. 3.7)

Viruses with (−) strand RNA genomes are found in seven families. Viruses of this type that can infect vertebrates include members of the *Bornaviridae, Filoviridae, Orthomyxoviridae, Paramyxoviridae*, and *Rhabdoviridae* families. Unlike (+) strand RNA, (−) strand RNA genomes cannot be translated directly into protein, but must be first copied to make (+) strand mRNA. There are no enzymes in the cell that can make mRNAs from the RNA genomes of (−) strand RNA viruses. These virus particles therefore contain virus-encoded RNA-dependent RNA polymerases. The genome is also the template for the synthesis of full-length (+) strands, which in turn are copied to produce (−) strand genomes. Such RNA viral genomes can be either single molecules (nonsegmented; some viruses with this configuration have been classified in the order *Mononegavirales*) or segmented.

The genomes of certain (−) strand RNA viruses (e.g., members of the *Arenaviridae* and *Bunyaviridae*) are ambisense: they contain both (+) and (−) strand information on a single strand of RNA (Fig. 3.7C). The (+) sense information in the genome is translated upon entry of the viral RNA into cells. Replication of the RNA genome yields additional (+) sense information, which is then translated.

What Do Viral Genomes Look Like?

Some small RNA and DNA genomes enter cells from virus particles as naked molecules of nucleic acid, whereas others are always associated with specialized nucleic acid-binding proteins. A fundamental difference between the genomes of viruses and those of hosts is that although viral genomes are often covered with proteins, they are usually not bound by histones (polyomaviral and papillomaviral genomes are an exception).

A ss (–) RNA: *Orthomyxoviridae, Paramyxoviridae, Rhabdoviridae*

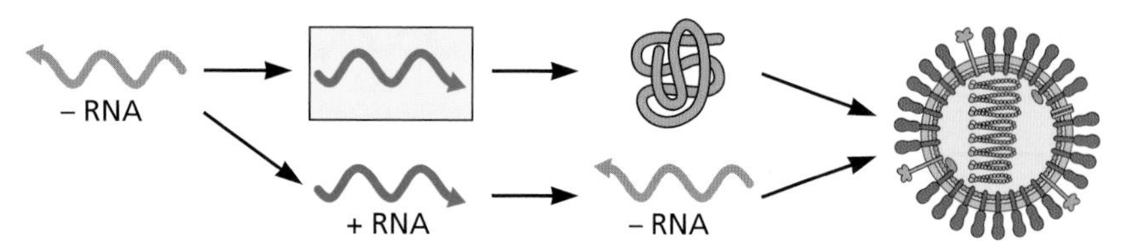

B Segmented genomes: *Orthomyxoviridae* (10–15 kb in 6–8 RNAs)

(–) strand RNA segments

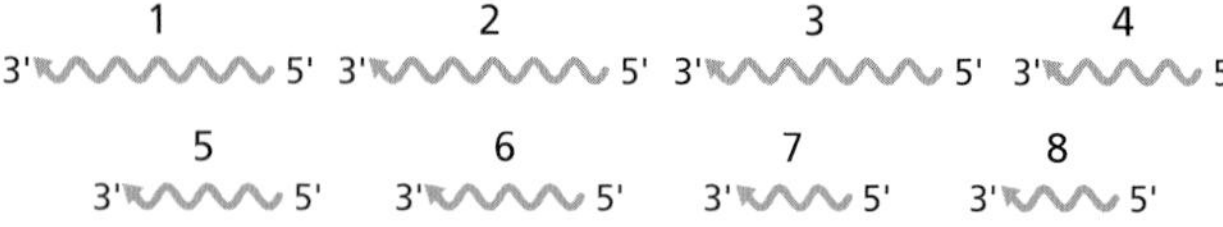

Nonsegmented genomes: *Paramyxoviridae* (15–16 kb)

3' 5'

***Rhabdoviridae* (13–16 kb)**

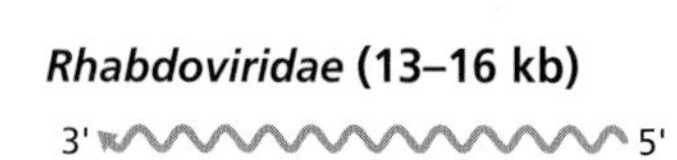

C Ambisense (–) strand RNA

***Arenaviridae* (11 kb in 2 RNAs)**
***Bunyaviridae* (12–23 kb in 3 RNAs)**

L RNA 5' C 3'

S RNA 5' C 3'

Figure 3.7 Structure and expression of viral single-stranded (−) RNA genomes. **(A)** Synthesis of genomes, mRNA, and protein. The icon represents an orthomyxovirus particle. **(B, C)** Genome configurations.

While viral genomes are all nucleic acids, they should **not** be thought of as one-dimensional structures. Virology textbooks (this one included) often draw genomes as straight, one-dimensional lines, but this notation is for illustrative purposes only; physical reality is certain to be dramatically different. Genomes have the potential to adopt amazing secondary and tertiary structures in which nucleotides may engage in long-distance interactions (Fig. 3.8).

The sequences and structures near the ends of viral genomes are often indispensable for viral replication (Fig. 3.9). For example, the DNA sequences at the ends of parvovirus genomes form T-shaped structures that are required for priming during DNA synthesis. Proteins covalently attached to 5′ ends, inverted and tandem repeats, and tRNAs may also participate in the replication of RNA and DNA genomes. Secondary RNA structures may facilitate translation (the internal ribosome entry site [IRES] of picornavirus genomes) and genome packaging (the structured packaging signal of retroviral genomes).

Coding Strategies

The compact genome of most viruses renders the "one gene, one mRNA" dogma inaccurate. Extraordinary tactics for information retrieval, such as the production of multiple subgenomic mRNAs, mRNA splicing, RNA editing, and nested transcription units (Fig. 3.10), allow the production of multiple proteins from a single viral genome. Further expansion of the coding capacity of the viral genome is achieved by posttranscriptional mechanisms, such as polyprotein synthesis, leaky scanning, suppression of termination, and ribosomal frameshifting. In general, the smaller the genome, the greater the compression of genetic information.

What Can Viral Sequences Tell Us?

Knowledge about the physical nature of genomes and coding strategies was first obtained by study of the nucleic acids of viruses. Indeed, DNA sequencing technology was perfected on viral genomes. The first genome of any kind to be sequenced was that of the *Escherichia coli* bacteriophage MS2, a linear ssRNA of 3,569 nucleotides. dsDNA genomes of larger viruses, such as herpesviruses and poxviruses (vaccinia virus), were sequenced completely by the 1990s. Since then, the complete sequences of >3,600 different viral genomes have been determined. Published viral genome sequences can be found at http://www.ncbi.nlm.nih.gov/genome/viruses/.

Viral genome sequences have many uses, including classification of viruses. Furthermore, sequence analysis has identified many relationships among diverse viral genomes, providing considerable insight into the origin of viruses. A great deal can also be learned from the lack of such relationships: >93% of the >2,500 genes of Pandoravirus salinus resemble nothing known. Consequently, their origin cannot be traced to any known cellular lineage, leading to the controversial suggestion that these giant viruses are derived from a now-extinct fourth domain of life. In outbreaks or epidemics of viral disease, even partial genome sequences can provide information about the identity of the infecting virus and its spread in different populations. New viral nucleic acid sequences can be associated with disease and characterized even in the absence of standard virological techniques (Volume II, Chapter 10). For example, human herpesvirus 8 was identified by comparing sequences present in diseased and nondiseased tissues.

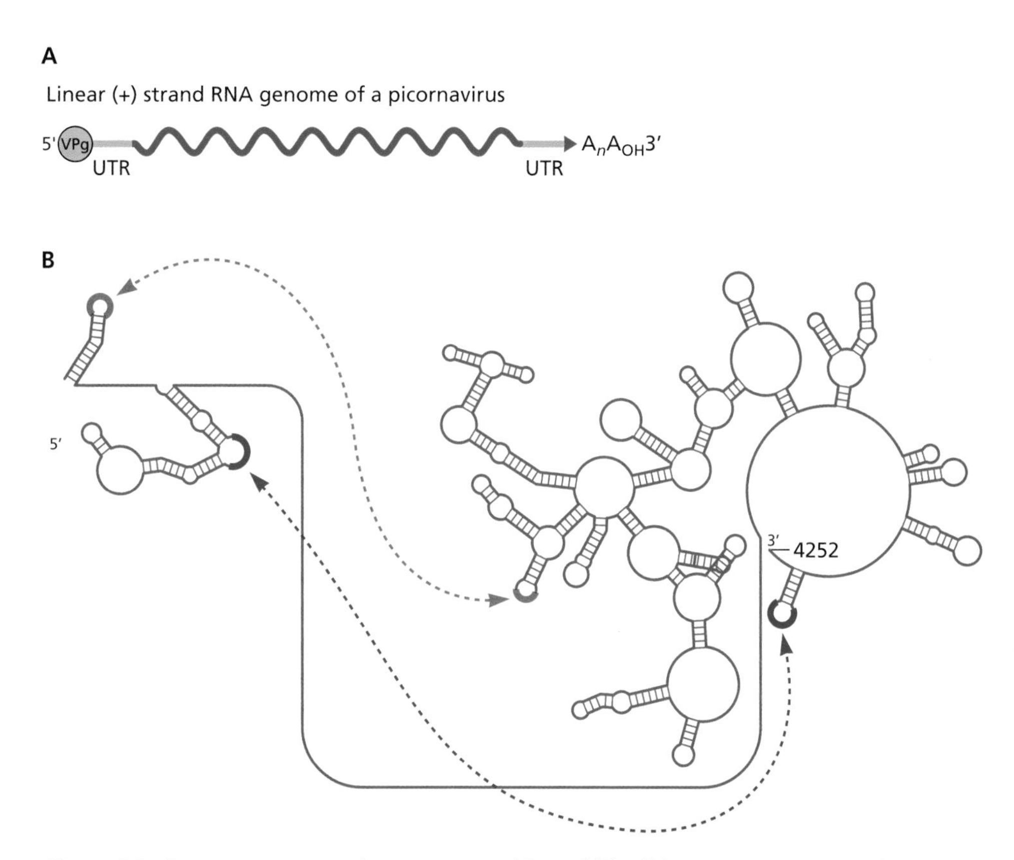

Figure 3.8 Genome structures in cartoons and in real life. (A) Linear representation of a picornavirus RNA genome. UTR, untranslated region. **(B)** Long-distance RNA-RNA interactions in a plant virus RNA genome. The 4,252-nucleotide viral genome is shown with secondary RNA structures at the 5′ and 3′ ends. Sequences that base pair are shown in blue (required for RNA frameshifting) and red (required to bring ribosomes from the 3′ end to the 5′ end). Courtesy of Anne Simon, University of Maryland.

Despite its utility, a complete understanding of how viruses reproduce cannot be obtained solely from the genome sequence or structure. The genome sequence of a virus is at best a biological "parts list": it provides some information about the intrinsic properties of a virus (e.g., predicted sequences of viral proteins and particle composition), but says little or nothing about how the virus interacts with cells, hosts, and populations. This limitation is best illustrated by the results of environmental metagenomic analyses, which reveal that the number of viruses around us (especially in the sea) is astronomical. Most are uncharacterized and, because their hosts are also unknown, cannot be studied. A reductionist study of individual components in isolation provides few answers. Although the reductionist approach is often experimentally the simplest, it is also important to understand how the genome behaves among others (population biology) and how the genome changes with time (evolution). Nevertheless, reductionism has provided much-needed detailed information for tractable virus-host systems. These systems allow genetic and biochemical analyses and provide models of infection *in vivo* and *in vitro*. Unfortunately, viruses and hosts that are difficult or impossible to manipulate in the laboratory remain understudied or ignored.

The Origin of Viral Genomes

The absence of bona fide viral fossils, i.e. ancient material from which viral nucleic acids can be recovered, might appear to make the origin of viral genomes an impenetrable mystery. However, the discovery of fragments of viral nucleic acids integrated into host genomes, coupled with an explosion in the determination of viral genome sequences, has allowed speculation on the evolutionary history of viruses. The origin of viruses is discussed in depth in Volume II, Chapter 10.

How viruses with DNA or RNA genomes arose is a compelling question. A predominant hypothesis is that RNA viruses are relics of the "RNA world," a period during which RNA was both genome and catalyst (no proteins yet existed). During this time, billions of years ago, life could have evolved from RNA, and the earliest organisms might have had RNA genomes. Viruses with RNA genomes might have evolved during this time. Later, DNA

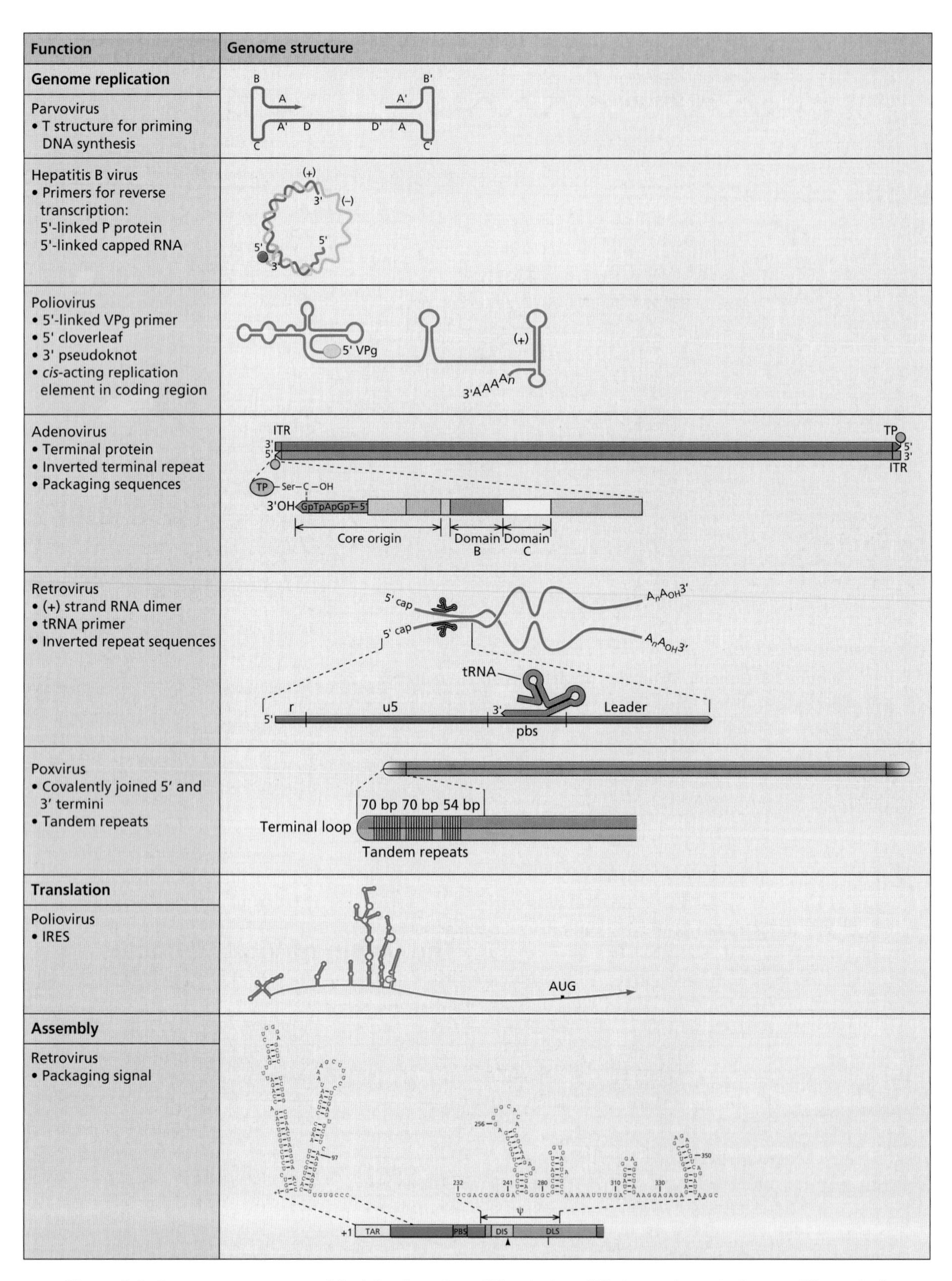

Figure 3.9 Genome structures critical for function. Abbreviations: ITR, inverted terminal repeat; TP, terminal protein; pbs and PBS, primer-binding site; IRES, internal ribosome entry site; TAR, trans-activating response element; DIS, dimerization initiation site; DLS, dimer linkage structure.

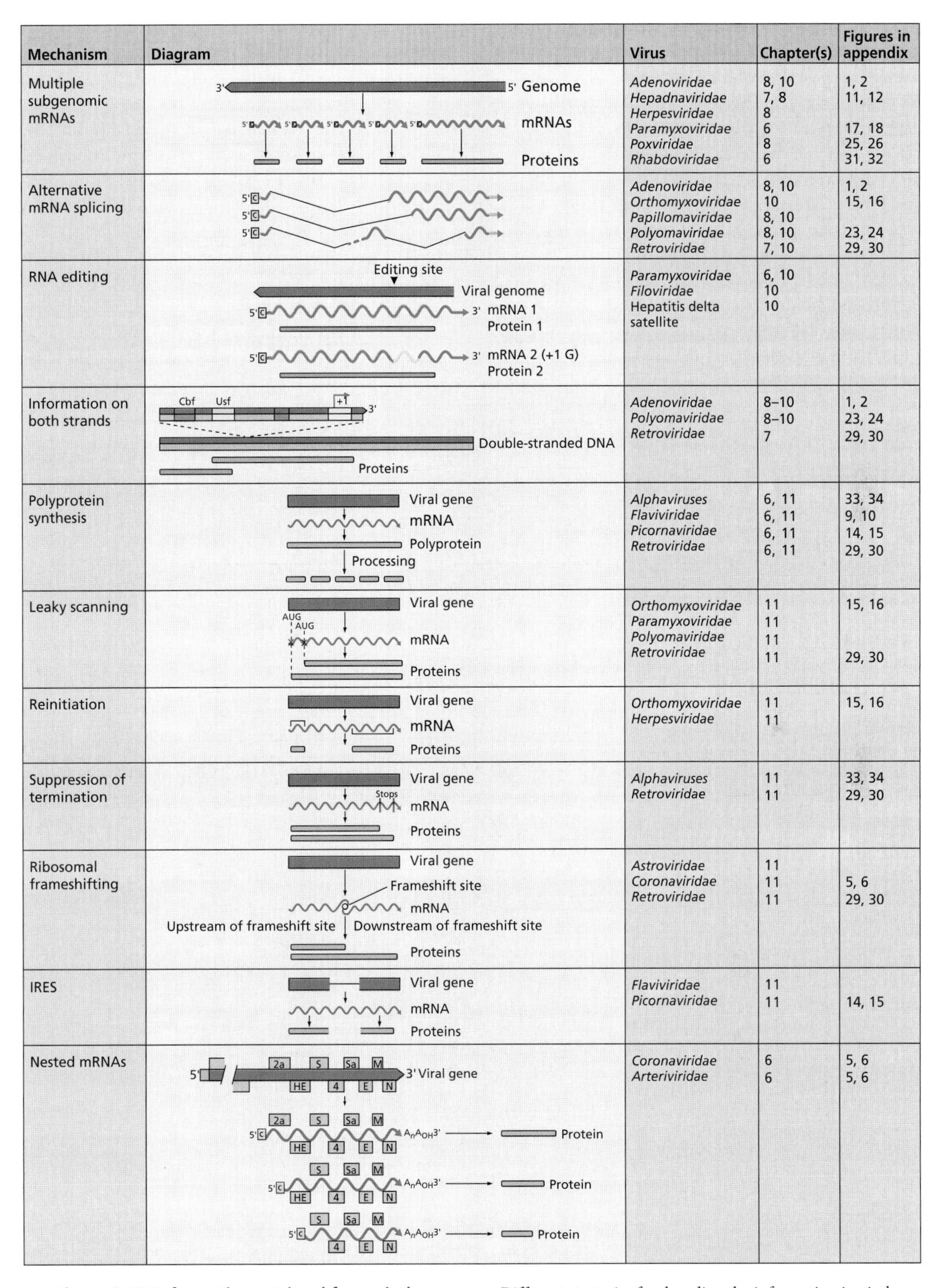

Mechanism	Diagram	Virus	Chapter(s)	Figures in appendix
Multiple subgenomic mRNAs	3' 5' Genome; mRNAs; Proteins	*Adenoviridae*	8, 10	1, 2
		Hepadnaviridae	7, 8	11, 12
		Herpesviridae	8	
		Paramyxoviridae	6	17, 18
		Poxviridae	8	25, 26
		Rhabdoviridae	6	31, 32
Alternative mRNA splicing	5'C	*Adenoviridae*	8, 10	1, 2
		Orthomyxoviridae	10	15, 16
		Papillomaviridae	8, 10	
		Polyomaviridae	8, 10	23, 24
		Retroviridae	7, 10	29, 30
RNA editing	Editing site; Viral genome; 5'C 3' mRNA 1; Protein 1; 5'C 3' mRNA 2 (+1 G); Protein 2	*Paramyxoviridae*	6, 10	
		Filoviridae	10	
		Hepatitis delta satellite	10	
Information on both strands	Cbf; Usf; +1; 3'; Double-stranded DNA; Proteins	*Adenoviridae*	8–10	1, 2
		Polyomaviridae	8–10	23, 24
		Retroviridae	7	29, 30
Polyprotein synthesis	Viral gene; mRNA; Polyprotein; Processing	*Alphaviruses*	6, 11	33, 34
		Flaviviridae	6, 11	9, 10
		Picornaviridae	6, 11	14, 15
		Retroviridae	6, 11	29, 30
Leaky scanning	Viral gene; AUG; AUG; mRNA; Proteins	*Orthomyxoviridae*	11	15, 16
		Paramyxoviridae	11	
		Polyomaviridae	11	
		Retroviridae	11	29, 30
Reinitiation	Viral gene; mRNA; Proteins	*Orthomyxoviridae*	11	15, 16
		Herpesviridae	11	
Suppression of termination	Viral gene; Stops; mRNA; Proteins	*Alphaviruses*	11	33, 34
		Retroviridae	11	29, 30
Ribosomal frameshifting	Viral gene; Frameshift site; mRNA; Upstream of frameshift site; Downstream of frameshift site; Proteins	*Astroviridae*	11	
		Coronaviridae	11	5, 6
		Retroviridae	11	29, 30
IRES	Viral gene; mRNA; Proteins	*Flaviviridae*	11	
		Picornaviridae	11	14, 15
Nested mRNAs	5' 2a S Sa M 3' Viral gene; HE 4 E N; 5'C $A_nA_{OH}3'$ Protein (×3)	*Coronaviridae*	6	5, 6
		Arteriviridae	6	5, 6

Figure 3.10 Information retrieval from viral genomes. Different strategies for decoding the information in viral genomes are depicted. Cbf, CCAAT-binding factor; Usf, upstream stimulatory factor; IRES, internal ribosome entry site.

replaced RNA as the genetic material, perhaps through the action of reverse transcriptases. With the emergence of DNA genomes came the evolution of DNA viruses. However, those with RNA genomes were and remain evolutionarily competitive, and hence they continue to survive to this day.

There is no evidence that viruses are monophyletic, i.e., descended from a common ancestor: there is no single gene shared by all viruses. Nevertheless, viruses with different genomes and replication strategies do share a small set of viral hallmark genes that encode icosahedral capsid proteins, nucleic acid polymerases, helicases, integrases, and other enzymes. There are only distant homologs of these hallmark genes in cellular genomes. It seems likely that the widespread presence of viral hallmark genes implies their ancient, possibly precellular origin.

Viral genomes display a greater diversity of genome composition, structure, and reproduction than any other organism. Understanding the function of such diversity is an intriguing problem. As viral genomes are survivors of constant selective pressure, all configurations must provide advantages. One possibility is that different genome configurations allow unique mechanisms for control over gene expression. These mechanisms include synthesis of a polyprotein from (+) strand RNA genomes or production of subgenomic mRNAs from (−) strand RNA genomes. There is some evidence that segmented RNA genomes might have arisen from monopartite genomes, perhaps to allow regulation of the production of individual proteins (Box 3.4). Segmentation probably did not emerge to increase genome size, as the largest RNA genomes are monopartite.

BOX 3.4

EXPERIMENTS

Origin of segmented RNA virus genomes

Segmented genomes are plentiful in the RNA virus world. They are found in virus particles from different families, and can be double stranded (*Reoviridae*) or single stranded, with (+) (*Closteroviridae*) or (−) (*Orthomyxoviridae*) polarity. Some experimental findings suggest that monopartite viral genomes emerged first, then later fragmented to form segmented genomes.

Insight into how a monopartite RNA genome might have fragmented to form a segmented genome comes from studies with the picornavirus foot-and-mouth disease virus (FMDV). The genome of this virus is a single molecule of (+) strand RNA. Serial passage of the virus in baby hamster kidney cells led to the emergence of genomes with two different large deletions (417 and 999 nucleotides) in the coding region. Neither mutant genome is infectious, but when they are introduced together into cells, an infectious virus population is produced. This population comprises a mixture of each of the two mutant genomes packaged separately into viral particles. Infection is successful because of complementation: when a host cell is infected with both particles, each genome provides the proteins missing in the other.

Further study of the deleted FMDV genomes revealed the presence of point mutations in other regions of the genome. These mutations had accumulated before the deletions appeared, and increased the fitness of the deleted genome compared with the wild-type genome.

These results show how monopartite viral RNAs may be divided, possibly a pathway to a segmented genome. It is interesting that the point mutations that gave the RNAs a fitness advantage over the standard RNA arose before fragmentation occurred—implying that the changes needed to occur in a specific sequence. The authors of the study conclude: "Thus, exploration of sequence space by a viral genome (in this case an unsegmented RNA) can reach a point of the space in which a totally different genome structure (in this case, a segmented RNA) is favored over the form that performed the exploration." While the fragmentation of the FMDV genome may represent a step on the path to segmentation, its relevance to what occurs in nature is unclear, because the results were obtained in cell culture.

A compelling picture of the genesis of a segmented RNA genome comes from the discovery of a new tick-borne virus in China, Jingmen tick virus (JMTV). The genome of this virus comprises four segments of (+) strand RNA. Two of the RNA segments have no known sequence homologs, while the other two are related to sequences of flaviviruses. The RNA genome of flaviviruses is not segmented: it is a single strand of (+) sense RNA. The proteins encoded by RNA segments 1 and 3 of JMTV are nonstructural proteins that are clearly related to the flavivirus NS5 and NS3 proteins (see the figure).

The genome structure of JMTV suggests that at some point in the past a flavivirus genome fragmented to produce the RNA segments encoding the NS3- and NS5-like proteins. This fragmentation might have initially taken place as shown for FMDV in cell culture, by fixing of deletion mutations that complemented one another. Next, coinfection of this segmented flavivirus with another unidentified virus took place to produce the precursor of JMTV.

The results provide new clues about the origins of segmented RNA viruses.

Moreno E, Ojosnegros S, García-Arriaza J, Escarmís C, Domingo E, Perales C. 2014. Exploration of sequence space as the basis of viral RNA genome segmentation. *Proc Natl Acad Sci U S A* **111**:6678–6683.

Qin XC, Shi M, Tian JH, Lin XD, Gao DY, He JR, Wang JB, Li CX, Kang YJ, Yu B, Zhou DJ, Xu J, Plyusnin A, Holmes EC, Zhang YZ. 2014. A tick-borne segmented RNA virus contains genome segments derived from unsegmented viral ancestors. *Proc Natl Acad Sci U S A* **111**:6744–6749.

RNA genome of JMTV virus. The viral genome comprises four segments of single-stranded, (+) sense RNA. Proteins encoded by each RNA are indicated. RNA segments 1 and 3 encode flavivirus-like proteins.

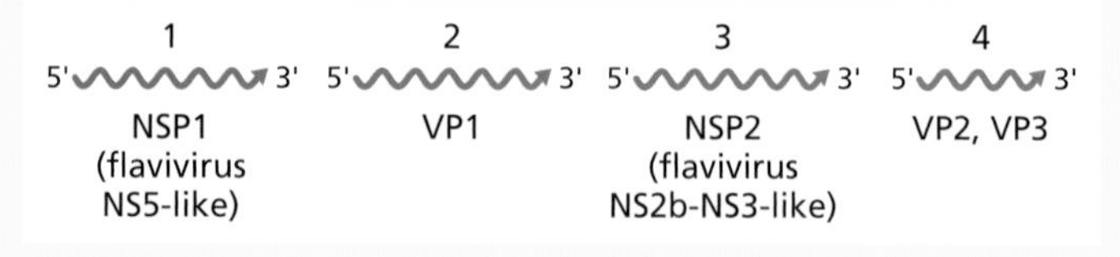

The "Big and Small" of Viral Genomes: Does Size Matter?

Currently, the prize for the smallest nondefective animal virus genome goes to members of the *Circoviridae* and *Anelloviridae*, which possess circular, ssDNA genomes of 1.7 to 2.2 kb and 2 to 4 kb, respectively (Fig. 3.3B). Members of the *Circoviridae* include agriculturally important pathogens of chickens and pigs; anelloviruses such as torque teno (TT) virus infect >90% of humans with no known consequence. The consolation prize goes to the *Hepadnaviridae*, such as hepatitis B virus, which causes hepatitis and liver cancer in millions of people. Its genome comprises 3.2 kb of gapped DNA (Fig. 3.2).

The largest known virus genome, a DNA molecule of 2,500 kbp, is that of Pandoravirus salinus, which infects amoebae. The largest RNA virus genome, 31 kb, is characteristic of some coronaviruses (Fig. 3.5). Despite detailed analyses, there is no evidence that one size is more advantageous than another. All viral genomes have evolved under relentless selection, so extremes of size must provide particular advantages. One feature distinguishing large genomes from smaller ones is the presence of many genes that encode proteins for viral genome replication, nucleic acid metabolism, and countering of host defense systems. In other words, these large viruses have sufficient coding capacity to escape some restrictions imposed by host cell biochemistry. The smallest genome of a free-living cell is predicted to comprise <300 genes (based on bacterial genome sequences). Remarkably, this number is smaller than the genetic content of large viral DNA genomes. Nevertheless, the big viruses are **not** cells: their replication absolutely requires the cellular translation machinery, as well as host cell systems to make membranes and generate energy.

The parameters that limit the size of viral genomes are largely unknown. There are cellular DNA and RNA molecules that are much longer than those found in virus particles. Consequently, the rate of nucleic acid synthesis is not likely to be limiting. In some cases, the capsid volume might limit genome size. There is a penalty inherent in having a large genome: a huge particle must be provided, and this is not a simple matter. In the case of the 150-kb herpes simplex virus genome, 50 to 60 gene products are needed to build the icosahedral nucleocapsid that houses the genome. This large number of protein products is encoded by 75% of the viral genome. Two of the largest known viral DNA genomes, those of Megavirus chilensis (1,259 kbp) and mamavirus (1,192 kbp), are housed in the biggest known capsids constructed with icosahedral symmetry. Although the principles of icosahedral symmetry are quite flexible in allowing a wide range of capsid sizes, it is possible that building a very large and stable capsid that can also come apart to release the viral genome is beyond the intrinsic properties of macromolecules.

One solution to the capsid size problem is to abandon icosahedral symmetry. Particles built with helical symmetry can in principle accommodate very large genomes, for example, baculoviruses with DNA genomes up to 180 kbp. The pandoraviruses, with the largest known DNA viral genomes (2,500 kbp), are housed in decidedly nonisometric ovoid particles 1 μm in length and 0.5 μm in diameter.

There is no reason to believe that the upper limit in viral particle and genome size has been reached. The core compartment of a mimivirus particle is larger than needed to accommodate the 1,200-kbp DNA genome. A particle of this size could, in theory, house a genome of 6 million bp if the DNA were packed at the same density as in polyomaviruses. Indeed, if the genome were packed into the particle at the same density reached in bacteriophages, it could be >12 million bp, the size of that of the smallest free-living unicellular eukaryote.

In cells, DNAs are much longer than RNA molecules. RNA is less stable than DNA, but in the cell, much of the RNA is used for the synthesis of proteins and therefore need not exceed the size needed to specify the largest polypeptide. However, this constraint does not apply to viral genomes. Yet the largest viral single-molecule RNA genomes, the 27- to 31-kb (+) strand RNAs of the coronaviruses, are dwarfed by the largest (2,500-kbp) DNA virus genomes. Susceptibility of RNA to nuclease attack might limit the size of viral RNA genomes, but there is little direct support for this hypothesis. The most likely explanation is that there are few known enzymes that can correct errors introduced during RNA synthesis. An exonuclease encoded in the coronavirus genome is one exception: its presence could explain the large size of these RNAs. DNA polymerases can eliminate errors during polymerization, a process known as proofreading, and remaining errors can also be corrected after synthesis is complete. The average error frequencies for RNA genomes are about 1 misincorporation in 10^4 or 10^5 nucleotides polymerized. In an RNA viral genome of 10 kb, a mutation frequency of 1 in 10^4 would produce about 1 mutation in every replicated genome. Hence, very long viral RNA genomes, perhaps longer than 32 kb, would sustain too many lethal mutations. Even the 7.5-kb genome of poliovirus exists at the edge of viability: treatment of the virus with the RNA mutagen ribavirin causes a >99% loss in infectivity after a single round of replication.

Genetic Analysis of Viruses

The application of genetic methods to study the structure and function of animal viral genes and proteins began with development of the plaque assay by Renato Dulbecco in 1952. This assay permitted the preparation of clonal stocks of virus, the measurement of virus titers, and a convenient system for studying viruses with conditional lethal mutations. Although a limited repertoire of classical genetic methods was available, the mutants that were isolated (Box 3.5) were invaluable in elucidating many aspects of infectious cycles and cell transformation. Contemporary methods of genetic analysis based

BOX 3.5

METHODS

Spontaneous and induced mutations

In the early days of experimental virology, mutant viruses could be isolated only by screening stocks for interesting phenotypes, for none of the tools that we now take for granted, such as restriction endonucleases, efficient DNA sequencing methods, and molecular cloning procedures, were developed until the mid- to late 1970s. RNA virus stocks usually contain a high proportion of mutants, and it is only a matter of devising the appropriate selection conditions (e.g., high or low temperature or exposure to drugs that inhibit viral reproduction) to select mutants with the desired phenotype from the total population. For example, the live attenuated poliovirus vaccine strains developed by Albert Sabin are mutants that were selected from a virulent virus stock (Volume II, Fig. 8.7).

The low spontaneous mutation rate of DNA viruses necessitated random mutagenesis by exposure to a chemical mutagen. Mutagens such as nitrous acid, hydroxylamine, and alkylating agents chemically modify the nucleic acid in preparations of virus particles, resulting in changes in base pairing during subsequent genome replication. Mutagens such as base analogs, intercalating agents, or UV light are applied to the infected cell to cause changes in the viral genome during replication. Such agents introduce mutations more or less at random. Some mutations are lethal under all conditions, while others have no effect and are said to be silent.

To facilitate identification of mutants, the population must be screened for a phenotype that can be identified easily in a plaque assay. One such phenotype is temperature-sensitive viability of the virus. Virus mutants with this phenotype reproduce well at low temperatures, but poorly or not at all at high temperatures. The permissive and nonpermissive temperatures are typically 33 and 39°C, respectively, for viruses that replicate in mammalian cells. Other commonly sought phenotypes are changes in plaque size or morphology, drug resistance, antibody resistance, and host range (that is, loss of the ability to reproduce in certain hosts or host cells).

on recombinant DNA technology confer an essentially unlimited scope for genetic manipulation; in principle, any viral gene of interest can be mutated, and the precise nature of the mutation can be predetermined by the investigator. Much of the large body of information about viruses and their lifestyles that we now possess can be attributed to the power of these methods.

Classical Genetic Methods

Mapping Mutations

Before the advent of recombinant DNA technology, it was extremely difficult for investigators to determine the locations of mutations in viral genomes. The **marker rescue** technique (described in "Introducing Mutations into the Viral Genome" below) was a solution to this problem, but before it was developed, other, less satisfactory approaches were exploited.

Recombination mapping can be applied to both DNA and RNA viruses. Recombination results in genetic exchange between genomes within the infected cell. The frequency of recombination between two mutations on a linear genome increases with the physical distance separating them. In practice, cells are coinfected with two mutants, and the frequency of recombination is calculated by dividing the titer of phenotypically wild-type virus (Box 3.6) obtained under restrictive conditions (e.g., high temperature) by the titer measured under permissive conditions (e.g., low temperature).

BOX 3.6

TERMINOLOGY

What is wild type?

Terminology can be confusing. Virologists often use terms such as "strains," "variants," and "mutants" to designate a virus that differs in some heritable way from a parental or wild-type virus. In conventional usage, the **wild type** is defined as the original (often laboratory-adapted) virus from which mutants are selected and which is used as the basis for comparison. A wild-type virus may **not** be identical to a virus isolated from nature. In fact, the genome of a wild-type virus may include numerous mutations accumulated during propagation in the laboratory. For example, the genome of the first isolate of poliovirus obtained in 1909 undoubtedly is very different from that of the virus we call wild type today. We distinguish carefully between laboratory wild types and new virus isolates from the natural host. The latter are called **field isolates** or **clinical isolates**.

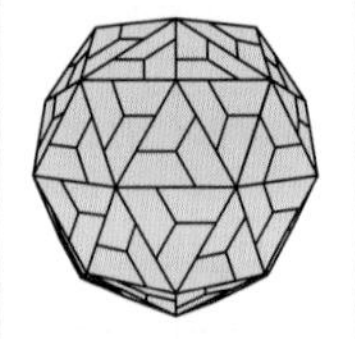

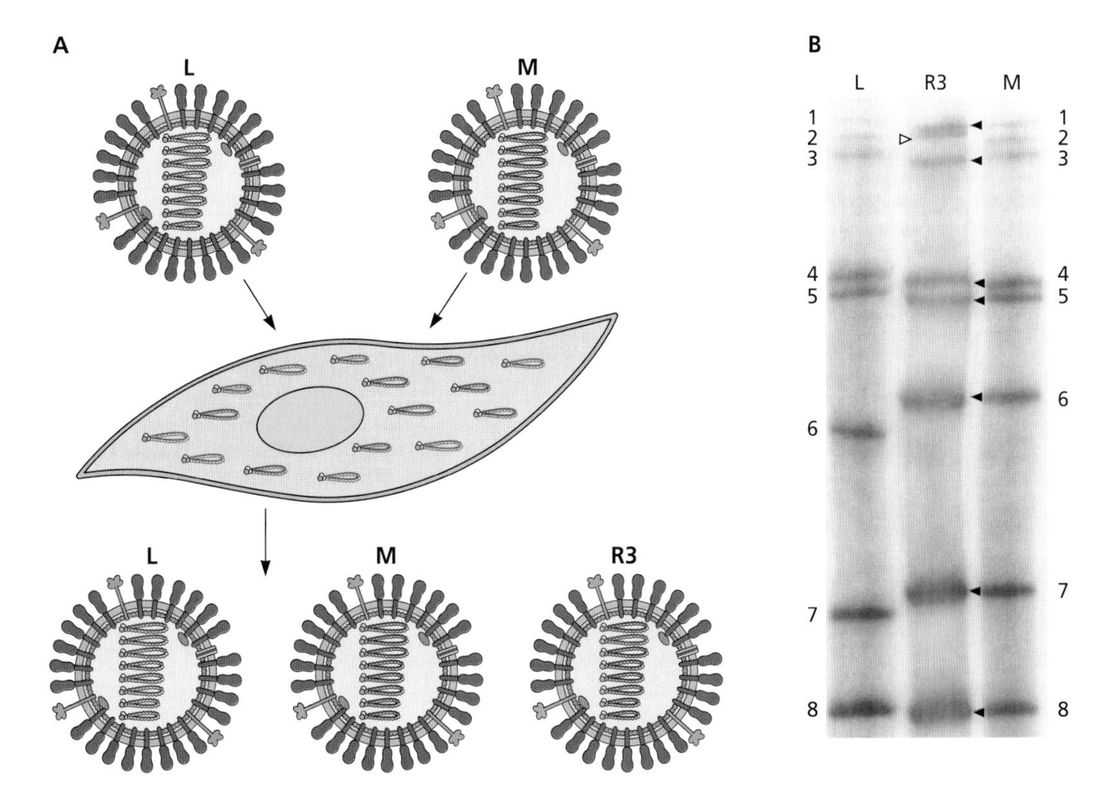

Figure 3.11 Reassortment of influenza virus RNA segments. (A) Progeny viruses of cells that are coinfected with two influenza virus strains, L and M, include both parents and viruses that derive RNA segments from them. Recombinant R3 has inherited segment 2 from the L strain and the remaining seven segments from the M strain. **(B)** ^{32}P-labeled influenza virus RNAs were fractionated in a polyacrylamide gel and detected by autoradiography. Migration differences of parental viral RNAs (M and L) permitted identification of the origin of RNA segments in the progeny virus R3. Panel B reprinted from V. R. Racaniello and P. Palese, *J Virol* **29**:361–373, 1979.

The recombination frequency between pairs of mutants is determined, allowing the mutations to be placed on a contiguous map. Although a location can be assigned for each mutation relative to others, this approach does not result in a physical map of the actual location of the base change in the genome.

In the case of RNA viruses with segmented genomes, the technique of **reassortment** allows the assignment of mutations to specific genome segments. When cells are coinfected with both mutant and wild-type viruses, the progeny includes **reassortants** that inherit RNA segments from either parent. The origins of the RNA segments can be deduced from their migration patterns during gel electrophoresis (Fig. 3.11) or by nucleic acid hybridization. By analyzing a panel of such reassortants, the segment responsible for the phenotype can be identified.

Functional Analysis

Complementation describes the ability of gene products from two different mutant viruses to interact functionally in the same cell, permitting viral reproduction. It can be distinguished from recombination or reassortment by examining the progeny produced by coinfected cells. True complementation yields only the two parental mutants, while wild-type genomes result from recombination or reassortment. If the mutations being tested are in separate genes, each virus is able to supply a functional gene product, allowing both viruses to be reproduced. If the two viruses carry mutations in the same gene, no reproduction will occur. In this way, the members of collections of mutants obtained by chemical mutagenesis were initially organized into complementation groups defining separate viral functions. In principle, there can be as many complementation groups as genes.

Engineering Mutations into Viral Genomes

Infectious DNA Clones

Recombinant DNA techniques have made it possible to introduce any kind of mutation anywhere in the genome of most animal viruses, whether that genome comprises DNA or RNA. The quintessential tool in virology today is the **infectious DNA clone**, a dsDNA copy of the viral genome that is carried on a bacterial vector such as a plasmid. Infectious DNA clones, or *in vitro* transcripts derived from them, can be introduced into cultured cells by **transfection** (Box 3.7) to

BOX 3.7

TERMINOLOGY

DNA-mediated transformation and transfection

The introduction of foreign DNA into cells is called DNA-mediated transformation to distinguish it from the oncogenic transformation of cells caused by tumor viruses and other insults. The term "transfection" (transformation-infection) was coined to describe the production of infectious virus after transformation of cells by viral DNA, first demonstrated with bacteriophage lambda. Unfortunately, the term "transfection" is now routinely used to describe the introduction of any DNA or RNA into cells. In this textbook, we use the correct nomenclature: the term "transfection" is restricted to the introduction of viral DNA or RNA into cells with the goal of obtaining virus reproduction.

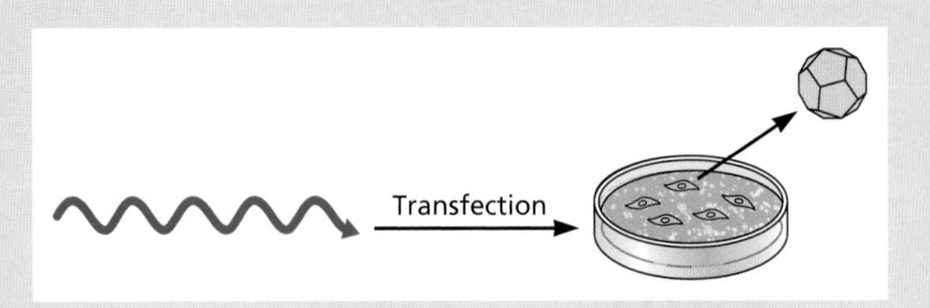

recover infectious virus. This approach is a modern validation of the Hershey-Chase experiment described in Chapter 1. The availability of site-specific bacterial restriction endonucleases, DNA ligases, and an array of methods for mutagenesis has made it possible to manipulate these infectious clones at will. Infectious DNA clones also provide a stable repository of the viral genome, a particularly important advantage for vaccine strains.

DNA viruses. Current genetic methods for the study of most viruses with DNA genomes are based on the infectivity of viral DNA. When deproteinized viral DNA molecules are introduced into permissive cells by transfection, they generally initiate a complete infectious cycle, although the infectivity (number of plaques per microgram of DNA) may be low. For example, the infectivity of deproteinized human adenoviral DNA is between 10 and 100 PFU per μg. When the genome is isolated by procedures that do not degrade the covalently attached terminal protein, infectivity is increased by 2 orders of magnitude, probably because this protein facilitates the assembly of initiation complexes on the viral origins of replication.

The complete genomes of polyomaviruses, papillomaviruses, and adenoviruses can be cloned in plasmid vectors, and such DNA is infectious under appropriate conditions. The DNA genomes of herpesviruses and poxviruses are too large to insert into conventional bacterial plasmid vectors, but they can be cloned in vectors that accept larger insertions (e.g., cosmids and bacterial artificial chromosomes). The plasmids containing such cloned herpesvirus genomes are infectious. In contrast, poxvirus DNA is not infectious, because the viral promoters cannot be recognized by cellular DNA-dependent RNA polymerase. Poxvirus DNA is infectious when early functions (viral DNA-dependent RNA polymerase and transcription proteins) are provided by a helper virus.

RNA viruses. *(+) strand RNA viruses.* The genomic RNA of retroviruses is copied into dsDNA by reverse transcriptase early during infection, a process described in Chapter 7. Such DNA is infectious when introduced into cells, as are molecularly cloned forms inserted into bacterial plasmids.

Introduction of a plasmid containing cloned poliovirus DNA into cultured mammalian cells results in the production of progeny virus (Fig. 3.12A). The mechanism by which cloned poliovirus DNA initiates infection is not known, but it has been suggested that the DNA enters the nucleus, where it is transcribed by cellular DNA-dependent RNA polymerase from cryptic, promoter-like sequences on the plasmid. The resulting (+) strand RNA transcripts initiate an infectious cycle. During genome replication, the extra terminal nucleotide sequences transcribed from the vector must be removed or ignored, because the virus particles that are produced contain RNA with the authentic 5′ and 3′ termini.

By incorporating promoters for bacteriophage T7 DNA-dependent RNA polymerase in plasmids containing poliovirus DNA, full-length (+) strand RNA transcripts can be synthesized *in vitro*. The specific infectivity of such RNA transcripts resembles that of genomic RNA (10^6 PFU per μg), which is higher than that of cloned DNA (10^3 PFU per μg). Infectious DNA clones have been constructed for many (+) strand RNA viruses.

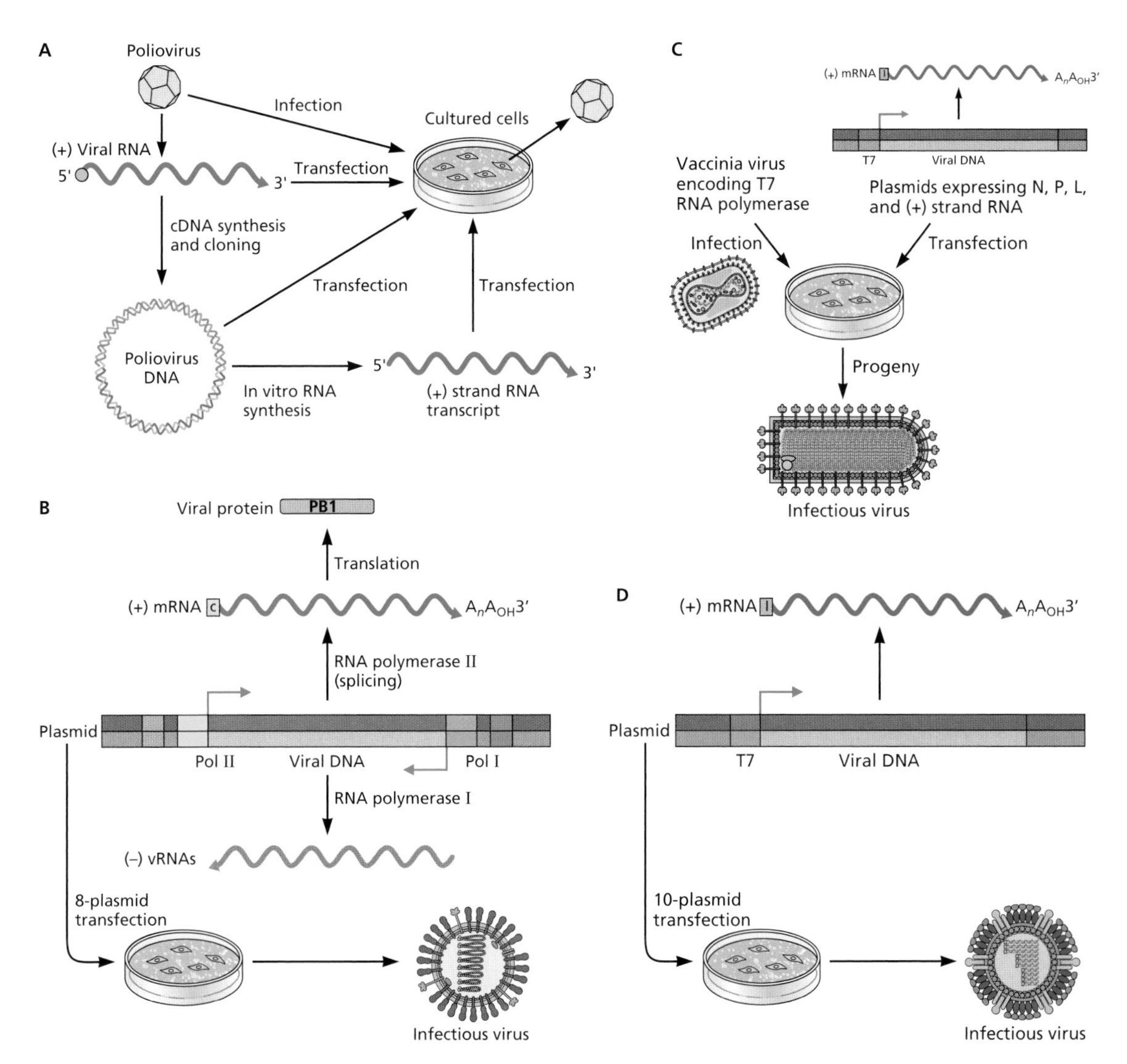

Figure 3.12 Genetic manipulation of RNA viruses. (A) Recovery of infectivity from cloned DNA of (+) strand RNA genomes as exemplified by genomic RNA of poliovirus, which is infectious when introduced into cultured cells by transfection. A complete DNA clone of the viral RNA (blue strands), carried in a plasmid, is also infectious, as are RNAs derived by *in vitro* transcription of the full-length DNA. **(B)** Recovery of influenza viruses by transfection of cells with eight plasmids. Cloned DNA of each of the eight influenza virus RNA segments is inserted between an RNA polymerase I promoter (Pol I, green) and terminator (brown), and an RNA polymerase II promoter (Pol II, yellow) and a polyadenylation signal (red). When the plasmids are introduced into mammalian cells, (−) strand viral RNA (vRNA) molecules are synthesized from the RNA polymerase I promoter, and mRNAs are produced by transcription from the RNA polymerase II promoter. The mRNAs are translated into viral proteins, and infectious virus is produced from the transfected cells. For clarity, only one cloned viral RNA segment is shown. Adapted from E. G. Hoffmann et al., *Proc Natl Acad Sci U S A* **97:**6108–6113, 2000, with permission. **(C)** Recovery of infectious virus from cloned DNA of viruses with a (−) strand RNA genome. Cells are infected with a vaccinia virus recombinant that synthesizes T7 RNA polymerase and transformed with plasmids that encode a full-length (+) strand copy of the viral genome RNA and proteins required for viral RNA synthesis (N, P, and L proteins). Production of RNA from these plasmids is under the control of the bacteriophage T7 RNA polymerase promoter (brown). Because bacteriophage T7 RNA transcripts are uncapped, an internal ribosome entry site (I) is included so the mRNAs will be translated. After the plasmids are transfected into cells, the (+) strand RNA is copied into (−) strands, which in turn are used as templates for mRNA synthesis and genome replication. The example shown is for viruses with a single (−) strand RNA genome (e.g., rhabdoviruses and paramyxoviruses). A similar approach has been demonstrated for Bunyamwera virus, with a genome comprising three (−) strand RNAs. **(D)** Recovery of infectious virus from cloned DNA of dsRNA viruses. Cloned DNA of each of the 10 reovirus dsRNA segments is inserted under the control of a bacteriophage T7 RNA polymerase promoter (brown). Because bacteriophage T7 RNA transcripts are uncapped, an internal ribosome entry site (I) is included so the mRNAs will be translated. Cells are infected with a vaccinia virus recombinant that synthesizes T7 RNA polymerase and transformed with all 10 plasmids. For clarity, only one cloned viral RNA segment is shown.

(−) strand RNA viruses. Genomic RNA of (−) strand RNA viruses is not infectious, because it can be neither translated nor copied into (+) strand RNA by host cell RNA polymerases, as discussed in Chapter 6. Two different experimental approaches have been used to develop infectious DNA clones of these viral genomes (Fig. 3.12B and C).

The recovery of influenza virus from cloned DNA is achieved by an expression system in which cloned DNA copies of the eight RNA segments of the viral genome are inserted between two promoters, so that complementary RNA strands can be synthesized (Fig. 3.12B). When the eight plasmids carrying DNA for each viral RNA segment are introduced into cells, infectious influenza virus is produced.

The full-length (−) strand RNA of viruses with a nonsegmented genome, such as vesicular stomatitis virus (a rhabdovirus), is not infectious, because it cannot be translated into protein or copied into mRNA by the host cell. When the full-length (−) strand is introduced into cells containing plasmids that produce viral proteins required for production of mRNA, no infectious virus is recovered. Unexpectedly, when a full-length (+) strand RNA is transfected into cells that synthesize the vesicular stomatitis virus nucleocapsid protein, phosphoprotein, and polymerase, the (+) strand RNA is copied into (−) strand RNAs. These RNAs initiate an infectious cycle, leading to the production of new virus particles.

dsRNA viruses. Genomic RNA of dsRNA viruses is not infectious because the (+) strand cannot be translated. The recovery of reovirus from cloned DNA is achieved by an expression system in which cloned DNA copies of the 10 RNA segments of the viral genome are inserted under the control of an RNA polymerase promoter (Fig. 3.12D). When 10 plasmids carrying DNA for each viral dsRNA segment are introduced into cells, infectious reovirus is produced.

Types of Mutation

Recombinant DNA techniques allow the introduction of many kinds of mutation at any desired site in cloned DNA (Box 3.8). Indeed, provided that the sequence of the segment of the viral genome to be mutated is known, there is little restriction on the type of mutation that can be introduced. **Deletion mutations** can be used to remove an entire gene to assess its role in reproduction, to produce truncated gene products, or to assess the functions of specific segments of a coding sequence. Noncoding regions can be deleted to identify and characterize regulatory sequences such as promoters. **Insertion mutations** can be made by the addition of any desired sequences. **Substitution mutations**, which can correspond to one or more nucleotides, are often made in coding or noncoding regions. Included in the former class are **nonsense mutations**, in which a termination codon is introduced, and **missense mutations**, in which a single nucleotide or a codon is changed, resulting in the synthesis of a protein with a single amino acid substitution. The introduction of a termination codon is frequently exploited to cause truncation of a membrane protein so that it is secreted or to eliminate the synthesis of a protein without changing the size of the viral genome or mRNA. Substitutions are used to assess the roles of specific nucleotides in regulatory sequences or of amino acids in protein function, such as polymerase activity or binding of a viral protein to a cell receptor.

Introducing Mutations into the Viral Genome

Mutations can be introduced rapidly into a viral genome when it is cloned in its entirety. Mutagenesis is usually carried out on cloned subfragments, which are then substituted into full-length cloned DNA. The final step is introduction of the mutagenized DNA into cultured cells by transfection. This approach has been applied to cloned DNA copies of RNA and DNA viral genomes.

BOX 3.8

TERMINOLOGY

Operations on nucleic acids and on protein

A mutation is a change in DNA or RNA comprising base changes and nucleotide additions, deletions, and rearrangements. When mutations occur in open reading frames, they can be manifested as changes in the synthesized proteins. For example, one or more base changes in a specific codon may produce a single amino acid substitution, a truncated protein, or no protein. The terms "mutation" and "deletion" are often used incorrectly, or ambiguously to describe alterations in proteins. In this textbook, these terms are used to describe genetic changes, and the terms "amino acid substitution" and "truncation" are used to describe protein alterations.

Introduction of mutagenized viral nucleic acid into cultured cells by transfection may produce one of several possible results. The mutation may have no effect on virus production, it may have a subtle effect, or it may impart a readily detectable phenotype.

Reversion Analysis

The phenotypes caused by mutation can **revert** in one of two ways: by change of the mutation to the wild-type sequence or by acquisition of a mutation at a second site, either in the same gene or a different gene. Phenotypic reversion caused by second-site mutation is known as **suppression**, or **pseudoreversion**, to distinguish it from reversion at the original site of mutation. Reversion has been studied since the beginnings of classical genetic analysis (Box 3.9). In the modern era of genetics, cloning and sequencing techniques can be used to demonstrate suppression and to identify the nature of the suppressor mutation (see below). The identification of suppressor mutations is a powerful tool for studying protein-protein and protein-nucleic acid interactions. Some suppressor mutations complement changes made at several sites, whereas **allele-specific** suppressors complement only a specific change. The allele specificity of second-site mutations provides evidence for physical interactions among proteins and nucleic acids.

Phenotypic revertants can be isolated either by propagating the mutant virus under restrictive conditions or, in the case of mutants exhibiting phenotypes (e.g., small plaques), by searching for wild-type properties. Chemical mutagenesis may be required to produce revertants of DNA viruses, but is not necessary for RNA viruses, which spawn mutants at a higher frequency. Nucleotide sequence analysis is then used to determine if the original mutation is still present in the genome of the revertant. The presence of the original mutation indicates that reversion has occurred by second-site mutation. Nucleotide sequence analysis is done to identify the suppressor mutation. The final step is introduction of the suspected suppressor mutation into the genome of the original mutant virus to confirm its effect. Several specific examples of suppressor analysis are provided below.

BOX 3.9

DISCUSSION

Is the observed phenotype due to the mutation?

In genetic analysis of viruses, mutations are made *in vitro* by a variety of techniques, all of which can introduce unexpected changes. Errors can be introduced during cloning, from PCR, during sequencing, and when the viral DNA or plasmid DNA is introduced into the eukaryotic cell.

With these potential problems in mind, how can it be concluded that a phenotype arises from the planned mutation? Here are some possible solutions.

- Repeat the construction. It is unlikely that an unlinked mutation with the same phenotype would occur twice.
- Look for marker rescue. Replace the mutation and all adjacent DNA with parental DNA. If the mutation indeed causes the phenotype, the wild-type phenotype should be restored in the rescued virus.
- Allow synthesis of the wild-type protein in the mutant background. If the wild-type phenotype is restored (complemented), then the probability is high that the phenotype arises from the mutation. The merit of this method over marker rescue is that the latter shows only that unlinked mutations are probably not the cause of the phenotype.

Each of these approaches has limitations, and it is therefore prudent to use more than one.

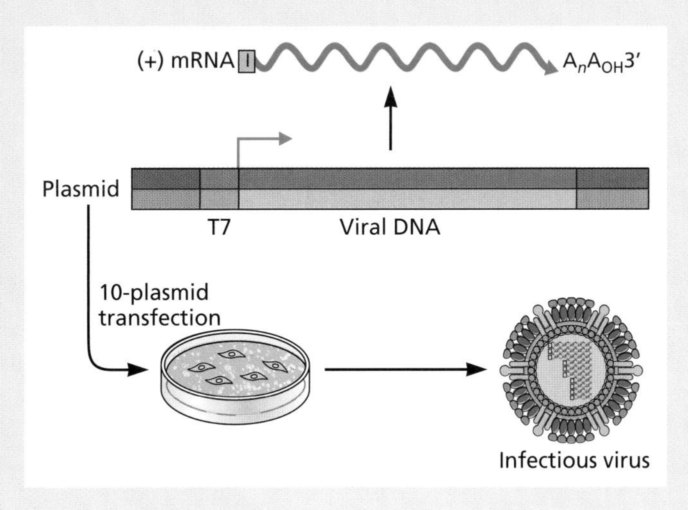

Some mutations within the origin of replication (Ori) of simian virus 40 reduce viral DNA replication and induce the formation of small plaques. Pseudorevertants of Ori mutants were isolated by random mutagenesis of mutant viral DNA followed by their introduction into cultured cells and screening for viruses that form large plaques. The second-site mutations that suppressed the replication defects were localized to a specific region within the gene for large T antigen. These results indicated that a specific domain of large T antigen interacts with the Ori sequence during viral genome replication.

The 5′ untranslated region of the poliovirus genome contains elaborate RNA secondary-structural features, which are important for RNA replication and translation, as discussed in Chapters 6 and 11, respectively. Disruption of such features by substitution of an 8-nucleotide sequence produces a virus that replicates poorly and readily gives rise to pseudorevertants that reproduce more efficiently (Fig. 3.13). Nucleotide sequence analysis of the genomes of two pseudorevertants demonstrated that they contain base changes that restore the disrupted secondary structure. These results confirm that the RNA secondary structure is important for the biological activity of this untranslated region.

RNA interference (RNAi)

RNA interference (Chapter 10) has become a powerful and widely used tool for analyzing gene function. In such analyses, duplexes of 21-nucleotide RNA molecules, called **small interfering RNAs (siRNAs)**, which are complementary to small regions of the mRNA, are synthesized chemically or by transcription reactions. siRNAs or plasmids that encode them are then introduced into cultured cells by transformation, and these small molecules block the production of specific proteins by inducing sequence-specific mRNA degradation or inhibition of translation. The functions of specific viral or cellular proteins during infection can therefore be studied by using this procedure (Fig. 3.14). In another application of this technology, thousands of siRNAs directed at all cellular mRNAs can be introduced into cells to identify genes that stimulate or block viral reproduction.

Targeted Gene Editing with CRISPR-Cas9

While the experimental use of RNAi can lead to reduced protein production, genomic manipulation by CRISPR-Cas9 (clustered regularly interspersed short palindromic repeat-CRISPR-associated nuclease 9) has advantages of complete depletion of the protein, and fewer effects on unintended targets. Though endogenous to bacteria and archaea (Box 10.11), the CRISPR-Cas9 system can be effectively and efficiently utilized to generate targeted gene disruptions in any genome. The specificity depends on the ability of the single-stranded guide RNAs (sgRNAs) to hybridize to the correct DNA sequence within the chromosome.

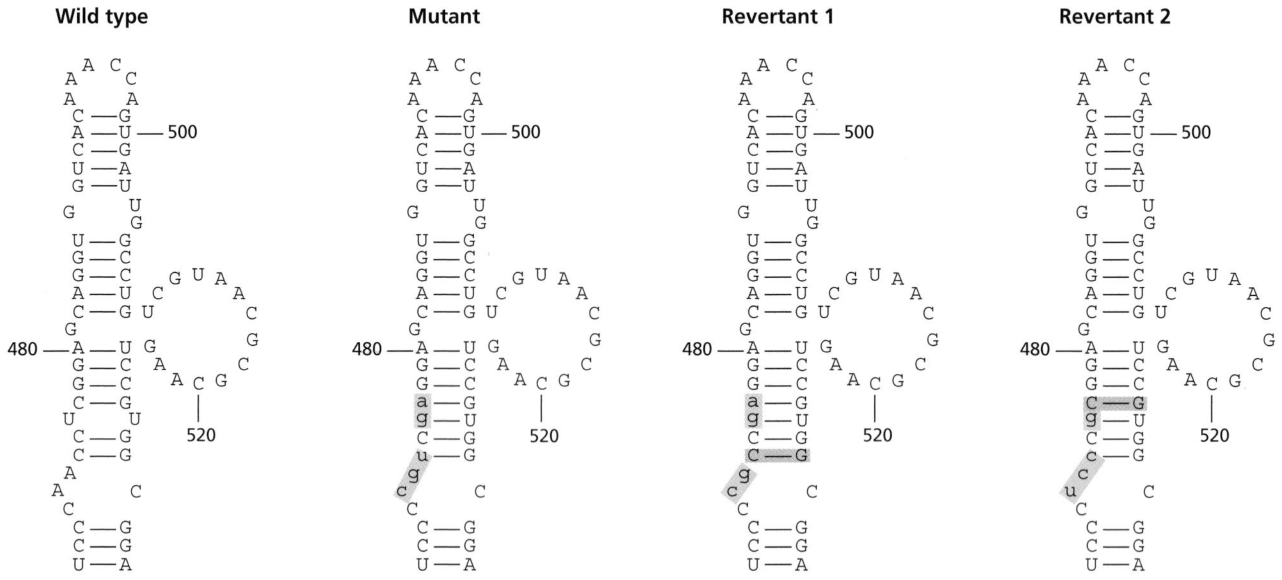

Figure 3.13 Effect of second-site suppressor mutations on predicted secondary structure in the 5′ untranslated region of poliovirus (+) strand RNA. Diagrams of the region between nucleotides 468 and 534, which corresponds to stem-loop V (Chapter 11), are shown. These include, from left to right, sequences of wild-type poliovirus type 1, a mutant containing the nucleotide changes highlighted in orange, and two phenotypic revertants. Two CG base pairs present in the wild-type parent and destroyed by the mutation are restored by second-site reversion (blue shading). Adapted from A. A. Haller et al., *J Virol* **70:**1467–1474, 1996, with permission.

Wild type **Mutant** **Revertant 1** **Revertant 2**

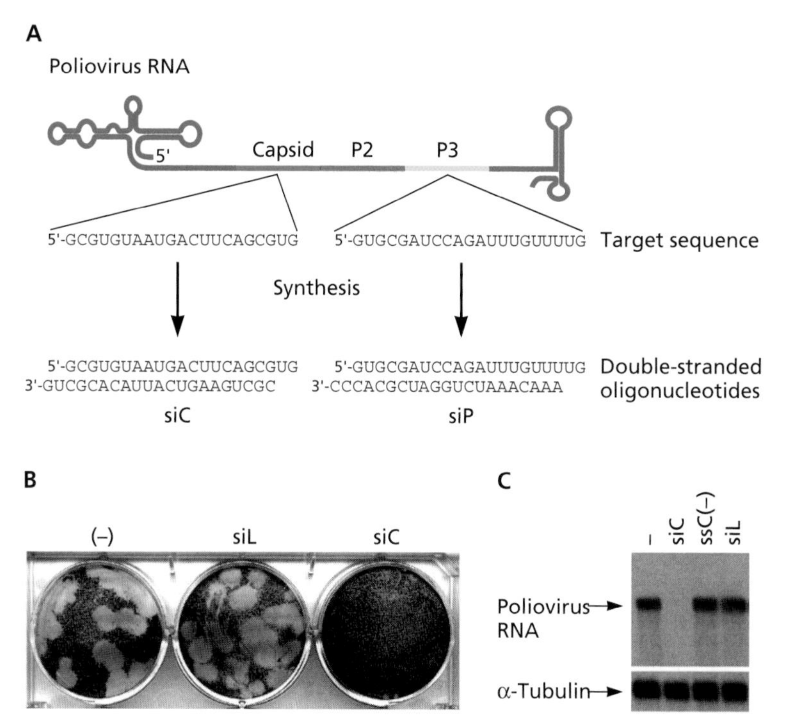

Figure 3.14 Inhibition of poliovirus replication by siRNA. siRNAs were introduced into cells by transformation, and the cells were then subjected to poliovirus infection. **(A)** Location of siRNAs siC and siP on a map of the poliovirus RNA genome. **(B)** Inhibition of plaque formation by siRNA siC. The number of plaques is not reduced in untreated cells (−) or when siRNA from *Renilla* luciferase is used (siL). Plaque formation was also inhibited with siP (not shown). **(C)** Northern blot analysis of RNA from poliovirus-infected cells 6 h after infection. Poliovirus RNA replication is blocked by siC but not by the (−) strand of siC RNA, ssC(−), or siL. The blot was rehybridized with a DNA probe directed against α-tubulin to ensure that all lanes contained equal amounts of RNA. Adapted from L. Gitlin et al., *Nature* **418:**430–434, 2002, with permission.

Once annealed, the endonuclease Cas9 catalyzes formation of a double-strand break, which is then repaired, creating frame-shifting insertion/deletion mutations within the gene. One advantage of using CRISPR-Cas9 methodology to genetically modify cell genomes is that the method can be applied to any cell type. Like siRNAs, CRISPR-Cas9 can be used to affect individual mRNAs or to identify cell genes that stimulate or block viral reproduction.

Engineering Viral Genomes: Viral Vectors

Naked DNA can be introduced into cultured animal cells as complexes with calcium phosphate or lipid-based reagents or directly by electroporation. Such DNA can direct synthesis of its gene products transiently or stably from integrated or episomal copies, respectively. Introduction of DNA into cells is a routine method in virological research and is also employed for certain clinical applications, such as the production of a therapeutic protein or a vaccine or the engineering of primary cells, progenitor cells, and stem cells for subsequent introduction into patients. However, this approach is not suitable for certain applications. For example, one goal of gene therapy is to deliver a gene to patients who either lack the gene or carry defective versions of it (Tables 3.1 and 3.2). The >7,000 monogenic human disorders, characterized by mutations in one gene, are especially amenable to viral gene therapy. In one application, DNA including the gene is introduced and expressed in cells recovered from the patient. After infusion into patients, the cells can become permanently established. If the primary cells to be used are limiting in a culture (e.g., stem cells), it is not practical to select and amplify the rare cells that receive naked DNA. Recombinant viruses carrying foreign genes can infect a greater percentage of cells and thus facilitate generation of the desired population. These viral vectors have also found widespread use in the research laboratory. A complete understanding of the structure and function of viral vectors requires knowledge of viral genome replication, a topic discussed in subsequent chapters for selected viruses and summarized in the Appendix.

Genetically engineered viruses are also being used to treat a wide variety of cancers, a field called viral oncotherapy (Box 3.10). Many tumor cells have defective innate immune signaling pathways and hence are susceptible to viral lysis. Viruses used for this purpose can be made more selective for tumor cells in a variety of ways. Another approach to viral oncotherapy is to utilize nonhuman viruses, such as the poxvirus myxoma virus and the picornavirus Seneca Valley virus, which can infect cells in human tumors but not normal tissues.

Design requirements for viral vectors include the use of an appropriate promoter; maintenance of genome size within the packaging limit of the particle; and elimination of viral virulence, the capacity of the virus to cause disease. Expression of foreign genes from viral vectors may be controlled by

Table 3.1 Clinical uses for viral vectors: some diseases being targeted in clinical trials of gene therapy with viral vectors

Disease	Defect	Incidence	Viral vector
Severe combined immunodeficiency	Adenosine deaminase (25% of patients)	Rare, <1 in 10^5 live births	Gammaretrovirus
	Common cytokine receptor γ chain (X-linked)	1 in 50,000–100,000 live births	Self-inactivating gammaretrovirus
Lipoprotein lipase deficiency	Lipoprotein lipase	Rare, 1–2 in 10^6 live births	AAV[a,b]
Hemophilia B	Factor IX deficiency	1 in 30,000 males	AAV
Hemoglobinopathies and thalassemias	Defects in α- or β-globin gene	1 in 600 in specific ethnic groups	Self-inactivating lentivirus
α_1-Antitrypsin deficiency (inherited emphysema, liver disease)	α_1-Antitrypsin not produced	1 in 3,500	AAV
Retinal degenerative disease, Leber's congenital amaurosis (LCA)	Retinal pigment epithelium-specific 65-kDa protein	<10% of LCA cases (LCA, ~1 in 80,000 live births)	AAV
X-linked adrenoleukodystrophy	ABCD1 transporter	1 in 20,000–50,000 live births	Self-inactivating lentivirus
Wiskott-Aldrich syndrome (eczema-thrombocytopenia-immunodeficiency syndrome)	Was protein	1–10 in 10^6 males	Self-inactivating lentivirus

[a]AAV, adenovirus-associated virus.

[b]Lipoprotein lipase gene therapy is approved for clinical use in Europe.

homologous or heterologous promoters and enhancers chosen to support efficient (e.g., the human cytomegalovirus immediate-early transcriptional control region) or cell-type-specific transcription, depending on the goals of the experiment. Such genes can be built directly into the viral genome or introduced by recombination in cells, as described above (see "Introducing Mutations into the Viral Genome"). The viral vector genome generally carries deletions and sometimes additional mutations. Deletion of some viral sequences is often required to overcome the limitations on the size of viral genomes that can be packaged in virus particles. For example, adenoviral DNA molecules more than 105% of the normal length are packaged very poorly. As this limitation would allow only 1.8 kbp of exogenous DNA to be inserted, adenovirus vectors often include deletions of the E3 gene (which is not essential for reproduction in cells in culture) and of the E1A and E1B transcription units, which encode proteins that can be provided by complementing cell lines.

Table 3.2 Clinical uses for viral vectors: some oncolytic viruses tested in clinical trials

Virus	Modification(s)	Delivery	Outcomes
Human adenovirus type 5 (e.g., ONYX-015, H101)	Deletion of E1B gene (increases virus reproduction in, and lysis of, tumor cells)	Intratumoral inoculation of tumors of head and neck	Decreased tumor volume in some patients when combined with chemotherapy; H101 in clinical use in China
Herpes simplex virus 1 (e.g., talimogene laherparepvec, aka OncoVEX)	Deletions in viral genes to confer tumor selectivity (ICP34.5, US11) or allow antigen presentation (ICP47); addition of cellular GM-CSF gene to stimulate tumor-specific immune responses	Intratumoral inoculation of malignant gliomas	Complete remission in 8 of 50 patients; improved overall survival
Vaccinia virus (JX-594)	Disruption of viral gene for ribonucleotide reductase (tumor selectivity); addition of human GM-CSF gene to stimulate tumor-specific immune responses	Intratumoral inoculation into primary and metastatic liver tumors	Decreased tumor volume in ~30% of patients; dose-dependent increase in survival time
Parvovirus (ParvOryx)	None	Myeloma	Phase 1 recruitment
Measles virus	Edmonton vaccine strain of measles virus; cannot block Stat1 and Mda5; addition of human gene for sodium-iodide symporter	Myeloma	2 of 2 patients resolved bone marrow plasmacytosis; 1 in complete remission
Poliovirus	Sabin vaccine strain with IRES from rhinovirus	Glioma	Phase 1 recruiting
Vesicular stomatitis virus	Addition of human interferon β gene	Hepatocellular carcinoma	Phase 1 recruiting
Murine leukemia virus	Amphotropic env gene added; addition of cytosine deaminase	Glioma	Phase 1/2

BOX 3.10

BACKGROUND

Viral oncotherapy

The use of viruses to treat cancer depends upon the ability of these agents to specifically infect and lyse cancer cells while not harming normal cells. These properties are made possible by a variety of tumor-specific abnormalities, including preferential production of certain proteins on the tumor cell surface that can serve as viral entry receptors; the enhanced activity of specific promoters and enhancers to drive expression of viral genes governing reproduction; the use of tumor-specific micro-RNAs to make viral gene expression cell specific; and the increased immunogenicity of tumor-specific antigens caused by the immune response to virus infection and the expression of immunostimulatory genes delivered by the vector.

Viruses from nine different families (*Adenoviridae*, *Picornaviridae*, *Herpesviridae*, *Paramyxoviridae*, *Parvoviridae*, *Reoviridae*, *Poxviridae*, *Retroviridae*, and *Rhabdoviridae*) are currently in clinical trials to test their safety and anticancer properties. The genomes of many viruses have been modified to confer greater efficacy and specificity for tumor cells. Oncolytic virotherapy has not been free of serious toxicities (see Volume II, Box 5.15), but in general, the treatments have been well tolerated after local or systemic injection.

A challenge to the development of oncolytic viruses is the host antiviral immune response, which can blunt therapeutic efficacy. Several approaches have been used to address this problem, including the substitution of structural proteins from different human or animal serotypes and the production of novel serotypes by chemical modification of virus particles. Different serotypes can be used when the patient is immune to the original vector, due to either previous infection or treatment.

Viral structural proteins may also be modified to bind proteins that are specific to the target cells, conferring greater specificity for tumor lysis. Such targeting may also involve postentry steps. For example, many tumor genes are expressed at aberrantly high levels; the promoters and enhancers responsible for such high expression have been identified and used to drive synthesis of viral genes encoding proteins that mediate in cell killing. Another approach to conferring specificity for tumor cells is to insert in the viral genome targets of micro-RNAs that are produced in nontumor cells.

Enhanced killing of tumor cells has also been achieved by inserting a gene in the viral vector that makes the cell more susceptible to destruction by drugs or immune therapies. An example is the insertion of the herpes simplex virus thymidine kinase gene, which converts prodrugs such as ganciclovir to a nucleoside analog that halts DNA synthesis. Insertion of the human sodium-iodide symporter gene into measles virus allows tumor cells to concentrate lethal beta-emitting isotopes. Oncolytic viruses have also been produced that carry the gene encoding granulocyte-macrophage colony-stimulating factor (GM-CSF). The synthesis of this protein stimulates proliferation of the eponymous cells that turn the adaptive immune system against the tumor cells.

From the first use of a vaccine strain of rabies virus to treat melanomatosis in the 1950s, our progress in understanding the biology of cancer, combined with the ability to genetically modify viruses by manipulation of infectious DNA clones, has led to the development of many rationally designed oncolytic viruses with greater clinical safety and efficacy.

Miest TS, Cattaneo R. 2014. New viruses for cancer therapy: meeting clinical needs. *Nat Rev Microbiol* **12**:23–34.

When viral vectors are designed for therapeutic purposes, it is essential to prevent their reproduction as well as destruction of target host cells. The deletions necessary to accommodate a foreign gene may contribute to such disabling of the vector. For example, the E1A protein-coding sequences that are invariably deleted from adenovirus vectors are necessary for efficient transcription of viral early genes; in their absence, viral yields from cells in culture are reduced by about 3 to 6 orders of magnitude (depending on the cell type). Removal of E1A coding sequences from adenovirus vectors is therefore doubly beneficial, although it is not sufficient to ensure that the vector cannot reproduce or induce damage in a host animal. Adenovirus-associated virus vectors are not lytic, obviating the need for such manipulations. As discussed in detail in Volume II, Chapter 8, production of virus vectors that do not cause disease can be more difficult to achieve.

As of this writing, >1,800 approved gene therapy clinical

Table 3.3 Some viral vectors

Virus	Insert size	Integration	Duration of expression	Advantages	Potential disadvantages
Adeno-associated virus	~5 kb	No	Long	Nonpathogenic, episomal, infects nondividing and dividing cells, broad tropism, low immunogenicity	Small packaging limit, helper virus needed for vector production
Adenovirus	~8–38 kb	No	Short	Efficient gene delivery, infects nondividing and dividing cells	Transient, immunogenic, high levels of preexisting immunity
Gammaretrovirus	8 kb	Yes	Short	Stable integration, broad tropism, low immunogenicity, low preexisting immunity	Risk of insertional mutagenesis, requires cell division
Herpes simplex virus	~50 kb	No	Long in central nervous system, short elsewhere	Infects nondividing cells, neurotropic, large capacity, broad tropism	Virulence, persistence in neurons, high levels of preexisting immunity, may recombine with genomes in latently infected cells
Lentivirus	9 kb	Yes	Long	Stable integration, transduces nondividing and dividing cells	Potential insertional mutagenesis; none detected in clinical trials
Rhabdovirus	~4.5 kb	No	Short	High-level expression, rapid cell killing, broad tropism, lack of preexisting immunity	Virulence, highly cytopathic, neurotropism, immunogenic
Vaccinia virus	~30 kb	No	Short	Wide host range, ease of isolation, large capacity, high-level expression, low preexisting immunity	Transient, immunogenic

trials have either been conducted or are in progress. These most often utilize adenovirus and retrovirus vectors, although poxvirus, adenovirus-associated virus, and herpes simplex virus vectors are also used. Cancer is the most common disease treated, followed by monogenetic and cardiovascular diseases.

A summary of viral vectors is presented in Table 3.3, and examples are discussed below.

DNA Virus Vectors

One goal of gene therapy is to introduce genes into terminally differentiated cells. Such cells normally do not divide, and they cannot be propagated in culture. Moreover, the organs they comprise cannot be populated with virus-infected cells. DNA virus vectors have been developed to overcome some of these problems.

Adenovirus vectors were originally developed for the treatment of cystic fibrosis because of the tropism of the virus for the respiratory epithelium. Adenovirus can infect terminally differentiated cells, but only transient gene expression is achieved, as this viral DNA is not integrated into host cell DNA. Adenoviruses carrying the cystic fibrosis transmembrane conductance regulator gene, which is defective in patients with this disease, have been used in clinical trials. Many other gene products with therapeutic potential have been produced from adenovirus vectors in a wide variety of cell types. In the earliest vectors that were designed, foreign genes were inserted into the E1 and/or E3 regions. As these vectors had limited capacity, genomes with minimal adenovirus sequences have been designed (Fig. 3.15). This strategy allows up to 38 kb of foreign sequence to be introduced into the vector. In addition, elimination of most viral genes reduces the host immune response to viral proteins, simplifying multiple immunizations. Considerable efforts have been made to modify the adenovirus capsid to target the vectors to different cell types. The fiber protein, which mediates adenovirus binding to cells, has been altered by insertion of ligands that bind particular cell surface receptors. Such alter-

Figure 3.15 Adenovirus vectors. High-capacity adenovirus vectors are produced by inserting a foreign gene and promoter into the viral E1 region, which has been deleted. The E3 region also has been deleted. Two *loxP* sites for cleavage by the Cre recombinase have been introduced into the adenoviral genome (black arrowheads). Infection of cells that produce Cre leads to excision of sequences flanked by the *loxP* sites. The result is a "gutless" vector that contains only the origin-of-replication-containing inverted terminal repeats (ITR), the packaging signal (yellow), the viral E4 transcription unit (orange), and the transgene with its promoter (green). Additional DNA flanking the foreign gene must be inserted to allow packaging of the viral genome (not shown). Adapted from A. Pfeifer and I. M. Verma, *in* D. M. Knipe et al. (ed.), *Fields Virology*, 4th ed. (Lippincott Williams & Wilkins, Philadelphia, PA, 2001), with permission.

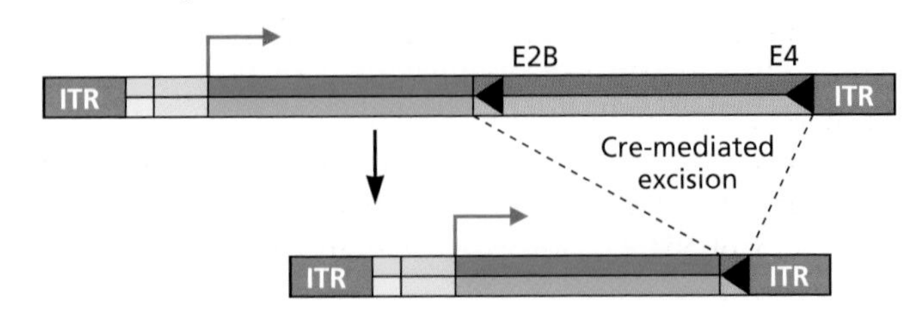

BOX 3.11

EXPERIMENTS

Restoring vision with viral gene therapy

Leber's congenital amaurosis is an autosomal recessive disease characterized by blindness as a consequence of retinal degeneration. Subretinal injection of adenovirus-associated virus vectors carrying the *RPE65* gene (figure) has restored patient vision is several clinical trials.

Mutations in the *RPE65* gene, which encodes a protein required for photoreceptor function in the retinal epithelium, account for ~10% of Leber's congenital amaurosis cases. Consequently, gene replacement has been studied as a therapeutic strategy for treatment of this disease. Human *RPE65* cDNA was packaged into an adenovirus-associated virus vector under the control of a chicken β-actin promoter. Infection of cells in culture with this virus, AAV2.hRPE65, leads to production of RPE65 protein. Introduction of this vector behind the retina of affected dogs led to sustained reversal of the visual deficit.

The safety and efficacy of AAV2.hRPE65 was assessed in three independent clinical trials. The results indicate that the vector is safe and in many cases leads to visual improvement for up to 1.5 years. These successful trials are likely to lead to licensure of this therapy to treat Leber's congenital amaurosis.

(See also the interview with Dr. Katherine High: http://bit.ly/Virology_High)

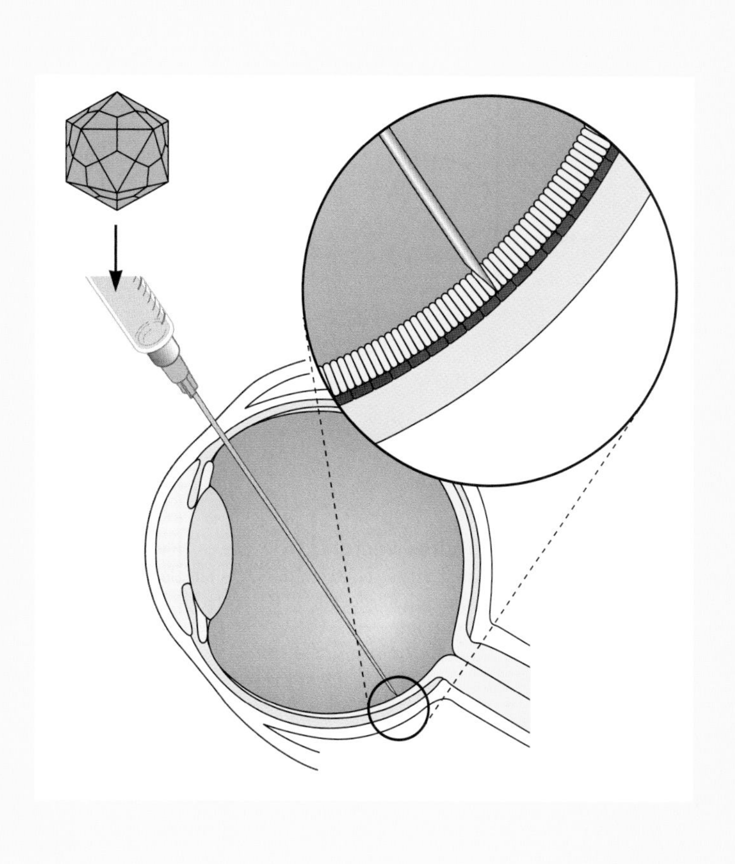

Pierce EA, Bennett J. 2015. The status of *RPE65* gene therapy trials: safety and efficacy. *Cold Spring Harb Perspect Med* doi:10.1101/cshperspect.a017285.

ations could increase the cell specificity of adenovirus attachment and the efficiency of gene transfer and thereby decrease the dose of virus that need be administered.

Adenovirus-associated virus has attracted much attention as a vector for gene therapy (Box 3.11). Genomes packaged into recombinant viruses replicate as an episome and persist, in some cases with high levels of expression, in many different tissues. There has been increasing interest in these vectors to target therapeutic genes to smooth muscle and other differentiated tissues, which are highly susceptible and support sustained high-level expression of foreign genes. Although the first-generation adenovirus-associated virus vectors were limited in the size of inserts that could be transferred, other systems have been developed to overcome the limited genetic capacity (Fig. 3.16). The cell specificity of adenovirus-associated virus vectors has been altered by inserting receptor-specific ligands into the capsid. In addition, many new viral serotypes that vary in their tropism and ability to trigger immune responses have been identified.

Vaccinia virus and other animal poxvirus vectors offer the advantages of a wide host range, a genome that accepts very large fragments, high expression of foreign genes, and relative ease of preparation. Foreign DNA is usually inserted into the viral genome by homologous recombination, using an approach similar to that described for marker transfer. Because of the relatively low pathogenicity of the virus, vaccinia virus recombinants have been considered candidates for human and animal vaccines.

RNA Virus Vectors

A number of RNA viruses have also been developed as vectors for foreign gene expression (Table 3.3). Vesicular stomatitis virus, a (−) strand RNA virus, has emerged as a

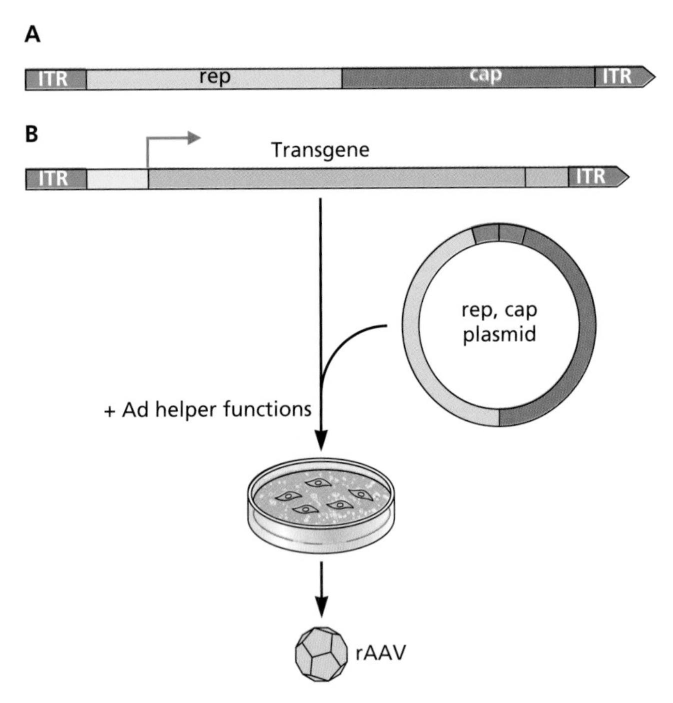

Figure 3.16 Adeno-associated virus vectors. (A) Map of the genome of wild-type adeno-associated virus. The viral DNA is single stranded and flanked by two inverted terminal repeats (ITR); it encodes capsid (blue) and nonstructural (orange) proteins. **(B)** In one type of vector, the viral genes are replaced with the transgene (pink) and its promoter (yellow) and a poly(A) addition signal (green). These DNAs are introduced into cells that have been engineered to produce capsid proteins, and the vector genome is encapsidated into virus particles. A limitation of this vector structure is that only 4.1 to 4.9 kb of foreign DNA can be packaged efficiently. Ad, adenovirus; rAAV, recombinant adenovirus-associated virus. Adapted from A. Pfeifer and I. M. Verma, *in* D. M. Knipe et al. (ed.), *Fields Virology*, 4th ed. (Lippincott Williams & Wilkins, Philadelphia, PA, 2001) with permission.

candidate for vaccine delivery (e.g., *Ebolavirus* vaccines) and for viral oncotherapy. The virus is well suited for the latter application because it reproduces preferentially in tumor cells, and recombinant vesicular stomatitis viruses have been engineered to improve tumor selectivity.

Retroviruses have enjoyed great popularity as vectors (Fig. 3.17) because their infectious cycles include the integration of a dsDNA copy of viral RNA into the cell genome, a topic of Chapter 7. The integrated provirus remains permanently in the cell's genome and is passed on to progeny during cell division. This feature of retroviral vectors results in permanent modification of the genome of the infected cell. The choice of the envelope glycoprotein carried by retroviral vectors has a significant impact on their tropism. The vesicular stomatitis virus G glycoprotein is often used because it confers a wide tissue tropism. Retrovirus vectors can be targeted to specific cell types by using other viral envelope proteins.

An initial problem encountered with the use of retroviruses in correcting genetic deficiencies is that only a few cell types can be infected by the commonly used murine retroviral vectors, and the DNA of these viruses can be integrated efficiently only in actively dividing cells. Often the cells that are targets of gene therapy, such as hepatocytes and muscle cells, do not divide. This problem can be circumvented if ways can be found to induce such cells to divide before being infected with the retrovirus. Another important limitation of the murine retrovirus vectors is the phenomenon of gene silencing, which represses foreign gene expression in many cells. An alternative approach is to use viral vectors that contain sequences from human immunodeficiency virus type 1 or other lentiviruses, which can infect nondividing cells and are less severely affected by gene silencing.

Perspectives

The information presented in this chapter can be used as a "road map" for navigating this book and for planning a virology course. Figures 3.1 to 3.7 serve as the points of departure for detailed analyses of the principles of virology. They illustrate seven strategies based on viral mRNA synthesis and genome replication. The material in this chapter can be used to structure individual reading or to design a virology course based on specific viruses or groups of viruses while adhering to the overall organization of this textbook by function. Refer to this chapter and the figures to find answers to questions about specific viruses. For example, Fig. 3.5 provides information about (+) strand RNA viruses and Fig. 3.10 indicates specific chapters in which these viruses are discussed.

Since the earliest days of experimental virology, genetic analysis has proven invaluable for studying the viral genome. Initially, methods were developed to produce viral mutants by chemical or UV mutagenesis, followed by screening for readily identifiable phenotypes. Because it was not possible to identify the genetic changes in such mutants, it was difficult to associate proteins with virus-specific processes. This limitation vanished with the development of infectious DNA clones of viral genomes, an achievement that enabled the introduction of defined mutations into any region of the viral genome. This complete genetic toolbox provides countless possibilities for studying the viral genome, limited only by the creativity and enthusiasm of the investigator. The ability to manipulate cloned DNA copies of viral genomes has also ushered in a new era of virus-based therapies. It may soon be possible to use viruses to treat genetic diseases and cancer and to deliver vaccines to prevent infectious diseases.

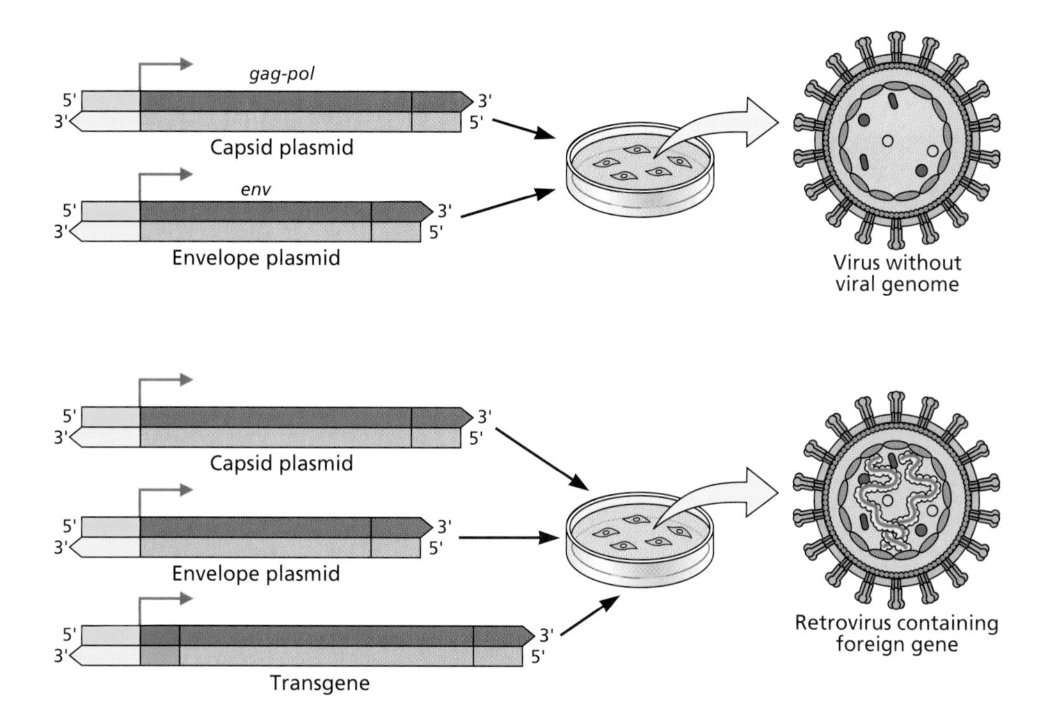

Figure 3.17 Retroviral vectors. The minimal viral sequences required for retroviral vectors are 5′- and 3′-terminal sequences (yellow and blue, respectively) that control gene expression and packaging of the RNA genome. The foreign gene (blue) and promoter (green) are inserted between the viral sequences. To package this DNA into viral particles, it is introduced into cultured cells with plasmids that encode viral proteins required for encapsidation, under the control of a heterologous promoter and containing no viral regulatory sequences. No wild-type viral RNA is present in these cells. If these plasmids alone are introduced into cells, virus particles that do not contain viral genomes are produced. When all three plasmids are introduced into cells, retrovirus particles that contain only the recombinant vector genome are formed. The host range of the recombinant vector can be controlled by the type of envelope protein. Envelope protein from amphotropic retroviruses allows the recombinant virus to infect human and mouse cells. The vesicular stomatitis virus glycoprotein G allows infection of a broad range of cell types in many species and also permits concentration with simple methods. Adapted from A. Pfeifer and I. M. Verma, *in* D. M. Knipe et al. (ed.), *Fields Virology*, 4th ed. (Lippincott Williams & Wilkins, Philadelphia, PA, 2001), with permission.

References

Books

Griffiths AJF, Gelbart WM, Lewontin RC, Miller JH. 2002. *Modern Genetic Analysis*. W. H. Freeman & Co, New York, NY.

Review Articles

Andino R, Domingo E. 2015. Viral quasispecies. *Virology* **479–480:**46–51.

Baltimore D. 1971. Expression of animal virus genomes. *Bacteriol Rev* **35:**235–241.

Kennedy EM, Cullen BR. 2015. Bacterial CRISPR/Cas DNA endonucleases: a revolutionary technology that could dramatically impact viral research and treatment. *Virology* **479–480:**213–220.

Koonin EV, Dolja VV. 2013. A virocentric perspective on the evolution of life. *Curr Opin Virol* **5:**546–557. (For discussion of this topic, see **Koonin EV.** 9 March 2014. Episode 275, *This Week in Virology*. http://www.twiv.tv/2014/03/09/twiv-275-virocentricity-with-eugene-koonin/.)

Koonin EV, Dolja VV, Krupovic M. 2015. Origins and evolution of viruses of eukaryotes: the ultimate modularity. *Virology* **479–480:**2–25.

Papers of Special Interest

Bridgen A, Elliott RM. 1996. Rescue of a segmented negative-strand RNA virus entirely from cloned complementary DNAs. *Proc Natl Acad Sci U S A* **93:**15400–15404.

Crotty S, Cameron CE, Andino R. 2001. RNA virus error catastrophe: direct molecular test by using ribavirin. *Proc Natl Acad Sci U S A* **98:**6895–6900.

Elbashir SM, Harborth J, Lendeckel W, Yalcin A, Weber K, Tuschl T. 2001. Duplexes of 21-nucleotide RNAs mediate RNA interference in cultured mammalian cells. *Nature* **411:**494–498.

Goff SP, Berg P. 1976. Construction of hybrid viruses containing SV40 and λ phage DNA segments and their propagation in cultured monkey cells. *Cell* **9:**695–705.

Hoffmann E, Neumann G, Kawaoka Y, Hobom G, Webster RG. 2000. A DNA transfection system for generation of influenza A virus from eight plasmids. *Proc Natl Acad Sci U S A* **97:**6108–6113.

Kobayashi T, Antar AA, Boehme KW, Danthi P, Eby EA, Guglielmi KM, Holm GH, Johnson EM, Maginnis MS, Naik S, Skelton WB, Wetzel JD, Wilson GJ, Chappell JD, Dermody TS. 2007. A plasmid-based reverse genetics system for animal double-stranded RNA viruses. *Cell Host Microbe* **1:**147–157.

Racaniello VR, Baltimore D. 1981. Cloned poliovirus complementary DNA is infectious in mammalian cells. *Science* **214:**916–919.

Saiki RK, Gelfand DH, Stoffel S, Scharf SJ, Higuchi R, Horn GT, Mullis KB, Erlich HA. 1988. Primer-directed enzymatic amplification of DNA with a thermostable DNA polymerase. *Science* **239:**487–491.

Sanger F, Coulson AR, Friedmann T, Air GM, Barrell BG, Brown NL, Fiddes JC, Hutchison CA, III, Slocombe PM, Smith M. 1978. The nucleotide sequence of bacteriophage ϕX174. *J Mol Biol* **125:**225–246.

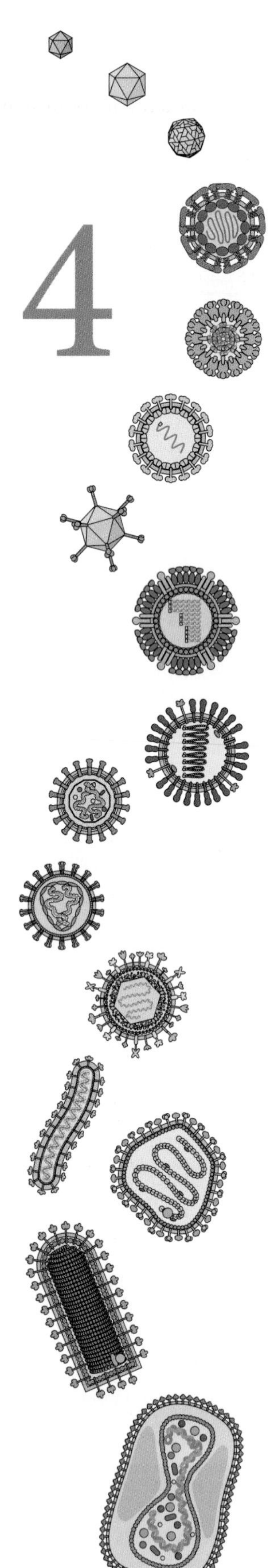

4 Structure

Introduction
- Functions of the Virion
- Nomenclature
- Methods for Studying Virus Structure

Building a Protective Coat
- Helical Structures
- Capsids with Icosahedral Symmetry
- Other Capsid Architectures

Packaging the Nucleic Acid Genome
- Direct Contact of the Genome with a Protein Shell
- Packaging by Specialized Viral Proteins
- Packaging by Cellular Proteins

Viruses with Envelopes
- Viral Envelope Components
- Simple Enveloped Viruses: Direct Contact of External Proteins with the Capsid or Nucleocapsid
- Enveloped Viruses with an Additional Protein Layer

Large Viruses with Multiple Structural Elements
- Bacteriophage T4
- Herpesviruses
- Poxviruses
- Giant Viruses

Other Components of Virions
- Enzymes
- Other Viral Proteins
- Nongenomic Viral Nucleic Acid
- Cellular Macromolecules

Perspectives

References

LINKS FOR CHAPTER 4

- *Video: Interview with Dr. Michael Rossmann*
  http://bit.ly/Virology_Rossmann
- *Movie 4.1: Virus-based Piezoelectric Generator*
  http://bit.ly/Virology_V1_Movie4-1
- *Movie 4.2: Cryo-EM reconstruction of the adenovirus type 5 capsid*
  http://bit.ly/Virology_V1_Movie4-2
- *Sizing up adenovirus*
  http://bit.ly/Virology_Twiv101
- *The Big Picture Book of Viruses*
  http://www.virology.net/Big_Virology/BVHomePage.html
- *ViralZone*
  http://viralzone.expasy.org/
- *Viruses in the extreme*
  http://bit.ly/Virology_5-28-15
- *Virus particle explorer*
  http://viperdb.scripps.edu/

In order to create something that functions properly—a container, a chair, a house—its essence has to be explored, for it should serve its purpose to perfection; i.e., it should fulfill its function practically and should be durable, inexpensive and beautiful.

WALTER GROPIUS
Neue Arbeiten der Bauhauswerkstätten, Bauhaus Book no. 7 1925

Introduction

Virus particles are elegant assemblies of viral, and occasionally cellular, macromolecules. They are marvelous examples of architecture on the molecular scale, with forms perfectly adapted to their functions. Virus particles come in many sizes and shapes (Fig. 4.1; also see Fig. 1.7) and vary enormously in the number and nature of the molecules from which they are built. Nevertheless, they fulfill common functions and are constructed according to general principles that apply to them all. These properties are described in subsequent sections, in which we also discuss some examples of the architectural detail characteristic of members of different virus families.

Functions of the Virion

Virus particles are designed for effective transmission of the nucleic acid genome from one host cell to another within a single organism or among host organisms (Table 4.1). A primary function of the **virion**, an infectious virus particle, is protection of the genome, which can be damaged irreversibly by a break in the nucleic acid or by mutation during passage through hostile environments. During its travels, a virus particle may encounter a variety of potentially lethal chemical and physical agents, including proteolytic and nucleolytic enzymes, extremes of pH or temperature, and various forms of natural radiation. In all virus particles, the nucleic acid is sequestered within a sturdy barrier formed by extensive interactions among the viral proteins that comprise the protein coat. Such protein-protein interactions maintain surprisingly stable capsids: many virus particles composed of only protein and nucleic acid survive exposure to large variations in the temperature, pH, or chemical composition of their environment. For example, when dried onto a solid surface, human rotavirus (a major cause of gastroenteritis) loses $<20\%$ of its infectivity in 30 days at room temperature, whereas the infectivity of poliovirus (a picornavirus) is reduced by some 5 orders of magnitude within 2 days. This same reduction in infectivity requires >250 days when poliovirus particles suspended in spring water are incubated at room temperature at neutral pH. Certain picornaviruses are even resistant to very strong detergents. The highly folded nature of coat proteins and their dense packing to form shells render them largely inaccessible to proteolytic enzymes. Some viruses also possess an **envelope**, typically derived from cellular membranes, into which viral glycoproteins have been inserted. The envelope adds not only a protective lipid membrane but also an external layer of protein and sugars formed by the glycoproteins. Like the cellular membranes from which they are derived, viral envelopes are impermeable to many molecules and block entry of chemicals or enzymes in aqueous solution.

To protect the nucleic acid genome, virus particles must be stable structures. However, they must also attach to an appropriate host cell and deliver the genome to the interior of that cell, where the particle is at least partially disassembled. The protective function of virus particles depends on stable intermolecular interactions among their components during assembly, egress from the virus-producing cell, and transmission. On the other hand, these interactions must be reversed readily during entry and uncoating in a new host cell. In only a few cases do we understand the molecular mechanisms by which these apparently paradoxical requirements are met. Nevertheless, it is clear that contact of a virion with the appropriate cell surface receptor or exposure to a specific intracellular environment can trigger substantial conformational changes. Virus particles are therefore **metastable structures** that have not yet attained the minimum free energy conformation. The latter state can be attained only once an unfavorable energy barrier has been surmounted, following induction of the irreversible conformational transitions that are associated

PRINCIPLES ***Structure***

- Virus particles are designed for protection and delivery of the genome.
- Virus structure can be studied at an atomic level of resolution.
- Genetic economy dictates construction of capsids from a small number of subunits.
- Rod-like viruses are built with helical symmetry and spherical viruses are built with icosahedral symmetry.
- The primary determinant of capsid size is the number of subunits: the more subunits, the larger the capsid.
- There are multiple ways to achieve icosahedral symmetry, even among small viruses.
- While ordered RNA can be observed, how genomes are condensed and organized within virus particles is largely obscure.
- The elaborate capsids of larger viruses contain viral proteins dedicated to stabilizing the capsid shell. In some cases, viruses may have multiple shells.
- Some large viruses are built with structural elements recognizable from simpler viruses.
- Virus particles contain nonstructural components, including enzymes, small RNAs, and cellular macromolecules.

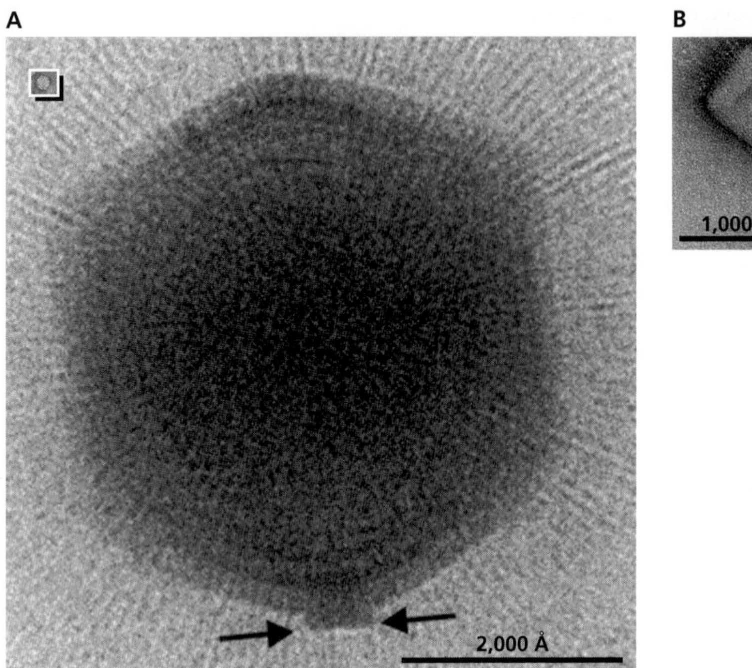

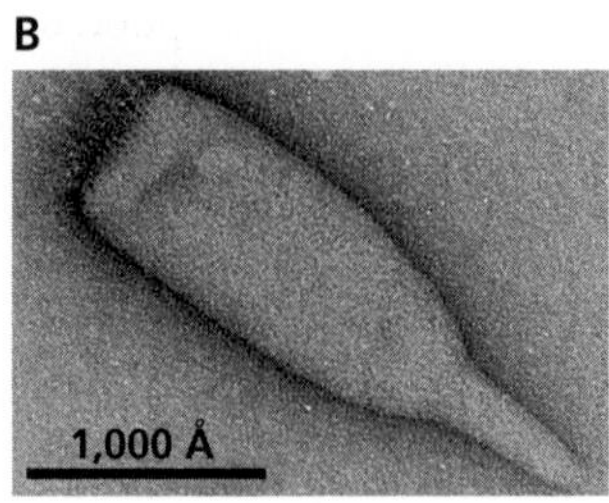

Figure 4.1 Variation in the size and shape of virus particles. (A) Cryo-electron micrographs of mimivirus and, in the inset (upper left), the parvovirus adeno-associated virus type 4, shown to scale relative to one another to illustrate the ~50-fold range in diameter among viruses that appear roughly spherical. Rod-shaped viruses also exhibit considerable variation in size, ranging in length from <200 nm to >2,000 nm. Adapted from C. Xiao et al., *J Mol Biol* **353:**493–496, 2005, and E. Pardon et al., *J Virol* **79:**5047–5058, 2005, respectively, with permission. Courtesy of M. G. Rossmann, Purdue University, and M. Agbandje-McKenna, University of Florida, Gainesville. **(B)** Non-symmetric shape of acidianus bottle virus isolated from a hot spring in Italy. The mimivirus particle (A) is also structurally complex: a large number of long, closely packed filaments project from its surface; and one vertex of the capsid carries a unique structure called the stargate, which opens in infected cells to release the viral genome. Adapted from M. Häring et al., *J Virol* **79:**9904–9911, 2005, with permission. Courtesy of D. Prangishvili, Institut Pasteur.

with attachment and entry. Virions are **not** simply inert structures. Rather, they are molecular machines (nanomachines) that play an active role in delivery of the nucleic acid genome to the appropriate host cell and initiation of the reproductive cycle.

Table 4.1 Functions of virion proteins

Protection of the genome
Assembly of a stable protective protein shell
Specific recognition and packaging of the nucleic acid genome
Interaction with host cell membranes to form the envelope
Delivery of the genome
Binding to external receptors of the host cell
Transmission of signals that induce uncoating of the genome
Induction of fusion with host cell membranes
Interaction with internal components of the infected cell to direct transport of the genome to the appropriate site
Other functions
Interactions with cellular components for transport to intracellular sites of assembly
Interactions with cellular components to ensure an efficient infectious cycle

As might be anticipated, elucidation of the structures of virus particles and individual structural proteins has illuminated the mechanisms of both assembly of viral nanomachines in the final stages of an infectious cycle and their entry into a new host cell. High-resolution structural information can also facilitate identification of targets for antiviral drugs, as well as the design of such drugs (Volume II, Chapter 9), and provide insights into the dynamic interplay between important viral pathogens and host adaptive immune responses (Volume II, Chapter 4). As we shall see, cataloguing of virus architecture has also revealed completely unanticipated relationships among viruses of different families that infect evolutionarily divergent hosts and has suggested new principles of virus classification.

Nomenclature

Virus architecture is described in terms of **structural units** of increasing complexity, from the smallest biochemical unit (the polypeptide chain) to the infectious particle (or virion). These terms, which are used throughout this text, are defined in Table 4.2. Although virus particles are complex assemblies of macromolecules exquisitely suited for protection and

Table 4.2 Nomenclature used in description of virus structure

Term	Synonym	Definition
Subunit (protein subunit)		Single, folded polypeptide chain
Structural unit	Asymmetric unit	Unit from which capsids or nucleocapsids are built; may comprise one protein subunit or multiple, different protein subunits
Capsid	Coat	The protein shell surrounding the nucleic acid genome
Nucleocapsid	Core	The nucleic acid-protein assembly packaged within the virion; used when this assembly is a discrete substructure of a particle
Envelope	Viral membrane	The host cell-derived lipid bilayer carrying viral glycoproteins
Virion		The infectious virus particle

delivery of viral genomes, they are constructed according to the general principles of biochemistry and protein structure.

Methods for Studying Virus Structure

Electron microscopy is the most widely used method for the examination of structure and morphology of virus particles. This technique, which has been applied to viruses since the 1940s, traditionally relied on staining of purified virus particles (or of sections of infected cells) with an electron-dense material. It can yield quite detailed and often beautiful images (Fig. 1.7; see the Appendix) and provided the first rational basis for the classification of viruses.

The greatest contrast between virus particle and stain (negative contrast) occurs where portions of the folded protein chain protrude from the surface. Consequently, surface knobs or projections, termed morphological units, are the main features identified by this method of electron microscopy. However, these structures are often formed by multiple proteins and so their organization does not necessarily correspond to that of the individual proteins that make up the capsid shell. Even when structure is well preserved and a high degree of contrast can be achieved, the minimal size of an object that can be distinguished by classical electron microscopy, its **resolution**, is limited to 50 to 75 Å. This resolution is far too poor to permit molecular interpretation: for example, the diameter of an α-helix in a protein is on the order of 10 Å. Cryo-electron microscopy (cryo-EM), in which samples are rapidly frozen and examined at very low temperatures in a hydrated, vitrified (noncrystalline, glass-like) state, preserves native structure. Because samples are not stained, this technique allows direct visualization of the contrast inherent in the virus particle. When combined with computerized mathematical methods of image analysis of single particles and three-dimensional reconstruction (Fig. 4.2), cryo-EM can increase resolution to the atomic level. As described in subsequent sections, the continual improvements in this method have provided unprecedented views of virus particles not amenable to other methods of structural analysis. Indeed, structures of even quite large and structurally sophisticated viruses like human adenovirus can now be determined at a resolution directly comparable to that achieved by X-ray crystallography (Box 4.1).

The inherent symmetry of most virus particles facilitates analysis of images obtained by cryo-EM for reconstruction of three-dimensional structure. This approach can be complemented by cryo-electron tomography, in which two-dimensional images are recorded as the vitrified sample is tilted at different angles to the electron beam and subsequently combined into a three-dimensional density map (Fig. 4.2). Within the past decade, cryo-electron microscopy and tomography have become standard tools of structural biology. Their application to virus particles has provided a wealth of previously inaccessible information about the external and internal structures of multiple members of at least 20 virus families.

The first descriptions of the molecular interactions that dictate the structure of virus particles were obtained by X-ray crystallography (Fig. 4.3) (see the interview with Dr. Michael Rossmann: http://bit.ly/Virology_Rossmann). A plant virus (tobacco mosaic virus) was the first to be crystallized, and the first high-resolution virus structure determined was that of tomato bushy stunt virus. Since this feat was accomplished in 1978, high-resolution structures of increasingly larger animal viruses have been determined, placing our understanding of the principles of capsid architecture on a firm foundation.

Not all viruses can be examined directly by X-ray crystallography: some do not form suitable crystals, and the larger viruses lie beyond the power of the current procedures by which X-ray diffraction spots are converted into a structural model. However, their architectures can be determined by using a combination of structural methods. Individual viral proteins can be examined by X-ray crystallography and by multidimensional nuclear magnetic resonance techniques. The latter methods, which allow structural models to be constructed from knowledge of the distances between specific atoms in a polypeptide chain, can be applied to proteins in solution, a significant advantage.

High-resolution structures of individual proteins have been important in deciphering mechanisms of attachment

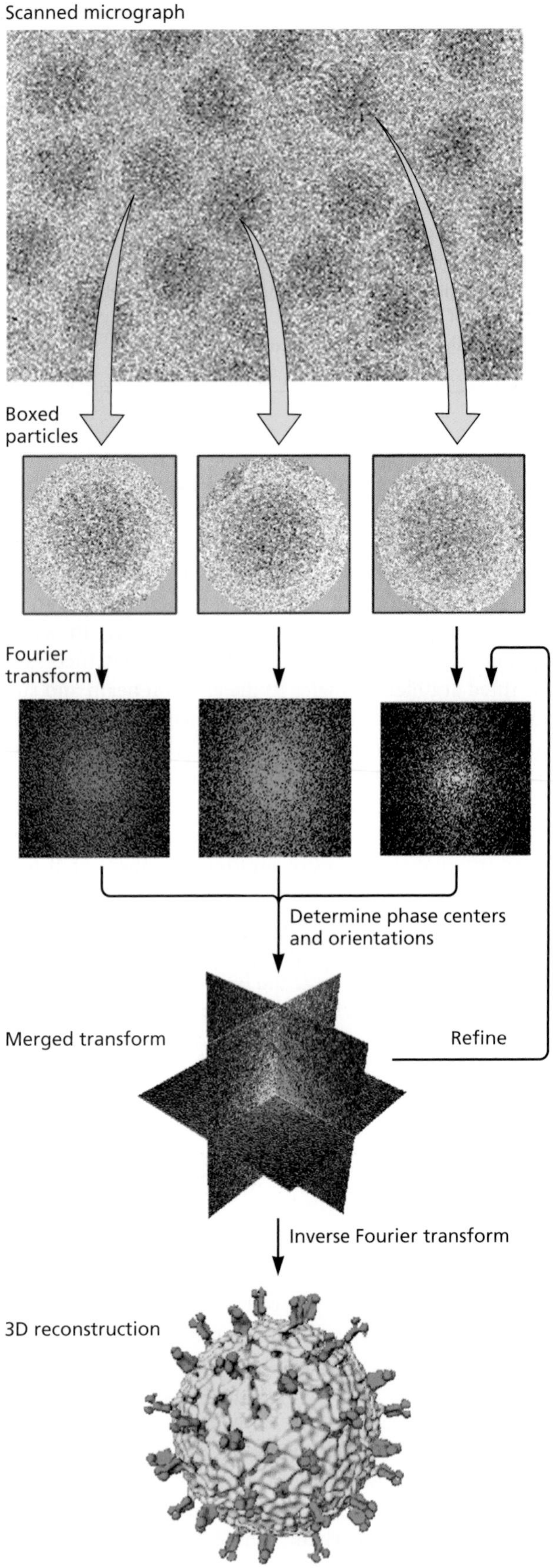

Figure 4.2 Cryo-EM and image reconstruction illustrated with images of rotavirus.

Concentrated preparations of purified virus particles are prepared for cryo-electron microscopy by rapid freezing on an electron microscope grid so that a glasslike, noncrystalline water layer is produced. This procedure avoids sample damage that can be caused by crystallization of the water or by chemical modification or dehydration during conventional negative-contrast electron microscopy. The sample is maintained at or below -160°C during all subsequent operations. Fields containing sufficient numbers of vitrified virus particles are identified by transmission electron microscopy at low magnification (to minimize sample damage from the electron beam) and photographed at high resolution (top).

These electron micrographs can be treated as two-dimensional projections (Fourier transforms) of the particles. Three-dimensional structures can be reconstructed from such two-dimensional projections by mathematically combining the information given by different views of the particles. For the purpose of reconstruction, the images of different particles are treated as different views of the same structure.

For reconstruction, micrographs are digitized for computer processing. Each particle to be analyzed is then centered inside a box, and its orientation is determined by application of programs that orient the particle on the basis of its icosahedral symmetry. In cryo-electron tomography, a series of images is collected with the sample at different angles to the electron beam and combined computationally to reconstruct a three-dimensional structure. The advantage of this approach is that no assumptions about the symmetry of the structure are required. The parameters that define the orientation of the particle must be determined with a high degree of accuracy, for example, to within 1° for even a low-resolution reconstruction (~40 Å). These parameters are improved in accuracy (**refined**) by comparison of different views (particles) to identify common data.

Once the orientations of a number of particles sufficient to represent all parts of the asymmetric unit have been determined, a low-resolution three-dimensional reconstruction is calculated from the initial set of two-dimensional projections by using computerized algorithms.

This reconstruction is refined by including data from additional views (particles). The number of views required depends on the size of the particle and the resolution sought. The reconstruction is initially interpreted in terms of the external features of the virus particle. Various computational and computer graphics procedures have been developed to facilitate interpretation of internal features. Courtesy of B. V. V. Prasad, Baylor College of Medicine.

And is it not true that even the small step of a glimpse through the microscope reveals to us images that we should deem fantastic and over-imaginative if we were to see them somewhere accidentally, and lacked the sense to understand them.

Paul Klee, *On Modern Art*, translated by Paul Findlay (London, United Kingdom, 1948)

BOX 4.1

METHODS

Structures of human adenovirus: technical tours de force

The nonenveloped adenovirus particle is quite large, ~900 Å in diameter excluding the fibers that project from each vertex, and built from multiple proteins. Structural models of this particle were obtained initially by combining the high-resolution structures of individual viral proteins with lower-resolution images obtained by cryo-EM (see Fig. 4.4). In 2010, two papers published in the same issue of *Science* described atomic-level-resolution structures of adenovirus.

A 3.5-Å-resolution structure was obtained by X-ray crystallography of a derivative of human adenovirus type 5, one of the largest and most complicated to be determined by this method. Cryo-EM and single-particle analysis of unmodified adenovirus type 5 generated a structure of comparable resolution, 3.6 Å. This reconstruction represented the most complex structure in which polypeptide chains could be traced directly (see figure). Both models revealed protein-protein interactions important for stabilizing the capsid. However, the cryo-EM-derived structure also included segments of the major capsid protein not evident in X-ray crystal structures. Furthermore, the cryo-EM density map was judged to be clearer than that obtained by X-ray crystallography.

By allowing a direct comparison, these studies showed that current methods of cryo-EM and single-particle analysis can be as powerful as X-ray crystallography, at least when applied to highly symmetric structures like the particles of icosahedral viruses.

Harrison S. 2010. Looking inside adenovirus. *Science* **329:**1026–1027.

Liu H, Jin L, Koh SB, Atanasov I, Schein S, Wu L, Zhou ZH. 2010. Atomic structure of human adenovirus by cryo-EM reveals interactions among protein networks. *Science* **329:**1038–1043.

Reddy VS, Natchiar SK, Stewart PL, Nemerow GR. 2010. Crystal structure of human adenovirus at 3.5 Å resolution. *Science* **329:**1071–1075.

The atomic model (sticks) of an α-helix present in the adenovirus type 5 major capsid protein (hexon), with amino acid side chains in red and the polypeptide backbone in blue, is shown superimposed on its cryo-EM electron density map (gray mesh). The identities of some of the side chains are labeled. Adapted from H. Liu et al., *Science* **329:**1038–1043, 2010, with permission. Courtesy of Z. H. Zhou, University of California, Los Angeles.

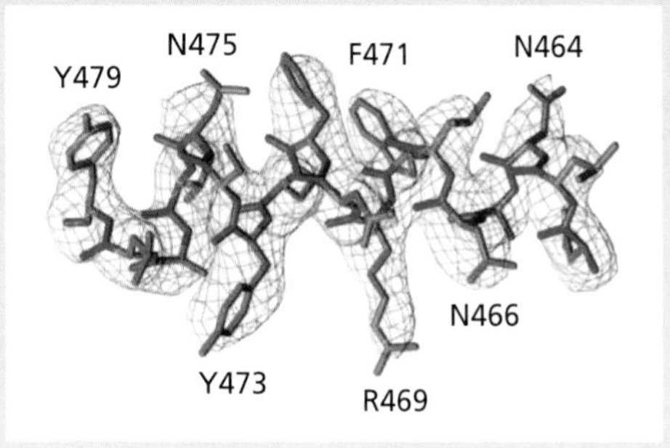

Figure 4.3 Determination of virus structure by X-ray diffraction. A virus crystal is composed of virus particles arranged in a well-ordered three-dimensional lattice. When the crystal is bombarded with a monochromatic X-ray beam traveling through the pinhole, each atom within the virus particle scatters the radiation. Interactions of the scattered rays with one another form a diffraction pattern that is recorded. Each spot contains information about the position and the identity of the atoms in the crystal. The locations and intensities of the spots are stored electronically. Determination of the three-dimensional structure of the virus from the diffraction pattern requires information that is lost in the X-ray diffraction experiment. This missing information (the phases of the diffracted rays) can be retrieved by collecting the diffraction information from otherwise identical (isomorphous) crystals in which the phases have been systematically perturbed by the introduction of heavy metal atoms at known positions. Comparison of the two diffraction patterns yields the phases. This process is called **multiple isomorphous replacement**. Alternatively, if the structure of a related molecule is known, the diffraction pattern collected from the crystal can be interpreted by using the phases from the known structure as a starting point and subsequently using computer algorithms to calculate the actual values of the phases. This method is known as **molecular replacement**. Once the phases are known, the intensities and spot positions from the diffraction pattern are used to calculate the locations of the atoms within the crystal.

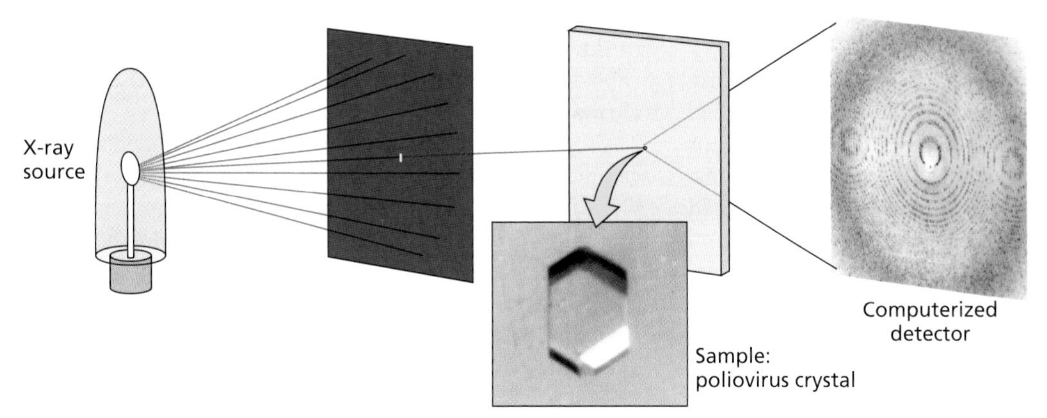

A

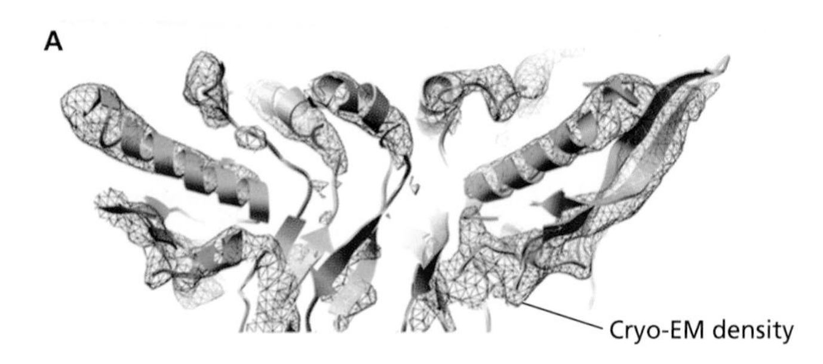

B

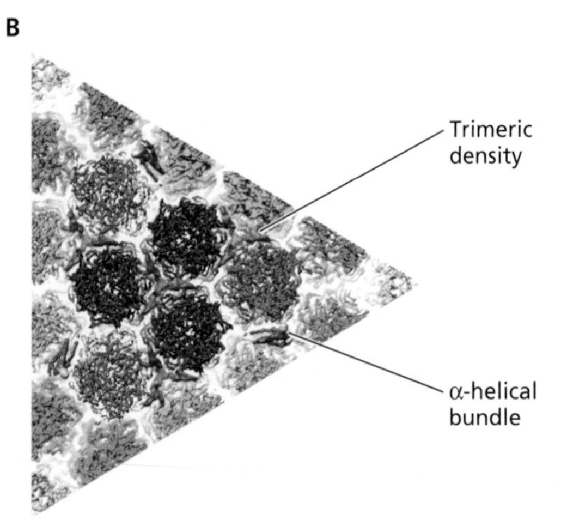

Figure 4.4 Difference mapping illustrated by a 6-Å-resolution reconstruction of adenovirus. (A) Comparison of α-helices of the penton base in the cryo-electron microscopic (cryo-EM) density and the crystal structure of this protein bound to a fiber peptide (ribbon). The excellent agreement established that α-helices could be reliably discerned in the 6-Å cryo-EM reconstruction. **(B)** Portion of the cryo-EM difference map corresponding to the surface of one icosahedral face of the capsid. The crystal structures of the penton base (yellow) and the hexons (green, cyan, blue, and magenta at different positions) at appropriate resolution were docked within the cryo-EM density at 6-Å resolution. The cryo-EM density that does not correspond to these structural units (the difference map) is shown in red. At this resolution, the difference map revealed four trimeric structures located between neighboring hexons and three bundles of coiled-coiled α-helices. The former were previously assigned to protein IX. Adapted from S. D. Saban et al., *J Virol* **80:**12049–12059, 2006, with permission. Courtesy of Phoebe Stewart, Vanderbilt University Medical Center.

and entry of enveloped viruses. Even more valuable are methods in which high-resolution structures of individual viral proteins are combined with cryo-EM reconstructions of intact virus particles. For example, in difference imaging, the structures of individual proteins are in essence subtracted from the reconstruction of the particle to yield new structural insight (Fig. 4.4). This powerful approach has provided fascinating views of interactions of viral envelope proteins embedded in lipid bilayers and even of internal surfaces and components of virus particles.

Atomic-resolution structures of individual proteins or domains can also be modeled into lower-resolution views (currently ~15 Å) obtained by small-angle X-ray scattering. This technique, which is applied to proteins in solution, provides information about the overall size and shape of flexible, asymmetric proteins and has provided valuable information about viral proteins with multiple functional domains (see, e.g., Chapter 7).

Building a Protective Coat

Regardless of their structural complexity, all virions contain at least one protein coat, the **capsid** or nucleocapsid, which encases and protects the nucleic acid genome (Table 4.2). As first pointed out by Francis Crick and James Watson in 1956, most virus particles appear to be rod shaped or spherical under the electron microscope. Because the coding capacities of viral genomes are limited, these authors proposed that construction of capsids from a small number of subunits would minimize the genetic cost of encoding structural proteins. Such genetic economy dictates that capsids be built from identical copies of a small number of viral proteins with structural properties that permit regular and repetitive interactions among them. These protein molecules are arranged to provide maximal contact and noncovalent bonding among subunits and structural units. The repetition of such interactions among a limited number of proteins results in a regular structure, with symmetry that is determined by the spatial patterns of the interactions. In fact, the protein coats of many viruses **do** display **helical** or **icosahedral symmetry**. Such well-defined symmetry has considerable practical value (Box 4.2).

Helical Structures

The **nucleocapsids** of some enveloped animal viruses, as well as certain plant viruses and bacteriophages, are rod-like or filamentous structures with helical symmetry. Helical symmetry is described by the number of structural units per turn of the helix, μ, the axial rise per unit, ρ, and the pitch of the helix, P (Fig. 4.5A). A characteristic feature of a helical structure is that any volume can be enclosed simply by varying the length of the helix. Such a structure is said to be **open**. In contrast, capsids with icosahedral symmetry (described below) are **closed** structures with fixed internal volume.

From a structural point of view, the best-understood helical nucleocapsid is that of tobacco mosaic virus, the very first virus to be identified. The virus particle comprises a single molecule of (+) strand RNA, about 6.4 kb in length, enclosed within a helical protein coat (Fig. 4.5B; see also Fig. 1.7). The coat is built from a single protein that folds into an extended structure shaped like a Dutch clog. Repetitive interactions among coat protein subunits form disks that have been

BOX 4.2

METHODS

Nanoconstruction with virus particles

Nanochemistry is the synthesis and study of well-defined structures with dimensions of 1 to 100 nm. Nano-building blocks span the size range between molecules and materials such as nylon. Molecular biologists study nanochemistry, nanostructures, and molecular machines including the ribosome and membrane-bound signaling complexes. Icosahedral viruses are proving to be precision building blocks for nanochemistry. The icosahedral cowpea mosaic virus particle is 30 nm in diameter, and its atomic structure is known in detail. Grams of particles can be prepared easily from kilograms of infected leaves, insertional mutagenesis is straightforward, and precise amino acid changes can be introduced. As illustrated in panel A of the figure, cysteine residues inserted in the capsid protein provide functional groups for chemical attachment of 60 precisely placed molecules, in this case, gold particles.

High local concentrations of attached chemical agents, coupled with precise placement, and the propensity of virus-like particles for self-organization into two- and three-dimensional lattices of well-ordered arrays of particles enable rather remarkable nanoconstruction. For example, the surface of the filamentous bacteriophage M13 can be patterned to carry separate binding sites for gold and cobalt oxide and assembled into nanowires to form the anodes of small lithium ion batteries. Remarkably, this bacteriophage also displays intrinsic piezoelectric properties, that is, the ability to generate an electric charge in response to mechanical deformation, and vice versa. The basis of this property is not fully understood, but modification of the sequence of the major protein to increase its dipole moment (figure, panel B) augmented the piezoelectric strength of the bacteriophage. Assembly of the modified M13 into thin films was exploited to build a piezoelectric generator that produced up to 6 mA of current and 400 mV of potential, sufficient to operate a liquid crystal display (see Movie 4.1: http://bit.ly/Virology_V1_Movie4-1, Box 4.2).

Virus particles also have considerable potential for the delivery of drugs and other medically relevant molecules.

Viruses are not just for infections anymore! They will provide a rich source of building blocks for applications spanning the worlds of molecular biology, materials science, and medicine.

Lee BY, Zhang J, Zueger C, Chung WJ, Yoo SY, Wang E, Meyer J, Ramesh R, Lee SW. 2012. Virus-based piezoelectric energy generation. *Nat Nanotechnol* 7:351–356.

Nam KT, Kim DW, Yoo PJ, Chiang CY, Meethong N, Hammond PT, Chiang YM, Belcher AM. 2006. Virus-enabled synthesis and assembly of nanowires for lithium ion battery electrodes. *Science* **312:** 885–888.

Tarascon JM. 2009. Nanomaterials: viruses electrify battery research. *Nat Nanotechnol* **4:**341–342.

Wang Q, Lin T, Tang L, Johnson JE, Finn MG. 2002. Icosahedral virus particles as addressable nanoscale building blocks. *Angew Chem Int Ed Engl* **41:**459–462.

Gold particles attached to cowpea mosaic virus. (A) Cryo-EM was performed on derivatized cowpea mosaic virus with a cysteine residue inserted on the surface of each of the 60 subunits and to which nanogold particles with a diameter of 1.4 nm were chemically linked. (Left) Difference electron density map obtained by subtracting the density of unaltered cowpea mosaic virus at 29 Å from the density map of the derivatized virus. This procedure reveals both the genome (green) and the gold nanoparticles. (Right) A section of the difference map imposed on the atomic model of cowpea mosaic virus. The positions of the gold indicate that it is attached at the sites of the introduced cysteine residues. Courtesy of M. G. Finn and J. Johnson, The Scripps Research Institute. **(B)** Increasing the piezoelectric strength of phage M13. Side view of a segment of M13 containing 10 copies of the helical major coat protein modified to contain four glutamine residues at its N terminus. The dipole moments (yellow arrows) are directed from the N terminus (blue) to the C terminus (red). Adapted from B. Y. Lee et al., *Nat Nanotechnol* 7:351–356, 2012, with permission. Courtesy of S.-W. Lee, University of California, Berkeley.

A

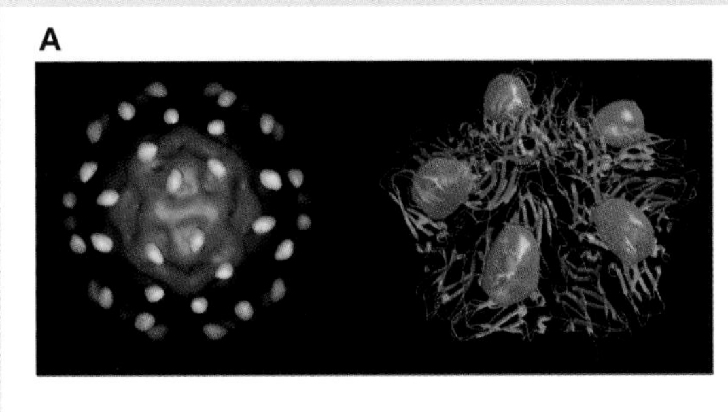

B

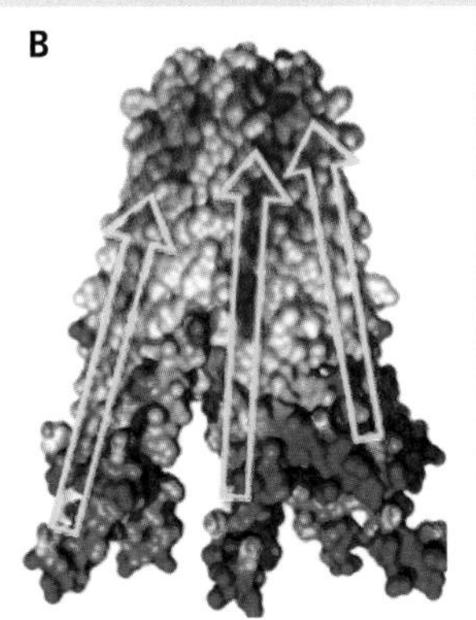

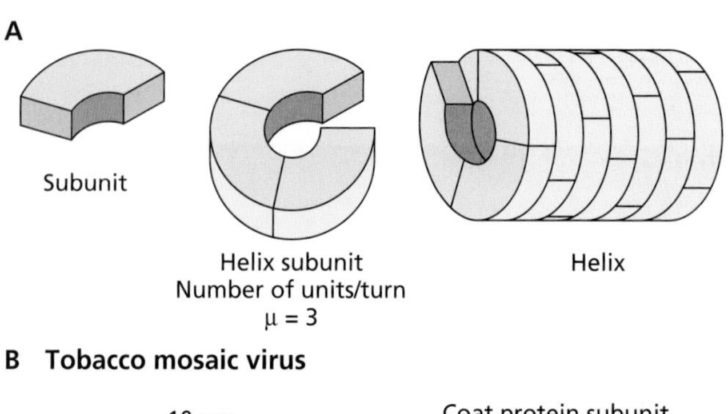

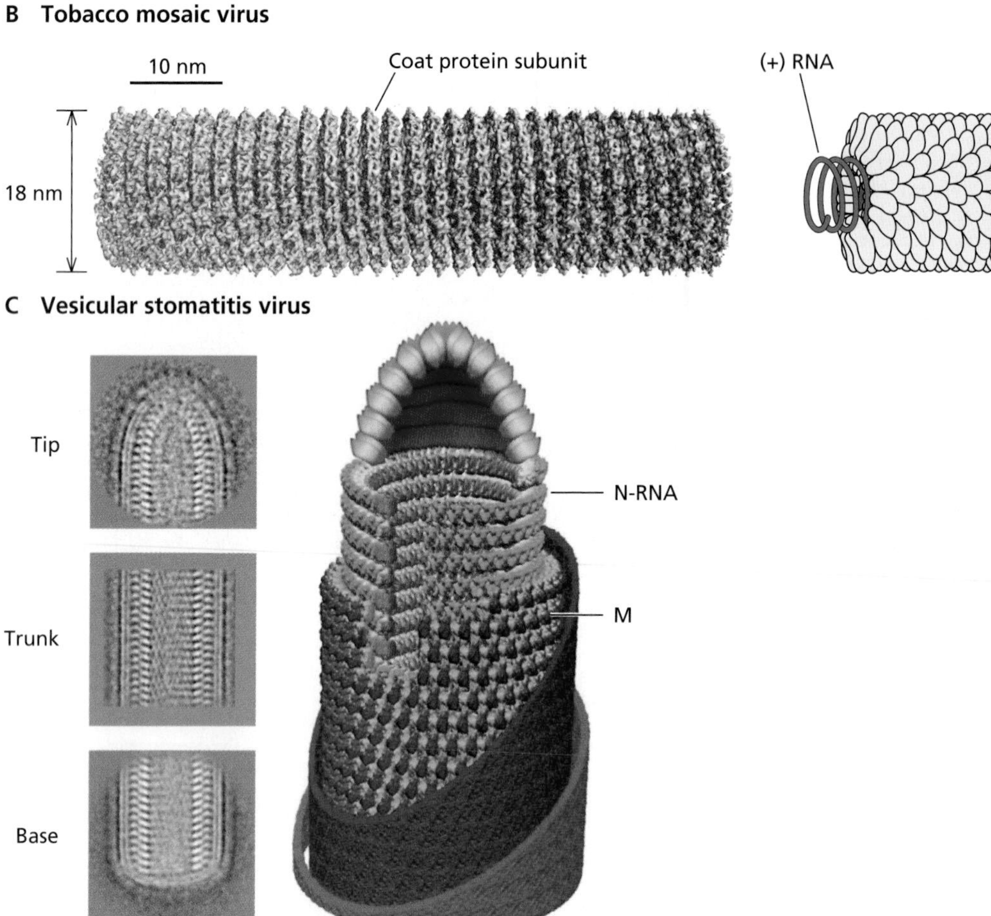

Figure 4.5 Virus structures with helical symmetry. (A) Schematic illustration of a helical particle, indicating the individual subunits, their interaction to form a helical turn, the helix, and the helical parameters ρ (axial rise per subunit) and μ (the number of subunits per turn). The pitch of the helix, P, is given by the formula $P = \rho \times \mu$. **(B)** Tobacco mosaic virus. (Left) A cryo-EM reconstruction at <5-Å resolution of a 70-nm segment of this particle. Each helical turn contains 16.3 protein molecules. Adapted from J. Sachse et al., *J Mol Biol* **371:**812–835, 2007, with permission. Courtesy of N. Grigorieff, Leibniz-Institut für Alterforschung, Jena, Germany. (Right) The regular interaction of the (+) strand RNA genome with coat protein subunits is illustrated in the model based on a 2.9-Å X-ray fiber diffraction structure. Adapted from K. Namba et al., *J Mol Biol* **208:**307–325, 1989, with permission. **(C)** Vesicular stomatitis virus. Representative averages of cryo-EM images of the central trunk, conical tip, and flat base of this bullet-shaped virus particle are shown at the left. The trunk and tip were analyzed and reconstructed separately to form the montage model shown on the right, with N and M proteins in green and blue, respectively, and the membrane in purple and pink. The N protein packages the (−) strand RNA genome in a left-handed helix. The crystal structure of N determined in an N-RNA complex (Fig. 4.6) fits unambiguously with the cryo-EM density of trunk N subunits. The turns of the N protein helix are not closely associated with one another, a property that accounts for the unwinding of the nucleoprotein in the absence of M (see text), which forms an outer, left-handed helix. In the trunk, the N helix contains 37.5 subunits per turn. Comparison of N-N interactions in such a turn and in rings of 10 N molecules determined by X-ray crystallography, as well as the results of mutational analysis, are consistent with formation of rings containing increasing numbers of N molecules from the apex of the tip via different modes of N-N interaction induced by association with long genomic RNA. Once a second turn of the N-RNA is stacked on the first, the M protein can bind to add rigidity. Adapted from P. Ge et al., *Science* **327:**689–693, 2010, with permission. Courtesy of Z. H. Zhou, University of California, Los Angeles.

likened to lock washers, which in turn assemble as a long, rod-like, right-handed helix. In the interior of the helix, each coat protein molecule binds three nucleotides of the RNA genome. The coat protein subunits therefore engage in **identical** interactions with one another and with the genome, allowing the construction of a large, stable structure from multiple copies of a single protein.

The particles of several families of animal viruses with (−) strand RNA genomes, including filoviruses, paramyxoviruses, rhabdoviruses, and orthomyxoviruses, contain internal structures with helical symmetry that are encased within an envelope. In all cases, these structures contain an RNA molecule, many copies of an RNA-packaging protein (designated NP or N), and the viral RNA polymerase and associated enzymes responsible for synthesis of mRNA. Despite common helical symmetry and similar composition, the internal components of these (−) strand RNA viruses exhibit considerable diversity in morphology and organization. For example, the nucleocapsids of the filovirus ebolavirus and the paramyxovirus Sendai virus are long, filamentous structures in which the NP proteins make regular interactions with the single molecule of the RNA genome. In contrast, those of rhabdoviruses such as vesicular stomatitis virus are bullet-shaped structures (Fig. 4.5C). Furthermore, an additional viral protein is essential to maintain their organization: vesicular stomatitis virus nucleocapsids released from within the envelope retain the dimensions and morphology observed in intact particles, but become highly extended and filamentous once the matrix (M) protein is also removed (Fig. 12.23). X-ray crystallography of a ring-like N protein-RNA complex containing 10 molecules of the N protein bound to RNA has revealed that each N protein molecule binds to 9 nucleotides of RNA, which is largely sequestered within cavities within the N proteins (Fig. 4.6). Furthermore, each N subunit makes extensive and regular contacts with neighboring N molecules, exactly as predicted from first principles by Crick and Watson.

The internal components of influenza A virus particles differ more radically. In the first place, they comprise not a single nucleocapsid but, rather, multiple ribonucleoproteins, one for each molecule of the segmented RNA genome present in the virus particle (Appendix, Fig. 15). Furthermore, with the exception of terminal sequences, the RNA in these ribonucleoproteins is fully accessible to solvent. This property suggests that the RNA is not sequestered in the interior of the ribonucleoprotein. The structure of ribonucleoproteins released from influenza A virus particles determined by cryo-EM is consistent with such a model: the ribonucleoprotein comprises a double helix of NP molecules connected at one end by an NP loop (Fig. 4.7A) (currently an unusual architecture for helical viral ribonucleoproteins) with the RNA bound along the exposed surface of each NP strand (Fig. 4.7B).

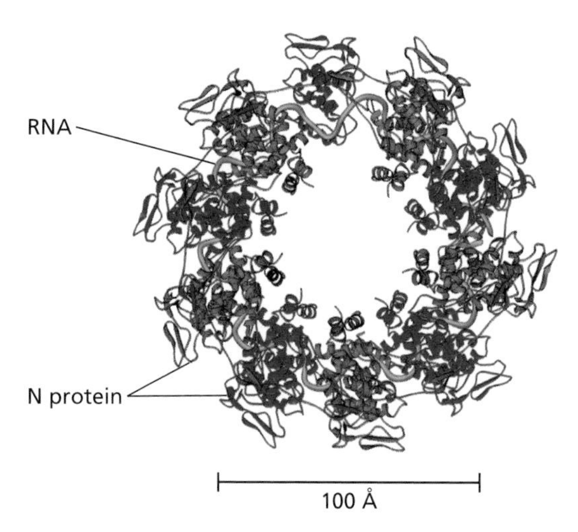

Figure 4.6 Structure of a ribonucleoprotein-like complex of vesicular stomatitis virus. Shown is the structure of the decamer of the N protein bound to RNA, determined by X-ray crystallography, with alternating monomers in the ring colored red and blue and the RNA ribose-phosphate backbone depicted as a green tube. To allow visualization of the RNA, the C-terminal domain of the monomer at the top center is not shown. The decamer was isolated by dissociation of the viral P protein from RNA-bound oligomers formed when the N and P proteins were synthesized in *Escherichia coli*. The N-terminal extension and the extended loop in the C-terminal lobe contribute to the extensive interactions among neighboring N monomers. Adapted from T. J. Green et al., *Science* **313**:357–360, 2006, with permission. Courtesy of M. Luo, University of Alabama at Birmingham.

Capsids with Icosahedral Symmetry

General Principles

Icosahedral symmetry. An icosahedron is a solid with 20 triangular faces and 12 vertices related by two-, three-, and fivefold axes of rotational symmetry (Fig. 4.8A). In a few cases, virus particles can be readily seen to be icosahedral (e.g., see Fig. 4.15 and 4.27. However, most closed capsids **look** spherical, and they often possess prominent surface structures or viral glycoproteins in the envelope that do not conform to the underlying icosahedral symmetry of the capsid shell. Nevertheless, the symmetry with which the structural units interact is that of an icosahedron.

In solid geometry, each of the 20 faces of an icosahedron is an equilateral triangle, and five such triangles interact at each of the 12 vertices (Fig. 4.8A). In the simplest protein shells, a trimer of a single viral protein (the **subunit**) corresponds to each triangular face of the icosahedron: as shown in Fig. 4.8B, such trimers interact with one another at the five-, three-, and twofold axes of rotational symmetry that define an icosahedron. As an icosahedron has 20 faces, 60 identical subunits (3 per face × 20 faces) is the minimal number needed to build a capsid with icosahedral symmetry.

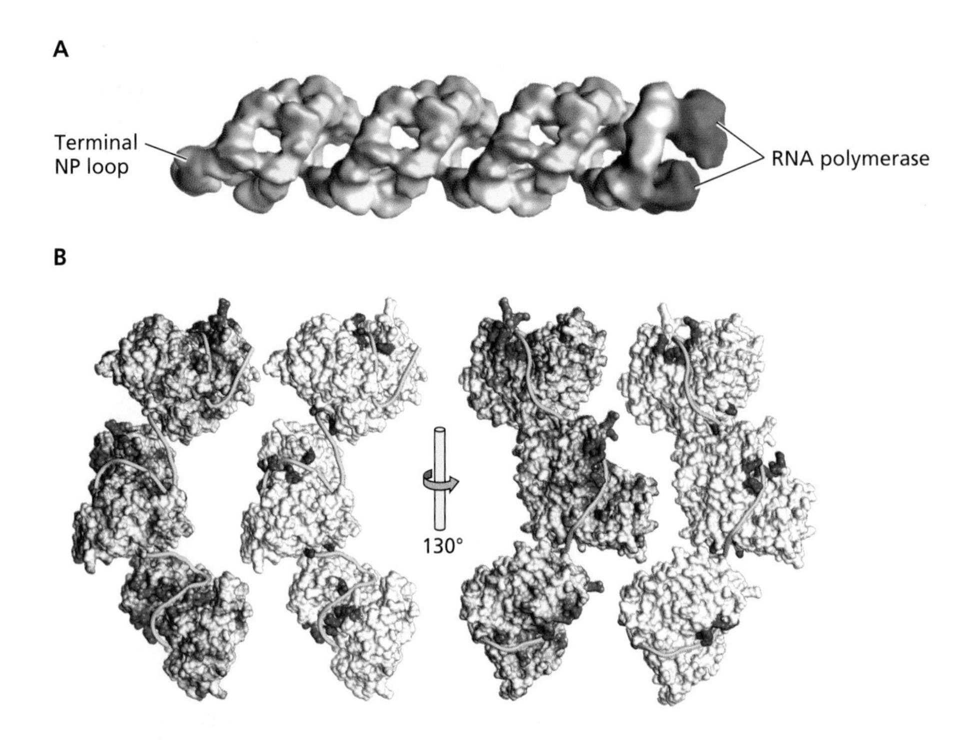

Figure 4.7 Structure of an influenza A virus ribonucleoprotein. (A) Ribonucleoproteins were isolated from purified influenza A virus particles and the central and terminal regions analyzed separately following cryo-EM. This procedure was adopted to overcome the heterogeneity in length of individual ribonucleoproteins and their flexibility. Class averaging of images of straight segments of central regions and three-dimensional reconstruction revealed that the RNA-binding NP protein forms a double helix closed by a loop at one end. In the model, the NP strands of opposite polarity are shown in blue and pink, with the NP loop in yellow and the RNA polymerase subunits at the other end in gray, green, and tan. **(B)** Four views of a single NP strand, indicating the likely localization of the (−) strand genome RNA (yellow ribbon). This localization was deduced from the surface electrostatic potential (models on the left, with positive and negative charge shown blue and red, respectively) and the positions of substitutions that impair binding of NP to RNA (blue in the models on the right). Adapted from R. Arranz et al., *Science* **338:**1634–1637, 2012, with permission. Courtesy of J. Martin-Benito, Centro Nacional de Biotecnologia, Madrid, Spain.

Large capsids and quasiequivalent bonding. In the simple icosahedral packing arrangement, each of the 60 subunits (**structural** or **asymmetric units**) consists of a single molecule in a structurally identical environment (Fig. 4.8B). Consequently, all subunits interact with their neighbors in an identical (or **equivalent**) manner, just like the subunits of helical particles such as that of tobacco mosaic virus. As the viral proteins that form such closed shells are generally $<\sim$100 kDa in molecular mass, the size of the viral genome that can be accommodated in this simplest type of particle is restricted severely. To make larger capsids, additional subunits must be included. Indeed, the capsids of the majority of animal viruses are built from many more than 60 subunits and can house quite large genomes. In 1962, Donald Caspar and Aaron Klug developed a theoretical framework accounting for the structural properties of larger particles with icosahedral symmetry. This theory has had enormous influence on the way virus architecture is described and interpreted.

The triangulation number, *T*. A crucial idea introduced by Caspar and Klug was that of **triangulation,** the description of the triangular face of a large icosahedral structure in terms of its subdivision into smaller triangles, termed **facets** (Fig. 4.9). This process is described by the triangulation number, *T*, which gives the number of structural units (small "triangles") per face (Box 4.3). Because the minimum number of subunits required is 60, the total number of subunits in the structure is $60T$.

Quasiequivalence. A second cornerstone of the theory developed by Caspar and Klug was the proposition that when a capsid contains >60 subunits, each occupies a quasiequivalent position; that is, the noncovalent bonding

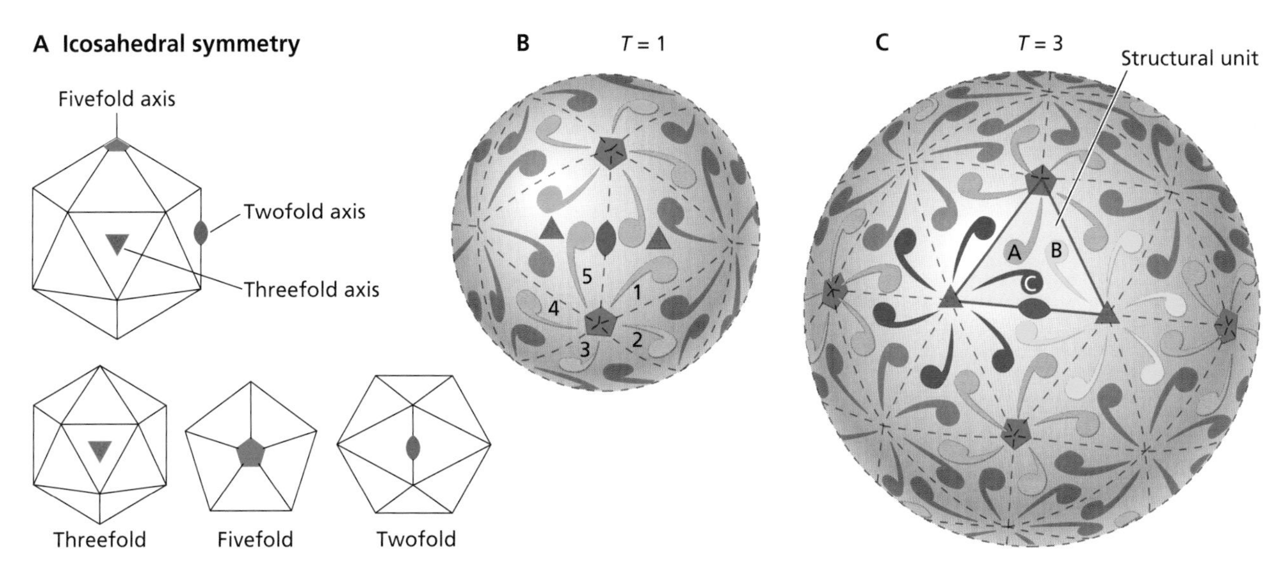

Figure 4.8 Icosahedral packing in simple structures. (A) An icosahedron, which comprises 20 equilateral triangular faces characterized by positions of five-, three-, and twofold rotational symmetry. The three views at the bottom illustrate these positions. **(B and C)** A comma represents a single protein molecule, and axes of rotational symmetry are indicated as in panel A. In the simplest case, $T = 1$ **(B)**, the protein molecule forms the structural unit, and each of the 60 molecules is related to its neighbors by the two-, three-, and fivefold rotational axes that define a structure with icosahedral symmetry. In such a simple icosahedral structure, the interactions of all molecules with their neighbors are identical. In the $T = 3$ structure **(C)** with 180 identical protein subunits, there are three modes of packing of a subunit (shown in orange, yellow, and purple): the structural unit (outlined in blue) is now the asymmetric unit, which, when replicated according to 60-fold icosahedral symmetry, generates the complete structure. The orange subunits are present in pentamers, formed by tail-to-tail interactions, and interact in rings of three (head to head) with purple and yellow subunits, and in pairs (head to head) with a purple or a yellow subunit. The purple and yellow subunits are arranged in rings of six molecules (by tail-to-tail interactions) that alternate in the particle. Despite these packing differences, the bonding interactions in which each subunit engages are similar, that is, quasiequivalent: for example, all engage in tail-to-tail and head-to-head interactions. Adapted from S. C. Harrison et al., *in* B. N. Fields et al. (ed.), *Fundamental Virology* (Lippincott-Raven, New York, NY, 1995), with permission.

properties of subunits in different structural environments are **similar**, but not identical, as is the case for the simplest, 60-subunit capsids. This property is illustrated in Fig. 4.8C for a particle with 180 identical subunits. In the small, 60-subunit structure, 5 subunits make fivefold symmetric contact at each of the 12 vertices (Fig. 4.8B). In the larger assembly with 180 subunits, this arrangement is retained at the 12 vertices, but the additional subunits are interposed to form clusters with sixfold symmetry. In such a capsid, each subunit can be present in one of three **different** structural environments (designated A, B, or C in Fig. 4.8C). Nevertheless, all subunits bond to their neighbors in similar **(quasi-equivalent)** ways, for example, via head-to-head and tail-to-tail interactions.

Capsid architectures corresponding to various values of *T*, some very large, have been described. The triangulation number and quasiequivalent bonding among subunits describe the structural properties of many simple viruses with icosahedral symmetry. However, it is now clear that the structures adopted by specific segments of capsid proteins can govern the packing interactions of identical subunits. Such large conformational differences between small regions of chemically identical subunits were not anticipated in early considerations of virus structure. This omission is not surprising, for these principles were formulated when little was known about the conformational flexibility of proteins. As we discuss in the next sections, the architectures of both small and more-complex viruses can, in fact, depart radically from the constraints imposed by quasiequivalent bonding. For example, the capsid of the small polyomavirus simian virus 40 is built from 360 subunits, corresponding to the $T = 6$ triangulation number excluded by the rules formulated by Caspar and Klug (Box 4.3). Moreover, a capsid stabilized by **covalent** joining of subunits to form viral "chain mail" has been described (Box 4.4). Our current view of icosahedrally symmetric virus structures is therefore one that includes greater diversity in the mechanisms by which stable capsids can be formed than was anticipated by the pioneers in this field.

Structurally Simple Capsids

Several nonenveloped animal viruses are small enough to be amenable to high-resolution analysis by X-ray

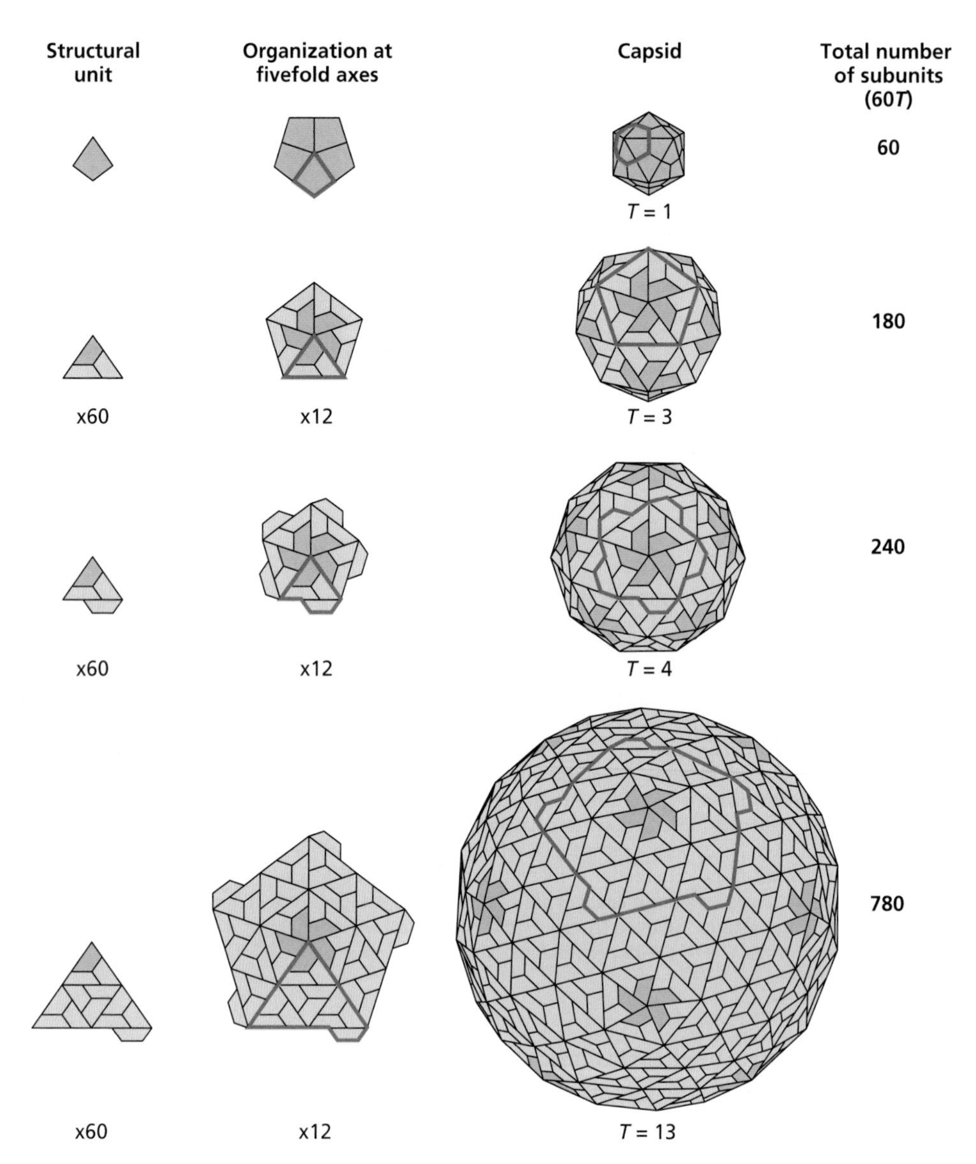

Figure 4.9 The principle of triangulation: formation of large capsids with icosahedral symmetry. The formation of faces of icosahedral particles by triangulation is illustrated by comparison of structural units, organization of structural units at fivefold axes of icosahedral symmetry, and capsids with the T number indicated. In each case, the protein subunits are represented by trapezoids, with those that interact at the vertices colored purple and all others tan. It is important to appreciate that protein subunits are **not**, in fact, flat, as shown here for simplicity, but highly structured (see, for examples, Fig. 4.10 and 4.12). Both the interaction of subunits around the fivefold axes of symmetry and the capsid, with an individual face outlined in red, are shown for each value of T, to illustrate the increase in face and particle size with increasing T.

crystallography. We have chosen three examples, the parvovirus adenovirus-associated virus 2, the picornavirus poliovirus, and the polyomavirus simian virus 40, to illustrate the molecular foundations of icosahedral architecture.

Structure of adeno-associated virus 2: classic $T = 1$ icosahedral design. The parvoviruses are very small animal viruses, with particles of ~25 nm in diameter that encase single-stranded DNA genomes of <5 kb. These small, naked capsids are built from 60 copies of a single subunit organized according to $T = 1$ icosahedral symmetry. The subunits of adenovirus-associated virus type 2, a member of the dependovirus subgroup of parvovirus (Appendix, Fig. 19), contain a core domain commonly found in viral

BOX 4.3

BACKGROUND

The triangulation number, T, *and how it is determined*

In developing their theories about virus structure, Caspar and Klug used graphic illustrations of capsid subunits, such as the net of flat hexagons shown at the top left of panel A in the figure. Each hexagon represents a hexamer, with identical subunits shown as equilateral triangles. When all subunits assemble into such hexamers, the result is a flat sheet, or lattice, which can never form a closed structure. To introduce curvature, and hence form three-dimensional structures, one triangle is removed from a hexamer to form a pentamer in which the vertex and faces project above the plane of the original lattice (A, far right). As an icosahedron has 12 axes of fivefold symmetry, 12 pentamers must be introduced to form a closed structure with icosahedral symmetry. If 12 adjacent hexamers are converted to pentamers, an icosahedron of the minimal size possible for the net is formed. This structure is built from 60 equilateral-triangle asymmetric units and corresponds to a $T = 1$ icosahedron (panel B, left). Larger structures with icosahedral symmetry are built by including a larger number of equilateral triangles (subunits) per face (B, right). In the hexagonal lattice, this is equivalent to converting 12 **nonadjacent** hexamers to pentamers at precisely spaced and regular intervals.

To illustrate this operation, we use nets in which an origin (O) is fixed and the positions of all other hexamers are defined by the coordinates along the axes labeled h and k, where h and k are any positive integer (B, left). The hexamer (h, k) is therefore defined as that reached from the origin (O) by h steps in the direction of the h axis and k steps in the direction of the k axis. In the $T = 1$ structure, $h = 1$ and $k = 0$ (or $h = 0$ and $k = 1$), and adjacent hexamers are converted to pentamers (B, left. When $h = 1$ and $k = 1$, pentamers are separated by one step in the h and one step in the k direction (B, right). Similarly, when $h = 2$ and $k = 0$ (or vice versa), two steps in a single direction separate the pentamers.

The triangulation number, T, is the number of asymmetric units per face of the icosahedron constructed in this way. It can be shown, for example by geometry, that

$$T = h^2 + hk + k^2$$

Therefore, when both h and k are 1, $T = 3$,

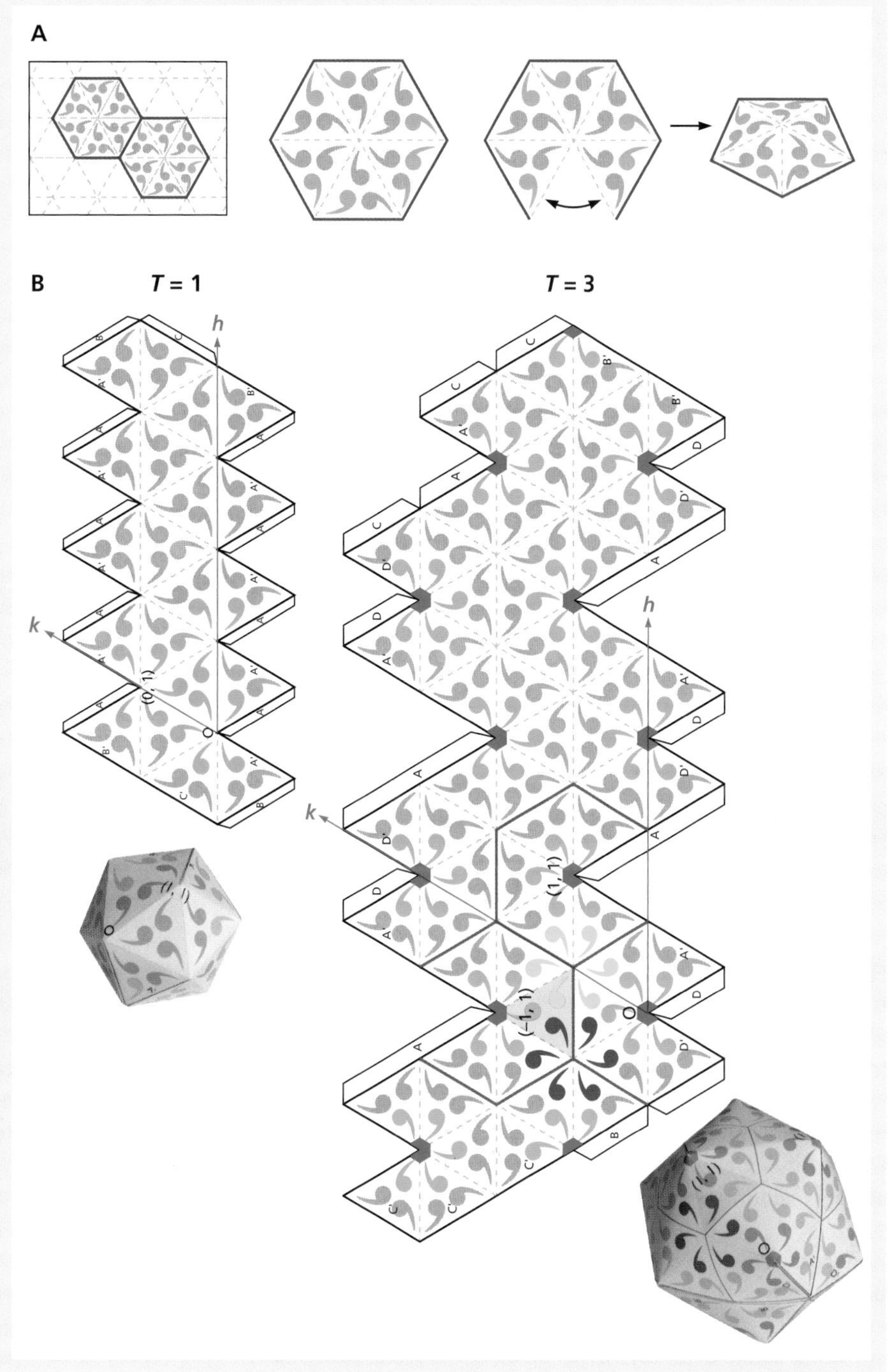

(continued)

BOX 4.3

BACKGROUND

The triangulation number, T, and how it is determined (continued)

and each face of the icosahedron contains three asymmetric units. The total number of units, which must be $60T$, is 180. When $T = 4$, there are four asymmetric units per face and a total of 240 units (Fig. 4.9).

As the integers h and k describe the spacing and spatial relationships of pentamers, that is, of fivefold vertices in the corresponding icosahedra, their values can be determined by inspection of electron micrographs of virus particles or their constituents (C). For example, in the bacteriophage p22 capsid (C, top), one pentamer is separated from another by two steps along the h axis and one step along the k axis, as illustrated for the bottom left pentamer shown. Hence, $h = 2$, $k = 1$, and $T = 7$. In contrast, pentamers of the herpes simplex virus type 1 (HSV-1) nucleocapsid (bottom) are separated by four and zero steps along the directions of the h and k axes, respectively. Therefore, $h = 4$, $k = 0$, and $T = 16$.

Panels A and B adapted from Fig. 2 of J. E. Johnson and A. J. Fisher, *in* R. G. Webster and A. Granoff (ed.), *Encyclopedia of Virology*, 3rd ed. (Academic Press, London, United Kingdom, 1994), with permission. Cryo-electron micrographs of bacteriophage p22 and HSV-1 courtesy of B. V. V. Prasad and W. Chiu, Baylor College of Medicine, respectively.

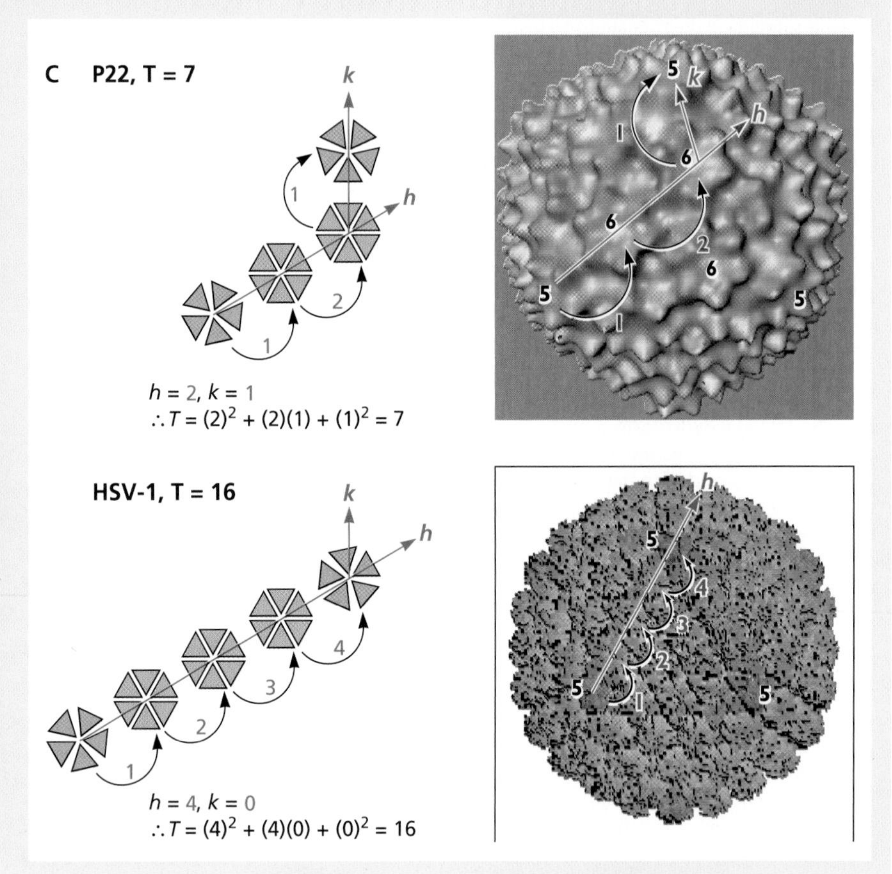

capsid proteins (the β-barrel jelly roll; see next section) in which β-strands are connected by loops (Fig. 4.10A). Interactions among neighboring subunits are mediated by these loops. The prominent projections near the threefold axes of rotational symmetry (Fig. 4.10B), which have been implicated in receptor binding of adenovirus-associated virus type 2, are formed by extensive interdigitation among the loops from adjacent subunits.

Structure of poliovirus: a $T = 3$ structure. As their name implies, the picornaviruses are among the smallest of animal viruses. In contrast to the $T = 1$ parvoviruses, the ~30-nm-diameter poliovirus particle is composed of 60 copies of a **multimeric** structural unit. It contains a (+) strand RNA genome of ~7.5 kb and its covalently attached 5′-terminal protein, VPg (Appendix, Fig. 21). Our understanding of the architecture of the *Picornaviridae* took a quantum leap in 1985 with the determination of high-resolution structures of human rhinovirus 14 (genus *Rhinovirus*) and poliovirus (genus *Enterovirus*).

The heteromeric structural unit of the poliovirus capsid contains one copy each of VP1, VP2, VP3, and VP4. The VP4 protein is synthesized as an N-terminal extension of VP2 and restricted to the inner surface of the particle. The poliovirus capsid is built from asymmetric units that contain one copy of each of three different proteins (VP1, VP2, and VP3), and is therefore described as a pseudo $T = 3$ structure (Fig. 4.11A). Although these three proteins are not related in amino acid sequence, all contain a central β-sheet structure termed a **β-barrel jelly roll**. The arrangement of β-strands in these β-barrel proteins is illustrated schematically in Fig. 4.11B, for comparison with the actual structures of VP1, VP2, and VP3. As can be seen in the schematic, two antiparallel β-sheets form a wedge-shaped structure. One of the β-sheets comprises one wall of the wedge, while the second, sharply twisted β-sheet forms both the second wall and the floor.

BOX 4.4

EXPERIMENTS

Viral chain mail: not the electronic kind

The mature capsid of the tailed, double-stranded DNA bacteriophage HK97 is a $T = 7$ structure built from hexamers and pentamers of a single viral protein, Gp5. The first hints of the remarkable and unprecedented mechanism of stabilization of this particle came from biochemical experiments, which showed the following:

- A previously unknown covalent protein-protein linkage forms in the final reaction in the assembly of the capsid: the side chain of a lysine (K) in every Gp5 subunit forms a covalent isopeptide bond with an asparagine (N) in an adjacent subunit. Consequently, **all** subunits are joined covalently to each other.
- This reaction is **autocatalytic**, depending only on Gp5 subunits organized in a particular conformational state: the capsid is enzyme, substrate, and product.
- HK97 mature particles are extraordinarily stable and cannot be disassembled into individual subunits by boiling in strong ionic detergent: it was therefore proposed that the cross-linking also interlinks the subunits from adjacent structural units to catenate rings of hexamers and pentamers.

The determination of the structure of the HK97 capsid to 3.6-Å resolution by X-ray crystallography has confirmed the formation of such capsid "chain mail" (figure, panel A), akin to that widely used in armor (B) until the development of the crossbow. The HK97 capsid is the first example of a protein catenane (an interlocked ring). This unique structure has been shown to increase the stability of the virus particle, and it may be of particular advantage as the capsid shell is very thin. The delivery of the DNA genome to host cells via the tail of the particle obviates the need for capsid disassembly.

Duda RL. 1998. Protein chainmail: catenated protein in viral capsids. *Cell* **94:**55–60.

Wikoff WR, Liljas L, Duda RL, Tsuruta H, Hendrix RW, Johnson JE. 2000. Topologically linked protein rings in the bacteriophage HK97 capsid. *Science* **289:**2129–2133.

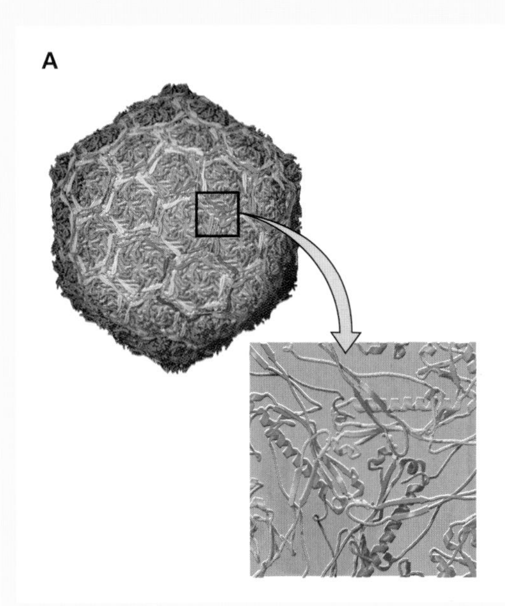

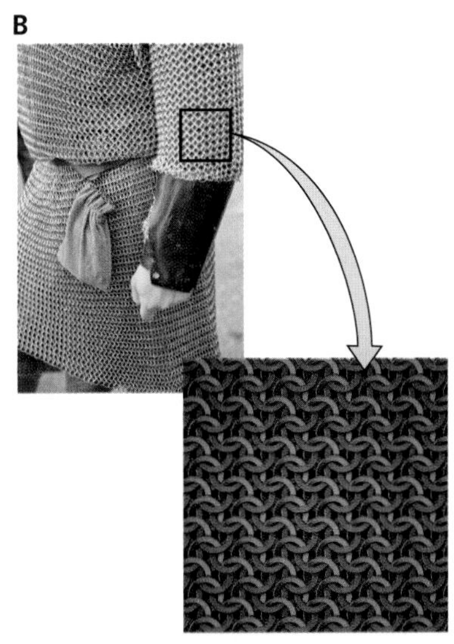

Chain mail in the bacteriophage HK97 capsid. (A) The exterior of the HK97 capsid is shown at the top, with structural units of the Gp5 protein in cyan. The segments of subunits that are cross-linked into rings are colored the same, to illustrate the formation of catenated rings of subunits. The cross-linking is shown in the more detailed view below, down a quasithreefold axis with three pairs of cross-linked subunits. The K-N isopeptide bonds are shown in yellow. The cross-linked monomers (shown in blue) loop over a second pair of covalently joined subunits (green), which in turn cross over a third pair (magenta). Adapted from W. R. Wikoff et al., *Science* **289:**2129–2133, 2000, with permission. Courtesy of J. Johnson, The Scripps Research Institute. **(B)** Chain mail armor and schematic illustration of the rings that form the chain mail.

The protein backbones in β-barrel domains of VP1, VP2, and VP3 are folded in the same way; that is, they possess the same **topology**, and the differences among these proteins are restricted largely to the loops that connect β-strands and to the N- and C-terminal segments that extend from the central β-barrel domains.

The β-barrel jelly roll conformation of these picornaviral proteins is also seen in the core domains of capsid proteins of a number of plant, insect, and vertebrate (+) strand RNA viruses, such as tomato bushy stunt virus and Nodamura virus. This structural conservation was entirely unanticipated. Even more remarkably, this relationship is not restricted to small RNA viruses: the major capsid proteins of the DNA-containing parvoviruses and polyomaviruses also contain such β-barrel domains. It is well established that the three-dimensional structures of cellular proteins have been highly conserved during evolution, even though there may be very little amino acid sequence identity. For example, all globins possess a common three-dimensional architecture based on a particular arrangement of eight α-helices, even though their amino acid sequences are different. One interpretation of the common occurrence of the β-barrel jelly roll domain in viral capsid proteins is therefore that seemingly unrelated modern viruses (e.g., picornaviruses and parvoviruses) share some portion of their evolutionary history. It is also possible that this domain topology represents one of a limited number commensurate

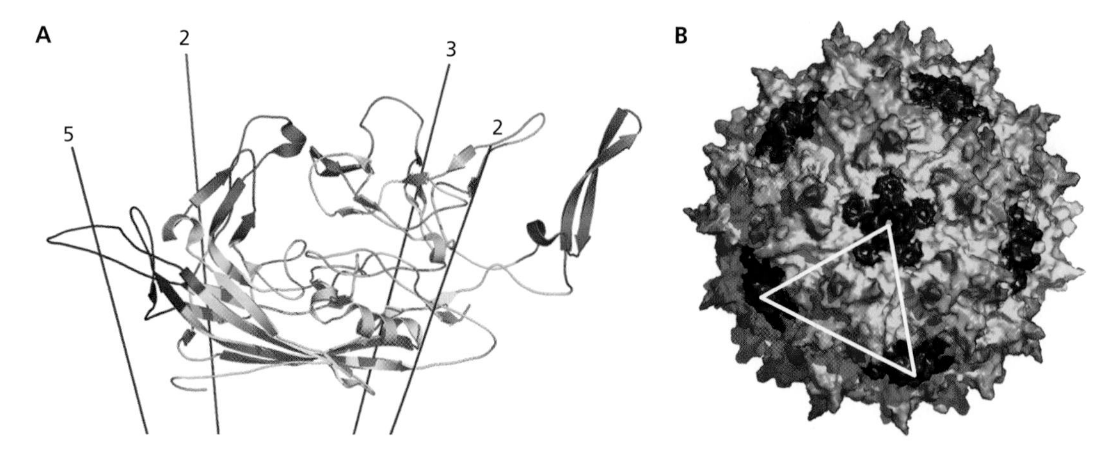

Figure 4.10 Structure of the parvovirus adeno-associated virus 2. (A) Ribbon diagram of the single coat subunit of the $T = 1$ particle. The regions of the subunit that interact around the five-, three-, and twofold axes (indicated) of icosahedral symmetry are shown in blue, green, and yellow, respectively. The red segments form peaks that cluster around the threefold axes. **(B)** Surface view of the 3-Å-resolution structure determined by X-ray crystallography of purified virions. The regions of the single subunits from which the capsid is built are colored as in panel A, and the face formed by three subunits is outlined in black. Adapted from Q. Xie et al., *Proc Natl Acad Sci U S A* **99:**10405–10410, 2002, with permission. Courtesy of Michael Chapman, Florida State University.

Figure 4.11 Packing and structures of poliovirus proteins. (A) The packing of the 60 VP1-VP2-VP3 structural units, represented by wedge-shaped blocks corresponding to their β-barrel domains. Note that the structural unit (outlined in black) contributes to two adjacent faces of an icosahedron rather than corresponding to a facet. When virus particles are assembled, VP4 is covalently joined to the N terminus of VP2. It is located on the inner surface of the capsid shell (see Fig. 4.12A). **(B)** The topology of the polypeptide chain in a β-barrel jelly roll is shown at the top left. The β-strands, indicated by arrows, form two antiparallel sheets juxtaposed in a wedge-like structure. The two α-helices (purple cylinders) that surround the open end of the wedge are also conserved in location and orientation in these proteins. As shown, the VP1, VP2, and VP3 proteins each contain a central β-barrel jelly roll domain. However, the loops that connect the β-strands in this domain of the three proteins vary considerably in length and conformation, particularly at the top of the β-barrel, which, as represented here, corresponds to the outer surface of the capsid. The N- and C-terminal segments of the protein also vary in length and structure. The very long N-terminal extension of VP3 has been truncated in this representation. Adapted from J. M. Hogle et al., *Science* **229:**1358–1365, 1985, with permission.

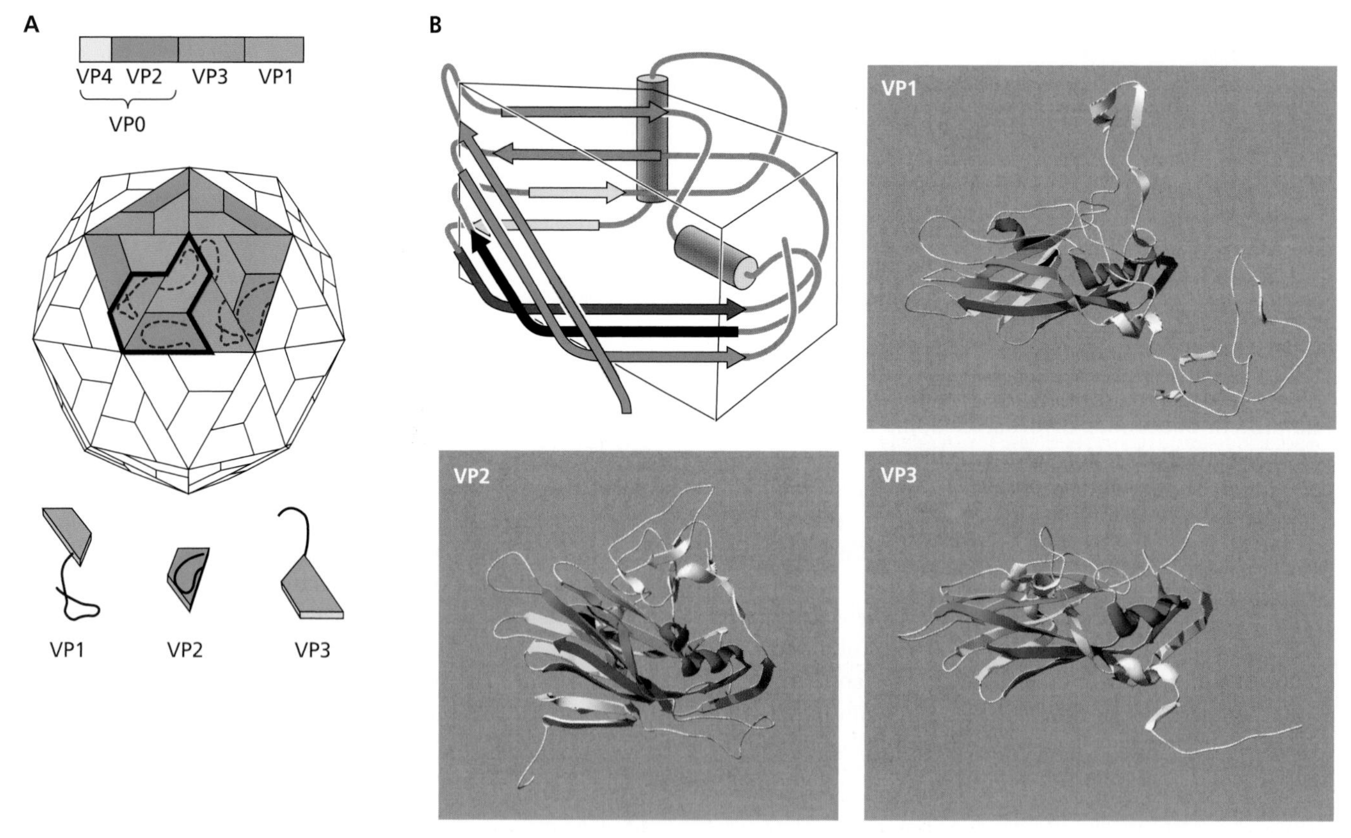

BOX 4.5

DISCUSSION

Remarkable architectural relationships among viruses with double-stranded DNA genomes

Viruses with double-stranded DNA genomes are currently classified by the International Committee on the Taxonomy of Viruses into 28 families (with some unassigned) on the basis of the criteria described in Chapter 1. As might be expected, these viruses exhibit different morphologies and infect diverse organisms representing all three domains of life. They span a large size range, with genomes from a few kilobase pairs (members of the *Polyomaviridae*) to >2,500 kbp (*Pandoravirus*). Nevertheless, consideration of structural properties indicates that these very disparate virus families in fact represent a limited number of architectural types.

Structural information is now available for the major capsid proteins of representatives of some 20 of the 28 families of double-stranded DNA viruses. Based on the fold of the proteins, most of these families can be assigned to one of just five structural classes. It is noteworthy that the two most common major capsid protein folds, the double jelly roll and the HK97-like, are found in viruses that infect *Bacteria*, *Archaea*, and *Eukarya* (including mammals), as summarized in the figure.

The small number of building blocks seen in the major capsid proteins of these viruses might indicate convergent evolution, the compatibility of only a tiny fraction of the >1,000 distinct protein folds described to date with assembly of an infectious virus particle. However, viruses that infect hosts as divergent as bacteria and humans share more than the architectural elements of their major capsid proteins. This property is exemplified by the bacteriophage PRD1 and human species C adenoviruses, in which the major structural unit comprises a trimer of monomers each with two jelly roll domains and hence exhibits pseudohexagonal symmetry. These icosahedral capsids also share a structural unit built from different proteins at the positions of fivefold symmetry, from which project proteins that attach to the host cell receptors project; features of their linear double-stranded DNA genomes, such as the presence of inverted terminal repetitions; and mechanisms of viral DNA synthesis. Extensive similarities in morphology and the mechanisms of particle assembly and active genome packaging are also shared by tailed, double-stranded DNA viruses that infect bacteria, e.g., phage T4, and herpesviruses. It is therefore difficult to escape the conclusion that these modern viruses evolved from an ancient common ancestor (see also Volume II, Chapter 10).

Abrescia NG, Branford DH, Grimes JM, Stuart DI. 2012. Structure unifies the viral universe. *Ann Rev Biochem* **81:**795–822.

Benson SD, Bamford JK, Bamford DH, Burnett RM. 1999. Viral evolution revealed by bacteriophage PRD1 and human adenovirus coat protein structures. *Cell* **98:**825–833.

Kropovic M, Banford DH. 2011. Double-stranded DNA viruses: 20 families and only five different architectural principles for virion assembly. *Curr Opin Virol* **1:**118–124.

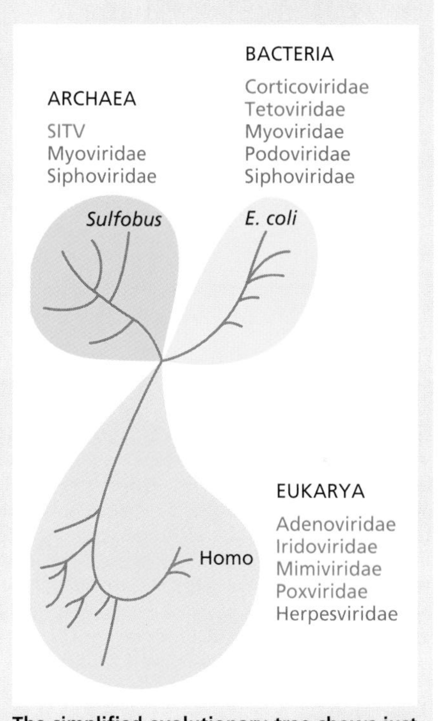

The simplified evolutionary tree shows just some of the branches within each domain of life, with archaeal, bacterial, and eukaryote hosts of viruses described in this chapter indicated. Viruses with major capsid proteins with the double jelly roll and HK97-like folds are listed in red and blue, respectively.

with packing of proteins to form a sphere, and is an example of convergent evolution. The structural (and other) properties of viruses with double-stranded DNA genomes provide compelling support for the first hypothesis (Box 4.5).

The overall similarity in shape of the β-barrel domains of poliovirus VP1, VP2, and VP3 facilitates both their interaction with one another to form the 60 structural units of the capsid and the packing of these structural units. How well these interactions are tailored to form a protective shell is illustrated by the model of the capsid shown in Fig. 4.12: the extensive interactions among the β-barrel domains of adjacent proteins form a dense, rigid protein shell around a central cavity in which the genome resides. The packing of the β-barrel domains is reinforced by a network of protein-protein contacts on the inside of the capsid. These interactions are particularly extensive about the fivefold axes (Fig. 4.12C). The interaction of five VP1 molecules, which is unique to the fivefold axes, results in a prominent protrusion extending to about 25 Å from the capsid shell (Fig. 4.12A). The protrusion appears as a steep-walled plateau encircled by a valley or cleft. In the capsids of many picornaviruses, these depressions, which contain the receptor-binding sites, are so deep that they have been termed **canyons**.

One of several important lessons learned from high-resolution analysis of picornavirus capsids is that their design does not conform strictly to the principle of **quasiequivalence**. For example, despite the topological identity and geometric similarity of the jelly roll domains of the proteins that form the capsid shell, the subunits do not engage in quasiequivalent bonding: interactions among VP1 molecules around the fivefold axes are neither chemically nor structurally equivalent to those in which VP2 or VP3 engage.

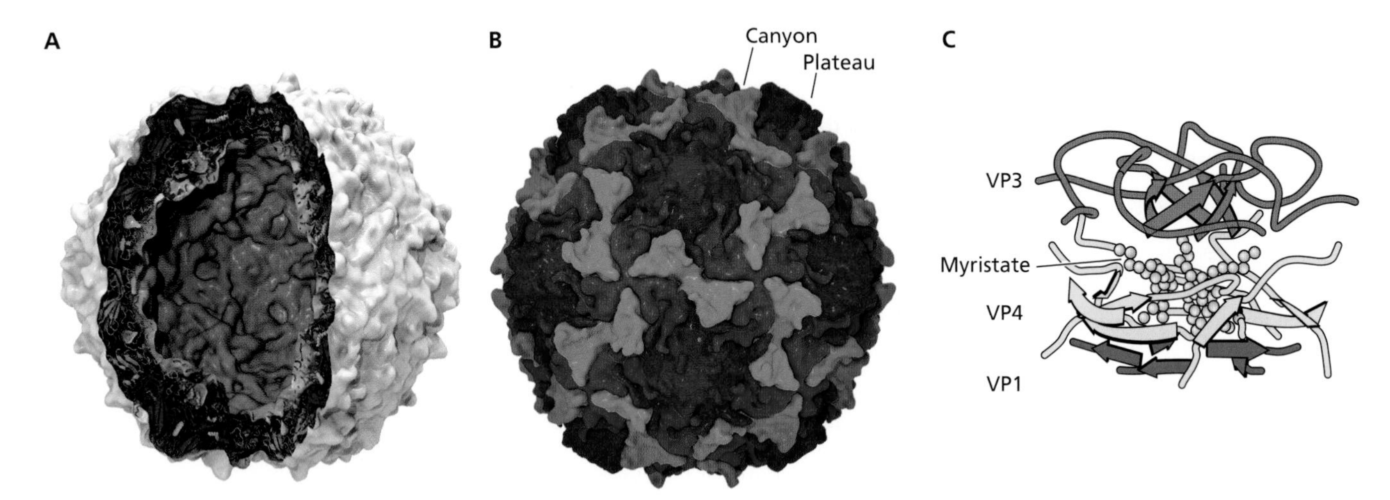

Figure 4.12 Interactions among the proteins of the poliovirus capsid. (A) Space-filling representation of the particle, with four pentamers removed from the capsid shell and VP1 in blue, VP2 in green, VP3 in red, and VP4 in yellow, as in Fig. 4.11A. Note the large central cavity in which the RNA genome resides; the dense protein shell formed by packing of the VP1, VP2, and VP3 β-barrel domains; and the interior location of VP4, which decorates the inner surface of the capsid shell. **(B)** Space-filling representation of the exterior surface showing the packing of the β-barrel domains of VP1, VP2, and VP3. Interactions among the loops connecting the upper surface of the β-barrel domains of these proteins create the surface features of the virion, such as the plateaus at the fivefold axes, which are encircled by a deep cleft or canyon. The particle is also stabilized by numerous interactions among the proteins on the inner side of the capsid. **(C)** These internal contacts are most extensive around the fivefold axes, where the N termini of five VP3 molecules are arranged in a tube-like, parallel β-sheet. The N termini of VP4 molecules carry chains of the fatty acid myristate, which are added to the protein posttranslationally. The lipids mediate interaction of the β-sheet formed by VP3 N termini with a second β-sheet structure, containing strands contributed by both VP4 and VP1 molecules. This internal structure is not completed until the final stages of, or after, assembly of virus particles, when proteolytic processing liberates VP2 and VP4 from their precursor, VP0. This reaction therefore stabilizes the capsid. Panels A and B were created by Jason Roberts, Doherty Institute, Melbourne, Australia.

An alternative icosahedral design: structure of simian virus 40. The capsids of the small DNA polyomaviruses simian virus 40 and mouse polyomavirus, ~50 nm in diameter, are organized according to a rather different design that is not based on quasiequivalent interactions. The structural unit is a pentamer of the major structural protein, VP1. The capsid is built from 72 such pentamers engaged in one of two kinds of interaction. Twelve structural units occupy the 12 positions of fivefold rotational symmetry, in which each is surrounded by five neighbors. Each of the remaining 60 pentamers is surrounded by 6 neighbors at positions of sixfold rotational symmetry in the capsid (Fig. 4.13A). Consequently, the 72 pentamers of simian virus 40 occupy a number of different local environments in the capsid, because of differences in packing around the five- and sixfold axes.

Like the three poliovirus proteins that form the capsid shell, simian virus 40 VP1 contains a large central β-barrel jelly roll domain, in this case with an N-terminal arm and a long C-terminal extension (Fig. 4.13B). However, the arrangement and packing of VP1 molecules bear little resemblance to the organization of poliovirus capsid proteins. In the first place, the VP1 β-barrels in each pentamer project outward from the surface of the capsid to a distance of about 50 Å, in sharp contrast to those of the poliovirus capsid proteins, which tilt along the surface of the capsid shell. As a result, the surface of simian virus 40 is much more "bristly" than that of poliovirus (compare Fig. 4.12A and 4.13A). Furthermore, the VP1 molecules present in adjacent pentamers in the simian virus 40 capsid do not make extensive contacts via the surfaces of their β-barrel domains. Rather, stable interactions among pentamers are mediated by their N- and C-terminal arms. The packing of VP1 pentamers in both pentameric and hexameric arrays requires different contacts among these structural units, depending on their local environment. In fact, there are just three kinds of interpentamer contact, which are the result of alternative conformations and noncovalent interactions of the long C-terminal arms of VP1 molecules. The same capsid design is also exhibited by human papillomaviruses.

Simian virus 40 and poliovirus capsids differ in their surface appearance, in the number of structural units, and in the ways in which these structural units interact. Nevertheless, they share important features, including modular organization of the proteins that form the capsid shell and a common β-barrel domain as the capsid building block. Neither poliovirus nor simian virus 40 capsids conform to strict quasiequivalent construction: all contacts made by all protein subunits are not similar, and in the case of simian virus 40, the majority of VP1 **pentamers** are packed in **hexameric** arrays. Nevertheless, close packing with icosahedral symmetry is achieved by limited variations of the contacts, either

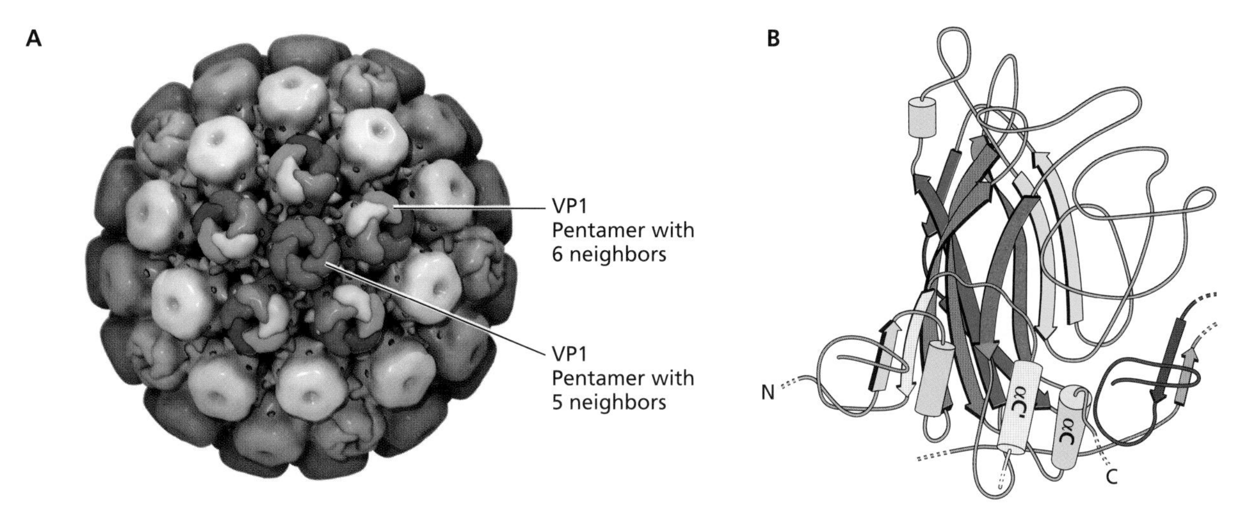

Figure 4.13 Structural features of simian virus 40. (A) View of the simian virus 40 virion showing the organization of VP1 pentamers. One of the 12 5-coordinated pentamers is shown in purple and 10 of the 60 pentamers present in hexameric arrays are in light gray. The individual VP1 molecules in the pentamers surrounding a pentamer with five neighbors are colored red, blue, green, yellow, and orange. The image was created by Jason Roberts, Doherty Institute, Melbourne, Australia. **(B)** The topology of the VP1 protein shown in a ribbon diagram, with the strands of the β-barrel jelly roll colored as in Fig. 4.11B. This β-barrel domain is perpendicular to the capsid surface. The C-terminal arm and α-helix (αC; orange) of the VP1 subunit invades a neighboring pentamer (not shown). The C-terminal arm and C α-helix shown in gray (αC′) is the invading arm from a different neighboring pentamer (not shown), which is clamped in place by extensive interactions of its β-strand with the N-terminal segment of the subunit shown. The subunit shown also interacts with the N-terminal arm from its anticlockwise neighbor in the same pentamer (dark gray). This subunit also interacts extensively with other segments from subunits in the same pentamer or in neighboring pentamers, shown in gray or black. Adapted from R. C. Liddington et al., *Nature* **354:**278–284, 1991, with permission.

among topologically similar, but chemically distinct, surfaces (poliovirus) or made by a flexible arm (simian virus 40).

Structurally simple icosahedral capsids in more complex particles. Several viruses that are structurally more sophisticated than those described in the previous sections nevertheless possess simple protein coats built from one or a few structural proteins. The complexity comes from the additional protein and lipid layers in which the capsid is enclosed (see "Viruses with Envelopes" below).

Structurally Complex Capsids

Some naked viruses are considerably larger and more elaborate than the small RNA and DNA viruses described in the previous section. The characteristic feature of such virus particles is the presence of proteins devoted to specialized structural or functional roles. Despite such complexity, detailed pictures of the organization of this type of virus particle can be constructed by using combinations of biochemical, structural, and genetic methods. Well-studied human adenoviruses and members of the *Reoviridae* exemplify these approaches.

Adenovirus. The most striking morphological features of the adenovirus particle (maximum diameter, 150 nm) are the well-defined icosahedral appearance of the capsid and the presence of long fibers at the 12 vertices (Fig. 4.14A). A fiber, which terminates in a distal knob that binds to the adenoviral receptor, is attached to each of the 12 penton bases located at positions of fivefold symmetry in the capsid. The remainder of the shell is built from 240 additional subunits, the hexons, each of which is a trimer of viral protein II (Fig. 4.14B). Formation of this capsid depends on nonequivalent interactions among subunits: the hexons that surround pentons occupy a different bonding environment than those surrounded entirely by other hexons. The X-ray crystal structures of the trimeric hexon (the major capsid protein) established that each protein II monomer contains two β-barrel domains, each with the topology of the β-barrels of the simpler RNA and DNA viruses described in the previous section (Fig. 4.14B). The very similar topologies of the two β-barrel domains of the three monomers facilitate their close packing to form the hollow base of the trimeric hexon. Interactions among the monomers are very extensive, particularly in the towers that rise above the hexon base and are formed by intertwining loops from each monomer. Consequently, the trimeric hexon is extremely stable.

The adenovirus particle contains seven additional structural proteins (Fig. 4.14A). The presence of so many proteins and the large size of the particle made elucidation of adenovirus architecture a challenging problem. One approach

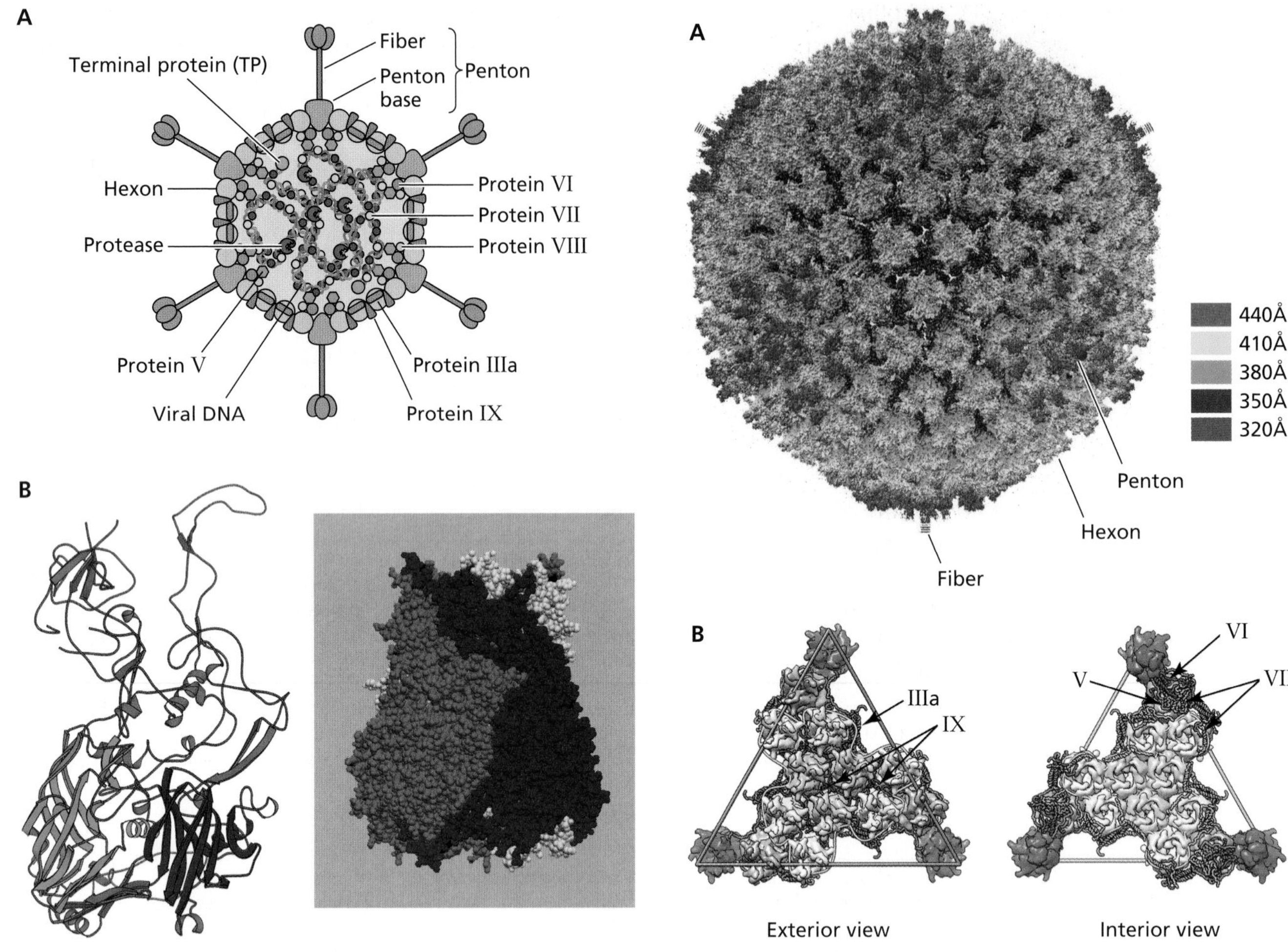

Figure 4.14 Structural features of adenovirus particles. (A) The organization of human adenovirus type 5 is shown schematically to indicate the locations of the major (hexon, penton base, and fiber) and minor (IIIa, VI, VIII, and IX) capsid proteins and of the internal core proteins, V, VII, and μ. These locations and the interactions between proteins shown were deduced from the composition of the products of controlled dissociation of viral particles and the results of cross-linking studies and confirmed in high-resolution structures of adenovirus particles. **(B)** Structure of the hexon homotrimer. The monomer (left) is shown as a ribbon diagram, with gaps indicating regions that were not defined in the X-ray crystal structure at 2.9-Å resolution, and the trimer (right) is shown as a space-filling model with each monomer in a different color. The monomer contains two β-barrel jelly roll domains colored green and blue in the left panel. The trimers are stabilized by extensive interactions within both the base and the towers. From M. M. Roberts et al., *Science* **232:**1148–1151, 1986, and F. K. Athappilly et al., *J Mol Biol* **242:**430–455, 1994, with permission. Courtesy of J. Rux, S. Benson, and R. M. Burnett, The Wistar Institute.

Figure 4.15 Interactions among major and minor proteins of the adenoviral capsid. (A) Cryo-EM reconstruction of the adenovirus type 5 capsid at 3.6-Å resolution radially colored by distance from the center, as indicated. This view is centered on a threefold axis of icosahedral symmetry. Only short stubs of the fibers are evident, as these structures are bent. For other views, see Movie 4.2 (http://bit.ly/Virology_V1_Movie4-2). **(B)** Views of the outer (left) and inner (right) surfaces indicating the locations of the minor capsid proteins IX (blue), IIIa (mauve), V (green), VI (red), and VIII (orange) with respect to hexons (light blue, pink, light green, and khaki) and penton base (magenta). The boundary of a group-of-nine hexon is shown by a white tube. Adapted from V. Reddy and G. Nemerow, *Proc Natl Acad Sci U S A* **111:**11715–11720, 2014, with permission. Courtesy of V. Reddy, The Scripps Research Institute.

that has proved generally useful in the study of complex viruses is the isolation and characterization of discrete subviral particles. For example, adenovirus particles can be dissociated into a core structure that contains the DNA genome, groups of nine hexons, and pentons. Analysis of the composition of such subassemblies identified two classes of proteins in addition to the major capsid proteins described above. One comprises the proteins present in the core, such as protein VII, the major DNA-binding protein. The remaining proteins are associated with either individual hexons or the groups of hexons that form an icosahedral face of the capsid, suggesting that they stabilize the structure. Protein IX has been clearly identified as capsid "cement": a mutant virus that lacks the protein IX coding sequence produces the typical yield of virions, but these particles are much less heat stable than the wild type.

The interactions of protein IX and other minor proteins with hexons and/or pentons were deduced initially by difference imaging (Fig. 4.4) and subsequently by X-ray crystallography or cryo-EM (Fig. 4.15A; Box 4.1). The minor capsid proteins make numerous contacts with the major structural units. For example, on the outer surface of the capsid, a network formed by extensive interactions among the extended molecules of protein IX knit together the hexons that form the groups of nine (Fig. 4.15B). Other minor capsid proteins are restricted to the inner surface, where they reinforce the groups of nine hexons, or weld the penton base to its surrounding hexons. During assembly, interactions among hexons and other major structural proteins must be relatively weak, so that incorrect associations can be reversed and corrected. However, the assembled particle must be stable enough to survive passage from one host to another. It has been proposed that the incorporation of stabilizing proteins like protein IX allows these paradoxical requirements to be met.

Reoviruses. Reovirus particles exhibit an unusual architecture: they contain multiple protein shells. They are naked particles, 70 to 90 nm in diameter with an outer $T = 13$ icosahedral protein coat, containing the 10 to 12 segments of the double-stranded genome and the enzymatic machinery to synthesize viral mRNA. The particles of human reovirus (genus Orthoreovirus) contain eight proteins organized in two concentric shells, with spikes projecting from the inner layer through and beyond the outer layer at each of the 12 vertices (Fig. 4.16A). Members of the genus Rotavirus, which includes the leading causes of severe infantile gastroenteritis in humans, contain three nested protein layers, with 60 projecting spikes (Fig. 4.16B). Although differing in architectural detail, reovirus particles have common structural features, including an unusual design of the innermost protein shell.

Removal of the outermost protein layer, a process thought to occur during entry into a host cell, yields an inner core structure, comprising one shell (orthoreoviruses) or two (rotaviruses and members of the genus *Orbivirus*, such as bluetongue virus). These subviral particles also contain the genome and virion enzymes and synthesize viral mRNAs under appropriate conditions *in vitro*. High-resolution structures have been obtained for bluetongue virus and human reovirus cores, some of the largest viral assemblies that have been examined by X-ray crystallography. The thin inner layer contains 120 copies of a single protein (termed VP3 in bluetongue virus). These proteins are not related in their primary sequences, but they nevertheless have similar topological features and the same plate-like shape. Moreover, in both cases, the dimeric proteins occupy one of two different structural environments, and to do so, they adopt one of two distinct conformational states, indicated as green and red in Fig. 4.16C (right). Because of this arrangement, the green and

Figure 4.16 Structures of members of the *Reoviridae*. The organization of mammalian reovirus **(A)** and rotavirus **(B)** particles is shown schematically to indicate the locations of proteins, deduced from the protein composition of intact particles and of subviral particles that can be readily isolated from them. **(C)** X-ray crystal structure of the core of bluetongue virus, a member of the *Orbivirus* genus of the *Reoviridae*, showing the core particle and the inner scaffold. Trimers of VP7 (VP6 in rotaviruses; panel B) project radially from the outer layer of the core particle (left). Each icosahedral asymmetric unit, two of which are indicated by the white lines, contains 13 copies of VP7 arranged as five trimers colored red, orange, green, yellow, and blue, respectively. This layer is organized with classical $T = 13$ icosahedral symmetry. As shown on the right, the inner layer is built from VP3 dimers that occupy one of two completely different structural environments, colored green and red. Green monomers span the icosahedral twofold axes and interact in rings of five around the icosahedral fivefold axes. In contrast, red monomers are organized as triangular "plugs" around the threefold axes. Differences in the interactions among monomers at different positions allow close packing to form the closed shell, the equivalent of a $T = 2$ lattice. As might be anticipated, VP7 trimers in pentameric or hexameric arrays in the outer layer make different contacts with the two classes of VP3 monomer in the inner layer. Nevertheless, each type of interaction is extensive, and in total, these contacts compensate for the symmetry mismatch between the two layers of the core. The details of these contacts suggest that the inner shell both defines the size of the virus particle and provides a template for assembly of the outer $T = 13$ structure. From J. M. Grimes et al., *Nature* **395**:470–478, 1998, with permission. Courtesy of D. I. Stuart, University of Oxford.

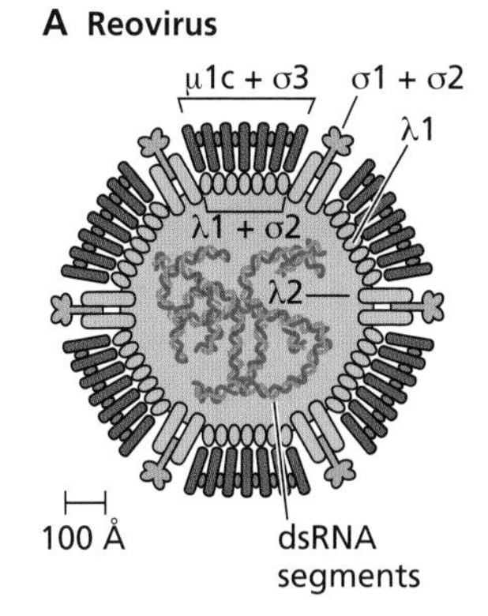

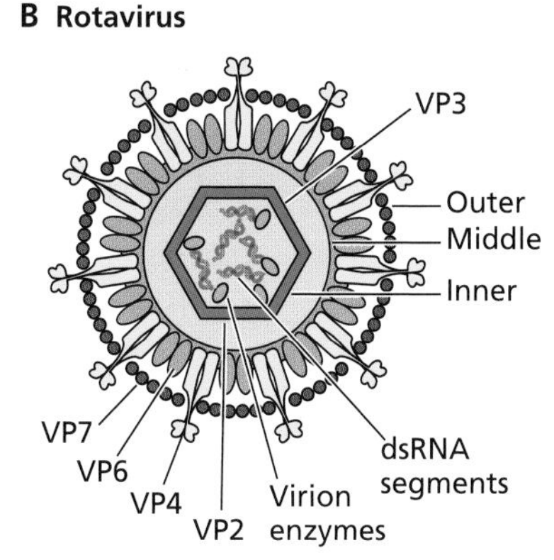

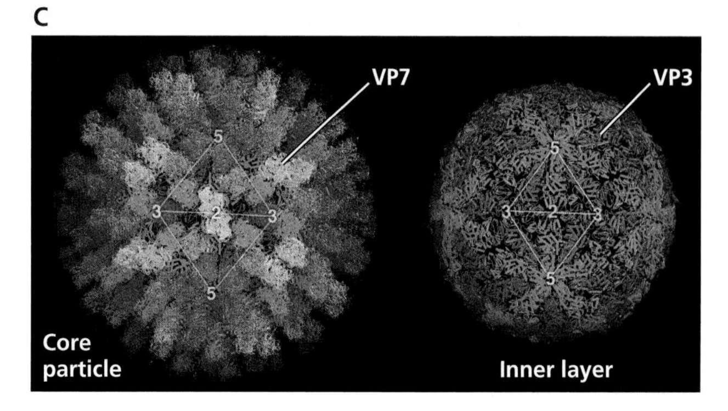

red dimers are **not** quasiequivalent, and virtually all contacts in which the two types of monomers engage are very different. However, these differences allow the formation of VP3 assemblies with either five- or threefold rotational symmetry and hence of an icosahedral shell ($T = 1$). This VP3 shell of bluetongue virus abuts directly on the inner surface of the middle layer, which comprises trimers of a single protein organized into a classical $T = 13$ lattice (Fig. 4.16C, left). A large number of different (nonequivalent) contacts between these trimers and VP3 weld the two layers together. These properties of reoviruses illustrate that a quasiequivalent structure is not the **only** solution to the problem of building large viral particles: viral proteins that interact with each other and with other proteins in multiple ways can provide an effective alternative. The architecture of the two protein shells described above appears to be conserved in all viruses with double-stranded RNA genomes. However, it is not yet known whether symmetry mismatch is also a feature of other large viruses that contain multiple protein layers.

Other Capsid Architectures

As noted previously, capsids with icosahedral or helical symmetry are characteristic of the great majority of virus particles described to date. Nevertheless, these architectures are not universal. As discussed subsequently, the composition and organization of internal structures that enclose the genome of some complex viruses, like the poxvirus vaccinia virus, are not well understood. However, the capsids of even some relatively small viruses can be constructed according to an alternative design. This property is exemplified by the capsids of retroviruses.

In the enveloped particles of all retroviruses, the capsid surrounds a nucleoprotein containing the diploid (+) strand RNA genome and is in turn encased by the viral matrix (M) protein. However, the capsid may be spherical, cylindrical, or conical, the shape exhibited by capsids of human immunodeficiency virus type 1 and other lentiviruses (Fig. 4.17A). These capsids are built from a single capsid (CA) protein, which can form both pentamers and hexamers (Fig. 4.17B). The details of the organization of the capsids of most retroviruses are not yet available. The odd appearance of the human immunodeficiency virus type 1 capsid might suggest that it represents an exception to the geometric rules that dictate the viral architectures described in previous sections. However, this is not the case: this capsid can be described by a fullerone cone model that combines principles of both icosahedral and helical symmetry. In this model, which has been confirmed by cryo-EM and modeling methods, a closed structure is formed using 12 pentamers, just as in an icosahedral capsid (Box 4.6). However, pentamers are not spaced at regular intervals throughout the

Figure 4.17 Asymmetric capsids of retroviruses. (A) Variation in the morphology of retroviruses shown schematically. Although all retrovirus particles are assembled from the same components (see the text), the cores are primarily spherical, cylindrical, or conical in the case of gammaretroviruses (e.g., Moloney murine leukemia viruses), betaretroviruses (e.g., Mason-Pfizer monkey virus), and lentiviruses (e.g., human immunodeficiency virus type 1), respectively. **(B)** Cryo-electron tomographic reconstruction of human immunodeficiency virus type 1 showing the conical core and the glycoprotein spikes projecting from the surface of the particle. Adapted from J. Liu et al., *Methods Enzymol* **483:**267–290, 2010, with permission. Courtesy of H. Winkler, Florida State University. **(C)** Structures of human immunodeficiency virus type 1 capsid (CA) pentamers and hexamers, determined by X-ray crystallography of these structures assembled *in vitro* and stabilized by disulfide bonds. Each subunit is shown in a different color with the additional subunit present in the hexamer in magenta. In each structural unit, the N-terminal domains of the CA subunits (saturated colors) interact in an inner ring surrounded by an outer belt formed by the C-terminal domains (pale colors). The interactions among N-terminal domains of individual subunits and between the N-terminal domain of one subunit and the C-terminal domain of its clockwise neighbor are very similar in pentamers and hexamers; that is, they are quasiequivalent. Reprinted from O. Pornillos et al., *Nature* **469:**424–427, 2011, with permission. Courtesy of O. Pornillos, University of Virginia.

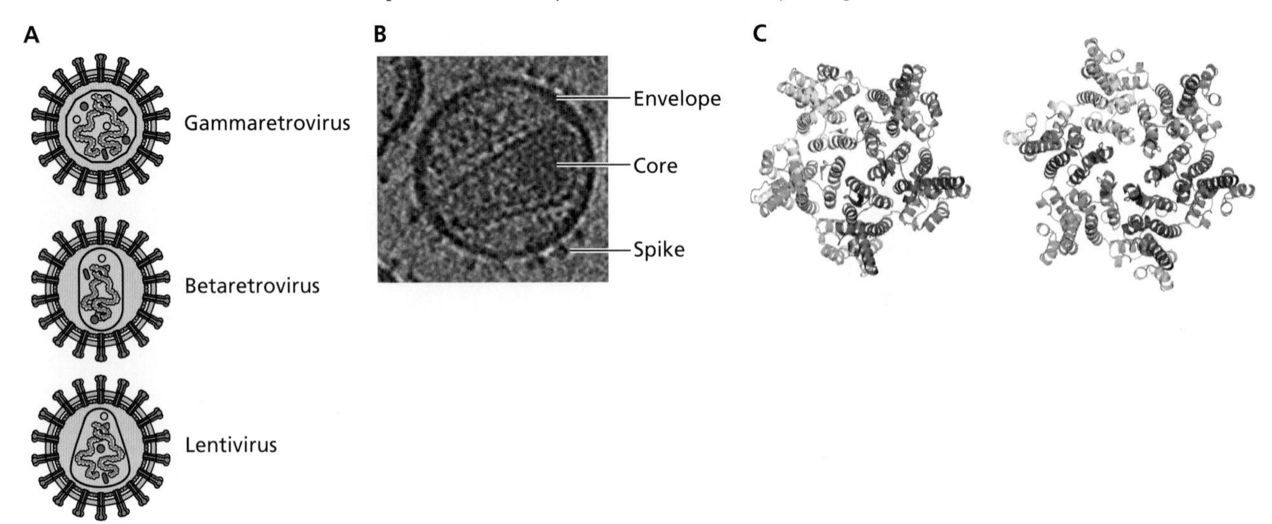

BOX 4.6

EXPERIMENTS

A fullerene cone model of the human immunodeficiency virus type 1 capsid

Diverse lines of evidence support a fullerone cone model of this capsid based on principles that underlie the formation of icosahedral and helical structures.

(A) A purified human immunodeficiency virus type 1 protein comprising the capsid linked to the nucleocapsid proteins, CA-NC self-assembles into cylinders and cones when incubated with a segment of the viral RNA genome *in vitro*. The cones assembled *in vitro* are capped at both ends, and many appear very similar in dimensions and morphology to cores isolated from viral particles (compare the two panels). From B. K. Ganser et al., *Science* **283:**80–83, 1999, with permission. Courtesy of W. Sundquist, University of Utah. **(B)** The very regular appearance of the synthetic CA-NC cones suggested that, despite their asymmetry, they are constructed from a regular, underlying lattice analogous to the lattices that describe structures with icosahedral symmetry discussed in Box 4.3. In fact, the human immunodeficiency virus type 1 cores can be modeled using the geometric principles that describe cones formed from carbon. Such elemental carbon cones comprise helices of hexamers closed at each end by caps of buckminsterfullerene, which are structures that contain pentamers surrounded by hexamers. As in structures with icosahedral symmetry, the positions of pentamers determine the geometry of cones. However, in cones, pentamers are present **only** in the terminal caps. The human immunodeficiency virus type 1 cones formed *in vitro* and isolated from mature virions can be modeled as a fullerene cone assembling on a curved hexagonal lattice with five pentamers (red) at the narrow end of the cone, as shown in the expanded view. The wide end would be closed by an additional 7 pentamers (because 12 pentamers are required to form a closed structure from a hexagonal lattice). In this type of structure, the cone angle at the narrow end can adopt only one of five allowed angles, determined by the number of pentamers. A narrow cap with five pentamers, as in the model shown in panel B, should exhibit a cone angle of 19.2°. Approximately 90% of all cones assembled *in vitro* examined met this prediction, consistent with the fullerene cone model. **(C)** A combination of cryo-EM of helical tubes of CA at higher resolution (8.6 Å), molecular dynamics simulation, and cryo-electron tomography of mature viral particles recently produced an atomic-level model of the human immunodeficiency virus type 1 capsid. The two top panels show representative sections through a three-dimensional reconstruction of isolated human immunodeficiency virus type 1 cores obtained by cryo-electron tomography. Red arrows indicate arrays of CA hexamers, and yellow stars indicate locations of sharp changes in curvature, i.e., the positions of pentamers. Stereoviews of the final model for a cone comprising 216 CA hexamers with N- and C-terminal domains in blue and red, respectively, and 12 pentamers (green) are shown below. Adapted from G. Zhao et al., *Nature* **497:**643–646, 2013, with permission. Courtesy of P. Zhang, University of Pittsburgh School of Medicine.

Briggs JA, Wilk T, Welker R, Kräusslich HG, Fuller SD. 2003. Structural organization of authentic, mature HIV-1 virions and cores. *EMBO J* **22:**1707–1715.

Ganser BK, Li S, Klishko VY, Finch JT, Sundquist WI. 1999. Assembly and analysis of conical models for the HIV-1 core. *Science* **283:**80–83.

Li S, Hill CP, Sundquist WI, Finch JT. 2000. Image constructions of helical assemblies of the HIV-1 CA protein. *Nature* **407:**409–413.

Zhao G, Perilla JR, Yufenyuy EL, Meng X, Chen B, Ning J, Ahn J, Gronenborn AM, Schulten K, Aiken C, Zhang P. 2013. Mature HIV-1 capsid structure by cryo-electron microscopy and all-atom molecular dynamics. *Nature* **497:**643–646.

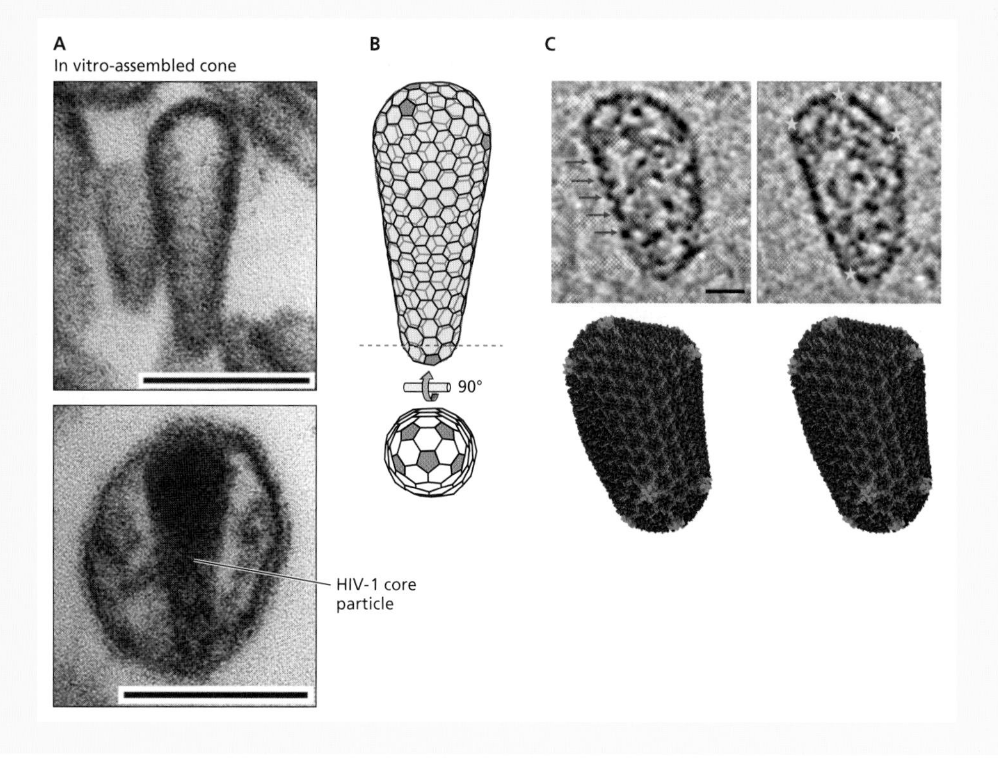

structure. Rather, they are restricted to the terminal caps and separated by spirals (a variant of helical symmetry) of CA hexamers that form the body of the cone.

Packaging the Nucleic Acid Genome

A definitive property of a virion is the presence of a nucleic acid genome. Incorporation of the genome requires its discrimination from a large population of cellular nucleic acid for packaging into virus particles. These processes are described in Chapter 13. The volumes of closed capsids are finite. Consequently, accommodation of viral genomes necessitates a high degree of condensation and compaction. A simple analogy illustrates vividly the scale of this problem; packing of the ~150-kbp DNA genome of herpes simplex virus type 1 into the viral capsid is equivalent to stuffing some 10 ft of 22 American gauge wire (diameter, 0.644 mm) into a tennis ball. Such confinement of the genome can result in high internal pressure, for example, some 18 and 25 atm within herpes simplex virus type 1 and phage capsids, respectively, and provides the force that powers ejection of DNA genomes. Packaging of nucleic acids is an intrinsically unfavorable process because of the highly constrained conformation imposed on the genome. In some cases, the force required to achieve packaging is provided, at least in part, by specialized viral proteins that harness the energy released by hydrolysis of ATP to drive the insertion of DNA. In many others, the binding of viral nucleic acids to capsid proteins appears to provide sufficient energy. The latter interactions also help to neutralize the negative charge of the sugar-phosphate backbone, a prerequisite for close juxtaposition of genome sequences.

We possess relatively little information about the organization of genomes within viral particles: nucleic acids or more-complex internal assemblies are not visible in the majority of high-resolution structural studies reported. This property indicates that the genomes or internal structures lack the symmetry of the capsid, do not adopt the same conformation in every viral particle, or both. Nevertheless, three mechanisms for condensing and organizing nucleic acid molecules within capsids can be distinguished and are described in the following sections.

Direct Contact of the Genome with a Protein Shell

In the simplest arrangement, the nucleic acid makes direct contact with the protein(s) that forms the protective shell of the virus particle. Proteins on the inner surfaces of the icosahedral capsids of many small RNA viruses interact with the viral genome. As we have seen, the interior surface of the poliovirus capsid can be described in detail. Nevertheless, we possess no structural information about the arrangement of the RNA genome, for the nucleic acid is not visible in the X-ray structure. However, the genome of the porcine picornavirus Seneca Valley virus has been visualized by this method (albeit at low resolution) (Fig. 4.18A). Much of the RNA genome forms an outer layer in which it makes extensive contact with the inner surface of the capsid. Highly ordered RNA genomes are also present in $T = 3$ nodaviruses, such as Flock house virus, in which an outer decahedral cage of ordered RNA surrounds additional rings.

Use of the same protein or proteins both to condense the genome and to build a capsid allows efficient utilization

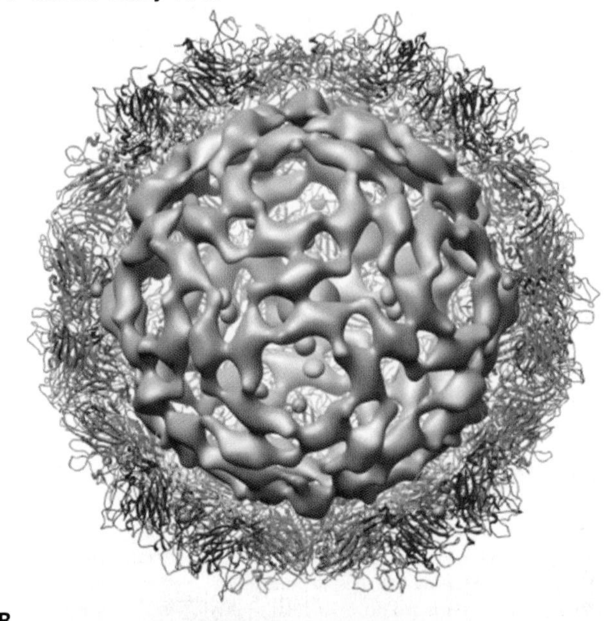

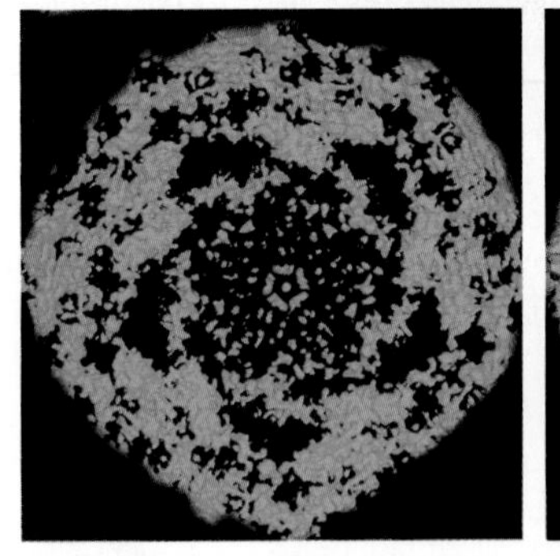

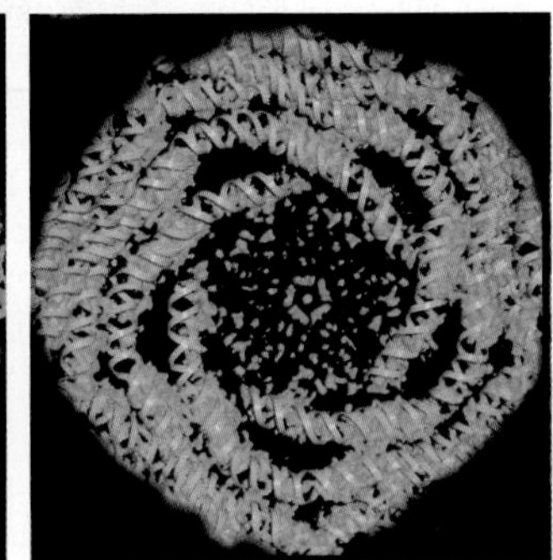

Figure 4.18: Ordered RNA genome in small and large icosahedral virus particles. (A) The 20-Å X-ray crystal structure of the picornavirus Seneca Valley virus viewed down a twofold axis of icosahedral symmetry, showing the density ascribed to the RNA genome (brown). The structural proteins are colored as in Fig. 4.11 and 4.12: VP1 (blue), VP2 (green), VP3 (red), and VP4 (yellow). **(B)** Outer layer of the double-stranded, segmented RNA genome of the rotavirus bluetongue virus observed at 6.5-Å resolution by X-ray crystallography of viral cores. The electron density of this layer of RNA from maps averaged between two closely related serotypes is shown alone (left) or with A-form duplex RNA modeled into the rods of density (right). These RNA spirals represent some 80% of the >19-kbp genome. Adapted from P. Gouet et al., *Cell* **97**:481–490, 1999, with permission. Courtesy of D. I. Stuart, University of Oxford.

of limited genetic capacity. It is therefore an advantageous arrangement for viruses with small genomes. However, this mode of genome packing is also characteristic of some more-complex viruses, notably rotaviruses and herpesviruses. The genome of rotaviruses comprises 11 segments of double-stranded RNA located within the innermost of the three protein shells of the particle. Remarkably, as much as 80% of the RNA genome appears highly ordered within the core, with strong elements of icosahedral symmetry (Fig. 4.18B).

One of the most surprising properties of the large herpesviral capsid is the absence of internal proteins associated with viral DNA: despite intense efforts, no such core proteins have been identified, and the viral genome has not yet been visualized. In contrast, cryo-EM has allowed visualization of the large, double-stranded DNA genome of bacteriophage T4, which is organized in closely apposed, concentric layers (Fig. 4.19). This arrangement illustrates graphically the remarkably dense packing needed to accommodate such large viral DNA genomes in closed structures of fixed dimensions. This type of organization must require neutralization of the negative charges of the sugar-phosphate backbone.

Figure 4.19 Dense packing of the double-stranded DNA genome in the head of bacteriophage T4 DNA. The central section of a 22-Å cryo-EM reconstruction of the head of bacteriophage T4 viewed perpendicular to the fivefold axis is shown. The concentric layers seen underneath the capsid shell have been attributed to the viral DNA genome. The connector, which is derived from the portal structure by which the DNA genome enters the head during assembly, connects the head to the tail. Adapted from A. Fokine et al., *Proc Natl Acad Sci U S A* **101:**6003–6008, 2004, with permission. Courtesy of M. Rossmann, Purdue University.

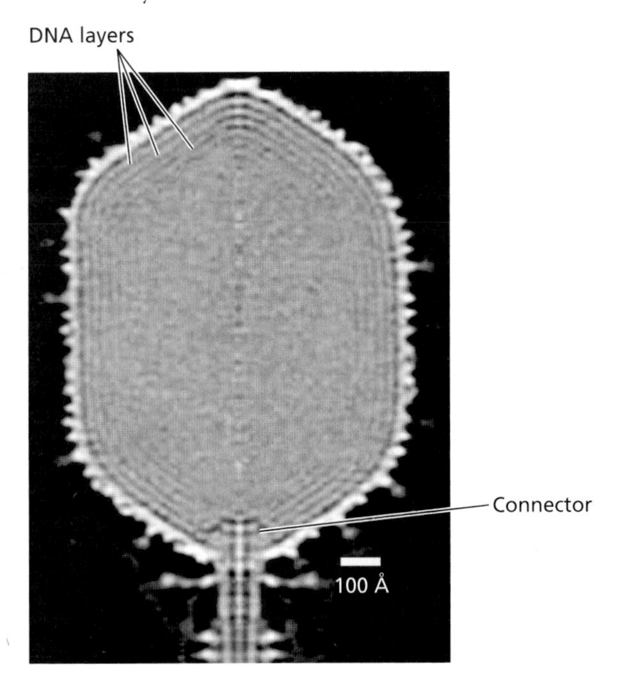

Neutralization might be accomplished by proteins that form the inner surface of the capsid, or by the incorporation of small, basic peptides made by the host cell, such as spermine and spermidine.

Packaging by Specialized Viral Proteins

In many virus particles, the genome is associated with specialized nucleic acid-binding proteins, such as the nucleocapsid proteins of (−) strand RNA viruses and retroviruses, or the core proteins of adenoviruses. An important function of such proteins is to condense and protect viral genomes. Consequently, they do not recognize specific nucleic acid sequences but rather bind nonspecifically to RNA or DNA genomes. This mode of binding is exemplified by the structure of the vesicular stomatitis virus N protein, in which 9 nucleotides of RNA are tightly but nonspecifically bound in a cavity formed between the two domains of each N protein molecule (Fig. 4.6). These protein-RNA interactions both sequester the RNA genome and organize it into a helical structure. Formation of helical ribonucleoproteins by two-domain RNA-binding proteins is a packaging mechanism common among (−) strand RNA viruses in the order *Mononegavirales*: the N proteins of representatives of other families in the order exhibit the same two-lobed structure and mode of RNA binding (Fig. 4.20).

Electron microscopy of cores released from adenovirus particles suggested that the internal nucleoprotein is also arranged in some regular fashion. However, how the viral DNA genome is organized and condensed by the core proteins is not known: the nucleoprotein was not observed in the high-resolution structures of adenovirus particles described previously, and the structures of core proteins have not been determined. The fundamental DNA packaging unit is a multimer of protein VII, which appears as beads on a string of adenoviral DNA when other core proteins are removed. Protein VII and the other core proteins are basic, as would be expected for proteins that bind to a negatively charged DNA molecule without sequence specificity.

Packaging by Cellular Proteins

The final mechanism for condensing the viral genome, by cellular proteins, is unique to polyomaviruses, such as simian virus 40, and papillomaviruses. The circular, double-stranded DNA genomes of these viruses are organized into nucleosomes that contain the four core histones, H2A, H2B, H3, and H4. These genomes are organized within the particle (and in infected cells) like cellular DNA in chromatin to form a minichromosome. The 20 or so nucleosomes that are associated with the viral genome condense the DNA by a factor of ~7. This packaging mechanism is elegant, with two major advantages: none of the limited viral genetic information needs to be devoted to DNA-binding proteins, and the viral genome, which is transcribed by cellular RNA

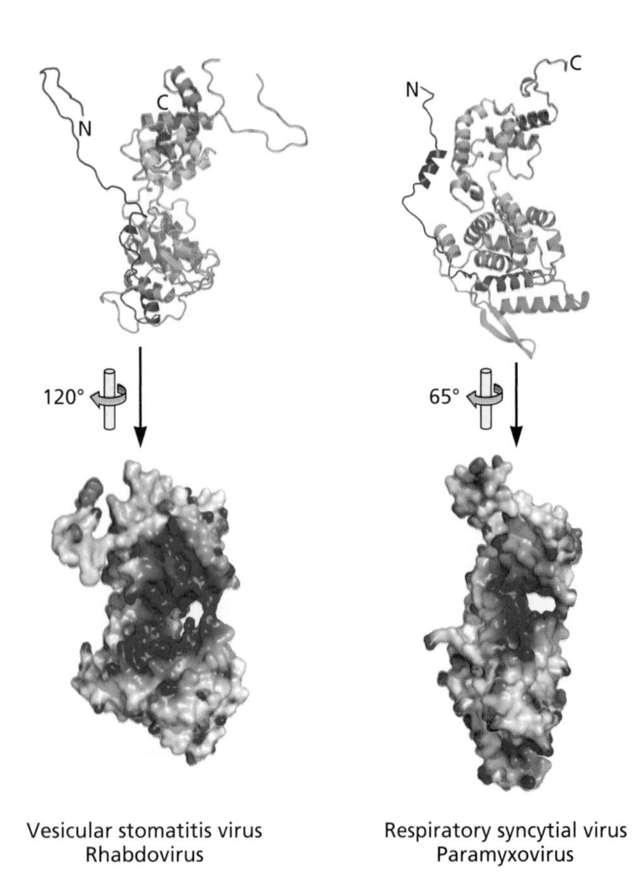

Figure 4.20 Conserved organization of the RNA-packaging proteins of nonsegmented (−) strand RNA viruses. Ribbon diagrams of the N proteins indicated are shown at the top, colored from purple at the N terminus to red at the C terminus. Their electrostatic surfaces from negative (red) to positive (blue) are shown in the space-filling models below, with the molecules rotated as indicated to show the RNA-binding cleft (blue) most clearly. Although differing in structural details, these N proteins share a two-lobed structure (top) and an RNA-binding cleft between the two lobes. Adapted from R. W. Ruigrok et al., *Curr Opin Microbiol* **14:**504–510, 2011, with permission. Courtesy of D. Kolakofsky, University of Geneva.

polymerase II, enters the infected cell nucleus as a nucleoprotein closely resembling the cellular templates for this enzyme.

Viruses with Envelopes

Many viruses contain structural elements in addition to the capsids described previously. Such virus particles possess an envelope formed by a viral protein-containing membrane that is derived from the host cell, but they vary considerably in size, morphology, and complexity. Furthermore, viral membranes differ in lipid composition, the number of proteins they contain, and their location. The envelopes form the outermost layer of enveloped animal viruses, but in bacteriophages and archaeal viruses of the PRD1 family the membrane lies **beneath** an icosahedral capsid (Box 4.7). Typical features of viral envelopes and their proteins are described in the next section, to set the stage for consideration of the structures of envelope proteins and the various ways in which they interact with internal components of the virion (Fig. 4.21).

Viral Envelope Components

The foundation of the envelopes of all animal viruses is a lipid membrane acquired from the host cell during assembly. The precise lipid composition is variable, for viral envelopes can be derived from different kinds of cellular membranes. Embedded in the membrane are viral proteins, the great majority of which are **glycoproteins** that carry covalently linked sugar chains, or **oligosaccharides**. Sugars are almost always added to the proteins posttranslationally, during transport to the cellular membrane at which progeny virus particles assemble. Intra- or interchain disulfide bonds, another common chemical feature of these proteins, are also acquired during transport to assembly sites. These covalent bonds stabilize the tertiary or quaternary structures of viral glycoproteins.

Envelope Glycoproteins

Viral glycoproteins are **integral membrane proteins** firmly embedded in the lipid bilayer by a short **membrane-spanning domain** (Fig. 4.22). The membrane-spanning domains of viral proteins are hydrophobic α-helices of sufficient length to span the lipid bilayer. They generally separate large external domains that are decorated with oligosaccharides from smaller internal segments. The former contain binding sites for cell surface virus receptors, major antigenic determinants, and sequences that mediate fusion of viral with cellular membranes during entry. Internal domains, which make contact with other components of the virion, are often essential for virus assembly.

With few if any exceptions, viral membrane glycoproteins form oligomers, which can comprise multiple copies of a single protein or may contain two or more protein chains. The subunits are held together by noncovalent interactions and disulfide bonds. On the exterior of particles, these oligomers can form surface projections, often called spikes. Because of their critical roles in initiating infection, the structures of many viral glycoproteins have been determined.

The hemagglutinin (HA) protein of human influenza A virus is a trimer that contains a globular head with a top surface that is projected ~135 Å from the viral membrane by a long stem (Fig. 4.23A). The latter is formed and stabilized by the coiling of α-helices present in each monomer. The membrane-distal globular domain contains the binding site for the host cell receptor. This important functional region is located >100 Å away from the lipid membrane of influenza virus particles. Other viral glycoproteins that mediate cell attachment and entry, such as the E protein of the flavivirus tick-borne encephalitis virus, adopt a quite different orientation (and structure); the external domain of E protein is a flat, elongated dimer that lies on the surface of the viral membrane

BOX 4.7

DISCUSSION

A viral membrane directly surrounding the genome

The membranes present in particles of animal viruses are external structures separated from the genome by at least one protein layer. As we have seen, internal protein layers contribute to condensation and organization of the genome via interactions of the nucleic acid with specialized nucleic acid-binding proteins or the internal surfaces of capsids. Nevertheless, this arrangement is not universal: in the particles of some archaeal and bacterial viruses, a host cell-derived membrane directly abuts the genome.

This property is exemplified by *Sulfolobus* turreted icosahedral virus, which infects a hyperthermophilic archaeon. This virus has a double-stranded DNA genome, a major capsid protein containing two β-barrel jelly roll domains, and pentons built from dedicated viral proteins. The capsid of *Sulfolobus* turreted icosahedral virus encases a lipid membrane rather than an internal nucleoprotein core. As shown in panel A of the figure, a large space separates the capsid and the membrane, with contact between the capsid and the membrane limited to the fivefold axes of icosahedral symmetry, where the most internal domain of the penton base protein contacts a viral transmembrane protein. Particles purified from *Sulfolobus* turreted icosahedral virus-infected cells include forms that lack the capsid and exhibit the size and morphology of lipid cores alone. These observations suggest that the membrane, rather than the capsid, is the major determinant of particle stability.

The unusual internal membrane of the *Sulfolobus* turreted icosahedral virus is built from membrane-forming lipids synthesized specifically in thermophilic and hyperthermophilic archaea: they comprise long chains (e.g., C_{40}, compared to C_{16} to C_{18} typical of mammalian cells) of branched, isoprenoid-like units ether linked at either end to various polar head groups. Because of the latter property, these lipids can form monolayer membranes, in contrast to the lipid bilayers formed in animal cells (panel B). The ether linkages and branched acyl chains considerably increase the stability of membranes formed from these specialized archaeal lipids. This property would appear to be essential for life in the extreme conditions (e.g., pH 3 and temperature of 80°C) inhabited by the host of *Solfolobus* turreted icosahedral virus and provides an effective mechanism to protect the viral genome during transit through such harsh environments.

Khayat R, Fu CY, Ortmann AC, Young MJ, Johnson JE. 2010. The architecture and chemical stability of the archaeal *Sulfolobus* turreted icosahedral virus. *J Virol* **84:**9575–9583.

Vessler D, Ng TS, Sendamarai AK, Eilers BJ, Lawrence CM, Lok SM, Young MJ, Johnson JE, Fu CY. 2013. Atomic structure of the 75 MDa extremophile *Sulfolobus* turreted icosahedral virus determined by CryoEM and X-ray crystallography. *Proc Natl Acad Sci U S A* **110:**5504–5509.

(A) Cross section through a near-atomic-resolution reconstruction of *Sulfolobus* turreted icosahedral virus, showing the unique pentonal structures (turrets) and the separation of the capsid shell from the membrane. The internal surface of the membrane (yellow) is in direct contact with the double-stranded DNA genome (red). Adapted from D. Vessler et al., *Proc Natl Acad Sci U S A* **110:**5504–5509, 2013, with permission. Courtesy of C.-Y. Fu, The Scripps Research Institute. **(B)** Schematic comparison of archaeal monolayer membrane-forming and eukaryotic bilayer membrane-forming lipids.

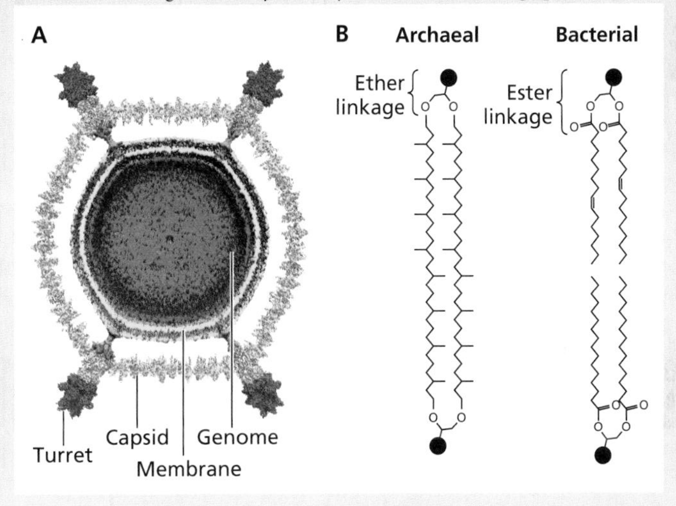

Figure 4.21 Schematic illustration of three modes of interaction of capsids or nucleocapsids with envelopes of virus particles.

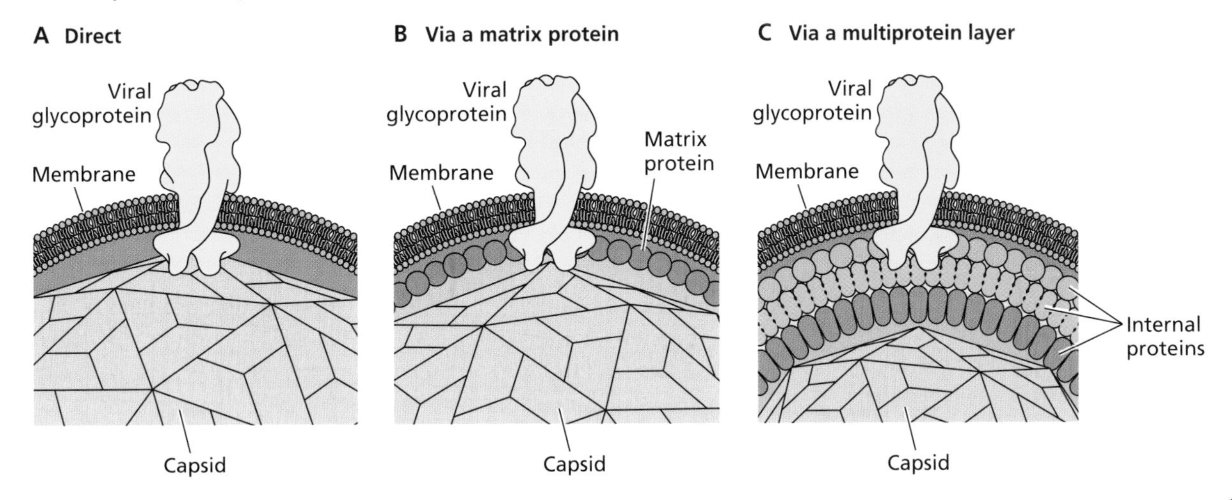

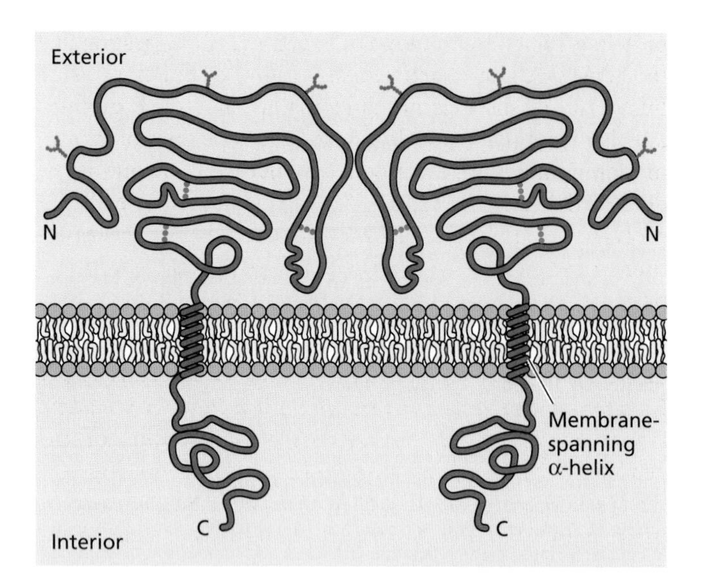

Figure 4.22 Structural and chemical features of a typical viral envelope glycoprotein shown schematically. The protein is inserted into the lipid bilayer via a single membrane-spanning domain. This segment separates a larger external domain, which is decorated with N-linked oligosaccharides (purple) and contains disulfide bonds (green), from a smaller internal domain.

rather than projecting from it (Fig. 4.23B). Despite their lack of common structural features, both the HA protein and the E protein are primed for dramatic conformational change to allow entry of internal virion components into a host cell.

The high-resolution viral glycoprotein structures mentioned above are those of the large external domains of the proteins that had been cleaved from the viral envelope by proteases. This treatment facilitated crystallization but, of course, precluded analysis of membrane-spanning or internal segments of the proteins, both of which may operate structurally or functionally. Membrane-spanning domains can contribute to the stability of oligomeric glycoproteins, as in influenza virus hemagglutinin (HA), while internal domains can participate in anchoring the envelope to internal structures (Fig. 4.21). Improvements in resolution achieved by application of cryo-electron microscopy or tomography have allowed visualization of these segments of glycoproteins of some enveloped viruses.

Other Envelope Proteins

The envelopes of some viruses, including orthomyxoviruses, herpesviruses, and poxviruses, contain integral membrane proteins that lack large external domains or possess multiple

Figure 4.23 Structures of extracellular domains of viral glycoproteins. These extracellular domains are depicted as they are oriented with respect to the membrane of the viral envelope. **(A)** X-ray crystal structure of the influenza virus HA glycoprotein trimer. Each monomer comprises HA1 (blue) and HA2 (red) subunits covalently linked by a disulfide bond. Adapted from J. Chen et al., *Cell* **95:**409–417, 1998, with permission. **(B)** X-ray structure of the tick-borne encephalitis virus (a flavivirus) E protein dimer, with the subunits shown in orange and yellow. PDB ID: 1SVB F. A. Rey and S. C Harrison, *Nature* **375:**291–298, 1995.

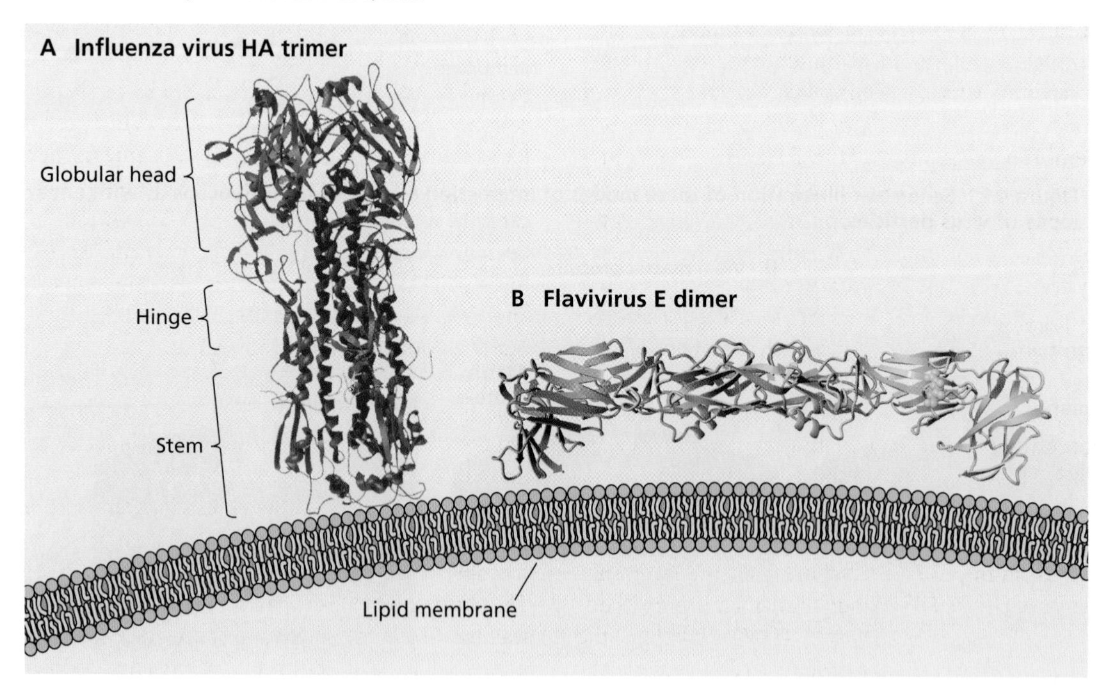

membrane-spanning segments. Among the best characterized of these is the influenza A virus M2 protein. This small (97-amino-acid) protein is a minor component of virus particles, estimated to be present at 14 to 68 copies per particle. In the viral membrane, two disulfide-linked M2 dimers associate to form a noncovalent tetramer that functions as an ion channel. The M2 ion channel is the target of the influenza virus inhibitor drug amantadine (Volume II, Fig. 9.13). The effects of this drug, as well as of mutations in the M2 coding sequence, indicate that M2 plays an important role during entry by controlling the pH of the virion interior.

Simple Enveloped Viruses: Direct Contact of External Proteins with the Capsid or Nucleocapsid

In the simplest enveloped viruses, exemplified by (+) strand RNA alphaviruses such as Semliki Forest, Sindbis, and Ross River viruses, the envelope directly abuts an inner nucleocapsid containing the (+) strand RNA genome. This inner protein layer is a $T = 4$ icosahedral shell built from 240 copies of a single capsid (C) protein arranged as hexamers and pentamers. The outer glycoprotein layer also contains 240 copies of the envelope proteins E1 and E2, which form heterodimers. They cover the surface of the particle, such that the lipid membrane is not exposed on the exterior. Strikingly, the glycoproteins are also organized into a $T = 4$ icosahedral shell (Fig. 4.24A).

The structure of Sindbis virus has been determined by cryo-EM and image reconstruction to some 9-Å resolution (Fig. 4.24A and B), while the structures of the E1 and C proteins of the related Semliki Forest virus have been solved at high resolution. The organization of the alphavirus envelope, including the transmembrane anchoring of the outer glycoprotein layer to structural units of the nucleocapsid, can therefore be described with unprecedented precision. The transmembrane segments of the E1 and E2 glycoproteins form a pair of tightly associated α-helices, with the cytoplasmic domain of E2 in close opposition to a cleft in the capsid protein (Fig. 4.24C and D). This interaction accounts for the 1:1 symmetry match between the internal capsid and exterior glycoproteins. On the outer surface of the membrane, the external portions of these glycoproteins, together with the E3 protein, form an unexpectedly elaborate structure: a thin $T = 4$ icosahedral protein layer (called the skirt) covers most of the membrane (Fig. 4.24B and C) and supports the spikes, which are hollow, three-lobed projections (Fig. 4.24D).

The structures formed by external domains of membrane proteins of the important human pathogens West Nile virus and dengue virus (family *Flaviviridae*) are quite different: they lie flat on the particle surface rather than forming protruding spikes (Fig. 4.25A; see also Box 4.8). Nevertheless, the alphavirus E1 protein and the single flavivirus envelope (E) protein exhibit the same topology (Fig. 4.25B), suggesting that the genes encoding them evolved from a common ancestor. Furthermore, the external domains of flaviviral E proteins are also icosahedrally ordered, and the envelopes of viruses of these families are described as **structured**. In contrast, as described in the next section, the arrangement of membrane proteins generally exhibits little relationship to the structure of the capsid when virus particles contain additional protein layers.

Enveloped Viruses with an Additional Protein Layer

Enveloped viruses of several families contain an additional protein layer that mediates interactions of the genome-containing structure with the viral envelope. In the simplest case, a single viral structural protein, termed the matrix protein, welds an internal ribonucleoprotein to the envelope (Fig. 4.21B). This arrangement is found in members of several groups of (−) strand RNA viruses (Fig. 4.5C; Appendix, Fig. 17 and 31). Retrovirus particles also contain an analogous, membrane-associated matrix protein (MA), which makes contact with an internal capsid in which the viral ribonucleoprotein is encased.

Because the internal capsids or nucleocapsids of these more complex enveloped viruses are not in direct contact with the envelope, the organization and symmetry of internal structures are not necessarily evident from the external appearance of the surface glycoprotein layer. Nor does the organization of these proteins reflect the symmetry of the capsid. For example, the outer surface of all retroviruses appears roughly spherical with an array of projecting knobs or spikes, regardless of whether the internal core is spherical, cylindrical, or cone shaped. Likewise, influenza virus particles, which contain helical nucleocapsids, are generally roughly spherical particles but are highly pleomorphic with long, filamentous forms common in clinical isolates (Box 4.9). Although the interior architecture of these enveloped viruses cannot be described in detail, high-resolution structures have been obtained for several matrix proteins. In conjunction with the results of *in vitro* assays for lipid binding and mutational analyses, such information allows molecular modeling of matrix protein-envelope interactions.

Internal proteins that mediate contact with the viral envelope are not embedded within the lipid bilayer but rather bind to its inner face. Such viral proteins are targeted to, and interact with, membranes by means of specific signals, which are described in more detail in Chapter 12. For example, a posttranslationally added fatty acid chain is important for membrane binding of the MA proteins of most retroviruses. The human immunodeficiency virus type 1 MA protein was the first viral peripheral membrane protein for which a high-resolution structure was determined, initially by nuclear

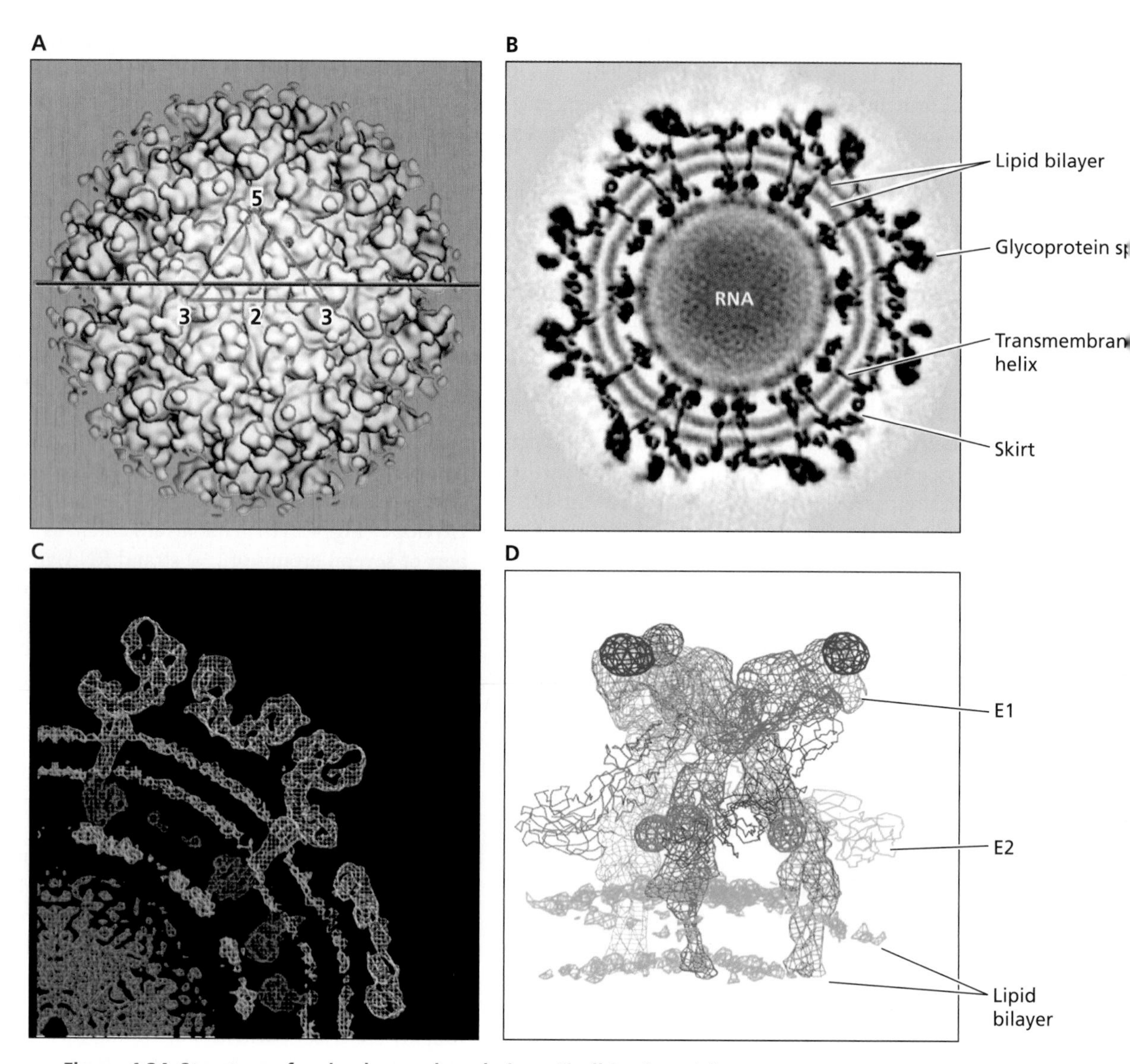

Figure 4.24 Structure of a simple enveloped virus, Sindbis virus. (A) The surface structure of Sindbis virus, a member of the alphavirus genus of the *Togaviridae*, at 20-Å resolution determined by cryo-EM. The boundaries of the structural (asymmetric) unit are demarcated by the red triangle, on which the icosahedral five-, three-, and twofold axes of rotational symmetry are indicated. This outer surface is organized as a $T = 4$ icosahedral shell studded with 80 spikes, each built from three copies of each of the transmembrane glycoproteins E1 and E2. These spikes are connected by a thin, external protein layer, termed the skirt. **(B)** Cross section through the density map at 11-Å resolution along the black line shown in panel A. The lipid bilayer of the viral envelope is clearly defined at this resolution, as are the transmembrane domains of the glycoproteins. **(C)** Different layers of the particle, based on the fitting of a high-resolution structure of the E1 glycoprotein into a 9-Å reconstruction of the virus particle. The nucleocapsid (red) surrounds the genomic (+) strand RNA. The RNA is the least well-ordered feature in the reconstruction, although segments (orange) lying just below the capsid protein appear to be ordered by interaction with this protein. The C protein penetrates the inner leaflet of the lipid membrane, where it interacts with the cytoplasmic domain of the E2 glycoprotein (blue). The membrane is spanned by rod-like structures that are connected to the skirt by short stems. **(D)** The structure of the E1 and E2 glycoproteins, obtained by fitting the crystal structure of the closely related Semliki Forest virus E1 glycoprotein into the 11-Å density map and assigning density unaccounted for to the E2 glycoprotein. The view shown is around a quasithreefold symmetry axis, with the three E2 glycoprotein molecules in a trimeric spike colored light blue, dark blue, and brown and the E1 molecules shown as backbone traces colored red, green, and magenta. The portions of the proteins that cross the lipid bilayer are helical, twisting around one another in a left-handed coiled coil. Adapted from W. Zhang et al., *J Virol* **76:**11645–11658, 2002, with permission. Courtesy of Michael Rossmann, Purdue University

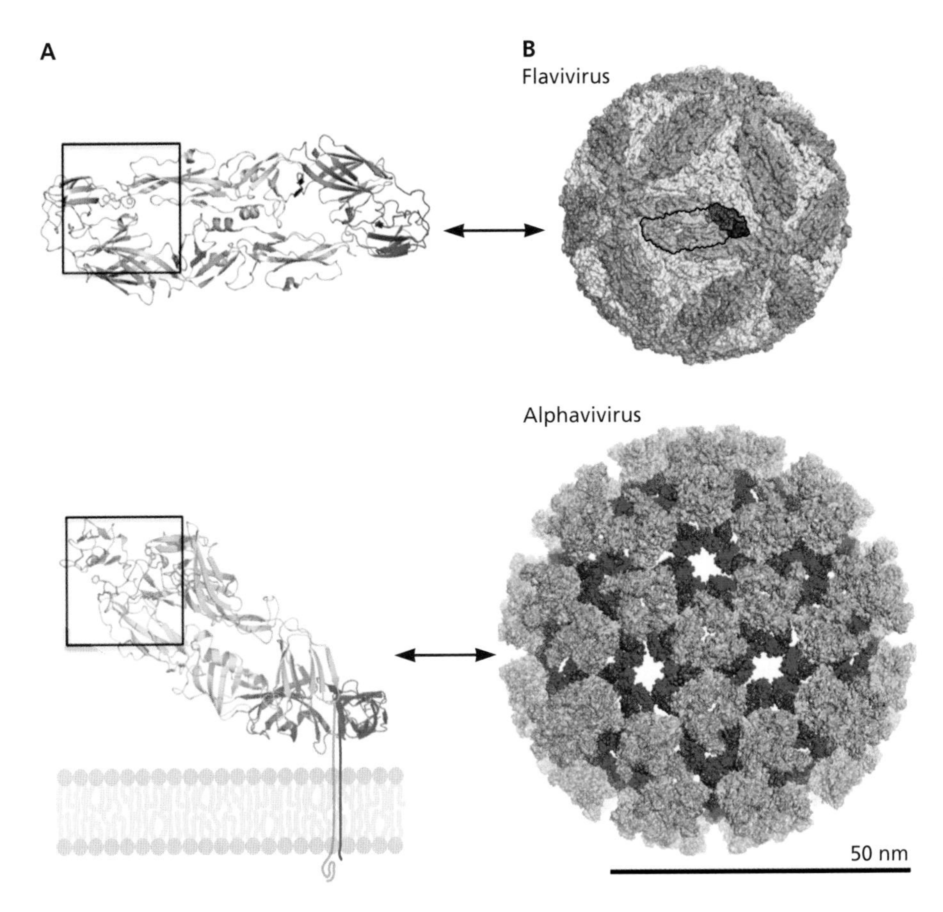

Figure 4.25 Conserved topology and regular packing of envelope proteins of small, (1) strand RNA viruses. (A) Ribbon diagrams of the flavivirus envelope (E) protein dimer (top) and the alphavirus E1/E2 heterodimer (bottom), with one E and the E2 subunit shown in gray. Conserved domains of E and E1 are colored red, yellow, and blue with the fusion loops required for entry in orange. The parallel and perpendicular orientations to the membrane of the flavivirus and alphavirus envelope proteins, respectively, result in the very different appearances of these particles shown in panel B. **(B)** Surface renderings on the same scale, showing the regular packing of flavivirus and alphavirus envelope protein dimers of flavivirus. The dimers related by two-, three-, and fivefold axes of icosahedral symmetry are colored blue, pale yellow, and mauve, respectively, except for the central dimer depicted, which is colored as in panel A. In the 80 spikes of the alphavirus envelope, E2 is shown gray and E1 colored by domain as in panel A. Adapted from M.-C. Vaney and F. A. Rey, *Cell Microbiol* **13:**1451–1459, 2011, with permission. Courtesy of F. A. Rey, Institut Pasteur.

magnetic resonance methods. Subsequent analysis by X-ray crystallography established that MA is a trimer. Each MA molecule comprises a compact, globular domain of α-helices capped by a β-sheet that contains positively charged amino acids necessary for membrane binding. As illustrated in the model of MA oriented on a membrane shown in Fig. 4.26, the basic residues form a positively charged surface, positioned for interaction with phospholipid head groups on the inner surface of the envelope. The matrix proteins of (−) strand RNA viruses such as vesicular stomatitis virus and influenza virus also contain positively charged domains required for membrane binding, despite having three-dimensional folds that are quite different from those of retroviral MA proteins (and from one another).

Large Viruses with Multiple Structural Elements

Virus particles that house large DNA genomes are structurally far more complex than any considered in previous sections. Such particles comprise obviously distinct components with different symmetries and/or multiple layers. In this section, we illustrate various ways in which multiple structural elements can be combined, using as examples bacteriophage T4, herpes simplex virus type 1, the poxvirus vaccinia virus, and giant viruses such as mimivirus. As we shall see, some of these elements are dedicated to specific functions.

Bacteriophage T4

Bacteriophage T4, which has been studied for more than 50 years, is the classic example of an architecturally elaborate virus that contains parts that exhibit both icosahedral and helical symmetry. The T4 particle, which is built from ~50 of the proteins encoded in the ~170-kbp double-stranded DNA genome, is a structurally elegant machine tailored for active delivery of the genome to host cells. The most striking feature is the presence of morphologically distinct and functionally specialized structures, notably the head containing the genome and a long tail that terminates in a baseplate from which six long tail fibers protrude (Fig. 4.27A).

The head of the mature T4 particle, an elongated icosahedron, is built from hexamers of a single viral protein (gp23*). In contrast to the other capsids considered so far, two T numbers are needed to describe the organization of gp23* in the two end structures ($T = 13$) and in the elongated midsection ($T = 20$). As in adenoviral capsids, the pentamers that occupy the vertices contain a different viral protein, and additional proteins reside on the outer or inner surfaces of the icosahedral shell (Fig. 4.27B). One of the 12 vertices is

BOX 4.8

DISCUSSION

A virus particle with different structures in different hosts

Throughout this chapter, we describe mature virus particles in terms of a single structure: "the" structure. However, it is important to appreciate that the architectures reported are those of particles isolated and examined under a single set of specific conditions. Recent studies of the flavivirus dengue virus, an important human pathogen, illustrate the conformational plasticity of some mature virus particles.

The organization of the single dengue virus envelope glycoprotein, E, described in the text (Fig. 4.25) is that observed in particles propagated in cells of the mosquito vector maintained at 28°C. As noted previously, the E protein dimers are tightly packed and icosahedrally ordered. However, the epitopes for binding of antibodies that neutralize the virus at 37°C are either partially or entirely buried, suggesting that the virus particle might undergo temperature-dependent conformational transitions. Indeed, when particles are exposed to temperatures encountered in the mammalian host (e.g., 37°C), they do expand significantly, exposing segments of the underlying membrane, and the E protein interactions are altered (compare the left and right panels in the figure). In fact, particles exposed to higher temperatures are heterogeneous, and the example shown in the figure (right) represents but one of multiple forms, identified during selection of particles for three-dimensional reconstruction. Because a heterogeneous population of particles with less well-ordered E protein dimers represent the form of dengue virus recognized by the human immune system, these observations have important implications for the design of dengue virus vaccines.

Fibriansah G, Ng TS, Kostyuchenko VA, Lee J, Lee S, Wang J, Lok SM. 2013. Structural changes in dengue virus when exposed to a temperature of 37°C. *J Virol* **87:**7585–7592.

Rey FA. 2013. Dengue virus: two hosts, two structures. *Nature* **497:**443–444.

Zhang X, Sheng J, Plevka P, Kuhn RJ, Diamond MS, Rossmann MG. 2013. Dengue structure differs at the temperatures of its human and mosquito hosts. *Proc Natl Acad Sci U S A* **110:**6795–6799.

Structures of dengue virus particles at 28°C (left) and at ≥34°C (right), with the axes of five-, three-, and twofold rotational symmetry indicated by a pentagon, triangle, and ellipse, respectively. The E protein dimers that lie at the twofold axes are shown in gray and the other dimers with one subunit in green and one in cyan. The particles exposed to higher temperatures are characterized by exposed patches of membrane (purple) and significant reduction of dimer contacts at the threefold axes of icosahedral symmetry. Adapted from F. A. Rey, *Nature* **497:**443–444, 2013. Courtesy of F. A. Rey, Institut Pasteur, with permission.

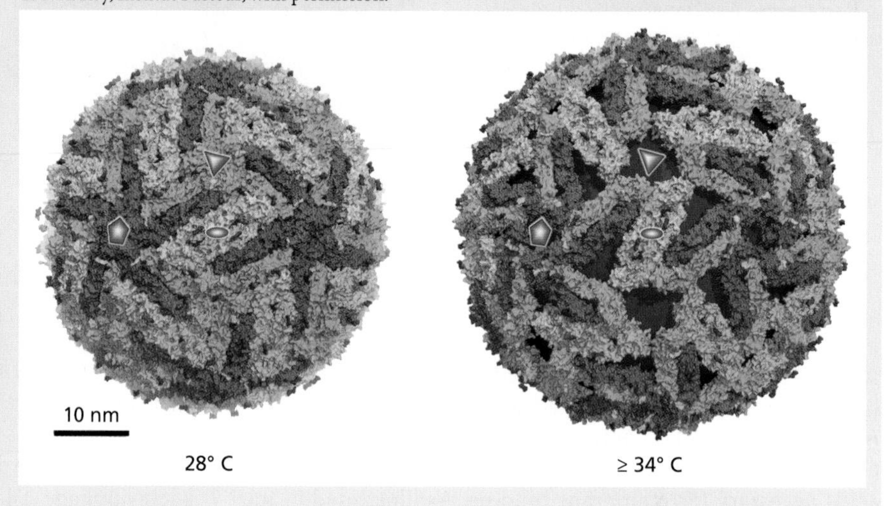

occupied by a unique structure termed the connector, which joins the head to the tail. Such structures are derived from the nanomachine that pulls DNA into immature heads termed the **portal**. Portals are a characteristic feature of the capsids of other families of DNA-containing bacteriophages, as well as of herpesviruses.

In contrast to the head, the ~100-nm-long tail, which comprises two protein layers, exhibits helical symmetry (Fig. 4.27A). The outer layer is a contractile sheath that functions in injection of the viral genome into host cells. The tail is connected to the head via a hexameric ring and at its other end to a complex, dome-shaped structure termed the baseplate, where it carries the cell-puncturing spike. Both long and short tail fibers project from the baseplate. The former, which are bent, are the primary receptor-binding structures of bacteriophage T4. As discussed in Chapter 5, remarkable conformational changes induced upon receptor binding by the tips of the long fibers are transmitted via the baseplate to initiate injection of the DNA genome.

Herpesviruses

Members of the *Herpesviridae* exhibit a number of unusual architectural features. More than half of the >80 genes of herpes simplex virus type 1 encode proteins found in the large (~200-nm-diameter) virus particles. These proteins are components of the envelope from which glycoprotein spikes project or of two distinct internal structures. The latter are the capsid surrounding the DNA genome and the protein layer encasing this structure, called the **tegument** (Fig. 4.28A).

A single protein (VP5) forms both the hexons and the pentons of the $T = 16$ icosahedral capsid of herpes simplex virus type 1 (Fig. 4.28B). Like the structural units of the smaller simian virus 40 capsid, these VP5-containing assemblies make direct contact with one another. However, the larger

BOX 4.9

DISCUSSION

The extreme pleomorphism of influenza A virus, a genetically determined trait of unknown function

Some enveloped viruses vary considerably in size and shape. For example, the particles of paramyxoviruses, such as measles and Sendai viruses, range in size from 120 to up to 540 nm in diameter and may contain multiple copies of the (−) strand RNA genome in helical nucleocapsids of different pitch. Influenza A virus particles exhibit even more extreme pleomorphism: they appear spherical, elliptical, or filamentous, and all forms come in a wide range of sizes (see the figure). Laboratory isolates are primarily filamentous but adopt the other morphologies when adapted to propagation in the laboratory.

Several lines of evidence indicate that the filamentous phenotype is genetically determined. For example, the particles of some influenza A virus isolates are primarily filamentous, whereas those of other isolates are not. Furthermore, genetic experiments have demonstrated that the viral matrix (M1 or M2) proteins, which are encoded within the same segment of the (−) strand, segmented RNA genome, govern formation of filamentous particles. However, what determines the choice between assembly of filamentous versus spherical particles is not understood. Nor is the physiological significance of the filamentous particles, despite their predominance in clinical isolates. It has been speculated that these forms might facilitate cell-to-cell transmission of virus particles through the respiratory mucosa of infected hosts.

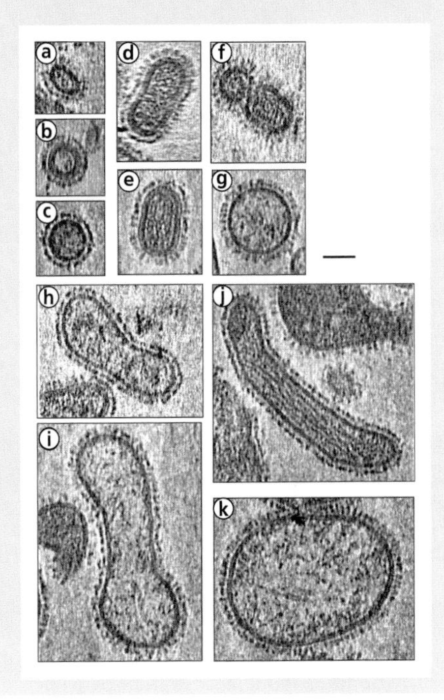

Cryo-electron tomogram sections of influenza A virus particles (strain PR8). Bar = 50 nm. Adapted from D. B. Nayak et al., *Virus Res* **143:**147–161, 2000. Courtesy of D. B. Nayak, University of California, Los Angeles, with permission.

herpesviral capsid is stabilized by additional proteins, VP19C and VP23, which form triplexes that link the major structural units. A second property shared with polyomaviruses (and papillomaviruses) is stabilization of the particle by disulfide bonds, which covalently link VP5, VP19C, and VP23 to one another and to specific tegument proteins. Although apparently a typical and quite simple icosahedral shell, this viral capsid is in fact an asymmetric structure: 1 of the 12 vertices is occupied not by a VP5 penton but by a unique structure termed the portal. The portal comprises 12 copies of the UL6 protein and is a squat, hollow cylinder that is wider at one end and surrounded by a two-tiered ring at the wider end (Fig. 4.28C). The asymmetry of the herpesviral capsid and the incorporation of the portal have important implications for the mechanism of assembly (see Chapter 13).

The tegument contains >20 viral proteins, viral RNAs, and cellular components. A few tegument proteins are icosahedrally ordered, as a result of direct contacts with the structural units of the capsid (Fig. 4.28D). However, some tegument proteins are **not** uniformly distributed around the capsid. Rather, they are concentrated on one side, where they form a well-defined cap-like structure (Fig. 4.28A). As this unanticipated asymmetry of herpesviral particles has been viewed only at low resolution, the molecular organization of the cap is not yet understood.

Poxviruses

Particles of poxviruses such as vaccinia virus also comprise multiple, distinct structural elements. However, none of these exhibit obvious icosahedral or helical symmetry, in contrast to components of bacteriophage T4 or herpesvirus particles. A second distinctive feature is that two forms of infectious particles are produced in vaccinia virus-infected cells (see Chapter 13), termed mature virions and enveloped extracellular virions, which differ in the number and origin of membranes. Mature virions are large, enveloped structures (~350 to 370 × 250 × 270 nm) comprising at least 75 proteins that appear in the electron microscope as brick or barrel shaped (depending on the orientation) (Fig. 4.29A). Whether one or two envelopes are present has been the subject of long-standing debate. However, there is now a growing consensus for the presence of just a single membrane. A number of internal structures have been observed by examination of thin sections through purified particles or by cryo-electron tomography (Fig. 4.29B). These features include the core wall, which surrounds the central core that contains the ~200-kbp DNA genome, and lateral bodies. Remarkably, the core contains some 20 enzymes with many different activities. Although viral proteins that contribute to these various structures have been identified, our understanding of vaccinia virus architecture remains at low resolution.

Figure 4.26 Model of the interaction of human immunodeficiency virus type 1 MA protein with the membrane. The membrane is shown with the polar head groups of membrane lipids in beige. This model is based on the X-ray crystal structure of recombinant MA protein synthesized in *E. coli* and consequently lacking the N-terminal (myristate) 14-carbon fatty acid normally added in human cells. The three monomers in the MA trimer are shown in different colors. Basic residues in the β-sheet that caps the globular α-helical domain are magenta or green. Substitution of those shown in magenta impairs reproduction of the virus in cells in culture. The positions of the N-terminal myristate (red), of MA, and of phosphatidylinositol 4,5-bisphosphate (orange) from the membrane were modeled schematically. From C. P. Hill et al., *Proc Natl Acad Sci U S A* **93:**3099–3104, 1996, with permission. Courtesy of C. P. Hill and W. I. Sundquist, University of Utah.

Giant Viruses

Since the discovery of mimivirus (first reported in 1992), a number of so-called "giant" viruses have been identified (Box 1.10). As might be anticipated, such viruses include previously unknown architectural details.

Despite their very large size (vertex-to-vertex diameter of ~5,000 Å), mimivirus particles exhibit some familiar structural features, notably icosahedral symmetry and a capsid built from a major capsid protein with the β-barrel jelly roll topology. Distinctive features include the dense coat of long fibers that cover the entire external surface with the exception of one vertex (Fig. 4.1). This vertex comprises a unique starfish-shaped structure, termed the stargate, the most distinctive structural element of this virus. The stargate opens within host *Acanthamoeba polyphaga* cells to facilitate release of the double-stranded DNA genome and also nucleates assembly of progeny virus particles.

The even larger pandoravirus and pithovirus, with double-stranded DNA genomes of 2.8 and 0.6 Mbp and particle lengths of ~1 and ~1.5 μm, respectively, bear little resemblance to any virus described previously. They share an amphora-like shape; a dense, striated outer layer surrounding an internal lipid membrane; and a rather featureless internal compartment. The apex of pithovirus is closed by a protruding "cork" with a hexagonal, grid-like appearance (Fig. 4.30). This unusual structure is expelled following uptake of virus particles into host cells by phagocytosis to allow fusion of the viral membrane with that of the cellular vacuole. Unprecedented assemblies specialized for release of the viral genome in host cells may prove to be a characteristic property of the very large viruses.

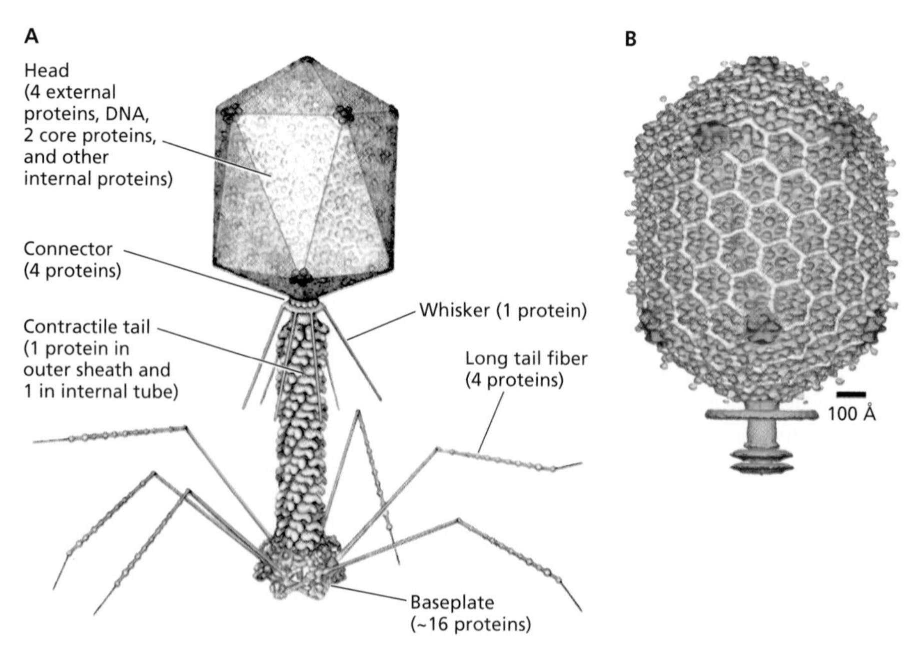

Figure 4.27 Morphological complexity of bacteriophage T4. (A) A model of the virus particle. Adapted from P. G. Leiman et al., *Cell Mol Life Sci* **60:**2356–2370, 2003, with permission. **(B)** Structure of the head (22-Å resolution) determined by cryo-EM, with the major capsid proteins shown in blue (gp23*) and magenta (gp24*), the protein that protrudes from the capsid surface in yellow, the protein that binds between gp23* subunits in white, and the beginning of the tail in green. Adapted from A. Fokine et al., *Proc Natl Acad Sci U S A* **101:**6003–6008, 2004, with permission. Courtesy of M. Rossmann, Purdue University.

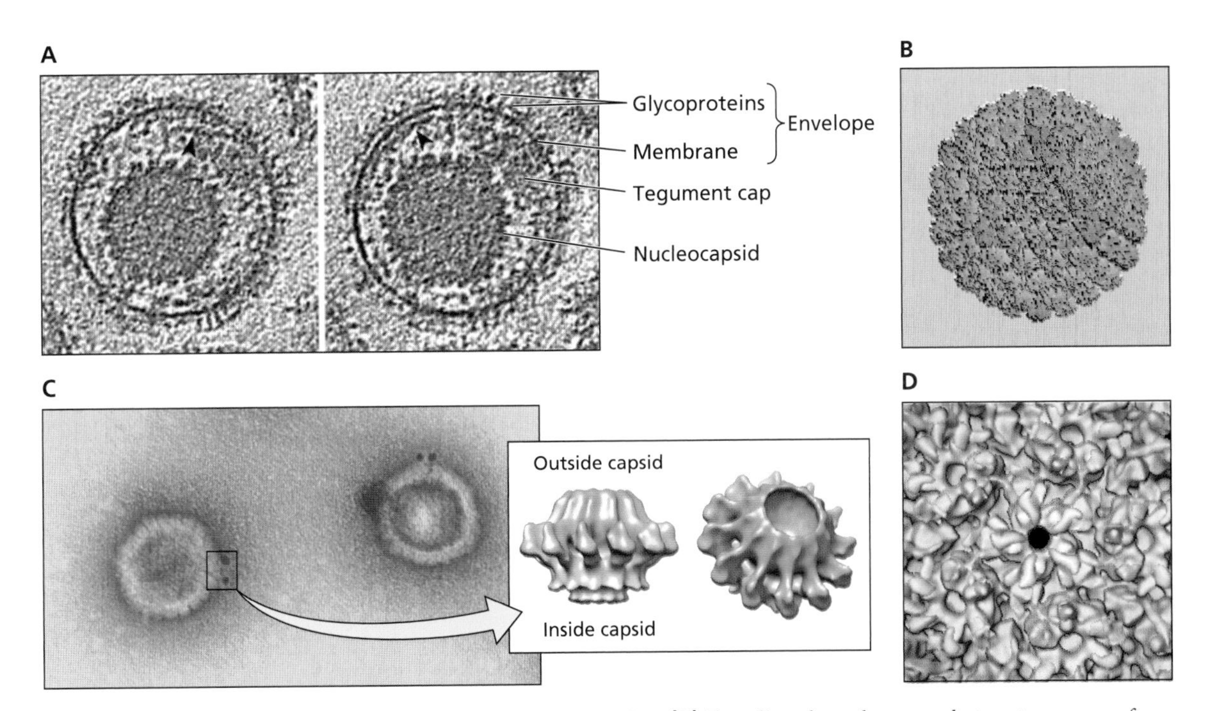

Figure 4.28 Structural features of herpesvirus particles. **(A)** Two slices through a cryo-electron tomogram of a single herpes simplex virus type 1 particle, showing the eccentric tegument cap. Adapted from K. Grunewald et al., *Science* **302:**1396–1398, 2003, with permission. **(B)** Reconstruction of the herpes simplex virus type 1 nucleocapsid (8.5-Å resolution), with VP5 hexamers and pentamers colored blue and red, respectively, and the triplexes that reinforce the connections among these structural units in green. VP5 hexamers, but not pentamers, are capped by a hexameric ring of VP26 protein molecules (not shown). Adapted from Z. H. Zhou et al., *Science* **288:**877–880, 2000, with permission. **(C)** The single portal of herpes simplex virus type 1 nucleocapsids visualized by staining with an antibody specific for the viral UL6 protein conjugated to gold beads is shown to the left. The gold beads are electron dense and appear as dark spots in the electron micrograph. They are present at a single vertex in each nucleocapsid, which therefore contains one portal. A 16-Å reconstruction of the UL6 protein portal based on cryo-EM is shown on the right. Adapted from B. L. Trus et al., *J Virol* **78:**12668–12671, 2004, with permission. **(D)** Interactions of two tegument proteins with the simian cytomegalovirus nucleocapsid. Tegument proteins that bind to hexons plus pentons and to triplexes are shown in blue and red, respectively. These proteins were visualized by cryo-EM, image reconstruction (to 22-Å resolution), and difference mapping of nucleocapsids purified from the nucleus and cytoplasm of virus-infected cells. The latter carry the tegument, but the former do not. Adapted from W. W. Newcomb et al., *J Virol* **75:**10923–10932, 2001, with permission. Courtesy of A. C. Steven, National Institutes of Health (A, C, and D) and W. Chiu, Baylor College of Medicine (B).

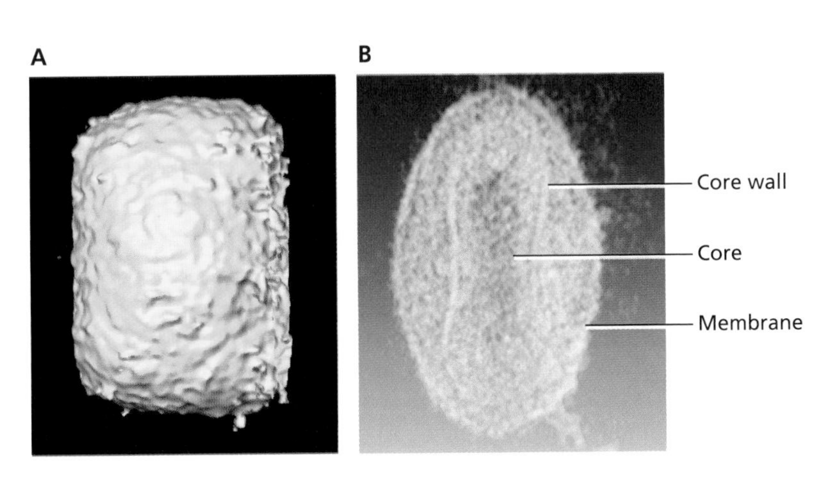

Figure 4.29 Structural features of the poxvirus vaccinia virus. **(A)** Surface rendering of intracellular mature particles of vaccinia virus reconstructed from cryo-electron tomograms showing the brick shape and irregular protrusions from the surface. **(B)** Translucent visualization of the reconstructed particle volume showing the dumbbell-shaped core and external membrane. Adapted from M. Cyrklaff et al., *Proc Natl Acad Sci U S A* **102:**2772–2777, 2005, with permission. Courtesy of J. L. Carrascosa, Universida Autonoma de Madrid. See also http://www.vacciniamodel.com.

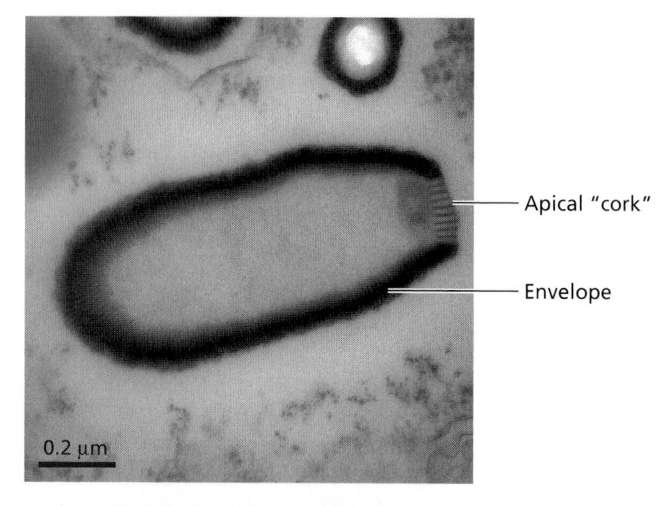

Figure 4.30 Morphology of pithovirus. The virus *Pithovirus sibericum* was isolated following culture of a suspension of soil from a sample of permafrost collected in 2000 in Siberia with the ameba *Acanthamoeba castellani*. Shown is an electron micrograph of a particle observed in infected ameba late in the infectious cycle following ultrathin sectioning of fixed cells and negative staining. Courtesy of Chantal Abergel and Jean-Michel Claverie, Aix-Marseille Université.

Other Components of Virions

Some virus particles comprise only the nucleic acid genome and structural proteins necessary for protection and delivery into a host cell. However, many contain additional viral proteins or other components, which are generally present at much lower concentrations but are essential or important for establishing an efficient infectious cycle (Table 4.3).

Enzymes

Many types of virus particles contain enzymes necessary for synthesis of viral nucleic acids. These enzymes generally catalyze reactions unique to virus-infected cells, such as synthesis of viral mRNA from an RNA template or of viral DNA from an RNA template. However, virions of vaccinia virus contain a DNA-dependent RNA polymerase, analogous to cellular RNA polymerases, as well as several enzymes that modify viral RNA transcripts (Table 4.3). This complement of enzymes is necessary because transcription of the viral double-stranded DNA genome takes place in the cytoplasm of infected cells, whereas cellular DNA-dependent RNA polymerases and the RNA-processing machinery are restricted to the nucleus. Other types of enzymes found in virus particles include integrase, cap-dependent endonuclease, and proteases. The proteases sever covalent connections within polyproteins or precursor proteins from which some virus particles assemble, a reaction that is necessary for the production of infectious particles.

Table 4.3 Some virion enzymes

Virus	Protein	Function(s)
Adenovirus		
Human adenovirus type 2	L3 23k	Protease; production of infectious particles
Herpesvirus		
Herpes simplex virus type 1	VP24	Protease; capsid maturation for genome encapsidation
	UL13	Protein kinase
	Vhs	RNase
Orthomyxovirus		
Influenza A virus	P proteins	RNA-dependent RNA polymerase; synthesis of viral mRNA and vRNA; cap-dependent endonuclease
Poxvirus		
Vaccinia virus[a]	DNA-dependent RNA polymerase (8 subunits)	Synthesis of viral mRNA
	Poly(A) polymerase (2 subunits)	Synthesis of poly(A) on viral mRNA
	Capping enzyme (2 subunits)	Addition of 5′ caps to viral pre-mRNA
	DNA topoisomerase	Sequence-specific nicking of viral DNA
	Proteases 1 and 2	Virus particle morphogenesis
Reovirus		
Reovirus type 1	λ2	Guanylyltransferase
	λ3	Double-stranded RNA-dependent RNA polymerase
Retrovirus		
Human immunodeficiency virus type 1	Pol	Reverse transcriptase; proviral DNA synthesis
	IN	Integrase; integration of proviral DNA into the cellular genome
	PR	Protease; production of infectious particles
Rhabdovirus		
Vesicular stomatitis virus	L	RNA-dependent RNA polymerase; synthesis of viral mRNA and vRNA

[a]Vaccinia virions contain some 20 enzymes, only a few of which are listed.

Other Viral Proteins

More-complex particles may also contain additional viral proteins that are not enzymes but nonetheless are important for an efficient infectious cycle. Among the best characterized are the protein primers for viral genome replication that are covalently linked to the genomes of picornaviruses such as poliovirus and adenoviruses. Others include several tegument proteins of herpesviruses, such as the VP16 protein, which activates transcription of viral immediate-early genes to initiate the viral program of gene expression. The cores of vaccinia virus also contain proteins that are essential for transcription of viral genes, as they allow recognition of viral early promoters. Other herpesvirus tegument proteins induce the degradation of cellular mRNA or block cellular mechanisms by which viral proteins are presented to the host's immune system. Retroviruses with complex genomes, such as human immunodeficiency virus type 1, contain additional proteins required for efficient viral reproduction in certain cell types, for example, Nef and Vpr. These proteins are discussed in Volume II, Chapter 6.

Nongenomic Viral Nucleic Acid

The presence of a viral nucleic acid genome has long been recognized as a definitive feature of virions. However, it is now clear that adenovirus, herpesvirus, and retrovirus particles also contain viral mRNAs. This property was first described for the mRNA that encodes the viral Env protein in avian sarcoma virus particles. A limited set of viral mRNAs, as well as some cellular and artificial reporter mRNAs, are packaged into particles of human cytomegalovirus, a betaherpesvirus that is an important human pathogen (Volume II, Appendix, Fig. 11), in proportion to their intracellular concentrations during the period of assembly of progeny particles. It is therefore difficult to exclude the possibility that their presence is a functionally irrelevant and secondary consequence of nonspecific nucleic acid binding by viral structural proteins. However, the viral mRNAs are translated soon after delivery to the host cell, and one has been demonstrated to encode a chemokine decoy that could modulate host immune responses.

Cellular Macromolecules

Virus particles can also contain cellular macromolecules that play important roles during the infectious cycle, such as the cellular histones that condense and organize polyomaviral and papillomaviral DNAs. Because they are formed by budding, enveloped viruses can readily incorporate cellular proteins and other macromolecules. For example, cellular glycoproteins may not be excluded from the membrane from which the viral envelope is derived. Furthermore, as a bud enlarges and pinches off during virus assembly, internal cellular components may be trapped within it. Enveloped viruses are also generally more difficult to purify than naked viruses. As a result, preparations of these viruses may be contaminated with vesicles formed from cellular membranes. Indeed, analysis by the sensitive proteomic methods provided by mass spectrometry has identified from 50 to 100 cellular proteins in purified, enveloped particles of various herpesviruses, filoviruses, and rhabdoviruses. Consequently, it can be difficult to distinguish cellular components specifically incorporated into enveloped virus particles from those trapped randomly or copurifying with the virus. Nevertheless, in some cases it is clear that cellular molecules are important components of virus particles: these molecules are reproducibly observed at a specific stoichiometry and can be shown to be essential or play important roles in the infectious cycle (Box 4.10). The cellular components captured in retrovirus particles have been particularly well characterized.

The primer for initiation of synthesis of the (−) strand DNA during reverse transcription in retroviral genomes is a specific cellular transfer RNA (tRNA). This RNA is incorporated into virus particles by virtue of its binding to a specific sequence in the RNA genome and to reverse transcriptase. A variety of cellular proteins are also present in some retroviral particles. One of the most unusual properties of human immunodeficiency virus type 1 is the presence of cellular cyclophilin A, a **chaperone** that assists or catalyzes protein folding. This protein is the major cytoplasmic member of a ubiquitous family of peptidyl-prolyl isomerases. It is incorporated within human immunodeficiency virus type 1 particles via specific interactions with the central portion of the capsid (CA) protein, and it catalyzes isomerization of a single Gly-Pro bond in this protein. Although incorporation of cyclophilin A is not a prerequisite for assembly, particles that lack this cellular chaperone have reduced infectivity. It is thought that in human cells, cyclophilin provides protection against an intrinsic antiviral defense mechanism (see Volume II, Chapter 7). Cellular membrane proteins, such as Icam-1 and Lfa1 (see Chapter 5), can also be incorporated in the viral envelope and can contribute to attachment and entry of retroviral particles. They may also influence pathogenesis (see Volume II, Chapter 6). Other cellular proteins assembled into viral particles, such as ADP-ribosylation factor 1 (Arf1) found in herpesviral particles, may facilitate intracellular transport (Box 4.10).

Cellular components present in virus particles may serve to facilitate virus reproduction, a property exemplified by the cellular tRNA primers for retroviral reverse transcription. However, incorporation of cellular components can also provide antiviral defense. As discussed in Volume II, Chapters 3 and 6, packaging of a cellular enzyme that converts cytosine to uracil (Apobec3) into retrovirus particles at the end of one infectious cycle leads to degradation and hypermutation of viral DNA synthesized early in the next cycle of infection.

It is clear from these examples that virus particles contain a surprisingly broad repertoire of biologically active molecules that are delivered to their host cells. This repertoire is undoubtedly larger than we presently appreciate, and the contributions of many components of virus particles to the infectious cycles of many viruses have yet to be established.

BOX 4.10

EXPERIMENTS

Cellular proteins in herpes simplex virus type 1 particles: distinguishing passengers from the crew

Powerful and sensitive methods of mass spectrometry have revolutionized cataloguing of the components of virus particles, particularly the structural proteins of viruses with large genomes. This approach has also been invaluable in identifying cellular proteins that are also present, with considerable numbers detected in the particles of several families of enveloped viruses. The sizeable populations (50 to 100) of such cellular proteins emphasize the importance of distinguishing those proteins that contribute to viral reproduction from those incorporated by chance.

Purified extracellular herpes simplex virus type 1 particles were found by mass spectrometry to contain 49 cellular proteins. This set included proteins reported to be present in the particles of other herpesviruses, such as cyclophilin A and actin, and many not detected previously. A two-step RNA interference screen was developed to assess the contributions to viral reproduction of cellular proteins introduced into cells via virus particles. In these experiments, a phenotypically wild-type virus with a capsid protein fused to the green fluorescent protein was exploited to allow rapid and accurate measurement of yields of extracellular virus particles. This assay was validated by the demonstration that small interfering RNA (siRNA)-mediated knockdown of the viral protein VP16, which is required for efficient expression of viral immediate-early genes, significantly decreased yield, whereas a scrambled version of this siRNA did not. Inhibition of synthesis of 24 of the 49 cellular proteins by RNA interference reduced virus yield to a statistically significant degree but did not impair viability of human cells used as host.

Particles depleted of 15 of the proteins were then prepared by infection of siRNA-treated cells (see the figure) and used to infect new cells, in which production of the same cellular protein was or was not inhibited. Removal of 13 of these proteins from particles reduced virus yield significantly, even in cells that continued to synthesize the proteins. These observations established unequivocally that some cellular proteins incorporated into herpesviral particles promote the next cycle of reproduction. This result is surprising, as a viral particle seems likely to contain many fewer molecules of these cellular proteins than the host cell. Perhaps viral reproduction is facilitated by delivery of particular cellular proteins already associated with components of viral particles or delivery of the proteins to specific sites during entry. Functions of such proteins include intracellular transport (e.g., Arf1 and Rab5A) and signaling (e.g., Mif and Cd59).

Loret S, Guay G, Lippé R. 2008. Comprehensive characterization of extracellular herpes simplex virus type 1 virions. *J Virol* **82:**8605–8618.

Stegen C, Yakova Y, Henaff D, Nadjar J, Duron J, Lippé R. 2013. Analysis of virion-incorporated host proteins required for herpes simplex virus type 1 infection through a RNA interference screen. *PLoS One* **8:**e53276. doi:10/1371/journal.pone.0053276.

Two-step method to assess the importance of cellular proteins incorporated into herpes simplex virus type 1 particles. In the first step, particles were isolated from human cells treated with siRNA against 1 of 15 cellular proteins or, as a control, treated with transduction agent (Lipofectamine) alone. As shown, control and siRNA-treated cells were then infected by depleted and control particles, and viral yield measured. The results obtained with particles depleted for VP16 (positive control) or the cellular proteins Mif and Cd59 are summarized. MOI, multiplicity of infection; HPI, hours postinfection. Adapted from C. Stegen et al., *PLos One* **8:**e53276, 2013, with permission.

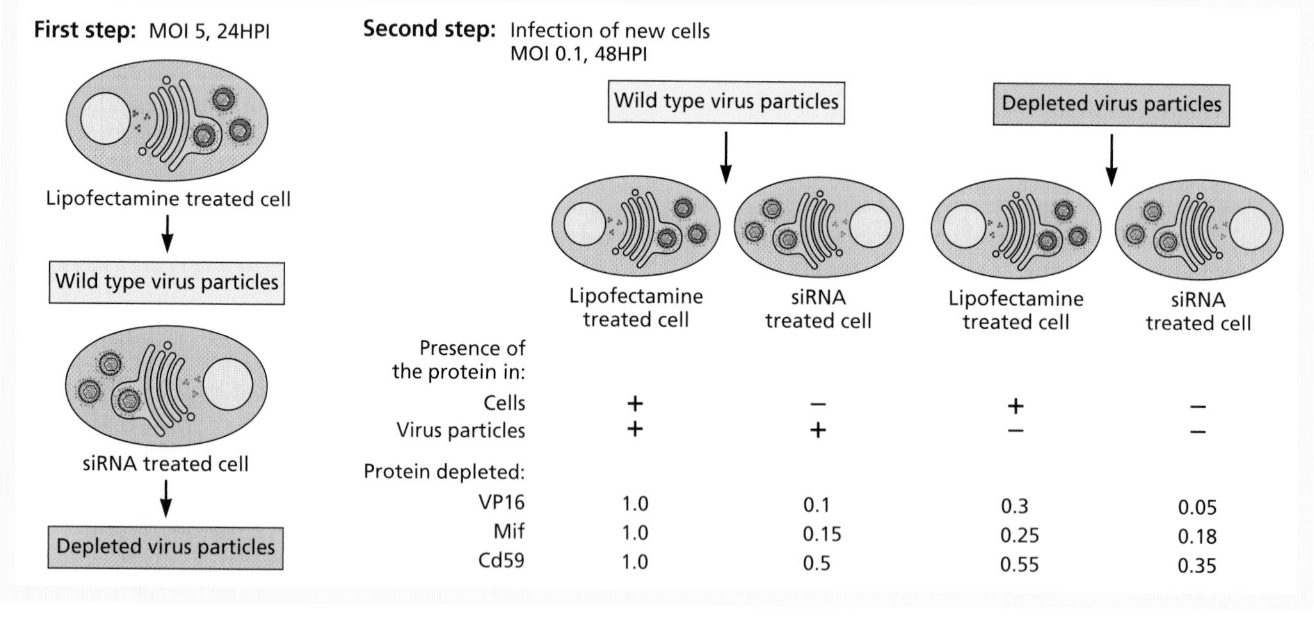

Perspectives

Virus particles are among the most elegant and visually pleasing structures found in nature, as illustrated by the images presented in this chapter. Now that many structures of particles or their components have been examined, we can appreciate the surprisingly diverse architectures they exhibit. Nevertheless, the simple principles of their construction proposed more than 50 years ago remain pertinent: with few exceptions, the capsid shells that encase and protect nucleic acid genomes are built from a small number of proteins arranged with helical or icosahedral symmetry.

The detailed views of nonenveloped virus particles provided by X-ray crystallography emphasize just how well these protein shells provide protection of the genome during passage from one host cell or organism to another. They have also identified several mechanisms by which identical or nonidentical subunits can interact to form icosahedrally symmetric structures. More-elaborate virus particles, which may contain additional protein layers, a lipid envelope carrying viral proteins, and enzymes or other proteins necessary to initiate the infectious cycle, pose greater challenges to the structural biologist. Indeed, for many years we possessed only schematic views of these structures, deduced from negative-contrast electron microscopy and biochemical or genetic methods of analysis. In the previous edition, we noted the power and promise of continuing refinements in methods of cryo-EM (or cryo-electron tomography), image reconstruction, and difference imaging. In the intervening period of just 5 years, these techniques have attained atomic-level resolution, providing remarkable views of large viruses with multiple components, viral envelopes, and, in some cases, the organization of genomes within particles. The structural descriptions of ever-increasing numbers of viruses representing diverse families have also allowed unique insights into evolutionary relationships among seemingly disparate viruses or viral proteins.

These extraordinary advances notwithstanding, important challenges remain, most obviously the visualization of structures that do not exhibit simple symmetry (or are not constructed from components that do). These structures include many genomes and the particles of some large viruses (e.g., poxviruses). The more recently described giant viruses, such as pandoravirus, with particles so large that they can be seen by light microscopy, also pose new technical challenges and suggest that unanticipated structural principles remain to be elucidated.

References

Book Chapters

Baker TS, Johnson JE. 1997. Principles of virus structure determination, p 38–79. *In* Chiu W, Burnett RM, Garcea RL (ed), *Structural Biology of Viruses*. Oxford University Press, New York, NY.

Reviews

Chang J, Liu X, Rochat RH, Baker ML, Chiu W. 2012. Reconstructing virus structures from nanometer to near-atomic resolutions with cryo-electron microscopy and tomography. *Adv Exp Med Biol* **726:**49–90.

Chapman MS, Giranda VL, Rossmann MG. 1990. The structures of human rhinovirus and Mengo virus: relevance to function and drug design. *Semin Virol* **1:**413–427.

Condit RC, Moussatche N, Traktman P. 2006. In a nutshell: structure and assembly of the vaccinia virion. *Adv Virus Res* **66:**31–124.

Engelman A, Cherepanov P. 2012. The structural biology of HIV-1: mechanistic and therapeutic insights. *Nat Rev Microbiol* **10:**279–290.

Ganser-Pornillos BK, Yeager M, Pornillos O. 2012. Assembly and architecture of HIV. *Adv Exp Med Biol* **726:**441–465.

Hryc CF, Chen DH, Chiu W. 2011. Near-atomic-resolution cryo-EM for molecular virology. *Curr Opin Virol* **1:**110–117.

Klose T, Rossmann MG. 2014. Structure of large dsDNA viruses. *Biol Chem* **395:**711–719.

Leiman PG, Kanamara S, Mesyanzhinov VV, Arisaka F, Rossmann MG. 2003. Structure and morphogenesis of bacteriophage T4. *Cell Mol Life Sci* **60:**2356–2370.

Nemerow GR, Stewart PL, Reddy VS. 2012. Structure of human adenovirus. *Curr Opin Virol* **2:**115–121.

Prasad BV, Schmid MF. 2012. Principles of virus structural organization. *Adv Exp Med Biol* **726:**17–47.

Ruigrok RW, Crépin T, Kolakofsky D. 2011 Nucleoproteins and nucleocapsids of negative-strand RNA viruses. *Curr Opin Microbiol* **14:**504–510.

Strauss JH, Strauss EG. 2001. Virus evolution: how does an enveloped virus make a regular structure? *Cell* **105:**5–8.

Stubbs G. 1990. Molecular structures of viruses from the tobacco mosaic virus group. *Semin Virol* **1:**405–412.

Vaney MC, Rey FA. 2011. Class II enveloped viruses. *Cell Microbiol* **13:** 1451–1459.

Zheng W, Tao YJ. 2013. Structure and assembly of the influenza A virus ribonucleoprotein complex. *FEBS Lett* **587:**1206–1214.

Papers of Special Interest

Theoretical Foundations

Caspar DL, Klug A. 1962. Physical principles in the construction of regular viruses. *Cold Spring Harbor Symp Quant Biol* **27:**1–22.

Crick FH, Watson JD. 1956. Structure of small viruses. *Nature* **177:** 473–475.

Structures of Nonenveloped Viruses

Brenner S, Horne RW. 1959. A negative staining method for high resolution electron microscopy of viruses. *Biochim Biophys Acta* **34:**103–110.

Harrison SC, Olson AJ, Schutt CE, Winkler FK, Bricogne G. 1978. Tomato bushy stunt virus at 2.9 Å resolution. *Nature* **276:**368–373.

Hogle JM, Chow M, Filman DJ. 1985. Three-dimensional structure of poliovirus at 2.9 Å resolution. *Science* **229:**1358–1365.

Liddington RC, Yan Y, Moulai J, Sahli R, Benjamin TL, Harrison SC. 1991. Structure of simian virus 40 at 3.8-Å resolution. *Nature* **354:**278–284.

Liu H, Jin L, Koh SB, Atanasov I, Schein S, Wu L, Zhou ZH. 2010. Atomic structure of human adenovirus by cryo-EM reveals interactions among protein networks. *Science* **329:**1038–1043.

Prasad BV, Rothnagel R, Zeng CQ, Jakana J, Lawton JA, Chiu W, Estes MK. 1996. Visualization of ordered genomic RNA and localization of transcriptional complexes in rotavirus. *Nature* **382:**471–473.

Reddy VS, Natchiar SK, Stewart PL, Nemerow GR. 2010. Crystal structure of human adenovirus at 3.5 Å resolution. *Science* **329:**1071–1075.

Rossmann MG, Arnold E, Erickson JW, Frankenberger EA, Griffith JP, Hecht HJ, Johnson JE, Kamer G, Luo M, Mosser AG, et al. 1985. Structure of a common cold virus and functional relationship to other picornaviruses. *Nature* **317:**145–153.

Settembre EC, Chen JZ, Dormitzer PR, Grigorieff N, Harrison SC. 2011. Atomic model of an infectious rotavirus particle. *EMBO J* **30:**408–416.

Structures of Enveloped Viruses

Arranz R, Coloma R, Chichón FJ, Conesa JJ, Carrascosa JL, Valpuesta JM, Ortín J, Martín-Benito J. 2012. The structure of native influenza virus ribonucleoproteins. *Science* **338:**1634–1637.

Bauer DW, Huffman JB, Homa FL, Evilevitch A. 2013. Herpes virus genome, the pressure is on. *J Am Chem Soc* **135:**11216–11221.

Briggs JA, Wilk T, Welker R, Kräusslich HG, Fuller SD. 2003. Structural organization of authentic, mature HIV-1 virions and cores. *EMBO J* **22:** 1707–1715.

Cheng RH, Kuhn RJ, Olson NH, Rossmann MG, Choi HK, Smith TJ, Baker TS. 1995. Nucleocapsid and glycoprotein organization in an enveloped virus. *Cell* **80:**621–630.

Cyrklaff M, Risco C, Fernández JJ, Jiménez MV, Estéban M, Baumeister W, Carrascosa JL. 2005. Cryo-electron tomography of vaccinia virus. *Proc Natl Acad Sci U S A* **102:**2772–2777.

Ge P, Tsao J, Schein S, Green TJ, Luo M, Zhou ZH. 2010. Cryo-EM model of the bullet-shaped vesicular stomatitis virus. *Science* **327:**689–693.

Lescar J, Roussel A, Wien MW, Navaza J, Fuller SD, Wengler G, Wengler G, Rey FA. 2001. The fusion glycoprotein shell of Semliki Forest virus: an icosahedral assembly primed for fusogenic activation at endosomal pH. *Cell* **105:**137–148.

Mancini EJ, Clarke M, Gowen BE, Rutten T, Fuller SD. 2000. Cryo-electron microscopy reveals the functional organization of an enveloped virus, Semliki Forest virus. *Mol Cell* **5:**255–266.

Newcomb WW, Juhas RM, Thomsen DR, Homa FL, Burch AD, Weller SK, Brown JC. 2001. The UL6 gene product forms the portal for entry of DNA into the herpes simplex virus capsid. *J Virol* **75:**10923–10932.

Pornillos O, Ganser-Pornillos BK, Yeager M. 2011. Atomic-level modelling of the HIV capsid. *Nature* **469:**424–427.

Wynne SA, Crowther RA, Leslie AG. 1999. The crystal structure of the human hepatitis B virus capsid. *Mol Cell* **3:**771–780.

Zhao G, Perilla JR, Yufenyuy EL, Meng X, Chen B, Ning J, Ahn J, Gronenborn AM, Schulten K, Aiken C, Zhang P. 2013. Mature HIV-1 capsid structure by cryo-electron microscopy and all-atom molecular dynamics. *Nature* **497:**643–646.

Zhou ZH, Dougherty M, Jakana J, He J, Rixon FJ, Chiu W. 2000. Seeing the herpesvirus capsid at 8.5 Å resolution. *Science* **288:**877–880.

Structures of Individual Proteins

Malashkevich VN, Schneider BJ, McNally ML, Milhollen MA, Pang JX, Kim PS. 1999. Core structure of the envelope glycoprotein GP2 from Ebola virus at 1.9-Å resolution. *Proc Natl Acad Sci U S A* **96:**2662–2667.

Massiah MA, Starich MR, Paschall C, Summers MF, Christensen AM, Sundquist WI. 1994. Three-dimensional structure of the human immunodeficiency virus type 1 matrix protein. *J Mol Biol* **244:**198–223.

Rey FA, Heinz FX, Mandl C, Kunz C, Harrison SC. 1995. The envelope glycoprotein from tick-borne encephalitis virus at 2 Å resolution. *Nature* **375:**291–298.

Sharma M, Yi M, Dong H, Qin H, Peterson E, Busath DD, Zhou HX, Cross TA. 2010. Insight into the mechanism of the influenza A proton channel from a structure in a lipid bilayer. *Science* **330:**509–512.

Wilson IA, Skehel JJ, Wiley DC. 1981. Structure of the haemagglutinin membrane glycoprotein of influenza virus at 3 Å resolution. *Nature* **289:** 366–373.

Zubieta C, Schoehn G, Chroboczek J, CusaCk S. 2005. The structure of the human adenovirus 2 penton. *Mol Cell* **17:**121–135.

Other Components of Virus Particles

Chertova E, Chertov O, Coren LV, Roser JD, Trubey CM, Bess JW, Jr, Sowder RC, II, Barsov E, Hood BL, Fisher RJ, Nagashima K, Conrads TP, Veenstra TD, Lifson JD, Ott DE. 2006. Proteomic and biochemical analysis of purified human immunodeficiency virus type 1 produced from infected monocyte-derived macrophages. *J Virol* **80:**9039–9052.

Kramer T, Greco TM, Enquist LW, Cristea IM. 2011. Proteomic characterization of pseudorabies virus extracellular virions. *J Virol* **85:**6427–6441.

Terhune SS, Schröer J, Shenk T. 2004. RNAs are packaged into human cytomegalovirus virions in proportion to their intracellular concentration. *J Virol* **78:**10390–10398.

Thali M, Bukovsky A, Kondo E, Rosenwirth B, Walsh CT, Sodroski J, Göttlinger HG. 1994. Functional association of cyclophilin A with HIV-1 virions. *Nature* **372:**363–365.

Websites

http://viperdb.scripps.edu/ *Virus Particle Explorer*

http://viralzone.expasy.org/ *ViralZone*

http://www.virology.net/Big_Virology/BVHomePage.html *The Big Picture Book of Viruses*

http://virology.wisc.edu/virusworld/ *Virusworld*

5 Attachment and Entry

Introduction

Attachment of Virus Particles to Cells

General Principles

Identification of Receptors for Virus Particles

Virus-Receptor Interactions

Entry into Cells

Uncoating at the Plasma Membrane

Uncoating during Endocytosis

Membrane Fusion

Movement of Viral and Subviral Particles within Cells

Virus-Induced Signaling via Cell Receptors

Import of Viral Genomes into the Nucleus

Nuclear Localization Signals

The Nuclear Pore Complex

The Nuclear Import Pathway

Import of Influenza Virus Ribonucleoprotein

Import of DNA Genomes

Import of Retroviral Genomes

Perspectives

References

LINKS FOR CHAPTER 5

- *Video: Interview with Dr. Jeffrey M. Bergelson*
  http://bit.ly/Virology_Bergelson
- *Video: Interview with Dr. Carolyn Coyne*
  http://bit.ly/Virology_Coyne
- *Bond, covalent bond*
  http://bit.ly/Virology_Twiv210
- *Breaking and entering*
  http://bit.ly/Virology_Twiv166
- *A new cell receptor for rhinovirus*
  http://bit.ly/Virology_4-30-15
- *Blocking HIV infection with two soluble cell receptors*
  http://bit.ly/Virology_2-26-15
- *Changing influenza virus neuraminidase into a receptor binding protein*
  http://bit.ly/Virology_11-21-13

Who hath deceived thee so often as thyself?
BENJAMIN FRANKLIN

Introduction

Because viruses are obligate intracellular parasites, the genome must enter a cell for the viral reproduction cycle to occur. The physical properties of the virion are obstacles to this seemingly simple goal. Virus particles are too large to diffuse passively across the plasma membrane. Furthermore, the viral genome is encapsidated in a stable coat that shields the nucleic acid as it travels through the harsh extracellular environment. These impediments must all be overcome during the process of viral entry into cells. Encounter of a virus particle with the surface of a susceptible host cell induces a series of events that lead to entry of the viral genome into the cytoplasm or nucleus. The first step in entry is adherence of virus particles to the plasma membrane, an interaction mediated by binding to a specific **receptor** molecule on the cell surface.

The receptor plays an important role in **uncoating**, the process by which the viral genome is exposed, so that gene expression and genome replication can begin. Interaction of the virus particle with its receptor may initiate conformational changes that prime the capsid for uncoating. Alternatively, the receptor may direct the virus particle into endocytic pathways, where uncoating may be triggered by low pH or by the action of proteases. These steps bring the genome into the cytoplasm, which is the site of replication of most RNA-containing viruses. The genomes of viruses that replicate in the nucleus are moved to that location by cellular transport pathways. Viruses that replicate in the nucleus include most DNA-containing viruses (exceptions include poxviruses and giant viruses), RNA-containing retroviruses, influenza viruses, and Borna disease virus.

Virus entry into cells is **not** a passive process but relies on viral usurpation of normal cellular processes, including endocytosis, membrane fusion, vesicular trafficking, and transport into the nucleus. Because of the limited functions encoded by viral genomes, virus entry into cells depends absolutely on cellular processes.

Attachment of Virus Particles to Cells

General Principles

Infection of cells by many, but not all, viruses requires binding to a receptor on the cell surface. Exceptions include viruses of yeasts and fungi, which have no extracellular phases, and plant viruses, which are thought to enter cells in which the cell wall has been physically damaged, for example by insects or farm machinery. A receptor is a cell surface molecule that binds the virus particle and participates in entry. It may induce conformational changes in the virus particle that lead to membrane fusion or penetration, and it may also transmit signals that cause uptake. The receptor may also bring the bound particle into endocytic pathways.

Receptors for viruses comprise a variety of cell surface proteins, carbohydrates, and lipids, all with functions in the cell unrelated to virus entry. Many virus receptors have been identified in the past 30 years and include immunoglobulin-like proteins, ligand-binding receptors, glycoproteins, ion channels, gangliosides, carbohydrates, proteoglycans, and integrins. The receptor may be the only cell surface molecule required for entry into cells, or an additional cell surface molecule, or **coreceptor**, may be needed (Box 5.1). Different receptors may serve for virus entry in diverse cell types, and unrelated viruses may bind to the same receptor (e.g., the Coxsackievirus and adenovirus receptor).

The receptor may determine the **host range** of a virus, i.e., its ability to infect a particular animal or cell culture. For example, poliovirus infects primates and primate cell cultures but not mice or mouse cell cultures. Mouse cells synthesize a protein that is homologous to the poliovirus receptor,

PRINCIPLES ***Attachment and entry***

- Virus particles are too large to diffuse across the plasma membrane, and thus entry must be an active process.
- Virus particles bind to receptors on their host cells to initiate entry.
- The cell receptor may determine the host range and tissue tropism of the virus.
- Viruses may bind multiple distinct receptors, and individual cellular proteins may be receptors for multiple viruses.
- Enveloped virus particles bind via their transmembrane glycoproteins; nonenveloped virus particles bind via the capsid surface or projections from the capsid.
- Attachment proteins may not lead to internalization and viral reproduction but may still be important for dissemination in the host.
- Some viruses uncoat at the plasma membrane, while others do so from intracellular vesicles.
- Many viruses enter host cells by the same cellular pathways used to take up macromolecules.
- The entry mechanism used by a particular virus may differ depending on the nature of the target cell.
- Viral particles and subviral particles depend on the cytoskeleton to move within an infected cell.
- Binding of virions to cell receptors may activate signaling pathways that facilitate virus entry and movement, or produce cellular responses that enhance virus propagation and/or affect pathogenesis.
- For viruses that undergo replication in the nucleus, import can occur either through use of the nuclear pore complex or during cell division, when the nuclear membrane breaks down.

BOX 5.1

TERMINOLOGY

Receptors and coreceptors

By convention, the first cell surface molecule that is found to be essential for virus binding is called its **receptor**. Sometimes, such binding is not sufficient for entry into the cell. When binding to another cell surface molecule is needed, that protein is called a **coreceptor**. For example, human immunodeficiency virus binds to cells via a receptor, CD4, and then requires interaction with a second cell surface protein such as CXCR4, the coreceptor.

In practice, the use of receptor and coreceptor can be confusing and inaccurate. A particular cell surface molecule that is a coreceptor for one virus may be a receptor for another. Distinguishing receptors and coreceptors by the order in which they are bound is difficult to determine experimentally and is likely to be influenced by cell type and multiplicity of infection. Furthermore, as is the case for the human immunodeficiency viruses, binding only to the coreceptor may be sufficient for entry of some members. Usage of the terms "receptor" and "coreceptor" is convenient when describing virus entry, but the appellations may not be entirely precise.

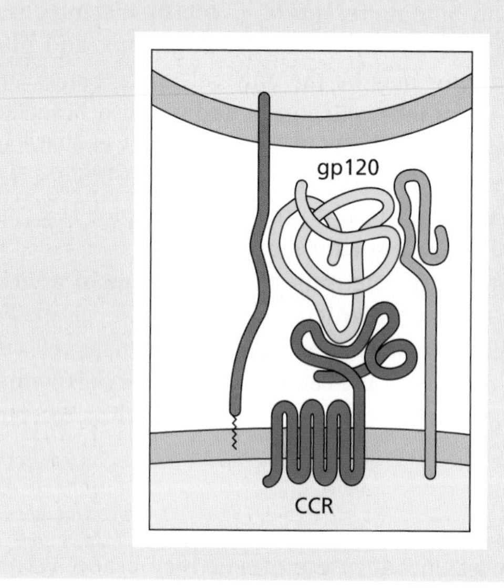

but sufficiently different that poliovirus cannot attach to it. In this example, the poliovirus receptor is **the** determinant of poliovirus host range. However, production of the receptor in a particular cell type does **not** ensure that virus reproduction will occur. Some primate cell cultures produce the poliovirus receptor but cannot be infected. The restriction of viral reproduction in these cells is most probably due to a block in viral reproduction beyond the attachment step. Receptors can also be determinants of tissue **tropism**, the predilection of a virus to invade and reproduce in a particular cell type. However, there are many other determinants of tissue tropism. For example, the sialic acid residues on membrane glycoproteins or glycolipids, which are receptors for influenza virus, are found on many tissues, yet viral reproduction in the host is restricted. The basis of such restriction is discussed in Volume II, Chapter 2.

Our understanding of the earliest interactions of virus particles with cells comes almost exclusively from analysis of synchronously infected cells in culture. The initial association with cells is probably via electrostatic forces, as it is sensitive to low pH or high concentrations of salt, but higher affinity binding relies mainly on hydrophobic and other short-range forces between the viral and cellular surfaces. Although the **affinity** of a receptor for a single virus particle is low, the presence of multiple receptor-binding sites on the virion and the fluid nature of the plasma membrane allow engagement of multiple receptors. Consequently, the **avidity** (the strength conferred by multiple interactions) of virus particle binding to cells is usually very high. Binding can usually occur at 4°C (even though entry does not) as well as at body temperature (e.g., 37°C). Infection of cultured cells can therefore be synchronized by allowing binding to take place at a low temperature and then shifting the cells to a physiological temperature to allow the initiation of subsequent steps.

The first steps in virus attachment are governed largely by the probability that a virus particle and a cell will collide, and therefore by the concentrations of free particles and host cells. The rate of attachment can be described by the equation

$$dA/dt = k[V][H]$$

where A is attachment, t is time, and $[V]$ and $[H]$ are the concentrations of virus particles and host cells, respectively, and k is a constant that defines the rate of the reaction. It can be seen from this equation that if a mixture of viruses and cells is diluted after an adsorption period, subsequent binding of particles is greatly reduced. For example, a 100-fold dilution of the virus and cell mixture reduces the attachment rate 10,000-fold (i.e., $1/100 \times 1/100$). Dilution can be used to prevent subsequent virus adsorption and hence to synchronize an infection.

Many receptor molecules can move in the plasma membrane, leading to the formation of microdomains that differ in composition. Bound virus may therefore localize to specialized areas of the membrane such as lipid rafts, **caveolae**, or coated pits. Localization of virus particle-receptor complexes can also cause transmembrane signaling, changes in the cytoskeleton, and recruitment of clathrin.

Identification of Receptors for Virus Particles

The development of three crucial technologies in the past 30 years has enabled identification of many receptors for viruses. Production of monoclonal antibodies provided a powerful means of isolating and characterizing individual cell surface proteins. Hybridoma cell lines that secrete monoclonal antibodies that block virus attachment are obtained after immunizing mice with intact cells. Such antibodies can be used to purify the receptor protein by affinity chromatography.

A second technology that facilitated the identification of receptors was the development of DNA-mediated transformation. This method was crucial for isolating genes that encode receptors, following introduction of DNA from susceptible cells into nonsusceptible cells (Fig. 5.1). Cells that acquire DNA encoding the receptor and carry the corresponding protein on their surface are able to bind virus specifically. Clones of such cells are recognized and selected, for example, by the binding of receptor-specific monoclonal antibodies. The receptor genes can then be isolated from these selected cells by using a third technology, molecular cloning. The power of these different technologies can lead to rapid progress: the receptor for a newly identified Middle Eastern respiratory syndrome coronavirus was identified just 4 months after the first description of the virus (Box 5.2). Although these technologies have led to the identification of many cell receptors for viruses, each method has associated uncertainties (Box 5.3).

The availability of receptor genes has made it possible to investigate the details of receptor interaction with viruses by site-directed mutagenesis. Receptor proteins can be synthesized in heterologous systems and purified, and their properties can be studied *in vitro*, while animal cells producing altered receptor proteins can be used to test the effects of alterations on virus attachment. Because of their hydrophobic membrane-spanning domains, many of these cell surface proteins are relatively insoluble and difficult to work with.

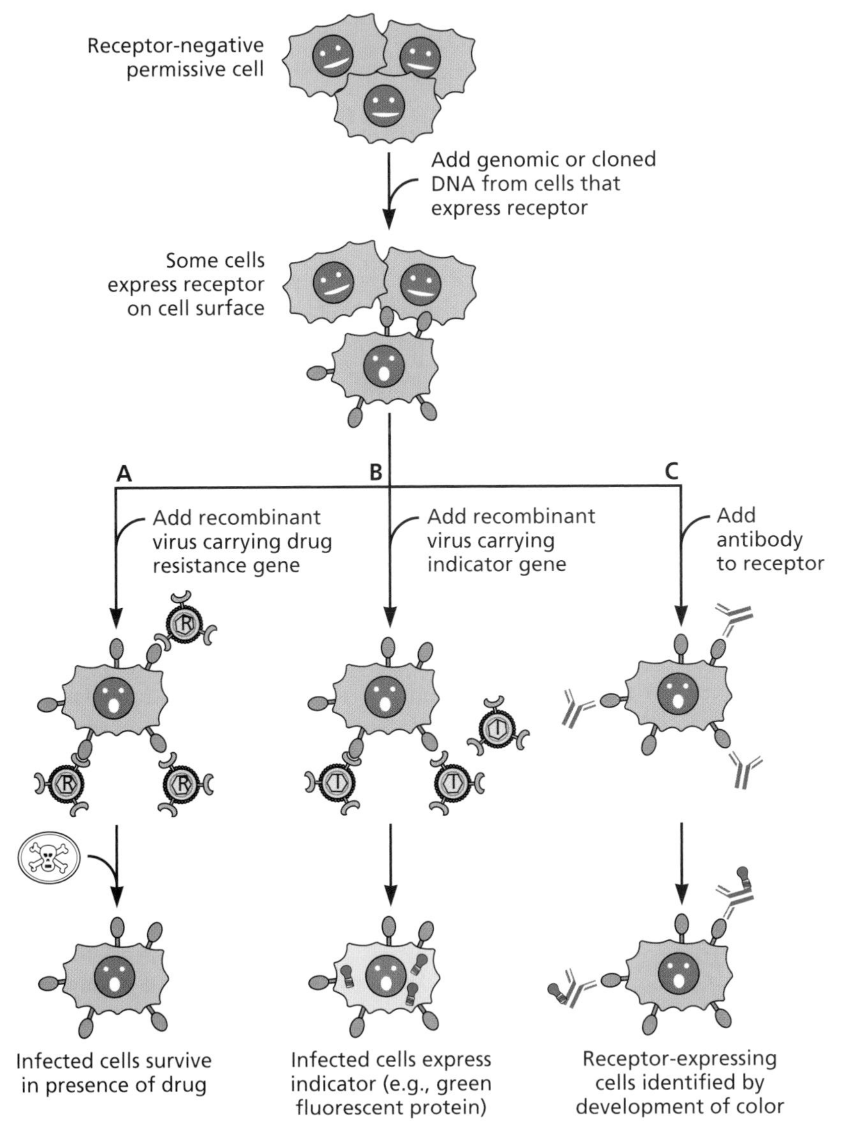

Figure 5.1 Experimental strategies for identification and isolation of genes encoding cell receptors for viruses. Genomic DNA or pools of DNA clones from cells known to synthesize the receptor are introduced into receptor-negative permissive cells. A small number of recipient cells produce the receptor. Three different strategies for identifying such rare receptor-expressing cells are outlined. **(A)** The cells are infected with a virus that has been engineered so that it carries a gene encoding drug resistance. Cells that express the receptor will become resistant to the drug. This strategy works only for viruses that persist in cells without killing them. **(B)** For lytic viruses, an alternative is to engineer the virus to express an indicator, such as green fluorescent protein or β-galactosidase. Cells that make the correct receptor and become infected with such viruses can be distinguished by a color change, such as green in the case of green fluorescent protein. **(C)** The third approach depends on the availability of an antibody directed against the receptor, which binds to cells that express the receptor gene. Bound antibodies can be detected by an indicator molecule. When complementary DNA (cDNA) cloned in a plasmid is used as the donor DNA, pools of individual clones (usually 10,000 clones per pool) are prepared and introduced individually into cells. The specific DNA pool that yields receptor-expressing cells is then subdivided, and the screening process is repeated until a single receptor-encoding DNA is identified.

BOX 5.2

METHODS

Affinity isolation

To identify the receptor for the newly emerged Middle Eastern respiratory syndrome coronavirus, the gene encoding the viral spike glycoprotein gene was fused with sequences encoding the Fc domain of human IgG. The fusion protein was produced in cells and incubated with lysates of cells known to be susceptible to the virus, and the resulting complexes were fractionated by native polyacrylamide gel electrophoresis. A single polypeptide of ~110 kDa was obtained by this procedure. This polypeptide was excised from the polyacrylamide gel, and its amino acid sequence was determined by mass spectrometric analysis, identifying it as dipeptidyl peptidase 4. When this protein was subsequently synthesized in nonsusceptible Cos-7 cells by DNA-mediated transformation, the cells became susceptible to Middle Eastern respiratory syndrome coronavirus infection. That a single protein was identified by this procedure is remarkable: typically, this approach identifies many nonspecific binding proteins.

Raj VS, Mou H, Smits SL, Dekkers DH, Müller MA, Dijkman R, Muth D, Demmers JA, Zaki A, Fouchier RA, Thiel V, Drosten C, Rottier PJ, Osterhaus AD, Bosch BJ, Haagmans BL. 2013. Dipeptidyl peptidase 4 is a functional receptor for the emerging human coronavirus-EMC. *Nature* **495:**251–256.

Identification of MERS-coronavirus cell receptor.

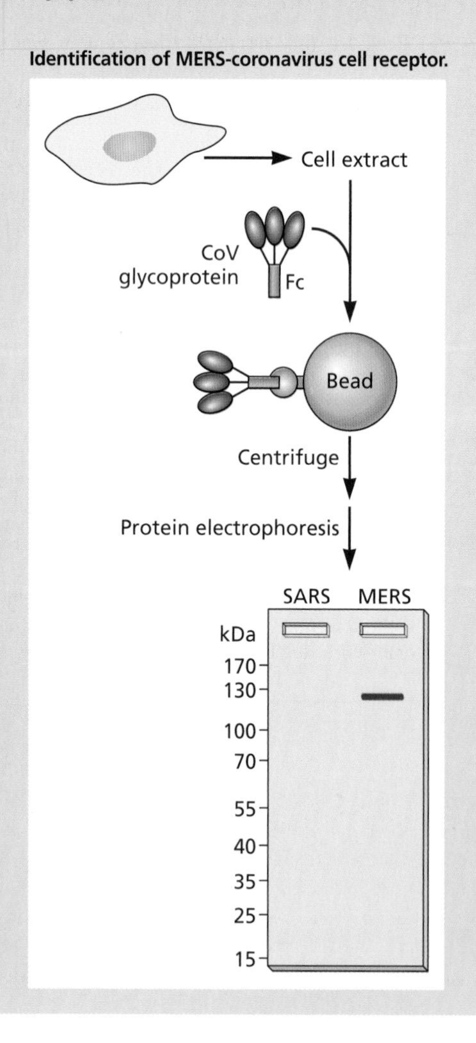

Soluble extracellular protein domains (with the virus binding sites) have been essential for structural studies of receptor-virus interactions. Receptor genes have also been used to produce transgenic mice that synthesize receptor proteins. Such transgenic animals can serve as useful models in the study of human viral diseases.

Virus-Receptor Interactions

Animal viruses have multiple receptor-binding sites on their surfaces. Of necessity, one or more of the capsid proteins of nonenveloped viruses specifically interact with the cell receptor. Typically, these form projections from or indentations in the surface. Receptor-binding sites for enveloped viruses are provided by oligomeric type 1 integral membrane glycoproteins encoded by the viral genome that have been incorporated into the cell-derived membranes of virus particles. Although the details vary among viruses, most virus-receptor interactions follow one of several mechanisms illustrated by the best-studied examples described below.

Nonenveloped Viruses Bind via the Capsid Surface or Projections

Attachment via surface features: canyons and loops. Members of the enterovirus genus of the *Picornaviridae* include human polioviruses, coxsackieviruses, echoviruses, enteroviruses, and rhinoviruses. The receptor for poliovirus, CD155, was identified by using a DNA transformation and cloning strategy (Fig. 5.1). It was known that mouse cells cannot be infected with poliovirus, because they do not produce the receptor. Transfection of poliovirus RNA into mouse cells in culture leads to poliovirus reproduction, indicating that there is no intracellular block to virus multiplication. Introduction of human DNA into mouse cells confers susceptibility to poliovirus infection. The human gene recovered from receptor-positive mouse cells proved to encode CD155, a glycoprotein that is a member of the immunoglobulin (Ig) superfamily (Fig. 5.2).

Mouse cells are permissive for poliovirus reproduction, and susceptibility is limited **only** by the absence of CD155. Consequently, it was possible to develop a small-animal model for poliomyelitis by producing transgenic mice that synthesize this receptor. Inoculation of CD155 transgenic mice with poliovirus by various routes produces paralysis, as is observed in human poliomyelitis. These CD155-synthesizing mice were the first new animal model created by transgenic technology for the study of viral disease. Similar approaches have subsequently led to animal models for viral diseases caused by measles virus and echoviruses.

Rhinoviruses multiply primarily in the upper respiratory tract and are responsible for causing up to 50% of all common colds. Over 150 rhinovirus genotypes have been identified and classified on the basis of genome sequence into three species, A, B, and C. Rhinoviruses bind to at least three

BOX 5.3

BACKGROUND

Criteria for identifying cell receptors for viruses

The use of monoclonal antibodies, molecular cloning, and DNA-mediated transformation provides a powerful approach for identifying cellular receptors for viruses, but each method has associated uncertainties. A monoclonal antibody that blocks virus attachment might recognize not the receptor but a closely associated membrane protein. To prove that the protein recognized by the monoclonal antibody **is** a receptor, DNA encoding the protein must be introduced into nonsusceptible cells to demonstrate that it can confer virus-binding activity. Any of the approaches outlined in Fig. 5.1 can result in identification of a cellular gene that encodes a putative receptor. However, the encoded protein might not be a receptor but may modify another cellular protein so that it can bind virus particles. One proof that the DNA codes for a receptor could come from the identification of a monoclonal antibody that blocks virus attachment and is directed against the encoded protein.

For some viruses, synthesis of the receptor on cells leads to binding but not infection. In such cases a coreceptor is required, either for internalization or for membrane fusion. The techniques of molecular cloning also can be used to identify coreceptors. For example, production of CD4 on mouse cells leads to binding of human immunodeficiency virus type 1 but not infection, because fusion of viral and cell membranes does not occur. To identify the coreceptor, a DNA clone was isolated from human cells that allowed membrane fusion catalyzed by the viral attachment protein in mouse cells synthesizing CD4.

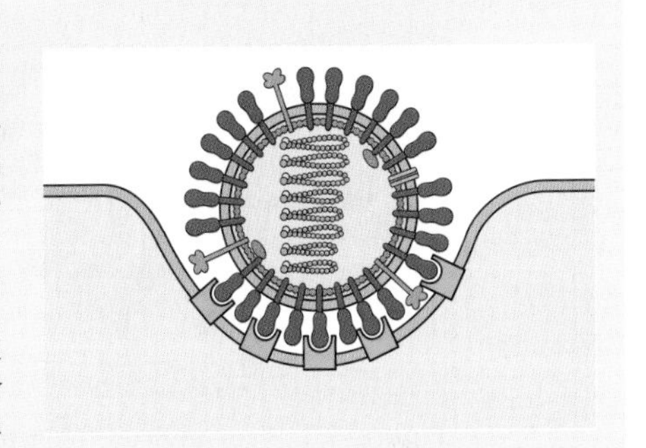

different receptor molecules. The cell surface receptor bound by most A and B species rhinoviruses was identified by using a monoclonal antibody that blocks rhinovirus infection and that recognizes a cell surface protein. This monoclonal antibody was used to isolate a 95-kDa cell surface glycoprotein by affinity chromatography. Amino acid sequence analysis of the purified protein, which bound to rhinovirus *in vitro*, identified it as the integral membrane protein intercellular adhesion molecule 1 (Icam-1). Cell receptors for other rhinoviruses are the low-density lipoprotein receptor and cadherin-related family member 3.

The RNA genomes of picornaviruses are protected by capsids made up of four virus-encoded proteins, VP1, VP2, VP3, and VP4, arranged with icosahedral symmetry (see Fig. 4.12). The capsids of rhinoviruses and polioviruses have deep canyons surrounding the 12 5-fold axes of symmetry (Fig. 5.3),

Figure 5.2 Some cell attachment factors and receptors for viruses. Schematic diagrams of cell molecules that function during virus entry. GlcNAc, N-acetylglucosamine; GalNAc, N-acetylgalactosamine; Ldlr, low-density lipoprotein receptor; DC-SIGN, dendritic cell-specific intercellular adhesion molecule-3-grabbing non-integrin; Car, coxsackievirus-adenovirus receptor.

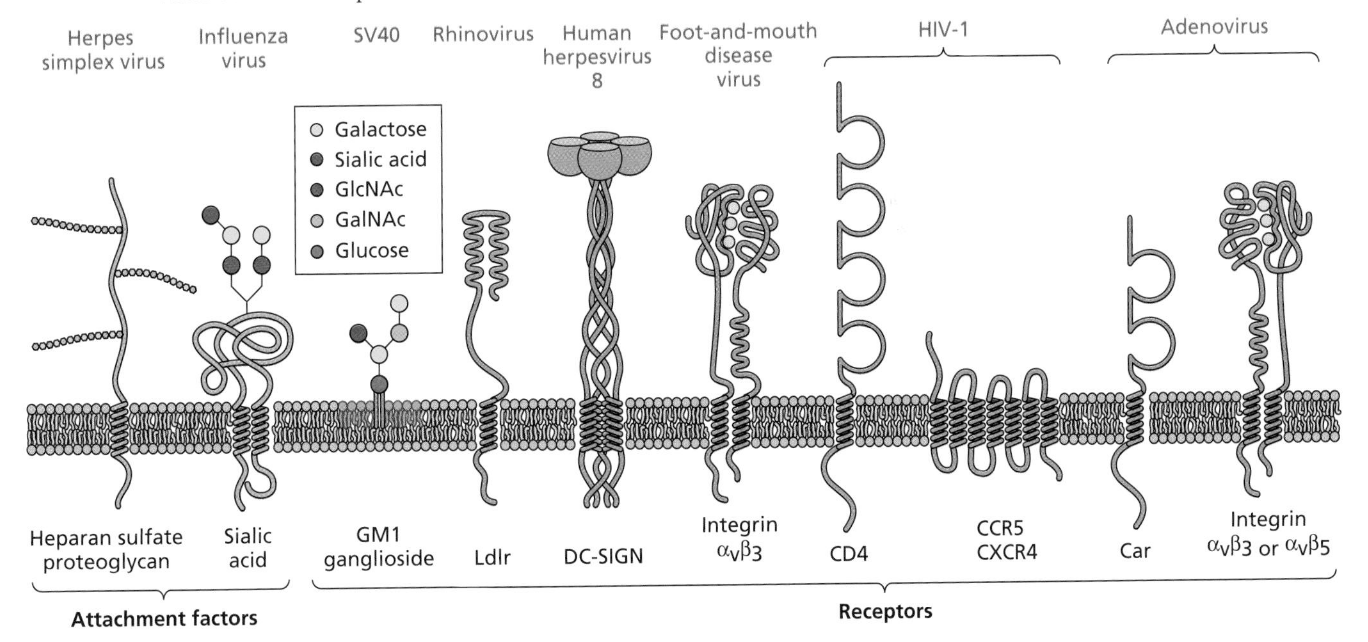

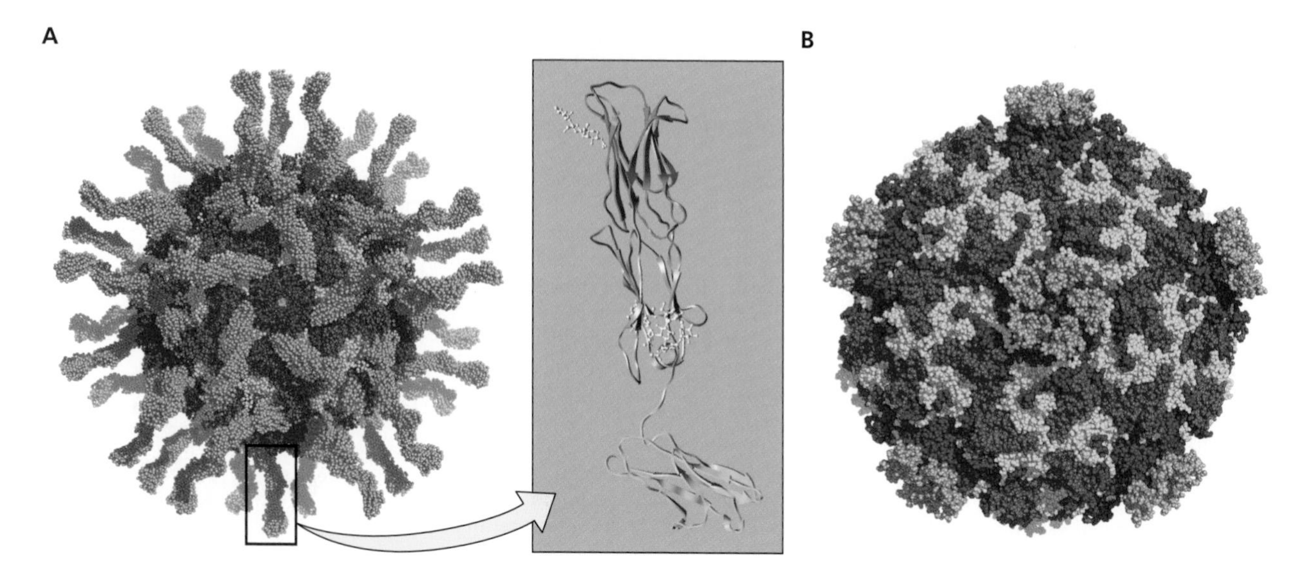

Figure 5.3 Picornavirus-receptor interactions. (A) Structure of poliovirus bound to a soluble form of CD155 (gray), derived by cryo-electron microscopy and image reconstruction. Capsid proteins are color coded (VP1, blue; VP2, yellow; VP3, red). One CD155 molecule is shown as a ribbon model in the panel to the right, with each Ig-like domain in a different color. The first Ig-like domain of CD155 (magenta) binds in the canyon of the viral capsid. **(B)** Structure of human rhinovirus type 2 bound to a soluble form of low-density lipoprotein receptor (gray). The receptor binds on the plateau at the 5-fold axis of symmetry of the capsid.

whereas cardioviruses and aphthoviruses lack this feature. The canyons in the capsids of some rhinoviruses and enteroviruses are the sites of interaction with cell surface receptors. Amino acids that line the canyons are more highly conserved than any others on the viral surface, and their substitution can alter the affinity of binding to cells. Poliovirus bound to a receptor fragment comprising CD155 domains 1 and 2 has been visualized in reconstructed images from cryo-electron microscopy. The results indicate that the first domain of CD155 binds to the central portion of the canyon in an orientation oblique to the surface of the virus particle (Fig. 5.3A).

Although canyons are present in the capsid of rhinovirus type 2, they are not the binding sites for the receptor, low-density lipoprotein receptor. Rather, this site on the capsid is located on the star-shaped plateau at the 5-fold axis of symmetry (Fig. 5.3B). Sequence and structural comparisons have revealed why different rhinovirus serotypes bind distinct receptors. A key VP1 amino acid, lysine, is conserved in all rhinoviruses that bind this receptor and interacts with a negatively charged region of low-density lipoprotein receptor. This lysine is not found in VP1 of rhinoviruses that bind Icam-1.

For picornaviruses with capsids that do not have prominent canyons, including coxsackievirus group A and foot-and-mouth disease virus, attachment is to VP1 surface loops that include amino acid sequence motifs recognized by their integrin receptors (Fig. 5.2).

Attachment via protruding fibers. The results of competition experiments indicated that members of two different virus families, group B coxsackieviruses and most human adenoviruses, share a cell receptor. This receptor is a 46-kDa member of the Ig superfamily called Car (coxsackievirus and adenovirus receptor). Binding to this receptor is not sufficient for infection by most adenoviruses. Interaction with a coreceptor, the α_v integrin $\alpha_v\beta_3$ or $\alpha_v\beta_5$, is required for uptake of the capsid into the cell by receptor-mediated endocytosis. An exception is adenovirus type 9, which can infect hematopoietic cells after binding directly to α_v integrins. Adenoviruses of subgroup B bind CD46, which is also a cell receptor for some strains of measles virus, an enveloped member of the *Paramyxoviridae*.

The nonenveloped DNA-containing adenoviruses are much larger than picornaviruses, and their icosahedral capsids are more complex, comprising at least 10 different proteins. Electron microscopy shows that fibers protrude from each adenovirus pentamer (Fig. 5.4; see the appendix in this volume, Fig. 1A). The fibers are composed of homotrimers of the adenovirus fiber protein and are anchored in the pentameric penton base; both proteins have roles to play in virus attachment and uptake.

For many adenovirus serotypes, attachment via the fibers is necessary but not sufficient for infection. A region comprising the N-terminal 40 amino acids of each subunit of the fiber protein is bound noncovalently to the penton base. The central shaft region is composed of repeating motifs of approximately 15 amino acids; the length of the shaft in different serotypes is determined by the number of these repeats. The three constituent shaft regions appear to form a rigid triple-helical structure in the trimeric fiber. The C-terminal 180 amino acids of each subunit interact to form a terminal knob. Genetic analyses and competition experiments indicate that determinants

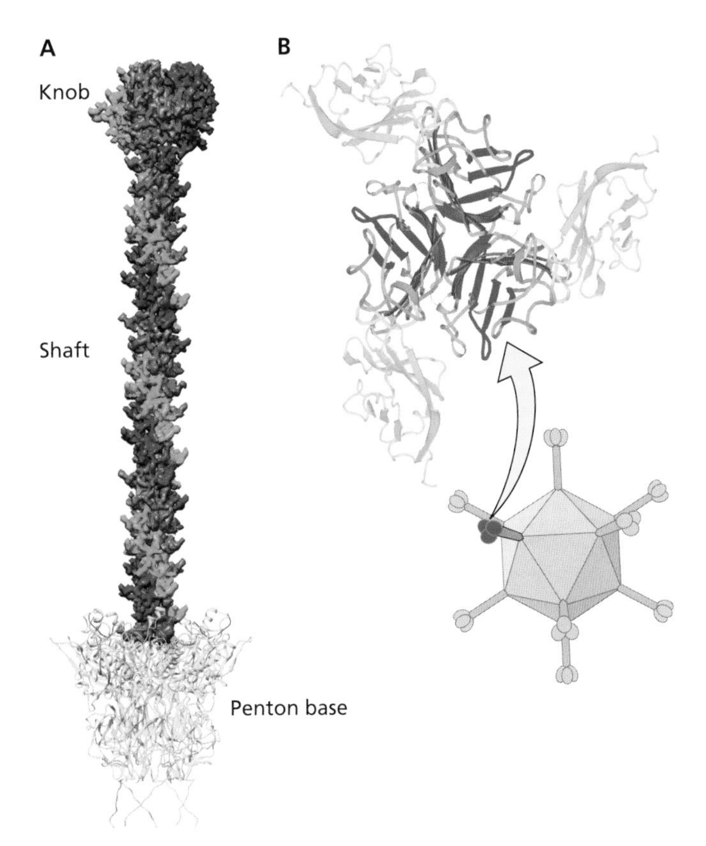

Figure 5.4 Structure of the adenovirus 12 knob bound to the Car receptor. **(A)** Structure of fiber protein, with knob, shaft, and tail domains labeled. Figure provided by Hong Zhou, University of California, Los Angeles, and Hongrong Liu, Hunan Normal University. **(B)** Ribbon diagram of the knob-Car complex as viewed down the axis of the viral fiber. The trimeric knob is in the center. The AB loop of the knob protein, which contacts Car, is in yellow. The first Ig-like domains of three Car molecules bound to the knob are colored blue. The binding sites of both molecules require trimer formation.

for the initial, specific attachment to host cell receptors reside in this knob. The structure of this receptor-binding domain bound to Car reveals that surface loops of the knob contact one face (Fig. 5.4).

Glycolipids, unusual cell receptors for polyomaviruses. The family *Polyomaviridae* includes simian virus 40 (SV40), mouse polyomavirus, and human BK virus. These viruses are unusual because they bind to ganglioside rather than protein receptors. Gangliosides are glycosphingolipids with one or more sialic acids linked to a sugar chain. There are over 40 known gangliosides, which differ in the position and number of sialic acid residues and are critical for virus binding. Simian virus 40, polyomavirus, and BK virus bind to three different types of ganglioside. Structural studies have revealed that sialic acid linked to galactose by an α(2,3) linkage binds to a pocket on the surface of the polyomavirus capsid (Fig. 5.5). Gangliosides are highly concentrated in lipid rafts (Chapter 2, Box 2.1) and participate in signal transduction, two properties

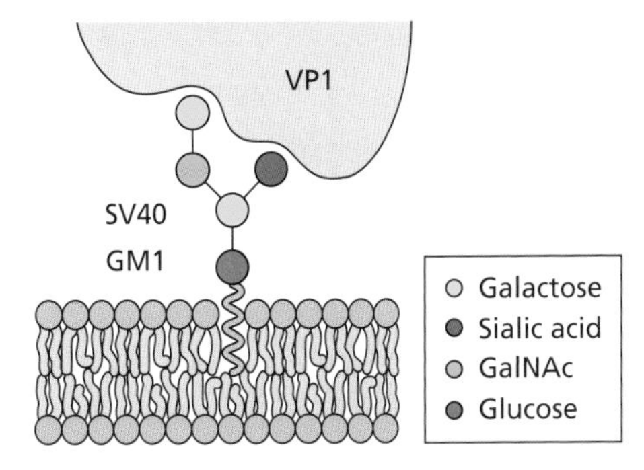

Figure 5.5 Interaction of polyomaviruses with ganglioside receptors. JC polyomavirus binds to a pentasaccharide with a terminal sialic acid linked by an α(2,6) bond to the penultimate galactose. The linear nature of this receptor differs from other branched polyomavirus ganglioside receptors. JC polyomavirus also appears to bind to a serotonin receptor for cell entry.

that are important during polyomavirus entry into cells. After binding a ganglioside, mouse polyomavirus interacts with α4β1 integrin to allow virus entry.

Enveloped Viruses Bind via Transmembrane Glycoproteins

The lipid membranes of enveloped viruses originate from those of the host cells. Membrane spanning viral proteins are inserted into these by the same mechanisms as cellular integral membrane proteins. Attachment sites on one or more of these envelope proteins bind to specific receptors. The two best-studied examples of enveloped virus attachment and its consequences are provided by the interactions of influenza A virus and the retrovirus human immunodeficiency virus type 1 with their receptors.

Influenza virus. The family *Orthomyxoviridae* comprises the three genera of influenza viruses, A, B, and C. These viruses bind to negatively charged, terminal sialic acid moieties present in oligosaccharide chains that are covalently attached to cell surface glycoproteins or glycolipids. The presence of sialic acid on most cell surfaces accounts for the ability of influenza virus particles to attach to many types of cell. The interaction of influenza virus with individual sialic acid moieties is of low affinity. However, the opportunity for multiple interactions among the numerous hemagglutinin (HA) molecules on the surface of the virus particle and multiple sialic acid residues on cellular glycoproteins and glycolipids results in a high overall avidity of the virus particle for the cell surface. The surfaces of influenza viruses were shown in the early 1940s to contain an enzyme that, paradoxically, removes the receptors for attachment from the surface of cells. Later, this enzyme was identified as the virus-encoded envelope glycoprotein neuraminidase, which cleaves the glycoside linkages of sialic acids (Fig. 5.6B). This enzyme is required for release of virus particles bound to the surfaces of

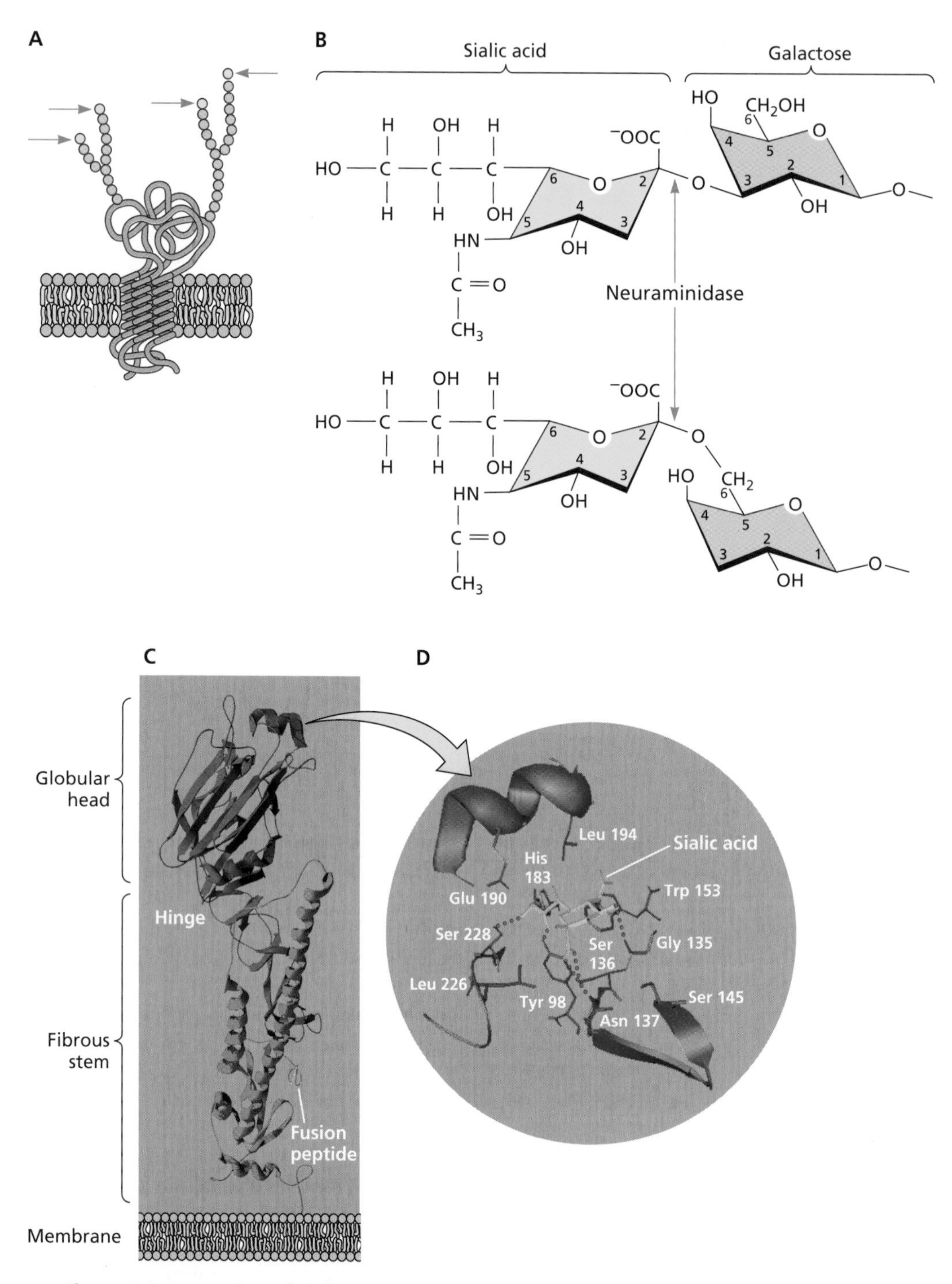

Figure 5.6 Interaction of sialic acid receptors with the hemagglutinin of influenza viruses. (A) An integral membrane glycoprotein; the arrows point to terminal sialic acid units that are attachment sites for influenza virus. **(B)** The structure of a terminal sialic acid moiety that is recognized by the viral envelope protein hemagglutinin. Sialic acid is attached to galactose by an α(2,3) (top) or an α(2,6) (bottom) linkage. The site of cleavage by the influenza virus envelope glycoprotein neuraminidase is indicated. The sialic acid shown is *N*-acetylneuraminic acid, which is the preferred receptor for influenza A and B viruses. These viruses do not bind to 9-*O*-acetyl-*N*-neuraminic acid, the receptor for influenza C viruses. **(C)** HA monomer modeled from the X-ray crystal structure of the natural trimer. HA1 (blue) and HA2 (red) subunits are held together by a disulfide bridge as well as by many noncovalent interactions. The fusion peptide at the N terminus of HA2 is indicated (yellow). **(D)** Close-up of the receptor-binding site with a bound sialic acid molecule. Side chains of the conserved amino acids that form the site and hydrogen-bond with the receptor are included.

infected cells, facilitating virus spread through the respiratory tract (Volume II, Chapter 9).

Influenza virus HA is the viral glycoprotein that binds to the cell receptor sialic acid. The HA monomer is synthesized as a precursor that is glycosylated and subsequently cleaved to form HA1 and HA2 subunits. Each HA monomer consists of a long, helical stalk anchored in the membrane by HA2 and topped by a large HA1 globule, which includes the sialic acid-binding pocket (Fig. 5.6C, D). While attachment of all influenza A virus strains requires sialic acid, strains vary in their affinities for different sialyloligosaccharides. For example, human virus strains bind preferentially sialic acids attached to galactose via an $\alpha(2,6)$ linkage, the major sialic acid present on human respiratory epithelium (Fig. 5.6B). Avian virus strains bind preferentially to sialic acids attached to galactose via an $\alpha(2,3)$ linkage, the major sialic acid in the duck gut epithelium. Amino acids in the sialic acid-binding pocket of HA (Fig. 5.6D) determine which sialic acid is preferred and can therefore influence viral host range. It is thought that an amino acid change in the sialic acid-binding pocket of the 1918 influenza virus, which may have evolved from an avian virus, allowed it to recognize the $\alpha(2,6)$-linked sialic acids that predominate in human cells.

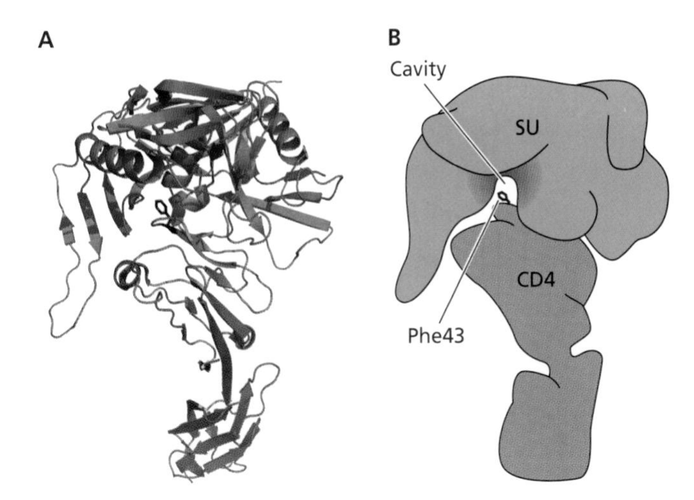

Figure 5.7 Interaction of human immunodeficiency virus type 1 SU with its cell receptor, CD4. (A) Ribbon diagram of SU (red) bound to CD4 (brown), derived from X-ray crystallographic data. The side chain of CD4 Phe43 is shown. **(B)** Cartoon of the CD4-SU complex. Mutagenesis has identified CD4 Phe43 as a residue critical for binding to SU. Phe43 is shown penetrating the hydrophobic cavity of SU. This amino acid, which makes 23% of the interatomic contacts between CD4 and SU, is at the center of the interface and appears to stabilize the entire complex.

Human immunodeficiency virus type 1. Animal retroviruses have long been of interest because of their ability to cause a variety of serious diseases, especially cancers (caused by oncogenic retroviruses) and neurological disorders (caused by lentiviruses). The acquired immunodeficiency syndrome (AIDS) pandemic has focused great attention on the lentivirus human immunodeficiency virus type 1 and its close relatives. The cell surface receptors of this virus have been among the most intensively studied and currently are the best understood.

When examined by electron microscopy, the envelopes of human immunodeficiency virus type 1 and other retroviruses appear to be studded with "spikes" (see Fig. 4.19). These structures are composed of trimers of the single viral envelope glycoprotein, which bind the cell receptor (Fig. 5.7). The monomers of the spike protein are synthesized as heavily glycosylated precursors that are cleaved by a cellular protease to form SU and TM. The latter is anchored in the envelope by a single membrane-spanning domain and remains bound to SU by numerous noncovalent bonds.

The cell receptor for human immunodeficiency virus type 1 is CD4 protein, a 55-kDa rodlike molecule that is a member of the Ig superfamily and has four Ig-like domains (Fig. 5.2). A variety of techniques have been used to identify the site of interaction with human immunodeficiency virus type 1, including site-directed mutagenesis and X-ray crystallographic studies of a complex of CD4 bound to the viral attachment protein SU (Fig. 5.7). The interaction site for SU in domain 1 of CD4 is in a region analogous to the site in CD155 that binds to poliovirus. Remarkably, two viruses with entirely different architectures bind to analogous surfaces of these Ig-like domains.

The atomic structure of a complex of human immunodeficiency virus type 1 SU, a two-domain fragment of CD4, and a neutralizing antibody against SU has been determined by X-ray crystallography (Fig. 5.7). The CD4-binding site in SU is a deep cavity, and the opening of this cavity is occupied by CD4 amino acid Phe43, which is critical for SU binding. Comparison with the structure of SU in the absence of CD4 indicates that receptor binding induces conformational changes in SU. These changes expose binding sites on SU for the chemokine receptors, which are required for fusion of viral and cell membranes (see "Uncoating at the Plasma Membrane" below).

Alphaherpesviruses. The alphaherpesvirus subfamily of the *Herpesviridae* includes herpes simplex virus types 1 and 2, pseudorabies virus, and bovine herpesvirus. Initial contact of these viruses with the cell surface is made by low-affinity binding of two viral glycoproteins, gC and gB, to glycosaminoglycans (preferentially heparan sulfate), abundant components of the extracellular matrix (Fig. 5.8). This interaction concentrates virus particles near the cell surface and facilitates subsequent attachment of the viral glycoprotein gD to an integral membrane protein, which is required for entry into the cell (Fig. 5.8). Members of at least two different protein families serve as entry receptors for alphaherpesviruses. One of these families, the nectins, comprises the poliovirus receptor CD155 and related proteins, yet another example of

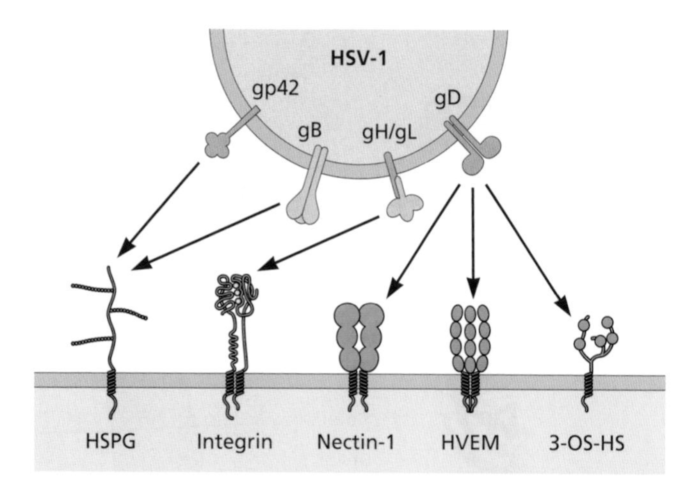

Figure 5.8 Cell receptors for herpes simplex virus type 1. Four viral glycoproteins, gp42, gB, gH/gL, and gD are shown binding with attachment molecule HSPG (heparin sulfate proteoglycan) or cell receptors (integrin, nectin-1, HVEM [herpesvirus entry mediator], 3-O-sulfate heparin sulfate). Virus entry does not require all interactions.

receptors shared by different viruses. When members of these two protein families are not present, 3-O-sulfated heparan sulfate can serve as an entry receptor for these viruses.

Multiple and Alternative Receptors

One type of receptor is not sufficient for infection by some viruses. Decay-accelerating protein (CD55), a regulator of the complement cascade, is the cell receptor for many enteroviruses, but infection also requires the presence of a coreceptor. Coxsackievirus A21 can bind to cell surface decay-accelerating protein, but this interaction does not lead to infection unless Icam-1 is also present. It is thought that Icam-1 inserts into the canyon where it triggers capsid uncoating. Some viruses bind to different cell receptors, depending on the nature of the virus isolate or the cell line. Often passage of viruses in cell culture selects variants that bind heparin sulfate. Infection of cells with foot-and-mouth disease virus type A12 requires the RGD-binding integrin $\alpha_v\beta_3$. However, the receptor for the O strain of this virus, which has been extensively passaged in cell culture, is not integrin $\alpha_v\beta_3$ but cell surface heparan sulfate. On the other hand, the type A12 strain cannot infect cells that lack integrin $\alpha_v\beta_3$, even if heparan sulfate is present. In a similar way, adaptation of Sindbis virus to cultured cells has led to the selection of variants that bind heparan sulfate. When receptors are rare, viruses that can bind to the more abundant glycosaminoglycan are readily selected.

Cell Surface Lectins and Spread of Infection

Cell surface **lectins** may bind to glycans present in viral glycoproteins, leading to dissemination within the host. An example is the lectin Dc-sign (dendritic cell-specific intercellular adhesion molecule-3-grabbing non-integrin), a tetrameric C-type lectin present on the surface of dendritic cells. This lectin binds high-mannose, N-linked glycans, such as those produced in insect cells. Viruses that reproduce in insects are delivered to the human skin via a bite and may bind and sometimes infect dendritic cells. These cells then carry the viruses to other parts of the body, particularly lymph nodes. However, not all viruses that bind Dc-sign replicate in insect cells. In humans, Dc-sign on the surface of dendritic cells binds human immunodeficiency virus type 1 virus particles, but cell entry does not take place. When the dendritic cells migrate to the lymph node, infectious virus is released where it can enter and reproduce in T cells. While the interaction of human immunodeficiency virus type 1 with Dc-sign is nonproductive, it leads to viral dissemination in the host.

Entry into Cells

Uncoating at the Plasma Membrane

The particles of many enveloped viruses, including members of the family *Paramyxoviridae* such as Sendai virus and measles virus, fuse directly with the plasma membrane at neutral pH. These virions bind to cell surface receptors via a viral integral membrane protein (Fig. 5.9). Once the viral and cell membranes have been closely juxtaposed by this receptor-ligand interaction, fusion is induced by a second viral glycoprotein known as fusion (F) protein, and the viral nucleocapsid is released into the cell cytoplasm (Fig. 5.10).

F protein is a type I integral membrane glycoprotein (the N terminus lies outside the viral membrane) with similarities to influenza virus HA in its synthesis and structure. It is a homotrimer that is synthesized as a precursor called F0 and cleaved during transit to the cell surface by a host cell protease to produce two subunits, F1 and F2, held together by disulfide bonds. The newly formed N-terminal 20 amino acids of the F1 subunit, which are highly hydrophobic, form a region called the **fusion peptide** because it inserts into target membranes to initiate fusion. Viruses with the uncleaved F0 precursor can be produced in cells that lack the protease responsible for its cleavage. Such virus particles are noninfectious; they bind to target cells but the viral genome does not enter. Cleavage of the F0 precursor is necessary for fusion, not only because the fusion peptide is made available for insertion into the plasma membrane, but also to generate the metastable state of the protein that can undergo the conformational rearrangements needed for fusion.

Because cleaved F-protein-mediated fusion can occur at neutral pH, it must be controlled, both to ensure that virus particles fuse with only the appropriate cell and to prevent aggregation of newly assembled virions. The fusion peptide of F1 is buried between two subunits of the trimer in the pre-fusion protein. Conformational changes in F protein lead to refolding of the protein, assembly of an α-helical coiled

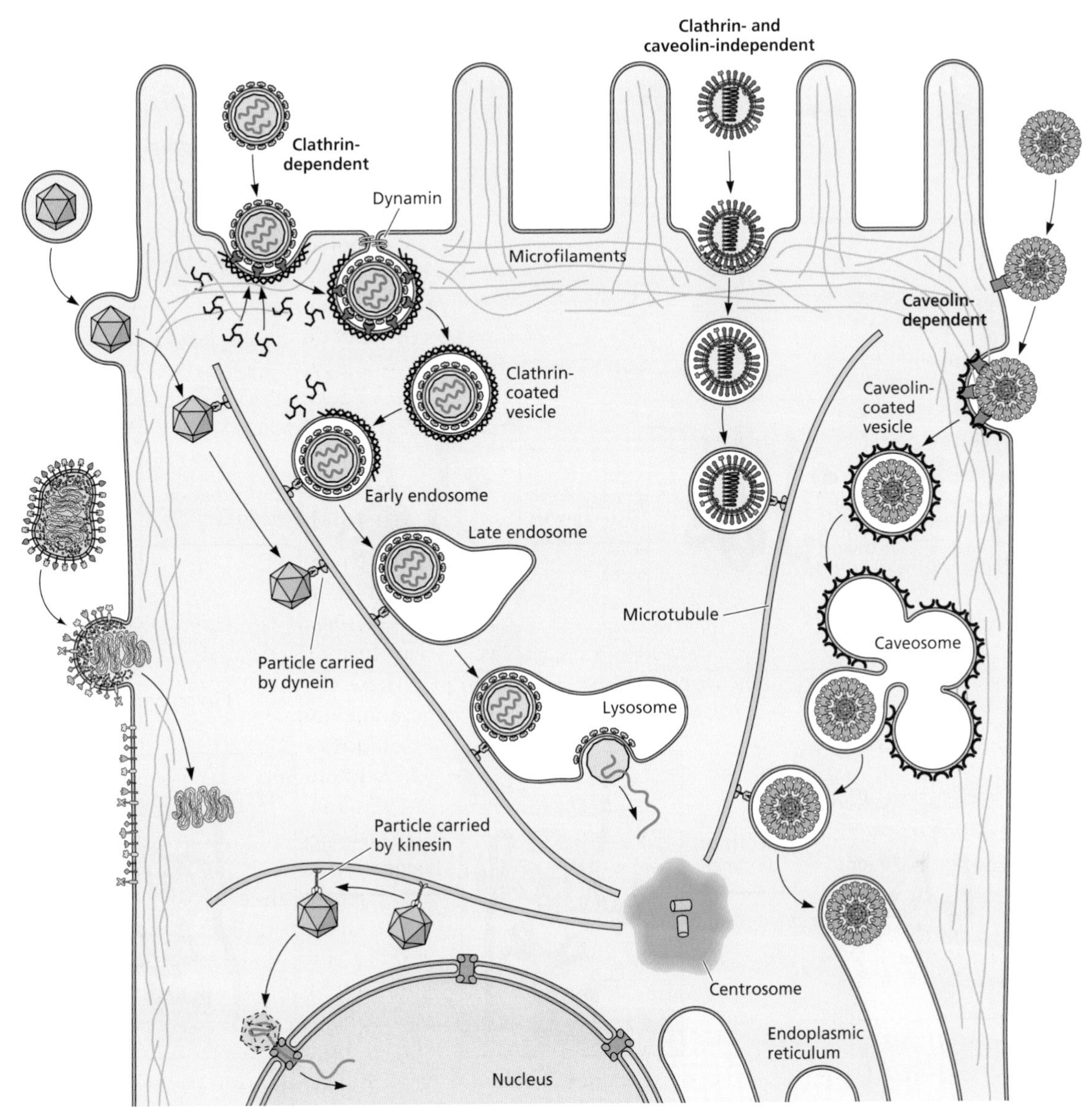

Figure 5.9 Virus entry and movement in cells. Examples of genome uncoating at the plasma membrane are shown on the left side of the cell. Fusion at the plasma membrane releases the nucleocapsid into the cytoplasm. In some cases, the subviral particle is transported on microtubules toward the nucleus, where the nucleic acid is released. Uptake of virions by clathrin-dependent endocytosis commences with binding to a specific cell surface receptor. The ligand-receptor complex diffuses into an invagination of the plasma membrane coated with the protein clathrin on the cytosolic side (clathrin-coated pits). The coated pit further invaginates and pinches off, a process that is facilitated by the GTPase dynamin. The resulting coated vesicle then fuses with an early endosome. Endosomes are acidic, as a result of the activity of vacuolar proton ATPases. Particle uncoating usually occurs from early or late endosomes. Late endosomes then fuse with lysosomes. Virus particles may enter cells by a dynamin- and caveolin-dependent endocytic pathway (right side of the cell). Three types of caveolar endocytosis have been identified. Dynamin 2-dependent endocytosis by caveolin 1-containing **caveolae** is observed in cells infected with simian virus 40 and polyomavirus. Dynamin 2-dependent, noncaveolar, lipid raft-mediated endocytosis occurs during echovirus and rotavirus infection, while dynamin-independent, noncaveolar, raft-mediated endocytosis is also observed during simian virus 40 and polyomavirus infection. This pathway brings virions to the endoplasmic reticulum via the caveosome, a pH-neutral compartment. Clathrin- and caveolin-independent endocytic pathways of viral entry have also been described (center of cell). Movement of endocytic vesicles within cells occurs on microfilaments or microtubules, components of the cytoskeleton. Microfilaments are two-stranded helical polymers of the ATPase actin. They are dispersed throughout the cell but are most highly concentrated beneath the plasma membrane, where they are connected via integrins and other proteins to the extracellular matrix. Transport along microfilaments is accomplished by myosin motors. Microtubules are 25-nm hollow cylinders made of the GTPase tubulin. They radiate from the **centrosome** to the cell periphery. Movement on microtubules is carried out by kinesin and dynein motors.

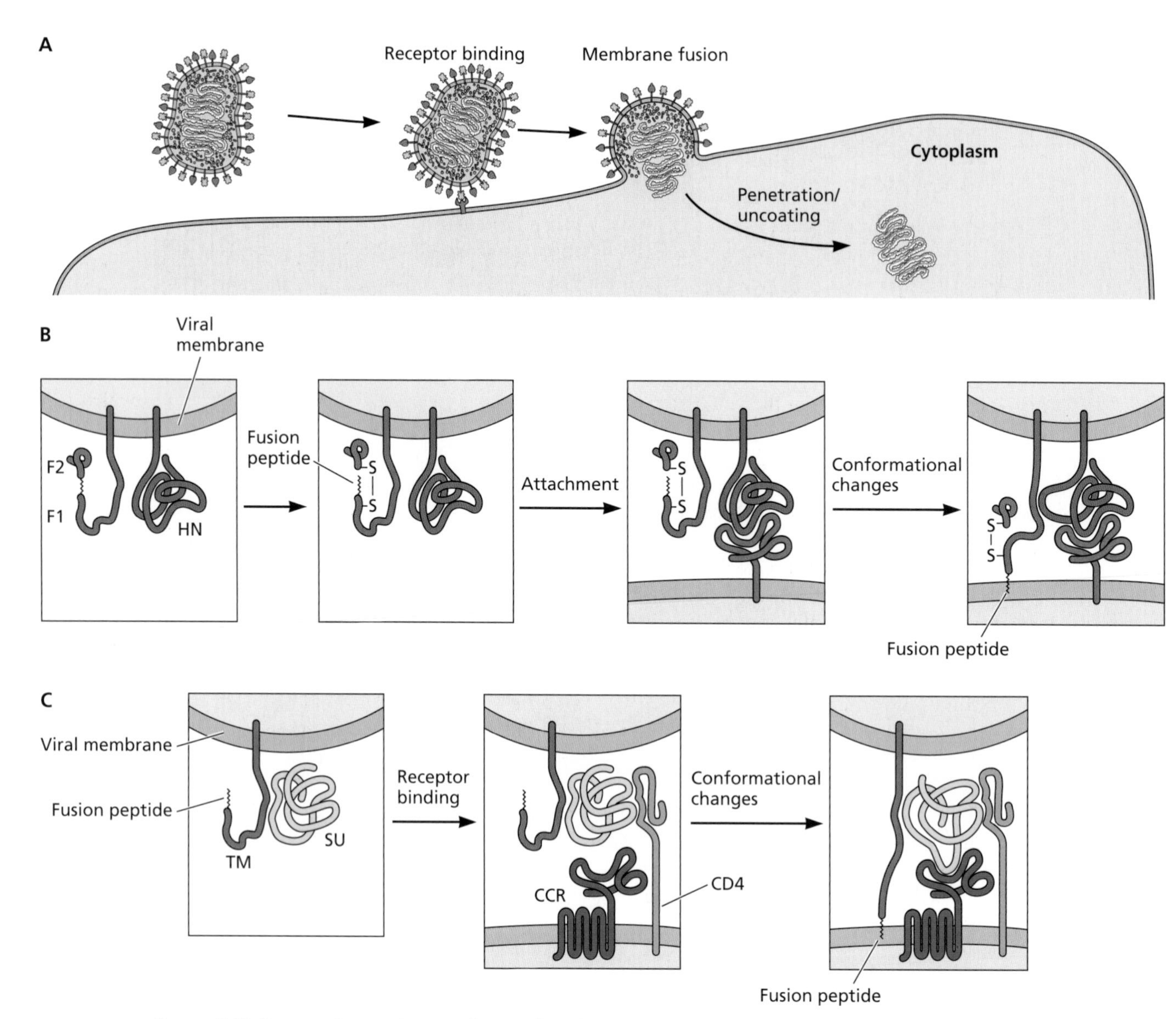

Figure 5.10 Penetration and uncoating at the plasma membrane. (A) Overview. Entry of a member of the *Paramyxoviridae*, which bind to cell surface receptors via the HN, H, or G glycoprotein. The fusion protein (F) then catalyzes membrane fusion at the cell surface at neutral pH. The viral nucleocapsid, as RNP, is released into the cytoplasm, where RNA synthesis begins. **(B)** Model for F-protein-mediated membrane fusion. Binding of HN to the cell receptor (red) induces conformational changes in HN that in turn induce conformational changes in the F protein, moving the fusion peptide from a buried position nearer to the cell membrane. **(C)** Model of the role of chemokine receptors in human immunodeficiency virus type 1 fusion at the plasma membrane. For simplicity, the envelope glycoprotein is shown as a monomer. Binding of SU to CD4 exposes a high-affinity chemokine receptor-binding site on SU. The SU-chemokine receptor interaction leads to conformational changes in TM that expose the fusion peptide and permit it to insert into the cell membrane, catalyzing fusion in a manner similar to that proposed for influenza virus (cf. Fig. 5.12 and 5.13).

coil, and movement of the fusion peptide toward the cell membrane (Fig. 5.10). Such movement of the fusion peptide has been described in atomic detail by comparing structures of the F protein before and after fusion.

The trigger that initiates conformational changes in the F protein is not known. The results of experiments in which hemagglutinin-neuraminidase (HN) and F glycoproteins are synthesized in cultured mammalian cells indicate that the fusion activity of F protein is absent or inefficient if HN is not present. It has therefore been hypothesized that an interaction between HN and F proteins is essential for fusion. It is thought that binding of HN protein to its cellular receptor induces conformational changes, which in turn trigger conformational change in the F protein, exposing the fusion peptide and making the protein fusion competent (Fig. 5.10). The requirement for HN protein in F fusion activity has been observed only with certain paramyxoviruses, including human parainfluenza virus type 3 and mumps virus.

As a result of fusion of the viral and plasma membranes, the viral nucleocapsid, which is a ribonucleoprotein (RNP) consisting of the (−) strand viral RNA genome and the viral proteins L, NP, and P, is released into the cytoplasm (Fig. 5.10). Once in the cytoplasm, the L, NP, and P proteins begin the synthesis of viral messenger RNAs (mRNAs), a process discussed in Chapter 6. Because members of the *Paramyxoviridae* replicate in the cytoplasm, fusion of the viral and plasma membranes achieves uncoating and delivery of the viral genome to this cellular compartment in a single step.

Fusion of human immunodeficiency virus type 1 with the plasma membrane requires participation not only of the cell receptor CD4 but also of an additional cellular protein. These proteins are cell surface receptors for small molecules produced by many cells to attract and stimulate cells of the immune defense system at sites of infection; hence, these small molecules are called **chemotactic cytokines** or **chemokines**. The chemokine receptors on such cells comprise a large family of proteins with seven membrane-spanning domains and are coupled to intracellular signal transduction pathways. There are two major coreceptors for human immunodeficiency virus type 1 infection. CXCr4 (a member of a family of chemokines characterized by having their first two cysteines separated by a single amino acid) appears to be a specific coreceptor for virus strains that infect T cells preferentially. The second is CCr5, a coreceptor for the macrophage-tropic strains of the virus. The chemokines that bind to this receptor activate both T cells and macrophages, and the receptor is found on both types of cell. Individuals who are homozygous for deletions in the CCr5 gene and produce nonfunctional coreceptors have no discernible immune function abnormality, but they appear to be resistant to infection with human immunodeficiency virus type 1. Even heterozygous individuals seem to be somewhat resistant to the virus. Other members of the CC chemokine receptor family (CCr2b and CCr3) were subsequently found to serve as coreceptors for the virus.

Attachment to CD4 appears to create a high-affinity binding site on SU for CCr5. The atomic structure of SU bound to CD4 revealed that binding of CD4 induces conformational changes that expose binding sites for chemokine receptors (Fig. 5.10). Studies of CCr5 have shown that the first N-terminal extracellular domain is crucial for coreceptor function, suggesting that this sequence might interact with SU. An antibody molecule fused to both the CD4 and CCr5 binding sites is being explored as a therapeutic compound to block infection (Box 5.4).

Human immunodeficiency virus type 1 TM mediates envelope fusion with the cell membrane. The high-affinity SU-CCr5 interaction may induce conformational changes in TM to expose the fusion peptide, placing it near the cell membrane, where it can catalyze fusion (Fig. 5.10). Such changes are similar to those that influenza virus HA undergoes upon exposure to low pH. X-ray crystallographic analysis of fusion-active human immunodeficiency virus type 1 TM revealed that its structure is strikingly similar to that of the low-pH fusogenic form of HA (see "Acid-Catalyzed Membrane Fusion" below).

Uncoating during Endocytosis

Many viruses enter cells by the same pathways by which cells take up macromolecules. The plasma membrane, the limiting membrane of the cell, permits nutrient molecules to enter and waste molecules to leave, thereby ensuring an appropriate internal environment. Water, gases, and small hydrophobic molecules such as ethanol can freely traverse the lipid bilayer, but most metabolites and ions cannot. These essential components enter the cell by specific transport processes. Integral membrane proteins are responsible for the transport of ions, sugars, and amino acids, while proteins and large particles are taken into the cell by phagocytosis or endocytosis. The former process (Fig. 5.11) is nonspecific, which means that any particle or molecule can be taken into the cell, and only occurs in specialized cell types such as dendritic cells and macrophages.

Clathrin-Mediated Endocytosis

A wide range of ligands, fluid, membrane proteins, and lipids are selectively taken into cells from the extracellular milieu by **clathrin-mediated endocytosis** (Fig. 5.9 and 5.11), also the mechanism of entry of many viruses. Ligands in the extracellular medium bind to cells via specific plasma membrane receptor proteins. The receptor-ligand assembly diffuses along the membrane until it reaches an invagination that is coated on its cytoplasmic surface by a cage-like lattice composed of the fibrous protein clathrin (Fig. 5.9). Such clathrin-coated pits can comprise as much as 2% of the surface area of a cell, and some receptors are clustered over these areas even in the absence of their ligands. Following the accumulation of receptor-ligand complexes, the clathrin-coated pit invaginates and then pinches off to form a clathrin-coated vesicle. Within a few seconds, the clathrin coat is lost and the vesicles fuse with small, smooth-walled vesicles located near the cell surface, called early **endosomes**. The lumen of early endosomes is mildly acidic (pH 6.5 to 6.0), a result of energy-dependent transport of protons into the interior of the vesicles by a membrane proton pump. The contents of the early endosome are then transported via endosomal carrier vesicles to late endosomes located close to the nucleus. The lumen of late endosomes is more acidic (pH 6.0 to 5.0). Late endosomes in turn

BOX 5.4

EXPERIMENTS

Blocking human immunodeficiency virus infection with two soluble cell receptors

Because viruses must bind to cell surface molecules to initiate replication, the use of soluble receptors to block virus infection has long been an attractive therapeutic option. Soluble CD4 receptors that block infection with human immunodeficiency virus type 1 (HIV-1) have been developed, but these have not been licensed because of their suboptimal potency. A newly designed soluble receptor for HIV-1 overcomes this problem and provides broad and effective protection against infection of cells and of nonhuman primates.

A soluble form of CD4 fused to an antibody molecule can block infection of most HIV-1 isolates and has been shown to be safe in humans, but its affinity for gp120 is low. Furthermore, human immunodeficiency virus can also be spread from cell to cell by fusion, a process that is not blocked by circulating, soluble CD4. Similarly, peptide mimics of the CCR5 coreceptor have been shown to block infection, but their affinity for gp120 is also low.

Combining the two gp120-binding molecules solved the problem of low affinity and in addition provided protection against a wide range of virus isolates. The entry inhibitor, called eCD4-Ig, is a fusion of the first two domains of CD4 to the Fc domain of an antibody molecule, with the CCR5-mimicking peptide at the carboxy terminus (illustrated). It binds strongly to gp120 and blocks infection with many different isolates of HIV-1, HIV-2, simian immunodeficiency virus (SIV), and HIV-1 resistant to broadly neutralizing monoclonal antibodies. The molecule blocks viral infection at concentrations that might be achieved in humans (1.5 to 5.2 micrograms per milliliter).

When administered to mice, eCD4-Ig protected the animals from HIV-1. Rhesus macaques inoculated with an adenovirus-associated virus (AAV) recombinant containing the gene for eCD4-Ig were protected from infection with large quantities of virus for up to 34 weeks after immunization. Concentrations of eCD4-Ig in the sera of these animals ranged from 17 to 77 micrograms per milliliter.

These results show that eCD4-Ig blocks HIV infection with a wide range of isolates more effectively than previously studied broadly neutralizing antibodies. Emergence of HIV variants resistant to neutralization with eCD4-Ig would likely produce viruses that infect cells less efficiently, reducing their transmission. eCD4-Ig is therefore an attractive candidate for therapy of HIV-1 infections. Whether sustained production of the protein in humans will cause disease remains to be determined. Because expression of the AAV genome persists for long periods, it might be advantageous to include a kill-switch in the vector: a way of turning it off if something should go wrong.

Gardner MR, Kattenhorn LM, Kondur HR, von Schaewen M, Dorfman T, Chiang JJ, Haworth KG, Decker JM, Alpert MD, Bailey CC, Neale ES, Jr, Fellinger CH, Joshi VR, Fuchs SP, Martinez-Navio JM, Quinlan BD, Yao AY, Mouquet H, Gorman J, Zhang B, Poignard P, Nussenzweig MC, Burton DR, Kwong PD, Piatak M, Jr, Lifson JD, Gao G, Desrosiers RC, Evans DT, Hahn BH, Ploss A, Cannon PM, Seaman MS, Farzan M. 2015. AAV-expressed eCD4-Ig provides durable protection from multiple SHIV challenges. *Nature* **519:**87–91.

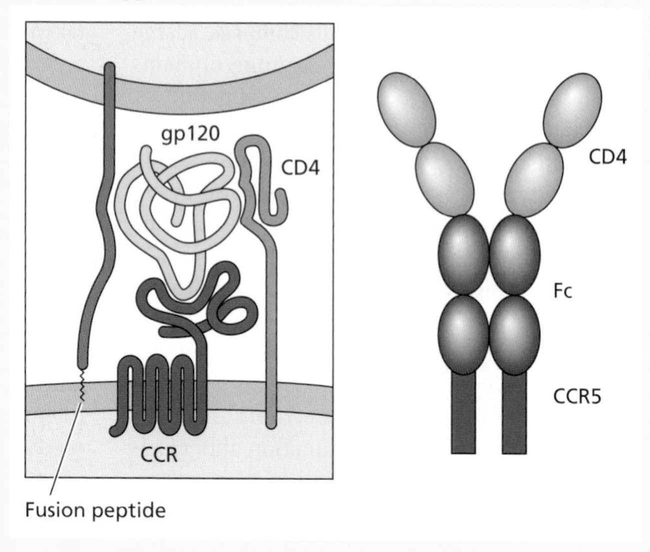

Left, binding of HIV-1 SU (gp120) to CD4 and a chemokine receptor, CCR. Right, illustration of soluble eCD4-Ig. The Fab domains of the antibody molecule are replaced with the first two Ig-like domains of CD4, and the gp120-binding part of CCR5 is added to the C terminus of the Fc domain.

fuse with **lysosomes**, which are vesicles containing a variety of enzymes that degrade sugars, proteins, nucleic acids, and lipids. Viruses with a high pH threshold for fusion, such as vesicular stomatitis virus, enter from early endosomes; most enter the cytoplasm from late endosomes, and a few enter from lysosomes.

Clathrin-mediated endocytosis is a continuous but regulated process. For example, the uptake of vesicular stomatitis virus into cells may be influenced by over 90 different cellular protein kinases. Influenza virus, vesicular stomatitis virus, and reovirus particles are taken into cells, not into preexisting pits but mainly by clathrin-coated pits that form after virus binds to the cell surface. It is not known how virus binding to the plasma membrane induces the formation of the clathrin-coated pit.

Caveolar and Lipid Raft-Mediated Endocytosis

Although uptake of most viruses occurs by the clathrin-mediated endocytic pathway, some viruses enter by caveolin- or raft-mediated endocytosis (Fig. 5.9). The caveolar pathway

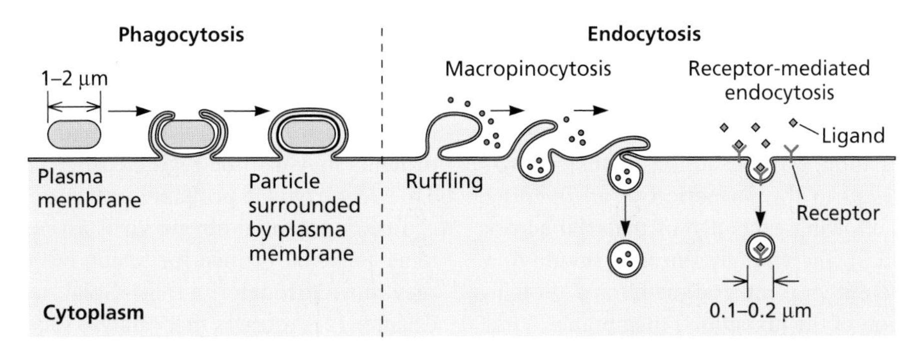

Figure 5.11 Mechanisms for the uptake of macromolecules from extracellular fluid. During phagocytosis, large particles such as bacteria or cell fragments that come in contact with the cell surface are engulfed by extensions of the plasma membrane. Phagosomes ultimately fuse with lysosomes, resulting in degradation of the material within the vesicle. Macrophages use phagocytosis to ingest bacteria and destroy them. Endocytosis comprises the invagination and pinching off of small regions of the plasma membrane, resulting in the nonspecific internalization of molecules (macropinocytosis) or the specific uptake of molecules bound to cell surface receptors (receptor-mediated endocytosis). Macropinocytosis is a mechanism for the uptake of extracellular fluid. It is triggered by ligand binding which initiates formation of plasma membrane ruffling, which traps material in large vacuoles. Adapted from J. Darnell et al., *Molecular Cell Biology* (Scientific American Books, New York, NY, 1986), with permission.

requires cholesterol (a major component of lipid rafts). Caveolae are distinguished from clathrin-coated vesicles by their flask-like shape, their smaller size, the absence of a clathrin coat, and the presence of a marker protein called caveolin. In the uninfected cell, caveolae participate in transcytosis, signal transduction, and uptake of membrane components and extracellular ligands. Binding of a virus particle to the cell surface activates signal transduction pathways required for pinching off of the vesicle, which then moves within the cytoplasm. Disassembly of filamentous actin also occurs, presumably to facilitate movement of the vesicle deeper into the cytoplasm. There it fuses with the **caveosome**, a larger membranous organelle that contains caveolin (Fig. 5.9). In contrast to endosomes, the pH of the caveosome lumen is neutral. Some viruses (e.g., echovirus type 1) penetrate the cytoplasm from the caveosome. Others (simian virus 40, polyomavirus, coxsackievirus B3) are sorted to the endoplasmic reticulum (ER) by a transport vesicle that lacks caveolin. These viruses enter the cytoplasm by a process mediated by thiol oxidases present in the lumen of the endoplasmic reticulum and by a component of the protein degradation pathway present in the membrane.

The study of virus entry by endocytosis can be confusing because some viruses may enter cells by multiple routes, depending on cell type and multiplicity of infection. For example, herpes simplex virus can enter cells by three different routes and influenza A virus may enter cells by both clathrin-dependent and clathrin-independent pathways.

Macropinocytosis

Macropinocytosis is a process by which extracellular fluid is taken into cells via large vacuoles. It is triggered by ligands and dependent on actin and a signaling pathway. It differs from phagocytosis by the signaling pathways needed and can take place in many cell types. This process serves as a pathway of entry for many viruses, including vaccinia virus, herpesviruses, and ebolaviruses. Upon receptor binding, viruses that enter cells via macropinocytosis trigger a signaling cascade that leads to changes in cortical actin and ruffling of the plasma membrane (Fig. 5.11). When these plasma membrane extensions retract, the viruses are brought into macropinosomes and eventually leave these vesicles via membrane fusion.

Membrane Fusion

The membranes of enveloped viruses fuse with those of the cell as a first step in delivery of the viral nucleic acid. Membrane fusion takes place during many other cellular processes, such as cell division, myoblast fusion, and exocytosis.

Membrane fusion must be regulated in order to maintain the integrity of the cell and its intracellular compartments. Consequently, membrane fusion does not occur spontaneously but proceeds by specialized mechanisms mediated by proteins. The two membranes must first come into close proximity. In cells, this reaction is mediated by interactions of integral membrane proteins that protrude from the lipid bilayers, a targeting protein on one membrane and a docking protein on the other. During entry of enveloped viruses, the virus and cell membranes are first brought into close contact by interaction of a viral glycoprotein with a cell receptor. The next step, fusion, requires an even closer approach of the membranes, to within 1.5 nm of each other. This step depends on the removal of water molecules from the membrane surfaces, an energetically unfavorable process. This step is hypothesized to occur when the viral glycoprotein undergoes a structural rearrangement called "hairpinning" (Fig. 5.12).

The precise mechanism by which lipid bilayers fuse is not completely understood, but the action of fusion proteins is thought to result in the formation of an opening called a **fusion pore**, allowing exchange of material across the membranes (Box 5.5). The viral glycoprotein bound to a cell receptor, or a different viral integral membrane protein, then catalyzes the fusion of the juxtaposed membranes. Viral fusion proteins are integral membrane proteins, often glycoproteins, that form homo- or hetero-oligomers.

Virus-mediated fusion must be regulated to prevent viruses from aggregating or to ensure that fusion does not occur in the incorrect cellular compartment. In some cases, fusogenic potential is masked until the fusion protein interacts with other integral membrane proteins. In others, low pH is required to expose fusion domains. The activity of fusion proteins may also be regulated by cleavage of a precursor. This requirement probably prevents premature activation of fusion potential during virus assembly. Viral fusion proteins are often primed for fusion by proteolytic cleavage as they move through the trans-Golgi network as described in Chapter 12. Proteases that catalyze such cleavage are typically furin family convertases that either cleave the fusion proteins directly (orthomyxoviruses, retroviruses, paramyxoviruses) or cleave a protein that masks the fusion protein (alphaviruses, flaviviruses).

Figure 5.12 Influenza virus entry. The globular heads of native HA mediate binding of the virus to sialic acid-containing cell receptors. The virus-receptor complex is endocytosed, and import of H^+ ions into the endosome acidifies the interior. Upon acidification, the viral HA undergoes a conformational rearrangement that produces a fusogenic protein. The loop region of native HA (yellow) becomes a coiled coil, moving the fusion peptides (red) to the top of the molecule near the cell membrane. At the viral membrane, the long α-helix (purple) packs against the trimer core, pulling the globular heads to the side. The long coiled coil bends, or hairpins, bringing the fusion peptides and the transmembrane domains together. This movement moves the cell and viral membranes close together so that fusion can occur. To allow release of vRNP into the cytoplasm, the H^+ ions in the acidic endosome are pumped into the particle interior by the M2 ion channel. As a result, vRNP is primed to dissociate from M1 after fusion of the viral and endosomal membranes. The released vRNPs are imported into the nucleus through the nuclear pore complex via a nuclear localization signal-dependent mechanism (see "Import of Influenza Virus Ribonucleoprotein" below). Adapted from C. M. Carr and P. S. Kim, *Science* **266:**234–236, 1994, with permission.

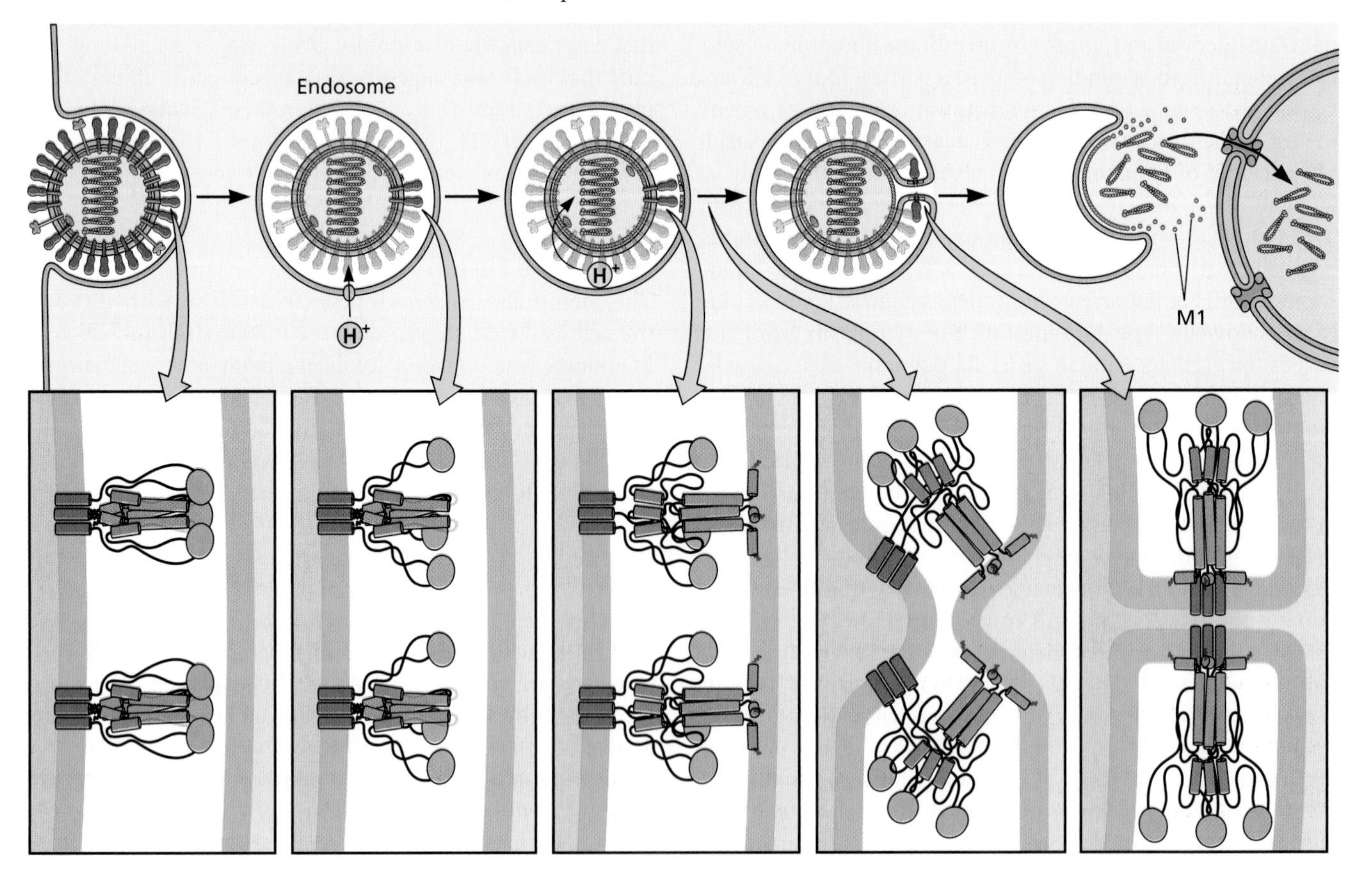

BOX 5.5

EXPERIMENTS

Membrane fusion proceeds through a hemifusion intermediate

Fusion is thought to proceed through a hemifusion intermediate in which the outer leaflets of two opposing bilayers fuse (see figure), followed by fusion of the inner leaflets and the formation of a fusion pore. Direct evidence that fusion proceeds via a hemifusion intermediate has been obtained with influenza virus HA (see figure). **(Left)** Cultured mammalian cells expressing wild-type HA are fused with erythrocytes containing two different types of fluorescent dye, one in the cytoplasm and one in the lipid membrane. Upon exposure to low pH, HA undergoes conformational change and the fusion peptide is inserted into the erythrocyte membrane. The green dye is transferred from the lipid bilayer of the erythrocyte to the bilayer of the cultured cell. The HA trimers tilt, causing reorientation of the transmembrane domain and generating stress within the hemifusion diaphragm. Fusion pore formation relieves the stress. The red dye within the cytoplasm of the erythrocyte is then transferred to the cytoplasm of the cultured cell. **(Right)** An altered form of HA was produced, lacking the transmembrane and cytoplasmic domains and with membrane anchoring provided by linkage to a glycosylphosphatidylinositol (GPI) moiety. Upon exposure to low pH, the HA fusion peptide is inserted into the erythrocyte membrane, and green dye is transferred to the membranes of the mammalian cell. When the HA trimers tilt, no stress is transmitted to the hemifusion diaphragm because no transmembrane domain is present, and the diaphragm becomes larger. Fusion pores do not form, and there is no mixing of the contents of the cytoplasm, indicating that complete membrane fusion has not occurred. These results prove that hemifusion, or fusion of only the inner leaflet of the bilayer, can occur among whole cells. The findings also demonstrate that the transmembrane domain of the HA polypeptide plays a role in the fusion process.

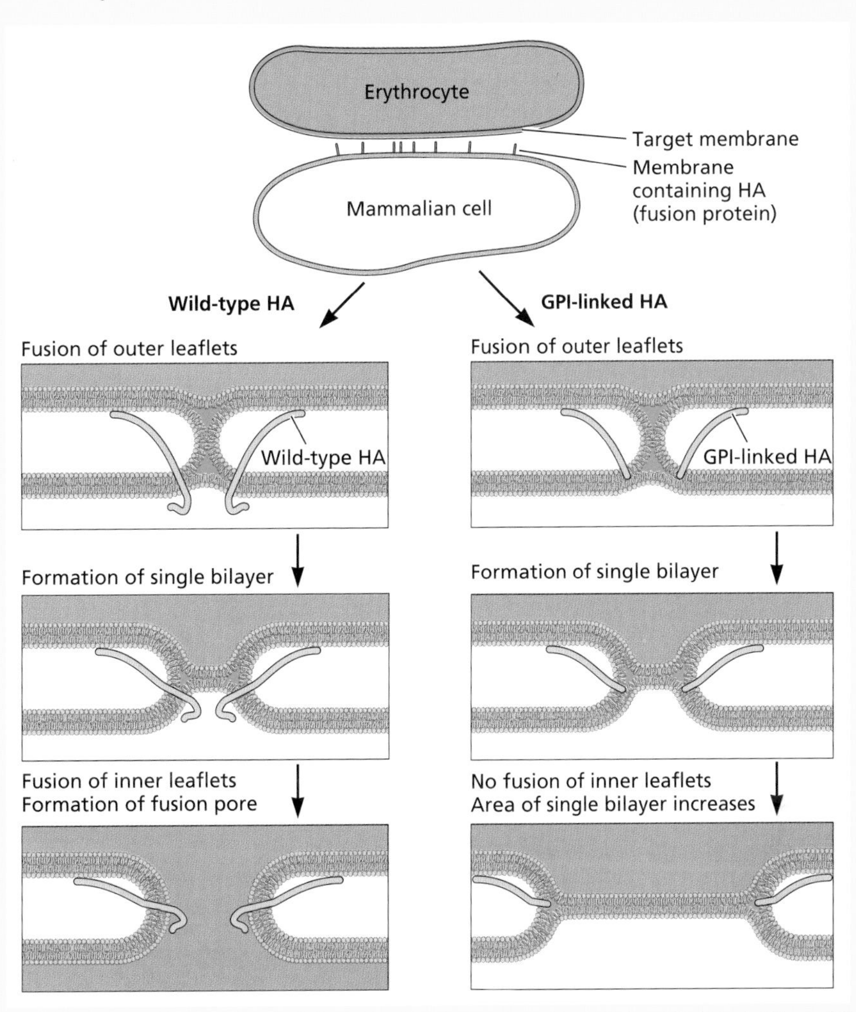

Glycosylphosphatidylinositol-anchored influenza virus HA induces hemifusion. (Left) Model of the steps of fusion mediated by wild-type HA. **(Right)** Effect on fusion by an altered form of HA lacking the transmembrane and cytoplasmic domains. Adapted from G. B. Melikyan et al., *J. Cell Biol* **131:**679–691, 1995, with permission.

Proteolytic cleavage not only is a mechanism to regulate where fusion occurs, but also generates the metastable states of viral glycoproteins that can subsequently undergo the conformational rearrangements required for fusion activity without additional energy. These structural changes expose the hidden fusion peptide so that it can insert into the target membrane and likely provide the energy needed to overcome the hydration force that prevents spontaneous membrane fusion. As a consequence the fusion protein is anchored in both viral and cellular membranes.

Acid-Catalyzed Membrane Fusion

The entry of influenza virus from the endosomal pathway is one of the best-understood viral entry mechanisms. At the cell surface, the virus attaches to sialic acid-containing receptors via the viral HA glycoprotein (Fig. 5.12). The virus-receptor complex is then internalized into the clathrin-dependent receptor-mediated endocytic pathway. When the endosomal pH reaches approximately 5.0, HA undergoes an acid-catalyzed conformational rearrangement, exposing a fusion peptide. The viral and endosomal membranes then fuse, allowing penetration of the viral RNP (vRNP) into the cytoplasm. Because influenza virus particles have a low pH threshold for fusion, uncoating occurs in late endosomes.

The fusion reaction mediated by the influenza virus HA protein is a remarkable event when viewed at atomic resolution (Fig. 5.13). In native HA, the fusion peptide is joined to the three-stranded coiled-coil core by which the HA monomers interact via a 28-amino-acid sequence that forms an extended loop structure buried deep inside the molecule, about 100 Å from the globular head. In contrast, in the low-pH HA structure, this loop region is transformed into a three-stranded coiled coil. In addition, the long α-helices of the coiled coil bend upward and away from the viral membrane. The result is that the fusion peptide is moved a great distance toward the endosomal membrane (Fig. 5.13). Despite these dramatic changes, HA remains trimeric and the globular heads can still bind sialic acid. In this conformation, HA holds the viral and endosome membranes 100 Å apart, too distant for fusion to occur. To bring the viral and cellular membranes closer, it is thought that the HA molecule bends, bringing together the fusion peptide and the transmembrane segment (Fig. 5.12). This movement brings the two membranes close enough to fuse.

In contrast to cleaved HA, the precursor HA0 is stable at low pH and cannot undergo structural changes. Cleavage of the covalent bond between HA1 and HA2 might simply allow movement of the fusion peptide, which is restricted in the uncleaved molecule. Another possibility is suggested by the observation that cleavage of HA is accompanied by movement of the fusion peptide into a cavity in the protein (Fig. 5.13). This movement buries ionizable residues of the fusion peptide, perhaps setting the low-pH "trigger." It should be emphasized that after cleavage, the N terminus of HA2 is tucked into the hydrophobic interior of the trimer (Fig. 5.13). This rearrangement presumably buries the fusion peptide so that newly synthesized virions do not aggregate and lose infectivity.

Viral fusion proteins have been placed into three structural classes called I, II, and III. Common characteristics of all three classes include insertion of a fusion peptide into the target membrane and refolding of the fusion protein so that cell and viral membranes are brought together. In class I fusion proteins (Fig. 5.14), the fusion peptides are presented to membranes on top of a three-stranded α-coiled coil.

The envelope proteins of alphaviruses and flaviviruses exemplify class II viral fusion proteins. Defining characteristics include a three-domain subunit, a fusion loop, and tight association with a second viral protein. Proteolytic cleavage of the second protein converts the fusion protein to a metastable state that can undergo structural rearrangements at low pH to promote fusion. In contrast, the fusion peptide of the influenza virus HA is adjacent to the cleavage point and becomes the N terminus of the mature fusion protein. The envelope proteins of alphaviruses (E1) and flaviviruses (E) do not form coiled coils, as do type I fusion proteins. Rather, they contain predominantly β-barrels that tile the surface of the virus particles as dimers. At low pH they extend toward the endosome membrane, allowing insertion of the fusion peptide (Fig. 5.15).

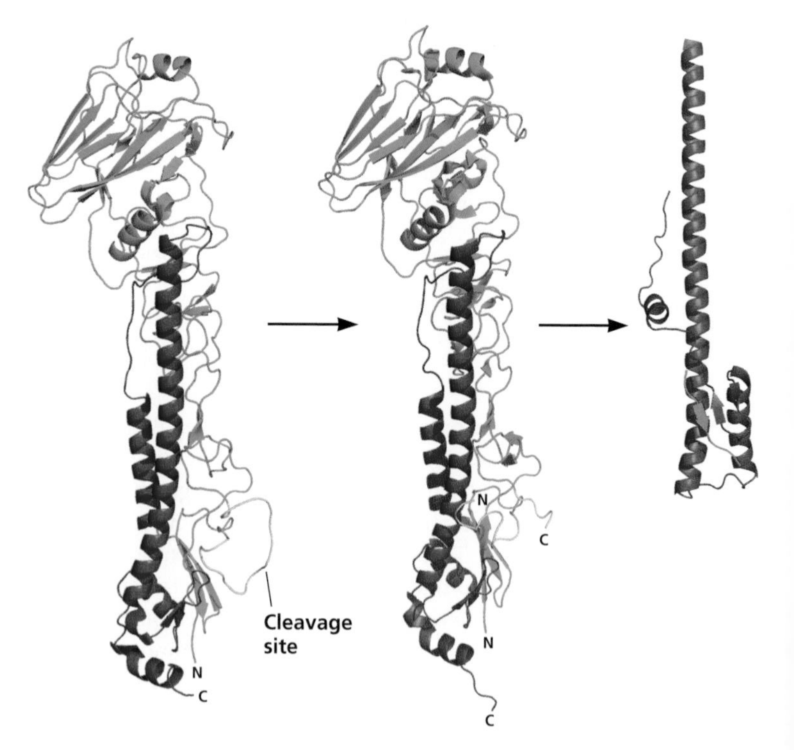

Figure 5.13 Cleavage- and low-pH-induced structural changes in the extracellular domains of influenza virus HA. **(Left)** Structure of the uncleaved HA0 precursor extracellular domain at neutral pH. HA1 subunits are green, HA2 subunits are red, residues 323 of HA1 to 12 of HA2 are yellow, and the locations of some of the N and C termini are indicated. The viral membrane is at the bottom, and the globular heads are at the top. The cleavage site between HA1 and HA2 is in a loop adjacent to a deep cavity. **(Middle)** Structure of the cleaved HA trimer at neutral pH. Cleavage of HA0 generates new N and C termini, which are separated by 20 Å. The N and C termini visible in this model are labeled. The cavity is now filled with residues 1 to 10 of HA2, part of the fusion peptide. **(Right)** Structure of the low-pH trimer. The protein used for crystallization was treated with proteases, and therefore the HA1 subunit and the fusion peptide are not present. This treatment is necessary to prevent aggregation of HA at low pH. At neutral pH the fusion peptide is close to the viral membrane, linked to a short α-helix, and at acidic pH this α-helix is reoriented toward the cell membrane, carrying with it the fusion peptide. The structures are aligned on a central α-helix that is unaffected by the conformational change. Only monomers of all three structures are shown. Adapted from J. Chen et al., *Cell* **95:**409–417, 1998, with permission.

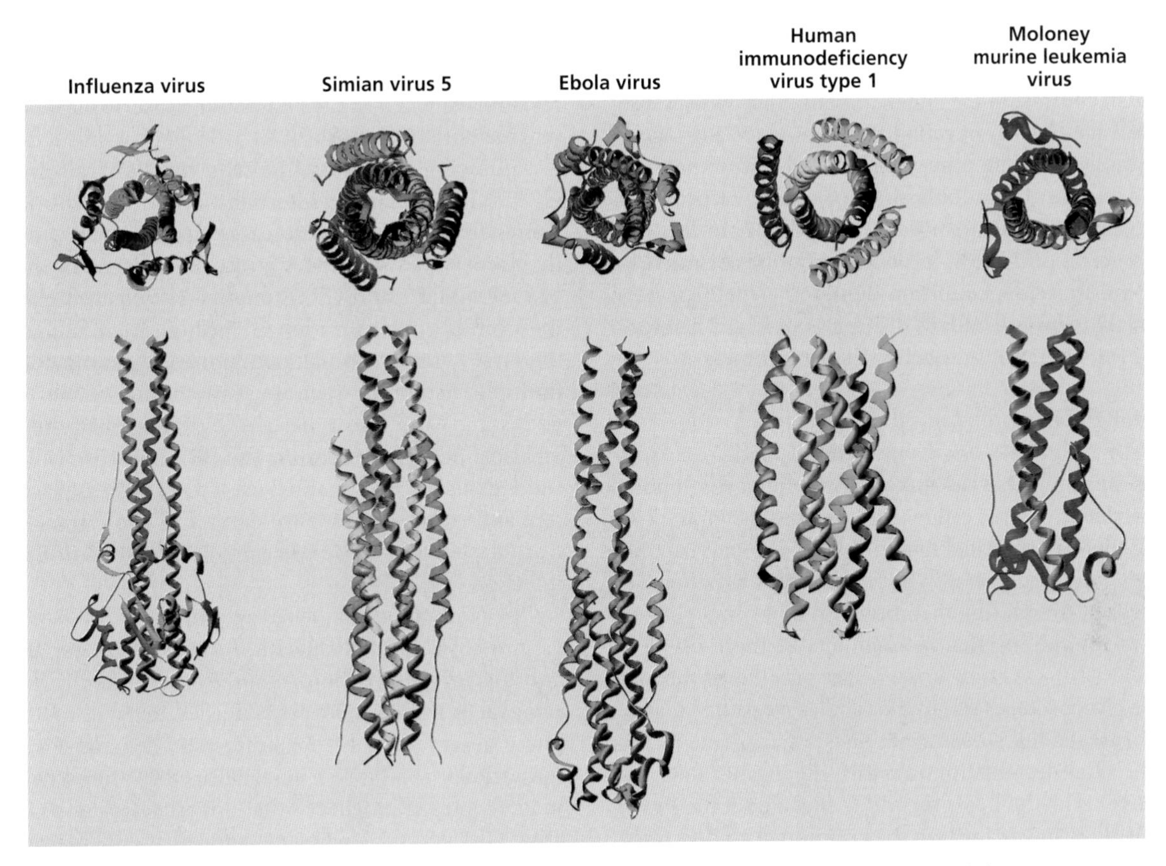

Figure 5.14 Similarities among five viral fusion proteins. **(Top)** View from the top of the structures. **(Bottom)** Side view. The structure shown for HA is the low-pH, or fusogenic, form. The structure of simian virus 5 F protein is of peptides from the N- and C-terminal heptad repeats. Structures of retroviral TM proteins are derived from interacting human immunodeficiency virus type 1 peptides and a peptide from Moloney murine leukemia virus and are presumed to represent the fusogenic forms because of structural similarity to HA. In all three molecules, fusion peptides would be located at the membrane-distal portion (the tops of the molecules in the bottom view). All present fusion peptides to cells on top of a central three-stranded coiled coil supported by C-terminal structures. Adapted from K. A. Baker et al., *Mol. Cell* **3**:309–319, 1999, with permission.

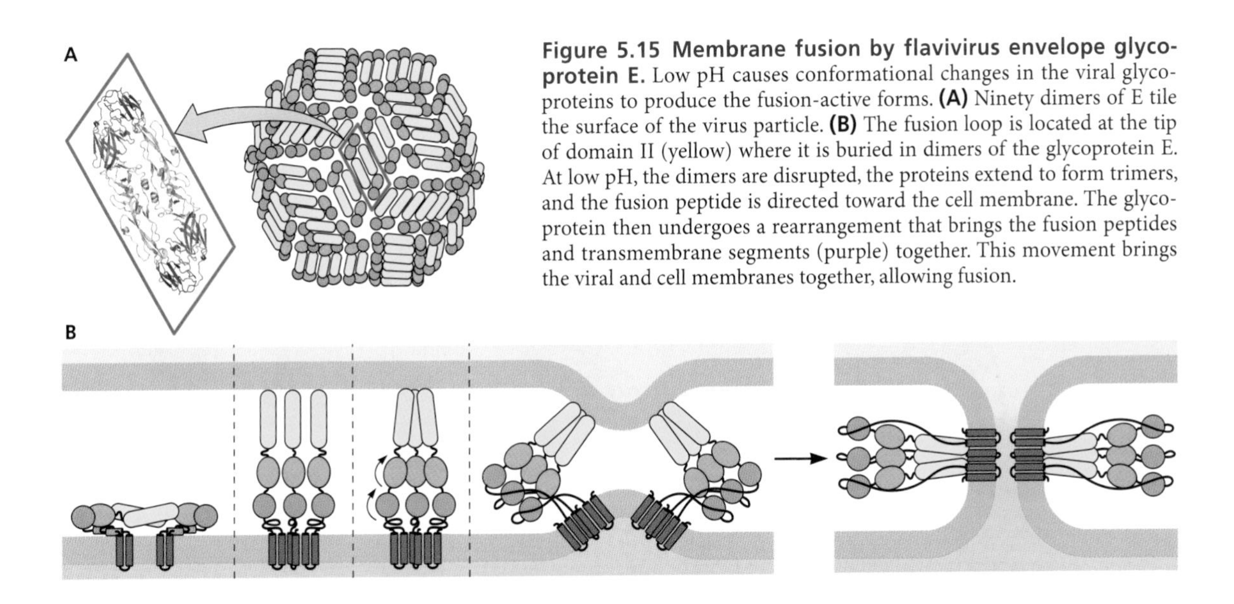

Figure 5.15 Membrane fusion by flavivirus envelope glycoprotein E. Low pH causes conformational changes in the viral glycoproteins to produce the fusion-active forms. **(A)** Ninety dimers of E tile the surface of the virus particle. **(B)** The fusion loop is located at the tip of domain II (yellow) where it is buried in dimers of the glycoprotein E. At low pH, the dimers are disrupted, the proteins extend to form trimers, and the fusion peptide is directed toward the cell membrane. The glycoprotein then undergoes a rearrangement that brings the fusion peptides and transmembrane segments (purple) together. This movement brings the viral and cell membranes together, allowing fusion.

In the alphavirus particle, the second viral protein, E2, acts as a clamp to hold the fusion protein in place; at low pH the clamp is released. In contrast to the situation with other viruses, proteolytic cleavage of E1 is not required to produce a fusogenic protein. However, protein processing controls fusion potential in another way: in the endoplasmic reticulum, E1 protein is associated with the precursor of E2, called p62. In this heterodimeric form, p62-E1, E1 protein cannot be activated for fusion by mildly acidic conditions. Only after p62 has been cleaved to E2 can low pH induce disruption of E1-E2 heterodimers and formation of fusion-active E1 homotrimers.

Receptor Priming for Low-pH Fusion: Two Entry Mechanisms Combined

During the entry of avian leukosis virus into cells, binding of the virus particle to the cell receptor primes the viral fusion protein for low-pH-activated fusion. Avian leukosis virus, like many other retroviruses with simple genomes, was believed to enter cells at the plasma membrane in a pH-independent mechanism resembling that of members of the *Paramyxoviridae* (Fig. 5.10). It is now known that binding of the viral membrane glycoprotein (SU) to the cellular receptors of avian leukosis viruses induces conformational rearrangements that convert the viral protein from a native metastable state that is insensitive to low pH to a second metastable state. In this state, exposure to low pH within the endosomal compartment leads to membrane fusion and release of the viral capsid.

An Endosomal Fusion Receptor

The study of Ebolavirus entry into cells has revealed a new mechanism in which the viral fusion protein binds to a specific fusion receptor in the endosome membrane. Previously, all known cases of fusion catalyzed by viral glycoproteins have taken place when the fusion peptide inserts into the endosomal membrane by virtue of its hydrophobic properties.

Following attachment to cells via the viral glycoprotein GP, viral particles are internalized and move to late endosomes. There, cysteine proteases cleave GP to remove heavily glycosylated sequences at the C terminus of the protein, producing GP1 and GP2 subunits. The cleaved glycoprotein then binds Niemann-Pick C1 protein, catalyzing fusion of the viral and endosomal membranes. Niemann-Pick C1 is a multiple membrane-spanning protein that resides in the late endosomes and lysosomes and participates in the transport of lysosomal cholesterol to the ER and other cellular sites. Individuals with Niemann-Pick type C1 disease lack the protein and consequently have defects in cholesterol transport; fibroblasts from these patients are resistant to Ebola virus infection (Fig. 5.16).

The binding site on the viral glycoprotein for Niemann-Pick C1 protein is located beneath the heavily glycosylated mucin and glycan cap of the protein, explaining why proteolytic removal of these sequences is needed for binding of viral GP. These observations demonstrate that Niemann-Pick C1 is an intracellular receptor for Ebola virus that promotes a late step in viral entry. It is believed that this receptor assists in dissociating GP1 and GP2, allowing conformational rearrangements needed for membrane fusion.

Release of Viral Ribonucleoprotein

The genomes of many enveloped RNA viruses are present as vRNP in the virus particle. One mechanism for release of

Figure 5.16 Entry of Ebolavirus into cells. Virus particles bind cells via an unidentified attachment receptor and enter by endocytosis. The mucin and glycan cap on the viral glycoprotein is removed by cellular cysteine proteases, exposing binding sites for NPC1. The latter is required for fusion of the viral and cell membranes, releasing the nucleocapsid into the cytoplasm. Courtesy of Kartik Chandran, Albert Einstein College of Medicine.

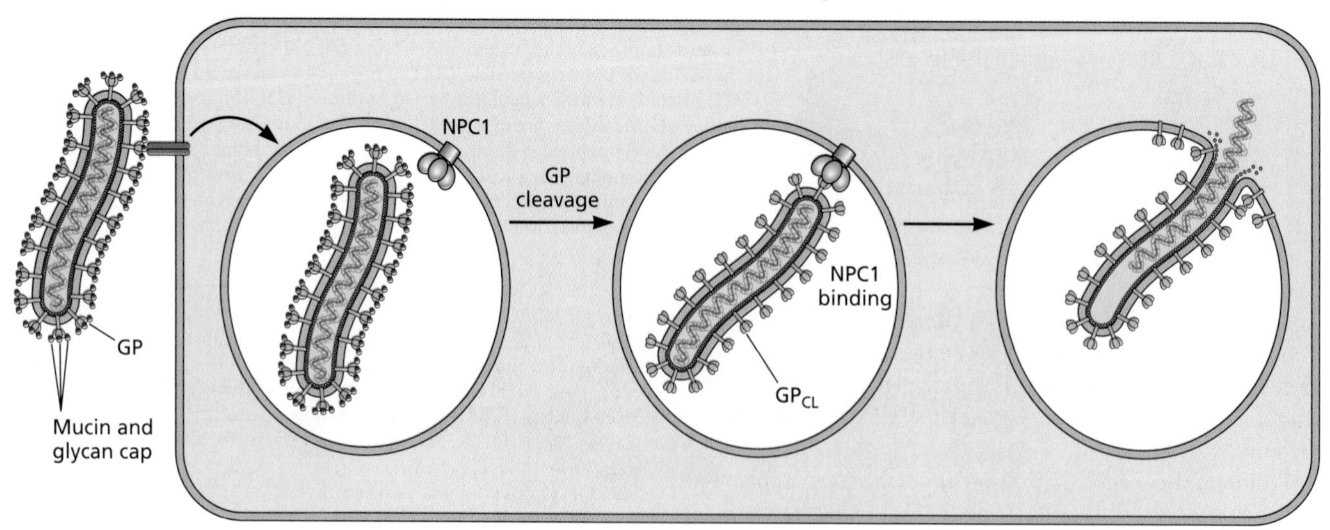

vRNP during virus entry has been identified by studies of influenza virus. Each influenza virus vRNP is composed of a segment of the RNA genome bound by nucleoprotein (NP) molecules and the viral RNA polymerase. This structure interacts with viral M1 protein, an abundant protein in virus particles that underlies the envelope and provides rigidity (Fig. 5.12). The M1 protein also contacts the internal tails of HA and neuraminidase proteins in the viral envelope. This arrangement presents two problems. Unless M1-vRNP interactions are disrupted, vRNPs might not be released into the cytoplasm. Furthermore, the vRNPs must enter the nucleus, where mRNA synthesis takes place. However, vRNP cannot enter the nucleus if M1 protein remains bound, because this protein masks a nuclear localization signal (see "Import of Influenza Virus Ribonucleoprotein" below).

The influenza virus M2 protein, the first viral protein discovered to be an ion channel, provides the solution to both problems. The envelope of the virus particle contains a small number of molecules of M2 protein, which form a homotetramer. When purified M2 was reconstituted into synthetic lipid bilayers, ion channel activity was observed, indicating that this property requires only M2 protein. The M2 protein channel is structurally much simpler than other ion channels and is the smallest channel discovered to date.

The M2 ion channel is activated by the low pH of the endosome before HA-catalyzed membrane fusion occurs. As a result, protons enter the interior of the virus particle. It has been suggested that the reduced pH of the particle interior leads to conformational changes in the M1 protein, thereby disrupting M1-vRNP interactions. When fusion between the viral envelope and the endosomal membrane takes place, vRNPs are released into the cytoplasm free of M1 and can then be imported into the nucleus (Fig. 5.12). Support for this model comes from studies with the anti-influenza virus drug amantadine, which specifically inhibits M2 ion channel activity (Volume II, Fig. 9.11). In the presence of this drug, influenza virus particles can bind to cells, enter endosomes, and undergo HA-mediated membrane fusion, but vRNPs are not released from endosomes.

Uncoating in the Cytoplasm by Ribosomes

Some enveloped RNA-containing viruses, such as Semliki Forest virus, contain nucleocapsids that are disassembled in the cytoplasm by pH-independent mechanisms. The icosahedral nucleocapsid of this virus is composed of a single viral protein, C protein, which encloses the (+) strand viral RNA. This structure is surrounded by an envelope containing viral glycoproteins E1 and E2, which are arranged as heterodimers clustered into groups of three, each cluster forming a spike on the virus surface.

Fusion of the viral and endosomal membranes exposes the nucleocapsid to the cytoplasm (Fig. 5.17). To begin translation

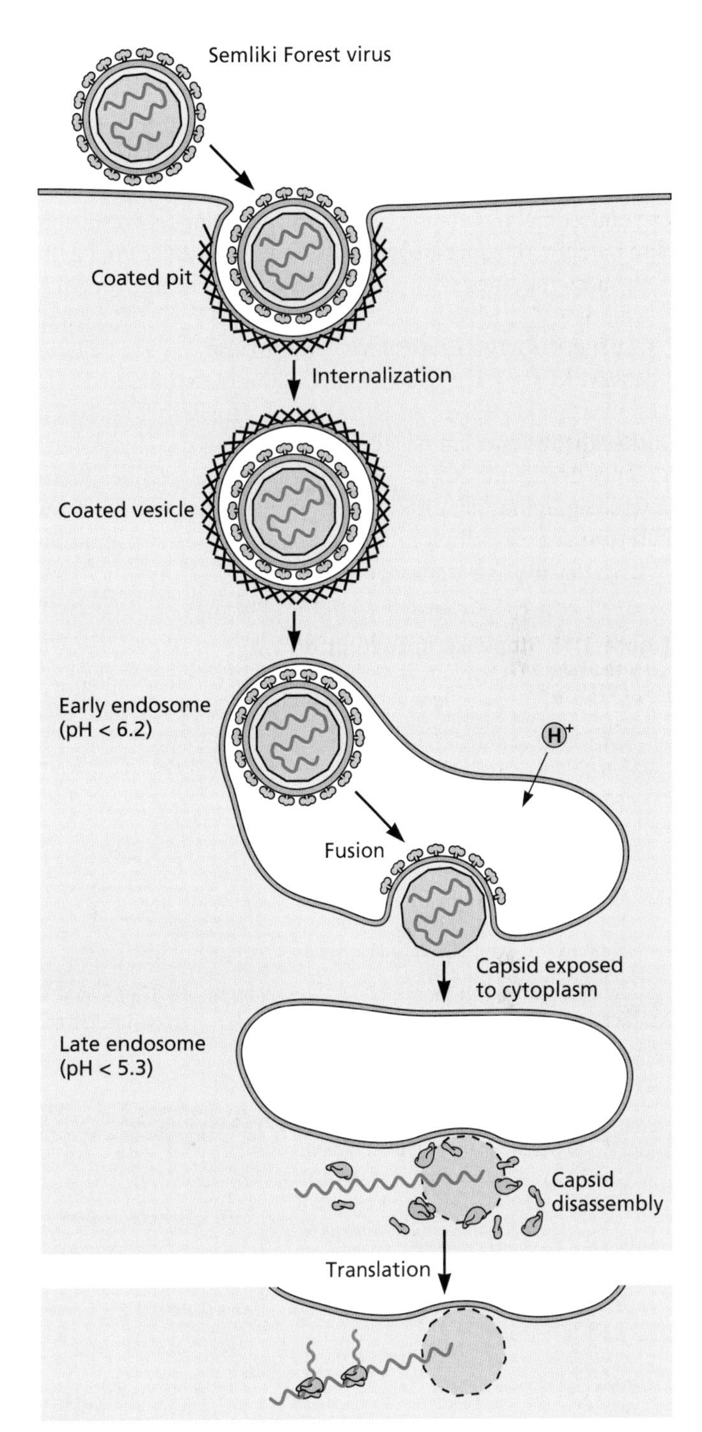

Figure 5.17 Entry of Semliki Forest virus into cells. Semliki Forest virus enters cells by clathrin-dependent receptor-mediated endocytosis, and membrane fusion is catalyzed by acidification of endosomes. Fusion results in exposure of the viral nucleocapsid to the cytoplasm, although the nucleocapsid remains attached to the cytosolic side of the endosome membrane. Cellular ribosomes then bind the capsid, disassembling it and distributing the capsid protein throughout the cytoplasm. The viral RNA is then accessible to ribosomes, which initiate translation. Adapted from M. Marsh and A. Helenius, *Adv. Virus Res.* **36:**107–151, 1989, with permission.

of (+) strand viral RNA, the nucleocapsid must be disassembled, a process mediated by an abundant cellular component—the ribosome. Each ribosome binds three to six molecules of C protein, disrupting the nucleocapsid. This process occurs while the nucleocapsid is attached to the cytoplasmic side of the endosomal membrane (Fig. 5.17) and ultimately results in disassembly. The uncoated viral RNA remains associated with cellular membranes, where translation and replication begin.

Disrupting the Endosomal Membrane

Adenoviruses are composed of a double-stranded DNA genome packaged in an icosahedral capsid (Chapter 4). Internalization of most adenovirus serotypes by receptor-mediated endocytosis requires attachment of fiber to an integrin or Ig-like cell surface receptor and binding of the penton base to a second cell receptor, the cellular vitronectin-binding integrins $\alpha_v\beta_3$ and $\alpha_v\beta_5$. Attachment is mediated by amino acid sequences in each of the five subunits of the adenovirus penton base that mimic the normal ligands of cell surface integrins. As the virus particle is transported via the endosomes from the cell surface toward the nuclear membrane, it undergoes multiple uncoating steps as structural proteins are removed sequentially (Fig. 5.18). As the endosome becomes acidified, the viral capsid is destabilized, leading to release of proteins from the capsid. Among these is protein VI, which causes disruption of the endosomal membrane, thereby delivering the remainder of the particle into the cytoplasm. An N-terminal amphipathic α-helix of protein VI is probably responsible for its pH-dependent membrane disruption activity. This region of the protein appears to be masked in the native capsid by the hexon protein. The liberated subviral particle then docks onto the nuclear pore complex (see "Import of DNA Genomes" below).

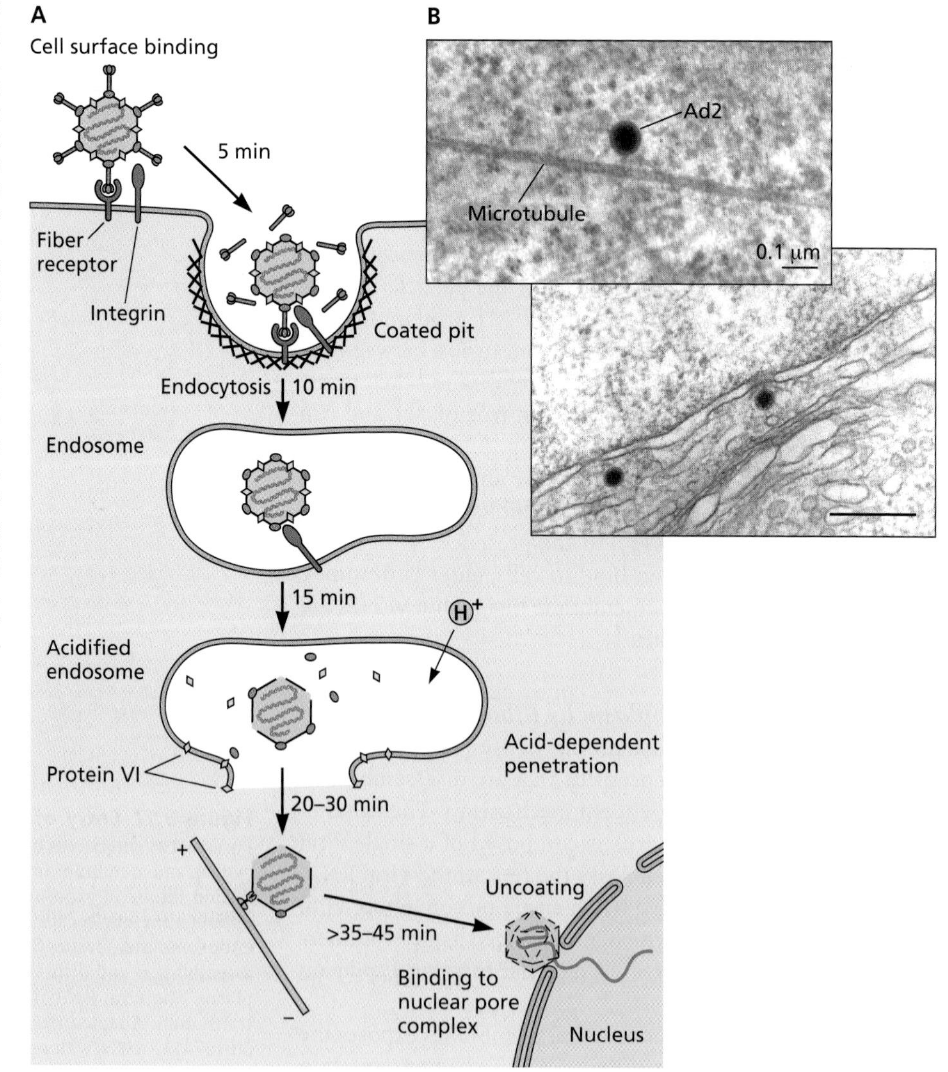

Figure 5.18 Stepwise uncoating of adenovirus. (A) Adenoviruses bind the cell receptor via the fiber protein. Interaction of the penton base with an integrin receptor leads to internalization by endocytosis. Fibers are released from the capsid during uptake. Low pH in the endosome causes destabilization of the capsid and release of protein VI (yellow diamonds). The hydrophobic N terminus of protein VI disrupts the endosome membrane, leading to release of a subviral particle into the cytoplasm. The capsid is transported in the cytoplasm along microtubules and docks onto the nuclear pore complex. **(B)** Electron micrograph of adenovirus type 2 particles bound to a microtubule (top) and bound to the cytoplasmic face of the nuclear pore complex (bottom). Bar in bottom panel, 200 nm. (A) Data from U. F. Greber et al., *Cell* **75**:477–486, 1993, and L. C. Trotman et al., *Nat. Cell Biol.* **3**:1092–1100, 2001. (B) Reprinted from U. F. Greber et al., *Trends Microbiol.* **2**:52–56, 1994, with permission. Courtesy of Ari Helenius, Urs Greber, and Paul Webster, University of Zurich.

Forming a Pore in the Endosomal Membrane

The genomes of nonenveloped picornaviruses are transferred across the cell membrane by a different mechanism, as determined by structural information at the atomic level and complementary genetic and biochemical data obtained from studies of cell entry. The interaction of poliovirus with its Ig-like cell receptor, CD155, leads to major conformational rearrangements in the virus particle and the production of an expanded form called an altered (A) particle (Fig. 5.19A). Portions of two capsid proteins, VP1 and VP4, move from the inner surface of the capsid to the exterior. These polypeptides are thought to form a pore in the cell membrane that allows transport of viral RNA into the cytoplasm (Fig. 5.19B). In support of this model, ion channel activity can be detected when A particles are added to lipid bilayers.

The properties of a virus with an amino acid change in VP4 indicate that this protein is required for an early stage of cell entry. Virus particles with such amino acid alterations can bind to target cells and convert to A particles but are blocked at a subsequent, unidentified step. During poliovirus assembly, VP4 and VP2 are part of the precursor VP0, which remains uncleaved until the viral RNA has been encapsidated. The cleavage of VP0 during poliovirus assembly therefore primes the capsid for uncoating by separating VP4 from VP2.

In cells in culture, release of the poliovirus genome occurs from within early endosomes located close (within 100 to 200 nm) to the plasma membrane (Fig. 5.19A). Uncoating is dependent upon actin and tyrosine kinases, possibly for movement of the capsid through the network of actin filaments, but not on dynamin, clathrin, caveolin, or flotillin (a marker protein for clathrin- and caveolin-independent endocytosis), endosome acidification, or microtubules. The trigger for RNA release from early endosomes is not known but is clearly dependent on prior interaction with CD155. This conclusion derives from the finding that antibody-poliovirus complexes can bind to cells that produce Fc receptors but cannot infect them. As the Fc receptor is known to be endocytosed, these results suggest that interaction of poliovirus with CD155 is required to induce conformational changes in the particle that are required for uncoating.

A critical regulator of the receptor-induced structural transitions of poliovirus particles appears to be a hydrophobic tunnel located below the surface of each structural unit (Fig. 5.19C). The tunnel opens at the base of the canyon and extends toward the 5-fold axis of symmetry. In poliovirus type 1, each tunnel is occupied by a molecule of sphingosine. Similar lipids have been observed in the capsids of other picornaviruses. Because of the symmetry of the capsid, each virus particle may contain up to 60 lipid molecules.

The lipids are thought to contribute to the stability of the native virus particle by locking the capsid in a stable conformation. Consequently, removal of the lipid is probably necessary to endow the particle with sufficient flexibility to permit the RNA to leave the shell. These conclusions come from the study of antiviral drugs known as WIN compounds (named after Sterling-Winthrop, the pharmaceutical company at which they were discovered). These compounds displace the lipid and fit tightly in the hydrophobic tunnel (Fig. 5.19C). Polioviruses containing bound WIN compounds can bind to the cell receptor, but A particles are not produced. WIN compounds may therefore inhibit poliovirus infectivity by preventing the receptor-mediated conformational alterations required for uncoating. The properties of poliovirus mutants that cannot replicate in the absence of WIN compounds underscore the role of the lipids in uncoating. These drug-dependent mutants spontaneously convert to altered particles at 37°C, in the absence of the cell receptor, probably because they do not contain lipid in the hydrophobic pocket. The lipids are therefore viewed as switches, because their presence or absence determines whether the virus is stable or will be uncoated. The interaction of the virus particle with its receptor probably initiates structural changes in the virion that lead to the release of lipid. Consistent with this hypothesis is the observation that CD155 docks onto the poliovirus capsid just above the hydrophobic pocket.

It is usually assumed that the 5′-end of (+) strand RNAs is the first to leave the capsid, to allow immediate translation by ribosomes. This assumption is incorrect for rhinovirus type 2: exit of viral RNA starts from the 3′-end. This directionality is a consequence of how the viral RNA is packaged in the virion, with the 3′-end near the location of pore formation in the altered particle. Whether such directionality is a general feature of nonenveloped (+) strand RNA viruses is unknown.

Uncoating in the Lysosome

Most virus particles that enter cells by receptor-mediated endocytosis leave the pathway before the vesicles reach the lysosomal compartment. This departure is not surprising, for lysosomes contain proteases and nucleases that would degrade virus particles. However, these enzymes play an important role during the uncoating of members of the *Reoviridae.*

Orthoreoviruses are naked icosahedral viruses containing a double-stranded RNA genome of 10 segments. The viral capsid is a double-shelled structure composed of eight different proteins. These virus particles bind to cell receptors via protein σ1 and are internalized into cells by endocytosis (Fig. 5.20A). Infection of cells by reoviruses is sensitive to bafilomycin A1, an inhibitor of the endosomal proton pump, indicating that acidification is required for entry. Low pH activates lysosomal proteases, which then modify several capsid proteins, enabling the virus to cross the vesicle membrane. One viral outer capsid protein is cleaved and another is removed from the particle, producing an infectious subviral particle. These particles have the viral μ1 protein, a myristoylated protein

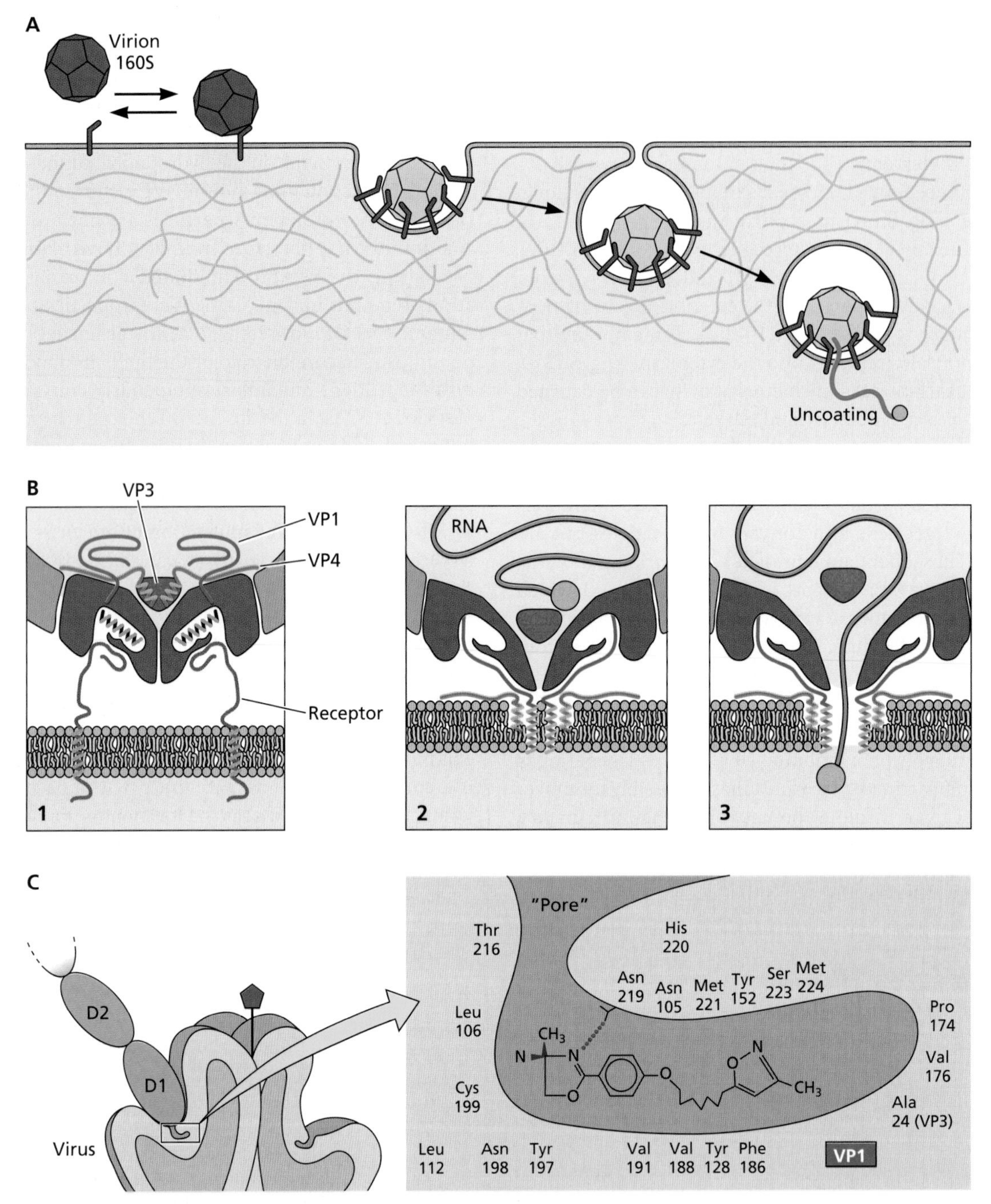

Figure 5.19 Model for poliovirus entry into cells. (A) Overview. The native virion (160S) binds to its cell receptor, CD155, and undergoes a receptor-mediated conformational transition resulting in the formation of altered (A) particles. The viral RNA, shown as a curved green line, leaves the capsid from within early endosomes close to the plasma membrane. **(B)** Model of the formation of a pore in the cell membrane after poliovirus binding. 1, poliovirus (shown in cross section, with capsid proteins purple) binds to CD155 (brown). 2, a conformational change leads to displacement of the pocket lipid (black). The pocket may be occupied by sphingosine in the capsid of poliovirus type 1. The hydrophobic N termini of VP1 (blue) are extruded and insert into the plasma membrane. 3, a pore is formed in the membrane by VP4 and the VP1 N termini, through which the RNA is released from the capsid into the cytosol. **(C)** Schematic diagram of the hydrophobic pocket below the canyon floor. Inset shows a WIN compound in the hydrophobic pocket. Adapted from J. M. Hogle and V. R. Racaniello, p. 71–83, *in* B. L. Semler and E. Wimmer (ed.), *Molecular Biology of Picornaviruses* (ASM Press, Washington, DC, 2002).

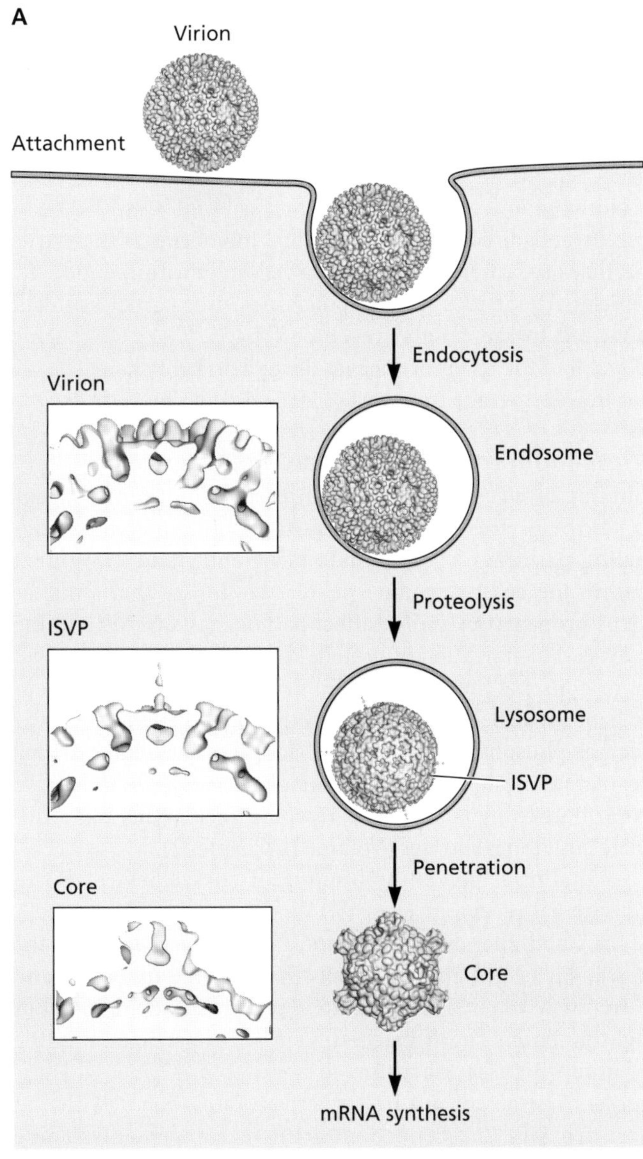

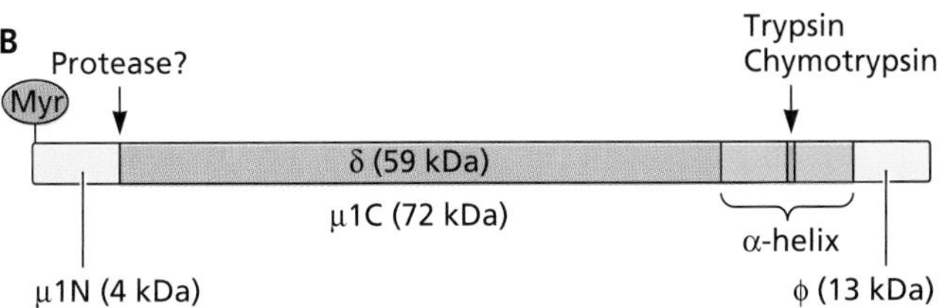

Figure 5.20 Entry of reovirus into cells. (A) The different stages in cell entry of reovirus. After the attachment of $\sigma 1$ protein to the cell receptor, the virus particle enters the cell by receptor-mediated endocytosis. Proteolysis in the late endosome produces the infectious subviral particle (ISVP), which may then cross the lysosomal membrane and enter the cytoplasm as a core particle. The intact virion is composed of two concentric, icosahedrally organized protein capsids. The outer capsid is made up largely of $\sigma 3$ and $\mu 1$. The dense core shell is formed mainly by $\lambda 1$ and $\sigma 2$. In the ISVP, 600 $\sigma 3$ subunits have been released by proteolysis, and the $\sigma 1$ protein changes from a compact form to an extended flexible fiber. The $\mu 1$ protein, which is thought to mediate interaction of the ISVP with membranes, is present as two cleaved fragments, μ1N and μ1C (see schematic of $\mu 1$ in panel B). The N terminus of μ1N is modified with myristate, suggesting that the protein functions in the penetration of membranes. A pair of amphipathic α-helices flank a C-terminal trypsin/chymotrypsin cleavage site at which μ1C is cleaved by lysosomal proteases. Such cleavage may release the helices to facilitate membrane penetration. The membrane-penetrating potential of μ1C in the virion may be masked by $\sigma 3$; release of the $\sigma 3$ in ISVPs might then allow μ1C to interact with membranes. The core is produced by the release of 12 $\sigma 1$ fibers and 600 $\mu 1$ subunits. In the transition from ISVP to core, domains of $\lambda 2$ rotate upward and outward to form a turretlike structure. (Insets) Close-up views of the emerging turretlike structure as the virus progresses through the ISVP and core stages. This structure may facilitate the entry of nucleotides into the core and the exit of newly synthesized viral mRNAs. **(B)** Schematic of the $\mu 1$ protein, showing locations of myristate, protease cleavage sites, and amphipathic α-helices. Virus images reprinted from K. A. Dryden et al., *J. Cell Biol.* **122:**1023–1041, 1993, with permission. Courtesy of Norm Olson and Tim Baker, Purdue University.

that interacts with membranes, on the surface. Consequently, subviral particles penetrate the lysosomal membrane and escape into the cytosol. Isolated infectious subviral particles cause cell membranes to become permeable to toxins and produce pores in artificial membranes. These particles can initiate an infection by penetrating the plasma membrane, entering the cytoplasm directly. Their infectivity is not sensitive to bafilomycin A1, further supporting the idea that these particles are primed for membrane entry and do not require further acidification for this process. The core particles generated from infectious subviral particles after penetration into the cytoplasm carry out viral mRNA synthesis.

Movement of Viral and Subviral Particles within Cells

Viral and subviral particles move within the host cell during entry and egress (Chapters 12 and 13). However, movement of molecules larger than 500 kDa does not occur by passive diffusion, because the cytoplasm is crowded with organelles, high concentrations of proteins, and the cytoskeleton. Rather, viral particles and their components are transported via the actin and microtubule cytoskeletons. Such movement can be visualized in live cells by using fluorescently labeled viral proteins (Chapter 2).

The cytoskeleton is a dynamic network of protein filaments that extends throughout the cytoplasm. It is composed of three types of filament—microtubules, intermediate filaments, and microfilaments. Microtubules are organized in a polarized manner, with minus ends situated at the microtubule-organizing center near the nucleus, and plus ends located at the cell periphery. This arrangement permits directed movement of cellular and viral components over long distances. Actin filaments (microfilaments) typically assist in virus movement close to the plasma membrane (Fig. 5.9).

Transport along actin filaments is accomplished by myosin motors, and movement on microtubules is carried out by kinesin and dynein motors. Hydrolysis of adenosine triphosphate (ATP) provides the energy for the motors to move their cargo along cytoskeletal tracks. Dyneins and kinesins participate in movement of viral components during both entry and egress (Chapters 12 and 13). In some cases, the actin cytoskeleton is remodeled during these processes, for example, when viruses bud from the plasma membrane.

There are two basic ways for viral or subviral particles to travel within the cell—within a membrane vesicle such as an endosome, which interacts with the cytoskeletal transport machinery, or directly in the cytoplasm (Fig. 5.9). In the latter case, some form of the virus particle must bind directly to the transport machinery. After leaving endosomes, the subviral particles derived from adenoviruses and parvoviruses are transported along microtubules to the nucleus. Although adenovirus particles have an overall net movement toward the nucleus, they exhibit bidirectional plus- and minus-end-directed microtubule movement. Adenovirus binding to cells activates two different signal transduction pathways that increase the net velocity of minus-end-directed motility. The signaling pathways are therefore required for efficient delivery of the viral genome to the nucleus. Adenovirus subviral particles are loaded onto microtubules through interaction of the capsid protein, hexon, with dynein. The particles move towards the **centrosome** and are then released and dock onto the nuclear pore complex, prior to viral genome entry into the nucleus.

Some virus particles move along the surfaces of cells prior to entry until a clathrin-coated pit is encountered. If the cell receptor is rare or inaccessible, particles may first bind to more abundant or accessible receptors, such as carbohydrates, and then migrate to receptors that allow entry into the cell. For example, after binding, polyomavirus particles move laterally ("surf") on the plasma membrane for 5 to 10 s before they are internalized. They can be visualized moving along the plasma membrane toward the cell body on **filopodia**, thin extensions of the plasma membrane (Fig. 5.11). Movement along filopodia occurs by an actin-dependent mechanism. Filopodial bridges mediate cell-to-cell spread of a retrovirus in cells in culture. The filopodia originate from uninfected cells and contact infected cells with their tips. The interaction of the viral envelope glycoprotein on the surface of infected cells with the receptor on uninfected cells stabilizes the interaction. Particles move along the outside of the filipodial bridge to the uninfected cell. Such transport is a consequence of actin-based movement of the viral receptor toward the uninfected cell.

A number of different viruses enter the peripheral nervous system and spread to the central nervous system via axons. As no viral genome encodes the molecular motors or cytoskeletal structures needed for long-distance axonal transport, viral adapter proteins are required to allow movement within nerves. An example is axonal transport of alphaherpesviruses. After fusion at the plasma membrane, the viral nucleocapsid is carried by retrograde transport to the neuronal cell body. Such transport is accomplished by the interaction of a major component of the tegument, viral protein VP1/2, with minus-end-directed dynein motors. In contrast, other virus particles are carried to the nerve cell body within endocytic vesicles. After endocytosis of poliovirus, virus particles remain attached to the cellular receptor CD155. The cytoplasmic domain of the receptor engages the dynein light chain Tctex-1 to allow retrograde transport of virus-containing vesicles.

Virus-Induced Signaling via Cell Receptors

Binding of virus particles to cell receptors not only concentrates the particles on the cell surface, but also activates signaling pathways that facilitate virus entry and movement within the cell or produce cellular responses that enhance virus propagation and/or affect pathogenesis. Virus binding may lead to activation of protein kinases that trigger cascades of responses at the plasma membrane, cytoplasm, and nucleus (Chapter 14). Second messengers that participate in signaling include phosphatidylinositides, diacylglycerides, and calcium; regulators of membrane trafficking and actin dynamics also contribute to signaling. Virus-receptor interactions also stimulate antiviral responses (Volume II, Chapter 3).

Signaling triggered by binding of coxsackievirus B3 to cells makes receptors accessible for virus entry. The coxsackievirus and adenovirus receptor, Car, is not present on the apical surface of epithelial cells that line the intestinal and respiratory tracts. This membrane protein is a component of tight junctions and is inaccessible to virus particles. Binding of group B coxsackieviruses to its receptor, CD55, which is present on the apical surface, activates Abl kinase, which in turn triggers Rac-dependent actin rearrangements. These changes allow movement of virus particles to the tight junction, where they can bind Car and enter cells.

Signaling is essential for the entry of simian virus 40 into cells. Binding of this virus particle to its glycolipid cell receptor, GM1 ganglioside, causes activation of tyrosine kinases. The signaling that ensues induces reorganization of actin filaments, internalization of the virus in caveolae, and transport of the caveolar vesicles to the endoplasmic reticulum. The activities of more than 50 cellular protein kinases regulate the entry of this virus into cells.

Import of Viral Genomes into the Nucleus

The reproduction of most DNA viruses, and some RNA viruses including retroviruses and influenza viruses, begins in the cell nucleus. The genomes of these viruses must therefore

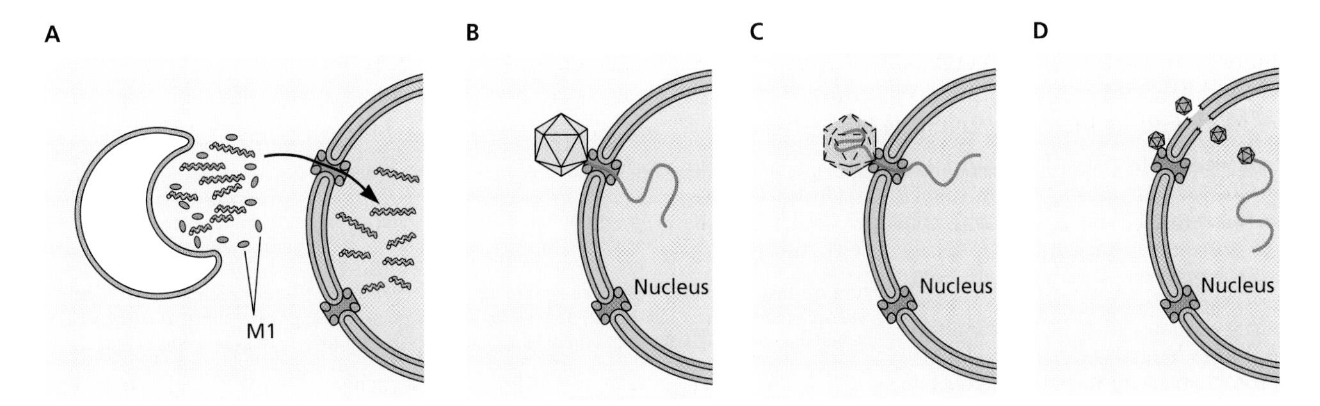

Figure 5.21 Different strategies for entering the nucleus. (A) Each segment of the influenza virus genome is small enough to be transported through the pore complex. **(B)** The herpes simplex virus type 1 capsid docks onto the nuclear pore complex and is minimally disassembled to allow transit of the viral DNA into the nucleus. **(C)** The adenovirus subviral particle is dismantled at the nuclear pore, allowing transport of the viral DNA into the nucleus. **(D)** The capsids of some viruses (parvovirus and hepadnavirus) are small enough to enter the nuclear pore complex without disassembly but do not enter by this route. These virus particles bind the nuclear pore complex, which causes disruption of the nuclear envelope followed by nuclear entry.

be imported from the cytoplasm. One way to accomplish this movement is via the cellular pathway for protein import into the nucleus. An alternative, observed in cells infected by some retroviruses, is to enter the nucleus after the nuclear envelope breaks down during cell division. When the nuclear envelope is reformed, the viral DNA is incorporated into the nucleus together with cellular chromatin. This strategy restricts infection to cells that undergo mitosis.

Many subviral particles are too large to pass through the nuclear pore complex. There are several strategies to overcome this limitation (Fig. 5.21). The influenza virus genome, which consists of eight segments that are each small enough to pass through the nuclear pore complex, is uncoated in the cytoplasm. Adenovirus subviral particles dock onto the nuclear pore complex and are disassembled by the import machinery, allowing the viral DNA to pass into the nucleus. Herpes simplex virus capsids also dock onto the nuclear pore but remain largely intact, and the nucleic acid is injected into the nucleus through a portal in the nucleocapsid. The DNA of some bacteriophages is packaged in virus particles at high pressure, which provides sufficient force to insert the viral DNA genome into the bacterial cell. A similar mechanism may allow injection of herpesviral DNA (Box 5.6). This mechanism would overcome the problem that transport through the nuclear pore complex depends upon hydrophobic interactions with nucleoporins: the charged and hydrophilic viral nucleic acids would have difficulty passing through the pore.

Nuclear Localization Signals

Proteins that reside within the nucleus are characterized by the presence of specific nuclear targeting sequences. Such **nuclear localization signals** are both necessary for nuclear entry of the proteins in which they are present and sufficient to direct heterologous, nonnuclear proteins to enter this organelle. Nuclear localization signals identified by these criteria share a number of common properties: they are generally fewer than 20 amino acids in length, and they are usually rich in basic amino acids. Despite these similarities, no consensus nuclear localization sequence can be defined.

Most nuclear localization signals belong to one of two classes, simple or bipartite sequences (Fig. 5.22). A particularly well characterized example of a simple nuclear localization signal is that of simian virus 40 large T antigen, which comprises five contiguous basic residues flanked by hydrophobic amino acids. This sequence is sufficient to relocate normally cytoplasmic proteins to the nucleus. Many other viral and cellular nuclear proteins contain short, basic nuclear localization signals, but these signals are not identical in primary sequence to that of T-antigen. The presence of a nuclear localization signal is all that is needed to target a macromolecular substrate for import into the nucleus.

The Nuclear Pore Complex

The nuclear envelope is composed of two typical lipid bilayers separated by a luminal space (Fig. 5.23). Like all other cellular membranes, it is impermeable to macromolecules such as proteins. However, the nuclear pore complexes that stud the nuclear envelopes of all eukaryotic cells provide aqueous channels that span both the inner and outer nuclear membranes for exchange of small molecules, macromolecules, and macromolecular assemblies between nuclear and cytoplasmic compartments. Numerous experimental techniques, including direct visualization of gold particles attached to proteins or RNA molecules as they are transported, have established that

BOX 5.6

DISCUSSION

The bacteriophage DNA injection machine

The mechanisms by which the bacteriophage genome enters the bacterial host are unlike those for viruses of eukaryotic cells. One major difference is that the bacteriophage particle remains on the surface of the bacterium as the nucleic acid passes into the cell. The DNA genome of some bacteriophages is packaged under high pressure (up to 870 lb/in^2) in the capsid and is injected into the cell in a process that has no counterpart in the entry process of eukaryotic viruses. The complete structure of bacteriophage T4 illustrates this remarkable process (see figure). To initiate infection, the tail fibers attach to receptors on the surface of *Escherichia coli*. Binding causes a conformational change in the baseplate, which leads to contraction of the sheath. This movement drives the rigid tail tube through the outer membrane, using a needle at the tip. When the needle touches the peptidoglycan layer in the periplasm, the needle dissolves and three lysozyme domains in the baseplate are activated. These disrupt the peptidoglycan layer of the bacterium, allowing DNA to enter.

Browning C, Shneider MM, Bowman VD, Schwarzer D, Leiman PG. 2011. Phage pierces the host cell membrane with the iron-loaded spike. *Structure* **20:**236–339.

Structure of bacteriophage T4. (A) A model of the 2,000-Å bacteriophage as produced from electron microscopy and X-ray crystallography. Components of the virion are color coded: virion head (beige), tail tube (pink), contractile sheath around the tail tube (green), baseplate (multicolored), and tail fibers (white and magenta). In the illustration, the virion contacts the cell surface, and the tail sheath is contracted prior to DNA release into the cell. Courtesy of Michael Rossmann, Purdue University. **Structure of bacteriophage membrane-piercing spike. (B)** CryoEM reconstruction of phi92 basplate. The spike is shown in red. **(C, D)** Trimers of bacteriophage phi92 gp138, shown as surface **(C)** and ribbon diagrams **(D)**. From P.G. Leiman et al., *Cell* **118:**419-430, 2004, with permission.

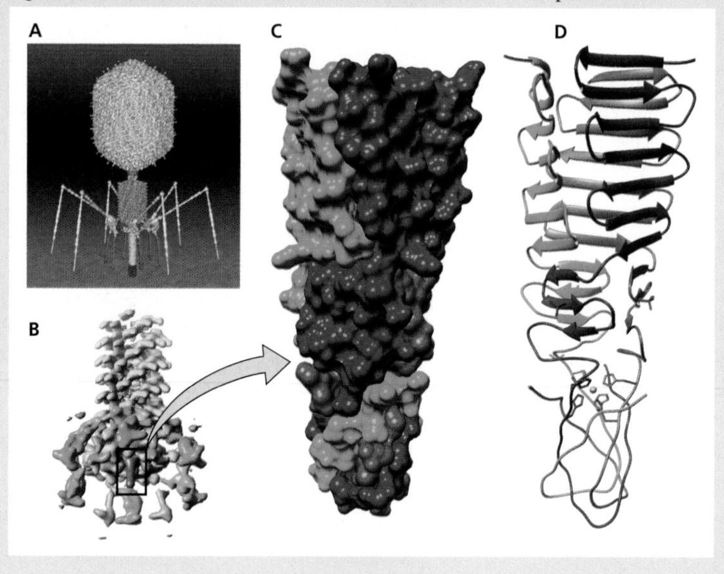

nuclear proteins enter and RNA molecules exit the nucleus by transport through the nuclear pore complex. The functions of the nuclear pore complex in both protein import and RNA export are not completely understood, not least because this important cellular machine is large (molecular mass, approximately 125 × 10^6 kDa in vertebrates), built from many different proteins, and architecturally complex (Fig. 5.23).

The nuclear pore complex allows passage of cargo in and out of the nucleus by either passive diffusion or facilitated translocation. Passive diffusion does not require interaction between the cargo and components of the nuclear pore complex and becomes inefficient as molecules approach 9 nm in diameter. Objects as large as 39 nm in diameter can pass through nuclear pore complexes by facilitated translocation. This process requires specific interactions between the cargo and components of the nuclear pore complex.

The Nuclear Import Pathway

Import of a protein into the nucleus via nuclear localization signals occurs in two distinct, and experimentally separable, steps (Fig. 5.23C). A protein containing such a signal first binds to a soluble cytoplasmic receptor protein. This complex then engages with the cytoplasmic surface of the nuclear pore complex, a reaction often called docking, and is translocated through the nuclear pore complex. In the nucleus, the complex is disassembled, releasing the protein cargo.

Different groups of proteins are imported by specific receptor systems. In what is known as the "classical system" of import, cargo proteins containing basic nuclear localization signals bind to the cytoplasmic nuclear localization signal receptor protein importin-α (Fig. 5.23C). This complex then binds importin-β, which mediates docking with the nuclear pore complex by binding to members of a family of nucleoporins. Some of these nucleoporins are found in the cytoplasmic filaments of the nuclear pore complex (Fig. 5.23), which associate with import substrates. The complex is translocated to the opposite side of the nuclear envelope, where the cargo is released. Other importins can bind cargo proteins directly without the need for an adapter protein.

A single translocation through the nuclear pore complex does not require energy consumption. However, maintenance of a gradient of the guanosine nucleotide-bound forms of Ran,

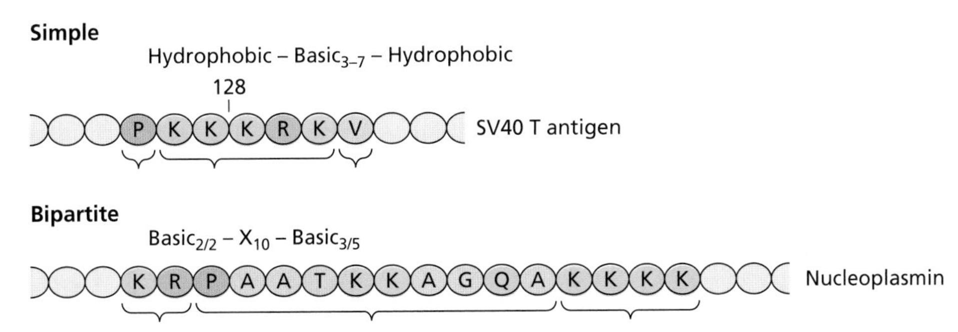

Figure 5.22 Nuclear localization signals. The general form and a specific example of simple and bipartite nuclear localization signals are shown in the one-letter amino acid code, where X is any residue. Bipartite nuclear targeting signals are defined by the presence of two clusters of positively charged amino acids separated by a spacer region of variable sequence. Both clusters of basic residues, which often resemble the simple targeting sequences of proteins like simian virus 40 T antigen, are required for efficient import of the proteins in which they are found. The subscript indicates either length (3–7) or composition (e.g., 3/5 means at least 3 residues out of 5 are basic). Gold particles with diameters as large as 26 nm are readily imported following their microinjection into the cytoplasm, as long as they are coated with proteins or peptides containing a nuclear localization signal.

with Ran-GDP and Ran-GTP concentrated in the cytoplasm and nucleus, respectively, is absolutely essential for continued transport. For example, conversion of Ran-GDP to Ran-GTP in the nucleus, catalyzed by the guanine nucleotide exchange protein Rcc-1, promotes dissociation of imported proteins from importins (Fig. 5.23).

Import of Influenza Virus Ribonucleoprotein

Influenza virus is among the few RNA-containing viruses with genomes that replicate in the cell nucleus. After vRNPs separate from M1 and are released into the cytosol, they are rapidly imported into the nucleus (Fig. 5.12). Such import depends on the presence of a nuclear localization signal in the NP protein, a component of vRNP: naked viral RNA does not dock onto the nuclear pore complex, nor is it taken up into the nucleus.

Import of DNA Genomes

The capsids of many DNA-containing viruses are larger than 39 nm and cannot be imported into the nucleus from the cytoplasm. One mechanism for crossing the nuclear membrane comprises docking of a capsid onto the nuclear pore complex, followed by delivery of the viral DNA into the nucleus. Adenoviral and herpesviral DNAs are transported into the nucleus via this mechanism. However, the strategies for DNA import are distinct: adenovirus DNA is covered with proteins and is recognized by the import system, while HSV DNA is naked and is injected.

Partially disassembled adenovirus capsids dock onto the nuclear pore complex by interaction with Nup214 (Fig. 5.24). Release of the viral genome requires capsid protein binding to kinesin-1, the motor protein that mediates transport on microtubules from the nucleus to the cell periphery. As the capsid is held on the nuclear pore, movement of kinesin-1 towards the plasma membrane is thought to pull apart the capsid. The released protein VII-associated viral DNA is then imported into the nucleus, where viral transcription begins. Herpesvirus capsids also dock onto the nuclear pore complex, and interaction with nucleoporins destabilizes a viral protein, pUL25, which locks the genome inside the capsid. This event causes the naked viral DNA, which is packaged in the nucleocapsid under very high pressure, to exit through the portal.

The 26-nm capsid of parvoviruses is small enough to fit through the nuclear pore (39 nm), and it has been assumed that these virus particles enter by this route. However, there is no experimental evidence that parvovirus capsids pass intact through the nuclear pore. Instead, virus particles bind to the nuclear pore complex, followed by disruption of the nuclear envelope and the nuclear lamina, leading to entry of virus particles (Fig. 5.21). After release from the endoplasmic reticulum, the 45-nm capsid of SV40 also docks onto the nuclear pore, initiating disruption of the nuclear envelope and lamina. Such nuclear disruption appears to involve cell proteins that also participate in the increased nuclear permeability that takes place during mitosis, raising the possibility that nuclear entry of these viral genomes is a consequence of remodeling a cell process.

Import of Retroviral Genomes

Fusion of retroviral and plasma membranes releases the viral core into the cytoplasm. The retroviral core consists of the viral RNA genome, coated with NC protein, and the enzymes reverse transcriptase (RT) and integrase (IN), enclosed by capsid (CA) protein. Retroviral DNA synthesis commences in the cytoplasm, within the nucleocapsid core, and after 4 to 8 h of DNA synthesis the preintegration complex, comprising

A

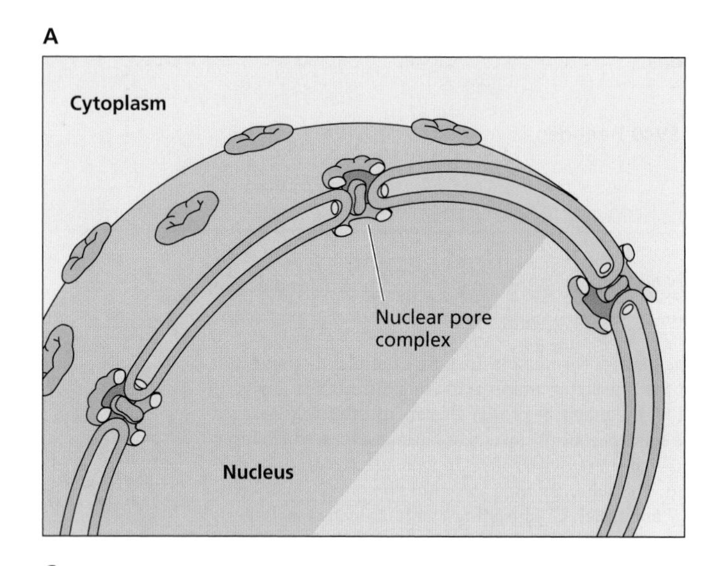

B

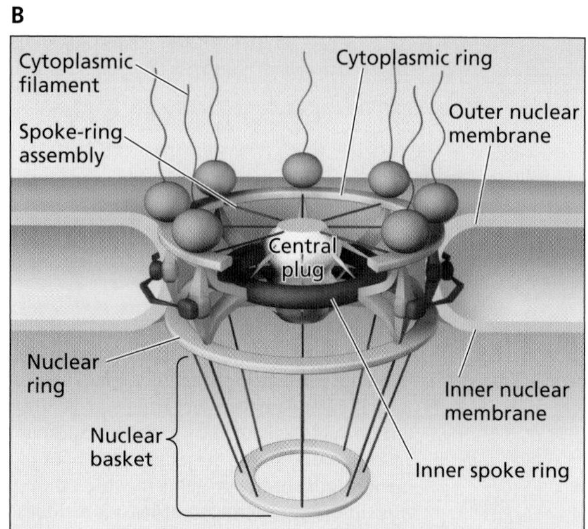

C

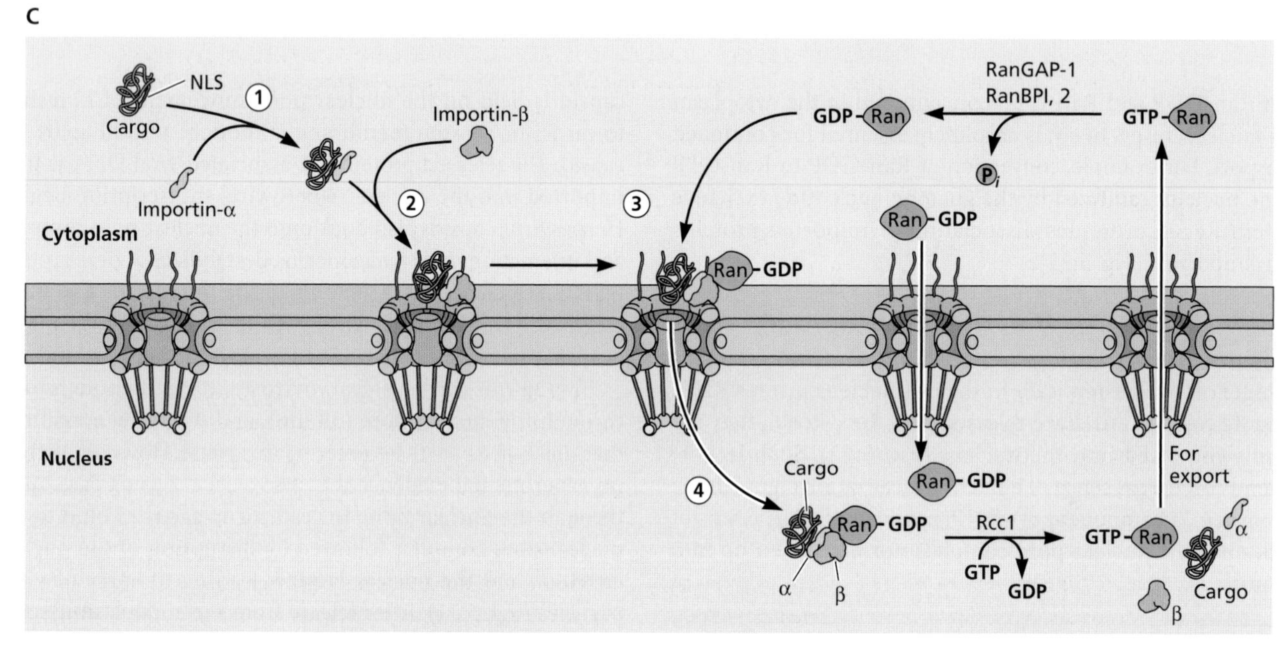

Figure 5.23 Structure and function of the nuclear pore complex. (A) Overview of the nuclear membrane, showing the topology of the nuclear pore complexes. **(B)** Schematic drawing of the nuclear pore complex, showing the spoke-ring assembly at its waist and its attachment to cytoplasmic filaments and the nuclear basket. The latter comprises eight filaments, extending 50 to 100 nm from the central structure and terminating in a distal annulus. The nuclear pore channel is shown containing the transporter. **(C)** An example of the classical protein import pathway for proteins with a simple nuclear localization signal (NLS). This pathway is illustrated schematically from left to right. Cytoplasmic and nuclear compartments are shown separated by the nuclear envelope studded with nuclear pore complexes. In step 1, a nuclear localization signal on the cargo (red) is recognized by importin-α. In step 2, importin-β binds the cargo–importin-α complex and docks onto the nucleus, probably by associating initially with nucleoporins present in the cytoplasmic filaments of the nuclear pore complex. Translocation of the substrate into the nucleus (step 4) requires additional soluble proteins, including the small guanine nucleotide-binding protein Ran (step 3). A Ran-specific guanine nucleotide exchange protein (Rcc1) and a Ran-GTPase-activating protein (RanGap-1) are localized in the nucleus and cytoplasm, respectively. The action of RanGAP-1, with the accessory proteins RanBp1 and RanBp2, maintains cytoplasmic Ran in the GDP-bound form. When Ran is in the GTP-bound form, nuclear import cannot occur. Following import, the complexes are dissociated when Ran-GDP is converted to Ran-GTP by Rcc1. Ran-GTP participates in export from the nucleus. The nuclear pool of Ran-GDP is replenished by the action of the transporter Ntf2/p10, which efficiently transports Ran-GDP from the cytoplasm to the nucleus. Hydrolysis of Ran-GTP in the cytoplasm and GTP-GDP exchange in the nucleus therefore maintain a gradient of Ran-GTP/Ran-GDP. The asymmetric distribution of RanGap-1 and Rcc1 allows for the formation of such a gradient. This gradient provides the driving force and directionality for nuclear transport. A monomeric receptor called transportin mediates the import of heterogeneous nuclear RNA-binding proteins that contain glycine- and arginine-rich nuclear localization signals. Transportin is related to importin-β, as are other monomeric receptors that mediate nuclear import of ribosomal proteins. (B) Adapted from Q. Yang, M. P. Rout, and C. W. Akey, *Mol. Cell* **1**:223–234, 1998, with permission.

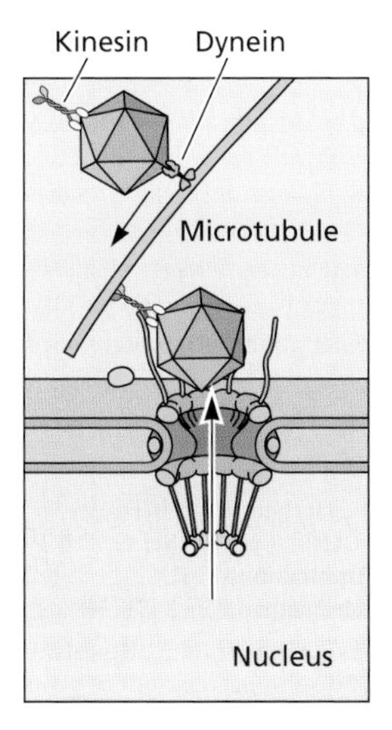

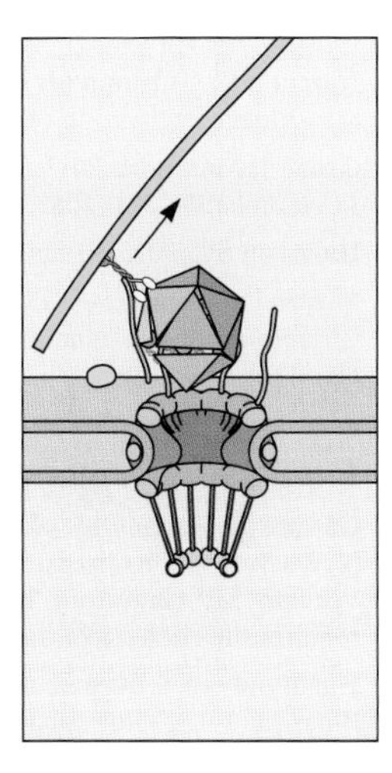
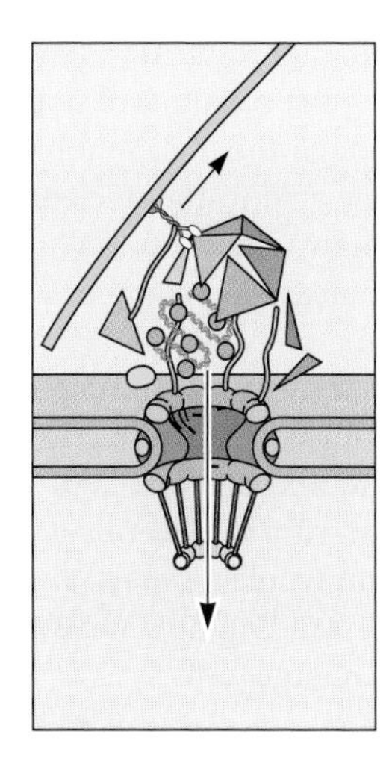
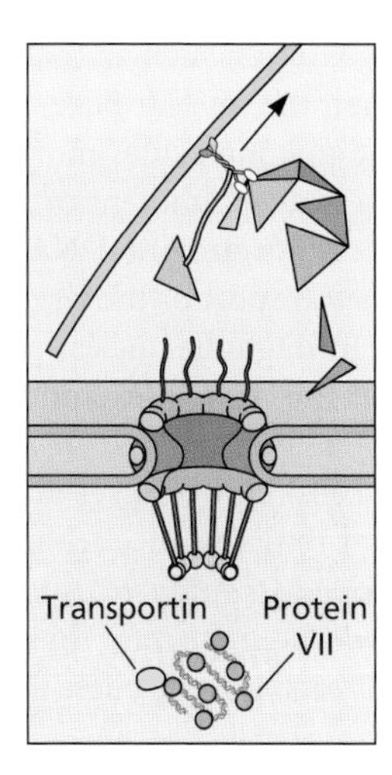

Figure 5.24 Uncoating of adenovirus at the nuclear pore complex. After release from the endosome, the partially disassembled capsid moves towards the nucleus by dynein-driven transport on microtubules. The particle docks onto the nuclear pore complex protein Nup214. The capsid also binds kinesin-1 light chains, which move away from the nucleus, pulling the capsid apart. The viral DNA, bound to protein VII, is delivered into the nucleus by the import protein transportin and other nuclear import proteins.

viral DNA, IN, and other proteins, localizes to the nucleus. There the viral DNA is integrated into the cellular genome, and viral transcription begins. The mechanism of nuclear import of the preintegration complex is poorly understood, but it is quite clear that this structure is too large to pass through the nuclear pore complex. The betaretrovirus Moloney murine leukemia virus can efficiently infect only dividing cells. The viral preintegration complex is tethered to chromatin when the nuclear membrane is broken down during mitosis and remains associated with chromatin as the nuclear membrane reforms in daughter cells, avoiding the need for active transport.

In contrast to Moloney murine leukemia virus, other retroviruses can reproduce in nondividing cells. The preintegration complex of these viruses must therefore be transported into an intact nucleus. The exact mechanism for nuclear entry is still unclear, but for the lentivirus human immunodeficiency virus type 1, there is evidence that CA-mediated attachment of the preintegration complex to the nuclear pore is required. In addition, various viral proteins that contain nuclear localization signals may facilitate the process (e.g., Vpr, MA, and IN).

Perspectives

The study of how viruses enter cells provides critical insight into the very first steps of virus reproduction. Virus entry comprises binding to receptors, transport within the cell, dismantling of the virus particle, and release of the genome. It has become clear that there are many pathways for virus entry into cells. Clathrin- and dynamin-dependent endocytosis is no longer the sole entry pathway known; other routes are caveolin-dependent endocytosis and clathrin- and caveolin-independent endocytosis. The road used seems to depend on the virus, the cell type, and the conditions of infection. As most of our current knowledge has been derived from studies with cells in culture, a crucial unanswered question is whether these same pathways of viral entry are utilized during infection of the heterogeneous tissues of living animals.

Virus binding to the cell surface leads to major alterations in cell activities, effects mediated by signal transduction. These responses include providing access to coreceptors, formation of pits, pinching off of vesicles, and rearrangement of actin filaments to facilitate vesicle movement. The precise signaling pathways required need to be elucidated. Such efforts may identify specific targets for inhibiting virus movement in cells.

The development of single-particle tracking methods has advanced considerably in the past 5 years. As a consequence, our understanding of the routes that viruses travel once they are inside the cell has improved markedly. The role of cellular transport pathways in bringing viral genomes to the site appropriate for their replication within the cell is beginning to be clarified. Yet many questions remain. How are viruses or subviral particles transported on the cytoskeletal network? What are the signals for their attachment and release from microtubules and filaments?

The genomes of many viruses are replicated in the nucleus. Some viral genomes enter this cellular compartment by transport through the nuclear pore complex. Studies of adenoviral DNA import into the nucleus have revealed an active role for components of both the nuclear pore complex and the microtubule network in subviral particle disassembly. Other viral DNA genomes, such as that of the herpesviruses, pass naked through the nuclear pore, raising the question of how these hydrophilic molecules move through the hydrophobic pore, against a steep gradient of nucleic acid in the nucleus. Nuclear import of retroviral DNA is barely understood. What signal allows transport of the large preintegration complex of retroviruses through the nuclear pore?

Nearly all the principles and specific features discussed in this chapter were derived from studies of viral infection in cultured cells. How viruses attach to and enter cells of a living animal remains an uncharted territory. Methods are being developed to study virus entry in whole animals, and the results will be important for understanding how viruses spread and breach host defenses to reach target cells.

References

Book

Pohlmann S, Simmons G. 2013. *Viral Entry into Host Cells*. Landes Bioscience, Austin, TX.

Reviews

Ambrose Z, Aiken C. 2014. HIV-1 uncoating: connection to nuclear entry and regulation by host proteins. *Virology* **454-455**:371–379.

Cosset F-L, Lavillette D. 2011. Cell entry of enveloped viruses. *Adv Genet* **73**:121–183.

Fay N, Panté N. 2015. Old foes, new understandings: nuclear entry of small non-enveloped DNA viruses. *Curr Opin Virol* **12**:59–65.

Garcia NK, Guttman M, Ebner JL, Lee KK. 2015. Dynamic changes during acid-induced activation of influenza hemagglutinin. *Structure* **23**:665–676.

Grove J, Marsh M. 2011. The cell biology of receptor-mediated virus entry. *J Cell Biol* **195**:1071–1082.

Harrison SC. 2015. Viral membrane fusion. *Virology* pii:S0042-6822.

Moyer CL, Nemerow GR. 2011. Viral weapons of membrane destruction: variable modes of membrane penetration by non-enveloped viruses. *Curr Opin Virol* **1**:44–49.

Papers of Special Interest

Carette JE, Raaben M, Wong AC, Herbert AS, Obernosterer G, Mulherkar N, Kuehne AI, Kranzusch PJ, Griffin AM, Ruthel G, Dal P Cin, Dye JM, Whelan SP, Chandran K, Brummelkamp TR. 2011. Ebola virus entry requires the cholesterol transporter Niemann-Pick C1. *Nature* **477**:340–346.

Harutyunyan S, Kumar M, Sedivy A, Subirats X, Kowalski H, Köhler G, Blaas D. 2013. Viral uncoating is directional: exit of the genomic RNA in a common cold virus starts with the poly-(A) tail at the 3′-end. *PLoS Pathog* **9**(4):e1003270.

Lukic Z, Dharan A, Fricke T, Diaz-Griffero F, Campbell EM. 2014. HIV-1 uncoating is facilitated by dynein and kinesin 1. *J Virol* **88**:13613–13625.

Rizopoulos Z, Balistreri G, Kilcher S, Martin CK, Syedbasha M, Helenius A, Mercer J. 2015. Vaccinia virus infection requires maturation of macropinosomes. *Traffic* doi:10.1111/tra.12290.

Scherer J, Yi J, Vallee RB. 2014. PKA-dependent dynein switching from lysosomes to adenovirus: a novel form of host-virus competition. *J Cell Biol* **205**:163–177.

Strauss M, Filman DJ, Belnap DM, Cheng N, Noel RT, Hogle JM. 2015. Nectin-like interactions between poliovirus and its receptor trigger conformational changes associated with cell entry. *J Virol* **89**(8):4143–4157.

6 Synthesis of RNA from RNA Templates

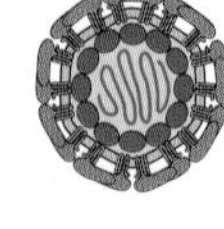

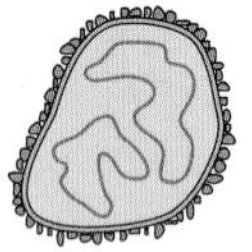

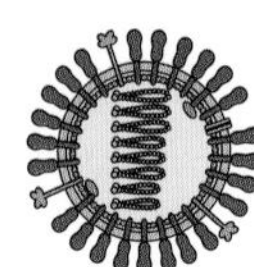

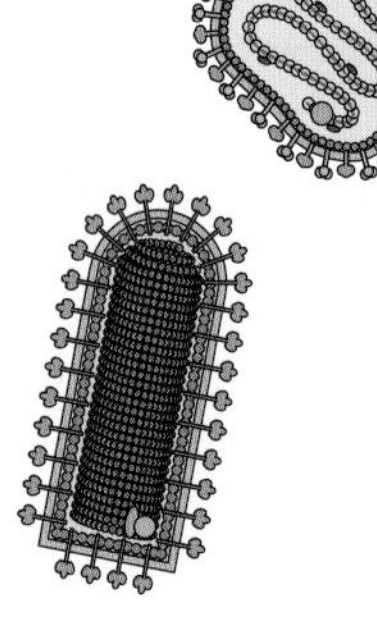

Introduction

The Nature of the RNA Template
- Secondary Structures in Viral RNA
- Naked or Nucleocapsid RNA

The RNA Synthesis Machinery
- Identification of RNA-Dependent RNA Polymerases
- Sequence Relationships among RNA Polymerases
- Three-Dimensional Structure of RNA-Dependent RNA Polymerases

Mechanisms of RNA Synthesis
- Initiation
- Capping
- Elongation
- Template Specificity
- Unwinding the RNA Template
- Role of Cellular Proteins

Paradigms for Viral RNA Synthesis
- (+) Strand RNA
- Synthesis of Nested Subgenomic mRNAs
- (−) Strand RNA
- Ambisense RNA
- Double-Stranded RNA
- Unique Mechanisms of mRNA and Genome Synthesis of Hepatitis Delta Satellite Virus
- Why Are (−) and (+) Strands Made in Unequal Quantities?
- Do Ribosomes and RNA Polymerases Collide?

Cellular Sites of Viral RNA Synthesis

Origins of Diversity in RNA Virus Genomes
- Misincorporation of Nucleotides
- Segment Reassortment and RNA Recombination
- RNA Editing

Perspectives

References

LINKS FOR CHAPTER 6

⏭ *Video: Interview with Dr. Karla Kirkegaard*
http://bit.ly/Virology_Kirkegaard

⏭ *A swinging gate*
http://bit.ly/Virology_Twiv330

> *Truth is ever to be found in the simplicity, and not in the multiplicity and confusion of things.*
> SIR ISAAC NEWTON

Introduction

The genomes of RNA viruses may be unimolecular or segmented, single stranded of (+), (−), or ambisense polarity, double stranded, or circular. These structurally diverse viral RNA genomes share a common requirement: they must be efficiently copied within the infected cell to provide both genomes for assembly into progeny virus particles and messenger RNAs (mRNAs) for the synthesis of viral proteins. The synthesis of these RNA molecules is a unique process that has no parallel in the cell. The genomes of all RNA viruses except retroviruses encode an **RNA-dependent RNA polymerase** (Box 6.1) to catalyze the synthesis of new genomes and mRNAs.

Virus particles that contain (−) strand or double-stranded RNA genomes must contain the RNA polymerase, because the incoming viral RNA can be neither translated nor copied by the cellular machinery. Consequently, the deproteinized genomes of (−) strand and double-stranded RNA viruses are noninfectious. In contrast, viral particles containing a (+) strand RNA genome lack a viral polymerase; the deproteinized RNAs of these viruses **are** infectious because they are translated in cells to produce, among other viral proteins, the viral RNA polymerase. An exception is the retrovirus particle, which contains a (+) stranded RNA genome that is not translated but rather copied to DNA by reverse transcriptase (Chapter 7).

The mechanisms by which viral mRNA is made and the RNA genome is replicated in cells infected by RNA viruses appear even more diverse than the structure and organization of viral RNA genomes (Fig. 6.1). For example, the genomes of both picornaviruses and alphaviruses are single molecules of (+) strand genomic RNA, but the strategies for the production of viral RNA are quite different. Nevertheless, each mechanism of viral RNA synthesis meets two essential requirements common to all infectious cycles: (i) during replication the RNA genome must be copied from one end to the other with no loss of nucleotide sequence; and (ii) viral mRNAs that can be translated efficiently by the cellular protein synthetic machinery must be made.

In this chapter we consider the mechanisms of viral RNA synthesis, the mechanism for switching from mRNA production to genome replication, and how the process of RNA-directed RNA synthesis leads to genetic diversity. Much of our understanding of viral RNA synthesis comes from experiments with purified components. Because it is possible that events proceed differently in infected cells, the results of such *in vitro* studies are used to build models for the different steps in RNA synthesis. While many models exist for each reaction, those presented in this chapter were selected because they are consistent with experimental results obtained in different laboratories or have been validated with simplified systems in cells in culture. The general principles of RNA synthesis deduced from such studies are illustrated with a few viruses as examples.

The Nature of the RNA Template

Secondary Structures in Viral RNA

RNA molecules are not simple linear chains but can form secondary structures that are important for RNA synthesis, translation, and assembly (Fig. 6.2). The first step in identifying a structural feature in RNA is to scan the nucleotide sequence with software designed to fold the nucleic acid into energetically stable structures. Comparative sequence analysis

PRINCIPLES *Synthesis of RNA from RNA templates*

- Viral RNA genomes must be copied to provide both genomes for assembly into progeny virus particles, and mRNAs for the synthesis of viral proteins.
- Viral RNA genomes may be naked in the virus particle (typically (+) strand RNAs) or organized into nucleocapsids in which proteins are bound to the genomic RNAs.
- Some viral RNA-dependent RNA polymerases can initiate RNA synthesis without a primer, while others are primer dependent.
- Viral RNA-dependent RNA polymerases, like the other three types of nucleic acid enzymes, resemble a right hand consisting of palm, fingers, and thumb domains, with the active site located in the palm.
- Viral RNA polymerases that initiate RNA synthesis without a primer appear to have an extra protein domain in the active site that acts as a "protein priming platform" to provide support for the initiating NTP.
- Primers for RNA polymerases may be capped fragments of cellular mRNAs or protein-linked nucleotides.
- Specificity of RNA polymerases for viral RNAs is conferred by the recognition of RNA sequences or structures.
- Host cell proteins are required for the activity of viral RNA polymerases.
- The single-stranded RNA genome of hepatitis delta virus is copied by host cell DNA-dependent RNA polymerase.
- Viral RNA synthesis takes place on specific structures in the cell, either nucleocapsids, subviral particles, or membrane-bound replication complexes.
- RNA synthesis is error prone, and this process, together with reassortment and recombination, yields diversity that is required for viral evolution.

BOX 6.1

TERMINOLOGY

What should we call RNA polymerases and the processes they catalyze?

Historically, viral RNA-dependent RNA polymerases were given two different names depending on their activities during infection. The term **replicase** was used to describe the enzyme that copies the viral RNA to produce additional genomes, while the enzyme that synthesizes mRNA was called **transcriptase.** In some cases, this terminology indicates true differences in the enzymes that carry out synthesis of functionally different RNAs, but for other RNA viruses, genomic replication and mRNA synthesis are the **same** reaction (Figure). For double-stranded RNA viruses, mRNA synthesis produces templates that can also be used for genomic replication. As these terms can therefore be inaccurate and misleading, they are not used here.

The production of mRNAs from viral RNA templates is often designated **transcription.** However, this term refers to a specific process, the copying of genetic information carried in DNA into RNA. Consequently, it is not used here to describe synthesis of the mRNAs of viruses with RNA genomes. Similarly, use of the term **promoter** is reserved to designate sequences controlling transcription of DNA templates.

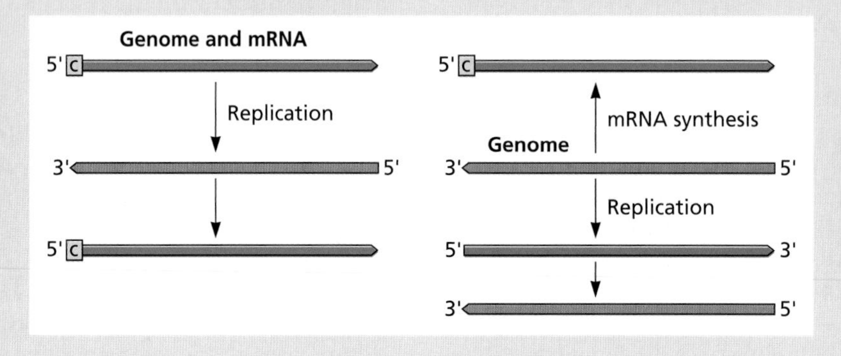

can predict RNA secondary structures. For example, comparison of the RNA sequences of several related viruses might establish that the structure, but not the sequence, of a stem-loop is conserved. Direct evidence for RNA structures comes from experiments in which RNAs are treated with enzymes or chemicals that attack single- or double-stranded regions specifically. The results of such analyses can confirm that predicted stem regions are base paired while loops are unpaired. Structures of RNA hairpins and pseudoknots have been determined by X-ray crystallography or nuclear magnetic resonance (Fig. 6.2C).

Naked or Nucleocapsid RNA

The genomes of (−) strand viruses are organized into nucleocapsids in which protein molecules, including the RNA-dependent RNA polymerase and accessory proteins, are bound to the genomic RNAs at regular intervals (Fig. 6.3). These tightly wound ribonucleoprotein complexes are very stable and generally resistant to RNase. The RNA polymerases of (−) strand viruses copy viral RNAs **only** when they are present in the nucleocapsid, such as that formed by the N protein of vesicular stomatitis virus bound to genomic RNA. In contrast, the genomes of (+) strand RNA viruses are not coated with proteins in the virus particle (exceptions are the (+) strand RNA genomes of members of the *Coronaviridae, Arteriviridae*, and *Retroviridae*). This difference is consistent with the fact that mRNAs are produced from the genomes of (−) strand RNA viruses upon cell entry, whereas the genomes of (+) strand RNA viruses are translated.

The viral nucleoproteins (NP) are cooperative, single-stranded RNA-binding proteins, as are the single-stranded nucleic acid-binding proteins required during DNA-directed DNA and RNA synthesis. Their function during replication is to keep the RNA single stranded and prevent base pairing between the template and product, so that additional rounds of RNA synthesis can occur. The nucleoproteins of nonsegmented (−) strand RNA viruses have a two-lobe architecture that forms a positively charged groove that binds and shields the genomic RNA (Fig. 6.3). Interactions between nucleoproteins lock monomers tightly, resulting in rigid NP-RNA complexes. The NP structures from segmented (−) strand RNA viruses are more varied and display less coordinated contacts between nucleoprotein subunits. These differences may explain why the NP-RNAs of these viruses are more susceptible to RNase digestion than those of nonsegmented (−) RNA viruses. The varied structures of the NP-RNA complexes also influence the access of the viral RNA polymerase to the template. The RNA polymerase of

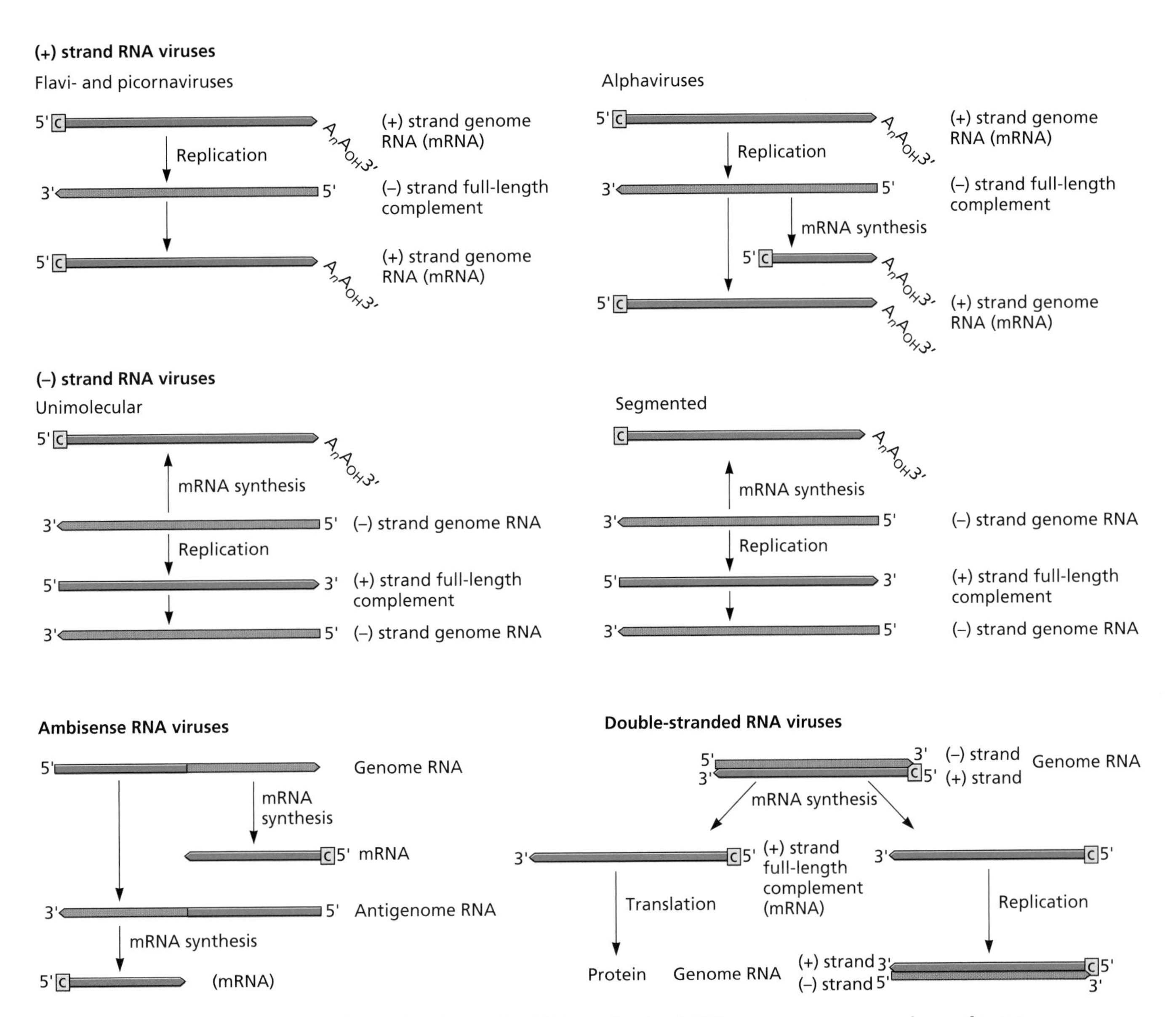

Figure 6.1 Strategies for replication and mRNA synthesis of RNA virus genomes are shown for representative virus families. Picornaviral genomic RNA is linked to VPg at the 5′ end. The (+) genomic RNA of some flaviviruses does not contain poly(A). Only one RNA segment is shown for segmented (−) strand RNA viruses.

segmented (−) strand RNA viruses can bind the NP-RNA template directly, whereas those of nonsegmented (−) strand RNA viruses cannot: a phosphoprotein (P) is required to recruit the RNA polymerase to the NP-RNA.

The genomes of many (+) strand RNA viruses encode helicases that serve a similar function as the nucleoproteins of (−) strand RNA viruses (see "Unwinding the RNA Template" below). In addition to its enzymatic activity, the poliovirus RNA polymerase ($3D^{pol}$) is a cooperative single-stranded RNA-binding protein and can unwind RNA duplexes without the hydrolysis of ATP that is characteristic of helicase-mediated unwinding. Polioviral RNA polymerase is therefore functionally similar to the RNA-binding nucleoproteins of (−) strand viruses.

The RNA Synthesis Machinery

Identification of RNA-Dependent RNA Polymerases

The first evidence for a viral RNA-dependent RNA polymerase emerged in the early 1960s from studies of mengovirus and poliovirus, both (+) strand RNA viruses. In these experiments, extracts were prepared from virus-infected cells and incubated with the four ribonucleoside

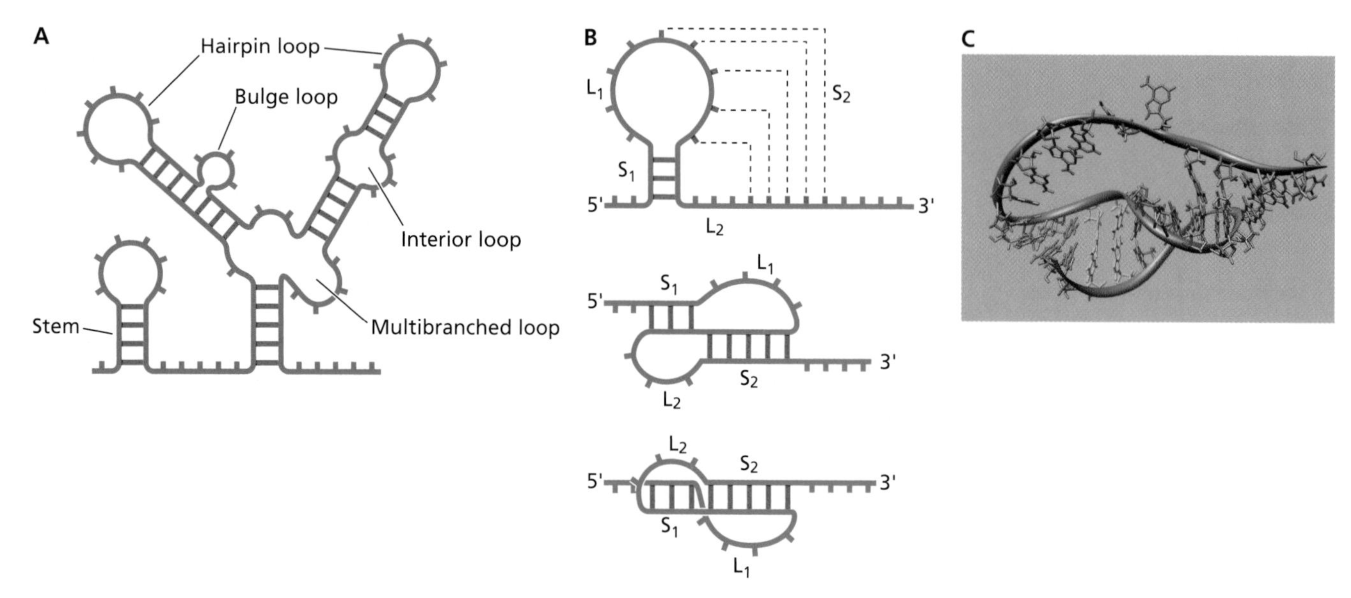

Figure 6.2 RNA secondary structure. (A) Schematic of different structural motifs in RNA. Red bars indicate base pairs; green bars indicate unpaired nucleotides. **(B)** Schematic of a pseudoknot. (Top) Stem 1 (S_1) is formed by base pairing in the stem-loop structure, and stem 2 (S_2) is formed by base pairing of nucleotides in the loop with nucleotides outside the loop. (Middle) A different view of the formation of stems S_1 and S_2. (Bottom) Coaxial stacking of S_1 and S_2 resulting in a quasicontinuous double helix. **(C)** Structure of a pseudoknot as determined by X-ray crystallography. The sugar backbone is highlighted with a green tube. Stacking of the bases in the areas of S_1 and S_2 can be seen. From Protein Data Bank file 1l2x. Adapted from C. W. Pleij, *Trends Biochem. Sci* **15:**143–147, 1990, with permission.

triphosphates (adenosine triphosphate [ATP], uridine triphosphate [UTP], cytosine triphosphate [CTP], and guanosine triphosphate [GTP]), one of which was radioactively labeled. The incorporation of nucleoside monophosphate into RNA was then measured. Infection with mengovirus or poliovirus led to the appearance of a cytoplasmic enzyme that could synthesize viral RNA in the presence of actinomycin D, a drug that was known to inhibit cellular DNA-directed RNA synthesis by intercalation into the double-stranded template. Lack of sensitivity to the drug suggested that the enzyme was virus specific and could copy RNA from an RNA template and not from a DNA template. This enzyme was presumed

Figure 6.3 Structure of viral ribonucleoproteins. (A) Ribbon diagram of vesicular stomatitis N protein molecule bound to RNA. The ribose-phosphate backbone of the RNA is shown as a green tube, and is bound in a groove located between N- and C-terminal lobes of the protein. From Protein Data Bank file 2qvj. **(B)** Ribbon diagram of Lassa virus NP bound to RNA. From Protein Data Bank file 3q7b. **(C)** Ribbon diagram of influenza virus NP bound to RNA. From Protein Data Bank file 2wfs. **(D)** Model of influenza A virus RNP. The polymerase complex is bound to a short NP-RNA template. Atomic structures of the NP are placed inside the model determined by cryo-electron microscopy. From Protein Data Bank file 2wfs and Electron Microscopy Databank entry 1603.

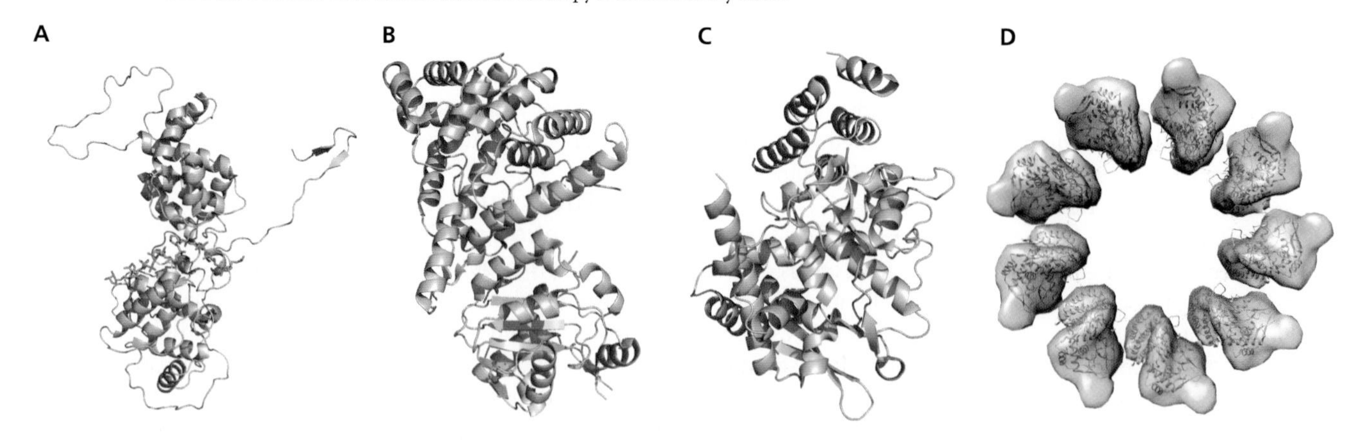

to be an RNA-dependent RNA polymerase. Similar assays were later used to demonstrate that the virions of (−) strand viruses and of double-stranded RNA viruses contain an RNA-dependent RNA polymerase that synthesizes mRNAs from the (−) strand RNA present in the particles.

The initial discovery of a putative RNA polymerase in poliovirus-infected cells was followed by attempts to purify the enzyme and show that it can copy viral RNA. Because polioviral genomic RNA contains a 3′-poly(A) sequence, polymerase activity was measured with a poly(A) template and an oligo(U) **primer**. After several fractionation steps, a poly(U) polymerase that could copy polioviral genomic RNA in the presence of this primer was purified from infected cells. Poly(U) polymerase activity coincided with a single polypeptide, now known to be the polioviral RNA polymerase $3D^{pol}$ (see Appendix, Fig. 21, for a description of this nomenclature). Purified $3D^{pol}$ RNA polymerase cannot copy polioviral genomic RNA in the absence of a primer.

Assays for RNA polymerase activity have been used to demonstrate virus-specific enzymes in virus particles or in extracts of cells infected with a wide variety of RNA viruses. Amino acid sequence alignments can be used to identify viral proteins with motifs characteristic of RNA-dependent RNA polymerases (see "Sequence Relationships among RNA Polymerases" below). These approaches were used to identify the L proteins of paramyxoviruses and bunyaviruses, the PB1 protein of influenza viruses, and the nsP4 protein of alphaviruses as candidate RNA polymerases. When the genes encoding these polymerases are expressed in cells, the proteins that are produced can copy viral RNA templates.

RNA-directed RNA synthesis follows a set of universal rules that differ slightly from those followed by DNA-dependent DNA polymerases. RNA synthesis initiates and terminates at specific sites in the template and is catalyzed by virus-encoded polymerases, but viral accessory proteins and even host cell proteins may also be required. Like cellular DNA-dependent RNA polymerases, some RNA-dependent RNA polymerases can initiate RNA synthesis *de novo*. Others require a primer with a free 3′-OH end to which nucleotides complementary to the template are added. Some RNA primers are protein linked, while others bear a 5′-cap structure (the cap structure is described in Chapter 10). A comparison of the structures and sequences of polynucleotide polymerases has led to the hypothesis that all polymerases catalyze synthesis by a mechanism that requires two metals (Box 6.2). RNA is usually synthesized by template-directed, stepwise incorporation of ribodeoxynucleoside monophosphates (NMPs) into the 3′-OH end of the growing RNA chain, which undergoes **elongation** in the 5′ → 3′ direction.

Sequence Relationships among RNA Polymerases

The amino acid sequences of viral RNA polymerases have been compared to identify conserved regions and to provide information about their evolution. Although polymerases have very different amino acid sequences, four conserved sequence motifs (A to D) have been identified in all RNA-dependent RNA polymerases (Fig. 6.4). Motif A and motif C contain the Asp residues that bind metal ions in the active site (Box 6.2), and motif C includes a Gly-Asp-Asp sequence conserved in the RNA polymerases of most (+) strand RNA viruses. It was suggested that this sequence is part of the active site of the enzyme. In support of this hypothesis, alterations in this sequence in polioviral $3D^{pol}$ and many other viral polymerases inactivate the enzyme. Evidence that a viral protein is an RNA polymerase is considerably strengthened when this 3-amino-acid sequence is found (Box 6.3).

Motifs A and C are also present in the sequences of RNA-dependent DNA polymerases that copy RNA templates, while all DNA-dependent polymerases share conserved sequence motifs A, B, and C (Fig. 6.4). These sequence comparisons indicate that all four classes of nucleic acid polymerases have a similar core catalytic domain (the palm domain) and most probably evolved from a common ancestor.

Three-Dimensional Structure of RNA-Dependent RNA Polymerases

The crystal structures of RNA-dependent RNA polymerases have confirmed the hypothesis that all polynucleotide polymerases are structurally similar. The shapes of all four types of polymerases resemble a right hand consisting of a palm, fingers, and a thumb, with the active site of the enzyme located in the palm (Fig. 6.4B). This shape supports the correct arrangement of substrates and metal ions at the catalytic site optimal for catalysis, and allows the dynamic changes needed during RNA synthesis. The structures of RNA-dependent RNA polymerases differ in detail from those of other polymerases, presumably to accommodate different templates and priming mechanisms.

All known RNA-dependent RNA polymerases adopt closed structures in which the active site is completely encircled (Fig. 6.4). In contrast, structures of other polynucleotide polymerases resemble an open hand. The closed structure, which is formed by interactions between the fingers and thumb domains, creates a nucleoside triphosphate (NTP) entry tunnel on one face of the enzyme and a template-binding site on the other. Residues within motif F, a conserved region unique to RNA-dependent RNA polymerases (Fig. 6.4), form the NTP entry tunnel.

The palm domain of RNA-dependent RNA polymerases is structurally similar to that of other polymerases, and contains the four motifs (A to D) that are conserved in all these enzymes (Fig. 6.4). The motifs confer specific functions, such as nucleotide recognition and binding (A and B), phosphoryl transfer (A and C), and determine the structure of the palm domain (D). The fifth motif, E, which is present in

BOX 6.2

BACKGROUND

Two-metal mechanism of catalysis by polymerases

All polynucleotide polymerases are thought to catalyze synthesis by a two-metal mechanism that requires two conserved aspartic acid residues (see figure). The carboxylate groups of these amino acids coordinate two divalent metal ions, shown as Mg^{2+} in the figure. One metal ion promotes deprotonation of the 3′-OH group of the nascent strand, and the other ion stabilizes the transition state at the α-phosphate of NTP and facilitates the release of pyrophosphate (PP_i).

Two-metal mechanism of polymerase catalysis. Red arrows indicate the net movement of electrons.

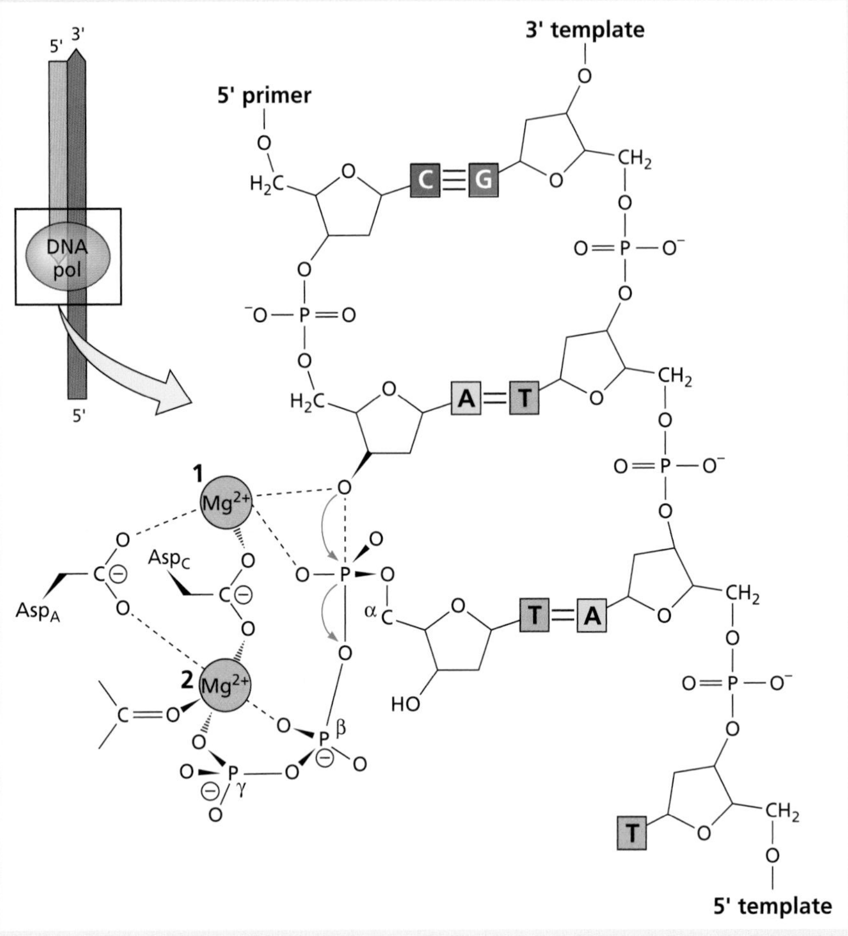

RNA-dependent, but not in DNA-dependent, polymerases, lies between the palm and thumb domains and binds the primer.

RNA-dependent RNA polymerases prefer to incorporate NTPs rather than deoxyribonucleoside triphosphates (dNTPs). NTP recognition by poliovirus $3D^{pol}$ is regulated by Asp238, which forms a hydrogen bond with the ribose 2′-OH (Fig. 6.5). dNTPs are not bound because Asp238 cannot form a hydrogen bond with 2′-deoxyribose. An Asp is present at this position in all RNA-dependent RNA polymerases. A Tyr at this position in RNA-dependent DNA polymerases is responsible for discriminating against NTPs and selecting dNTPs. Motif C of $3D^{pol}$ contains the Asp-Asp sequence conserved in RNA-dependent polymerases; the first Asp is also conserved in DNA-dependent polymerases. The two Asp residues of motif C and the conserved Asp238 of motif A form a cluster that coordinates the triphosphate moiety of the NTP and the metal ions required for catalysis (Fig. 6.5).

The interaction of RNA polymerase and the RNA template has been revealed by structural studies of poliovirus $3D^{pol}$ together with elongation complexes produced after several rounds of nucleotide incorporation (Fig. 6.6). In contrast to that by other types of nucleic acid polymerase, catalysis by $3D^{pol}$ does not depend on repositioning by the fingers domain of the nascent template-NTP from a preinsertion site to the active site. Closure of the active site is accomplished by base pairing of the initial NTP to a template nucleotide, leading to structural changes in the palm domain that cause Mg^{2+} binding and catalysis.

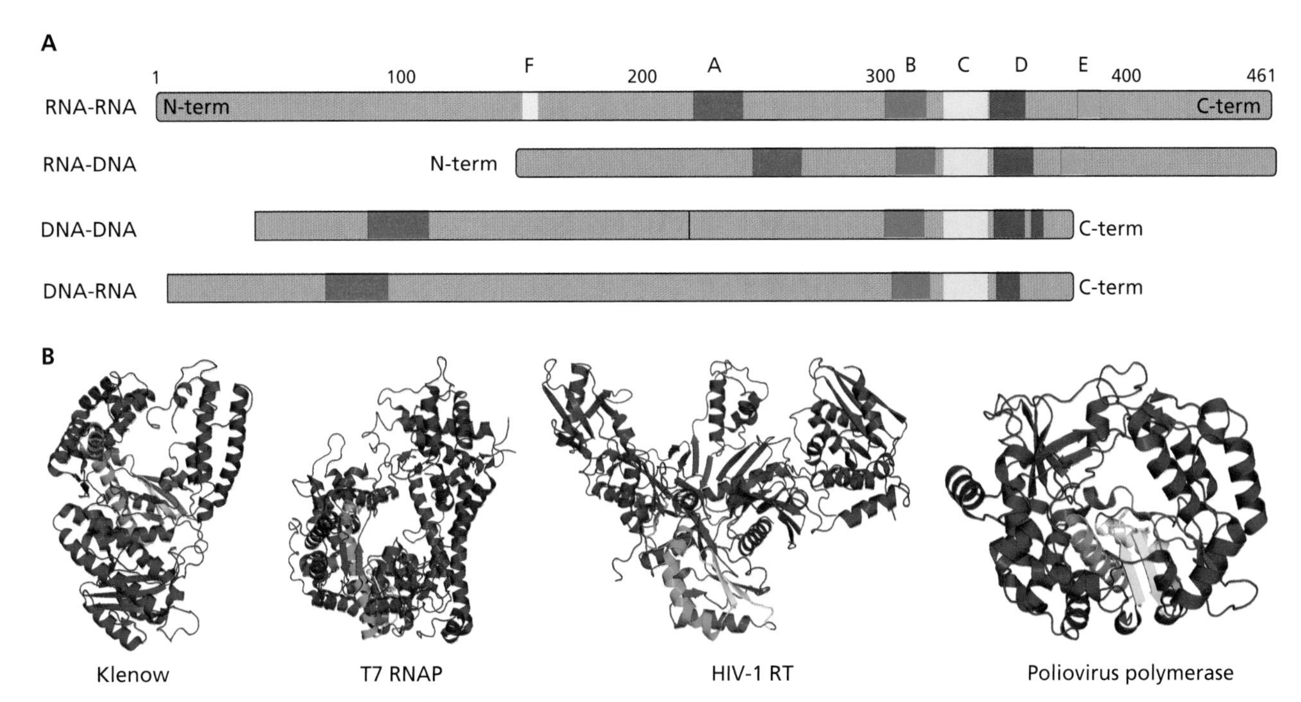

Figure 6.4 Protein domain alignments for the four categories of nucleic acid polymerases. (A) Schematic diagrams of polymerases. Numbers at the top are from the 3D^{pol} amino acid sequence. Sequence and structure motifs in each polymerase category are colored. Motif F is found only in RNA-dependent RNA polymerases. **(B)** Representative structures of each of the four types of nucleic acid polymerases. Ribbon diagrams of the polymerase domain of the large (Klenow) fragment of *Escherichia coli* DNA polymerase I, a DNA-dependent DNA polymerase; T7 RNA polymerase (T7 RNAP), a DNA-dependent RNA polymerase; human immunodeficiency virus type 1 reverse transcriptase (HIV-1 RT), an RNA-dependent DNA polymerase; and polioviral 3D^{pol}, an RNA-dependent RNA polymerase. The thumb domain is at the right, and the fingers domain is at the left. The conserved structure/sequence motifs A, B, C, D, and E are red, green, yellow, cyan, and purple, respectively. From Protein Data Bank files 1qsl, 1s77, 3hvt, and 1ra6.

BOX 6.3

BACKGROUND

The Gly-Asp-Asp sequence of RNA polymerase motif C

The Asp-Asp sequence of motif C is also conserved in RNA-dependent DNA polymerases of retroviruses and in RNA polymerases of double-stranded RNA and segmented (−) strand viruses. The RNA polymerases of nonsegmented (−) strand viruses contain Gly-Asp-Asn instead of Gly-Asp-Asp. Mutational studies have shown that this sequence in the RNA polymerase (L protein) of vesicular stomatitis virus is essential for RNA synthesis. The RNA polymerase of birnavirus, an insect virus with a double-stranded RNA genome, has Ala-Asp-Asn instead of Gly-Asp-Asp. An RNA polymerase with Ala-Asp-Asn substituted with Gly-Asp-Asp has increased enzymatic activity. This observation has led to the suggestion that Ala-Asp-Asn may have been selected during the evolution of these birnaviruses to reduce pathogenicity and facilitate virus spread.

RNA-dependent RNA polymerase is depicted as a brown rectangle, with motifs A to F colored. Conserved motif C amino acids are shown for viral RNA-dependent RNA polymerases and reverse transcriptase of HIV-1.

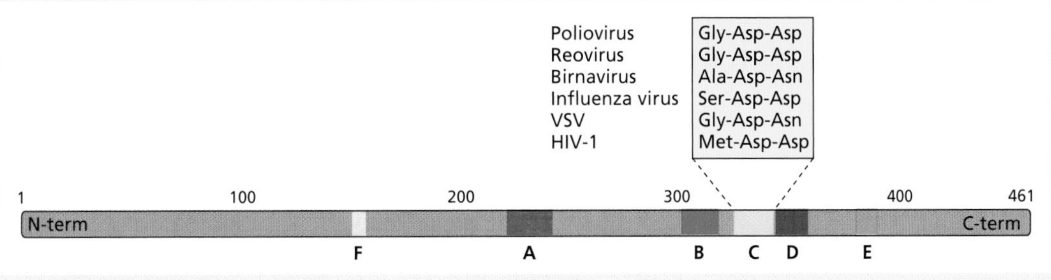

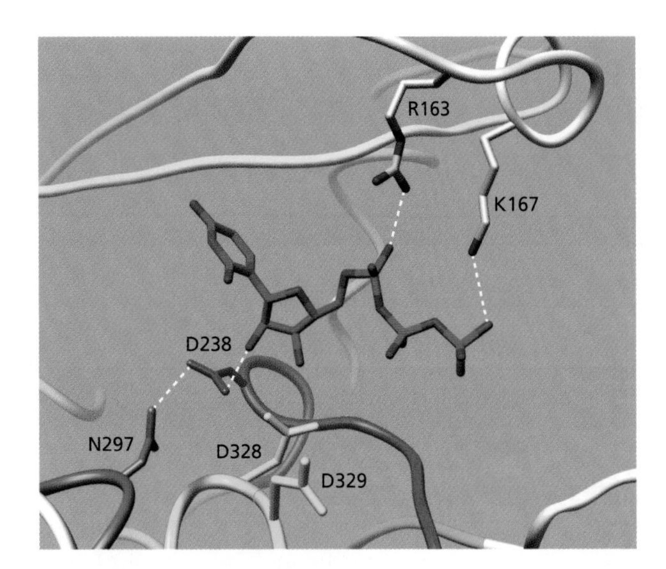

Figure 6.5 Structure of UTP bound to poliovirus 3D^{pol}. The NTP bridges the fingers (top) and palm (bottom) domains. The base is stacked with Arg174 from the fingers. Hydrogen bonds are shown as dashed lines. The Asp238 of motif A, which is conserved in all RNA-dependent RNA polymerases, hydrogen bonds with the 2′-OH of the ribose moiety; this interaction discriminates NTPs from dNTPs. Asp328 and Asp329, which coordinate Mg^{2+}, are also labeled. From Protein Data Bank file 2im2.

Mechanisms of RNA Synthesis

Initiation

As polymerases synthesize nucleic acid in a 5′ to 3′ direction, the nucleotidyl transfer reaction is initiated at the 3′ end of the template strand. The requirement for a primer for initiation of nucleic acid synthesis varies among the different classes of polymerases. All DNA polymerases are primer-dependent enzymes, while DNA-dependent RNA polymerases initiate RNA synthesis *de novo*. Some RNA-dependent RNA polymerases (e.g., those of flaviviruses and rhabdoviruses) can also initiate RNA synthesis *de novo*, while others require a primer (Fig. 6.7). Nucleic acid synthesis by these RNA polymerases is initiated by a protein-linked primer (picornaviruses) or an oligonucleotide cleaved from the 5′ end of cellular mRNA (influenza viruses).

***De novo* initiation**

3′-terminal initiation — 3′- N1 N2 ▬ 5′ / NTP NTP / OH

Internal initiation — 5′-pppG / 3′-AUC AUC AUC UG ▬ 5′

Elongation

5′-pppG UAG AC / 3′-AUC AUC AUC UG ▬ 5′

Slip back

5′-pppG UAG AC / 3′-AUC AUC AUC UG ▬ 5′

Primer-dependent initiation

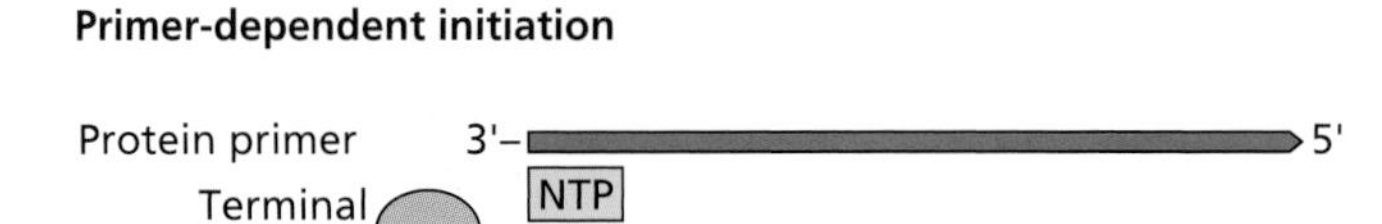

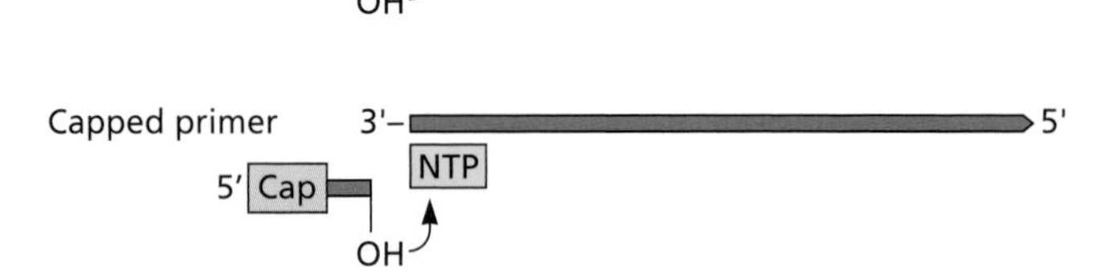

Figure 6.7 Mechanisms of initiation of RNA synthesis. *De novo* initiation may occur at the 3′ end of the viral RNA or from an internal base. When a primer is required, it may be a capped or protein-linked oligonucleotide.

De Novo Initiation

In this process the first phosphodiester bond is made between the 3′-OH of the initiating NTP and the second NTP (Fig. 6.7). The viral polymerase then copies the entire viral genome without dissociating. Initiation takes place at the exact 3′ end of the template, except during replication

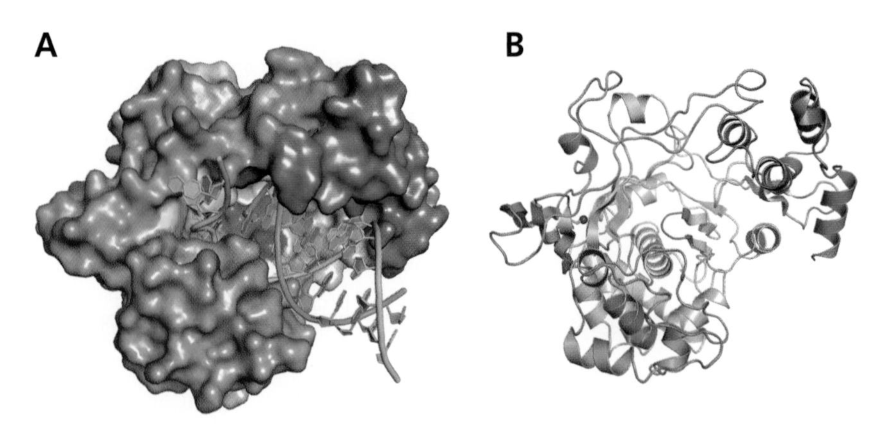

Figure 6.6 Structure of RNA polymerase with template and primer. (A, B) Views from the top of poliovirus 3D^{pol} polymerase looking down into the active site. **(A)** Surface representation of the elongation complex is shown with bound template (cyan) and product (green) RNAs showing how the duplex RNA is clamped in place between the pinky finger and thumb structures. From Protein Data Bank file 1ra6.

of the genomes of some (−) strand RNA viruses, such as bunyaviruses and arenaviruses (Fig. 6.7). Initiation begins at an internal C, and after extension of a few nucleotides, the daughter strand is shifted in the 3′ direction so that the 5′-terminal G residue is not base paired with the template strand. Because the daughter strand slips, this mechanism is called "prime and realign."

The structure of the RNA-dependent RNA polymerase of hepatitis C virus reveals how a primer-independent RNA polymerase positions the first nucleotide on the RNA template: a dinucleotide primer is synthesized by the polymerase using a beta-loop insertion in the thumb domain as a "protein platform" in the active site (Fig. 6.8). After the product reaches a certain length, the polymerase undergoes a conformational change that moves the priming platform out of the way and allows the newly synthesized complementary RNA to exit as the enzyme moves along the template strand.

A protein platform also appears to be involved in *de novo* priming by the reovirus RNA polymerase, a cubelike structure with a catalytic site in the center that is accessible by four tunnels. One tunnel allows template entry, one serves for the exit of newly synthesized double-stranded RNA, a third permits exit of mRNA, and a fourth is for substrate entry. A priming loop that is not observed in this region of other RNA polymerases is present in the palm domain. The loop supports the initiating NTP, then retracts into the palm and fits into the minor groove of the double-stranded RNA product. This movement assists in the transition between initiation and elongation, and also allows the newly synthesized RNA to exit the polymerase.

Protein platforms also appear to be involved in the *de novo* priming of RNA synthesis by other flaviviruses (dengue and West Nile viruses), influenza virus (genome RNA synthesis is primer independent), and bacteriophage Φ6.

Primer-Dependent Initiation

Protein priming. A protein-linked oligonucleotide serves as a primer for RNA synthesis by RNA polymerases of members of the *Picornaviridae* and *Caliciviridae*. Protein priming also occurs during DNA replication of adenoviruses, certain DNA-containing bacteriophages (Chapter 9), and hepatitis B virus (Chapter 7). A terminal protein provides a hydroxyl group (in a tyrosine or serine residue) to which the first oligonucleotide can be linked, by viral polymerases, via a phosphodiester bond. The protein-linked primer is then used for elongation.

Polioviral genomic RNA, as well as newly synthesized (+) and (−) strand RNAs, are covalently linked at their 5′ ends to the 22-amino-acid protein VPg (Fig. 6.9A), initially suggesting that VPg might function as a primer for RNA synthesis. This hypothesis was supported by the discovery of a uridylylated form of the protein, VPg-pUpU, in infected cells.

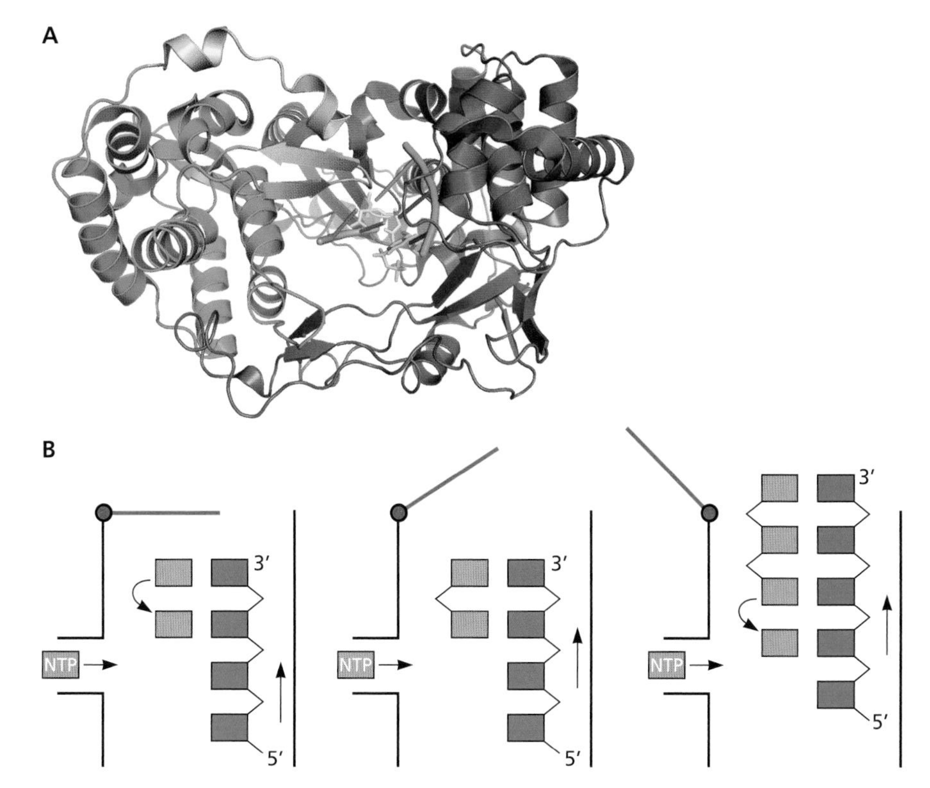

Figure 6.8 Mechanism of *de novo* initiation. (A) Ribbon diagram of RNA polymerase of hepatitis C virus. Fingers, palm, and thumb domain are colored blue, green, and magenta. The C-terminal loop that blocks the active site is shown in brown. Active site residues are yellow. Produced from Protein Data Bank file 4wtm. **(B)** Swinging-gate model of initiation. With the RNA template (green) in the active site of the enzyme, a short beta-loop (red) provides a platform on which the first complementary nucleotide (light green) is added to the template (left). The second nucleotide is then added, producing a dinucleotide primer for RNA synthesis (middle). At this point nothing further can happen because the priming platform blocks the exit of the RNA product from the enzyme. The solution to this problem is that the polymerase undergoes a conformational change that moves the priming platform out of the way and allows the newly synthesized complementary RNA (right) to exit as the enzyme moves along the template strand.

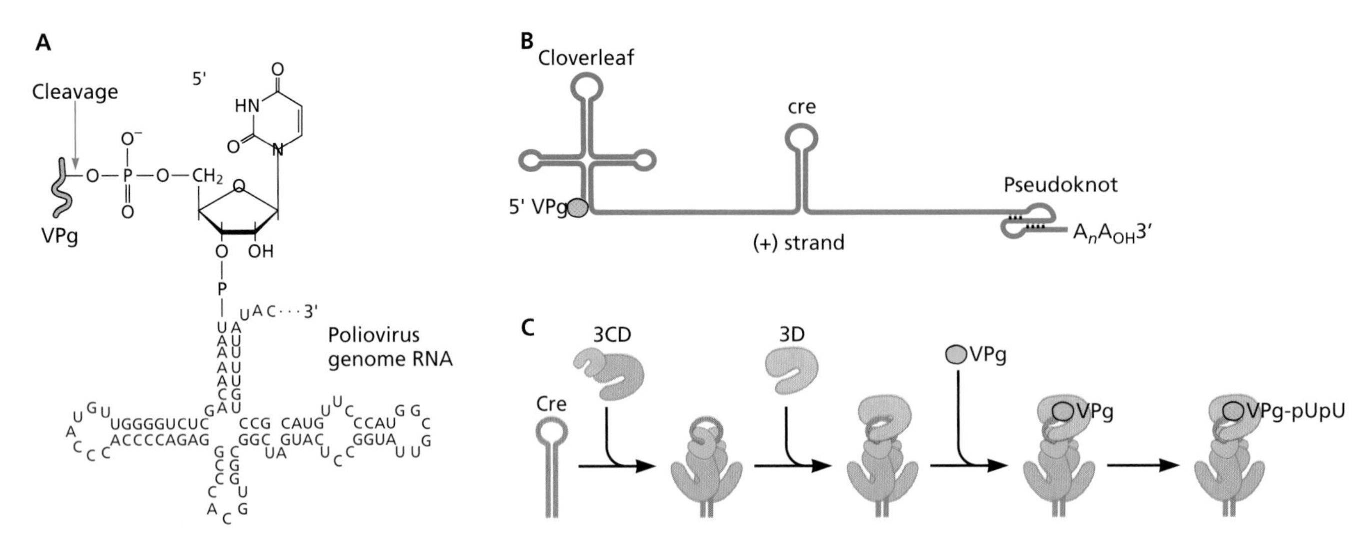

Figure 6.9 Uridylylation of VPg. (A) Linkage of VPg to polioviral genomic RNA. Polioviral RNA is linked to the 22-amino-acid VPg (orange) via an *O*4-(5′-uridylyl)-tyrosine linkage. This phosphodiester bond is cleaved at the indicated site by a cellular enzyme to produce the viral mRNA containing a 5′-terminal pU. **(B)** Structure of the poliovirus (+) strand RNA template, showing the 5′-cloverleaf structure, the internal cre (*cis*-acting replication element) sequence, and the 3′ pseudoknot. **(C)** Model for assembly of the VPg uridylylation complex. Two molecules of 3CD bind to cre. The 3C dimer melts part of the stem. $3D^{pol}$ binds to the complex by interactions between the back of the thumb domain and the surface of 3C. VPg then binds the complex and is linked to two U moieties in a reaction templated by the cre sequence.

VPg can be uridylylated *in vitro* by $3D^{pol}$ and can then prime the synthesis of VPg-linked poly(U) from a poly(A) template. The template for uridylylation of VPg is either the 3′-poly(A) on (+) strand RNA (during synthesis of (−) strand RNA, Fig. 6.10), or an RNA hairpin, the *cis*-acting replication element (cre), located in the coding region (during synthesis of (+) strand RNA (Fig. 6.9B and C).

Structures of the RNA polymerases of different picornaviruses and caliciviruses indicate that the active site is more accessible than in polymerases with a *de novo* mechanism of initiation. The small thumb domains of these polymerases leave a wide central cavity that can accommodate the template primer and the protein primer.

Uridylylation of VPg can be achieved in a reaction containing $3D^{pol}$, a template $(rA)_{10}$, UTP, and Mg^{2+} and Mn^{2+}. Crystallographic analysis of this structure reveals that VPg-pU is bound in the template-binding channel, with the N terminus of VPg in the NTP entry channel and the C terminus pointing toward the template-binding channel. The hydroxyl group of a tyrosine in VPg is covalently linked to the α-phosphate of UMP and interacts with a divalent metal ion that binds an Asp of the Gly-Asp-Asp motif in the active site. This arrangement of VPg is similar to that of the primer terminus in the nucleotidyl transfer reaction, demonstrating that $3D^{pol}$ catalyzes VPg uridylylation using the same two-metal mechanism as the nucleotidyl transfer reaction.

When VPg uridylylation begins at the 3′-poly(A) tail of the (+) strand template, the polymerase continues nucleotidyl transfer reactions and replicates the entire genome. However, when uridylylation of VPg takes place on the CRE, the protein must dissociate and transfer to the 3′ end of the RNA. How this process is accomplished is not known.

Protein priming by the birnavirus RNA polymerase VP1 is unusual because the primer **is** the polymerase, not a separate protein. Even in the absence of a template, VP1 has self-guanylylation activity that is dependent on divalent metal ions. The guanylylation site is a serine located approximately 23 Å from the catalytic site of the polymerase. The long distance between these sites suggests that guanylylation may be carried out at a second active site. The finding that some altered polymerases that are inactive in RNA synthesis retain self-guanylylation activity supports this hypothesis. After two G residues are added to VP1, it binds to a conserved CC sequence at the terminus of the viral RNA template to initiate RNA synthesis. The 5′ ends of mRNAs and genomic double-stranded RNAs produced by this reaction are therefore linked to a VP1 molecule.

Priming by capped RNA fragments. Influenza virus mRNA synthesis is blocked by treatment of cells with the fungal toxin α-amanitin at concentrations that inhibit cellular DNA-dependent RNA polymerase II. This surprising finding demonstrated that the viral RNA polymerase is dependent on a host cell RNA polymerase II. Inhibition by α-amanitin is explained by a requirement for newly synthesized cellular transcripts made by this enzyme to provide primers for viral mRNA synthesis. Presumably, these transcripts must be

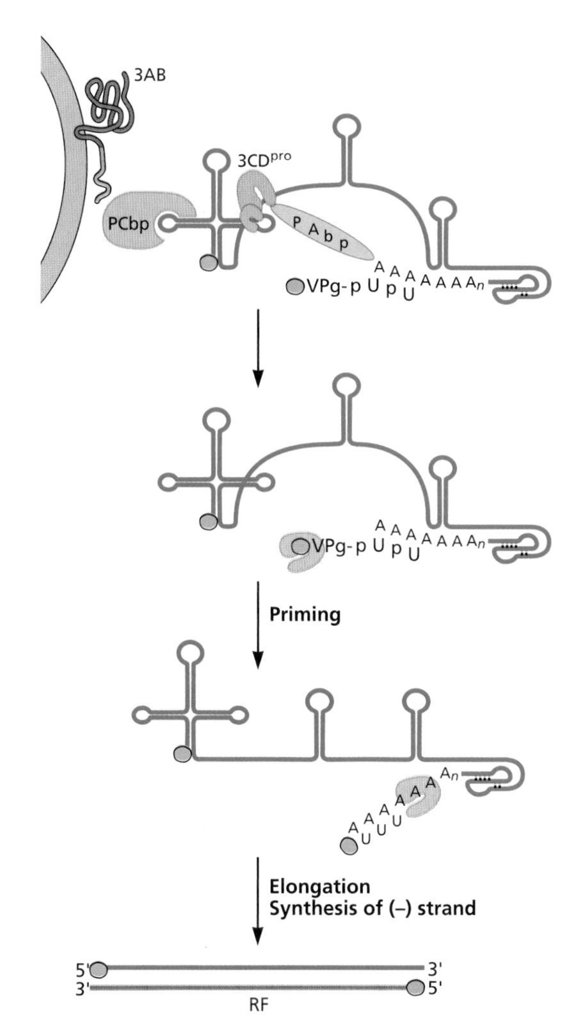

Figure 6.10 Poliovirus (−) strand RNA synthesis. The precursor of VPg, 3AB, contains a hydrophobic domain and is a membrane-bound donor of VPg. A ribonucleoprotein complex is formed when poly(rC)-binding protein 2 (PCbp2) and 3CD^{pro} bind the cloverleaf structure located within the first 108 nucleotides of (+) strand RNA. The ribonucleoprotein complex interacts with poly(A)-binding protein 1 (PAbp1), which is bound to the 3′-poly(A) sequence, bringing the ends of the genome into close proximity. Protease 3CD^{pro} cleaves membrane-bound 3AB, releasing VPg and 3A. VPg-pUpU is synthesized by 3D^{pol} using the 3′-poly(A) sequence as a template, and used by 3D^{pol} as a primer for RNA synthesis. Modified from A. V. Paul, p. 227–246, *in* B. L. Semler and E. Wimmer (ed), *Molecular Biology of Picornaviruses* (ASM Press, Washington, DC, 2002).

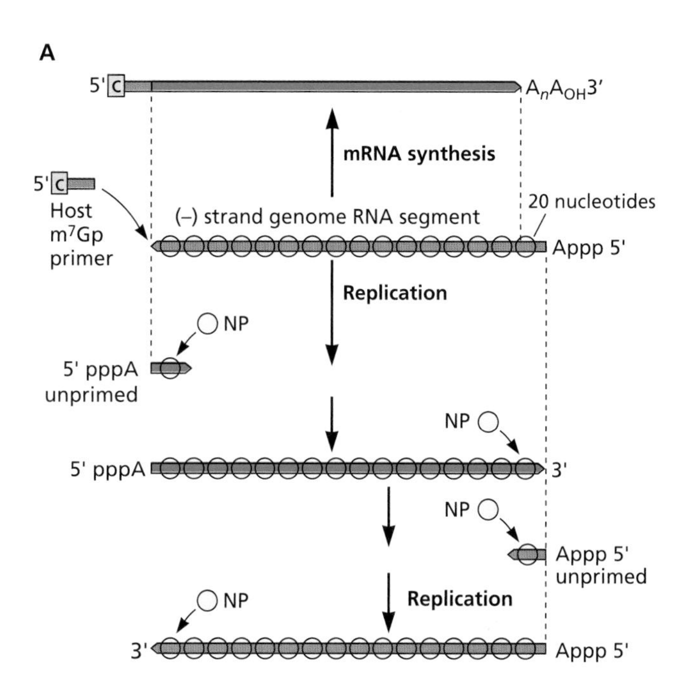

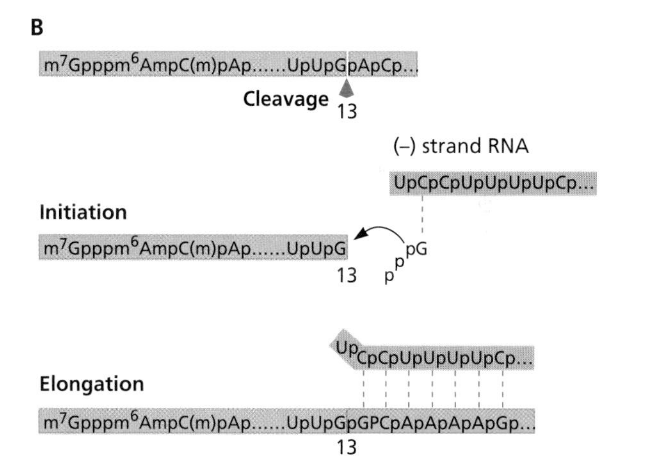

Figure 6.11 Influenza virus RNA synthesis. (A) Viral (−) strand genomes are templates for the production of either subgenomic mRNAs or full-length (+) strand RNAs. The switch from viral mRNA synthesis to genomic RNA replication is regulated by both the number of nucleocapsid (NP) protein molecules and the acquisition by the viral RNA polymerase of the ability to catalyze initiation without a primer. Binding of the NP protein to elongating (+) strands enables the polymerase to read to the 5′ end of genomic RNA. **(B)** Capped RNA-primed initiation of influenza virus mRNA synthesis. Capped RNA fragments cleaved from the 5′ ends of cellular nuclear RNAs serve as primers for viral mRNA synthesis. The 10 to 13 nucleotides in these primers do not need to hydrogen bond to the common sequence found at the 3′ ends of the influenza virus genomic RNA segments. The first nucleotide added to the primer is a G residue templated by the penultimate C residue of the genomic RNA segment; this is followed by elongation of the mRNA chains. The terminal U residue of the genomic RNA segment does not direct the incorporation of an A residue. The 5′ ends of the viral mRNAs therefore comprise 10 to 13 nucleotides plus a cap structure snatched from host nuclear pre-mRNAs. Adapted from S. J. Plotch et al., *Cell* **23:**847–858, 1981, with permission.

made continuously because they are exported rapidly from the nucleus once processed. Such transcripts are cleaved by a virus-encoded, cap-dependent endonuclease that is part of the RNA polymerase (Fig. 6.11). The resulting 10- to 13-nucleotide capped fragments serve as primers for the initiation of viral mRNA synthesis.

Bunyaviral mRNA synthesis is also primed with capped fragments of cellular RNAs. In contrast to that of influenza virus, bunyaviral mRNA synthesis is not inhibited by α-amanitin because it occurs in the cytoplasm, where capped cellular RNAs are abundant.

The influenza virus RNA polymerase is a heterotrimer composed of PA, PB1, and PB2 proteins (Fig. 6.12). The PB1 protein is the RNA polymerase, the PB2 subunit binds capped host mRNAs, and the PA protein harbors endonuclease activity. In contrast, the bunyavirus RNA polymerase is a single protein (L). The N-terminal domains of influenza PA and bunyavirus L have endonuclease activity that participates in cap snatching. The structures of endonuclease domains from these viruses reveal the presence of a common nuclease fold.

Figure 6.12 Activation of the influenza virus RNA polymerase by specific virion RNA sequences. The three P proteins form a multisubunit assembly that can neither bind to capped primers nor synthesize mRNAs. Addition of a sequence corresponding to the 5′-terminal 11 nucleotides of the viral RNA, which is highly conserved in all eight genome segments, activates the cap-binding activity of the P proteins. The PB1 protein binds this RNA sequence and activates the cap-binding PB2 subunit, probably by conformational change. Concomitantly with activation of cap binding, the P proteins acquire the ability to bind to a conserved sequence at the 3′ ends of genomic RNA segments. This second interaction activates the endonuclease that cleaves host cell RNAs 10 to 13 nucleotides from the cap, producing the primers for viral mRNA synthesis. The RNA polymerase can then carry out initiation and elongation of mRNAs. p, polymerase active site. 5′ and 3′ indicate the binding sites for the 5′ and 3′ ends, respectively, of (−) strand genomic RNA. Blue indicates an inactive site, and red indicates an active site. The polymerase is bound to both the 5′ and 3′ ends of the genomic RNA, with the capped RNA primer associated with the PB2 protein. Adapted from D. M. Knipe et al. (ed), *Fields Virology*, 4th ed. (Lippincott Williams & Wilkins, Philadelphia, PA, 2001), with permission.

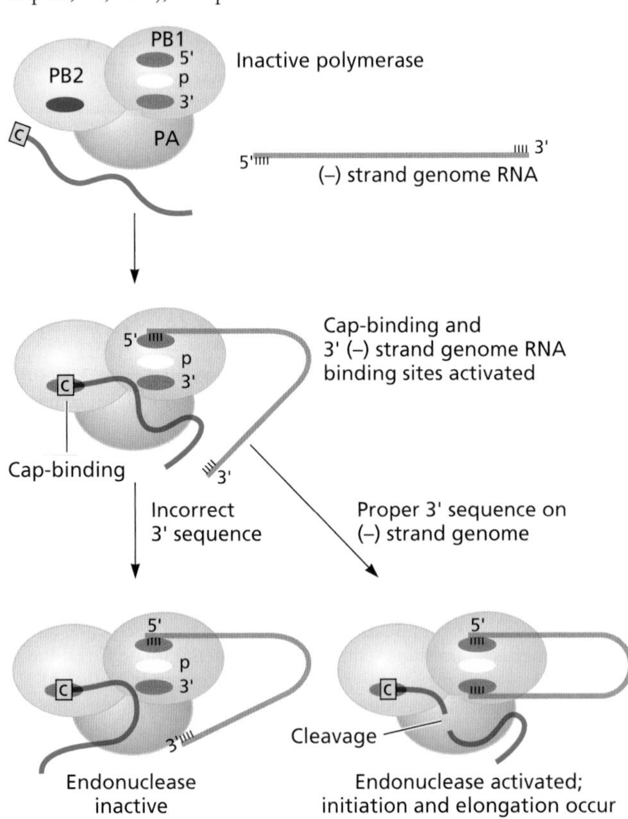

Capping

Most viral mRNAs carry a 5′-terminal cap structure (exceptions include picornaviruses and the flavivirus hepatitis C virus), but the modification is made in different ways. Three mechanisms can be distinguished: acquisition of preformed 5′ cap structures from cellular pre-mRNAs or mRNAs, or during priming of mRNA synthesis as described in the previous section. Details of the latter processes can be found in Chapter 10.

Elongation

After a polymerase has associated stably with the nucleic acid template, the enzyme then adds nucleotides without dissociating from the template. Most polymerases are highly **processive**; that is, they can add thousands of nucleotides before dissociating. The poliovirus RNA polymerase $3D^{pol}$ can add 5,000 and 18,000 nucleotides in the absence or presence, respectively, of the accessory protein 3AB. The vesicular stomatitis virus P protein enhances the processivity of the RNA polymerase L, possibly as a result of conformational changes that occur upon binding of P. The increased processivity induced by P protein is enhanced in the presence of N, possibly because the template must be kept unstructured so as not to impede the progress of L. Full processivity of the influenza virus RNA polymerase also requires the presence of NP.

All nucleic acid synthesis begins with the formation of a complex of polymerase, template-primer, and initiating NTP. This complex becomes activated, a process that includes a conformational change from an open to a closed form. The nucleotidyl transfer reaction then takes place, pyrophosphate is released, and the template-primer moves by one base. RNA-dependent RNA polymerases have been visualized in template-primer, open NTP, and closed NTP forms. Interactions between the fingertips and the thumb domains of RNA-dependent RNA polymerases precede the large-scale conformational changes upon binding of template, primer, or NTP in the template-binding channel, and are characteristic of other polymerases.

The template-primer double-stranded RNA binds to the template-binding channel of picornavirus $3D^{pol}$ with the 3′ end of the template leaving the active site through one face of the enzyme. The template strand interacts mainly with the fingers domain, and the primer strand interacts with the thumb and palm domains. Seven nucleotides can fit into the template-binding channel. The 3′-hydroxyl of the primer forms a hydrogen bond with the catalytic Asp in motif C (Asp338) in the active site. This Asp also binds a metal ion, the only one observed in the active site in this state.

Catalysis begins with the template above the active site, followed by loading of the first NTP and closure of the active site to produce a precatalytic state. The phosphate group of the incoming NTP binds a metal ion near Asp338 in motif C, and the 3′-OH of this NTP forms a hydrogen bond with Asp245 in motif A. This structure cannot carry out nucleotidyl transfer reactions until the 3′-hydroxyl of the primer terminus binds the second metal ion. After catalysis, the active site is opened, followed by translocation of the template-product. Such translocation requires conformational flexibility of a loop sequence in the B motif of the palm domain.

The closed conformation has been visualized in an elongation assembly composed of the norovirus RNA polymerase, self-complementary RNA, incoming NTP, and Mg^{2+} and Mn^{2+}. This structure is ready to carry out the nucleotidyl transfer reaction. Binding of the template-primer-NTP complex is associated with rotation of helices in the thumb domain. This movement creates a binding groove for a primer strand. The incoming NTP base pairs with the template, and the phosphates bind two Mn^{2+} ions in the active site. Motif A Asp242, and Asp343 and Asp344 of the Asp-Gly-Asp motif also coordinate the divalent metal ions. The interaction of polymerase with the 2′-hydroxyl of the rNTP also participates in addition of NTP to the nascent strand. The 2′-hydroxyl of the rNTP forms hydrogen bonds with Ser and Asn residues in motif B of the fingers domain. In the "open" conformation this interaction does not occur and the incoming NTP is not correctly assembled for catalysis. This conformation therefore represents a closed complex trapped immediately prior to catalysis.

Template Specificity

Viral RNA-dependent RNA polymerases must select viral templates from among a vast excess of cellular mRNAs and then initiate correctly to ensure accurate RNA synthesis. Different mechanisms that contribute to template specificity have been identified. Initiation specificity may be regulated by the affinity of the RNA polymerase for the initiating nucleotide. For example, the RNA polymerases of bovine viral diarrhea virus and bacteriophage ϕ6 prefers 3′-terminal C. Reovirus RNA polymerase prefers a G at the second position of the template RNA. This preference is controlled by hydrogen bonding of carbonyl and amino groups of the G with two amino acids of the enzyme. Both preferences would exclude initiation on cellular mRNAs that end in poly(A) (the great majority).

Template specificity may also be conferred by the recognition of RNA sequences or structures at the 5′ and 3′ ends of viral RNAs by viral proteins. RNA synthesis initiates specifically within a polypyrimidine tract in the 3′-untranslated region of hepatitis C virus RNA. The 3′-noncoding region of polioviral genomic RNA contains an **RNA pseudoknot** structure that is conserved among picornaviruses (Fig. 6.10). A viral protein (3AB-3CD) binds this structure and may direct the polymerase to that site for the initiation of (−) strand RNA synthesis. The precursor to the poliovirus RNA polymerase ($3CD^{pro}$) plays an important role in viral RNA synthesis by participating in the formation of a ribonucleoprotein at the 5′ end of the (+) strand RNA. This protein, together with cellular poly(rC)-binding protein 2, binds to a cloverleaf structure in the viral RNA (Fig. 6.10). Alterations within the RNA-binding domain of 3CD inhibit binding to the cloverleaf and RNA synthesis.

Internal RNA sequences may confer initiation specificity to RNA polymerases. The *cis*-acting replication elements (cre) in the coding sequence of poliovirus protein 2C and rhinovirus capsid protein VP1 contain short RNA sequences that are required for RNA synthesis. These sequences are binding sites for $3CD^{pro}$ and, as discussed previously, serve as a template for uridylylation of the VPg protein (Fig. 6.10).

During mRNA synthesis by influenza virus polymerase, sequences at the RNA termini ensure that the 5′ ends of newly synthesized viral mRNAs are not cleaved and used as primers (Fig. 6.12). If such cleavage were to occur, there would be no net synthesis of viral mRNAs. Such binding to two sites in the genomic RNA blocks access of a second P protein and protects newly synthesized viral mRNA from endonucleolytic cleavage by P proteins.

Protein-protein interactions can also direct RNA polymerases to the RNA template. The vesicular stomatitis virus RNA polymerase for mRNA synthesis consists of the P protein and the L protein, the catalytic subunit. The P protein binds both the L protein and the ribonucleoprotein containing N and the (−) strand RNA. In this way the P protein brings the L polymerase to the RNA template (See "(−) Strand RNA" below). Cellular general initiation proteins have a similar function in bringing RNA polymerase II to the correct site to initiate transcription of DNA templates.

While viral RNA polymerases copy only viral RNAs in the infected cell, purified polymerases often lack template specificity. The replication complex in the infected cell may contribute to template specificity by concentrating reaction components to create an environment that copies viral RNAs selectively. Replication of viral RNAs on membranous structures might contribute to such specificity (see "Cellular Sites of Viral RNA Synthesis").

Unwinding the RNA Template

Base-paired regions in viral RNA must be disrupted to permit copying by RNA-dependent RNA polymerase. RNA helicases, which are encoded in the genomes of many RNA viruses, are thought to unwind the genomes of double-stranded RNA viruses, as well as secondary structures in template RNAs. They also prevent extensive base pairing between template RNA and the nascent complementary strand. The RNA helicases of several viruses that are important human pathogens, including the flaviviruses hepatitis C virus and dengue virus, have been studied extensively because these proteins are potential targets for therapeutic intervention. To facilitate the development of new agents that inhibit these helicases, their three-dimensional structures have been determined by X-ray crystallography. These molecules comprise three domains that mediate hydrolysis of NTPs and RNA binding (Fig. 6.13). Between the domains is a cleft that is large enough to accommodate single-stranded but not double-stranded RNA. Unwinding of double-stranded RNA probably occurs as one strand of RNA passes through the cleft and the other is excluded.

The bacteriophage ϕ6 RNA polymerase can separate the strands of double-stranded RNA without the activity of a helicase. Examination of the structure of the enzyme suggests how such melting might be accomplished. This RNA polymerase has a plowlike protuberance around the entrance to the template channel that is thought to separate the two strands, allowing only one to enter the channel.

Role of Cellular Proteins

Host cell components required for viral RNA synthesis were initially called "host factors," because nothing was known about their chemical composition. Evidence that cellular proteins are essential components of a viral RNA polymerase first came from studies of the bacteriophage Qβ enzyme. This viral RNA-dependent RNA polymerase is a multisubunit enzyme, consisting of a 65-kDa virus-encoded protein and four host proteins: ribosomal protein S1, translation elongation proteins (EF-Tu and EF-Ts), and an RNA-binding protein. Proteins S1 and EF-Tu contain RNA-binding sites that enable the RNA polymerase to recognize the viral RNA template. The 65-kDa viral protein exhibits no RNA polymerase activity in the absence of the host proteins, but has sequence and structural similarity to known RNA-dependent RNA polymerases.

Figure 6.13 Structure of a viral RNA helicase. The RNA helicase of yellow fever virus is shown in surface representation, colored red, white, or blue depending on the distance of the amino acid from the center of the molecule. A model for melting of double-stranded RNA is shown. From Protein Data Bank file 1yks.

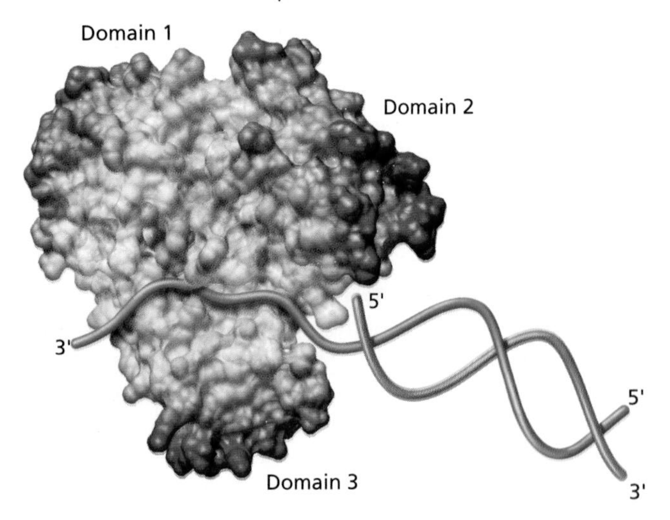

Polioviral RNA synthesis also requires host cell proteins. When purified polioviral RNA is incubated with a cytoplasmic extract prepared from uninfected permissive cells, the genomic RNA is translated and the viral RNA polymerase is made. If guanidine hydrochloride is included in the reaction mixture, the polymerase assembles on the viral genome, but RNA synthesis is not initiated. The RNA polymerase-template assembly can be isolated free of guanidine, but RNA synthesis does not occur unless a new cytoplasmic extract is added, indicating that soluble cellular proteins are required for initiation. A similar conclusion comes from studies in which polioviral RNA was injected into oocytes derived from the African clawed toad *Xenopus laevis*: the viral RNA cannot replicate in *Xenopus* oocytes unless it is coinjected with a cytoplasmic extract from human cells. These observations can be explained by the requirement of the viral RNA polymerase for one or more mammalian proteins that are absent in toad oocytes.

One of these host cell proteins is poly(rC)-binding protein, which binds to a cloverleaf structure that forms in the first 108 nucleotides of (+) strand RNA (Fig. 6.10). Formation of a ribonucleoprotein composed of the 5′ cloverleaf, 3CD, and poly(rC)-binding protein is essential for the initiation of viral RNA synthesis. Interaction of poly(rC)-binding protein with the cloverleaf facilitates the binding of viral protein 3CD to the opposite side of the same cloverleaf.

Another host protein that is essential for polioviral RNA synthesis is poly(A)-binding protein 1. This protein brings together the ends of the viral genome by interacting with poly(rC)-binding protein 2, $3CD^{pro}$, and the 3′-poly(A) tail of poliovirus RNA. (Fig. 6.10). Formation of this circular ribonucleoprotein complex is required for (−) strand RNA synthesis.

Recent technical advances have facilitated identification of host cell proteins required for viral RNA synthesis. Interactions among cellular and viral proteins can be identified readily by mass spectrometry, and their function in viral replication can be determined by silencing their production by RNA interference or other methods. This approach has been used to show that the cellular RNA helicase A participates in influenza viral RNA synthesis, and heat shock protein 70 associates with the replication complex of Japanese encephalitis virus and positively regulates RNA synthesis.

Paradigms for Viral RNA Synthesis

Exact replicas of the RNA genome must be made for assembly of infectious viral particles. However, the mRNAs of most RNA viruses are **not** complete copies of the viral genome.

The replication cycle of these viruses must therefore include a switch from mRNA synthesis to the production of full-length genomes. The majority of mechanisms for this switch regulate either the initiation or the termination of RNA synthesis.

(+) Strand RNA

For some (+) strand RNA viruses, the genome and mRNA are identical. The genome RNA of the *Picornaviridae* and *Flaviviridae* is translated upon entry into the cytoplasm to produce viral proteins, including the RNA-dependent RNA polymerase and accessory proteins. The (+) strand RNA genome is copied to a (−) strand, which in turn is used as a template for the synthesis of additional (+) strands (Fig. 6.1). Newly synthesized (+) strand RNA molecules can serve as templates for further genomic replication, as mRNAs for the synthesis of viral proteins, or as genomic RNAs to be packaged into progeny virions. Because picornaviral mRNA is identical in sequence to the viral RNA genome, all RNAs needed for the reproduction of these viruses can be made by a simple set of RNA synthesis reactions (Fig. 6.1). Such simplicity comes at a price, because synthesis of individual viral proteins cannot be regulated. However, polioviral gene expression can be controlled by the rate and extent of polyprotein processing. For example, the precursor of the viral RNA polymerase, $3D^{pol}$, cannot polymerize RNA. Rather, this protein is a protease that cleaves at certain Gln-Gly amino acid pairs in the polyprotein. Regulating the processing of the precursor $3CD^{pro}$ controls the concentration of RNA polymerase.

The mRNAs synthesized during infection by most RNA viruses contain a 3′-poly(A) sequence, as do the vast majority of cellular mRNAs (exceptions are mRNAs of arenaviruses and reoviruses). The poly(A) sequence is encoded in the genome of (+) strand viruses. For example, polioviral (+) strand RNAs contain a 3′ stretch of poly(A), approximately 62 nucleotides in length, which is required for infectivity. The (−) strand RNA contains a 5′ stretch of poly(U), which is copied to form this poly(A).

The mechanisms of mRNA synthesis of other (+) strand RNA viruses allow structural and nonstructural proteins (generally needed in greater and lesser quantities, respectively) to be made separately. The latter are synthesized from full-length (+) strand (genomic) RNA, while structural proteins are translated from subgenomic mRNA(s). This strategy is a feature of the replication cycles of coronaviruses, caliciviruses, and alphaviruses. Translation of the Sindbis virus (+) strand RNA genome yields the nonstructural proteins that synthesize a full-length (−) strand (Fig. 6.14). Such RNA molecules contain not only a 3′-terminal sequence for initiation of (+) strand RNA synthesis, but also an internal initiation site, used for production of a 26S subgenomic mRNA.

Alphaviral genome and mRNA synthesis is regulated by the sequential production of three RNA polymerases with different template specificities. All three enzymes are derived from the nonstructural polyprotein P1234 and contain the complete amino acid sequence of this precursor (Fig. 6.14). The covalent connections among the segments of the polyprotein are successively broken, with ensuing alterations in the specificity of the enzyme (Fig. 6.15). It seems likely that each proteolytic cleavage induces a conformational change in the polymerase that alters its template specificity.

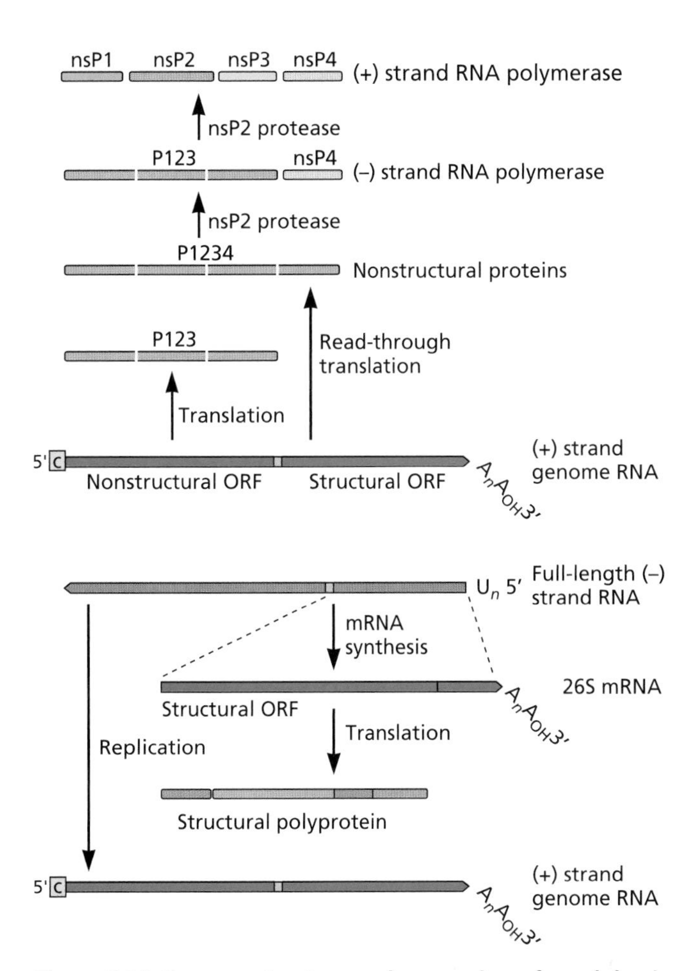

Figure 6.14 Genome structure and expression of an alphavirus, Sindbis virus. The 11,703-nucleotide Sindbis virus genome contains a 5′-terminal cap structure and a 3′-poly(A) tail. A conserved RNA secondary structure at the 3′ end of (+) strand genomic RNA is thought to control the initiation of (−) strand RNA synthesis. At early times after infection, the 5′ region of the genomic RNA (nonstructural open reading frame [ORF]) is translated to produce two nonstructural polyproteins: P123, whose synthesis is terminated at the first translational stop codon (indicated by the box), and P1234, produced by an occasional (15%) readthrough of this stop codon. The P1234 polyprotein is proteolytically cleaved to produce the enzymes that catalyze the various steps in genomic RNA replication: the synthesis of a full-length (−) strand RNA, which serves as the template for (+) strand synthesis, and either full-length genomic RNA or subgenomic 26S mRNA. The 26S mRNA, shown in expanded form, is translated into a structural polyprotein (p130) that undergoes proteolytic cleavage to produce the virion structural proteins. The 26S RNA is not copied into a (−) strand because a functional initiation site fails to form at the 3′ end.

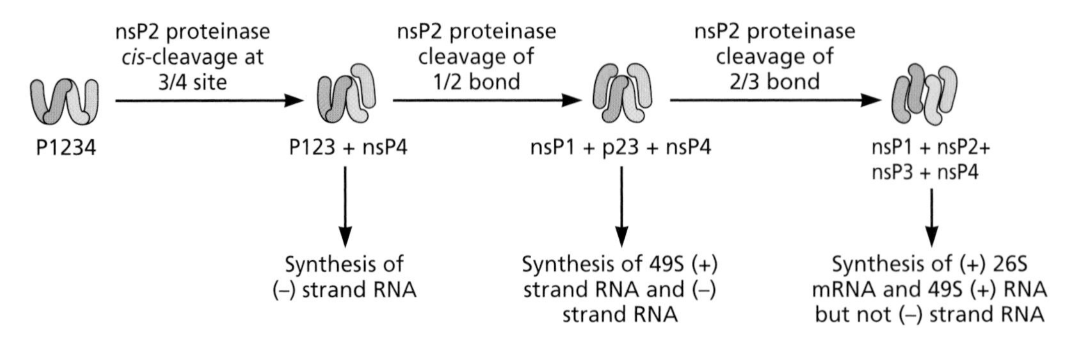

Figure 6.15 Three RNA polymerases with distinct specificities in alphavirus-infected cells. These RNA polymerases contain the entire sequence of the P1234 polyprotein and differ only in the number of proteolytic cleavages in this sequence.

Synthesis of Nested Subgenomic mRNAs

An unusual pattern of mRNA synthesis occurs in cells infected with members of the families *Coronaviridae* and *Arteriviridae*, in which subgenomic mRNAs that form a 3′-coterminal nested set with the viral genome are synthesized (Fig. 6.16). These viral families were combined into the order *Nidovirales* to denote this shared property (*nidus* is Latin for nest).

The subgenomic mRNAs of these viruses comprise a leader and a body that are synthesized from noncontiguous sequences at the 5′ and 3′ ends, respectively, of the viral (+) strand genome (Fig. 6.16A). The leader and body are separated by a conserved junction sequence encoded both at the 3′ end of the leader and at the 5′ end of the mRNA body. Subgenome-length (−) strands are produced when

Figure 6.16 Nidoviral genome organization and expression. (A) Organization of open reading frames. The (+) strand viral RNA is shown at the top, with open reading frames as boxes. The genomic RNA is translated to form polyproteins 1a and 1ab, which are processed to form the RNA polymerase. Structural proteins are encoded by nested mRNAs. **(B)** Model of the synthesis of nested mRNAs. Discontinuous transcription occurs during (−) strand RNA synthesis. Most of the (+) strand template is not copied, probably because it loops out as the polymerase completes synthesis of the leader RNA (orange). The resulting (−) strand RNAs, with leader sequences at the 3′ ends, are then copied to form mRNAs.

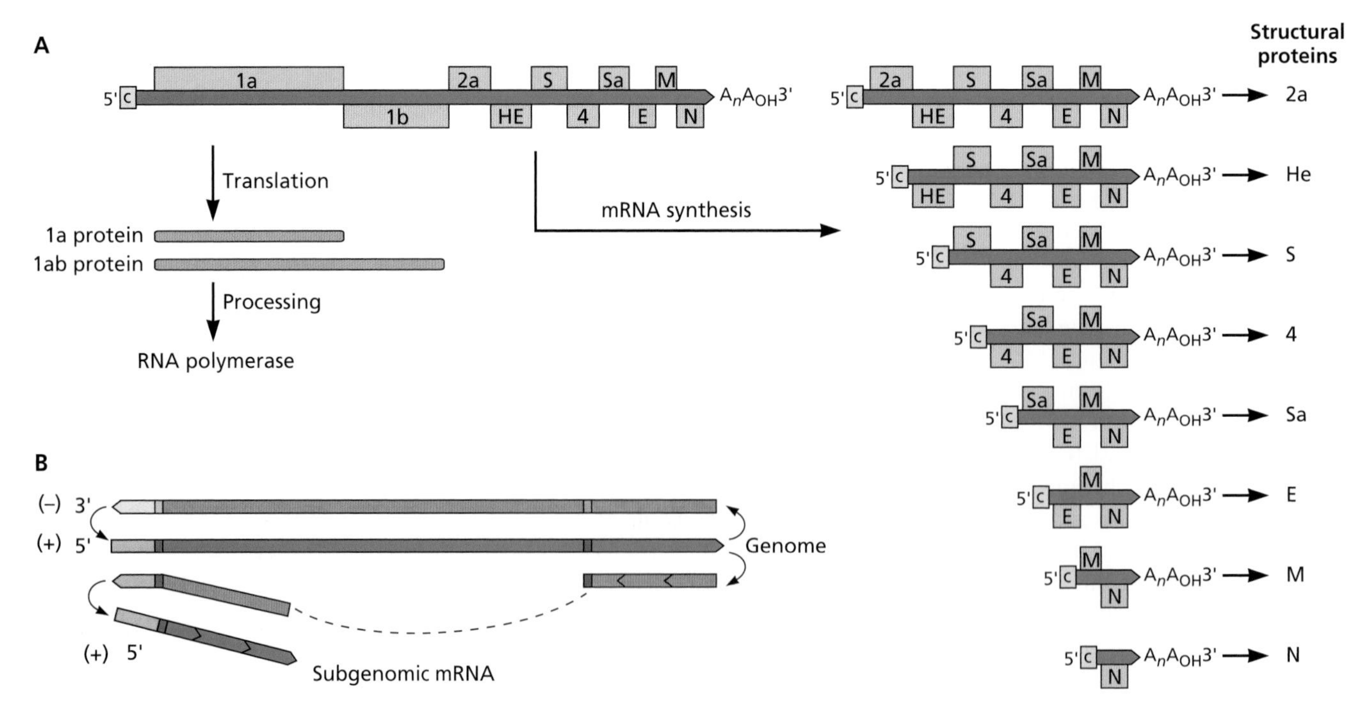

the template loops out as the polymerase completes synthesis of the leader RNA (Fig. 6.16B). These (−) strand subgenome-length RNAs then serve as templates for mRNA synthesis.

(−) Strand RNA

The genes of RNA viruses with a nonsegmented (−) strand RNA genome are expressed by the production of subgenomic mRNAs in infected cells (Fig. 6.17). An RNA polymerase composed of one molecule of L protein associated with four molecules of P protein is thought to carry out vesicular stomatitis virus mRNA synthesis. Individual mRNAs are produced by a series of initiation and termination reactions as the RNA polymerase moves down the viral genome (Fig. 6.18). This start-stop mechanism accounts for the observation that 3′-proximal genes must be copied before downstream genes (Box 6.4). The viral RNA polymerase is unable to initiate synthesis of each mRNA independently.

Vesicular stomatitis virus mRNA synthesis illustrates a second mechanism for poly(A) addition, reiterative copying of, or "stuttering" at, a short U sequence in the (−) strand template. After initiation, vesicular stomatitis virus mRNAs are elongated until the RNA polymerase reaches a conserved stop-polyadenylation signal [3′-AUACU$_7$-5′] located in each intergenic region (Fig. 6.19). Poly(A) (approximately 150 nucleotides) is added by reiterative copying of the U stretch, followed by termination.

The transition from mRNA to genome RNA synthesis in cells infected with vesicular stomatitis virus is dependent on the viral nucleocapsid (N) protein. To produce a full-length (+) strand RNA, the stop-start reactions at intergenic regions must be suppressed, a process that depends on the synthesis of the N and P proteins. The P protein maintains the N protein in a soluble form so that it can encapsidate the newly synthesized RNA. N-P assemblies bind to leader RNA and cause antitermination, signaling the polymerase to begin processive RNA synthesis. Additional N protein molecules then associate with the (+) strand RNA as it is elongated, and eventually bind to the seven A bases in the intergenic region. This interaction blocks reiterative copying of the seven U bases in the genome because the A bases cannot slip backward along the genomic RNA template. Consequently, RNA synthesis continues through the intergenic regions. The number of N-P protein complexes in infected cells therefore regulates

Figure 6.17 Vesicular stomatitis viral RNA synthesis. Viral (−) strand genomes are templates for the production of either subgenomic mRNAs or full-length (+) strand RNAs. The switch from mRNA synthesis to genomic RNA replication is mediated by two RNA polymerases and by the N protein. mRNA synthesis initiates at the beginning of the N gene, near the 3′ end of the viral genome. Poly(A) addition is a result of reiterative copying of a sequence of seven U residues present in each intergenic region. Chain termination and release occur after approximately 150 A residues have been added to the mRNA. The RNA polymerase then initiates synthesis of the next mRNA at the conserved start site 3′UUGUC . . . 5′. This process is repeated for all five viral genes. Synthesis of the full-length (+) strand begins at the exact 3′ end of the viral genome and is carried out by the RNA polymerase L-N-(P)4. The (+) strand RNA is bound by the viral nucleocapsid (N) protein, which is associated with the P protein in a 1:1 molar ratio. The N-P complexes bind to the nascent (+) strand RNA, allowing the RNA polymerase to read through the intergenic junctions at which polyadenylation and termination take place during mRNA synthesis.

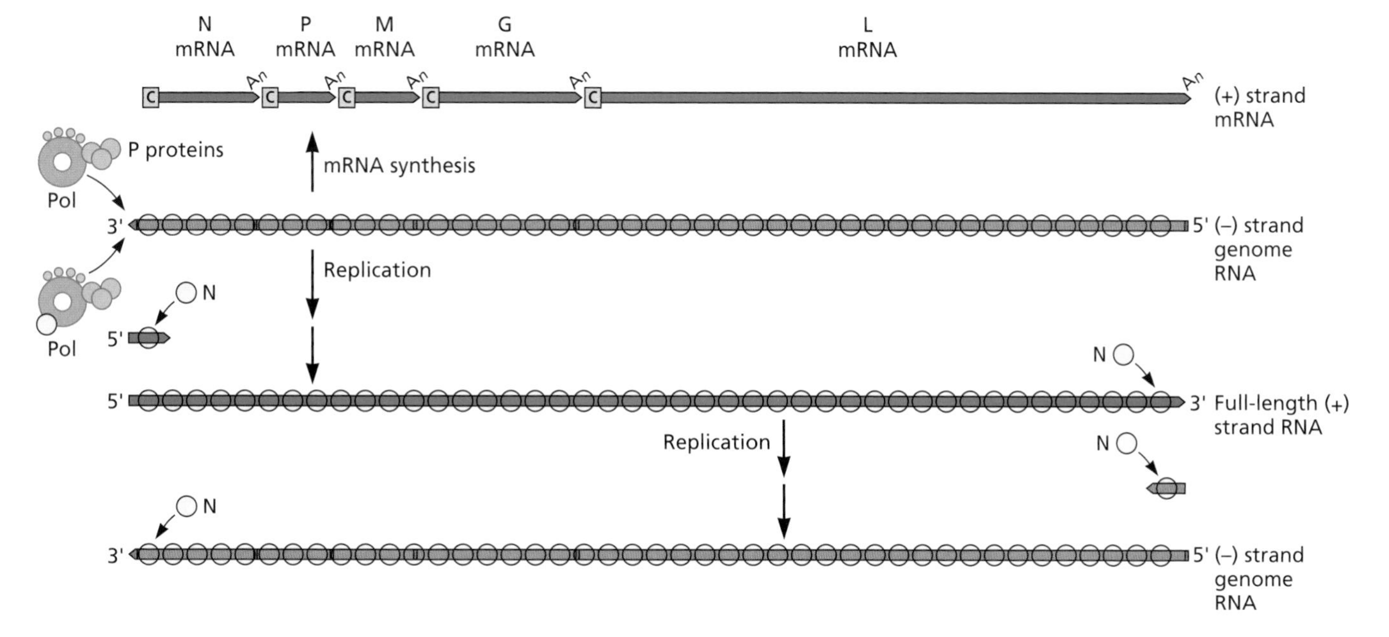

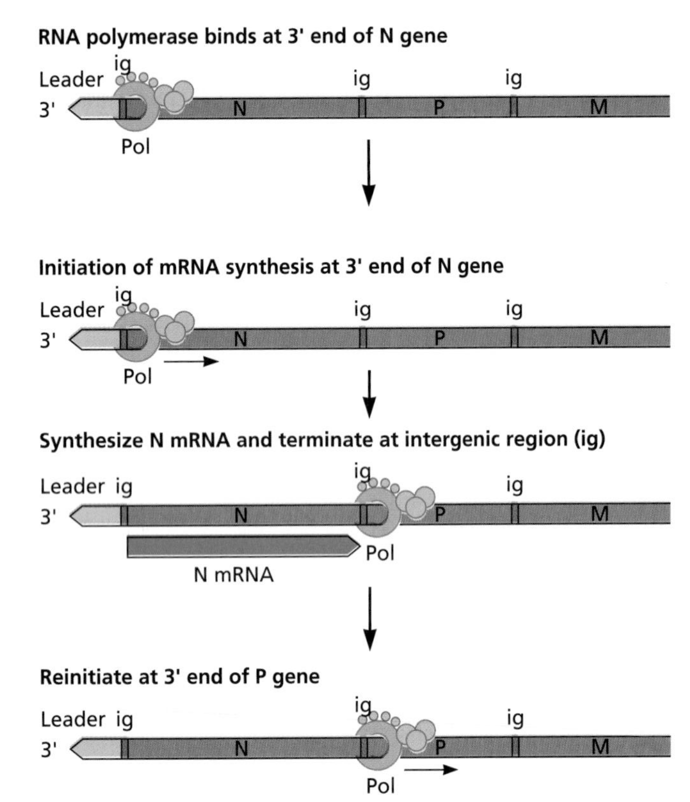

Figure 6.18 Stop-start model of vesicular stomatitis mRNA synthesis. The RNA polymerase (Pol) initiates RNA synthesis at the 3′ end of the N gene. After synthesis of the N mRNA, RNA synthesis terminates at the intergenic region, followed by reinitiation at the 3′ end of the P gene. This process continues until all five mRNAs are synthesized. Reinitiation does not occur after the last mRNA (the L mRNA) is synthesized, and, as a consequence, the 59 5′-terminal nucleotides of the vesicular stomatitis virus genomic RNA are not copied. Only a fraction of the polymerase molecules successfully make the transition from termination to reinitiation of mRNA synthesis at each intergenic region.

the relative efficiencies of mRNA synthesis and genome RNA replication. The copying of full-length (+) strand RNAs to (−) strand genomic RNAs also requires the binding of N-P protein complexes to elongating RNA molecules. Newly synthesized (−) strand RNAs are produced as nucleocapsids that can be readily packaged into progeny viral particles.

The (−) strand RNA genome of paramyxoviruses is copied efficiently only when its length in nucleotides is a multiple of 6. This requirement, called the **rule of six**, is probably a consequence of the association of each N monomer with exactly six nucleotides. Assembly of the nucleocapsid begins with the first nucleotide at the 5′ end of the RNA and continues until the 3′ end is reached. If the genome length is not a multiple of 6, then the 3′ end of the genome will not be precisely aligned with the last N monomer. Such misalignment reduces the efficiency of initiation of RNA synthesis at the 3′ end. Curiously, although the rhabdovirus N protein binds nine nucleotides of RNA, the genome length need not be a multiple of this number for efficient copying.

The segmented (−) strand RNA genome of influenza virus is expressed by the synthesis of subgenomic mRNAs in infected cells by a heterotrimeric RNA polymerase described previously (Fig. 6.12). Individual mRNAs are initiated with a capped primer derived from host cell mRNA, and terminate 20 nucleotides short of the template 3′ end. Polyadenylation of these mRNAs is achieved by a similar mechanism to that observed during vesicular stomatitis virus mRNA synthesis, reiterative copying of a short U sequence in the (−) strand template. Such copying is thought to be a consequence of the RNA polymerase specifically binding the 5′ end of (−) strand RNA and remaining at this site throughout mRNA synthesis. The genomic RNAs are threaded through the polymerase in a 3′ → 5′ direction as mRNA synthesis proceeds (Fig. 6.20). Eventually the template is unable to move, leading to reiterative copying of the U residues.

The influenza virus NP protein also regulates the switch from viral mRNA to full-length (+) strand synthesis (Fig. 6.11). The RNA polymerase for genome replication reads through the polyadenylation and termination signals for mRNA production only if NP is present. This protein is thought to bind nascent (+) strand transcripts and block poly(A) addition by a mechanism analogous to that described for vesicular stomatitis virus N protein. Copying of (+) strand RNAs into (−) strand RNAs also requires NP protein. Intracellular concentrations of NP protein are therefore an important determinant of whether mRNAs or full-length (+) strands are synthesized.

Ambisense RNA

Although arenaviruses are considered (−) strand RNA viruses, their genomic RNA is in fact **ambisense**: mRNAs are produced both from (−) strand genomic RNA and from complementary full-length (+) strands. The arenavirus genome comprises two RNA segments, S (small) and L (large) (Fig. 6.21). Shortly after infection, RNA polymerase that enters from viral particles synthesizes mRNAs from the 3′ region of both RNA segments. Synthesis of each mRNA terminates at a stem-loop structure. These mRNAs, which are translated to produce the nucleocapsid (NP) protein and RNA polymerase (L) protein, respectively, are the only viral RNAs made during the first several hours of infection. Later in infection, the block imposed by the stem-loop structure is overcome, permitting the synthesis of full-length S and L (+) RNAs. It was initially thought that melting of the stem-loop structure by the NP protein allowed the transcription termination signal to be bypassed. It now seems more likely that two different RNA polymerases are made in infected cells, one that produces mRNAs and a second that synthesizes full-length copies of the genome. The finding that viral mRNAs are capped while genomes are not is consistent with this hypothesis.

BOX 6.4

EXPERIMENTS

Mapping gene order by UV irradiation

The effects of ultraviolet (UV) irradiation provided insight into the mechanism of vesicular stomatitis virus mRNA synthesis. In these experiments, virus particles were irradiated with UV light, and the effect on the synthesis of individual mRNAs was assessed. UV light causes the formation of pyrimidine dimers that block passage of the RNA polymerase. In principle, larger genes require less UV irradiation to inactivate mRNA synthesis and have a larger **target size.** The dose of UV irradiation needed to inactivate synthesis of the N mRNA corresponded to the predicted size of the N gene, but this was not the case for the other viral mRNAs. The target size of each other mRNA was the sum of its size plus the size of other genes located 3′ to it. For example, the UV target size of the L mRNA is the size of the entire genome. These results indicate that these mRNAs are synthesized sequentially, in the 3′ ⊠ 5′ order in which their genes are arranged in the viral genome: N-P-M-G-L.

Ball LA, White CN. 1976. Order of transcription of genes of vesicular stomatitis virus. *Proc Natl Acad Sci. U S A* **73:**442–446.

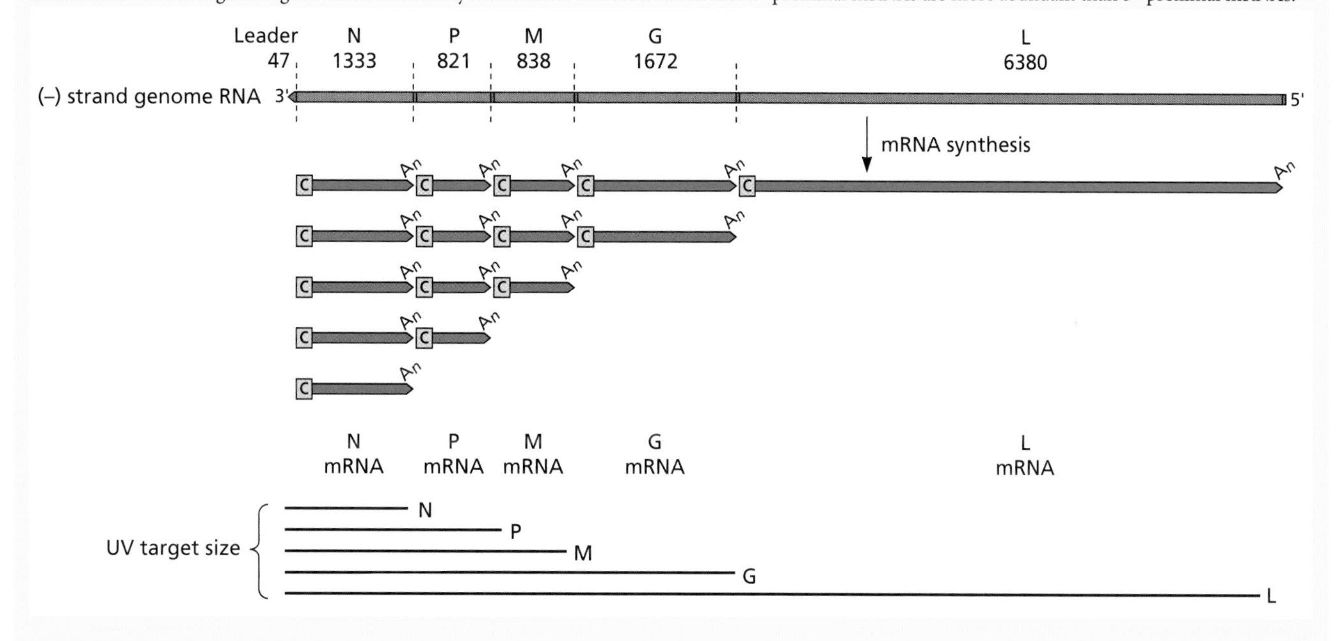

Vesicular stomatitis virus mRNA map and UV map. The genome is shown as a dark green line at the top, and the N, P, M, G, and L genes and their relative sizes are indicated. The 47-nucleotide leader RNA is encoded at the 3′ end of the genomic RNA. The leader and intergenic regions are shown in orange. The RNAs encoded at the 3′ end of the genome are made in larger quantities than the RNAs encoded at the 5′ end of the genome. UV irradiation experiments determined the size of the vesicular stomatitis virus genome (UV target size) required for synthesis of each of the viral mRNAs. The UV target size of each viral mRNA corresponded to the size of the genomic RNA sequence encoding the mRNA plus all of the genomic sequence 3′ to this coding sequence. The transition from reiterative copying and termination to initiation is not perfect, and only about 70 to 80% of the polymerase molecules accomplish this transition at each intergenic region. Such inefficiency accounts for the observation that 3′-proximal mRNAs are more abundant than 5′-proximal mRNAs.

Double-Stranded RNA

A distinctive feature of the infectious cycle of double-stranded RNA viruses is the production of mRNAs and genomic RNAs from distinct templates in different viral particles. Because the viral genomes are double stranded, they cannot be translated. Therefore, the first step in infection is the production of mRNAs from each viral RNA segment by the virion-associated RNA polymerase (Fig. 6.22). Reoviral mRNAs carry 5′-cap structures but lack 3′-poly(A) sequences.

In the reovirus core, the λ3 polymerase molecules are attached to the inner shell at each fivefold axis, below an RNA exit pore. Viral mRNAs are synthesized by the polymerase inside the viral particle and then extruded into the cytoplasm through this pore. Attachment of the polymerase molecules to the pores ensures that mRNAs are actively threaded out of the particle, without depending upon diffusion, which would be very inefficient. Examination of the structure of actively transcribing rotavirus, a member of the *Reoviridae*, has allowed a three-dimensional visualization of how mRNAs are released from the particle (Box 6.5). Viral (+) strand RNAs that will serve as templates for (−) strand RNA synthesis are first packaged into newly assembled subviral particles (Fig. 6.22). Each (+) strand RNA is then copied just once within this particle to produce double-stranded RNA.

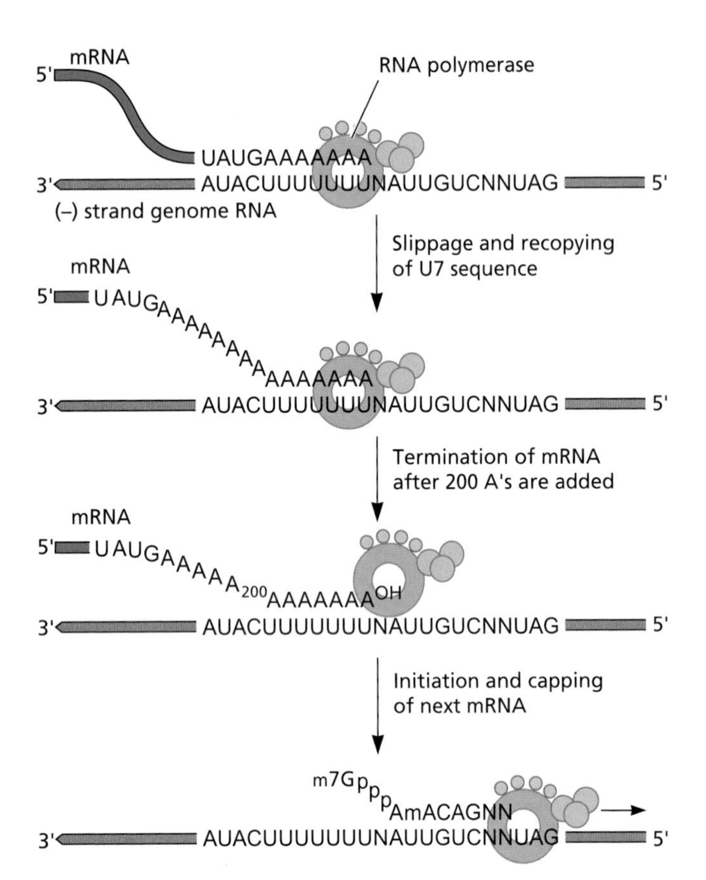

Figure 6.19 Poly(A) addition and termination at an intergenic region during vesicular stomatitis virus mRNA synthesis. Copying of the last seven U residues of an mRNA-encoding sequence is followed by slipping of the resulting seven A residues in the mRNA off the genomic sequence, which is then recopied. This process continues until approximately 200 A residues are added to the 3′ end of the mRNA. Termination then occurs, followed by initiation and capping of the next mRNA. The dinucleotide NA in the genomic RNA is not copied.

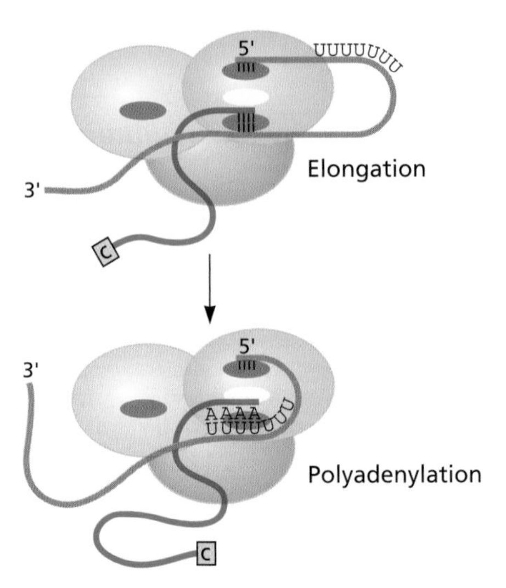

Figure 6.20 Moving-template model for influenza virus mRNA synthesis. During RNA synthesis, the polymerase remains bound to the 5′ end of the genomic RNA, and the 3′ end of the genomic RNA is threaded through (or along the surface of) the polymerase as the PB1 protein catalyzes each nucleotide addition to the growing mRNA chain. This threading process continues until the mRNA reaches a position on the genomic RNA that is close to the binding site of the polymerase. At this point the polymerase itself blocks further mRNA synthesis, and reiterative copying of the adjacent U_7 tract occurs. After about 150 A residues are added to the 3′ end of the mRNA, mRNA synthesis terminates. Adapted from D. M. Knipe et al. (ed), *Fields Virology*, 4th ed. (Lippincott Williams & Wilkins, Philadelphia, PA, 2001), with permission.

Members of different families of double-stranded RNA viruses carry out RNA synthesis in diverse ways. Replication of the genome of bacteriophage ϕ6 (3 double-stranded RNA segments) and birnaviruses (2 double-stranded RNA segments) is semiconservative, whereas that of reoviruses (10 to 12 double-stranded RNA segments) is conservative (Fig. 6.23). During conservative replication, the double-stranded RNA that exits the polymerase must be melted, so that the newly synthesized (+) strand is released and the template (−) strand reanneals with the original (+) strand. In reovirus particles, each double-stranded RNA segment is attached to a polymerase molecule, by interaction of the 5′-cap structure with a cap-binding site on the RNA polymerase. Attachment of the 5′ cap to the polymerase facilitates insertion of the 3′ end of the (−) strand into the template channel. This arrangement allows very efficient reinitiation of RNA synthesis in the crowded core of the particle. The RNA polymerase of bacteriophage ϕ6 and birnaviruses do not have such a cap-binding site, as would be expected for enzymes that copy both strands of the double-stranded RNA segments. This strategy appears less efficient, but may be sufficient when the genome consists of only two or three double-stranded RNA segments.

Unique Mechanisms of mRNA and Genome Synthesis of Hepatitis Delta Satellite Virus

The strategy for synthesis of the (+) strand RNA genome of hepatitis delta satellite virus is apparently unique among animal viruses (Fig. 6.24). The genome does not encode an RNA polymerase: viral RNAs are produced by host cell RNA polymerase II (Box 6.6), and the hepatitis delta virus RNAs are RNA catalysts, or **ribozymes** (Box 6.7). The genome of hepatitis delta virus is a 1,700-nucleotide (−) strand circular RNA, the only RNA with this structure that has been found in animal cells. As approximately 70% of the nucleotides are base paired, the viral RNA is folded into a rodlike structure.

All hepatitis delta satellite virus RNAs are synthesized in the nucleus. The switch from mRNA synthesis to the production of full-length (+) RNA is controlled by suppression of a poly(A) signal. Full-length (−) and (+) strand RNAs are copied by a rolling-circle mechanism, and ribozyme self-cleavage

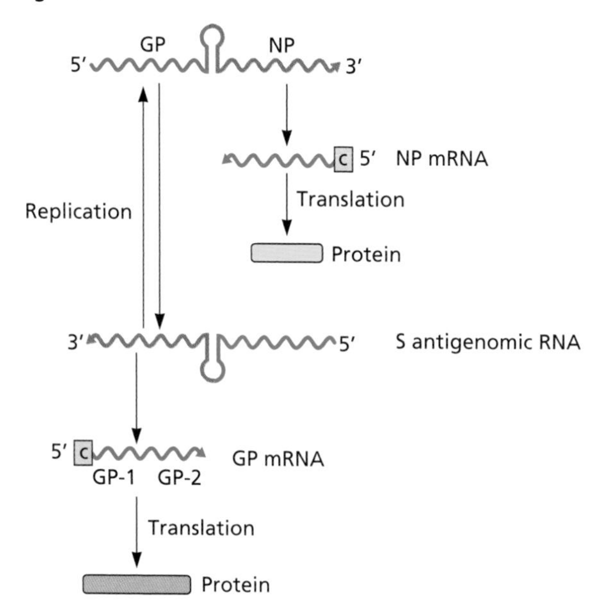

Figure 6.21 Arenavirus RNA synthesis. Arenaviruses contain two genomic RNA segments, L (large) and S (small) (top). At early times after infection, only the 3′ region of each of these segments is copied to form mRNA: the N mRNA from the S genomic RNA and the L mRNA from the L genomic RNA. Copying of the remainder of the S and L genomic RNAs may be blocked by a stem-loop structure in the genomic RNAs. After the S and L genomic RNAs are copied into full-length strands, their 3′ regions are copied to produce mRNAs: the glycoprotein precursor (GP) mRNA from S RNA and the Z mRNA (encoding an inhibitor of viral RNA synthesis) from the L RNA. Only RNA synthesis from the S RNA is shown in detail.

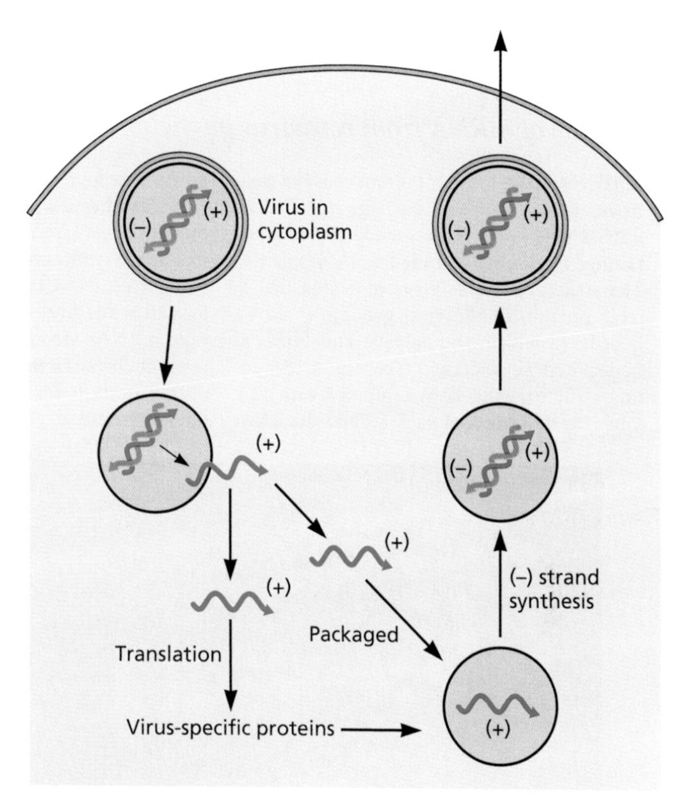

Figure 6.22 mRNA synthesis and replication of double-stranded RNA genomes. These processes occur in subviral particles containing the RNA templates and necessary enzymes. During cell entry, the virion passes through the lysosomal compartment, and proteolysis of viral capsid proteins activates the RNA synthetic machinery. Single-stranded (+) viral mRNAs, which are synthesized in parental subviral particles, are extruded into the cytoplasm, where they serve either as mRNAs or as templates for the synthesis of (−) RNA strands. In the latter case, viral mRNAs are first packaged into newly assembled subviral particles in which the synthesis of (−) RNAs to produce double-stranded RNAs occurs. These subviral particles become infectious particles. Only 1 of the 10 to 12 double-stranded RNA segments of the reoviral genome is shown.

releases linear monomers. Subsequent ligation of the two termini by the same ribozyme produces a monomeric circular RNA. The hepatitis delta virus ribozymes are therefore needed to process the intermediates of rolling-circle RNA replication. This enzyme initiates viral mRNA synthesis at a position on the genome near the beginning of the delta antigen-coding region. Once the polymerase has moved past a polyadenylation signal and the self-cleavage domain (Fig 6.24), the 3′-poly(A) of the mRNA is made by host cell enzymes. The RNA downstream of the poly(A) site is not degraded, in contrast to that of other mRNA precursors made by RNA polymerase II, but is elongated until a complete full-length (+) strand is made. The poly(A) addition site in this full-length (+) RNA is not used. The delta antigen bound to the rodlike RNA may block access of cellular enzymes to the poly(A) signal, thereby inhibiting polyadenylation.

Why Are (−) and (+) Strands Made in Unequal Quantities?

Different concentrations of (+) and (−) strands are produced in infected cells. For example, in cells infected with poliovirus, genomic RNA is produced at 100-fold higher concentrations than its complement. There are different explanations for these observations. RNA genomes and their complementary strands might have different stabilities, or the two strands might be synthesized by mechanisms with different efficiencies.

Viral (−) strand RNA is approximately 20 to 50 times more abundant than (+) strand RNA in cells infected with vesicular stomatitis virus. It was suggested that such asymmetry is

BOX 6.5

EXPERIMENTS

Release of mRNA from rotavirus particles

Rotaviruses, the most important cause of gastroenteritis in children, are large icosahedral viruses made of a three-shelled capsid containing 11 double-stranded RNA segments. The structure of this virus indicated that a large portion of the viral genome (~25%) is ordered within the particle and forms a dodecahedral structure (see Fig. 4.18). In this structure, the RNAs interact with the inner capsid layer and pack around the RNA polymerase located at the fivefold axis of symmetry. Further analysis of rotavirus particles in the process of synthesizing mRNA has shown that newly synthesized molecules are extruded from the capsid through several channels located at the fivefold axes (see figure). Multiple mRNAs are released at the same time from such particles. On the basis of these observations, it has been suggested that each double-stranded genomic RNA segment is copied by an RNA polymerase located at a fivefold axis of symmetry. This model may explain why no double-stranded RNA virus with more than 12 genomic segments, the maximum number of fivefold axes, has been found.

Lawton JA, Estes MK, Prasad BV. 1997.Three-dimensional visualization of mRNA release from actively transcribing rotavirus particles. *Nat Struct Biol* **4**:118–121.

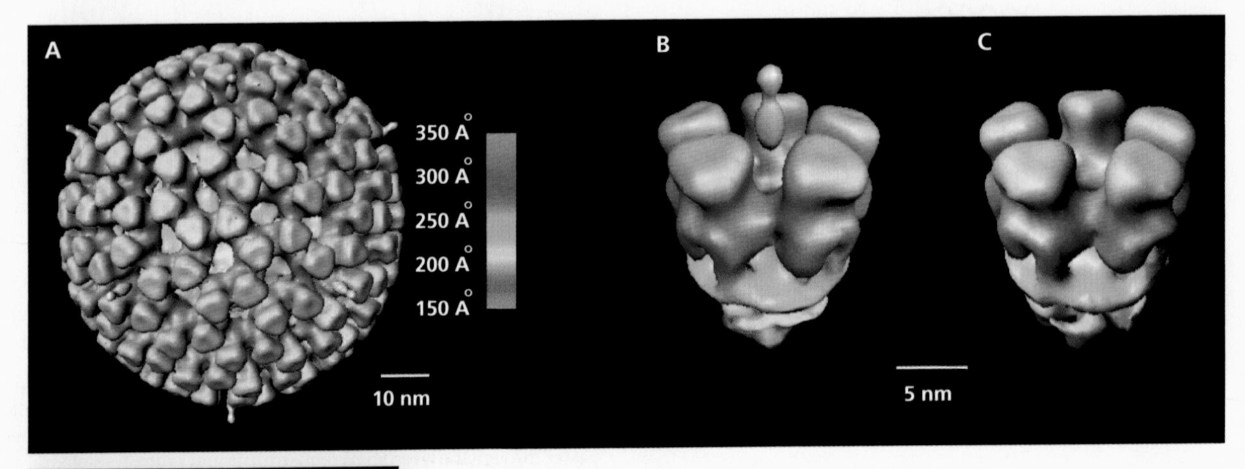

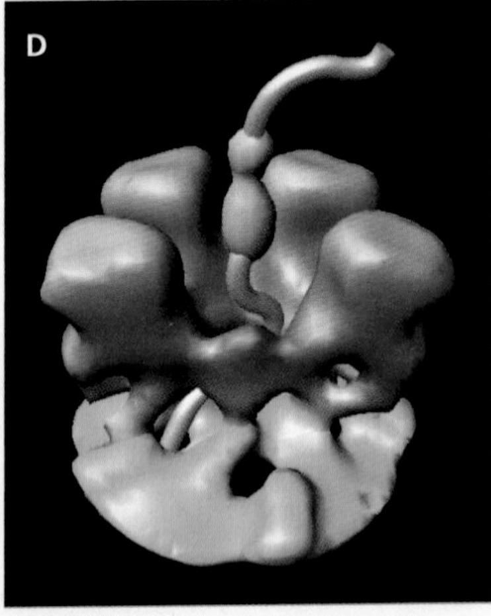

Three-dimensional visualization of mRNA release from rotavirus particles synthesizing mRNA. (A) Structure of a rotavirus particle in the process of synthesizing mRNA. The capsid is depth cued according to the color chart. Parts of newly synthesized mRNA that are ordered, and therefore structurally visible, are shown in magenta at the fivefold axes of symmetry. **(B)** Close-up view of the channel at the fivefold axis and the visible mRNA. The mRNA is surrounded by five trimers of capsid protein VP6. **(C)** Close-up view of the channel at the fivefold axis of a particle not in the process of synthesizing mRNA. **(D)** Model of the pathway of mRNA transit through the capsid. One VP6 trimer has been omitted for clarity. The green protein is VP2, and the mRNA visible in the structure is shown in pink. The gray tube represents the possible path of an mRNA molecule passing through the VP2 and VP6 layers through the channel. Courtesy of B. V. V. Prasad, Baylor College of Medicine. Reprinted from J. A. Lawton et al., *Nat Struct Biol* **4**:118–121, 1997, with permission.

a consequence of more efficient initiation of RNA synthesis at the 3′ end of (+) strand RNA than of (−) strand RNA. An elegant proof of this hypothesis came from the construction and study of a second rhabdovirus genome, that of rabies virus, with identical initiation sites at the 3′ ends of both (−) and (+) strand RNAs. In cells infected with this virus, the ratio of (−) to (+) strands is 1:1.

In alphavirus-infected cells, the abundance of genomic RNA is explained by the fact that (−) strand RNAs are synthesized for but a short time early in infection. The RNA polymerase that catalyzes (−) strand RNA synthesis is present only during this period. The synthesis of (+) strands continues for much longer, leading to accumulation of mRNA and (+) strand genomic RNA.

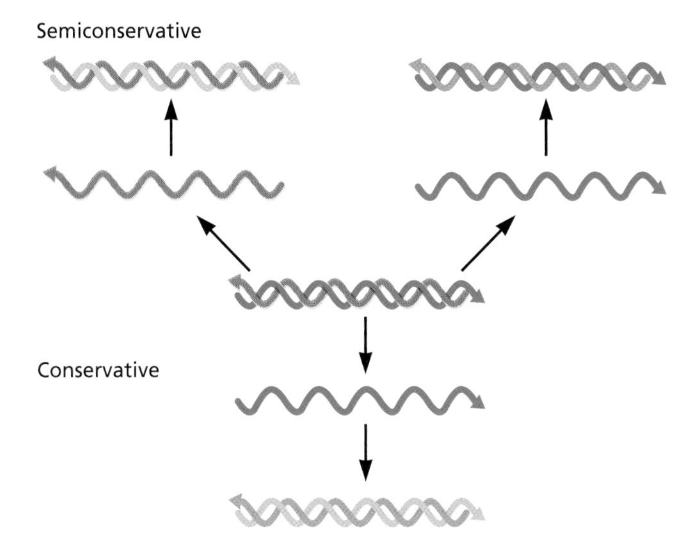

Figure 6.23 Two mechanisms for copying nucleic acids. During semiconservative replication, both strands of nucleic acid serve as templates for the synthesis of new strands (shown in red). In contrast, only one strand is copied during conservative replication.

Do Ribosomes and RNA Polymerases Collide?

The genomic RNA of (+) strand viruses can be translated in the cell, and the translation products include the viral RNA polymerase. At a certain point in infection, the RNA polymerase copies the RNA in a 3′ ⊠ 5′ direction while ribosomes traverse it in an opposite direction (Fig. 6.25), raising the question of whether the viral polymerase avoids collisions with ribosomes. When ribosomes are frozen on polioviral RNA by using inhibitors of protein synthesis; replication is blocked. In contrast, when ribosomes are run off the template, replication of the RNA increases. These results suggest that ribosomes must be cleared from viral RNA before it can serve as a template for (−) strand RNA synthesis; in other words, replication and translation cannot occur simultaneously.

The interactions of viral and cellular proteins with the polioviral 5′-untranslated region might determine whether the genome is translated or replicated. In this model, binding of cellular poly(rC)-binding protein 2 within the 5′-untranslated region initially stimulates translation. Once the viral protease has been synthesized, it cleaves poly(rC)-binding protein; consequently, binding of the cellular protein is reduced. However, cleaved poly(rC)-binding protein can still bind to a different segment of the 5′-untranslated region (the cloverleaf) (Fig. 6.10) and promote viral genome synthesis.

Restricting translation and RNA synthesis to distinct compartments may prevent collisions of ribosomes and polymerases. Viral mRNA synthesis takes place in the reovirus capsid, where the enzymes responsible for this process are located. The viral mRNAs are exported to the cytoplasm for translation. Retroviral RNAs are synthesized in the cell nucleus, where translation does not take place. The architecture of membranous replication complexes of (+) strand RNA viruses may favor RNA synthesis and exclude translation.

Even if mechanisms exist for controlling whether the genomes of RNA viruses are translated or replicated, some ribosome-RNA polymerase collisions are likely to occur. The isolation of a polioviral mutant with a genome that contains an insertion of a 15-nucleotide sequence from 28S ribosomal RNA (rRNA) is consistent with this hypothesis. After colliding with a ribosome, the RNA polymerase apparently copied 15 nucleotides of rRNA.

Cellular Sites of Viral RNA Synthesis

Genomes and mRNAs of most RNA viruses are made in the cytoplasm, invariably in specific structures such as the nucleocapsids of (−) strand RNA viruses, subviral particles of double-stranded RNA viruses, and membrane-bound replication complexes in the case of (+) strand RNA viruses. Such replication complexes are morphologically diverse, and the membranes originate from various cellular compartments (Chapter 14). Alphaviral RNA synthesis occurs on the cytoplasmic surface of endosomes and lysosomes, and polioviral RNA polymerase is located on the surfaces of small, membranous vesicles. Different viral proteins have been implicated in the formation of these viral replicative organelles.

The membrane vesicles observed early in poliovirus-infected cells are thought to originate from two sources. One appears to be the endoplasmic reticulum (ER), specifically vesicles whose production is regulated by proteins of coat protein complex II (CopII) (Chapter 12). Unlike vesicles produced from the ER in uninfected cells, those in poliovirus-infected cells do not fuse with the Golgi and therefore accumulate in the cytoplasm. The vesicles produced later during poliovirus infection bear several hallmarks of autophagosomes, including a double membrane and colocalization with protein markers of these vesicles. Synthesis of poliovirus 2BC and 3A proteins in uninfected cells leads to production of such autophagosomes. Similar double-membrane vesicles are observed during infection with a variety of (+) strand RNA viruses, indicating that they may serve as a general replication platform.

Flavivirus RNA replication takes place on perinuclear, double-membraned vesicles derived from the ER (Fig. 6.26). These vesicles are connected to the cytoplasm through a pore and are often near sites of virus assembly. Coronaviral proteins remodel cellular membranes to produce the replication complex, which consists of a network of convoluted membranes, double-membrane vesicles, and vesicle packets, all of which are continuous with the ER (Fig. 6.26).

It is thought that membrane association of viral replication assemblies ensures high local concentrations of relevant components, and hence increases the rates or efficiencies of

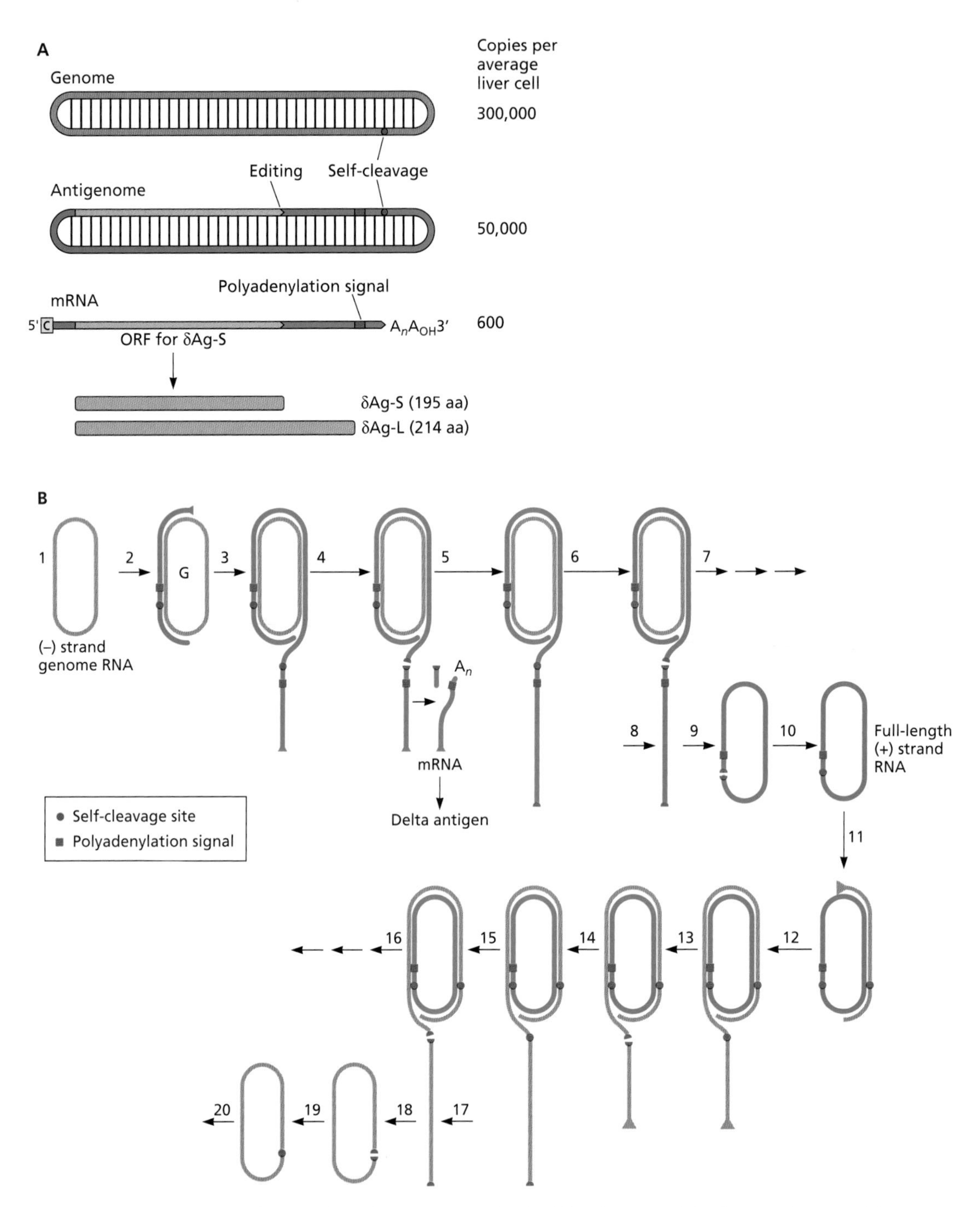

Figure 6.24 Hepatitis delta satellite virus RNA synthesis. (A) Schematic of the forms of hepatitis delta virus RNA and δ antigen found in infected cells. aa, amino acids; ORF, open reading frame. **(B)** Overview of hepatitis delta virus mRNA and genomic RNA synthesis. In steps 1 to 3, RNA synthesis is initiated, most probably by host RNA polymerase II, at the indicated position on the (−) strand genomic RNA. The polymerase passes the poly(A) signal (purple box) and the self-cleavage domain (red circle). In steps 4 and 5, the 5′ portion of this RNA is processed by cellular enzymes to produce delta mRNA with a 3′-poly(A) tail, while RNA synthesis continues beyond the cleavage site and the RNA undergoes self-cleavage (step 6). RNA synthesis continues until at least one unit of the (−) strand genomic RNA template is copied. The poly(A) signal is ignored in this full-length (+) strand. In steps 7 to 10, after self-cleavage to release a full-length (+) strand, self-ligation produces a (+) strand circular RNA. In steps 11 to 20, mRNA synthesis initiates on the full-length (+) strand to produce (−) strands by a rolling-circle mechanism. Unit-length genomes are released by the viral ribozyme (step 15) and self-ligated to form (−) strand circular genomic RNAs. Adapted from J. M. Taylor, *Curr Top Microbiol Immunol* **239:**107–122, 1999, with permission.

BOX 6.6

DISCUSSION

RNA-dependent RNA polymerase II

The mRNAs produced during hepatitis delta satellite virus infection of cells have typical properties of DNA-dependent RNA polymerase II products, including a 5′ cap and 3′-poly(A) tail. Production of these satellite mRNAs is also sensitive to α-amanitin, an inhibitor of DNA-dependent RNA polymerase II. Furthermore, the RNA genome of plant viroids can be copied by plant DNA-dependent RNA polymerase II. Based on these observations, it was suggested that the RNA genome of hepatitis delta satellite virus is copied by RNA polymerase II. Experimental support for this hypothesis has been obtained. When purified mammalian RNA polymerase II was incubated with NTPs and an RNA template-primer, an RNA product was produced. Similar results were obtained when the antigenome of hepatitis delta satellite virus was used in the reaction. Structural studies revealed that the RNA template-product duplex occupies the same site on the enzyme as the DNA-RNA hybrid during transcription. When transcription protein IIS was added to the reaction mixture, the satellite genome was cleaved, and the new 3′ end was used as a primer. Compared with DNA-dependent RNA synthesis, RNA-dependent RNA synthesis by RNA polymerase II was slower and less processive. These properties may explain why the enzyme can copy only short RNA templates.

The ability of DNA-dependent RNA polymerase II to copy an RNA template provides a missing link in molecular evolution. This activity supports the hypothesis that an ancestor of RNA polymerase II copied RNA genomes that are thought to have existed during the ancient RNA world. During the transition from RNA to DNA genomes, this enzyme evolved to copy DNA templates. Today these enzymes can still copy small RNAs such as the genome of hepatitis delta satellite virus.

Chang J, Nie X, Chang HE, Han Z, Taylor J. 2008. Transcription of hepatitis delta virus RNA by RNA polymerase II. *J Virol* **82:**1118–1127.

Lehmann E, Brueckner F, Cramer P. 2007. Molecular basis of RNA-dependent RNA polymerase II activity. *Nature* **450:**445–449.

Rackwitz HR, Rohde W, Sanger HL. 1981. DNA-dependent RNA polymerase II of plant origin transcribes viroid RNA into full-length copies. *Nature* **291:**297–301.

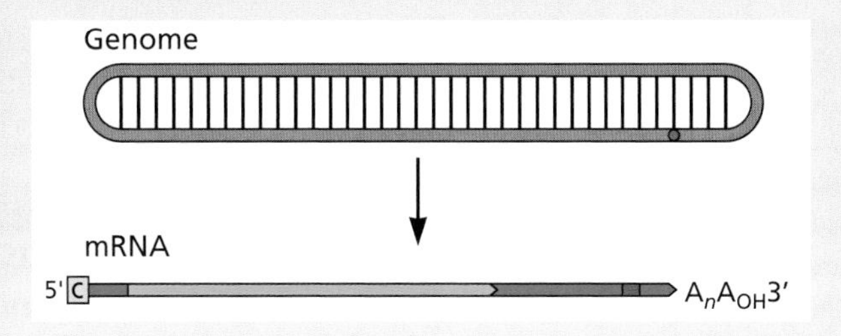

BOX 6.7

BACKGROUND

Ribozymes

A **ribozyme** is an enzyme in which RNA, not protein, carries out catalysis. The first ribozyme discovered was the group I intron of the ciliate *Tetrahymena thermophila*. Other ribozymes have since been discovered, including RNase P of bacteria, group II self-splicing introns, hammerhead RNAs of viroids and satellite RNAs, and the ribozyme of hepatitis delta virus. These catalytic RNAs are very diverse in size, sequence, and the mechanism of catalysis. For example, the hepatitis delta satellite virus ribozyme (see figure) catalyzes a transesterification reaction that yields products with 2′,3′-cyclic phosphate and 5′-OH termini. Only an 85-nucleotide sequence is required for activity of this ribozyme, and can cleave optimally with as little as a single nucleotide 5′ to the site of cleavage.

Ribozymes have been essential for producing infectious RNAs from cloned DNA copies of the genomes of (−) strand RNA viruses. Such transcripts often have extra sequences at the 3′ end. By joining the 85-nucleotide ribozyme fragment to upstream sequences, accurate 3′ ends of heterologous RNA transcripts synthesized *in vitro* can be obtained.

Kruger K, Grabowski PJ, Zaug AJ, Sands J, Gottschling DE, Cech TR. 1982. Self-splicing RNA: autoexcision and autocyclization of the ribosomal RNA intervening sequence of Tetrahymena. *Cell* **31:**147–157.

Westhof E, Michel F. 1998. Ribozyme architectural diversity made visible. *Science* **282:**251–252.

Whelan SP, Ball LA, Barr JN, Wertz GT. 1995. Efficient recovery of infectious vesicular stomatitis virus entirely from cDNA clones. *Proc Natl Acad. Sci U S A* **92:**8388–8392.

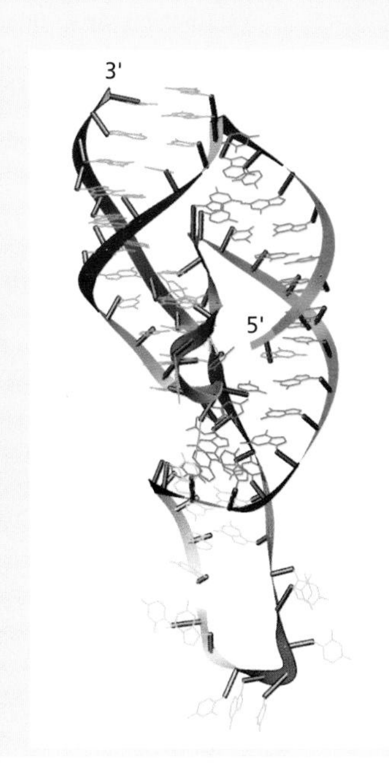

Crystal structure of the hepatitis delta satellite virus ribozyme. The RNA backbone is shown as a ribbon. The two helical stacks are shown in red and blue, and unpaired nucleotides are gray. The 5′ nucleotide, which marks the active site, is green. Produced from Protein Data Bank file 1cx0.

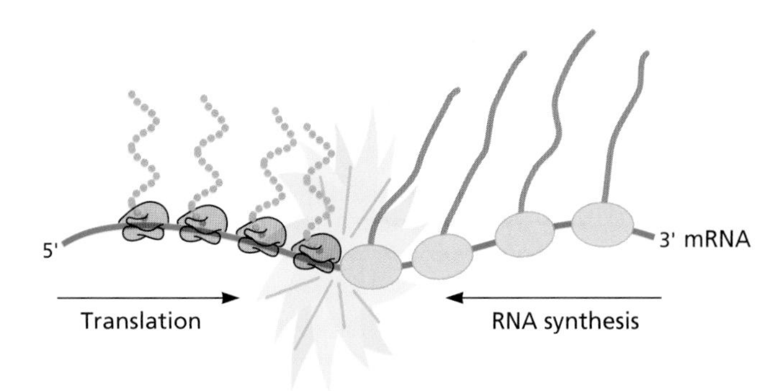

Figure 6.25 Ribosome-RNA polymerase collisions. A strand of viral RNA is shown, with ribosomes translating in the 5′ → 3′ direction and RNA polymerase copying the RNA chains in the 3′ → 5′ direction. Ribosome-polymerase collisions would occur in cells infected with (+) strand RNA viruses unless mechanisms exist to avoid simultaneous translation and replication.

these reactions. As we have seen, this property may contribute to the specificity of polioviral 3D^{pol} for viral RNA templates. Membrane association may also have other functions, such as allowing efficient packaging of progeny RNA into virus particles, or providing lipid components or physical support to the replication machinery. The surfaces of membranous replication complexes isolated from poliovirus-infected cells appear to be coated with two-dimensional arrays of polymerase. These arrays are formed by interaction of 3D^{pol} molecules in a head-to-tail fashion. Surface catalysis is known to confer several advantages, including a higher probability of collision among reactants, an increase in substrate affinity from clustering of multiple binding sites, and retention of reaction products. The last property would facilitate multiple rounds of copying (+) and (−) strand RNA templates.

In addition to providing a location for efficient viral RNA synthesis, membranous replication structures may also protect nucleic acids from nucleases and shield RNAs that activate host intrinsic defenses (Volume II, Chapter 3).

Viral RNA polymerases are recruited to membranous replication compartments in different ways. A C-terminal, transmembrane segment of the hepatitis C virus RNA polymerase, NS5b, is responsible for attachment of the enzyme to cellular membrane replication complexes. Polioviral 3D^{pol} cannot by itself associate with membranes but is brought to the replication complex by binding to protein 3AB. When the membrane association of this protein is disrupted by amino acid changes, viral RNA synthesis is inhibited. The hydrophobic domain of 3AB can be substituted for the C-terminal transmembrane segment of NS5b with little effect on RNA polymerase activity, indicating that membrane association is the sole function of this sequence.

Origins of Diversity in RNA Virus Genomes

Misincorporation of Nucleotides

All nucleic acid polymerases insert incorrect nucleotides during chain elongation. DNA-directed DNA polymerases have **proofreading** capabilities in the form of exonuclease activities that can correct such mistakes. Most RNA-dependent RNA polymerases do not possess

Figure 6.26 Membranous sites of viral RNA synthesis. (A) Tomogram of dengue virus-infected cell showing virus-induced vesicles as invaginations of the ER **(B)** Three-dimensional rendering of dengue virus-induced vesicles in the ER. Virus particles are colored red. **(C)** Model of SARS-coronavirus induced convoluted membranes (CM), double-membrane vesicles (DMV), and vesicle packets (VP). The cluster of large and small double-membrane vesicles (outer membrane, gold; inner membrane, silver) are connected to a vesicle packet and convoluted membrane structure (bronze). A, B from S. Welsch et al., *Cell Host Microbe* **23:**365–375, 2009; C from K. Knoops et al. *PLoS Biol* **6:**e226, 2008, with permission.

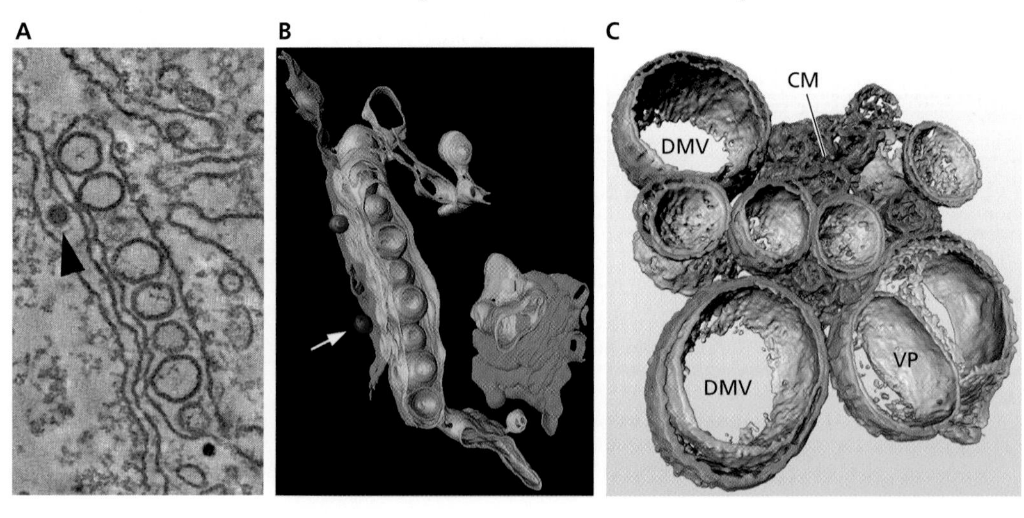

this capability. The result is that error frequencies in RNA replication can be as high as one misincorporation per 10^3 to 10^5 nucleotides polymerized, whereas the frequency of errors in DNA replication is about 10,000-fold lower. Many of these polymerization errors cause lethal amino acid changes, while other mutations may appear in the genomes of infectious virus particles. This phenomenon has led to the realization that RNA virus populations are **quasispecies**, or mixtures of many different genome sequences. The errors introduced during RNA replication have important consequences for viral pathogenesis and evolution, as discussed in Volume II, Chapter 10. Because RNA viruses exist as mixtures of genotypically different viruses, viral mutants may be isolated readily. For example, live attenuated poliovirus vaccine strains are viral mutants that were isolated from an unmutagenized stock of wild-type virus.

Fidelity of copying by RNA-dependent RNA polymerases is determined by how the template, primer, and NTP interact at the active site. Nucleotide binding occurs in two steps: first, the NTP is bound in such a way that the ribose cannot interact properly with the Asp of motif A and the Asn of motif B (Fig. 6.5). Next, if the NTP is correctly base paired with the template, there is a conformational change in the enzyme, which reorients the triphosphate and allows phosphoryl transfer to occur. This change requires reorientation of the Asp and Asn residues, which would stabilize the position of the ribose in the binding pocket, and is thought to be a key fidelity checkpoint for the picornaviral RNA polymerase. This model is based on the structures of $3D^{pol}$ bound to a template primer and NTP (Fig. 6.6) and the study of an altered poliovirus $3D^{pol}$ with higher fidelity than the wild-type enzyme. The increased fidelity of this enzyme, which has a single amino acid substitution in the fingers domain, is the result of a change in the equilibrium constant for the conformational transition. Although this amino acid is remote from the active site, it participates in hydrogen bonding to motif A, which, as discussed above, is important in holding the NTP in a catalytically appropriate conformation. Of great interest is the observation that a similar interaction between fingers and motif A can be observed in RNA polymerases from a wide variety of viruses. This mechanism of enhancing fidelity may therefore be conserved in all RNA-dependent RNA polymerases.

These studies also provide mechanistic information on how ribavirin, an antiviral compound, causes lethal mutagenesis. The structure of foot-and-mouth disease virus RNA polymerase $3D^{pol}$ bound to ribavirin shows the compound positioned in the active site of the enzyme, adjacent to the 3′ end of the primer. The ribose of ribavirin is bound in the pocket, indicating that it has bypassed the fidelity checkpoint and induced the conformational change that holds the analog in a position ready for catalysis. Therefore, ribavirin is a mutagen because it can be substituted for any of the four NTPs in newly synthesized RNA molecules.

The RNA polymerase of members of the *Nidovirales* (Fig. 6.16) allows faithful replication of the large (up to 32 kb) RNA genomes. The RNA synthesis machinery includes proteins not found in other RNA viruses, such as ExoN, a 3′-5′ exonuclease. Inactivation of this enzyme does not impair viral replication, but leads to 15- to 20-fold increases in mutation rates. This observation suggests that ExoN provides a proofreading function for the viral RNA polymerase, similar to the activity associated with DNA synthesis. Viruses lacking the ExoN gene display attenuated virulence in mice, and are being considered as vaccine candidates.

Segment Reassortment and RNA Recombination

Reassortment is the exchange of entire RNA molecules between genetically related viruses with segmented genomes. In cells coinfected with two different influenza viruses, the eight genome segments of each virus replicate. When new progeny virus particles are assembled, they can package RNA segments from **either** parental virus. Because reassortment is the simple exchange of RNA segments, it can occur at high frequencies.

In contrast to reassortment, recombination is the exchange of nucleotide sequences among different genomic RNA molecules (Fig. 6.27). Recombination, a feature of many RNA viruses, is an important mechanism for producing new genomes with selective growth advantages. This process has shaped the RNA virus world by rearranging genomes, or creating new ones. RNA recombination was first discovered in

Figure 6.27 RNA recombination. Schematic representation of RNA recombination occurring during template switching by RNA polymerase, or copy choice. Two parental genomes are shown. The RNA polymerase (purple oval) has copied the 3′ end of the donor genome and is switching to the acceptor genome. The resulting recombinant molecule is shown.

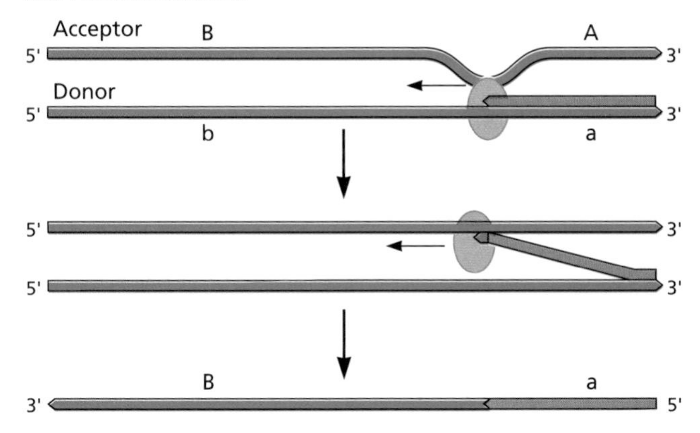

cells infected with poliovirus, and was subsequently observed for other (+) strand RNA viruses. The frequency of recombination can be relatively high: it has been estimated that 10 to 20% of polioviral genomic RNA molecules recombine in a single growth cycle. Recombinant polioviruses are readily isolated from the feces of individuals immunized with the three serotypes of Sabin vaccine. The genome of such viruses, which are recombinants of the vaccine strains with other enteroviruses found in the human intestine, may possess an improved ability to reproduce in the human alimentary tract and have a selective advantage over the parental viruses. Recombination can also lead to the production of pathogenic viruses (Box 6.8).

Polioviral recombination is predominantly **base pairing dependent**: it occurs between nucleotide sequences that have a high percentage of nucleotide identity. Other viral genomes undergo **base-pairing-independent** recombination between very different nucleotide sequences. RNA recombination is coupled with the process of genomic RNA replication: it occurs by template exchange during (−) strand synthesis, as first demonstrated in poliovirus-infected cells. The RNA polymerase first copies the 3′ end of one parental (+) strand and then exchanges one template for another at the corresponding position on a second parental (+) strand. Template exchange in poliovirus-infected cells occurs predominantly during (−) strand synthesis, presumably because the concentration of (+) strands is 100-fold higher than that of (−) strands. This template exchange mechanism of recombination is also known as **copy choice**. The exact mechanism of template exchange is not known, but it might be triggered by pausing of the polymerase during chain elongation or damage to the template.

If the RNA polymerase skips sequences during template switching, deletions will occur. Such RNAs will replicate if

BOX 6.8

DISCUSSION

RNA recombination leading to the production of pathogenic viruses

A remarkable property of pestiviruses, members of the *Flaviviridae*, is that RNA recombination generates viruses that cause disease. Bovine viral diarrhea virus causes a usually fatal gastrointestinal disease. Infection of a fetus with this virus during the first trimester is noncytopathic, but RNA recombination produces a cytopathic virus that causes severe gastrointestinal disease after the animal is born.

Pathogenicity of bovine viral diarrhea virus is associated with the synthesis of a nonstructural protein, NS3, encoded by the recombinant cytopathic virus (see figure). The NS3 protein cannot be made in cells infected by the noncytopathic parental virus because its precursor, the NS2-3 protein, is not proteolytically processed. In contrast, NS3 is synthesized in cells infected by the cytopathic virus because RNA recombination adds an extra protease cleavage site in the viral polyprotein, precisely at the N terminus of the NS3 protein (see figure). This cleavage site can be created in several ways. One of the most frequent is insertion of a cellular RNA sequence coding for ubiquitin, which targets cellular proteins for degradation. Insertion of ubiquitin at the N terminus of NS3 permits cleavage of NS2-3 by any member of a widespread family of cellular proteases. This recombination event provides a selective advantage, because pathogenic viruses outgrow nonpathogenic ones. Why cytopathogenicity is associated with release of the NS3 protein, which is thought to be part of the machinery for genomic RNA replication, is not known.

Retroviruses acquire cellular genes by recombination, and the resulting viruses can have lethal disease potential, as described in Volume II, Chapters 6 and 11.

Pathogenicity of bovine viral diarrhea virus is associated with production of the NS3 protein. Two cytopathic viruses, Osloss and CP1, in which the ubiquitin sequence (UCH) has been inserted at different sites, are shown. In Osloss, UCH has been inserted into the NS2-3 precursor, and NS3 is produced. In CP1, a duplication has also occurred such that an additional copy of NS3 is present after the UCH sequence. Adapted from D. M. Knipe et al. (ed), *Fields Virology*, 4th ed. (Lippincott Williams & Wilkins, Philadelphia, PA, 2001), with permission.

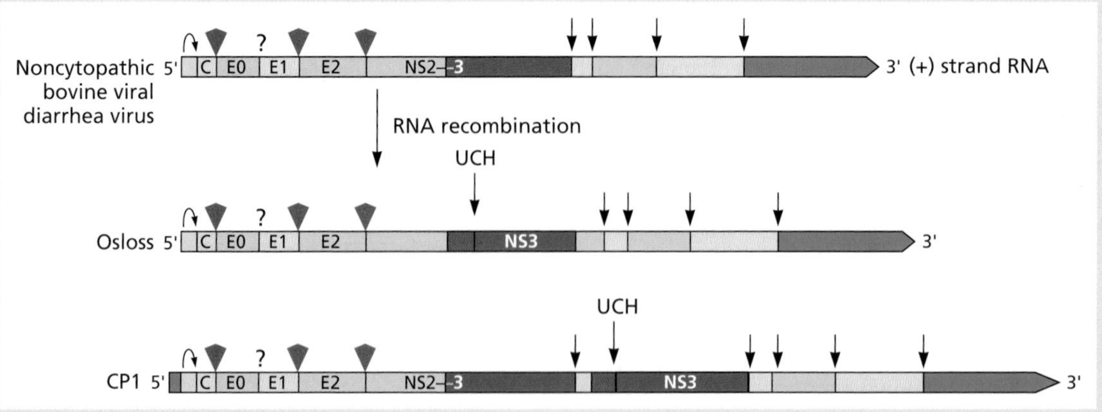

they contain the appropriate signals for the initiation of RNA synthesis. Because of their smaller size, subgenomic RNAs replicate more rapidly than full-length RNA, and ultimately compete for the components of the RNA synthesis machinery. Because of these properties, they are called **defective interfering RNAs**. Such RNAs can be packaged into viral particles only in the presence of a **helper virus** that provides viral proteins.

Defective interfering particles accumulate during the replication of both (+) and (−) strand RNA viruses. Production of these particles requires either a high multiplicity of infection or serial passaging, conditions that are achieved readily in the laboratory but rarely in nature. It is not known whether defective interfering viruses generally play a role in viral pathogenesis. However, some recombination reactions that lead to the appearance of cytopathic bovine viral diarrhea viruses delete viral RNA sequences rather than inserting cellular sequences (Box 6.8). Such deletions create a new protease cleavage site at the N terminus of the NS3 protein, and the defective interfering viruses also cause severe gastrointestinal disease in livestock.

RNA Editing

Diversity in RNA viral genomes is also achieved by RNA editing. Viral mRNAs can be edited either by insertion of a nontemplated nucleotide during synthesis or by alteration of the base after synthesis. Examples of RNA editing have been documented in members of the *Paramyxoviridae* and *Filoviridae* and in hepatitis delta satellite virus. This process is described in Chapter 10.

Perspectives

Structural biology has made important contributions toward understanding the mechanisms of RNA synthesis carried out by RNA-dependent RNA polymerases of (+) and double-stranded RNA viruses. These structures underscore the relationship of these enzymes to other nucleic acid polymerases, but also highlight differences that accommodate the wide diversity of RNA genome configurations. The first structure of an RNA polymerase of a (−) strand RNA virus, influenza virus, reveals how the endonuclease and cap-binding proteins are arranged with respect to the catalytic subunit. This structure should be the basis for an understanding at the atomic level of the different functions of this enzyme. We expect that forthcoming structures from other (−) strand RNA viruses will detail the conformational movements that take place during the switch between initiation and elongation, and the changes that occur as the polymerase moves from an open to a closed conformation.

RNA viral genetic diversity, and the ability to undergo rapid evolution, is made possible by errors made during nucleic acid synthesis, as well as recombination and reassortment. The importance of polymerase errors is underscored by the dramatic decrease in poliovirus fitness caused by a single amino acid change in the polymerase that decreases error rate. A different amino acid change in the RNA polymerase, which increases error frequency, has a similar effect. These observations demonstrate that the mutational diversity of RNA viruses is almost precisely where it must be, a place that is determined in large part by the error frequency of the RNA polymerase. A recent suggestion that not all mutations in the poliovirus genome are attributable to the viral RNA polymerase merits further investigation.

The function of host proteins that are required for viral RNA synthesis remain largely obscure, as does the reason why RNA synthesis is restricted to certain cellular compartments and structures induced by viral proteins. Some kinds of membranes are the sites of RNA replication for nearly all eukaryotic (+) strand RNA viruses; their architectures vary within viral families or orders. How these membranes are assembled, and their contribution to RNA synthesis, remain obscure. Prokaryotic (+) strand RNA viral replication does not occur on membranes and therefore this property cannot be a fundamental characteristic of RNA-directed RNA synthesis. Perhaps membrane-bound RNA synthesis was an adaptation required for colonization of eukaryotes. If so, it would important to determine if membranes are a requirement for RNA-directed RNA synthesis in *Archaea*.

References

Book

Rossmann MG, Rao VB (ed). 2012. *Viral Molecular Machines.* Springer Science+Business Media LLC.

Reviews

Belov GA. 2014. Modulation of lipid synthesis and trafficking pathways by picornaviruses. *Curr Opin Virol* **9:**19–23.

Flores R, Grubb D, Elleuch A, Nohales MÁ, Delgado S, Gago S. 2011. Rolling-circle replication of viroids, viroid-like satellite RNAs and hepatitis delta virus: variations on a theme. *RNA Biol* **8:**200–206.

Fodor E. 2013. The RNA polymerase of influenza A virus: mechanisms of viral transcription and replication. *Acta Virol* **57:**113–122.

Kok CC, McMinn PC. 2009. Picornavirus RNA-dependent RNA polymerase. *Int J Biochem Cell Biol* **41:**498–502.

McDonald SM, Tao YJ, Patton JT. 2009. The ins and outs of four-tunneled *Reoviridae* RNA-dependent RNA polymerases. *Curr Opin Struct Biol* **19:**775–782.

Morin B, Kranzusch PJ, Rahmeh AA, Whelan SPJ. 2013. The polymerase of negative-stranded RNA viruses. *Curr Opin Virol* **3:**103–110.

Neuman B, Angelini MM, Buchmeier MJ. 2014. Does form meet function in the coronavirus replicative organelle? *Trends Microbiol* **22:**642–647.

Sholders AJ, Peersen OB. 2014. Distinct conformations of a putative translocation element in poliovirus polymerase. *J Mol Biol* **426:**1407–1409.

Steil BP, Barton DJ. 2009. Cis-active RNA elements (CREs) and picornavirus RNA replication. *Virus Res* **139:**240–252.

Papers of Special Interest

Appleby TC, Perry JK, Murakami E, Barauskas O, Fend J, Cho A, Fox D, Wetmore DR, McGrath ME, Ray AS, Sofia MJ, Swaminathan S, Edwards TE. 2015. Structural basis for RNA replication by the hepatitis C virus polymerase. *Science* **347:**771–775.

Gong P, Peersen OB. 2010. Structural basis for active site closure by the poliovirus RNA-dependent RNA polymerase. *Proc Natl Acad Sci U S A* **107:**22505–22510.

Gong P, Kortus MG, Nix JC, Davis RE, Peersen OB. 2013. Structures of Coxsackievirus, rhinovirus, and poliovirus polymerase elongation complexes solved by engineering RNA mediated crystal contacts. *PLoS One* **8:**e60272.

Heinrich BS, Morin B, Rahmeh AA, Whelan SP. 2012. Structural properties of the C terminus of vesicular stomatitis virus N protein dictate N-RNA complex assembly, encapsidation, and RNA synthesis. *J Virol* **86:**8720–8729.

Iwasaki M, Ngo N, Cubitt B, de la Torre JC. 2015. Efficient interaction between arenavirus nucleoprotein (NP) and RNA-dependent RNA polymerase (L) is mediated by the virus nucleocapsid (NP-RNA) template. *J Virol* **89:**5734–5738.

Pan J, Lin L, Tao YJ. 2009. Self-guanylylation of birnavirus VP1 does not require an intact polymerase active site. *Virology* **395:**87–96.

Pflug A, Guilligay D, Reich S, Cusack S. 2014. Structure of influenza A polymerase bound to the viral RNA promoter. *Nature* **516:**355–361.

Richards AL, Soares-Martins JAP, Riddell GT, Jackson WT. 2014. Generation of unique poliovirus RNA replication organelles. *MBio* **5:**e00833-13.

Subissi L, Posthuma CC, Collet A, Zevenhoven-Dobbe JC, Gorbalenya AE, Decroly E, Snijder EJ, Canard B, Imbert I. 2014. One severe acute respiratory syndrome coronavirus protein complex integrates processive RNA polymerase and exonuclease activities. *Proc Natl Acad Sci U S A* **111:**E3900–E3909.

7 Reverse Transcription and Integration

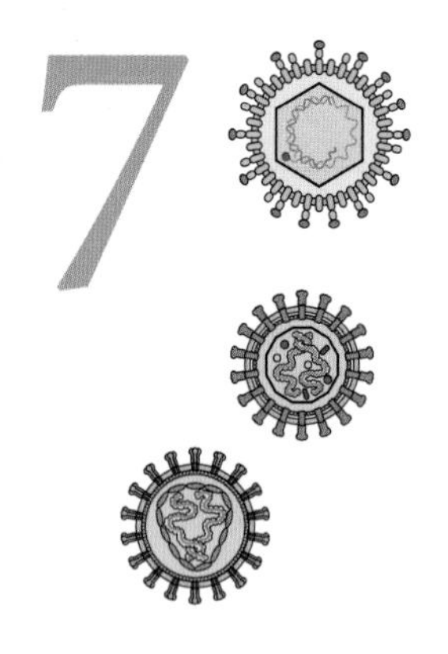

Retroviral Reverse Transcription
- Discovery
- Impact
- The Process of Reverse Transcription
- General Properties and Structure of Retroviral Reverse Transcriptases
- Other Examples of Reverse Transcription

Retroviral DNA Integration Is a Unique Process
- The Pathway of Integration: Integrase-Catalyzed Steps
- Integrase Structure and Mechanism

Hepadnaviral Reverse Transcription
- A DNA Virus with Reverse Transcriptase
- The Process of Reverse Transcription

Perspectives

References

LINKS FOR CHAPTER 7

⏭ *Video: Interview with Dr. David Baltimore*
http://bit.ly/Virology_Baltimore

⏭ *Movie 7.1: Crystal structure of the prototype foamy virus integrase tetramer bound to viral DNA ends and a target sequence*
http://bit.ly/Virology_V1_Movie7-1

⏭ *Retroviruses and cranberries*
http://bit.ly/Virology_Twiv320

⏭ *Retroviral influence on human embryonic development*
http://bit.ly/Virology_4-23-15

⏭ *A retrovirus makes chicken eggshells blue*
http://bit.ly/Virology_9-11-13

⏭ *Museum pelts help date the Koala retrovirus*
http://bit.ly/Virology_10-11-12

⏭ *Unexpected viral DNA in RNA virus-infected cells*
http://bit.ly/Virology_6-5-14

> *"One can't believe impossible things," said Alice.*
> *"I dare say you haven't had much practice," said the Queen.*
> *"Why, sometimes I've believed as many as six impossible things before breakfast."*
>
> LEWIS CARROLL
> *Alice in Wonderland*

Retroviral Reverse Transcription

Discovery

Back-to-back reports from the laboratories of Howard Temin and David Baltimore in 1970 provided the first concrete evidence for the existence of an RNA-directed DNA polymerase activity in retrovirus particles. The pathways that led to this unexpected finding were quite different in the two laboratories. In Temin's case, the discovery came about through attempts to understand how infection with this group of viruses, which have (+) strand RNA genomes, could alter the heredity of cells permanently, as they do in the process of oncogenic transformation. Temin believed that retroviral RNA genomes became integrated into the host cell's chromatin in a DNA form: studies of bacterial viruses such as bacteriophage lambda had established a precedent for viral DNA integration into host DNA (Box 7.1). However, it was a difficult hypothesis to test with the technology available at the time, and attempts by Temin and others to demonstrate the existence of such a phenomenon in infected cells were generally met with skepticism. Baltimore's entrée into the problem of reverse transcription came from his interest in virus-associated polymerases, in particular one that he had found to be present in particles of vesicular stomatitis virus, which contain a (−) strand RNA genome. It occurred to Baltimore and Temin independently that retrovirus particles might also contain such an enzyme, in this case the sought-after RNA-dependent DNA polymerase. Subsequent experiments confirmed this prediction; the enzyme that had earlier eluded Temin was discovered to be an integral component of these virus particles. Five years later, Temin and Baltimore were awarded the Nobel Prize in physiology or medicine for their independent discoveries of retroviral reverse transcriptase (RT).

Impact

The immediate impact of the discovery of RT was to amend the accepted central dogma of molecular biology, that the transfer of genetic information is unidirectional: DNA → RNA → protein. It was now apparent that there could also be a "retrograde" flow of information from RNA to DNA, and the name **retroviruses** eventually came to replace the earlier designation of RNA tumor viruses. In the years following this revision of dogma, many additional reverse transcription reactions were discovered. Furthermore, as Temin hypothesized, study of oncogenic retroviruses has provided a framework for current concepts of the genetic basis of cancer (Volume II, Chapter 6). Analysis of the reverse transcription and integration processes has enhanced our understanding of how retroviral infections persist and clarified aspects of the pathogenesis of acquired immunodeficiency syndrome (AIDS), caused by the human immunodeficiency virus. Finally, RT itself, first purified from virus particles and now produced in bacteria, has become an indispensable tool in molecular biology, for example, allowing experimentalists to capture cellular messenger RNAs (mRNAs) as complementary DNAs (cDNAs), which can then be converted into double-stranded DNAs, cloned, and expressed by well-established methodologies. Furthermore, the high efficiency of DNA integration mediated by the retroviral RT partner enzyme, integrase (IN), has been widely exploited for construction of viral vectors for gene transfer. For such reasons, we devote an entire chapter to these very important reactions (see interview with Dr. David Baltimore for background and personal account: http://bit.ly/Virology_Baltimore).

The Process of Reverse Transcription

Insight into the mechanism of reverse transcription can be obtained by comparing the amino acid sequences of RTs

PRINCIPLES *Reverse transcription and integration*

- Reverse transcriptases (RTs) are enzymes that synthesize DNA from both RNA and DNA templates.
- One immediate consequence of the discovery of the first (retroviral) RT was amendment of the central dogma, DNA → RNA → protein.
- Retrotransposons are elements in cellular DNAs that are copied from an RNA intermediate by RTs and inserted at other loci. Such intracellular transposable elements are present in the genomes of most, if not all, members of the tree of life.
- Retroviral RT is an indispensable tool in molecular biology, allowing capture of cellular mRNAs as complementary DNAs (cDNAs).
- As with all DNA polymerases, DNA synthesis by RTs requires a primer, either a fragment of RNA or a protein.
- Reverse transcription is an error-prone process, as RTs lack proofreading activity.
- Reverse transcription of retroviral and hepadnaviral RNAs is facilitated by the presence of terminal repeat sequences, and the dynamic and multifunctional properties of their RTs.
- Retroviruses are the only animal viruses with genomes that encode integrase proteins.
- Integrase is a specialized recombinase that mediates the insertion of retroviral DNA into its host genome, where it is called a provirus.
- Genomic RNA templates are packaged by all viruses that replicate via reverse transcription, but the RTs of some of these viruses convert the RNA to a DNA copy before infection of a new cell.

BOX 7.1

TRAILBLAZER

Bacteriophage lambda, a paradigm for the joining of viral and host DNAs

In 1962, Allan Campbell proposed an elegant, but at the time revolutionary, model for site-specific integration of DNA of the bacteriophage lambda into the chromosome of its host, *Escherichia coli*. The model was deduced from the fact that different linkage maps could be constructed for viral genomes at different stages in its life cycle. One linkage map, that of the integrated prophage, was obtained from the study of lysogenic bacteria. A different linkage map was obtained by measuring recombination frequencies of phage progeny (see part A of figure).

Campbell proposed that these unique features could be explained by a model for integration in which the incoming, linear, double-stranded DNA phage genome must first circularize. Subsequent recombination between a specific, internal sequence in the phage genome (called *attP*) and a specific sequence in the bacterial chromosome (called *attB*) would produce an integrated viral genome, with a linkage map that was a circular permutation of that of the linear phage genome, as had been observed (see part B of figure).

Although this model seems obvious today, it was not so in the 1960s. An alternative, in which the linear viral DNA was attached by a partial binding or "synapse" with the bacterial chromosome, was favored by a number of investigators. However, shortly after Campbell's elaboration of his model, circular molecules of lambda DNA were detected in infected cells, and the linear DNA extracted from purified phage particles was found to possess short, complementary single-strand extensions, "cohesive ends," that could promote circle formation. Other predictions of the model were also validated in several laboratories, and viral and cellular proteins that catalyzed integration were identified.

Lambda DNA integration remains an important paradigm for understanding the molecular mechanisms of DNA recombination and the parameters that influence the joining of viral and host DNAs.

Campbell AM. 1962. Episomes. *Adv Genet* **11:**101–145.

Distinct orientations of the bacteriophage lambda genetic map. (A) Comparison of the integrated prophage map, in which genes N and J are flanked by bacterial DNA and genes R and A are located centrally, with DNA extracted from virus particles, in which genes N and J are in the center of the genetic map and viral genes A and R are at the termini. Complementary single-strand ends in the viral DNA are shown in expanded scale. **(B)** Organization of viral genes in the circular form of DNA produced by annealing and ligation of the single-strand ends of viral DNA following infection of host cells. The *attP* site (yellow box) lies between genes N and J. The bacterial insertion site, *attB* (orange box), is shown below the circle, flanked by genes that encode enzymes required for galactose metabolism (*gal*) and biosynthesis of the vitamin biotin (*bio*). **(C)** Map of the integrated prophage, flanked by hybrid *att* sites. Upon induction, recombination at the hybrid *att* sites, catalyzed by another viral enzyme, leads to excision of a circular viral genome and viral DNA replication.

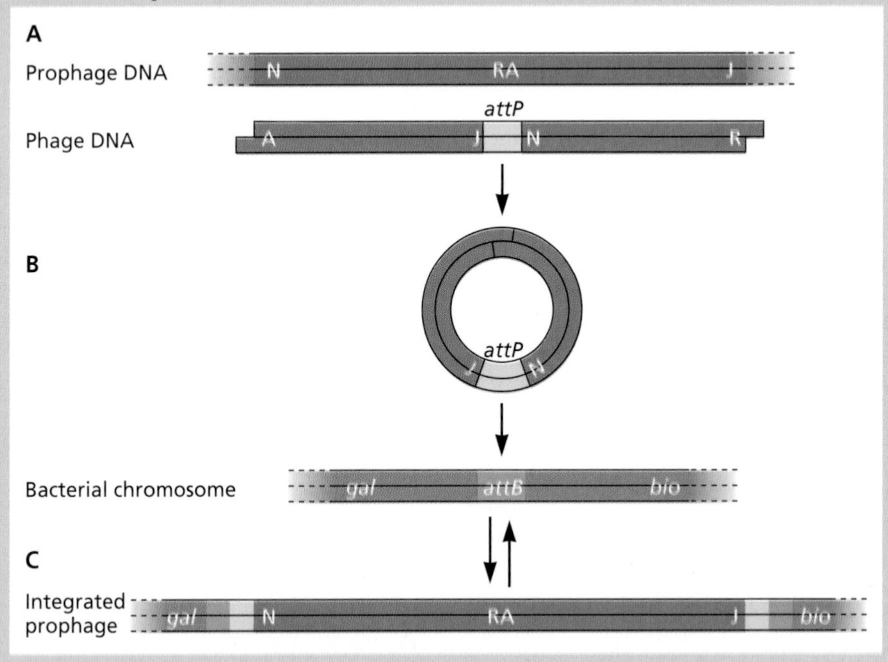

with those of other enzymes that catalyze similar reactions. For example, RTs share certain sequence motifs with RNA and DNA polymerases of bacteria, archaea, and eukaryotes, in regions known to include critical active-site residues (Fig. 6.4). Consequently, it is not surprising that these enzymes employ similar catalytic mechanisms for nucleic acid polymerization reactions. Like DNA polymerases, viral RTs cannot initiate DNA synthesis *de novo*, but require a specific primer. It should be noted that even as arcane and distinct as the viral systems may appear, they do not exhaust the repertoire for reverse transcription reactions that exist in nature. Wide varieties of primers, as well as sites and modes of initiation, are used by other RTs.

Much of what has been learned about reverse transcription in retroviruses comes from the identification of intermediates in the reaction pathway that are formed in infected cells. Reverse transcription intermediates have also been detected in **endogenous reactions**, which take place within purified virus particles, using the encapsidated viral RNA template. It was amazing to discover that intermediates and products virtually identical to those made in infected cells can actually be synthesized in purified virus particles; all that is required is treatment with a mild detergent to permeabilize the envelope and addition of the metal cofactor and deoxynucleoside triphosphate (dNTP) substrates. The fidelity and robustness of the endogenous reaction suggests that the reverse transcription system is poised for action as soon as the virus particle enters the cell. Retroviral reverse transcription intermediates have also been analyzed in fully reconstituted reactions with purified enzymes and model RNA templates.

Retroviral RT is the only protein required to accomplish all the diverse steps in the pathway described below. However, as the reactions that take place inside cells are more efficient than those observed in either endogenous or reconstituted systems, it is unlikely that all of the significant molecular interactions have been reproduced *in vitro*.

Essential Components

Genomic RNA. Retrovirus particles contain two copies of the RNA genome held together by multiple regions of base pairing. (See Box 7.2 for labeling conventions.) When purified from virus particles, this RNA sediments at 70S, as expected for a dimer of 35S genomes. Partial denaturation and electron microscopic analyses of the 70S RNA indicate that the most stable pairing is via sequences located near the 5′ ends of the two genomes (Fig. 7.1A). Sequence interactions that promote dimerization have been identified in human immunodeficiency virus type 1 RNA (Fig. 7.1B). The 70S RNA also includes two molecules of a specific cellular transfer RNA (tRNA) that serves as a primer for the initiation of reverse transcription, discussed in the following section.

Despite the fact that two genomes are encapsidated, only one copy of integrated retroviral DNA is typically detected after infection with a single particle. Therefore, retroviral virus particles are said to be **pseudodiploid**. The availability of two RNA templates could help retroviruses survive extensive damage to their genomes. At least parts of both genomes can be, and typically are, used as templates during the reverse transcription process, accounting for the high rates of genetic recombination in these viruses. Presumably, being able to patch together one complete DNA copy from two randomly interrupted or mutated RNA genomes would provide survival value. Nevertheless, genetic experiments have shown that the use of two RNA templates is not an essential feature of the reverse transcription process. Therefore, all of the known steps in reverse transcription can take place on a single genome.

Like the genomes of (−) strand RNA viruses, the retroviral genome is coated along its length by a viral nucleocapsid protein (NC), with approximately one molecule for every 10 nucleotides. This small, basic protein can bind nonspecifically to both RNA and DNA and promote the annealing of nucleic acids. Biochemical experiments suggest that NC may facilitate template exchanges and function in reverse transcription like the bacterial single-stranded-DNA-binding proteins. In the synthesis of DNA catalyzed by bacterial DNA polymerases, the single-stranded-DNA-binding proteins enhance **processivity** (ability to continue synthesis without dissociating from the template). The ability of NC to organize RNA genomes within the virus particle and to facilitate reverse transcription within the infected cell may account for some of the differences in efficiency observed when comparing reactions reconstituted *in vitro* with those that take place in a natural infection.

Primer tRNA. In addition to the viral genome, retrovirus particles contain a collection of cellular RNAs. These include ~100 copies of a nonrandom sampling of tRNAs, some 5S rRNA, 7S RNA, and traces of cellular mRNAs. We do not know how most of these cellular RNAs become incorporated into virus particles, and most have no obvious function. However, one particular tRNA molecule is critical: it serves as a primer for the initiation of reverse transcription. The tRNA primer is positioned on the template genome during virus assembly, in a reaction that is facilitated by interactions with the viral polyprotein precursors (Gag and Gag-Pol; see Appendix, Fig. 29 and 30) during particle assembly. The primer tRNA

BOX 7.2

TERMINOLOGY

Conventions for designating sequences in nucleic acids

For clarity, lowercase designations are used throughout this chapter to refer to RNA sequences; uppercase designations identify the same or complementary sequences in DNA (e.g., pbs in RNA; PBS in DNA).

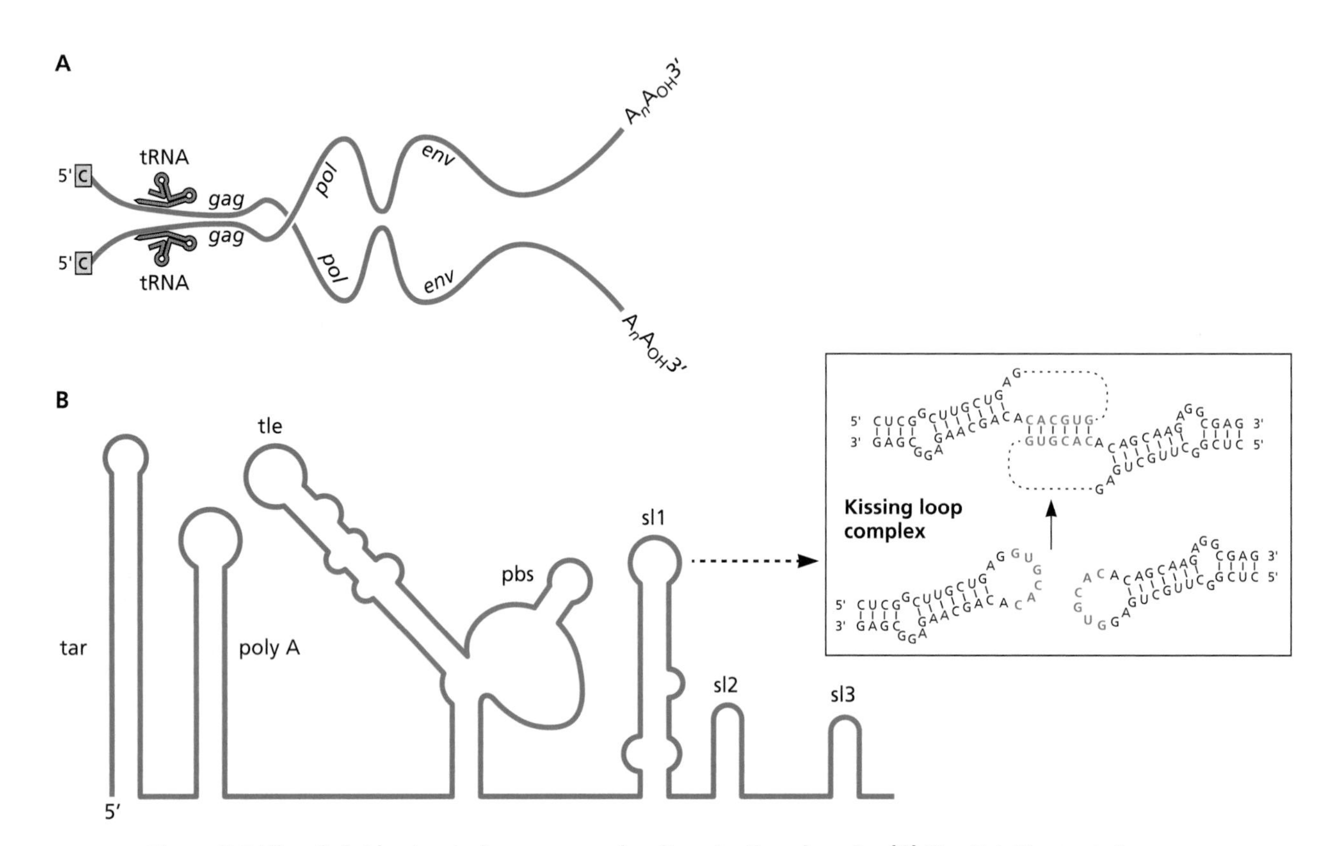

Figure 7.1 The diploid retroviral genome and a dimerization domain. (A) The diploid retroviral genome includes the following, from 5′ to 3′: the m^7Gppp cap; the coding regions for viral structural proteins and enzymes; *gag*, *pol*, and *env*; and the 3′-poly(A) sequence. The cell-derived primer tRNA is also shown. Points of contact represent multiple short regions of complementary base pairing. From J. M. Coffin, p 1767–1848, *in* B. N. Fields et al (ed), *Fields Virology*, 3rd ed (Lippincott-Raven Publishers, Philadelphia, PA, 1996), with permission. **(B)** Structural elements in the 5′ end of genomic RNA comprise distinct stem-loop structures. In the human immunodeficiency virus RNA, these elements include the Tat-binding site (tar), a poly(A) stem-loop, and a section that resembles a tRNA anticodon loop called the tle (for tRNA-like element). The adjacent primer-binding site (pbs), comprising a sequence complementary to the 3′ end of the tRNA primer, is followed by a stem-loop structure, sl1, that initiates genome dimerization by hybridizing with sl1 in a second viral RNA molecule to form a "kissing loop," as illustrated in the box. The sl1, sl2, and sl3 elements are required for efficient viral RNA packaging.

is partially unwound and hydrogen-bonded to complementary sequences near the 5′ end of each RNA genome in a region called the **primer-binding site (pbs)** (Fig. 7.2). The RTs of all retroviruses studied to date are primed by one of only a few classes of cellular tRNAs. Most mammalian retroviral RTs rely on $tRNA^{Pro}$, $tRNA^{Lys3}$, or $tRNA^{Lys1,2}$ for this function, and the relevant primer RNAs are packaged selectively into virus particles. Lysyl-tRNA synthetase is also packaged in human immunodeficiency virus type 1 particles: this cellular protein binds to viral RNA and facilitates positioning of the $tRNA^{Lys3}$ primer on the pbs (Box 7.3). It seems possible that a similar mechanism promotes primer binding on other retroviral genomes, but the generality of this process has not yet been tested.

In addition to the 3′-terminal 18 nucleotides that anneal to the pbs, other regions in the tRNA primer contact the RNA template and modulate reverse transcription. The template-primer interaction has been studied extensively in reconstituted reactions with RNA and RT of avian sarcoma/leukosis virus. In these *in vitro* analyses, the ability of the viral RNA to form stem-loop structures, and specific interactions between the primer $tRNA^{Trp}$ and one of these loops, appear to be critical for reverse transcription (Fig. 7.2). Similar interactions have been reported for human immunodeficiency virus RNA and its primer. Although the interactions are likely to be significant biologically, we do not yet know how RTs recognize structural features in these template-primer complexes.

Reverse transcriptase. Each retrovirus particle contains 50 to 100 molecules of RT. Reducing the number of enzymatically active copies of RT by more than 2- to 3-fold dramatically inhibits the process of reverse transcription in cells infected in culture. However, the number of molecules that are actually engaged in reverse transcription in each virus particle is not known. Reverse transcription can be initiated in

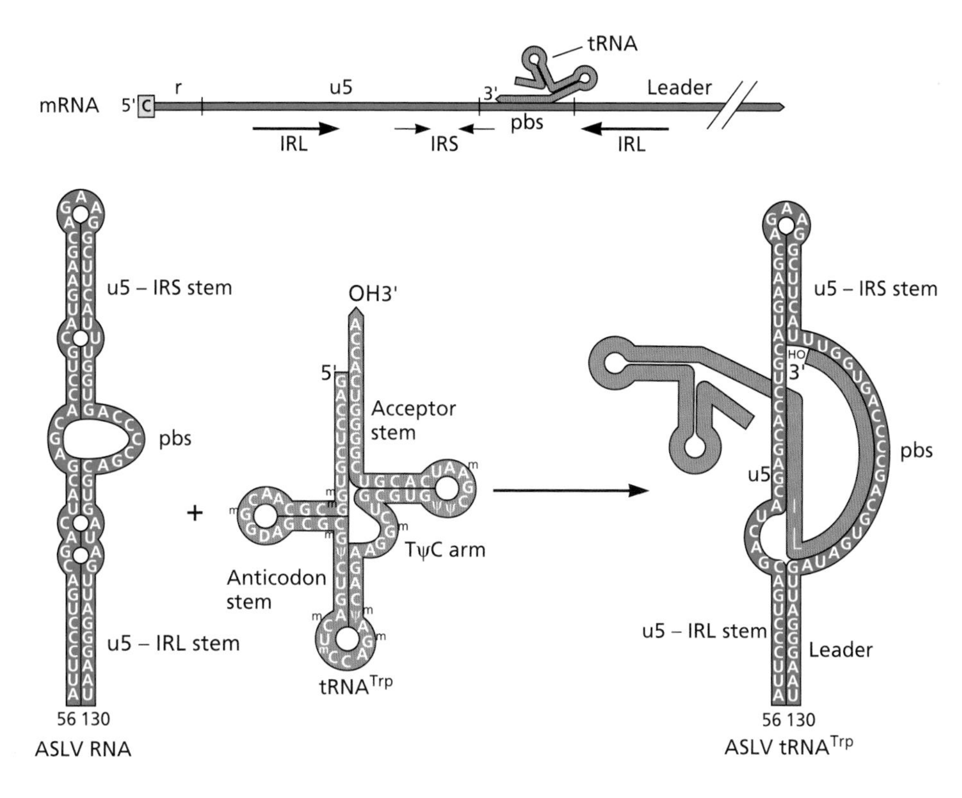

Figure 7.2 Primer tRNA binding to the retroviral RNA genome. (Top) Linear representation of the 5′ terminus of retroviral RNA, indicating locations of the r, u5, and leader regions. A tRNA primer is annealed schematically to the pbs. Two inverted-repeat (IRS and IRL) sequences that flank the pbs are represented by arrows. (Bottom) The avian sarcoma/leukosis virus (ASLV) RNA can form an extended hairpin structure around the pbs in the absence of primer tRNA (left). Primer tRNA^{Trp} is shown in the cloverleaf structure (middle). Modified bases are indicated. Viral RNA annealed to tRNA^{Trp}, with flanking u5-leader and u5-IR stem structures (right). The TΨC arm of the primer and u5 RNA also form hydrogen bonds. Bottom diagram is from J. Leis et al, p 33–47, *in* A. M. Skalka and S. P. Goff (ed), *Reverse Transcriptase* (Cold Spring Harbor Laboratory Press, Cold Spring Harbor, NY, 1993), with permission.

virus particles as soon as the viral envelope is made permeable to dNTP substrates, and it has been established that DNA synthesis takes place in the cytoplasm shortly after entry, within a subviral nucleoprotein structure that retains a partially dissociated capsid. Enzymes of the three retroviruses that have been studied most extensively, avian sarcoma/leukosis virus, murine leukemia virus, and human immunodeficiency virus type 1, are used as examples throughout this chapter.

Retroviral RTs are intricate molecular machines with moving parts and multiple activities. The distinct catalytic activities brought into play at various stages in the pathway of reverse transcription include RNA-directed and DNA-directed DNA polymerization, DNA unwinding, and the hydrolysis of RNA in RNA-DNA hybrids (RNase H). The first three activities reside within the polymerase domain, while the RNase H is in a separate domain. The RNase H of RT functions as an endonuclease, producing fragments of 2 to 15 nucleotides from the genomic RNA after it has been copied into cDNA. RNase H activity also produces the primer for (+) strand DNA synthesis from the genomic RNA, and removes this primer and the tRNA primer from the 5′ ends of the nascent viral DNA strands at specific steps in the reaction.

Distinct Steps in Reverse Transcription

Initiation of (−) strand DNA synthesis. Our understanding of DNA synthesis indicates that the simplest way of copying an RNA template to produce full-length complementary DNA would be to start at its 3′ end and finish at its 5′ end. It was therefore somewhat of a shock for early researchers to discover that retroviral reverse transcription in fact starts near the 5′ end of the viral genome—only to run out of template after little more than ~100 nucleotides (Fig. 7.3). However, this counterintuitive strategy for initiation of DNA synthesis allows the duplication and translocation of essential transcription and integration signals encoded in both the 5′ and 3′ ends of the genomic RNA, called u5 and u3, respectively.

The 5′ end of the genome RNA is degraded by the RNase H domain of RT, after (or as) it is copied to form (−) strand DNA. The short (ca. 100-nucleotide) DNA product of this first reaction, attached to the tRNA primer, accumulates in large quantities in the endogenous and reconstituted systems and is called **(−) strong-stop DNA**. For simplicity, the reactions illustrated in Fig. 7.3 to 7.6 are shown as taking place on a single RNA genome.

The first template exchange. In the next distinct step (Fig. 7.4), the 3′ end of the RNA genome is engaged as a template via hydrogen bonding between the R sequence in the (−) strong-stop DNA and the complementary r sequence upstream of the poly(A) tail. This reaction corresponds to the substitution of one end of the RNA for another to be copied by the RT "machine." As (−) strong-stop DNA is barely detectable in infected cells, this first template exchange must be efficient. Once the 3′ end of the genome RNA is engaged, the RNA-dependent DNA polymerase activity of RT can continue

BOX 7.3

EXPERIMENTS

tRNA mimicry and the primer-binding site of human immunodeficiency virus type 1

A highly conserved region in the 5′ end of human immunodeficiency virus type 1 genomic RNA contains multiple sequences that are critical for control of viral transcription, genome replication, and genome packaging (Fig. 7.1 and 7.2). Not only are specific nucleotides important, but numerous regions of base pairing mediate formation of particular structures that are also critical for function. Recent analysis of the three-dimensional structures of functional sections of this region has shown that molecular mimicry may explain how the tRNA primer is positioned selectively on one of these sections, the primer-binding site (pbs).

Human $tRNA^{Lys3}$ is the primer for synthesis of cDNA by the RT of human immunodeficiency virus type 1. While only one primer (or perhaps two) is needed for the two copies of genomic RNA in a virus particle, approximately 20 to 25 $tRNA^{Lys}$ molecules are encapsidated, along with equal amounts of their major binding protein, human lysyl-tRNA synthetase (hLysRS). The enrichment of $tRNA^{Lys}$ in virus particles is due, in part, to the interaction of hLysRS with the viral Gag and Gag-Pol proteins during particle assembly. It has also been established that hLysRS binds tightly to viral RNA, and that such binding depends on sequences in a tRNA-like element (tle) proximal to the pbs, which resembles the anticodon loop of $tRNA^{Lys}$. Annealing of the tRNA primer has also been proposed to promote a conformation required for genome dimerization via the downstream initiation site.

Structural analysis of a 99-nucleotide fragment corresponding to the tle-pbs region of human immunodeficiency virus type 1 RNA shows that this region, either alone or with an annealed 18-deoxynucleotide primer sequence, adopts a bent conformation in solution that resembles the shape of a tRNA (see figure, bottom). This capacity for molecular mimicry and analysis of the relative affinities of hLysRS for the tle and tRNA have suggested the model illustrated at the top of the figure. In the model, competition by the tle for binding to hLysRS leads to release of $tRNA^{Lys3}$, which can now be positioned selectively on the adjacent pbs.

Jones CP, Cantara WA, Olson ED, Musier-Forsyth K. 2014. Small-angle X-ray scattering-derived structure of the HIV-1 5′ UTR reveals 3D tRNA mimicry. *Proc Natl Acad Sci U S A* **111:**3395–3400.

Jones CP, Saadatmand J, Kleiman L, Musier-Forsyth K. 2013. Molecular mimicry of human $tRNA^{Lys}$ anti-codon domain by HIV-1 RNA genome facilitates tRNA primer annealing. *RNA* **19:** 219–229.

Seif E, Niu M, Kleiman L. 2013. Annealing to sequences within the primer binding site loop promotes an HIV-1 RNA conformation favoring RNA dimerization and packaging. *RNA* **19:**1384–1393.

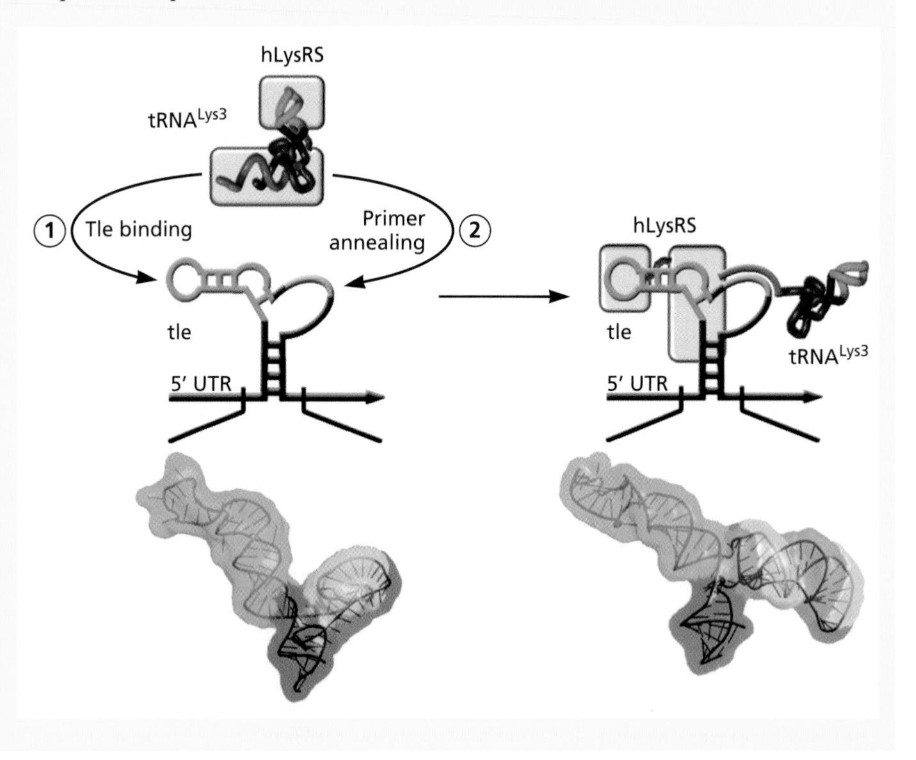

Model for $tRNA^{Lys3}$ primer placement onto the pbs. The $tRNA^{Lys3}$ primer (top left) shares 18 nucleotides of complementarity (red) to human immunodeficiency virus type 1 RNA pbs (blue), to which the primer must be annealed for reverse transcription to begin from its CCA-3′-OH end. The tle is part of an hLysRS-binding domain that effectively competes with tRNA for binding to hLysRS (left, step 1). Such competition is believed to facilitate release of bound $tRNA^{Lys3}$ from the synthetase, the 3′ end of which can then be annealed to the pbs in viral RNA (left, step 2). The final annealed complex with hLysRS bound to the tle is shown at the right. The three-dimensional models at the bottom of the figure are derived from small-angle X-ray scattering in conjunction with molecular dynamics and simulated annealing of a 99-nucleotide RNA that includes the tle-pbs domain alone (left) or annealed to an 18-nucleotide complementary DNA fragment, anti-PBS, that represents the priming, 3′ end of $tRNA^{Lys3}$ (right). The color scheme is the same as at the top, with the tle shown in green, the pbs in blue, and anti-PBS in red. Figure courtesy of W. A. Cantara, E. D. Olson, C. P. Jones, and K. Musier-Forsyth, Ohio State University.

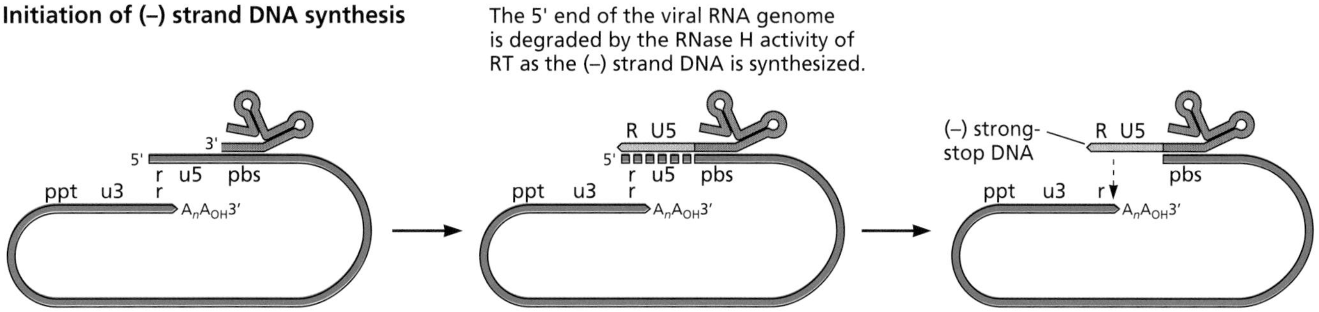

Initiation of (–) strand DNA synthesis

The 5' end of the viral RNA genome is degraded by the RNase H activity of RT as the (–) strand DNA is synthesized.

(–) strong-stop DNA

Figure 7.3 Retroviral reverse transcription: initiation of (−) strand DNA synthesis from the tRNA primer. Retroviral DNA synthesis begins with copying of the 5′ end of the viral RNA genome, using the 3′ end of a tRNA as the primer.

copying all the way to the 5′ end of the template, with the RNase H activity digesting the RNA template in its wake.

Initiation of (+) strand DNA synthesis. Among the early products of RNase H degradation of genomic RNA is a fragment comprising a **polypurine tract (ppt)** of ~13 to 15 nucleotides. This RNA fragment is especially important as it serves as the primer for (+) strand DNA synthesis, which begins even before (−) strand DNA synthesis is completed (Fig. 7.4). Following initiation from the ppt, synthesis of (+) strand DNA proceeds to the nearby end of the (−) strand DNA template and terminates after copying the first 18 nucleotides of the primer tRNA, when it encounters a modified base that cannot be copied. This product is called (+) **strong-stop DNA** (Fig. 7.5, left). The (−) strand DNA synthesis continues to the end of the viral DNA template, which includes the pbs sequence that had been annealed to the tRNA primer. The production of (+) strong-stop DNA and the converging (−) strand DNA synthesis disengages the template ends. The product is a (−) strand of viral DNA comprising the equivalent of an entire genome (but in permuted order) annealed to the (+) strong-stop DNA. The ppt and tRNA primer are removed by the RNase H, probably via recognition of structural features in these RNA-DNA junctions. The single-stranded 3′ end of the (−) strong-stop DNA then becomes available for annealing to complementary sequences in the single-stranded PBS at the 3′ end of the (+) strand DNA (Fig. 7.5, right).

The second template exchange. The next steps in the pathway of reverse transcription begin with a **second template exchange** in which annealing of the complementary PBS sequences provides a circular DNA template for polymerization by RT (Fig. 7.6, top). Synthesis of the (+) strand DNA can now continue, using (−) strand DNA as a template. The (−) strand DNA synthesis also continues to the end of U3, displacing the 5′ end of the (+) strand, a reaction that opens the DNA circle. Synthesis stops when RT reaches the terminus of each template strand (Fig. 7.6, left). The final product is a linear, DNA duplex copy of the viral genome with **long terminal repeats (LTRs)**, containing critical *cis*-acting signals at either end. This linear form of viral DNA is the major product of reverse transcription found in the nucleus of infected cells.

Figure 7.4 Retroviral reverse transcription: first template exchange, mediated by annealing of short terminal repeat sequences. Although a template exchange of the 5′ end of one RNA genome for the 3′ end of the second RNA genome can also occur, the principles illustrated and the final end products would be the same.

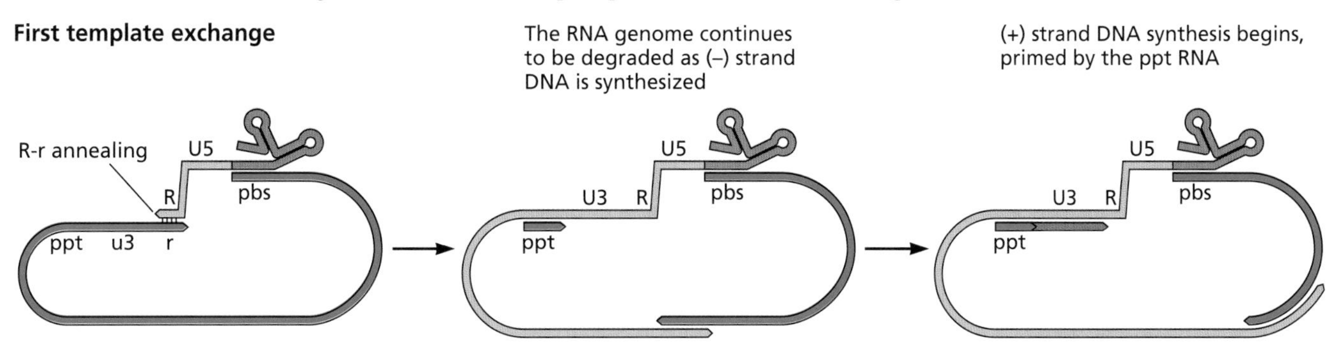

(+) strand DNA synthesis

The pbs sequence is copied twice:
- once from the RNA genome
- once from the tRNA primer

RNase H endonuclease activity of RT removes both primer RNAs

DNA ends are juxtaposed by annealing at complementary PBS sequences

U5
R
U3
Modified base
PPT
PBS
(+) strong-stop DNA
PBS
PPT U3 R U5 PBS

Figure 7.5 Retroviral reverse transcription: (+) strand DNA synthesis primed from ppt RNA.

Figure 7.6 Retroviral reverse transcription: the second template exchange and formation of the final linear DNA product. The second exchange is facilitated by annealing of PBS sequences in (+) and (−) strands of retroviral DNA. Formation of two minor (normally <1%), circular products is shown in the shaded box on the right. The smaller of the two circles contains only one LTR, and can arise either from a failure of strand displacement synthesis of RT or by recombination between the terminal LTR sequences in the linear molecule. The circle with two LTRs is presumed to arise by ligation of the ends of the linear viral DNA. Formation of this product requires a nuclear enzyme, DNA ligase, and it is easy to detect by PCR techniques. Consequently, the two-LTR circles are typically used as convenient markers for the transport of viral DNA into the nucleus.

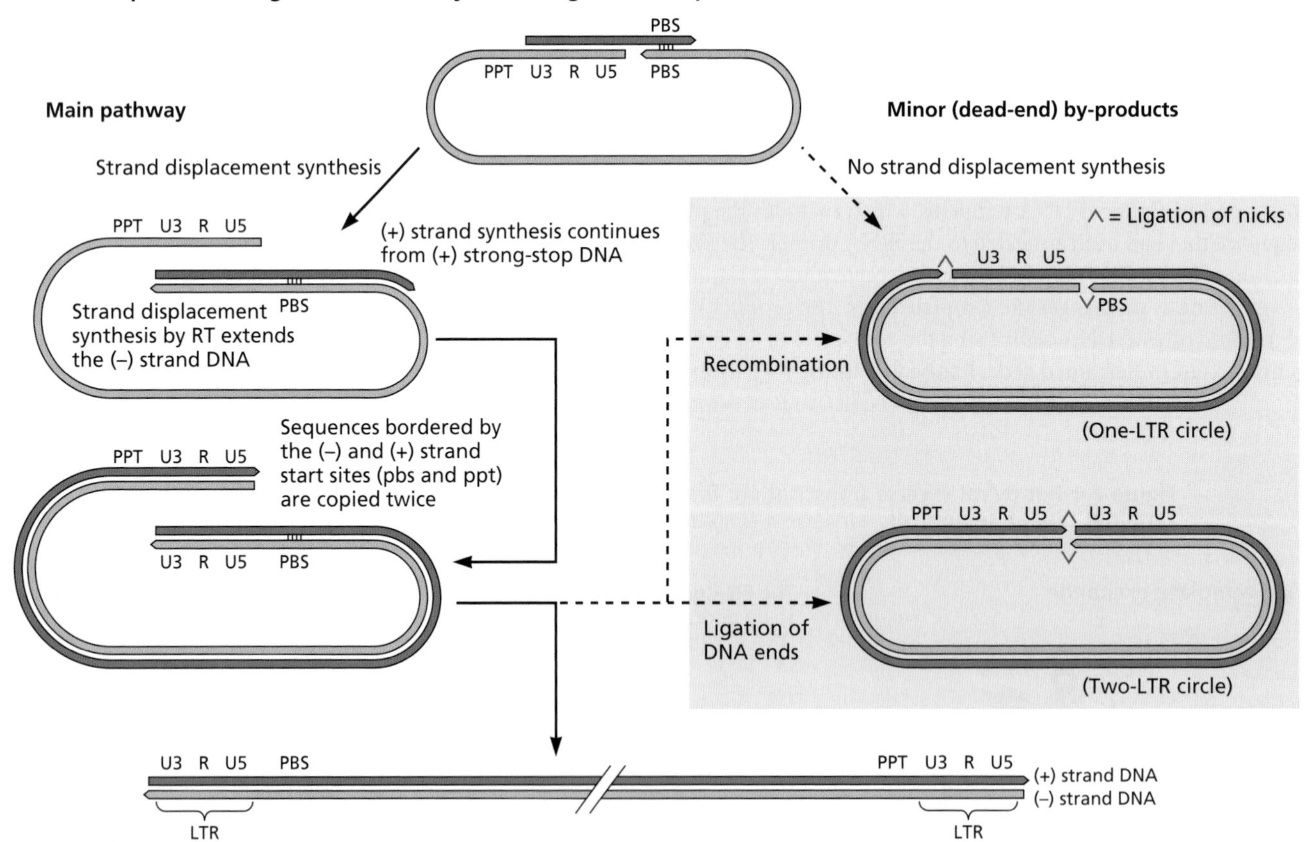

Small quantities of two circular DNA products are also invariably present in the nucleus. These are nonfunctional, dead-end products; their presumed origin is illustrated in Fig. 7.6 (right).

Retroviral reverse transcription has been called "destructive replication," as there is no net gain of genomes, but rather a substitution of one double-stranded DNA for two molecules of single, (+) strand RNA. However, by this rather intricate but elegant pathway, RT not only makes a linear DNA copy of the retroviral genome to be integrated, but also produces the LTRs that contain signals necessary for transcription of integrated DNA, which is called the **provirus**. The promoter in the upstream LTR is now in the appropriate location for synthesis of progeny RNA genomes and viral mRNAs by host cell RNA polymerase II (Chapter 8). Integration also ensures subsequent replication of the provirus via the host's DNA synthesis machinery as the cell divides.

Reverse transcription promotes recombination. The high rate of genetic recombination, a hallmark of retrovirus reproduction, is facilitated by the presence of two RNA templates within the reverse transcriptase complex. Although only one viral DNA molecule is normally produced by each infecting virion, recombination can occur during reverse transcription. The incorporation of two distinct RNA templates in a single virus particle can lead to new combinations of sequences (Box 7.4).

The above description of reverse transcription has been idealized for clarity. Analyses of reaction intermediates show that

BOX 7.4

WARNING

Retroviral recombination and the rise and fall of XMRV

Because a mutation in a human gene encoding a viral defense protein (RNase L) is a known risk factor for prostate cancer, a 2006 report of the isolation of a new retrovirus from tissue samples of individuals homozygous for this mutation attracted considerable attention. The excitement was compounded by a report in 2009 that the same virus could be isolated from the blood of patients suffering from chronic fatigue syndrome (CFS). While the association with CFS was controversial and never confirmed (indeed, it was later retracted), some desperate CFS patients were nevertheless treated with RT inhibitors.

The new virus was called XMRV (for xenotropic murine leukemia virus-related virus) because its sequence is closely related to well-known murine virus strains. Such strains are called xenotropic because they can infect foreign cells, such as human cells, in culture but are unable to reinfect mouse cells. The discovery was also noteworthy because XMRV is a gammaretrovirus, and this genus was not previously known to include human pathogens. Numerous investigators took up the study of this new virus, but the scientific literature was soon filled with contradictory reports concerning its association with prostate cancer, most based on the results from extremely sensitive PCR assays. Many investigators began to wonder if XMRV was indeed a human virus, and questioned its association with cancer.

These issues were addressed in a careful study, published in 2011, which showed that XMRV was derived from the recombination of two previously unknown defective murine endogenous retroviruses, and that the event probably occurred between 1993 and 1996 when a human tumor was implanted into nude mice, a process that is necessary to establish prostate cancer cell lines. These findings suggested that the reported PCR-based evidence of XMRV in clinical specimens could be explained by laboratory contamination. This explanation was supported by results from another team of investigators, which included some of the initial "discoverers" of XMRV, who established that the original archived prostate cancer tissue was indeed negative for XMRV, but the archival extracted RNA from the original study was positive for the viral genome. They also found that the source of XMRV contamination in the archival extracted RNA was the XMRV-infected cell line used in the laboratory. The contradictory results reported by numerous laboratories can be attributed therefore to the superb sensitivity of the PCR methods used to detect XMRV sequences and the ubiquitous presence of mouse DNA and/or sources of likely contamination.

Lee D, Das Gupta J, Gaughan C, Steffan I, Tang N, Luk KC, Qiu X, Urisman A, Fischer N, Molinaro R, Broz M, Schochetman G, Klein EA, Ganem D, DeRisi JL, Simmons G, Hackett J, Jr, Silverman RH, Chiu CY. 2012. In-depth investigation of archival and prospectively collected samples reveals no evidence for XMRV infection in prostate cancer. *PLoS One* 7:e44954. doi:10.1371/journal.pone.0044954.

Lombardi VC, Ruscetti FW, Das Gupta J, Pfost MA, Hagen KS, Peterson DL, Ruscetti SK, Bagni RK, Petrow-Sadowski C, Gold B, Dean M, Silverman RH, Mikovits JA. 2009. Detection of an infectious retrovirus, XMRV, in blood cells of patients with chronic fatigue syndrome. *Science* **326:**585–589.

Paprotka T, Delviks-Frankenberry KA, Cingoz O, Martinez A, Kung HJ, Tepper CG, Hu WS, Fivash MJ, Jr, Coffin JM, Pathak VK. 2011. Recombinant origin of the retrovirus XMRV. *Science* **333:**97–101.

Urisman A, Molinaro RJ, Fisher N, Plummer SJ, Casey G, Klein EA, Malathi K, Magi-Galluzzi C, Tubbs RR, Ganem D, Silverman RH, DeRisi JL. 2006. Identification of a novel gammaretrovirus in prostate tumors of patients homozygous for R462Q *RNASEL* variant. *PLoS Pathog* **2:**e25. doi:10.1371/journal.ppat.0020025.

Origin of XMRV. XMRV arose via recombination between two previously unknown defective murine endogenous retroviruses (arbitrarily called Pre XMRV-1 and -2), which included six separate crossover events. For discussion, see http://www.virology.ws/2011/05/31/xmrv-is-a-recombinant-virus-from-mice/.

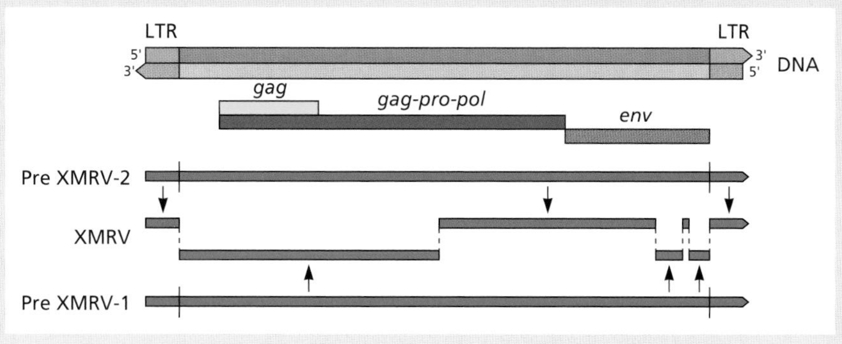

RT pauses periodically during synthesis, presumably at some sequences, structures, or breaks that impede copying. If a break is encountered in one RNA template, synthesis can be completed by utilization of the second RNA genome. Such internal template exchanges (known to occur even in the absence of breaks) probably proceed at regions of homology via the same steps outlined for the first template exchange. Internal exchanges that take place during RNA-directed DNA synthesis are estimated to be the main source of genetic recombination, a mechanism known as **copy choice**. Exchange of single-stranded ends from one DNA template to another during (+) strand synthesis can also lead to recombination via a mechanism known as **strand displacement synthesis** (Fig. 7.7).

Figure 7.7 Two models for recombination during reverse transcription. Virtually all retroviral recombination occurs between coencapsidated genomes at the time of reverse transcription. The copy choice model **(A)** postulates a mechanism for genetic recombination during RNA-directed (−) strand DNA synthesis. This mechanism predicts that a homoduplex DNA product is formed as the recombined (−) strand of DNA is copied to form a (+) complementary strand. The strand displacement synthesis model **(B)** postulates a mechanism for genetic recombination that can occur when (+) strand DNA synthesis is initiated at internal sites on the (−) strand DNA template. Such internal initiations are known to arise frequently during reverse transcription by avian sarcoma/leukosis virus and human immunodeficiency virus RTs. Recombination can occur if (−) strand DNA has been synthesized from both RNA genomes in the particle. This mechanism can be distinguished from copy choice because a heteroduplex DNA product will be formed, i.e., A′/A, in which only the (+) strand (light blue in the figure) is a recombinant. The two mechanisms are not mutually exclusive, and while copy choice is most frequent, there is experimental support for both. Viral genetic markers, arbitrarily labeled *a*, *b*, and *c*, are indicated to illustrate recombination. While multiple crossovers are frequently observed, single recombination events are shown for simplicity, focusing on the hypothetical *a* allele, with the mutant form in red. For more details, see R. Katz and A. M. Skalka, *Annu Rev Genet* **24**:409–445, 1990.

A Copy Choice

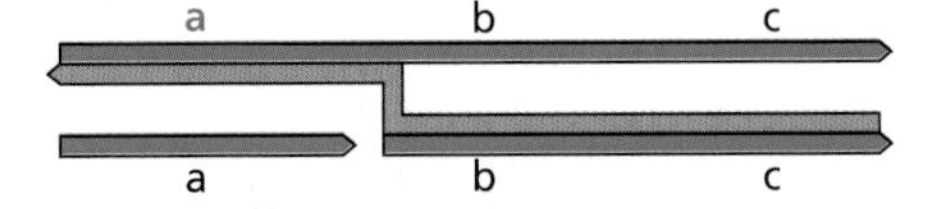

(–) (first)-strand DNA synthesis starts on one genome and switches to the second at a break point, pause site, or random location

B Strand displacement synthesis

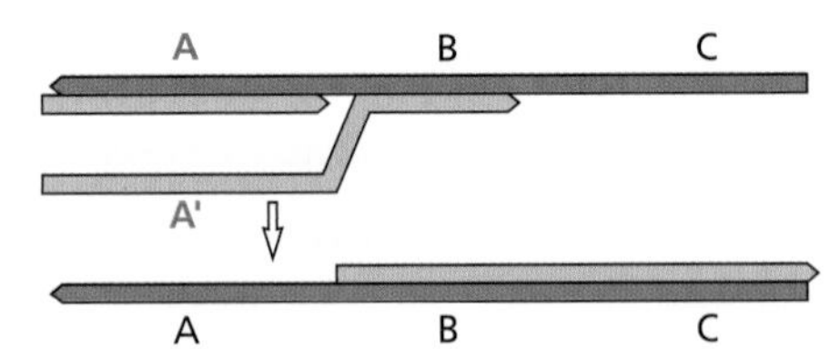

(+) (second)-strand DNA synthesis accompanied by strand displacement and assimilation of single-strand DNA tails onto DNA from the second genome (white arrow)

General Properties and Structure of Retroviral Reverse Transcriptases

Domain Structure and Variable Subunit Organization

The RTs of retroviruses are encoded in the *pol* genes. Despite the sequence homologies and similar organization of coding sequences, retroviral species-specific differences in proteolytic processing of the Gag-Pol polyprotein precursors leads to the inclusion of additional sequences or domains in the RTs. For example, the avian sarcoma/leukosis virus RT incudes a C-terminal integrase domain, and that of the prototype foamy virus includes an N-terminal protease domain (Fig. 7.8). Furthermore, although most retroviral RTs function as monomers, the enzymes of the avian sarcoma/leukosis and human immunodeficiency viruses function as heterodimers. It is difficult to gauge the significance of this structural diversity, which may simply be the result of different evolutionary histories.

Catalytic Properties

DNA polymerization is slow. The biochemical properties of retroviral RTs have been studied with enzymes purified from virus particles or synthesized in bacteria, using model templates and primers. Kinetic analyses have identified an ordered reaction pathway for DNA polymerization similar to that of other polymerases. Like cellular polymerases and nucleases, RTs require divalent cations as cofactors (most likely Mg^{2+} in the infected cell). The rate of elongation by RT on natural RNA templates *in vitro* is 1 to 1.5 nucleotides per s, approximately 1/10 the rate of other eukaryotic DNA polymerases. Assuming that DNA synthesis is initiated promptly upon viral entry, the long time period required to produce a complete copy of retroviral RNA after infection (~4 h for ~9,000 nucleotides) supports the view that reverse transcription is also a relatively slow process *in vivo*.

In reactions *in vitro*, the rate of dissociation of the enzyme from the template-primer increases considerably after addition of the first nucleotide, suggesting that initiation and elongation are distinct steps in reverse transcription, as is the case during DNA synthesis by DNA-dependent DNA polymerases. In contrast to most other DNA polymerases, retroviral RTs dissociate from their template-primers frequently *in vitro*, a property described as "poor processivity." This may not be a limitation *in vivo*, where genomic RNA is reverse transcribed within the confines of a subviral particle.

Fidelity is low. Retroviral genomes, like those of other RNA viruses, accumulate mutations at much higher rates than do cellular genomes. RTs not only are error prone, but also lack an endonuclease capable of excising mispaired nucleotides. These properties contribute to the high mutation rate of retroviruses in infected cells.

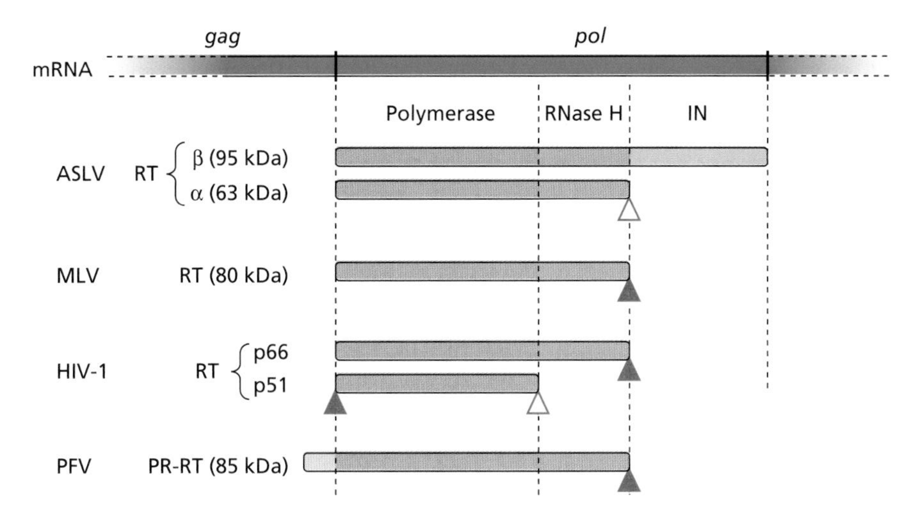

Figure 7.8 Domain and subunit relationships of the RTs of different retroviruses. The organization of open reading frames in the mRNAs of all but the spumavirus prototype foamy virus (PFV) is indicated at the top. PFV RT is expressed from a spliced *pro-pol* mRNA. Protein products (not to scale) are shown below, with arrows pointing to the sites of proteolytic processing that produce the diversity of RT subunit composition. Open red arrows indicate partial (asymmetric) processing, and solid red arrows indicate complete processing. ASLV, the alpharetrovirus avian sarcoma/leukosis virus; MLV, the gammaretrovirus murine leukemia virus; HIV-1, the lentivirus human immunodeficiency virus type 1; PFV, the spumavirus, prototype foamy virus.

Errors introduced by purified enzymes include not only misincorporations but also rearrangements such as deletions and additions (Fig. 7.9). Misincorporations by human immunodeficiency virus type 1 RT can occur as frequently as 1 per 70 copies at some template positions, and as infrequently as 1 per 10^6 copies at others. Both deletions and insertions are also known to occur during reverse transcription within an infected cell, apparently because template exchanges can take place within short sequence repeats (e.g., 4 or 5 nucleotides) that are not in homologous locations on the two RNA templates. Many types of genetic experiments have been conducted in attempts to determine the error rates of RTs in a single infectious cycle within a cell (see Volume II, Box 10.2). The general conclusion is that such rates are also quite high, with reported misincorporations in the range of 1 per 10^4 to 1 per 10^6 nucleotides polymerized, in contrast to 1 per 10^7 to 1 per 10^{11} for cellular DNA replication. As retroviral genomes are $\sim10^4$ nucleotides in length, ~1 lesion per retroviral genome per replication cycle can be expected, simply by misincorporation. This high mutation rate explains, in part, the difficulties inherent in treating AIDS patients with inhibitors of RT or other viral proteins; a large population of mutant viruses preexist in every chronically infected individual, some encoding drug-resistant proteins. These mutants can propagate in the presence of a drug and quickly comprise the bulk of the population (see Volume II, Chapters 7 and 9).

Figure 7.9 Mutational intermediates for base substitution and frameshift errors. Several unique activities of RTs are likely to contribute to their high error rates. The avian sarcoma/ leukosis and human immunodeficiency virus enzymes are both proficient at extending mismatched terminal base pairs, such as those that result from nontemplated addition **(A)**. This process facilitates incorporation of mismatched (red) nucleotides into the RT product. A certain type of slippage within homopolymeric runs in which one or more bases are extruded on the template strand can also happen during reverse transcription **(B, C)**; mispairing occurs after the next deoxyribonucleotide is added and the product strands attempt to realign with the template. Deletions can also be produced by this mechanism **(D)**. Slippage and dislocations are assumed to be mediated by looping out of nucleotides in the template. Only single-nucleotide dislocations are shown here, but large dislocations leading to deletions are also possible. From K. Bebenek and T. A. Kunkel, p 85–102, *in* A. M. Skalka and S. P. Goff (ed), *Reverse Transcriptase* (Cold Spring Harbor Laboratory Press, Cold Spring Harbor, NY, 1993), with permission.

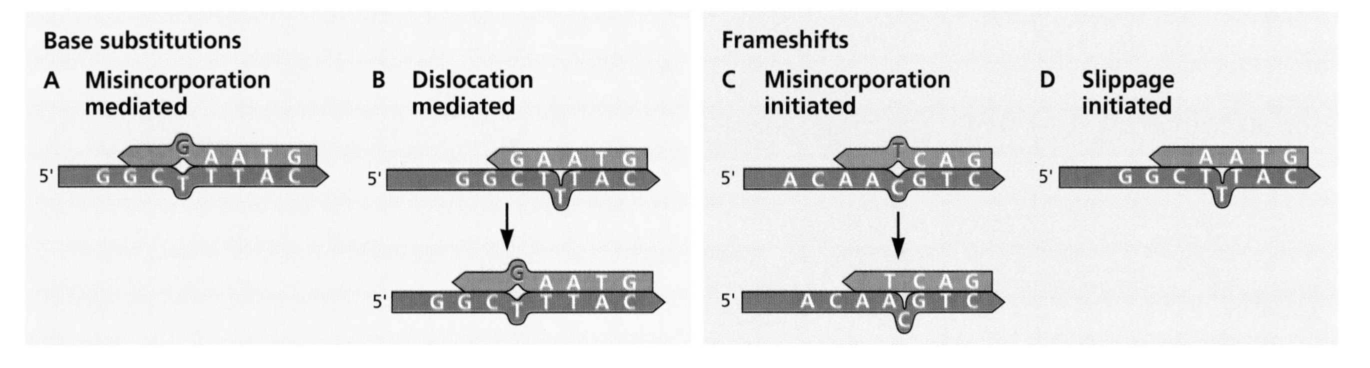

RNase H. The RNase H of RT also requires a divalent cation, most likely Mg^{2+}, which is abundant in cells. Like other RNase H enzymes, present in all bacterial, archaeal, and eukaryotic cells, the RNase H of RT digests only RNA that is annealed to DNA. Three activities of the RNase H of human immunodeficiency virus type 1 and murine leukemia virus RTs have been distinguished: endonucleases that are directed to the ends of hybrid duplexes in which the 3′ terminus of the DNA is either extended or recessed, and an internal endonuclease that is not end directed. A loose consensus site for all three activities has also been observed.

Structure of RT

Although RTs from avian and murine retroviruses have been studied extensively *in vitro*, the importance of human immunodeficiency virus type 1 RT as a target for drugs to treat AIDS has focused intense interest and resources on this enzyme. As a consequence, we know more about this RT than any other. Three aspartic acid residues in the polymerase region are included in conserved motifs in a large number of polymerases (see Fig. 6.4). These residues coordinate the required metal ions and contribute to binding deoxyribonucleoside triphosphates and subsequent catalysis.

The primary sequence of the smaller (p51) subunit of human immunodeficiency virus RT is the same as that of the larger (p66), minus the RNase H domain (Fig. 7.8). Consequently, it was somewhat surprising when the first crystallographic studies of this RT revealed **structural asymmetry** of these subunits in the heterodimer. Not only are analogous portions arranged quite differently (Fig. 7.10, bottom); they also perform different functions in the enzyme. All catalytic functions are contributed by p66. Nevertheless, p51 is required for enzyme activity and may perform a unique function in the RT heterodimer, that of binding the tRNA primer. In the heterodimer, these two subunits are nestled on top of each other, with an extensive interface (Fig. 7.10, top). The p66 polymerase domain is divided into three subdomains denoted "finger," "palm," and "thumb" by analogy to the convention used for describing the topology of the *Escherichia coli* DNA polymerase I Klenow fragment, described in Chapter 6 (Fig. 6.4). A fourth subdomain called the "connection" lies between the remainder of the polymerase and the RNase H domain. This subdomain contains the major contacts between the two subunits. The extended thumb of p51 contacts the RNase H domain of p66, an interaction that appears to be required for RNase H activity. Not only are human immunodeficiency virus type 1 RT and *E. coli* DNA polymerase similar topologically, but this retroviral RT can actually substitute for the bacterial enzyme in *E. coli* cells that lack a functional DNA polymerase I.

Highly dynamic interactions between template-primer, dNTP substrates, and RT must occur during reverse transcription. A schematic rendition (Fig. 7.11) of an RNA-DNA heteroduplex bound in the cleft region of the human immunodeficiency virus type 1 RT illustrates how the RNA strand is fed into the polymerizing site. The substrates are bound in a defined order: the template-primer first, and then the complementary dNTP to be added. The dNTP substrate interacts directly with two fingertip residues, and this contact induces closure of the binding pocket. This conformational change facilitates attack of the 3′-OH of the primer on the α-phosphate of the dNTP. Release of the diphosphate product and reopening of the fingers allows the template-primer to **translocate** by one nucleotide in preparation for the binding and addition of the next dNTP. As a new DNA chain is synthesized by addition of deoxyribonucleotides to the primer, the template RNA is moved in stepwise fashion toward the RNase H.

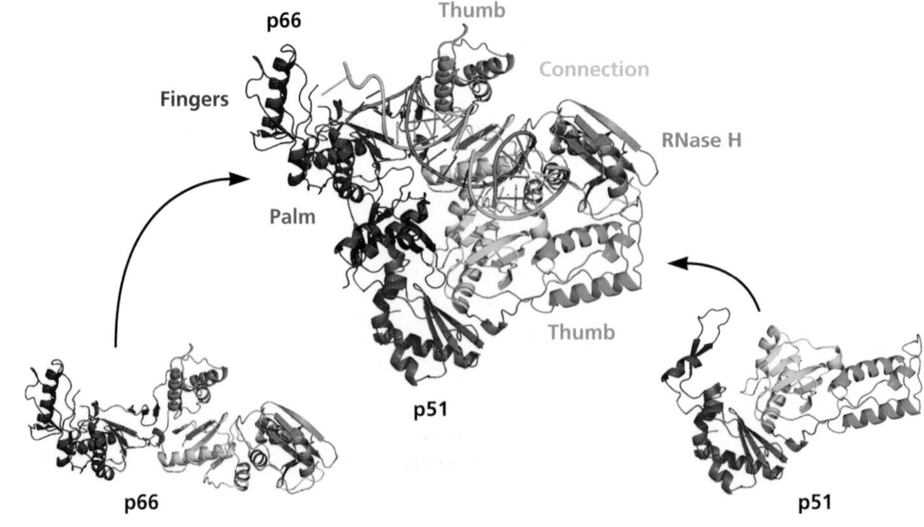

Figure 7.10 Ribbon representation of human immunodeficiency virus type 1 RT in complex with a model RNA template-DNA primer. The p61-p51 heterodimer is shown at the top, with subdomains in the catalytic subunit, p66, identified. The light blue RNA strand in the template-primer model is nicked; the DNA strand is shown in purple. The p61 and p51 subunits are shown separated at the bottom to emphasize the distinct organization of subdomains in each. For additional details, see M. Lapkouski et al., *Nat Struct Mol Biol* **20**:230–236, 2013. Courtesy of S. F. Le Grice, National Cancer Institute, Frederick, MD.

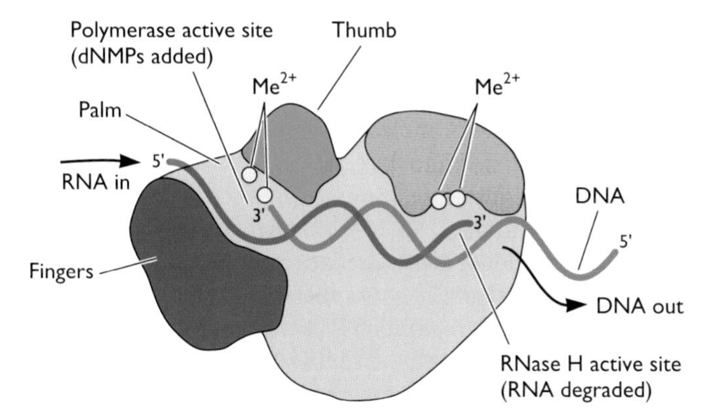

Figure 7.11 Model for a DNA-RNA hybrid bound to human immunodeficiency virus type 1 RT. The RNA template-DNA product duplex is shown lying in a cleft. The polymerase active site and the putative RNase H active site are indicated. Me^{2+} signifies a divalent metal ion. As illustrated, the RNA template enters at the polymerizing site and is degraded at the RNase H active site after being copied into DNA, which then exits from the RNase H site. Adapted from L. A. Kohlstaedt et al., p 223–250, *in* A. M. Skalka and S. P. Goff (ed), *Reverse Transcriptase* (Cold Spring Harbor Laboratory Press, Cold Spring Harbor, NY, 1993), with permission.

Remarkable dynamic capabilities of human immunodeficiency virus RT have been revealed in studies of the purified enzyme. The RNase H and polymerase sites do not act simultaneously; conformational changes are required to optimize each function. Furthermore, while synthesizing DNA from an RNA template, the enzyme is able to bind the primer in a position poised for polymerization or one that is "flipped" 180°; sliding and flipping can occur without RT disengaging from the DNA (Box 7.5). Such molecular contortions may explain the observation that polymerase-coupled RNase H activity results in cleavage of the RNA template about once for every 50 to 100 dNTPs polymerized.

Production of two protein subunits that possess identical amino acid sequences, but have structures and functions that are distinct, is an excellent example of viral genetic economy. The C terminus of the p51 subunit, at the end of the connection domain, is buried within the N-terminal β-sheet of the RNase H domain of p66. This organization suggests a model for proteolytic processing in which a p66 homodimer intermediate is arranged asymmetrically and the RNase H domain of the subunit destined to become p51 is unfolded. Such an

BOX 7.5

DISCUSSION

Reverse transcriptase can reverse direction

The exchange of one template for another to be copied by either DNA or RNA polymerases is sometimes referred to as enzyme "jumping." This inappropriate term comes from a too literal reading of simplified illustrations of the process, in which the templates to be exchanged may be opposite ends of the nucleic acid or different nucleic acid molecules. In actuality, such enzyme movement is quite improbable, and use of this terminology can cloud thinking about these processes. In almost all cases, these polymerases are components of large assemblies with architecture designed to bring different parts of the template, or different templates, close to each other. Consequently, it is likely that most of the "movement" is made by the flexible nucleic acid templates. Nevertheless, some dynamic changes in protein conformation must occur to accommodate template exchanges, as implied by another dynamic property ascribed to RT, called "flipping."

Application of a single-molecule assay to measure enzyme-substrate interactions has shown that RT of the human immunodeficiency virus type 1 can switch rapidly from one orientation to another on a single primer-template. The assay made use of surface-immobilized template-primer oligonucleotide substrate molecules: the protein and a nucleic acid end were labeled with donor and acceptor fluorophores. The position of the enzyme relative to the substrate was then measured by fluorescence resonance energy transfer. The results showed that a single RT heteroduplex can switch from one orientation to another without dissociating from the substrate (see figure).

Abbondanzieri EA, Bokinsky G, Rausch JW, Zhang JX, Le Grice SF, Zhuang X. 2008. Dynamic binding orientations direct activity of HIV reverse transcriptase. *Nature* **453:**184–189.

Liu S, Abbondanzieri EA, Rausch JW, Le Grice SF, Zhuang X. 2008. Slide into action: dynamic shuttling of HIV reverse transcriptase on nucleic acid substrates. *Science* **322:**1092–1097.

RT dynamics. (Top) During reverse transcription, DNA is synthesized and the RNA template is degraded by RT. (Bottom) RT can adopt an alternative orientation and switch from RNA-directed to DNA-directed synthesis without disengaging from the substrate. Polymerase-coupled RNase H degradation of the viral RNA template may also be facilitated by such flipping. P indicates the polymerase domain and H the RNase H domain of RT.

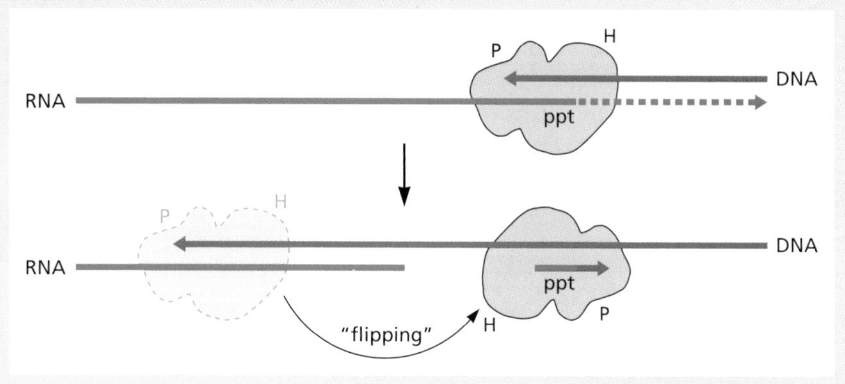

arrangement could account for asymmetric cleavage by the viral protease.

Other Examples of Reverse Transcription

When it was first discovered, RT was thought to be a peculiarity of retroviruses. We now know that other animal viruses, hepadnaviruses, and some plant viruses, such as the caulimoviruses, synthesize genomic DNA via an RNA intermediate. All are therefore classified as **retroid viruses**. In fact, the discovery of RT activity in some strains of myxobacteria and *E. coli* places the evolutionary origin of this enzyme before the separation of bacteria and eukaryotes. It is now widely held that the biological world was initially based on RNA molecules, functioning as both catalysts and genomes. Consequently, the development of the modern (DNA) stage of evolution would have required an RT activity. If so, the retroid viruses may also be viewed as living fossils, shining the first dim light into an ancient evolutionary passageway from the primordial world (Volume II, Chapter 10).

During the retroviral life cycle (Appendix, Fig. 30), the double-stranded DNA molecule synthesized by reverse transcription is integrated into the genomes of infected animal cells by the retroviral integrase. In some cases, retroviral DNA may be integrated into the DNA of germ line cells in a host organism. These integrated DNAs are then passed on to future generations in Mendelian fashion as **endogenous proviruses**. Such proviruses are often replication defective, a property that may facilitate coexistence with their hosts. Some 8% of the human genome comprises endogenous proviruses (Fig. 7.12). While many of these proviruses were established in the primate lineage millions of years ago, a present-day example of this phenomenon can be observed in another mammalian species (Box 7.6).

Since the discovery of RT in retroviruses, additional RT-related sequences have been found in **retroelements**,

Figure 7.12 Retroelements resident in eukaryotic genomes and their representation in the human genome. (A) Gene arrangements of retroelements in eukaryotic genomes. The genetic content and organization of endogenous proviruses and LTR-containing retrotransposons are similar (Appendix, Fig. 29), but most retrotransposons lack an *env* gene. LINEs (for long interspersed nuclear elements) are a distinct class of retrotranposon; they lack LTRs but contain untranslated sequences (UTRs) that include an internal promoter for transcription by cellular RNA polymerase II. The LINE *orf1* gene encodes a protein chaperone, and *orf2* encodes a protein with endonuclease and RT activities, which catalyzes reverse transcription of mRNA intermediates and integration of the DNA product. As with the other retrotransposons, the presence of flanking duplications of cellular DNA (represented by arrows below the maps) is a hallmark of transposition by LINEs. SINEs (for short interspersed nuclear elements) are classified as retroposons. They have no known open reading frames, and RNA from these sequences is retrotransposed in *trans* by the RTs of active LINEs. Processed pseudogenes comprise a less abundant group of such non-autonomously transposed retroelements. They have no introns (hence "processed"), and their sequences are related to exons in functional genes that map elsewhere in the genome. Processed pseudogenes include long, A-rich stretches. However, they contain no promoter for transcription and no RT, and are thought to arise from reverse transcription of cellular mRNAs catalyzed by the RTs of retroviruses or nondefective LINEs. Genetic maps of the retroelements are not to scale. **(B)** Retroelements in the human genome. The percentage of the human genome that each element represents and the total number of retroelements in each major class are indicated in the boxes. The percentage of the human genome that is represented by each type of element is shown beneath the boxes. Data are excerpted from N. Bannert and R. Kurth, *Proc Natl Acad Sci U S A* **101:**14572–14579, 2004.

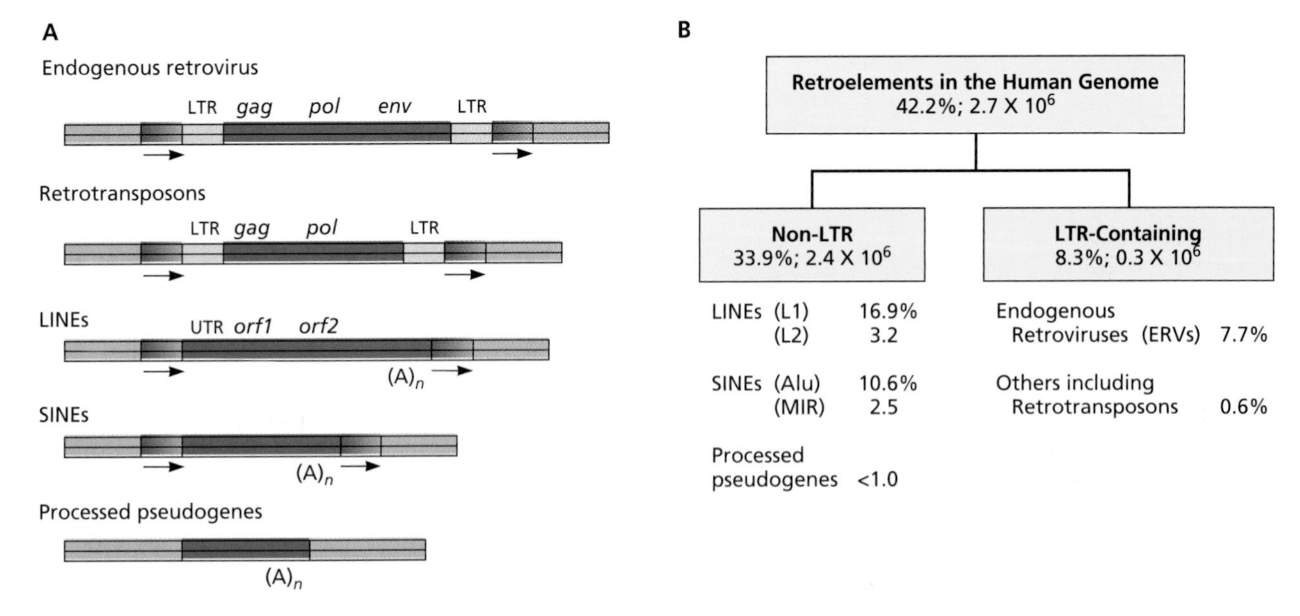

BOX 7.6

BACKGROUND

Present-day establishment of an endogenous retrovirus

Endogenous retroviruses are common in the genomes of humans and other mammals; most are ancient, defective relics of germ line infections that occurred millions of years ago. The koala retrovirus recently isolated from wild and captive animals in Australia, and in koala populations in zoos in other countries, appears to be a contemporary exception.

The koala virus is a gammaretrovirus, related to the gibbon ape leukemia and feline leukemia viruses. The first subtype to be isolated, KoRV-A, uses the same cell surface receptor for entry as the gibbon ape leukemia virus (the sodium-dependent phosphate transporter PiT1) and is thought to have been acquired by koalas as a result of cross-species transmission from rodents or bats. This viral subtype is widely distributed among koalas in northern Australia and, based on analysis of museum specimens, was circulating in this population more than 150 years ago. KoRV-A proviruses have been found in multiple copies in the genomes of every cell (including germ cells) in koalas in northern Australia. A smaller fraction of koalas in southern Australia were found to carry the virus, but genomic studies have identified several independent germ line insertions in animals from this region as well. In contrast, KoRV-A infection has been detected in only ~14% of the animals on an island off the southern coast of Australia to which koalas were introduced in the early 1900s. It was suggested, therefore, that both exogenous infection and the establishment of endogenous KoRV-A proviruses are probably still occurring in this host species as the virus spreads southward from a focus in northern Australia.

Denner J, Young PR. 2013. Koala retroviruses: characterization and impact on the life of koalas. *Retrovirology* **10:**108. doi:10.1186/1742-4690-10-108.

Ishida Y, Zhao K, Greenwood AD, Roca AL. 2015. Proliferation of endogenous retroviruses in the early stages of a host germ line invasion. *Mol Biol Evol* **32:**109–120.

Tarlinton RE, Meers J, Young PR. 2006. Retroviral invasion of the koala genome. *Nature* **442:**79–81.

Xu W, Stadler CK, Gorman K, Jensen N, Kim D, Zheng H, Tang S, Switzer WM, Pye GW, Eiden MV. 2013. An exogenous retrovirus isolated from koalas with malignant neoplasias in a US zoo. *Proc Natl Acad Sci U S A* **110:**11547–11552.

sequences that can be propagated from one locus to others in a cellular genome via reverse transcription (Fig. 7.12). One class of such transposable elements, the LTR-containing **retrotransposons**, is widely dispersed in nature. The gene contents and arrangements of these retrotransposons are similar to those of retroviruses. Most are distinguished from retroviruses by lack of an extracellular phase. They have no *env* gene, and hence the virus-like particles formed within the cell are not infectious. As with retroviruses, such particles contain RNA copies of the integrated sequences, and element-specific RT and integrase. DNA is synthesized and inserted into additional loci following entry into the nucleus of the same cell in which the particles are produced. However, members of one genus in the family *Metaviridae* do include open reading frames corresponding to *env*, and at least one of these elements, the *Drosophila* gypsy, produces infectious, extracellular particles. Phylogenetic comparisons of LTR-retrotransposons from invertebrates provide evidence that several have acquired *env* sequences via genetic recombination with both RNA and DNA viruses. These results support the view that LTR-containing retrotransposons were retroviral progenitors. An alternative possibility, but with less phylogenetic support, is that they are degenerate forms of retroviruses.

While there are distinct structural and biochemical differences among the RTs of retroviruses and LTR-containing retrotransposons, X-ray crystallographic comparisons have revealed striking topological similarities. For example, in the monomeric RT of a murine leukemia-related virus, the catalytic subunits of the heterodimeric RT of human immunodeficiency virus (p66), and the homodimeric RT of the yeast retrotransposon Ty3 (subunit A), the fingers-palm-thumb subdomains and the position of bound nucleic acids are almost superimposable (Fig. 7.13). Topologies of the two protein dimers are also superimposable despite the fact that the arrangements of their subdomains are quite distinct, having been modified and repositioned through evolution to provide analogous functions.

A second class of retrotransposons that is dispersed widely in the human genome is called **LINEs** (for long interspersed nuclear elements). LINEs can be up to 6 kbp in length; they lack LTRs, but contain internal promoters for transcription by cellular RNA polymerase II. All have A-rich stretches at one terminus, presumed to be derived by reverse transcription of the 3′-poly(A) tails in their RNA intermediates. Most LINEs encode RT-related sequences, but these often contain large deletions and translational stop codons; such elements cannot mediate their own retrotransposition and are considered to be "dead." However, ~80 to 100 human LINEs encode functional RTs, and a number of disease-causing genetic lesions resulting from

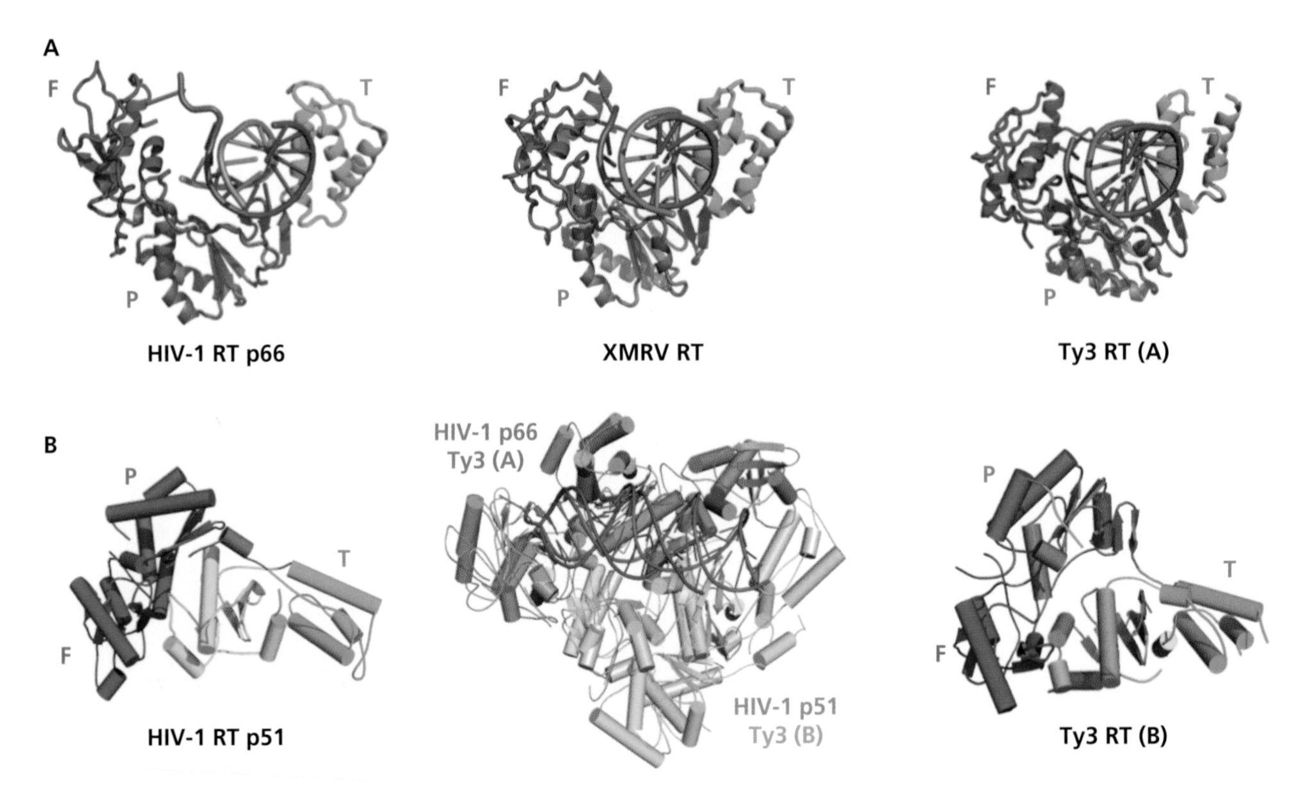

Figure 7.13 Comparison of the structures of three RTs. (A) The DNA polymerase domains of the lentivirus human immunodeficiency virus type 1 (HIV-1 p66), a gammaretrovirus related to murine leukemia virus (XMRV), and the yeast LTR-retrotransposon Ty3 (subunit A) RTs. Fingers, palm, and thumb subdomains are designated F, P, and T, respectively. RNA is shown in magenta and DNA in blue. **(B)** Architectures of the noncatalytic subunits of the dimeric RTs: HIV-1 p51 and Ty3 subunit B. Both subunits contain F, P, and T subdomains in analogous positions. Unexpectedly, the p51 connection and Ty3 (subunit B) RNase H domain, shown in yellow and brown, respectively, are also in similar positions. Superposition of the asymmetric p66-p51 HIV-1 RT heterodimer and the symmetric Ty3 (A)-(B) homodimer, together with bound nucleic acids (blue and orange strands), is shown in the center. HIV RT subunits are in orange and gray and Ty3 subunits in green and yellow. For additional details, see E. Nowak et al., *Nat Struct Mol Biol* **21**:389–396, 2014. Illustration prepared by Jason Rausch and Stuart Le Grice, National Cancer Institute, Frederick, MD, and Marcin Nowotny, International Institute of Molecular and Cell Biology, Poland.

LINE-mediated retrotransposition events have been documented. Reverse transcription by LINE RTs accounts for the wide distribution of genetic elements that lack this enzyme, including short interspersed nuclear elements (**SINEs**) and processed pseudogenes that are related to exons in functional genes but map elsewhere in the genome. Some 40% of the human genome is now known to comprise retroelements (Fig. 7.12).

Retroviral DNA Integration Is a Unique Process

The **integrase (IN)** of retroviruses and the related retrotransposons catalyzes specific and efficient insertion of the DNA product of RT into host cell DNA. This activity is unique in the eukaryotic virus world. Establishment of an integrated copy of the genome is a critical step in the life cycle of retroviruses, as this reaction ensures stable association of viral DNA with the host cell genome. The integrated **proviral DNA** is transcribed by cellular RNA polymerase II to produce the viral RNA genome and the mRNAs required to complete the infectious cycle.

IN is encoded in the 3′ region of the retroviral *pol* gene (Fig. 7.8), and the mature protein is produced by viral protease (PR)-mediated processing of the Gag-Pol polyprotein precursor. During progeny virus assembly, all three viral enzymes (PR, RT, and IN) are incorporated into the viral capsid. Virus particles contain equimolar quantities of RT and IN (some 50 to 100 molecules per particle). The viral DNA product of RT is the direct substrate for IN, and genetic and biochemical studies indicate that these enzymes function in concert within infecting particles. As already noted, IN sequences actually are present in one of the subunits of avian sarcoma/leukosis virus RT, and gentle extraction of murine leukemia virus particles yields RT-IN complexes. However, as with RT, virtually nothing is known about the molecular organization of IN within virus particles.

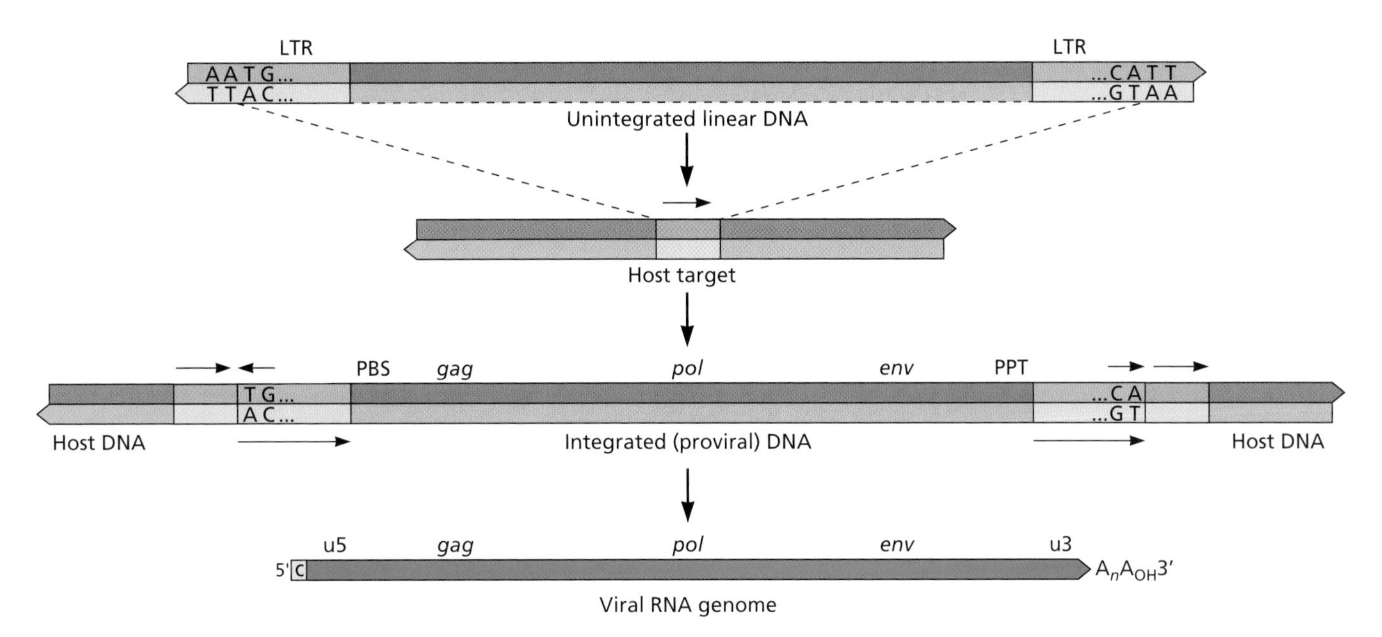

Figure 7.14 Characteristic features of retroviral integration. Unintegrated linear DNA of the avian retrovirus avian sarcoma/leukosis virus (top) after reverse transcription has produced blunt-ended LTRs (Fig. 7.6). The dashes under the bottom (+) strand indicate that this strand may include discontinuities, whereas the top (−) strand must be continuous (Fig. 7.7). Two base pairs (AA·TT) are lost from both termini upon completion of the integration process, and a 6-bp "target site" in host DNA (pink, indicated by an arrow) is duplicated on either side of the proviral DNA. The integrated proviral DNA (middle) includes short, imperfect inverted repeats at its termini, which end with the conserved 5′-TG...CA-3′ sequence; these repeats are embedded in the LTR, which is itself a direct repeat. The gene order is identical in unintegrated and proviral DNA, and is colinear with that in the viral RNA genome (bottom), for which a provirus serves as a template (described more fully in Chapter 8).

The first insights into the mechanism of the integration process came in the early 1980s, when it was established that proviral DNA is flanked by LTRs and the coding sequences are colinear in the unintegrated viral DNA and the RNA genome (Fig. 7.14). Nucleotide sequencing of cloned retroviral DNAs and host-virus DNA junctions revealed several unique features of the process. Both viral and cellular DNAs undergo characteristic changes. Viral DNA is cropped, usually by 2 bp from each end, and a short duplication of host DNA flanks the provirus on either end. Finally, the proviral ends of all retroviruses include the same dinucleotide. This dinucleotide is often embedded in an extended, imperfect inverted repeat that can be as long as 20 bp for some viral genomes. The fact that the length of the host cell DNA duplication is characteristic of the virus provided the first clue that a viral protein must play a critical role in the integration process.

The inverted terminal repeat, conserved terminal dinucleotide sequence, and flanking direct repeats of host DNA are strikingly reminiscent of features observed earlier in a number of bacterial transposons and the *E. coli* bacteriophage Mu (for "mutator"). Homologies to the predicted amino acid sequences of a portion of the retroviral IN were also found in the transposases of certain bacterial transposable elements such as Tn5. This observation suggested that, like RT, IN probably evolved before the divergence of bacteria and eukaryotes. These similarities predicted what is now known to be a common mechanism for retroviral DNA integration and DNA transposition.

The Pathway of Integration: Integrase-Catalyzed Steps

A generally accepted model for the IN-catalyzed reactions has been developed from the results of many different types of experiment, including studies of infected cells and reconstituted systems (Box 7.7). IN functions as a multimer, and two reactions occur in biochemically and temporally distinct steps (Fig. 7.15). In the first step, nucleotides (usually two) are removed from each 3′ end of the viral DNA. This "processing" step requires a virus-specific nucleotide sequence and duplex DNA ends. As this reaction can only occur when RT has completed synthesis of the viral DNA ends, the probability that defective molecules with imperfect ends will be integrated is limited. It has been shown that processing can take place in the cytoplasm of an infected cell before viral DNA enters the nucleus, within a subviral structure commonly referred to as the **preintegration complex** (described below). Although there is strong genetic evidence for sequence specificity for processing, only limited sequence-specific binding

BOX 7.7

BACKGROUND

Model in vitro *reactions elucidate catalytic mechanisms*

The development of simple *in vitro* assays for the processing and joining steps catalyzed by IN marked an important turning point for investigation of the biochemistry of these reactions. With such assays, it was discovered that the retroviral IN protein is both necessary and sufficient for catalysis; that no exogenous source of energy (ATP or an ATP-generating system) is needed; and that the only required cofactor is a divalent cation, Mn^{2+} or Mg^{2+}. Use of simple substrates with purified IN protein helped to delineate the sequence and structural requirements for DNA recognition. Such *in vitro* assays also formed the basis for drug screening, enabling development of FDA-approved IN inhibitors for the treatment of AIDS.

In the simplest assays, substrates comprise short duplex DNAs (ca. 25 bp), with sequences corresponding to one retroviral DNA terminus, with labeled terminal nucleotides (red asterisk in the figure). IN can also catalyze an apparent reversal of this joining reaction *in vitro*, which has been called disintegration. While the processing, joining, and disintegration reactions produce different products, their underlying chemistry is the same: all comprise a nucleophilic attack on a phosphorus atom by the oxygen in an OH group, and result in cleavage of a phosphodiester bond in the DNA backbone. In **processing (panel A of the figure)**, the -OH comes from a water molecule. In **joining (B)**, the -OH is derived from the processed 3′ end of the viral DNA, and the result is a direct transesterification. In **disintegration (C)**, also a direct transesterification, a 3′-OH end in the interrupted duplex attacks an adjacent phosphorus atom, forming a new phosphodiester bond and releasing the overlapping DNA. The products of all these reactions can be distinguished by gel electrophoresis.

Although assays with short, single-viral-end model substrates have been invaluable in elucidation of the catalytic mechanisms of IN, they are limited in that the major products represent "half reactions" in which only one viral end is processed and joined to a target. Subsequently, conditions for efficient, **concerted processing and joining (D)** of two viral DNA ends to a target DNA were described, with a variety of specially designed "miniviral" model DNA substrates. After preincubation, excess plasmid DNA is added as target, and the concerted joining of two donor fragments produces a linear DNA product.

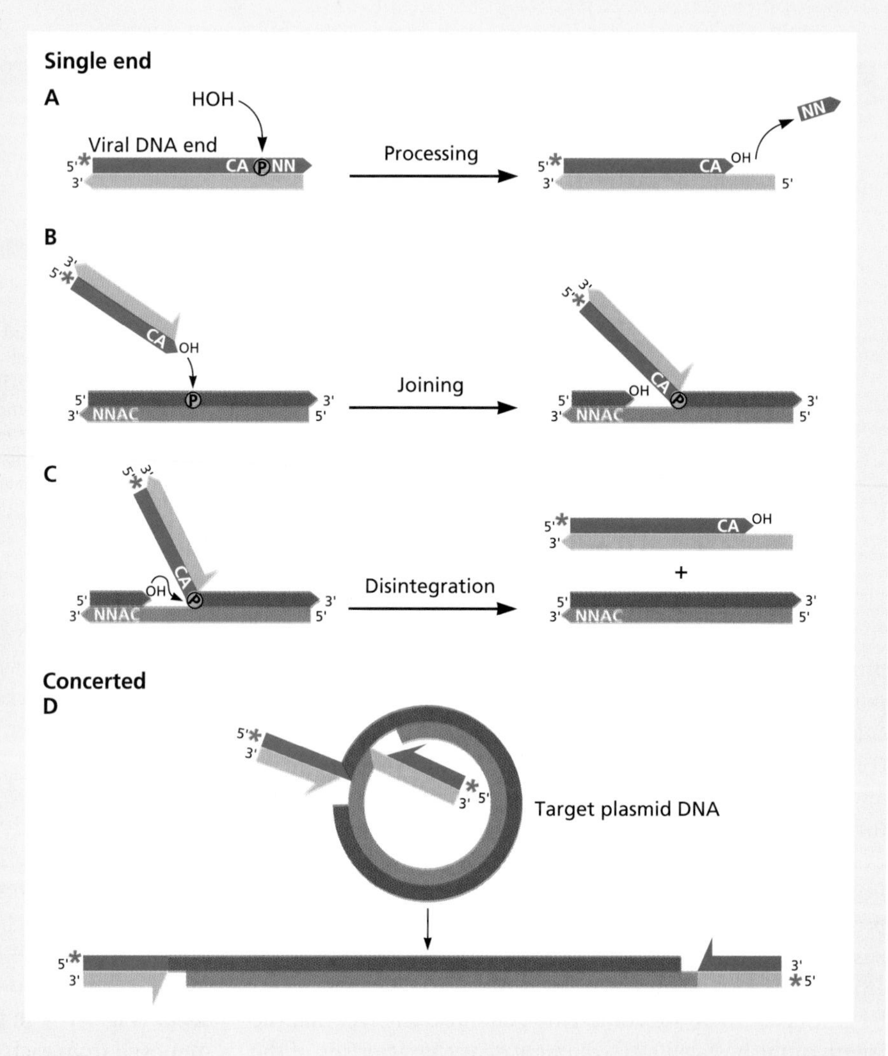

of purified IN protein to retroviral DNA has been detected in reconstituted systems. It seems likely, therefore, that some structural features or interactions among components within the preintegration complex help to place IN at its site of action near the viral DNA termini. The second step catalyzed by IN is a concerted cleavage and ligation reaction in which the two newly processed 3′ viral DNA ends are joined to staggered (4- to 6-bp) phosphates at the target site in host DNA (Fig. 7.15). The product of the joining step is a **gapped intermediate** in which the 5′-PO_4 ends of the viral DNA are not linked to the 3′-OH ends of host DNA.

Host Proteins Are Recruited for Repair of the Integration Intermediate

Retroviral DNA integration creates a discontinuity in the host cell chromatin, and repair of this damage is required

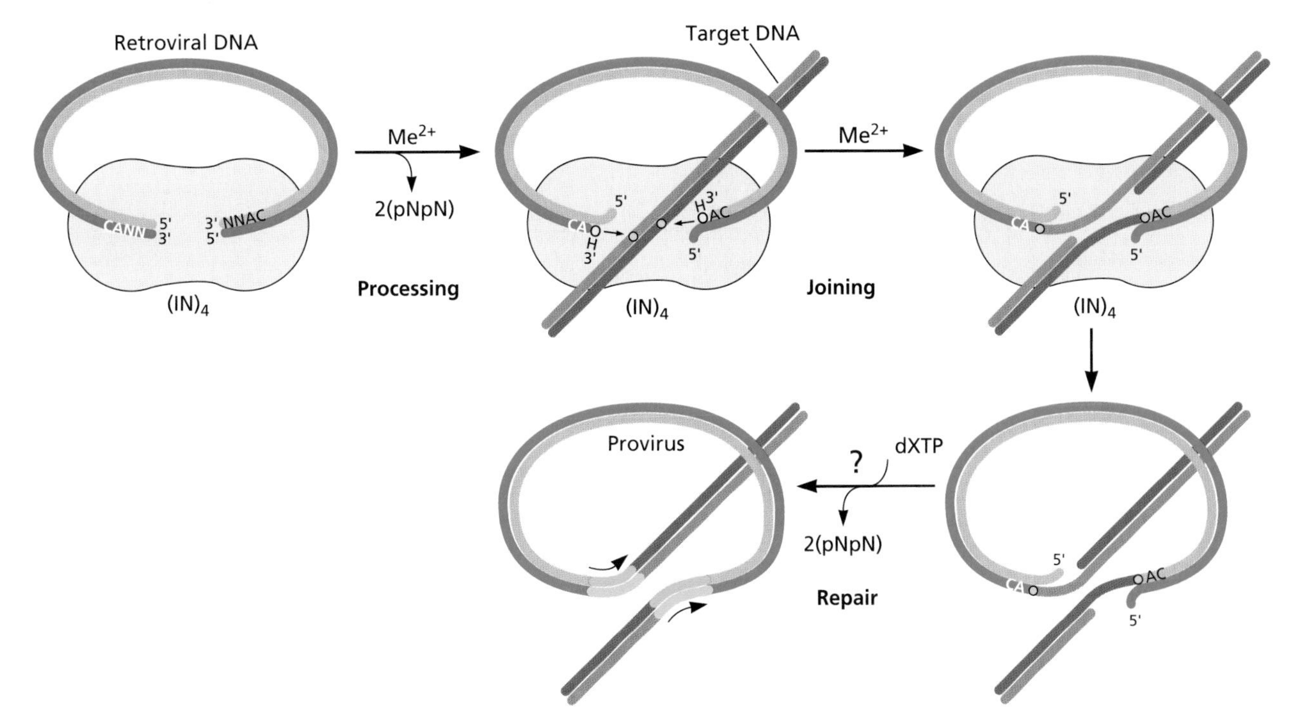

Figure 7.15 Three steps in the retroviral DNA integration process. Endonucleolytic nicking adjacent to the conserved dinucleotide near each DNA end results in the removal of a terminal dinucleotide, and formation of a new, recessed CA_{OH}-3′ end that will be joined to target DNA in the second step of the IN-catalyzed reaction. Both processing and joining reactions require a divalent metal, Mg^{2+} or Mn^{2+}. The viral DNA ends are bound by a tetramer of IN protein, $(IN)_4$, and the complex is called an intasome. Results of site-directed mutagenesis of viral DNA ends established that the conserved CA_{OH}-3′ dinucleotide is essential for correct and efficient integration. The small gold circles represent the phosphodiester bonds cleaved and re-formed in the joining reaction. The final step in the integration process is a host cell-mediated repair process.

to complete the integration process (Fig. 7.15). As with double-strand breaks produced by ionizing radiation or genotoxic drugs, retroviral DNA integration promotes recruitment of proteins of the DNA damage-sensing pathways. Components of the nonhomologous end-joining DNA repair pathway (DNA-dependent protein kinase, ligase IV, and Xrcc4) are required for postintegration repair. Retroviral DNA integration can trigger either cell cycle arrest or programmed cell death in cells that are defective in any of these proteins. It is likely that other host proteins play a role in both postintegration repair and reconstitution of chromatin structure at the site of integration, among them cellular DNA polymerases to fill any gaps, and histones and chromatin remodeling proteins to position nucleosomes on the proviral DNA.

Rapidly sedimenting preintegration complexes have been isolated from the cytoplasm of cells infected with several retroviruses. These nucleoprotein assemblies contain IN and viral DNA in a form that can be joined to exogenously provided plasmid or bacteriophage DNA. Such *ex vivo* reactions exhibit all the features expected for products of authentic integration. The mechanisms by which the preintegration complexes of different retroviruses gain access to the nucleus are likely to vary, but the details are still unclear (see Chapter 5).

Multiple Parameters Affect Selection of Host DNA Target Sites

DNA sequence features. A preference for retroviral integration into DNA sequences that are intrinsically bent, or underwound as a consequence of being wrapped around a nucleosome, was established in early analyses, but only limited sequence specificity was detected. Subsequently, advances in bioinformatics and high-throughput sequencing methodologies provided a wealth of information concerning the selection of integration sites. Weak consensus sequences for host target sites have been identified by a number of investigators, who, collectively, have mapped thousands of integration sites in human and other cell lines. As illustrated for the two cases in Fig. 7.16, the preferred sequence patterns for different retroviral genera are distinct. All target sequences studied to date form imperfect palindromes. The combinations of adjacent purine and pyrimidine nucleotides at the center of the palindromes possess different base stacking properties and hence different flexibility. The compositions of the palindromes

Random

Base	−5	−4	−3	−2	−1	1	2	3	4	5	6	7	8	9	10
A	0.31	0.29	0.35	0.30	0.28	0.28	0.30	0.28	0.30	0.32	0.30	0.30	0.27	0.31	0.30
C	0.18	0.20	0.20	0.21	0.21	0.22	0.22	0.21	0.23	0.19	0.19	0.22	0.23	0.18	0.25
G	0.21	0.22	0.19	0.22	0.21	0.22	0.22	0.22	0.19	0.21	0.20	0.18	0.19	0.22	0.20
T	0.30	0.29	0.26	0.28	0.29	0.27	0.26	0.30	0.28	0.28	0.31	0.31	0.31	0.29	0.25

ASLV

Base	−4	−3	−2	−1	1	2	3	4	5	6	7	8	9	10
A	0.21	0.22	0.27	0.34	0.18	0.28	0.29	0.32	0.28	0.13	0.26	0.34	0.43	0.23
C	0.28	0.13	0.21	0.14	0.32	0.20	0.23	0.24	0.21	0.32	0.24	0.23	0.17	0.29
G	0.24	0.18	0.24	0.20	0.39	0.19	0.21	0.22	0.23	0.30	0.13	0.21	0.15	0.26
T	0.36	0.47	0.28	0.32	0.11	0.33	0.27	0.23	0.28	0.25	0.31	0.22	0.25	0.22
Preferred base		T			G/C					G/C			A	

HIV-1

Base	−5	−4	−3	−2	−1	1	2	3	4	5	6	7	8	9	10
A	0.34	0.25	0.22	0.29	0.34	0.22	0.24	0.32	0.46	0.10	0.27	0.35	0.46	0.34	0.31
C	0.19	0.25	0.11	0.08	0.13	0.29	0.10	0.16	0.14	0.41	0.34	0.23	0.22	0.16	0.19
G	0.19	0.19	0.24	0.26	0.31	0.40	0.12	0.16	0.10	0.25	0.13	0.10	0.13	0.19	0.21
T	0.28	0.32	0.43	0.37	0.22	0.09	0.54	0.36	0.30	0.24	0.26	0.32	0.19	0.31	0.29
Preferred base			T		G	G	T		A	C	C		A		

Figure 7.16 Palindromic consensus sequences at retroviral integration sites. The frequency of each base at each position around the integration sites was calculated, where 1 equals 100%. Integration occurs between positions −1 and 1 on the top strand, shown in the figure. Colored positions have statistically different frequencies of bases from those of randomly generated sequences shown at the top. Bases with a >10% increase of frequency at a position are blue, and bases with a >10% decrease of frequency at a position are yellow. The preferred bases are listed below. Inferred duplicated target sites are in the blue box, and joining to the 3′ ends of viral DNA occurs at positions labeled by arrows. The symmetry of the palindromic patterns is centered on the duplicated target sites. Adapted from X. Wu et al., *J Virol* **79:**5211–5214, 2005.

determine the degree of "bendability" of the target sequences that are optimal for interaction with specific integrases. For example, more bending is required by IN proteins that join viral DNA ends to their target 4 bp apart than those that join 6 bp apart. The symmetry in the patterns is consistent with the idea that IN complexes function as symmetrical multimers in the preintegration complex (discussed below).

These same large-scale, global analyses have also shown that all human chromosomes are targets for integration, but that different retroviruses display distinct preferences for particular chromosomal features (Table 7.1). For example, human immunodeficiency virus type 1 DNA is integrated preferentially inside genes, especially in those that are highly transcribed, whereas murine leukemia virus DNA is integrated preferentially in and near transcription start sites. These observations suggest that the interaction of preintegration complexes with different chromatin-bound proteins promotes integration into distinct chromosomal locations.

Cellular tethers. Cellular tethers, proteins that bind to both cellular chromatin and IN proteins, were first described for yeast retrotransposons: Pol III components for Ty3 IN, and heterochromatin proteins for Ty5 IN. The mechanism for tethering of retroviral IN protein is understood most fully for the lentivirus proteins, which bind directly to the transcriptional coactivator Ledgf (lens epithelium-derived growth factor/p75 protein) (Fig. 7.17A). The efficiency of integration of human immunodeficiency virus type 1 is greatly reduced in cells in which Ledgf has been depleted by treatment with small interfering RNA (siRNA) or in which the gene encoding Ledgf has been deleted. Furthermore, the pattern

Table 7.1 Comparison of retroviral integration site preferences

	% Integration[a]					
	Human cells[b]				Mouse cells[c]	
Site or region	**Random**	**ASLV**	**MLV**	**HIV**	**HIV$^{LEDGF+/-}$**	**HIV$^{LEDGF-/-}$**
Within genes	26	42	40	60–70	62	44
Transcription start sites	5	8	20	10	6	17

[a]HIV, human immunodeficiency virus; ASLV, avian sarcoma/leukosis virus; MLV, murine leukemia virus.

[b]Percentages are approximates for integration into human cells and are from A. Narezkina et al., *J Virol* **78:** 11656–11663, 2004.

[c]Percentages for mouse embryo fibroblasts are from M. C. Shun et al., *Genes Dev* **21:**1767–1768, 2007. Calculations performed in this study indicated HIV gene usage in mouse LEDGF$^{-/-}$ cells at ~8% above random, which was ~3% less than ASLV/MLV gene usage in human cells.

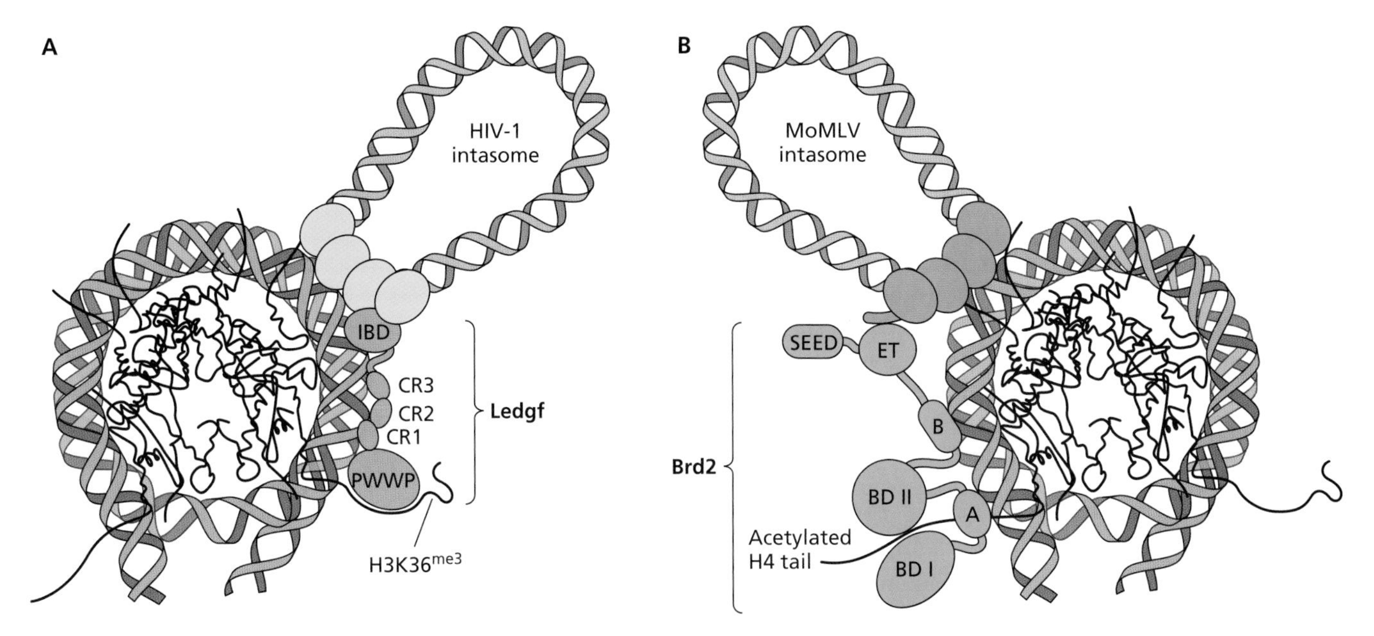

Figure 7.17 Models for chromatin tethering of intasomes by cellular proteins. (A) The intasome of human immunodeficiency type 1 virus, comprising a tetramer of IN bound to processed ends of viral DNA, is tethered to chromatin by attachment to the cellular protein Ledgf/p75. Histones are represented by black lines. The IN-binding domain (IBD) at the C-terminal end of Ledgf/p75 binds in the dimer interface(s) of the IN tetramer, and is anchored to nucleosomes by cooperative binding of the PWWP domain in Ledgf/p75 with the H3K36^{me3} modified histone tails associated with transcribed genes and by the three charge regions (CR1 to -3) that bind DNA. **(B)** The intasome of Moloney murine leukemia virus (MoMLV) is tethered to chromatin by Brd proteins. The ET domain in the C-terminal region of these proteins binds to the C-terminal tail of the IN protein(s) of this virus, and is anchored to nucleosomes by the interaction of two N-terminal bromodomains with acetylated H3 and H4 histone tails and by binding of motifs A and B to the host DNA. For more detailed information, see M. Kvaratskhelia et al., *Nucleic Acids Res* **42:**10209–10225, 2014.

of preference for various chromosomal features is altered in the small percentage of Ledgf-deficient cells in which integration does occur (Table 7.1). Target site selection is also correlated with histone modifications. Integration of human immunodeficiency virus type 1 is disfavored in regions of human chromosomes that contain modifications associated with transcriptional repression but is favored in regions with modifications that are associated with active transcription. X-ray crystallographic studies of the IN-binding domain of Ledgf bound to the catalytic core dimer interface of human immunodeficiency virus IN proteins have provided sufficient atomic detail to develop small-molecule inhibitors that reduce integration efficiency in cultured cells (see Volume II, Box 9.1).

Studies with the gammaretroviruses murine and feline leukemia viruses have identified the cellular bromodomain-containing proteins Brd2, -3, and -4 as specific tethers of these retroviral IN proteins. Conserved extraterminal (ET) domains in the Brd proteins bind to the 27-amino-acid C-terminal extension in gammaretroviral IN proteins and direct integration of the associated viral DNA to transcription start sites, which are rich in acetylated histones H3 and H4. A chromatin-tethering function for these cellular proteins has been established for papillomaviral genomes, via interaction with the DNA-binding viral E protein. In this case, viral genomes are tethered to condensed mitotic chromosomes, thereby ensuring their distribution to daughter cells following cell division. As Brd proteins promote assembly of transcriptional activators at transcription start sites, their tethering of the gammaretroviral IN provides a satisfying explanation for the integration preferences of these proviral genomes.

Although cellular tethers have not yet been identified for other retrovirus IN proteins, the examples described above support a general two-step mechanism: (i) binding to chromatin-associated proteins brings preintegration complexes in close proximity to host DNA (Fig. 7.17B); (ii) IN-catalyzed joining of viral DNA ends then occurs at viral-specific, preferred host DNA sequences nearby (Fig. 7.16).

Other Host Proteins May Affect Integration

Close to 100 cellular proteins have been identified as possible participants in the integration reactions of murine leukemia viruses or human immunodeficiency viruses, based on their association with preintegration complexes and/or their ability to bind to IN protein. Roles for candidate host proteins with DNA-binding properties, such as transcriptional regulators, chromatin components, and DNA repair enzymes, are not unexpected. The tethering functions of the Ledgf and Brd proteins are two validated examples. The possible contributions of other candidates are less apparent, and only a few have been characterized extensively.

One of the first of the candidate host proteins to be investigated was the 89-amino-acid cellular protein called barrier-to-autointegration factor (Baf), detected as a component of the preintegration complex of murine leukemia virus. Baf was shown to prevent integration into the newly synthesized viral DNA (autointegration), a reaction that would be suicidal for the virus. Purified Baf forms dimers in solution, binds to DNA, and can produce intermolecular bridges that compact the DNA, a reaction that prevents autointegration. The homologous human protein has been shown to block autointegration in isolated human immunodeficiency virus preintegration complexes. As purified virus particles do not contain this cellular protein, it must be acquired from the cytoplasm of a newly infected host cell (Fig. 7.18).

Another cellular protein thought initially to affect the integration process directly is the human immunodeficiency virus type 1 IN-interacting protein 1 (INi-1), which is a core component of the Swi/Snf chromatin-remodeling complex. Although INi-1 stimulates IN catalysis *in vitro*, this does not appear to be a physiologically important activity. Interestingly, IN mutants that cannot bind to INi-1 exhibit defects in virus particle morphology and reverse transcription and may provide clues to noncatalytic functions of IN.

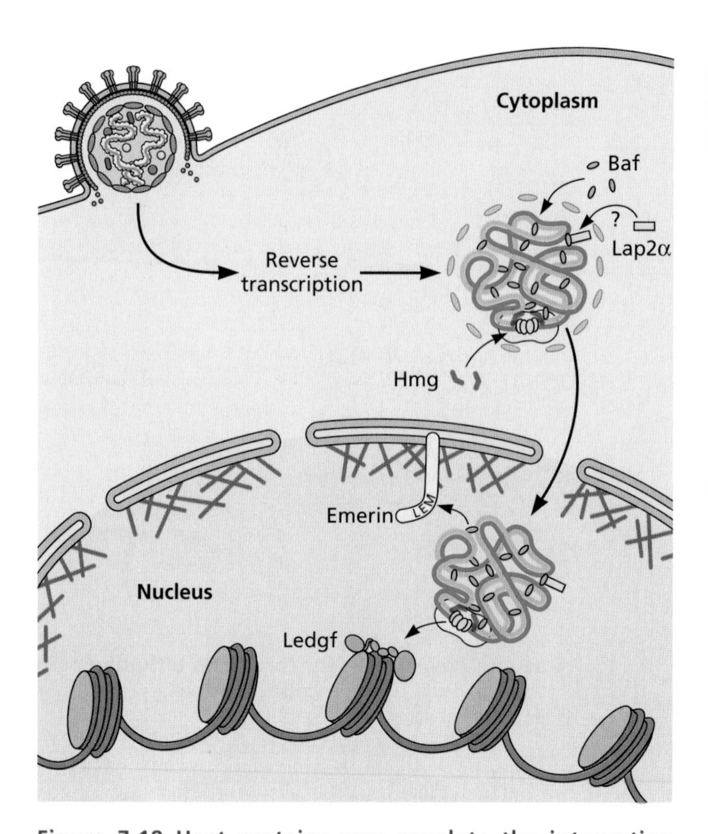

Figure 7.18 Host proteins may regulate the integration process. The barrier-to-autointegration factor protein (Baf binds to newly synthesized murine leukemia virus and human immunodeficiency virus type 1 DNAs, causing them to condense. Such DNA compaction prevents autointegration. The high mobility group 1a protein (Hmg), present in preintegration complexes of human immunodeficiency virus type 1, was originally thought to target integration to chromatin. Because cells that lack this protein have no obvious defect in virus reproduction, the protein is either redundant or not essential for integration. As this Hmg protein has been found to enhance transcription of the integrated provirus by recruitment of chromatin-remodeling complexes, it may facilitate a postintegration step. Lap2α proteins accumulate in the preintegration complexes of the human and murine viruses, respectively, but their functions, if any, are still uncertain. Once inside the nucleus, binding of Baf to emerin is proposed to facilitate access of the IN-DNA complex to chromatin, but this interaction does not appear to be essential for integration. IN binding to cellular tethers (e.g., Ledgf/p75 for lentiviruses or Brd family proteins for gammaretroviruses) anchors the intasomes to chromatin, thereby increasing integration efficiency. Fluorescence *in situ* hybridization studies with human immunodeficiency virus type 1–infected primary CD4 T cells indicate that viral DNA is integrated into actively transcribed regions of the host chromatin that are in close proximity to nuclear pores, and that the process is facilitated by particular nuclear pore proteins (nucleoporins).

Integrase Structure and Mechanism

IN Proteins Are Composed of Three Structural Domains

Retroviral IN proteins are ~300 amino acids in length and include three common domains connected by linkers of varying length (Fig. 7.19). The N-terminal domain is characterized by two pairs of invariant, Zn^{2+}-chelating histidine and cysteine residues (HHCC motif). The catalytic core domain contains a constellation of three invariant acidic amino acids, the last two separated by 35 residues [D,D(35)E motif]. These acidic amino acids chelate the two Mg^{2+} cofactors that are required for both processing and joining. Topologies of the catalytic core domains of the human and avian viral proteins established a relationship of IN to a large superfamily of nucleases and recombinases that includes the RNase H domain of RT. The amino acid sequence of the C-terminal domain is the least conserved among IN proteins from different retroviral genera, but the three-dimensional structures of this domain are quite similar in all examples analyzed to date. Some retroviral IN proteins (e.g., murine leukemia virus and prototype foamy virus) have an additional domain at their N termini, called the N-terminal extension domain (Fig. 7.19).

A Multimeric Form of IN Is Required for Catalysis

Properties of the integration reaction were first delineated by analysis of purified proteins. While a dimeric form appears to be sufficient to perform the processing reaction *in vitro*, a tetramer is required for the concerted integration of two viral DNA ends into a target DNA. The IN tetramer is stabilized by interaction with a pair of viral DNA ends, and each end is held mainly through contacts with C-terminal domain residues in one IN monomer, but acted upon by the catalytic core domain of another. The viral DNA ends do not remain double-stranded when bound at the active site of the enzyme, but the strands are partially unwound and distorted.

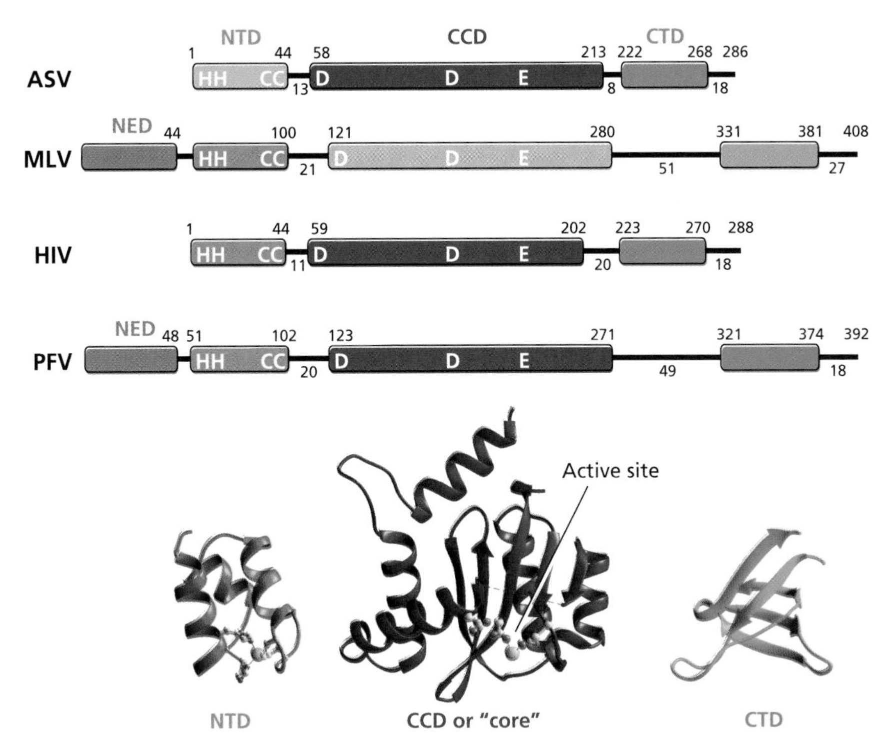

Figure 7.19 Domain maps of integrase proteins from different retroviral genera, and the structures of conserved domains in human immunodeficiency virus type 1 protein. (Top) Numbers above the maps indicate amino acid residues, starting with 1 at the N termini. Flexible linkers between the conserved domains, or C-terminal "tails," are represented by straight lines. Domain color coding is as follows: red, N-terminal domain (NTD); blue, catalytic core domain (CCD) or "core"; green, C-terminal domain (CTD); purple, N-terminal extension domain (NED). Domains for which atomic detail structures are not yet available are shown in faded colors. Evolutionarily conserved amino acids are indicated in the single-letter code within the domains. ASV, the alpharetrovirus avian sarcoma/leukosis virus; MLV, the gammaretrovirus murine leukemia virus; HIV, the lentivirus human immunodeficiency virus type 1; PFV, the spumavirus prototype foamy virus. (Bottom) The domain models are from crystal structures of the HIV-1 NTD, CCD, and CTD (PDB codes 1K6Y, 1BIU, and 1EX4, respectively). The Zn^{2+} ion in the NTD is shown as an aqua sphere, and in this structure of the HIV IN CCD with metal, only one of the two Mg^{2+} ions is bound in the active site, as indicated by the green sphere. The conserved Glu residue of the D,D(35)E motif is presumed to chelate the second metal ion together with the first conserved Asp residue.

These and other results suggested a model in which the core domains of only two of the four subunits in the IN tetramer provide catalytic function.

The retroviral INs are unusual enzymes with very low turnover rates *in vitro* (ca. 0.1 sec^{-1}). This low rate may not be a limitation *in vivo*, as only one concerted joining reaction is required to attach viral to host DNA in an infected cell.

Characterization of an Intasome

Since the last edition of this textbook was prepared, solutions of crystal structures of the prototype foamy virus IN-DNA complexes have confirmed many of the predictions from biochemical studies. In the presence of short duplex DNA fragments, with sequence corresponding to a viral DNA end, the prototype foamy virus IN protein assembles into a tetramer (a "dimer of dimers"). It was somewhat surprising to find that in this structure, called an **intasome** (Fig. 7.20A), the two "inner" subunits of the tetramer not only perform catalysis but also make **all** of the contacts with the viral DNA. The N-terminal domains in each of these inner subunits span the structure to bind the opposing viral DNA ends. The common N-terminal domains also reach over to interact with the opposing catalytic core domains. The C-terminal domains of the inner subunits are in the center, positioned to promote fraying of the unprocessed viral DNA ends and, following processing, to interact with a 30-bp target DNA fragment that contains the preferred palindromic sequence of the prototype foamy virus. Analysis of crystals with and without target DNA showed that the overall topology of the IN-DNA complex does not change upon target binding. However, as predicted from earlier biochemical studies, the target DNA must bend to fit into the active site. Crystals of IN-containing processed viral DNA ends and the DNA target were catalytically active in the presence of the Mg^{2+} cofactor, performing a concerted joining reaction 4 bp apart on the target, as expected for this viral IN protein (Fig. 7.20B).

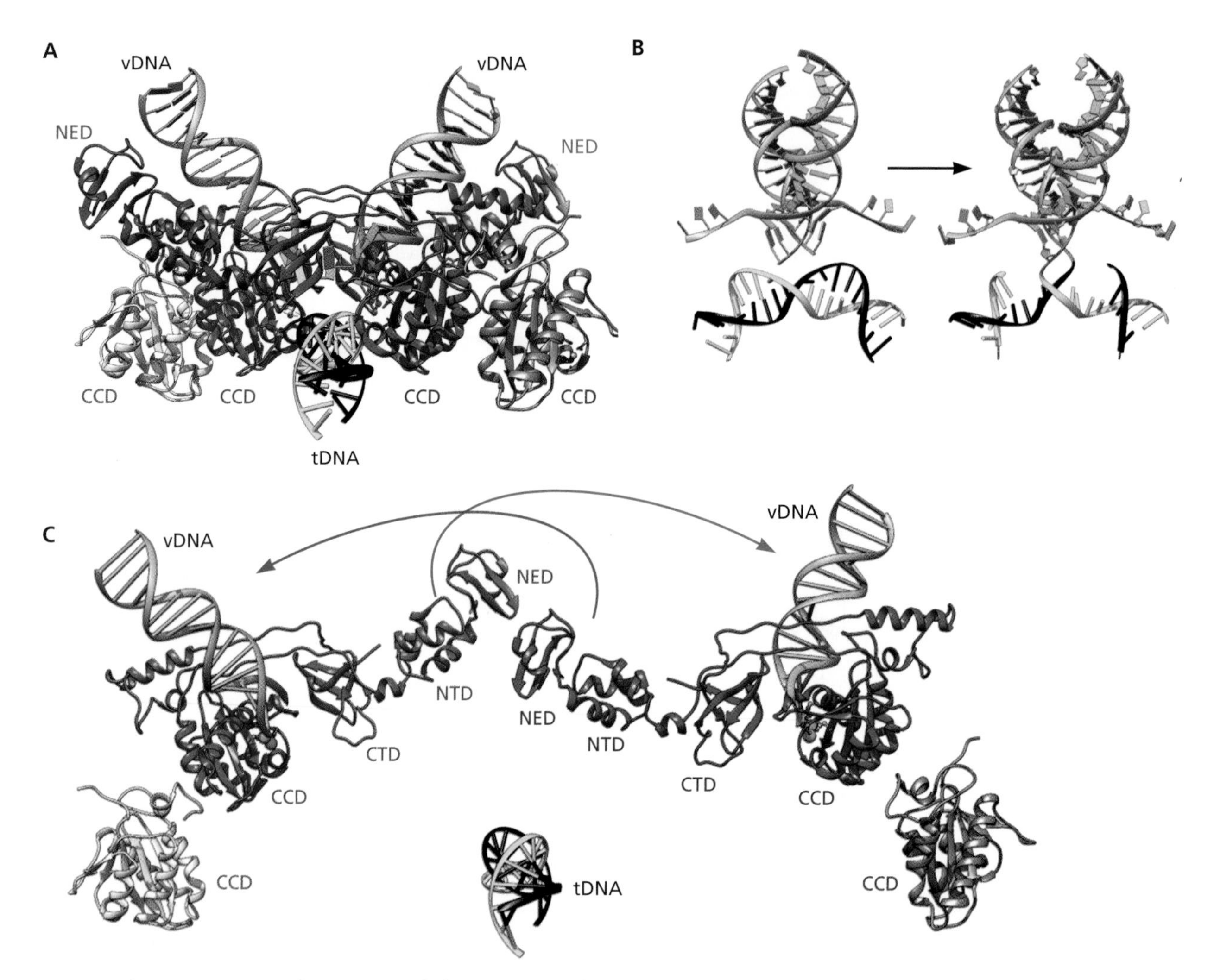

Figure 7.20 Crystal structure of the prototype foamy virus integrase tetramer bound to viral DNA ends and a target sequence. (A) The assembled complex (PDB code 4E7K) is shown in ribbon representation with the inner subunits in red and blue. Only the catalytic core domains (CCDs) of the outer subunits (gray) were resolved. Viral DNA (vDNA) oligonucleotides are in orange ribbon-ladder representation and the target DNA fragment (tDNA) in yellow and black. The locations of the N-terminal extension domains (NEDs) and CCDs are indicated. **(B)** DNA components of the complex portrayed before and after joining, rotated 90° about the *y* axis from panel A. The left view shows processed vDNA ends prior to joining, and the right view after joining to the target DNA. **(C)** The complex shown in panel A is pulled apart to show the positions of all domains in the inner subunits. Interactions between the distal N-terminal domain (NTD) and NED of one inner subunit and vDNA held in the CCD of the other inner subunit are indicated by the arrows. CTD, C-terminal domain. Assembly of the complex is shown in Movie 7.1 (http://bit.ly/Virology_V1_Movie7-1). For more detail on the prototype foamy virus structures, see G. N. Maertens et al., *Nature* **468:**326–329, 2010, and K. Gupta et al., *Structure* **20:**1918–1928, 2012. Image and movie courtesy of Mark Andrake, Fox Chase Cancer Center, Philadelphia, PA.

Analysis of the prototype foamy virus IN crystal structures allowed elucidation of the roles of the metal cofactors in catalyzing the reaction, and resolved details of interactions with metal-chelating, Food and Drug Administration (FDA)-approved inhibitors of human immunodeficiency virus IN. Furthermore, despite the differences in amino acid sequence, domain composition, and linker length, the prototype foamy virus IN-DNA structures have also been valuable in modeling similar assemblies for other retroviruses. While the terminal domains of the outer subunits in the prototype foamy virus IN crystals were not resolved, subsequent analysis of this complex in solution showed that these domains are extended outward from the complex and have no contact with the substrate DNAs (Fig. 7.20C).

With no obvious role for the outer subunits *in vitro*, it has been proposed that they may stabilize the tetrameric structure or promote its assembly (Box 7.8). It is also possible that the outer subunits facilitate integration *in vivo* by interacting

BOX 7.8

DISCUSSION

Intasome assembly

The concerted integration of two viral DNA ends into a target DNA requires assembly of an IN tetramer. The multimerization properties of purified retroviral IN proteins differ substantially. For example, at a concentration in which human immunodeficiency virus type 1 IN is mainly tetrameric, avian sarcoma/leukosis viral IN is a dimer, while prototype foamy virus IN is a monomer at even twice the concentration of the other two proteins. In solution, human immunodeficiency virus IN can form two apparently equally stable dimers: one (called a core dimer) resembles the outer dimer of the foamy virus intasome, and the other (called a reaching dimer) resembles the inner dimer of the intasome. However, the tetrameric form, a dimer of dimers, has no room for the DNA substrates without undergoing substantial conformational change (ergo, closed tetramer) (panel A in the figure).

The tetramer in the foamy virus intasome assembles from IN monomers *in vitro*, with substrate DNA imparting stability to the complex. Some evidence suggests that human immunodeficiency virus intasomes (panel B) are also assembled from monomers *in vitro*, but whether this is the case for other retroviral INs and how intasome assembly is accomplished in infected cells are currently unknown.

Bojja RS, Andrake MD, Merkel G, Weigand S, Dunbrack RL Jr, Skalka AM. 2013. Architecture and assembly of HIV integrase multimers in the absence of DNA substrates. *J Biol Chem* **288:**7373–7386.

Bojja RS, Andrake MD, Weigand S, Merkel G, Yarychkivska O, Henderson A, Kummerling M, Skalka AM. 2011. Architecture of a full-length retroviral integrase monomer and dimer, revealed by small angle X-ray scattering and chemical cross-linking. *J Biol Chem* **286:**17047–17059.

Gupta K, Curtis JE, Krueger S, Hwang Y, Cherepanov P, Bushman FD, Van Duyne GD. 2012. Solution conformations of prototype foamy virus integrase and its stable synaptic complex with U5 viral DNA. *Structure* **20:**1918–1928.

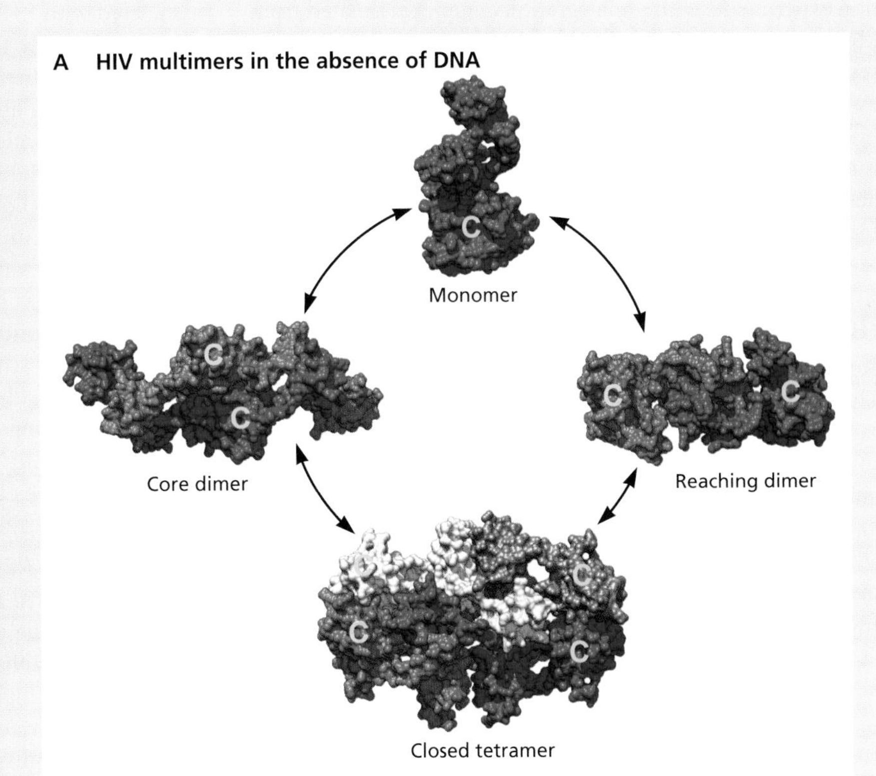

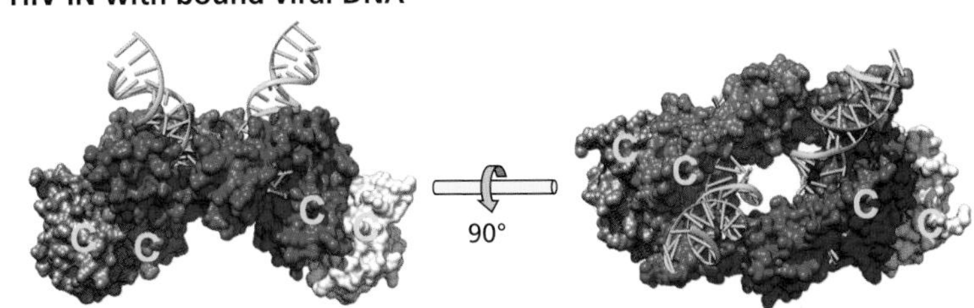

Multimers of human immunodeficiency virus type 1 (HIV-1) IN. (A) Models for the forms of HIV-1 IN that are at equilibrium in solution. Surface representations of the monomer, dimers, and a "closed" tetramer structure are based on architectures determined by small-angle X-ray scattering analysis, chemical cross-linking, and mass spectrometry with avian sarcoma/leukosis virus IN and HIV IN proteins in the absence of viral DNA substrates. In the dimers, one monomer is blue and the second is red. In the closed tetramer, additional subunits are dark and light gray. **(B)** Surface representations of HIV-1 IN with bound viral DNA end oligonucleotides, modeled from the prototype foamy virus crystal structure depicted in Fig. 7.20, in which only the catalytic core domains of the outer dimers were resolved (L. Krishnan et al., *Proc Natl Acad Sci U S A* **107:**15010–15915, 2010). The inner dimer comprises red and blue monomers as in panel A. The core domains of the outer dimers are shown in dark and light gray. Viral DNA substrate oligonucleotides are in gold ribbon-ladder representation. For aid in orientation, positions of the catalytic core domains are marked "C." Courtesy of Mark Andrake, Fox Chase Cancer Center, Philadelphia, PA.

with nucleosomes when the target is chromatin. While only 4 IN molecules comprise the intasome, virus particles contain some 50 to 100 molecules of this viral protein. DNA protection experiments with preintegration complexes isolated from Moloney murine leukemia virus-infected cells suggest that several hundred base pairs of the viral DNA ends are protected by association with IN, and consequently, many monomers may be bound. IN may also provide a noncatalytic structural function, as the C-terminal domain of the human immunodeficiency virus protein has been reported to interact with the capsid protein, and virus particles that lack IN possess empty capsids and displaced genomes.

Hepadnaviral Reverse Transcription

A DNA Virus with Reverse Transcriptase

The revolutionary concept that a virus with an RNA genome can replicate by means of a DNA intermediate was followed, about a decade later, by another big surprise: RNA as an intermediate in the replication of a virus with a DNA genome. Early hints that a mechanism other than semiconservative DNA synthesis was responsible for hepadnaviral replication came from the discovery of asymmetries in the genomic DNA and in the product of an endogenous DNA polymerase reaction in isolated virus particles. The viral DNA comprises one full-length (−) strand and an incomplete, complementary (+) strand (Fig. 7.21), but the endogenous polymerase reaction could extend only the (+) strand. The replication intermediates isolated from infected cells were also unusual, comprising mainly (−) strands of less than unit length, few of which were associated with (+) strands. All this seemed suspiciously like retroviral reverse transcription, and landmark studies published in 1982 disclosed the unique features of hepadnaviral replication with duck hepatitis B virus. Unlike the endogenous reaction typical of extracellular virus particles, newly formed intracellular "core" particles were found to incorporate dNTPs into both strands. As with RNA-dependent DNA polymerization in retroviral particles, synthesis of the (−) strand was resistant to the DNA-intercalating drug actinomycin D, whereas synthesis of the (+) strand was inhibited by this compound. Furthermore, a portion of the newly synthesized (−) strand DNA sedimented with the density of RNA-DNA hybrids. These and related findings marked an important turning point in our understanding of hepadnaviruses and greatly extended our knowledge of reverse transcription. (See Box 7.9 for a subsequent surprise.)

Figure 7.21 Hepadnaviral DNA. The DNA in extracellular hepadnavirus particles is a partially duplex molecule of ~3 kb with circularity that is maintained by overlapping 5′ ends. The (−) strand is slightly longer than unit length, and the polymerase, shown as a blue ball, is attached to its 5′ end. The (+) strand has a capped RNA of 18 nucleotides at its 5′ end and is less than unit length. The 5′ ends are near or in (10- to 12-bp) direct repeats called DR1 and DR2 (colored purple and yellow, respectively). As in retroviruses, these repeat sequences play the critical role of facilitating template transfers during reverse transcription. In mammalian hepadnavirus genomes, the (+) strand is shorter than the (−) strand and has heterogeneous ends. In avian hepadnavirus genomes, the (+) strand is almost the same length as the (−) strand. Details of the genetic content are provided in the Appendix, Fig. 11.

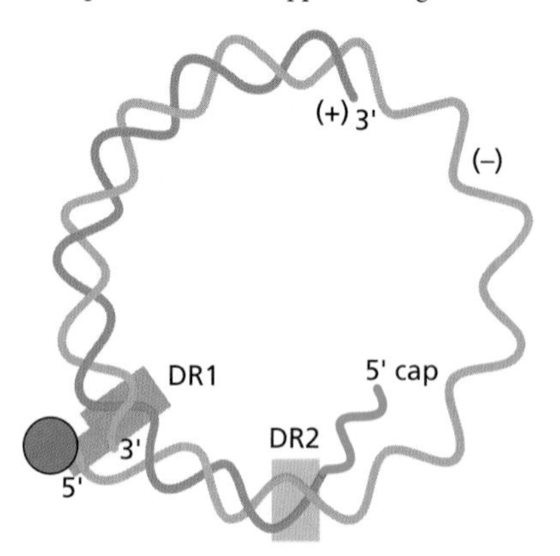

Function of Reverse Transcription in the Hepadnaviral Life Cycle

Analyses of the single-cell reproduction cycle of hepadnaviruses have established that the gapped DNA of an entering virus particle is imported into the nucleus, where it is repaired to produce a covalently closed circular molecule (Fig. 7.22). The exact mechanism of formation of these circular molecules is not yet known. One attractive hypothesis is that, just as with retroviral DNA integration, the incoming hepadnaviral genome is seen as "damaged" by the cell. Enzymes of cellular DNA repair pathways that normally excise damaged bases or DNA adducts might then remove the bound P protein and capped RNA (Fig. 7.21) so that the viral DNA ends can then be filled in and ligated. The hepadnaviral genomes encode no integrase, and hepadnaviral DNA is not normally integrated into the host's genome. However, the covalently closed circular DNA, with acquired cellular histones, persists in the nucleus as a nonreplicating minichromosome from which cellular RNA polymerase II transcribes viral RNAs.

The 3.5-kb pregenomic mRNA is exported to the cytoplasm, where it serves as the template for reverse transcription. This process takes place in a newly formed subviral "core" particle that includes the pregenomic mRNA, plus capsid and polymerase proteins (products of the C and P genes). P protein provides all the activities required for reverse transcription. The DNA-containing, nascent core particles can then follow one of two pathways. Late in infection, when the cisternae of the endoplasmic reticulum contain an abundance of viral envelope glycoprotein, they can bud into the endoplasmic reticulum and eventually be secreted as progeny virus particles (Chapter 13). Alternatively, if they do not become enveloped, the core particles can be directed to the nucleus, where their

BOX 7.9

BACKGROUND

A retrovirus with a DNA genome?

The *Spumavirinae* comprise a subfamily of retroviruses isolated from primate, feline, and bovine species, among others. Spumaviruses are commonly called **foamy viruses**, because they cause vacuolization and formation of syncytia in cultured cells. These viruses exhibit no known pathogenesis and received little attention from virologists until recently. However, it is now clear that the foamy viruses are most unconventional retroviruses, with many properties that seem more similar to those of hepadnaviruses than of other retroviral family members, in the following respects.

- Reverse transcription is a late event in foamy virus production, and is largely complete before extracellular virus particles infect new host cells. Furthermore, although they contain both RNA and DNA, the genome-length DNA extracted from foamy virus particles can account entirely for viral infectivity. Like other retroviral family members, foamy virus genome replication requires an RNA intermediate, but as with hepadnaviruses, the functional nucleic acid in extracellular foamy virus particles appears to be DNA.
- Although the arrangement of genes and the mechanism of reverse transcription are the same as those in other retroviruses, the prototype foamy virus RT is not synthesized as part of a Gag-Pol precursor, but rather by translation of a separate *pol* mRNA, as is also the case for hepadnaviral RT.
- Mature foamy virus particles do not include the usual processed retroviral structural proteins (MA, CA, and NC), but instead contain two large Gag proteins that differ only by a 3-kDa extension at the C terminus. These Gag proteins contain glycine-arginine-rich domains that bind with equal affinity to RNA and DNA, much like the hepadnaviral core (C) protein.
- As with the hepadnaviruses, foamy virus budding requires both Gag and Env proteins, and most budding occurs into the endoplasmic reticulum.
- Most foamy virus particles remain within the infected cell. This property probably accounts for the large quantities of intracellular viral DNA, and might explain why persistently infected cells contain numerous integrated proviruses. It is possible that some foamy virus DNA integration occurs via an intracellular recycling pathway of progeny genomes, similar to that which occurs with hepadnaviruses.

Linial MA. 1999. Foamy viruses are unconventional retroviruses. *J Virol* **73:**1747–1755.

Cells infected with primate foamy virus (A) show large syncytia and numerous vacuoles. Uninfected cells **(B)** lack such vacuoles and have only single nuclei. Nuclei are stained blue, α-tubulin is red, and viral Gag protein is green. Micrographs were obtained by Alison Yu and generously provided by Maxine Linial, Fred Hutchinson Cancer Research Center.

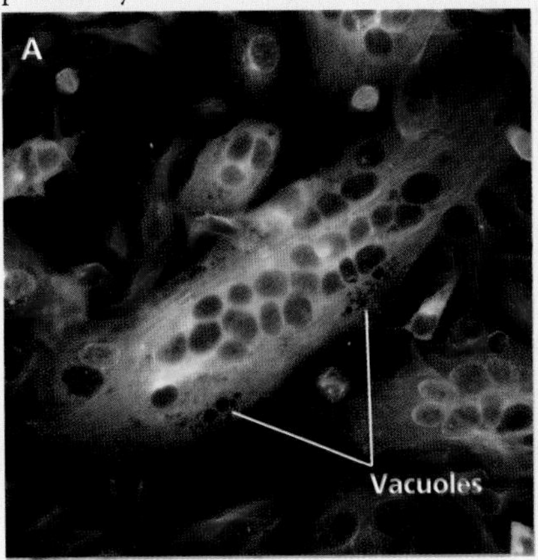

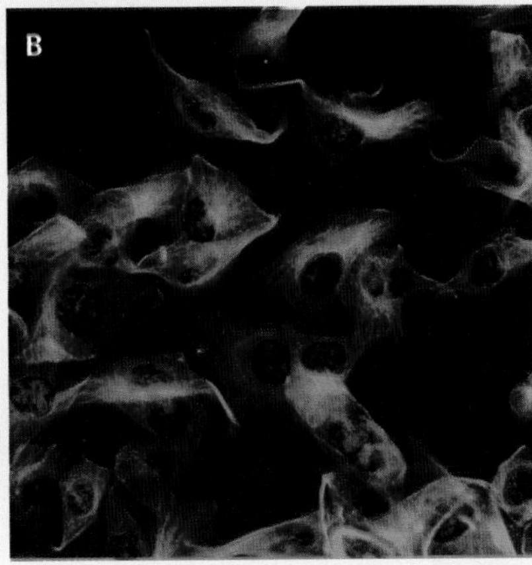

DNA is converted to additional copies of the covalently closed circular molecules. This pathway predominates at early times after infection, when little envelope protein is available. Eventually, as many as 30 hepadnaviral episomes can accumulate in the nucleus.

The DNA in hepadnaviral episomes is not replicated by the host's DNA synthesis machinery; **all** hepadnaviral DNA is produced by reverse transcription. This situation contrasts with that of retroviruses, in which the integrated nuclear form, the provirus, is replicated along with the host DNA. Consequently, both retroviral and hepadnaviral DNAs are maintained in infected cells for the life of those cells, but in quite different ways.

Analysis of hepadnaviral reverse transcription has been difficult for a number of technical reasons. Suitable tissue culture systems were not available until hepatoma cell lines were identified in which virus reproduction could take place following transfection with cloned viral DNA. Furthermore, mutational studies are confounded by the compact coding organization of the DNA. The tiny genome (~3 kb) is organized very efficiently, with more than half of its nucleotides translated in more than one reading frame. This arrangement makes it more difficult to produce mutations that change only one gene product. Finally, although reverse transcription takes place in newly assembled core particles, it was not possible initially to prepare enzymatically active P protein to study the reaction. Nevertheless, currently available details reveal fascinating analogies, but also striking differences, in the reverse transcription of hepadnaviruses and retroviruses.

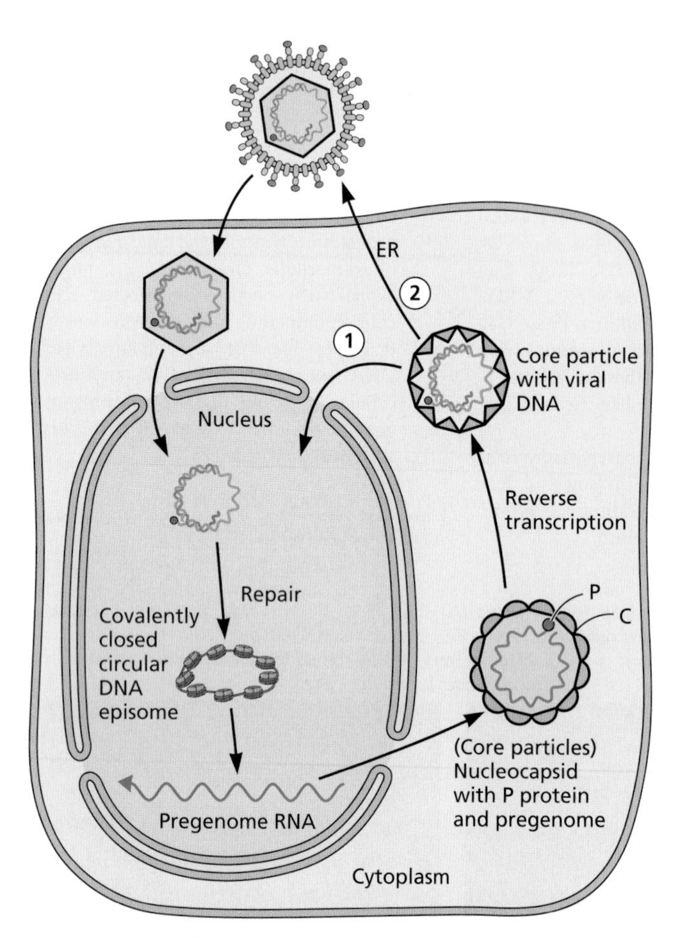

Figure 7.22 Single-cell replication cycle for hepadnaviruses. Pathway 1 provides additional copies of covalently closed circular minichromosomes. Pathway 2 represents exit of enveloped particles through the endoplasmic reticulum (ER). Additional details of the single-cell reproductive cycle are provided in the Appendix, Fig. 12.

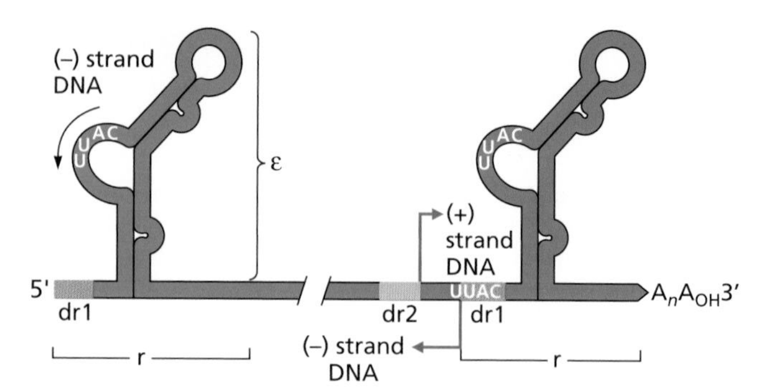

Figure 7.23 Essential *cis*-acting signals in pregenomic mRNA. The viral pregenomic mRNA bears terminal repetitions of ca. 200 nucleotides (r) that contain copies of the packaging signal (ε), but only the 5′ copy has functional activity *in vivo*. Indicated are positions for initiation of the 5′ ends of (−) and (+) strand DNAs, and the 5′-UUAC-3′ motifs in duck hepatitis B virus within ε and at dr1, that are important for (−) strand DNA synthesis. Both the structural features of ε and the specific sequence in the loop are critical for its function. Adapted from C. Seeger and W. S. Mason, p 815–832, *in* M. L. DePamphilis (ed), *DNA Replication in Eukaryotic Cells* (Cold Spring Harbor Laboratory Press, Cold Spring Harbor, NY, 1996), with permission.

The Process of Reverse Transcription

Essential Components

Pregenomic mRNA. The pregenomic mRNA that provides the template for production of hepadnaviral genomic DNA is capped and polyadenylated, and also serves as the mRNA for both capsid (C) and reverse transcription (P) proteins. The transcription of pregenomic mRNA from covalently closed circular DNA in the nucleus is initiated at a position ~6 bp upstream of one copy of a short direct repeat called DR1 (Fig. 7.21). Transcription then proceeds along the entire DNA molecule, past the initiation site, to terminate after a polyadenylation signal just downstream of DR1. Consequently, pregenomic mRNA is longer than its template DNA. Because the region from the transcription initiation site to the polyadenylation site is copied twice, there is a long direct repeat (~200 nucleotides) (r) at either end of the RNA. This long repeat includes dr1 and a structural element of about 100 nucleotides called **epsilon** (ε) (Fig. 7.23). Deletion of ε within the 3′ copy of r has no impact on genome replication. In contrast, ε at the 5′ end is essential as it provides both the site for initiation of (−) strand synthesis and the signal for encapsidation of DNA into core particles. Although all viral transcripts have ε at their 3′ ends, only the pregenomic mRNA has this important copy of ε at its 5′ end.

There is a marked preference for reverse transcription of the pregenomic mRNA molecules from which P protein is translated. The basis of such *cis*-selectivity is unknown; C protein, which is also translated from this RNA and has nucleic acid-binding properties, appears to function perfectly well in *trans*. It is possible that the nascent P polypeptide binds to its own mRNA cotranslationally. An attendant benefit from such a mechanism would be the selection for genomes that express functional P protein. Analysis of cytoplasmic core particles suggests that there is one molecule of P protein per molecule of DNA, implying that hepadnaviruses contain one copy of the viral genome per virus particle (Table 7.2). This selectivity would be determined, in part, by the presence of the encapsidation signal(s) at the 5′ end of the pregenomic mRNA.

Primers. The primers for hepadnaviral RT remain attached to the 5′ ends of the viral DNA strands. They are, for (−) strand synthesis, the P protein itself, and, for (+) strand synthesis, a capped RNA fragment derived from the 5′ end of pregenomic RNA. A protein-priming mechanism (Chapter 9) was first described for adenovirus DNA replication and later for the bacteriophage ϕ29. Priming by a viral protein, VPg, is also a feature of poliovirus RNA synthesis. Hepadnaviral reverse transcription is distinguished by the fact that the primer and the polymerase are within a single protein.

Table 7.2 Comparison of retroviral and hepadnaviral reverse transcription

Parameter	Retroviruses[a]	Hepadnaviruses[a]
Viral genome	RNA (pseudodiploid)	DNA (incomplete duplex)
Template RNA also serves as:	Genomic *RNA mRNA* (*gag*, *pol*)	Pregenomic *RNA mRNA* (C and P proteins)
DNA intermediate	*Circular DNA with 5′ overlaps*	*Circular DNA with 5′ overlaps*
Virus-encoded enzyme	RT	P protein
No. of molecules/core	50–100	1
Functions	*DNA polymerase*, *RNase H*, helicase (strand displacement)	*DNA polymerase*, *RNase H*, protein priming, template RNA encapsidation
Primer, first (−) DNA strand	tRNA (host)	Viral P protein (TP domain)
Site of initiation	*Near 5′ end of genome*	*Near 5′ end of pregenome*
First DNA product	(−) strong-stop DNA, ca. 100 nucleotides	4 nucleotides copied from bulge in 5′ ε
First template exchange	*To complementary sequence in repeated sequence, r, at 3′ end of template RNA*	*To complementary sequence in repeated sequence, R, at 3′ end of template RNA*
Primer, second (+) DNA strand	*Derived from template RNA*, internal RNase H product (ppt)	*Derived from template RNA*, 5′ cap, terminal RNase H product
Site of initiation	*Near 5′ end of (−) DNA*	*Near 5′ end of (−) DNA*
Time of initiation	Before completion of (−) strand	After completion of (−) strand
Type of priming	Priming *in situ*	Primer translocated
Second template exchange	*To the 3′ end of (−) strand DNA via complementary* pbs *sequence*	*To the 3′ end of (−) strand DNA via complementary sequence*
Reverse transcribing nucleoprotein complex	*Subviral "core"* particles, deposited in the *cytoplasm* upon viral entry	Nascent *subviral "cores"*; *cytoplasmic* intermediates in viral assembly
Final product(s)	Double-stranded linear DNA	Circular viral DNA or covalently closed episomal DNA
DNA maintained in the nucleus	Integrated into host genome, proviral DNA	Nonintegrated episome in host nucleus

[a]Italics indicate similarities.

P protein is a self-priming reverse transcriptase. P protein has C-terminal enzymatic domains that were first identified by amino acid sequence alignment with the retroviral RTs (Fig. 7.24). The highly conserved residues in the homologous domains are essential for hepadnaviral reverse transcription. Hepadnaviral P protein also contains an N-terminal domain separated from the RT region by a spacer, believed to provide a flexible hinge between these two regions of the protein. The N-terminal domain, referred to as the terminal protein region, includes a tyrosine residue utilized for priming (−) strand DNA synthesis. In addition to its other functions, P protein is required for encapsidation of viral RNA, a process

Figure 7.24 Comparison of hepadnaviral and retroviral RTs. Linear maps of the duck hepatitis B virus (DHBV) and human immunodeficiency virus type 1 (HIV-1) *pol* gene products. The maps were aligned relative to amino acids that are generally conserved among all RTs. Approximate locations of motifs (T3 and RT1) in the hepadnaviral P proteins that interact with epsilon (ε) in pregenome mRNA are indicated.

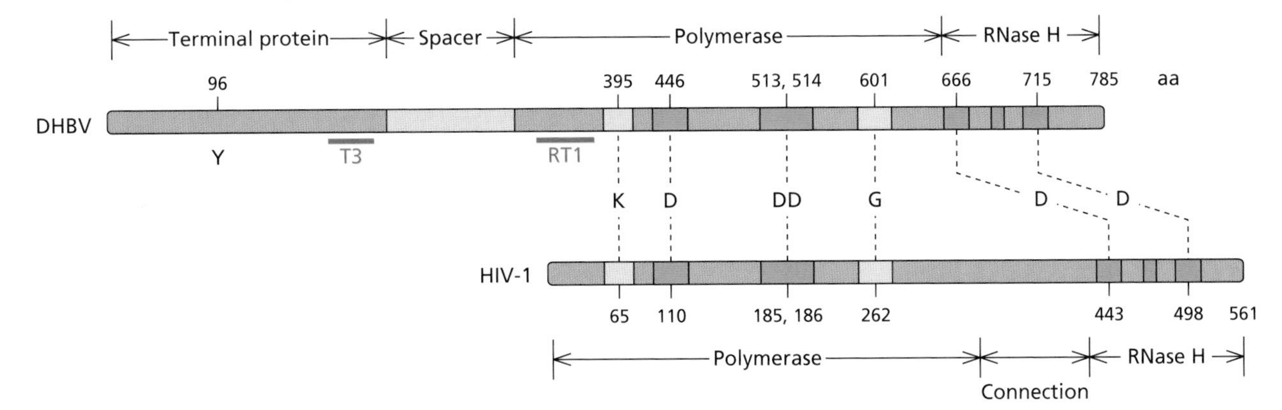

that depends on the interaction of both the RT and terminal protein domains with the 5′ ε structure. This mechanism represents a departure from the retroviral scheme, in which the NC protein sequences in the Gag polyprotein serve this purpose (Table 7.2). Indeed, the requirement for a DNA polymerase in hepadnaviral RNA encapsidation is unique among retroelements.

Host proteins may facilitate P-protein folding. An important breakthrough in the study of hepadnaviral reverse transcription was achieved with the demonstration that enzymatically active P protein can be produced upon translation of P mRNA from duck hepatitis B virus in a cell-free rabbit reticulocyte lysate. P protein is the only viral protein required for initiation of hepadnaviral DNA synthesis. If ε is not present during synthesis of P protein in yeast, the enzyme is inactive, even if ε is supplied later. Furthermore, P protein that is synthesized when ε is present in the mRNA is more resistant to proteolysis. Consequently, binding to ε may be required for the P protein to fold into an active conformation. Host cell proteins also appear to affect P-protein folding; synthesis of active P protein in the cell-free system requires the presence of cellular chaperone proteins and a source of energy (ATP). Furthermore, incorporation of these host cell proteins into viral capsids appears to require the polymerase activity of P protein. It has been proposed that chaperones are needed to maintain this viral protein in a conformation that is competent to bind to ε and prime DNA synthesis, and also to interact with assembling capsid subunits (Fig. 7.25).

Critical Steps in Reverse Transcription

Initiation and the first template exchange. Synthesis of the (−) strand of duck hepatitis B virus DNA is initiated by the polymerization of three or four nucleotides primed by the -OH group of a tyrosine residue located in the terminal protein domain of the single P-protein molecule present in the capsid. This single protein molecule acts both as primer and catalyst for all subsequent steps in reverse transcription (Box 7.10). This initial synthesis is followed by a template exchange in which the enzyme-bound, 4-nucleotide product anneals to a complementary sequence at the edge of dr1 at the 3′ end of the pregenomic RNA (Fig. 7.23 and 7.26, step 2). Although the sequence at this end is complementary to the short, initial product, it is not unique in the pregenomic mRNA (Fig. 7.23). In human hepadnavirus genome replication, appropriate positioning of the nascent DNA strand is promoted by a *cis*-acting sequence (ϕ), which anneals to the upper stem of the 5′ ε to which P protein is bound. It seems likely that selection of the normal site is also guided by the specific organization of pregenomic RNA in core particles. P protein remains covalently attached to the 5′ end of the (−) strand during the first template exchange and, as noted previously, through **all** subsequent steps.

Elongation and RNase H degradation of the RNA template. Following the first template exchange, (−) strand DNA synthesis continues all the way to the 5′ end of the pregenomic RNA template (Fig. 7.26, steps 3 and 4). Because synthesis is initiated in the 3′ dr1, a short repeat of 7 to

Figure 7.25 Model for the assembly of hepadnavirus nucleocapsids. P protein is synthesized in an inactive conformation (labeled "closed complex"). Interaction with a chaperone assembly (heat shock protein 90 [Hsp90], together with four cochaperones, Hsp70, Hop, Hsp40, and p23) induces a conformational change (open complex) that allows binding of P protein to ε RNA, facilitated by interaction with the T3 and RT1 motifs in the terminal protein (TP) and RT domains of P, respectively (productive complex). Such binding provides the signal for nucleocapsid assembly and initiation of viral DNA synthesis (priming complex) at the active site of the RT (indicated by a pale oval). TP provides the primer tyrosine residue for the initiation of reverse transcription. For additional details, see C. Seeger et al., p 2185–2221, *in* D. M. Knipe and P. M. Howley (ed), *Fields Virology*, 6th ed, vol 2 (Lippincott Williams & Wilkins, Philadelphia, PA, 2013).

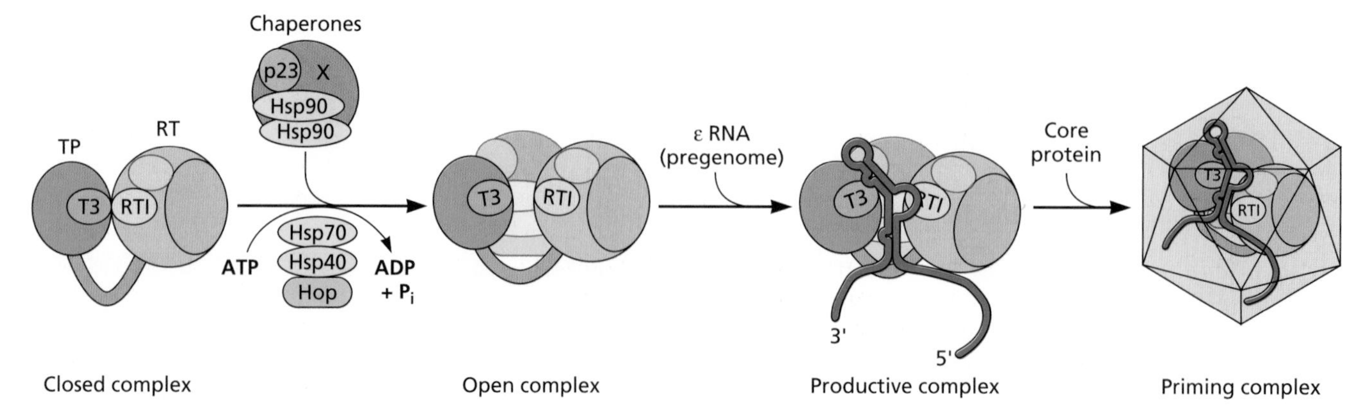

BOX 7.10

DISCUSSION

A single P-protein molecule does it all?

It is quite difficult to envision how a single protein can perform all of the gymnastics required for synthesis of a double-stranded circular DNA product while remaining attached to one viral DNA end. Nevertheless, it is widely believed among hepadnavirus researchers that there is only **one** P molecule in an infectious particle. Several observations support this view.

- Hepadnavirus assembly requires binding of the P protein to the ε stem-loop structure in a single pregenome mRNA molecule. In the absence of ε, there is no RNA packaging or assembly of core particles with RT activity.
- P protein in virus particles, and immature cores that contain nascent (−) DNA strands, do not use exogenous templates. Because such templates are copied by RT in permeabilized retroviral particles, one might expect that if one or more additional P proteins (not covalently attached to the genome) were present, they would bind and copy an exogenous template.
- P protein synthesized *in vitro* does not form dimers. Furthermore, even though the two functional domains are separated by a hinge, a P variant with a mutation in the terminal protein cannot complement a variant with a mutation in the polymerase domain.
- Single image particle reconstructions of RNA-filled cores reveal a structure consistent with a single P protein anchored in a unique position, touching the pregenome mRNA, which is aligned along the inner surface of the core (see the figure).

It seems likely that hepadnaviral core architecture and components help to ensure that the required interactions between P protein and nucleic acid templates can occur in such a way that exchanges are facilitated and templates can transit to the active site as product strands are synthesized.

Hirsch RC, Lavine JE, Chang LJ, Varmus HE, Ganem D. 1990. Polymerase gene products of hepatitis B viruses are required for genomic RNA packaging as well as for reverse transcription. *Nature* **344:**552–555.

Junker-Niepmann M, Bartenschlager R, Schaller H. 1990. A short *cis*-acting sequence is required for hepatitis B virus pregenome encapsidation and sufficient for packaging of foreign RNA. *EMBO J* **9:**3389–3396.

Kaplan PM, Greenman RL, Gerin JL, Purcell RH, Robinson WS. 1973. DNA polymerase associated with human hepatitis B antigen. *J Virol* **12:**995–1005.

Radziwill G, Tucker W, Schaller H. 1990. Mutational analysis of the hepatitis B virus P gene product: domain structure and RNase H activity. *J Virol* **64:**613–620.

Electron microscopic reconstruction of a hepadnavirus RNA-filled core showing possible location of P protein. In a cross section, the outer portion of the core is assembled from C protein dimers (gray). Pregenome mRNA density (pgRNA, yellow) coats the inner surface of the core. A uniquely positioned density (red) is tentatively assigned to the P protein. Additional internal (blue) density represents unidentified encapsidated proteins and/or misaligned capsid. A close-up shows that the homologous polymerase domain of RT from human immunodeficiency virus type 1 in blue ribbon representation (PDB ID code 1RTD) can fit neatly into the right-hand doughnut-shaped red density of the putative P protein. The TP and RNase H domains of P protein are unresolved. From Fig. 4 in J. C.-Y. Wang et al., *Proc Natl Acad Sci U S A* **111:**11329–11334, 2014, with permission.

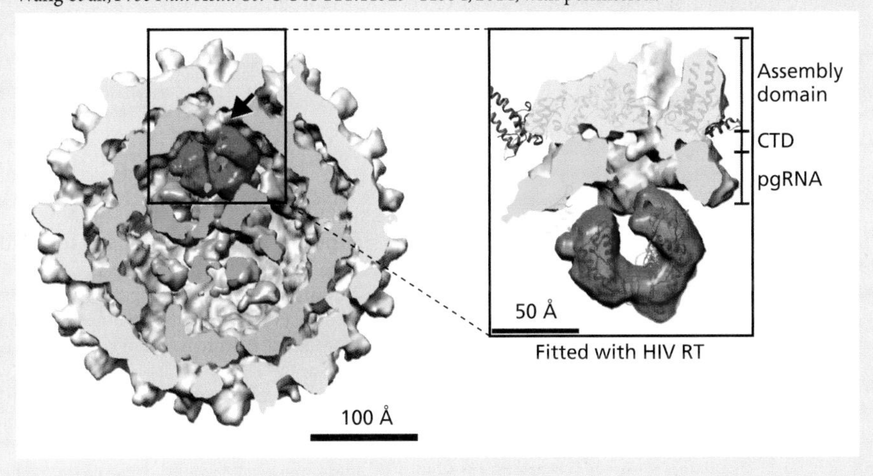

8 nucleotides (3′R) is produced at the end of this elongation step when the 5′ dr1 sequence is copied (Fig. 7.26, step 4). The RNA template is degraded by the RNase H activity of P protein as (−) strand synthesis proceeds. Unlike the retroviral RNase H products, none of these hepadnaviral RNA fragments are used as primers for (+) strand DNA synthesis (Table 7.2). The final product of RNase H digestion is a short RNA molecule, corresponding to the capped end of the pregenomic RNA, which includes the 5′ dr1 and serves as a primer for (+) strand DNA synthesis. It is noteworthy that (+) strand DNA synthesis can begin only after completion of (−) strand DNA synthesis, because such completion is required for formation of this primer.

Translocation of the primer for (+) strand DNA synthesis. Translocation of the primer for (+) strand DNA synthesis is likely to be facilitated by the homology between DR1 and DR2 (Fig. 7.26, step 5): the capped RNA primer, which includes dr1 sequences, can anneal to both. How the primer is induced to dissociate from DR1 and associate with DR2 is unclear. A small hairpin structure that includes the 5′ end of DR1 in the (−) strand of duck hepatitis B virus DNA appears to contribute to the translocation by inhibiting *in situ* priming and, perhaps, facilitating annealing of the capped RNA fragment with the complementary sequence in DR2 (Fig. 7.27). As in the first template exchange, a particular organization of the template in the core particles is thought to facilitate the process.

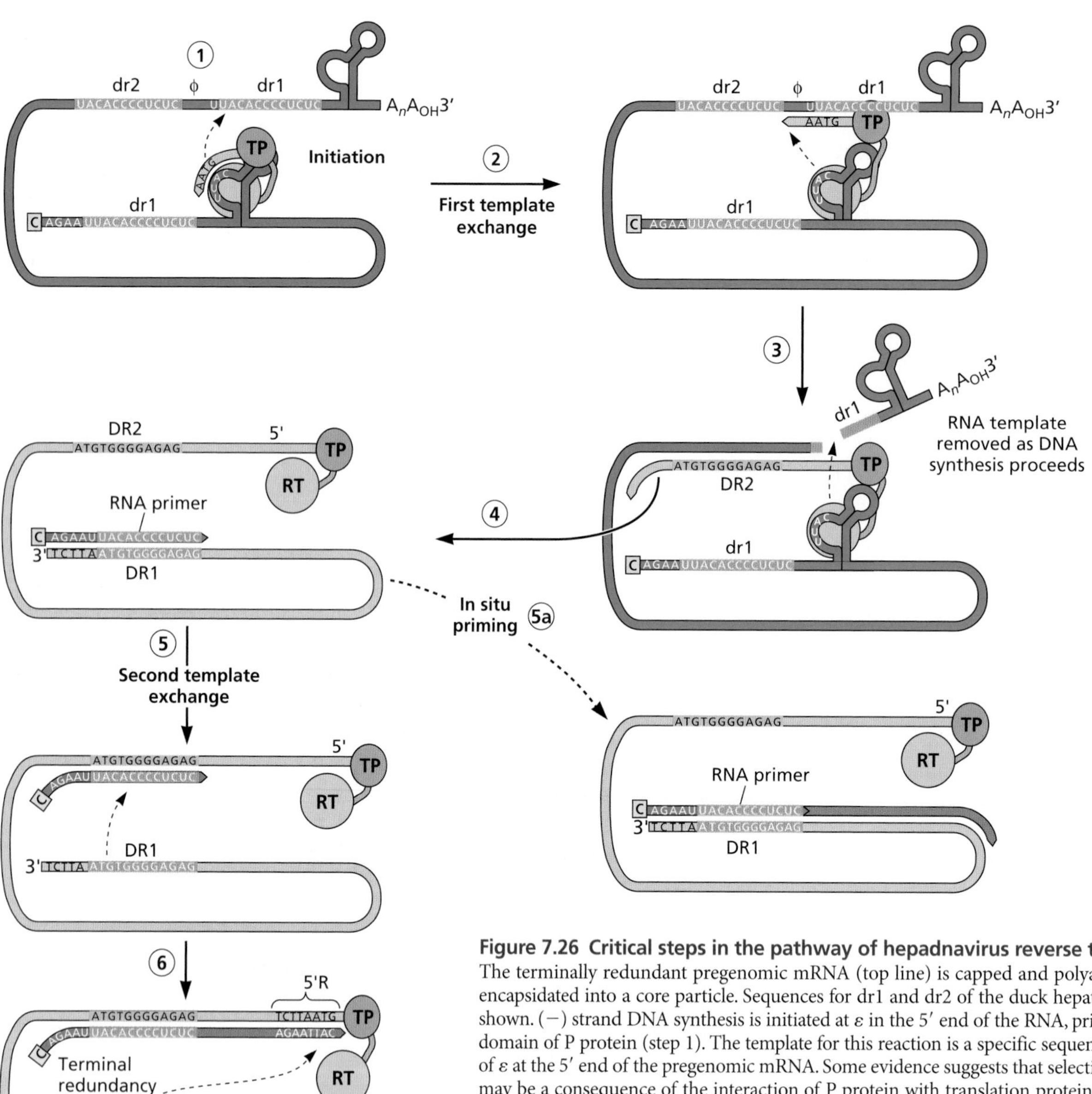

Figure 7.26 Critical steps in the pathway of hepadnavirus reverse transcription. The terminally redundant pregenomic mRNA (top line) is capped and polyadenylated and encapsidated into a core particle. Sequences for dr1 and dr2 of the duck hepatitis B virus are shown. (−) strand DNA synthesis is initiated at ε in the 5′ end of the RNA, primed by the TP domain of P protein (step 1). The template for this reaction is a specific sequence in the bulge of ε at the 5′ end of the pregenomic mRNA. Some evidence suggests that selection of this copy may be a consequence of the interaction of P protein with translation proteins at this end of the pregenomic mRNA. The first template exchange (step 2) is promoted by interaction with a sequence (ϕ) that lies between dr1 and dr2 at the 3′ end of the pregenome. DNA synthesis continues, using the 3′ copy of dr1 as the template (step 3). Mutation of the normal acceptor sequence leads to the synthesis of (−) strands with 5′ ends that map to other sites in the vicinity of the 3′ dr1, which apparently can serve as alternative acceptors. A deletion analysis with the woodchuck virus has suggested that a region 1 kb upstream of the 3′ dr1 includes a signal that specifies the acceptor site. As (−) strand DNA synthesis proceeds, the RNA template is degraded by the RNase H domain of P protein (step 4). The entire pregenome mRNA template is copied by the RT, producing a terminally redundant, complete (−) strand DNA species with short redundancies (7 or 8 nucleotides) that are denoted 3′R (see step 6). The primer for (+) strand synthesis is generated from the 5′-terminal 15 to 18 nucleotides of the pregenomic mRNA, which remains as the limit product of RNase H digestion. The primer is capped and includes the short sequence 3′ of dr1. At a low frequency (5 to 10%), the (+) strand primer is extended *in situ* instead of being translocated (the structure set off by a dashed arrow) (step 5a); elongation of this (+) strand results in a duplex linear genome. In the majority of cases, the primer is translocated to base pair with the DR2 sequence near the 5′ end of (−) strand DNA (step 5), facilitated as illustrated in Fig. 7.27: if the potential for the primer to hybridize with DR2 is disrupted by mutation, the pathway leading to formation of linear duplex DNA molecules predominates. After (+) strand synthesis is initiated, elongation proceeds (step 6). On reaching the 5′ end of (−) strand DNA, an intramolecular template exchange occurs, resulting in a circular DNA genome (step 7). This exchange is promoted by the short terminal redundancy, 5′R, in (−) strand DNA. (+) strand DNA synthesis then continues for a variable distance, resulting in the circular form of the genome found in mature virus particles. Adapted from Fig. 1 of J. W. Habig and D. D. Loeb, *J Virol* **76:**980–989, 2002, with permission.

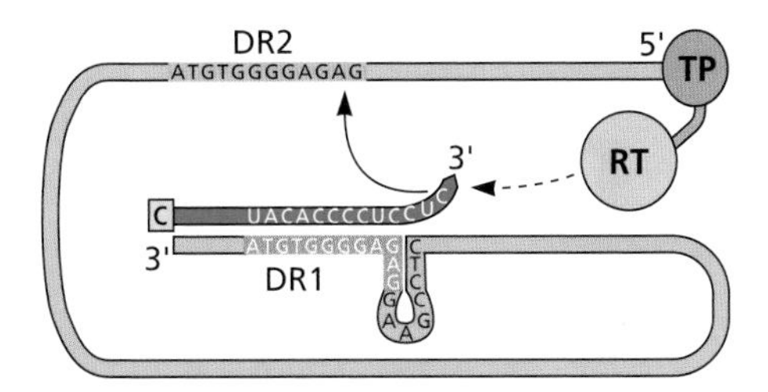

Figure 7.27 Model for (+) strand priming. Formation of a hairpin in the (−) strand DNA template displaces the 3′ end of the capped RNA fragment, preventing *in situ* priming and facilitating annealing with the homologous sequence in DR2. The ensuing translocation of the RNA primer allows initiation of (+) strand DNA synthesis. Adapted from Fig. 2 of J. W. Habig and D. D. Loeb, *J Virol* **76:**980–989, 2002.

The (+) strand synthesis primed by the translocated capped hepadnaviral RNA primer is similar to that which produces the strong-stop DNAs in retroviral reverse transcription. The (+) strand DNA synthesis begins near the 5′ end at DR2 and soon runs out of (−) strand template. As in the retroviral case, this problem is solved by a second template exchange, in this instance facilitated by the short repeat, 5′R, produced during synthesis of the (−) strand (Fig. 7.26, step 6).

The second template exchange creates a noncovalent circle. The structural requirements for the next step in hepadnaviral reverse transcription must allow displacement of the 5′ end of the (−) strand while still attached to the DNA. In addition to DR1 and DR2, *cis* interactions among other sequences at the ends and in a central region of (−) strand DNA have been implicated in this final step. It has been suggested that the simultaneous interaction of the central region with both ends may hold the termini in a position that facilitates both (+) strand primer translocation and the second template exchange. However, even with such "help," it is difficult to envision how a single protein accommodates all three DNA ends at once and catalyzes polymerization while still attached to one of them. Nevertheless, this exchange does occur with high efficiency in infected cells, and subsequent incomplete elongation of the (+) strand produces the partially duplex, noncovalent circle with variable (+) strand ends that comprises virion DNA (Fig. 7.26, step 7).

It is not clear what causes premature termination during synthesis of the (+) strand of hepadnaviral DNA. This synthesis is affected by mutations in the C protein. It has been proposed that DNA synthesis induces a change in the outer surface of the core, and that envelopment is regulated by interaction of the envelope proteins with this altered structure. Once these cores (capsids) are enveloped, DNA synthesis stops, presumably because dNTP substrates can no longer enter the particle.

Perspectives

The description of the reactions in hepadnaviral reverse transcription reveals interesting points of similarity to, and contrast with, retroviral systems (Table 7.2). Amino acid sequences and functions are conserved among retroviral RT and hepadnaviral P proteins, and both enzymes use terminal nucleic acid repeats to mediate template exchanges. However, the mechanisms by which their templates are reverse transcribed are quite distinct. Differences in the form and function of the final products of the two pathways are especially striking. A DNA circle with overlapping 5′ ends is an intermediate in the formation of a linear duplex DNA, the final product of retroviral reverse transcription. Repair of the circular intermediate is an unusual reaction, and covalently closed circle forms are dead-end products. In contrast, linear DNA is an aberrant product of hepadnaviral reverse transcription, and the covalently closed circle is the functional form for transcription.

The single-cell reproduction cycles of retroviruses and hepadnaviruses are, in a sense, permutations of one another. In comparing them to each other and to the unconventional foamy viruses (Box 7.9), it is instructive to include the cauliflower mosaic virus, a plant retroid virus that seems to combine some features of the animal viruses during reverse transcription (Fig. 7.28). This plant virus has a circular DNA genome and directs synthesis of a covalently closed episomal form, but its reverse transcription and priming mechanisms are quite analogous to those of retroviruses and retrotransposons. On the other hand, as with hepadnaviruses, RNA primers remain attached to the 5′ ends of cauliflower mosaic virus DNA. Retroid viruses appear to represent a continuum in evolution, and remind us of the varied combinations of strategies that exist in nature for replicating viral genomes and related genetic elements.

Biochemical and structural analyses reported since the last edition of this textbook have expanded our knowledge substantially and provided new insight into some of the remarkable properties of the viral RTs and the retroviral IN protein. It is now clear that these protein molecules can be scaffolds as well as catalysts, and their multiple functions appear to be enabled by a remarkable capacity for dynamic conformational change. Functional versatility is perhaps most striking in the hepadnaviral P protein, which performs all of the reactions necessary to synthesize a duplex circular DNA product from a linear RNA template while remaining covalently attached to one viral DNA end. Models derived from X-ray crystal structures of the retroviral RT and IN proteins have not only illuminated mechanistic details but have also informed efforts to develop new inhibitors that can be used in the clinic, while also providing insight into the molecular basis of drug resistance (Volume II, Chapter 9). Furthermore, identification of

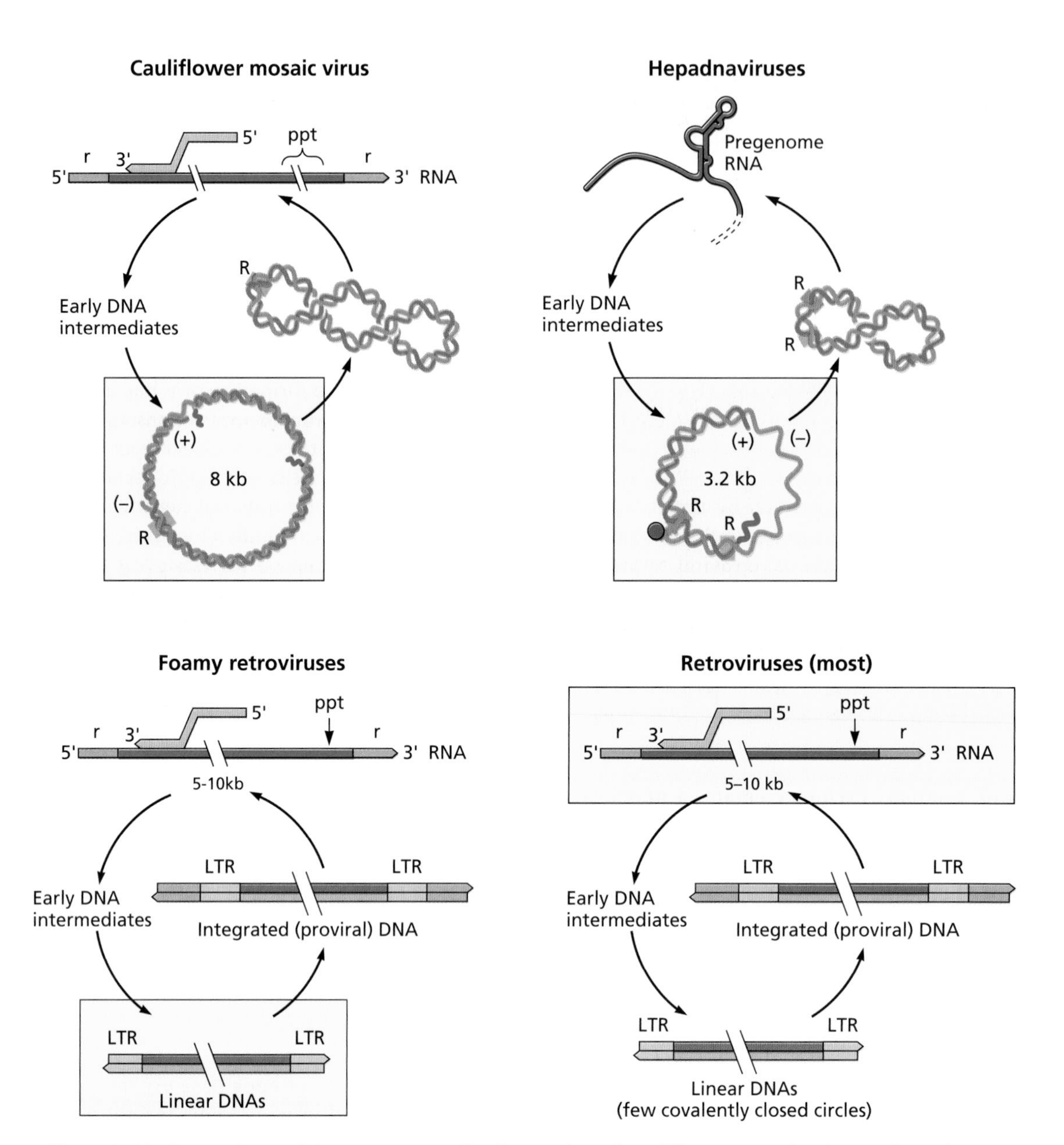

Figure 7.28 Comparison of the genome replication cycles of cauliflower mosaic viruses, hepadnaviruses, and retroviruses. The double-stranded DNA circle found in cauliflower mosaic virus particles contains three interruptions. At each interruption there is a short 5′ overlap of DNA as if formed by strand displacement synthesis. Ribonucleotides are often found attached to the 5′ ends. The (−) strand starts with either a ribo- or a deoxyriboadenosine. The 5′ ends of the (+) strand each contain 8 to 10 purine-rich matches to the viral DNA at the same location, suggesting a primer function. r, short sequence at both ends of viral RNA; R, same sequence in DNA. The shaded boxes indicate that the nucleic acids (genomes) encapsidated in particles of each virus represent different components in analogous pathways.

cellular tethering proteins and elucidation of their roles in retroviral DNA integration have led to the design of chimeric tethers that can direct integration to predetermined sites in host chromosomes. As might be expected, such progress has elicited important new questions to be addressed in the future.

References

Books

Cooper GM, Temin RG, Sugden W (ed). 1995. *The DNA Provirus: Howard Temin's Scientific Legacy.* ASM Press, Washington, DC.

Neamati N (ed). 2011. *HIV-1 Integrase.* John Wiley & Sons, Inc, Hoboken, NJ.

Recent Reviews

Bushman F, Craigie R. 2015. Host factors in retroviral integration and selection of target sites, p 1035–1050. *In* Craig N (ed), *Mobile DNA*. ASM Press, Washington, DC.

Engelman A, Cherapanov P. 2015. Retroviral integrase structure and DNA recombination mechanism, p 1011–1032. *In* Craig N (ed), *Mobile DNA*. ASM Press, Washington, DC.

Goff SP. 2013. *Retroviridae*, p 1424–1473. *In* Knipe DM, Howley PM (ed), *Fields Virology*, 6th ed, vol 2. Lippincott Williams & Wilkins, Philadelphia, PA.

Hughes S. 2015. Reverse transcription of retroviruses and LTR retrotransposons, p 1051–1077. *In* Craig N (ed), *Mobile DNA*. ASM Press, Washington, DC.

Krishnan L, Engelman A. 2012. Retroviral integrase proteins and HIV-1 DNA integration. *J Biol Chem* **287**:40858–40866.

Le Grice SF. 2012. Human immunodeficiency virus reverse transcriptase: 25 years of research, drug discovery, and promise. *J Biol Chem* **287**: 40850–40857.

Seeger C, Zoulim F, Mason WS. 2013. Hepadnaviruses, p 2185–2221. *In* Knipe DM, Howley PM (ed), *Fields Virology*, 6th ed, vol 2. Lippincott Williams & Wilkins, Philadelphia, PA.

Skalka AM. 2015. Retroviral DNA transposition: themes and variations, p 1101–1123. *In* Craig N (ed), *Mobile DNA*. ASM Press, Washington, DC.

Skalka AM. Retroviral integrase. *Annu Rev Virol*, in press.

Suzuki Y, Chew ML, Suzuki Y. 2012. Role of host-encoded proteins in restriction of retroviral integration. *Front Microbiol* **3**:227. doi:10.3389/fmicb.2012.00227.

Some Landmark Papers

Retroviral Reverse Transcription and DNA Integration

Baltimore D. 1970. RNA-dependent DNA polymerase in virions of RNA tumour viruses. *Nature* **226**:1209–1211.

Temin HM, Mizutani S. 1970. RNA-dependent DNA polymerase in virions of Rous sarcoma virus. *Nature* **226**:1211–1213.

Brown PO, Bowerman B, Varmus HE, Bishop JM. 1987. Correct integration of retroviral DNA in vitro. *Cell* **49**:347–356.

Craigie R, Fujiwara T, Bushman F. 1990. The IN protein of Moloney murine leukemia virus processes the viral DNA ends and accomplishes their integration in vitro. *Cell* **62**:829–837.

Grandgenett DP, Vora AC, Schiff RD. 1978. A 32,000-dalton nucleic acid-binding protein from avian retravirus cores possesses DNA endonuclease activity. *Virology* **89**:119–132.

Katz RA, Merkel G, Kulkosky J, Leis J, Skalka AM. 1990. The avian retroviral IN protein is both necessary and sufficient for integrative recombination in vitro. *Cell* **63**:87–95.

Hepadnaviral Reverse Transcription

Summers J, Mason WS. 1982. Replication of the genome of a hepatitis B–like virus by reverse transcription of an RNA intermediate. *Cell* **29**:403–415.

Wang GH, Seeger C. 1992. The reverse transcriptase of hepatitis B virus acts as a protein primer for viral DNA synthesis. *Cell* **71**:663–670.

8 Synthesis of RNA from DNA Templates

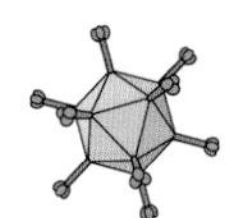

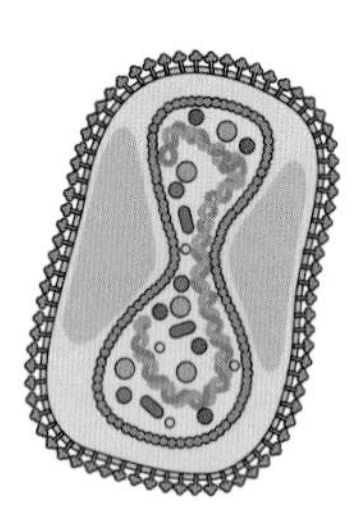

Introduction
- Properties of Cellular RNA Polymerases That Transcribe Viral DNA
- Some Viral Genomes Must Be Converted to Templates Suitable for Transcription

Transcription by RNA Polymerase II
- Regulation of RNA Polymerase II Transcription
- Common Properties of Proteins That Regulate Transcription

The Cellular Machinery Alone Can Transcribe Viral DNA Templates

Viral Proteins That Govern Transcription of Viral DNA Templates
- Patterns of Regulation
- The Human Immunodeficiency Virus Type 1 Tat Protein Autoregulates Transcription
- The Transcriptional Cascades of DNA Viruses
- Entry into One of Two Alternative Transcriptional Programs

Transcription of Viral Genes by RNA Polymerase III
- The VA-RNA I Promoter
- Regulation of VA-RNA Gene Transcription

Inhibition of the Cellular Transcriptional Machinery

Unusual Functions of Cellular Transcription Components

A Viral DNA-Dependent RNA Polymerase

Perspectives

References

LINKS FOR CHAPTER 8

⏭ *Video: Interview with Dr. Arnold Berk.*
http://bit.ly/Virology_Berk

⏭ *Movie 8.1: Initiation of transcription by RNA polymerase II.*
http://bit.ly/Virology_V1_Movie8-1

> *It is possible that nature invented DNA for the purpose of achieving regulation at the transcriptional rather than at the translational level.*
>
> A. Campbell, 1967

Introduction

During the infectious cycles of viruses with DNA genomes, viral messenger RNA (mRNA) synthesis must precede production of proteins. In most cases, this step is accomplished by the host cell enzyme that produces cellular mRNA, RNA polymerase II (Table 8.1). This enzyme also transcribes the proviral DNA of retroviruses. The signals that control expression of the genes of these viruses are similar to those of cellular genes. In fact, much of our understanding of the mechanisms of cellular transcription stems from study of viral DNA templates. In contrast, viral RNA polymerases transcribe the large DNA genomes of viruses that replicate in the cytoplasm, such as poxviruses. These enzymes resemble their host cell counterparts in several respects.

The expression of viral genes in a strictly defined, reproducible sequence is a hallmark of cells infected by DNA viruses. In general, enzymes and regulatory proteins needed in smaller quantities are made during the initial period of infection, whereas structural proteins of virus particles are made only after viral DNA synthesis begins. Such orderly gene expression is primarily the result of transcriptional regulation by viral proteins. This pattern is quite different from the continual expression of all viral genes that is characteristic of the infectious cycles of many RNA viruses (Chapter 6). As discussed in this chapter, the elucidation of the molecular strategies that ensure sequential transcription of the genes of DNA viruses has identified a number of common mechanisms executed in virus-specific fashion. As a collateral dividend, we have gained insights into the cellular mechanisms that control progression through the cell cycle.

Properties of Cellular RNA Polymerases That Transcribe Viral DNA

Eukaryotes Have Three Transcriptional Systems

A general feature of eukaryotic cells is the division of transcriptional labor among three DNA-dependent RNA polymerases. These enzymes, designated RNA polymerases I, II, and III, synthesize different kinds of cellular RNA (Table 8.2). RNA polymerase II makes precursors to mRNA, as well as the precursors of small, regulatory RNA molecules (microRNAs, miRNAs; see Chapter 10). The other two enzymes produce stable RNAs, such as ribosomal RNAs (rRNAs) and transfer RNAs (tRNAs). Synthesis of these stable "housekeeping" RNAs must be adjusted to match the rates of cell growth and division. But regulation of mRNA synthesis is crucial for orderly development and differentiation in eukaryotes, as well as for the responses of cells to their environment. The evolution of RNA polymerases with distinct transcriptional responsibilities appears to be a device for maximizing opportunities for control of mRNA synthesis, while maintaining a constant and abundant supply of the RNA species essential for the metabolism of all cells.

Despite their different functions, several of the 12 to 16 subunits of the large eukaryotic RNA polymerases are identical, while others are related in sequence to one another or to subunits of bacterial RNA polymerases. Such conservation of sequence can be attributed to the common biochemical capabilities of the enzymes. These activities include binding of ribonucleoside triphosphate substrates, binding to template DNA and to product RNA, and catalysis of phosphodiester bond formation. The structure of yeast RNA polymerase II revealed that the organization of its active center is similar to that of smaller DNA-dependent RNA polymerases, as well as of enzymes that make DNA from DNA or RNA templates (Fig. 6.4).

PRINCIPLES *Synthesis of RNA from DNA templates*

- Transcription is the first biosynthetic reaction to occur in cells infected by double-stranded DNA viruses.
- To form a template suitable for transcription, gapped, double-stranded or single-stranded DNA genomes are converted to double-stranded DNA molecules by cellular enzymes; retroviral RNA genomes are converted to double-stranded proviral DNA that is integrated into the cellular genome by viral enzymes.
- Studies in virology led to the identification of elements in DNA that direct pre-mRNA or mRNA synthesis, including promoters and enhancers that are binding sites for components of the transcriptional machinery.
- The cellular transcriptional machinery alone is sufficient to transcribe some viral DNA templates.
- Viral proteins can stimulate transcription of their own transcriptional unit to establish a positive autoregulatory loop or activate transcription of different viral genes.
- Transcription of subsets of viral genes in distinct temporal periods (phases) is a characteristic feature of the reproductive cycles of all viruses with DNA genomes, including bacteriophages. Transitions from one phase to the next depend on viral activators and synthesis of progeny viral genomes.
- Viral proteins that regulate transcription may bind directly to viral promoter sequences or indirectly in association with cellular proteins.
- Some viruses, including the herpesviruses, establish latent infections in which transcription of lytic genes is inhibited and, in some cases, unique latency-associated transcription units are expressed.
- Most viral genes are transcribed by the cellular RNA polymerase II, but some small viral RNAs are produced by RNA polymerase III.
- Suppression of cellular transcription by viral components diverts limited cellular resources to aid viral transcription.

Table 8.1 Strategies of transcription of viral DNA templates

Origin of transcriptional components	Virus
Host only	Retroviruses with simple genomes, caulimoviruses
Host plus one viral protein	
The viral protein transcribes late genes	Bacteriophages T3 and T7
The viral protein regulates transcription	Parvoviruses, papillomaviruses, polyomaviruses, retroviruses with complex genomes, geminiviruses
Host plus several viral proteins	Adenoviruses, bacteriophage T4, herpesviruses
Viral	Poxviruses

Transcription of cellular and viral genes requires not only template-directed synthesis of RNA but also correct interpretation of DNA punctuation signals that mark the sites at which transcription must start and stop. Initiation of transcription comprises recognition of the point at which copying of the DNA should begin, the **initiation site**, and synthesis of the first few phosphodiester bonds in the RNA. During the elongation phase, nucleotides are added rapidly to the 3′ end of the nascent RNA, as the transcriptional machinery reads the sequence of a gene. When termination sites are encountered, both the RNA product and the RNA polymerase are released from the DNA template. Purified RNA polymerases I, II, and III perform the elongation reactions *in vitro* but are incapable of specific initiation of transcription without the assistance of additional proteins.

Cellular RNA Polymerases II and III Transcribe Viral Templates

Viral mRNAs or their precursors (pre-mRNAs) are made by RNA polymerase II in cells infected by DNA viruses with both small and large genomes, such as polyomaviruses and herpesviruses, respectively. This enzyme also synthesizes the precursors to viral, as well as cellular, miRNAs. However, it can also carry out at least one reaction unique to virus-infected cells,

Table 8.2 Eukaryotic RNA polymerases synthesize different classes of cellular and viral RNA

	RNAs synthesized[a]	
Enzyme	**Cellular**	**Viral**
RNA polymerase I	Pre-rRNA	None known
RNA polymerase II	Pre-mRNA	Pre-mRNA and mRNA
	Pri-miRNA	Pri-miRNA
	snRNAs	HDV genome RNA and mRNA
RNA polymerase III	Pre-tRNAs	Ad2 VA-RNAs
	5S rRNA	EBV EBER RNAs
	U6 snRNA	MHV68 pre-miRNA

[a]Ad2, adenovirus type 2; EBER, Epstein-Barr virus-encoded small RNA; EBV, Epstein-Barr virus; HDV, hepatitis delta virus; MHV68, murine gammaherpesvirus 68; pri-miRNA, primary transcripts containing precursor to miRNAs; snRNA, small nuclear RNA.

the transcription of an RNA template by RNA polymerase II to produce hepatitis delta satellite virus genomes and mRNA (Table 8.2).

Some animal viral DNA genomes also encode small, noncoding RNAs that are made by RNA polymerase III. This phenomenon was initially observed in human cells infected by adenovirus, but RNA polymerase III transcription units are present in the genomes of other viruses (Table 8.2).

Some Viral Genomes Must Be Converted to Templates Suitable for Transcription

All viral DNA molecules must enter the infected cell nucleus to be transcribed by cellular RNA polymerases, but there is considerable variation in the reactions needed to produce templates that can be recognized by the cellular machinery. Some viral genomes are double-stranded DNA molecules that can be transcribed as soon as they reach the nucleus. Transcription of specific genes is therefore the first biosynthetic reaction in cells infected by adenoviruses, herpesviruses, papillomaviruses, and polyomaviruses. Other viral DNA genomes must be converted from the form in which they enter the cell to double-stranded molecules that serve as transcriptional templates (Fig. 8.1). The hepadnaviral genome is an incomplete circular DNA molecule with a large gap in one strand that is repaired by cellular enzymes to form a fully double-stranded DNA molecule. Similarly, single-stranded genomes such as that of the adenovirus-associated virus, a parvovirus, are converted to double-stranded molecules by a cellular DNA polymerase (Chapter 9). The prerequisites for expression of retroviral genetic information are even more demanding, for the (+) strand RNA genome must be both converted into viral DNA and integrated into the cellular genome. Reverse transcription creates an appropriate double-stranded DNA template that includes the signals needed for its recognition by components of the cellular transcriptional machinery (Chapter 7).

The cellular templates for transcription by RNA polymerase II are DNA sequences packaged in chromatin, which contains the conserved histones and many other proteins. The fundamental structural unit of chromatin is the nucleosome, which comprises ~140 bp of DNA wrapped around an octamer containing two copies each of histones H2A, H2B, H3, and H4. The posttranslational modifications of the histones help distinguish highly condensed, transcriptionally silent heterochromatin from transcriptionally active genes. As the organization of DNA into nucleosomes can both block recognition of regulatory sequences and impose barriers to transcriptional elongation, numerous proteins that regulate transcription function by overcoming such obstacles.

Many viral DNA genomes transcribed by RNA polymerase II are also organized by cellular nucleosomes. Because they are integrated into the cellular genome, the proviral DNA templates for retroviral transcription are organized into chromatin indistinguishable from that of the host cell. The DNA genomes of papillomaviruses and polyomaviruses enter cells as "minichromosomes" in which the viral DNA is bound to

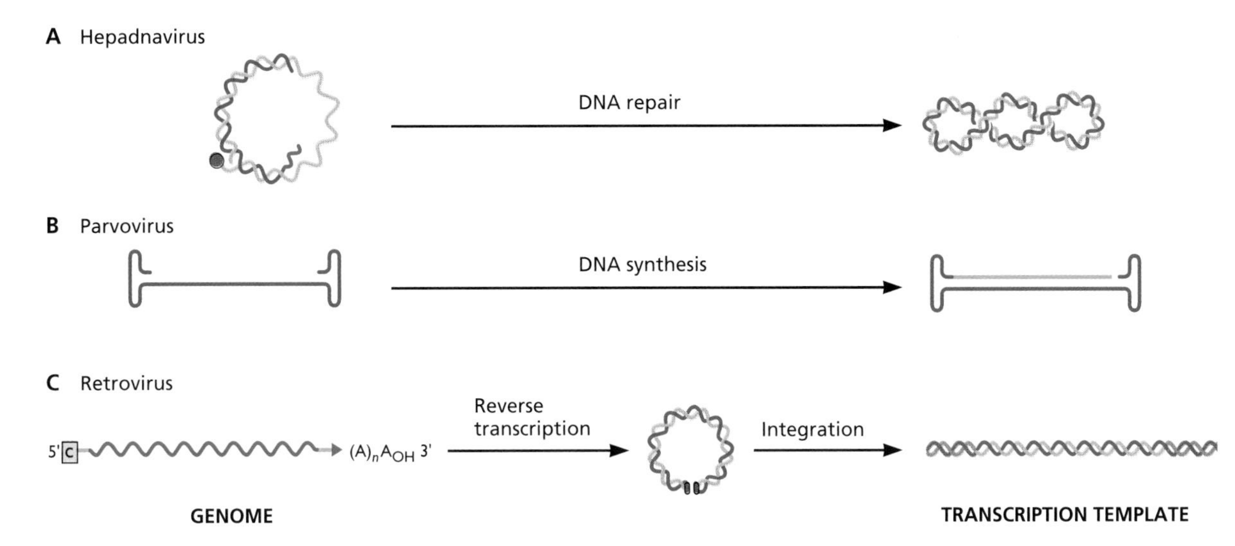

Figure 8.1 Conversion of viral genomes to templates for transcription by RNA polymerase. (A) Hepadnaviral templates for transcription are closed, circular, double-stranded DNA molecules. The mechanisms by which such DNA is formed by repair of the partially double-stranded, gapped DNA genomes are not well understood. **(B)** The single-stranded DNA genomes of parvoviruses such as adenovirus-associated virus carry an inverted terminal repetition with a free 3′ OH end. Copying of the viral genome from this primer by cellular DNA polymerase produces a double-stranded template for transcription. **(C)** Viral enzymes catalyze the conversion of retroviral (+) RNA genomes to double-stranded DNA and its subsequent integration into the host cell genome (proviral DNA) (see Chapter 7).

BOX 8.1

METHODS

Association of histones and other proteins with DNA in vivo: the *chromatin immunoprecipitation assay*

The chromatin immunoprecipitation (ChIP) assay is widely used to investigate the association of specific proteins with viral or cellular DNA within cells. Proteins are initially cross-linked to DNA by exposure of intact cells to formaldehyde. Following cell lysis and fragmentation of the DNA by sonication or enzymatic digestion, DNA fragments bound to the protein of interest are isolated by immunoprecipitation with antibodies against that protein. Enrichment for the DNA sequences of interest in the immunoprecipitate is then assessed by PCR amplification of specific sequences or by high-throughput DNA sequencing.

Application of this assay to herpes simplex virus type 1-infected cells, as illustrated, has demonstrated that histone H3 binds to immediate-early, early, and late genes in entering, but not in newly replicated, viral genomes. However, nucleosomes do not organize viral DNA into a regular structure like that of cellular chromatin, in which the histone octamers are spaced at regular intervals on DNA.

Kent JR, Zeng PY, Atanasice D, Fraser NW, Berger SL. 2004. During lytic infection herpes simplex virus type 1 is associated with histones bearing modifications that correlate with active transcription. *J Virol* **78:**10178–10186.

Kwiatkowski DL, Thompson HW, Bloom DC. 2009. The polycomb group protein Bmi1 binds to the herpes simplex virus 1 latent genome and maintains repressive histone marks during latency. *J Virol* **83:** 8173–8181.

Oh J, Fraser NW. 2008. Temporal association of the herpes simplex virus genome with histone proteins during a lytic infection. *J Virol* **82:**3503–3537.

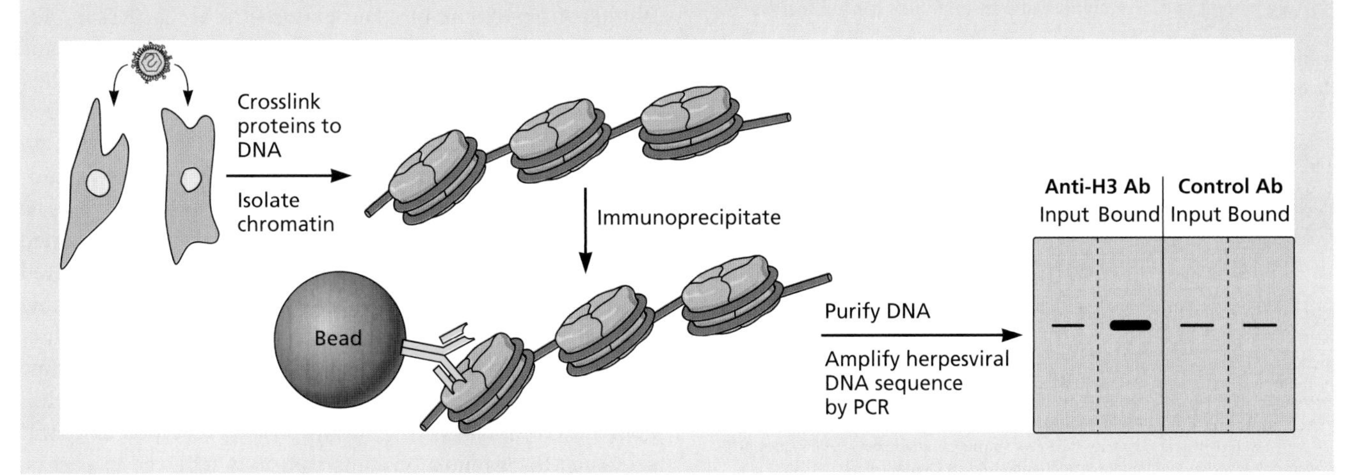

nucleosomes, whereas histones become associated with herpesviral genomes soon after their entry into infected cell nuclei (Box 8.1). Such nucleosomal organization suggests that mechanisms analogous to those regulating transcription of cellular chromatin are likely to operate on these viral templates. Indeed, as we shall see, the properties of viral "chromatin" can result in transcriptional silencing and prevent transcription of the majority of viral genes in cells latently infected by some herpesviruses. Although transcription of viral DNA templates associated with histones is a common phenomenon, it is not universal: the initial templates for adenoviral gene expression are nucleoproteins comprising the linear, double-stranded DNA genome and the major core protein of the viral particle, protein VII.

Transcription by RNA Polymerase II

Accurate initiation of transcription by RNA polymerase II is directed by specific DNA sequences located near the site of initiation and called the **promoter** (Fig. 8.2). The promoter and the additional DNA sequences that govern transcription make up the **transcriptional control region.** These sequences of DNA viruses and retroviruses were among the first to be examined experimentally. For example, the human adenovirus type 2 major late promoter was the first from which accurate initiation of transcription was reconstituted *in vitro* (Box 8.2). Subsequently, the study of viral transcription yielded fundamental information about the mechanisms by which RNA polymerase II transcription is initiated and regulated.

Figure 8.2 RNA polymerase II transcriptional control elements. The site of initiation is represented by the red arrow drawn in the direction of transcription on the nontranscribed DNA strand, a convention used throughout this text. The core promoter comprises the minimal sequence necessary to specify accurate initiation of transcription. The TATA sequence is the binding site for TfIId (Box 8.3), and the initiator is a sequence sufficient to specify initiation at a unique site. The activity of the core promoter is modulated by local regulatory sequences typically found within a few hundred base pairs of the initiation site. The location of these sequences upstream of the TATA sequence as shown is common, but such sequences can also lie downstream of the initiation site. Distant regulatory sequences that stimulate (enhancers) or repress (silencers) transcription are present in a large number of transcriptional control regions.

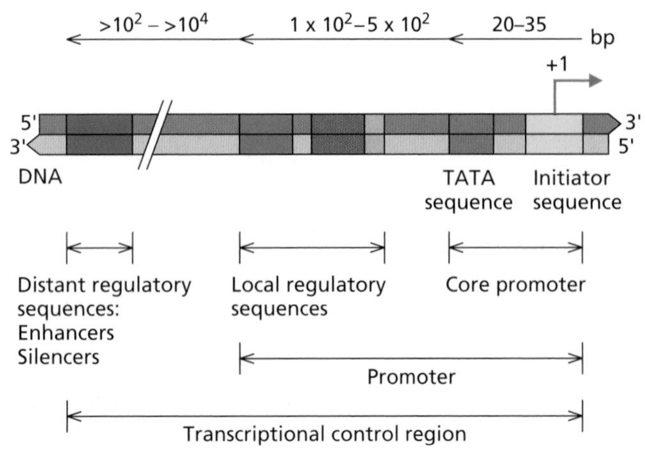

Biochemical studies using model transcriptional control regions, such as the adenoviral major late promoter, established that initiation of transcription is a multistep process. The initiation reactions include promoter recognition, unwinding of the duplex DNA around the initiation site to form an open initiation complex, and movement of the transcribing complex away from the promoter (promoter clearance) (Fig. 8.3 and Movie 8.1: http://bit.ly/Virology_V1_Movie8-1). At least 40 proteins, which comprise RNA polymerase II itself and auxiliary initiation proteins, are needed to complete the intricate process of initiation. Our understanding of the functions of these proteins and of the DNA sequences that control initiation is based largely on *in vitro* systems or simple assays for detecting gene expression within cells. Application of these methods has identified a very large number of transcriptional control sequences. Fortunately, all of them can be assigned to one of the three functionally distinct regions identified in Fig. 8.2.

Core promoters of viral and cellular genes contain all the information necessary for recognition of the site of initiation and assembly of precisely organized **preinitiation complexes.** These assemblies contain RNA polymerase II and a common set of general initiation proteins required for accurate and precise initiation. A hallmark of many core RNA polymerase II promoters is the presence of a TA-rich **TATA** sequence 20 to 35 bp upstream of the site of initiation (Fig. 8.2 and 8.4), which is recognized by the TATA-binding protein (Tbp) (Box 8.3). Short sequences, termed **initiators,** which specify accurate (but inefficient) initiation of transcription in the absence of any other promoter sequences, are also commonly found (Fig. 8.4).

Many of the interactions among components of the transcriptional machinery take place before a promoter is encountered: RNA polymerase II is present in cells in extremely large assemblies that contain the initiation proteins, as well as others that are essential for transcription or its regulation. Such assemblies, termed **holoenzymes,** appear to be poised to initiate transcription as soon as they are recruited to a promoter.

Regulation of RNA Polymerase II Transcription

Numerous patterns of gene expression are necessary for eukaryotic life: some RNA polymerase II transcription units must be expressed in all cells, whereas others are transcribed only during specific developmental stages or in specialized differentiated cells. Many others must be maintained in an almost silent state, from which they can be activated rapidly in response to specific stimuli, and to which they can be returned readily. Transcription of viral genes is also regulated during the infectious cycles of most of the viruses considered in this chapter. Large quantities of viral proteins for assembly of progeny virions must be made within a finite (and often short) infectious cycle. Consequently, some viral genes must be transcribed at higher rates than others. In many cases, viral genes are transcribed in a specific and stereotyped temporal sequence. Such regulated transcription is achieved in part by

BOX 8.2

EXPERIMENTS

Mapping of a human adenovirus type 2 initiation site and accurate transcription in vitro

When cellular RNA polymerase II was identified in 1969, investigators had access only to preparations of total cellular DNA, and nothing was known about the organization of eukaryotic transcription units. Consequently, the genomes of the DNA viruses simian virus 40 and human adenovirus type 2 served as valuable resources for investigation of mechanisms of transcription. Indeed, it was detailed information about a particular adenoviral transcription unit that finally allowed biochemical studies of the mechanism of initiation. In 1978, the site at which major late transcription begins was mapped precisely, by determining the sequence of the 5′ end of the RNA transcript. This knowledge was exploited to develop a simple assay for accurate initiation of transcription, the "runoff" assay, using a linear template that includes a transcription initiation site shown in the figure. Purified RNA polymerase II produced no specific transcripts in the runoff assay, but unfractionated nuclear extracts of human cells were shown to contain all the components necessary for accurate initiation of transcription.

Weil PA, Luse DS, Segall J, Roeder RG. 1979. Selective and accurate initiation of transcription at the Ad2 major late promoter in a soluble system dependent on purified RNA polymerase II and DNA. *Cell* **18:**469–484.

Ziff EB, Evans RM. 1978. Coincidence of the promoter and capped 5′ terminus of RNA from the adenovirus 2 major late transcription unit. *Cell* **15:**1463–1475.

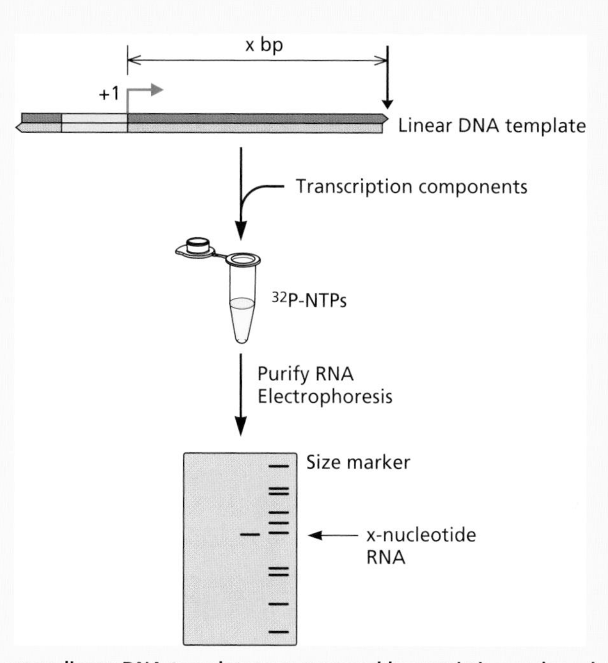

In this simple assay, linear DNA templates are prepared by restriction endonuclease cleavage (black arrow), a known distance, *x* bp, downstream of the initiation site (+1). When the template is incubated with the transcriptional machinery and nucleoside triphosphate (NTP) substrates, transcription initiated at position +1 continues until the transcribing complex "runs off" the linear template. Specific transcription is therefore assayed as the production of ^{32}P-labeled RNA *x* nucleotides in length. This runoff transcription assay is convenient and has been used to assess both specificity and efficiency of transcription.

means of cellular control mechanisms, for example, cellular proteins that repress transcription. In general, however, viral proteins are critical components of the circuits that establish orderly transcription of viral genes.

Recognition of Local and Distant Regulatory Sequences

Both local and distant sequences can control transcription from core promoters. However, local sequences are often sufficient for proper transcriptional regulation. These local regulatory sequences are recognized by sequence-specific DNA-binding proteins (Fig. 8.5), a property first demonstrated with the simian virus 40 early promoter. An enormous number of sequence-specific proteins that regulate transcription are now known, many first identified through analyses of viral promoters. Unfortunately, the nomenclature applied to these regulatory proteins presents serious difficulties for both writer and reader, for it is unsystematic and idiosyncratic (Box 8.4).

Efficient transcription of many viral and cellular genes also requires more distant regulatory sequences in the DNA template, which possess properties that were entirely unanticipated. The first such **enhancer**, so named because it stimulated transcription to a large degree, was discovered in the genome of simian virus 40. Enhancers are defined by their position- and orientation-independent stimulation of transcription of homologous and heterologous genes over distances as great as 10,000 bp in the genome. Despite these unusual properties, enhancers are built with binding sites for the proteins that recognize local promoter sequences.

The Simian Virus 40 Enhancer: a Model for Viral and Cellular Enhancers

The majority of viral DNA templates described in this chapter contain enhancers that are recognized by cellular DNA-binding proteins. The simian virus 40 enhancer has been studied intensively, and its properties and mechanism of action are characteristic of many enhancers, whether of viral or cellular origin.

The simian virus 40 enhancer is built from three units, termed enhancer elements, which are subdivided into

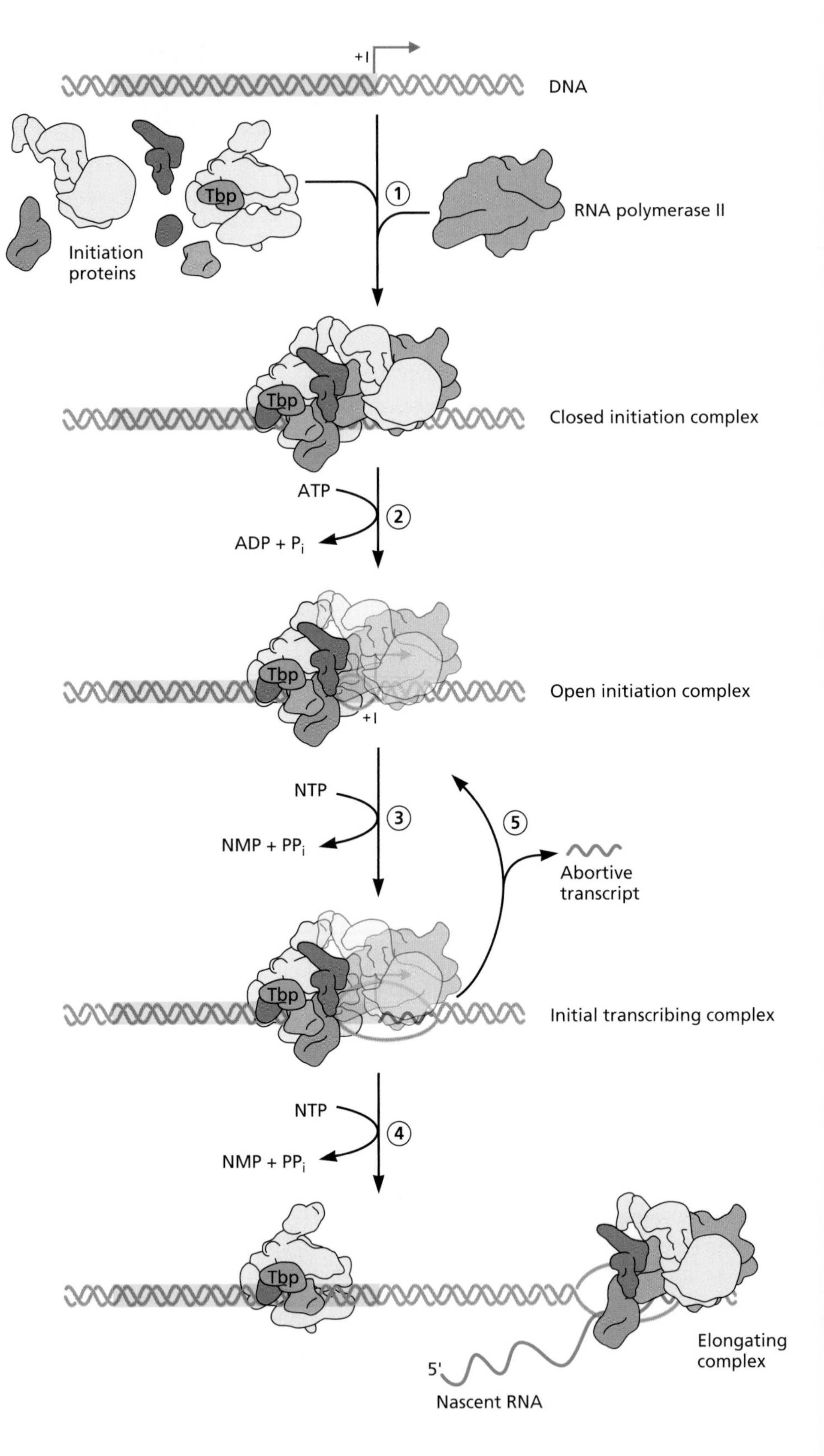

Figure 8.3 Initiation of transcription by RNA polymerase II. Assembly of the closed initiation complex (step 1) is followed by unwinding of the DNA template in the region spanning the site of initiation (step 2). RNA polymerase II then synthesizes short transcripts (less than 10 to 15 nucleotides) by template-directed incorporation of nucleotides (step 3). The initial transcribing complex is thought to be conformationally strained, because RNA polymerase II remains in contact with promoter-bound initiation proteins as it synthesizes short RNAs. The severing of these contacts allows the transcribing complex to escape from the promoter and proceed with elongation (step 4). This promoter clearance step is often inefficient, with abortive initiation (step 5) predominating. In the latter process, initial transcripts are released, reforming the open initiation complex. The initial elongating transcriptional complex contains some but not all of the proteins that form the preinitiation complex, as well as proteins that stimulate elongation (not shown). Structural data collected for RNA polymerase II and initiation proteins associated with nucleic acid has been used to produce a movie of initiation and elongation (Movie 8.1: http://bit.ly/Virology_V1_Movie8-1), adapted from A. C. Cheung and P. Cramer, *Cell* **149:**1431–1437, 2012, with permission.

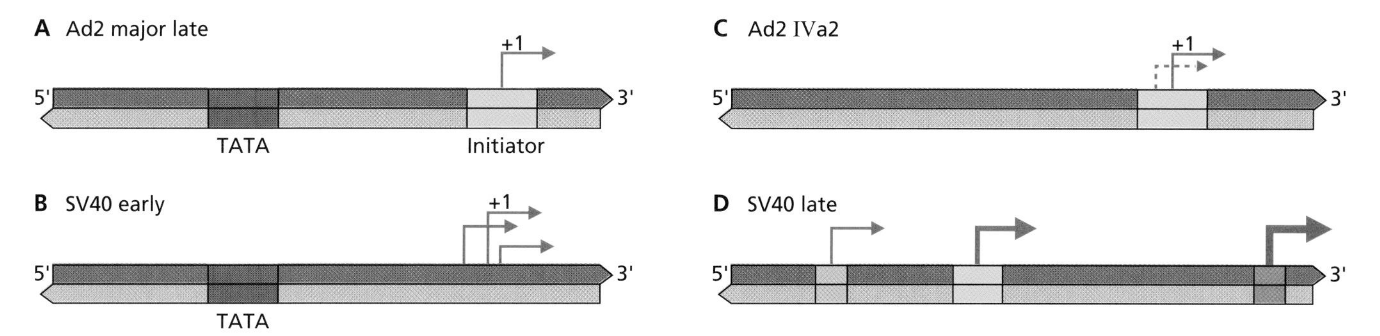

Figure 8.4 Variations in core RNA polymerase II promoter architecture. Variations in promoter architecture are illustrated using four viral promoters represented as in Fig. 8.2. The TATA or initiator sequences of the different promoters are not identical in DNA sequence. In the case of the simian virus 40 (SV40) late transcription unit **(D)**, each of the sites of initiation is included within a DNA sequence resembling those of initiators. It has not been shown experimentally that all actually function as autonomous initiator sequences. The relative frequencies with which different initiation sites in a single promoter are used are indicated by the thickness of the red arrows. Ad2, adenovirus type 2.

BOX 8.3

BACKGROUND

The RNA polymerase II closed initiation complex

The closed initiation complex is shown on a promoter that contains both a TATA and an initiator sequence (e.g., the adenovirus major late promoter). The TfIId protein contains a subunit that recognizes the TATA sequence (Tbp) and 8 to 10 additional subunits. The X-ray crystal structures of DNA-bound Tbp, such as that of *Arabidopsis thaliana*, bound to the adenoviral major late TATA sequence shown in the inset (courtesy of S. K. Burley, The Rockefeller University), revealed that this protein induces sharp bending of the DNA. One popular hypothesis is that such bending facilitates interactions among proteins bound to regulatory sequences located upstream of the TATA sequence and the basal transcriptional machinery. TfIId is required for transcription from all RNA polymerase II promoters. It can recognize those that lack TATA sequences by various mechanisms. TfIIh supplies DNA-dependent ATPase and helicase activities essential for transcription and a kinase that phosphorylates the C-terminal segment of the largest subunit of RNA polymerase II. The depictions of the transcription initiation proteins are based on visualization of initiation by cryo-electron microscopy.

Buratowski S, Hahn S, Guarente L, Sharp PA. 1989. Five intermediate complexes in transcription initiation by RNA polymerase II. *Cell* **56:**549–561.

He Y, Fang J, Taatjes DJ, Nogales E. 2013. Structural visualization of key steps in human transcription initiation. *Nature* **495:**481–486.

Kim JL, Nikolov DB, Burley SK. 1993. Co-crystal structure of TBP recognizing the minor groove of a TATA element. *Nature* **365:**520–527.

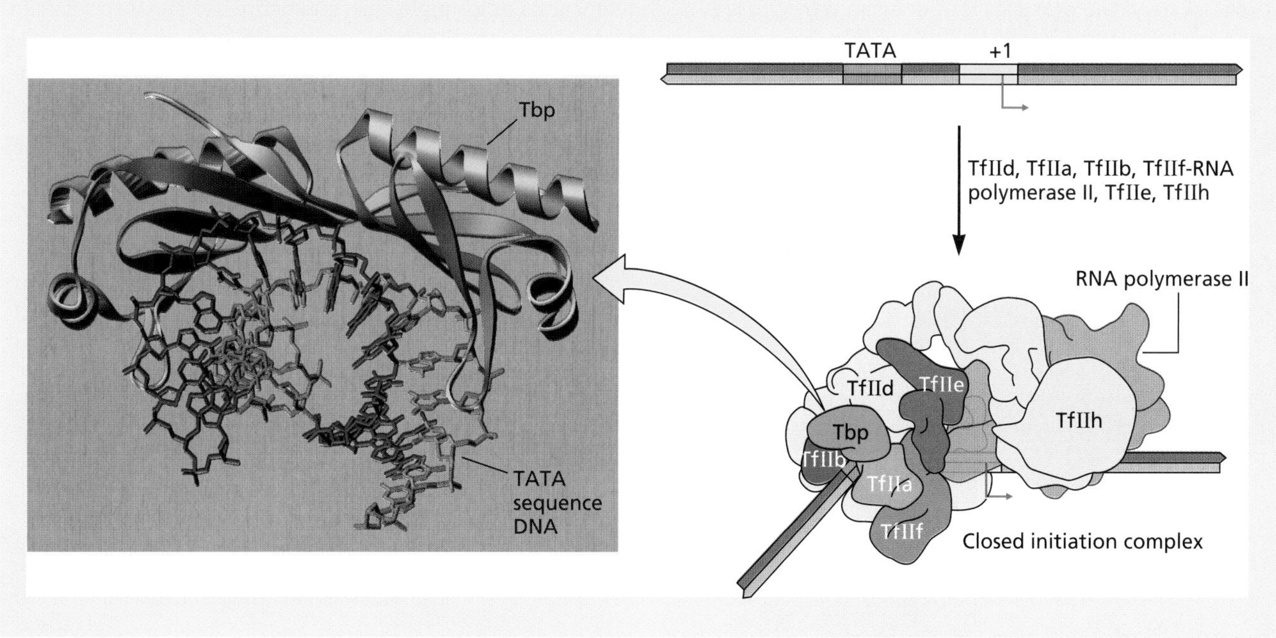

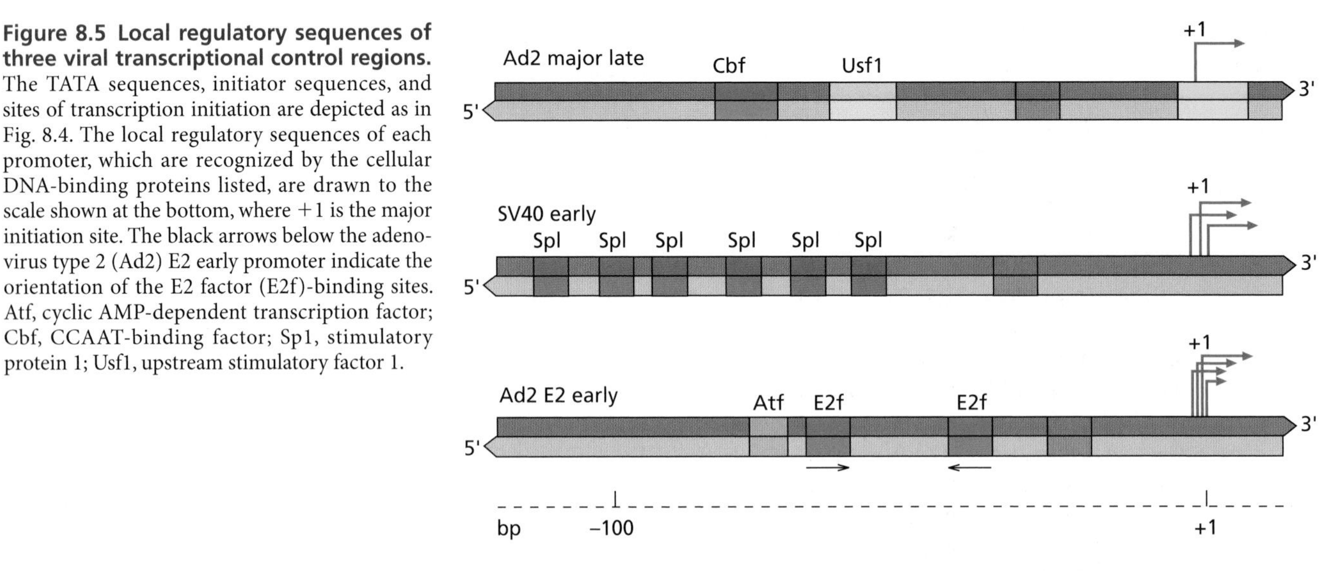

Figure 8.5 Local regulatory sequences of three viral transcriptional control regions. The TATA sequences, initiator sequences, and sites of transcription initiation are depicted as in Fig. 8.4. The local regulatory sequences of each promoter, which are recognized by the cellular DNA-binding proteins listed, are drawn to the scale shown at the bottom, where +1 is the major initiation site. The black arrows below the adenovirus type 2 (Ad2) E2 early promoter indicate the orientation of the E2 factor (E2f)-binding sites. Atf, cyclic AMP-dependent transcription factor; Cbf, CCAAT-binding factor; Spl, stimulatory protein 1; Usf1, upstream stimulatory factor 1.

smaller sequence motifs recognized by DNA-binding proteins (Fig. 8.6). The DNA-binding proteins that interact with this viral enhancer are differentially produced in different cell types. For example, nuclear factor κb (Nf-κb) and certain members of the octamer-binding protein (Oct) family are enriched in cells of lymphoid origin, and their binding sites are necessary for enhancer activity in these cells. Other elements of the enhancer, such as the activator protein 1 (Ap1)-binding sites, confer responsiveness to cellular signaling pathways. This constellation of enhancer elements ensures transcription of the viral early gene and initiation of the viral infectious cycle in many different cellular environments. This property is exhibited by several other viral enhancers, including those present in adenoviral and herpesviral genomes and the proviral DNA of avian retroviruses. In contrast, some viral templates for RNA polymerase II transcription contain enhancers that are active only in a specific cell type, only in the presence of viral proteins, or only under particular metabolic conditions (Box 8.5).

The simian virus 40 enhancer is located within 200 bp of the transcription initiation site, but enhancers of cellular genes are typically found thousands or tens of thousands of base pairs up- or downstream of the promoters that they regulate. The most popular model of the mechanism by which these sequences exert remote control of transcription, the DNA-looping model, invokes interactions among enhancer-bound proteins and the transcriptional components assembled at the promoter, with the intervening DNA looped out. Compelling evidence in favor of this model has been collected by using the simian virus 40 enhancer (Box 8.6). These regulatory sequences can also facilitate access of the transcriptional machinery to chromatin templates. For example, the simian virus 40 enhancer contains DNA sequences that induce formation of a nucleosome-free region of the viral genome in infected cells. Enhancers can,

BOX 8.4

TERMINOLOGY

The idiosyncratic nomenclature for sequence-specific DNA-binding proteins that regulate transcription

When proteins that bind to specific promoter sequences to regulate transcription by RNA polymerase II were first identified, no rules for naming mammalian proteins (or the genes encoding them) were in place. Consequently, the names given by individual investigators were based on different properties of the protein.

- Some names indicate the function of the regulator, e.g., the glucocortoid receptor (Gr) and serum response factor (Srf).
- Some names indicate the promoter sequence to which the protein binds, e.g., cyclic AMP response element (CRE)-binding protein (Creb) and octamer-binding protein 1 (Oct-1).
- Some names are based on the promoter in which binding sites for the regulator were first identified, e.g., adenovirus E2 transcription factor (E2f).
- Some names report some very general property of the regulator, e.g., stimulatory protein 1 (Sp1 [the first sequence-specific activator to be identified]), upstream stimulatory factor (Usf), and host cell factor (Hcf).

Such inconsistency, coupled with the universal use of acronyms, can mystify rather than inform: the historical origins of the names of transcriptional regulators are not known to most readers. The subsequent recognition that many "factors" are members of families of closely related proteins compounds such difficulties.

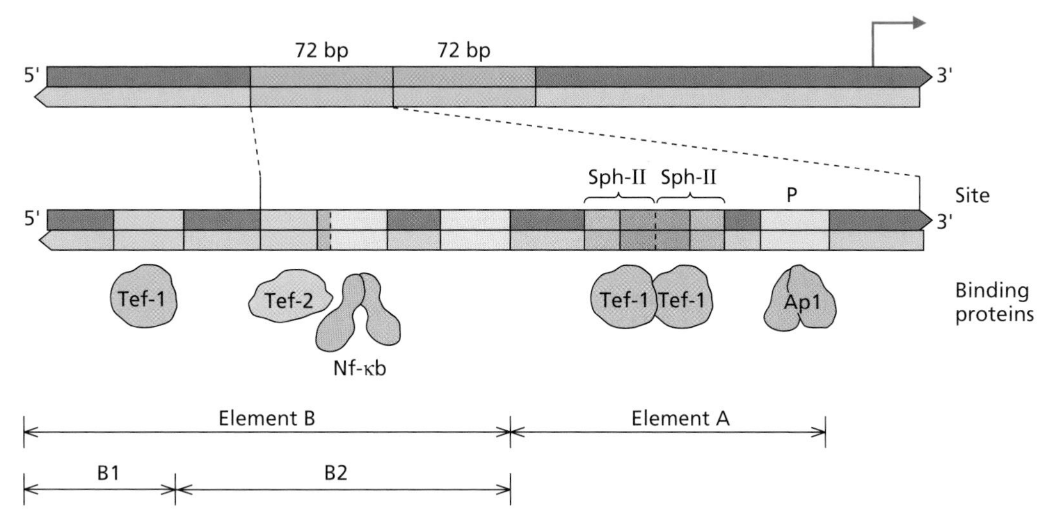

Figure 8.6 Organization of the archetypal simian virus 40 enhancer. The positions of the 72-bp repeat region containing the enhancer elements are shown relative to the early promoter at the top. Shown to scale below are functional DNA sequence units of the early promoter-distal 72-bp repeat and its 5′ flanking sequence, which forms part of enhancer element B, and the proteins that bind to them. All the protein-binding sites shown between the expansion lines are repeated in the promoter-proximal 72-bp repeat. The complete enhancer contains one copy of the enhancer element B1 and two directly repeated copies of the enhancer elements B2 and A. Some enhancer elements are built from repeated binding sites for a single sequence-specific protein. For example, cooperative binding of transcriptional enhancer factor (Tef)-1 to the two Sph-II sequences forms a functional enhancer element. Such cooperative binding renders enhancer activity sensitive to small changes in the concentration of a single protein. A second class of enhancer elements comprises sequences bound by two different proteins, as illustrated by the sequences bound by Tef-1 and Tef-2: binding is not cooperative, but these proteins interact once bound to DNA to form an active enhancer element. Ap1, activator protein 1.

BOX 8.5

DISCUSSION

Host cell metabolism can regulate viral enhancers

The hepatitis B virus genome contains two enhancers (I and II) that control transcription of viral genes (see figure). Enhancer I is bound by several ubiquitous transcriptional activators, as well as by activators that are present only in hepatocytes or enriched in these cells. Activation of enhancer II, which controls synthesis of the pregenome RNA and pre-C mRNA (Appendix, Fig. 11), requires the prior function of enhancer I. Enhancer II is recognized by multiple liver-enriched transcriptional activators including hepatocyte nuclear factor (Hnf) 3 family members, Hnf4, and the farnesol X and peroxisome proliferator-activated receptors (Fxr and Ppar-α, respectively).

The constellation of hepatocyte-specific or -enriched proteins that confer activity upon enhancers I and II accounts for the tropism of hepatitis B virus for the liver. However, this organ is a major metabolic hub: its many metabolic functions include synthesis of glucose in response to low concentrations of this sugar in the blood, synthesis of cholesterol and bile salts, and deamination of amino acids. Several of the proteins that govern hepatitis B virus transcription are also important regulators of metabolism, and hence sensitive to the metabolic status of the host. For example, Ppar-α is activated under conditions of fasting or starvation to promote gluconeogenesis and capture of energy by fatty acid catabolism. Consequently, this metabolic state also stimulates transcription from hepatitis B virus promoters and viral reproduction. Poor nutritional status may therefore directly promote disease caused by hepatitis B virus.

Bar-Yishay I, Shaul Y, Shlomai A. 2011. Hepatocyte metabolic signalling pathways and regulation of hepatitis B virus expression. *Liver Int* **31:**282–290.

Shlomai A, Paran N, Shaul Y. 2006. PGC-1α controls hepatitis B virus through nutritional signals. *Proc Natl Acad Sci U S A* **103:**16003–16008.

The segment of the hepatitis B virus genome containing enhancers I and II, the coding sequence for protein X, and the 59 end of the pre-C coding sequence are shown to scale. Hepatocyte-specific or -enriched proteins that bind the enhancers are listed below, as are metabolic signals that induce their activity (green arrows).

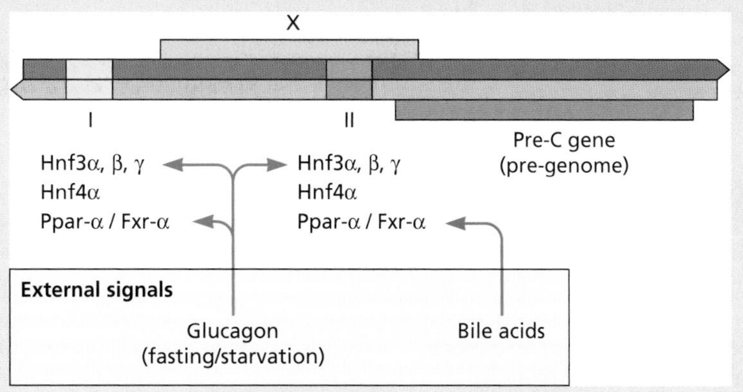

BOX 8.6

EXPERIMENTS

Mechanisms of enhancer action

(A) The DNA-looping model postulates that proteins bound to a distant enhancer (orange), here shown upstream of a gene, interact directly with the components of the transcription initiation complex, with the intervening DNA looped out. Such interactions might stabilize the initiation complex and therefore stimulate transcription. **(B)** An enhancer noncovalently linked to a promoter via a protein bridge is functional. When placed upstream of the rabbit β-globin gene promoter in a circular plasmid, the simian virus 40 enhancer stimulates specific transcription *in vitro* by a factor of 100. In the experiment summarized here, the enhancer and promoter were separated by restriction endonuclease cleavage. Under this condition, the enhancer cannot stimulate transcription. Biotin was added to the ends of each DNA fragment by incorporation of biotinylated UTP. Biotin binds the protein streptavidin noncovalently, but with extremely high affinity (K_d, 10^{-15} M). Because streptavidin can bind four molecules of biotin, its addition to the biotinylated DNA fragments allows formation of a noncovalent protein "bridge" linking the enhancer and the promoter. Under these conditions, the viral enhancer stimulates *in vitro* transcription almost as efficiently as when present in the same DNA molecule, as summarized in the column on the right. This result indicated that an enhancer can stimulate transcription when present in a separate DNA molecule (i.e., in *trans*) and ruled out models in which enhancers are proposed to serve as entry sites for RNA polymerase II. The results of this experiment are therefore consistent with the looping model shown in panel A.

Müller HP, Sogo JM, Schaffner W. 1989. An enhancer stimulates transcription in *trans* when attached to the promoter via a protein bridge. *Cell* **58**:767–777.

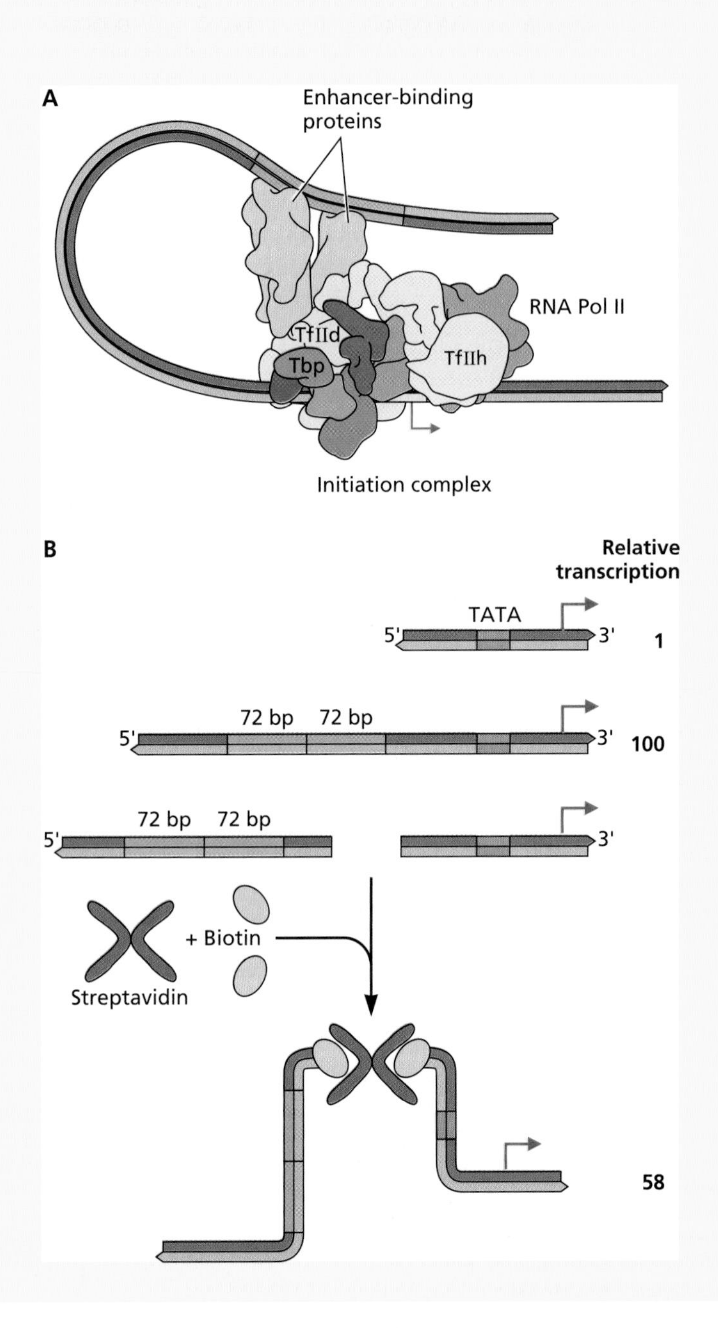

therefore, stimulate RNA polymerase II transcription by multiple molecular mechanisms. The primary effect of these mechanisms is to increase the probability that the gene to which an enhancer is linked will be transcribed.

Common Properties of Proteins That Regulate Transcription

Cellular, sequence-specific transcriptional regulators play pivotal roles in expression of viral genes. However, the genomes of many viruses also encode additional regulatory proteins. The cellular and viral DNA-binding proteins necessary for transcription from viral DNA templates share a number of common properties. Their most characteristic feature is modular organization: they are built from discrete structural and functional domains. The basic modules are a DNA-binding domain and an activation domain, which function as independent units. Other common properties include binding to DNA as dimers (Fig. 8.7).

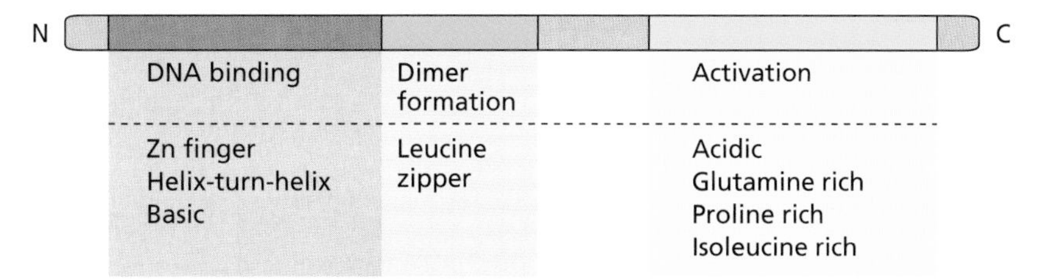

Figure 8.7 Modular organization of sequence-specific transcriptional activators. Common functional domains of eukaryotic transcriptional regulators are shown at the top, with some of the types of each domain listed below. DNA-binding and activation domains are defined by their structure (e.g., Zn finger or helix-turn-helix) and chemical makeup (e.g., acidic, glutamine rich), respectively. Transcriptional activators are often more complex than illustrated here. They can contain two activation domains, as well as regulatory domains, such as ligand-binding domains, and the various domains may be located at different positions with respect to the N and C termini of the protein.

Regulation of transcription by sequence-specific DNA-binding proteins usually requires additional proteins termed **coactivators** or **corepressors**. In general, these proteins cannot bind specifically to DNA, nor can they modulate transcription on their own. However, once recruited to a promoter by interaction with a DNA-bound protein, they dramatically augment (or damp) transcriptional responses. Coactivators can cooperate with multiple, sequence-specific activators and stimulate transcription from many promoters. A common property of many coregulators is their ability to alter the structure of nucleosomal templates, including viral templates, either directly or by interaction with appropriate enzymes. Several coactivators are histone acetyltransferases that catalyze the addition of acetyl groups to specific lysine residues in histones. This class includes p300, which was first identified by virtue of its interaction with adenoviral E1A proteins. Such histone acetyltransferases, and the deacetylases associated with corepressors, help establish the patterns of histone posttranscriptional modifications that distinguish transcriptionally active chromatin (Box 8.7). A second class of coactivators, exemplified by members of the Swi/Snf family, contain ATP-dependent chromatin-remodeling enzymes that alter the way in which DNA is bound to nucleosomes. It is thought that the coordinated action of these two classes of enzymes makes nucleosomal DNA accessible for both transcription initiation and elongation.

The ability of the RNA polymerase II system to mediate many patterns of transcription stems, in part, from the variety in both the nature of core promoters and the constellations of sequence-specific proteins and coactivators that govern their activity. Equally important is the power of the transcriptional machinery to integrate signals from multiple, promoter-bound regulators. This machinery must also be able respond to environmental cues, such as those provided by circulating hormones or growth factors. The proteins that control transcription are therefore frequently regulated by mechanisms that govern their activity, availability, or intracellular concentration. These mechanisms include modulation of the phosphorylation (or other modification) of specific amino acids, which can determine how well a protein binds to DNA, its **oligomerization** state, or the properties of its regulatory domain(s). In some cases, the intracellular location of a sequence-specific DNA-binding protein, or its association with inhibitory proteins, is modulated. Autoregulation of expression of the genes encoding transcriptional regulators is also common. This brief summary illustrates the varied repertoire of mechanisms available for regulation of transcription of viral templates by RNA polymerase II. Not surprisingly, virus-infected cells provide examples of all items on this menu, with the added zest of virus-specific mechanisms.

The Cellular Machinery Alone Can Transcribe Viral DNA Templates

In cells infected by many retroviruses, the components of the cellular transcriptional machinery described in the previous section complete the viral transcriptional program without the assistance of **any** viral proteins. The proviral DNA created by reverse transcription of retroviral RNA genomes and integration comprises a single RNA polymerase II transcription unit organized into chromatin, exactly like the cellular templates for transcription. Its transcription therefore produces a single viral RNA, which serves as both the genome and the source of viral mRNA species. Because the genomes of these retroviruses do not encode transcriptional regulators, the rate at which proviral DNA is transcribed is determined by the constellation of cellular transcription proteins present in an infected cell. This rate may be influenced by the nature and growth state of the infected cell, as well as by the organization of cellular chromatin containing the proviral DNA. Nevertheless, transcription of viral genetic information can occur throughout the lifetime of the host cell, indeed even in descendants of the cell initially infected. This strategy for transcription of viral DNA is exemplified by avian sarcoma and leukosis viruses, such as Rous-associated viruses. The long terminal repeat (LTR) of

BOX 8.7

DISCUSSION

The histone code hypothesis

In eukaryotic cells, genomic DNA is organized and highly compacted by histones and many other proteins in chromatin. Transcriptionally active DNA is present in a less condensed form called **euchromatin**. It has been known for decades that the nucleosomal histones present in euchromatin are enriched in acetylated residues, and it is now clear that complex patterns of histone posttranslational modification govern the properties of chromatin.

Residues in the N-terminal tails of the four core histones of the nucleosome (H2A, H2B, H3, and H4) are subject to a variety of posttranslational modifications, including acetylation, methylation, phosphorylation, ubiquitinylation, and sumoylation of specific residues. Panel A of the figure shows the positions of some of the known modifications of this segment of histone H3. The large number of possible modifications of numerous residues results in a far greater number of combinations. For example, mass spectrometry has identified more than 200 combinations present in different molecules of human histone H3, just 2 of which are shown in panel B.

It was initially proposed that particular combinations of posttranscriptionally modified histones could identify transcriptionally active or inactive DNA. Consistent with this "histone code" hypothesis, some combinations are characteristic of transcriptionally active genes. Furthermore, modified amino acids carrying posttranslational modifications serve as recognition sites for proteins that further modify histones, modify DNA, remodel nucleosomes, or facilitate transcription by other mechanisms. Although the idea of a simple code of histone posttranslational modifications has great appeal, it is now clear that it may be more appropriate to consider this a complex "language": for example, the same modification can recruit either activators or repressors of transcription, probably depending on the cellular or local context. Furthermore, histone modifications are dynamic, changing, for example, during transcriptional elongation or from one transcriptional cycle to another.

Berger SL. 2007. The complex language of chromatin regulation during transcription. *Nature* **447:**407–412.

Garcia BA, Pesavento JJ, Mizzen CA, Kelleher NL. 2007. Pervasive combinational modification of histone H3 in human cells. *Nat Methods* **4:**487–489.

Jenuwein T, Allis CD. 2001. Translating the histone code. *Science* **293:**1074–1079.

Young NL, DiMaggio PA, Plazas-Mayorca MD, Baliban RC, Floudas CA, Garcia BA. 2009. High throughput characterization of combinatorial histone codes. *Mol Cell Proteomics* **8:**2266–2284.

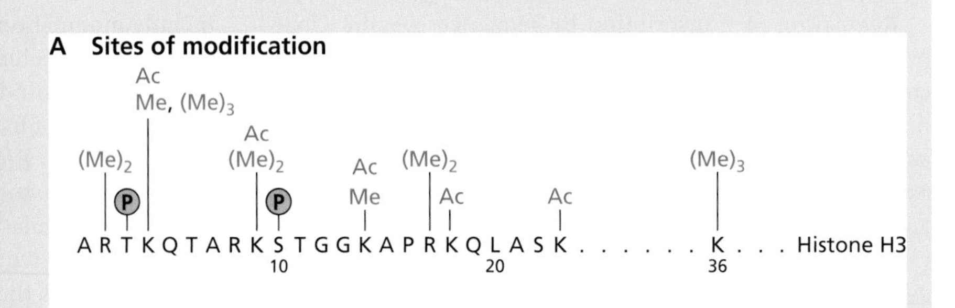

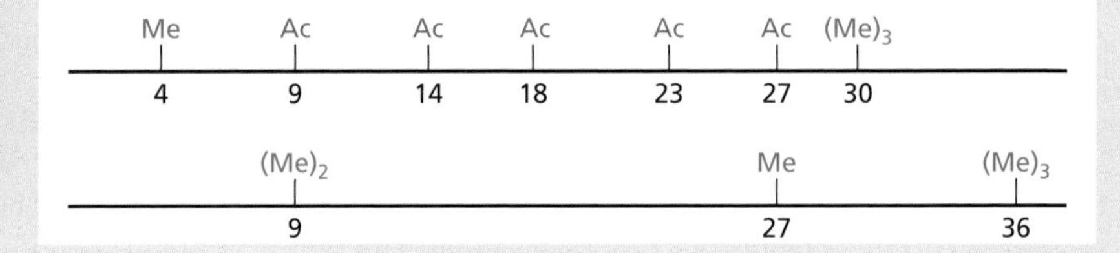

these proviral DNAs contains a compact enhancer located immediately upstream of the viral promoter (Fig. 8.8). The avian and mammalian serum response proteins that bind to the enhancer also recognize a specific sequence in the promoter. The other proteins that bind to this enhancer are all members of a family defined by a "leucine zipper" motif responsible for dimerization (Fig. 8.7).

The most remarkable property of the avian retroviral transcriptional control region is that it is active in many different cell types of both the natural avian hosts and mammals. This unusual feature can be explained by the widespread distribution of the cellular proteins that bind to it. Nevertheless, transcription of proviral DNA is not an inevitable consequence of integration. Rather, it can be blocked or impaired by specific cellular proteins that induce repressive histone (and DNA) modifications, and hence epigenetic silencing of proviral transcription (Box 8.8). As discussed in Volume II, Chapter 3, such inhibition of proviral transcription is but one example of intrinsic antiviral defense mechanisms.

Because the LTRs are direct repeats of one another (Fig. 8.8), transcription directed by the 3′ LTR extends into cellular DNA and cannot contribute to the expression of retroviral genetic information. In fact, the transcriptional control region of the 3′ LTR is normally inactivated by a process called **promoter occlusion**: the passage of transcribing complexes initiating at the 5′ LTR through the 3′ LTR prevents recognition of the latter by enhancer- and promoter-binding proteins. Occasionally, transcription from the 3′ LTR **does** occur, with profound consequences for the host cell (see Volume II, Chapter 6).

Absolute dependence on cellular components for the production of viral transcripts avoids the need to devote limited viral genetic capacity to transcriptional regulatory proteins. Nevertheless, such a strategy is the exception, not the rule.

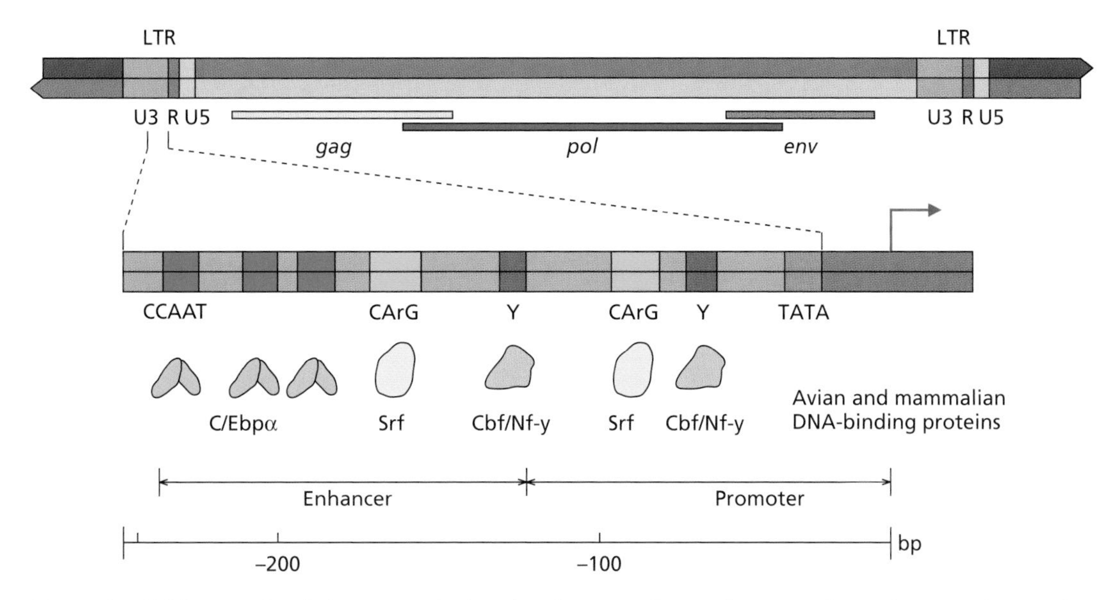

Figure 8.8 Widespread cellular transcriptional activators of an avian retrovirus. The proviral DNA of an avian leukosis virus is shown at the top. The enhancer and promoter present in the U3 regions of the LTRs are drawn to scale below. Each of the multiple CCAAT, CArG, and Y box sequences, which are required for maximally efficient transcription, is recognized by the proteins listed below, which are present in both avian and mammalian cells. Nf-y, nuclear transcription factor Y; Srf, serum response factor.

Viral Proteins That Govern Transcription of Viral DNA Templates

Patterns of Regulation

Transcription of many viral DNA templates by the RNA polymerase II machinery results in the synthesis of large quantities of viral transcripts (in some cases, more than 10^5 copies of individual mRNA species per cell) in relatively short periods. Such bursts of transcription are elicited by viral proteins that stimulate RNA polymerase II transcription and establish one of two kinds of regulatory circuits (Fig. 8.9). The first is a **positive autoregulatory loop**, epitomized by transcription of human immunodeficiency virus type 1 proviral DNA (Fig. 8.9A). A viral activating protein stimulates the rate of transcription but does not alter the complement of viral proteins made in infected cells. The second is a **transcriptional cascade**, in which different viral transcription units are activated in an ordered sequence (Fig. 8.9B). This mechanism, which ensures that different classes of viral proteins are made during different periods of the infectious cycle, is characteristic of viruses with DNA genomes. The participation of viral proteins confers a measure of control lacking when the transcriptional program is executed solely by cellular components. The following sections describe some well-studied examples of the regulatory circuits established by viral proteins.

The Human Immunodeficiency Virus Type 1 Tat Protein Autoregulates Transcription

The proteins of retroviruses with complex genomes are encoded in a single proviral transcription unit controlled by an LTR enhancer and promoter. However, in addition to the common structural proteins and enzymes, these genomes encode auxiliary proteins, including transcriptional regulators. Some of these proteins, such as the Tax protein of human T-lymphotropic virus type 1, resemble activators of other virus families, and stimulate transcription from a wide variety of viral and cellular promoters. Others, exemplified by the transactivator of transcription (Tat) of human immunodeficiency virus type 1, are unique: they recognize an RNA element in nascent transcripts.

In principle, the positive feedback loop that is established once a sufficient concentration of Tat has accumulated in an infected cell is simplicity itself (Fig. 8.9A). Cellular proteins initially direct transcription of the proviral DNA in infected cells at some basal rate: among the processed products of the primary viral transcript are the spliced mRNAs from which the Tat protein is synthesized; this protein is imported into the nucleus, where it stimulates transcription of the proviral template upon binding to its RNA recognition site in nascent viral transcripts. However, the molecular mechanisms that establish this autostimulatory loop are sophisticated and unusual. Their elucidation has been an important area of research, because the Tat protein is essential for virus propagation and represents a valid target for antiviral therapy.

Cellular Proteins Recognize the Human Immunodeficiency Virus Type 1 LTR

Cellular proteins that bind to the LTR enhancer and promoter proteins support a low rate of proviral transcription

BOX 8.8

EXPERIMENTS

Epigenetic silencing of integrated proviral DNAs

It is well established that expression of exogenous genes introduced into cells in culture via retroviral vectors can gradually become inhibited, and that human immunodeficiency virus type I proviral DNA is maintained in a transcriptionally latent state in resting T lymphocytes. These phenomena illustrate the fact that integration of a proviral DNA into the host cell genome does **not** guarantee transcription of viral genetic information. Such repression is mediated, at least in part, by epigenetic mechanisms that are important for silencing of expression of cellular genes during differentiation and development. These include the addition of repressive posttranslational modifications to nucleosomal histones (see Box 8.7) and methylation of cytosine in DNA to form 5^{Me}CpG. For example, avian sarcoma proviral DNAs are subjected to rapid epigenetic silencing in mammalian cells but not in natural avian host cells. This observation indicated that one or more mammalian proteins promote an antiviral defense that represses expression of avian proviral DNAs. One such candidate protein, human death domain-associated protein 6 (Daxx), was identified by virtue of its binding to the viral integrase in a yeast two-hybrid screen and in infected cells.

Subsequent studies exploited avian sarcoma viruses that carried a green fluorescent protein (GFP) reporter gene to facilitate analysis of proviral expression and repression. It was observed that

- Daxx was not required for early events in avian sarcoma virus replication, but viral reporter gene expression was increased significantly when synthesis of Daxx in human cells was inhibited by RNA interference (RNAi), and in murine $Daxx^{-/-}$ cells
- the histone deacetylases Hdac1 and Hdac2 were associated with viral DNA in Daxx-producing but not in $Daxx^{-/-}$ cells, as assessed by chromatin immunoprecipitation
- in populations of human cells in which proviral LTR promoters were silenced and heavily methylated, knockdown of Daxx by RNAi induced expression of GFP reporter genes (panel A), as did inhibition of synthesis of specific DNA methyltransferases (Dnmts)
- in such silenced cells, Daxx and Dnmts were associated with one another (coimmunoprecipitation) and with proviral promoters (chromatin immunoprecipitation), and Daxx knockdown also substantially reduced methylation of proviral promoter DNA

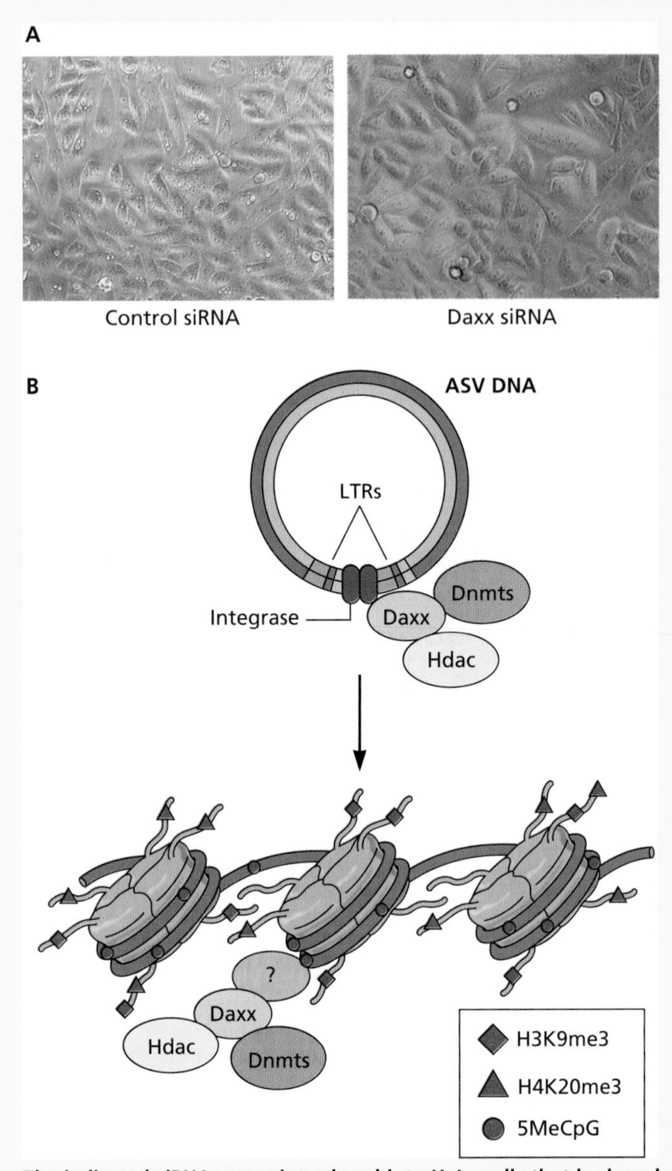

The indicated siRNAs were introduced into HeLa cells that harbored avian sarcoma viral (SV) DNA that carried a silent GFP reporter gene. (A) Images of transfected cells were taken using a fluorescent microscope 96 h thereafter. siRNA-mediated knockdown of cellular Daxx protein resulted in release of epigenetic gene silencing, as shown by reactivation of expression of the reporter gene. Courtesy of Andrey Poleshko and A. Skalka, Fox Chase Cancer Center. **(B)** Models for the initiation and maintenance of retroviral silencing by Daxx via recruitment of Hdacs and Dnmts (top and bottom, respectively). Adapted from N. Shalginskikh et al., *J Virol* **87:** 2137–2150, 2013, with permission.

Based on these and other observations, it has been proposed that Daxx associates with the viral integrase prior to proviral integration to recruit Dnmts and enzymes that catalyze formation of repressive chromatin (panel B).

Transcription of human immunodeficiency virus type 1 proviruses is also silenced in resting CD4⁺ T lymphocytes, a reservoir of "invisible" infected cells that complicates treatment (see Volume II, Chapter 7). As discussed in the text, the sequestration in the cytoplasm of Nf-κb, which is necessary for efficient initiation of transcription, contributes to the silencing of LTR-dependent transcription, as does DNA methylation. The establishment of such latent human immunodeficiency virus proviruses has also been studied using viruses carrying genes encoding fluorescent reporter proteins. The LTRs of latent proviral DNAs were shown to be associated with histone deacetylases and nucleosomes carrying repressive posttranslational modifications. Proviral reporter gene expression upon activation of the T cells correlated not only with reversal of such repressive chromatin modification and concomitant association of RNA polymerase II with the LTR, but also with substantial increases in the nuclear concentration of p-Tefb.

Greger JG, Katz RA, Ishov AM, Maul GG, Skalka AM. 2005. The cellular protein Daxx interacts with avian sarcoma virus integrase and viral DNA to repress viral transcription. *J Virol* **79:**4610–4618.

Shalginskikh N, Poleshko A, Skalka AM, Katz RA. 2013. Retroviral DNA methylation and epigenetic repression are mediated by the antiviral host protein Daxx. *J Virol* **87:**2137–2150.

Tyagi M, Pearson RJ, Karn J. 2010. Establishment of HIV latency in primary CD4⁺ cells is due to epigenetic transcriptional silencing and P-TEFb restriction. *J Virol* **84:**6425–6437.

before Tat is made in infected cells. In contrast to avian retroviruses, human immunodeficiency virus type 1 propagates efficiently in only a few cell types, notably CD4⁺ T lymphocytes and cells of the macrophage/monocyte lineage. Viral reproduction (i.e., transcription) in infected T cells in culture is stimulated by T cell growth factors, indicating that viral transcription requires cellular components available only in such stimulated T cells. Indeed, the failure of the virus to propagate efficiently in unstimulated T cells correlates with the absence of active forms of particular enhancer-binding proteins. The distribution of cellular enhancer-binding proteins is therefore an important determinant of the host range of retroviruses with both simple and complex genomes. However, the transcription of the provirus of retroviruses with

Figure 8.9 Mechanisms of stimulation of transcription by viral proteins. Cellular transcriptional components acting alone transcribe the viral gene encoding protein X. Once synthesized and returned to the nucleus, viral protein X can stimulate transcription either of the same transcription unit **(A)** or of a different one **(B)**. In either case, viral protein X acts in concert with components of the cellular transcriptional machinery.

A

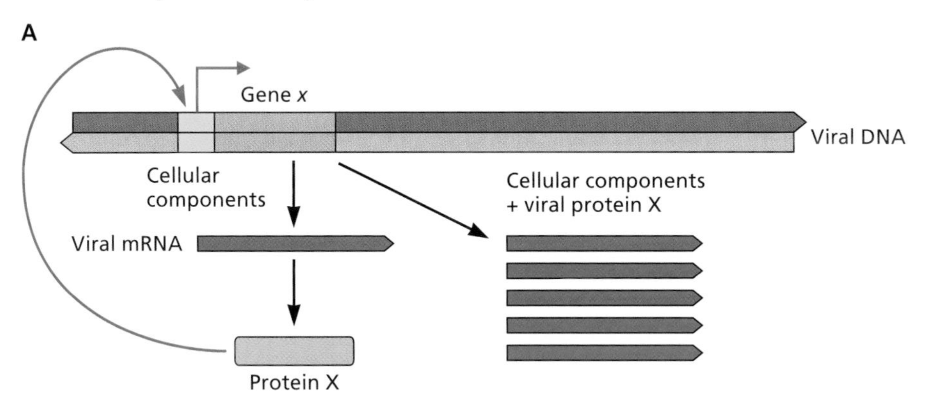

B

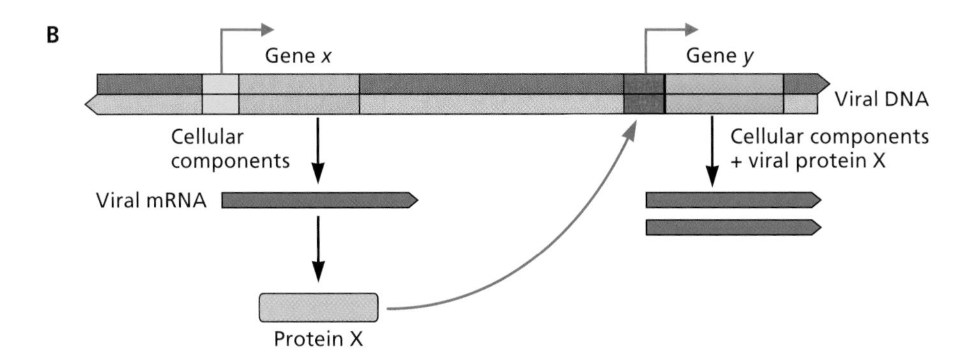

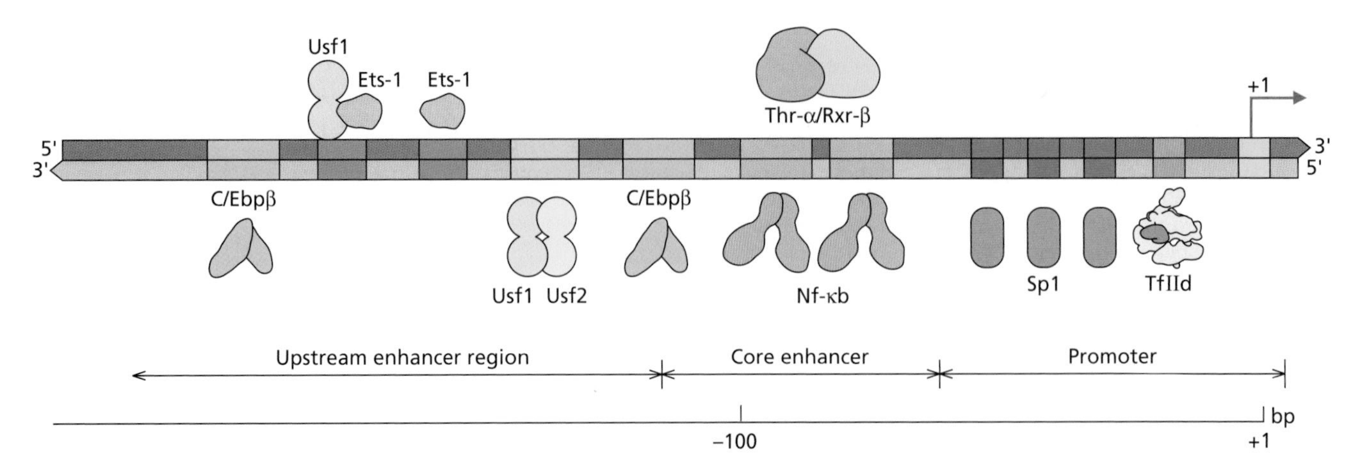

Figure 8.10 Cell-type-specific regulators bind to the transcriptional control region of human immunodeficiency virus type 1. The organization of the U3 region of the proviral LTR is shown to scale, with proteins that bind to promoter or enhancer sequences indicated above or below the DNA. Activation of C/Ebpβ (also known as NfIL6) stimulates viral gene expression in macrophages, as does T3rα-1/Rxr-β, while the other enhancer-binding proteins shown stimulate transcription in T cells. Not all the binding sites shown are well conserved in different viral isolates. Ets-1, protein C-ets1; Rxr-β, retinoic acid receptor-β; Thrα, thyroid hormone receptor-α.

simple genomes, such as the avian retroviruses, depends on proteins that are widely distributed, whereas human immunodeficiency virus type 1 transcription requires proteins that are found in only a few cell types or active only under certain conditions.

Within the human immunodeficiency virus type 1 LTR, the promoter is immediately preceded by two important regulatory regions (Fig. 8.10), termed the core and upstream enhancers, that are necessary for efficient viral transcription in both peripheral blood lymphocytes and certain T cell lines. Both the core and the upstream enhancers are densely packed with binding sites for cellular proteins, many of which are enriched in the types of cell in which the virus can reproduce. These proteins were typically identified because they stimulated transcription from the LTR in transient-expression assays. Because these assays do not reproduce physiological conditions (Box 8.9), a positive result establishes only that a certain protein **can** stimulate transcription, not that it normally does so. However, it is now clear that many of these proteins **do** stimulate viral transcription and replication in the types of cell in which human immunodeficiency virus type I can reproduce, for example, Ets-1 and C/Ebpβ in T lymphocytes and monocytes/macrophages, respectively.

Regulation of viral transcription by cellular pathways is exemplified by the critical role of the transcriptional activator Nf-κb in replication of human immunodeficiency virus type 1 in T cells (Fig. 8.11). Unstimulated T cells display no Nf-κb activity, because the protein is retained in an inactive form in the cytoplasm by binding of inhibitory proteins of the Iκb family. Growth factors that activate T cells trigger signal transduction cascades that lead to phosphorylation and subsequent degradation of these inhibitors by the cytoplasmic multiprotease complex (the **proteasome**). Consequently, Nf-κb is freed for transit to the nucleus, where it can bind to its recognition sites within the viral LTR core enhancer (Fig. 8.10). This pathway can account for the induction of human immunodeficiency virus type 1 transcription observed when T cells are stimulated. The severe, or complete, inhibition of virus reproduction (transcription) in normal human $CD4^+$ T lymphocytes caused by mutations in the Nf-κb-binding sites emphasizes the importance of activation of this cellular protein in the infectious cycle of the virus. Nevertheless, Nf-κb and the other cellular proteins that act via LTR enhancer- or promoter-binding sites do not support efficient expression of viral genes: this process depends on synthesis of the viral Tat protein, as discussed in the next section.

A characteristic feature of human immunodeficiency virus type 1 infection of individuals is a period of clinical latency in which few symptoms are manifested. The provirus can be considered dormant in infected cells that lack the constellation of active enhancer-binding proteins necessary to allow synthesis of small quantities of Tat mRNA and protein, and hence induction of the positive autoregulatory loop (Fig. 8.9A). Nevertheless, during clinical latency, virus is produced continuously in cells that contain the necessary cellular proteins, because the positive autoregulatory circuit is triggered whenever infected cells can support LTR enhancer-dependent transcription of proviral DNA: symptoms develop when the patient's immune system can no longer maintain effective countermeasures (Volume II, Chapter 7).

BOX 8.9

WARNING

Caution: transient-expression assays do not reproduce conditions within virus-infected cells

Transient-expression assays (see figure) provide a powerful, efficient way to investigate regulation of transcription. Advantages include the following:

- simplicity and sensitivity of assays for reporter gene activity
- ready analysis of mutated promoters to identify DNA sequences needed for the action of the regulatory protein
- application with chimeric fusion proteins and synthetic promoters to avoid transcriptional responses due to endogenous cellular proteins
- simplification of complex regulatory circuits to focus on the activity of a single protein

Despite these advantages, transient-expression assays do not necessarily tell us how transcription is regulated in virus-infected cells, because they do not reproduce normal intracellular conditions. Important differences include the following:

- abnormally high concentrations of exogenous template DNA: concentrations of reporter genes as high as 10^6 copies per cell are not unusual. This value is significantly greater than even the maximal concentrations of viral DNA molecules attained toward the end of an infectious cycle, up to 10^3 and 10^4 copies/cells in the case of alphaherpesviruses and adenovirus, respectively
- abnormally high concentrations of the regulatory protein as a result of its deliberate overproduction
- the potential for spurious interactions of the viral protein with template, or cellular components, because of these high concentrations of template and protein
- the absence of viral components that might negatively or positively modulate the activity of the protein under study

The last three caveats apply to any experiment in which a viral protein is overproduced, for example, for investigation of its interactions with other proteins.

Because of their inherent limitations, models of regulation of viral transcription based on results obtained by exploiting the advantages of transient-expression assays require validation in infected cells.

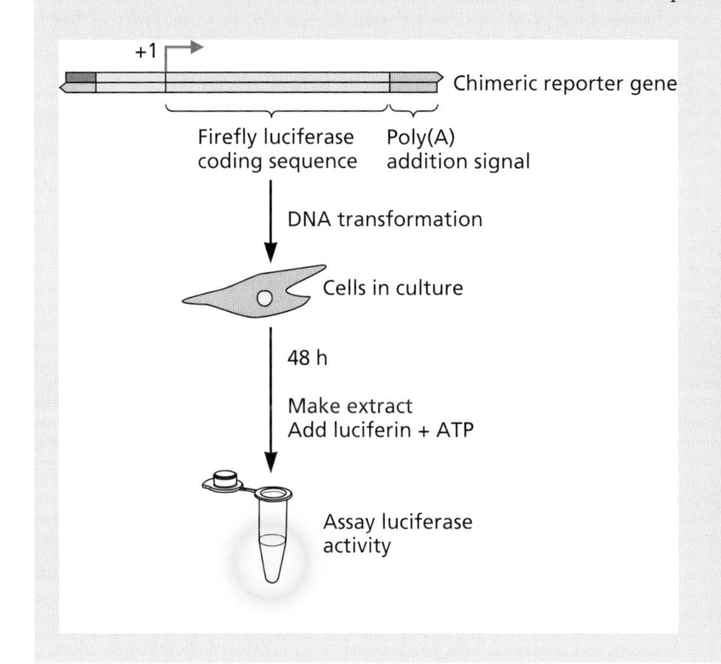

The transient-expression assay. A segment of DNA containing the transcriptional control region of interest (yellow) is ligated to the coding sequence (orange) of an enzyme not synthesized in the recipient cells to be used, luciferase in this example, and RNA-processing signals such as those specifying polyadenylation (green box). Plasmids containing such chimeric reporter genes are introduced into cells in culture by any one of several methods, including electroporation and incubation with synthetic vesicles containing the plasmid DNA. Within a cell that takes up the plasmid, the DNA enters the nucleus, where the transcriptional control region directs transcription of chimeric RNA. The RNA is processed, exported from the nucleus, and translated by cytoplasmic polyribosomes. The activity of the luciferase enzyme is then assayed, generally 48 h after introduction of the reporter gene. This indirect measure of transcription assumes that it is **only** the activity of the transcriptional control region that determines the concentration of the enzyme. Alternatively, the concentration of the chimeric reporter RNA can be measured.

The Tat Protein Regulates Transcription by Unique Mechanisms

Tat recognizes an RNA structure. Stimulation of human immunodeficiency virus type 1 transcription by Tat requires an LTR sequence, termed the transactivation response (TAR) element, which lies within the transcription unit (Fig. 8.12A). The observation that mutations that disrupted the predicted secondary structure of TAR RNA inhibited Tat-dependent transcription suggested that the TAR element is recognized as RNA. Indeed, the Tat protein binds specifically to a trinucleotide bulge and adjacent base pairs in the stem of the TAR RNA stem-loop structure (Fig. 8.12B). Binding of Tat to this region of TAR induces a local rearrangement in the RNA, resulting in formation of a more stable, compact, and energetically favorable structure (Fig. 8.12D). Recognition of a viral transcriptional control sequence as RNA remains unique to Tat proteins.

Tat stimulates transcriptional elongation. Binding of Tat to TAR RNA stimulates production of viral RNA by as much

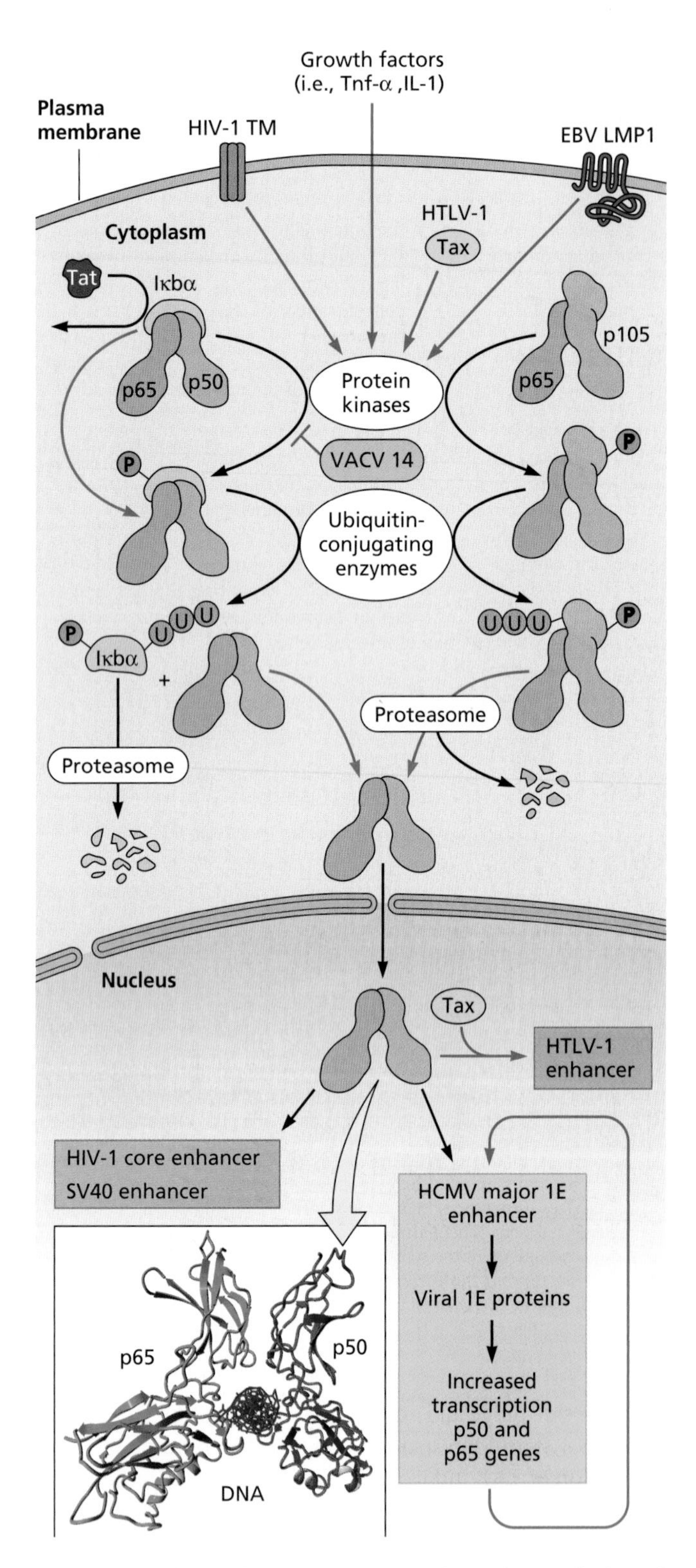

Figure 8.11 The cellular transcriptional regulator Nf-κb and its participation in viral transcription. The members of the Nf-κb–cRel protein family (p50-p65) are defined by the presence of the Rel homology region, which contains DNA-binding and dimerization motifs, and a nuclear localization signal. The p65 (Rel) protein of the p50-p65 heterodimer (left) also contains an acidic activation domain at its C terminus. p50 is synthesized as an inactive precursor, p105 (right). The p105-p65 heterodimer is one of two forms of inactive Nf-κb found in the cytoplasm (e.g., of unstimulated T cells). The second consists of mature p50-p65 heterodimers associated with an inhibitory protein such as Iκbα (left), which blocks the nuclear localization signals of the p50 and p65 proteins. The C-terminal segment of p105 functions like Iκb, with which it shares sequences, to block nuclear localization signals and retain this heterodimer in the cytoplasm. Exposure of the cells to any of several growth factors results in activation (green arrows) of protein kinases that phosphorylate specific residues of Iκb or p105. Upon phosphorylation, Iκb dissociates and is recognized by the system of enzymes that adds branched chains of ubiquitin (Ub) to proteins, a modification that targets them for degradation by the proteasome. Specific p105 cleavage by the proteasome also produces the p50-p65 dimer. Unencumbered Nf-κb dimers produced by either mechanism can translocate to the nucleus, because nuclear localization signals are now accessible. In the nucleus of uninfected cells, Nf-κb binds to specific promoter sequences to stimulate transcription via the p65 activation domain. Viral transcriptional control regions to which Nf-κb binds and some viral proteins that induce activation (green arrows) of Nf-κb are indicated. The X-ray crystal structure of a p50-p65 heterodimer bound specifically to DNA is shown in the inset. The structure is viewed down the helical axis of DNA with the two strands in blue and with the p50 and p65 subunits in red and green, respectively. The dimer makes extensive contact with DNA via protein loops. EBV, Epstein-Barr virus, HCMV, human cytomegalovirus; HIV-1, human immunodeficiency virus type 1; HTLV-1, human T-lymphotropic virus type 1; SV40, simian virus 40; VACV, vaccinia virus. NDB ID: RDR0333 F. E. Chen, D.D. Humag, Y.Q. Chen and G. Ghosh *Nature* **391:**410–413,

as 100-fold. In contrast to many cellular and viral proteins that stimulate transcription, the Tat protein has little effect on initiation. Rather, it greatly improves elongation. Complexes that initiate transcription in the absence of Tat elongate poorly, and many terminate transcription within 60 bp of the initiation site (Fig. 8.13A). The Tat protein overcomes such poor **processivity** of elongating complexes, thereby allowing efficient production of full-length viral transcripts. Consequently, in the absence of Tat, full-length transcripts of proviral DNA account for no more than 10% of the total. This property resolves the paradox of why the human immunodeficiency virus type 1 LTR enhancer and promoter are not sufficient to support efficient viral RNA synthesis.

How Tat stimulates transcriptional elongation. A search for cellular proteins that stimulate viral transcription when bound to the N-terminal region of Tat (Fig. 8.12C) identified the human Ser/Thr kinase p-Tefb (positive-acting transcription factor b), which was known to stimulate elongation of cellular transcripts. This cellular protein is essential for Tat-dependent stimulation of processive viral transcription both *in vitro* and in infected cells. One subunit of the p-Tefb heterodimer is a **cyclin**, cyclin T. Cyclins are so named because members of the family accumulate during specific periods of the cell cycle. Cyclin T regulates the activity of the second subunit of p-Tefb, cyclin-dependent kinase 9 (Cdk9). Tat and

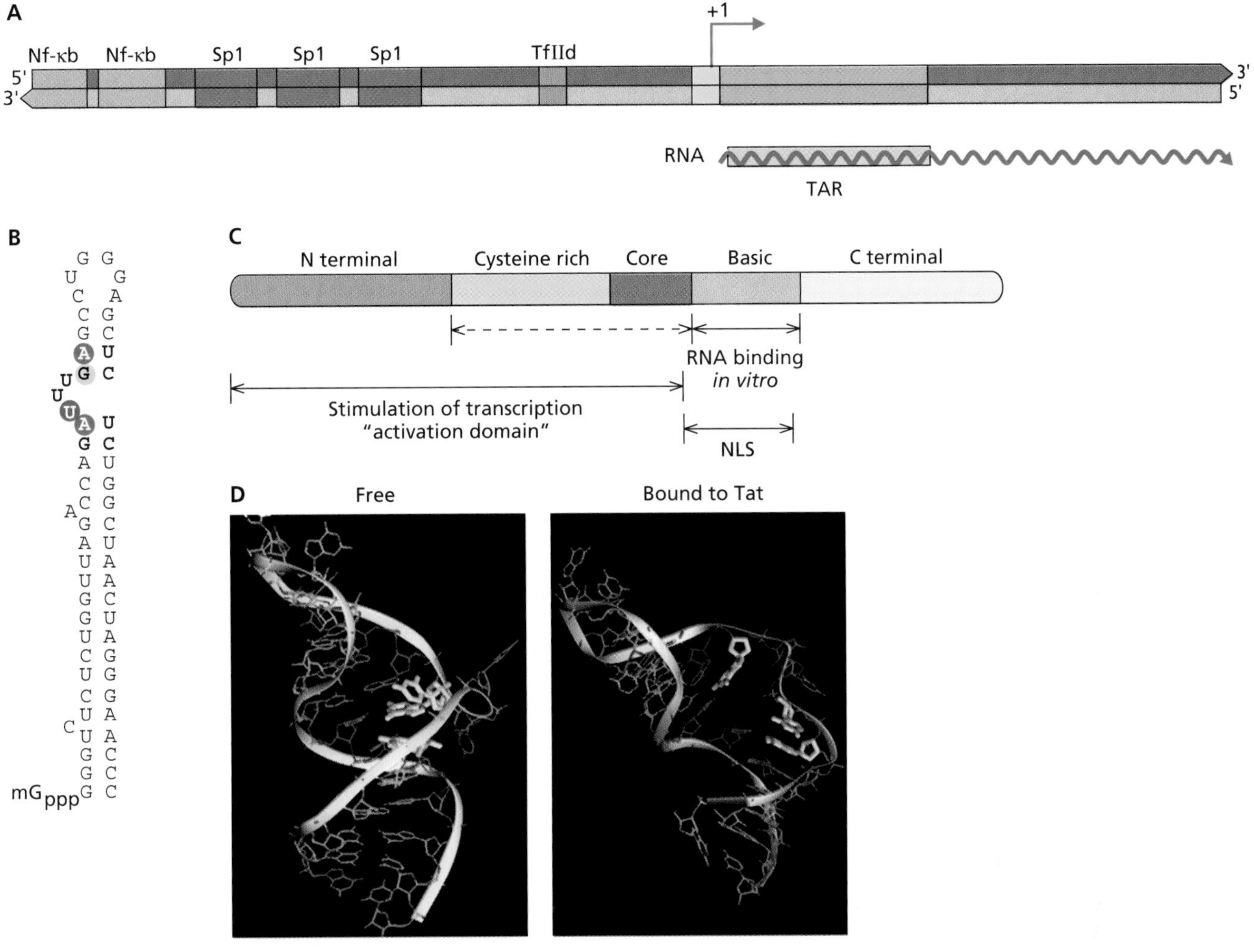

Figure 8.12 Human immunodeficiency virus type 1 TAR and the Tat protein. (A) The region of the viral genome spanning the site of transcription initiation is drawn to scale, with the core enhancer and promoter depicted as in Fig. 8.10. The DNA sequence lying just downstream of the initiation site (pink) negatively regulates transcription. Transcription of the proviral DNA produces nascent transcripts that contain the TAR sequence (tan box). **(B)** The TAR RNA hairpin extends from position +1 to position +59 in nascent viral RNA. Sequences important for recognition of TAR RNA by the Tat protein are colored. Optimal stimulation of transcription by Tat requires not only this binding site in TAR but also the terminal loop. **(C)** The Tat protein is made from several different, multiply spliced mRNAs (Appendix, Fig. 29B) and therefore varies in length at its C terminus. The regions of the protein are named for the nature of their sequences (basic, cysteine rich) or greatest conservation among lentiviral Tat proteins (core). Experiments with fusion proteins containing various segments of Tat and a heterologous RNA-binding domain identified the N-terminal segment indicated as sufficient to stimulate transcription. The basic region, which contains the nuclear localization signal (NLS), can bind specifically to RNA containing the bulge characteristic of TAR RNA. However, high-affinity binding, effective discrimination of wild-type TAR from mutated sequences *in vitro*, and RNA-dependent stimulation of transcription within cells require additional N-terminal regions of the protein, shown by the dashed arrow. **(D)** Major groove views of structures of a free TAR RNA corresponding to the apical stem and loop regions (but with a truncated stem) (left) and of the same RNA when bound to the Tat peptide (right) were determined by nuclear magnetic resonance methods. The bases shown in yellow are A22, U23, and G26, which are colored red, blue, and yellow, respectively, in panel B. Note the energetically favorable change in the conformation of the trinucleotide bulge region to a more stable, compact structure on binding of the Tat peptide. From F. Aboul-ela et al., *J Mol Biol* **253**:313–332, 1995, and F. Aboul-ela et al., *Nucleic Acids Res* **24**:3974–3981, 1996, with permission. Courtesy of M. Afshar, RiboTargets, and J. Karn, MRC Laboratory of Molecular Biology.

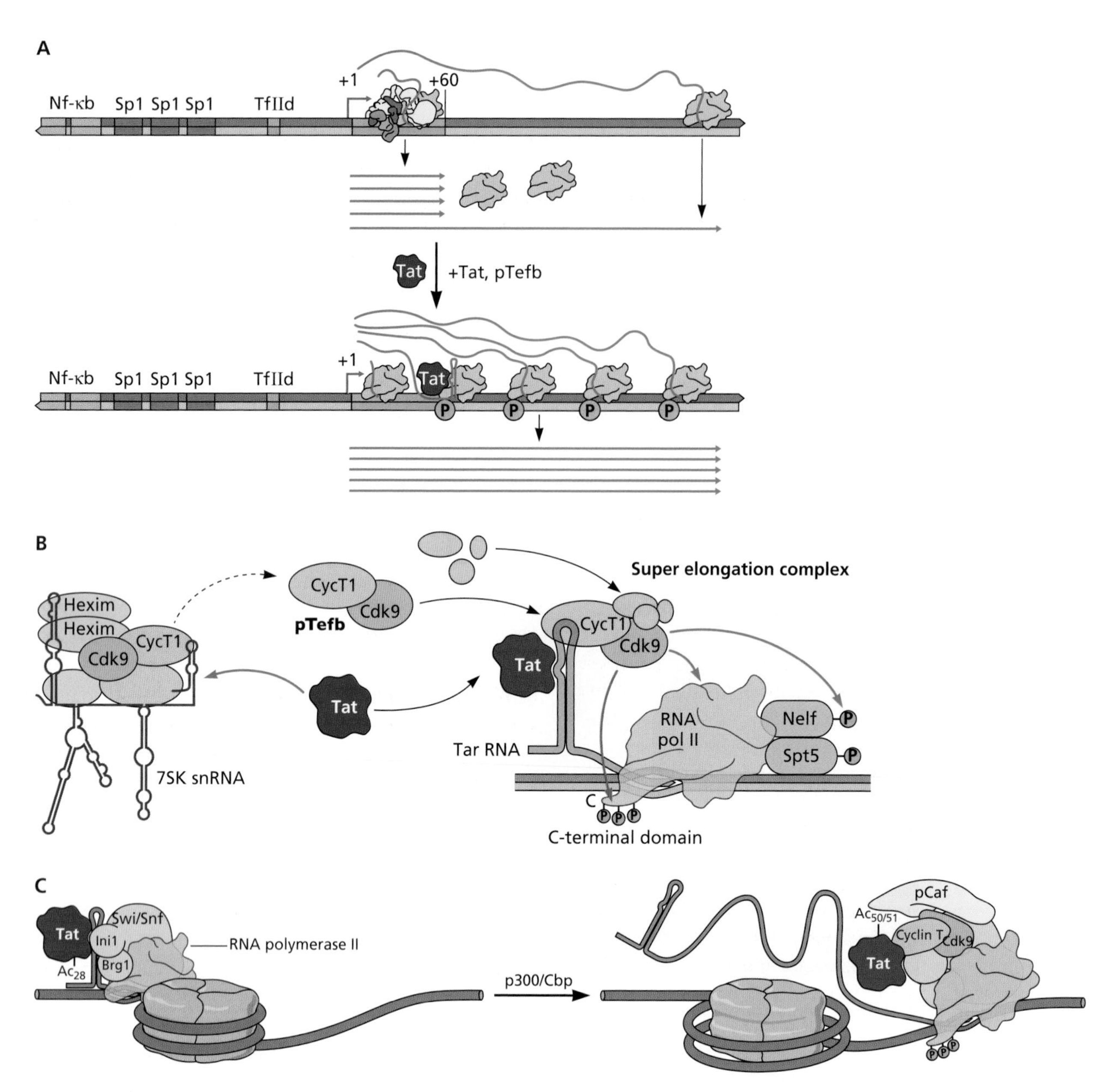

Figure 8.13 Mechanisms of stimulation of transcription by the human immunodeficiency virus type 1 Tat protein. (A) Model for the stimulation of elongation. The regulatory sequences flanking the site of initiation of transcription are depicted as in Fig. 8.10. In the absence of Tat, transcription complexes are poorly processive, and the great majority (9 of 10) terminate within 60 bp of the initiation site, releasing transcription components and short transcripts. Production of the Tat protein upon translation of mRNAs spliced from rare, full-length transcripts and its recruitment of p-Tefb and other regulators of elongation to nascent RNA allow transcriptional complexes to pass through the elongation block and synthesis of full-length viral RNA. **(B)** Cooperative binding to TAR of Tat and p-Tefb (via its cyclin T subunit) leads to phosphorylation (P) of the C-terminal domain of the largest subunit of RNA polymerase II by the Cdk9 kinase subunit of p-Tefb. This enzyme also phosphorylates and inactivates negative regulators of transcriptional elongation (e.g., transcription elongation factor [Spt5] and negative elongator factor complex [Nelf]). Positive regulators of elongation, such as RNA polymerase II elongation factor 2 (Ell2), are also recruited to form a super elongation complex. The net result is that transcriptional complexes become competent to carry out highly processive transcription. A second function of Tat (left) is to increase the concentration of p-Tefb available to bind to TAR in infected cells by inducing dissociation of a 7SK small nuclear RNA (snRNA)-containing ribonucleoprotein that sequesters p-Tefb from the transcriptional machinery. Hexim, hexamethylene bis-acetamide-inducible protein. Adapted from M. Ott et al., *Cell Host Microbe* **10:**426–435, 2011, with permission. **(C)** Model for nucleosome remodeling. The initial transcript (green) of proviral DNA (blue line) is depicted with Tat bound to the TAR sequence. Acetylation of Tat at Lys28 is critical for high-affinity binding to TAR and p-Tefb. The nucleosome located a short distance downstream of the initiation site blocks transcriptional elongation, and nucleosome remodeling by the Swi/Snf complex is required for efficient elongation of transcription. Specific subunits of this remodeling complex (e.g., Ini-1 and Brg-1) bind to Tat, but only once this protein is acetylated on Lys50 and Lys51 by the histone acetyltransferases Crebbp/p300 and Gcn5. These modifications also induce dissocation of Tat from TAR, presumably by neutralizing positive charge in the RNA-binding region of the protein.

p-Tefb bind **cooperatively** to the TAR RNA stem-loop (that is, with higher affinity than either protein alone), and with greater specificity. This property is the result of the interaction of the cyclin T1 subunit of p-Tefb with nucleotides within the TAR RNA loop (Fig. 8.12B) that are not contacted by Tat but are nevertheless crucial for stimulation of transcription.

Assembly of the ternary complex containing TAR RNA, Tat, and p-Tefb promotes elongation of human immunodeficiency virus type 1 transcription in several different ways (Fig. 8.13). This process induces conformational changes that activate the Cdk9 kinase subunit of p-Tefb. Once associated with transcription complexes, the active kinase phosphorylates Ser residues within an unusual domain at the C terminus of the largest subunit of RNA polymerase II, which is hypophosphorylated when RNA polymerase II is present in preinitiation complexes (Fig 8.13B). These modifications are essential for Tat-dependent stimulation of elongation and hence complete proviral transcription. The Cdk9 kinase also modifies and inactivates negative regulators of elongation to facilitate release of paused transcriptional complexes (Fig. 8.13B). Furthermore, via its interaction with p-Tefb, Tat recruits additional proteins that increase the rate of elongation to establish a super elongation complex for viral transcription. The results of experiments in which p-Tefb was inhibited in infected cells, as well as genetic analyses, have established that p-Tefb is essential for Tat-dependent stimulation of viral transcription *in vivo*.

The Tat protein also facilitates human immunodeficiency virus transcription indirectly, by inducing release of p-Tefb from a complex in which it is sequestered in an inactive form (Fig. 8.13B). Such inhibition is mediated by Hexim1 (hexamethylene bis-acetamide-inducible protein) and requires a scaffold provided by a small cellular RNA. The mechanism by which Tat disrupts this complex is not well understood but is thought to include competition with Hexim1 for binding to a segment of the cellular RNA that is structurally similar to TAR RNA.

Tat also facilitates nucleosome remodeling. As discussed previously, integrated proviral DNA templates for transcription are organized in chromatin. Although human immunodeficiency virus type 1 proviral DNA is integrated preferentially into or near transcriptionally active genes of the host cell (Chapter 7), efficient transcription requires reorganization of nucleosomes, which are located at specific positions on this LTR. The promoter and enhancers are nucleosome free, and hence accessible to the transcriptional activators described above. In contrast, a nucleosome is located immediately downstream of the site of initiation of transcription and must be repositioned to allow transcriptional elongation. Cellular transcriptional activators, such as Nf-κb, are important for such remodeling, but this process is facilitated by Tat.

In addition to binding to the cyclin T1 subunit of p-Tefb, Tat can bind to specific subunits of ATP-dependent chromatin-remodeling enzymes of the Swi/Snf family, as well as to several histone acetyltransferases. The data currently available are consistent with a model in which binding of Tat to TAR RNA recruits not only p-Tefb but also Swi/Snf enzymes, which then alter the position or structure of the downstream nucleosome to promote elongation of viral transcription (Fig. 8.13C). Although some details are not yet clear, the inhibition of Tat-dependent transcription induced by small interfering RNA (siRNA)-mediated knockdown of specific subunits of Swi/Snf enzymes provides compelling evidence for the important contribution of this function of Tat to transcription of integrated proviral DNA.

The ability of Tat to bind to cellular proteins is governed by posttranslational modification. For example, acetylation at Lys28 promotes high-affinity binding to TAR and to p-Tefb, whereas acetylation at Lys50, within the RNA-binding region of Tat, prevents this interaction. The latter modification also provides a site for association of Tat with specific Swi/Snf subunits and a histone acetyltransferase (Fig. 8.13C).

Why such unusual transcriptional regulation? At this juncture, it is difficult to appreciate the value of the intricate transcriptional program of human immunodeficiency virus type 1 and related viruses. Those of many other viruses are executed successfully by proteins that operate, directly or indirectly, via specific DNA sequences. Binding of Tat to nascent viral RNA close to the site at which many transcriptional complexes pause or stall (Fig. 8.13A) could provide a particularly effective way to recruit the cellular proteins that stimulate processive transcription. Alternatively, it may be that regulation of transcription via an RNA sequence is a legacy from some ancestral virus-host cell interaction in an RNA world.

The Transcriptional Cascades of DNA Viruses

Common Strategies Are Executed by Virus-Specific Mechanisms

An overview of three DNA virus transcriptional programs is presented in this section to illustrate both their diversity and the common themes of the central role of virus-encoded transcriptional regulators and the coordination of transcriptional control with viral DNA synthesis.

The transcriptional strategies characteristic of the infectious cycles of viruses with DNA genomes exhibit a number of common features. The most striking is the transcription of viral genes in a reproducible and precise temporal sequence (Fig. 8.14). Prior to initiation of genome replication, during **early** phases, infected cells synthesize viral proteins necessary for efficient viral gene expression, viral DNA synthesis, or other regulatory functions. Transcription of the **late** genes, most of which encode structural proteins, requires genome replication. This property ensures coordinated production of the DNA genomes and the structural proteins from which progeny virus

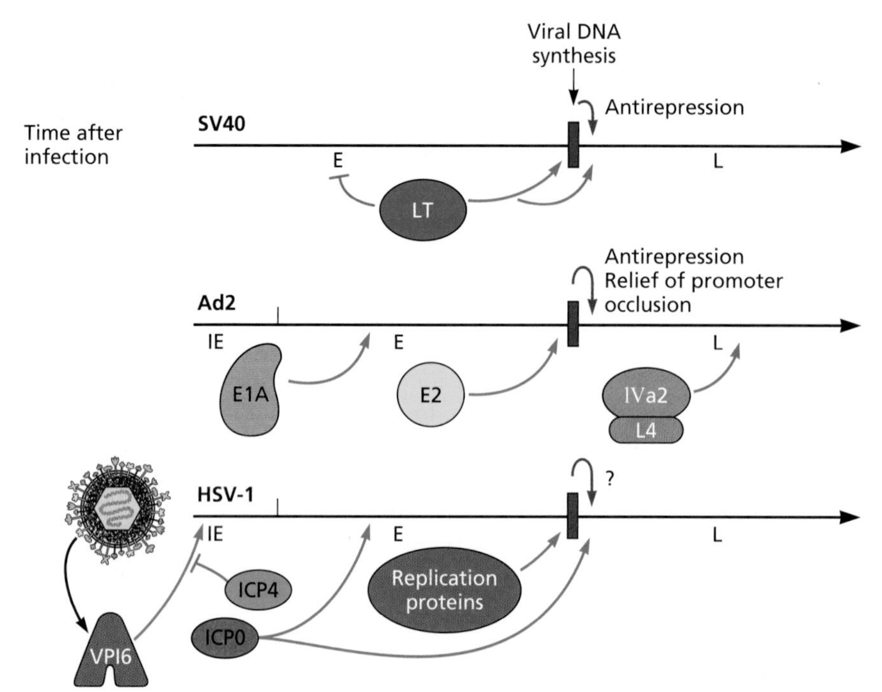

Figure 8.14 Important features of the simian virus 40 (SV40), human adenovirus type 2 (Ad2), and herpes simplex virus type 1 (HSV-1) transcriptional programs. The transcriptional programs of these three viruses are depicted by the horizontal time lines, on which the onset of viral DNA synthesis is indicated by the purple boxes. For comparative purposes **only**, the three reproductive cycles are represented by lines of equal length. The immediate-early (IE), early (E), and late (L) transcriptional phases are indicated, as are viral proteins that participate in regulation of transcription. Stimulation of transcription by these proteins and effects contingent on viral DNA synthesis in infected cells are indicated by green and purple arrows, respectively. Red bars indicate negative regulation of transcription.

particles are assembled. Another common feature is the control of the transitions from one transcriptional stage to the next by both viral proteins and genome replication. Such viral programs closely resemble those that regulate many developmental processes in animals, in both the transcription of individual genes in a predetermined sequence and the sequential action of proteins that regulate the transcription of different sets of genes.

The simplest transcriptional programs comprise only two phases. For example, the genome of simian virus 40 contains one early and one late transcription unit (Appendix, Fig. 23B), each of which encodes more than one protein. Although significantly larger, the genomes of human adenoviruses also encode multiple proteins within each of a limited number of transcription units (Appendix, Fig. 1B). This type of organization reduces the genetic information that must be devoted to transcription punctuation marks and regulatory sequences, a significant advantage when genome size is limited by packaging constraints. The price for such a transcriptional strategy is heavy dependence on the host cell's RNA-processing systems to generate multiple mRNAs by differential polyadenylation and/or splicing of a single primary transcript (Chapter 10). In contrast, the more than 80 known protein-coding sequences of herpes simplex virus type 1 are, with few exceptions, expressed as individual transcription units. Furthermore, splicing of primary transcripts is the exception. The basic distinction of early and late phases is maintained in the herpesviral transcriptional program, but temporal control of the activity of more than 80 viral promoters is obviously more complicated. In fact, the potential for finely tuned regulation is much greater when the viral genome contains a large number of independent transcription units.

Cellular enhancer- and promoter-binding proteins are sufficient to initiate the polyomaviral and adenoviral transcription programs and the synthesis in infected cells of crucial viral transcriptional regulators, large T antigen and E1A proteins, respectively. In contrast, a viral activating protein imported into cells as the virion structural component VP16 is necessary for efficient transcription of the first herpesviral genes to be expressed (called immediate-early genes) (Fig. 8.14). This simple strategy might seem to guarantee transcription of these genes in **all** infected cells. Surprisingly, however, this is not the case, because VP16 functions only in conjunction with specific cellular proteins (see next section).

As might be anticipated, the sophistication of the regulatory circuits that govern the transcriptional cascade of these DNA viruses increases with genome size. Synthesis of a single viral protein, large T antigen, in simian virus 40-infected cells leads inevitably to entry into the late phase of infection: T antigen both induces initiation of viral DNA synthesis and activates late transcription. In contrast, several transcriptional regulators control the transitions from one phase to the next during the adenoviral infectious cycle. The immediate-early E1A proteins, which regulate transcription by multiple mechanisms, are necessary for efficient transcription of all early transcription units. Among this set is the E2 gene, which encodes the proteins required for viral DNA synthesis. Accumulation of progeny viral genomes leads to relief of repression of transcription of the gene that encodes the sequence-specific DNA-binding protein IVa_2 and subsequent activation of transcription from the major late promoter (Fig. 8.14). This promoter controls synthesis of the majority of structural proteins (Appendix, Fig. 1B).

The synthesis of progeny adenoviral DNA molecules is therefore indirectly coordinated with production of the protein components that will encapsidate them in virus particles.

As noted above, the large number of individual transcription units suggests that the regulatory scheme of herpes simplex virus type 1 may be even more elaborate. Indeed, the immediate-early viral gene products include two transcriptional regulators, ICP4 and ICP0. Like the adenoviral E1A proteins, the ICP4 protein is necessary for efficient progression beyond the immediate-early phase of infection and is regarded as the major transcriptional activator. It stimulates transcription of both early and late genes and also acts as a repressor of immediate-early transcription. Some herpesviral late genes are transcribed only following synthesis of progeny genomes, the pattern exemplified by simian virus 40 late transcription, but others attain their maximal rates of transcription during the late phase. More subtle distinctions among the large number of late genes may be made as their transcriptional regulation becomes better understood.

Examples of Viral Proteins That Stimulate Transcription

In this section, we focus on a few well-characterized viral regulators to illustrate general principles of their operation, or fundamental insights into cellular processes that have been gained through their study. These proteins all promote progression through the infectious cycle, but differ in the mechanisms by which they become associated with viral promoters, and consequently in the specificity with which they stimulate transcription. The viral protein may itself bind to a specific viral DNA sequence (the Epstein-Barr virus Zta protein) or may be recruited to promoters indirectly, either via a single cellular DNA-binding protein (herpes simplex virus type 1 VP16) or by association with several cellular activators to stimulate transcription from most promoters in the viral genome (adenovirus E1A protein).

Some viral transcriptional regulators are close relatives of cellular proteins that bind to specific DNA sequences in promoters or enhancers (Table 8.3). These viral proteins possess

Table 8.3 Properties and functions of some viral transcriptional regulators

Virus	Protein	Sequence-specific DNA binding	Properties	Function[a]
Adenovirus				
Species C human adenovirus	IVa_2	Yes	Operates with a viral L4 protein	Stimulates ML transcription
	E1A 289R, 243R	No	Bind to multiple cellular regulators of transcription	289R stimulates E gene transcription; both overcome sequestration of E2f by Rb
Herpesviruses				
Herpes simplex virus type 1	VP16	No	Binds a specific promoter sequence via cellular Oct-1 and Hcf proteins	Stimulates transcription from IE promoters
	ICP4	Yes	Typical domain organization	Stimulates transcription from E and L promoters; represses IE transcription
Epstein-Barr virus	Zta	Yes	Basic-leucine zipper protein	Activates E gene transcription; commits to lytic infection
Papillomavirus				
Bovine papillomavirus type I	E2	Yes	Typical domain organization	Stimulates transcription from viral promoters; required for genome replication
Polyomavirus				
Simian virus 40	Large T antigen	No[b]	Can bind to several transcription initiation proteins	Stimulates L gene transcription; required for genome replication
Poxvirus				
Vaccinia virus	VETF	Yes	Binds as heterodimer; DNA-dependent ATPase	Essential for recognition of E promoters by the viral RNA polymerase
Retrovirus				
Human T cell lymphotropic virus type I	Tax	No	Modulates cellular basic-leucine zipper proteins; reverses cytoplasmic sequestration of Nf-κb	Stimulates transcription from viral LTR and cellular promoters

[a]E, early; IE, immediate early; L, late; ML, major late.

[b]The sequence-specific DNA-binding activity of large T antigen (see Fig. 9.3) is not required for stimulation of transcription by this protein.

discrete DNA-binding domains, some with sequence motifs characteristic of cellular DNA-binding proteins, and activation domains that interact with cellular initiation proteins. These properties are described in more detail for one such protein, the Epstein-Barr virus Zta protein, in the next section.

Sequence-specific DNA-binding proteins play ubiquitous roles in the transcription of cellular genes by RNA polymerase II, so it is not surprising that viral DNA genomes transcribed by this enzyme encode analogous proteins. However, viral transcriptional regulators that possess no intrinsic ability to bind specifically to DNA are equally, if not more, common (Table 8.3). Two examples, the herpes simplex virus type 1 VP16 and adenovirus E1A proteins, illustrate the diversity of mechanisms by which these viral proteins regulate transcription. The preponderance of such viral proteins, quite unexpected when they were first characterized, was a strong indication that host cells also contain proteins that modulate transcription without themselves binding to DNA. Many such proteins (e.g., coactivators) have now been recognized.

The Epstein-Barr virus Zta protein: a sequence-specific DNA-binding protein that induces entry into the productive cycle. When the gammaherpesvirus Epstein-Barr virus infects B lymphocytes, only a few viral genes are transcribed and a latent state described in "Entry into One of Two Alternative Transcriptional Programs" below is established. The products of these genes maintain the viral genome via replication from a latent phase-specific origin of replication (OriP) (Fig. 8.15A), modulate the immune system, and alter the growth properties of the cells. Virus reproduction begins with synthesis of three viral proteins that regulate gene expression. However, just one of these, the transcriptional regulator Zta (also known as ZEBRA, Z, or EB-1), is sufficient to interrupt latency and induce entry into the productive cycle.

The Zta protein exhibits many properties characteristic of the cellular proteins that recognize promoter sequences: it is a modular, sequence-specific DNA-binding protein that belongs to the basic-leucine zipper family (Table 8.3 and Fig. 8.7). Dimerization of Zta via this domain is required for its direct binding to viral promoters. The discrete activation domain, which can bind directly to cellular initiation proteins, such as subunits of transcription factor IID (TfIId), is thought to facilitate the assembly of preinitiation complexes, and hence initiation of transcription from these promoters.

The availability or activity of Zta is regulated by numerous mechanisms. In latently infected cells, transcription from the Zta promoter is blocked by binding of cellular transcriptional repressors to several sites (Fig. 8.15B). Upon B cell activation and induction of signal transduction cascades in response to external stimuli, such as binding of antigens to B cell receptors, several cellular regulators, including members of the Sp1 and Atf families, bind to and activate transcription from the Zta promoter. Synthesis of Zta augments transcription from this promoter, as the protein is a positive autoregulator. The availability of Zta mRNA for translation is also regulated, in part, by annealing of Zta pre-mRNA to the complementary transcripts of the viral EBNA-1 gene (Fig. 8.15A). The net effect of these regulatory mechanisms, which depends on the type and the proliferation and differentiation states of the Epstein-Barr virus-infected cell, determines whether active Zta protein is available. Entry into the infectious cycle appears to be an inevitable consequence of production of active Zta: this protein not only stimulates transcription from the promoters of its own gene and other early genes but also plays an important role in replication from the lytic origins.

The herpes simplex virus type 1 VP16 protein: sequence-specific activation of transcription via a cellular DNA-binding protein. The herpesviral VP16 protein, which enters infected cells in the virus particle (Fig. 8.14), has taught us much about mechanisms by which transcription by RNA polymerase II can be stimulated. Furthermore, its unusual mode of promoter recognition illustrates the importance of conformational change in proteins during formation of DNA-bound protein assemblies.

The VP16 protein lacks a DNA-binding domain. Its acidic activation domain is one of the most potent known and has been exploited to investigate mechanisms of stimulation of transcription. Chimeric proteins in which this domain is fused to heterologous DNA-binding domains strongly stimulate transcription from promoters that contain the appropriate binding sites. When part of such fusion proteins, the VP16 acidic activation domain can stimulate several reactions required for initiation of transcription (Fig. 8.16A). It can also increase the rate of transcriptional elongation and promote transcription from chromatin templates (Fig. 8.16B). These properties established that a single protein can regulate RNA polymerase II transcription by multiple molecular mechanisms.

The VP16 protein is the founding member of a class of viral regulators that possess no sequence-specific DNA-binding activity, yet activate transcription from promoters that contain a specific consensus sequence. The 5′ flanking regions of viral immediate-early genes contain at least one copy of the consensus sequence that is necessary for VP16-dependent activation of their transcription, 5′TAATGARAT3′ (where R is a purine). This sequence is bound by VP16 only in association with at least two cellular proteins, Oct-1 and host cell factor (Hcf) (Fig. 8.17). The Oct-1 protein is a ubiquitous transcriptional activator named for its recognition of a DNA sequence termed the octamer motif. This protein and VP16 can associate to form a ternary (three-component) complex on the 5′TAATGARAT3′ sequence, but the second cellular protein, Hcf, is necessary for stable, high-affinity binding. The VP16

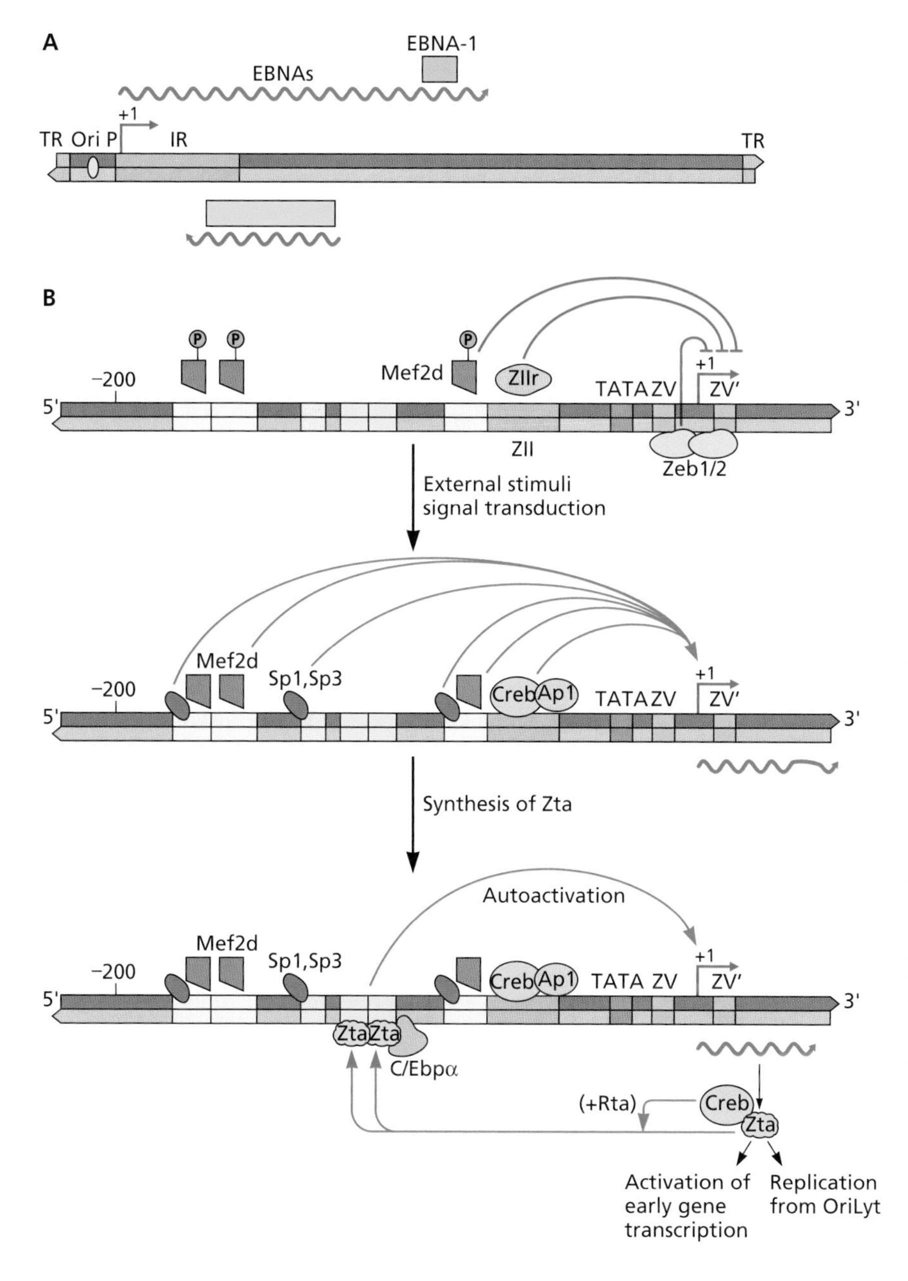

Figure 8.15 Organization and regulation of the Epstein-Barr virus Zta gene promoter. **(A)** Organization of the transcription units that contain the coding sequence for the Epstein-Barr virus nuclear antigen (EBNA) proteins (an ~100-kb transcription unit) and Zta. The locations of the genomic terminal (TR) and internal (IR) repeat sequences, the origin of replication for plasmid maintenance (OriP), and the coding sequences for the EBNA-1 and Zta proteins are indicated. The Zta protein is synthesized from spliced mRNAs processed from the primary transcripts shown. **(B)** (Top) Sequences that regulate transcription from the Zp promoter of the Zta gene are shown to scale and in the conventional 5′→3′ direction. In primary B lymphocytes, transcription from this promoter is repressed by synergistic binding of Zn-finger E-box-binding homeobox (Zeb) 1 or 2 proteins to the ZV and ZV′ sequences and of an as yet unidentified protein (designated ZII repressor, or ZIIr) to the ZII sequence. Binding of a phosphorylated form of myocyte-specific enhancer factor 2D (Mef2d) also contributes to repression by recruiting Hdacs. (Middle) Activation of B cells, for example, by reagents that induce cross-linking and activation of B cell surface receptor, induces signal transduction pathways that lead to reversal of the inhibitory modification of Mef2D, allowing recruitment of histone acetyltransferases and activation of positive regulators, such as Creb and Ap1. These proteins, in conjunction with ubiquitous activators of the Sp1 family, stimulate transcription from the Zta promoter. (Bottom) Synthesis of Zta activates transcription of early genes by reversing repressive modifications of nucleosome associated with early promoters and also promotes viral genome replication. This viral protein also establishes a positive autoregulatory circuit by binding to specific promoter sequences and cooperating with cellular C/Ebpα.

protein and Hcf form a heteromeric complex in the absence of Oct-1 or DNA. An important function of Hcf appears to be stabilization of conformational change in VP16, to allow its high-affinity binding to Oct-1 on the immediate-early promoters (Fig. 8.17). The VP16 protein interacts with Hcf and Oct-1 proteins via its N-terminal region. Its C-terminal region contains the acidic activation domain described previously. The results of chromatin immunoprecipitation experiments suggest that stimulation of immediate-early gene transcription in infected cells is mediated by several of the biochemical activities exhibited by the acidic activation domain in simplified experimental systems (Box 8.10).

One of the most remarkable features of the mechanism by which the VP16 protein is recruited to immediate-early promoters is its specificity for Oct-1. This protein is a member of a family of related transcriptional regulators defined by a common DNA-binding motif called the POU-homeodomain. The VP16 protein distinguishes Oct-1 from all other members of this family, including Oct-2, which binds to exactly the same DNA sequence as Oct-1. In fact, VP16 detects a **single** amino acid difference in the exposed surfaces of DNA-bound Oct-1 and Oct-2 homeodomains.

The incorporation of the VP16 protein into virus particles at the end of one infectious cycle appears to be an effective

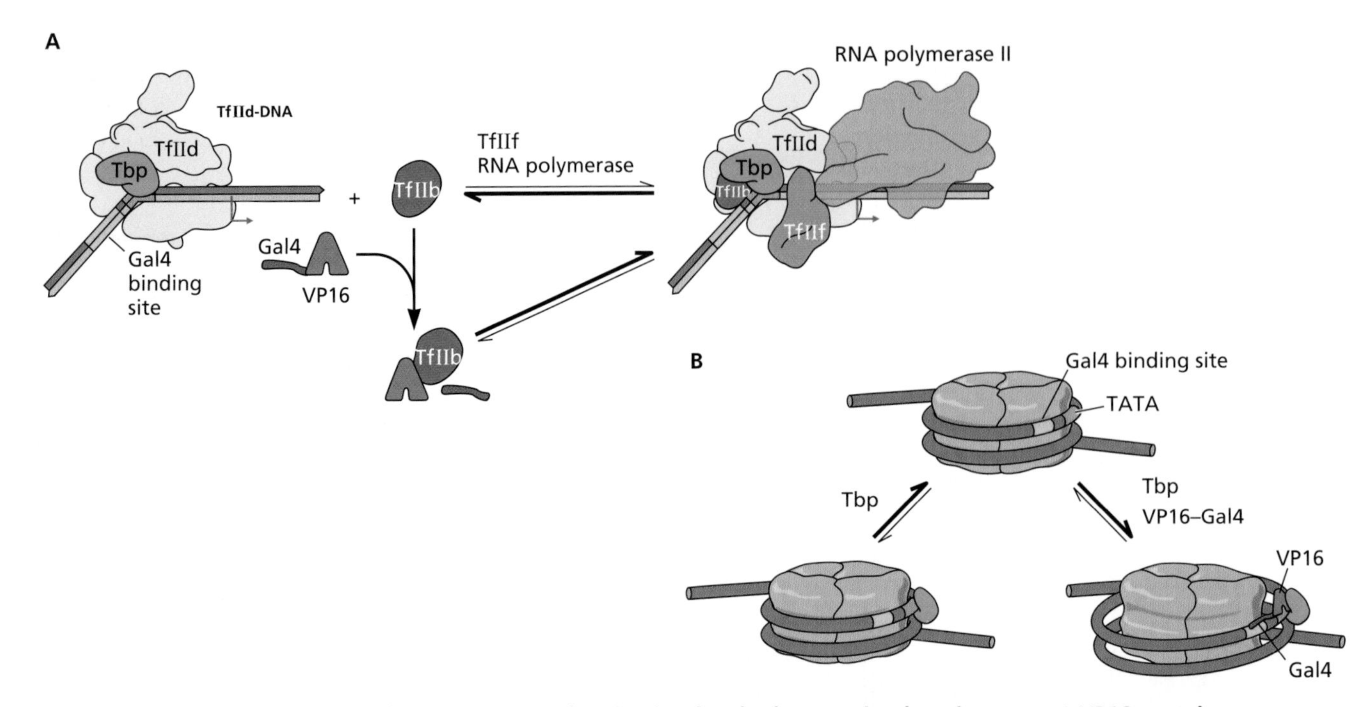

Figure 8.16 Models for transcriptional activation by the herpes simplex virus type 1 VP16 protein. **(A)** Induction of conformational change in TfIIb. In native TfIIb, the N- and C-terminal domains associate with one another such that internal segments of the protein that interact with the TfIIf-RNA polymerase II complex are inaccessible. Binding of the acidic activation domain of VP16, for example, as a chimera with the DNA-binding domain of the yeast protein Gal4, disrupts this intermolecular association of TfIIb domains, exposing its binding sites for TfIIf and RNA polymerase II. Consequently, formation of the preinitiation complex that contains TfIIb, TfIIf, and RNA polymerase II is now a more favorable reaction. **(B)** Alleviation of transcriptional repression by nucleosomes. Many activators, including the acidic activation domain of VP16, stimulate transcription from nucleosomal DNA templates to a much greater degree than they do transcription from naked DNA. This property is the result of their ability to alleviate repression of transcription by nucleosomes. Organization of DNA into a nucleosome can block access of proteins to their DNA-binding sites, as illustrated for binding of Tbp to a TATA sequence (left). Association of the acidic activation domain of VP16 with the template alters the interaction of the DNA with the nucleosome to allow Tbp access to the TATA sequence (right), presumably as a result of recruitment of ATP-dependent chromatin-remodeling enzymes and/or histone acetyltransferases (Box 8.10).

way to ensure transcription of viral genes and initiation of viral reproduction in a new host cell. Nevertheless, some features of this mechanism are not fully appreciated, in particular the benefits conferred by the indirect mechanism by which VP16 recognizes viral promoters. One advantage over direct DNA binding may be the opportunity to monitor the growth state of the host cell that is provided by the requirement for binding to Hcf: this protein regulates transcription during the cell cycle and is important for proliferation of uninfected cells. Furthermore, Hcf is a component of several chromatin-modifying complexes, and its recruitment to immediate-early promoters is required for replacement of repressive with activating modifications of the nucleosomal histones associated with these promoters. The dependence on Hcf may also contribute to the establishment of latent infections in neurons (see "Entry into One of Two Alternative Transcriptional Programs" below).

Adenoviral E1A proteins: regulation of transcription by multiple mechanisms. Two E1A proteins are synthesized from differentially spliced mRNAs during the immediate-early phase of adenovirus infection (Fig. 8.18). These two proteins share all sequences except for an internal segment (conserved region 3 [CR3]) that is unique to the larger protein. Nevertheless, they differ considerably in their regulatory potential, because the CR3 segment is primarily responsible for stimulation of transcription of viral early genes. As the larger E1A protein neither binds specifically to DNA nor depends on a specific promoter sequence, it is often considered the prototypical example of viral proteins that stimulate transcription by indirect mechanisms.

The CR3 segment of the larger E1A protein comprises an N-terminal zinc finger motif followed by 10 amino acids that are highly conserved among human adenoviruses (Fig. 8.19A). The latter region mediates binding of the E1A

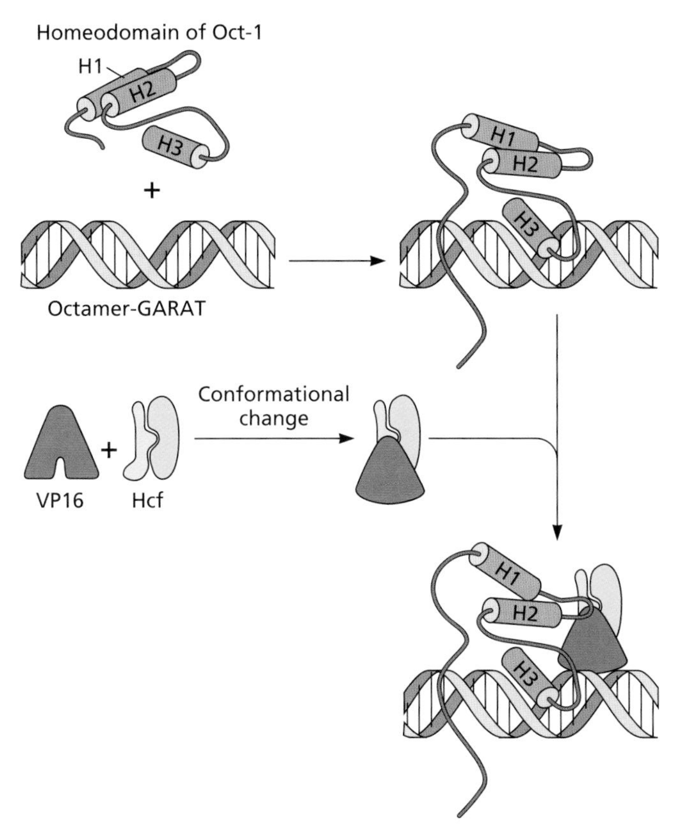

Figure 8.17 Conformational changes and recruitment of VP16 to herpes simplex virus type 1 promoters. Binding of the octamer-binding protein 1 (Oct-1) homeodomain to DNA containing the GARAT sequence and of VP16 to host cell factor (Hcf) induces conformational changes that allow specific recognition of GARAT-bound Oct-1 by VP16. This mechanism ensures that the VP16 protein is recruited only to promoters that contain the GARAT sequence, that is, viral immediate-early promoters.

protein to cellular, sequence-specific activators such as Atf-2 and Sp1, and hence association with the viral promoters. The zinc finger motif is essential for stimulation of transcription by the E1A protein in infected cells. It binds with exceptionally high affinity to a single component (Med23) of the human mediator complex (see the interview with Dr. Arnold Berk: http://bit.ly/Virology_Berk), which contains at least 20 different subunits and is essential for regulation of transcription by RNA polymerase II. The Med23-E1A interaction recruits the mediator complex to promoters at which E1A is bound to DNA-binding cellular activators. This association leads to activation of transcription in infected cells, by stimulation of preinitiation complex assembly and recruitment of the super elongation complex described previously (see "The Tat Protein Regulates Transcription by Unique Mechanisms" above). Like other viral activators we have discussed, the short E1A CR3 segment appears to stimulate multiple reactions in the transcription cycle.

Adenoviral E1A proteins activate transcription by a second mechanism, which is mediated by the conserved N-terminal regions CR1 and CR2. The CR1 or CR2 segments interact with several cellular proteins, including Rb and p300 (Fig. 8.18). The Rb protein is the product of the cellular retinoblastoma susceptibility gene, a tumor suppressor that plays a crucial role in cell cycle progression (Volume II, Chapter 6). In uninfected cells, Rb binds to cellular E2f proteins, which are sequence-specific transcriptional activators originally discovered because they bind to the human adenovirus type 2 E2 early promoter (Fig. 8.5). Such E2f-Rb complexes possess the specific DNA-binding activity characteristic of E2f, but Rb represses transcription (Fig. 8.19B). Competition for Rb by the E1A proteins disrupts the Rb-E2f association and allows transcription from E2f-dependent promoters. During the early phase of infection, E2f proteins are essential for efficient transcription of the gene that encodes the proteins required for viral DNA synthesis. Sequestration of Rb by the E1A proteins therefore ensures synthesis of replication proteins and progression into the late phase of the infectious cycle (Fig. 8.14).

The N-terminal sequences common to the two E1A proteins also bind directly to the cellular coactivators p300 and Creb-binding protein (Crebbp) (Fig. 8.18). As noted previously, these proteins are histone acetyltransferases and bind to other such enzymes to modify histones and alter the structure of transcriptionally active chromatin. The E1A proteins disrupt the interactions of p300 and Crebbp with specific activators and compete for their binding to other histone acetylases. This activity of E1A proteins has been implicated in repression of enhancer-dependent transcription and is required for induction of cell proliferation (Chapter 9).

The multiplicity of mechanisms by which the E1A proteins engage with components of the cellular transcriptional machinery is one of their most interesting features. Regulation by multiple mechanisms may prove to be a general property of viral proteins that cannot bind directly to DNA. For example, the human T-lymphotropic virus type 1 Tax protein stimulates transcription by binding to specific cellular members of the basic-leucine zipper family, and also by activating Nf-κb.

Coordination of Transcription of Late Genes with Viral DNA Synthesis

In cells infected by the viruses under consideration in this chapter, synthesis of the large quantities of structural proteins needed for assembly of progeny virus particles is restricted to the late phase of infection, following the onset of viral genome replication. This pattern, first characterized in studies of bacteriophages such as T7 and T4, is a general, if not universal, feature of the reproductive cycles of viruses with

BOX 8.10

EXPERIMENTS

In vivo *functions of the VP16 acidic activation domain*

The acidic activation of domain of VP16 has been studied extensively as a model for transcriptional activation. It was shown to stimulate transcription by multiple mechanisms in simplified experimental systems (see text). The functions of VP16 in vivo were investigated using the chromatin immunoprecipitation assay (Box 8.1) to compare the proteins associated with herpes simplex virus type 1 immediate-early promoters in cells infected by the wild-type virus or a mutant encoding VP16 that lacks the acidic activation domain. Cross-linked DNA was immunoprecipitated with antibodies to VP16, RNA polymerase, or several other cellular proteins. The concentrations of viral promoter DNA present in such immunoprecipitates were then assessed by using PCR.

The results of these experiments provide validation for mechanisms of activation of the VP16 activation domain deduced using simplified experimental systems, notably stimulation of initiation complex assembly and induction of chromatin remodeling (see text). As summarized in the table, association of RNA polymerase II and Tbp with the viral promoters (initiation complex assembly) depended on synthesis of VP16 containing an activation domain, as did efficient recruitment of histone acetyltransferases (Crebbp and p300) and ATP-dependent remodeling proteins (Brg-1), as well as loss of histone H3 (chromatin remodeling).

Herrera FJ, Triezenberg SJ. 2004. VP16-dependent association of chromatin-modifying coactivators and underrepresentation of histones at immediate-early gene promoters during herpes simplex virus infection. *J Virol* **78:**9689–9696.

	Viral promoter DNA	
	VP16 acidic activation domain	
Promoter-bound proteins	**Not present**	**Present**
VP16	++	++
Oct-1	++	++
RNA polymerase II	–	++
Tbp	–	++
Crebbp	+	++
Brg-1	–	++
Histone H3	++	–

DNA genomes, and offers a number of potential advantages (Box 8.11). The restriction of synthesis of structural proteins to the end of a cycle of viral reproduction results from the dependence of late gene transcription on viral DNA replication: drugs or mutations that inhibit viral DNA synthesis in infected cells block efficient expression of late genes. Indeed, late genes are defined experimentally as those that are not transcribed, or are transcribed much less efficiently, when viral DNA synthesis is blocked. Despite their importance, the mechanisms by which activation of transcription can be integrated with viral DNA synthesis remain incompletely understood.

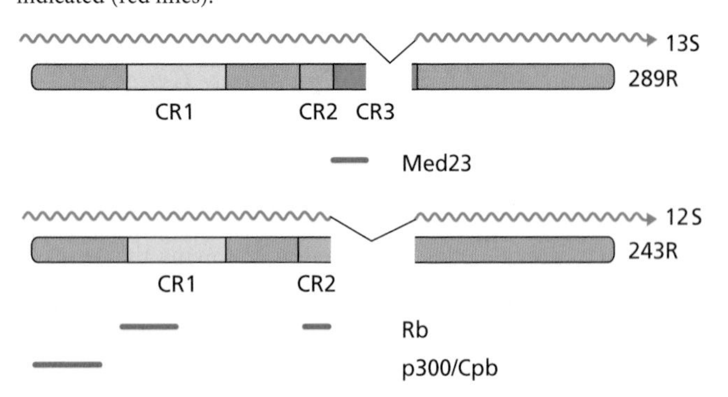

Figure 8.18 The adenoviral E1A proteins bind to multiple transcriptional regulators. Primary transcripts of the immediate-early E1A gene are alternatively spliced to produce the abundant 13S and 12S mRNAs. As such splicing does not change the translational reading frame, the E1A proteins are identical, except for an internal segment of 46 amino acids unique to the larger protein. The three most highly conserved regions are designated CR1, CR2, and CR3. The regions of the E1A proteins necessary for interaction with the Rb protein, the histone acetyltransferases p300 and Crebbp, and the mediator subunit Med23 are indicated (red lines).

Titration of cellular repressors. The most obvious consequence of genome replication in cells infected by DNA viruses is the large increase in concentration of viral DNA molecules. Even in experimental situations, infected cells contain a relatively small number of copies of the viral genome during the early phase of infection, typically 1 to 100 copies per cell depending on the multiplicity of infection (defined in Chapter 2). As soon as viral DNA synthesis begins, this number increases rapidly to values as high as hundreds of thousands of viral DNA molecules per infected cell nucleus. At such high concentrations, viral promoters can compete effectively for components of the cellular transcription machinery.

The increase in DNA template concentration also titrates cellular transcriptional repressors that bind to specific sequences of certain viral late promoters. For example, the simian virus 40 major late promoter remains inactive, because of the binding to it of a cellular repressor that belongs to the steroid/thyroid hormone receptor superfamily. Viral DNA replication increases the concentration of the late promoter

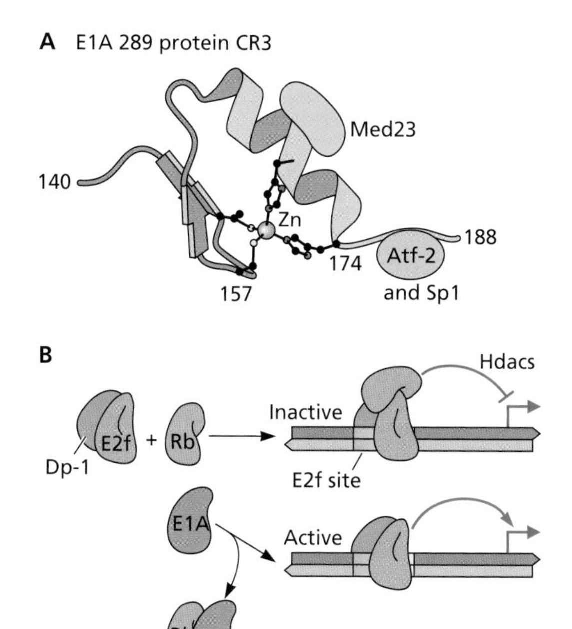

Figure 8.19 Indirect stimulation of transcription by adenoviral E1A proteins. (A) Interactions of the E1A CR3 sequences with components of the RNA polymerase II transcriptional machinery. The C-terminal segment of CR3 interacts with several cellular activators that bind to specific DNA sequences, as indicated by Sp1 and cyclic AMP-dependent transcription factor Atf-2, as well as with particular Taf subunits of TfIId. The Zn finger motif is required for tight binding to the Med23 subunit of the mediator complex. CR3-dependent stimulation of transcription of viral early genes is impaired in mutant cells homozygous for deletion of the *Med23* gene and when Med23 is depleted from permissive human cells by RNA interference. Such depletion also reduces association of Tbp with viral early promoters. This interaction is required for stimulation of viral early gene transcription by the super elongation complex component Cdk9. The exceptionally high-affinity binding of CR3 to Med23 may also facilitate reinitiation. **(B)** Model of competition between E1A proteins and E2f for binding to Rb protein. The E2f transcriptional activators are heterodimers of a member of the E2f protein family (described in Chapter 9) and E2f dimerization partner 1 (Dp-1). The binding of E2f to its recognition sites in specific promoters is not inhibited by association with the Rb protein, but Rb represses transcription from E2f-dependent promoters via recruitment of Hdacs (top). The CR1 and CR2 regions of the adenoviral E1A proteins made in infected (or transformed) cells bind to Rb and disrupt the E2f-Rb interaction. They also induce proteasomal degradation of Rb. Consequently, Rb is removed from association with E2f, which can then stimulate transcription.

until it exceeds that of repressor, and therefore allows this promoter to become active (Fig. 8.20). This "antirepression" mechanism directly coordinates activation of transcription of late genes with viral genome replication and is highly efficient. Consequently, it is not surprising that the same mechanism regulates transcription of the adenoviral IVa_2 gene (Fig. 8.14). The IVa_2 protein is itself a sequence-specific activator of transcription: it cooperates with a second viral protein that also binds to specific DNA sequences to stimulate the rate of initiation of transcription from the major late promoter at least 20-fold. Activation of this promoter is therefore coupled indirectly to adenovirus DNA synthesis: this process initiates a transcriptional cascade in which late promoters are activated sequentially (Fig. 8.14).

Although viral DNA synthesis is sufficient for activation of transcription of some viral late genes (e.g., the adenoviral IVa_2 gene), this process is usually facilitated by one

BOX 8.11

DISCUSSION

Some potential advantages of temporal regulation of viral gene expression

The genomes of DNA viruses come in a variety of conformations and an enormous range of sizes (Chapter 3). Nevertheless, temporal regulation of viral gene expression appears to be a universal feature of their reproductive cycles: genes encoding viral structural proteins are expressed only during the late phase, following the onset of genome replication, with earlier periods devoted to synthesis of viral enzymes and regulatory proteins. This pattern, which is also characteristic of the infectious cycles of some RNA viruses (Chapter 6), must therefore facilitate reproduction of these viruses. Possible advantages of sequential expression of viral genes and synthesis of the large quantities of structural proteins required for assembly of progeny virus particles only later in the infectious cycle may include

- the availability of viral proteins (products of early genes) that mediate efficient production of late mRNAs posttranscriptionally via effects on splicing or mRNA export from the nucleus (Chapter 10)
- reorganization of infected cell components orchestrated by early proteins, and their assembly with replicated viral DNA molecules at specialized sites for optimal transcription of late genes (Chapter 9)
- coordination of synthesis of structural proteins and viral DNA to facilitate genome encapsidation and assembly of progeny virus particles
- prevention of premature cessation of viral genome replication and transcription (these processes and encapsidation are mutually exclusive)
- postponement of competition for finite cellular resources (e.g., substrates for DNA and RNA synthesis, amino acids), potentially deleterious for both the host cell and reproduction of the virus until late in infection; by this time, the only reactions required to complete the infectious cycle are assembly and release of progeny virus particles
- restriction of synthesis of cytotoxic viral proteins that facilitate release of viral particles (Chapter 13) to the end of the infectious cycle

This strategy might therefore be considered analogous to the "just in time" inventory control method widely used in industry. This approach is defined by Wikipedia as "a production strategy that strives to improve a business return on investment by reducing in-process inventory and associated carrying costs."

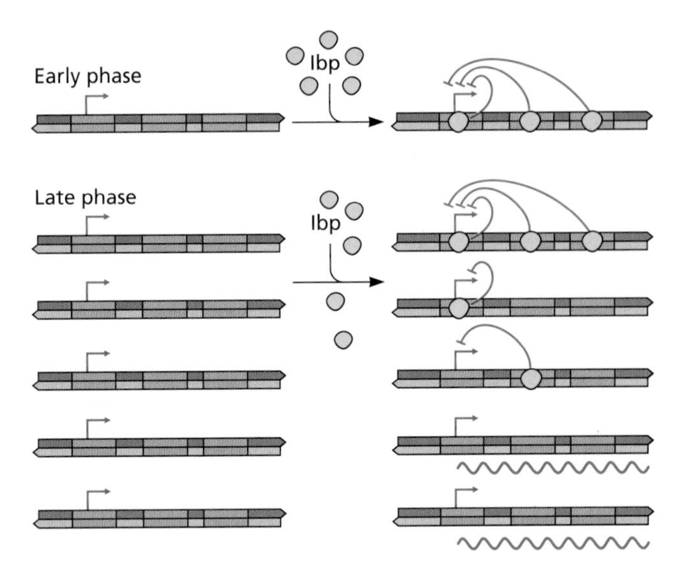

Figure 8.20 Cellular repressors regulate the activity of the simian virus 40 late promoter. The sequence surrounding the simian virus 40 major late initiation site (the thickest arrow in Fig. 8.4D) contains three binding sites for the cellular repressor termed initiator-binding protein (Ibp), which contains members of the steroid/thyroid receptor superfamily. During the early phase of infection, the concentration of Ibp relative to that of the viral major late promoter is sufficiently high to allow all Ibp-binding sites in the viral genomes to be occupied. The concentration of Ibp does not change during the course of infection. However, as viral DNA synthesis takes place in the infected cell, the concentration of the major late promoter becomes sufficiently high that not all Ibp-binding sites can be occupied. Consequently, the major late promoter becomes accessible to cellular transcription components. Although we generally speak of "activation" of late gene transcription, this DNA replication-dependent mechanism is, in fact, one of escape from repression.

or more viral proteins. For example, maximally efficient transcription from the simian virus 40 major late promoter depends on the viral early gene product large T antigen. This protein controls simian virus 40 late transcription both directly, as an activating protein (Fig. 8.14), and indirectly, as a result of its essential functions in viral DNA synthesis (Chapter 9).

Transcription of herpesviral late genes also requires viral DNA replication and synthesis of viral activators, such as ICP4 and ICP0 in the case of herpes simplex virus type 1. Transcription of viral late genes is generally regulated by inhibition or activation of initiation. However, the synthesis of progeny viral genomes can also alter termination, a regulatory mechanism illustrated by the species C human adenovirus major late transcription unit. During the early phase of infection, major late transcription terminates within a region in the middle of the transcription unit. As discussed in Chapter 10, such restricted transcription is coupled with preferential utilization of specific RNA-processing signals to produce a single major late mRNA and protein during the early phase. Viral DNA synthesis is necessary to induce full-length transcription to a termination site close to the right-hand end of the viral genome, and therefore expression of the many other major late coding sequences (Appendix, Fig. 1). The fact that only replicated viral DNA molecules can support such complete transcription suggests an unusual regulatory mechanism. One hypothesis is that alterations in template structure upon viral DNA synthesis may contribute to this process.

Availability and structure of templates. Newly replicated viral DNA molecules can enter into additional replication cycles, serve as templates for transcription, or become assembled into virus particles, with different fates predominating at different times in the infectious cycle. Transcription of all viral DNA molecules made in infected cells would seem to be a simple mechanism to ensure efficient transcription of late genes. Amazingly, however, no more than 5 to 10% of the large numbers that accumulate are transcriptionally active. In the case of simian virus 40, synthesis of viral DNA molecules is coordinated with assembly into nucleosomes, and transcriptional activity can be ascribed to establishment of an open chromatin region spanning the viral promoters and the enhancer in minichromosomes. It is not clear whether subsets of adenoviral and herpesviral DNA molecules are also marked in some way for transcriptional activity. However, one important parameter governing the concentration of transcriptional templates must be the relative concentrations of viral DNA molecules and the proteins that package them during assembly of virus particles, because packaging and transcription of genomes are mutually exclusive.

Entry into One of Two Alternative Transcriptional Programs

Studies of bacteriophage lambda led to the discovery that some viral infections result in maintenance of a quiescent viral genome for long periods in infected cells (lysogeny) rather than in viral replication (Chapter 1). Whether lambda enters this lysogenic state or the lytic cycle is determined by the outcome of the opposing actions of two viral proteins that repress transcription (Box 8.12). This regulatory mechanism, which was among the first to be elucidated in detail, emphasized the importance of repression of transcription of specific genes and established a general paradigm for transcriptional switches. Several animal viruses can establish a similar pattern of infection. For example, **latent infection** is a characteristic feature of herpesvirus infection of specific types of host cells. As in bacteriophage lambda lysogeny, latent infections are characterized by both lack of efficient expression of many viral genes and activation of a unique, latent-phase transcriptional program. Whether a herpesvirus infection is latent or lytic, as well as reentry into the productive cycle from latency (**reactivation**), is governed by mechanisms that regulate transcription.

BOX 8.12

DISCUSSION

Two bacteriophage lambda repressors govern the outcome of infection

Infection of *Escherichia coli* by bacteriophage lambda leads to either synthesis of progeny virions and lysis of the host cell (lytic infection) or stable integration of the viral genome into that of the host cell (lysogenic infection) (panel A of the figure). During lysogeny, lytic genes are not expressed. Remarkably, the actions of two repressors of transcription encoded within the viral genome, the cI repressor and Cro, make a major contribution to the lytic/lysogeny "decision." When first encountered, the regulatory circuits by which these proteins govern expression of lytic and lysogenic genes can be difficult to understand: they include several promoters and multiple binding sites for the repressors. However, these circuits are crucial for survival of the bacteriophage and were among the first to be understood in detail.

The region of the lambda genome containing the *cI* repressor and *cro* genes is illustrated at the top of panel B. These coding sequences are flanked by genes encoding proteins that regulate transcription during lytic infection (e.g., N) or that are required during establishment of lysogeny (e.g., *int*, which encodes an integrase). Although both repressors bind to the operator sequences O_R and O_L adjacent to the right (P_R) and left (P_L) promoters, respectively, events at O_R are critical in determining the outcome of infection. The expanded view of the region of the genome containing O_R and P_R indicates the three binding sites for the repressors and the two promoters from which the *cI* gene is expressed, the promoters for repressor establishment and for repressor maintenance, P_{RE} and P_{RM}, respectively.

When the lambda genome enters a host cell, transcription from the P_R and P_{RE} promoters by the bacterial RNA polymerase leads to synthesis of the cI repressor and Cro. The highest-affinity binding site for the cI repressor in O_R is O_{R1}, but this dimeric protein binds cooperatively to O_{R1} and O_{R2}. As these sites overlap sequences of the P_R promoter essential for binding of *E. coli* RNA polymerase, transcription of *cro* (and other rightward lytic genes) is repressed (red bar). Transcription from P_L is blocked in the same way by binding of the cI repressor to O_{L1} and O_{L2}. The N-terminal domain of cI repressor bound to O_{R2} contacts the subunit of RNA polymerase that

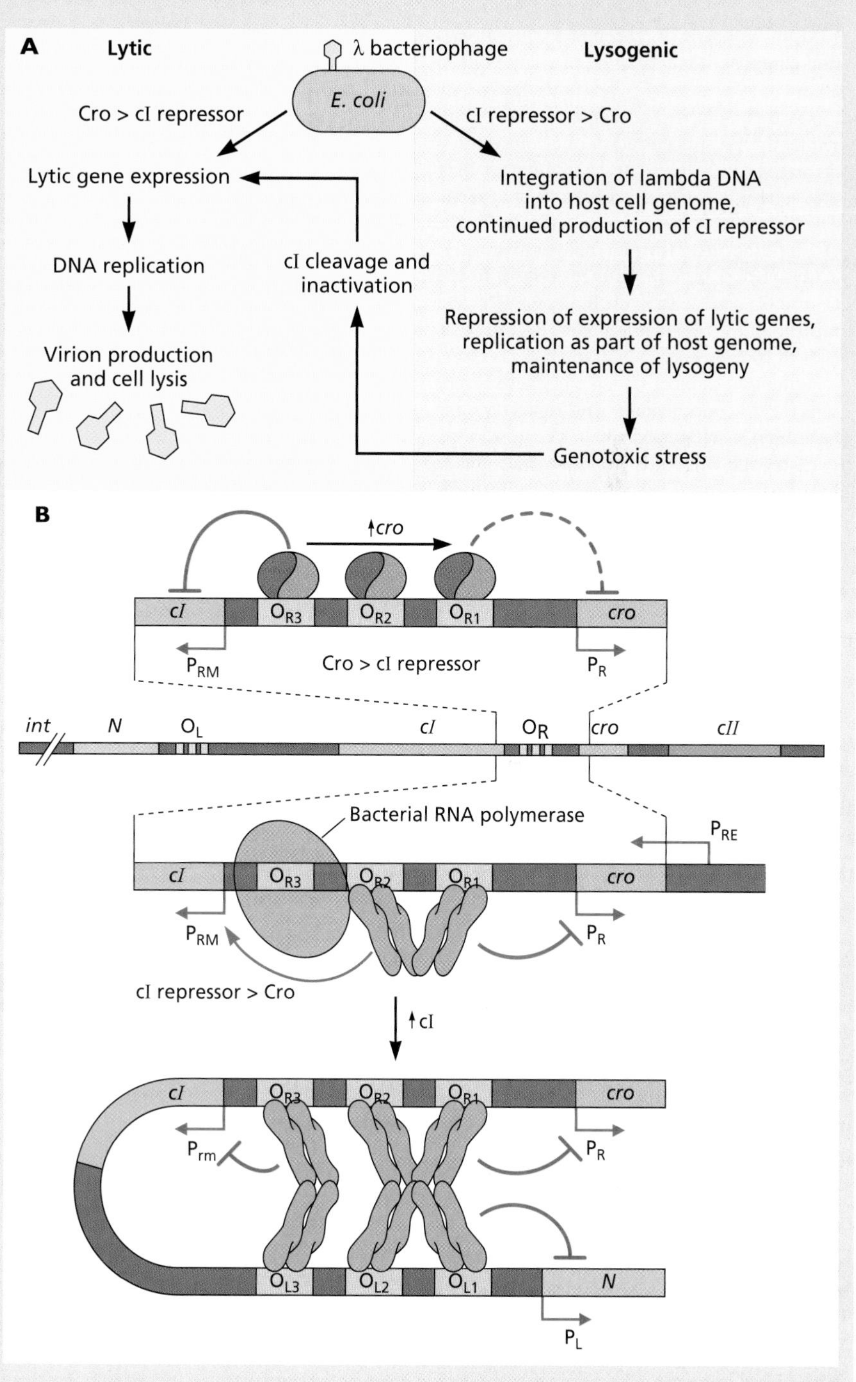

(continued)

BOX 8.12

DISCUSSION

Two bacteriophage lambda repressors govern the outcome of infection (continued)

binds to the nearby P_{RM} promoter. This interaction stimulates the formation of an open initiation complex at the P_{RM} promoter, and hence transcription of the *cI* gene (green arrow). Consequently, the concentration of cI repressor is increased to a value some 10-fold higher than that compatible with expression of lytic genes. The cI repressor has only low affinity for the O_{R3}-binding site. However, cooperative interactions occur between dimers bound to the O_L and O_R sites, to facilitate binding to O_{R3} and repression of transcription from P_{RM}. Because of such cooperative binding, whether cI repressor stimulates or blocks its own synthesis is very sensitive to concentration, and repressor concentration is maintained within a narrow range.

Although Cro binds to the same O_R sites as the cI repressors, it has the highest affinity for O_{R3}. It therefore occupies this site preferentially, and then binds to O_{R2}, to block association of RNA polymerase with the P_{RM} promoter. Consequently, the cI repressor does not attain the concentrations necessary for establishment (and maintenance) of lysogeny. Binding of Cro to O_{R2} and O_{R1} leads to weak repression of transcription from P_R (and from P_L by an analogous mechanism). This function of Cro favors lytic infection, for example, by reducing production of the cII transcriptional regulator, which promotes lysogeny by activating transcription of the *cI* gene from P_{RE}, and of the integrase gene.

It has been known for many years that environmental conditions and the activities of particular host cell gene products influence the outcome of lambda infection. The lysis/lysogeny decision was one of the first to be analyzed using a statistical-thermodynamic model of regulation of promoter activity. The results indicated that random thermal fluctuations in the rates of the reactions that comprise the regulatory circuits can lead to random phenotypic variation ("choice" between lytic and lysogenic infection) among the infected cells in a population. This conclusion is consistent with experimental observations.

Arkin A, Ross J, McAdams HC. 1998. Stochastic kinetic analysis of developmental pathway bifunction in phage λ-infected *Escherichia coli*. *Genetics* **149:** 1633–1648.

Dodd B, Shearwin KE, Egan JB. 2005. Revisited gene regulation in bacteriophage λ. *Curr Opin Genet Dev* **15:**145–152.

Ptashne M, Jeffrey A, Johnson AD, Maurer R, Meyer BJ, Pabo CO, Roberts TM, Sauer RT. 1980. How λ repressor and Cro work. *Cell* **19:**1–11.

As described in a previous section, the availability and activity of a single viral protein, Zta, determine whether Epstein-Barr virus infection is latent or lytic in B cells. This protein is necessary for transcription of viral early genes, as well as for viral DNA replication during the lytic cycle. Consequently, a latent infection ensues until the infected cell is exposed to conditions that activate transcription of the Zta gene. As Zta also represses transcription of the genes expressed in latently infected cells, it can be viewed as a simple regulatory switch. In contrast, more complex mechanisms appear to determine the outcome of infection by the alphaherpesviruses, which establish latent infections in neurons.

During latent infection of neurons by herpes simplex virus type 1, transcription of lytic genes is blocked and only a single transcription unit is expressed efficiently as latency-associated transcripts (LATs) (Fig. 8.21). As noted previously, the viral genome becomes circularized and associated with cellular nucleosomes upon entry into infected cell nuclei. In latently infected neurons, lytic genes are organized by nucleosomes that carry repressive posttranslational modifications and are associated with cellular repressors of transcription. In contrast, the LAT gene is associated with nucleosomes containing histones with modifications characteristic of actively transcribed genes. The mechanisms that lead to the silencing of lytic gene expression, the establishment of such distinct domains of "chromatin" on the viral genome, and why this process is specific to neurons are not fully understood. However, one important parameter is likely to be limited stimulation of expression of the immediate-early lytic genes by VP16: in neurons, the essential VP16 cofactor Hcf is localized largely in the cytoplasm, sequestered from viral genomes and VP16 that enter infected cell nuclei. In addition, Hcf binds to Zhangfei, a cellular protein that is a strong repressor of transcription. Zhangfei is synthesized in sensory neurons (a natural site of latency) but not in most other cell types. The synthesis of the LAT RNAs may also facilitate the establishment and maintenance of latency.

The major 2.0-kb (and 1.5-kb) LATs (Fig. 8.21), which accumulate to 40,000 to 100,000 copies in nuclei of latently infected neurons, lack poly(A) tails and are not linear molecules. Indeed, all properties observed to date indicate that they are stable introns produced by splicing of precursor RNA. The primary LAT also serves as the precursor for production of several viral miRNAs (Fig. 8.21) present at high concentrations in latently but not lytically infected neurons. Studies of properties of the RNAs synthesized from the LAT region are consistent with roles in the establishment or maintenance of latency. For example, when stably produced in neuronal cells, the LAT introns suppress replication of the viral genome and the synthesis of the immediate-early gene products that are needed for progression through the infectious cycle. They can also inhibit apoptosis and interfere with expression of interferon genes, functions that could promote the survival of latently infected neurons. Indeed, the LAT locus is required for maintenance of latently infected neurons that can support reentry into the lytic cycle. Similarly, several of the LAT-encoded miRNAs inhibit synthesis of viral transcriptional regulators such as ICP0 and ICP4 in transient-expression assays, suggesting that they might contribute to repression

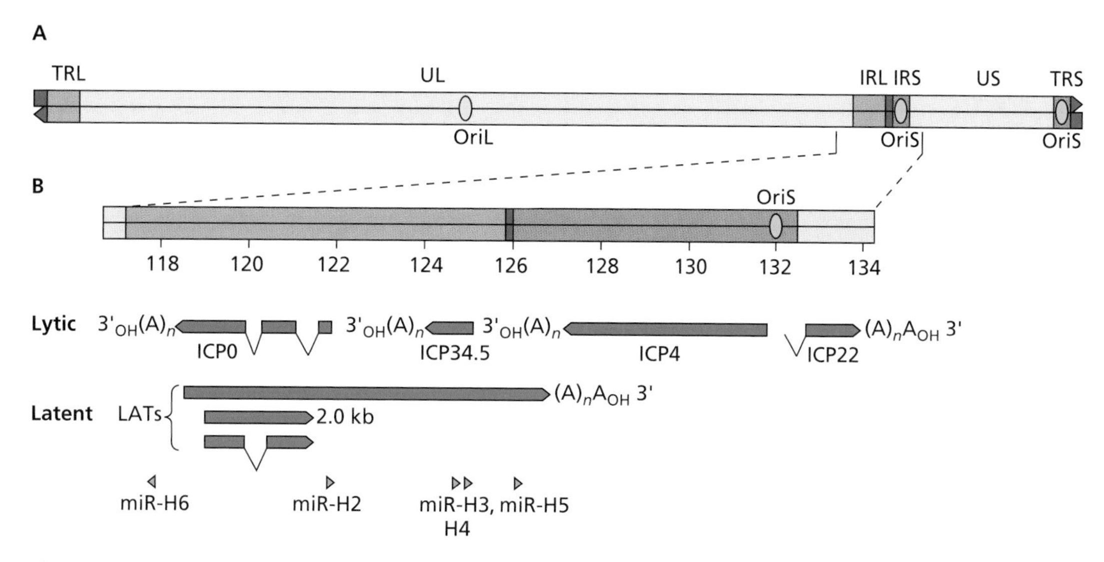

Figure 8.21 The latency-associated transcripts of herpes simplex virus type 1. (A) Diagram of the herpes simplex virus type 1 genome, showing the unique long and short segments, UL and US, respectively; the terminal repeat (TRL and TRS) and internal repeat (IRL and IRS) sequences; and the origins of replication, OriL and OriS. **(B)** Expanded map of the region shown, with the scale in kilobase pairs. This region encodes immediate-early proteins ICP0, ICP4, and ICP22, which play important roles in establishing a productive infection. Below are shown the locations of sequences encoding the major LATs. The arrows indicate the direction of transcription, and $(A)_nA_{OH}$ and $_{OH}A(A)_n$ 3′ poly(A) sequences. Below are shown positions of coding sequences for miRNAs miR-H2 to miR-H6, which are synthesized in high concentrations in latently infected murine trigeminal ganglion neurons. Deletions of the LAT promoter only or the promoter followed by 1.8 kbp of downstream sequence reduce the concentrations of all these miRNAs by some 2 orders of magnitude. The maintenance of the LAT region in active chromatin while lytic genes become associated with repressive nucleosomes may be facilitated by specialized DNA sequences (insulators) that flank the LAT region and demarcate different types of chromatin domains.

of lytic gene expression in latently infected cells. However, it has proved difficult to establish the specific contributions of the individual RNAs synthesized from the latency-associated region to the establishment of, maintenance of, or reactivation from latency. For example, the phenotypes exhibited by mutants carrying LAT gene deletions, such as failure to establish repressive chromatin on lytic gene promoters, were initially ascribed to the absence of LAT stable introns. However, these deletions remove the LAT promoter and hence also prevent production of the LAT-associated miRNAs. Furthermore, LAT deletions reduce the efficiency with which latency is established in some types of murine neurons (e.g., those of the trigeminal ganglia) but not in others.

How the lytic cycle transcriptional program is initiated during reactivation from latency presents a conundrum: VP16, the critical activator of this program, is a structural protein made only during the late phase of infection, which is never attained in latently infected neurons. For many years, it was thought that the need for VP16 must be circumvented. However, it is now clear that *de novo* synthesis of VP16 is induced when latently infected neurons are exposed to stresses that result in reactivation, and various lines of evidence implicate stimulation of transcription by VP16 in efficient reactivation (Box 8.13).

Transcription of Viral Genes by RNA Polymerase III

As noted previously, RNA polymerase III is dedicated to synthesis of small RNAs (typically comprising <200 nucleotides) that are made in large quantities (Table 8.2). The genomes of several of the viruses considered in this chapter contain genes that are transcribed by RNA polymerase III genomes (Table 8.4). The first, and still best-understood, example is the gene encoding human species C adenovirus virus-associated RNA I (VA-RNA I). The VA-RNA I gene specifies an RNA product that ameliorates the effects of a host cell defense mechanism (Volume II, Chapter 3) and also serves as a precursor for production of viral miRNAs. It contains a typical intragenic promoter that has been widely used in studies of initiation of transcription by RNA polymerase III.

The VA-RNA I Promoter

The human adenovirus type 5 genome contains two VA-RNA genes located very close to one another (Appendix, Fig. 1B).

BOX 8.13

EXPERIMENTS

New insights into herpes simplex virus type 1 reactivation from studies *in vivo*

As we have seen, efficient transcription from herpes simplex virus type 1 immediate-early promoters requires VP16 brought into infected cells as a structural component of virus particles, as well as the cellular proteins Oct-1 and Hcf. In latently infected neurons, viral late genes, including that encoding VP16, are not expressed. How then can transcription of immediate-early genes occur to initiate entry into the lytic cycle during reactivation from latency? Since initial studies of this issue performed more than 25 years ago, it has been thought that VP16 does not contribute to this process. Rather, it was proposed that a different viral activator (ICP0) initiates transcription of late genes or that early gene expression and hence genome replication and late gene transcription occur prior to transcription of immediate-early genes during reactivation. These initial studies relied on ganglia explanted from mice. During preparation of these explanted neurons, their axons are cut. It is now clear that neurons damaged in this way exhibit large-scale changes in gene expression and, within 2 to 3 h, characteristic features of neuronal degeneration. These observations indicated that reactivation from latent infection in explanted neurons might not reproduce the mechanisms by which herpes simplex virus type 1 replication is reactivated *in vivo*.

More recently, the mechanism of reactivation in latently infected trigeminal ganglion neurons of herpes simplex virus type 1-infected mice has been examined by exploiting viruses carrying reporters for expression of specific viral genes (e.g., VP16) and sensitive methods for quantification of viral genomes in individual neurons. In such experiments, it was observed that

- mutations that prevent synthesis of ICP0 or viral genome replication do not impair synthesis of VP16 in neurons during reactivation in response to heat shock (mimicking fever).
- mutations that impair activation of transcription by VP16, for example, because of deletion of its C-terminal activation domain (see figure), prevent reactivation *in vivo*, but not in neurons explanted into culture
- deletion of the ICP0 TAATGARAT promoter sequence recognized by the VP16–Oct-1–Hcf complex (see text) greatly reduced the efficiency of reactivation *in vivo*, but reactivation and viral replication were restored when this sequence was reintroduced into the promoter

It has therefore been proposed that VP16 in fact induces the lytic transcriptional program during reactivation of latent infection, as it does at the beginning of a lytic infection. The mechanism(s) that induce VP16 synthesis in latently infected neurons remains to be investigated.

Thompson RL, Preston CM, Sawtell NM. 2009. De novo synthesis of VP16 coordinates the exit from HSV latency in vivo. *PLoS Pathog* **5:**e1000352. doi:10.1371/journal.ppat.1000352.

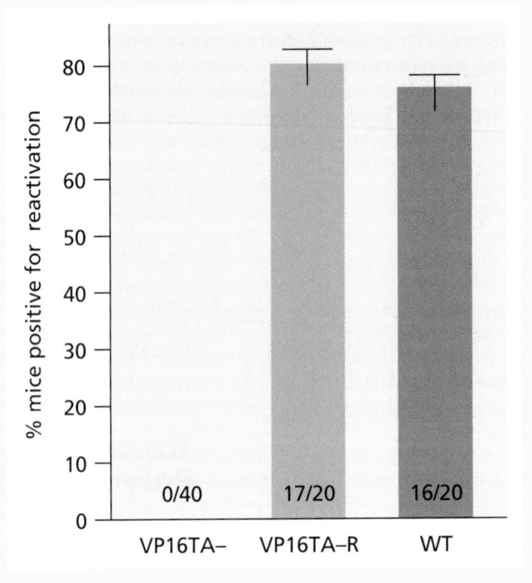

Mice were infected with a wild-type strain of herpes simplex virus type 1 (WT), a mutant defective for the transcriptional activation function of VP16 (VP16TA–), or a derivative of the mutant in which the sequences removed from the VP16 gene were restored (VP16TA–R). When latent infections had been established (40 days later), the mice were subject to hyperthermic stress to induce reactivation. Trigeminal ganglia were removed 22 h thereafter, and the concentrations of infectious virus particles were measured by plaque assay on cells in culture. The ratios on the histograms show the number of mice positive versus the number tested. Adapted from Thompson et al., *PLoS Pathog* 5:e100352, 2009, with permission.

Table 8.4 Viral RNA polymerase III transcription units

Virus	RNA polymerase III transcript	Function
Adenovirus		
Human adenovirus type 5	VA-RNA I	Blocks activation of RNA-dependent protein kinase; pre-miRNA
	VA-RNA II	Pre-miRNA
Herpesviruses		
Epstein-Barr virus	EBER-1, EBER-2	Made in latently infected cells; implicated in transformation and oncogenesis
Herpesvirus saimiri	HSVR 1–5	Degradation of certain cellular mRNAs
Murine gammaherpesvirus 68	Pre-miRNAs	Not known
Retrovirus		
Moloney murine leukemia virus	Let	Stimulation of transcription of specific cellular genes

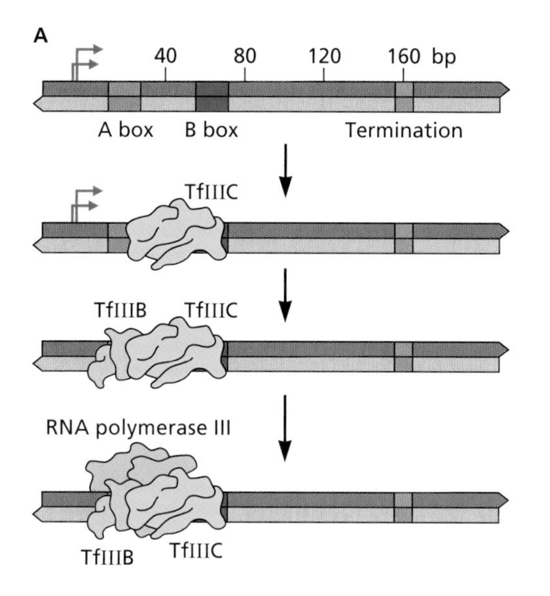

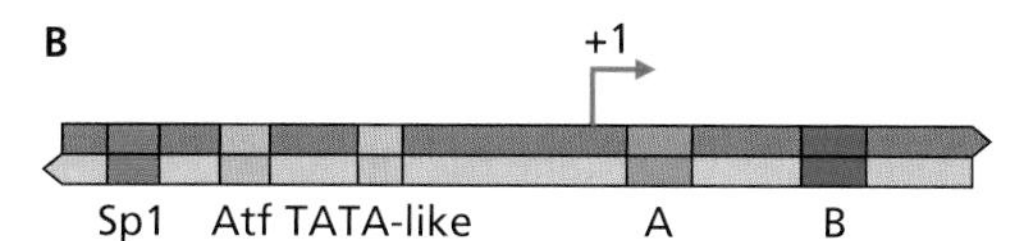

Figure 8.22 Organization of viral RNA polymerase III promoters. (A) The human adenovirus type 5 promoter. The VA-RNA I gene is depicted to scale, in base pairs. The intragenic A and B box sequences are essential for efficient VA-RNA I transcription and are closely related to the consensus A and B sequences of cellular tRNA genes. The VA-RNA termination site sequences are also typical of those of cellular genes transcribed by RNA polymerase III. **(B)** The Epstein-Barr virus EBER-2 promoter. The 5′ end of the Epstein-Barr virus EBER-2 transcription unit is shown to scale. This gene contains typical intragenic A and B box sequences. However, efficient transcription by RNA polymerase III also depends on the 5′ flanking sequence, which includes binding sites for the RNA polymerase II stimulatory proteins Sp1 and Atf. The TATA-like sequence is also important for efficient transcription and essential for specifying transcription by RNA polymerase III.

The VA-RNA I promoter is described here, for it is the more thoroughly characterized. Transcription of this gene depends on two intragenic sequences, the A and B boxes (Fig. 8.22A). As in the RNA polymerase II system, the essential promoter sequences are binding sites for accessory proteins necessary for promoter recognition. The internal sequences are recognized by the RNA polymerase III-specific initiation protein TfIIIc, which binds to the promoter to seed assembly of an initiation complex that also contains TfIIIb and the enzyme. This pathway of initiation was elucidated by using *in vitro* assays. We can be confident that this same mechanism operates in adenovirus-infected cells, because there is excellent agreement between the effects of A and B box mutations on VA-RNA I synthesis *in vitro* and in mutant virus-infected cells.

Regulation of VA-RNA Gene Transcription

The two VA-RNA genes are initially transcribed at similar rates, but during the late phase of infection, production of VA-RNA I is accelerated greatly. Such preferential transcription is the result of competition between the strong VA-RNA I and the intrinsically much weaker VA-RNA II promoters for a limiting component of the RNA polymerase III transcriptional machinery. Repression of VA-RNA I transcription may account for the similar rates at which the two genes are transcribed during the early phase. The control of transcription of VA-RNA genes emphasizes the fact that transcription by RNA polymerase III can, and must, be regulated, although the mechanisms are less elaborate than those that govern transcription by RNA polymerase II. Other viral RNA polymerase III transcription units include upstream promoter elements and hence illustrate the kinship of the RNA polymerase II and III systems (Fig. 8.22B).

Inhibition of the Cellular Transcriptional Machinery

Inhibition of cellular transcription in virus-infected cells offers several advantages. Cellular resources, such as substrates for RNA synthesis, can be devoted exclusively to the production of viral mRNAs (and, in many cases, RNA genomes), and competition between viral and cellular mRNAs for components of the translational machinery is minimized. The essential participation of cellular transcriptional systems in the infectious cycles of most viruses considered in this chapter precludes inactivation of this machinery. However, posttranscriptional mechanisms allow selective expression of adenoviral and herpesviral genes (Chapter 10). Furthermore, transcription of many cellular genes is inhibited following infection with herpes simplex virus type 1. Selective transcription of viral genes is accompanied by loss of RNA polymerase II phosphorylated at a specific amino acid, induced by the viral ICP22 protein, and proteasomal degradation of the hypophosphorylated form of the enzyme correlates with inhibition of transcription of cellular genes. Infection by poxviruses, with genomes that encode all components of a viral transcription machine, leads to rapid inhibition of synthesis of all classes of cellular RNA. Such inhibition requires viral proteins, but these have not been identified.

Reproduction of the majority of viruses with RNA genomes requires neither the cellular transcriptional machinery nor its RNA products, and is often accompanied by inhibition of cellular mRNA synthesis. Among the best-characterized examples is the inhibition of transcription by RNA polymerase II that is characteristic of poliovirus-infected cells. Such inhibition can be explained by the fact that the viral 3C^{pro} protease cleaves the Tbp subunit of TfIId at several sites. This modification eliminates the DNA-binding activity of Tbp and hence transcription by RNA polymerase II. The TATA-binding protein is also a subunit of initiation proteins that function with RNA polymerase III (TfIIIb) and RNA polymerase I. Consequently, its cleavage by 3C^{pro} in poliovirus-infected cells appears to be a very efficient way to prevent transcription of all cellular genes. As poliovirus yields are reduced in cells that synthesize an altered form of Tbp that is resistant to cleavage by 3C^{pro}, it is clear that inhibition of cellular transcription is necessary for optimal virus reproduction. The RNA genomes of alphaviruses such as Sindbis virus also encode a protein that induces degradation of an essential component of the cellular transcriptional machinery, in this case one of the catalytic subunits of RNA polymerase II.

Two gene products of the rhabdovirus vesicular stomatitis virus have been implicated in inhibition of cellular transcription. Following synthesis in the cytoplasm, the leader RNA described in Chapter 6 enters the nucleus, and is primarily responsible for the rapid reduction in cellular RNA synthesis in infected cells. The question of how short RNA molecules impair DNA-dependent RNA transcription cannot yet be answered, although *in vitro* experiments suggest that binding of a cellular protein to specific sequences within the RNA may be important. The viral M protein is also a potent inhibitor of transcription by RNA polymerase II, even in the absence of other viral gene products. This activity may become important later in infection, when replication of genome RNA predominates over mRNA synthesis and less leader RNA is produced.

Unusual Functions of Cellular Transcription Components

In the preceding sections, we concentrated on the similarities among the mechanisms by which viral and cellular DNA are transcribed. Even though all mechanisms of regulation of expression of viral genes by the host cell's RNA polymerase II or RNA polymerase III cannot be described in detail, the majority are not unique to viral systems. It is therefore an axiom of molecular virology that **every** mechanism by which viral transcription units are expressed by cellular components, or by which their activity is regulated, will prove to have a normal cellular counterpart. However, virus-infected cells also provide examples of functions or activities of cellular transcription proteins that have no known cellular counterparts.

One example of such a virus-specific function is the production of hepatitis delta satellite virus RNA from an **RNA** template by RNA polymerase II, described in Chapter 6. The RNA of viroids, infectious agents of plants, is synthesized in the same manner (Volume II, Chapter 12). Such RNA-dependent RNA synthesis by RNA polymerase II is one of the most remarkable interactions of a viral genome with the cellular transcriptional machinery. No cellular analog of this reaction is yet known. Even more divergent functions of cellular transcriptional components in virus-infected cells are illustrated by the participation of the RNA polymerase III initiation proteins TfIIIb and TfIIIc in integration of the yeast retrotransposon Ty3 (see Chapter 7). Given the large repertoire of molecular and biochemical activities displayed by components of the cellular transcriptional machinery, it seems likely that other unusual activities of these cellular proteins will be discovered in virus-infected cells.

A Viral DNA-Dependent RNA Polymerase

The DNA genomes of viruses considered in preceding sections replicate in the nucleus of infected cells, where the cellular transcriptional machinery resides. In contrast, poxviruses such as vaccinia virus are reproduced exclusively in the cytoplasm of their host cells. This feat is possible because the genomes of these viruses encode the components of transcription and RNA-processing systems that produce viral mRNAs with the hallmarks of cellular mRNA, such as 5′ caps and 3′ poly(A) tails. These components, which are carried into infected cells within virus particles, include a DNA-dependent RNA polymerase with striking structural and functional resemblance to cellular RNA polymerases.

Like those of other DNA viruses, vaccinia virus genes are expressed at different times in the infectious cycle (early, intermediate, and late). Distinguishing intermediate from late genes has been difficult: both are transcribed only after viral genome replication and their promoters share sequence similarities. In fact, construction of a genome-scale map of these transcription units has been achieved only recently (Box 8.14). All viral genes are transcribed by the viral RNA polymerase. This enzyme, like the cellular RNA polymerases, is a large, multisubunit enzyme built from the products of at least eight genes. The amino acid sequences of several of these subunits (including the two largest and the smallest) are clearly related to subunits of RNA polymerase II. Like its cellular counterparts, the vaccinia viral RNA polymerase recognizes promoters by cooperation with additional proteins. For example, formation of initiation complexes on vaccinia virus early promoters is mediated by the viral

BOX 8.14

EXPERIMENTS

The challenges of mapping vaccinia virus transcripts

The vaccinia virus genome contains more than 200 closely spaced open reading frames that are expressed sequentially in infected cells. As discussed in the text, synthesis of early, intermediate, and late viral transcripts depends on cooperation of the viral RNA polymerase with different sets of initiation proteins. By definition, early genes are those transcribed prior to viral genome replication. Early transcripts are therefore the first to be synthesized during synchronous infection and the only viral RNAs made when infected cells are maintained in the presence of inhibitors of DNA synthesis. Consequently, 118 early genes and 93 expressed only after genome replication were readily distinguished by high-throughput sequencing and mapping of polyadenylated RNA isolated from cells infected for increasing periods or maintained in the absence or presence of an inhibitor of viral DNA synthesis. In contrast, intermediate and late genes could not be distinguished using this approach alone, for several reasons. For example, only a short period separates the onset of expression of intermediate and late genes following viral DNA synthesis. This problem is compounded by the close spacing of open reading frames and extensive read-through transcription from one gene into neighboring downstream genes. Such read-through transcription is particularly pronounced after genome replication and results in representation of virtually every nucleotide in the viral genome in the infected cell RNA population.

The dependence of late gene expression on dedicated initiation proteins was used to solve this problem and to allow intermediate and late genes to be distinguished. These experiments relied on tightly inducible expression from the viral genome of the G8R gene, which encodes the protein necessary for late transcription, under the control of the *E. coli lac* operator: in cells infected by this vaccinia virus derivative, expression of the G8R genes is 99% inhibited unless the infected cells are exposed to the inducer isopropyl-β-D-thiogalactopyranoside (IPTG). High-throughput sequencing of polyadenylated RNA isolated from cells infected by the recombinant virus in the presence or absence of IPTG identified a large number of intermediate genes (see figure). Subsequent experiments took advantage of the fact that viral genes with an intermediate promoter are expressed from plasmids when introduced into vaccinia virus-infected cells maintained in the presence of a DNA synthesis inhibitor, whereas viral late promoters are not active. These studies confirmed the identity of 53 intermediate and 38 late genes.

Yang Z, Bruno DP, Martens CA, Porcella SF, Moss B. 2010. Simultaneous high resolution analysis of vaccinia virus and host cell transcriptomes by deep RNA sequencing. *Proc Natl Acad Sci U S A* **107:** 11513–11518.

Yang Z, Reynolds SE, Martens CA, Bruno DP, Porcella SF, Moss B. 2011. Expression profiling of the intermediate and late stages of poxvirus replication. *J Virol* **85:**9899–9908.

The results of RNA sequencing are plotted as the number of reads per nucleotide along the viral genome, with read counts above and below the line representing RNAs transcribed in the rightward and leftward directions, respectively. Reads obtained for RNA made in the presence and absence of IPTG are shown in red and green, respectively. Yellow = superimposed reads. Adapted from Z. Yang et al., *J Virol* **85:**9899–9908, 2011, with permission.

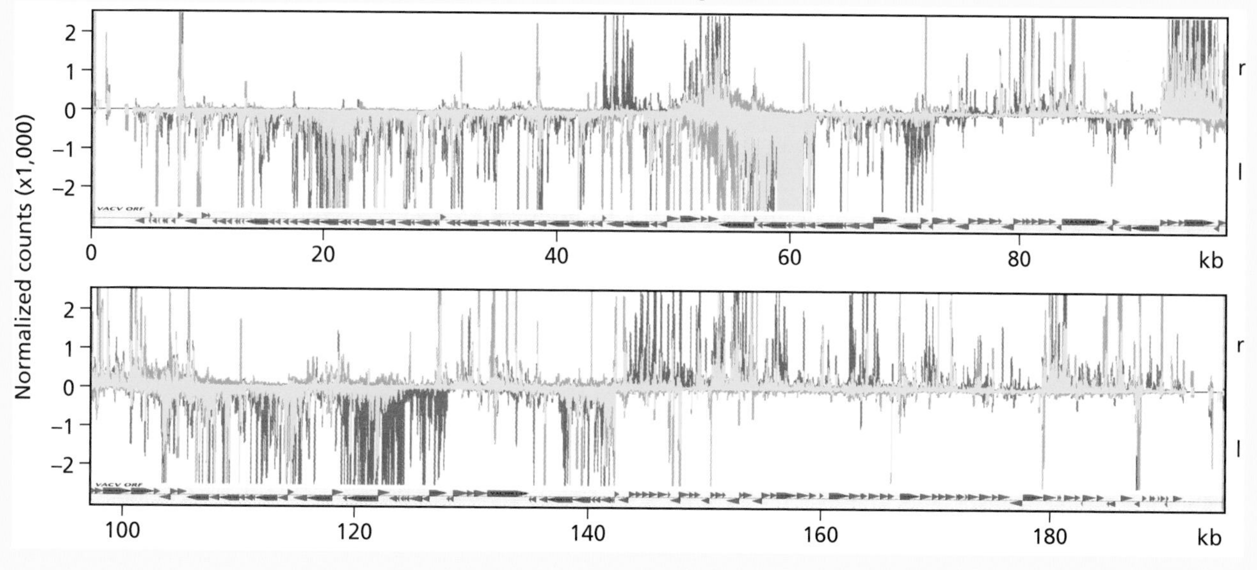

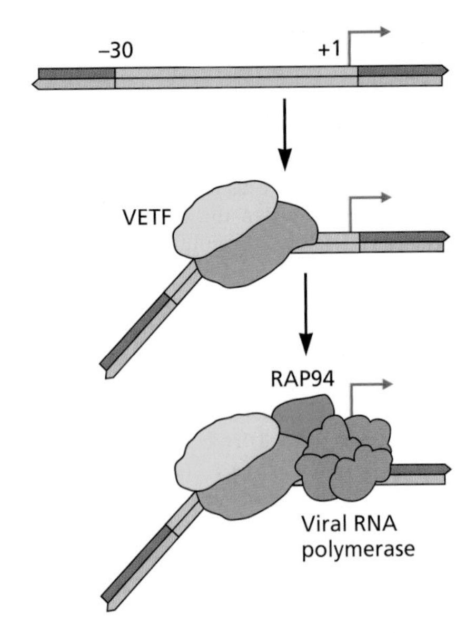

Figure 8.23 Assembly of an initiation complex on a vaccinia virus early promoter. Vaccinia virus early promoters contain an AT-rich sequence (tan) immediately upstream of the site of initiation. Vaccinia virus RNA polymerase cannot recognize these (or any other) viral promoters in the absence of other viral proteins. VETF is necessary for early promoter recognition and must bind before the viral RNA polymerase. This heteromeric protein associates specifically with early promoters and induces DNA bending. It also possesses DNA-dependent ATPase activity. VETF and the second protein necessary for early promoter specificity, RAP94, enter infected cells in virus particles. The RAP94-RNA polymerase complex associates with early promoter-bound VETF to form a functional initiation complex. Assembly of these vaccinia virus initiation complexes is therefore analogous to, although simpler than, formation of RNA polymerase II initiation complexes (Box 8.3).

proteins vaccinia virus early transcription protein (VETF) and RAP94, which are responsible for the recognition of promoter sequences and recruitment of the RNA polymerase, respectively (Fig. 8.23). These viral proteins are functional analogs of the cellular RNA polymerase II initiation proteins TfIId and TfIIf (Box 8.3). However, the vaccinia virus transcriptional machine is not analogous to its cellular counterpart in every respect. Cellular RNA polymerase II generally transcribes far beyond the sites at which the 3′ ends of mature cellular or viral mRNAs are produced by processing of the primary transcript, and does not terminate transcription at simple sequences. In contrast, transcription of the majority of vaccinia virus early genes **does** terminate at discrete sites, 20 to 50 bp downstream of specific T-rich sequences in the template. Termination requires the viral termination protein, which is also the viral mRNA-capping enzyme (see Chapter 10). The 3′ ends of the viral mRNAs correspond to sites of transcription termination. This viral mechanism is considerably simpler than the cellular counterpart.

In addition to the viral RNA polymerase, the several other proteins necessary for transcription of early genes enter host cells within vaccinia virus particles. Subsequent viral gene expression depends on viral genome replication and the ordered synthesis of viral proteins that permit sequential recognition of intermediate and late promoters. For example, transcription of intermediate genes requires synthesis of the viral RPO30 gene product (a subunit of the viral polymerase) and a second viral protein, while late transcription depends on production of several intermediate gene products. The viral genome also encodes several proteins that regulate elongation during transcription of late genes. Transcription of vaccinia virus genetic information is therefore regulated by mechanisms similar to those operating in cells infected by other DNA viruses, even though the transcriptional machinery is viral in origin.

Surprisingly, the vaccinia virus transcription system is not entirely self-contained: a cellular protein is necessary for transcription of viral intermediate genes. This protein (Vitf2) is located in the nucleus of uninfected cells but is present in both the cytoplasm and the nucleus of infected cells. As a significant number of vaccinia virus genes encode proteins necessary for transcription, such dependence on a cellular protein must confer some advantage. An attractive possibility is that interaction of the viral transcriptional machinery with a cellular protein serves to integrate the viral reproductive cycle with the growth state of its host cell. The identification of Vitf2 as a heterodimer of proteins that are produced in greatest quantities in proliferating cells is consistent with this hypothesis.

Perspectives

It is difficult to exaggerate the contributions of viral systems to the elucidation of mechanisms of transcription and its regulation in eukaryotic cells. The organization of RNA polymerase II promoters considered typical was first described for viral transcriptional control regions, enhancers were first discovered in viral genomes, and many important cellular regulators of transcription were identified by virtue of their specific binding to viral promoters. Perhaps even more importantly, efforts to elucidate the molecular basis of regulatory circuits that are crucial to viral infectious cycles have established general principles of transcriptional control. These include the importance of proteins that do not recognize DNA sequences directly and the ability of a single transcriptional regulator to modulate multiple components of the machinery. The insights into regulation of elongation by RNA polymerase II gained from studies of the human immunodeficiency virus type 1 Tat protein emphasize the intimate relationship of viral proteins with cellular components that make viral systems such rich resources for the investigation of eukaryotic transcription.

The identification of cellular and viral proteins necessary for transcription of specific viral genes has allowed many regulatory circuits to be traced. For example, the tissue distribution or the availability of particular cellular activators that bind to specific viral DNA sequences can account for the tropism of individual viruses, or conditions under which different transcriptional programs (latent or lytic) can be established. Furthermore, the mechanisms that allow sequential expression of viral genes are quite well established. Regardless of whether regulatory circuits are constructed of largely cellular or mostly viral proteins, these transcriptional cascades share such mechanistic features as sequential production of viral activators and integration of transcription of late genes with synthesis of viral DNA.

The models for the individual regulatory processes described in this chapter were developed initially by using convenient and powerful experimental systems. Such simplified systems (e.g., *in vitro* transcription reactions and transient-expression assays) do not reproduce the features characteristic of infected cells. Nor can they address such issues as how transcription of specific genes can be coupled with replication of the viral genome. It is therefore crucial that models be tested in virus-infected cells, even though it is more difficult to elucidate the molecular functions and mechanisms of action of transcriptional components. Many viral regulatory proteins perform multiple functions, a property that can confound genetic analysis, and the study of individual intracellular reactions, such as binding of a protein to a specific promoter sequence, is technically demanding. Nevertheless, viral *cis*-acting sequences and regulatory proteins remain more amenable to genetic analyses of their function in the natural context than do their cellular counterparts. In conjunction with increasingly powerful and sensitive methods for examining intracellular processes, continued efforts to exploit such genetic malleability will eventually establish how transcription of viral DNA templates is mediated and regulated within infected cells.

References

Book Chapters

Dasgupta A, Yalamanchilli P, Clark M, Kliewer S, Fradkin L, Rubinstein S, Das S, Sken Y, Weidman MK, Banerjee R, Datta U, Igo M, Barat P, Berk AJ. 2002. Effects of picornavirus proteinases on host cell transcription, p 321–336. *In* Semler BL, Wimmer E (ed), *Molecular Biology of Picornaviruses*. ASM Press, Washington, DC.

Reviews

Broyles SS. 2003. Vaccinia virus transcription. *J Gen Virol* **84:**2293–2303.

DeCaprio JA. 2009. How the Rb tumor suppressor structure and function was revealed by the study of Adenovirus and SV40. *Virology* **384:** 274–284.

Flint J, Shenk T. 1997. Viral transactivating proteins. *Annu Rev Genet* **31:**177–212.

Jenuwein T, Allis CD. 2002. Translating the histone code. *Science* **293:** 1074–1080.

Knipe DM, Lieberman PM, Jung JU, McBride AA, Morris KV, Ott M, Margolis D, Victo A, Nevels M, Parks RJ, Kristie TM. 2013. Snapshots: chromatin control of viral infection. *Virology* **435:**141–156.

Li B, Carey M, Workman JL. 2007. The role of chromatin during transcription. *Cell* **128:**707–719.

Liu X, Bushnell DA, Kornberg RD. 2013. RNA polymerase II transcription: structure and mechanism. *Biochim Biophys Acta* **1829:**2–8.

Malik S, Roeder RG. 2010. The metazoan Mediator co-activator complex as an integrative hub for transcriptional regulation. *Nat Rev Genet* **11:** 761–772.

Marsman J, Horsfield JA. 2012. Long distance relationships: enhancer-promoter communication and dynamic gene transcription. *Biochim Biophys Acta* **1819:**1217–1227.

Mbonye V, Karn J. 2011. Control of HIV latency by epigenetic and non-epigenetic mechanisms. *Curr HIV Res* **9:**555–567.

Naar AM, Lemon BD, Tjian R. 2001. Transcriptional coactivator complexes. *Annu Rev Biochem* **70:**475–501.

Ott M, Greyer M, Zhou Q. 2011. The control of HIV transcription: keeping RNA polymerase II on track. *Cell Host Microbe* **10:**426–435.

Roeder RG. 1996. The role of general initiation factors in transcription by RNA polymerase II. *Trends Biochem Sci* **21:**327–335.

Ruddell A. 1995. Transcription regulatory elements of the avian retroviral long terminal repeat. *Virology* **206:**1–7.

Sinclair AJ. 2003. bZIP proteins of gammaherpesviruses. *J Gen Virol* **84:**1941–1949.

Weake VM, Workman JL. 2010. Inducible gene expression: diverse regulatory mechanisms. *Nat Rev Genet* **11:**426–437.

Wysocka J, Herr W. 2003. The herpes simplex virus VP16-induced complex: the makings of a regulatory switch. *Trends Biochem Sci* **28:**294–304.

Papers of Special Interest

Viral RNA Polymerase II Promoters and the Cellular Transcriptional Machinery

Dynan WS, Tjian R. 1983. The promoter-specific transcription factor Sp1 binds to upstream sequences in the SV40 early promoter. *Cell* **35:**79–87.

Grüss P, Dhar R, Khoury G. 1981. Simian virus 40 tandem repeated sequences as an element of the early promoter. *Proc Natl Acad Sci U S A* **78:**943–947.

Hu SL, Manley JL. 1981. DNA sequences required for initiation of transcription *in vitro* from the major late promoter of adenovirus 2. *Proc Natl Acad Sci U S A* **78:**820–824.

Kovesdi I, Reichel R, Nevins JR. 1987. Role of an adenovirus E2 promoter binding factor in E1A-mediated coordinate gene control. *Proc Natl Acad Sci U S A* **84:**2180–2184.

Matsui T, Segall J, Weil PA, Roeder RG. 1980. Multiple factors required for accurate initiation of transcription by purified RNA polymerase II. *J Biol Chem* **255:**11992–11996.

Yamamoto T, de Crombrugghe B, Pastan I. 1980. Identification of a functional promoter in the long terminal repeat of Rous sarcoma virus. *Cell* **22:**787–797.

Viral Enhancers

Banerji J, Rusconi S, Schaffner W. 1981. Expression of the β-globin gene is enhanced by remote SV40 DNA sequences. *Cell* **27:**299–308.

Fromental C, Kanno M, Nomiyama H, Chambon P. 1988. Cooperativity and hierarchical levels of functional organization in the SV40 enhancer. *Cell* **54:**943–953.

Moreau P, Hen R, Wasylyk B, Everett R, Gaub MP, Chambon P. 1981. The SV40 72 base repeat has a striking effect on gene expression both in SV40 and other chimeric recombinants. *Nucleic Acids Res* **9:**6047–6068.

Nabel GJ, Baltimore D. 1987. An inducible transcription factor that activates expression of human immunodeficiency virus in T cells. *Nature* **326:**711–713.

Spalholz BA, Yang YC, Howley PM. 1985. Transactivation of a bovine papillomavirus transcriptional regulatory element by the E2 gene product. *Cell* **42:**183–191.

Yee J. 1989. A liver-specific enhancer in the core promoter of human hepatitis B virus. *Science* **246:**658–670.

RNA-Dependent Stimulation of Human Immunodeficiency Virus Type 1 Transcription by the Tat Protein

Barboric M, Yik JH, Czudnochowski N, Yang Z, Chen R, Contteras X, Geyer M, Matjia Perterlin B, Zhou Q. 2007. Tat competes with HEXIM1 to increase the active pool of P-TEFb for HIV-1 transcription. *Nucleic Acids Res* **35:**2003–2012.

Brès V, Tagami H, Pèloponèse JM, Loret E, Jeang KT, Nakatani Y, Emiliani S, Benkirane M, Kiernan RE. 2002. Differential aceyltation of Tat coordinates its interactions with the coactivators cyclin T1 and VCAF. *EMBO J* **21:**6811–6819.

Dingwall C, Ernberg J, Gait MJ, Green SM, Heaphy S, Karn J, Singh M, Shinner JJ. 1990. HIV-1 Tat protein stimulates transcription by binding to a U-rich bulge in the stem of the TAR RNA structure. *EMBO J* **9:**4145–4153.

Feinberg MB, Baltimore D, Frankel AD. 1991. The role of Tat in the human immunodeficiency virus life cycle indicates a primary effect on transcriptional elongation. *Proc Natl Acad Sci U S A* **88:**4045–4049.

Mancebo HS, Lee G, Flygare J, Tomassini J, Luu P, Zhu Y, Peng J, Blau C, Hazuda D, Price D, Flores O. 1997. P-TEFb kinase is required for HIV Tat transcriptional activation in vivo and in vitro. *Genes Dev* **11:**2633–2644.

Parada CA, Roeder RG. 1996. Enhanced processivity of RNA polymerase II triggered by Tat-induced phosphorylation of its carboxy-terminal domain. *Nature* **384:**375–378.

Tréand C, du Chéné I, Brès V, Kiernan R, Benarous R, Benkirane M, Emiliani S. 2006. Requirement of SWI/SNF chromatin-remodeling complex in Tat-mediated activation of the HIV-1 promoter. *EMBO J* **25:**1690–1699.

Wei P, Garber MF, Fang SM, Fischer WH, Jones KA. 1998. A novel CDK9-associated C-type cyclin interacts directly with HIV-1 Tat and mediates its high-affinity, loop-specific binding to TAR RNA. *Cell* **92:**451–462.

Regulation of Transcription of DNA Genomes by Viral Proteins

Bandara LR, La Thangue NB. 1991. Adenovirus E1A prevents the retinoblastoma gene product from complexing with a cellular transcription factor. *Nature* **351:**494–497.

Batterson W, Roizman B. 1983. Characterization of the herpes simplex virion-associated factor responsible for the induction of α genes. *J Virol* **46:**371–377.

Berk AJ, Lee F, Harrison T, Williams JF, Sharp PA. 1979. Pre-early adenovirus 5 gene product regulates synthesis of early viral messenger RNAs. *Cell* **17:**935–944.

Chellappan SP, Hiebert S, Mudryj M, Horowitz JM, Nevins JR. 1991. The E2F transcription factor is a cellular target for the Rb protein. *Cell* **65:**1053–1061.

Chi T, Carey M. 1993. The ZEBRA activation domain: modular organization and mechanism of action. *Mol Cell Biol* **13:**7045–7055.

Jones N, Shenk T. 1979. An adenovirus type 5 early gene function regulates expression of other early viral genes. *Proc Natl Acad Sci U S A* **76:**3665–3669.

Keller JM, Alwine JC. 1984. Activation of the SV40 late promoter: direct effects of T antigen in the absence of viral DNA replication. *Cell* **36:**381–389.

Lai JS, Cleary MA, Herr W. 1992. A single amino acid exchange transfers VP16-induced positive control from the Oct-1 to the Oct-2 homeo domain. *Genes Dev* **6:**2058–2065.

Stevens JL, Cantin GT, Wang G, Shevchenko A, Shevchenko A, Berk AJ. 2002. Transcription control by E1A and MAP kinase pathway via Sur2 Mediator subunit. *Science* **296:**755–758.

Tribouley C, Lutz P, Staub A, Kedinger C. 1994. The product of the adenovirus intermediate gene IVa2 is a transcriptional activator of the major late promoter. *J Virol* **68:**4450–4457.

Wiley SR, Kraus RJ, Zuo F, Murray EE, Loritz K, Mertz JE. 1993. SV40 early-to-late switch involves titration of cellular transcriptional repressors. *Genes Dev* **7:**2206–2219.

Wilson AC, Freeman RW, Goto H, Nishimoto T, Herr W. 1997. VP16 targets an amino-terminal domain of HCF involved in cell cycle progression. *Mol Cell Biol* **17:**6139–6146.

Establishing Latent or Lytic Infections

Atanasia D, Kent JR, Gartner JJ, Fraser NW. 2006. The stable 2-kb LAT intron of herpes simplex stimulates the expression of heat shock proteins and protects cells from stress. *Virology* **350:**26–33.

Kramer MF, Jurak I, Pesola JM, Boise S, Knipe DM, Coen DM. 2011. Herpes simplex virus 1 microRNAs expressed abundantly during latent infection are not essential for latency in mouse trigeminal ganglia. *Virology* **417:**239–247.

Kraus RJ, Perrigoue JG, Mertz JE. 2003. ZEB negatively regulates the lytic switch BZLF1 gene promoter of Epstein-Barr virus. *J Virol* **77:**199–207.

Wang QY, Zhou C, Johnson KE, Colgrove RC, Coen DM, Knipe DM. 2005. Herpesviral latency-associated gene promotes assembly of heterochromatin on viral lytic-gene promoters in latent infection. *Proc Natl Acad Sci U S A* **102:**16055–16059.

Transcription of Viral Genes by RNA Polymerase III

Fowlkes DM, Shenk T. 1980. Transcriptional control regions of the adenovirus VA1 RNA gene. *Cell* **22:**405–413.

Howe JG, Shu MD. 1993. Upstream basal promoter element important for exclusive RNA polymerase III transcription of EBER 2 gene. *Mol Cell Biol* **13:**2655–2665.

Inhibition of the Cellular Transcriptional Machinery

Akhrymuk I, Kulemzin SV, Frolova EI. 2012. Evasion of the innate immune response: the Old World alphavirus nsP2 protein induces rapid degradation of Rpb1, a catalytic subunit of RNA polymerase II. *J Virol* **86:**7180–7191.

Clark ML, Lieverman PM, Berk AJ, Dasgupta A. 1993. Direct cleavage of TATA-binding protein by poliovirus protease 3C in vivo and in vitro. *Mol Cell Biol* **13:**1232–1237.

Spencer CA, Dahmus ME, Rice SA. 1997. Repression of host RNA polymerase II transcription by herpes simplex virus type 1. *J Virol* **71:**2031–2040.

The Poxviral Transcriptional System

Hagler J, Shuman S. 1992. A freeze-frame view of eukaryotic transcription during elongation and capping of nascent RNA. *Science* **255:**983–986.

Kates JR, McAuslan BR. 1967. Poxvirus DNA-dependent RNA polymerase. *Proc Natl Acad Sci U S A* **58:**134–141.

Passarelli AL, Kovacs GR, Moss B. 1996. Transcription of a vaccinia virus late promoter template: requirement for the product of the A2L intermediate-stage gene. *J Virol* **70:**4444–4450.

Rosales R, Sutter G, Moss B. 1994. A cellular factor is required for transcription of vaccinia viral intermediate-stage genes. *Proc Natl Acad Sci U S A* **91:**3794–3798.

9 Replication of DNA Genomes

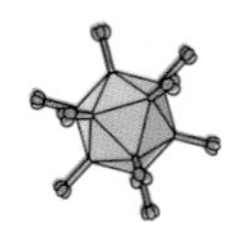

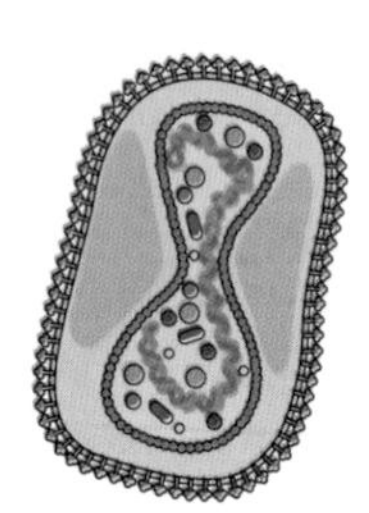

Introduction

DNA Synthesis by the Cellular Replication Machinery
- Eukaryotic Replicons
- Cellular Replication Proteins

Mechanisms of Viral DNA Synthesis
- Lessons from Simian Virus 40
- Replication of Other Viral DNA Genomes
- Properties of Viral Replication Origins
- Recognition of Viral Replication Origins
- Viral DNA Synthesis Machines
- Resolution and Processing of Viral Replication Products

Exponential Accumulation of Viral Genomes
- Viral Proteins Can Induce Synthesis of Cellular Replication Proteins
- Synthesis of Viral Replication Machines and Accessory Enzymes
- Viral DNA Replication Independent of Cellular Proteins
- Delayed Synthesis of Structural Proteins Prevents Premature Packaging of DNA Templates
- Inhibition of Cellular DNA Synthesis
- Viral DNAs Are Synthesized in Specialized Intracellular Compartments

Limited Replication of Viral DNA Genomes
- Integrated Parvoviral DNA Can Replicate as Part of the Cellular Genome
- Different Viral Origins Regulate Replication of Epstein-Barr Virus
- Limited and Amplifying Replication from a Single Origin: the Papillomaviruses

Origins of Genetic Diversity in DNA Viruses
- Fidelity of Replication by Viral DNA Polymerases
- Inhibition of Repair of Double-Strand Breaks in DNA
- Recombination of Viral Genomes

Perspectives

References

LINKS FOR CHAPTER 9

⏭ *Video: Interview with Dr. Sandra Weller*
http://bit.ly/Virology_Weller

⏭ *Mark Challberg, a cold room kind of guy*
http://bit.ly/Virology_Twiv203

The more the merrier.

ANONYMOUS

Introduction

The genomes of DNA viruses span a considerable size range, from some 1.7 kb (circoviruses) to >2.5 Mbp (pandoraviruses), and may be single- or double-stranded DNA molecules that are linear or circular (Fig. 9.1). Whatever their physical nature, viral DNA molecules must be replicated within an infected cell to provide genomes for assembly into progeny virus particles. Such replication invariably requires the synthesis of at least one, but usually several, viral proteins. Consequently, viral DNA synthesis cannot begin immediately upon arrival of the genome at the appropriate intracellular site, but rather is delayed until viral replication proteins have attained a sufficient concentration. Initiation of viral DNA synthesis typically leads to many cycles of replication and the accumulation of large numbers of newly synthesized DNA molecules. However, longer-lasting latent infections are also common, both in nature and in the laboratory. In these circumstances, the number of viral DNA molecules made is strictly controlled.

Replication of all DNA, from the genome of the simplest virus to that of the most complex vertebrate cell, follows a set of universal rules: (i) DNA is always synthesized by template-directed, stepwise incorporation of deoxynucleoside monophosphates (dNMPs) from deoxynucleoside triphosphate (dNTP) substrates into the 3′-OH end of the growing DNA chain; (ii) each parental strand of a duplex DNA template is copied by base pairing to produce two daughter molecules identical to one another and to their parent (**semiconservative replication**); (iii) replication of DNA begins and ends at specific sites in the template, termed **origins** and **termini**, respectively; and (iv) DNA synthesis is catalyzed by DNA-dependent DNA polymerases, but many accessory proteins are required for initiation or elongation. In contrast to all DNA-dependent, and many RNA-dependent, RNA polymerases, **no** DNA polymerase can initiate template-directed DNA synthesis *de novo*. All require a **primer** with a free 3′-OH end to which dNMPs complementary to those of the template strand are added.

The genomes of RNA viruses must encode enzymes that catalyze RNA-dependent RNA or DNA synthesis. In contrast, those of DNA viruses can be replicated by the cellular machinery. Indeed, replication of the smaller DNA viruses, such as parvoviruses and polyomaviruses, requires but a single viral replication protein, and the majority of reactions are carried out by cellular proteins (Fig. 9.1). This strategy avoids the need to devote limited viral genetic coding capacity to enzymes and other proteins required for DNA synthesis. In contrast, the genomes of all larger DNA viruses encode DNA polymerases and additional replication proteins. In the extreme case, exemplified by poxviruses, the viral genome encodes a complete DNA synthesis system and is replicated in the cytoplasm of host cells.

PRINCIPLES *Replication of DNA genomes*

- As during cellular DNA replication, viral DNA is always synthesized by template-directed, stepwise incorporation of deoxynucleoside monophosphates (dNMPs) from deoxynucleoside triphosphate (dNTP) substrates into the 3′-OH end of the growing DNA chain.
  - Each parental strand of a duplex DNA template is copied by base pairing to produce two daughter molecules identical to one another and to their parent (**semiconservative replication**).
  - Replication of DNA begins and ends at specific sites in the template, termed **origins** and **termini**, respectively.
- In contrast to many RNA polymerases, no known DNA polymerase can initiate synthesis *de novo*: all require a primer with a free 3′-OH group.
- Priming of viral DNA synthesis can be via the 3′-OH terminus of RNA, a protein, or the ends of specialized structures in the genomic DNA.
- Viral DNA replication occurs either by copying of both strands at a replication fork or by copying of one strand and displacement of the other.
- The polyomavirus simian virus 40 was essential for elucidating crucial aspects of replication of viral DNA genomes, as well as identifying essential cellular replication proteins.
- Viral origins are assembly points for DNA replication machines and are recognized by dedicated origin-binding proteins.
- Viral DNA synthesis depends on a combination of viral and cellular replication proteins; in extreme cases, all replication proteins are encoded in the viral genomes.
- When viral DNA replication is carried out largely by viral proteins, cellular DNA synthesis is inhibited, probably in order to increase the pool of substrates for optimal viral replication.
- Viral DNA replication and transcription occur in discrete compartments within the cell, in which the viral proteins that participate in these processes are concentrated.
- During viral persistence, alternative replication mechanisms maintain viral genomes at low concentrations and partition them into daughter cells.
- DNA viruses are replicated with high fidelity because both cellular and viral DNA polymerases possess proofreading capability.
- Recombination drives viral diversity, and components of recombination systems may participate in viral DNA replication.

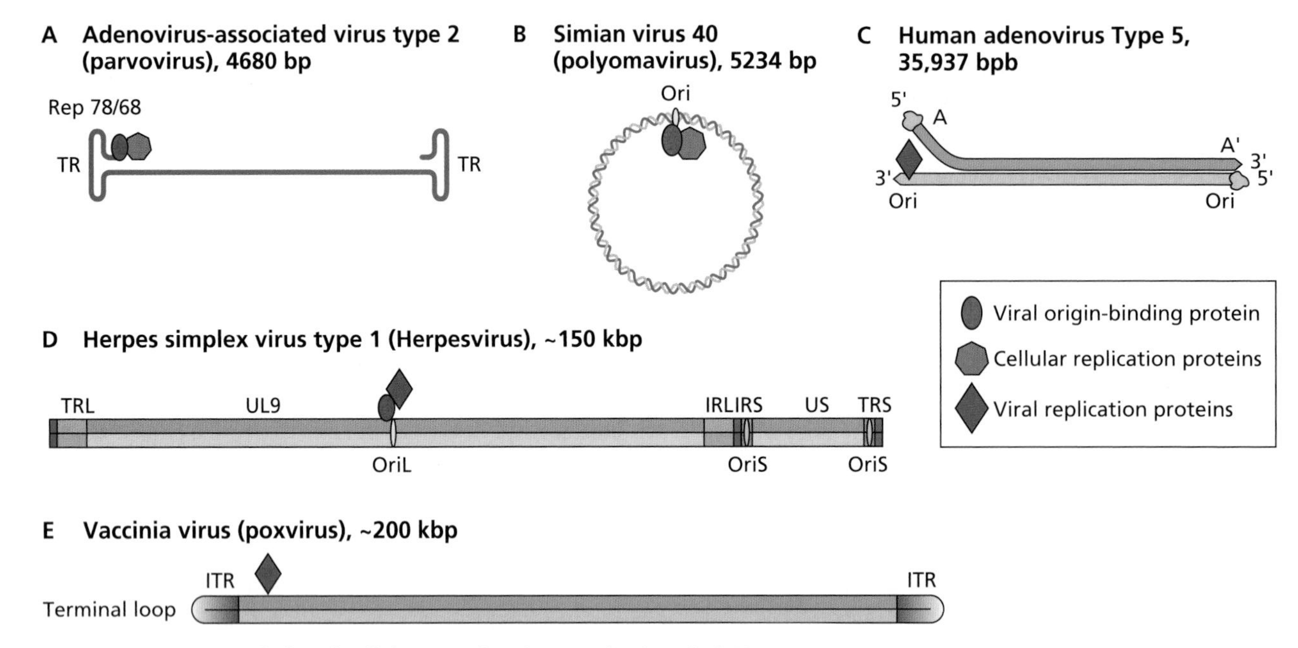

Figure 9.1 Viral and cellular proteins that synthesize viral DNA genomes. The genomes of the viruses listed are shown schematically and **not** to scale with respect to one another. The herpes simplex virus type 1 genome comprises long and short unique regions (UL and US) flanked by internal and terminal repeat sequences (IRL, IRS, TRL, TRS). When present, the positions of origins of replication (Ori) are indicated, as are the viral proteins that recognize origins, and the cellular or viral origin of the proteins that carry out DNA synthesis. ITR, inverted terminal repetition.

There is also variety in the mechanism of priming of viral DNA synthesis. In some cases, short RNA primers are first synthesized, as during replication of cellular genomes. In others, structural features of the genome or viral proteins provide primers. Despite such distinctions, the replication strategies of different viral DNAs are based on common molecular principles and one of only two mechanisms: copying of both strands of a double-stranded DNA template at a replication fork or copying of only one strand while its complement is displaced (Box 9.1). For example, the genomes of polyomaviruses and herpesviruses, which are quite different in size and structure, are replicated by the cellular replication machinery and viral replication proteins,

BOX 9.1

BACKGROUND

The two mechanisms of synthesis of double-stranded viral DNA molecules

Replication of double-stranded nucleic acids proceeds by **either** copying of both strands at a replication fork **or** copying of only one strand while its complement is displaced. No other replication mechanisms are known.

Among viral genomes, only those of certain double-stranded DNA viruses are synthesized via a replication fork. Replication of viral double-stranded RNAs **never** proceeds via this mechanism.

DNA synthesis via a replication fork is **always** initiated from an RNA primer. In contrast, strand displacement synthesis of viral DNA **never** requires an RNA primer.

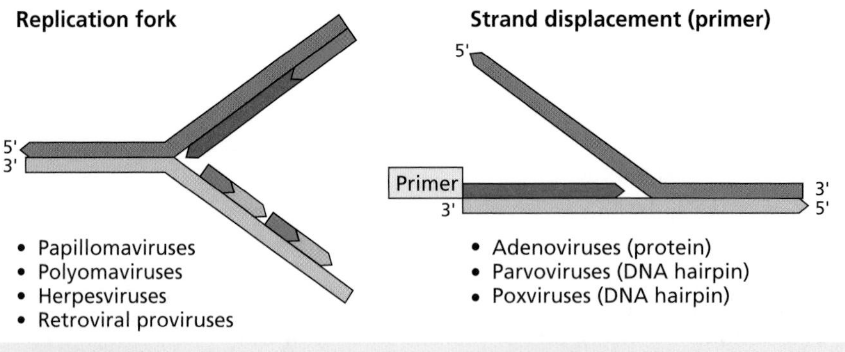

Parental DNA, RNA primers, and newly synthesized DNA are shown in blue, green, and red/pink, respectively. The primer indicated by the tan box can be a DNA structure or a protein.

respectively (Fig. 9.1). Nevertheless, synthesis of these two DNAs is initiated by the same priming mechanism, and the herpesviral replication machinery carries out the same biochemical reactions as the host proteins that mediate synthesis of polyomavirus DNA.

DNA Synthesis by the Cellular Replication Machinery

Our current understanding of the intricate reactions by which both strands of a typical double-stranded DNA template are copied in eukaryotic cells is based on *in vitro* studies of simian virus 40 DNA synthesis. In the next section, we discuss the cellular replication machinery that catalyzes these reactions and the molecular functions of its components that were established by such studies. Here, we briefly describe general features of eukaryotic DNA replication, and why simian virus 40 proved to be an invaluable resource for those seeking to understand this process.

Eukaryotic Replicons

General Features

The replication of large eukaryotic genomes within the lifetime of an actively growing cell depends on their organization into smaller units of replication termed **replicons** (Fig. 9.2). At the maximal rate of replication observed *in vivo*, a typical human chromosome could not be copied from a single origin as a single unit in less than 10 days! Each chromosome therefore contains many replicons, ranging in length from ~20 to 300 kbp. All but the smallest viral DNA genomes also contain two or three origins (see "Properties of Viral Replication Origins" below).

Each replicon contains an origin at which replication begins. The sites at which nascent DNA chains are being synthesized, the ends of "bubbles" seen in the electron microscope (Fig. 9.2A), are termed **replication forks**. In bidirectional replication, two replication forks are established at a single origin and move away from it as the new DNA strands are synthesized (Fig. 9.2B). However, as DNA must be synthesized in the 5′ → 3′ direction, only one of the two parental strands can be copied continuously from a primer deposited at the origin. The long-standing conundrum of how the second strand is synthesized was solved with the elucidation of the discontinuous mechanism of synthesis (Fig. 9.3A): RNA primers for DNA synthesis are synthesized at multiple sites, such that the second new DNA strand is made initially as short, discontinuous segments, termed **Okazaki fragments** in honor of the investigator who discovered them.

The discontinuous mechanism of DNA synthesis creates a special problem at the ends of linear DNAs, where excision of the terminal primer creates a gap at the 5′ end of the daughter DNA molecules (Fig. 9.3B). In the absence of a mechanism for completing synthesis of termini, discontinuous DNA

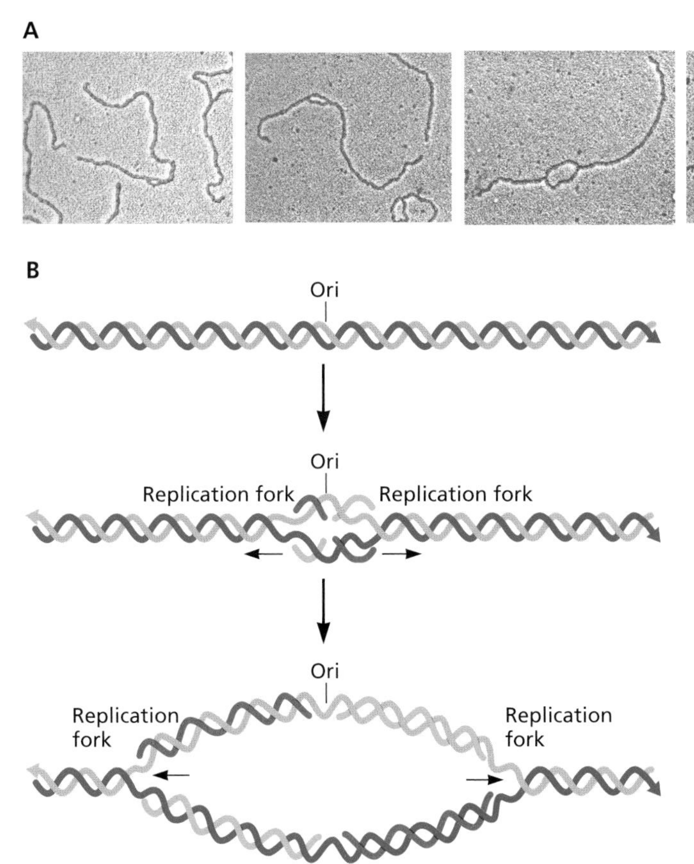

Figure 9.2 Properties of replicons. (A) Electron micrographs of replicating simian virus 40 DNA, showing the "bubbles" of replicating DNA, in which the two strands of the template are unwound. These linear DNA molecules were obtained by restriction endonuclease cleavage of viral DNA that had replicated to different degrees in infected cells. They are arranged in order of increasing degree of replication to illustrate the progressive movement of the two replication forks from a single origin of replication. From G. C. Fareed et al., *J Virol* **10**:484–491, 1972, with permission. **(B)** Bidirectional replication from an origin. Newly synthesized DNA is shown in red and pink, a convention used throughout the text.

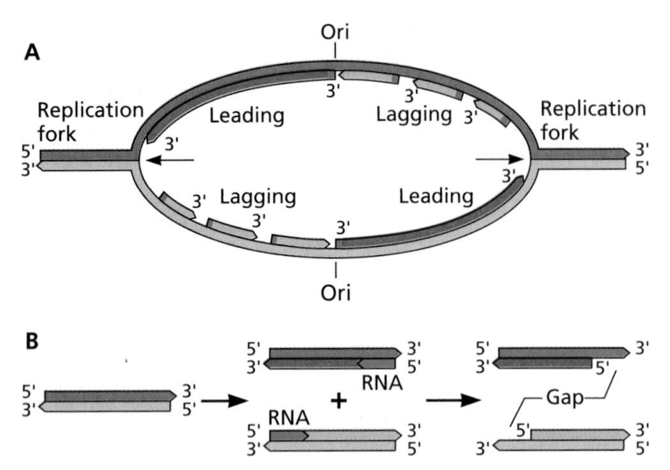

Figure 9.3 Semidiscontinuous DNA synthesis from a bidirectional origin. (A) Semidiscontinuous synthesis of the daughter strands. Synthesis of the RNA primers (green) at the origin allows initiation of continuous copying of one of the two strands on either side of the origin in the replication bubble. The second strand cannot be made in the same way (see the text). The nascent DNA population contains many small molecules termed **Okazaki fragments** in honor of the investigator who first described them. The presence of short segments of RNA at the 5′ ends of Okazaki fragments indicated that the primers necessary for DNA synthesis are molecules of RNA. With increasing time of replication, these small fragments are incorporated into long DNA molecules, indicating that they are precursors. It was therefore deduced that the second nascent DNA strand is synthesized discontinuously, also in the 5′ → 3′ direction. Because synthesis of this strand cannot begin until the replication fork has moved some distance from the origin, it is called the **lagging strand**, while the strand synthesized continuously is termed the **leading strand**. Complete replication of the lagging strand requires enzymes that can remove RNA primers, repair the gaps thus created, and ligate the individual DNA fragments to produce a continuous copy of the template strand. **(B)** Incomplete synthesis of the lagging strand. When a DNA molecule is linear, removal of the terminal RNA primer from the 5′ end of the lagging strand creates a gap that cannot be repaired by any DNA-dependent DNA polymerase.

synthesis would lead to an intolerable loss of genetic information. In chromosomal DNA, specialized elements, called **telomeres**, at the ends of each chromosome prevent loss of terminal sequences. These structures comprise simple, repeated sequences maintained by reverse transcription of an RNA template, which is an essential component of the ribonucleoprotein enzyme telomerase. As discussed subsequently, complete replication of all sequences of linear viral DNA genomes is achieved by a variety of elegant mechanisms.

Origins of Cellular Replication

It is well established that replication initiates at numerous, specific sites in eukaryotic genomes. The origins of the simple eukaryote *Saccharomyces cerevisiae* (budding yeast) can be characterized readily, because they support replication of small plasmids that are maintained as episomes. All yeast origins behave as such **autonomously replicating sequences** and can therefore be defined in detail. In contrast, this simple functional assay failed to identify analogous mammalian sequences, even when applied to DNA segments that contained origins mapped in mammalian chromosomes. Genome-wide analysis of sites of both initiation of DNA synthesis and binding of conserved replication proteins, such as the origin recognition complex (Orc), indicated that mammalian origins do not comprise specific consensus sequences (as in budding yeast). Rather, initiation sites are defined by a variety of parameters, including proximity to active promoters, presence of CG-rich sequences, and chromatin structure. The difficulties in identifying functional origins in mammalian genomes made compact viral genomes like that of simian virus 40 essential tools for elucidation of mechanisms of origin-dependent DNA synthesis.

Cellular Replication Proteins

Eukaryotic DNA Polymerases

It has been known for more than 50 years that eukaryotic cells contain DNA-dependent DNA polymerases. Mammalian cells contain several such nuclear enzymes, which are distinguished by their sensitivities to various inhibitors and their degree of **processivity**, the number of nucleotides incorporated into a nascent DNA chain per initiation reaction. These characteristics can be readily assayed in *in vitro* reactions with artificial template-primers, such as gapped or nicked DNA molecules. The requirements for viral DNA synthesis *in vitro* and genetic analyses (performed largely with yeasts) identified DNA polymerases α, δ, and ε as the enzymes that participate in genome replication. Other DNA polymerases are restricted to mitochondria or act only during repair of damaged DNA (e.g., DNA polymerase β). Only DNA polymerase α is associated with priming activity, because it is bound tightly to a heteromeric **primase**.

One of the most striking properties of these DNA polymerases is their obvious evolutionary relationships to prokaryotic and viral enzymes. All template-directed nucleic acid polymerases share several sequence motifs and probably a similar core architecture (Chapters 6 and 7), indicating that important features of the catalytic mechanisms are also common to all these enzymes.

Other Proteins Required for DNA Synthesis in Mammalian Cells

Analogy with well-characterized bacterial DNA replication machines indicated that several proteins in addition to DNA polymerase and primase would be required for mammalian DNA synthesis. Identification of such proteins awaited the development of cell-free systems for origin-dependent initiation. This feat was first accomplished for synthesis of adenoviral DNA, a breakthrough soon followed by origin-dependent replication of simian virus 40 DNA *in vitro*. Because cellular components are largely responsible for simian virus 40 DNA synthesis, development of this system proved to be the watershed in the investigation of eukaryotic DNA replication: it allowed the identification of previously unknown cellular replication proteins and elucidation of their mechanisms of action.

Mechanisms of Viral DNA Synthesis

In this section, we first describe the contribution of studies of simian virus 40 genome replication, for subsequent comparison to the variety of virus-specific solutions to the mechanistic problems associated with each step in DNA synthesis.

Lessons from Simian Virus 40

The Origin of Simian Virus 40 DNA Replication

The simian virus 40 (SV40) origin was the first viral control sequence to be located on a physical map of the viral genome, in which the reference points were restriction endonuclease cleavage sites (Box 9.2). We now possess a detailed picture of this viral origin (Fig. 9.4) and of the binding sites for the viral origin recognition protein, large T antigen (LT). A 64-bp sequence, the **core origin**, which lies between the sites at which early and late transcription begin, is sufficient for initiation of DNA synthesis in infected cells. This sequence contains four copies of a pentanucleotide-binding site for LT, flanked by an AT-rich element and a 10-bp imperfect palindrome (Fig. 9.4). Additional sequences within this busy control region of the viral genome increase the efficiency of initiation of DNA synthesis from this core origin.

Mechanism of Simian Virus 40 DNA Synthesis

Origin recognition and unwinding. The first step in SV40 DNA synthesis is the recognition of the origin by LT, the major early gene product of the virus. This viral protein can bind to pentanucleotide repeat sequences in the core origin to form a hexamer (Fig. 9.4). However, initiation of viral DNA synthesis also requires the sequences that flank the minimal origin. When bound to ATP, LT assembles to form a double hexamer on the origin and elicits structural distortions in the flanking sequences. In concert with cellular replication protein A (Rp-A), which possesses single-stranded-DNA-binding activity, the intrinsic $3' \rightarrow 5'$ helicase activity of LT then harnesses the energy of ATP hydrolysis to unwind DNA bidirectionally from the core

BOX 9.2

EXPERIMENTS

Mapping of the simian virus 40 origin of replication

As illustrated in panel **A** of the figure (left), exposure of simian virus 40-infected monkey cells to [^{3}H]thymidine ([^{3}H]dT) for a period less than the time required to complete one round of replication (e.g., 5 min) results in labeling of the growing points of replicating DNA. If replication proceeds from a specific origin (Ori) to a specific termination site (T), the DNA replicated last will be labeled preferentially in the population of completely replicated molecules (panel **A**, right). The distribution of [^{3}H]thymidine among the fragments of completely replicated viral DNA generated by digestion with restriction endonucleases HindII and HindIII is shown in panel **B**. The simian virus 40 genome is represented as cleaved within the G fragment, and relative distances are given with respect to the junction of the A and C fragments. The observation of two decreasing gradients of labeling that can be extrapolated (dashed lines) to the same region of the genome confirmed that simian virus 40 replication is bidirectional (Fig. 9.2B) and allowed location of the origin on the physical map of the viral genome. Modified from K. J. Danna and D. Nathans, *Proc Natl Acad Sci U S A* **69:**3097–3100, 1972, with permission.

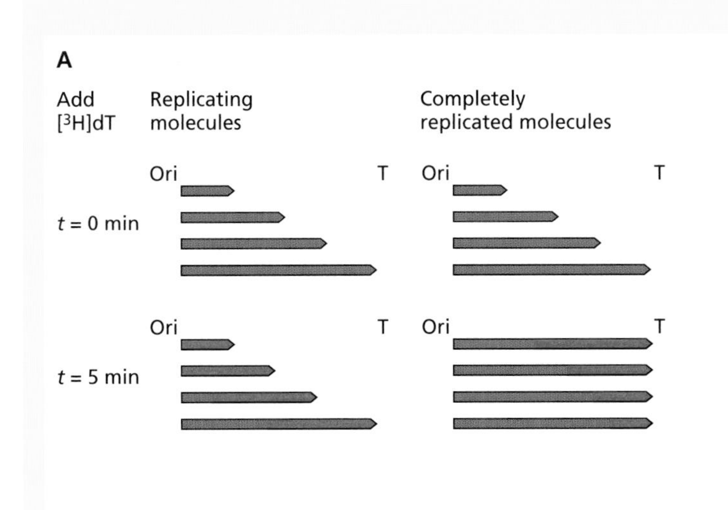

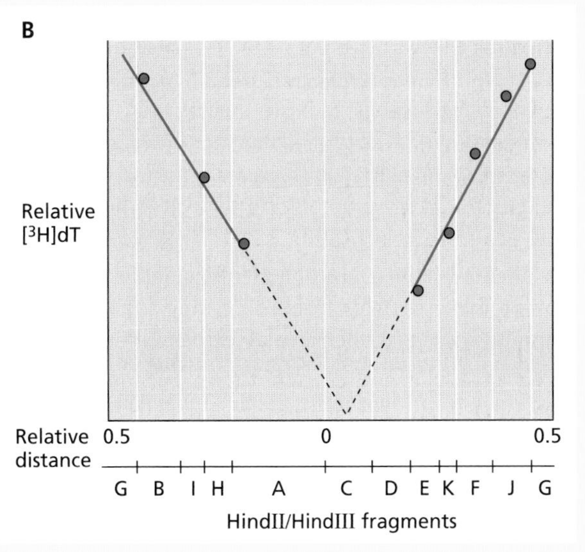

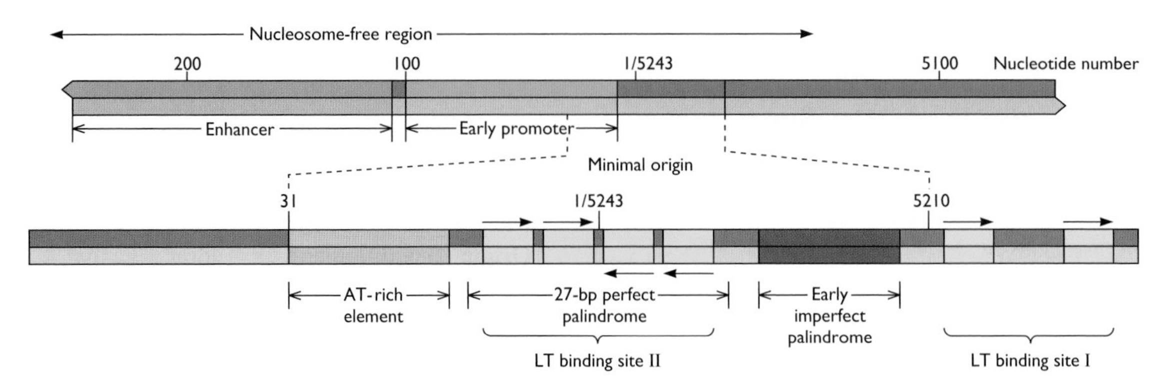

Figure 9.4 The origin of simian virus 40 DNA replication. The positions in the simian virus 40 genome of the minimal origin necessary for DNA replication *in vivo* and *in vitro* and of the enhancer and early promoter (see Chapter 8) are indicated. The pentameric LT-recognition sequences are shown in yellow. The AT-rich element and early imperfect palindrome, as well as LT-binding site II, are essential for replication. A second LT-binding site (site I) stimulates replication modestly *in vivo*. Other sequences, including the enhancer and Sp1-binding sites of the early promoter, increase the efficiency of viral DNA replication at least 10-fold. The activation domains (see Chapter 8) of transcriptional regulators that bind to these sequences might help recruit essential replication proteins to the origin. Alternatively, the binding of transcriptional activators might induce remodeling of chromatin in the vicinity of the origin. This possibility is consistent with the fact that, as indicated at the top, the region of the genome containing the origin and transcriptional control regions is nucleosome free in a significant fraction (~25%) of minichromosomes in infected cells.

origin (Fig. 9.5). Assembly of LT at the SV40 origin resembles assembly reactions at well-characterized bacterial origins, such as *Escherichia coli* OriC, or the origin of phage λ, in which multimeric protein structures assemble on AT-rich sequences. Furthermore, formation of hexamers around DNA is a property common to several viral and cellular replication proteins.

Leading-strand synthesis. Binding of DNA polymerase α-primase to both LT and Rp-A at the SV40 origin sets the stage for the initiation of leading-strand synthesis (Fig. 9.6). The primase synthesizes the RNA primers of the leading strand at each replication fork, while DNA polymerase α extends them to produce short fragments. The 3′-OH ends of these DNA fragments are then bound by cellular replication factor C (Rf-C), proliferating-cell nuclear antigen (Pcna), and DNA polymerase ε. Pcna is the processivity factor for DNA polymerase ε: it is required for synthesis of long DNA chains from a single primer. This mammalian protein is the functional analog of the β subunit of *E. coli* DNA polymerase III and phage T4 gene 45 product. These remarkable **sliding clamp** proteins form closed rings that track along the DNA template and serve as movable platforms for DNA polymerases. Subsequent binding of the replicative DNA polymerase completes assembly of a multiprotein assembly capable of leading-strand synthesis by continuous copying of the parental template strand.

Lagging-strand synthesis. The first Okazaki fragment of the lagging strand is synthesized by the DNA polymerase α-primase complex (Fig. 9.6, step 4). The lagging strand is synthesized by DNA polymerase δ, and transfer of the 3′ end of the first Okazaki fragment to this enzyme is thought to proceed as on the leading strand. The lagging-strand template is then copied **toward** the origin of replication. Consequently, synthesis of the lagging

Figure 9.5 Model of the recognition and unwinding of the simian virus 40 origin. In the presence of ATP, two hexamers bind to the origin via the pentanucleotide LT-binding sites (step 1). Binding of LT hexamers protects the flanking AT-rich (A/T) and early palindrome (EP) sequences of the minimal origin from DNase I digestion and induces conformational changes, for example, distortion of the early palindrome (step 2). Stable unwinding of the origin requires the cellular, single-stranded-DNA-binding protein replication protein A (Rp-A), which binds to LT. LT helicase activity, in concert with Rp-A and topoisomerase I, progressively unwinds the origin (step 3).

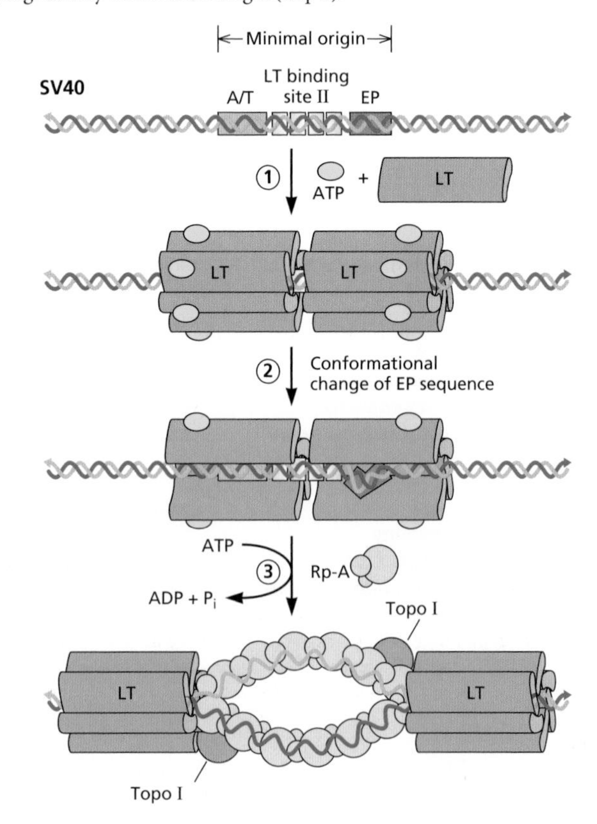

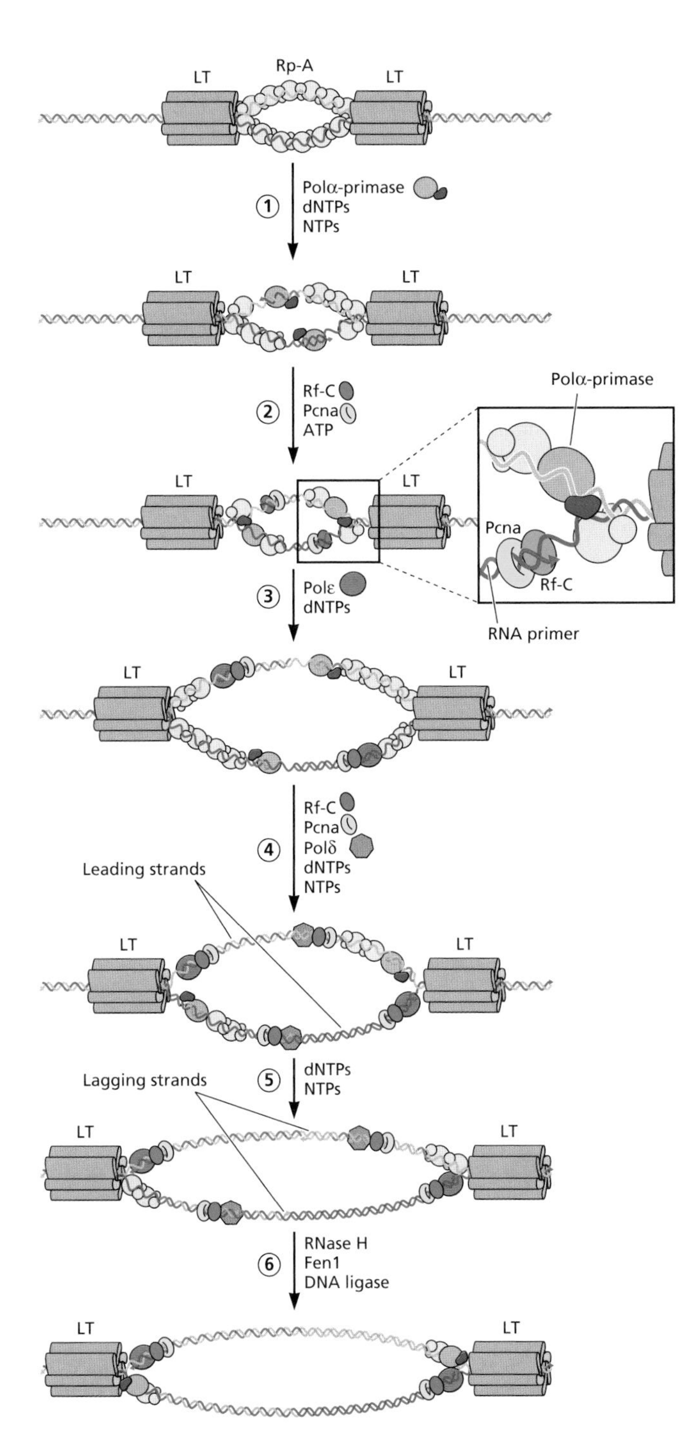

Figure 9.6 Synthesis of leading and lagging strands. The DNA polymerase (Pol) α-primase responsible for the synthesis of Okazaki fragments binds specifically to both replication protein A (Rp-A) and large T antigen (LT) assembled at the origin in the presynthesis complex. Once bound, the enzymes synthesize leading-strand RNA primers that are subsequently extended as DNA (step 1). The 3′-OH group of the nascent RNA-DNA fragment (~30 nucleotides in total length) is then bound by replication factor C (Rf-C) in a reaction that requires ATP but not its hydrolysis. Rf-C allows ATP-dependent opening of the proliferating-cell nuclear antigen (Pcna) ring and its loading onto the template (step 2). This reaction induces dissociation of DNA polymerase α-primase. Replicative DNA polymerase (usually ε) then binds to the Pcna/Rf-C complex (step 3). Because the clamp-loading protein Rf-C binds to the 5′-OH ends of the DNA fragments, it places the processivity protein at the replication forks. This replication complex is competent for continuous and highly processive synthesis of the leading strands (steps 4 and 5). Lagging-strand synthesis begins with synthesis of the first Okazaki fragment by DNA polymerase α-primase (step 3). Processive DNA polymerase (δ) is recruited as during leading-strand synthesis and produces a lagging-strand segment (step 5). The multiple DNA fragments produced by discontinuous lagging-strand synthesis are sealed by removal of the primers by RNase H (an enzyme that specifically degrades RNA hybridized to DNA) and the 5′ → 3′ exonuclease Fen1, repair of the resulting gaps by DNA polymerase δ, and joining of the DNA fragments by DNA ligase I (step 6).

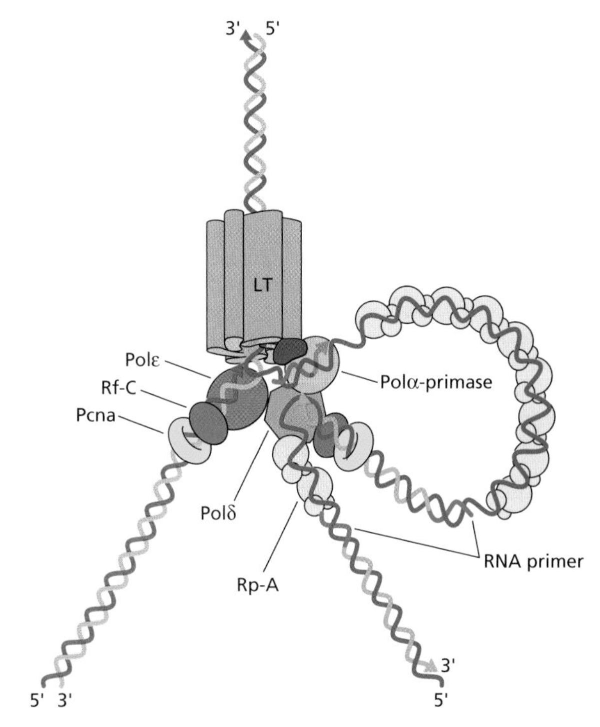

Figure 9.7 A model of the simian virus 40 replication machine. A replication machine containing all proteins necessary for both continuous synthesis of the leading strand and discontinuous synthesis of the lagging strand would assemble at each replication fork. Spooling of a loop of the template DNA strand for discontinuous synthesis would allow the single complex to copy the two strands in opposite directions. Pol, polymerase.

strand requires initiation by DNA polymerase α-primase at many sites progressively farther from the origin. The mechanisms by which leading- and lagging-strand synthesis are coordinated are not fully understood. If the replication machinery tracked along an immobile DNA template, the complexes responsible for leading- and lagging-strand synthesis would have to move in opposite directions. A more attractive alternative is that the DNA template is spooled through an immobile replication complex that contains all the proteins necessary for synthesis of both daughter strands. This mechanism would allow simultaneous copying of the template strands in opposite directions at each fork (Fig. 9.7). Consistent with this idea, replication

BOX 9.3

EXPERIMENTS

Unwinding of the simian virus 40 origin leads to spooling of DNA

Visualization by electron microscopy of structures formed during LT-dependent unwinding from the simian virus 40 origin *in vitro* suggested that the two hexamers remain in contact as DNA is unwound. LT was incubated with origin-containing DNA and ATP for 15 min in the presence of *E. coli* single-stranded binding protein (Ssb) to stabilize unwound DNA. Proteins were then cross-linked to DNA and samples processed for negative-contrast electron microscopy. **(A)** LT bound to the origin, as a characteristic bilobed structure (the double hexamer shown in Fig. 9.5); **(B)** unwinding intermediates; **(C)** the intermediate at the bottom right in panel B at higher magnification. This intermediate contains a bilobed LT complex connecting the two replication forks, so the single-stranded DNA (ssDNA), which is marked by the Ssb molecules bound to it, is looped out as "rabbit ears." The formation of such structures containing a dimer of the LT hexamer, in which each monomer is bound to a replication fork, stimulates the helicase activity of LT. This property supports the view that the DNA template is spooled through an immobile replication machine (see the text). dsDNA, double-stranded DNA. From R. Wessel et al., *J Virol* **66:**804–815, 1992, with permission. Courtesy of H. Stahl, Universität des Saarlandes.

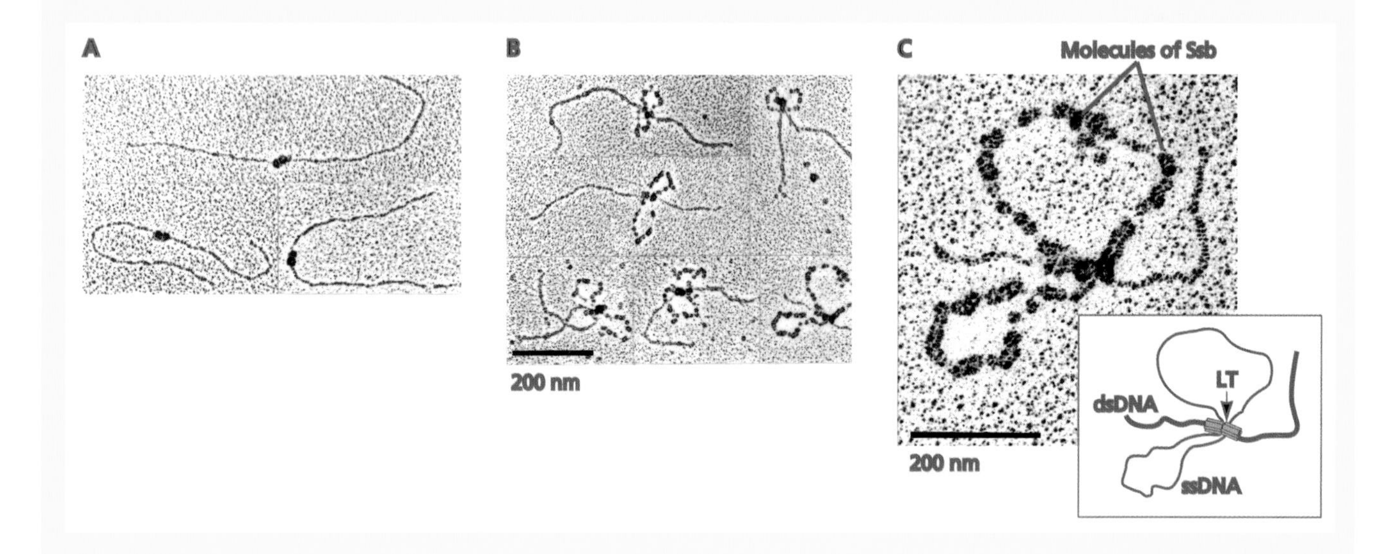

of chromosomal DNA occurs at fixed sites in the nucleus, and proteins that interact with the replicative helicase and both leading-strand and lagging-strand DNA polymerases have been identified. Furthermore, structures indicative of DNA spooling have been observed in the electron microscope during the initial, LT-dependent unwinding of the SV40 origin (Box 9.3).

Base pairing of dNTP substrates with template DNA requires the unwinding of double-stranded DNA genomes like that of SV40. LT is the helicase responsible for unwinding DNA at the origin, and remains associated with the replication forks, unwinding the template during elongation (Fig. 9.6).

Termination and resolution. Because the circular SV40 DNA genome possesses no termini, its replication does not lead to gaps in the strands made discontinuously. Nevertheless, additional cellular proteins are needed for the production of two daughter molecules from the circular template. These essential components are the cellular enzymes topoisomerases I and II, which alter the topology of DNA. These enzymes, which differ in their catalytic mechanisms and functions in the cell, reverse the winding of one duplex DNA strand around another (**supercoiling**). Because they remove supercoils, topoisomerases are said to **relax** DNA. In a closed circular DNA molecule, the unwinding of duplex DNA at the origin and subsequently at the replication forks is necessarily accompanied by supercoiling of the remainder of the DNA (Fig. 9.8A). If not released, the torsional stress so introduced would act as a brake on movement of the replication forks, eventually bringing them to a complete halt. Topoisomerase I associates with LT and is required for progression of SV40 replication forks and for viral reproduction. A single cycle of SV40 DNA synthesis produces two interlocked (catenated) circular DNA molecules that can be separated only when one DNA molecule is passed through a double-strand break in the other. The break is then resealed. Topoisomerase II catalyzes this series of reactions (Fig. 9.8B).

Replication of chromatin templates. The SV40 genome is associated with cellular nucleosomes both in virus particles and

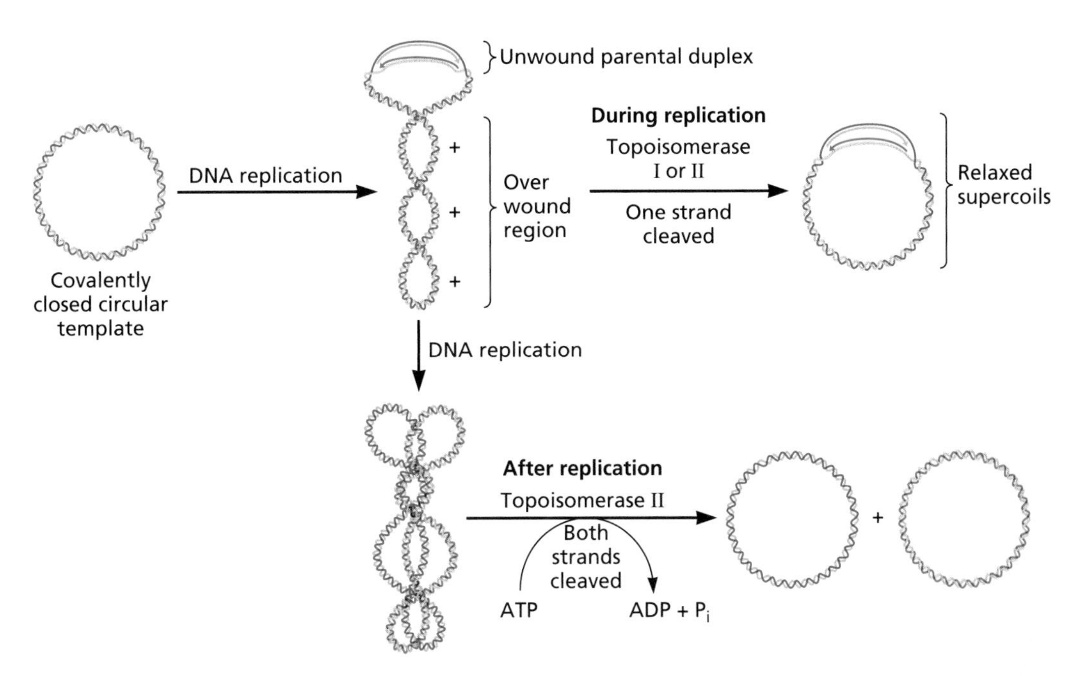

Figure 9.8 Function of topoisomerases during simian virus 40 DNA replication. Unwinding of the template DNA at the origin and two replication forks leads to overwinding (positive supercoiling) of the DNA ahead of the replication forks (middle). Either topoisomerase I or topoisomerase II can remove the supercoils to relieve such overwinding and allow continued movement of the replication fork. However, LT binds to topoisomerase I, and substitutions that impair this interaction inhibit LT-dependent DNA synthesis *in vitro* and viral reproduction in infected cells. The products of genome replication are interlocked daughter molecules (below). Their separation requires topoisomerase II, which makes a double-strand break in DNA, passes one double strand over the other to unwind one turn, and reseals the DNA in reactions that require hydrolysis of ATP.

in infected cell nuclei. It is therefore replicated as a minichromosome, in which the DNA is wrapped around nucleosomes. This arrangement raises the question of how the replication machinery can copy a DNA template that is bound to nucleosomal histones. A similar problem is encountered during the replication of many viral RNA genomes, when the template RNA is packaged by viral RNA-binding proteins in a large ribonucleoprotein. The mechanisms by which replication complexes circumvent such barriers to movement are not understood in detail. Nevertheless, numerous proteins that couple ATP hydrolysis to remodeling of nucleosomal DNA have been identified (see Chapter 8). The organization of the SV40 genome into a minichromosome also implies that viral DNA replication must be coordinated with binding of newly synthesized DNA to cellular nucleosomes. In fact, new nucleosomes are deposited at viral replication forks, a reaction that is catalyzed by the essential human protein chromatin assembly factor 1.

Summary. Analysis of simian virus 40 replication *in vitro* identified essential cellular replication proteins, led to molecular descriptions of crucial reactions in the complicated process of DNA synthesis, and provided new insights into chromatin assembly. The detailed understanding of the reactions completed by the cellular DNA replication machinery laid the foundation for elucidation of the mechanisms by which other animal viral DNA genomes are replicated, and of some of the intricate circuits that regulate DNA synthesis and its initiation.

Replication of Other Viral DNA Genomes

The replication of all viral DNA genomes within infected cells comprises reactions analogous to those necessary for simian virus 40 DNA synthesis, namely, origin recognition and assembly of a presynthesis complex, priming of DNA synthesis, elongation, termination, and often resolution of the replication products. However, the mechanistic problems associated with each of these reactions are solved by a variety of virus-specific mechanisms. Synthesis of viral DNA molecules is initiated not only by RNA priming, but also by unusual mechanisms in which DNA and even protein molecules function as primers. As we shall see, these latter priming strategies circumvent the need for discontinuous synthesis of daughter DNA molecules.

Synthesis of Viral RNA Primers by Cellular or Viral Enzymes

The standard method of priming is synthesis of a short RNA molecule by a specialized primase. As we have seen, cellular DNA polymerase α-primase synthesizes all RNA primers needed for replication of both template strands of polyomaviral

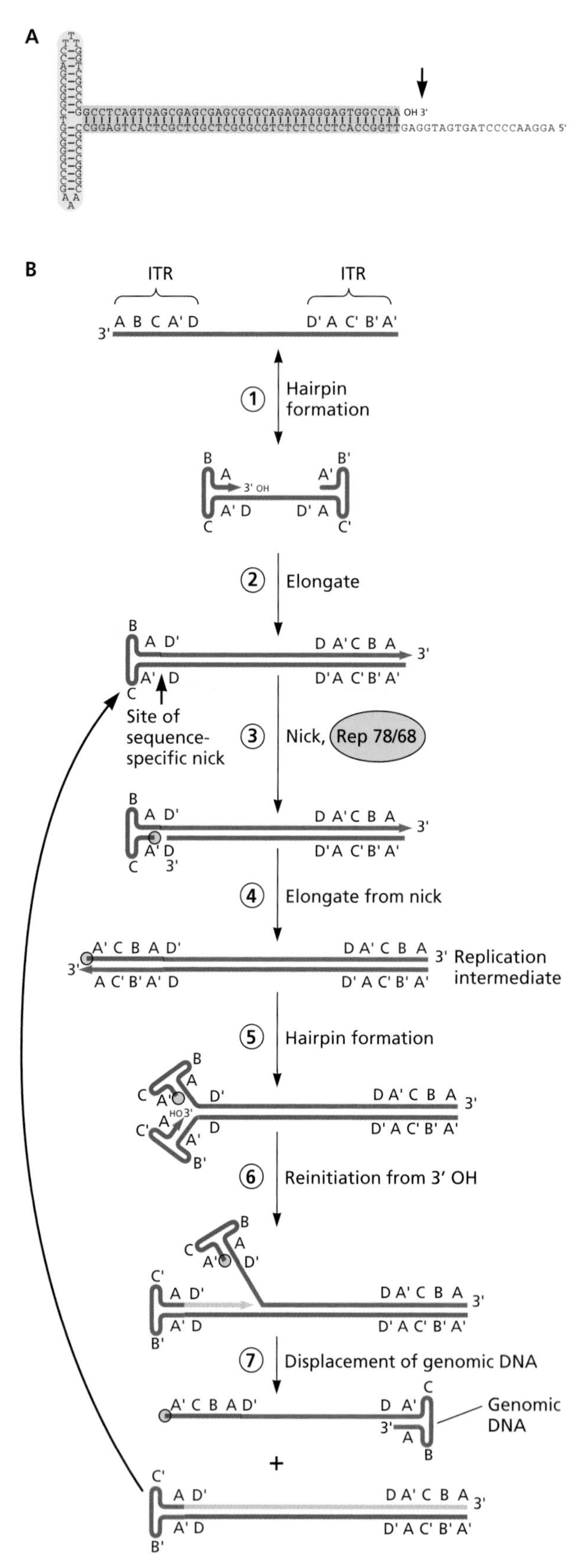

genomes. A similar mechanism operates at papillomavirus origins and those of some herpesviruses, such as that directing replication of the episomal Epstein-Barr viral genome in latently infected cells (see " Different Viral Origins Regulate Replication of Epstein-Barr Virus" below). The integrated proviral genomes of retroviruses are also replicated via RNA primers, which are synthesized by the cellular primase at the origin of the cellular replicon in which the provirus resides. In actively dividing cells, proviral DNA is therefore replicated once per cell cycle by the cellular replication machinery. Viral primases can also synthesize RNA primers. This mechanism is characteristic of genome replication during productive infection by herpesviruses, such as herpes simplex virus type 1.

Priming via DNA: Specialized Structures in Viral Genomes

Self-priming of viral DNA synthesis via specialized structures in the viral genome is a hallmark of all *Parvoviridae*, among the smallest DNA viruses that replicate in animal cells. This virus family includes the dependoviruses, such as adenovirus-associated viruses, and the autonomous parvoviruses, such as minute virus of mice. This mechanism is illustrated here with adenovirus-associated virus, which has a small (<5-kb) genome of single-stranded, linear DNA that carries **inverted terminal repetitions** (ITRs). Genomic DNA is of both (+) and (−) polarity, for both strands are encapsidated, but in separate virus particles. Palindromic sequences within the central 125 nucleotides of the ITR base pair to form T-shaped structures (Fig. 9.9A). Formation of this

Figure 9.9 Replication of parvoviral DNA. (A) Sequence and secondary structure of the adenovirus-associated virus type 2 inverted terminal repetition (ITR). A central palindrome (tan background) is flanked by a longer palindrome (light blue background) within the ITR. When these bases pair at the 3′ end of the genome, a T-shaped structure in which the internal duplex stem terminates in a free 3′-OH group (arrow) is formed. **(B)** Model of adenovirus-associated virus DNA replication. The ITRs are represented by 3′ABCA′D5′ and 5′A′B′C′AD′3′. Formation of the 3′-terminal hairpin provides a template-primer (step 1). Elongation from the 3′-OH group of the hairpin allows continuous synthesis (red) to the 5′ end of the parental strand (step 2). To complete copying of the parental strand, a nick to generate a new 3′ OH is introduced at the specific **terminal resolution site** (marked by the arrow) by the viral Rep 78/68 proteins (step 3). Elongation from the nick results in copying of sequences that initially formed the self-priming hairpin to form the double-stranded replication intermediate (step 4). However, the parental strand then contains newly replicated DNA (red) at its 3′ end. As a result, the ITR of the parental strand is no longer the initial sequence but rather its complement. This palindromic sequence is therefore present in populations of adenovirus-associated virus DNA molecules in one of two orientations. Such sequence heterogeneity provided an important clue for elucidation of the mechanism of viral DNA synthesis. The newly replicated 3′ end of the replication intermediate can form the same terminal hairpin structure (step 5) to prime a new cycle of DNA synthesis (step 6) with displacement of a molecule of single-stranded genomic DNA, and the formation of the incompletely replicated molecule initially produced (step 7). The latter molecule can undergo additional cycles of replication as in steps 3 and 4.

structure at the 3′ end of single-stranded viral DNA provides an ideal template-primer for initiation of viral DNA synthesis (Fig. 9.9B). Experimental evidence for such **self-priming** includes the dependence of adenovirus-associated virus DNA synthesis on self-complementary sequences within the ITR. Following recognition of the free 3′-OH end of the viral DNA primer, the single template strand of an infecting genome can be copied by a continuous mechanism, analogous to leading-strand synthesis during replication of double-stranded DNA templates. In subsequent cycles of replication, the same 3′-terminal priming structures form in the duplex replication intermediate produced in the initial round of synthesis (Fig. 9.9B). Adenovirus-associated virus DNA synthesis is therefore always continuous and requires cellular DNA polymerase δ, Rf-C, and Pcna, but not DNA polymerase α-primase.

On the other hand, a specialized mechanism **is** necessary to complete replication of each strand, because the initial product retains the priming hairpin and is largely duplex DNA in which parental and daughter strands are covalently connected (Fig. 9.9.B, step 2). Complete copying is initiated by nicking of this intermediate within the parental DNA strand at a specific site. The new 3′-OH end liberated in this way then primes continuous synthesis to the end of the DNA molecule (Fig. 9.9B, step 4). The nick is introduced by the related viral proteins Rep 78 and Rep 68 (Rep 78/68). These proteins are site- and strand-specific endonucleases, which bind to, and cut at, specific sequences within the ITR. During this terminal resolution process, Rep 78/68 becomes covalently linked to the cleaved DNA at the sites that will become the 5′ termini of the fully replicated molecule and single-stranded daughter genome. This covalent linkage is maintained during genome encapsidation and assembly of virus particles, but the subsequent fate of genome-linked Rep 78/68 is not known. Following the synthesis of a duplex of the genomic DNA molecule (the **replication intermediate**), formation of the 3′-terminal priming hairpin allows continuous synthesis of single-stranded genomes by a strand displacement mechanism, with re-formation of the replication intermediate (Fig. 9.9B, steps 6 and 7).

Rep 78 and Rep 68 are similar to simian virus 40 LT in several respects and can be considered origin recognition proteins (Table 9.1). They are the only viral gene products necessary for parvoviral DNA synthesis. In addition to recognizing and cleaving the terminal resolution site, these proteins provide the ATP-dependent, 3′ → 5′ helicase activity needed for unwinding of the replicated ITR and re-formation of the priming hairpin (Fig. 9.9B, step 5). However, the cellular helicase Mcm (minichromosome maintenance complex) is also required for adenovirus-associated virus DNA synthesis *in vitro* and in infected cells.

Whether priming of DNA synthesis via complementary sequences in the genome is a unique feature of parvoviral replication is not yet clear. Single-stranded, linear genomes of other viruses, such as the widespread geminivirus of plants, are not replicated in this way, but by a rolling-circle mechanism (Box 9.4). However, a self-priming and strand displacement mechanism is consistent with some, but not all, properties of replication of the large, double-stranded DNA genomes of poxviruses, such as vaccinia virus (Box 9.5).

Table 9.1 Viral origin recognition proteins

Virus	Protein(s)	Origin-binding properties	Other activities and functions
Parvovirus			
Adenovirus-associated virus	Rep 78/68	Binds to specific sequences in ITR as hexamer	Site- and strand-specific endonuclease; ATPase and helicase; transcriptional regulator
Papovaviruses			
Simian virus 40	LT	Binds cooperatively to origin site II to form double hexamer; distorts origin	ATPase and helicase; binds to cellular Rp-A and polymerase α-primase; represses early and activates late transcription; binds cellular Rb protein to induce progression through the cell cycle
Bovine papillomavirus type 5	E1	Binds strongly and cooperatively only in presence of E2 protein	ATPase and helicase
	E2	Binds to specific sequences in origin as dimer	Regulates transcription by binding to viral enhancers
Adenovirus			
Human adenovirus type 5	Pre-TP–DNA polymerase	Binds to minimal origins	Primes continuous synthesis of both strands of viral genome
Herpesviruses			
Herpes simplex virus type 1	UL9	Binds cooperatively to specific sites in viral origins; distorts DNA	ATPase and helicase; binds UL29 protein, UL8 subunit of viral primase, and UL42 processivity protein
Epstein-Barr virus	EBNA-1	Binds to multiple sites in OriP as dimer	Stimulates transcription from viral promoters

BOX 9.4

BACKGROUND

Rolling-circle replication

The rolling-circle replication mechanism of DNA synthesis was discovered during studies of the replication of the single-stranded DNA genome of bacteriophage ϕX174. However, it also operates during replication of double-stranded genomes, such as that of bacteriophage λ.

Rolling-circle replication is initiated by introduction of a nick that creates a 3′-OH end in one strand of a double-stranded, circular DNA. One strand of the template is copied continuously, and multiple times, while the displaced strand is copied discontinuously. As shown in the figure, this mechanism produces genome concatemers.

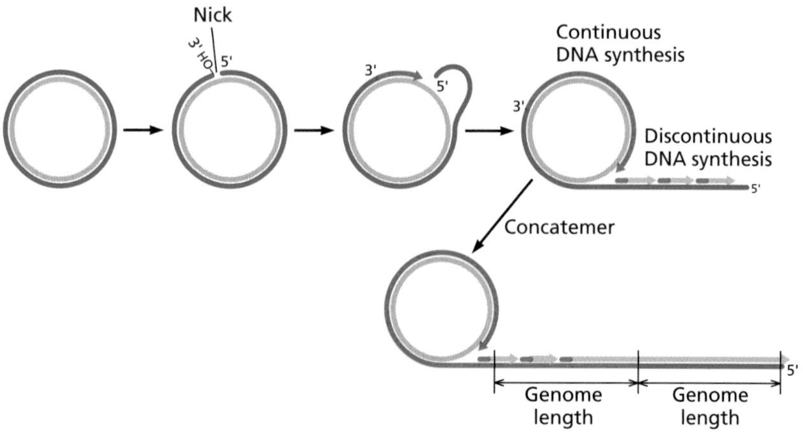

Protein Priming

Initiation of DNA synthesis via a protein primer is a relatively rare mechanism, restricted to some bacteriophages (e.g., ϕ29 and PRD1) and to hepadnaviruses and adenoviruses among those DNA viruses that infect animal cells. The replication of some viral RNA genomes is also initiated from a protein primer, notably the VPg protein of poliovirus discussed in Chapter 6. Here, we use adenoviral replication to illustrate the mechanism of protein priming.

The 5′ ends of adenoviral genomes that enter infected nuclei are covalently linked to the terminal protein (TP). The precursor to this protein (Pre-TP) serves as the primer for viral DNA synthesis. The adenoviral DNA polymerase covalently links the α-phosphoryl group of dCMP to the hydroxyl group of a specific serine residue in Pre-TP (Fig. 9.10). The 3′-OH group of the protein-linked dCMP then primes synthesis of daughter viral DNA strands by the viral DNA polymerase. Once the first few nucleotides have been incorporated, the DNA polymerase must disassociate from Pre-TP to allow elongation of the daughter DNA strand. The structure of the ϕ29 DNA polymerase bound to its priming terminal protein suggests that such dissociation is the result of conformational change induced by displacement of the priming domain from the catalytic site in the polymerase (Box 9.6). The nucleotide is added to Pre-TP **only** when this protein primer is assembled with the DNA polymerase into preinitiation complexes at the origins of replication. As the origins lie at the ends of the linear genome, each template strand is then copied continuously from one end to the other by strand displacement (Fig. 9.10). The parental template strand displaced initially is copied by the same mechanism, following annealing of an ITR sequence to re-form the duplex DNA sequence present at the ends of parental DNA. This unusual strand displacement mechanism therefore results in semiconservative replication, even though the two parental strands of viral DNA are not copied at the same replication fork.

Properties of Viral Replication Origins

Origins of replication contain the sites at which viral DNA synthesis begins and can be defined experimentally as the minimal DNA segment necessary for initiation of replication in cells or *in vitro* reactions. Viral origins of replication support initiation of DNA synthesis by a variety of mechanisms, including some with no counterpart in cellular DNA synthesis. Nevertheless, they are discrete DNA segments that contain sequences recognized by viral origin recognition proteins to seed assembly of multiprotein complexes, and they exhibit a number of common features.

Number of Origins

In contrast to papillomaviral and polyomaviral DNAs, the genomes of the larger DNA viruses contain not one, but two or three origins. As noted above, the two identical adenoviral origins at the ends of the linear genome are the sites of assembly of preinitiation complexes (Fig. 9.10). The genomes of herpesviruses, such as Epstein-Barr virus and herpes simplex virus type 1, contain three origins of replication. Different functions can be ascribed to the different Epstein-Barr virus origins: a single origin (OriP) allows maintenance of episomal genomes in latently infected cells (see "Different Viral Origins Regulate Replication of Epstein-Barr Virus" below), while the two others (OriLyt)

BOX 9.5

DISCUSSION

Self-priming or RNA priming of vaccinia viral DNA synthesis?

The two strands of the large DNA genome of vaccinia virus are covalently connected by terminal, unpaired loops at the ends of inverted terminal repeated sequences (Appendix, Fig. 25). The potential of these terminal sequences to form hairpins suggests the possibility for self-pairing and continuous synthesis of viral DNA, initiated following introduction of a nick near one or both termini to form a 3′-OH primer (see the figure). This model is consistent with the observation that the products of DNA synthesis are concatemers of the viral genome (see the figure), and with the changes in the sedimentation properties of viral DNA that are indicative of nicking that occur following entry into host cells. In addition, the terminal sequences of the viral genome confer optimal replication upon linear minichromosome templates. However, vaccinia virus-infected cells support replication of exogenous circular plasmids that contain **no** viral DNA. Such origin-independent replication takes place at the sites of viral genome replication, specialized cytoplasmic replication factories, and requires viral replication proteins. These properties suggest an alternative model for vaccinia viral DNA synthesis.

Indeed, early studies reported the detection of short fragments of newly synthesized vaccinia viral DNA that were covalently linked to RNA and became incorporated into larger DNA molecules. The production of these viral Okazaki fragments suggests a mechanism that includes RNA priming and discontinuous DNA synthesis during genome replication. Viral enzymes that catalyze reactions pivotal to this mechanism have been identified only quite recently. For example, the D5R protein, long known to be a helicase, contains an N-terminal primase that catalyzes DNA-dependent synthesis of RNA with little template specificity. Viral replication also requires a DNA ligase, the enzyme that joins DNA fragments following discontinuous DNA synthesis, encoded by either the viral A50R gene, or, in actively growing cells, cellular DNA ligase 1.

These apparently contradictory observations illustrate the difficulties of establishing how large viral DNA genomes are replicated. They have yet to be reconciled (or explained), and there is currently no generally accepted model of vaccinia viral DNA synthesis.

De Silva FS, Lewis W, Berglund P, Koonin EV, Moss B. 2007. Poxvirus DNA primase. *Proc Natl Acad Sci U S A* **104:**18724–18729.

De Silva FS, Moss B. 2005. Origin-independent plasmid replication occurs in vaccinia virus cytoplasmic factories and requires all five known poxvirus replication factors. *Virol J* **2:**23. doi:10.1186/1743-422X-2-23.

Du S, Traktman P. 1996. Vaccinia virus DNA replication: two hundred base pairs of telomeric sequence confer optimal replication efficiency on minichromosome templates. *Proc Natl Acad Sci U S A* **93:**9693–9698.

Moss B. 2013. Poxvirus DNA replication. *Cold Spring Harb Perspect Biol* **5:**a010199. doi:10.1101/cshperspect.a010199.

A self-priming model for vaccinia viral DNA synthesis. The double-stranded DNA is not depicted to scale, to emphasize the inverted terminal repetitions and terminal loops. Complementary sequences within these regions are indicated by upper- and lower-case letters. Viral DNA synthesis would be initiated by introduction of a nick that creates a 3′-OH end (step 1). Following synthesis of DNA to the end of the genome (step 2), re-formation of the terminal hairpins by base pairing (step 3) would allow continuous DNA synthesis (step 4) and production of concatemers (step 5).

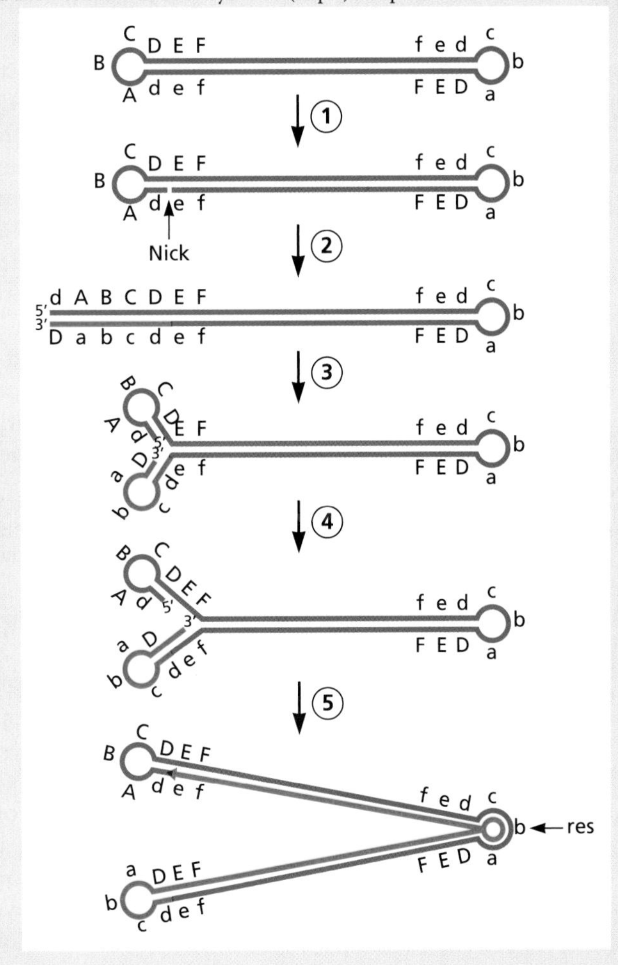

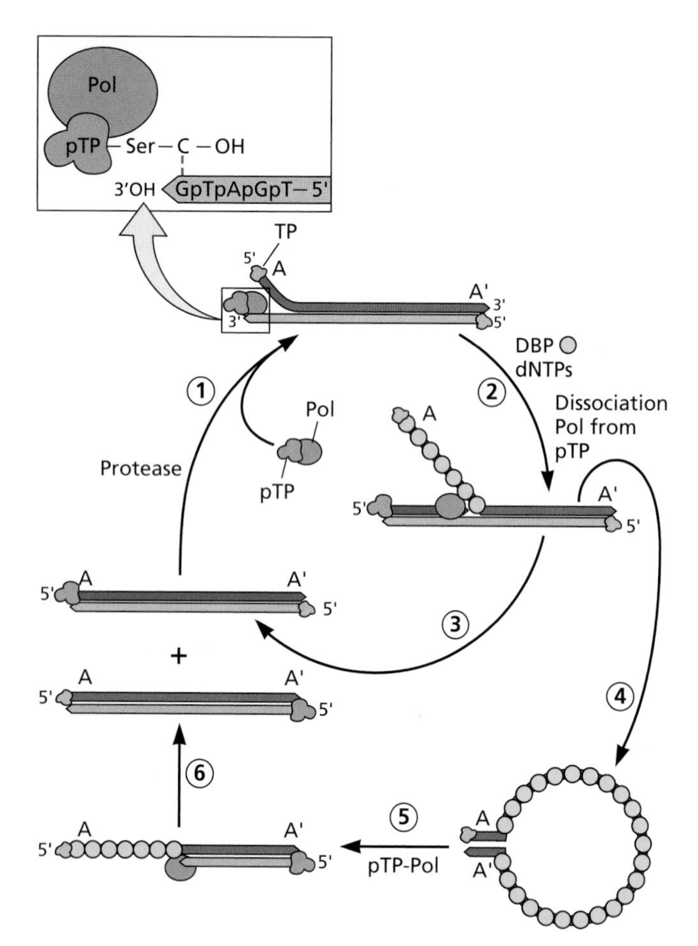

Figure 9.10 Replication of adenoviral DNA. Assembly of the viral preterminal protein (pTP) and DNA polymerase (Pol) into a preinitiation complex at each terminal origin of replication activates covalent linkage of dCMP to a specific serine residue in pTP by the DNA polymerase (step 1). The free 3′-OH group of pTP-dCMP primes continuous synthesis in the 5′ → 3′ direction by Pol (step 2). This reaction also requires the viral E2 single-stranded-DNA-binding protein (DBP), which coats the displaced second strand of the template DNA molecule, and a cellular topoisomerase. As the terminal segments of the viral genome comprise an inverted repeat sequence (A and A′), there is an origin at each end, and both parental strands can be replicated by this displacement mechanism (step 3). Reannealing of the complementary terminal sequences of the parental strand initially displaced forms a short duplex stem identical to the terminus of the double-stranded genome (step 4). The origin re-formed in this way directs a new cycle of protein priming and continuous DNA synthesis (steps 5 and 6). The pTP is cleaved by the viral protease to the terminal protein (TP) during maturation of viral particles.

support replication of the genome during productive infection. The herpes simplex virus type 1 genome contains two copies of OriS and one of OriL (Fig. 9.11). The two types of origin possess considerable nucleotide sequence similarity, but differ in their organization and can be distinguished functionally. For example, OriL is activated when differentiated neuronal cells are exposed to a glucocorticoid hormone, but OriS is repressed. As glucocorticoids are produced in response to stress, a condition reactivates latent herpes simplex virus type 1 infection, it has been suggested that replication from OriL may be particularly important during the transition to a productive infection.

Viral Replication Origins Share Common Features

Even though the origins of replication of double-stranded DNA viruses are recognized by different proteins and support different mechanisms of initiation, they exhibit a number of common features (Fig. 9.12). The most prominent of these is the presence of AT-rich sequences. In general, AT base pairs contain only two hydrogen bonds, whereas GC pairs interact via three such bonds. The less stable AT-rich sequences are thought to facilitate the unwinding of origins that is necessary for initiation of viral DNA synthesis on double-stranded templates. Another general feature is the close relationship between origin sequences and those that regulate transcription. For example, sequences adjacent to the polyomaviral and adenoviral core origins that increase replication efficiency include binding sites for transcriptional activators. Other viral origins, those of papillomaviruses and parvoviruses and OriLyt of Epstein-Barr virus, contain binding sites for viral proteins that are **both** transcriptional regulators and essential replication proteins. And all three herpes simplex virus type 1 origins lie between promoters for viral transcription. Assembly of viral preinitiation complexes on adenoviral origins is stimulated by direct interactions with cellular transcriptional activators that bind to adjacent sequences (Fig. 9.12). In other cases, such cellular proteins may promote viral DNA synthesis indirectly via alterations in the properties of nucleosomes with which the viral genomes are associated (see Chapter 8).

Recognition of Viral Replication Origins

The paradigm for viral origin recognition is the simian virus 40 LT protein. We therefore describe its properties as the prelude to discussion of other viral proteins with similar functions.

Properties of Simian Virus 40 LT

Functions and organization. The LT proteins of polyomaviruses provide functions essential for viral DNA synthesis, viral gene expression, and optimization of the intracellular environment (Table 9.1). As we have seen, simian virus 40 LT is both necessary and sufficient for recognition of the viral origin and also supplies the helicase activity that drives origin unwinding and perhaps movement of the replication fork. The LT proteins make a major contribution to the species specificity of polyomaviruses. Although the genomes of simian virus 40 and mouse polyomavirus are closely related in organization and sequence, they replicate only in simian and murine cells, respectively. Such host specificity is largely the result of species-specific binding of LT to the largest subunit of DNA polymerase α of the host cell in which the virus will replicate. Although the precise mechanism remains to be

BOX 9.6

DISCUSSION

Model for the transition between initiation and elongation during protein-primed DNA synthesis

Association of the adenoviral DNA polymerase with the Pre-TP primer is necessary for catalysis of covalent linkage of the priming dCMP to Pre-TP (see the text). However, this interaction must be reversed following initiation to allow processive elongation by the enzyme. Clues about how this transition occurs have come from structural studies of bacteriophage ϕ29 replication proteins.

Replication of the linear, double-stranded ϕ29 genome is initiated by protein priming from origins at the ends of the genome. The phage DNA polymerase (Pol) and priming terminal protein (TP) form a heterodimer and the enzyme catalyzes linkage of the priming nucleotide to TP, just as in adenoviral DNA synthesis (see Fig. 9.10). The structure of the ϕ29 Pol-TP heterodimer has been determined by X-ray crystallography. In this complex, the TP priming domain lies in the site occupied by the DNA template-primer in a model of the elongating enzyme. The loop that contains the serine to which the priming nucleotide is attached lies closest to the Pol active site. The priming domain is connected to a domain (the intermediate domain) that makes extensive contacts with the DNA polymerase via a hinge.

The results of modeling studies indicate that up to 6 or 7 nucleotides can be added to the nascent DNA while TP maintains close contacts with DNA polymerase: motion about the hinge allows displacement of the priming domain while the intermediate domain maintains contact with Pol (see figure). However, this mechanism cannot accommodate further translocation of the priming domain. Rather, the intermediate domain of TP and Pol must dissociate, presumably as a result of additional structural changes. Consequently, the DNA polymerase is released for elongation, as illustrated for incorporation of eight dNMPs in the figure.

Kamtekar S, Berman AJ, Wang J, Lazaro JM, de Vega M, Blanco L, Salas M, Steitz TA. 2006. The ϕ29 DNA polymerase: protein primer structure suggests a model for the initiation to elongation transition. *EMBO J* 25:1335–1343.

The 39-OH group of the priming nucleotide attached to TP and nascent DNA are shown in green and red, respectively, with newly incorporated dNMPs indicated by orange circles. Adapted from S. Kamtekar et al., *EMBO J* **25**:1335–1343, 2006, with permission.

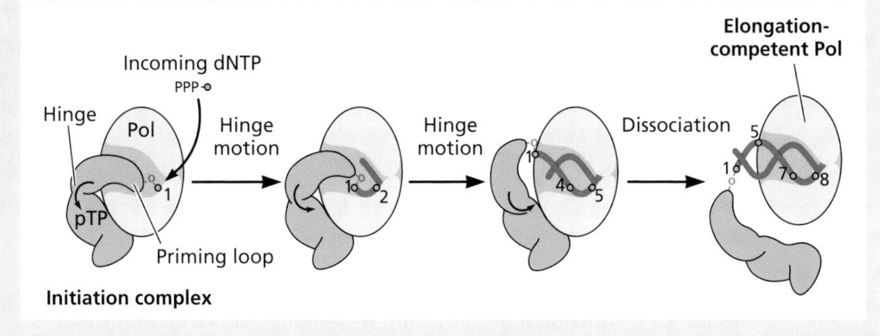

determined, assembly of preinitiation complexes competent for unwinding of the origin does not take place when the LT of one polyomavirus binds to the origin of another.

LT proteins also ensure that the cellular components needed for simian virus 40 DNA synthesis are available in the host cell. By binding and sequestering specific cellular proteins, LT perturbs mechanisms that control cell proliferation and can induce infected cells to enter S phase (see "Viral Proteins Can Induce Synthesis of Cellular Replication Proteins" below). LT also regulates its own synthesis and activates late gene expression.

Sequences of simian virus 40 LT that are necessary for its numerous activities have been mapped by analysis of the effects of specific alterations in the protein on virus replication in infected cells, DNA synthesis *in vitro*, or the individual biochemical activities of the protein. The properties of such altered proteins indicate that LT contains discrete structural and functional domains, such as the minimal domain for specific binding to the viral origin (Fig. 9.13). However, the activities of such functional regions defined by genetic and biochemical methods may be influenced by distant sites, as discussed in the next section.

Figure 9.11 Features of the herpes simplex virus type 1 genome. The long (L) and short (S) regions of the viral genome that are inverted with respect to one another in the four genome isomers are indicated at the top. Each segment comprises a unique sequence (UL or US) flanked by internal and terminal repeated sequences (IR and TR). The locations of the two identical copies of OriS, in repeated sequences, and of the single copy of OriL are indicated.

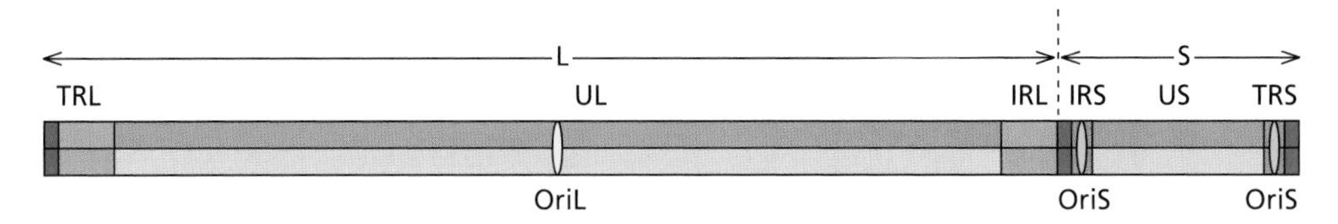

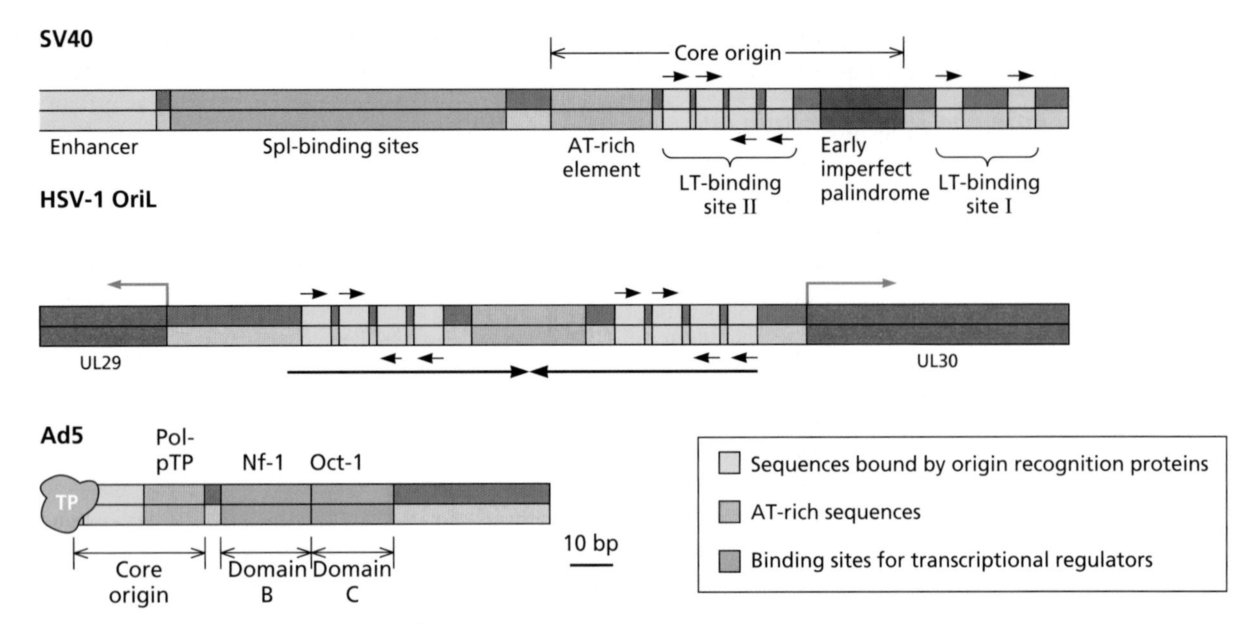

Figure 9.12 Common features of viral origins of DNA replication. The simian virus 40 (SV40) origin, herpes simplex virus type 1 (HSV-1) OriL (Fig. 9.11), and adenovirus type 5 (Ad5) origin (Fig. 9.10) are illustrated to scale, emphasizing the common features shown in the key. Sites of initiation of transcription are indicated by jointed red arrows and palindromic DNA sequences by black arrows. The two copies of herpesviral OriS (Fig. 9.11) are very similar in sequence to OriL. The terminal sequence of the adenoviral origin designated the core origin functions inefficiently in the absence of the adjacent binding site for the transcriptional activator nuclear factor 1 (Nf-1).

Figure 9.13 Functional organization of simian virus 40 LT. LT is represented to scale. Indicated are the sequences required for binding to the DNA polymerase α-primase complex (Polα), to the cellular chaperone Hsc70, to the cellular retinoblastoma (Rb) and p53 proteins, to the origin of replication (origin DNA binding), and to single-stranded (ss) DNA. Also shown are segments necessary for the helicase and ATPase activities, hexamer assembly at the origin, the nuclear localization signal (NLS), and a C-terminal sequence necessary for production of viral particles but not viral DNA synthesis (host range). The region that binds to Hsc70 lies within an N-terminal segment termed the J domain, because it shares sequences and functional properties with the *E. coli* protein DnaJ, a chaperone that assists the folding and assembly of proteins and is required during reproduction of bacteriophage λ. The chaperone functions of the J domain and Hsc70 are essential for replication in infected cells and seem likely to assist assembly or rearrangement of the preinitiation complex. Below are shown the two regions of the protein in which sites of phosphorylation are clustered, indicating modifications that have been shown to inhibit (red) or activate (green) the replication activity of LT.

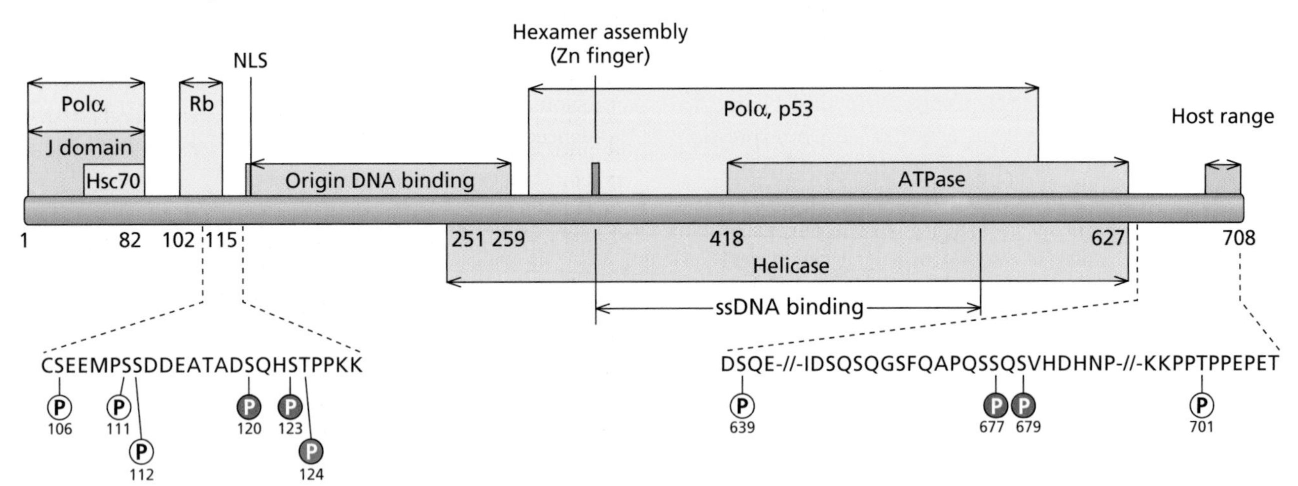

During initiation of viral DNA synthesis, LT first binds specifically to double-stranded pentanucleotide repeat sequences in the origin. It must then interact with single-stranded DNA nonspecifically during origin distortion and unwinding and when the protein couples the hydrolysis of ATP to translocate along DNA at replication forks. Structural studies of various forms of individual domains have provided important insights into the different interactions of LT with DNA. For example, X-ray crystallography of the origin-binding domain bound to the core origin identified numerous contacts between residues in LT and specific origin sequences. In the absence of DNA, the origin-binding domain forms a gapped, hexameric spiral with a large central channel lined by these residues, as well as the distinct but overlapping set that mediates binding to single-stranded DNA (Fig. 9.14). Comparison of the high-resolution structures of various forms of this domain, and of the helicase domain, indicates that both can undergo substantial conformational change. How such reorganizations within individual domains are integrated and coordinated during the intricate process of LT-dependent initiation of DNA synthesis is not clear, in part because the inherent flexibility of full-length LT has limited structural studies. Although important questions remain to be addressed, elegant studies of LT-dependent unwinding of single DNA molecules have established that LT translocates along single-stranded DNA in the 3′ → 5′ direction to unwind DNA by steric exclusion (Box 9.7).

Figure 9.14 Structure of origin-binding domain of simian virus 40 large T antigen, determined by X-ray crystallography. The structure of the origin-binding domain (amino acids 131 to 260) hexamer is shown in surface representation. In this model of the hexamer bound to DNA, the DNA is gray, with the palindromic LT-binding sequences (Fig. 9.4) in cyan and magenta. The DNA-binding regions of LT are colored red and purple. The results of mutational analysis indicate that in the double hexamer of the full-length protein, the origin-binding domains in the two hexamers interact with one another. Adapted from G. Meinke et al., *J Virol* **80:**4304–4312, 2006, with permission. Courtesy of Andrew Bohm, Tufts University School of Medicine.

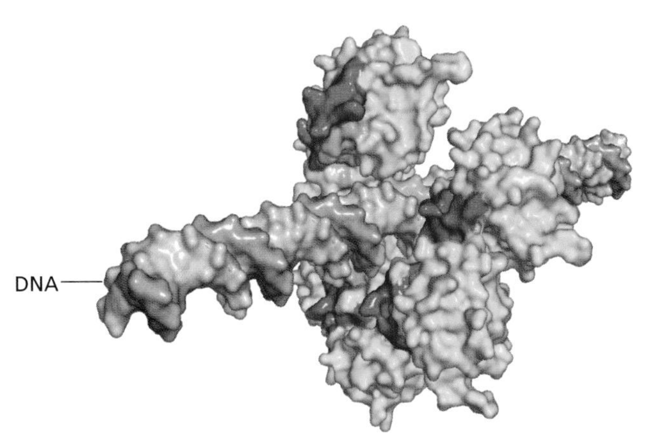

Regulation of LT activity. The viral early gene encoding LT is transcribed efficiently as soon as the viral chromosome enters the nucleus (Chapter 8). The spliced LT mRNA is the predominant product of processing of these early transcripts. Although production of LT is not regulated during the early phase of infection in simian cells in culture, its activity is tightly controlled.

Specific posttranslational modifications regulate the ability of LT to support viral DNA synthesis. For example, the combination of phosphorylation of Thr124 with lack of phosphorylation of Ser120 and Ser123 stimulates binding of LT to origin site II, promotes assembly of the double hexamer (Fig. 9.5), and is essential for unwinding of DNA from the origin. As Thr124 does not lie within the minimal origin-binding domain (Fig. 9.13), such regulation of DNA-binding activity is thought to be the result of conformational change induced by phosphorylation at this site. The best candidate for the protein kinase that phosphorylates Thr124 is cyclin-dependent kinase 2 (Cdk2) associated with cyclin A.

Viral Origin Recognition Proteins Share Several Properties

Other viral origin recognition proteins share with simian virus 40 LT the ability to bind specifically to DNA sequences within the cognate origin of replication. They also interact with other replication proteins (although these may be viral or cellular), and several possess the biochemical activities exhibited by LT (Table 9.1). For example, the herpes simplex virus type 1 protein UL9, which recruits viral rather than cellular replication proteins, binds cooperatively to specific origin sequences and distorts adjacent AT-rich sequences of the viral origins (Fig. 9.11). It also possesses an ATP-dependent helicase activity that unwinds DNA in the 3′ → 5′ direction. The adenovirus-associated virus Rep 78/68 protein possesses these same activities but is also the site-specific endonuclease that is essential for terminal resolution (Fig. 9.9). The domain that mediates sequence-specific binding adjacent to the terminal resolution site includes a large region very similar in architecture to the origin-binding domains of simian virus 40 LT and the papillomavirus E1 protein (Fig. 9.15). Such structural homology is remarkable, as there is no amino acid identity among the three viral proteins.

In many respects, the herpesviral UL9 protein is a typical origin-binding protein (Table 9.1). However, it is required only during the initial stage of viral DNA synthesis, as are the viral origins. The UL9 protein is cleaved by the cellular protease cathepsin B following the onset of viral DNA synthesis. Such cleavage may contribute to a switch from origin-dependent to origin-independent replication by preventing UL9-dependent initiation of DNA synthesis at the origins.

Although recognition of viral origins of replication by a single viral protein is common, it is not universal. The

BOX 9.7

EXPERIMENTS

The mechanism by which simian virus 40 LT unwinds and translocates along DNA

Despite decades of study, fundamental questions about the mechanism by which simian virus 40 LT unwinds DNA during genome replication remain unanswered. These include whether LT functions as a double hexamer throughout replication, as, for example, suggested by the studies described in Box 9.3, and how it translocates along DNA during unwinding. Structural studies of a hexamer of the LT helicase domain established that the central channel can expand sufficiently to accommodate double-stranded DNA, and identified side channels through which single-stranded DNA might be extruded, consistent with translocation on double-stranded DNA. However, the structurally related papillomavirus E1 helicase, as well as the cellular helicase minichromosome maintenance complex (Mcm), translocate in the 3′ → 5′ direction along single-stranded DNA to unwind double-stranded templates by steric exclusion.

In one approach to examine the first question (panel A of the figure, left), DNA containing the simian virus 40 origin was attached at both ends to the surface of a microfluidic flow cell. LT was drawn into the cell and allowed to assemble at the origin. The single-stranded-DNA-binding protein Rp-A fused to a green fluorescent-like protein (designated *-Rp-A) was then introduced, and fluorescent images were recorded for 60 min. Symmetrically growing linear tracks of the fluorescent Rp-A protein were observed (panel A, right), consistent with spatial separation of LT hexamers during unwinding from the origin. The observation that tethering both ends of the origin-containing template reduced neither the extent nor the rate of replication compared to tethering of just one end was inconsistent with an LT double hexamer drawing double-stranded DNA into its central chamber to unwind it: in this case, replication of doubly tethered templates would proceed more slowly. Why uncoupling of LT hexamers was observed in these but not in previous (Box 9.3) experiments remains to be explained and the form in which LT unwinds DNA in infected cells to be established.

To investigate the mechanism of translocation, a small, linear, origin-containing DNA template was modified by addition of a molecule of biotin to a specific site on the top or the bottom strand. As shown in panel B (left), biotin attached to the top strand can be displaced only if LT translocates along double-stranded DNA. Radioisotopically labeled versions of the templates were incubated with excess streptavidin, and then with LT and Rp-A for 30 min. Strand displacement was examined as the appearance of single-stranded DNA decreased in mobility by binding of streptavidin to the biotin tag. Control experiments established that denaturation of the templates resulted in complete release of both the biotin-tagged strands. LT-dependent displacement of a biotin-tagged single strand was observed **only** when biotin was attached to the bottom strand (panel B, right). These results indicate that, following assembly at the viral origin, simian virus 40 LT translocates in the 3′ → 5′ direction on the leading-strand template. This mechanism is therefore a common feature of viral and cellular helicases that operate during DNA synthesis in mammalian cells.

Yardimci H, Wang X, Lowland AB, Zudner DZ, Hurwitz J, van Oijen AM, Walter JC. 2012. Bypass of a protein barrier by a replicative DNA helicase. *Nature* **492:**205–209.

A

T-ag (45 min)

RPA^{mKikGR}

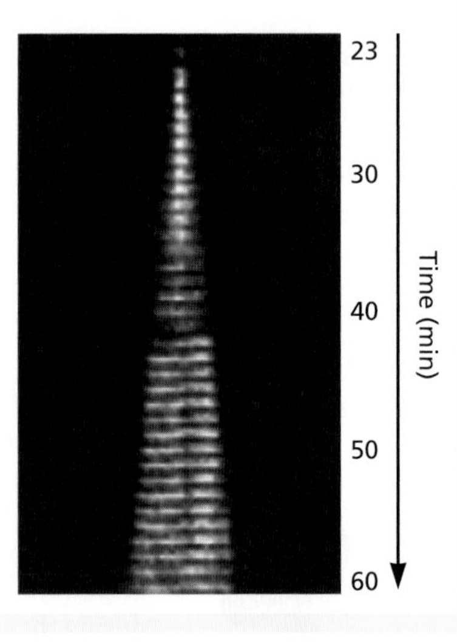

Visualization of formation of a replication bubble during DNA unwinding by LT. (A) The experimental strategy is depicted at the left, and the results obtained upon binding of fluorescent Rp-A during LT unwinding from the origin of a single template molecule are shown on the right. The symmetrical increase in length in the unwound DNA (bound by fluorescent Rp-A) as a function of time indicates that LT hexamers uncouple and move apart after initiation of unwinding. *(continued)*

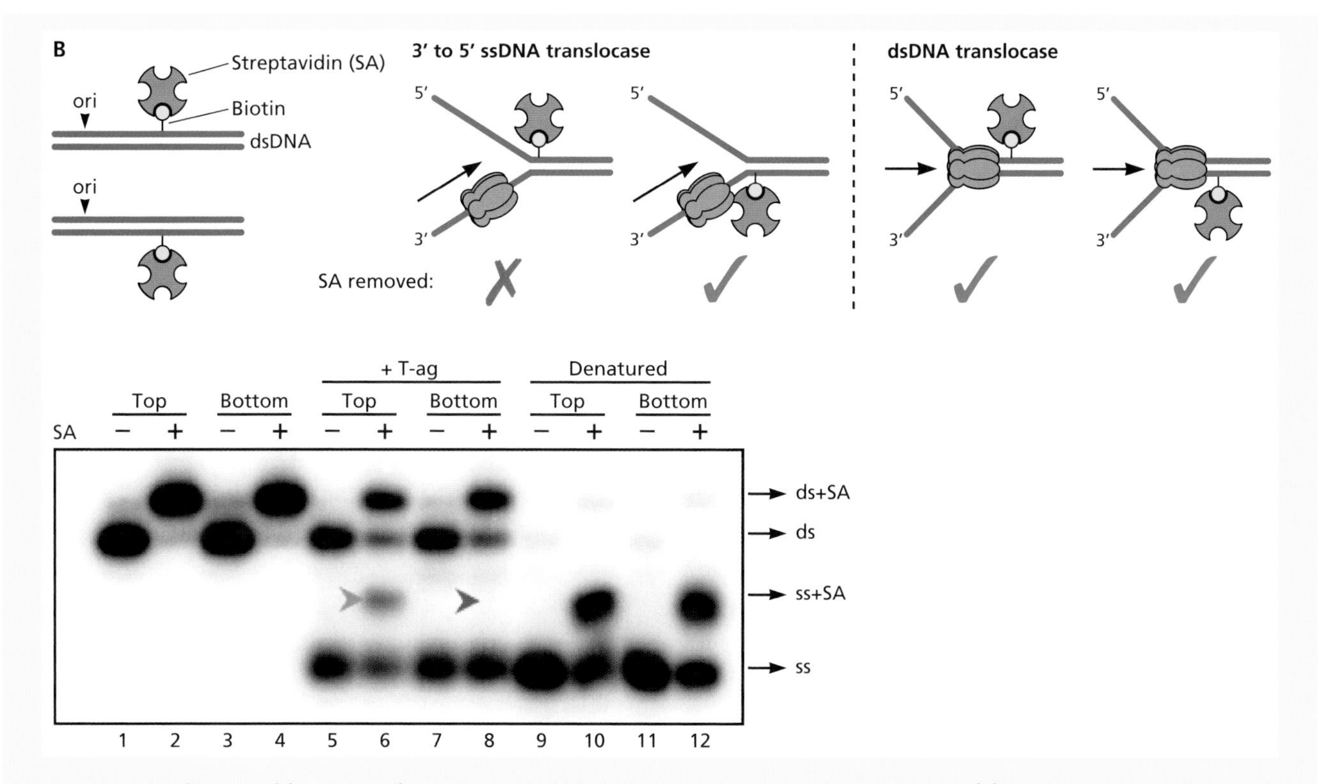

Visualization of formation of a replication bubble during DNA unwinding by LT. *(continued)* **(B)** As depicted at the top, LT can displace a biotin-tagged upper strand of an origin-containing template only if the protein translocates on double-stranded DNA. However, only the bottom biotin-tagged strand was released, as detected by the decrease in mobility after binding of streptavidin, when DNA unwinding was carried out by LT and Rp-A (right).

Figure 9.15 Structural homology among DNA-binding domains of viral origin recognition proteins. The X-ray crystal structures of the adenovirus-associated virus type 5 Rep 68 DNA-binding endonuclease domain and the bovine papillomavirus E1 and simian virus 40 LT origin-binding domains are shown in ribbon form. Each protein contains a central antiparallel β-sheet flanked by three α-helices. However, the Rep protein includes a cleft on one surface of the β-sheet that contains the endonuclease active site (residues shown in ball-and-stick). In the other two viral proteins, no cleft is present, as this region is occupied by N-terminal extensions (red) and helices shifted with respect to the position in Rep (orange). Adapted from A. Hickman et al., *Mol Cell* **10:**327–337, 2002, with permission. Courtesy of Alison Hickman, National Institutes of Health.

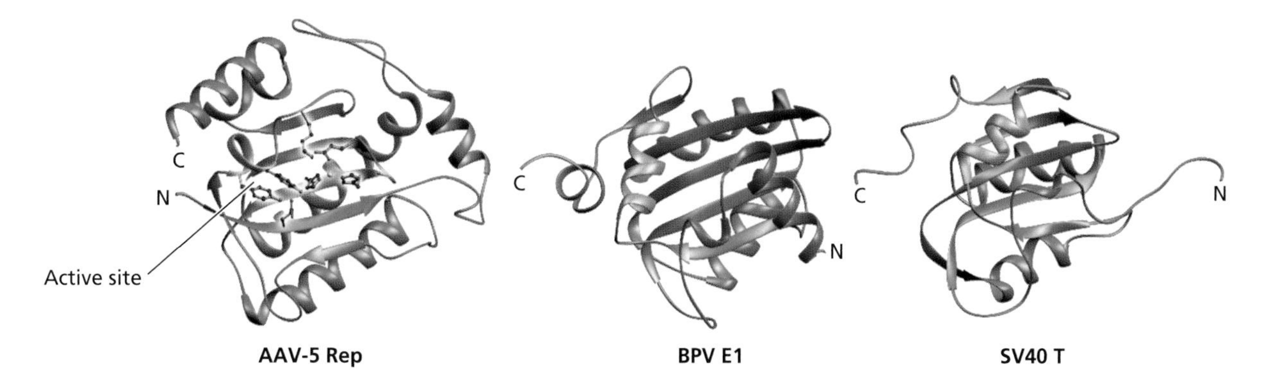

papillomavirus E1 proteins possess the same activities as simian virus 40 LT (Table 9.1), to which they are related in sequence, organization, and structure (Fig. 9.15). Nevertheless, the E1 protein cannot support papillomaviral DNA replication in infected cells: a second viral protein, the E2 transcriptional regulator, is also necessary. The minimal origin of replication of papillomaviral genomes includes adjacent binding sites for both the E1 and E2 proteins (Fig. 9.16A). The E1 protein binds to origin DNA with only low specificity. In contrast, when the E1 and E2 proteins bind cooperatively, the specificity and affinity of the E1-DNA interaction are increased significantly. Once a specific E1-E2 protein complex has assembled on the origin, hydrolysis of ATP to E1 appears to induce a conformational change that leads to dissociation of E2, allowing additional molecules of E1 to bind (Fig. 9.16B). The final product is an E1 double hexamer assembled on single-stranded DNA.

The adenoviral origins of replication are also recognized by two viral proteins, the preterminal protein and viral DNA polymerase. In this case, the proteins associate as they are synthesized in the cytoplasm and, once within the nucleus, bind specifically to a conserved sequence within the minimal origins of replication (Fig. 9.12).

Viral DNA Synthesis Machines

Larger viral DNA genomes encode DNA polymerases and other essential replication proteins. A particularly simple viral replication apparatus is that of adenoviruses, which comprises the Pre-TP primer and DNA polymerase and only one other protein, a single-stranded-DNA-binding protein. The latter protein stimulates initiation and is essential during elongation, when it coats the displaced strands of the template DNA molecule (Fig. 9.10). Cooperative binding of this protein to single-stranded DNA stimulates the activity of the viral DNA polymerase as much as 100-fold and induces highly processive DNA synthesis. Remarkably, no ATP hydrolysis is required. Rather, the DNA-binding protein multimerizes via a C-terminal hook (Fig. 9.17A), and the formation of long protein chains by cooperative, high-affinity binding to single-stranded DNA provides the driving force for ATP-independent unwinding of the duplex template (Fig. 9.17B). Other single-stranded-DNA-binding proteins, such as the herpes simplex virus type 1 UL8 protein and cellular replication protein A, may destabilize double-stranded DNA helices by a similar mechanism.

Other viral replication systems include a larger number of accessory replication proteins (Table 9.2). Herpes simplex virus type 1 genes that encode essential replication proteins have been discovered by both genetic methods and a DNA-mediated transformation assay that identifies the gene products necessary for plasmid replication directed by a viral origin (Fig. 9.18). Replication from a herpes simplex virus

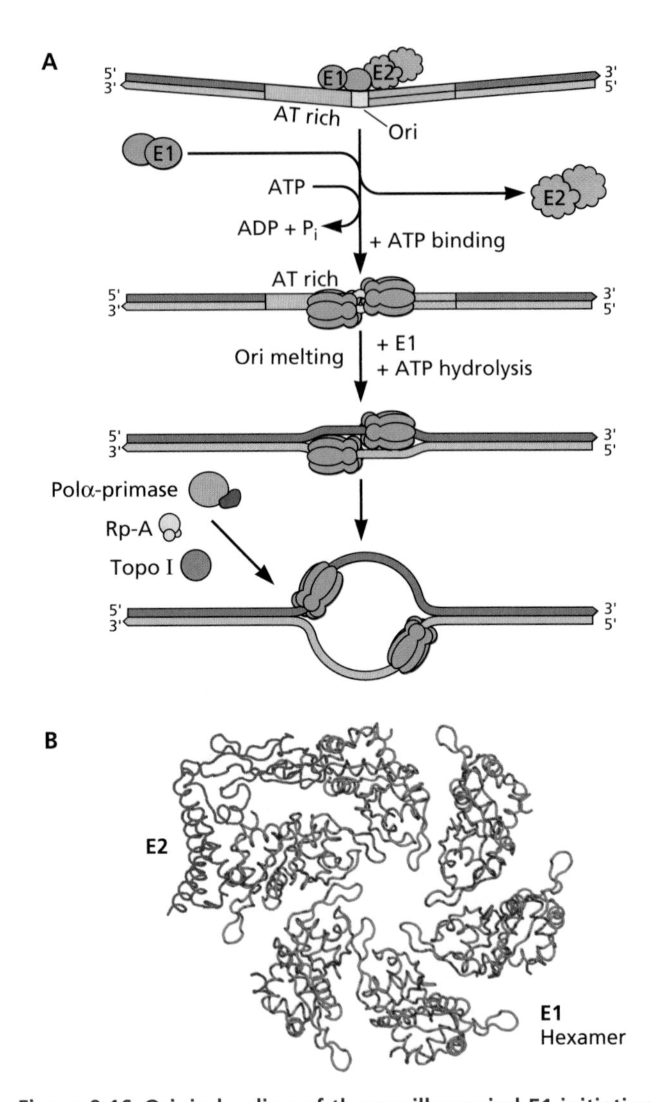

Figure 9.16 Origin loading of the papillomaviral E1 initiation protein by the viral E2 protein. (A) Schematic model. The sequence features of the minimal origin of replication of bovine papillomavirus type 1 are depicted as in Fig. 9.12. This origin contains an essential binding site for the viral E2 protein, a sequence-specific transcriptional regulator. The model of the origin loading of the viral E1 by the E2 protein is based on *in vitro* studies of the interactions of these proteins with the origin. The E1 and E2 proteins, which are both homodimers, bind cooperatively to the viral origin, with specificity and affinity far greater than that exhibited by the E1 protein alone. When ATP is hydrolyzed (presumably by the ATPase of the E1 protein), the $(E1)_2(E2)_2$-Ori complex is destabilized, the E2 dimers are displaced, and additional E1 molecules bind, initially forming double trimers. Upon further ATP hydrolysis and unwinding of origin DNA by an assembly intermediate, E1 double hexamers assemble, each encircling a single strand of DNA. **(B)** X-ray crystal structures of the E2 activation domain (red) and the E1 ATPase/helicase domain (blue) are shown in ribbon form. This overlay of E2 and the E1 hexamer illustrates how association with E2 blocks the E1 surface that mediates hexamer assembly. Hence, E2 must dissociate prior to E1 assembly. Consistent with this model, the E1 and E2 proteins form a 1:1 complex in the absence of ATP, but in the presence of ATP E1 assembles into a high-molecular-mass form that contains no E2. Adapted from E. Abbate et al., *Genes Dev* **18:**1981–1986, 2004, with permission. Courtesy of Eric Abbate and Michael Botchan, University of California, Berkeley.

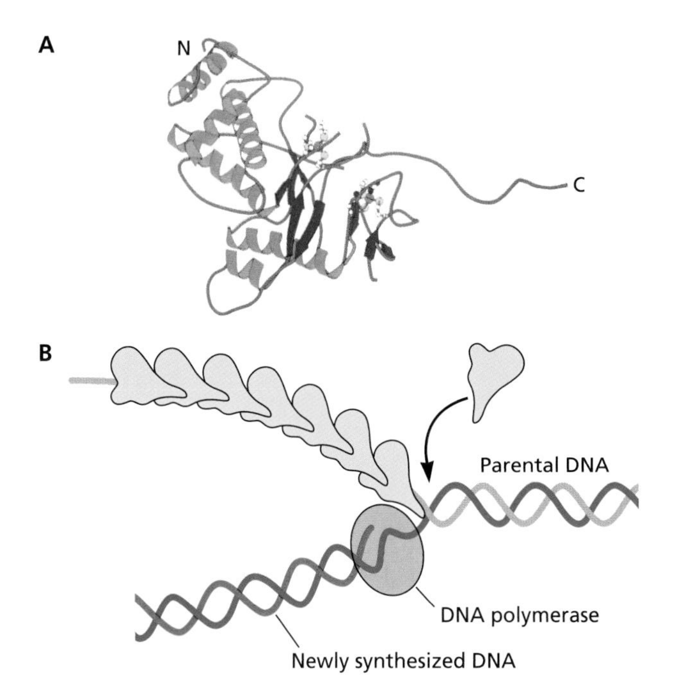

Figure 9.17 Crystal structure of the adenoviral single-stranded-DNA-binding protein. (A) Ribbon diagram of the C-terminal nucleic acid-binding domain (amino acids 176 to 529) of the human adenovirus type 5 protein, showing the two sites of $Zn^{2}+$ (red atom) coordination. The most prominent feature is the long (<40-Å) C-terminal extension. This C-terminal extension of one protein molecule invades a cleft between two α-helices in its neighbor in the protein array formed in the crystal. Deletion of the C-terminal 17 amino acids of the DNA-binding protein fragment eliminates cooperative binding of the protein to DNA, indicating that the interaction of one molecule with another via the C-terminal hook is responsible for cooperativity in DNA binding. From P. A. Tucker et al., *EMBO J* **13**:2994–3002, 1994, with permission. Courtesy of P. C. van der Vliet, Utrecht University. **(B)** Model of unwinding of double-stranded adenoviral DNA by cooperative interactions among the viral single-stranded-DNA-binding protein.

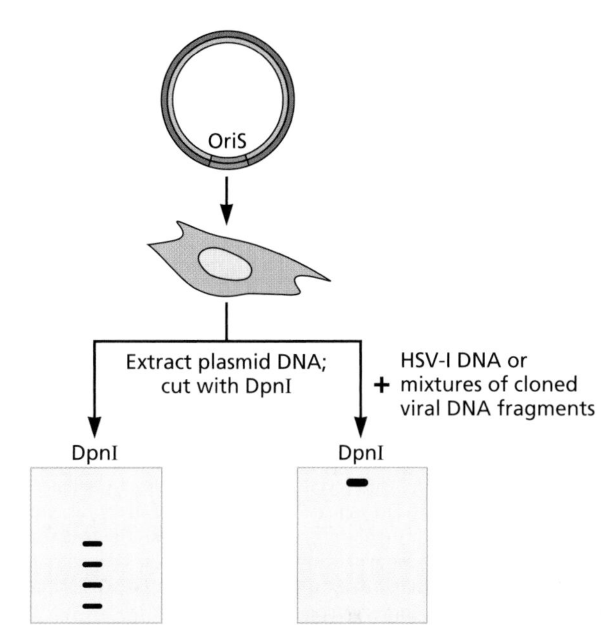

Figure 9.18 DNA-mediated transformation assay for essential herpes simplex virus type 1 replication proteins. A plasmid carrying a viral DNA fragment spanning OriS is introduced into monkey cells permissive for herpesvirus replication. In the absence of viral proteins (left), the plasmid DNA is not replicated and retains the methyl groups added to A residues in a specific sequence by the *E. coli dam* methylation system. As these sequences include the recognition site for the restriction endonuclease DpnI, which cleaves only such methylated DNA, the unreplicated plasmid DNA is sensitive to DpnI cleavage. When all viral genes encoding proteins required for OriS-dependent replication are also introduced into the cells (right), the plasmid is replicated. Because the newly replicated DNA is **not** methylated at DpnI sites, it cannot be cleaved by this enzyme. Resistance of the plasmid to DpnI cleavage therefore provides a simple assay for plasmid replication, and for the identification of viral proteins required for replication from OriS.

Table 9.2 Replication systems of large DNA viruses

Function	Viral protein(s)	
	Herpes simplex virus type 1	**Vaccinia virus**
Common components		
DNA polymerase and associated 3′→5′ exonuclease	UL30	E9L
Primase/helicase	UL5, UL8, and UL52 Heterotrimer	D5R
Processivity factor	UL42	A20R
Single-stranded-DNA-binding protein	UL2	I3L
Apparently unique components		
Origin recognition protein	UL9	–
Type I topoisomerase	–	H6R
DNA ligase	–	A50R

type 1 origin requires five proteins in addition to the viral DNA polymerase and origin recognition protein. These proteins carry out the same reactions as essential components of the cellular replication machinery, as do the proteins necessary for synthesis of vaccinia viral DNA (Table 9.2). Although these herpesviral proteins provide an extensive repertoire of replication functions, they are not sufficient for viral DNA synthesis *in vitro*. While all the additional viral and/or cellular proteins needed to reconstitute herpesviral DNA synthesis have not been identified, cellular topoisomerase II is essential for replication in infected cells.

Resolution and Processing of Viral Replication Products

Several of the viral DNA replication mechanisms described in preceding sections yield products that do not correspond to the parental viral genome. As we have seen, replication of simian virus 40 DNA yields two interlocked, double-stranded,

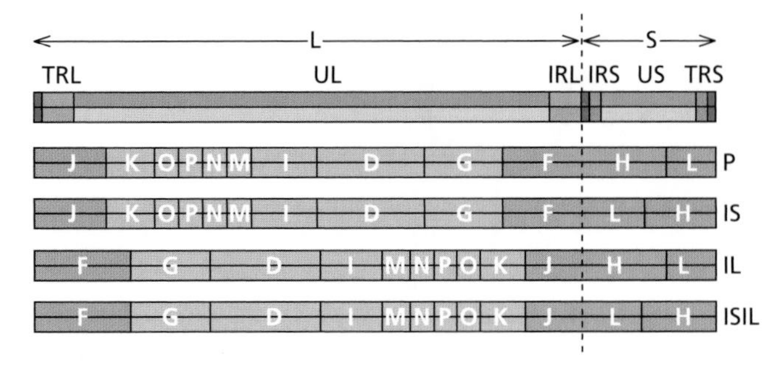

Figure 9.19 Isomers of the herpes simplex virus type 1 genome. The organization of the unique and repeated sequences of the viral genome are depicted at the top, as in Fig. 9.11. This orientation is defined as the prototype (P) genome isomer. The other three isomers differ, with respect to the P form, in the orientation of S (IS), in the orientation of L (IL), or in both S and L (ISIL). These differences are illustrated using HindIII fragments. The unusual isomerization of this viral genome was deduced from the presence of fragments that span the terminal or internal inverted repeat sequences at 0.5 and 0.25 molar concentrations, respectively, in such HindIII digests, and examination of partially denatured DNA in the electron microscope.

circular DNA molecules that must be separated by cellular topoisomerase II. Such resolution is required whenever circular templates (e.g., papillomavirus or episomal Epstein-Barr virus DNA) are replicated as monomers. In other cases, replication yields multimeric DNA molecules, from which linear genomes of fixed length must be processed for packaging into virus particles. This situation is exemplified by the herpes simplex virus type 1 genome (Fig. 9.19).

The products of herpesviral DNA synthesis are head-to-tail **concatemers** containing multiple copies of the viral genome. It is well established that the linear viral genomes that enter infected cell nuclei at the start of a productive infection are converted rapidly to "endless" molecules in which the DNA termini are joined together. This reaction requires cellular DNA ligase IV, which normally mediates joining of nonhomologous DNA ends during a cellular repair process, but it is not clear if unit-length circles (as found in latently infected cells) or linear concatemers are produced (Box 9.8). This distinction is of more than esoteric interest, as the configuration of the parental DNA has profound implications about the mechanism of viral DNA synthesis. When a genome is circular, concatemers can be synthesized by the rolling-circle mechanism (Box 9.4), as during initial replication of the double-stranded DNA genome of bacteriophage λ. In contrast, recombination is required to produce longer-than-unit-length genomes during replication of a linear template (see "Recombination of Viral Genomes" below).

Linear herpes simplex virus type 1 DNA molecules with termini identical to those of the infecting genome are liberated from concatemeric replication products by cleavage at specific sites within the *a* repeats (Fig. 9.11). Such cleavage is coupled with encapsidation of viral DNA molecules during assembly of virus particles (Chapter 13).

Exponential Accumulation of Viral Genomes

The details of the mechanisms by which DNA genomes are replicated vary considerably from one virus family to another. Nevertheless, each of these strategies results in efficient viral DNA synthesis. Production of 10^3 to 10^4 viral genomes, or more, per infected cell is not uncommon, as the products of one cycle of replication are recruited as templates for the next. Such exponential viral DNA synthesis sets the stage for assembly of a large burst of progeny virus particles. In this section, we discuss regulatory mechanisms that ensure efficient viral DNA synthesis.

Viral Proteins Can Induce Synthesis of Cellular Replication Proteins

With few exceptions, virus reproduction is studied by infecting established cell lines that are susceptible and permissive for the virus of interest. Such immortal or transformed cell lines proliferate indefinitely and differ markedly from the cells in which viruses reproduce in nature. For example, highly differentiated cells, such as neurons or the outer cells of an epithelium, do not divide and are permanently in a specialized resting state, termed the **G_0 state**. Many other cells in an organism divide only rarely, or only in response to specific stimuli, and therefore spend much of their lives in G_0. Such cells do not contain many of the components of the replication machinery and are also characterized by generally low rates of synthesis of RNAs and proteins. Virus reproduction entails the synthesis of large quantities of viral nucleic acids (and proteins), often at a high rate. Consequently, the resting state does not provide a hospitable environment. Nevertheless, viruses often reproduce successfully within cells infected when they are in G_0. In some cases, such as replication of the genomes of several herpesviruses in neurons, the DNA synthesis machinery is encoded within the viral genome. Infection by other viruses stimulates resting or slowly growing cells to abnormal activity, by disruption of cellular circuits that restrain cell proliferation. This strategy is characteristic of polyomaviruses and adenoviruses.

Functional Inactivation of the Rb Protein

Loss or mutation of both copies of the cellular retinoblastoma (*rb*) gene is associated with the development of tumors of the retina in children and young adults. Because it is the **loss** of normal function that leads to tumor formation, *rb* is defined as a **tumor suppressor gene**. The Rb protein is an important component of the regulatory program that ensures that cells grow, duplicate their DNA, and divide in an orderly manner (Volume II, Chapter 6). In particular, the Rb protein

BOX 9.8

DISCUSSION

Circularization or concatemerization of the herpes simplex virus type 1 genome in productively infected cells?

It has been known for some time that the linear herpes simplex virus type 1 DNA that enters nuclei of productively infected cells rapidly adopts a new conformation in which the termini are fused: restriction endonuclease cleavage of viral DNA recovered shortly after infection established that cleavage products generated from free ends decrease in concentration as those characteristic of joined ends increase. As shown in the figure, joined ends could arise by either circularization or concatemerization (as long as infected cell nuclei contain multiple copies of the genome). The origin of joined ends has been difficult to determine, for the following reasons:

- Formation of **either** unit-length circles or concatemers results in loss of free termini.
- The presence of an internal, inverted copy of the joined terminal repeats precludes the use of an assay based on detection of joined termini.
- Conventional methods for separation and identification of linear and circular DNA molecules by electrophoresis cannot be applied to the large herpesviral genome.
- The large size of the herpesviral genome renders it very sensitive to damage during extraction from infected cells.
- Under conditions that facilitate detection of entering viral DNA, high multiplicity of infection, the majority of infecting DNA molecules are neither transcribed nor replicated.

Consequently, application of different experimental approaches has led to reports that entering viral genomes become circular and that they form concatemers, and these divergent conclusions have yet to be reconciled.

Jackson SA, DeLuca NA. 2003. Relationship of herpes simplex virus genome configuration to productive and persistent infections. *Proc Natl Acad Sci U S A* **100:**7871–7876.

Strang BL, Stow ND. 2005. Circularization of the herpes simplex virus type 1 genome upon lytic infection. *J Virol* **79:**12487–12494.

The herpes simplex virus type 1 genome is depicted as in Fig. 9.11. As shown below, either circularization or formation of concatemers leads to loss of genome termini, and hence of restriction endonuclease fragments that contain them.

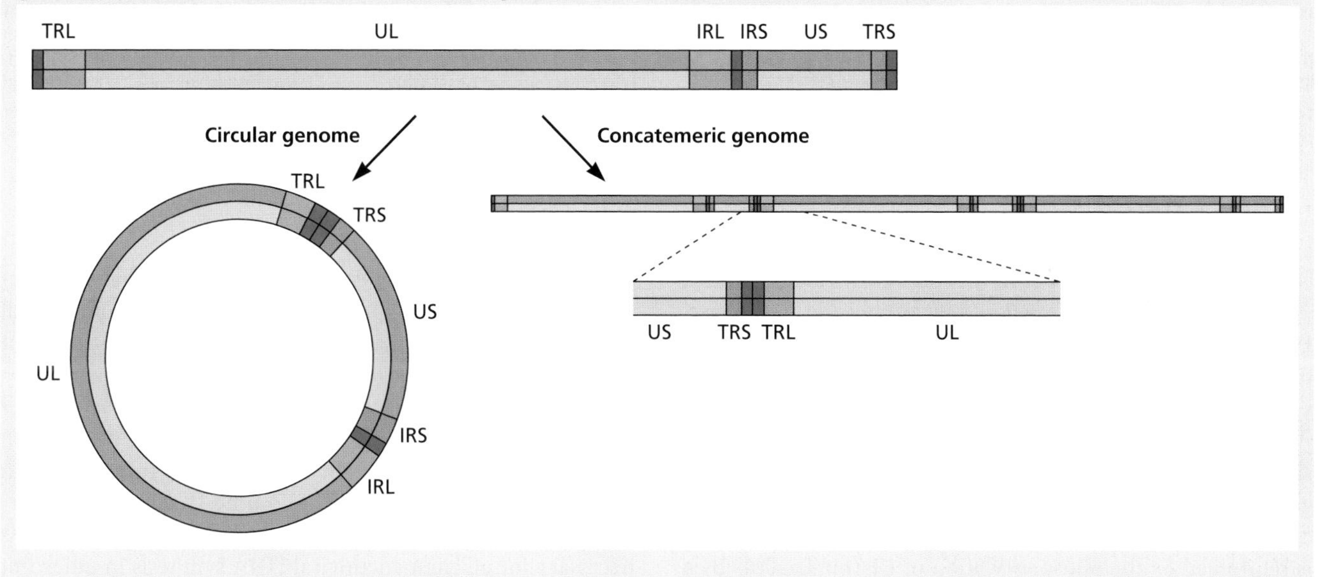

controls entry into the period of the cell cycle in which DNA is synthesized, the **S phase**, from the preceding (G_1) phase. Our current appreciation of the critical participation of this protein in the control of cell cycle progression, and of the mechanism by which it operates, stems from the discovery that Rb binds directly to the two adenoviral E1A proteins (see Chapter 8) and functionally analogous proteins of papillomaviruses and polyomaviruses.

In the G_1 phase of uninfected cells, the Rb protein is bound to transcriptional regulators of the E2f family. These complexes, which bind to specific promoters via the DNA-binding activity of E2f, function as repressors of transcription (Fig. 9.20A). Binding of adenoviral E1A proteins, simian virus 40 LT, or E7 proteins of highly oncogenic human papillomaviruses to Rb releases E2f from this association and sequesters Rb. The E2f proteins therefore become available to stimulate transcription of cellular genes for proteins that participate directly or indirectly in DNA synthesis or in control of cell cycle progression (Fig. 9.20A; Volume II, Chapter 6).

Many of the genes that encode DNA polymerases, accessory replication proteins, and enzymes that catalyze synthesis of dNTPs contain E2f-binding sites in their transcriptional

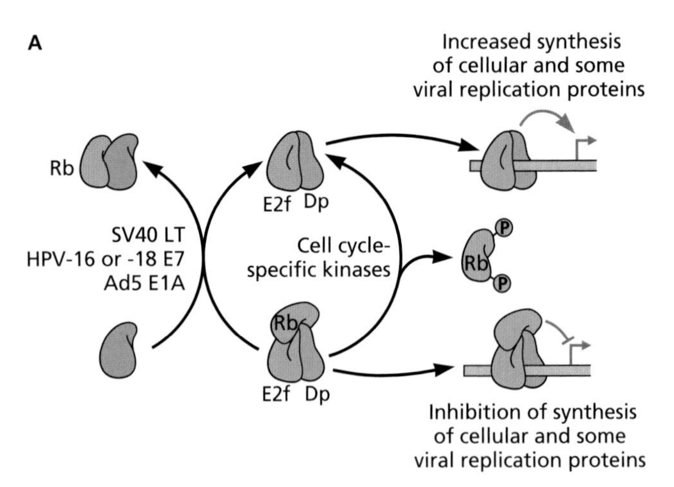

Figure 9.20 Regulation of production of cellular and viral replication proteins. (A) Model for the abrogation of the function of the Rb protein by viral proteins. E2f transcriptional regulators are heterodimeric proteins, each containing one E2f and one Dp (E2f dimerization partner) subunit. E2f dimers stimulate transcription of cellular genes encoding replication proteins, histones, and proteins that allow passage through the cell cycle (green arrow). Binding of Rb protein does not prevent promoter recognition by E2f. However, Rb protein represses transcription (red bar). Phosphorylation of Rb protein at specific sites induces its dissociation from E2f and activates transcription of cellular genes expressed in S phase. The adenoviral E1A proteins, simian virus 40 LT, and the E7 proteins of certain human papillomaviruses (for example, types 16 and 18) bind to the region of Rb protein that contacts E2f to disrupt Rb-E2f complexes and activate E2f-dependent transcription. **(B)** Stimulation of transcription from the adenoviral E2 early promoter by E1A proteins. The E2E promoter-binding sites for the cellular Atf (activating transcription factor), E2f, and TfIId (transcription factor IId) proteins are necessary for E2E transcription in infected cells. The inversion of the two E2f sites (arrows) and their precise spacing are essential for assembly of an E2f-DNA complex unique to adenovirus-infected cells, in which the viral E4 Orf6/7 protein is bound to each E2f heterodimer. Binding of the E4 protein promotes cooperative binding of E2f and increases the lifetime of E2f-DNA complexes. The availability of the cellular E2f and viral E4 Orf6/7 proteins is a result of the action of immediate-early E1A proteins: either the 243R or 289R protein can sequester unphosphorylated Rb to release active E2f from Rb-E2f complexes, and the 289R protein stimulates transcription from the E4 promoter. This larger E1A protein can also stimulate transcription from the E2E promoter directly.

control regions. This property normally restricts synthesis of these gene products to when they are needed in S phase. However, the sequestration of Rb by simian virus 40 LT allows production of the cellular proteins necessary for viral DNA synthesis, regardless of the proliferation state of the host cell. As LT is the only viral protein needed for viral DNA synthesis, its production seems likely to maximize the efficiency of genome replication, by coordinating initiation of this process with entry of the host cell into S phase. Such integration is reinforced by the phosphorylation of LT (on Thr124) by a kinase, Cdk2-cyclin A, which is present **only** during this period, because this modification is essential for initiation of viral DNA synthesis.

A major consequence of activation of E2f in adenovirus-infected cells is stimulation of production of the three viral replication proteins. The viral DNA polymerase, Pre-TP primer, and DNA-binding protein are encoded within the E2 gene, which is transcribed from an early promoter that contains two binding sites for E2f (Fig. 9.20B). In fact, these critical cellular regulators derive their name from the E2 promoter-binding sites, which are necessary for efficient E2 transcription during the early phase. As noted above, the viral E1A proteins disrupt Rb-E2f complexes to release E2f, but also stimulate transcription from the E2 promoter by two other mechanisms (Fig. 9.20B). The E1A-dependent regulatory mechanisms presumably operate synergistically, to allow synthesis of the viral mRNAs that encode replication proteins in quantities sufficient to support numerous cycles of viral DNA synthesis.

The mechanism by which LT and E1A proteins counter Rb (and Rb family members) to induce cell cycle progression are well established, as is their importance in transformation of nonpermissive rodent cells (Volume II, Chapter 6). Furthermore, it has been shown that the smaller E1A protein is necessary for efficient adenoviral DNA synthesis in quiescent human cells, and that this protein displaces Rb and related proteins from cellular promoters, including those that contain binding sites for E2f.

Synthesis of Viral Replication Machines and Accessory Enzymes

The DNA genomes of several viruses, exemplified by those of herpes simplex virus type 1 and the poxvirus vaccinia virus, encode large cohorts of proteins that participate in viral genome replication directly (the proteins that mediate viral DNA synthesis described previously) or indirectly (accessory enzymes). These enzymes are viral analogs of cellular proteins that catalyze synthesis of dNTP substrates, such as thymidine kinase and ribonucleotide reductase, or participate in repair

Table 9.3 Viral enzymes of nucleic acid metabolism

Virus	Protein	Functions
Herpesvirus		
Herpes simplex virus type 1	Thymidine kinase (UL23 protein, ICP36)	Phosphorylates thymidine and other nucleosides; essential for efficient reproduction in animal hosts
	Ribonucleotide reductase ($\alpha_2\beta_2$ dimer of UL39 and U40 proteins)	Reduces ribose to deoxyribose in ribonucleotides; essential in nondividing cells
	dUTPase (UL50 protein)	Hydrolyzes dUTP to dUMP, preventing incorporation of dUTP into DNA and providing dUMP for conversion to dTMP
	Uracil DNA glycosylase	Corrects insertion of dUTP or deamination of C in viral DNA
	Alkaline nuclease (UL12 protein)	Required for production of infectious DNA
Poxvirus		
Vaccinia virus	Thymidine kinase	Phosphorylates thymidine; required for efficient virus reproduction in animal hosts
	Thymidylate kinase	Phosphorylates TMP
	Ribonucleotide reductase, dimer	Reduces ribose to deoxyribose in ribonucleotides; essential in nondividing cells
	dUTPase	Hydrolyzes dUTP to dUMP (see above)
	DNase	Has nicking-joining activity; present in virion cores
	D4R protein	Uracil DNA glycosylase

of DNA (Table 9.3). In general, such proteins are dispensable for replication in proliferating cells in culture, because cellular enzymes supply the substrates for DNA synthesis. However, herpes simplex viruses that lack thymidine kinase or ribonucleotide reductase genes cannot reproduce in neurons: such terminally differentiated cells are permanently withdrawn from the cell cycle and do not make enzymes that produce substrates for DNA synthesis.

Timely synthesis of herpes simplex virus type 1 replication proteins is the result of the viral transcriptional cascade described in Chapter 8. Expression of the early genes that encode these viral proteins is regulated by immediate-early proteins. These regulatory proteins operate transcriptionally (e.g., ICP0 and ICP4) or posttranscriptionally (e.g., ICP27) to induce synthesis of viral replication proteins at concentrations sufficient to support efficient replication of viral genomes.

Viral DNA Replication Independent of Cellular Proteins

One method guaranteed to ensure replicative success of a DNA virus, regardless of the proliferation state of the host cell, is to encode **all** of the necessary proteins in the viral genome. On the other hand, this mechanism is genetically expensive, which may be the reason why it is restricted to the viruses with the largest DNA genomes, such as the poxvirus vaccinia virus. The genome of this virus, which is replicated in the cytoplasm, encodes a DNA polymerase, several accessory replication proteins, and enzymes for synthesis of dNTPs (Table 9.3). None of the latter appear to be essential for virus reproduction in actively growing cells. However, several of them, such as the thymidine kinase, are necessary for efficient virus propagation in quiescent cells or in animal hosts, where they presumably contribute to synthesis of nucleotide substrates for genome replication.

Delayed Synthesis of Structural Proteins Prevents Premature Packaging of DNA Templates

Provided that daughter viral DNA molecules are available to serve as templates, each cycle of genome replication increases the number of DNA molecules that can be copied in the subsequent cycle. One process that sequesters potential templates is encapsidation of genomes during assembly of new virus particles. However, particle assembly is delayed with respect to initiation of viral DNA synthesis, in part because transcription of late genes that encode structural proteins depends on genome replication (Chapter 8). The increase in the pool of replication templates therefore seems likely to make an important contribution to rapid amplification of genomes.

Inhibition of Cellular DNA Synthesis

When viral DNA replication is carried out largely by viral proteins, cellular DNA synthesis is often inhibited, presumably to increase the availability of substrates for viral genome replication. Indeed, infection by the larger DNA viruses (herpesviruses and poxviruses) induces severe inhibition of cellular DNA synthesis. This process is also blocked when adenoviruses infect proliferating cells in culture. Although inhibition of cellular DNA synthesis in cells infected by these DNA viruses was described in some of the earliest studies of their infectious cycles, very little is known about the mechanisms that shut down this cellular process.

There is some evidence that inhibition of cellular DNA synthesis is an active process rather than an indirect result of passive competition between viral and cellular DNA polymerases for the finite pools of dNTP substrates. For example, infection of proliferating cells by adenovirus or betaherpesviruses such as human cytomegalovirus induces cell cycle arrest, as does synthesis of the Epstein-Barr virus Zta protein, a sequence-specific transcriptional regulator and origin-binding protein. In the latter case, arrest is the result of **increased** concentrations of cellular proteins that negatively regulate progression through the cell cycle, such as the Rb protein.

Viral DNAs Are Synthesized in Specialized Intracellular Compartments

A common, probably universal, feature of cells infected by viruses with DNA genomes is the presence of virus-specific territories that are the sites of viral DNA synthesis. Vaccinia virus DNA is replicated in the cytoplasm, in discrete **viral factories** that contain viral genomes and lie near infected cell nuclei. Visualization of viral genomes in living cells and *in situ* hybridization suggests that each viral factory is established by a single infectious particle (Fig. 9.21A). These compartments contain the viral replication proteins, all the viral enzymes and other proteins necessary for synthesis of viral mRNAs, and cellular translation proteins. Consequently, the viral proteins that participate in replication of the vaccinia virus genome are produced within the specialized compartments in which they will operate.

The replication of viral DNA genomes within infected cell nuclei also takes place in specialized compartments, which can be visualized as distinctive, infected cell-specific foci containing viral proteins. Such structures, known as **replication centers** or **replication compartments,** have been best characterized in human cells infected by adenovirus or herpes simplex virus type 1 (Fig. 9.21B). They contain newly synthesized viral DNA and the viral proteins necessary for viral DNA

Figure 9.21 Discrete sites of viral replication. (A) Cytoplasmic vaccinia virus factories. Monkey cells stably synthesizing the DNA-binding bacteriophage λ Cro repressor fused to enhanced green fluorescent protein (Cro-EGFP) were infected with a 1:1 mixture of vaccinia viruses carrying in their genomes the coding sequence for either bacteriophage T7 RNA polymerase or *E. coli* LacZ. Direct fluorescent imaging of living cells indicated that Cro-EGFP labeled both cellular DNA in the cytoplasm and cytoplasmic viral DNA, via nonspecific binding to DNA. The top panel shows such as a fluorescent image recorded from 1 to 6.5 h after infection, by which time the initial structures have increased significantly in size. The infected cells were then examined by fluorescent *in situ* hybridization (FISH) with probes specific for the T7 RNA polymerase (red) or LacZ (green) genes, and stained with 4′,6-diamidino-2-phenylindole (blue). The processing necessary for FISH denatures Cro-EGFP and eliminates its fluorescence. As shown in the bottom panel, replication of individual incoming viral genomes encoding either T7 RNA polymerase or LacZ takes place in distinct factories. However, some genome mixing is evident (white arrow) as individual factories coalesce or fuse. Adapted from Y.-C. Lin and D. H. Evans, *J Virol* **84:**2432–2443, 2010,with permission. Courtesy of D. H. Evans, University of Alberta, Edmonton, Canada. **(B)** Adenoviral replication centers visualized in infected cells exposed to the dNTP analog bromodeoxyuridine (BrdU) for 1 h at the times postinfection (p.i.) indicated to mark newly synthesized DNA. Such DNA (green) and the E2 single-stranded-DNA-binding protein (DBP) (red) were detected by indirect immunofluorescence. As illustrated, replication centers develop from small foci (white arrows in top panel) to larger ring-like structures as the infectious cycle progresses. Adapted from D. Gautam and E. Bridge, *J Virol* **87:**8687–8696, 2013, with permission. Courtesy of E. Bridge, University of Miami.

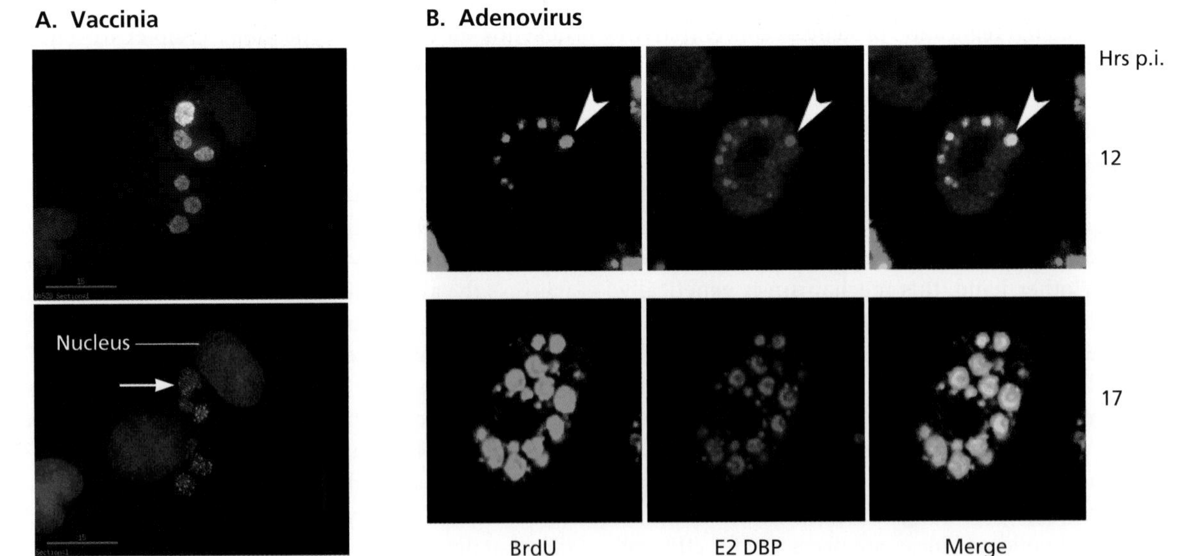

synthesis and increase in size as viral genome replication takes place. As is the case for vaccinia virus cytoplasmic factories, each replication center formed in alphaherpesvirus-infected cells originates from a single viral genome. However, the number of such genomes that can be expressed and replicated in a single infected cell is strictly limited (Box 9.9). A characteristic feature of the replication centers established in cells infected by herpesviruses is the recruitment to these sites of numerous cellular DNA repair and recombination proteins (Fig. 9.22). Several of the latter, including DNA mismatch

BOX 9.9

EXPERIMENTS

Counting the number of herpesviral genomes that can be expressed and replicated

Conventional methods for visualization of viral DNA molecules in infected cells, such as indirect immunofluorescence or *in situ* hybridization, detect **all** viral genomes and cannot distinguish functional genomes from those that are nonfunctional. Consequently, whether all viral genomes that enter permissive host cells can be expressed and replicated to produce progeny virus particles is a long-standing question. This issue has now been addressed for an alphaherpesvirus exploiting the properties of mixing of light of different wavelengths.

For these experiments, isogenic derivatives of pseudorabies virus that direct the synthesis of a red, cyan, or yellow fluorescent protein were constructed. Porcine kidney epithelial cells were infected with an equal mixture of the three viruses at increasing multiplicities of infection (MOIs), and the color profiles of infected cells visualized 6 h after infection by using epifluorescence microscopy. The color spectra of thousands of cells per condition were determined and plotted according to their position on a triangle plot, as shown in panel A of the figure. In this plot, each vertex of the triangle represents a pure color, each side represents a mixture of two colors, and mixtures of the three colors observed in individual cells are represented by the points within the triangle.

As the MOI is increased, each cell should be infected by an increasing number of genomes. Consequently, the number of cells with mixed colors should increase as those exhibiting a single color (red, cyan, or yellow) decrease. While this pattern was observed, a significant number of cells exhibited single or double colors even at the highest MOIs (panel A). The number of fluorescent proteins (0 to 3) per cell was determined from the colors of individual cells. This parameter was then used in conjunction with a mathematical model to estimate the average number of genomes expressed in an infected cell (l). Strikingly, when l was examined as a function of MOI, the number of expressed genomes expressed did not increase linearly. Rather, this value approached the low limit of <10 genomes per cell (panel B). This result was independent of the viral promoter from which the genes encoding fluorescent proteins were expressed and of whether the reporter proteins were made early or late after infection. It was also shown that the genomes that are expressed are also those that are replicated.

These experiments establish that the number of herpesviral genomes that support viral reproduction is strictly limited, presumably by properties of the host cell. The number of active genomes correlates closely with the number of viral replication centers that are established in infected cell nuclei and the number of genomes that are packaged into virus particles. Although the mechanisms responsible for this limitation of active genomes are not yet known, one possibility is that most infecting DNA genomes are repressed by intrinsic nuclear defense systems.

Kobiler O, Brodersen P, Taylor MP, Ludmir EB, Enquist LW. 2011. Herpesvirus replication compartments originate with single incoming viral genomes. *MBio* **2:**e00278-11. doi:10.1128/mbio.00278-11.

Kobiler O, Lipman Y, Therkelsen K, Daubechies I, Enquist LW. 2010. Herpesviruses carrying a Brainbow cassette reveal replication and expression of limited numbers of incoming genomes. *Nat Commun* **1:**146. doi:10.1038/ncomms1145.

(A) Confluent porcine kidney epithelial cells were infected with mixtures of equal concentrations of infectious particles of pseudorabies viruses that direct expression of red, cyan, or yellow fluorescent protein (RFP, CFP, and YFP, respectively) at the MOIs indicated. Representative color profiles visualized by epifluorescence microscopy are shown at the top with triangle plots for >3,000 cells per condition shown below. Bar = 100 μm. **(B)** The values of λ calculated from two experiments (each with three separate replicate wells) plotted as a function of MOI. The range of λ values among the replicates is represented for each point by the bar. Adapted from O. Kobiler et al., *Nat Commun* **1:**146, 2010, with permission.

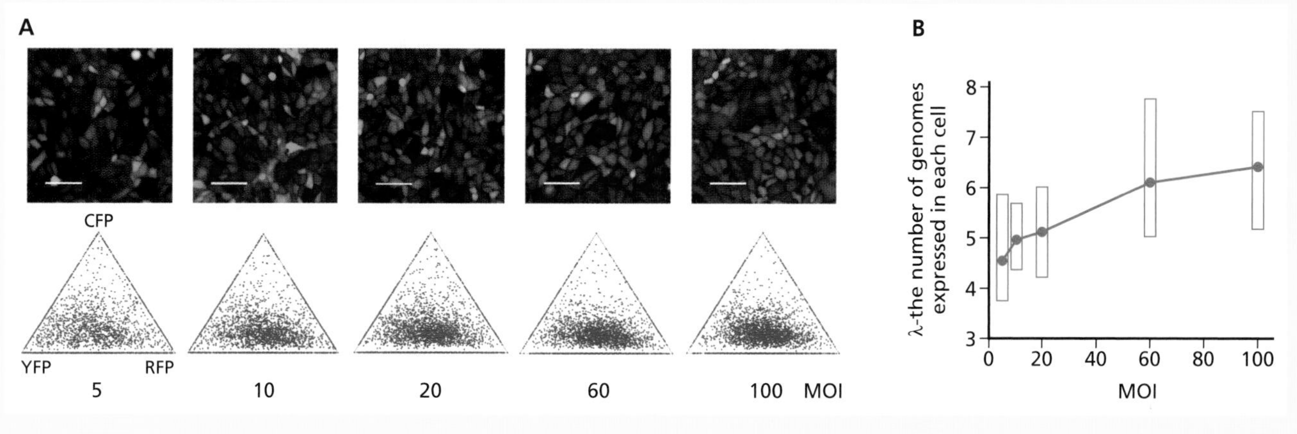

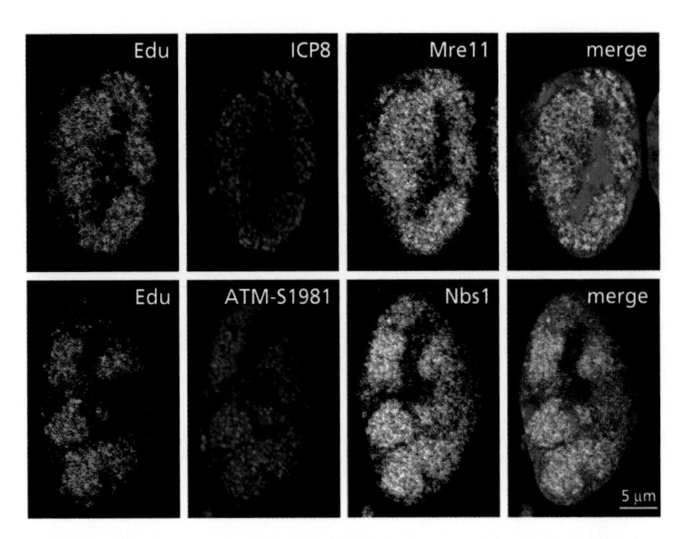

Figure 9.22 Association of cellular DNA damage response proteins with herpesviral replication centers. The cellular DNA damage response proteins Nbs1, Mre11, and activated (phosphorylated) Atm kinase (ATM-S1981) and the viral single-stranded-DNA-binding protein ICP8 (UL42) were detected by indirect immunofluorescence 8 h after herpes simplex virus type 1 infection of HeLa cells. Edu indicates viral DNA detected by incorporation of ethynyl deoxyuridine, biotinylation by "click" chemistry, and reaction with anti-biotin antibodies. Courtesy of M. D. Weitzman, University of Pennsylvania.

repair proteins and components of a signal transduction pathway activated in response to DNA damage, are required for maximally efficient viral replication.

The localization of the templates for viral DNA synthesis as well as the replication proteins at a limited number of sites undoubtedly facilitates efficient genome replication. This arrangement increases the local concentrations of proteins that must interact with one another, or with viral origin sequences or replication forks, favoring such intermolecular interactions by the law of mass action. In addition, the high local concentrations of replication templates and proteins are likely to allow efficient recruitment of the products of one replication cycle as templates for the next. Viral replication centers also serve as foci for viral gene expression, presumably in part by concentrating templates for transcription with the proteins that carry out or regulate this process. For example, the herpes simplex virus type 1 immediate-early ICP4 and ICP27 proteins, as well as the host cell's RNA polymerase II, are recruited to these nuclear sites.

Viral replication centers do not assemble at random sites, but rather are formed by viral colonization of specialized niches within mammalian cell nuclei. When they enter the nucleus, infecting adenoviral or herpes simplex virus type 1 genomes, and those of papillomaviruses and polyomaviruses, localize to preexisting nuclear bodies that contain the cellular promyelocytic leukemia proteins (Pmls). These are therefore called **Pml bodies**, or nuclear domains 10, a name derived from the average number present in most cells. Viral proteins then induce reorganization of Pml bodies as viral replication centers are established (Fig. 9.23). The human adenovirus type 5 E4 Orf3 protein induces disruption of these structures, with relocalization of some components, such as specific Pml isoforms, to viral replication centers and of others to the cytoplasm for degradation. The herpes simplex virus type 1 ICP0 protein is responsible for similar reorganization of Pml bodies and degradation of several Pml body proteins. This viral protein is an E3 ubiquitin ligase, which catalyzes addition of polyubiquitin chains to proteins, thereby targeting them for destruction by the proteasome (Box 9.10).

The association of replication centers of different DNA viruses with constituents of the same intranuclear bodies suggests that reorganization of host cell nuclei facilitates viral DNA synthesis. The discovery that the genomes of nuclear DNA viruses home to Pml bodies stimulated characterization of their components, but much remains to be learned about their molecular functions. There is evidence that Pml bodies

Figure 9.23 Reorganization of Pml bodies by the adenoviral E4 Orf3 protein. Monkey cells were infected with a wild-type adenovirus type 5 (Ad5) or a mutant that cannot direct synthesis of the E4 Orf3 protein (inORF3). This viral protein (red) and Pml protein (green) were examined by indirect immunofluorescence. In the presence of the E4 Orf3 protein, Pml foci are rearranged to track like structures that contain this viral protein. Adapted from A. J. Ullman et al., *J Virol* **81**:4744–4752, 2007, with permission. Courtesy of P. Hearing, Stony Brook University.

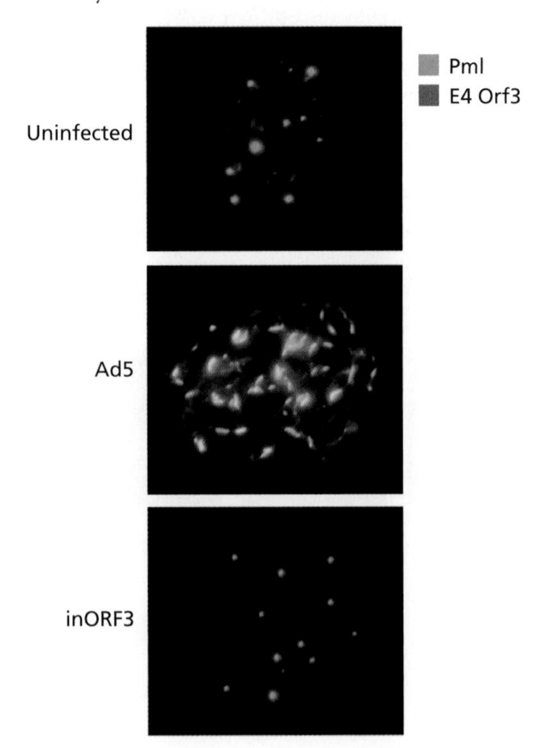

BOX 9.10

BACKGROUND

Ubiquitinylation of proteins: a posttranslational modification that can target for destruction

Covalent linkage of the small (76 amino acids) protein ubiquitin to Lys residues is a posttranslational modification that is ubiquitous (hence the protein name) and conserved in eukaryotes. Reversible addition to proteins of small chemical groups, e.g., during phosphorylation or acetylation, requires but a single enzyme, such as a protein kinase. In contrast, ubiquitinylation depends on the sequential activation of three enzymes, a ubiquitin-activating enzyme (E1), a ubiquitin-conjugating enzyme (E2), and an E3 ubiquitin ligase that catalyzes transfer of ubiquitin from the E2 enzyme to a Lys residue of the substrate. The human E1-activating enzyme Ubal cooperates with multiple E2s and a very large number of E3s, which determine substrate specificity. As summarized in the figure, these ubiquitin ligases are divided into two groups on the basis of the presence of a RING (really interesting new gene) or a HECT (homologous to E6-Ap carboxy terminus) domain.

Ubiquitin itself contains multiple Lys residues to which an additional molecule of the small protein modifier can be linked. Indeed, the substrates of E3 ubiquitin ligases may be polyubiquitinylated via different types of linkages among ubiquitin moieties, or monoubiquitinylated. As illustrated, the nature and site of the modification determines whether the substrate protein is targeted for degradation by the proteasome (polyubiquitinylation at K48 of ubiquitin molecules) or its activity regulated (e.g., monoubiquitinylated). The reversible addition of other small proteins discovered subsequently, such as Sumo (small ubiquitin-like modifier) proteins and ubiquitin-like protein-Nedd8, can also regulate the location or activity of proteins.

The genomes of members of various families encode proteins that are themselves E3 ubiquitin ligases or that form these enzymes with distinct specificities upon association with components of cellular E3 ubiquitin ligases. The former class includes herpes simplex virus type 1 ICP0, which induces polyubiquitinylation and degradation of Pml and other Pml body components described in the text, and human herpesvirus 8 proteins (K3 and K5) that target major histocompatibility class I proteins and other components of immune defenses for proteasomal degradation. Viral proteins that redirect the activities of cellular E3 ubiquitin ligases are more numerous. This set includes the human adenovirus type 5 E1B 55-kDa and E4 Orf6 proteins, which coopt the cellular proteins Cul5, EloB and C, and Rbx1 to mark components of the MRN complex (see the text) and the human tumor suppressor p53 (Volume II, Chapter 6) for degradation; retroviral Vif proteins, which cooperate with the same set of cellular proteins to block an innate host defense (Volume II, Chapter 7); and the human papillomavirus type 16 and 18 E6 proteins, which induce degradation of p53 (and other proteins) by recruiting the cellular E3 ubiquitin ligase E6-Ap.

Gustin JK, Moses AV, Früh K, Douglas JL. 2011. Viral takeover of the host ubiquitin system. *Front Microbiol* **2:**161. doi:10.3389/fmicb.2011.00161.

Kerscher O, Felberbaum R, Hockstrasser M. 2006. Modification of proteins by ubiquitin and ubiquitin-like proteins. *Annu Rev Cell Dev Biol* **22:**159–180.

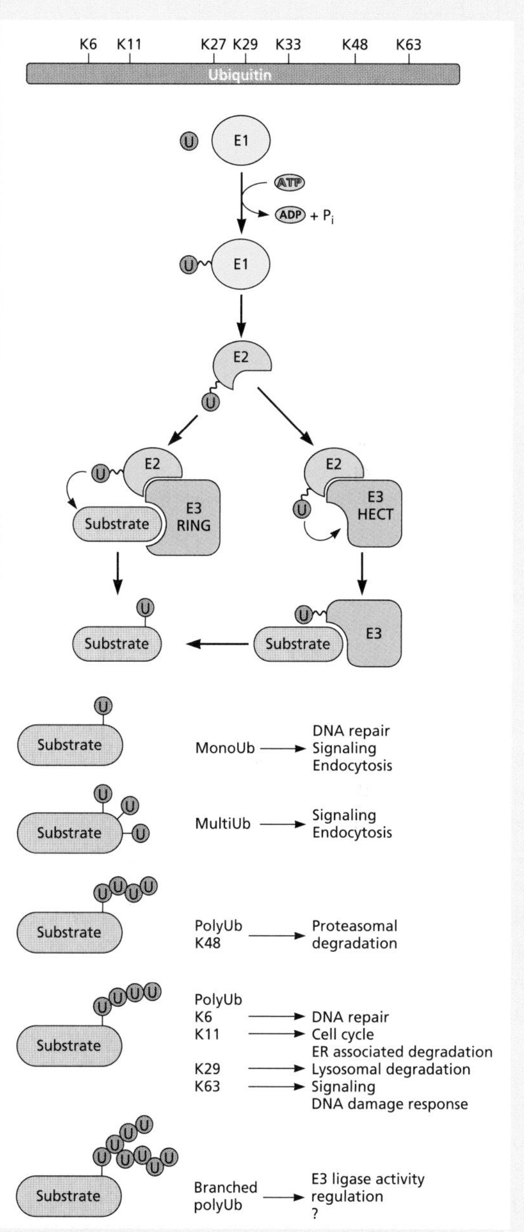

The sequential action of the enzymes required to covalently link ubiquitin to a Lys residue in a substrate protein and the two major classes of E3 ubiquitin ligases are shown. As indicated, the nature of the modification determines its impact on the target protein.

represent a form of intrinsic antiviral defense (Volume II, Chapter 3). For example, exposure of cells to antiviral cytokines (interferons) increases both the number and size of Pml bodies. However, other advantages conferred by the degradation or dispersal of Pml body proteins are likely to be virus specific. The human papillomavirus type 18 E6 protein induces proteasomal degradation of a Pml isoform (Pml-IV) that causes primary human cells to become senescent, a state in which cellular proteins required for replication of the viral genome are not made. In contrast, herpesviral DNA synthesis may require cellular repair and recombination proteins that become relocalized from Pml bodies to viral replication centers.

Limited Replication of Viral DNA Genomes

Synthesis of large numbers of genomes is the typical pattern when DNA viruses infect cells in culture. Nevertheless, several can establish long-term relationships with their hosts and host cells, in which the number of genomes produced is limited. Various mechanisms that achieve copy number control are described in this section.

Integrated Parvoviral DNA Can Replicate as Part of the Cellular Genome

The adenovirus-associated viruses reproduce only in cells coinfected with a helper adenovirus or herpesvirus. Although the latter viruses are widespread in hosts infected by adenovirus-associated viruses, the chances that a particular host cell will be infected simultaneously by two viruses are very low. The strategy of exploiting other viruses to provide essential functions would therefore appear to impose an obstacle to reproduction of individual adenovirus-associated virus particles. In fact, this is not the case, for this viral genome can survive in the absence of a helper virus by an alternative mechanism: its genome becomes integrated into that of the host cell and is replicated as part of a cellular replicon.

This program for long-term survival of the adenovirus-associated virus genome depends on expression of its regulatory region (Rep) (Appendix, Fig. 19). The two larger proteins encoded by this region, Rep 78/68, are multifunctional and control all phases of the viral life cycle (Table 9.1). When helper virus proteins, such as adenoviral E1A, E1B, and E4 proteins, allow synthesis of large quantities of Rep 78/68, adenovirus-associated virus DNA is replicated by the mechanism described previously. In the absence of helper functions, only small quantities of Rep 78/68 are made, there is little viral DNA synthesis, and the genome becomes integrated into that of the host cell. Integration is also mediated by Rep 78/68.

One of the most unusual features of the integration reaction is that it occurs preferentially near one end of human chromosome 19. It was believed for many years that integration required the recognition of the viral ITR origin (Fig. 9.9) by Rep 78/68. However, the observation that integration of DNA molecules containing only the ITR was exceedingly inefficient led to the identification of a viral sequence that increased the frequency of site-specific integration by up to 100-fold in established lines of human cells. This sequence, which can function as an origin, overlaps the p5 promoter (Fig. 9.24). The Rep 78/68 protein can bind simultaneously to both viral DNA and the related human chromosomal 19 DNA sequences that are required for integration, at least *in vitro*. The current model of integration therefore proposes that its specificity is the result of such simultaneous binding to the two DNA molecules by multimeric Rep 78/68.

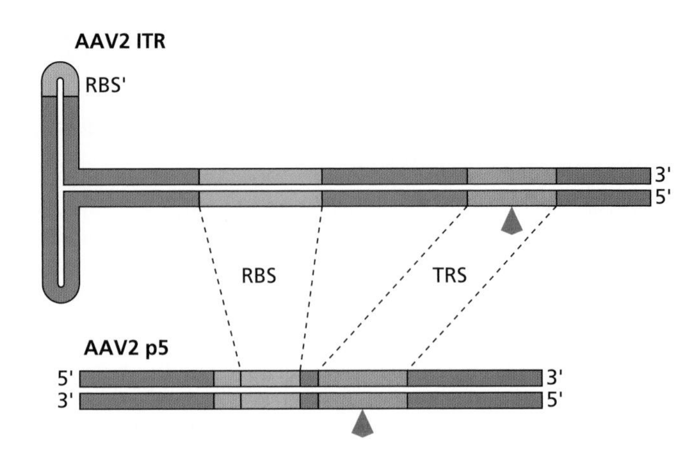

Figure 9.24 Common features of the adenovirus-associated virus type 2 ITR and the p5 sequences important for integration. The Rep 78/68-binding sites are shown in pink, and the terminal resolution sites (TRS) are shown in purple. The p5 origin TRS, like that of the ITR origin, has been shown to be cleaved by the viral protein (red arrowheads). Adapted from D. L. Glauser et al., *J Virol* **79:**12218–12230, 2005, with permission.

In the absence of Rep protein, as in cells infected by typical adenovirus-associated virus vectors (Chapter 3), viral genomes commonly persist as episomal concatemers. It is thought that double-stranded, circular genomes form initially, for example, upon annealing of complementary single-stranded genomes, and then undergo recombination to give rise to concatemers. The long-term persistence of these forms in cells that do not divide is likely to be an important reason for the therapeutic success of some adenovirus-associated virus vectors.

Site-specific integration of the viral genome also occurs in cells infected by human herpesvirus 6 and chicken Marek's disease virus, the result of recombination between the repeated sequences that comprise telomeres and related sequences in viral DNA (Box 9.11). Such integration accounts for persistence and congenital transmission of human herpesvirus 6 DNA and is important for transformation and tumor formation by the oncogenic Marek's disease virus.

BOX 9.11

DISCUSSION

Integration into host cell telomeres as a mechanism of herpesvirus latency?

A characteristic property of herpesviruses is the establishment of latent infections in specific cell types, for example, neurons and B cells in the case of alphaherpesviruses and Epstein-Barr virus, respectively. Although linear in virus particles, viral genomes persist in such latently infected cells as circular episomes, either because the cells do not divide (neurons) or as a result of coordination of replication and segregation of viral genomes with the host cell cycle (B cells). Studies of the betaherpesvirus human herpesvirus 6 have prompted consideration of an alternative mode of herpesviral latency.

Primary infection with human herpesvirus 6, which is widespread in the human population, occurs early in life, and in 25 to 35% of cases is associated with development of fever and a characteristic rash in babies (roseola infantum). The virus establishes latency following primary infection, and reactivation from this state can cause serious disease, particularly in immunocompromised individuals. The first indications of an unusual mechanism of persistence of this herpesviral genome were reports that ~1% of the human population in several different countries carry chromosomally integrated herpesvirus 6 DNA, often in multiple cell types including cells of the germ line.

The human herpesvirus 6 genome contains a unique sequence bounded by direct repeats. These direct repeats are in turn flanked by multiple copies of short sequences that are either identical to the 6-bp repeat sequence that comprises human telomeric DNA or imperfect copies of the telomere repeat sequence (see the figure). Analysis of integrated viral DNA by fluorescent *in situ* hybridization in cells recovered from patients revealed integration sites in host cell telomere sequences in all cases examined. This conclusion has been confirmed by direct sequencing of junctions between viral and host cell DNA recovered by PCR from both patients' cells and cells infected in culture. It is thought that integration is the result of homologous recombination between the telomere repeat sequences in the viral genome and present at the ends of human chromosomes. While the mechanism remains to be established, the presence of telomere repeat-related sequences in the viral genome is not sufficient: such sequences are also present in the genome of human herpesvirus 7, but integration of this herpesviral genome has never been observed. This difference could be explained if integration of human herpesvirus 6 DNA is mediated by the viral U94 protein (a unique gene product), which has sequence homology to the adenovirus-associated virus Rep 78/68 endonuclease/helicase and can complement a Rep 78/68 deletion mutant.

As yet, there is no consensus as to whether integration of herpesvirus 6 DNA into telomeres represents a form of latency or is an epiphenomenon of DNA replication and represents a dead end for the viral genome.

Arbuckle JH, Medveczky MM, Luka J, Hadley SH, Luegmayr A, Ablashi D, Lund TC, Tolar J, De Meirleir K, Montoya JG, Komaroff AL, Ambros PF, Medveczky PG. 2010. The latent human herpesvirus-6A genome specifically integrates in telomeres of human chromosomes in vivo and in vitro. *Proc Natl Acad Sci U S A* **107:**5563–5568.

Arbuckle JH, Pantry SN, Medveczky MM, Prichett J, Loomis KS, Ablashi D, Medveczky PG. 2013. Mapping the telomere integrated genome of human herpesvirus 6A and 6B. *Virology* **442:**3–11.

Morissette G, Flamand L. 2010. Herpesviruses and chromosomal integration. *J Virol* **84:**12100–12109.

Diagram of the human herpesvirus 6 (HHV-6) genome showing the position of the direct repeats (DR_R and DR_L), the perfect or imperfect telomere repeat sequences (TRS and hetTRS, respectively), and the coding sequence for the U94 protein. Adapted from G. Morissette and L. Flamand, *J Virol* **84:**12100–12109, 2010, with permission.

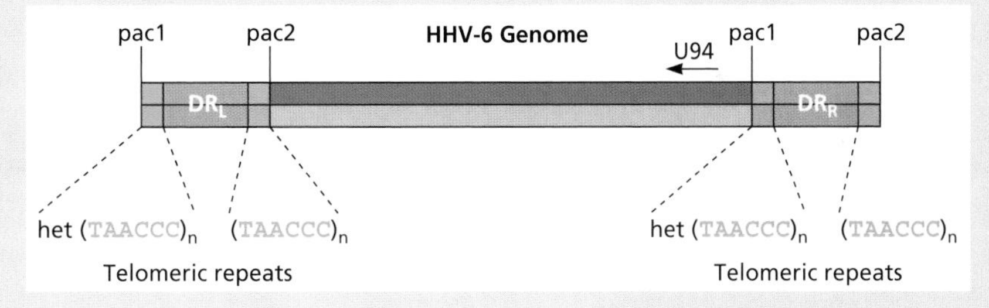

Different Viral Origins Regulate Replication of Epstein-Barr Virus

During herpesviral latent infections, the viral genome is stably maintained at low concentrations, often for long periods (Volume II, Chapter 5). Furthermore, replication of viral and cellular genomes can be coordinated. This pattern is characteristic of human B cells latently infected by Epstein-Barr virus. Many such cell lines have been established from patients with Burkitt's lymphoma, and this state is the usual outcome of infection of B cells in culture. Characteristic features of latent Epstein-Barr virus infection include expression of only a small number of viral genes, the presence of a finite number of viral genomes, and replication from a specialized origin. Because replication from this origin, which is not active in productively infected cells, is responsible for maintenance of episomal viral genomes, it is termed the **origin for plasmid maintenance** (OriP).

The Epstein-Barr virus genome is maintained in nuclei of latently infected cells as a stable circular **episome**, present at 10 to 50 copies per cell. For example, one immortal Burkitt's lymphoma cell line (Raji) has carried ~50 copies per cell of episomal viral DNA for more than 40 years and many passages. When Epstein-Barr virus infects a B cell, the linear viral genome circularizes by a mechanism that is not well understood. The circular viral DNA is then amplified during S phase of the host cell to the final concentration noted above. Such replication is by the cellular DNA polymerases and accessory proteins that synthesize simian virus 40 DNA. However, this process also requires OriP (Fig. 9.25A) and the

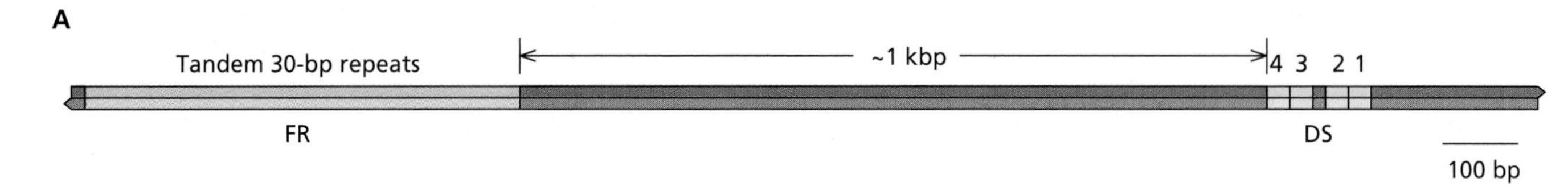

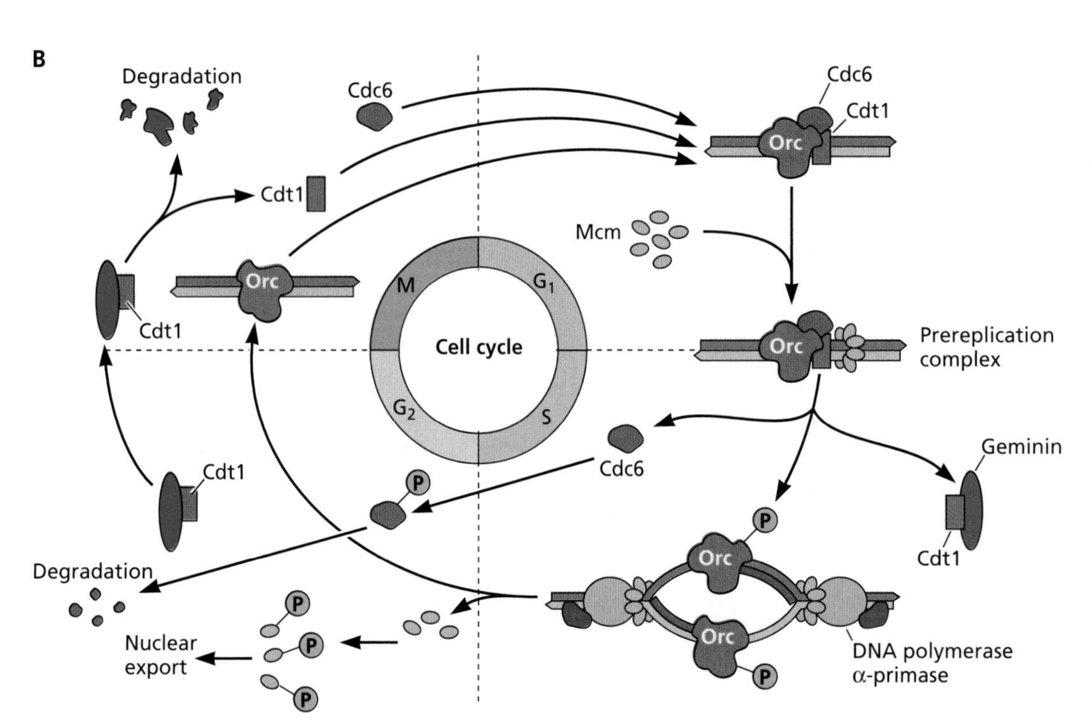

Figure 9.25 Licensing of replication from Epstein-Barr virus OriP. (A) Organization of EBNA-binding sites, shown to scale. The dyad symmetry (DS) sequence, which comprises two pairs of binding sites (1 to 4) for EBNA-1 dimers, is the site of initiation of DNA synthesis. The activity of the DS origin is regulated by sequences adjacent to the EBNA-1-binding sites that are recognized by cellular telomere-binding proteins (see the text) and stimulated by the family of repeat (FR) sequence. Proteins that bind to telomere repeat-related sequences include Trf2 (see text) and telomere-associated poly(ADP-ribose) polymerase. The latter, as well as a second poly(ADP-ribose) polymerase, interfere with preinitiation complex assembly at OriP by catalyzing addition of polymers of ADP-ribose to EBNA-1. Binding of EBNA-1 to multiple FR sequences is necessary for maintenance of episomal viral DNA in latently infected B cells. **(B)** The multiprotein origin recognition complex (Orc) is present throughout the cell cycle and is associated with replication origins. However, initiation of DNA synthesis requires loading of the hexameric minichromosome maintenance complex (Mcm), which provides helicase activity. It is the recruitment of Mcm that is regulated during the cell cycle to limit the initiation of DNA synthesis to S phase. This reaction requires two proteins, Cdc6 and Cdt1. The concentrations and activities of both are tightly controlled during the cell cycle. As cells complete mitosis and enter G_1, Cdc6 and Cdt1 accumulate in the nucleus, where they associate with DNA-bound Orc. These interactions permit loading of Mcm at the G_1-to-S-phase transition, and subsequently of components of the DNA synthesis machinery, such as Rp-A and DNA polymerase α-primase. The latter step requires phosphorylation of specific components of the prereplication complex by cyclin-dependent kinases that accumulate during the G_1-to-S-phase transition (Volume II, Chapter 6). Reinitiation of DNA synthesis is prevented by several mechanisms. A cyclin-dependent kinase that accumulates during the G_2 and M phases phosphorylates both Mcm proteins and Cdc6. This modification induces nuclear export of the former and degradation of the latter. In addition, the protein called geminin is present in the nucleus from S until M phase (when it is degraded). This protein binds to Cdt1, sequestering it from interaction with Cdc6 and Orc. As a consequence of such regulatory mechanisms, the prereplication complex can form **only** in the G_1 phase, ensuring firing of the origin once per cell cycle. The association of Mcm with OriP during G_1 and S but not during G_2 and the inhibition of OriP-dependent replication by overproduction of a protein that prevents recruitment of Mcm provide strong support for the conclusion that synthesis of viral DNA genomes in latently infected cells is governed by the mechanisms that ensure once-per-cell-cycle firing of cellular origins.

viral protein that binds specifically to it, Epstein-Barr virus nuclear antigen 1 (EBNA-1) (Table 9.1), which is always synthesized in latently infected cells. Amplification of the episomal viral genome is limited to a few cycles. Subsequently, viral DNA genomes are duplicated once per cell cycle during S phase and partitioned evenly to daughter cells during mitosis (Box 9.12). The EBNA-1 protein and OriP are sufficient for both once-per-cell-cycle replication and the orderly segregation of viral genomes when host lymphocytes divide.

The availability of cellular replication proteins only in late G_1 and S can account for the timing of Epstein-Barr virus replication in latently infected cells. However, this property **cannot** explain why each genome is replicated only once in each cell cycle, just as each cellular replicon: OriP and cellular origins fire **once and only once** in each S phase. The mechanisms that control once-per-cycle firing of eukaryotic origins, a process termed **replication licensing**, were initially elucidated in budding yeasts, which contain compact origins of replication. Mammalian homologs of the yeast origin recognition complex (Orc) and proteins that regulate initiation of DNA synthesis, such as Mcm, have been identified in all other eukaryotes examined. The human Orc proteins, which are associated with OriP and can bind to EBNA-1, are necessary for OriP-dependent replication, as is the hexameric Mcm helicase, which became associated with OriP during the G_1 phase. Several mechanisms ensure that the essential Mcm helicase is available **only** at the G_1-to-S-phase transition, and hence limit origin firing to once per cell cycle (Fig. 9.25B). For example, recruitment of Mcm to the origin requires cell division control protein 6 homolog (Cdc6) and DNA replication factor Cdt1. These proteins accumulate in the nucleus during S phase, but are subsequently degraded (Cdc6) or sequestered (Cdt1).

Initiation of Epstein-Barr virus DNA synthesis is also regulated temporally and takes place late during S phase. For reasons that are not yet clear, initiation early in S phase is detrimental to both replication efficiency and the maintenance of episomal viral genomes. One parameter important for such temporal regulation is phosphorylation of telomere repeat-binding protein 2 (Trf2) early during S phase by Chk2, a kinase implicated in control of replication timing in yeast. Trf2 binds to three copies of a telomere repeat-related sequence in OriP and interacts with both Orc and histone deacetylases. Consequently, it is thought to coordinate nucleosome remodeling at OriP with recruitment of the proteins essential for initiation of DNA synthesis from this origin. Phosphorylation of Trf2 by Chk2 inhibits these functions and is thought to contribute to preventing too-early initiation of OriP-dependent replication.

Orderly segregation of episomal viral DNA molecules during mitosis (Box 9.12) requires binding of EBNA-1 to its high-affinity sites in the family of repeat (FR) sequences of OriP (Fig. 9.25A). Direct observation of episomal viral genomes by *in situ* hybridization has established that these DNA molecules become tethered to the cellular sister chromatids that are separated during mitosis. Tethering of viral DNA chromosomes, and their subsequent partitioning, is mediated by an N-terminal EBNA-1 sequence that contains two domains that bind directly to AT-rich DNA. In metaphase chromosomes, regions of less condensed (that is, accessible) AT-rich DNA are found between segments that are highly condensed. Any derivative of EBNA-1 that contains two such AT-hook domains (even if these are derived from cellular proteins) binds to chromosomes and supports maintenance of OriP-containing episomes in a host cell population.

As a latent infection is established, the Epstein-Barr virus genome becomes increasingly methylated at C residues present in CG dinucleotides. Sequences that must function in latently infected cells, such as OriP, generally escape this modification, but how methylation specificity is established is not known. Such DNA methylation is associated with repression of transcription and contributes to inhibition of viral gene expression. The viral genome also becomes packaged by cellular nucleosomes and is therefore replicated as a circular minichromosome, much like that of simian virus 40. Replication of the Epstein-Barr virus genome once per cell cycle persists unless conditions that induce entry into the viral productive cycle are encountered. The critical step for this transition is activation of transcription of the viral genes that encode the transcriptional activators Zta and Rta (Chapter 8). These proteins induce expression of the early genes that encode the viral DNA polymerase and other proteins necessary for replication from OriLyt. In addition, Zta appears to be the viral OriLyt recognition protein. Consequently, once this protein is made in an Epstein-Barr virus-infected cell, its indirect and direct effects on viral DNA synthesis ensure a switch from OriP-dependent to OriLyt-dependent replication, and progression through the infectious cycle.

Limited and Amplifying Replication from a Single Origin: the Papillomaviruses

Papillomaviruses reproduce in the differentiating cells of an epithelium, with distinct modes of viral DNA synthesis associated with cells in various differentiation states (Fig. 9.26). Entry of a papillomaviral genome into the nucleus of an undifferentiated, proliferating basal cell initiates a period of amplification of the circular genome, just as during the early stages of latent infection by Epstein-Barr virus. Replication continues until a moderate number of viral genomes (~50 to 100) has accumulated. A maintenance replication pattern, in which the viral genomes are duplicated on average once per cell cycle, is then established as the cells differentiate and move toward the cell surface. The mechanism that governs the switch from amplification to maintenance replication is not known.

BOX 9.12

EXPERIMENTS

Visualization of duplication and partitioning of Epstein-Barr virus plasmid replicons in living cells

It is well established that episomal Epstein-Barr virus genomes are maintained in populations of proliferating host cells. However, the mechanisms responsible for such maintenance, which is governed by the partition of replicated genomes to daughter cells, are not well understood. In one approach to address this issue, duplication and partition of Epstein-Barr virus replicons were visualized in individual cells by live-cell imaging.

The plasmid replicon used in these experiments contained all sequences of OriP and many copies of the binding site for the Lac repressor (panel A of the figure). Its replication and partitioning were observed in HeLa cells that stably produce the viral EBNA-1 protein, and the Lac repressor fused to two copies of the red fluorescent protein and to a nuclear localization signal (Lac-RFP). This protein was prevented from binding to replicons until visualization was desired by inclusion in the medium of an inducer of the Lac repressor, isopropyl-β-D-1-thiogalactopyranoside. Two clones of the modified HeLa cells that maintained on average 3 to 4 copies/cell of the plasmid were identified for use in subsequent experiments.

To examine the fate of OriP-containing plasmids, the cells described above were synchronized by blocking cell cycle progression at the beginning of S phase. The plasmid replicons were then visualized via binding of Lac-RFP and live-cell imaging at various times during cell cycle progression after release from the block (panel B). Of 370 plasmids observed in this way,

- 84% were duplicated in S phase, with close colocalization of the daughter plasmids.
- 88% of colocalized pairs of replicated plasmids partitioned accurately (i.e., one plasmid per daughter cell) during the subsequent mitosis.
- 16% of the plasmids failed to replicate and partitioned randomly.

These observations revealed the previously unknown colocalization of newly synthesized OriP-containing episomes and the coupling of the dependence of subsequent, nonrandom partitioning of the episomes during mitosis upon such colocalization.

Nanbo A, Sugden A, Sugden B. 2007. The coupling of synthesis and partitioning of EBV's plasmid replicon is revealed in live cells. *EMBO J* **26:**4252–4262.

(A) Schematic of the replicon, indicating the presence of OriP, binding sites for the Lac repressor, and a neomycin resistance gene (neo^r). This gene allows selection of cells in which the plasmid is present. **(B)** Examples of plasmid segregation (bright dots) during mitosis. This cell contains two pairs of plasmids (as determined by the fluorescence intensity) that segregate as mitosis proceeds from metaphase and partition equally to the daughter cells. Adapted from A. Nanbo et al., *EMBO J* **26:**4252–4262, 2007, with permission. Panel B courtesy of B. Sugden, University of Wisconsin, Madison.

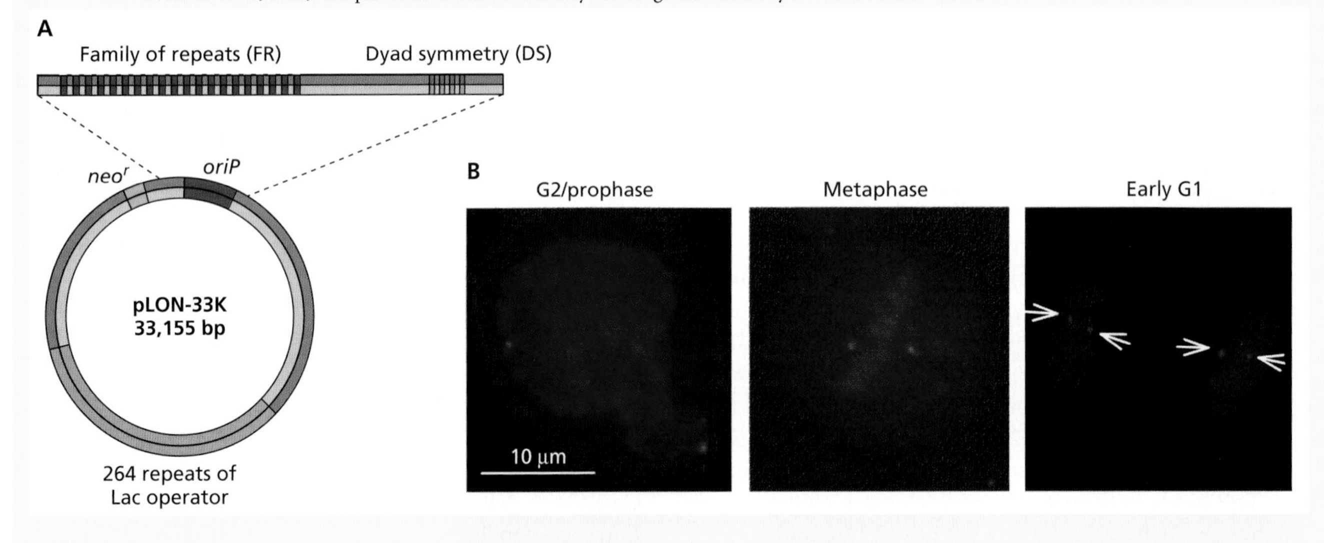

The single viral origin and the viral E1 and E2 proteins that bind to specific origin sequences (Fig. 9.16; Table 9.1) are necessary for both the initial amplification of the papillomavirus genome and its maintenance for long periods at a more or less constant concentration. Initial studies of bovine papillomavirus indicated that such maintenance replication is not the result of strict, once-per-cell-cycle replication of viral DNA. Rather, replication of individual viral episomes occurs at random, taking place on average once per cell cycle. Subsequent studies of human papillomavirus DNA replication in different epithelial cell lines, including those derived from naturally infected cervical epithelia, have established that the viral genome can be replicated by both random and strict, once-per-cell-cycle mechanisms. Which mode of replication

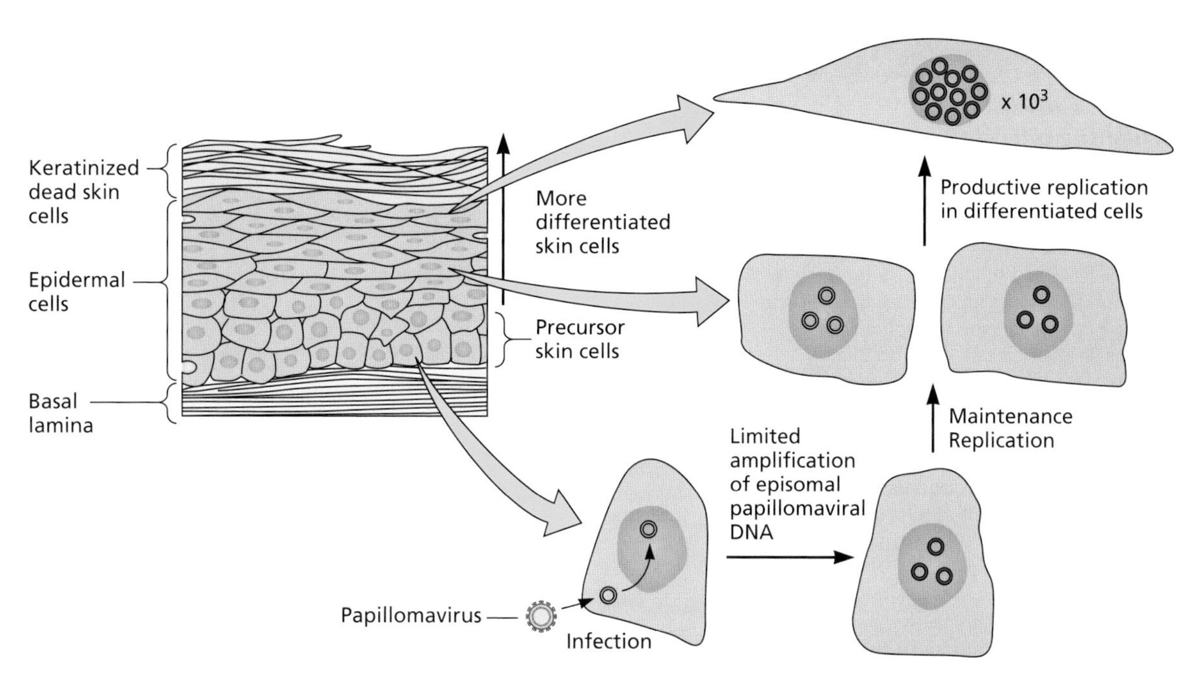

Figure 9.26 Regulation of papillomaviral DNA replication in epithelial cells. The outer layers of the skin are shown as depicted in Fig. 2.5. The virus infects proliferating basal epithelial cells, to which it probably gains access after wounding. The double-stranded, circular viral DNA genome is imported into the infected cell nucleus and initially amplified to a concentration of 50 to 100 copies per cell. This concentration of viral DNA episomes is maintained by further limited replication as the basal and parabasal cells of the epithelium divide **(maintenance replication)**. As cells move to the outer layers of the epidermis and differentiate, productive replication of the viral genome to thousands of copies per cell takes place.

prevails is determined by both the nature of the host cell and the concentration of the viral E1 protein (Box 9.13).

Stable maintenance of the viral genome requires an additional sequence, called the **minichromosome maintenance element**, which is composed of multiple binding sites for the E2 protein. When bound by the viral protein, the minichromosome maintenance element is attached to mitotic chromosomes and remains associated with them during all stages of mitosis. This association is mediated by binding of E2 to the cellular bromodomain-containing protein 4 (Brd 4), an acetylated histone H4-binding protein that interacts with mitotic chromosomes. Such tethering seems to be a mechanism that is shared among a number of viruses. The Brd 4 protein has also been implicated in binding to mitotic chromosomes of episomal DNA of human herpesvirus 8, a herpesvirus that is associated with various human tumors (Volume II, Chapter 6). Brd 4 and the related Brd 2 and Brd 3 proteins have been shown to facilitate integration of the DNA of the gammaretroviruses murine leukemia virus and feline leukemia virus by binding to the retroviral integrase protein and tethering the preintegration complex to host chromosomes.

Remarkably, the final stage of papillomaviral DNA replication, production of high concentrations of the viral genome for assembly into progeny virus particles, is restricted to nondividing, differentiated epithelial cells, such as terminally differentiated keratinocytes (Fig. 9.26). Induction of the DNA damage response mediated by the Atm (ataxia telangiectasia mutated) kinase is required for such genome amplification. The viral E1 and E7 proteins contribute to activating this response and induce the accumulation of Atm and other proteins that mark sites of DNA damage, such as Chk2 and components of the Mre11-Rad50-Nbs1 (MRN) complex, with the viral replication proteins at discrete nuclear foci. These sites are also associated with cellular proteins that participate in homologous recombination. However, it is not clear whether recombination is necessary for genome amplification or to resolve concatemeric replication products into unit-length circular genomes.

Origins of Genetic Diversity in DNA Viruses

Fidelity of Replication by Viral DNA Polymerases

Proofreading Mechanisms

Cellular DNA replication is a high-fidelity process with an error rate of only about one mistake in every 10^9 nucleotides incorporated. Such fidelity, which is essential to maintain the integrity of the genome, is based on accurate base pairing during genome replication and after this process is complete. Nonstandard base pairs between template and substrate deoxyribonucleotide bases can form quite readily, but DNA

BOX 9.13

EXPERIMENTS

Distinguishing once-per-cell-cycle from random replication of human papillomavirus DNA

In once-per-cell-cycle replication, each molecules of episomal viral DNA is replicated just once during S phase. In random replication, some DNA molecules are replicated several times in a single cell cycle, some are replicated once, and some do not replicate. As illustrated in the figure, these mechanisms can be distinguished by the densities of the DNA molecules synthesized when cells are incubated with the dense analog of thymidine bromodeoxyuridine (BUdR) for a period of less than the time required to complete one cell cycle. Newly synthesized DNA into which BUdR is incorporated is heavy (H), whereas parental DNA is lighter (L).

Results obtained when this method was applied to W12 cervical keratinocytes that contain human papillomavirus type 16 DNA are shown schematically in the figure. In these cells, viral DNA replication is by the once-per-cell-cycle mechanism: no HH DNA could be detected (left). When a vector for expression of the viral E1 protein was introduced, random replication of the viral DNA ensued (right).

How E1 induces this switch has not been established. However, it might override licensed once-per-cell-cycle replication mediated by cellular proteins such as Mcm. This proposal is consistent with reports that maintenance replication does not, in fact, require E1. Alternatively, overproduction of E1 might circumvent mechanisms that limit E1-dependent replication to once per cell cycle, such as regulation of its nuclear localization by phosphorylation by the S-phase-specific kinase cyclin E-Cdk2.

Hoffman R, Hirt B, Bechtold V, Beard P, Raj K. 2006. Different modes of human papillomavirus DNA replication during maintenance. *J Virol* **80:**4431–4439.

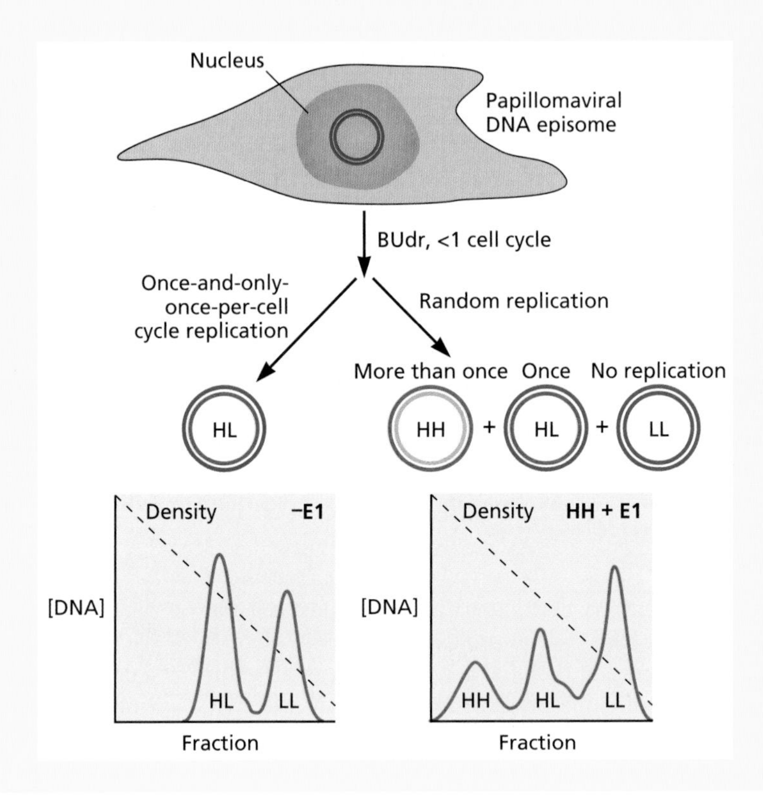

synthesis does not proceed if the terminal nucleotide, or the preceding region of the primer-template, is mismatched. In such circumstances, the mismatched base in the primer strand is excised by a 3′ → 5′ exonuclease present in all replicative DNA polymerases until a perfectly base-paired primer-template is created (Fig. 9.27). Replicative DNA polymerases are therefore self-correcting enzymes, removing errors made in newly synthesized DNA. Mispaired bases that are not eliminated by such proofreading activity are corrected subsequently by mismatch repair. During this process, errors presently in the newly synthesized strand are corrected using the information present in the template strand.

The cellular DNA polymerases that replicate small viral DNA genomes possess proofreading exonucleases. Infection by these viruses (e.g., papillomaviruses and polyomaviruses) does not result in inhibition of cellular protein synthesis, and indeed may induce expression of cellular replication proteins. As the cellular mechanisms of mismatch repair are available to operate on progeny viral genomes, replication of the genomes of these small DNA viruses is likely to be as accurate as that of the genomes of their host cells.

The small, single-stranded DNA genomes of viruses like the parvoviruses and circoviruses are also synthesized by cellular DNA polymerases with proofreading activity. Nevertheless, the rates of mutation of such genomes are considerably higher than those of double-stranded DNA genomes, on the order of 10^{-6} substitutions/site/genome, and these viruses evolve rapidly. Such lower fidelity may result from the inability of the mismatch repair system to detect and correct errors when newly synthesized DNA is single-stranded.

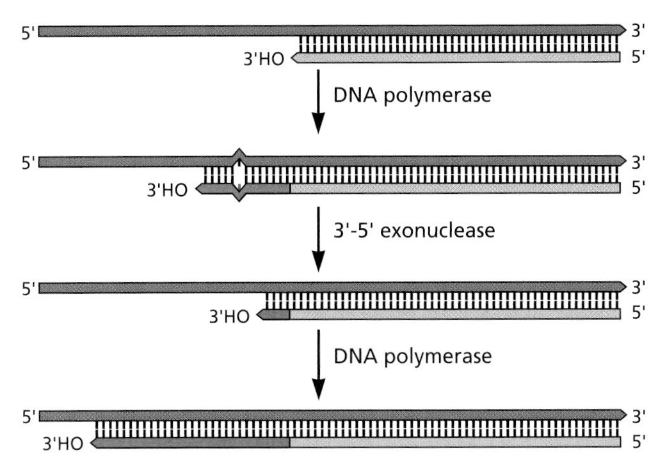

Figure 9.27 Proofreading during DNA synthesis. If permanently fixed into the genome, mispaired bases would result in mutation. However, the majority are removed by the proofreading activity of replicative DNA polymerases. A mismatch at the 3′-OH terminus of the primer-template during DNA synthesis activates the 3′ → 5′ exonuclease of all replicative DNA polymerases, which excises the mismatched region to create a perfect duplex for further extension. In the best-characterized case, DNA polymerase I of *E. coli*, the rate of extension from a mismatched nucleotide is much lower than when a correct base pair is formed at the 3′ terminus of the nascent strand. This low rate of extension allows time for spontaneous unwinding (breathing) of the new duplex region of the DNA and transfer of the 3′ end to the 3′ → 5′ exonuclease site for removal of the mismatched nucleotide. Because preferential excision of mismatched nucleotides is the result of differences in the **rate** at which the polymerase can add the next nucleotide, this mechanism is called **kinetic proofreading**.

Proofreading by Viral DNA Polymerases

The question of how accurately viral DNA is replicated by viral DNA polymerases, such as those of adenoviruses, herpesviruses, and poxviruses, has received relatively little attention. However, each of these viral enzymes possesses an intrinsic 3′ → 5′ exonuclease that preferentially excises mismatched nucleotides from duplex DNAs *in vitro*, and mutations that impair the exonuclease activity of the herpes simplex virus type 1 DNA polymerase greatly increase the mutation rate.

Relatively little is known about the effects of infection by the larger DNA viruses on the production or function of cellular mismatch repair proteins that normally back up proofreading. Because expression of cellular genes and cellular DNA synthesis are generally inhibited in cells infected by these viruses, it is possible that mismatch repair proteins are not present in the concentrations necessary for effective surveillance and repair of newly synthesized viral DNA. Indeed, infection of primary human fibroblasts by human cytomegalovirus (a betaherpesvirus) reduces the activity of an enzyme important for excision of alkylated bases. More detailed information about the rates at which viral DNA polymerases introduce errors during DNA synthesis *in vitro*, and the rates of mutation of viral DNA genomes during productive infection, would help to establish whether cellular repair systems help to maintain the integrity of these viral genomes. Similarly, the contributions of viral enzymes that could prevent or repair DNA damage, such as the dUTPase and uracil DNA glycosylase of herpesviruses and poxviruses (Table 9.3), remain to be established.

Inhibition of Repair of Double-Strand Breaks in DNA

Exposure of mammalian cells to ultraviolet (UV) or infrared light, as well as stalling or collapse of replication forks, can produce double-strand breaks in the DNA genome. Such lesions are potentially lethal, so it is not surprising that they elicit powerful and sensitive damage-sensing and response systems. Proteins that recognize double-stranded DNA ends initiate signaling to effector proteins that both halt progression through the cell cycle (to allow time for repair) and repair the broken ends. The DNA ends are sealed by either nonhomologous end joining or homologous recombination repair (Fig. 9.28). Nonhomologous end joining is an error-prone process in which broken DNA ends are simply joined together after trimming. This important repair pathway is blocked in cells infected by several DNA viruses.

The products of adenoviral DNA synthesis are unit-length copies of the linear viral genome that require no processing prior to packaging (Fig. 9.10). However, accumulation of these viral DNA molecules requires inactivation of nonhomologous end joining. In the absence of the viral E4 Orf3 and Orf6 proteins, newly synthesized viral DNA molecules are joined end to end to form concatemers far too large to be packaged into progeny virus particles. Accumulation of such multimeric DNA molecules depends on cellular proteins that function in nonhomologous end joining, including DNA ligase IV and the MRN complex, which normally accumulates at sites of DNA damage (Fig. 9.28). In adenovirus-infected cells, the protein components of this complex become redistributed within nuclei by the E4 Orf3 protein and are then degraded. The E4 Orf6 and the viral E1B 55-kDa proteins assemble with several cellular proteins to form an infected cell-specific E3 ubiquitin ligase that modifies both MRN components and DNA ligase IV, thereby targeting them for proteasomal degradation. When neither this virus-specific E3 ligase nor the E4 Orf3 protein is present, the cellular repair proteins accumulate in viral replication centers and viral DNA synthesis is inhibited.

Inhibition of nonhomologous end joining is an obvious prerequisite for replication and packaging of linear adenoviral DNA molecules. However, proteasomal degradation of MRN complex components is also induced by LT and increases the yield of simian virus 40. As this virus possesses a circular, double-stranded DNA genome, the reason for this increase is not obvious.

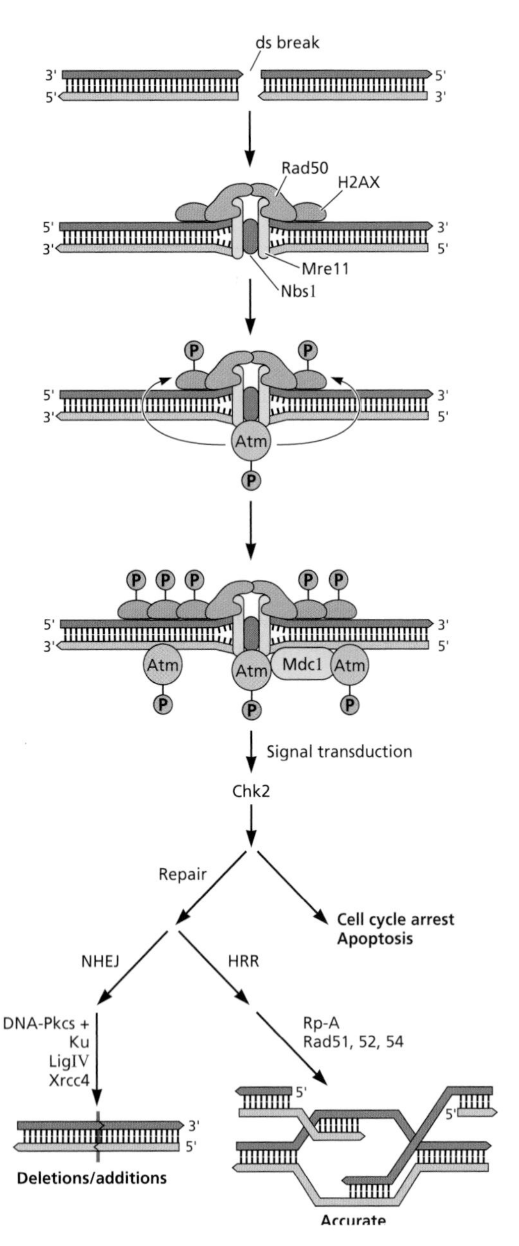

Figure 9.28 Detection of double-strand breaks in DNA. Induction of a double-strand break in the DNA genome triggers rapid accumulation of the MRN complex at the break. This complex contains two copies each of the Mre11 and Rad50 proteins, which move from the cytoplasm into the nucleus, and one of the Nbs1 protein. Mre11 possesses 3′ → 5′ exonuclease, single-stranded DNA endonuclease, and helicase activities. It is thought that these activities unwind the DNA ends at the site of the break, allowing recruitment of the large protein kinase ataxia telangiectasia mutated, Atm. This kinase then becomes activated, perhaps by conformational change and autophosphorylation, and phosphorylates substrates such as the variant histone H2aX. This modification allows amplification of the signal via binding of additional MRN complexes and of mediator of DNA damage checkpoint protein 1, Mdc1. Both this protein and Nsb1 bind phosphorylated H2aX. The Mdc1 protein transduces the signal via additional protein kinases (e.g., Chk2) and other proteins to induce such responses as cell cycle arrest and DNA repair. The two major repair pathways are nonhomologous end joining (NHEJ) and homologous recombination repair (HRR).

Recombination of Viral Genomes

General Mechanisms of Recombination

Genetic recombination is an important source of genetic variation in populations. It also makes a major contribution to repair of breaks in a DNA genome (Fig. 9.28) and can rescue replication when this process has stalled at unfavorable sequences in the template. Much of our understanding of the mechanisms of recombination is based on studies of bacterial viruses, such as bacteriophage λ. Similar principles apply to recombination of DNA genomes of animal viruses.

Two types of recombination are generally recognized: site specific and homologous. In **site-specific recombination**, exchange of DNA takes place at short DNA sequences that are specifically recognized by proteins that catalyze recombination, such as the λ and retroviral integrases. These sequences may be present in only one or both of the DNA sequences that are recombined in this way. Much more common during reproduction of DNA viruses is **homologous recombination**, the exchange of genetic information between **any** pair of related DNA sequences.

Origin-Independent, Recombination-Dependent Replication

In previous sections, we focused on viral genome replication initiated by binding of specialized proteins to origins of replication to induce unwinding of the template and establishment of replication forks. However, early studies of the replication of the genomes of bacteriophages T4 and λ identified an alternative mechanism (Box 9.14). This replication mechanism does not require recognition of viral origin sequences, but rather depends on viral recombination proteins. For example, mutations in the bacteriophage T4 genes that encode such proteins lead to arrest of viral DNA synthesis. In such recombination-dependent replication, recombination proteins catalyze the invasion of double-stranded DNA by a single DNA strand with a 3′-OH terminus, hence providing a primer for DNA synthesis (Fig. 9.29).

As we have seen, the replication of herpes simplex virus genomes exhibits several properties consistent with such a recombination-dependent replication mechanism: viral DNA synthesis becomes independent of the origins and origin-binding protein late in infection; certain cellular DNA repair and recombination proteins become associated with viral replication centers, and inhibition of their synthesis impairs virus reproduction; and the viral genome encodes proteins like those that form the bacteriophage λ recombinase (see next section). Furthermore, herpes simplex virus DNA replication is accompanied by a high degree of recombination between repeated sequences in the genome. Indeed, conversion of the genome from one of its four isomers to another (Fig. 9.19) occurs by the time that newly replicated DNA can first be detected in infected cells. These properties suggest that

BOX 9.14

DISCUSSION

Replication and recombination/repair are two sides of the same coin: earliest insights from bacteriophage λ

In the early 1970s, studies of the replication of bacteriophage λ showed that mutants defective in viral recombination genes (*redα*⁻ or *redβ*⁻, *gam*⁻) synthesize DNA at only half to one-third the wild-type rate. Furthermore, the concatemers typical of late DNA synthesis were on average shorter than usual, and viral bursts were only 30 to 40% of wild-type values. The role of Gam was explained by its inhibition of the cellular RecBCD nuclease, which would be expected to destroy free concatemer ends. However, the role of Red proteins was not so readily apparent. Furthermore, the fact that viral *red*⁻ mutants failed to plate at all on certain cells, for example, those that were deficient in host DNA polymerase I or ligase, suggested a critical role for recombination and repair functions in λ DNA replication.

An elegant series of genetic and biochemical experiments led to a model (shown here) for the transition from circle to rolling-circle replication, which proposed a mechanism by which viral recombination or host DNA repair proteins might produce new replication forks when encountering damage induced by a single-strand break.

It was suggested at the time that the principles illustrated in this model might be applicable to cellular DNA metabolism. The idea that recombination could generate a replication origin was novel at the time, but current schemes for the repair of stalled replication forks in both bacterial and eukaryotic cells incorporate the very same ideas elaborated from studies of λ more than 30 years ago. Furthermore, it has been reported recently that homologous recombination alone can support efficient replication of the 2.85-Mbp genome of the archaeon *Haloferax volcanic*.

Enquist LW, Skalka A. 1973. Replication of bacteriophage lambda DNA dependent on the function of host and viral genes. I. Interaction of *red*, *gam*, and *rec*. *J Mol Biol* **75:**185–212.

Hawkins M, Malla S, Blythe MJ, Nieduszynski CA, Allers T. 2013. Accelerated growth in the absence of DNA replication origins. *Nature* **503:**544–547.

Skalka A. 1974. A replicator's view of recombination (and repair), p 421–432. *In* Grell RF (ed), *Mechanisms in Recombination*. Plenum Press, New York, NY.

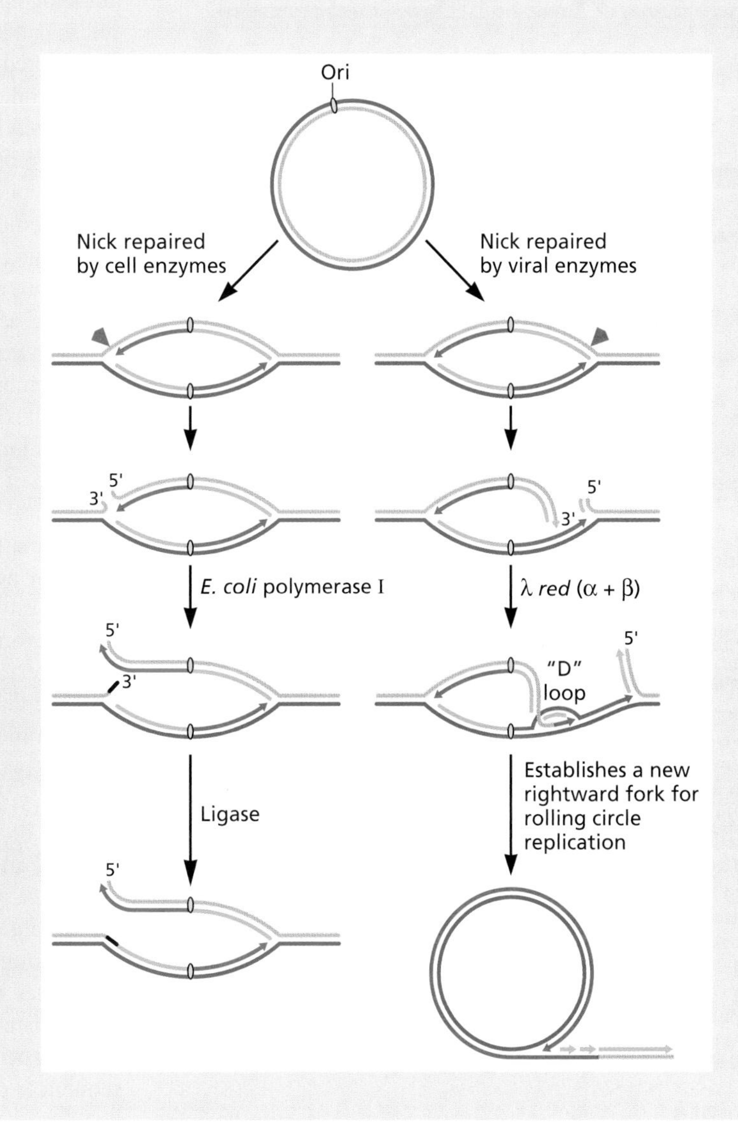

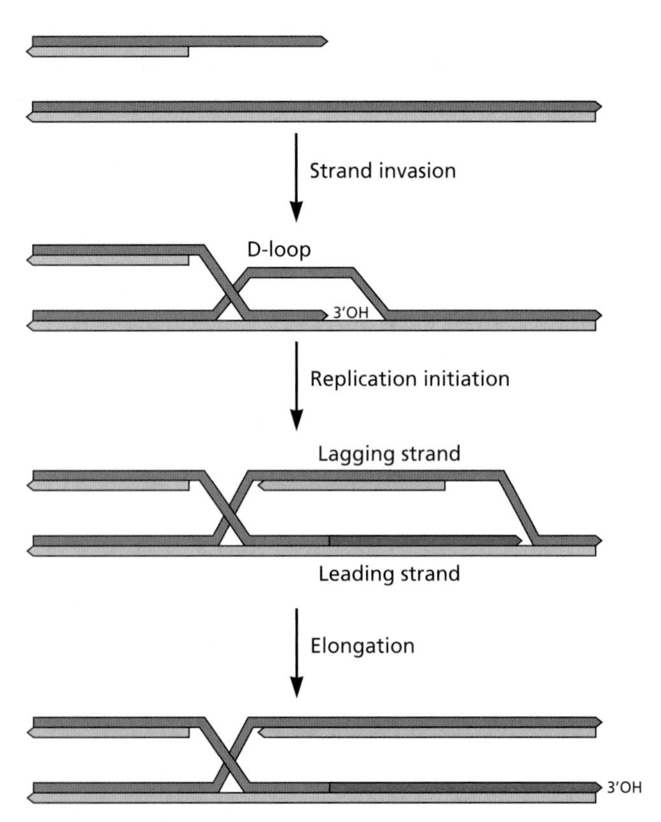

Figure 9.29 General model for initiation of recombination-dependent replication. Replication of linear viral DNA genomes by the standard mechanism of leading- and lagging-strand synthesis from RNA primers produces DNA molecules with single-stranded ends carrying a 3′-OH terminus (see Fig. 9.2B). Such a process is thought to occur during one or a few initial cycles of bacteriophage T4 DNA synthesis. The action of cellular repair proteins that are recruited to double-strand breaks in DNA can also create such single-stranded 3′ ends. The crucial reaction for initiation of recombination-dependent replication is invasion of a homologous sequence in another viral DNA molecule to form a D-loop. This process is exactly analogous to the initial step of general homologous recombination and is catalyzed by repair proteins. The 3′-OH terminus of the invading strand provides a primer for initiation of leading-strand synthesis. Once the D-loop is sufficiently enlarged (equivalent to a replication bubble), lagging-strand synthesis takes place (via synthesis of RNA primers). Continued elongation of the two daughter strands leads to replication to the end of a linear viral DNA template. The newly synthesized strand with the terminating 3′OH at the end can invade homolgous sequences in another viral DNA molecule repeat the process and generate large, branched concatemeric viral DNA molecules.

recombination may be an essential reaction during herpes simplex virus type 1 DNA synthesis. However, it remains to be established whether recombination promotes initiation of viral DNA synthesis or stimulates replication indirectly, for example, by processing of replication intermediates.

Viral Genome Recombination

The integration of adenovirus-associated virus DNA into a specific region of chromosome 19, and its excision when conditions are appropriate, is the result of **site-specific** recombination reactions mediated by the Rep 78/68 viral proteins, which bind to specific sequences in both viral and host DNA molecules. In contrast, although there are some host sequence preferences, integration of retroviral DNA (Chapter 7) is sequence specific only for the viral DNA.

All viral DNA genomes undergo homologous recombination. Because the initial step in recombination, pairing of homologous sequences with one another, depends on random collision, it is concentration dependent. Recombination is therefore favored by the large numbers of viral DNA molecules present in productively infected cells, and their concentration within specialized replication compartments. Furthermore, the structures of replication intermediates, or the nicking of viral DNA during replication or packaging that yields DNA ends, can facilitate recombination. The formation of nuclear replication compartments can also result in the concentration of cellular proteins that participate in recombination (and repair) with viral genomes, as, for example, observed in cells infected by herpes simplex virus type 1. The ease with which viral DNA sequences can recombine is an important factor in the evolution of these viruses. It is also of great benefit to the experimenter, facilitating introduction of specific mutations into the viral genome or construction of viral vectors (see Chapter 3).

As viral genomes do not generally encode homologous recombination proteins, it is thought that this process is catalyzed by host cell enzymes. One exception is the herpes simplex virus type 1 alkaline nuclease (Table 9.3). This enzyme is a 5′ →3′ exonuclease with homology to the Red α component of the bacteriophage λ recombinase (Box 9.14). In conjunction with the viral single-stranded-DNA-binding protein (ICP8), the alkaline nuclease can mediate the exchange of strands between two DNA molecules *in vitro*. Recombination mediated by these viral proteins during infection is important for production of normal, infectious genomes: the viral DNA synthesized and packaged in cells infected by mutants that do not direct synthesis of the active nuclease contains structural abnormalities and is poorly infectious.

Although recombination among animal viral DNA sequences has been widely exploited in the laboratory, the mechanisms have not received much attention. One important exception is the homologous recombination of DNA sequences of some herpesviruses, including herpes simplex virus type 1, that is responsible for isomerization of the genome. Populations of viral DNA molecules purified from herpes simplex virus type 1 virus particles contain four isomers of the genome, defined by the relative orientations of the two unique sequence segments (L and S) with respect to one another (Fig. 9.19). All four isomers are present at equimolar concentrations when viral DNA is isolated from a single plaque, suggesting that a single virus particle containing just one genome isomer gives rise to all four by recombination between repeated DNA sequences. Recombination between the inverted repeats that flank unique sequences in the viral genome promotes inversion of the L and S segments. Such homologous recombination takes place during viral DNA synthesis and requires the viral replication machinery.

Despite some 30 years of study, the function of the unusual isomerization of the genome of herpes simplex virus type 1 and certain other herpesviruses remains enigmatic. Isomerization is not absolutely essential for virus reproduction in cells in culture, because viruses "frozen" as a single isomer by deletion of internal inverted repeats are viable. On the other hand, the reduced yield of such mutants, and the presence of the inverted repeat sequences in **all** strains of herpes simplex virus type 1 examined, emphasize the importance of the repeated sequences in natural infections. It may be that these sequences themselves fulfill some beneficial function (as yet unknown). Recombinational isomerization would then be a secondary result of the presence of multiple, inverted copies of these sequences in the viral genome. Alternatively, as discussed previously, isomerization might be a consequence of an important role for recombination in replication of the viral genome. The value of the unusual isomerization of the herpes simplex virus type 1 genome may become clearer as understanding of the mechanism of viral DNA synthesis improves.

Perspectives

Our understanding of mammalian replication proteins and the intricate reactions they carry out during DNA synthesis would still be rudimentary were it not for pioneering studies that focused on the simian virus 40 origin of replication, and the discovery that this relatively simple viral DNA sequence could support origin-dependent replication *in vitro* when cellular proteins are supplemented with a single viral protein, LT.

Knowledge of the mechanism of synthesis of this small viral DNA genome also provided the conceptual framework within which to appraise the diversity in replication of other viral DNA genomes. One parameter that varies considerably is the degree of dependence on the host cell's replication machinery. In contrast to those of papillomaviruses, parvoviruses, and polyomaviruses, the genomes of the larger DNA viruses (herpesviruses and poxviruses) encode the components of a complete DNA synthesis system, as well as accessory enzymes responsible for the production of dNTP substrates. Nevertheless, replication of **all** viral DNA genomes requires proteins that carry out the reactions first described for simian virus 40 DNA synthesis, namely, an origin recognition protein(s), one or more DNA polymerases, proteins that promote processive DNA synthesis, origin-unwinding and helicase proteins, and, usually, proteins that synthesize, or serve as, primers.

Strategies for replication of viral DNA genomes range from simple, continuous synthesis of both strands of a double-stranded DNA template (adenovirus) to baroque (and not well-understood) mechanisms that produce DNA concatemers (herpesviruses). These diverse strategies represent alternative mechanisms for circumventing the inability of all known DNA-dependent DNA polymerases to initiate DNA synthesis *de novo*. In some cases, initiation of viral DNA synthesis requires RNA primers and the lagging strand is synthesized discontinuously, but in others, all daughter DNA strands are synthesized continuously from protein or DNA sequence primers.

Efficient reproduction of DNA viruses requires the production of large numbers of progeny viral DNA molecules for assembly of viral particles in relatively short periods. One parameter important for such genome amplification is the efficient production of the proteins that mediate or support DNA synthesis, be they viral or cellular in origin. Viral DNA replication at specialized intracellular sites, a common feature of cells infected by these viruses, is also likely to contribute. Further exploration of this incompletely understood phenomenon should shed new light on host cell biology, in particular the structural and functional compartmentalization of the nucleus. The cues that set the stage for alternative modes of limited replication that are characteristic of some DNA viruses also remain incompletely understood. Elucidation of the mechanisms that result in close integration of viral DNA synthesis with the physiological state of the host cell seems certain to continue to provide important insights into both host cell control mechanisms and the long-term relationships these viruses can establish with their hosts.

References

Books

Kornberg A, Baker T. 1992. *DNA Replication*, 2nd ed. W. H. Freeman and Company, New York, NY.

Book Chapters

Berns, K, Parrish CR. 2007. Parvoviridae, p 2437–2477. *In* Knipe DM, Howley PM, Griffin DE, Lamb RA, Martin MA, Roizman B, Straus SE (ed), *Fields Virology*, 5th ed. Lippincott Williams & Wilkins, Philadelphia, PA.

Reviews

Bell SP. 2002. The origin recognition complex: from simple origins to complex functions. *Genes Dev* **16:**659–672.

Everett RD. 2013. The spatial organization of DNA virus genomes in the nucleus. *PLoS Pathog* **9:**e1003386. doi:10.1371/journal.ppat.1003386.

Frappier L. 2012. EBNA1 and host factors in Epstein-Barr virus latent DNA replication. *Curr Opin Virol* **2:**733–739.

Kreuzer KN. 2000. Recombination-dependent DNA replication in phage T4. *Trends Biochem Sci* **25:**165–173.

Lindahl T, Wood RD. 1999. Quality control by DNA repair. *Science* **286:** 1897–1905.

Machida YJ, Hamlin JL, Dulta A. 2005. Right place, right time, and only once: replication initiation in metazoans. *Cell* **123:**13–24.

Moss B. 2013. Poxvirus DNA replication. *Cold Spring Harb Perspect Biol* **5:**a010199. doi:10.1101/cshperspect.a010199.

Sakakibara N, Chen D, McBride AA. 2013. Papillomaviruses use recombination-dependent replication to vegetatively amplify their genomes in differentiated cells. *PLoS Pathog* **9:**e1003321. doi:10.1371/journal.ppat.1003321.

Sowd GA, Fanning E. 2012. A wolf in sheep's clothing: SV40 co-opts host genome maintenance proteins to replicate viral DNA. *PLoS Pathog* **8:**e1002994. doi:10.1371/journal.ppat.1002994.

Stenlund A. 2003. Initiation of DNA replication: lessons from viral initiation proteins. *Nat Rev Mol Cell Biol* **4:**777–785.

Sugden B. 2002. In the beginning: a viral origin exploits the cell. *Trends Biochem Sci* **27:**1–3.

Van der Vliet PC. 1995. Adenovirus DNA replication. *Curr Top Microbiol Immunol* **199**(Pt 2)**:**1–30.

Waga S, Stillman B. 1998. The DNA replication fork in eukaryotic cells. *Annu Rev Biochem* **67:**721–751.

Weinberg RA. 1995. The retinoblastoma protein and cell cycle control. *Cell* **81:**323–330.

Weitzman MD, Lilley CE, Chaurushiya MS. 2010. Genomes in conflict: maintaining genome integrity during virus infection. *Annu Rev Microbiol* **64:**61–81.

Weller SK, Coen DM. 2012. Herpes simplex viruses: mechanisms of DNA replication. *Cold Spring Harb Perspect Biol* **4:**a0130111. doi:10.1101/cshperspect.a013011.

Zhang Y, Zhou J, Lim CU. 2006. The role of NBS1 in DNA double strand break repair, telomere stability, and cell cycle checkpoint control. *Cell Res* **16:**45–54.

Papers of Special Interest

Viral Replication Mechanisms and Replication Proteins

Challberg MD. 1986. A method for identifying the viral genes required for herpesvirus DNA replication. *Proc Natl Acad Sci U S A* **83:**9094–9098.

Challberg MD, Desiderio SV, Kelly TJ, Jr. 1980. Adenovirus DNA replication *in vitro*: characterization of a protein covalently linked to nascent DNA strands. *Proc Natl Acad Sci U S A* **77:**5105–5109.

De Silva FS, Lewis W, Bergland P, Koonin EV, Moss B. 2007. Poxvirus DNA primase. *Proc Natl Acad Sci U S A* **104:**18724–18729.

Im DS, Muzyczka N. 1990. The AAV origin binding protein Rep68 is an ATP-dependent site-specific endonuclease with DNA helicase activity. *Cell* **61:**447–457.

Li JJ, Kelly TJ. 1984. Simian virus 40 DNA replication *in vitro*. *Proc Natl Acad Sci U S A* **81:**6973–6977.

Li JJ, Peden KW, Dixon RA, Kelly T. 1986. Functional organization of the simian virus 40 origin of DNA replication. *Mol Cell Biol* **6:**1117–1128.

Link MA, Silva LA, Schaffer PA. 2007. Cathepsin B mediates cleavage of herpes simplex virus type 1 origin binding protein (OBP) to yield OBPC-1, and cleavage is dependent upon viral DNA replication. *J Virol* **81:**9175–9182.

Mohni KN, Dee AR, Smith S, Schumacher AJ, Weller SK. 2013. Efficient herpes simplex virus 1 replication requires cellular ATR pathway proteins. *J Virol* **87:**531–542.

Rekosh DM, Russell WC, Bellett AJ, Robinson AJ. 1977. Identification of a protein linked to the ends of adenovirus DNA. *Cell* **11:**283–295.

Sanders CM, Stenlund A. 1998. Recruitment and loading of the E1 initiator protein: an ATP-dependent process catalysed by a transcription factor. *EMBO J* **17:**7044–7055.

Regulation of Viral DNA Replication

Burkham J, Coen DM, Hwang CB, Weller SK. 2001. Interactions of herpes simplex virus type 1 with ND10 and recruitment of PML to replication compartments. *J Virol* **75:**2353–2367.

Doncas V, Ishov AM, Romo A, Juguilon H, Weitzman MD, Evans RM, Maul GG. 1996. Adenovirus replication is coupled with the dynamic properties of the PML nuclear structure. *Genes Dev* **10:**196–207.

Moarefi IF, Small D, Gilbert I, Höpfner M, Randall SK, Schneider C, Russo AA, Ramsperger U, Arthur AK, Stahl H, Kelly TJ, Fanning E. 1993. Mutation of the cyclin-dependent kinase phosphorylation site in simian virus 40 (SV40) large T antigen specifically blocks SV40 origin DNA unwinding. *J Virol* **67:**4992–5002.

Taylor TJ, Knipe DM. 2004. Proteomics of herpes simplex virus replication compartments: association of cellular DNA replication, repair, recombination, and chromatin remodeling protein with ICP8. *J Virol* **78:**5856–5866.

Whyte P, Buchkovich KJ, Horowitz JM, Friend SH, Raybuck M, Weinberg RA, Harlow E. 1988. Association between an oncogene and an anti-oncogene: the adenovirus E1A proteins bind to the retinoblastoma gene product. *Nature* **334:**124–129.

Limited Replication of Viral Genomes

Dhar SK, Yoshida K, Machida Y, Khaira P, Chaudhuri B, Wohlschlegel JA, Leffak M, Yates J, Dutta A. 2001. Replication from oriP of Epstein-Barr virus requires human ORC and is inhibited by geminin. *Cell* **106:**287–296.

Flores ER, Lambert PF. 1997. Evidence for a switch in the mode of human papillomavirus type 16 DNA replication during the viral life cycle. *J Virol* **71:**7167–7179.

Kotin RM, Siniscalco M, Samulski RJ, Zhu XD, Hunter L, Laughlin CA, McLaughlin S, Muzyczka N, Rocchi M, Berns KI. 1990. Site-specific integration by adeno-associated virus. *Proc Natl Acad Sci U S A* **87:**2211–2215.

Philpott NJ, Gomos J, Berns KI, Falck-Pedersen E. 2002. A p5 integration efficiency element mediates Rep-dependent integration into AAVS1 at chromosome 19. *Proc Natl Acad Sci U S A* **99:**12381–12385.

Reisman D, Yates J, Sugden B. 1985. A putative origin of replication of plasmids derived from Epstein-Barr virus is composed of two *cis*-acting components. *Mol Cell Biol* **5:**1822–1832.

Schnepp BC, Jensen RL, Chen CL, Johnson PR, Clark KR. 2005. Characterization of adeno-associated virus genomes isolated from human tissues. *J Virol* **79:**14793–14803.

You J, Croyle JL, Nishimura A, Ozato K, Howley PM. 2004. Interaction of the bovine papillomavirus E2 protein with Brd4 tethers the viral DNA to host mitotic chromosomes. *Cell* **117:**349–360.

Zhou J, Deng Z, Norseen J, Lieberman PM. 2010. Regulation of the Epstein-Barr virus origin of plasmid replication (OriP) by the S-phase checkpoint kinase Chk2. *J Virol* **84:**4979–4987.

Repair and Recombination of Viral DNA

Evans JD, Hearing P. 2005. Relocalization of the Mre11-Rad50-Nbs1 complex by the adenovirus E4 ORF3 protein is required for viral replication. *J Virol* **79:**6207–6215.

Hayward GS, Jacob RJ, Wadsworth SC, Roizman B. 1975. Anatomy of herpes simplex virus DNA: evidence for four populations of molecules that differ in the relative orientations of their long and short components. *Proc Natl Acad Sci U S A* **72:**4243–4247.

Linden RM, Winocour E, Berns KI. 1996. The recombination signals for adeno-associated virus site-specific integration. *Proc Natl Acad Sci U S A* **93:**7966–7972.

Reuven NB, Staire AE, Myers RS, Weller SK. 2003. The herpes simplex virus type 1 alkaline nuclease and single-stranded DNA binding protein mediate strand exchange in vitro. *J Virol* **77:**7425–7433.

Sakakibara N, Mitra R, McBride AA. 2011. The papillomavirus E1 helicase activates a cellular DNA damage response in viral replication foci. *J Virol* **55:**8981–8995.

Stracker TH, Carson CT, Weitzman MD. 2002. Adenovirus oncoproteins inactivate the Mre11-Rad50-NBS1 DNA repair complex. *Nature* **418:** 348–352.

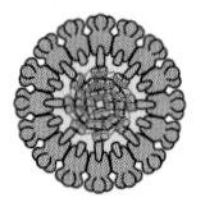

10 Processing of Viral Pre-mRNA

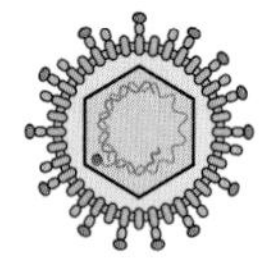

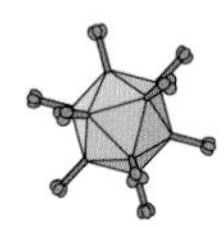

Introduction

Covalent Modification during Viral Pre-mRNA Processing

- Capping the 5′ Ends of Viral mRNA
- Synthesis of 3′ Poly(A) Segments of Viral mRNA
- Splicing of Viral Pre-mRNA
- Alternative Processing of Viral Pre-mRNA
- Editing of Viral mRNAs

Export of RNAs from the Nucleus

- The Cellular Export Machinery
- Export of Viral mRNA

Posttranscriptional Regulation of Viral or Cellular Gene Expression by Viral Proteins

- Temporal Control of Viral Gene Expression
- Viral Proteins Can Inhibit Cellular mRNA Production

Regulation of Turnover of Viral and Cellular mRNAs in the Cytoplasm

- Regulation of mRNA Stability by Viral Proteins
- mRNA Stabilization Can Facilitate Transformation

Production and Function of Small RNAs That Inhibit Gene Expression

- Small Interfering RNAs, Micro-RNAs, and Their Synthesis
- Viral Micro-RNAs
- Viral Gene Products That Block RNA Interference

Perspectives

References

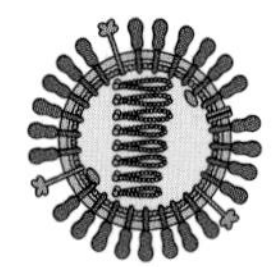

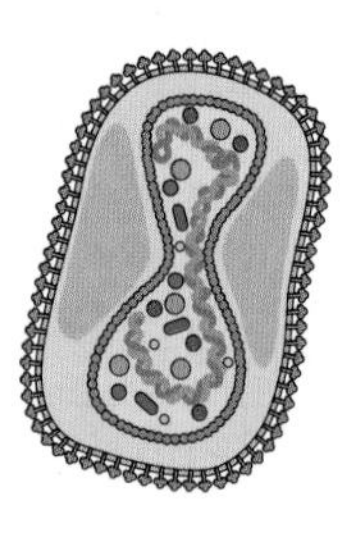

LINKS FOR CHAPTER 10

⏭ *Video: Interview with Dr. Phillip Sharp*
http://bit.ly/Virology_Sharp

All is flux.

HERACLITUS

Introduction

Viral messenger RNAs (mRNAs) are synthesized by either viral or cellular enzymes and may be made in the nucleus or the cytoplasm of an infected cell. Regardless of how and where they are made, all must be translated by the protein-synthesizing machinery of the host cell. A series of covalent modifications, collectively known as **RNA processing** (Fig. 10.1), facilitate recognition of mRNAs by the protein synthesis machinery and translation of the coding sequences by cellular ribosomes. Most RNA-processing reactions were discovered in viral systems, primarily because virus-infected cells provide large quantities of specific mRNAs for analysis.

Two modifications important for efficient translation are the addition of m^7GpppN to the 5′ end (**capping**) and the addition of multiple A nucleotides to the 3′ end (**polyadenylation**) (Fig. 10.1). The enzymes that perform these chemical additions may be encoded by viral or cellular genes. When an RNA is produced in the nucleus, another chemical rearrangement, called **splicing**, is possible. During splicing, short blocks of noncontiguous coding sequences (**exons**) are joined precisely to create a complete protein-coding sequence for translation, while the intervening sequences (**introns**) are discarded (Fig. 10.1). Splicing therefore dramatically alters the precursor mRNA (pre-mRNA) initially synthesized. As no viral genome is known to encode even part of the intricate machinery needed to catalyze splicing reactions, splicing of viral pre-mRNAs is accomplished by cellular gene products. Some viral pre-mRNAs undergo a different type of internal chemical change, in which a single base is replaced by another or one or more nucleotides are inserted at specific positions. Such **RNA editing** reactions introduce nucleotides that are not encoded in the genome, and consequently may change the sequence of the encoded protein.

When a viral RNA is produced in the nucleus, it must be exported to the cytoplasm for translation (Fig. 10.1). Such export of mature viral and cellular mRNAs is considered to be part of mRNA processing, even though the RNA is not known to undergo any chemical change during transport. Viral mRNAs invariably leave the nucleus by cellular pathways, but the cargo transported by nuclear export pathways may be altered in virus-infected cells. Once within the cytoplasm, an mRNA has a finite lifetime before it is recognized and degraded by ribonucleases. The susceptibilities of individual mRNA species to attack by these destructive enzymes vary greatly, and can be modified in virus-infected cells.

These RNA-processing reactions (Fig. 10.1) not only produce functional mRNAs but also provide numerous opportunities for posttranscriptional control of gene expression. Regulation of RNA processing can increase the coding capacity of the viral genome, determine when specific viral proteins are made during the infectious cycle, and facilitate selective expression of viral genetic information. An additional component of the varied repertoire of posttranscriptional mechanisms that regulate viral and cellular gene expression has been recognized more recently. Cellular and viral genomes encode small RNAs that induce mRNA degradation or inhibition of translation upon base pairing to an mRNA. This phenomenon is known as RNA silencing or **RNA interference**. RNA interference by cellular RNAs is an important component of antiviral defense, and both cellular and viral small RNAs can modulate virus-host cell interactions.

In this chapter, we focus on these RNA-processing reactions to illustrate both critical viral regulatory mechanisms and the seminal contributions of viral systems to the elucidation of essential cellular processes.

PRINCIPLES *Processing of viral pre-mRNA*

- Viral mRNAs must be translated by the host cell machinery.
- Addition of a modified nucleotide to the 5′ end of an mRNA, "capping," ensures efficient translation, protects the mRNA from exonucleases, and prevents activation of antiviral responses.
- Addition of poly(A) to the 3′ end of an mRNA enhances translation and mRNA stability.
- Viral mRNA splicing is mediated by host cell components; no viral genome encodes any splicing components.
- Alternative splicing and editing of viral pre-mRNAs expand coding capacity and can regulate viral gene expression.
- The export of viral mRNAs is mediated by the host cell machinery and, in most cases, is indistinguishable from export of analogous cellular RNAs.
- Reproduction of some viruses requires production of unspliced or partially spliced mRNAs; virus-encoded proteins or sequences allow the nuclear export of these molecules.
- The genomes of several viruses encode proteins that regulate one or more RNA-processing reactions and are important for temporal regulation of viral gene expression or inhibition of the production of cellular mRNAs.
- Inhibiting the production of cellular mRNAs can favor viral mRNA translation.
- Cellular micro-RNAs may inhibit or facilitate reproduction of a variety of viruses; virally encoded miRNAs may promote viral replication, persistence, or latency or inhibit the host response.

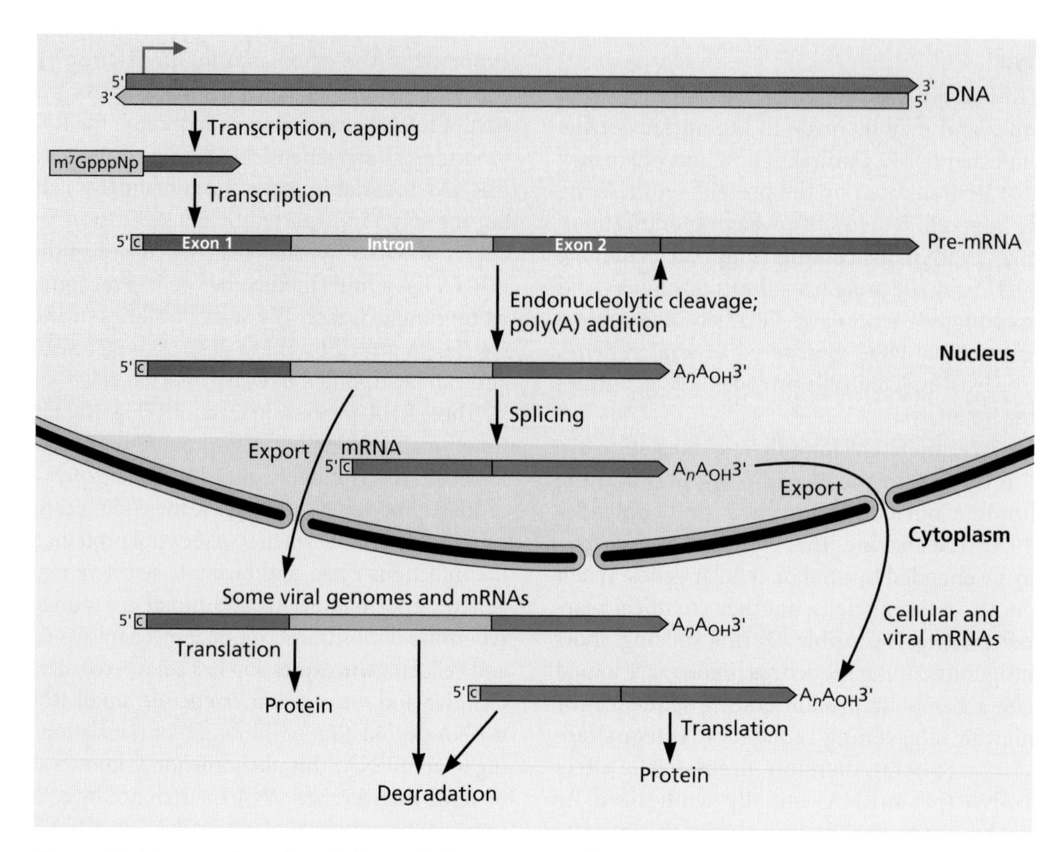

Figure 10.1 Processing of a viral or cellular mRNA synthesized by RNA polymerase II. The reactions by which mature mRNA is made from a typical RNA polymerase II transcript are shown. The first such reaction, capping, takes place cotranscriptionally. For clarity, the exons of a hypothetical, partially processed (i.e., polyadenylated but unspliced) pre-mRNA are depicted, even though polyadenylation and splicing are often coupled and many splicing reactions are cotranscriptional. Most cellular and viral pre-mRNAs synthesized by RNA polymerase II are processed by this pathway (right). However, some viral mRNAs that are polyadenylated but not spliced, or are incompletely spliced, are also exported to the cytoplasm (left).

Covalent Modification during Viral Pre-mRNA Processing

Capping the 5′ Ends of Viral mRNA

The first mRNAs shown to carry the 5′-terminal structure termed the **cap** were those of reovirus and vaccinia virus (Box 10.1). These viral mRNAs are made and processed by virus-encoded enzymes, but subsequent research established that the great majority of viral and cellular mRNAs possess the same cap structure, m^{7}GpppN, where N is any nucleotide (Fig. 10.2A). This structure protects mRNAs from 5′ exonucleolytic attack and is essential for the efficient translation of most mRNAs, as it is recognized by translation initiation proteins. The principal exceptions are the uncapped mRNAs of certain (+) strand viruses, notably picornaviruses and the flavivirus hepatitis C virus, which are translated by the cap-independent mechanism described in Chapter 11. Cytoplasmic RNA molecules with uncapped 5′-triphosphate termini can be recognized by components of intrinsic defense systems of the host cell (Volume II, Chapter 3). Capping of viral mRNAs blocks such recognition and consequently mitigates induction of cellular antiviral defenses.

Although most viral mRNAs carry a 5′-terminal cap, there is considerable variation in how this modification is made. Three mechanisms can be distinguished: *de novo* synthesis by cellular enzymes, synthesis by viral enzymes, and acquisition of preformed 5′ cap structures from cellular pre-mRNAs or mRNAs.

Synthesis of Viral 5′ Cap Structures by Cellular Enzymes

Viral pre-mRNA substrates for the cellular capping enzyme are invariably made in the infected cell nucleus by cellular RNA polymerase II. The formation of cap structures on the 5′ ends of such pre-mRNAs, the first step in their processing, is a cotranscriptional reaction that takes place when the nascent RNA is only 20 to 30 nucleotides in length. Phosphorylation of paused RNA polymerase II at specific serines in the C-terminal domain of the largest subunit is the signal for binding of the capping enzyme (see below) and capping of the

BOX 10.1

TRAILBLAZER

Identification of 5′ cap structures on viral mRNAs

The first clues that the termini of mRNAs made in eukaryotic cells possess special structures came when viral mRNAs did not behave as predicted from the known structure of bacterial mRNAs. The figure summarizes one of the experiments that identified 5′ cap structures. The 5′ end of reoviral (or vaccinia virus) mRNA, in contrast to that of a bacterial mRNA, could not be labeled by polynucleotide kinase and [γ-^{32}P]ATP (yellow) after alkaline phosphatase treatment. This property established that the 5′ end did not carry a simple phosphate group, but rather was blocked. The structure of the 5′ blocking group (termed the **cap**) was elucidated subsequently by differential labeling of specific chemical groups of the viral mRNA, such as methyl and terminal phosphate groups, followed by digestion of the mRNA with nucleases with different specificities.

Furuichi Y, Morgan M, Muthukrishnan S, Shatkin AJ. 1975. Reovirus messenger RNA contains a methylated, blocked 5′-terminal structure: m^{7}G(5′)ppp(5′)G^{m}pCp-. *Proc Natl Acad Sci U S A* **72:**362–366.

Wei CM, Moss B. 1975. Methylated nucleotides block 5′-terminus of vaccinia virus messenger RNA. *Proc Natl Acad Sci U S A* **72:**318–322.

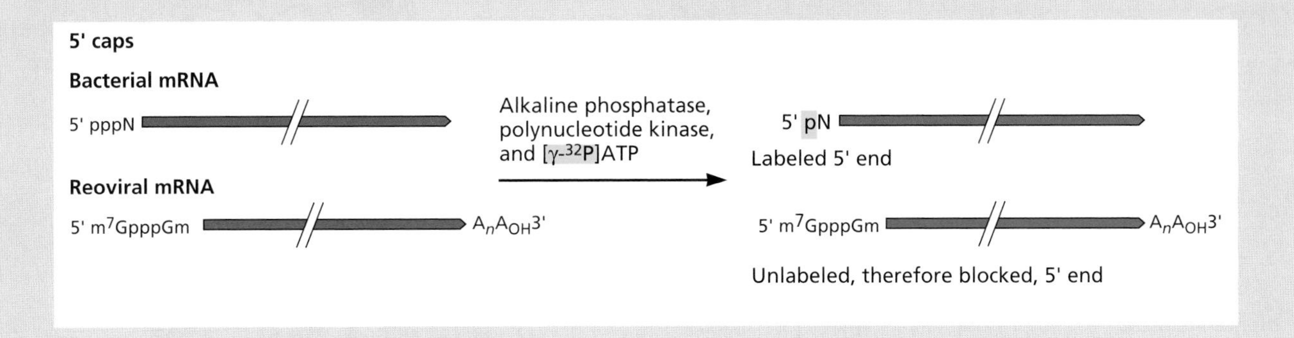

nascent RNA. The intimate relationship between the cellular capping enzyme and RNA polymerase II ensures that all transcripts made by this enzyme are capped at their 5′ ends.

The 5′ cap structure is assembled by the action of several enzymes. In mammalian cells, a single protein, commonly called **capping enzyme**, contains both activities required for synthesis of a 5′ cap (Fig. 10.2B). Following the action of capping enzyme, the terminal residues are modified by methylation at specific positions. The cap 1 structure, m^{7}GpppNm, is common in viral and mammalian mRNAs. However, the sugar of the second nucleotide can also be methylated by a cytoplasmic enzyme to form the cap 2 structure (Fig. 10.2B). Methylation of the guanine base added during capping is important for recognition of mRNA by the translation machinery, whereas 2′-0 methylation of the sugar(s) blocks the inhibition of translation induced by interferons, major components of the initial antiviral defenses (Volume II, Chapter 3).

Synthesis of Viral 5′ Cap Structures by Viral Enzymes

When viral mRNAs are made in the cytoplasm of infected cells, their 5′ cap structures are, of necessity, synthesized by viral enzymes. These enzymes form cap structures typical of those present on cellular mRNA, although with some variations; for example, alphaviral mRNAs carry the cap 0 structure (Fig. 10.2B). Like their cellular counterparts, viral capping enzymes are intimately associated with the RNA polymerases responsible for mRNA synthesis. In the simplest case, exemplified by the vesicular stomatitis virus L protein, the several enzymatic activities required for synthesis of the mRNA and a 5′ cap structure are supplied by a single viral protein. The large (>2,000-amino-acid) L protein contains discrete domains that catalyze RNA and cap synthesis and subsequent methylation. This arrangement presumably facilitates coordination of capping with RNA synthesis. More-complex viruses encode dedicated capping enzymes, such as the λ-2 protein of reovirus particles and the VP4 protein of bluetongue virus (both members of the *Reoviridae*). The latter protein catalyzes all of the four reactions required for synthesis of the cap 1 structure, and its active sites are organized as a capping "assembly line" (Fig. 10.3). One of the first capping enzymes to be analyzed in detail was the vaccinia virus enzyme, which displays striking functional similarities to its host cell counterpart: it binds directly to the viral RNA polymerase and adds 5′ cap structures cotranscriptionally to nascent viral transcripts that are ~30 nucleotides in length.

Most viral capping enzymes cooperate with viral RNA-dependent RNA polymerases that can synthesize both (−) and (+) strand RNAs, but cap only (+) strand RNAs. The mechanisms that coordinate capping activity with viral mRNA synthesis are not fully understood. In some cases, sequence or structural features of the (+) strand RNA may be recognized by capping enzymes. For example, the methyltransferase of

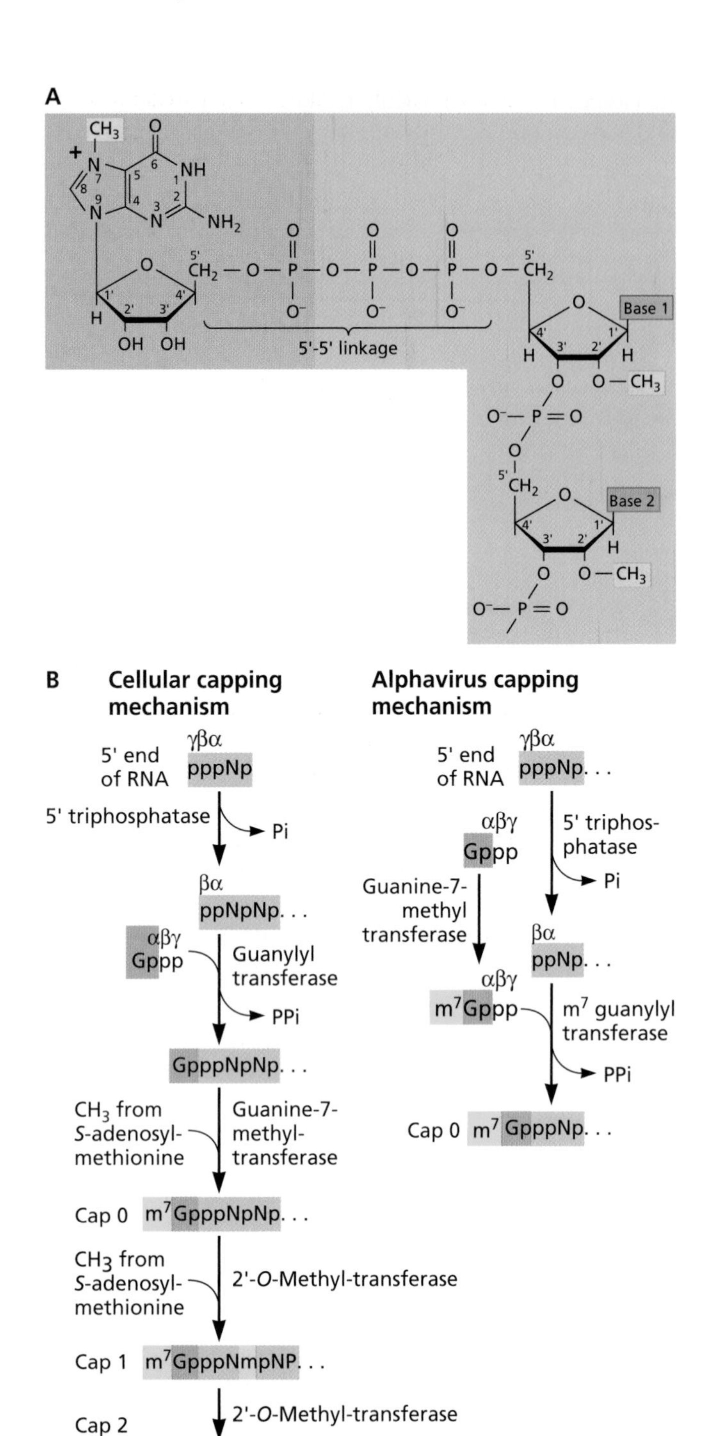

Figure 10.2 The 59 cap structure and its synthesis by cellular or viral enzymes. (A) In the cap structure shown, cap 2, the sugars of the two transcribed nucleotides (green) adjacent to the terminal m^{7}G (gray) contain 2′-*O*-methyl groups (yellow). The first and second nucleotides synthesized are methylated in the nucleus and in the cytoplasm, respectively. **(B)** The enzymes and reactions by which this cap is synthesized by cellular enzymes are listed (left) and compared to the synthesis of the caps of Semliki Forest virus (a togavirus) mRNAs by viral enzymes in the cytoplasm of infected cells (right).

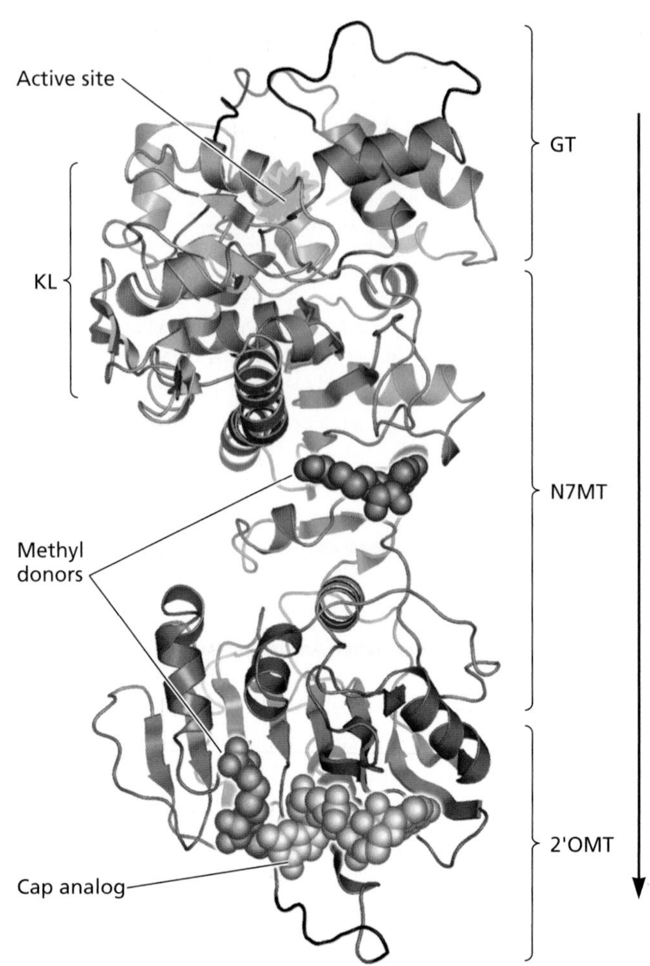

Figure 10.3 A unimolecular assembly line for capping. The structure of the bluetongue virus VP4 protein determined by X-ray crystallography is shown in ribbon form, with each of the four domains in a different color. Localization of the binding sites for substrates and products (e.g., a cap analog) identified the 2′-*O*-methyltransferase (2′OMT, purple), guanine-7-methyltransferase (N7MT, green), and guanylyltransferase (GT, blue) domains. The latter may also contain the RNA 5′-triphosphatase active site. The linear layout of the active sites in the sequence in which capping reactions take place (Fig. 10.2B) allows efficient coordination of these reactions. The KL domain (orange), which is located on one side of the otherwise linear protein, contains no active sites and is thought to mediate interactions with other proteins, such as the viral RNA-dependent RNA polymerase. Adapted from G. Sutton et al., *Nat Struct Mol Biol* **14:**449–451, 2007, with permission. Courtesy of Polly Roy, London School of Hygiene and Tropical Medicine.

the flavivirus West Nile virus binds specifically to a stem-loop structure at the 5′ end of (+) strand RNA. Substitutions of specific residues within this region inhibit cap methylation and viral replication. In other cases, such as the alphaviruses Sindbis virus and Semliki Forest virus, activation of capping enzymes may be the result of proteolytic processing. The viral P1234 polyprotein is responsible for the initial synthesis of (−)

strand RNA from the (+) strand viral genome (Chapter 6). This polyprotein includes the sequences of the RNA polymerase and the capping enzyme, but the latter is inactive. Cleavage of the polyprotein, which is necessary for synthesis of viral mRNAs (see Fig. 6.15), also releases the capping enzyme.

Acquisition of Viral 5′ Cap Structures from Cellular RNAs

The 5′ cap structures of orthomyxoviral and bunyaviral mRNAs are produced by cellular capping enzymes, but in a unique manner: the 5′ caps of these viral mRNAs are acquired when viral cap-dependent endonucleases cleave cellular transcripts to produce the primers needed for viral mRNA synthesis, a process called **cap snatching** (see Fig. 6.17). The 5′-terminal segments and caps of influenza virus mRNAs are obtained from cellular pre-mRNA in the nucleus. On the other hand, bunyaviral mRNA synthesis is primed with 5′-terminal fragments cleaved from mature cellular mRNAs in the cytoplasm.

Synthesis of 3′ Poly(A) Segments of Viral mRNA

Like the 5′ cap structure, a 3′ poly(A) segment was first identified in a viral mRNA (Box 10.2). This 3′-end modification was soon found to be a common feature of mRNAs made in eukaryotic cells, including most viral mRNAs. Like the 5′ cap, the 3′ poly(A) sequence stabilizes mRNA, and also increases the efficiency of translation. Those RNAs that are not endowed with a 3′ poly(A) tail, such as reoviral and arenaviral mRNAs, may survive by virtue of 3′-terminal stem-loop structures that block nucleolytic attack. Such structures are also present at the 3′ ends of cellular, poly(A)-lacking mRNAs that encode histones. The addition of 3′ poly(A) segments to viral pre-mRNAs, like capping of their 5′ ends, can be carried out by either cellular or viral enzymes. However, cellular and viral polyadenylation mechanisms can differ markedly.

Polyadenylation of Viral Pre-mRNA by Cellular Enzymes

Viral pre-mRNAs synthesized in infected cell nuclei by RNA polymerase II are invariably polyadenylated by cellular enzymes. Transcription of a viral or cellular gene by this enzyme proceeds beyond the site at which poly(A) will be added. The 3′ end of the mRNA is determined by endonucleolytic cleavage of its pre-mRNA at a specific position. Such cleavage is also required for termination of transcription. Poly(A) is then added to the new 3′ terminus, while the RNA downstream of the cleavage site is degraded (Fig. 10.4). Cleavage and polyadenylation sites are identified by specific sequences, first characterized in simian virus 40 and adenovirus pre-mRNAs, including the highly conserved and essential polyadenylation signal, 5′AAUAAA3′. The first reaction in polyadenylation is recognition of this sequence by the protein termed Cpsf (cleavage and polyadenylation specificity protein), an interaction that is stabilized by other proteins (Fig. 10.4). Poly(A) polymerase is then recruited and, following cleavage of the pre-mRNA, synthesizes a poly(A) segment of 200 to 250 nucleotides in a two-stage process. Like

BOX 10.2

TRAILBLAZER

Identification of poly(A) sequences on viral mRNAs

Polyadenylation of viral mRNAs was first indicated by the observation that an RNA chain resistant to digestion by RNase A, which cleaves after U and C, was produced when vaccinia virus mRNA, but not bacterial mRNA, was treated with this enzyme following labeling with [^{3}H]ATP (see the figure). The presence of a tract of poly(A) was confirmed by the specific binding of vaccinia virus mRNA, but not of bacterial mRNA, to poly(U)-Sepharose under conditions that allowed annealing of complementary nucleic acids. The position of the poly(A) sequence in viral mRNA was determined by analysis of the products of alkaline hydrolysis, when phosphodiester bonds are broken to produce nucleotides with 5′ hydroxyl and 3′ phosphate (Ap) groups. The liberation of A residues carrying 3′ hydroxyl groups by this treatment indicated that the poly(A) was located at the 3′ end of the mRNA.

Kates J. 1970. Transcription of the vaccinia virus genome and occurrence of polyriboadenylic acid sequences in messenger RNA. *Cold Spring Harb Symp Quant Biol* **35:**743–752.

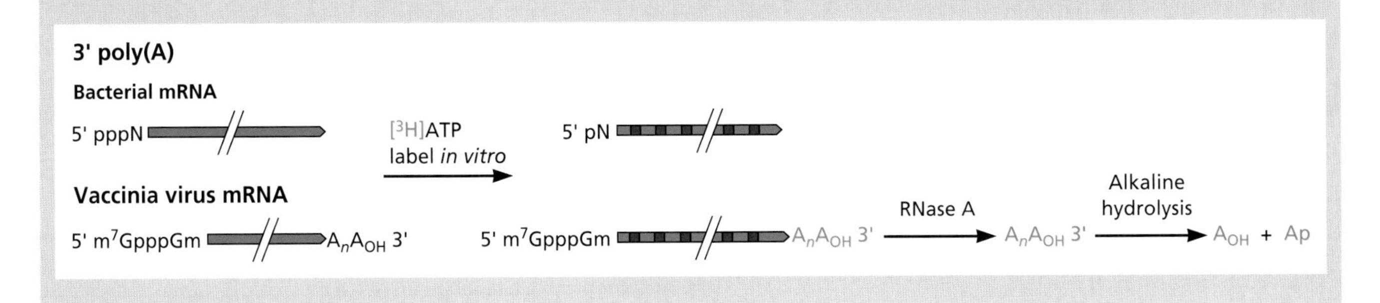

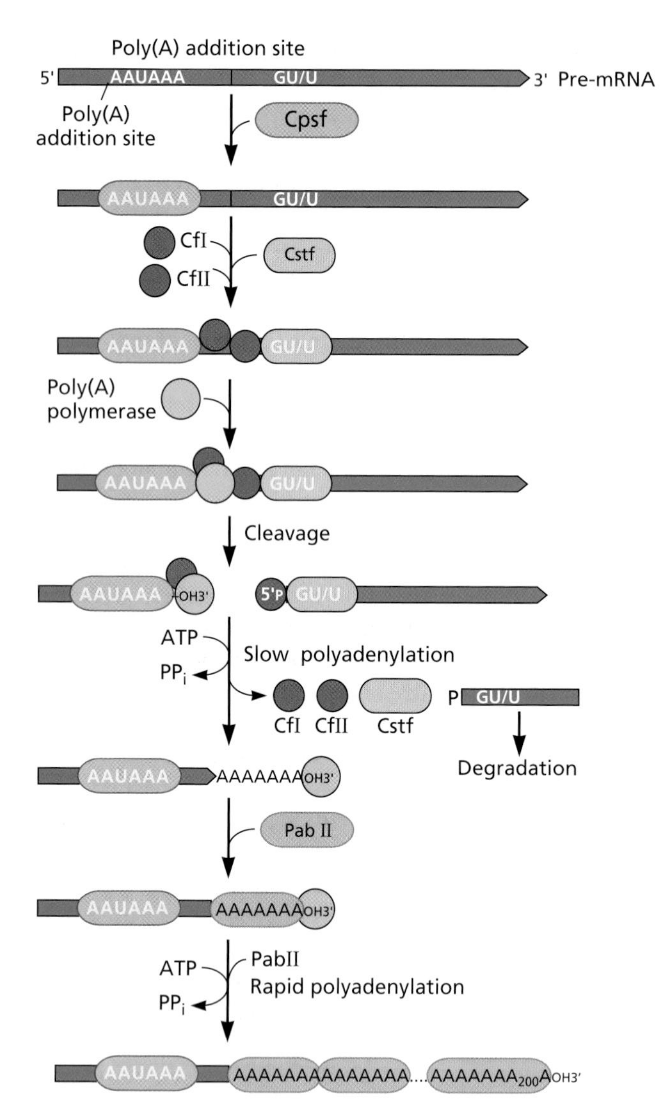

Figure 10.4 Cleavage and polyadenylation of vertebrate pre-mRNAs. The 3′ end of mature mRNA is formed 10 to 30 nucleotides downstream of the essential polyadenylation signal, 5′AAUAAA3′. However, this sequence is **not** sufficient to specify poly(A) addition. For example, it is found within mRNAs at internal positions that are never used as polyadenylation sites. Sequences at the 3′ side of the cleavage site, notably a U- or GU-rich sequence located 5 to 20 nucleotides downstream, are required. In many mRNAs (particularly viral mRNAs), additional sequences 5′ to the cleavage site are also important. The cleavage and polyadenylation specificity protein (Cpsf), which contains four subunits, binds to the 5′AAUAAA3′ poly(A) addition signal. Cleavage stimulatory protein (Cstf) then interacts with the downstream U/GU-rich sequence to stabilize a complex that also contains the two cleavage proteins, CfI and CfII. Binding of poly(A) polymerase is followed by cleavage at the poly(A) addition site by a subunit of Cpsf, and CfI, CfII, Cstf, and the downstream RNA cleavage product are then released. The polymerase slowly adds 10 to 15 A residues to the 3′-OH terminus produced by the cleavage reaction. Poly(A)-binding protein II (PabII) then binds to this short poly(A) sequence and, in conjunction with Cpsf, tethers poly(A) polymerase to the poly(A) sequence. This association facilitates rapid and processive addition of A residues until a poly(A) chain of ~200 residues has been synthesized.

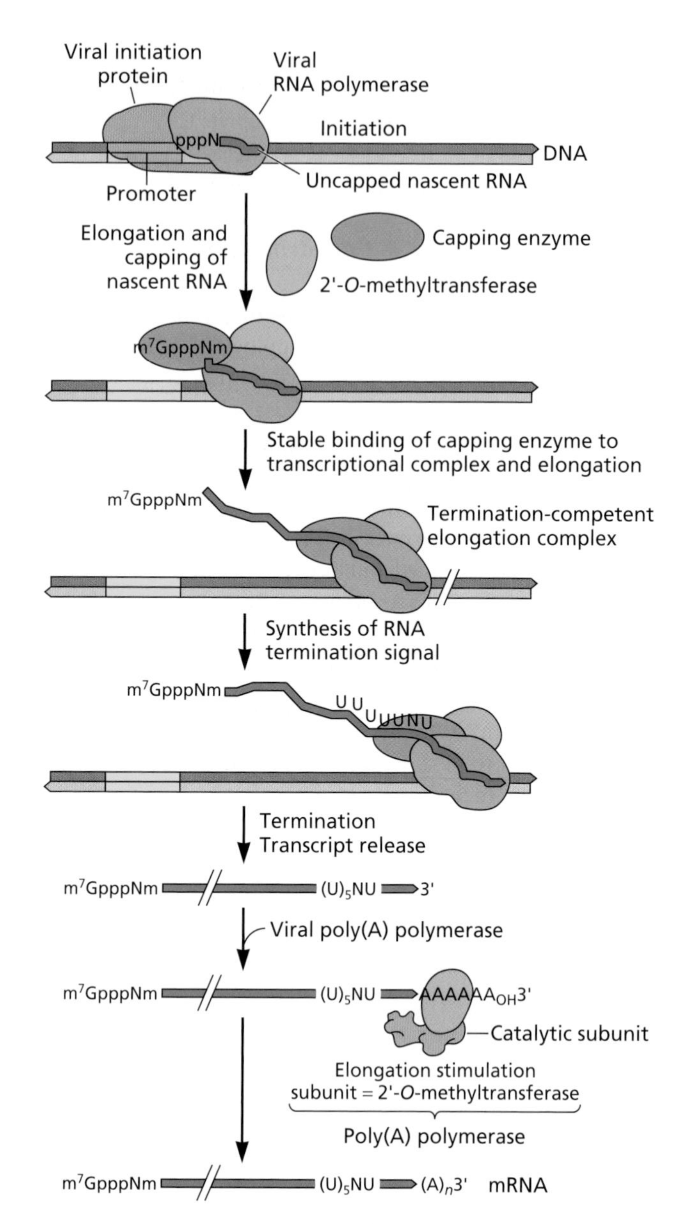

Figure 10.5 The vaccinia virus capping enzyme and 2′-*O*-methyltransferase process both the 5′ and 3′ ends of vaccinia virus mRNAs. After capping the 5′ ends of nascent viral mRNA chains ~30 nucleotides in length (step 1), the capping enzyme remains bound to the nascent RNA chain and to the RNA polymerase as the latter enzyme transcribes the template DNA. The viral 2′-*O*-methyltransferase, which produces a cap 1 structure, also binds to the viral RNA polymerase and stimulates elongation during transcription of viral intermediate and late genes (step 2). This protein is also a subunit of the viral poly(A) polymerase. Termination of transcription (step 3), which takes place 30 to 50 nucleotides downstream of the RNA sequence 5′UUUUNU3′, is mediated by the termination protein/capping enzyme and the viral nucleoside triphosphate phosphohydrolase I, which is a single-stranded-DNA-dependent ATPase. A fraction of the 2′-*O*-methyltransferase molecules act as an elongation stimulation protein for the viral poly(A) polymerase, analogous to cellular poly(A)-binding protein II. This viral enzyme, like its cellular counterpart (Fig. 10.4), adds poly(A) to the 3′ ends of the mRNA in a two-step process (steps 4 and 5).

capping, polyadenylation at the 3′ ends of an mRNA appears to be coordinated with synthesis of the pre-mRNA: binding of both Cpsf and Cstf (cleavage stimulatory protein) (Fig. 10.4) to the C-terminal domain of the largest subunit of RNA polymerase II is essential for polyadenylation *in vivo*.

Polyadenylation of Viral Pre-mRNAs by Viral Enzymes

Synthesis of poly(A) tails by viral enzymes can occur either posttranscriptionally, like polyadenylation of cellular mRNAs, or during viral mRNA synthesis.

Formation of 3′ ends by termination of transcription. A poly(A) polymerase synthesizes the 3′ poly(A) sequence of vaccinia viral early mRNAs in a two-step process remarkably like that catalyzed by the cellular enzyme (compare Fig. 10.4 and Fig. 10.5). Nevertheless, this viral system for formation of the 3′ ends of mRNA is distinctive in two major respects. The 3′ ends of vaccinia virus early mRNAs are formed by termination of transcription by the viral DNA-dependent RNA polymerase at specific sites (Fig. 10.5), a mechanism with no known counterpart in cellular mRNA synthesis systems. The vaccinia virus capping enzyme is one protein that is required for termination of transcription. Furthermore, all the proteins needed for such termination of transcription and synthesis of poly(A) are also components of the viral capping machinery (Fig. 10.5). These dual-function viral RNA-processing proteins seem likely to facilitate coordination of the reactions by which viral mRNAs are produced.

Polyadenylation during viral mRNA synthesis. The posttranscriptional synthesis of the poly(A) segments of vaccinia virus early mRNAs resembles the cellular mechanism in several respects. In contrast, the poly(A) sequences of other mRNAs made by viral RNA polymerases are produced during synthesis of the mRNA. In the simplest case, exemplified by (+) strand picornaviruses, a poly(U) sequence present at the 5′ end of the (−) strand RNA template is copied directly into a poly(A) sequence of equivalent length. The mRNAs of (−) strand RNA viruses like vesicular stomatitis virus and influenza virus are polyadenylated by reiterative copying of short stretches of U residues in the (−) strand RNA template, a mechanism described in Chapter 6.

Splicing of Viral Pre-mRNA

Discovery of Splicing

Between 1960 and the mid-1970s, the study of putative nuclear precursors of mammalian mRNAs established that these RNAs are larger than the mRNAs translated in the cytoplasm and heterogeneous in size. They were therefore named **heterogeneous nuclear RNAs** (hnRNAs). Such hnRNAs were shown to carry both 5′-terminal cap structures and 3′ poly(A) sequences, leading to the conclusion that both ends of the hnRNA were preserved in the smaller, mature mRNA. Investigators were faced with the conundrum of deducing how smaller mRNAs could be produced from larger hnRNAs while both ends of the hnRNA were retained.

The puzzle was solved by two groups of investigators, led by Phillip Sharp and Richard Roberts, who shared the 1993 Nobel Prize in physiology or medicine (See the interview with Dr. Phillip Sharp: http://bit.ly/Virology_Sharp). These investigators showed that adenoviral major late mRNAs are encoded by four **separate** genomic sequences (Box 10.3). The distribution of the mRNA-coding sequences into four separate blocks in the genome, in conjunction with the large size of major late mRNA precursors, implied that these mRNAs were produced by excision of noncoding sequences from primary transcripts (introns), with precise joining of coding sequences (exons). The demonstration that nuclear major late transcripts contain the introns confirmed that the mature mRNAs are formed by **splicing** of noncontiguous coding sequences in the pre-mRNA. This mechanism had great appeal, because it could account for the puzzling properties of hnRNA. Indeed, it was shown within a matter of months that splicing of pre-mRNA is not an obscure, virus-specific device: splicing occurs in all eukaryotic cells, and the great majority of mammalian pre-mRNAs, like the adenoviral major late mRNAs, comprise exons interspersed among introns.

The organization of protein-coding sequences into exons separated by introns has profound implications for the evolution of the genes of eukaryotes and their viruses. Introns are generally much longer than exons, and only short sequences at their ends are necessary for accurate splicing (see "Mechanism of Splicing" below). Consequently, introns provide numerous sites at which DNA sequences can be broken and rejoined without loss of coding information, and greatly increase the frequency with which random recombination reactions can create new functional genes by rearrangement of exons. Evidence of such "exon shuffling" can be seen in the modular organization of many modern proteins. Such proteins comprise combinations of a finite set of structural and functional domains or motifs, or multiple repeats of a single protein domain, each often encoded by a single exon. The presence of introns is also thought to have facilitated the recombination of viral and cellular genetic information.

Any viral transcript that is synthesized by cellular RNA polymerase II can potentially be spliced. Indeed, splicing is the rule for the transcripts of parvo-, papilloma-, polyoma-, and adenoviral genomes, as well as those of integrated proviral DNAs of retroviruses. Furthermore, alternative splicing of these viral transcripts is an important mechanism for expanding the coding capacity of such viral genomes. Although the (+) strand RNAs of influenza A virus are synthesized by

BOX 10.3

TRAILBLAZER

Discovery of the spliced structure of adenoviral major late mRNAs

(A) Digestion of adenoviral major late mRNAs with RNase T_1, which cleaves after G, and isolation of the capped 5′ oligonucleotides indicated that the **same** 11-nucleotide sequence was present at the 5′ ends of several different mRNAs. This observation was surprising and puzzling. Hybridization studies indicated that these 5′ ends were **not** encoded adjacent to the main segments of major late mRNAs. Direct visualization of such mRNAs hybridized to viral DNA provided convincing proof that their coding sequences are dispersed in the viral genome. **(B)** Schematic diagram of one major late mRNA (hexon mRNA) hybridized to a complementary adenoviral DNA fragment extending from the left end of the genome to a point within the hexon-coding sequence. Three loops of unhybridized DNA (thin lines), designated A, B, and C, bounded or separated by three short segments (1, 2, and 3) and one long segment (hexon mRNA) of DNA-RNA hybrid (thick lines) were observed. Other adenoviral late mRNAs yielded the same sets of hybridized and unhybridized viral DNA sequences at their 5′ ends, but differed in the length of loop C and the length and location of the 3′-terminal RNA-DNA hybrid. It was therefore concluded that the major late mRNAs contain a common 5′-terminal segment (segments 1, 2, and 3) built from sequences encoded at three different sites in the viral genome and termed the tripartite leader sequence. This sequence is joined to the mRNA body, a long sequence complementary to part of the hexon-coding sequence in the example shown. Panel B adapted from S. M. Berget et al., *Proc Natl Acad Sci U S A* **74:**3171–3175, 1977, with permission.

Berget SM, Moore C, Sharp PA. 1977. Spliced segments at the 5′ terminus of adenovirus 2 late mRNA. *Proc Natl Acad Sci U S A* **74:**3171–3175.

Chow LT, Gelinas RE, Booker TR, Roberts RJ. 1977. An amazing sequence arrangement at the 5′ ends of adenovirus 2 messenger RNA. *Cell* **12:**1–8.

Gelinas RE, Roberts RJ. 1977. One predominant undecanucleotide in adenovirus late messenger RNAs. *Cell* **11:**533–544.

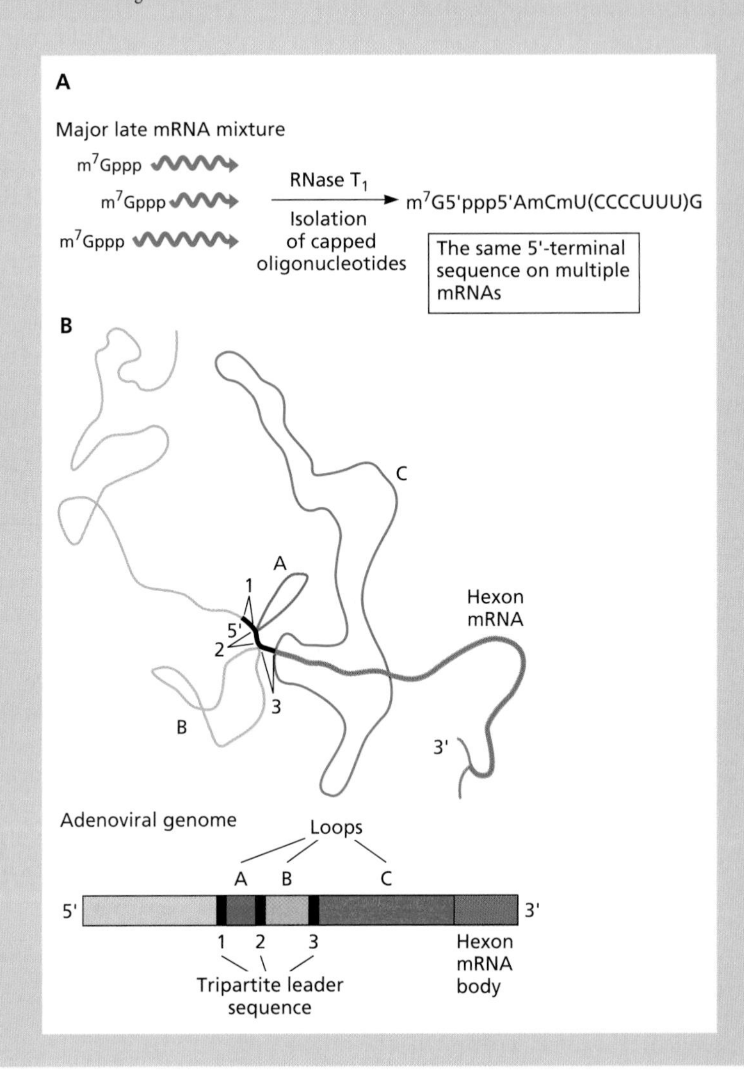

a viral RNA polymerase (Chapter 6), some are nevertheless spliced. For example, the (+) M RNA serves as the mRNA for the matrix protein, but is also spliced to produce the mRNA that specifies the M2 ion channel protein (Appendix, Fig. 15). Such splicing, which may account for the synthesis of influenza A virus RNAs in infected cell nuclei, is an exception to the coordination among the cellular components that synthesize and process pre-mRNAs. The production of a stable intron in neurons latently infected by herpes simplex virus type 1 (Box 10.4) is a second example of a virus-specific permutation of the RNA-processing reactions of the host cell.

Mechanism of Splicing

Sequencing of DNA copies of a large number of cellular and viral mRNAs and of the genes that encode them identified short consensus sequences at the 5′ and 3′ **splice sites**, which are joined to each other in mature mRNA (Fig. 10.6A). The conserved sequences lie largely within the introns. The dinucleotides GU and AG are found at the 5′ and 3′ ends, respectively, of almost all introns. Mutation of any one of these four nucleotides eliminates splicing, indicating that all are essential. Elucidation of the mechanism of splicing came with the development of *in vitro* systems in which model pre-mRNAs

BOX 10.4

DISCUSSION

Cell-type-specific production of a stable viral intron

The most abundant RNA detected in neurons latently infected with herpes simplex virus type 1, the 2.0-kb major latency-associated transcript (LAT), is an excised intron: it is not linear, lacks a 3′ poly(A) sequence, and contains a branch point like those of the intron lariats excised during mRNA splicing (Fig. 10.6B).

Processing of LAT precursor RNA is much more efficient in the sensory ganglia, in which herpes simplex virus type 1 establishes latent infections, than in other cell types (see the figure). Furthermore, in contrast to typical introns, which are degraded rapidly, the LAT intron is remarkably stable, with a half-life of >24 h.

These properties imply that the LAT RNA intron fulfills a beneficial function during establishment or maintenance of latency or during subsequent reactivation and entry into the viral lytic cycle. However, the functions of this RNA remain unknown.

Gussow AM, Giordani NV, Tran RK, Imai Y, Kwiatkowski DL, Rall GF, Margolis TP, Bloom DC. 2006. Tissue-specific splicing of the herpes simplex virus type latency-associated transcript (LAT) intron in LAT transgenic mice. *J Virol* **80:**9414–9423.

Zabolotny JM, Krummenacher C, Fraser NW. 1997. The herpes simplex virus type 1 2.0-kilobase latency-associated transcript is a stable intron which branches at a guanosine. *J Virol* **71:**4199–4208.

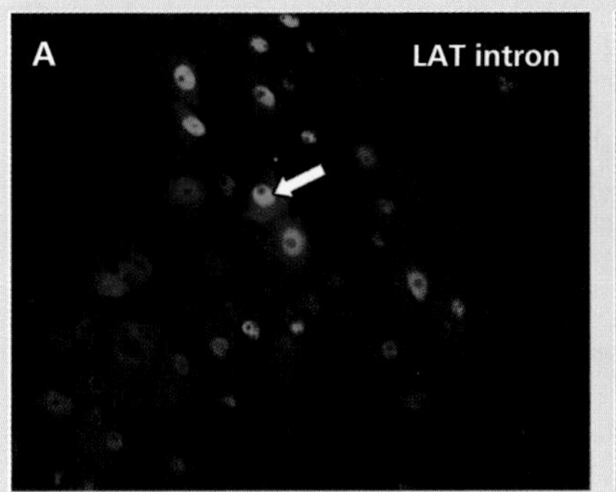

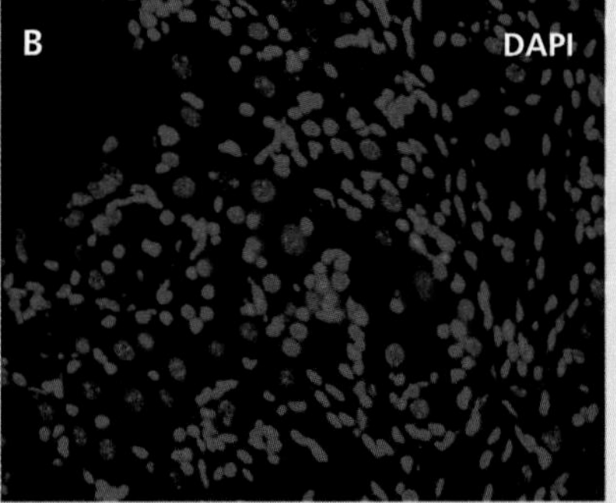

Trigeminal neurons isolated from LAT transgenic mice were hybridized to a digoxigenin-containing complementary RNA specific for spliced LAT RNA, incubated with fluorescein-labeled anti-digoxigenin antibodies **(A)**, stained with 4′,6-diamidino-2-phenylindole (DAPI) **(B)**, and examined by fluorescent microscopy. The white arrow indicates a neuron in which LAT RNA splicing was efficient. The many nuclei that do not contain LAT RNA are those of glial cells. Adapted from A. M. Gussow et al., *J Virol* 80:9414–9423, 2006, with permission.

Figure 10.6 Splicing of pre-mRNA. (A) Consensus splicing signals in cellular and viral pre-mRNAs. The most conserved sequences are found at the 5′ and 3′ splice sites at the junctions of exons (green) and introns (pink) and at the 3′ ends of introns. The intronic 5′GU3′ and 5′AG3′ dinucleotides at the 5′ and 3′ ends, respectively, of introns and branch point A (highlighted) are present in all but rare mRNAs made in higher eukaryotes. **(B)** The two transesterification reactions of pre-mRNA splicing. In the first reaction, the 2′ hydroxyl group of the conserved A residue in the intronic branch point sequence makes a nucleophilic attack on the phosphodiester bond at the 5′ side of the GU dinucleotide at the 5′ splice site to produce the intron-3′ exon lariat and the 5′ exon. A second nucleophilic attack by the newly formed 3′ hydroxyl group of the 5′ exon on the phosphodiester bond at the 3′ splice site then yields the spliced exons and the intron lariat.

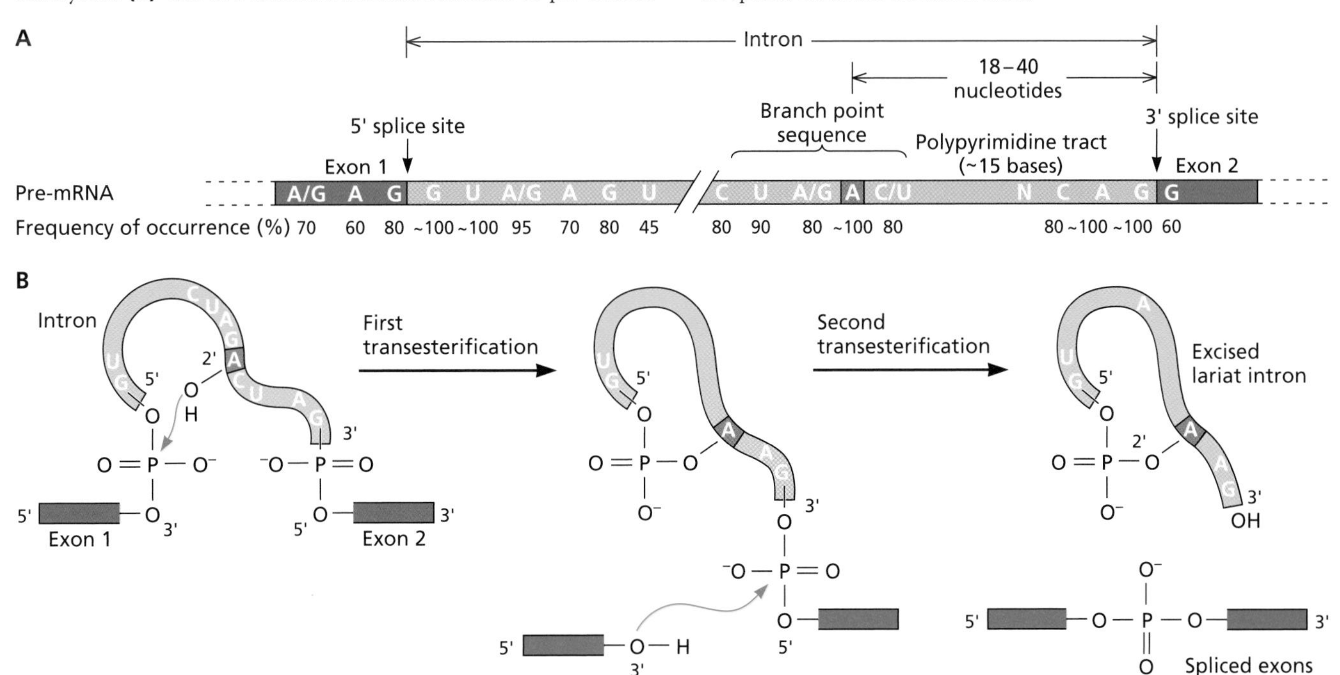

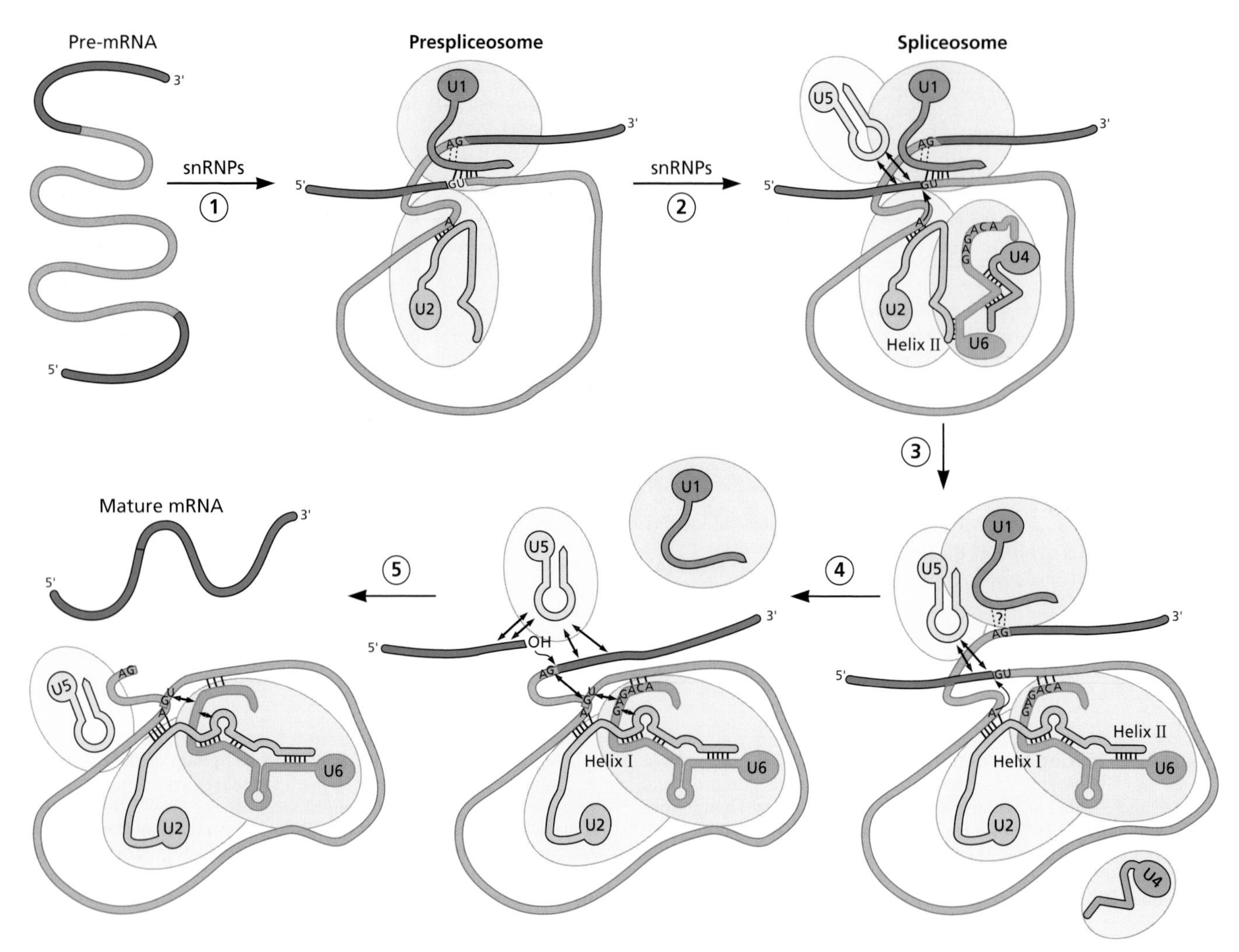

Figure 10.7 RNA-RNA interactions organize substrates and catalysts during splicing. Base pairs are indicated by dashes and experimentally observed or presumed contacts among the RNA molecules by the two-headed arrows. The U1 and U2 snRNAs initially base pair with the 5′ splice site and branch point sequence, respectively, in the pre-mRNA (step 1). The other snRNPs then enter the assembling spliceosome (step 2). The U4 and U6 snRNAs, which are present in a single snRNP, are base paired with one another over an extended complementary region. This snRNP binds to the U5 snRNP, and the snRNP complex associates with that containing the pre-mRNA and U1 and U2 snRNAs to form the spliceosome. RNA rearrangements then activate spliceosomes for catalysis of splicing (step 3). U4 snRNA dissociates from U6 snRNA, which forms hydrogen bonds with both U2 snRNA and the pre-mRNA. The interaction of U6 snRNA with the 5′ splice site displaces U1 snRNA (step 4). One of the U2 sequences hydrogen bonded to U6 snRNA (helix I) is adjacent to the U2 snRNA sequence that is base paired with the pre-mRNA branch point region. The interactions of U2 and U6 snRNAs with each other and with the pre-mRNA therefore juxtapose the branch point and 5′ splice site sequences for the first transesterification reaction (Fig. 10.6B). The U5 snRNA base pairs to sequences in both the 5′ and 3′ exons to align them for the second transesterification reaction (step 5). The many proteins that participate in spliceosome assembly and activation or that package the pre-mRNA and snRNAs are not shown. Adapted from T. W. Nilsen, *Cell* **78:**1–4, 1994, with permission.

(initially of viral origin) are accurately spliced. Pre-mRNA splicing occurs by two transesterification reactions, in which one phosphodiester bond is exchanged for another without the need for an external supply of energy. The first reaction yields two products, the 5′ exon and the intron-3′ exon **lariat**. In the second, the two exons are joined and the intron lariat is released (Fig. 10.6B).

From a chemical point of view, the splicing of pre-mRNA is a simple process. However, each splicing reaction must be completed with a high degree of accuracy to ensure that coding information is not lost or altered, and the chemically active hydroxyl groups must also be brought into close proximity to the phosphodiester bonds they will attack (Fig. 10.6B). Many genes contain a large number of introns separating multiple exons, which must be spliced in the correct order. It is presumably for such reasons that pre-mRNA splicing occurs in the large structure called the **spliceosome**, which contains both many proteins and several small RNAs.

Five small nuclear RNAs (snRNAs) participate in splicing: the U1, U2, U4, U5, and U6 snRNAs. In vertebrate cells, these RNAs vary in length from 100 to 200 nucleotides and are associated with proteins to form **small nuclear ribonucleoproteins** (snRNPs). The RNA components of the snRNPs recognize splice sites and other sequences in cellular and viral pre-mRNAs. Indeed, they participate in multiple dynamic interactions with the pre-mRNA and with each other both during the initial ordered assembly of the spliceosome and in splicing. These base-pairing interactions juxtapose first the 5′ splice site and the branch point for the first transesterification reaction and then the 5′ and the 3′ exons (Fig. 10.7). However, RNAs of the snRNPs do much more than simply organize the pre-mRNA sequences into a geometry suitable for transesterification. It has long been suspected that the spliceosome might be an RNA enzyme (or **ribozyme**), and compelling evidence for catalysis by U6 snRNAs during splicing has been reported recently (Box 10.5).

Although the snRNAs play essential roles in splicing as both guides and catalysts, the spliceosome also contains ~150 non-snRNP proteins. One class comprises proteins that package the pre-mRNA substrate. Many other splicing proteins contain both RNA-binding and protein-protein interaction domains. Such proteins bind to pre-mRNA sequences within or adjacent to exons to facilitate spliceosome assembly or regulate splicing or exon recognition. Other proteins important for splicing are RNA-dependent helicases, which are thought to catalyze the multiple rearrangements of hydrogen bonding among different snRNAs and the pre-mRNA substrate (Fig. 10.7). Such helicases are generally ATP dependent, and spliceosome assembly and rearrangement depend on energy supplied by ATP hydrolysis.

Splicing of pre-mRNAs is commonly cotranscriptional, and components of the splicing machinery associate with the hyperphosphorylated form of the C-terminal domain of the largest subunit of RNA polymerase II during elongation of

BOX 10.5

DISCUSSION

Catalysis of pre-mRNA splicing by RNA

The catalytic activity of RNA was established with the discovery of *Escherichia coli* RNase P, which contains a small RNA molecule essential for catalysis, and self-splicing RNAs. The first such RNA to be described was a short (414-nucleotide) intron in pre-rRNA of *Tetrahymena*. Specific nucleotides in the introns coordinate metal ions to catalyze the same phosphoryl transfer reactions that accomplish pre-mRNA splicing. This precedent, the formation of a lariat intermediate during both pre-mRNA splicing and that of some self-splicing introns, and the essential role of snRNAs in pre-mRNA splicing suggested that the latter process is also catalyzed by RNA. Consistent with this view, U2 and U6 snRNAs synthesized *in vitro* base pair to form a stable structure with the configuration thought to be present at the active site of the spliceosome. In the absence of any protein, these RNAs are sufficient to catalyze a reaction analogous to the first transesterification during pre-mRNA splicing. Nevertheless, direct evidence that U6 catalyzes pre-mRNA splicing has been reported only recently.

In these experiments, sulfur was substituted for single oxygen atoms at 20 positions in U6 snRNA that were known to be important for pre-mRNA splicing or analogous to catalytic residues in self-splicing introns. Splicing reactions require Mg^{2+}, which binds efficiently to oxygen but not to sulfur. These modified U6 snRNAs were then assembled into spliceosomes and model splicing substrates added. Five of the substitutions inhibited splicing in the presence of Mg^{2+} ions. However, splicing was restored when metal ions that bind sulfur, Mn^{2+} or Cd^{2+}, were supplied. Analysis of which reactions were blocked by individual substitutions and other experiments indicated that one of the two metal ions is coordinated to the nucleophilic hydroxyl group that initiates the first transesterification reaction, and the second is coordinated to the leaving group (M2 and M1, respectively, in the figure). These studies provide direct evidence for RNA-mediated catalysis during pre-mRNA splicing.

Fica SM, Tuttle N, Novak T, Li NS, Lu J, Koodathingal P, Dai Q, Staley JP, Piccirilli JA. 2013. RNA catalyses nuclear pre-mRNA splicing. *Nature* **503:**229-234.

Valadkhan S, Manley JL. 2001. Splicing-related catalysis by protein-free snRNAs. *Nature* **413:**701–707.

The hairpin structure of U6 snRNA and its base pairing to U2 snRNA are shown, with the phosphoryl groups that coordinate Mg^{2+} ions (M1 and M2) shown by the asterisks.

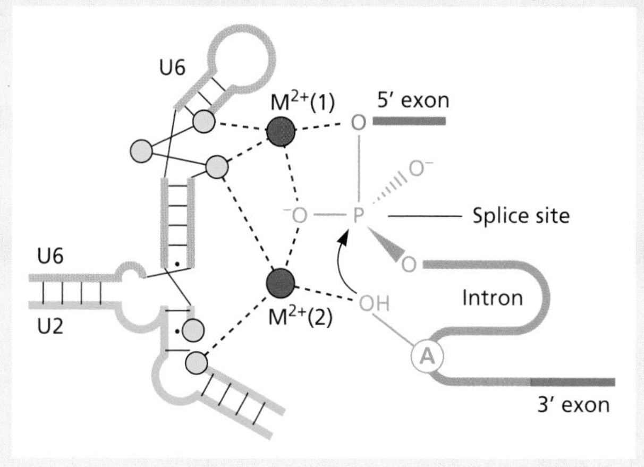

transcription. Peptide mimics of the C-terminal domain, or antibodies raised against it, inhibit pre-mRNA splicing in cells in culture or *in vitro*. Furthermore, nontranscribed sequences within RNA polymerase II promoters can dictate whether a particular exon is retained or removed during splicing. As we have seen, association of components of the 5′ capping and 3′ polyadenylation systems with this domain of RNA polymerase is necessary for these processing reactions. The synthesis of pre-mRNA and its complete processing are therefore coordinated as a result of association of specific proteins needed for each processing reaction with the C-terminal domain of RNA polymerase II. Such a transcription and RNA-processing machine is analogous to that of vaccinia virus, described above, but much more elaborate.

Alternative Splicing

Many viral and cellular pre-mRNAs contain multiple exons. Splicing of many such transcripts removes all introns and joins all exons in the order in which they are present (Fig. 10.8A). However, numerous cellular and many viral pre-mRNAs yield more than one mRNA as a result of the splicing of different combinations of exons, a process termed **alternative splicing**. Several different types of alternative splicing can be defined (Fig. 10.8B). Alternative splicing, which can comprise selection among large numbers of exons in a pre-mRNA, is governed by multiple parameters, including splice site and regulatory sequences in the pre-mRNA and the constellation of splicing proteins present in particular cell types (Box 10.6).

As alternative splicing generally leads to the synthesis of mRNAs that differ in their protein-coding sequences, its most obvious advantage is that it can expand the limited coding capacity of viral genomes. The early genes of polyomaviruses and adenoviruses each specify two or more proteins as a result of splicing of primary transcripts at alternative 5′ or 3′ splice sites. Alternative splicing can also be important for temporal regulation of viral gene expression or the control of a crucial balance in the production of spliced and unspliced mRNAs.

Although described separately in previous sections, capping, polyadenylation, and splicing of a pre-mRNA by cellular components are not independent. Rather, one processing reaction governs the efficiency or specificity of another. For example, interaction of the nuclear cap-binding protein with the 5′ end of a pre-mRNA facilitates both removal of the 5′-terminal intron and efficient cleavage at the 3′ poly(A) addition site. Similarly, the presence of a 3′ poly(A) addition signal generally stimulates removal of the intron closest to it.

Alternative Processing of Viral Pre-mRNA

Cellular Differentiation Regulates Production of Papillomaviral Late Pre-mRNAs

The late proteins of bovine papillomavirus type 1 are synthesized efficiently only in highly differentiated keratinocytes. Productive replication of viral DNA and assembly and release of virus particles are also restricted to these outer cells in an epithelium. Alterations in polyadenylation and splicing are crucial for production of the mRNA encoding the major capsid proteins, such as the L1 mRNA (Fig. 10.9A). *In situ* hybridization studies have shown that this mRNA is made only in fully differentiated cells (Fig. 10.9B). Production of the late mRNAs requires utilization of both polyadenylation and 3′ splice sites that are not recognized in undifferentiated cells. The 3′ splice sites for papillomavirus late mRNAs are suboptimal, and their recognition is governed by *cis*-acting suppressor or enhancer sequences in the pre-mRNA (Fig. 10.9C). It has therefore been proposed that terminal differentiation of keratinocytes is accompanied by changes in the activity or abundance of the cellular proteins that bind to these sequences. Similarly, binding of U1 snRNP to a pseudo-5′ splice site appears to suppress recognition of the poly(A) addition site for

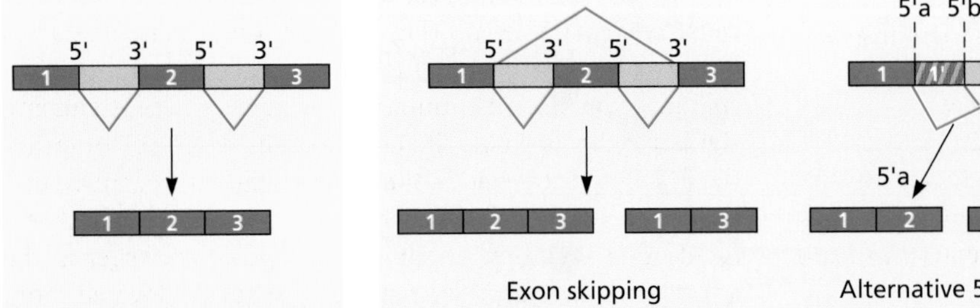

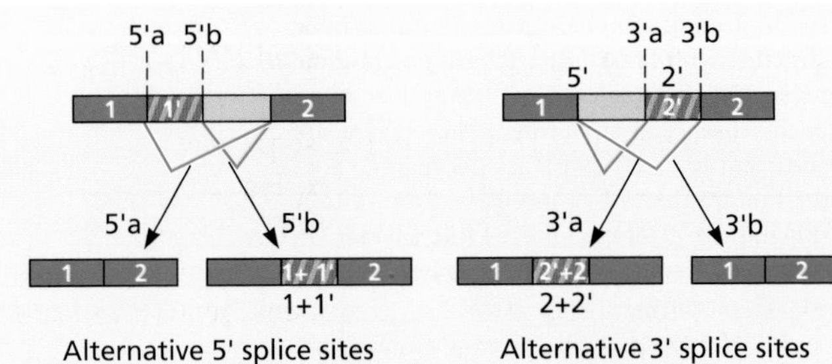

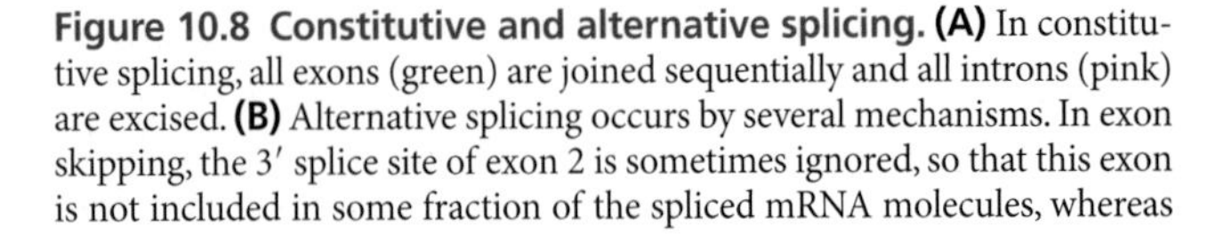

Figure 10.8 Constitutive and alternative splicing. (A) In constitutive splicing, all exons (green) are joined sequentially and all introns (pink) are excised. **(B)** Alternative splicing occurs by several mechanisms. In exon skipping, the 3′ splice site of exon 2 is sometimes ignored, so that this exon is not included in some fraction of the spliced mRNA molecules, whereas in intron retention, a set of splice sites is ignored. Alternatively, one of two 5′ splice sites (5′a and 5′b) in exon 1 or one of two 3′ splice sites (3′a and 3′b) in exon 2 are recognized. Recognition of different 5′ and 3′ splice sites produces alternatively spliced simian virus 40 early and adenoviral major late (Fig. 10.11) mRNAs, respectively.

BOX 10.6

BACKGROUND

Discrimination among splice sites for alternative splicing of pre-mRNA

Many mammalian, and some viral, pre-mRNAs are alternatively spliced. This process, which can yield large numbers of mRNAs from a primary transcript (see, for example, Fig. 10.11), requires the splicing machinery to discriminate between alternative 5′ or alternative 3′ splice sites, to skip an exon or to retain an intron (that is, ignore particular combinations of 5′ and 3′ splice sites) (Fig. 10.8). Detailed studies of alternative splicing of specific pre-mRNAs or model substrates and genome-wide analyses have identified multiple features of the substrate that cooperate to govern alternative splicing. These parameters include the following:

- Splice site "strength," that is, match to the consensus 5′ and 3′ splice site sequences (Fig. 10.6A)
- The presence of stable RNA secondary structures that interfere with recognition of individual splice sites
- The lengths of exons and introns. In mammalian cells, interactions among splicing components at a 5′ splice site and the upstream 3′ splice site are responsible for recognition of most exons. Such "exon definition" is more efficient when exons are short. Furthermore, exons flanked by long introns are much more likely to be alternatively spliced than those flanked by short introns.
- The presence in both exons and introns of binding sites for proteins that facilitate or suppress recognition of neighboring splice sites, and hence alternative splicing. Examples of these sequences, termed exonic or intronic splicing enhancers or suppressors, are described in the text. The impact of such a regulatory element can be context dependent, determined by distance from adjacent splice site(s), the number of copies of the sequence, and the constellation of other splicing regulatory elements in the vicinity (see the figure).

The concentration of proteins that recognize splicing regulatory sequences in particular cells or tissues is an important determinant of alternative splicing, resulting in tissue-specific synthesis of different isoforms of many mammalian mRNAs. The alternative splicing of pre-mRNAs can also be coupled to transcription, by recruitment of particular splicing proteins by binding to RNA polymerase II, and as a result of the rate at which splice sites and splicing regulatory sequences are made in a nascent pre-mRNA.

Hertel KJ. 2008. Combinatorial control of exon recognition. *J Biol Chem* **283:**1211–1215.

Kornblihtt AR, Schor IE, Alló M, Dujardin G, Petrillo E, Muñoz MJ. 2013. Alternative splicing: a pivotal step between eukaryotic transcription and translation. *Nat Rev Mol Cell Biol* **14:**153–165.

Wang Z, Burge CB. 2008. Splicing regulation: from a parts list of regulatory elements to an integrated splicing code. *RNA* **14:**802–813.

Schematic illustration of splicing regulatory sequences and some of the proteins that recognize them. ESE, exonic splicing enhancer; ESS, exonic splicing suppressor; ISE, intronic splicing enhancer; ISS, intronic splicing suppressor; SR, serine- and arginine-rich splicing proteins.

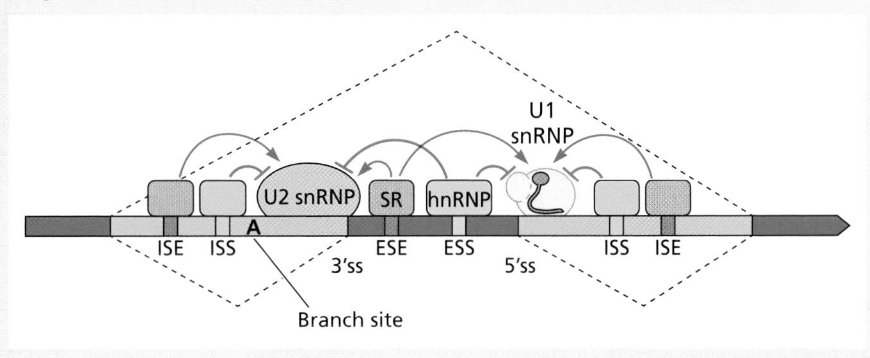

production of bovine papillomavirus type 1 late mRNAs until the keratinocyte host cell is fully differentiated (Fig. 10.9D).

Production of Spliced and Unspliced RNAs Essential for Virus Replication

The expression of certain coding sequences in retroviral genomes (Gag and Pol) (Appendix, Fig. 29) and orthomyxoviruses (M1 and NS1) (Appendix, Fig. 15) depends on an unusual form of alternative splicing that produces both spliced and unspliced mRNAs. This phenomenon has been well studied in retrovirus-infected cells.

In cells infected by retroviruses with simple genomes, such as avian leukosis virus, a full-length, unspliced transcript of proviral DNA serves as both the genome and the mRNA for the capsid proteins and viral enzymes, while a singly spliced mRNA specifies the viral envelope protein (Fig. 10.10A). Retrovirus production depends rather critically on the maintenance of a proper balance in the proportions of unspliced and spliced RNAs: modest changes in splicing efficiencies cause replication defects (Fig. 10.10A). This phenomenon has been used as a genetic tool to select for mutations that affect splicing control. Such mutations arise in different splicing signals at the 3′ splice site and alter the efficiency of either the first or second step in the splicing reaction. Features that maintain the proper splicing balance include suboptimal recognition of the 3′ splice site, and a splicing enhancer in the adjacent exon. A negative regulatory sequence located more than 4000 nucleotides upstream of the 3′ splice site is also important. This sequence, which is bound by both U1 snRNP and specific cellular proteins (Fig. 10.10A), has been proposed to act as a "decoy" 5′ splice site: it forms a complex with the 3′ splice site for production of Env mRNA, but one that does

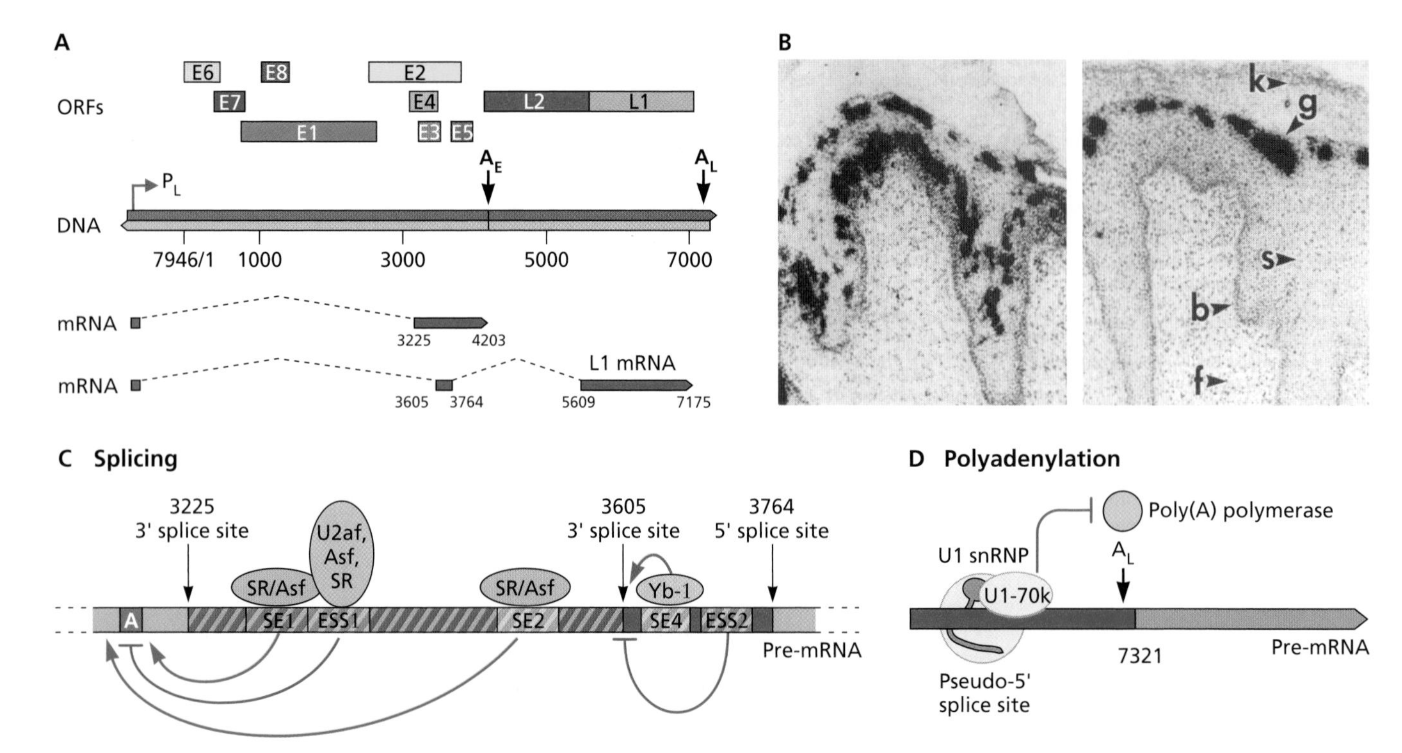

Figure 10.9 Alternative polyadenylation and splicing control the production of bovine papillomavirus type 1 late mRNAs. **(A)** The circular bovine papillomavirus type 1 genome is represented in linear form, with open reading frames (ORFs) shown above. Two of the many mRNAs made from transcripts from the late promoter (P_L) are shown to illustrate the changes in recognition of splice sites and of poly(A) addition sites necessary to produce the L1 mRNA. Synthesis of this mRNA depends on recognition of a 3′ splice site at position 3605, rather than that at 3225, which is used during the early phase of infection. Polyadenylation of pre-mRNAs must also switch from the early (A_E) to the late (A_L) polyadenylation site. **(B)** *In situ* hybridization of bovine fibropapillomas to probes that specifically detect mRNAs spliced at the 3225 3′ splice site (left) or at the 3605 site (right). The cell layers of the fibropapilloma are indicated in the right panel. Abbreviations: k, keratin horn; g, granular cell layer; s, spinous cell layer; b, basal cell layer; f, fibroma. Note the production of late mRNA spliced at the 3605 3′ splice site only in the outermost layer (g) of fully differentiated cells. From S. K. Barksdale and C. C. Baker, *J Virol* **69:**6553–6556, 1995, with permission. Courtesy of C. C. Baker, National Institutes of Health. **(C)** Mechanisms that regulate splicing to produce L1 mRNA, which are specific to highly differentiated keratinocytes of the granular cell layer. The sequences that control alternative splicing at the 3225 and 3605 3′ splice sites are located between these splice sites. The splicing enhancers, SE1, SE2, and SE4, are recognized by cellular SR (serine- and arginine-rich) and other splicing proteins such as Asf (alternative splicing factor). The SE1 enhancer and the adjacent sequence that inhibits splicing at the 3605 3′ splice site, termed exonic splicing suppressor (ESS1), are thought to facilitate recruitment of U2-associated protein (U2af) and recognition of the branch point sequence upstream of the 3225 3′ splice site. SE2 is located very close to the 3605 3′ splice site and may block access to the branch point for splicing at this site until keratinocytes differentiate. The hnRNP A1 protein that binds to a sequence that inhibits recognition of the downstream 3′ splice site of the late pre-mRNA is not present in differentiated keratinocytes. **(D)** Inhibition of polyadenylation at the A_E site by the binding of U1 snRNP to a pseudo-5′ splice site located nearby in the primary transcript (see the text). Such inhibition is the result of binding of the U1 snRNP 70k subunit to poly(A) polymerase.

not participate in splicing reactions. The splicing of human immunodeficiency virus type 1 pre-mRNA is necessarily much more complicated, as >40 alternatively spliced mRNAs are made in infected cells. Nevertheless, alternative splicing is also regulated by specific sequences that promote or repress recognition or utilization of splice sites and by the degree of conformity of 3′ splice sites to the optimal sequence.

The long terminal repeats at each end of proviral DNAs include a poly(A) addition signal (Fig. 10.10B). Transcription of some proviral DNAs, such as that of Rous sarcoma virus, initiates downstream of the polyadenylation signal in the 5′ long terminal repeat sequence so that a poly(A) addition site is present only at the 3′ ends of pre-mRNAs. However, many other retroviral transcripts carry complete signals for this modification at both their 5′ and 3′ ends, but poly(A) is added to only the 3′ ends of these pre-mRNAs. At least two mechanisms ensure that the correct poly(A) addition signal of human immunodeficiency virus type 1 is recognized. Sequences present only at the 3′ end of the pre-mRNA stimulate polyadenylation *in vitro* and in cells in culture, by facilitating binding of Cpsf to the nearby 5′AAUAAA3′ sequence. In addition, recognition of the 5′ poly(A) signal is suppressed by the 5′ splice site lying immediately downstream (Fig. 10.10B).

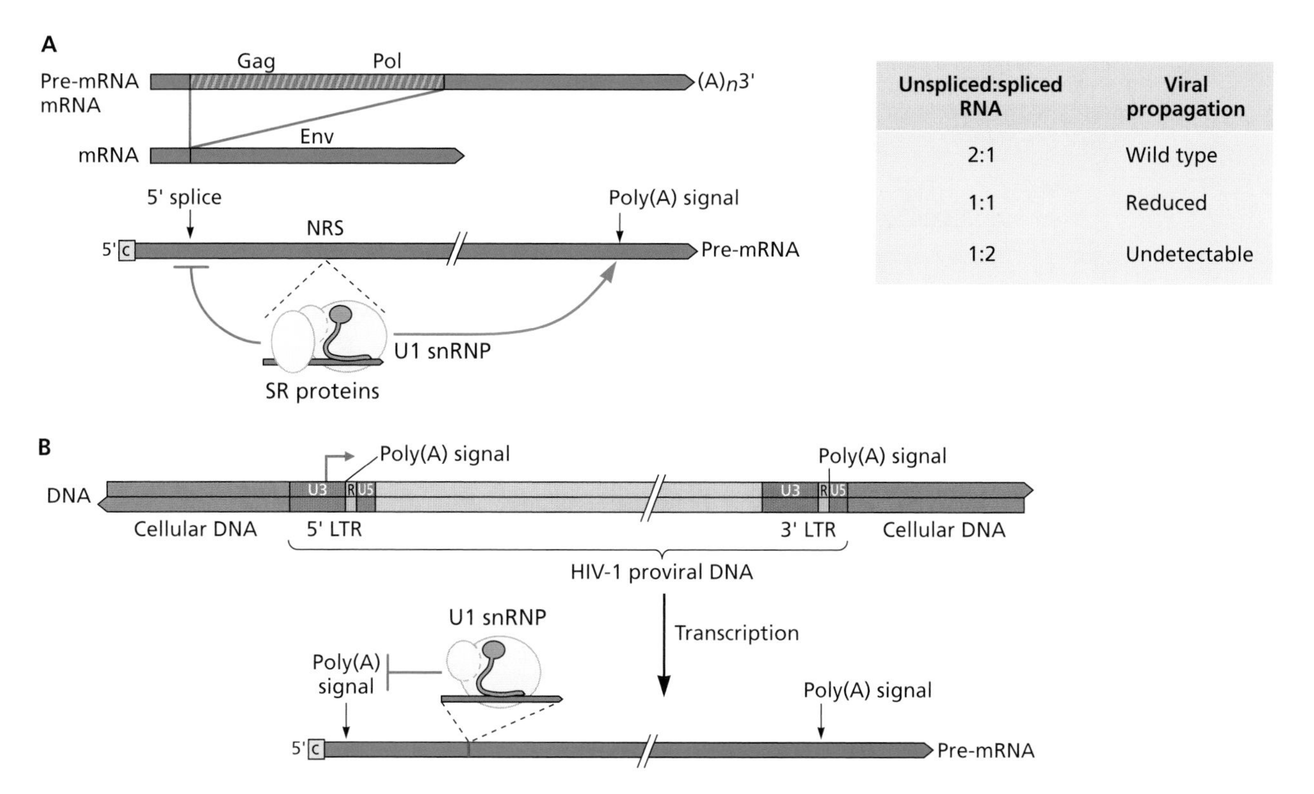

Figure 10.10 Control of RNA-processing reactions during retroviral gene expression. (A) Balanced production of spliced and unspliced mRNAs is illustrated for avian leukosis virus. A single 3′ splice site is recognized in about one-third of the primary transcripts to produce spliced mRNA encoding the Env protein. The Gag and Pol proteins are synthesized from unspliced transcripts. Even a 2-fold reduction in the ratio of unspliced to spliced mRNA impairs virus reproduction (right). Shown below is the negative regulatory sequence (NRS) located within Gag-coding sequences, which is bound by U1 snRNP and SR proteins. This sequence is believed to act as a "decoy" 5′ splice site to inhibit splicing (red bar). It also stimulates polyadenylation (green arrow) by an unknown mechanism. **(B)** Suppression of poly(A) site recognition. Utilization of the 5′ polyadenylation site in primary transcripts of human immunodeficiency virus type 1 proviral DNA is inhibited by binding of U1 snRNP to the major 5′ splice site located 195 nucleotides downstream. The ability of the U1 snRNP protein U1a to bind to both poly(A) polymerase and Cpsf suggests that the U1 snRNP might inhibit their activity.

Temporal Regulation of Synthesis of Adenoviral Major Late mRNAs

The production of adenoviral major late mRNAs epitomizes complex alternative splicing and polyadenylation at multiple sites in a pre-mRNA, which in this case can give rise to at least 15 different mRNAs. These mRNAs fall into five families (L1 to L5) defined by which of five polyadenylation sites is recognized (Fig. 10.11). The frequency with which each site is used must therefore be regulated to allow production of all major late mRNAs. High-efficiency polyadenylation at the L1 site during the early phase of infection prevents synthesis of L2 to L5 mRNAs. In contrast, during the late phase of adenovirus infection, each of the five polyadenylation sites directs 3′-end formation with approximately the same efficiency. The mechanism(s) responsible for such balanced recognition of multiple poly(A) addition sites is not fully understood, but alteration in the activities of cellular polyadenylation proteins as infection proceeds allows the switch from polyadenylation at only the L1 site (see "Posttranscriptional Regulation of Viral or Cellular Gene Expression by Viral Proteins" below).

All major late mRNAs contain the 5′-terminal tripartite leader sequence. The splicing reactions that produce this sequence from three small exons (Box 10.3) take place before polyadenylation of the primary transcript. The final splicing reaction joins the tripartite leader sequence to one of many mRNA sequences (Fig. 10.11). Each primary transcript therefore yields only a **single** mRNA, even though it contains the sequences for many, and most of its sequence is discarded. It remains a mystery why the majority of adenoviral late mRNAs are made by this bizarre mechanism. However, one contributing factor may be that it ensures that each major late mRNA molecule carries the 5′-terminal tripartite leader sequence, which is important for efficient translation late in the infectious cycle (Chapter 11).

Editing of Viral mRNAs

The term **RNA editing** describes the process by which nucleotides not specified in the genome are introduced into mRNAs, first reported in 1980 for a mitochondrial mRNA of

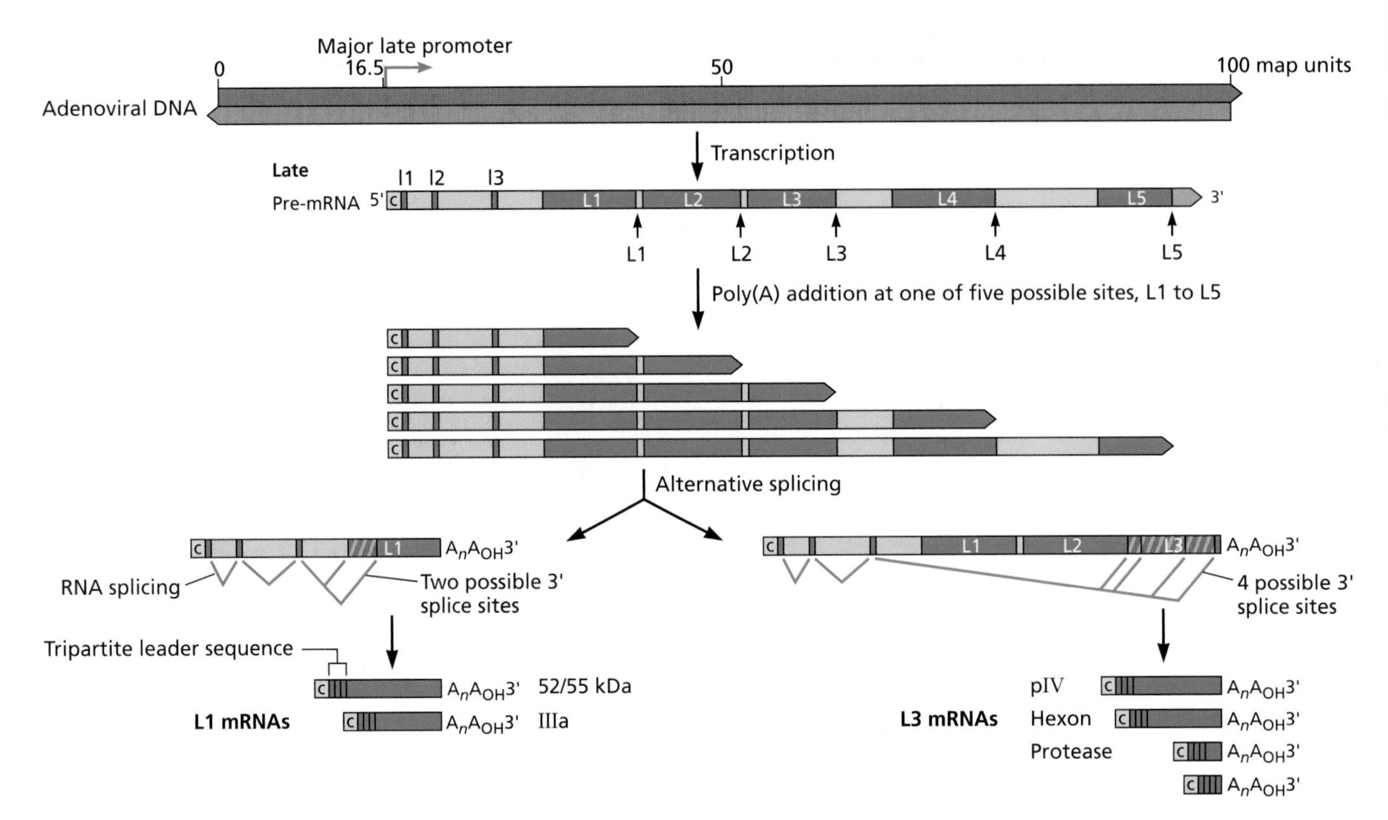

Figure 10.11 Alternative polyadenylation and splicing of adenoviral major late transcripts. During the late phase of adenovirus infection, major late primary transcripts extend from the major late promoter almost to the right end of the genome. They contain the sequences for at least 15 mRNAs and are polyadenylated at one of five sites, L1 to L5, as a result of decreased activity of Cstf (see the text). The tripartite leader sequence, present at the 5′ ends of all late mRNAs, is assembled by the splicing of three short exons, l1, l2, and l3. This sequence is then ligated to alternative 3′ splice sites. Such joining of the spliced tripartite leader sequence to an mRNA sequence has been reported to take place after polyadenylation of pre-mRNA. Polyadenylation therefore appears to determine which 3′ splice sites can be utilized during the final splicing reaction.

trypanosomes. Since this modification was discovered, RNA editing has been identified in many different eukaryotes, as well as in some viral systems.

Viral mRNAs are edited by either insertion of nucleotides not directly specified in the template during synthesis or alteration of a base *in situ*, changing the sequence and function of the protein specified by edited mRNA. Consequently, RNA editing has the potential to make an important contribution to regulation of viral gene expression.

For the most part, viral mRNAs are edited during or following their synthesis from viral RNA genomes. These reactions are described in Chapter 6. However, like their cellular counterparts, transcripts of viral DNA templates can also be edited by Adar1 (adenosine deaminase acting on RNA 1), which deaminates adenine bases to generate inosineI. Transcripts of the K12 region of the human herpesvirus 8 genome are edited efficiently at just one site, both in infected cells and by Adar1 *in vitro*. This modification appears to regulate the function of the kaposin protein that is encoded within the K12 region: this protein exhibits transforming activity only when it is made from the unedited coding sequence, and editing of this site predominates in productively infected cells.

Editing by Adar1 also increases the efficiency of reproduction of human immunodeficiency virus type 1. Edited A residues have been identified at several positions in the viral RNA genome, including the 5′ untranslated region and just downstream of the Rev-response element that directs export of unspliced and partially spliced viral transcripts from the nucleus (see "The Human Immunodeficiency Virus Type 1 Rev Protein Directs Export of Intron-Containing mRNAs" below). Mutational studies have shown that editing of the latter sequence stimulates production of unspliced viral RNA and specific viral proteins, but the consequences of editing of other genomic sequences are not yet known.

Editing as a Powerful Antiviral Defense Mechanism

Adar1 can facilitate reproduction of several viruses by editing-independent inhibition of Pkr, an important component of antiviral defenses mounted in response to exposure of cells to interferon (Volume II, Chapter 3). However, cellular editing enzymes can also inhibit virus reproduction. This phenomenon is exemplified by the enzymes known as Apobec3s, which edit RNA by deamination of cytidine to uridine. Inhibition of the activity of such enzymes is important for the successful reproduction of several viruses, including hepatitis B virus and human immunodeficiency virus type 1 (Volume II, Chapter 7).

Export of RNAs from the Nucleus

Any mRNA made in the nucleus must be transported to the cytoplasm for translation. Other classes of RNA, including small cellular and viral RNAs made by RNA polymerase III, also enter the cytoplasm permanently (e.g., transfer RNAs [tRNAs]) or transiently (snRNAs). The export of viral mRNAs is mediated by the host cell machinery and, in most cases, is indistinguishable from export of analogous cellular RNAs. In this section, we describe the cellular export machinery and the mechanisms that ensure export of some atypical viral mRNA substrates.

The Cellular Export Machinery

The substrates for mRNA export are not naked RNA molecules, but rather ribonucleoproteins. Indeed, export of RNA molecules (with the exception of tRNAs) is directed by sequences present in the proteins associated with them. Like proteins entering the nucleus, RNA molecules travel between nuclear and cytoplasmic compartments via the nuclear pore complexes described in Chapter 5. Numerous genetic, biochemical, and immunocytochemical studies have demonstrated that specific nucleoporins (the proteins from which nuclear pore complexes are built) participate in nuclear export. Export of RNA molecules also shares several mechanistic features with import of proteins into the nuclei: substrates for nuclear export or import are identified by specific protein signals, and some soluble proteins, including the small guanosine nucleotide-binding protein Ran, function in both import and export. And RNA export, like protein import, is mediated by receptors that recognize nuclear export signals and direct the proteins, and ribonucleoproteins that contain them, to and through nuclear pore complexes.

Export of Viral mRNA

All viral mRNAs made in infected cell nuclei carry the same 5′- and 3′-terminal modifications as cellular mRNAs that are exported. Furthermore, many viral mRNAs, like their cellular counterparts, are produced by splicing of intron-containing precursors. Cellular pre-mRNAs that contain introns and splice sites and have not been spliced ordinarily are retained in the nucleus, at least in part because they remain associated with spliceosomes. Furthermore, a protein complex that marks mature mRNAs for export is assembled on the RNA only during splicing, and efficient export requires cooperation among multiple adapter proteins that are deposited as a pre-mRNA is processed. However, reproduction of retroviruses, herpesviruses, and orthomyxoviruses requires production of mRNAs that are not spliced at all, because either the intron is retained or the mRNAs contain no introns or splice sites. Efforts to address the question of how these unusual mRNAs leave the nucleus provided important insights into the molecular mechanisms that mediate export of macromolecules, including the identification of RNA export receptors.

The Human Immunodeficiency Virus Type 1 Rev Protein Directs Export of Intron-Containing mRNAs

The human immunodeficiency virus type 1 Rev protein is the best understood of the viral proteins that modulate mRNA export from the nucleus. This protein and related proteins of other lentiviruses promote export of the unspliced (and partially spliced) viral mRNAs. Rev binds specifically to an RNA sequence termed the **Rev-responsive element** that lies within an alternatively spliced intron of viral pre-mRNA (Fig. 10.12). The Rev-responsive element is some 350 nucleotides in length and forms several stem-loops (Fig. 10.13A), one of which contains a high-affinity binding site for the arginine-rich RNA-binding domain of Rev (Fig. 10.13B). This site is formed by conformational change in the RNA following the initial interaction with Rev. Subsequently, Rev monomers oligomerize cooperatively on the RNA (Fig. 10.13C). Export of RNAs that contain the Rev-responsive element depends on the formation of these RNA-bound oligomers, and a leucine-rich nuclear export signal present in Rev.

When oligomeric Rev is assembled on the RNA, the nuclear export signals of the protein become organized on one surface (Fig. 10.13C). One cellular protein that binds to the nuclear export signal of Rev is exportin-1 (Xpo1, also known as Crm-1). This protein, which binds simultaneously to Rev and the GTP-bound form of Ran, is the **receptor** for Rev-dependent export of the human immunodeficiency virus type 1 RNAs bound to it. The viral protein functions as an **adapter**, directing viral, intron-containing mRNAs to a preexisting cellular export receptor. Translocation of the complex containing viral RNAs, Rev, and cellular proteins through the nuclear pore complex to the cytoplasm requires specific nucleoporins and other proteins (Fig. 10.14). In the cytoplasm, hydrolysis of GTP bound to Ran by a Ran-specific GTPase-activating protein present only in the cytoplasm induces dissociation of the export machinery. Rev then shuttles back into the nucleus via a typical nuclear localization signal, where it can pick up another cargo RNA molecule.

Perhaps the most interesting aspect of Rev-dependent RNA export is the exit of mRNAs by a pathway that normally does not handle such cargo, but rather exports small RNA species (and proteins) of the host cell. The Rev nuclear export

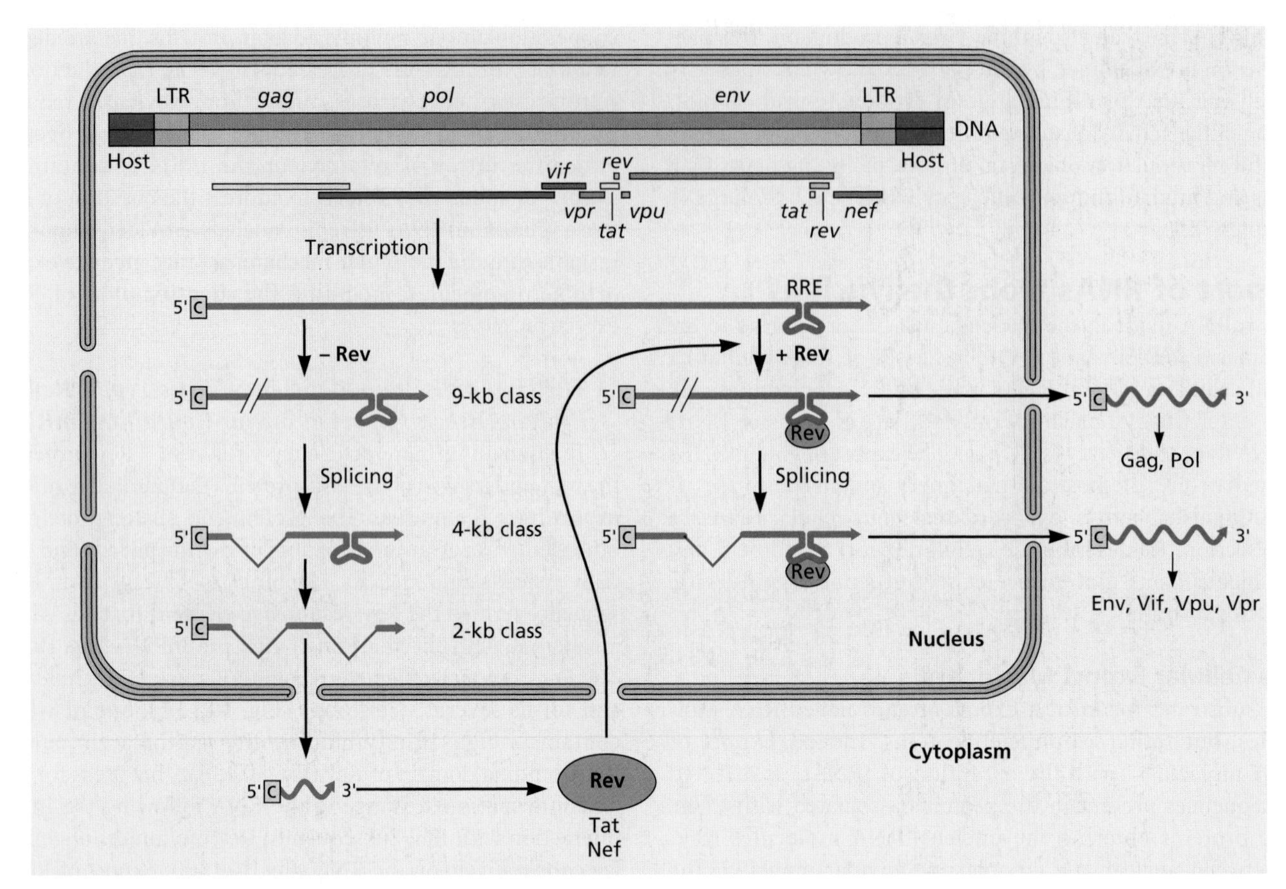

Figure 10.12 Regulation of export of human immunodeficiency virus type 1 mRNAs by the viral Rev protein. Before the synthesis of Rev protein in the infected cell, only fully spliced (2-kb class) viral mRNAs are exported to the cytoplasm (left). These mRNAs specify viral regulatory proteins, including Rev. The Rev protein enters the nucleus, where it binds to an RNA structure, the Rev-responsive element (RRE) present in unspliced (9-kb class) and singly spliced (4-kb class) viral mRNAs. This interaction induces export to the cytoplasm of the RRE-containing mRNAs, from which viral structural proteins and enzymes are made (right). The Rev protein therefore alters the pattern of viral gene expression as the infectious cycle progresses.

signal is similar to, and can be functionally replaced by, that of the cellular protein TfIIIa. This protein binds specifically to 5S rRNA and is required for export of this cellular RNA from the nucleus. Peptides containing the Rev nuclear export signal inhibit export of 5S rRNA (and other small RNAs), but not of mRNAs. The human immunodeficiency virus type 1 Rev protein therefore circumvents the normal restriction on the export of intron-containing pre-mRNAs from the host cell nucleus by diverting such viral mRNAs to a cellular pathway that handles intronless RNAs. Export of unspliced viral RNA via the Xpo1 pathway is required for efficient assembly of virus particles in the cytoplasm, for reasons that are not yet clear.

RNA Signals Can Mediate Export of Intron-Containing Viral mRNAs by Cellular Proteins

The genomes of retroviruses with simple genomes do not encode proteins analogous to Rev, even though unspliced viral RNAs must reach the cytoplasm. These unspliced viral mRNAs contain specific sequences that promote export. Because they must function by means of cellular proteins, such sequences were termed **constitutive transport elements** (CTEs). The first such sequence was found in the 3′ untranslated region of the genome of Mason-Pfizer monkey virus.

Even low concentrations of RNA containing the Mason-Pfizer monkey virus CTE inhibit export of mature mRNAs when microinjected into *Xenopus* oocyte nuclei, but CTE RNA does **not** compete with Rev-dependent export. This observation indicated that this retroviral RNA sequence is recognized by components of a cellular mRNA export pathway. A search for such proteins led to the first identification of a mammalian protein mediating mRNA export, the human nuclear export factor 1 (Nxf1, also known as Tap). This protein binds specifically to the CTE and is essential for export from the nucleus of the unspliced viral RNAs and spliced cellular mRNAs.

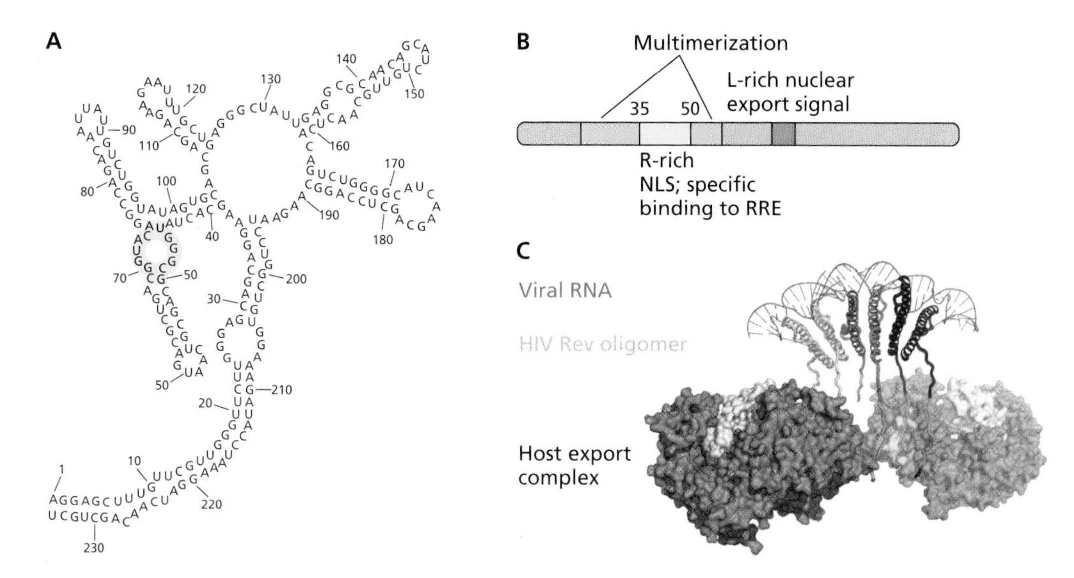

Figure 10.13 Features of the Rev-responsive element and Rev protein. (A) Predicted secondary structure of the 234-nucleotide Rev-responsive element, with the high-affinity binding site for Rev shaded in yellow. **(B)** The functional organization of the Rev protein. **(C)** Model of a Rev hexamer (three Rev dimers: light blue, green, and dark blue) assembled on the RRE, based on the crystal structure of a Rev dimer lacking the disordered C-terminal 46 amino acids and carrying alterations at positions 12 and 60 to prevent aggregation, and the observation that six Rev monomers assemble on the RRE. Modeling of the nuclear export signal-containing C termini not present in the structure indicated that they project away from the RNA-binding domain of Rev. They are shown modeled into the binding sites (cyan) for leucine-rich nuclear export signals in Xpo1 bound to Ran-GTP complex. Adapted from M. D. Daugherty et al., *Nat Struct Mol Biol* **11**:1337–1342, 2010, with permission. Courtesy of A. D. Frankel, University of California, San Francisco.

The pathway of Nxf1-dependent mRNA export has not yet been fully elucidated, but the Ran protein does **not** participate. The direct and specific binding of Nxf1 to the CTE of unspliced retroviral RNAs bypasses a cellular process that ensures that export is normally coupled with transcription and pre-mRNA processing (Fig. 10.15). This protein can bind only nonspecifically and with low affinity to cellular pre-mRNAs, but is recruited by export adapters, including several SR splicing proteins, a subunit of Cpsf, and a protein complex called Trex1 (transcription and export complex 1) that is deposited during splicing. Such indirect recruitment of Nxf1 to an mRNA, which couples export to RNA synthesis and processing, is circumvented in the case of retroviral pre-mRNAs containing CTEs: these unspliced RNAs are recognized directly by Nxf1, allowing their export from the nucleus.

Control of the Balance between Export and Splicing

The relative efficiencies of splicing and export maintain a finely tuned balance in the production of spliced and unspliced retroviral RNAs. This balance is of critical importance as even a 2-fold change prevents viral reproduction (Fig. 10.10A). On one hand, splicing of viral pre-mRNA must be inefficient to allow export of the essential, intron-containing mRNAs. Indeed, increasing the efficiency of splicing of human immunodeficiency virus type 1 pre-mRNA, by replacing the natural, suboptimal splice sites with efficient ones, leads to complete splicing of all pre-mRNA molecules before Rev can recognize and export the unspliced mRNA to the cytoplasm. On the other hand, when unspliced RNAs remain in the nucleus (e.g., before Rev is made in infected cells), they are eventually spliced to completion. Efficient export is therefore required to place unspliced mRNA into the cytoplasm for translation or incorporation into virus particles.

Export of Single-Exon Viral mRNAs

Most of the viral mRNAs made in nuclei of cells infected by hepadnaviruses, herpesviruses, or orthomyxoviruses are not spliced. In contrast to the retroviral mRNAs described in previous sections, these viral mRNAs do not contain introns. Rather, the viral genes that encode them contain no such sequences, and consequently the RNAs **cannot** be spliced. We therefore designate such mRNAs as **single-exon mRNAs** to distinguish them from those that retain introns.

Single-exon mRNAs are rare in uninfected mammalian cells, numbering only a few hundred. The majority encode regulatory proteins, such as signal components of signal transduction pathways and cytokines. Viral and cellular single-exon mRNAs cannot become associated with export adapters during spliceosome assembly and splicing, but nonetheless must be transported efficiently to the cytoplasm. The export of such viral mRNA is promoted by specific RNA sequences or viral proteins, analogous to the retroviral CTE and Rev protein, respectively.

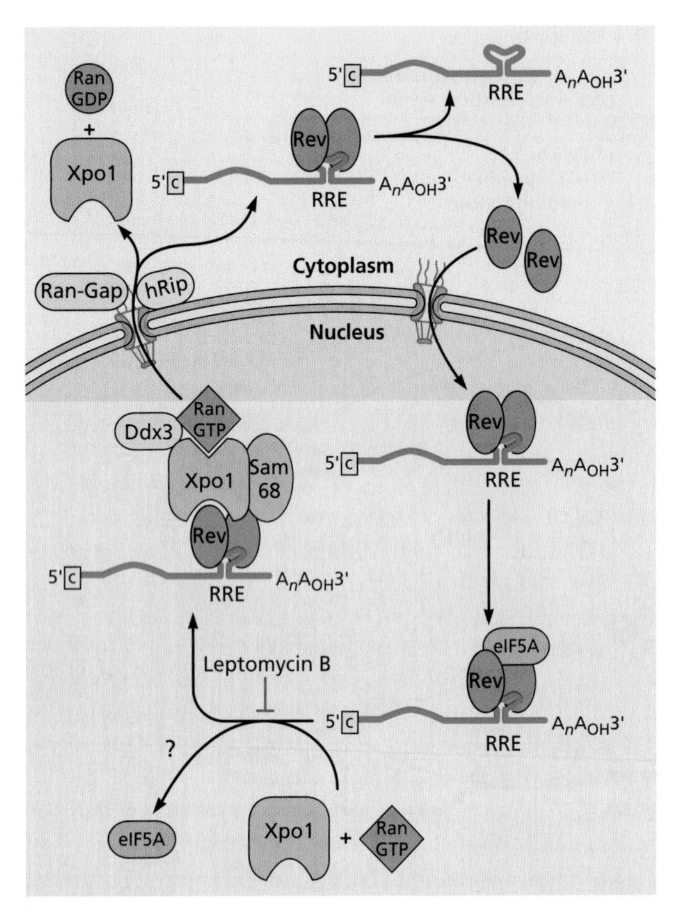

Figure 10.14 Mechanism of Rev protein-dependent export. The cellular nuclear proteins exportin-1 (Xpo1), the GTP-bound form of Ran (Ran-GTP), and the 68-kDa Src-associated protein in mitosis (Sam68) have been implicated in Rev-dependent mRNA export, for example, by analysis of the effects of dominant negative forms of the proteins. In the presence of Ran-GTP, Rev binds to Xpo1. This protein is related to the import receptors described in Chapter 5, and interacts with nucleoporins. The complex containing Rev, Xpo1, and Ran-GTP bound to the Rev-responsive element in RNA is translocated through the nuclear pore complex to the cytoplasm via interactions of Xpo1 with nucleoporins, such as Can/Nup14 and Nup98. Translocation may be facilitated by the action of Ddx3, an ATP-dependent RNA helicase. The Sam68 protein can bind to the Rev nuclear export signal, but does not appear to shuttle between nucleus and cytoplasm. It may therefore act prior to docking of the viral RRE-containing RNA complex at the nuclear pore. The human Rev-interacting protein (hRip) appears to act following translocation, as it is essential for efficient release of Rev-associated RNA into the cytoplasm. Hydrolysis of GTP bound to Ran to GDP induced by the cytoplasmic Ran GTPase-activating protein (Ran-Gap) is presumed to dissociate the export complex, releasing viral RNA for translation or assembly of virus particles, and Ran, Xpo1, and other proteins for reentry into the nucleus.

Specific RNA sequences that promote export of single-exon mRNAs are exemplified by the conserved posttranscriptional regulatory element (PRE) of hepadnaviral mRNAs. The PRE is recognized by export adapters such as components of the Trex1 complex and is analogous to export-promoting sequences subsequently identified in cellular single-exon mRNAs, including those encoding interferon $\alpha 1$ and interferon $\beta 1$. The viral sequence is sufficient to facilitate cytoplasmic accumulation of heterologous mRNAs, and has therefore been included in many vectors for gene expression in mammalian cells (Box 10.7).

Efficient export of herpesviral single-exon mRNAs depends on a viral protein, ICP27 in the case of herpes simplex virus type 1. Like the human immunodeficiency virus type 1 Rev protein, ICP27 shuttles between the nucleus and cytoplasm and binds to viral RNA, in this case via distinct N- and C-terminal RNA-binding domains. Although ICP27 contains a leucine-rich nuclear export signal, it also binds to Nxf1, and various lines of evidence indicate that it serves as a virus-specific adapter for export via the Nxf1 pathway. The influenza A virus NS1 protein, which can bind to both Nxf1 and viral mRNAs, may serve a similar function for export of viral mRNA made later in the infectious cycle.

Posttranscriptional Regulation of Viral or Cellular Gene Expression by Viral Proteins

The genomes of several viruses encode proteins that regulate one or more RNA-processing reactions. These proteins are critical for temporal regulation of viral gene expression, or inhibit the production of cellular mRNAs.

Temporal Control of Viral Gene Expression

Regulation of Alternative Polyadenylation by Viral Proteins

During infection by adenoviruses, herpesviruses, and papillomaviruses, the frequencies of utilization of alternative poly(A) addition sites within specific viral pre-mRNAs change. As discussed previously, the polyadenylation of bovine papillomavirus type 1 late mRNA is activated by a specific complement of cellular proteins found only in fully differentiated cells of the epidermis. In contrast, viral proteins have been implicated in regulation of polyadenylation in cells infected by the larger DNA viruses.

Despite its name, the adenoviral major late promoter is active during the early phase of infection, prior to the onset of viral DNA synthesis. The major late pre-mRNAs made during this period are polyadenylated predominantly at the L1 mRNA site, even though they also contain the L2 and L3 3′ processing sites (Fig. 10.11). Such selective recognition of this polyadenylation signal depends on Cstf, which binds to the U/GU-rich sequence 3′ to the cleavage site (Fig. 10.3). As infection continues, the activity of this cellular protein decreases. It has been shown experimentally that synthesis of the viral L4 33-kDa protein is essential for the switch to the late pattern of gene expression (Box 10.8). It is not yet known whether this protein modulates the activity of Cstf

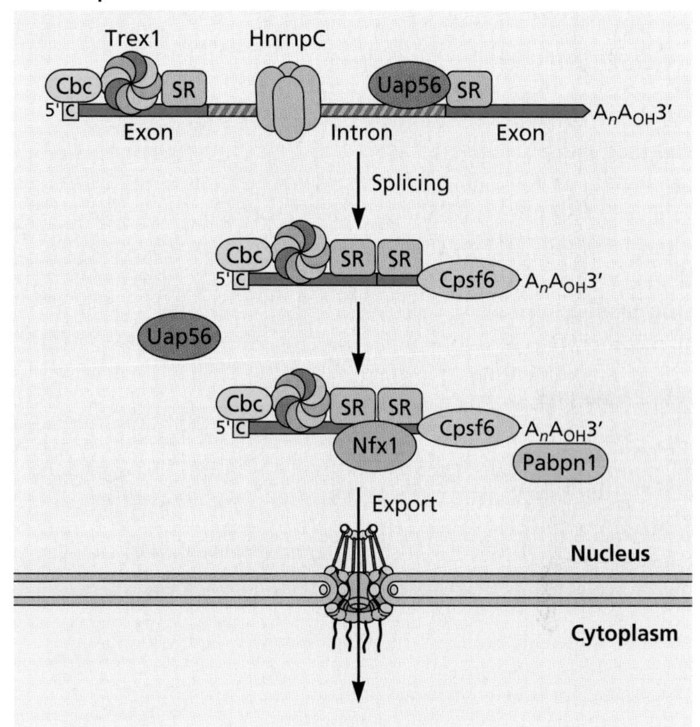

Figure 10.15 Export of unspliced RNA of retroviruses with simple genomes and cellular mRNAs from the nucleus. Export of unspliced, primary transcripts of many retroviruses depends on the constitutive transport element (CTE) in the RNA (left). This sequence is recognized by the cellular Nxf1 subunit of the export receptor dimer Nxf1-Nxt1, which is then bound by proteins that mark mRNAs as appropriate substrates for export, such as Ref. A variety of experimental approaches, including genetic studies in yeast, have indicated that Nxf1 is an essential component of the major pathway for export of cellular mRNAs. However, Nxf1 does not bind to cellular mRNAs with high affinity, but rather becomes associated with them indirectly via interactions with specific proteins, such as several SR proteins. Mature mRNAs are exported from the nucleus as associated with numerous proteins, i.e., as ribonucleoproteins, including those recognized by RNA export receptors. As indicated at the right, such export adapters become associated with mammalian pre-mRNAs, most of which contain introns and must be spliced, as primary transcripts are synthesized and processed. In this way, mRNA export is coupled to transcription and processing reactions. Several SR splicing proteins, the cap-binding complex (Cbc), a subunit of Cpsf, and the multiple-subunit protein Trex1, which is deposited on mammalian pre-mRNAs during splicing, interact with Nxf1. These adapters cooperate to direct efficient export via Nxf1-Nxt1. The export substrate is shown as a compact structure, in which the 5′ and 3′ ends are held in proximity by association of Cbc with the nuclear poly(A)-binding protein Pabn1 (see Chapter 11). Entry into this pathway also requires binding to nascent transcripts made by RNA polymerase II of HnRnpC. This interaction prevents export via the Xpo1 receptor. The direct interaction of Nxf1 with retroviral CTEs (left) therefore bypasses the mechanism(s) that couples splicing of cellular mRNAs with their export. Adapted from M. Müller-McNicoll and K. M. Neugebauer, *Nat Rev Genet* **14:**275–287, 2013, with permission.

or other components of the polyadenylation machinery. The recognition of the other four polyadenylation sites present in major late pre-mRNA synthesized during the late phase (Fig. 10.11) is much less dependent on Cstf. It is therefore likely that these poly(A) addition signals compete more effectively with the L1 site for components of the polyadenylation machinery later in the infectious cycle.

Viral Proteins Can Regulate Alternative Splicing

Some viral proteins that regulate pre-mRNA splicing alter the balance among alternative splicing reactions at specific points in the infectious cycle. For example, the ratios of alternatively spliced mRNA products of several adenoviral pre-mRNAs change with the transition into the late phase of infection. This phenomenon has been studied most extensively using the L1 mRNAs. The L1 pre-mRNA can be spliced at one of two alternative 3′ splice sites. However, only the L1 mRNA that specifies the 52/55-kDa protein is made prior to the onset of viral DNA synthesis, because binding of cellular SR proteins to a negative regulatory sequence located immediately upstream of the branch point for the L1 IIIa mRNA blocks its recognition (Fig. 10.16). Such inhibition is overcome by a viral early protein encoded within the E4 transcription unit, which induces dephosphorylation of the SR proteins. Overproduction of the SR protein Sf2 in adenovirus-infected cells impairs synthesis of the L1 IIIa mRNA, as well as viral reproduction.

BOX 10.7

METHODS

Increasing expression of transgenes in mammalian cells using the woodchuck hepatitis virus posttranscriptional regulatory element

Initial studies demonstrated that efficient expression of genes of the hepadnavirus hepatitis B virus depends on an RNA sequence. This sequence, termed the posttranscriptional regulatory element (PRE), acts in *cis* to allow transport to the cytoplasm of the single-exon viral mRNAs. It can be considered functionally equivalent to an intron: the PRE can stimulate expression of β-globin cDNA, normally very low because of the absence from the transcripts of an intron and splice site, and conversely, inclusion of an intron in the viral surface protein mRNA (in the absence of the PRE) stimulates production of this protein. Subsequently, a similar PRE in the genome of woodchuck hepatitis virus was identified and characterized. This sequence, called WPRE, shares two of its three elements with the hepatitis B virus PRE (see the figure) and is more effective in inducing export to the cytoplasm of a single-exon viral mRNA.

These properties prompted inclusion of the woodchuck hepatitis virus PRE in vectors for expression of transgenes in mammalian cells: to circumvent the large size of many genes of humans (and other mammals), as well as the prevalence of alternative splicing of primary transcripts, such transgenes are typically cDNAs that contain no introns. For example, the presence of WPRE in the 3′ untranslated regions of intronless reporter genes in either retroviral or lentiviral vectors increased synthesis of the reporter proteins 5- to 8-fold. Such stimulation was promoter independent, but was observed only when the WPRE was in the sense orientation.

Donello JE, Loeb JE, Hope TJ. 1998. Woodchuck hepatitis virus contains a tripartite posttranscriptional regulatory element. *J Virol* **72:**5085–5092.

Huang ZM, Yen TS. 1995. Role of the hepatitis B virus posttranscriptional regulatory element in export of intronless transcripts. *Mol Cell Biol* **15:**3864–3869.

Zufferey R, Donello JE, Trono D, Hope TJ. 1999. Woodchuck hepatitis virus posttranscriptional regulatory element enhances expression of transgenes delivered by retroviral vectors. *J Virol* **73:**2886–2892.

(A) The hepdnaviral genome and transcripts are depicted as in the Appendix, Fig. 11, with the position of the PRE indicated. (B) The hepatitis B virus (HBV) and woodchuck hepatitis virus (WHV) PREs are compared. The positions of the reading frames for the polymerase (Pol) and X proteins and the shared α and β subelements of the PREs are indicated. enh I, enhancer I.

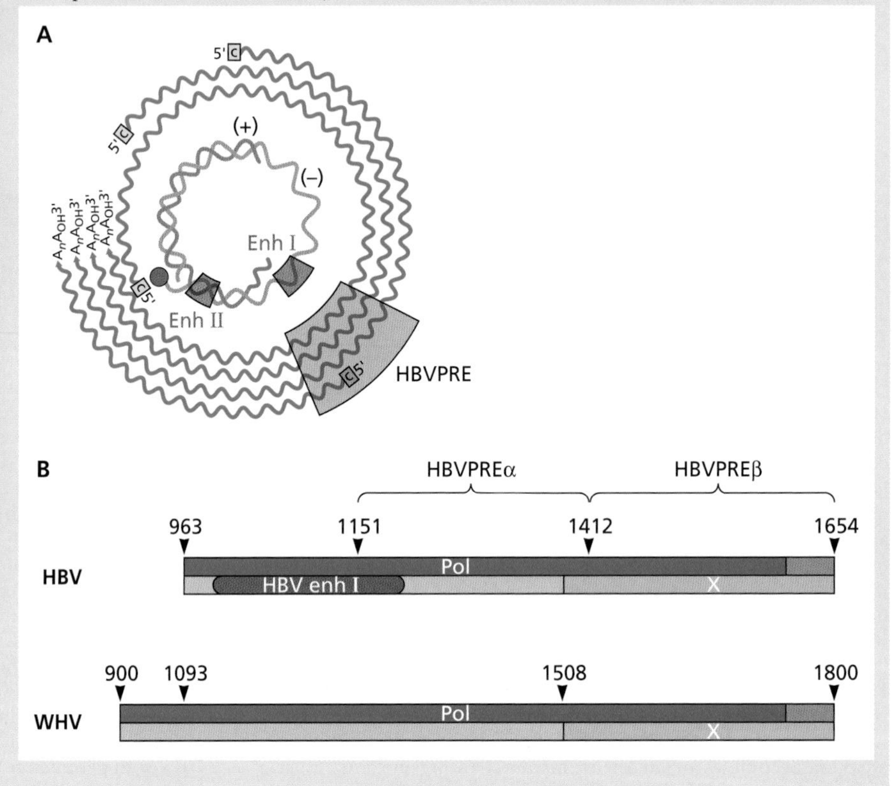

This observation indicates that dephosphorylation of cellular SR proteins makes a major contribution to posttranscriptional regulation of adenoviral gene expression. However, efficient production of the L1 IIIa mRNA depends on a splicing enhancer and the viral L4 33-kDa protein (Fig. 10.16).

Regulation of mRNA Export

Even though all are encoded within a single proviral transcription unit, the regulatory and structural proteins of human immunodeficiency virus type 1 are made sequentially in infected cells, as a result of regulation of mRNA export by the Rev protein: this protein regulates a switch in viral gene expression from early production of viral regulatory proteins to a later phase, in which components of virus particles are made (Fig. 10.12).

Viral proteins that modulate mRNA export may play secondary, but nonetheless crucial, roles in temporal regulation of viral gene expression. For example, the transcriptional program described in Chapter 8 results in efficient transcription of herpes simplex virus late genes only following initiation of viral DNA synthesis in infected cells. Because all but one of the late mRNAs contains a single exon, their entry into the cytoplasm and the synthesis of viral late proteins require ICP27. Consequently, this viral posttranscriptional regulator is essential for putting the viral transcriptional program into effect. Similarly, the complete panoply of adenoviral major

BOX 10.8

EXPERIMENTS

A single adenoviral protein controls the early-to-late switch in major late RNA processing

In efforts to develop cell lines that stably produce adenoviral late proteins, plasmids containing various segments of the major late (ML) transcription unit under the control of an inducible promoter were introduced into human cells. As summarized in the figure, the plasmid ML1-5 supported very efficient expression of all the ML-coding sequences and synthesis of the full set of the ML proteins. In contrast, **only** the L1 52/55-kDa protein was synthesized efficiently in cells containing the plasmid carrying the L1, L2, and L3 sequences. Examination of the cytoplasmic concentrations of processed ML mRNAs showed that only the L1 52/55-kDa mRNA was made efficiently in cells containing this truncated plasmid, as is also the case during the early phase of infection. These observations implied that one or more viral proteins encoded in the L4 or L5 region induce the early-to-late switch in processing of ML pre-mRNA. In fact, synthesis of the L4 33-kDa protein in cells containing the ML1-3 plasmid allowed production of the L1 IIIa and the L2 and L3 proteins. This viral protein stimulated the synthesis of fully processed hexon mRNA, but did not alter the nuclear concentration of the pre-mRNA. It was therefore concluded that the L4 33-kDa protein is necessary and sufficient to switch processing of the ML pre-mRNA from the early to the late pattern.

The subsequent discovery of a promoter that directs transcription of the L4 region of the adenoviral genome solved the puzzle of how the major late-encoded L4 33-kDa protein became available to induce the late pattern of viral gene expression.

Farley DC, Brown JL, Leppard KN. 2004. Activation of the early-late switch in adenovirus type 5 major late transcription unit expression by L4 gene products. *J Virol* **78:**1782–1791.

Wright J, Leppard KN. 2013. The human adenovirus 5 L4 promoter is activated by cellular stress response protein p53. *J Virol* **87:**11617–11625.

The major late (ML) coding regions (L1 to L5) of the adenovirus type 5 genome are shown to scale at the top, with the regions in the ML1-3 and ML1-5 plasmids introduced into human cells shown below. 1, 2, and 3 indicate the positions of the three segments of the tripartite leader sequence. The proteins made in cells containing these plasmids, and the ML1-3 plasmid plus a vector directing synthesis of the viral L4 33-kDa protein, are indicated below.

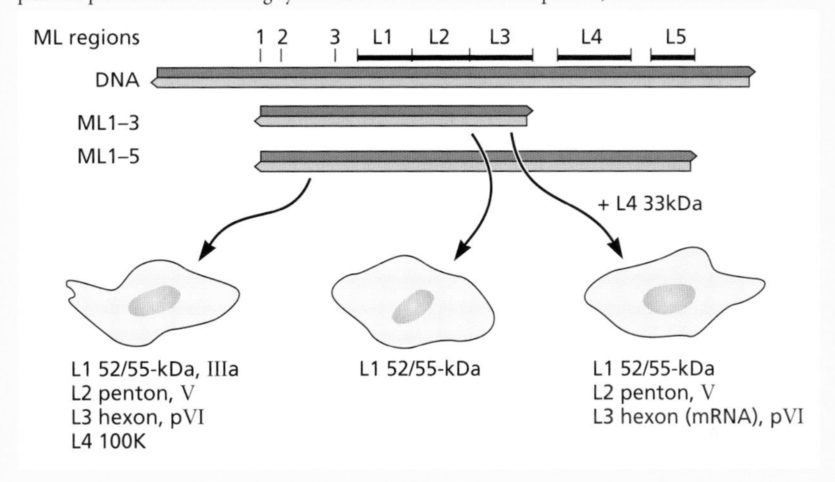

late gene products can be produced only when the viral L4 33-kDa protein induces the switch to the late pattern of processing of these pre-mRNAs (Box 10.8).

Viral Proteins Can Inhibit Cellular mRNA Production

All viral mRNAs are translated by the protein synthesis machinery of the host cell. Inhibition of production of cellular mRNAs can therefore favor this essential step in viral replication. Several mechanisms of selective inhibition of cellular RNA processing operate in virus-infected cells (Fig. 10.17).

Inhibition of Polyadenylation and Splicing

The influenza virus NS1 protein can inhibit both polyadenylation and splicing of cellular pre-mRNAs. A C-terminal segment of this viral protein is required for inhibition of polyadenylation and contains binding sites for both Cpsf and poly(A)-binding protein II (PabII) (Fig. 10.4). Its interaction with Cpsf inhibits polyadenylation of cellular mRNAs in experimental systems. When NS1 is not made, infected cells produce larger quantities of cellular mRNAs that encode interferons and other proteins with antiviral activities. Inhibition of processing of such cellular mRNAs may therefore contribute to the circumvention of host cellular defenses, a critical function of the NS1 protein (Volume II, Chapter 3).

In addition to its other activities, herpes simplex virus ICP27 inhibits splicing of cellular pre-mRNAs. This protein inhibits splicing in *in vitro* reactions, probably because its direct interaction with components of the spliceosome blocks an early step in spliceosome assembly. Genetic analyses have shown that disruption of cellular RNA processing by ICP27 leads to inhibition of cellular protein synthesis, and that this function is distinct from the requirement for the protein for efficient production of viral late mRNAs. Because herpesviral genes generally lack introns, inhibition of splicing is an effective strategy for the selective inhibition of cellular gene expression.

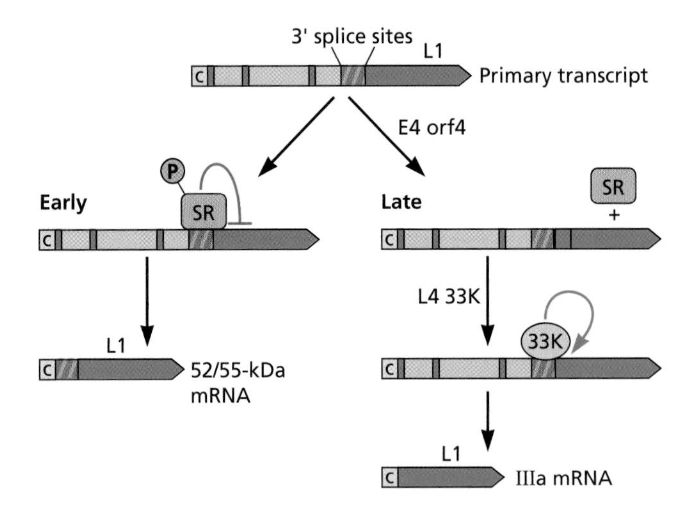

Figure 10.16 Regulation of alternative splicing of adenoviral major late L1 pre-mRNA. The polyadenylated L1 pre-mRNA contains alternative 3′ splice sites, for the 52/55-kDa protein and protein IIIa. (Left) During the early phase of infection, only the 3′ splice site for the 52/55-kDa protein is utilized, because binding of SR proteins to the pre-mRNA blocks recognition of the 3′ splice site for production of the mRNA for protein IIIa. (Right) An E4 protein induces dephosphorylation of these cellular proteins by protein phosphatase 2. This modification inhibits binding of the SR proteins to the pre-mRNA. However, efficient utilization of the IIIa mRNA 3′ splice site (during the late phase) requires the viral L4 33-kDa protein, which activates splicing via an infected cell-specific splicing enhancer. This L4 protein also stimulates splicing at other suboptimal 3′ splice sites in major late pre-mRNAs, such as those that produce the L2 mRNAs for proteins V and pre-VII.

Inhibition of Cellular mRNA Export

To facilitate production of viral mRNAs. In contrast to the other viruses considered in this section, adenovirus infection disrupts cellular gene expression by inhibition of export of cellular mRNAs from the nucleus. Synthesis and processing of cellular pre-mRNAs are unaffected, but these RNAs are not exported and are degraded within the nucleus (Fig. 10.17). Consequently, during the late phase of infection, the great majority of newly synthesized mRNAs entering the cytoplasm are viral in origin. When selective viral mRNA export is prevented by mutations in the viral genome, both the quantities of late proteins made in infected cells and virus yield are reduced substantially. These same phenotypes are seen in herpes simplex virus-infected cells when the ICP27 protein is defective for the inhibition of pre-mRNA splicing. These properties emphasize the importance of posttranscriptional inhibition of cellular mRNA production for efficient virus reproduction.

The preferential export of late mRNAs in adenovirus-infected cells requires two viral early proteins, the E1B 55-kDa and E4 Orf6 proteins, which associate with one another and with proteins present in cellular E3 ubiquitin ligases to form a virus-specific enzyme. E3 ubiquitin ligases typically add chains of ubiquitin to mark proteins for degradation by the proteasome. Although assembly of the adenovirus-specific enzyme is required, it is not known how it regulates mRNA export.

The selectivity of mRNA export in adenovirus-infected cells is especially puzzling, because the viral mRNAs possess all the characteristic features of cellular mRNAs, are made in the same way, and are exported via the Nxf1 pathway. One hypothesis is that the viral E1B-E4 protein complex recruits nuclear proteins needed for export of mRNA to the specialized sites within the nucleus at which the adenoviral genome is replicated and transcribed. As a result of such sequestration, viral mRNAs would be exported preferentially.

To block antiviral responses. The genomes of many RNA viruses encode all the enzymes necessary for synthesis and processing of viral mRNAs in the cytoplasm. Infection by some of these viruses results in inhibition of export of cellular RNAs from the nucleus. This response can reduce competition of cellular with viral mRNAs for components of the translational machinery. However, such inhibition can also facilitate virus reproduction indirectly, by impairing host antiviral responses, as observed in cells infected by rhabdoviruses and picornaviruses.

The vesicular stomatitis virus M protein inhibits export of cellular mRNAs (as well as small RNAs) by binding to a cellular nucleoporin (Nup98) and the cellular export protein Rae1, which normally shuttles between the nucleus and cytoplasm and binds to Nxf1 (Fig. 10.17). The consequent disruption of cellular mRNA export probably reduces host cell protein synthesis. This response also appears to block an important antiviral defense, as expression of the cellular Rae1 and Nup98 genes is induced by interferon, a potent antiviral cytokine (see Volume II, Chapter 3). The observation that specific alterations in the M protein that prevent inhibition of export of interferon β mRNA reduce viral reproduction is consistent with this interpretation.

Picornaviruses also disrupt trafficking from the nucleus to the cytoplasm. The poliovirus 2A protease induces relocation of particular nuclear proteins to the cytoplasm (Fig. 10.17). Such redistribution correlates with loss of structure from the central channel of the nuclear pore, and cleavage of specific nucleoporins (e.g., Nup153). The small leader (L) protein of encephalomyocarditis virus, a member of the cardiovirus group within the *Picornaviridae*, binds to Ran-GTPase, an essential component of Ran-dependent nuclear export and import pathways, and induces hyperphosphorylation of several of the nucleoporins that are cleaved by the poliovirus 2A protease. The phenotypes of mutants with deletions in the L gene suggest that inhibition of trafficking between the nucleus and cytoplasm both tempers the interferon antiviral response and contributes to inhibition of cellular protein synthesis.

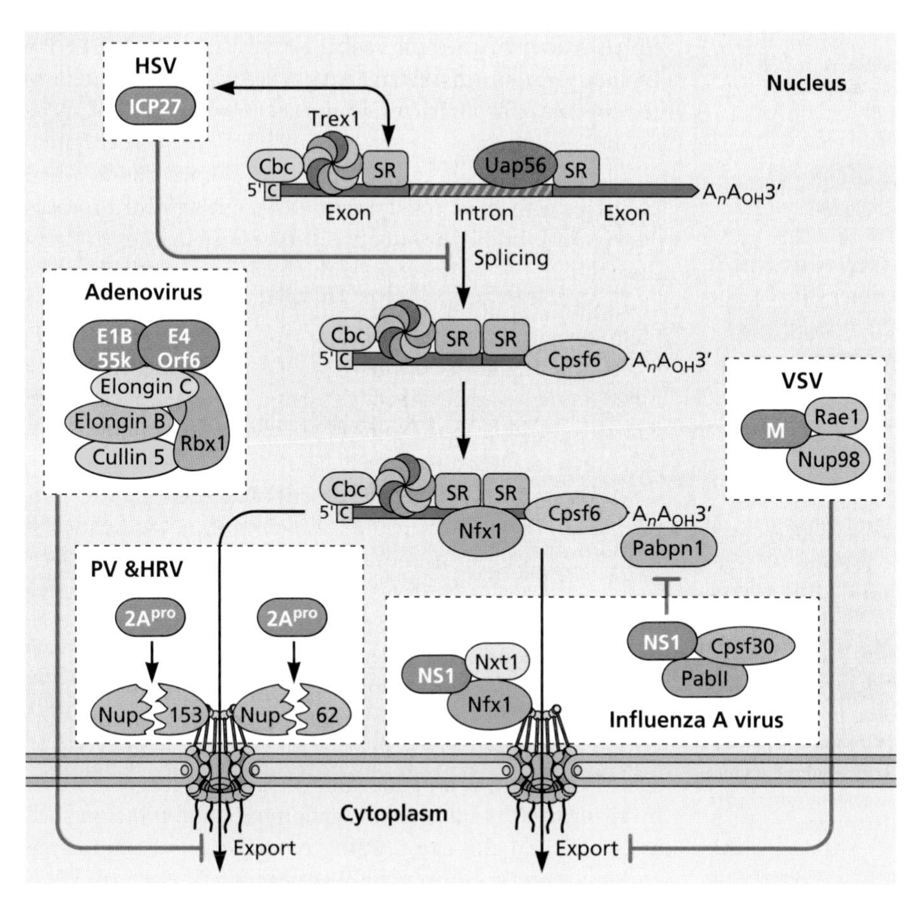

Figure 10.17 Inhibition of cellular pre-mRNA processing by viral proteins. The integration of synthesis and processing of cellular pre-mRNA with export of the mature mRNA to the cytoplasm is depicted as in Fig. 10.15 (right). Some viral proteins inhibit splicing (herpes simplex virus type 1 [HSV-1] ICP27), polyadenylation (influenza A virus NS1), or export (vesicular stomatitis virus [VSV] M and influenza A virus NS1) by interaction with the cellular proteins required for these processes. Inhibition of export of processed cellular mRNAs in cells infected by human adenovirus depends on assembly of the virus-specific E3 ubiquitin ligase that contains the viral E1B 55-kDa and E4 Orf6 proteins, but the mechanism of inhibition has not been elucidated. Export from the nucleus is inhibited in cells infected by picornaviruses as a result of degradation of specific nucleoporins by the viral 2A protease ($2A^{Pro}$), poliovirus (PV), and human rhinovirus (HRV) or hyperphosphorylation of these same nucleoproteins induced by the leader (L) protein of encephalomyocarditis virus (not shown).

Regulation of Turnover of Viral and Cellular mRNAs in the Cytoplasm

Individual mRNAs may differ in the rate at which they are translated, and also in such properties as cytoplasmic location and stability. Indeed, the intrinsic lifetime of an mRNA can be a critical parameter in the regulation of gene expression.

In the cytoplasm of mammalian cells, the lifetimes of specific mRNAs can differ by as much as 100-fold. This property is described in terms of the time required for 50% of the mRNA population to be degraded under conditions in which replenishment of the cytoplasmic pool is blocked, the **half-life** of the mRNA. Many mRNAs are very stable, with half-lives exceeding 12 h. As might be anticipated, these mRNAs encode proteins needed in large quantities throughout the lifetimes of all cells, such as ribosomal proteins. At the other extreme are unstable mRNAs with half-lives of <30 min. This class includes mRNAs specifying regulatory proteins that are synthesized in a strictly controlled manner in response to cues from external or internal environments of the cell, such as cytokines, and proteins that regulate cell cycle progression. The short lifetimes of these mRNAs ensure that synthesis of their products can be shut down effectively once they are no longer needed. Specific sequences that signal the rapid turnover of the mRNAs in which they reside have been identified, such as a 50- to 100-nucleotide AU-rich sequence within the 3′ untranslated region. Mammalian mRNAs are degraded by the pathways summarized in Fig. 10.18, in which deadenylation of the 3′ end of the mRNA triggers either removal of the 5′ cap for degradation by the 5′ → 3′ exoribonuclease Xrn1 or 3′ → 5′ degradation by the conserved, multiprotein exosome. These reactions take place in dynamic cytoplasmic foci, termed P (processing) bodies, that are enriched in proteins that mediate mRNA degradation or inhibit translation, mRNAs that are translationally silent (often deadenylated), and micro-RNAs (see next section).

The stabilities of viral mRNAs have not been examined in much detail, in part because many viral infectious cycles are completed within the normal range of mRNA half-lives and many viral mRNAs carry the 5′ caps and 3′ poly(A) tails that protect against degradation. When these features are absent, the 5′ and/or 3′ ends of viral mRNAs often form structures that block exonucleolytic attack,

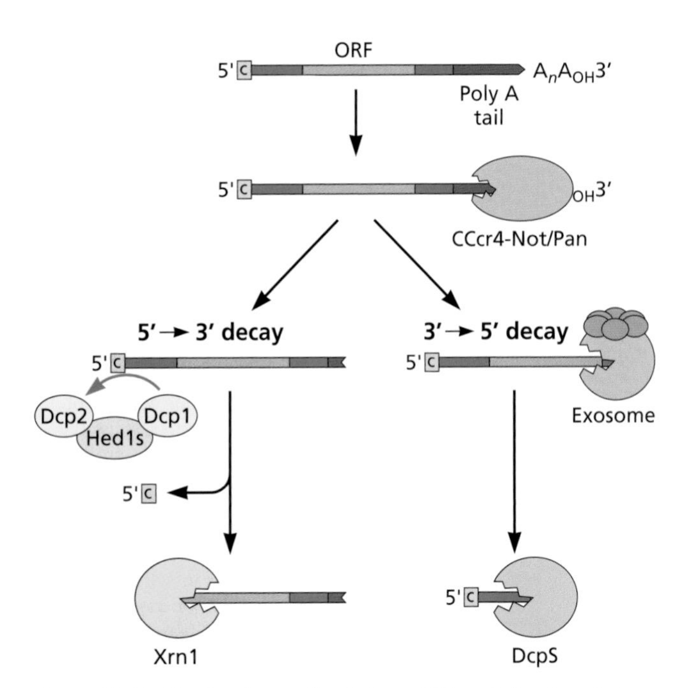

Figure 10.18 Mechanisms of cellular mRNAs of deadenylation-dependent degradation. A major pathway of mRNA degradation in the cytoplasm depends on initial deadenylation. In the case of short-lived mRNAs, this process can be initiated by binding of specific proteins to 5′AUUUUA3′ sequences. Regardless, shortening of the poly(A) tail is a two-step process catalyzed by different deadenylases, the Pan and Ccr4-Not complexes. In subsequent 5′ → 3′ decay (left), shortening of the poly(A) tail to <110 nucleotides triggers decapping by the enzyme-decapping protein 2 (Dcp2). This reaction is stimulated by Dcp1, which interacts with Dcp2 via the Hedls (human enhancer of decapping large subunit) protein. The exact way in which the decapping enzyme is recruited to target mRNAs, and the contribution of shortening of the poly(A) tail, are not fully understood. The decapped mRNA is then degraded by 5′ → 3′ exonucleases, such as Xrn1. Alternatively (right), the deadenylated mRNA can be degraded in the 3′ → 5′ direction by the exosome, which contains multiple exo- and endonucleases, including a processive 3′ → 5′ exoribonuclease, and decapping by Dcp5. Adapted from V. K. Nagarajan et al., *Biochim Biophys Acta* **1829:** 590–603, 2013, with permission.

such as stem-loop structures in the 3′ untranslated regions of arenavirus, bunyavirus, and flavivirus mRNAs. Viral RNAs can also include binding sites for cellular proteins that increase their stability. Viral proteins that induce RNA degradation make an important contribution to selective expression of viral genes in cells infected by large DNA viruses. Regulation of the stability of specific viral or cellular mRNAs has also been implicated in the permanent changes in cell growth properties (transformation) induced by some viruses. Furthermore, RNA-mediated induction of degradation of specific mRNAs, a widespread phenomenon known as RNA interference, is thought to contribute to host antiviral defense mechanisms, and degradation of genomic RNA even contributes to the pathogenesis of flaviviruses (Box 10.9).

Regulation of mRNA Stability by Viral Proteins

Cellular proteins that participate in mRNA degradation are removed or relocalized in cells infected by several viruses. For example, subunits of the deadenylation (Pan3) and decapping (Dcp1a) enzymes and Xrn1 (Fig. 10.18) are degraded in poliovirus-infected cells, most likely by the viral protease $3C^{pro}$, and P bodies are disrupted. These structures are also dismantled in cells infected by the flavivirus West Nile virus, and by the adenovirus E4 Orf3 protein. How these changes facilitate virus reproduction has not yet been established, although destruction of the 5′ → 3′ exonuclease might stabilize poliovirus mRNA, which lacks a protective 5′ cap (see Chapter 6). The genomes of other viruses encode proteins that accelerate RNA decay.

The first such protein to be described, the virion host shutoff protein (Vhs) of herpes simplex virus type 1, reduces the stability of mRNAs in infected cells. As its name implies, Vhs is a structural protein: it is present at low concentrations in the tegument and hence delivered to infected cells at the start of the infectious cycle. It remains in the cytoplasm, where it mediates degradation of some cellular mRNAs to facilitate viral gene expression, presumably by reducing or eliminating competition from cellular mRNAs during translation. The Vhs protein is an endoribonuclease that targets mRNA by virtue of its binding to translation initiation proteins, such as eIF4H and the cap-binding complex eIF4F. Following endonucleolytic cleavage by Vhs near the 5′ end, mRNA is degraded by Xrn1. Although recruited to mRNAs by different mechanisms, the human herpesvirus 8 SOX and severe acute respiratory syndrome coronavirus nsp1 proteins also induce cleavage of mRNA to allow exonucleolytic degradation (Fig. 10.19).

Vhs cannot distinguish viral mRNAs from their cellular counterparts, and induces degradation of both. Although more Vhs protein is made in infected cells once its coding sequence is expressed during the late phase of infection, the protein is sequestered in the tegument of assembling virus particles by interaction with the viral VP16 protein. As a result, the activity of Vhs decreases as the infection cycle progresses. This mechanism presumably contributes to the efficient synthesis of viral proteins characteristic of the late phase of infection. In contrast, the coronavirus nsp1 protein induces selective degradation of cellular mRNA, because viral mRNAs are protected from endonucleolytic cleavage by the common leader sequence present at their 5′ ends (Chapter 6).

The genomes of poxviruses, such as vaccinia virus, also contain the coding sequence for enzymes that induce degradation of viral and cellular mRNAs. These proteins, D9 and D10, are not, however, RNases, but rather decapping enzymes that share a motif with their cellular counterpart and hydrolyze the cap to release m^7GDP (Fig. 10.19). It is clear from the results of genetic experiments that the D10

BOX 10.9

DISCUSSION

Coopting a cellular mechanism of RNA degradation for viral pathogenesis

The family *Flaviviridae* includes important agents of human disease, many of which are spread by arthropod vectors, notably yellow fever virus, West Nile virus, Japanese encephalitis virus, and dengue virus. In infected cells, the (+) strand RNA genome is not only translated and replicated, but also serves as the precursor for subgenomic flaviviral RNAs (sfRNAs) 300 to 500 nucleotides in length. These RNAs correspond to most of the 3′ untranslated regions of the (+) strand RNAs and are produced by incomplete degradation of the full-length (+) strand RNA by the cellular 5′ → 3′ exonuclease Xrn1 (following removal of the cap by an unknown mechanism) (panel A of the figure). The introduction of mutations that prevent production of sfRNAs (by disrupting the RNA structures that block the progress of exonuclease described below) reduced the ability of West Nile virus to kill cells in culture. Such mutant viruses also failed to induce encephalitis, and death, in young mice. The essential role of sfRNAs in pathogenesis focused attention on the features of the 3′ untranslated regions of the genomic RNA that confer resistance to Xrn1.

A combination of phylogenetic studies, *in silico* prediction, and experimental analysis of RNA structure using chemical probes indicated that the 3′ untranslated regions of these flaviviruses are rich in secondary (and higher-order) structures (panel B of the figure). When present in short model RNAs, each of the stem-loops at the 5′ ends of the 3′ untranslated regions was shown to be resistant to Xrn1 digestion *in vitro*. They were also required for accumulation of specific sfRNAs in infected cells.

The tertiary structure of the first stem-loop inferred from the results of mutagenesis and RNA-folding experiments was confirmed by X-ray crystallography. This structure contains a three-way junction, and two of its three RNA helices form a ring-like structure with the 5′ end of the RNA passing through its center and "tied" in place by base pairing and other interactions with bases in the ring (panel B). Simple unwinding of RNA helices could not make the 5′ end of the RNA accessible, explaining how this structure blocks degradation by Xrn1, which proceeds in the 5′ → 3′ direction.

Because they are relatively short and highly conserved among the arthropod-borne flaviviruses, it has been suggested that it might be possible to develop therapeutic agents that target the structures that confer Xrn1 resistance,. Furthermore, mutations that prevent formation of sfRNAs should attenuate these viruses and may facilitate development of vaccines.

Chapman EG, Costantino DA, Rabe JL, Moon SL, Wilusz J, Nix JC, Kieft JS. 2014. The structural basis of pathogenic subgenomic flavivirus RNA (sfRNA) production. *Science* **344:**307–310.

Chapman EG, Moon SL, Wilusz J, Kieft JS. 2014. RNA structures that resist degradation by Xrn1 produce a pathogenic Dengue virus RNA. *eLife* **3:**e01892. doi:10.7554/eLife.01892.

Pijlman GP, Funk A, Kondratieva N, Leung J, Torres S, van der Aa L, Liu WJ, Palmenberg AC, Shi PY, Hall RA, Khromykh AA. 2008. A highly structured, nuclease-resistant, noncoding RNA produced by flaviviruses is required for pathogenicity. *Cell Host Microbe* **11:**579–591.

(A) Model for production of sfRNAs. (**B**) Organization and sequence of conserved RNA stem-loop 1, in this case of Murray Valley encephalitis virus, and structure (right) determined by X-ray crystallography at 2.5-Å resolution. In the structural model, the different elements are color coded as in the secondary structure representation shown at the left. Note the interaction between bases at the 5′ end of the RNA and those present in helices P1 and P3. ORF, open reading frame; UTR, untranslated region. Adapted from E. G. Chapman et al., *Science* **344:**307–310, 2014, with permission. Courtesy of J. S. Kieft, University of Colorado, Denver.

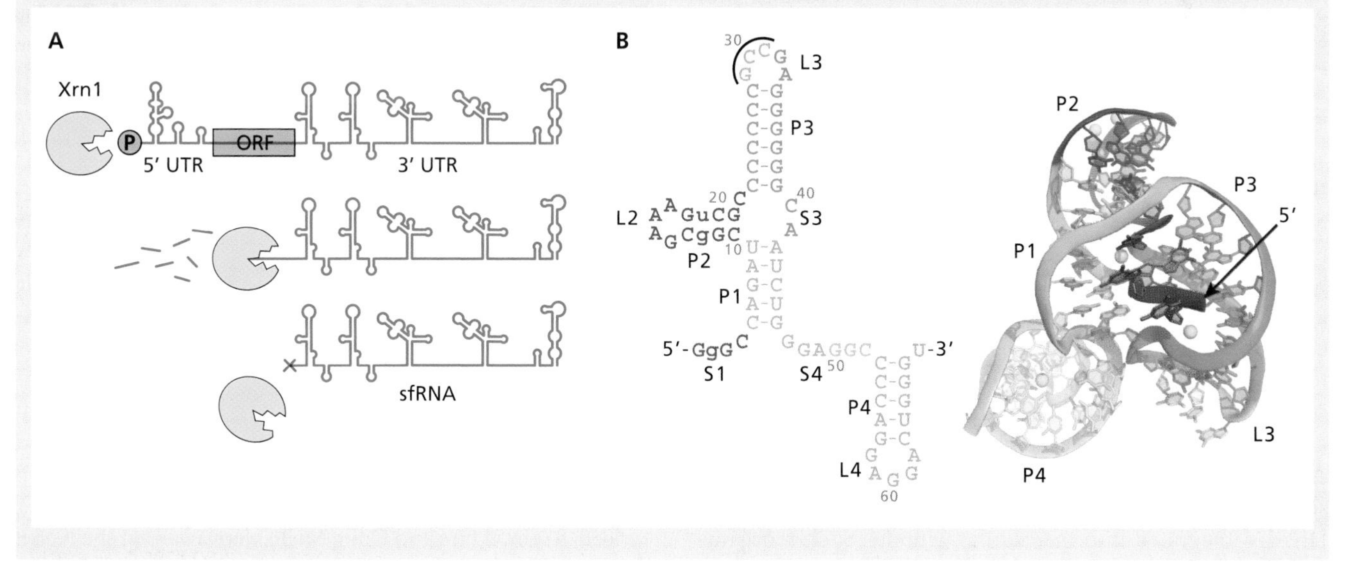

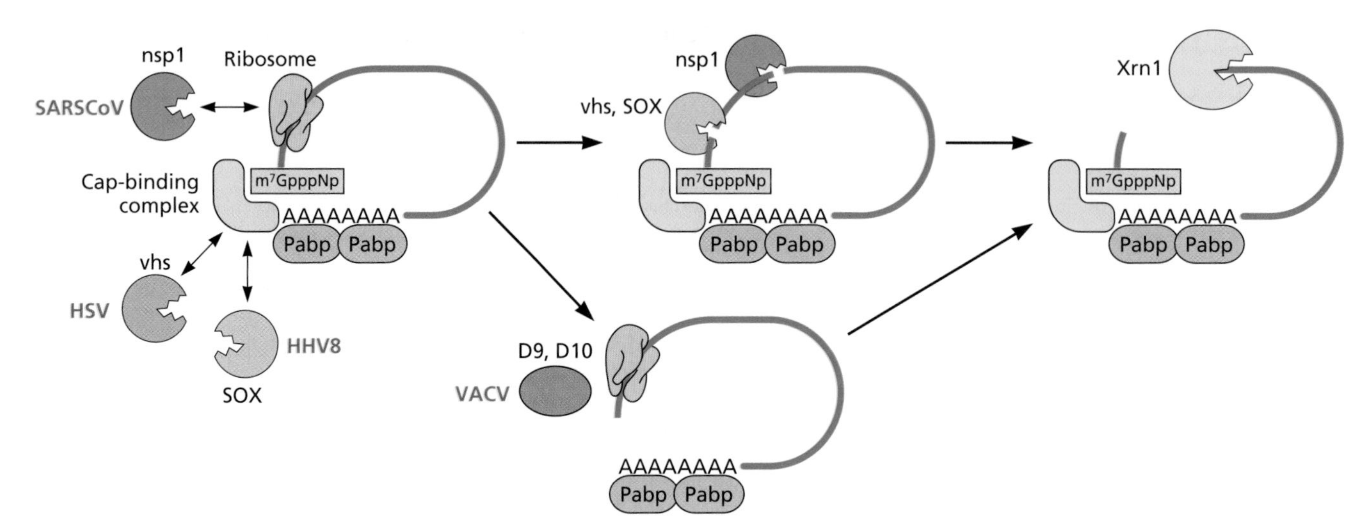

Figure 10.19 Viral proteins initiate mRNA degradation by different mechanisms. The genomes of alpha- and gammaherpesviruses encode endonucleases that initiate degradation of mRNA in infected cells, exemplified by the herpes simplex virus type 1 (HSV-1) vhs and human herpesvirus 8 (HHV8) SOX proteins. However, the former is a nuclease of the Fen1 family, whereas SOX is related to members of a different endonuclease family. Vhs is recruited to mRNA by interaction with the cap-binding complex, prior to ribosome binding. SOX also associates with mRNA before the ribosome, but the cellular components with which it interacts have not yet been identified. These viral proteins cleave the mRNA, within the 5′ untranslated region in the case of vhs, to allow subsequent 5′ → 3′ exonucleolytic degradation by Xrn1. The severe acute respiratory syndrome coronavirus (SARSCoV) nsp1, which is not obviously related to any viral or cellular nuclease, functions to cleave mRNA in a similar manner, but targets this substrate by interaction with the 40S ribosomal subunit. In contrast, the D9 and D10 proteins of the poxvirus vaccinia virus (VACV) share sequence motifs with cellular decapping enzymes and remove this 5′ protective structure to initiate mRNA degradation. Pabp, poly(A)-binding protein. Adapted from M. M. Gaglia et al., *J Virol* **86:**9527–9530, 2012.

protein induces rapid turnover of viral and cellular mRNAs, and hence facilitates inhibition of cellular protein synthesis in infected cells. It has been suggested that turnover of viral mRNAs may facilitate the production of specific sets of viral proteins during the successive phases of the infectious cycle, with the D9 and D10 enzymes acting early and late in infection, respectively.

mRNA Stabilization Can Facilitate Transformation

Stabilization of specific viral mRNAs appears to be important in the development of cervical carcinoma associated with infection by high-risk human papillomaviruses, such as types 16 and 18. The E6 and E7 proteins of these viruses induce abnormal cell proliferation (Volume II, Chapter 6). In benign lesions, the circular human papillomavirus genome is not integrated. The E6 and E7 mRNAs that are synthesized from such templates contain destabilizing, AU-rich sequences in their 3′ untranslated regions and possess short half-lives. In cervical carcinoma cells, the viral DNA is integrated into the cellular genome. Such reorganization of viral DNA frequently disrupts the sequences encoding the E6 and E7 mRNAs, such that their 3′ untranslated regions are copied from cellular DNA sequences (Fig. 10.20). These hybrid mRNAs therefore lack the destabilizing AU-rich sequences and are more stable. The increase in the stability of the viral mRNAs accounts, at least in part, for the higher concentrations of the papillomaviral transforming proteins in tumor cells.

Production and Function of Small RNAs That Inhibit Gene Expression

Small Interfering RNAs, Micro-RNAs, and Their Synthesis

In the early 1990s, attempts to produce more vividly purple petunias by creation of transgenic plants carrying an additional copy of the gene for the enzyme that makes the purple pigment often resulted in white flowers. It is now clear that this seemingly esoteric observation represented the first example of a previously unknown mechanism of post-transcriptional regulation of gene expression, called **RNA interference** or RNA silencing. We now know that RNA-based silencing of gene expression is widespread and ancient (Box 10.10). Our understanding of the mechanisms and functions of RNA interference, as well as its exploitation as an experimental tool, has advanced at a remarkably rapid pace. Indeed, Andrew Fire and Craig Mello were awarded the Nobel Prize in physiology or medicine in 2006, just 8 years after the publication of their groundbreaking study of the mechanism of RNA interference.

RNA interference is mediated by small RNA molecules (typically 19 to 25 nucleotides in length) that function in antiviral defense or regulate gene expression. The two main

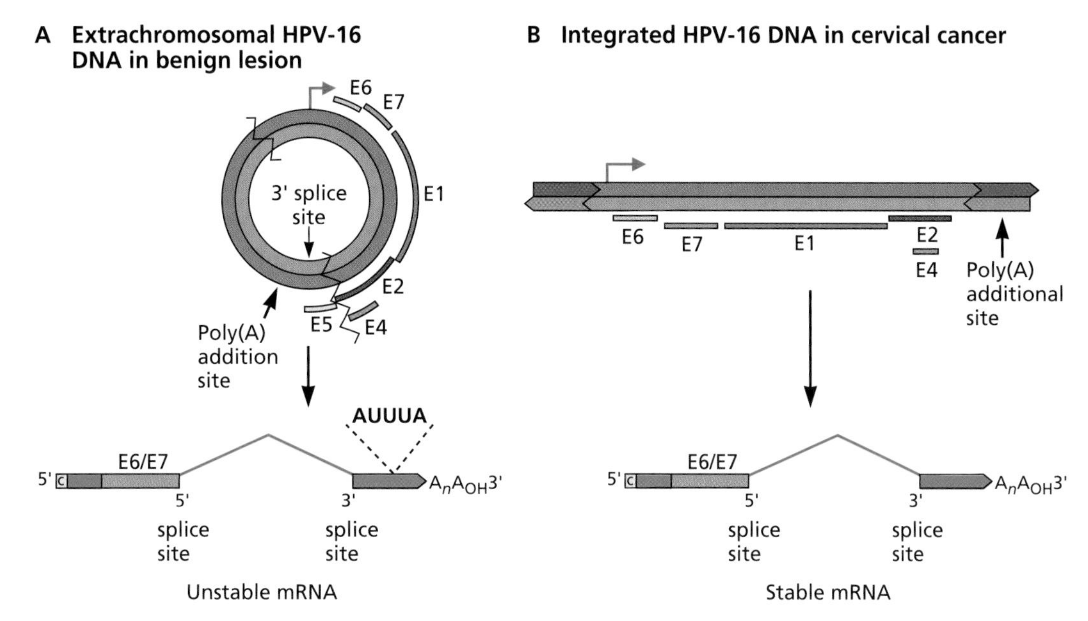

Figure 10.20 Stabilization of human papillomavirus type 16 (HPV-16) mRNAs upon integration of the viral genome into cellular DNA. (A) In benign lesions, the viral genome is maintained as an extrachromosomal, circular episome. Transcription of such viral DNA and pre-mRNA processing produce various alternatively spliced mRNAs containing the E6 and/or E7 protein-coding sequences, but, as illustrated, all contain destabilizing 5′AUUUA3′ sequences in their 3′ untranslated regions. **(B)** In cervical carcinoma cells, the viral genome is integrated into cellular DNA (purple) such that the viral genome is disrupted upstream of the E6/E7 mRNA 3′ splice site. The mRNAs encoding these viral proteins are therefore made by using 3′ splice and polyadenylation sites transcribed from adjacent cellular DNA, and they lack the destabilizing sequence.

types of these regulatory RNA molecules present in eukaryotes are distinguished by how they are synthesized. **Small interfering RNAs** (siRNAs), such as those first discovered in plants, are initially processed by endonucleolytic cleavage of double-stranded RNAs by cytoplasmic Dicer enzymes. The double-stranded RNA precursors are formed by base pairing of transcripts that contain complementary sequences, such as the (+) and (−) strand RNAs synthesized in cells infected by many viruses with RNA genomes. MicroRNAs (miRNAs) can be processed from RNAs synthesized by RNA polymerase III or from introns within pre-mRNAs. However, their precursors are generally capped and polyadenylated transcripts synthesized by RNA polymerase II, in which self-complementary regions form imperfect hairpin structures (Fig. 10.21). The sequences that encode miRNAs are often clustered, an arrangement that allows synthesis of transcripts containing multiple miRNA sequences. Such transcripts are initially processed by endonucleolytic cleavage in the nucleus to liberate pre-miRNAs, imperfect hairpins of 60 to 80 nucleotides. Further processing of pre-miRNAs occurs following export to the cytoplasm, where they are cleaved by Dicer enzymes.

In the case of both siRNAs and miRNAs, the products of Dicer cleavage are largely double-stranded, with two unpaired bases at the 3′ ends. These RNAs are then unwound from one 5′ end (Box 10.11), and one strand becomes tightly associated with a member of the argonaute (Ago) family of proteins in the effector ribonucleoprotein, termed the RNA-induced silencing complex, Risc. In these complexes, the small RNA acts as a "guide," identifying the target mRNA by base pairing to specific sequences within it prior to cleavage of the mRNA or inhibition of its translation. Perfect base pairing with the target mRNA usually results in mRNA cleavage. Such cleavage requires Ago2, the only one of the four human Ago proteins found in Risc that possesses endoribonuclease activity. However, inhibition of translation by miRNAs is often followed by deadenylation and decay of the mRNA.

The introduction of small, double-stranded RNAs analogous to the products formed by Dicer has proved to be a very valuable experimental tool. Such exogenous RNAs are incorporated into Riscs with high efficiency, allowing the experimenter to inhibit expression of particular genes by targeting siRNAs to degrade the corresponding mRNA. However, mammalian cells synthesize miRNAs rather than siRNAs. Some cellular miRNAs have a significant impact on virus-host cell interactions, and the genomes of several DNA viruses and retroviruses contain sequences coding for miRNAs.

BOX 10.10

BACKGROUND

An ancient antiviral defense guided by RNA: the CRISPR system

Bacteria (and archaea) are infected by numerous viruses with DNA genomes, and are also exposed to foreign DNA as a result of transduction and conjugation. It has been estimated that bacteriophages represent the most abundant biological entities on the planet and that they are responsible for destruction of 4 to 50% of bacterial populations. Not surprisingly, these organisms have developed a variety of defense mechanisms, such as the well-characterized bacterial restriction-modification systems that destroy foreign DNA (the sources of the restriction endonucleases so widely used in research). The most recent to be discovered, the clustered regularly interspersed short palindromic repeat (CRISPR) system, is a powerful mechanism that provides acquired immunity to exogenous DNAs via short RNAs.

As its name indicates, a definitive feature of this system is the presence in the genome of arrays of short repeated sequences interspersed with nonrepetitive spacers (panel A of the figure). Such an array was first described in 1987 in the *E. coli* genome, and the name CRISPR was coined in 2000 following identification of such arrays in the genomes of other bacteria and archaea. The number of CRISPR loci, the number of repeat-spacer units per locus, and the lengths of repeats and spacers vary considerably among bacterial and archaeal species. However, the presence of palindromic sequences, and hence the ability to form hairpins, is observed in most CRISPR repeat sequences. Such CRISPR loci are adjacent to a set of conserved protein-coding genes. These genes vary in number, position, and orientation with respect to CRISPR loci (e.g., panel A), but include a core set encoding the proteins necessary for defense against foreign DNAs.

The breakthrough in understanding the function of CRISPR arrays came from computational studies of the origin of the spacer sequences: among those represented in sequence databases, most matched sequences of bacteriophage genomes or plasmids. Furthermore, species containing a spacer derived from a particular invader proved to be resistant to that invader. These observations led to the hypothesis that CRISPR loci provide a form of acquired immunity. The demonstration that strains of *Streptococcus thermophilus* selected for resistance to specific bacteriophages carried new, phage-derived sequences in the 5′ end of their CRISPR locus provided direct experimental evidence for this hypothesis.

Subsequently, multiple experimental approaches confirmed that the CRISPR system provides defense against invading genetic elements, and elucidated the mechanism summarized in panel B. When a foreign DNA, such as a bacteriophage genome, enters a bacterial cell, some fraction is fragmented. Short fragments that match conserved sequence motifs adjacent to the CRISPR spacers, termed protospacer adjacent motifs (PAMs), then become integrated into the CRISPR locus. Proteins necessary for this process have been identified, but the molecular mechanisms are not well understood. Following transcription, CRISPR RNAs are processed by a multiprotein complex (Cascade) to produce CrRNAs (~60 nucleotides), each of which carries a repeat-derived sequence at its 5′ end, a 3′ hairpin, and a unique, spacer-derived internal sequence (panel B). When the spacer of a CrRNA base pairs with a complementary sequence in an invading DNA molecule, an R-loop in which one strand of the foreign DNA is single-stranded is formed. CRISPR-associated endonucleases, such as Cas1 in *E. coli*, cleave the DNA, which then becomes extensively degraded. The integration of the sequences of the invading DNA into the host cell genome, from which they can be mobilized in the form of CrRNAs, provides a form of "memory" and acquired immunity. Remarkably, bacteriophages that infect *Vibrio cholerae* were discovered subsequently to encode a CRISPR/Cas system that counteracts host chromosomal sequences that inhibit bacteriophage reproduction.

Karginov FV, Hannon GJ. 2010. The CRISPR system: small RNA-guided defense in bacteria and archaea. *Mol Cell* **37:**7–19.

Seed KD, Lazinski DW, Calderwood SB, Camilli A. 2013. A bacteriophage encodes its own CRISPR/Cas adaptive response to evade host innate immunity. *Nature* **494:**489–491.

Westra ER, Swarts DC, Staals RH, Jore MM, Brouns SJ, van der Oost J. 2012. The CRISPRs, they are a-changin': how prokaryotes generate adaptive immunity. *Annu Rev Genet* **46:**311–339.

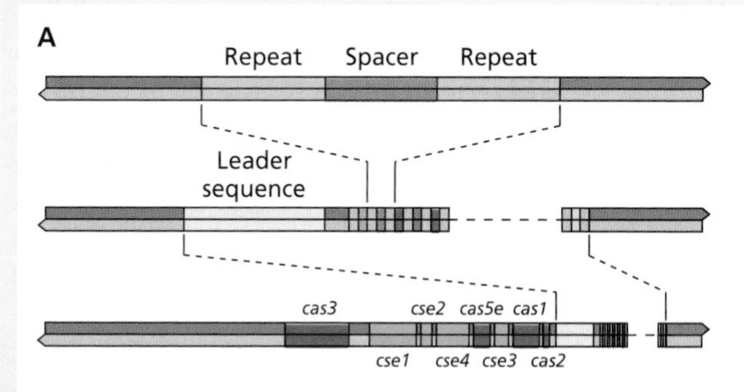

(A) The organization of CRISPR repeat and spacer sequences is illustrated at the top, and the position of the *E. coli* CRISPR locus with respect to CRISPR-associated genes (cas and cse genes) below. Adapted from F. V. Karginov and G. J. Hannon, *Mol Cell* **37:**7–19, 2010, with permission. **(B)** The structure of a CrRNA, following processing of CRISPR transcripts that contain multiple repeat-spacer units, is summarized at the top. As shown below, perfect base pairing of CrRNA with a complementary sequence adjacent to a PAM exposes a strand of the DNA for endonucleolytic attack by a CRISPR-associated endonuclease, such as Cas1. Adapted from E. R. Westra et al., *Annu Rev Genet* **46:**311–339, 2012, with permission.

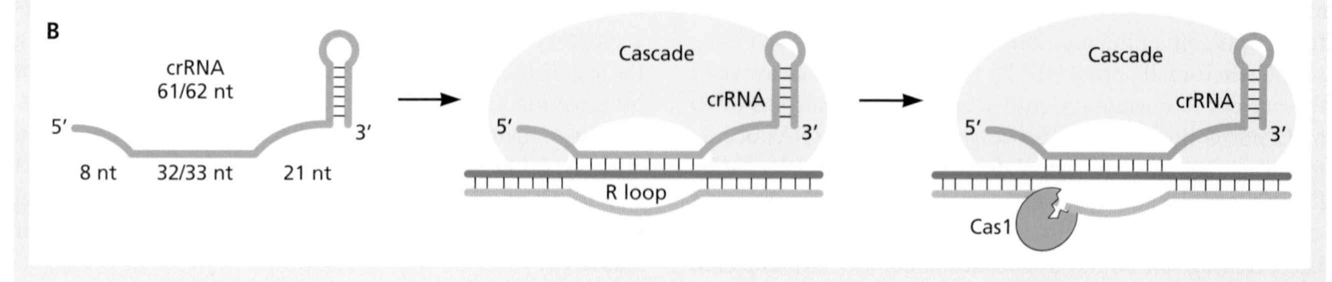

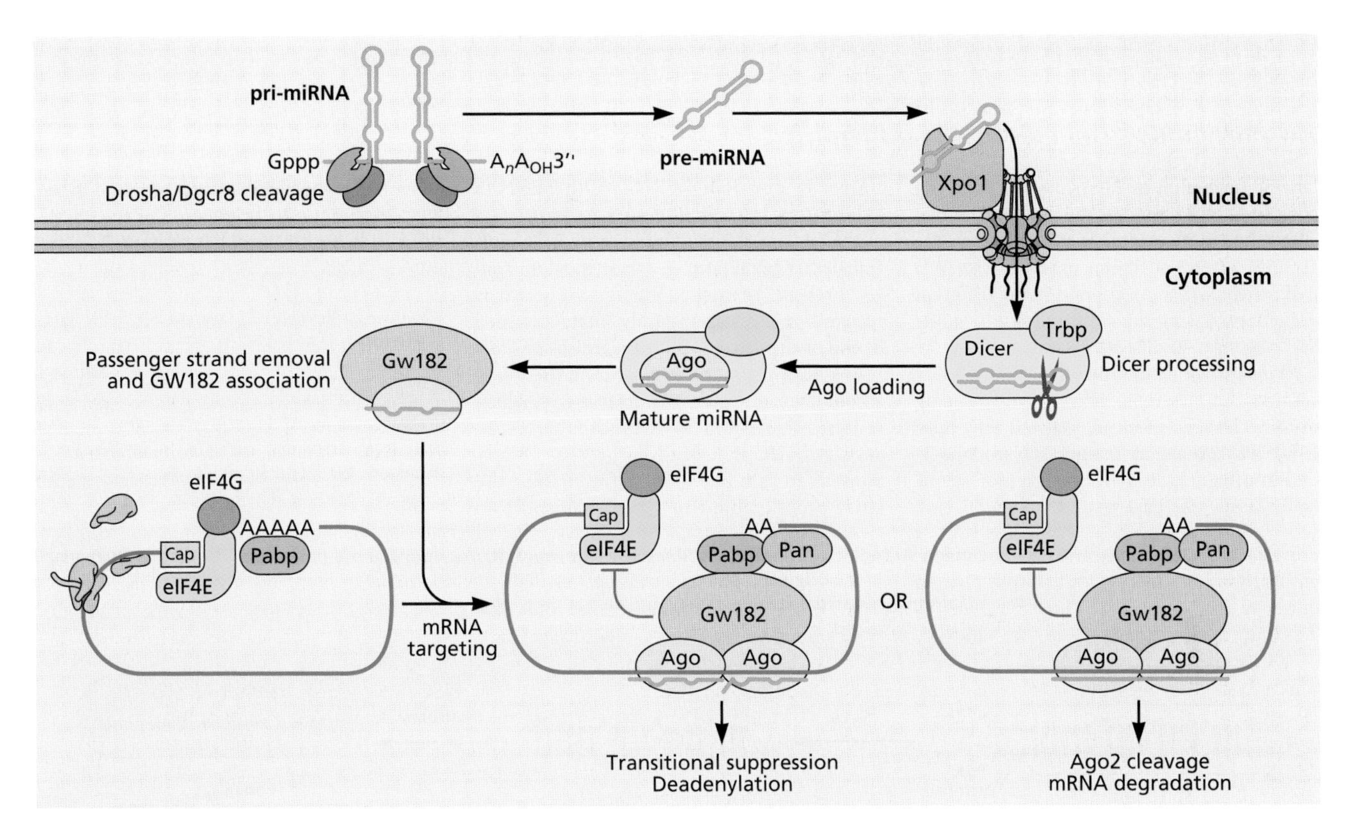

Figure 10.21 Synthesis and function of miRNAs. The precursors of miRNAs (pri-miRNAs), which are typically transcripts made by RNA polymerase II, undergo initial processing in the nucleus. Such transcripts are cleaved by the microprocessor, which comprises the ribonuclease Drosha and DiGeorge syndrome critical region 8 (Dgcr8) protein that is necessary for binding to the RNA substrate. The pre-miRNAs thus produced are then exported to the cytoplasm via the export receptor exportin-5. In the cytoplasm, further processing by the enzyme Dicer associated with a double-stranded-RNA-binding protein such as human TAR RNA-binding protein (Trbp) liberates 22-nucleotide, double-stranded RNAs with unpaired nucleotides at the 3′ ends. Upon unwinding of these duplexes, one RNA strand becomes tightly associated with an argonaute (Ago) protein (and others) in the RNA-induced silencing complex (Risc). The other strand is degraded. When human RISC contains Ago2, an mRNA to which its miRNA base pairs perfectly can be cleaved by this endonuclease (left). More generally (right), Ago proteins (Ago1-4 in human cells) interact with proteins of the Gw182 family that are required for RNA-mediated silencing, and induce inhibition of translation. Such proteins recruit deadenylases [e.g., poly(A)-binding protein (Pabp)-dependent poly(A)-specific ribonuclease (Pan)] that initiate degradation of mRNA by deadenylation, decapping, and 5′ → 3′ exonucleolytic degradation (Fig. 10.18).

Cellular miRNAs in Virus-Infected Cells

Micro-RNAs made in particular cell types have been reported to inhibit reproduction of a variety of viruses, including hepatitis B and C viruses, herpesviruses, human immunodeficiency virus type 1, influenza virus, and papillomaviruses. For example, at least six different cellular miRNAs can target human immunodeficiency virus type 1. However, cellular miRNAs can also dictate the outcome of virus infection and even facilitate virus reproduction. Examples of these phenomena are described below.

Cellular miRNA-155 Promotes Viral Oncogenesis

When Epstein-Barr virus infects primary B cells, it establishes a latent state characterized by limited expression of viral genetic information and maintenance replication of the viral genome (Chapters 8 and 9). The infected cells are immortalized and transformed and, *in vivo*, give rise to various B-cell malignancies. This process depends on viral gene products (Volume II, Chapter 6). However, induction of synthesis of cellular miRNA-155 is also important. This miRNA, which is present at high concentrations in human B-cell lymphomas and tumors, is encoded by a gene first identified as a common integration site of the retrovirus avian leukosis virus (Volume II, Chapter 6). The concentration of miRNA-155 is increased substantially in B cells transformed by Epstein-Barr virus in culture. Inhibition of its activity in such cells, by introduction of excess short, complementary RNA (an RNA “sponge”), inhibited cell proliferation and induced apoptosis. Although a number of mRNA targets of miRNA-155 have been identified, the mechanism by which this small RNA promotes oncogenic transformation has not yet been elucidated.

BOX 10.11

DISCUSSION

How the guide strand of siRNAs is identified

During formation of RNA-induced silencing complexes, one strand of the double-stranded siRNA, the guide strand, is retained while the second (often called the passenger strand) is destroyed. siRNAs contain many different sequences, raising the question of how guide and passenger strands are distinguished.

The answer came from efforts to identify siRNAs that are most effective in inducing mRNA cleavage when introduced into cells. It was observed that, in such siRNAs, the 5′ end of the guide RNA forms thermodynamically less stable base pairs than the 3′ end. Naturally occurring siRNAs and miRNAs exhibit this same asymmetry. As base pairs at the ends of double-stranded nucleic acids transiently break and re-form (they are said to "breathe"), this property might favor recognition of the transiently single-stranded 5′ end of the guide strand. Regardless, the less stable base pairs at the 5′ end of the guide strand favor unwinding from that end, which requires Dicer and RNA-binding proteins, such as TAR RNA-binding protein (Trbp). Subsequent studies demonstrated that the latter proteins bind to the siRNA end that contains the most-stable base pairs and therefore determines the orientation of the Dicer–RNA-binding protein heterodimer on the duplex siRNA.

Khvorova A, Reynolds A, Jayasena SD. 2003. Functional siRNAs and miRNAs exhibit strand bias. *Cell* **115**:209–216.

Schwarz DS, Hutvágner G, Du T, Xu Z, Aronin N, Zamore PD. 2003. Asymmetry in the assembly of the RNAi enzyme complex. *Cell* **115**:199–208.

Tomari Y, Matranga C, Haley B, Martinez N, Zamore PD. 2004. A protein sensor for siRNA asymmetry. *Science* **306**:1377–1380.

Processing and assembly into Risc of a guide strand of a pre-miRNA and release (for rapid degradation) of the passenger strand are illustrated. Which strand functions as the guide is independent of the original orientation in the pre-miRNA, but determined by properties of the small RNA duplex produced by Dicer, such as thermodynamic asymmetry, and the identity of the nucleotide at the 5′ end.

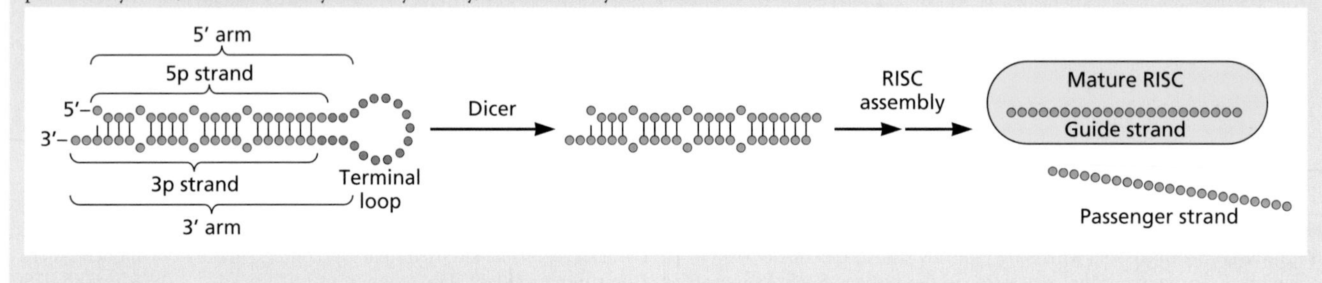

Cellular miRNA-122 Promotes Liver-Specific Reproduction of Hepatitis C Virus

The flavivirus hepatitis C virus is a widespread human pathogen that can establish chronic infection of the liver and is associated with the development of cirrhosis and hepatocellular carcinoma. The liver-specific reproduction of this virus is facilitated by cellular miRNA-122, the most abundant miRNA in hepatocytes: inhibition of the function of this miRNA impairs accumulation and expression of (+) strand viral RNA in hepatocytes in culture and in infected chimpanzees. The stimulation of hepatitis C virus reproduction by miRNA-122 is **not** the result of inhibition of synthesis of cellular proteins that directly or indirectly block the infectious cycle but rather of unusual virus-specific adaptations of base pairing with miRNAs: the miRNA both stabilizes (+) strand viral RNAs and impairs their translation, freeing them to serve as templates for genome replication (Box 10.12).

Viral Micro-RNAs

The first viral miRNAs were identified in 2004, by cloning and sequencing of small RNA molecules made in cells latently infected by Epstein-Barr virus. Subsequently, miRNAs of a number of other viruses have been described. Such RNAs are typically identified by combining computational methods that screen viral genomes for sequences with the properties of pre-miRNAs with assays for detection, such as high-throughput sequencing of low-molecular-weight RNAs isolated from infected cells. miRNA databases—e.g., miRBase (http://www.mirbase.org/index.shtml) and VIRmiRNA http://crdd.osdd.net/servers/virmirna/ hundreds of viral miRNAs, but the functions of the great majority are not yet known. We therefore describe a few well-characterized examples to illustrate what are likely to be general roles of such viral gene products.

Polyomavirus miRNAs That May Promote Persistence of Infected Cells

The genomes of simian virus 40 and the closely related human polyomaviruses JC virus and BK virus contain the sequence for a single pre-miRNA, which is transcribed as part of the late pre-mRNA and, unusually, processed to produce two miRNAs (Fig. 10.22). The miRNAs induce cleavage and degradation of the mRNA for the early gene product, large T antigen (LT). Mutations designed to disrupt the simian virus 40 pre-miRNA secondary structure prevented both viral miRNA synthesis and LT mRNA degradation and reduced the

BOX 10.12

DISCUSSION

A cellular miRNA that protects the hepatitis C virus genome from degradation and promotes its replication

miRNAs typically interact with target sequences in the 3′ untranslated regions of mRNAs. However, miRNA-122 base pairs with two complementary sequences present in the 5′ untranslated region of (+) strand hepatitis C virus, and in so doing protects the genome, which lacks a 5′ cap or 3′ poly(A) sequence, from degradation by Xrn1: mutated genomic RNA that lacks the miRNA-122 binding sites is less stable than the wild type, unless production of Xrn1 is also inhibited. Protection of the (+) strand RNA from degradation correlates with recruitment of a Risc-like complex to its 5′ end.

Removal of Xrn1 is not sufficient to restore replication of mutant viral genomes that lack the miRNA-122 binding sites, indicating that the cellular miRNA contributes to efficient reproduction of hepatitis C virus by one or more additional mechanisms. When the concentration of miRNA-122 was increased in infected cells, the steady-state concentrations of both viral mRNA and viral protein also increased, regardless of the presence or absence of Xrn1. Subsequent kinetic analyses of the accumulation of newly synthesized viral mRNA and protein established that the miRNA stimulated viral mRNA synthesis, but not its translation. Such stimulation is the result of competition for binding to the 5′ end of the (+) mRNA between miRNA-122 and a cellular protein that facilitates translation of this mRNA, Pcbp2 [poly(rC)-binding protein 2].

Antagonists of miRNA-122, such as antisense oligonucleotides, block virus reproduction with no harmful effects in animal models, and are currently in clinical trials in humans.

Gottwein E. 2013. Roles of microRNAs in the life cycles of mammalian viruses. *Curr Top Microbiol Immunol* **371:**201–227.

Jopling CL, Yi M, Lancaster AM, Lemon SM, Sarnow P. 2005. Modulation of hepatitis C virus RNA abundance by a liver-specific microRNA. *Science* **309:**1577–1581.

Masaki T, Arend KC, Li Y, Yamane D, McGivern DR, Kato T, Wakita T, Moorman NJ, Lemon SM. 2015. miR-122 stimulates hepatitis C virus RNA synthesis by altering the balance of viral RNAs engaged in replication versus translation. *Cell Host Microbe* **17:**217–228.

Shimakami T, Yamane D, Jangra RK, Kempf BJ, Spaniel C, Barton DJ, Lemon SM. 2012. Stabilization of hepatitis C virus RNA by an Ago2-miR-122 complex. *Proc Natl Acad Sci U S A* **109:**941–946.

The organization of the 5′ end of the (+) strand hepatitis C virus genome is summarized at the top, but not to scale. The secondary structures shown are consistent with the results of structural and mutational studies. The expansion illustrates the base pairing of cellular miRNA-122 (miR-122) with viral sequences on either side of stem-loop 1 (SL1), with the miRNA sequence shown in red. Adapted from E. Gottwein, *Curr Top Microbiol Immunol* **371:**201–227, 2013, with permission.

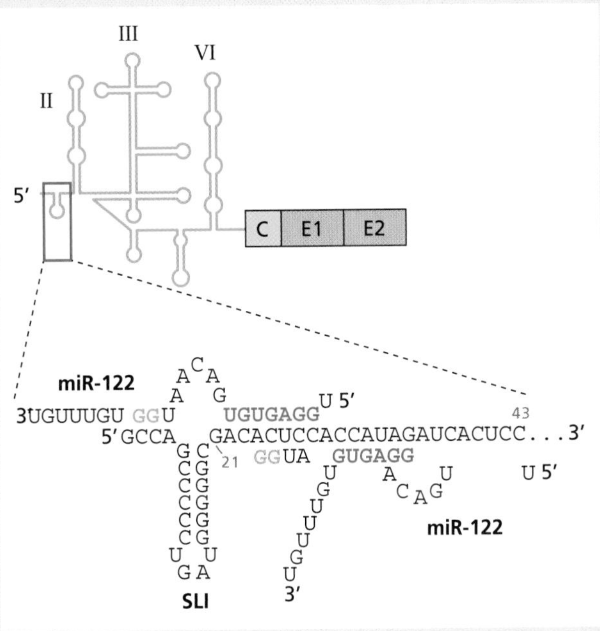

susceptibility of infected cells to killing by cytotoxic T cells specific for LT. However, no effects of such mutations on simian virus 40 reproduction in cells in culture or *in vivo* could be discerned. More-recent studies show that these miRNAs limit LT production and genome replication of an archetypical strain of BK virus that has not acquired rearrangement of sequences that control viral gene expression (Box 10.13). It is therefore possible that this function can promote the establishment of persistent BK virus infection in cells of the urinary tract.

Latency-Associated miRNAs of Herpesviruses

It is striking that pre-miRNA-coding sequences that are expressed in latently infected cells have been identified in regions of several alpha-, beta-, and gammaherpesviral genomes. For example, some 12 miRNAs are made in cells latently infected by human herpesvirus 8, which is a causative agent of Kaposi's sarcoma and B-cell lymphoma (Volume II, Chapter 6). The latency-associated miRNAs are processed from three overlapping transcripts synthesized from viral promoters active in latently infected cells. Ectopic expression of the viral miRNA coding region reduced substantially the concentrations of eight cellular mRNAs for proteins that participate in regulation of proliferation, immune responses, and apoptosis, such as that encoding the cyclin-dependent kinase inhibitor p21, which blocks cell cycle progression (Volume II, Chapter 6). Other human herpesvirus 8 miRNAs

BOX 10.13

WARNING

BK virus miRNAs and viral early gene expression: impact of sequence variations acquired during laboratory adaptation

Mutations that prevent synthesis of the viral miRNAs had no effect on the reproduction of simian virus 40 or mouse polyomavirus. However, subsequent experiments with BK virus suggest that these failures may be the result of noncoding sequence rearrangements present in the laboratory-adapted strains used in these studies.

The human polyomavirus BK virus is associated with persistent infection in healthy individuals and diseases of the kidney in immunosuppressed transplant patients. The organization of the coding and control sequences in the BK virus genome is like that of the closely related simian virus 40 (Appendix, Fig. 23). A strain of the virus isolated from healthy people, designated the archetype strain, does not produce progeny virus particles in primary renal proximal tubule epithelia cells in culture, and both LT and viral DNA synthesis are extremely inefficient. In contrast, variants with deletions and duplications within the noncoding control region reproduce efficiently in such cells. As the viral miRNAs restrict production of LT (Fig. 10.22), the origin recognition protein necessary for viral DNA synthesis, their role in regulating viral replication was compared in archetype and variant viruses.

Mutations that prevent processing of the miRNAs, but do not alter the overlapping coding sequence of LT (Fig. 10.22), were introduced into the genomes of these viruses, and the inhibition of miRNA synthesis was confirmed. In the absence of miRNAs, accumulation of LT (and its mRNA), viral DNA synthesis, and production of virions were increased significantly in primary renal epithelial cells infected by archetype BK virus (panel A of the figure). In contrast, although LT concentration was also increased somewhat and viral genome replication decreased, the mutations did not alter the yield of the variant BK virus. It was also established that the viral miRNAs can be made during the early phase of infection. Furthermore, their production is impaired, while early promoter activity is increased, by the rearrangements in the variant viral genome.

These observations led to a model in which the combination of the action of the viral miRNAs and a weak early promoter severely limit LT synthesis during the early phase of archetype BK virus infection (panel B). Consequently, the genome cannot replicate to a degree sufficient for assembly of progeny virus particles. These control parameters are reversed (strong early promoter, inefficient miRNA synthesis) by alterations in variant BK virus genomes, resulting in accumulation of high concentrations of LT and efficient genome replication and virus particle production. Limitation of LT synthesis by viral miRNAs during the early phase of infection *in vivo* is likely to be important for establishing the persistent infection characteristic of archetype BK virus.

Broekema NM, Imperiale MJ. 2013. miRNA regulation of BK polyomavirus replication during early infection. *Proc Natl Acad Sci U S A* **110:**8200–8205.

(A) Accumulation of LT (TAg) examined by immunoblotting, genome accumulation, and virus yield in viral DNA synthesis in primary renal proximal tubule epithelial cells infected by wild-type (wt) archetype BK virus (A) or a variant with a rearranged noncoding region (R) or derivations carrying mutations that prevent production of the viral miRNAs (mut). GAPDH, cellular glyceraldehyde-3-phosphate dehydrogenase (internal control); VP1, virion protein 1; ns, not significant; ‡, below the limit of detection; *, $P < 0.05$. **(B)** Model of control of archetype BK virus replication by miRNAs. Adapted from N. M. Broekema and M. J. Imperiale, *Proc Natl Acad Sci U S A* **110:**8200–8205, 2013, with permission. Panel A courtesy of M. Imperiale, University of Michigan Medical School.

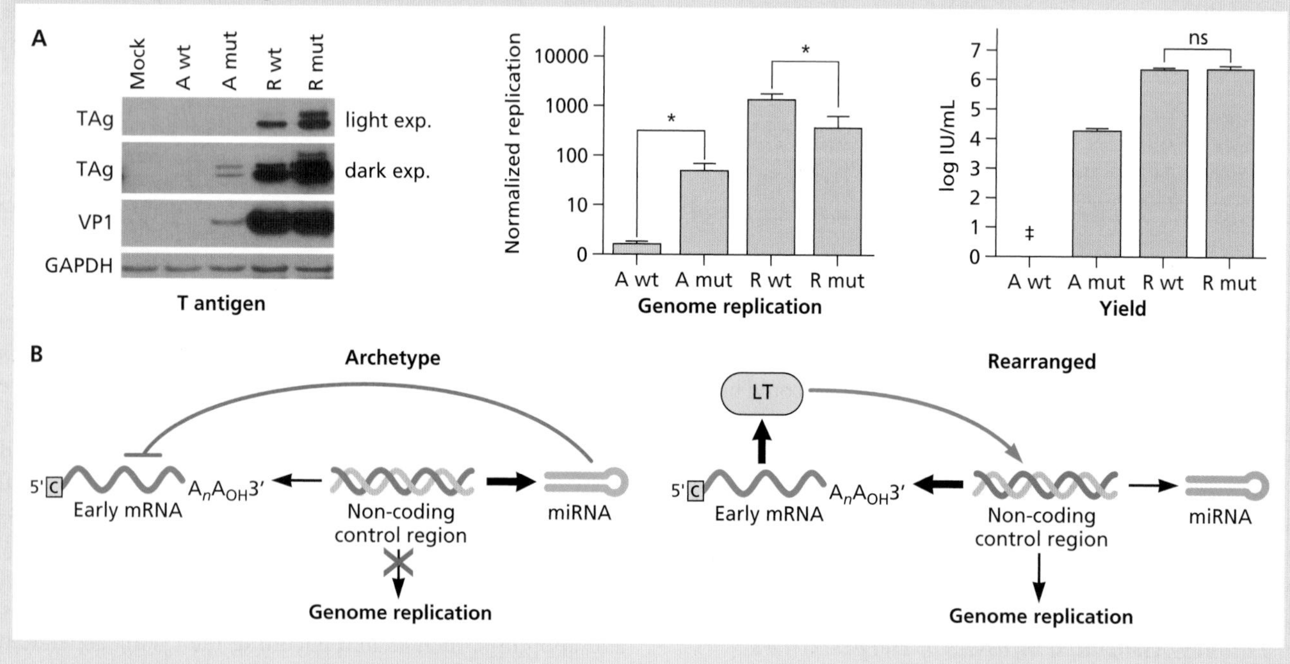

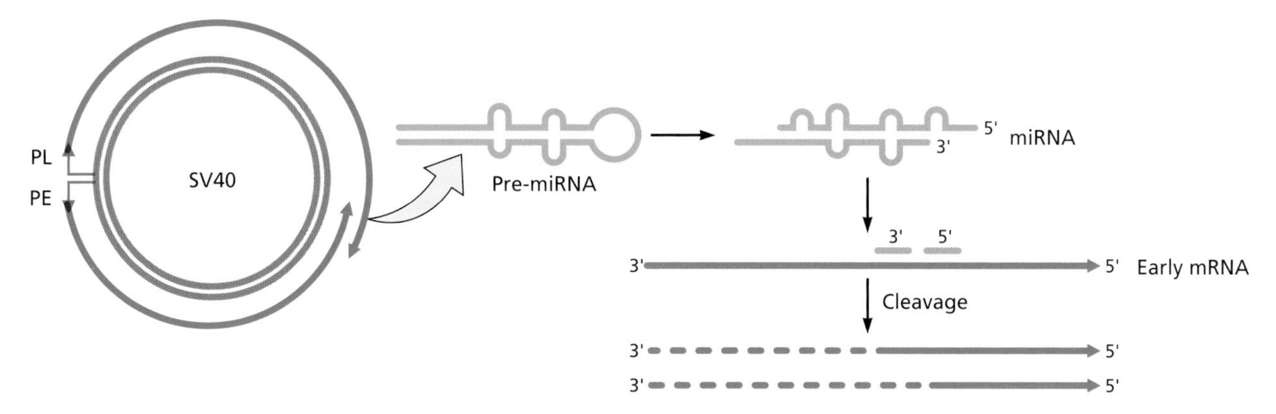

Figure 10.22 The miRNAs of simian virus 40. The circular simian virus 40 genome is shown at the left, with the positions of the early (P_E) and late (P_L) promoters and the primary transcripts indicated. As shown, the 3′ ends of early and late pre-mRNAs are encoded by opposite strands of the same sequence. Downstream of its polyadenylation site (arrowhead), the late pre-RNA contains a pri-miRNA sequence that is processed to a 57-nucleotide pre-miRNA and then to two miRNAs, designated 3′ and 5′. Both are perfectly complementary to specific sequences in the early mRNAs that encode LT and induce its cleavage.

act as functional homologs of cellular miRNAs that promote cell survival and proliferation or prevent synthesis of the protein that activates expression of lytic genes, thereby helping to maintain latent infection.

Inhibition of Antiviral Defenses

The genomes of several herpesviruses encode miRNAs that have been implicated in blocking intrinsic and immune antiviral defenses. Examples include four Epstein-Barr virus miRNAs that impair production of specific proapoptotic proteins (Volume II, Chapter 3); several beta- and gammaherpesviral miRNAs that protect infected cells against natural killer cells (Volume II, Chapter 3); and human herpesvirus 8 miRNA-K5 and -K9, which target components of signaling pathways that induce production of interferons and proinflammatory cytokines in response to infection.

Viral Gene Products That Block RNA Interference

siRNAs provide antiviral defenses in plants and invertebrates (Volume II, Chapter 3). As might therefore be anticipated, the genomes of viruses that replicate in these organisms encode proteins that suppress RNA silencing, such as the tomato bushy stunt virus and Flock house virus double-stranded-RNA-binding proteins p19 and B2, respectively. Viral inhibition of the production or function of miRNAs in mammalian cells appears to be rare, so far reported only for adenoviruses and poxviruses, such as vaccinia virus. The enormous quantities of the small virus-associated (VA) RNAs that accumulate in adenovirus-infected cells compete with cellular pre-miRNAs for binding to the active site of Dicer. Production of cellular miRNAs is impaired, as Dicer processes the viral RNAs to produce viral miRNAs (of unknown function). The VA RNAs, which are synthesized by RNA polymerase III in the nucleus, also block export of the Dicer mRNA to the cytoplasm by binding to Xpo5, a function that contributes to optimal reproduction of the virus. In contrast, degradation of cellular miRNAs is induced in vaccinia virus-infected cells. The viral poly(A) polymerase, which is both necessary and sufficient, targets cellular miRNAs for destruction by catalyzing the addition of short (<10 nucleotides) oligo(A) segments to their 3′ ends. This mechanism may represent a virus-specific variant of the recently described cellular system in which pre-miRNAs are targeted for destruction by addition of 3′ oligo(U).

Perspectives

Many of the molecular processes required for reproduction of animal viruses, including such virus-specific reactions as synthesis of genomic RNAs and mRNAs from an RNA template, were foretold by the properties of the bacteriophages that parasitize bacterial cells. In contrast, the covalent modifications necessary to produce functional mRNAs in eukaryotic cells were without precedent when discovered in viral systems. Study of the processing of viral RNAs has yielded much fundamental information about the mechanisms of capping, polyadenylation, and splicing. More recently, viral systems have provided equally important insights into export of mRNA from the nucleus to the cytoplasm. Perhaps the most significant lesson learned from the study of viral mRNA processing is the importance of these reactions in the regulation of gene expression.

Viral RNA processing can be regulated passively, by differences in the concentrations or activities of specific cellular components in different cell types, or actively by viral gene products. Several mechanisms by which viral gene products or RNAs can regulate or inhibit polyadenylation or splicing reactions, export of mRNA from the nucleus, or mRNA stability have been quite well characterized. However, our understanding of regulation of viral gene expression via RNA-processing reactions is far from complete: the mechanisms of action of several critical viral regulatory proteins have not been fully elucidated, and many of the specific mechanisms deduced by using experimental systems have yet to be confirmed in virus-infected cells. Similarly, much remains to be learned about the benefits for virus reproduction of the increasing numbers of viral proteins now known to destroy or relocalize components of the cellular mRNA degradation machinery, or that are analogs of such components.

Since we prepared the previous edition of this book, the catalog of viral miRNAs has grown enormously and noncanonical functions of viral miRNAs have been discovered. Furthermore, we can now document some fascinating instances of cellular miRNAs that shape the outcome of virus-host cell interactions. It therefore seems likely that the continued exploration of the molecular and physiological functions of viral miRNAs, currently unknown in most cases, will transform our understanding of the interplay among viruses and their hosts.

References

Reviews

Banks JD, Beemon KL, Linial MF. 1997. RNA regulatory elements in the genomes of simple retroviruses. *Semin Virol* **8:**194–204.

Colgan DF, Manley JL. 1997. Mechanism and regulation of mRNA polyadenylation. *Genes Dev* **11:**2755–2766.

DeCroly E, Ferron F, Lescar J, Canard B. 2012. Conventional and unconventional mechanisms for capping viral mRNA. *Nat Rev Microbiol* **10:**51–65.

Hsin JP, Manley JL. 2012. The RNA polymerase II CTD coordinates transcription and RNA processing. *Genes Dev* **26:**2119–2137.

Karn J, Stoltzfus CM. 2012. Transcriptional and posttranscriptional regulation of HIV-1 gene expression. *Cold Spring Harb Perspect Med* **2:**a006916. doi:10/1101/cshperpest.a006916.

Kincaid RP, Sullivan CS. 2012. Virus-encoded microRNAs: an overview and look to the future. *PLoS Pathog* **8:**e1003018. doi:10:1371/journal-ppat.1003018.

Müller-McNicoll M, Neugebauer KM. 2013. How cells get the message: dynamic assembly and function of mRNA-protein complexes. *Nat Rev Genet* **14:**275–287.

Narayan K, Makino S. 2013. Interplay between viruses and host mRNA degradation. *Biochim Biophys Acta* **1829:**732–741.

Rodriguez MS, Dargemont C, Stutz F. 2004. Nuclear export of RNA. *Biol Cell* **96:**639–655.

Samuel CE. 2011. Adenosine deaminases acting on RNA (ADARs) are both antiviral and proviral. *Virology* **411:**180–193.

Sandri-Goldin RM. 2008. The many roles of the regulatory protein ICP27 during herpes simplex virus infection. *Front Biosci* **13:**5241–5256.

Shatkin AJ, Manley JL. 2000. The ends of the affair: capping and polyadenylation. *Nat Struct Biol* **7:**838–842.

Staley JP, Guthrie C. 1998. Mechanical devices of the spliceosome: motors, clocks, springs, and things. *Cell* **92:**315–326.

Yarbrough ML, Mata MA, Sakthivel R, Fontoura BM. 2013. Viral subversion of nucleocytoplasmic trafficking. *Traffic* **15:**127–140.

Zhang ZM, Baker CC. 2006. Papillomavirus genome structure, expression and post-transcriptional regulation. *Front Biosci* **11:**2286–2302.

Papers of Special Interest

Capping of Viral mRNAs or Pre-mRNAs

Luo Y, Mao X, Deng L, Cong P, Shuman S. 1995. The D1 and D12 subunits are both essential for the transcription termination activity of vaccinia virus capping enzyme. *J Virol* **69:**3852–3856.

Plotch SJ, Bouloy M, Ulmanen I, Krug RM. 1981. A unique cap (m^7GpppXm)-dependent influenza virion endonuclease cleaves capped RNAs to generate the primers that initiate viral RNA transcription. *Cell* **23:**847–858.

Salditt-Georgieff M, Harpold M, Chen-Kiang S, Darnell JE, Jr. 1980. The addition of 5′ cap structures occurs early in hnRNA synthesis and prematurely terminated molecules are capped. *Cell* **19:**69–78.

Polyadenylation of Viral mRNA or Pre-mRNA

Fitzgerald M, Shenk T. 1981. The sequence 5′AAUAAA3′ forms part of the recognition site for polyadenylation of late SV40 mRNAs. *Cell* **24:**251–260.

Gilmartin GM, Hung SL, DeZazzo JD, Fleming ES, Imperiale MJ. 1996. Sequences regulating poly(A) site selection within the adenovirus major late transcription unit influence the interaction of constitutive processing factors with the pre-mRNA. *J Virol* **70:**1775–1783.

Schnierle BS, Gershon PD, Moss B. 1992. Cap-specific mRNA (nucleoside-O2′-)-methyltransferase and poly(A) polymerase stimulatory activities of vaccinia virus are mediated by a single protein. *Proc Natl Acad Sci U S A* **89:**2897–2901.

Wilusz JE, Beemon KL. 2006. The negative regulator of splicing element of Rous sarcoma virus promotes polyadenylation. *J Virol* **80:**9634–9640.

Splicing of Viral Pre-mRNA

Barksdale SK, Baker CC. 1995. Differentiation-specific alternative splicing of bovine papillomavirus late mRNAs. *J Virol* **69:**6553–6556.

Bouck J, Fu XD, Skalka AM, Katz RA. 1995. Genetic selection for balanced retroviral splicing: novel regulation involving the second step can be mediated by transitions in the polypyrimidine tract. *Mol Cell Biol* **15:**2663–2671.

Lamb RA, Lai CJ. 1980. Sequence of interrupted and uninterrupted mRNAs and cloned DNA coding for the two overlapping non-structural proteins of influenza virus. *Cell* **21:**475–485.

McPhillips MG, Veerapraditsin T, Cumming SA, Karali D, Milligan SG, Boner W, Morgan IM, Graham SV. 2004. SF2/ASF binds the human papillomavirus type 16 late RNA control element and is regulated during differentiation of virus-infected epithelial cells. *J Virol* **78:**10598–10605.

Export of Viral mRNAs from the Nucleus

Ernst RK, Bray M, Rekosh D, Hammarskjöld ML. 1997. A structured retroviral RNA element that mediates nucleocytoplasmic export of intron-containing RNA. *Mol Cell Biol* **17:**135–144.

Fischer U, Huber J, Boelens WC, Mattaj IW, Lührmann R. 1995. The HIV-1 Rev activation domain is a nuclear export signal that accesses an export pathway used by specific cellular RNAs. *Cell* **82:**475–483.

Kang Y, Cullen BR. 1999. The human Tap protein is a nuclear mRNA export factor that contains novel RNA-binding and nucleocytoplasmic transport sequences. *Genes Dev* **13:**1126–1139.

Pasquinelli AE, Ernst RK, Lund E, Grimm C, Zapp ML, Rekosh D, Hammarskjöld ML, Dahlberg JE. 1997. The constitutive transport element (CTE) of Mason-Pfizer monkey virus (MMPV) accesses a cellular mRNA export pathway. *EMBO J* **16:**7500–7510.

Regulation of RNA Processing by Viral Proteins

Castelló A, Izquierdo JM, Welnowska E, Carrasco L. 2009. RNA nuclear export is blocked by poliovirus 2A protease and is concomitant with nucleoporin cleavage. *J Cell Sci* **122**(Pt 20):3799–3809.

Faria PA, Chakraborty P, Levay A, Barber GN, Ezelle HJ, Enninga J, Arana C, van Deursen J, Fontoura BM. 2005. VSV disrupts the Rae1/mrnp14 mRNA export pathway. *Mol Cell* **17**:93–102.

Pilder S, Moore M, Logan J, Shenk T. 1986. The adenovirus E1B-55K transforming polypeptide modulates transport or cytoplasmic stabilization of viral and host cell mRNAs. *Mol Cell Biol* **6**:470–476.

Törmänen H, Backström E, Carlsson A, Akusjärvi G. 2006. L4-33K, an adenovirus-encoded alternative RNA splicing factor. *J Biol Chem* **281**:36510–36517.

Woo JL, Berk AJ. 2007. Adenovirus ubiquitin-protein ligase stimulates viral late mRNA nuclear export. *J Virol* **81**:575–587.

Stability of Viral and Cellular mRNA

Everly DN, Jr, Feng P, Mian IS, Read GS. 2002. mRNA degradation by the virion host shutoff (Vhs) protein of herpes simplex virus: genetic and biochemical evidence that Vhs is a nuclease. *J Virol* **76**:8560–8571.

Gaglia MM, Covarrubias S, Wong W, Glaunsinger BA. 2012. A common strategy for host RNA degradation by divergent viruses. *J Virol* **86**:9527–9530.

Jeon S, Lambert PF. 1995. Integration of human papillomavirus type 16 DNA into the human genome leads to increased stability of E6 and E7 mRNAs: implications for cervical carcinogenesis. *Proc Natl Acad Sci U S A* **92**:1654–1658.

Parrish S, Resch W, Moss B. 2007. Vaccinia virus D10 protein has mRNA decapping activity, providing a mechanism for control of host and viral gene expression. *Proc Natl Acad Sci U S A* **104**:2139–2144.

RNA Interference

Backes S, Shapiro JS, Sabin LR, Pham AM, Reyes I, Moss B, Cherry S, tenOever BR. 2012. Degradation of host microRNAs by poxvirus poly(A) polymerase reveals terminal RNA methylation as a protective antiviral mechanism. *Cell Host Microbe* **12**:200–210.

Bennasser Y, Chable-Bessia C, Triboulet R, Gibbings D, Gwizdek C, Dargemont C, Kremer EJ, Voinnet O, Benkirane M. 2011. Competition for XPO5 binding between Dicer mRNA, pre-miRNA and viral RNA regulates human Dicer levels. *Nat Struct Mol Biol* **18**:323–327.

Cui C, Griffiths A, Li G, Silva LM, Kramer MF, Gaasterland T, Wang XJ, Coen DM. 2006. Prediction and identification of herpes simplex virus 1-encoded mircoRNAs. *J Virol* **80**:5499–5508.

Linnstaedt SD, Gottwein E, Skalsky RL, Luftig MA, Cullen BR. 2010. Virally induced cellular microRNA miR-155 plays a key role in B-cell immortalization by Epstein-Barr virus. *J Virol* **84**:11670–11678.

Pfeffer S, Zavolan M, Grässer FA, Chien M, Russo JJ, Ju J, John B, Enright AJ, Marks D, Sander C, Tuschl T. 2004. Identification of virus-encoded microRNAs. *Science* **304**:734–736.

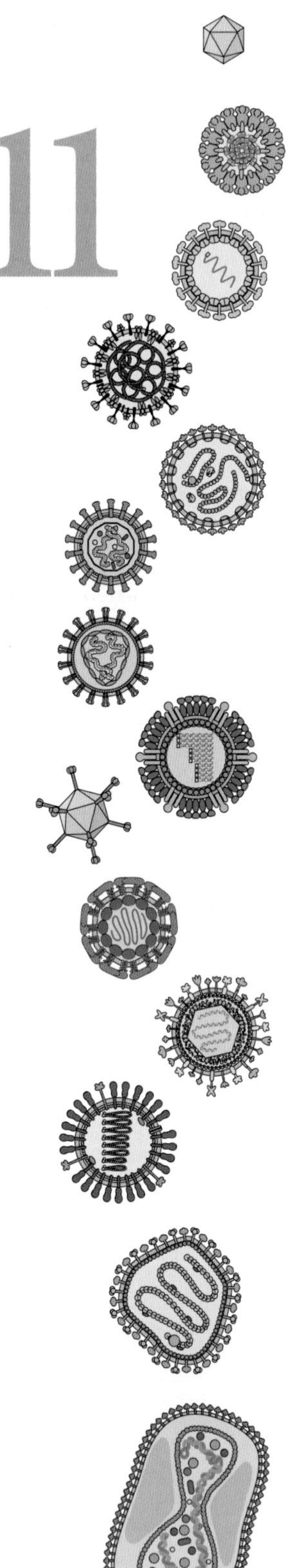

11 Protein Synthesis

Introduction

Mechanisms of Eukaryotic Protein Synthesis

General Structure of Eukaryotic mRNA

The Translation Machinery

Initiation

Elongation and Termination

The Diversity of Viral Translation Strategies

Polyprotein Synthesis

Leaky Scanning

Reinitiation

Suppression of Termination

Ribosomal Frameshifting

Bicistronic mRNAs

Regulation of Translation during Viral Infection

Inhibition of Translation Initiation after Viral Infection

Regulation of eIF4F

Regulation of Poly (A)-Binding Protein Activity

Regulation of eIF3

Interfering with RNA

Stress-Associated RNA Granules

Perspectives

References

LINKS FOR CHAPTER 11

Video: Interview with Dr. Ian Mohr
http://bit.ly/Virology_Mohr

California virology
http://bit.ly/Virology_Twiv97

Hantavirus protein replaces eIF4F
http://bit.ly/Virology_1-22-09

> *Translation is that which transforms everything so that nothing changes.*
>
> GÜNTER GRASS

Introduction

No viral genome encodes a complete translational apparatus (Box 11.1). Consequently, translation of viral messenger RNAs (mRNAs) is wholly dependent on the host cell. To allow efficient production of viral proteins, the translational machinery is usually modified to ensure that viral mRNAs are preferentially translated. Viral mRNAs are also translated in noncanonical ways to maximize their coding potential and allow the production of multiple proteins from a single mRNA.

Studies of virus-infected cells have contributed considerably to our understanding of protein synthesis and its regulation. Before the advent of recombinant DNA technology, infected cells were a rich source of large quantities of relatively pure mRNAs for *in vitro* studies of protein synthesis. The 5′ cap structure was identified on a viral RNA, and new translation initiation mechanisms, such as internal ribosomal entry, were discovered during studies of infected cells. Our understanding of how the activity of the multisubunit cap-binding complex can be regulated originated from the finding that one of its subunits is cleaved in infected cells.

Translation is a universal process in which proteins are synthesized from the amino to the carboxy terminus from mRNA templates read in the 5′ ⊠ 3′ direction. Each amino acid is specified by a genetic code consisting of three bases, a **codon**, in the mRNA. Translation takes place on **ribosomes**, and **transfer RNAs (tRNAs)** are the adapter molecules that link specific amino acids with individual codons in the mRNA. This chapter explores the basic mechanisms by which translation occurs in eukaryotic cells, the many ways by which viral mRNAs are translated to expand the limited coding capacity in genomes of limited size, and how translation is regulated in infected cells.

Mechanisms of Eukaryotic Protein Synthesis

General Structure of Eukaryotic mRNA

With the exception of organelle and certain viral mRNAs, eukaryotic mRNAs begin with a 5′ 7-methylguanosine (m^7G) **cap structure** (Fig. 11.1; see also Fig. 10.2). It is joined to the second nucleotide by a 5′-5′ phosphodiester linkage, in contrast to the 5′-3′ bonds found in the remainder of the mRNA. The unique cap structure directs pre-mRNAs to processing and transport pathways, regulates mRNA turnover, and is required for efficient translation by the 5′-end-dependent mechanism. Eukaryotic mRNAs contain **5′ untranslated regions**, which may vary in length from 3 to >1,000 nucleotides, although they are typically 50 to 70 nucleotides long. Such 5′ untranslated regions often contain secondary structures (e.g., hairpin loops [see Fig. 6.2]) formed by base pairing of the RNA. These double-helical regions must be unwound to allow passage of 40S ribosomal subunits during translation.

Translation begins and ends at **initiation codons** and **termination codons**, respectively. The termination codon is followed by a **3′ untranslated region**, which can regulate initiation, translation efficiency, and mRNA stability. At the very 3′ end of the mRNA is a stretch of adenylate residues known as the **poly(A) tail**, which is added to nascent pre-mRNA. The poly(A) tail is necessary for efficient translation, and for interactions among proteins that bind both ends of the mRNA.

Most bacterial and archaeal mRNAs are **polycistronic**: they encode several proteins, and each open reading frame is separated from the next by an untranslated spacer region. The vast majority of eukaryotic mRNAs are **monocistronic**; i.e., they encode only a single protein (Fig. 11.1). A small number of eukaryotic mRNAs are functionally polycistronic, and there are different strategies for synthesizing multiple proteins from a single mRNA. Members of the virus family *Dicistroviridae* are unique because the virus particles contain true bicistronic mRNAs.

PRINCIPLES ***Protein synthesis***

- No viral genome encodes the complete translational apparatus.
- The majority of viral mRNAs are translated by 5′-end-dependent mechanisms, but there is appreciable variation in this process, including mimicry of the initiator transfer RNA (tRNA) in the viral genome.
- Some viral RNAs are translated by a 5′-end-independent mechanism in which ribosomes bind internally to internal ribosome entry sites (IRESs).
- IRESs require RNA-binding proteins for activity.
- (+) strand RNA genomes that lack caps and poly(A) tails require a 3′-cap-independent translational enhancer for protein synthesis.
- A variety of unusual translation mechanisms expand the coding capacity of viral genomes and allow the synthesis of multiple polypeptides from a single RNA genome.
- Alterations in the cellular translational apparatus are commonplace in virus-infected cells.
- RNA granules are cytoplasmic aggregates that are assembled in response to viral infection to sequester RNAs, and many virus infections inhibit their formation or function.

BOX 11.1

TRAILBLAZER

Viral contributions to the translational machinery

Analysis of the nucleic acid of the largest DNA viruses challenges the belief that no viral genomes encode any part of the translational machinery. The 330- to 380-kbp DNA genome of viruses that infect the unicellular green alga *Chlorella* encode 10 to 15 tRNAs. These viral tRNAs are produced in infected cells, and some of them are aminoacylated, suggesting that they function during protein synthesis. These viral genomes also encode a homolog of elongation protein 3 that is synthesized in infected cells. DNA genomes of other giant viruses, including *Mimivirus*, *Pandoravirus*, and *Cafeteria roenbergensis* virus, encode multiple tRNAs, aminoacyl-tRNA synthetases, and a variety of initiation, elongation, and termination proteins, some of which have been shown to be functional.

These remarkable observations suggest that parts of the cellular translational machinery might be replaced by viral gene products. Support for this hypothesis comes from the observation that mimivirus-encoded translation termination proteins are synthesized by two recoding events: translational read-through and frameshifting. Although the amino acid sequences of these proteins are clearly eukaryotic, the regulatory features are specific to bacteria.

Why viral genomes encode gene products that participate in protein synthesis is not known. One possibility is that they modify the translation apparatus to favor the production of viral proteins. For example, the use of viral tRNAs may compensate for the low abundance of some tRNAs in host cells, allowing more efficient reproduction.

Jeudy S, Abergel C, Claverie JM, Legendre M. 2012. Translation in giant viruses: a unique mixture of bacterial and eukaryotic termination schemes. *PLoS Genet* **8:**e1003122. doi:10.1371/journal.pgen.1003122.

Nishida K, Kawasaki T, Fujie M, Usami S, Yamada T. 1999. Aminoacylation of tRNAs encoded by *Chlorella* virus CVK2. *Virology* **263:**220–229.

Raoult D, Audic S, Robert C, Abergel C, Renesto P, Ogata H, La Scola B, Suzan M, Claverie JM. 2005. The 1.2-megabase genome sequence of *Mimivirus*. *Science* **306:**1344–1350.

Yamada T, Fukuda T, Tamura K, Furukawa S, Songsri P. 1993. Expression of the gene encoding a translational elongation factor 3 homolog of *Chlorella* virus CVK2. *Virology* **197:**742–750.

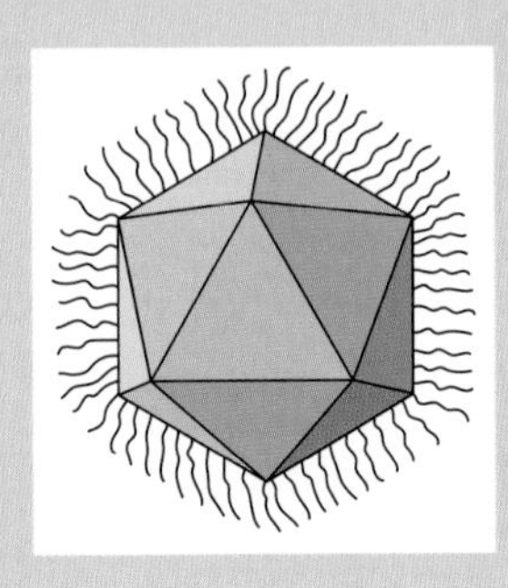

The Translation Machinery

Ribosomes

Mammalian ribosomes, the sites of protein synthesis, are composed of two subunits designated according to their sedimentation coefficients, 40S and 60S (Fig. 11.2A). The 40S subunit comprises an 18S rRNA molecule and 30 proteins, while the 60S subunit contains 3 rRNAs (5S, 5.8S, and 28S rRNAs) and 50 proteins. Actively growing mammalian cells may contain as many as 10 million ribosomes.

The mRNA moves past three sites on the ribosome, called A (aminoacyl or acceptor), P (peptidyl), and E (exit). The initiator tRNA enters at the P site, but all subsequent charged tRNAs enter the A site. The peptide bond is formed at the P site, while exit of the uncharged tRNA takes place at the E site.

Remarkably, the catalytic activity of ribosomes resides in RNA, not protein. After removal of 95% of the ribosomal proteins, the 60S ribosomal subunit can still catalyze the formation of peptide bonds; the peptidyltransferase center, where peptide bonds are formed, contains only RNA. The ribosome is the largest known RNA catalyst, providing evidence for an RNA world in which RNA, not proteins, carried out chemical reactions. The protein components of ribosomes help fold the rRNAs properly, so that they can fulfill their catalytic function, and to position the tRNAs.

Figure 11.1 Structure of eukaryotic and bacterial/archaeal mRNAs. UTR, untranslated region; AUG, initiation codon; ORF, open reading frame; Stop, termination codon. Adapted from G. M. Cooper, *The Cell: a Molecular Approach* (ASM Press, Washington, DC, and Sinauer Associates, Sunderland, MA, 1997).

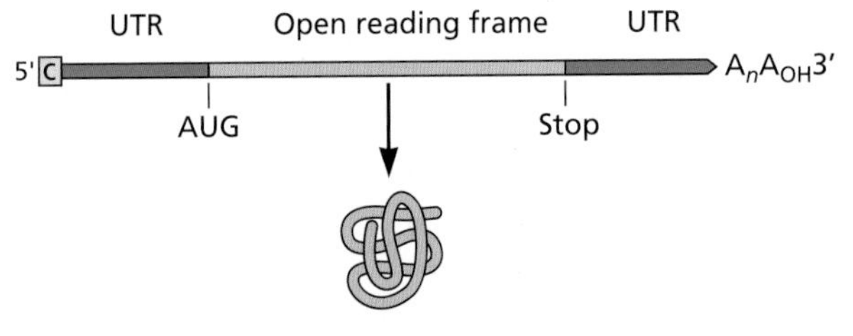

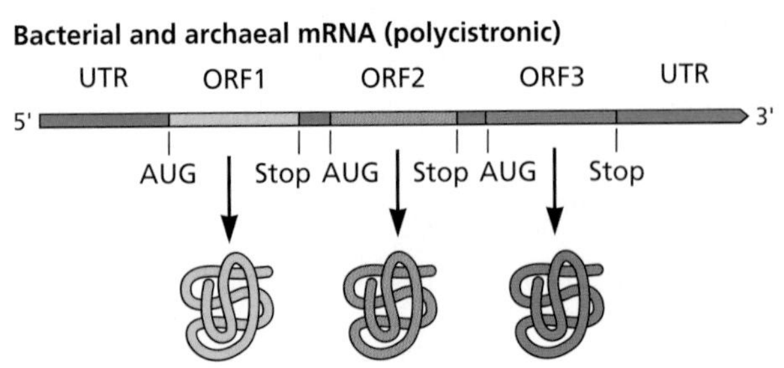

tRNAs

tRNAs are adapter molecules that align each amino acid with its corresponding codon on the mRNA. Each tRNA is 70 to 80 nucleotides in length and folds into a highly base-paired L-shaped structure (Fig. 11.2B). This shape is thought to be required for the appropriate interaction between tRNA and the ribosome during translation. The adapter function of tRNAs is carried out by two distinct regions of the molecule. At their 3′ ends, all tRNAs have the sequence 5′-CCA-3′, to

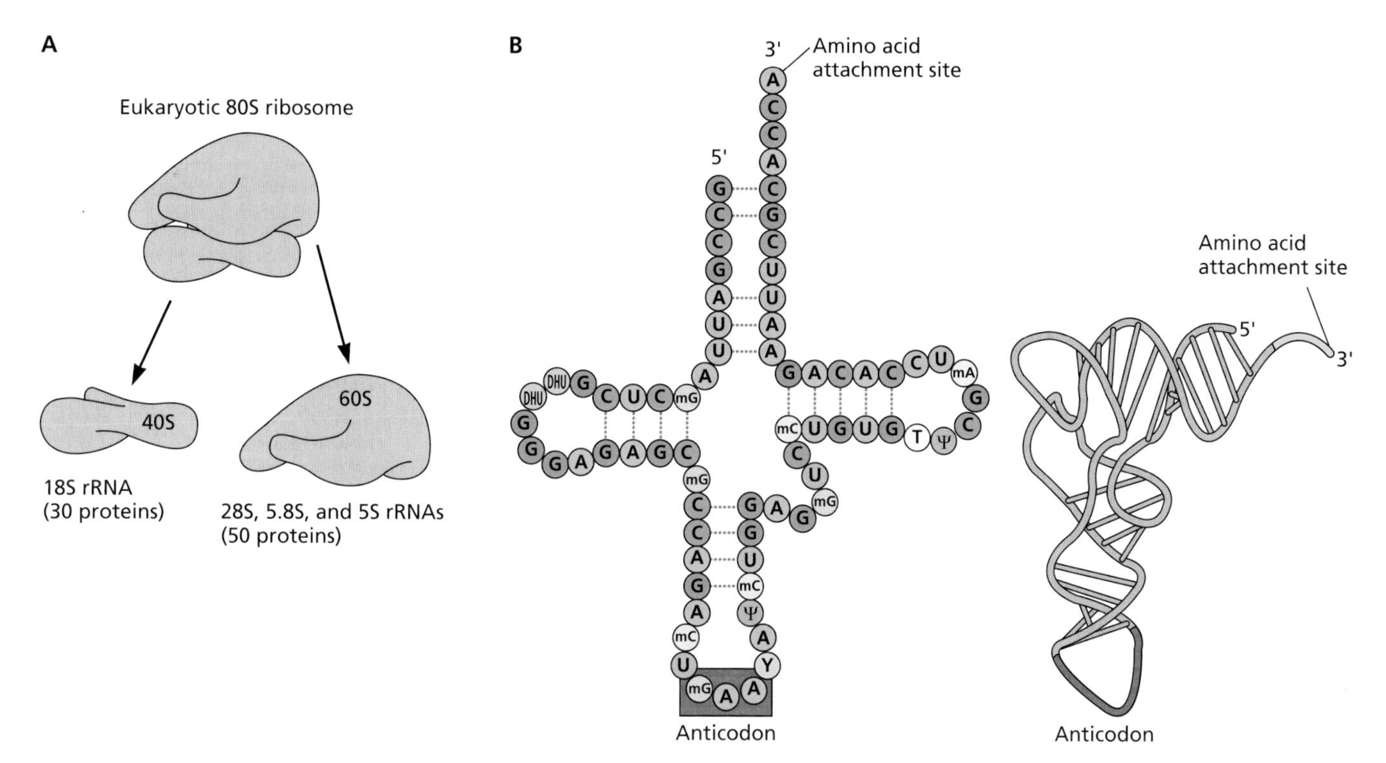

Figure 11.2 Ribosomes and tRNAs. (A) Model of a eukaryotic ribosome. The 80S ribosome consists of 60S and 40S subunits, which are made of ribosomal proteins and rRNAs. **(B)** Structure of tRNA. The model on the left shows how base pairing among the nucleotides of the tRNA results in a cloverleaf-like structure. Modified bases include methylguanosine (mG), methylcytosine (mC), dihydrouridine (DHU), ribothymidine (T), a modified purine (Y), and pseudouridine (Ψ). On the right is a folded representation showing the L-shaped structure. Adapted from G. M. Cooper, *The Cell: a Molecular Approach* (ASM Press, Washington, DC, and Sinauer Associates, Sunderland, MA, 1997).

which amino acids are covalently linked by **aminoacyl-tRNA synthetase**. Each of these enzymes recognizes a single amino acid and the correct tRNA. At the opposite end of the tRNA is the **anticodon loop**, which base pairs with the mRNA template. The accuracy of protein synthesis is maintained by two different proofreading mechanisms: faithful incorporation of amino acids depends on the specificity of codon-anticodon base pairing, as well as on the correct attachment of amino acids to tRNAs by aminoacyl-tRNA synthetases.

Translation Proteins

Many nonribosomal proteins are required for eukaryotic translation. Some form multisubunit assemblies containing as many as 11 different proteins, while others function as monomers. Translation can be separated experimentally into three distinct stages: initiation, elongation, and termination. The proteins that participate at each stage are named eukaryotic initiation, elongation, and termination proteins. These proteins are named in the same way as their bacterial and archaeal counterparts, with the prefix "e" to designate those of eukaryotic origin. The amino acid sequences of these proteins are conserved from yeasts to mammals, indicating that the mechanisms of translation are similar throughout eukaryotes.

Initiation

The majority of regulatory mechanisms function during initiation, because it is the rate-limiting step in the translation of most mRNAs (see "Regulation of Translation during Viral Infection" below). At least 11 initiation proteins participate in this energy-dependent process. The end result is formation of a complex containing the mRNA, the ribosome, and the initiator Met-$tRNA_i$, in which the reading frame of the mRNA has been set. The 80S ribosome, which is the predominant species in cells, must be dissociated, because it is the 40S subunit that participates in initiation. Three initiation proteins, eIF1A, eIF3, and eIF6, promote such dissociation.

There are two mechanisms by which ribosomes bind to mRNA in eukaryotes. In 5′-end-dependent initiation, by which the majority of mRNAs are translated, the initiation complex binds to the 5′ cap structure and moves, or scans, in a 3′ direction until the initiating AUG codon is encountered. In contrast, during 5′-end-independent initiation, the initiation complex binds at, or just upstream of, the initiation codon.

Internal ribosome entry sites were first discovered in picornavirus mRNAs, and are now known to be present in some cellular mRNAs.

5′-End-Dependent Initiation

How ribosomes assemble at the correct end of mRNA. The first step in the 5′-end-dependent initiation pathway is recognition of the m^7G cap by the cap-binding protein, eIF4E (Fig. 11.3). eIF4G acts as a scaffold between the cap structure and the 40S subunit, which associates with the mRNA via an interaction of eIF3 with the C-terminal domain of eIF4G. This important adapter molecule was first discovered as the target of proteolytic cleavage in poliovirus-infected cells, a modification that results in the inhibition of host protein synthesis. After binding near the cap, the 40S ribosomal subunit, which is part of a **preinitiation complex** that includes Met-$tRNA_i$ and other initiation proteins, moves in a 3′ direction on the mRNA in a process called **scanning**. Such movement depends upon a conformation of the 40S ribosomal subunit that allows processive motion, and unwinding of double-stranded structures to permit the RNA to thread through the ribosome and expose codon triplets (see "The role of mRNA secondary structure in translation" below). Scanning is a combination of a series of forward and backward movements with overall net movement in the 5′ → 3′ direction. When the preinitiation complex reaches the AUG initiation codon, an event detected by the second two bases of Met-$tRNA_i$ with the assistance of eIF1 and eIF1A, GTP is hydrolyzed and initiation proteins are released, allowing the 60S ribosomal subunit to associate with the 40S subunit to form the 80S initiation complex.

Role of the poly(A) tail in initiation. The presence of a poly(A) tail can stimulate mRNA translation. This effect is a consequence of interactions between proteins associated with the 5′ and 3′ ends of the mRNA, which promote 40S subunit

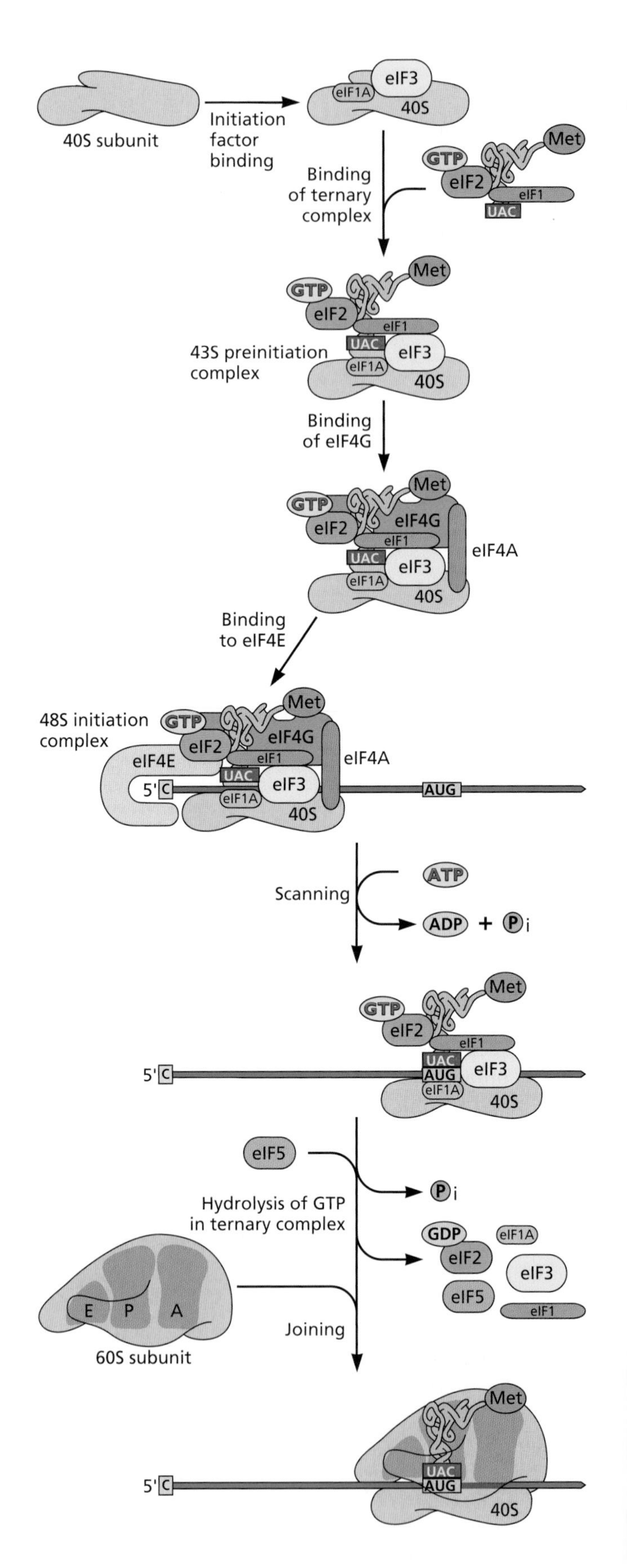

Figure 11.3 5′-cap-dependent assembly of the initiation complex. Initiation proteins eIF3 and eIF1A bind to free 40S subunits to prevent their association with the 60S subunit, while interaction of eIF6 (not shown) with the larger subunit prevents it from associating with the 40S subunit. eIF4F, which consists of three proteins, eIF4A, eIF4E, and eIF4G, binds the cap via the eIF4E subunit, and the ribosome binds a ternary complex containing eIF2, GTP, and Met-$tRNA_i$, forming a 43S preinitiation complex. The ribosome then binds eIF4G via eIF3. Alternatively, eIF4G may first join the 43S preinitiation complex and then bind the mRNA via eIF4E bound to the cap. The 40S subunit then scans down the mRNA until the AUG initiation codon is reached. eIF1 and eIF1A are required for selection of the correct AUG initiation codon. eIF5 triggers GTP hydrolysis, eIF2 bound to GDP is released along with other initiation proteins, and the 60S ribosomal subunit joins the complex. The ribosomal A site binds the aminoacylated tRNA; the P site binds the peptidyl-tRNA, and the uncharged tRNA leaves at the E site. Adapted from G. M. Cooper, *The Cell: a Molecular Approach* (ASM Press, Washington, DC, and Sinauer Associates, Sunderland, MA, 1997).

recruitment. Such interactions were first demonstrated in the yeast *Saccharomyces cerevisiae*, in which poly(A)-binding protein Pabp1 is required for efficient mRNA translation. Stimulation of translation by poly(A) occurs by enhancing the binding of 40S ribosomal subunits to mRNA. Pabp1 interacts with the N terminus of eIF4G (Fig. 11.4). Alteration of this binding site on eIF4G destroys stimulation of translation by poly(A). These results have led to a model in which Pabp1, bound to the poly(A) tail, associates with eIF4G bound to the 5′ cap, stabilizing the interaction and assisting in recruitment of 40S subunits (Fig. 11.4). A consequence of these interactions is that the 5′ and 3′ ends of the mRNA are brought into close proximity.

Figure 11.4 5′-end-dependent initiation. (Top) Schematic of eIF4G. Adapted from S. J. Morley et al., *RNA* **3:**1085–1104, 1997, with permission. (Middle) Model of initiation complex assembly. eIF4F is brought to the mRNA 5′ end by interaction of eIF4E with the cap structure (top) or genome-linked VPg (middle). The N terminus of eIF4G binds eIF4E, and the C terminus binds eIF4A. The 40S ribosomal subunit binds to eIF4G indirectly via eIF3. (Bottom) 5′-end-dependent initiation is stimulated by the poly(A)-binding protein Pabp1, which interacts with eIF4G. This interaction may bring the mRNA ends together and facilitate formation of the initiation complex at the 5′ end. Adapted from M. W. Hentze, *Science* **275:**500–501, 1997, with permission.

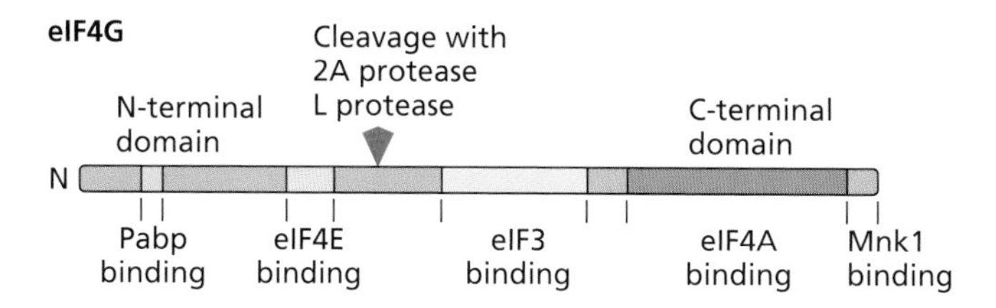

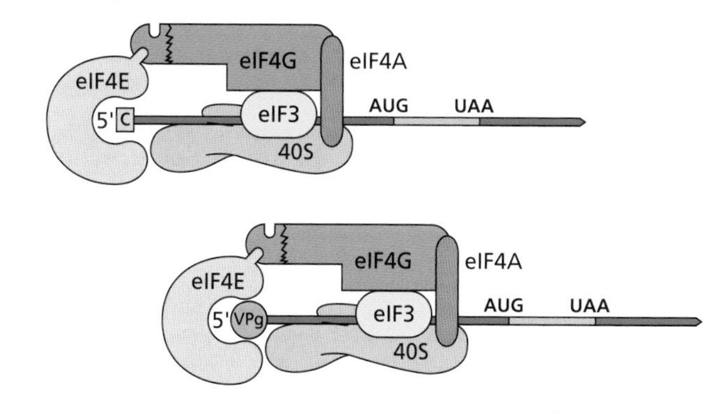

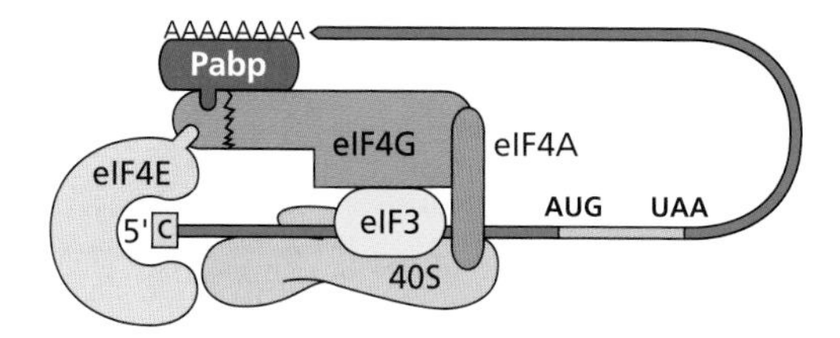

Some viral mRNAs, such as those of certain plant viruses, lack a 5′-terminal cap and 3′ poly(A) sequence. Nevertheless, the ends of these mRNAs are brought together by base pairing between discrete sequences in the 5′ and 3′ untranslated regions. Translation of mRNA of the flavivirus dengue virus, which has a 5′ cap structure but lacks a 3′ poly(A) sequence, may also depend on complementarity between sequences in the untranslated regions.

The juxtaposition of mRNA ends might be a mechanism to ensure that only intact mRNAs that contain a 5′ cap and 3′ poly(A) are translated. Such structures could also stabilize mRNA, by preserving the interaction among the translation initiation proteins associated with the ends, and hence sequestering them to attack by exonucleases. Translation reinitiation might also be stimulated by such an arrangement: once the ribosome terminates translation, it might be repositioned at the AUG initiation codon rather than dissociating from the mRNA template.

VPg-dependent ribosomal recruitment. The 40S ribosomal subunit appears to be brought to the mRNAs of members of the *Potyviridae* and the *Caliciviridae* via interactions with VPg, the small protein linked to the first base of the RNA (Fig. 6.11). VPg of the plant virus turnip mosaic virus (*Potyviridae*) binds eIF4E, thereby recruiting eIF4G, eIF3, and the 40S ribosomal subunit to the mRNA (Fig. 11.4). In cells infected with members of the *Caliciviridae*, VPg binds both eIF4E and eIF3. Such interactions may also facilitate selective translation of viral mRNAs, although the mechanisms involved have not been elucidated.

The role of mRNA secondary structure in translation. Translation efficiency is reduced by the presence of a stable secondary structure in the mRNA 5′ untranslated region. There are at least two reasons for this effect. If an RNA stem-loop structure is adjacent to the 5′ cap, it can inhibit binding of the 40S ribosomal subunit. In addition, the presence of secondary structure blocks ribosome movement toward the initiation codon.

The ATP-dependent RNA helicase activity of eIF4A, assisted by eIF4B, unwinds intramolecular regions of double-stranded RNA (dsRNA) near the 5′ end of the mRNA, allowing the 43S preinitiation complex to bind. The helicase may also migrate in a 3′ direction, unwinding dsRNA and enabling movement of ribosomes. mRNAs with less secondary structure in the 5′ untranslated region have a reduced requirement for RNA helicase activity during translation, and hence are less dependent on the cap structure, which brings the helicase to the mRNA. Dependence of translation on the cap can be measured experimentally by determining the effect on protein synthesis of cap analogs, such as m^7GDP and m^7GTP. These compounds inhibit 5′-end-dependent initiation competitively by binding to eIF4E. For example, the 5′ untranslated region of alfalfa mosaic virus (a plant bromovirus) RNA segment 4 is largely free of secondary structure, and translation of this mRNA is quite resistant to inhibition by cap analogs.

Choosing the initiation codon. The selection of the initiating AUG codon depends on both its position in the mRNA and the surrounding nucleotide sequence. For >90% of mRNAs, translation initiates at the 5′-proximal AUG codon. If the 5′-proximal AUG codon is mutated so that it cannot serve as an initiation codon, translation starts at the next downstream AUG. Insertion of an AUG codon upstream of the initiating codon leads to initiation at the more 5′-proximal site. The efficiency of initiation is influenced by the nucleotide sequence surrounding this codon. Studies of the effects of mutating these sequences have shown that the consensus sequence 5′-GCC**A**CCAUGG-3′ is recognized most efficiently in mammalian cells: the presence of a purine at the −3 position (boldface) is most important. However, only 5% of eukaryotic mRNAs contain this ideal consensus sequence: most have suboptimal sequences that result in less-efficient translation. This finding indicates that not all mRNAs must be translated at maximal efficiency, but rather only at levels appropriate for the function of the protein product. If a very poor match to this consensus sequence is present, the AUG codon may be passed over by the ribosome and initiation may occur farther downstream (see "The Diversity of Viral Translation Strategies" below).

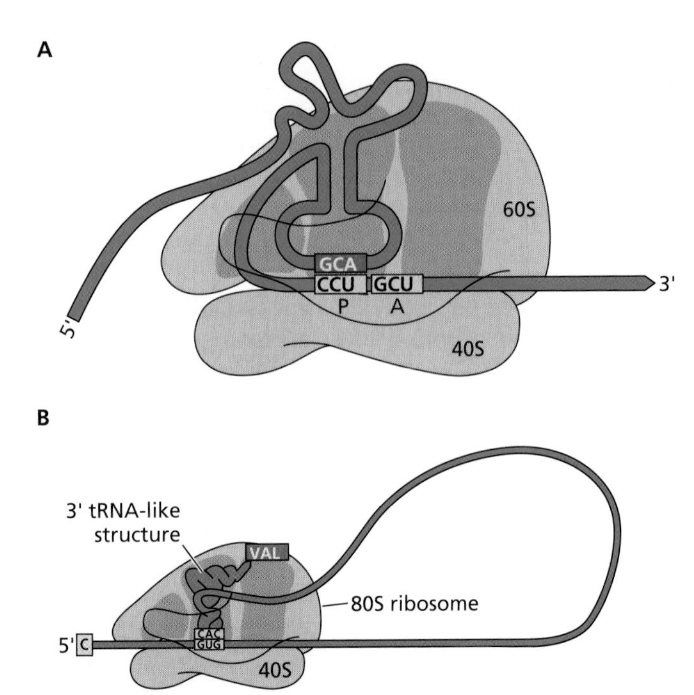

Figure 11.5 Two mechanisms of methionine-independent initiation. (A) The viral mRNA of picornavirus-like viruses of insects mimics the structure of tRNA, which occupies the P site of the ribosome, allowing initiation to take place within the A site. Adapted from M. Bushell and P. Sarnow, *J Cell Biol* **158:**395–399, 2002, with permission. **(B)** A tRNA-like structure in the 3′ untranslated region of turnip yellow mosaic virus RNA, aminoacylated with valine, occupies the P site of the ribosome. Adapted from S. Barends et al., *Cell* **112:**123–129, 2003, with permission.

Methionine-independent initiation. The structural proteins of some viruses begin not with methionine but with glutamine (CAA), proline (CCU), or alanine (GCU or GCA). Initiation of synthesis of these viral proteins does not require Met-tRNA$_i$ or the ternary complex, because the viral mRNA mimics the structure of tRNA (Fig. 11.5A). The tRNA-like structure occupies the P site of the ribosome, allowing initiation to take place within the A site. These mRNAs require no translation initiation proteins, and can bind ribosomes and induce them to enter the elongation phase of translation. Methionine-independent initiation of the mRNA of turnip yellow mosaic virus (a *Tymovirus*) is accomplished in a similar way, except that the tRNA-like structure is located in the 3′ untranslated region of the viral RNA (Fig. 11.5B). The tRNA-like structure is aminoacylated with valine, which is incorporated as the first amino acid of the viral polyprotein.

Ribosome shunting. Stable RNA secondary structures in 5′ untranslated regions may inhibit scanning of 40S ribosomes. In some RNAs, such hairpin structures are not inhibitory because ribosomes bypass them. This process, called **ribosome shunting**, may be dependent or independent of viral proteins. Shunting on the 35S cauliflower mosaic virus RNA requires translation of a very short upstream open reading frame on the same viral mRNA. Upon termination of translation of this open reading frame, ribosomes bypass a 480-nucleotide stem-loop structure that includes 8 AUG codons, and resume scanning just beyond the structure (Fig. 11.6). The ability to bypass this structure is thought to be a consequence of initiation proteins retained

Figure 11.6 Hypothetical model of ribosome shunting. The 40S ribosomal subunit binds to the mRNA by a cap-dependent mechanism and then bypasses regions of the mRNA with secondary structure to reach the AUG initiation codon. Shunting elements, such as the loops in the figure, and viral or cellular proteins may direct ribosome movement.

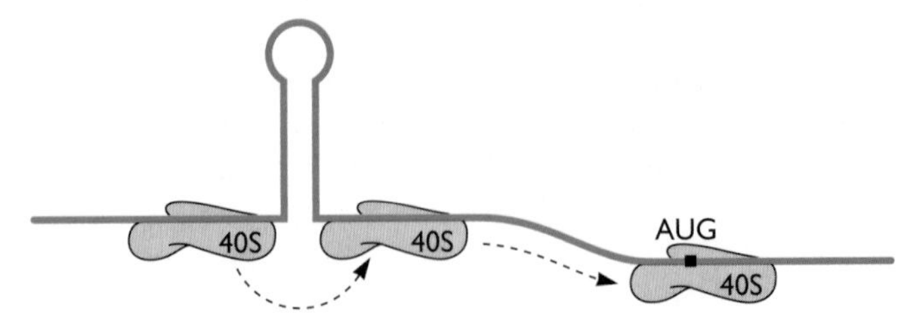

on the 40S subunit after translation, together with a temporary loss of other proteins that allow discontinuous scanning over the stem-loop base. In contrast, shunting on other viral mRNAs, including those of adenoviruses, paramyxoviruses, reoviruses, and hepadnaviruses, takes place in the absence of viral proteins.

5′-End-Independent Initiation

The internal ribosome entry site. The mRNAs of picornaviruses differ from most host cell mRNAs: they lack the 5′-terminal cap structure, and the 5′ untranslated regions are highly structured and contain multiple AUG codons. Infection of host cells by many picornaviruses results in the inhibition of translation of cellular mRNAs. These observations led to the hypothesis that translation of the mRNA of (+) strand picornaviruses was initiated by an unusual mechanism. It was suggested that the ribosome bound internally, rather than at the mRNA 5′ end. In an important experiment, the 5′ untranslated region of poliovirus mRNA was shown to promote internal binding of the 40S ribosomal subunit, and was termed the **internal ribosome entry site (IRES)** (Box 11.2).

An IRES has been identified in the mRNAs of all picornaviruses, in other viral mRNAs including those of pestiviruses and hepatitis C virus, and in some cellular mRNAs. Viral IRESs have been placed in five groups, depending on a variety of criteria, including primary sequence and secondary structure conservation, the location of the initiation codon, and activity in different cell types. There is very little nucleotide sequence conservation among members of the different groups, with the exception of a 25-nucleotide-long oligopyrimidine tract at the 3′ end of the IRES. Viral IRESs contain extensive regions of RNA secondary structure (Fig. 11.7). Although such secondary structure is not strictly conserved, it is of extreme importance for ribosome binding.

The discovery of the IRES makes even more puzzling the rarity of eukaryotic mRNAs that contain multiple long open reading frames (Fig. 11.1). In principle, all the open reading frames of a polycistronic mRNA can be translated in a eukaryotic cell as long as each frame is preceded by an IRES. Nevertheless, only one such naturally occurring polycistronic mRNA has been identified in eukaryotes, and it is not known if an IRES is present. Bicistronic mRNAs produced in the laboratory have been used in the expression of cloned genes (Box 11.3).

The mechanism of internal initiation. Different sets of translation initiation proteins are required for the function of various IRESs. Internal ribosome binding on the hepatitis A virus IRES requires all the initiation proteins, including eIF4E. At the other extreme, the intergenic IRES of cricket paralysis virus requires **none** of them. However, the activity of most IRESs depends on a subset of initiation proteins. Initiation on the type 1, 2, and 5 IRESs requires all except eIF4E, and either the carboxy-terminal two-thirds or the central one-third fragment of eIF4G. Both fragments contain binding sites for eIF3 and eIF4A (Fig. 11.8) and function better than the full-length protein in IRES-directed protein synthesis. In poliovirus-infected cells, eIF4G is cleaved, reducing the translation of most cellular mRNAs.

Initiation of translation via an IRES comprises binding of the 40S ribosomal subunit, followed by scanning to the initiation codon. Depending on the IRES, the 40S subunit may bind directly to the RNA or may be recruited to the IRES by means of interaction with translation initiation proteins (Fig. 11.8). For example, the cleavage products of eIF4G bind directly to the type 1 or type 2 IRES, and the 40S ribosomal subunit is recruited to the IRES via interaction with eIF3.

The IRESs of hepatitis C virus (Fig. 11.7C), some pestiviruses, and teschoviruses function very differently from those of picornaviruses. The formation of the 48S initiation complex on the mRNA is independent of eIF4A, eIF4B, and eIF4F. Purified 40S ribosomal subunits bind directly to stem-loop IIId of the hepatitis C virus IRES, and single point mutations in this structure abolish both the interaction and internal initiation. Translation requires formation of a 3-nucleotide base pair between a loop in the IRES and a helix in 18S rRNA. Addition of only Met-tRNA$_i$, eIF2, and GTP is required to assemble the 48S complexes. A dramatic conformational change in the 40S ribosomal subunit occurs when it binds the hepatitis C virus IRES, clamping the mRNA in place and setting the AUG initiation codon within the P site of the ribosome. The IRES also contacts the E site of the ribosome, where the deacylated tRNA is harbored after translocation of the 80S ribosome. Initiation of translation from the IRES of hepatitis C virus and related viruses therefore resembles initiation of translation of bacterial mRNAs.

The intergenic IRESs of picornavirus-like viruses of insects are bound by the 40S ribosome independent of initiation proteins, and translation does not begin at an AUG codon. The secondary structure of the IRES of these viruses mimics an uncharged tRNA, and mutations that destabilize the fold abrogate translation. The tRNA-like structure is recognized and bound by the 40S ribosomal subunit, placing the initiation codon within the A site instead of the P site (Fig. 11.5A). Initiation is therefore dependent on elongation proteins eEF1A and eEF2 and the appropriate aminoacylated tRNAs. Consequently, initiation from these IRESs is inhibited by the ternary complex (Met-tRNA$_i$–eIF2–GTP) and a high concentration of Met-tRNA$_i$. Furthermore, in cells infected with these viruses, recycling of eIF2-GDP is blocked and the concentration of the ternary complex is low, both consequences of eIF2α phosphorylation by the host as an antiviral defense. Cellular mRNA translation is inhibited, but the activity of the intergenic IRES is not reduced, because the ternary complex is not needed.

BOX 11.2

EXPERIMENTS

Discovery of the IRES

The hypothesis that poliovirus mRNA is translated by internal ribosome binding was first tested by examining the translation of mRNAs containing two open reading frames (ORFs) separated by the poliovirus 5′ untranslated region (figure, panel A). The second ORF was efficiently translated only if it was preceded by the picornavirus 5′ untranslated region. It was concluded that ribosomes bind within the viral 5′ untranslated region, thereby permitting translation of the second ORF. The segment of the 5′ untranslated region that directs internal ribosome entry was called the IRES.

It had long been known that covalently closed circular mRNAs cannot be translated by 5′-end-dependent initiation. Translation by internal ribosome binding, however, should not require a free 5′ end. To test this hypothesis, circular mRNAs with and without an IRES were created. The circular mRNA was translated only if an IRES was present (figure, panel B). This experiment formally proved that translation initiation directed by an IRES occurs by internal binding of ribosomes and does not require a free 5′ end.

Chen CY, Sarnow P. 1995. Initiation of protein synthesis by the eukaryotic translational apparatus on circular RNAs. *Science* **268:**415–417.

Jang SK, Kräusslich HG, Nicklin MJ, Duke GM, Palmenberg AC, Wimmer E. 1988. A segment of the 5′ nontranslated region of encephalomyocarditis virus RNA directs internal entry of ribosomes during in vitro translation. *J Virol* **62:**2636–2643.

Pelletier J, Sonenberg N. 1988. Internal initiation of translation of eukaryotic mRNA directed by a sequence derived from poliovirus RNA. *Nature* **334:**320–325.

Assays for an IRES. (A) Bicistronic mRNA assay. Plasmids were constructed that encode bicistronic mRNAs encoding the thymidine kinase (tk) and chloramphenicol acetyltransferase (cat) proteins separated by a spacer (light green) or a poliovirus IRES (dark green). Plasmids were introduced into mammalian cells by transformation. In uninfected cells containing either plasmid (top lines), both tk and cat proteins were detected, although without an IRES, cat synthesis was inefficient. Translation of cat from this plasmid probably occurs by reinitiation. In poliovirus-infected cells, 5′-end-dependent initiation is blocked (stop sign), and no proteins are observed without an IRES. cat protein is detected in infected cells when the IRES is present, demonstrating internal ribosome binding. Adapted from J. Pelletier and N. Sonenberg, *Nature* **344:**320–325, 1988, with permission. **(B)** Circular mRNA assay for an IRES. Circular mRNAs containing an ORF (yellow) were produced and translated *in vitro*. No protein product was observed unless an IRES was included in the circular mRNA. Adapted from C. Y. Chen and P. Sarnow, *Science* **268:**415–417, 1995, with permission.

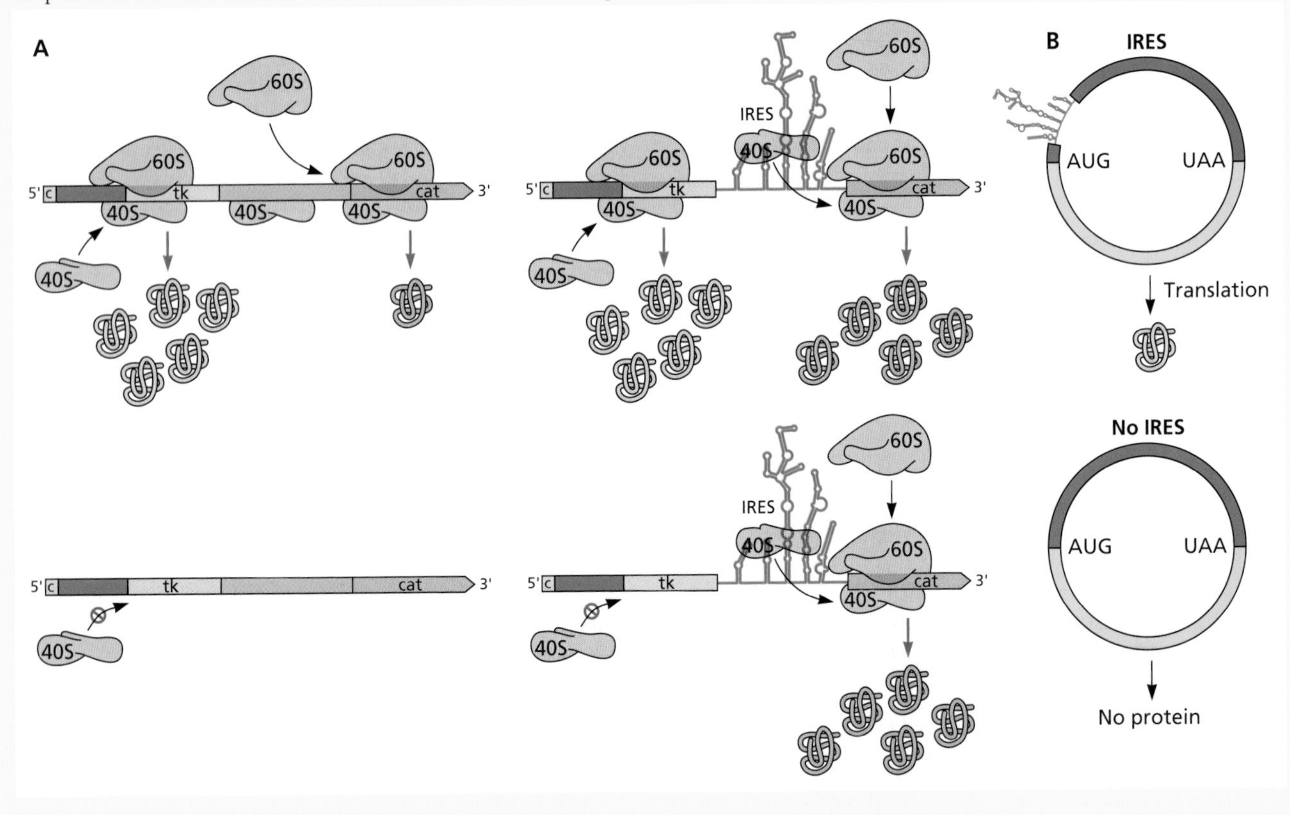

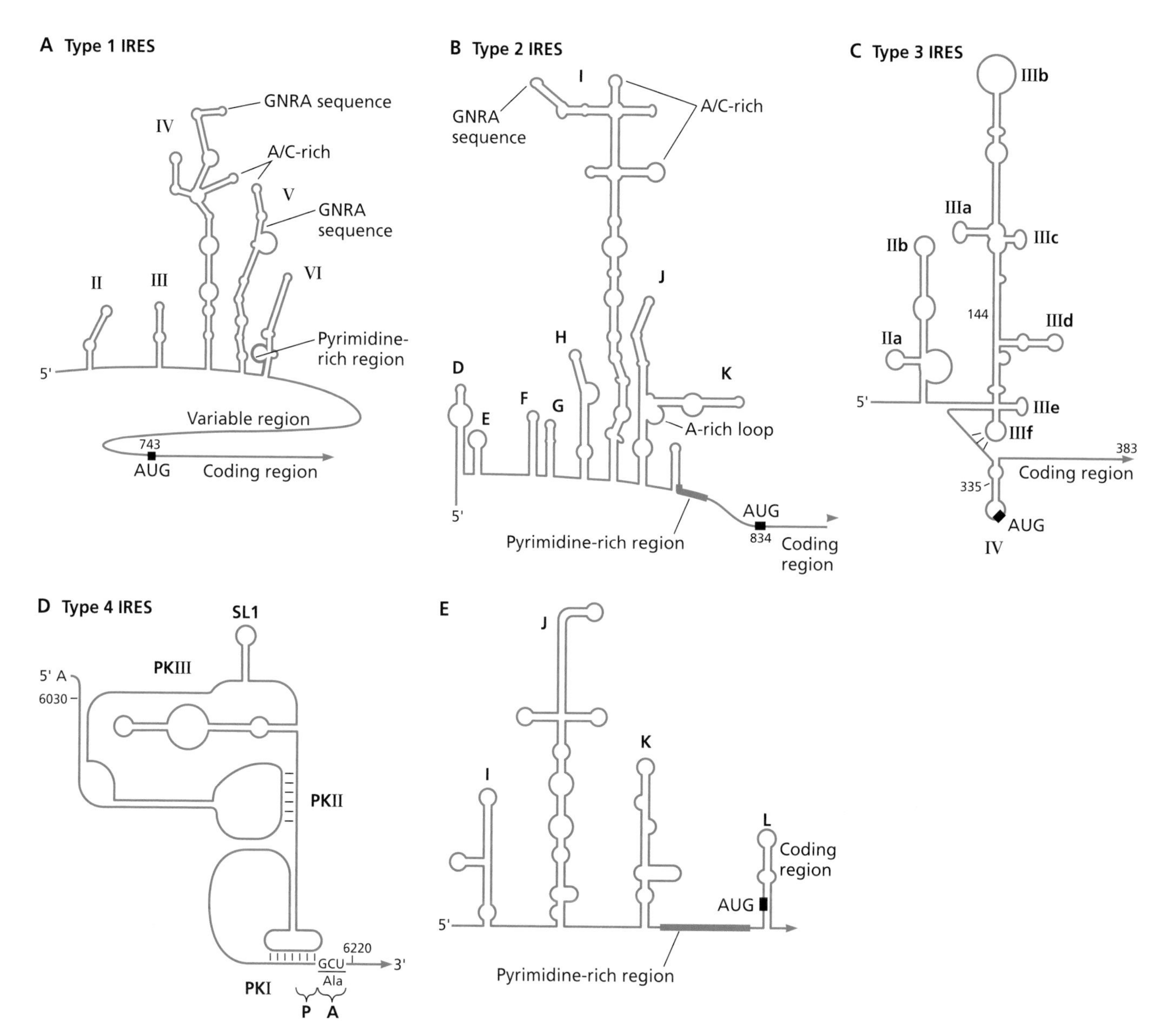

Figure 11.7 Five types of IRES. The 5′ untranslated regions from genome RNAs of poliovirus **(A)**, encephalomyocarditis virus **(B)**, hepatitis C virus **(C)**, cricket paralysis virus **(D)**, and Aichi virus **(E)** are shown. Predicted secondary and tertiary RNA structures (RNA pseudoknots) are shown. Nearly every picornavirus IRES can be placed into one type, with the exception of hepatitis A virus. The poliovirus IRES is a type 1 IRES, which is found in the genomes of enteroviruses and rhinoviruses. The ribosome probably enters the IRES at domains V and VI and scans to the AUG initiation codon, which is located 50 to 100 nucleotides past the 3′ end of the IRES. The type 2 IRES is found in the genomes of aphthoviruses and cardioviruses. The 3′ end of the hepatitis C virus IRES (type 3) extends beyond the AUG initiation codon (black box). The IRES of picornavirus-related viruses of insects (type 4), such as cricket paralysis virus, mimics a tRNA and occupies the P site in the 40S ribosomal subunit. Translation initiates with a non-AUG codon from the A site. The fifth class of IRES is exemplified by the 5′ untranslated region of Aichi virus, which comprises four domains. GNRA, a four-base hairpin loop sequence comprised of guanine, any base, a purine, and adenine; PK, pseudoknot; SL, stem-loop. (A and B) Adapted from S. R. Stewart and B. L. Semler, Semin. Virol. **8:**242–255, 1997, with permission. (C) Adapted from S. M. Lemon and M. Honda, Semin. Virol. **8:**274–288, 1997, with permission. (D) Adapted from E. Jan and P. Sarnow, J. Mol. Biol. **324:**889–902, 2002, with permission.

BOX 11.3

BACKGROUND

Use of the IRES in expression vectors

The IRES has been used widely in the expression of exogenous genes in eukaryotes. One strategy is to produce mRNAs in the cytoplasm by using a bacteriophage DNA-dependent RNA polymerase, such as T7 RNA polymerase. Such mRNAs are poorly translated because they are not capped; inclusion of an IRES in the 5′ untranslated region allows them to be translated efficiently.

Another application of IRESs is in gene therapy, where the ability to introduce multiple therapeutic genes is desirable. An example is the treatment of ischemic disease by coproduction of vascular endothelial growth factor and angiopoietin: synergistic effects are obtained. To accomplish this goal, an IRES is incorporated between two transgenes (figure, top panel). A single mRNA is produced from which two proteins are translated.

IRESs have also been used in the isolation of mutant mice by homologous recombination in embryonic stem cells. Bicistronic vectors are designed to produce mRNA encoding the altered protein and β-galactosidase, separated by an IRES (figure, bottom panel). Because β-galactosidase is encoded on the same mRNA as the targeted gene product, it serves as a marker for expression of the mutated gene.

Renaud-Gabardos E, Hantelys F, Morfoisse F, Chaufour X, Garmy-Susini B, Prats AC. 2015. Internal ribosome entry site-based vectors for combined gene therapy. *World J Exp Med* **5**:11–20.

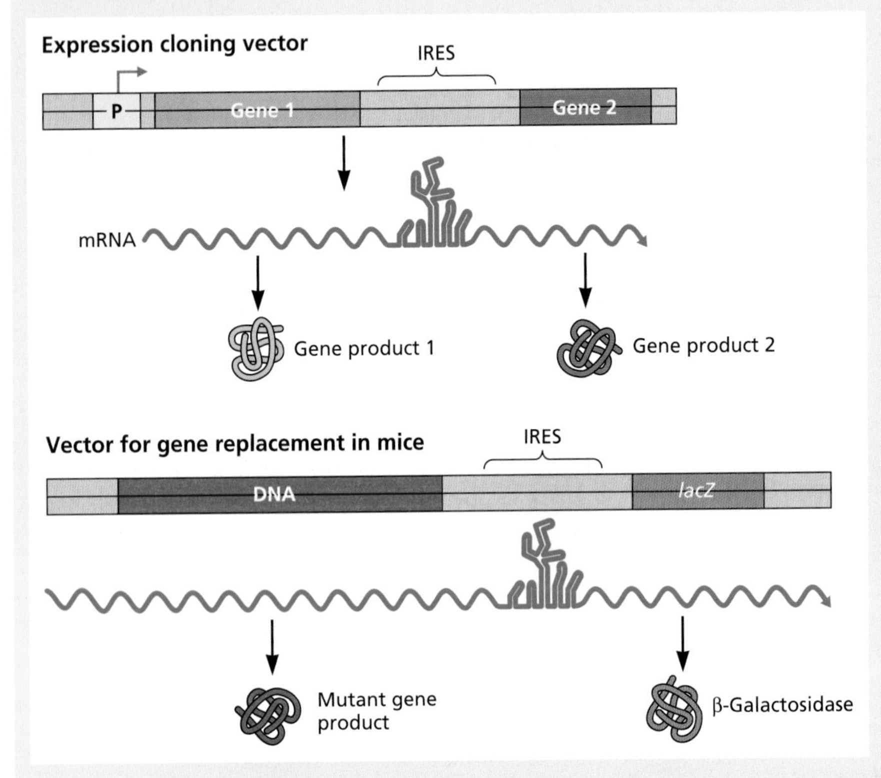

(Top) Design of plasmids for expression of two genes. DNA encoding the first gene is followed by an IRES and then a second gene. A single mRNA is produced from a promoter when this plasmid DNA is introduced into cells. The first gene is translated by 5′-end-dependent translation, and that of the second by internal ribosome entry. **(Bottom)** Vector for gene replacement in mice. In this example, the goal is to replace the gene with a mutant version. The targeting plasmid consists of mutant DNA followed by an IRES and the *lacZ* gene. The flanking light blue bars represent sequences from the mouse gene that mediate homologous recombination. After replacement of the endogenous gene with this synthetic version, mRNA that encodes the mutant gene product as well as the β-galactosidase protein will be produced. The latter can be detected in tissues by staining with the chromogenic substrate X-Gal (5-bromo-4-chloro-3-indolyl-β-D-galactopyranoside).

As discussed above, translation of cellular mRNAs is enhanced by the juxtaposition of mRNA ends (Fig. 11.4). Translation of viral mRNAs by internal initiation is also stimulated by this arrangement. An example is the 5′ and 3′ ends of the RNA genome of foot-and-mouth disease virus, which are brought together by RNA-RNA interactions (Fig. 11.9A). This mechanism is distinct from the protein-RNA interactions that bring together RNA ends of cellular mRNAs.

Other host cell proteins that contribute to IRES function. In addition to canonical translation proteins, activity of IRESs requires other cellular RNA-binding proteins. These were first discovered because the poliovirus IRES functions poorly in reticulocyte lysates, in which most capped mRNAs are translated efficiently (Box 11.4). Addition of a cytoplasmic extract from other cells to reticulocyte lysates restores efficient translation from this IRES. These observations led to the suggestion that ribosome binding to the IRES requires more than translation initiation proteins. Such proteins were first identified by their ability to bind to the IRES and to restore its function in the reticulocyte lysate.

The requirements for RNA-binding proteins differ among various IRESs, and no single host cell protein that is essential for the function of all of them has been identified.

Type 1 or 2 IRES

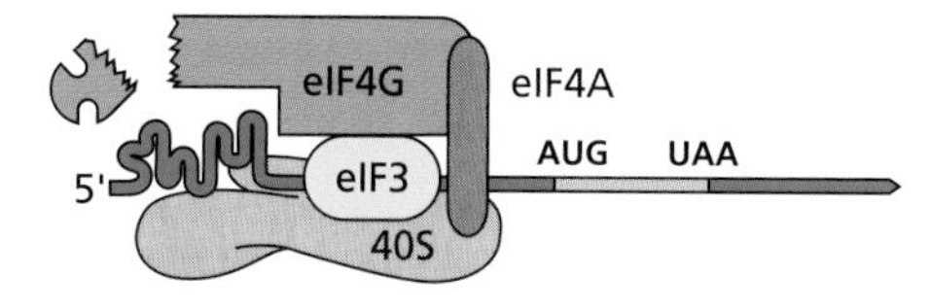

Hepatitis C virus IRES

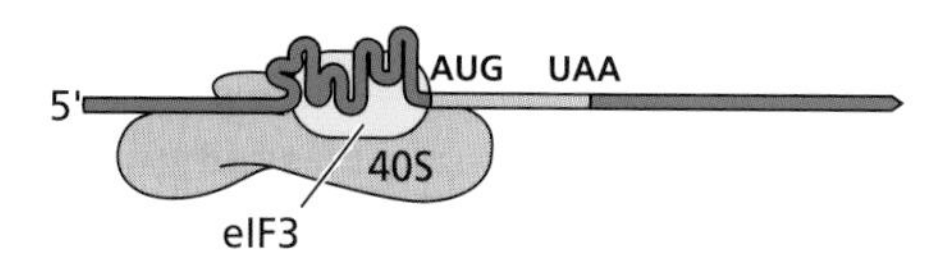

Figure 11.8 5′-end-independent initiation. (Top) Initiation on the type 1 or 2 IRES does not depend on the presence of a cap structure, but requires the C-terminal fragment of eIF4G to recruit the 40S ribosomal subunit via its interaction with eIF3. eIF4G probably binds directly to the IRES. (Bottom) The ribosomal 40S subunit binds to the hepatitis C virus IRES without the need for translation initiation proteins. eIF3 also binds the IRES and is thought to be necessary for recruitment of the 60S ribosomal subunit.

All type 1 IRESs require the cytoplasmic RNA-binding protein poly(rC)-binding protein 2 (Pcbp2) for activity. This protein was originally identified by its ability to bind stem-loop IV of the poliovirus IRES (Fig. 11.7A). Mutations in the poliovirus 5′ untranslated region that abolish binding of Pcbp2 lead to decreased translation *in vitro*. Depletion of Pcbp2 from human translation extracts inhibits translation dependent on the IRESs of poliovirus, Coxsackievirus B, and rhinovirus, but not on those of encephalomyocarditis virus or foot-and-mouth disease virus. Translation activity of the IRESs was restored by addition of purified Pcbp2. This protein binds to, and functions cooperatively during internal initiation with, serine/arginine-rich splicing factor 3 (SRp20), a protein that is essential for constitutive splicing and regulation of alternative splice site selection. Cleavage of Pcbp2 is thought to enable a switch from translation to replication during poliovirus infection (Chapter 6).

The encephalomyocarditis virus IRES is highly active in the absence of RNA-binding proteins, while the rhinovirus and foot-and-mouth disease virus IRESs require polypyrimidine-tract-binding protein (Ptb), also called heterogeneous nuclear ribonucleoprotein I (hnRnpI), a negative regulator of alternative pre-mRNA splicing. The poliovirus and Aichi virus IRESs also require Ptb for activity. This predominantly nuclear protein is redistributed to the cytoplasm during poliovirus infection. Ptb binds to sequences upstream of the pyrimidine-rich sequence of the poliovirus IRES, and to both the 5′ and 3′ untranslated regions of hepatitis C virus RNA. Yet other RNA-binding proteins are required for the activities of the foot-and-mouth disease virus, rhinovirus, and poliovirus IRESs.

There is no evidence that such RNA-binding proteins facilitate recruitment of 43S preinitiation complexes to the IRES. Rather, it appears that these proteins act as RNA chaperones to maintain the IRES in a secondary and tertiary structure that is appropriate for binding to ribosomes and translation initiation proteins. In support of this hypothesis is the observation that all are RNA-binding proteins that can form multimers that contact the IRES at multiple points; these proteins protect some IRESs from enzymatic degradation. They bind at numerous sites, consistent with a role in constraining three-dimensional flexibility. The binding site for Ptb on the poliovirus IRES overlaps that of eIF4G, leading to

Figure 11.9 Long-range RNA-RNA interactions aid translation. (A) Activity of the IRES of foot-and-mouth disease virus is enhanced by interactions with sequences at the 3′ end of the viral RNA. **(B)** The 3′-cap-independent translational enhancer (3′ CITE) found in some plant viral RNAs binds eIF4F, allowing recruitment of the 40S ribosomal subunit. A long-range interaction of this sequence with the 5′ end of the viral RNA positions the 40S ribosomal subunit at the initiation codon.

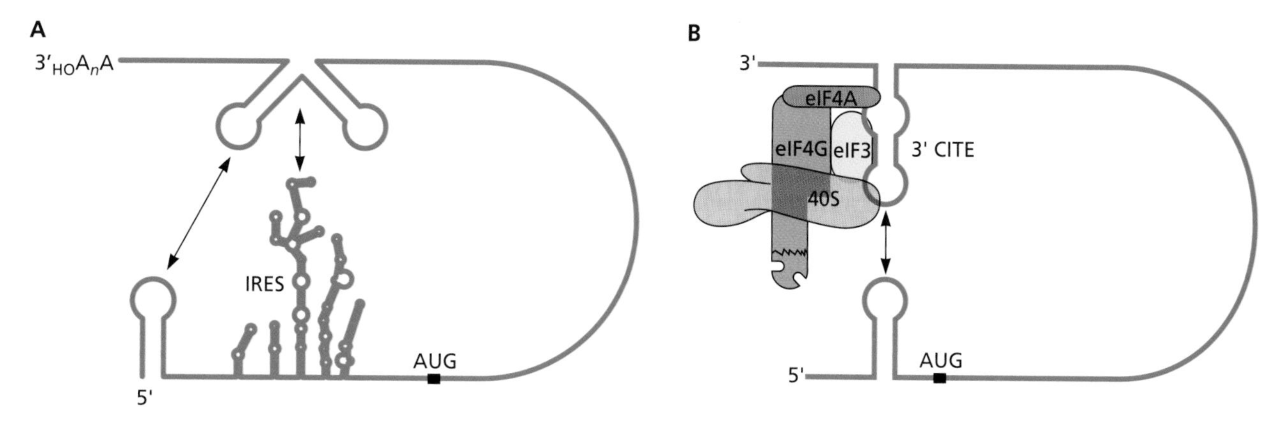

BOX 11.4

METHODS

Translation in vitro*: the reticulocyte lysate and wheat germ extract*

Our present understanding of the fundamentals of translation initiation, elongation, and termination, as well as viral translation strategies, would not be possible without the technique of *in vitro* translation in cell extracts. In this method, cells are lysed and the nuclei are removed by centrifugation. The mRNA is added to the lysate, and the mixture is incubated to allow translation to proceed.

The ideal extract for *in vitro* translation has two important properties: high translation efficiency and low protein synthesis in the absence of exogenous mRNA. By the early 1970s, cell extracts prepared from Krebs II ascites tumor cells or rabbit reticulocytes (immature red blood cells that primarily produce hemoglobin; they lack nuclei) were found to translate protein with high efficiency, but the presence of endogenous mRNAs that were also translated complicated the analysis of proteins made from added mRNA. In 1973, a cell extract from commercial wheat germ that had low background protein synthesis, and in which exogenous mRNAs were translated very efficiently, was developed. A few years later, the background in a reticulocyte lysate was eliminated by treatment with micrococcal nuclease, which destroyed the endogenous mRNA. This nuclease requires calcium for its activity, and it was therefore a simple matter of adding a calcium chelator, EGTA, to the reaction to prevent the degradation of exogenously added mRNA.

Wheat germ extract and reticulocyte lysate are still widely used in studies of translation, because the cells are abundant, inexpensive, and excellent sources of initiation proteins. Micrococcal nuclease followed by calcium chelation has been successfully used to make mRNA-dependent extracts from many mammalian cell types, although the translation efficiency of such systems does not approach that of wheat germ or reticulocyte lysates. Unfortunately, it has not been possible to prepare translation extracts from normal mammalian tissues consistently, a failure that has hampered the study of regulation of tissue-specific translation in virus-infected and uninfected cells.

Pelham HR, Jackson RJ. 1976. An efficient mRNA-dependent translation system from reticulocyte lysates. *Eur J Biochem* **67**:247–256.

Roberts BE, Patterson BM. 1973. Efficient translation of tobacco mosaic virus RNA and rabbit globin 9S RNA in a cell-free system from commercial wheat germ. *Proc Natl Acad Sci U S A* **70**:2330–2334.

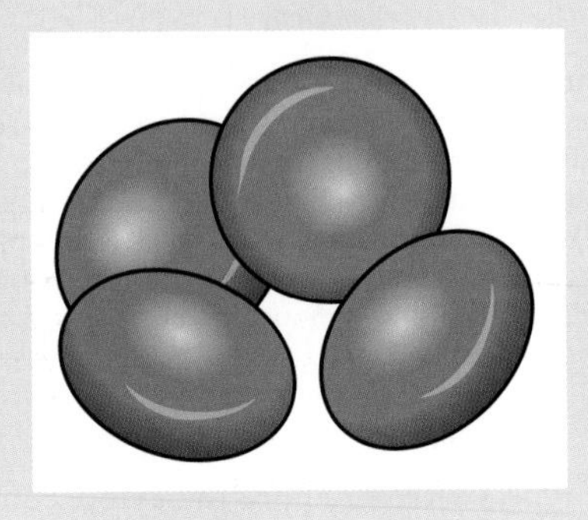

repositioning of the protein, which may explain the need for this chaperone.

3′-cap-independent translational enhancers. The (+) strand RNA genomes of a number of plant viruses that lack both 5′ caps and 3′ poly(A) tails require a 3′-cap-independent translational enhancer (3′ CITE) for protein synthesis. These structures have been placed into several different classes, but all recruit ribosomes by binding directly or via eIF4G or eIF4E. Because they are located in the 3′ noncoding region of the viral RNA, long range RNA-RNA interactions are required to place ribosomes or translation proteins at the 5′ end, where translation begins. An example is the genome of barley yellow dwarf virus, in which complementary sequences located in the 5′ untranslated region and 3′ CITE form an RNA-RNA bridge by a kissing-loop interaction (Fig. 11.9B). Simultaneous binding of the 3′ CITE to eIF4F and the 5′ untranslated region recruits the 40S ribosomal subunit to the RNA 5′ end.

Some viral genomes that lack caps and poly(A) tails utilize both a 5′ IRES and 3′ CITE for translation. For example, in the mRNA of blackcurrant reversion virus, a member of the *Picornavirales*, long-range RNA-RNA interactions between these elements are required for translation. This interaction might represent yet another way to maximize translation efficiency by juxtaposing the 5′ and 3′ ends of uncapped and unpolyadenylated RNAs.

Elongation and Termination

During elongation, the ribosome selects aminoacylated tRNA according to the sequence of the mRNA codon, and catalyzes the formation of a peptide bond between the nascent polypeptide and the incoming amino acid. The 40S ribosomal subunit is responsible for both decoding and selection of the cognate tRNA. The RNA of the 60S subunit catalyzes the peptidyltransferase reaction without any soluble nonribosomal proteins or a source of energy. Elongation is assisted by three proteins that maintain the speed and accuracy of translation. In the 80S initiation complex, the Met-tRNA$_i$ is bound to the P site of the ribosome (Fig. 11.10). Elongation of the peptide chain begins with addition of the next amino acid encoded by the triplet that occupies the A site. An important component of this process is elongation factor eEF1A, which is bound to aminoacylated tRNA, a molecule of GTP, and the nucleotide exchange protein eEF1B.

Interaction between the codon and the anticodon leads to a conformational change in the ribosome called **accommodation**, the hydrolysis of GTP and the release of eEF1A-GDP. Accommodation maintains the fidelity of translation, because it can occur only upon proper codon-anticodon base pairing and is required for GTP hydrolysis. If an incorrect tRNA enters the A site, accommodation does not occur and the aminoacylated tRNA is rejected. The large ribosomal subunit catalyzes the formation of a peptide bond between the amino acids

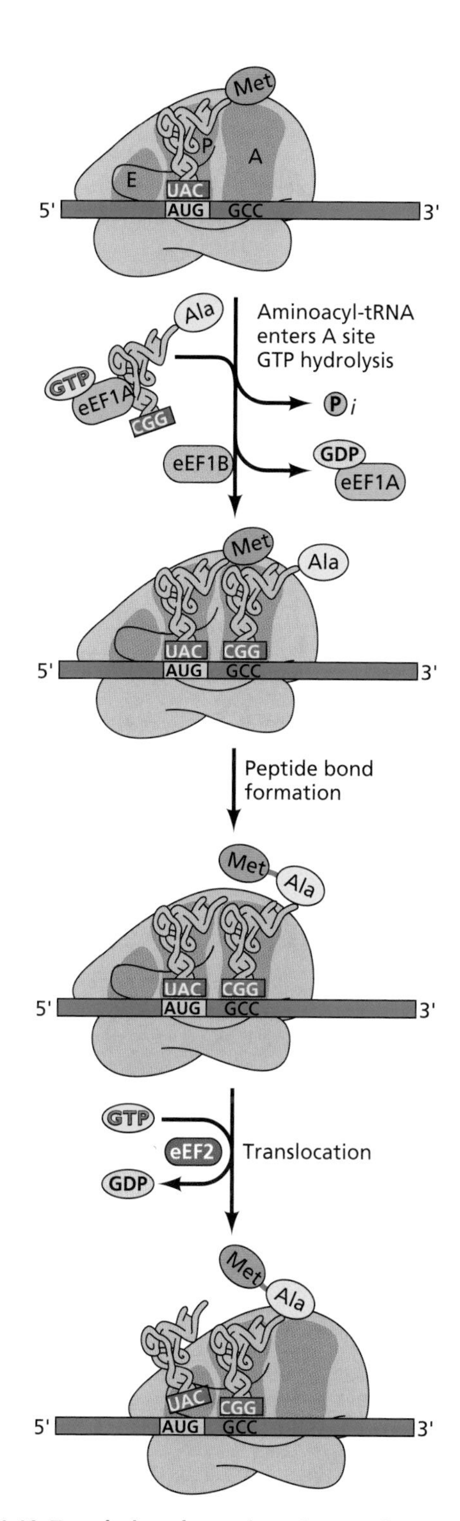

Figure 11.10 Translation elongation. There are three tRNA-binding sites on the ribosome, called peptidyl (P), aminoacyl or acceptor (A), and exit (E). After the initiating Met-$tRNA_i$ is positioned in the P site, the second aminoacyl-tRNA (alanyl-tRNA is shown) is brought to the A site by eEF1A bound to GTP. After GTP hydrolysis, eEF1A is released. The guanine nucleotide exchange protein eEF1B exchanges GDP of eEF1A-GDP with GTP, allowing eEF1A to interact with a tRNA synthetase and bind a newly aminoacylated tRNA. The peptide bond is then formed; this reaction is followed by movement of the ribosome 3 nucleotides along the mRNA, a step that requires GTP hydrolysis and eEF2. The peptidyl (Met-Ala) tRNA moves to the P site, and the uncharged tRNA moves to the E site. The A site is now empty, ready for another aminoacyl-tRNA. Adapted from G. M. Cooper, *The Cell: a Molecular Approach* (ASM Press, Washington, DC, and Sinauer Associates, Sunderland, MA, 1997).

occupying the P and A sites. The 80S ribosome then moves 3 nucleotides along the mRNA. Translocation is dependent upon eEF2 and hydrolysis of GTP. This motion moves the uncharged tRNA to the exit (E) site and the peptidyl-tRNA to the P site, enabling a new aminoacylated tRNA to enter the A site and subsequent release of the uncharged tRNA. This cycle is repeated until the ribosome encounters a stop codon. mRNAs are usually bound by many ribosomes (**polysomes**), with each ribosome separated from its neighbors by ~100 to 200 nucleotides.

Termination is a modification of the elongation process: once the stop codon enters the A site of the ribosome, it is recognized by the 40S subunit, and the 60S subunit cleaves the ester bond between the protein chain and the last tRNA. Recognition of the three stop codons (UAA, UAG, and UGA) by the 40S ribosomal subunit is facilitated by the release proteins eRF1 and eRF3 (Fig. 11.11). The structure of eRF1 mimics that of tRNA, allowing the release protein to occupy the A site of the ribosome. The N terminus of eRF1 recognizes all three stop codons. Once bound in the A site, eRF1 and eRF3 cooperate to induce a rearrangement of the 80S ribosome, translocation of the P-site codon, and release of the polypeptide. The interaction between eRF1 and the ribosome stimulates the GTPase activity of eRF3, which is bound to the C terminus of eRF1. GTP hydrolysis is required for release of the nascent polypeptide.

In addition to accommodation, the E site is an important determinant of the fidelity of protein synthesis. When the E site is occupied by a deacylated tRNA, the affinity of the A site for aminoacyl-tRNA is low. Consequently, incorrect tRNAs are readily rejected. When the E site is empty, the affinity of the A site for aminoacyl-tRNA is significantly higher, making rejection of incorrect tRNAs less likely. An occupied E site also prevents tRNA slippage; when this site is empty, increased ribosomal frameshifting occurs.

Although stop codons are the major determinants of translation termination, other sequences can affect the efficiency of this process. The nucleotide immediately downstream of the stop codon can influence chain termination and ribosome dissociation. In eukaryotes, the preferred termination signals are UAA(A/G) and UGA(A/G).

After release of the polypeptide chain, the 60S ribosomal subunit and tRNA are released from the mRNA by the cooperation of eIF1, eIF1A, and eIF3 (Fig. 11.12). It has

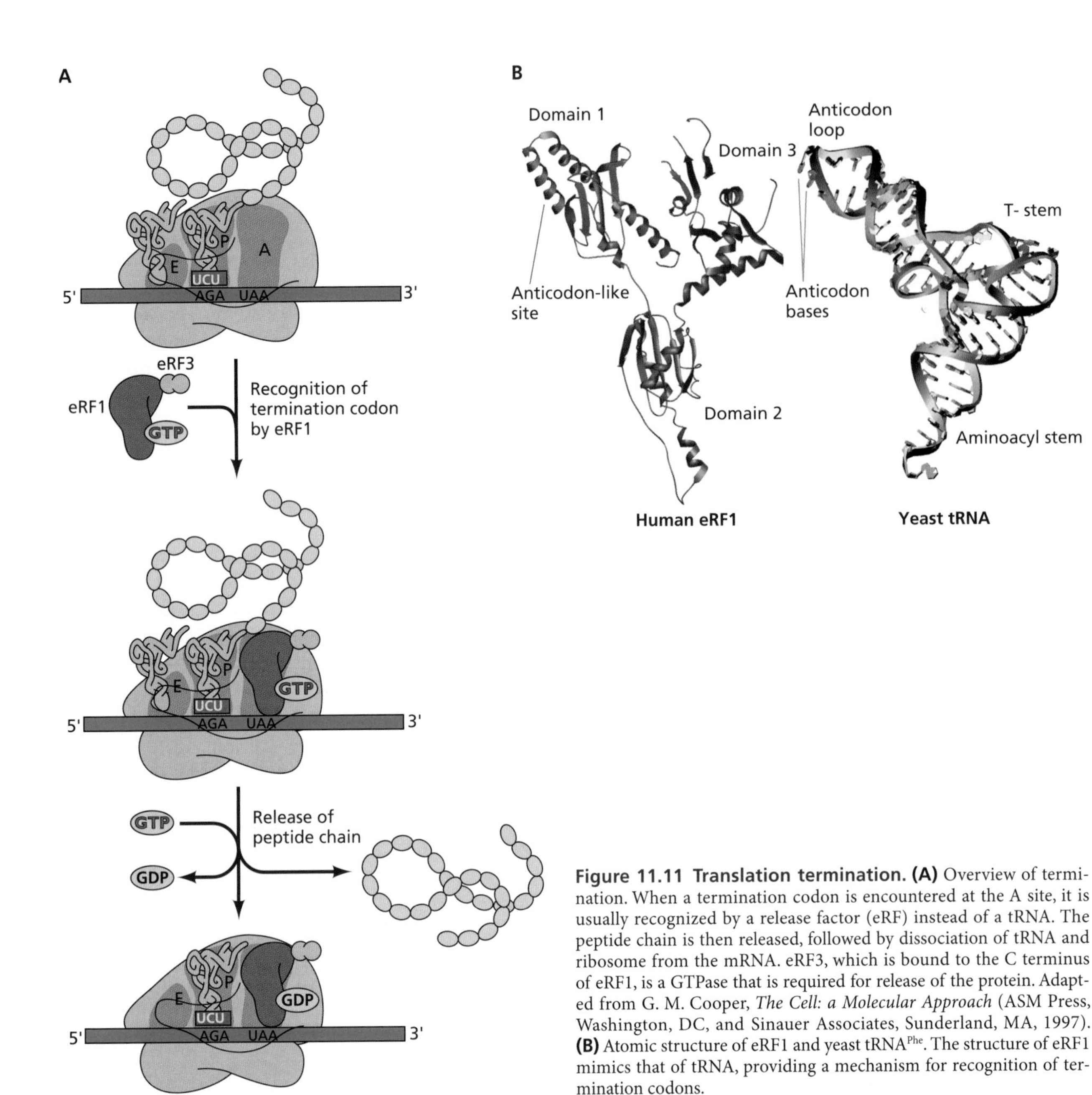

Figure 11.11 Translation termination. (A) Overview of termination. When a termination codon is encountered at the A site, it is usually recognized by a release factor (eRF) instead of a tRNA. The peptide chain is then released, followed by dissociation of tRNA and ribosome from the mRNA. eRF3, which is bound to the C terminus of eRF1, is a GTPase that is required for release of the protein. Adapted from G. M. Cooper, *The Cell: a Molecular Approach* (ASM Press, Washington, DC, and Sinauer Associates, Sunderland, MA, 1997). **(B)** Atomic structure of eRF1 and yeast $tRNA^{Phe}$. The structure of eRF1 mimics that of tRNA, providing a mechanism for recognition of termination codons.

been suggested that 40S ribosomal subunits preferentially engage in new rounds of translation initiation on the same mRNA. This hypothesis is supported by the finding that eIF3, which remains bound to the 40S ribosomal subunit after termination, also binds eIF4G (Fig. 11.13). Other observations that are consistent with this model include the ability of eRF3 to bind Pabp1 (Fig. 11.13) and the stimulation of 60S ribosomal subunit joining by this protein. As a result, ribosomes may shuttle from the 3*f* end of the mRNA back to the 5*f* end, beginning the synthesis of another molecule of the protein.

The Diversity of Viral Translation Strategies

A variety of unusual translation mechanisms expand the coding capacity of viral genomes and allow the synthesis of multiple polypeptides from a single RNA (Fig. 11.14). All were discovered in virus-infected cells and subsequently shown to operate during translation of cellular mRNAs. Nontranslational solutions for maximizing the number of proteins encoded in viral genomes are discussed in other chapters and include the synthesis of multiple subgenomic mRNAs, mRNA splicing, and RNA editing.

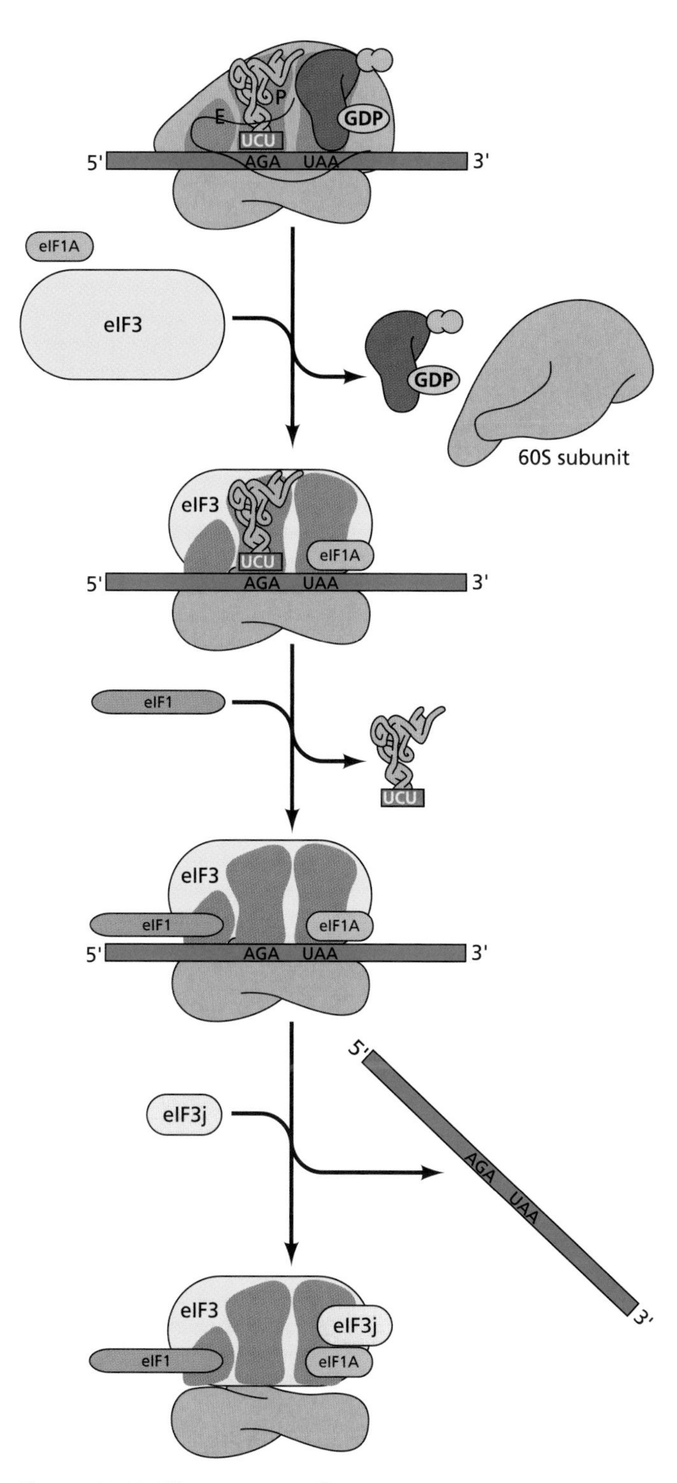

Figure 11.12 Ribosome recycling. After peptide release, eIF1A and eIF3 cause dissociation and release of the 60S ribosomal subunit. Release of the P-site deacylated tRNA is promoted by eIF1 and is followed by dissociation of mRNA mediated by eIF3j binding.

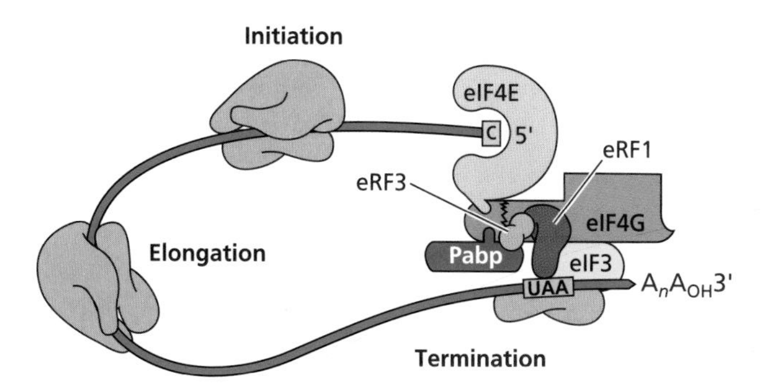

Figure 11.13 Juxtaposition of mRNA ends. Shown is a juxtaposition of mRNA ends by interactions of termination and initiation proteins, Pabp1, and the mRNA 5′ and 3′ ends. eRF3 binds both eRF1 and Pabp1. Adapted from N. Uchida et al., *J Biol Chem* **277**:50286–50292, 2002, with permission.

Polyprotein Synthesis

One strategy allowing for the production of multiple proteins from an RNA genome is to synthesize from a single mRNA a polyprotein precursor, which is then proteolytically processed to form functional viral proteins. A dramatic example of protein processing occurs in picornavirus-infected cells: nearly the entire (+) strand RNA is translated into a large polyprotein (Fig. 11.15A). Processing of this precursor is carried out by two virus-encoded proteases, $2A^{pro}$ and $3C^{pro}$, which cleave between Tyr and Gly and between Gln and Gly, respectively. In both cases, flanking amino acids control the efficiency of cleavage so that not all Tyr-Gly and Gln-Gly pairs in the polyprotein are processed. These two proteases are active in the nascent polypeptide and release themselves by self-cleavage. Consequently, the polyprotein is not observed in infected cells because it is processed as soon as the protease-coding sequences have been translated. After the proteases have been released, they cleave other polyprotein molecules.

Protein production can be controlled by the rate and extent of polyprotein processing. In addition, alternative utilization of cleavage sites can produce proteins with different activities. For example, the poliovirus protease $3C^{pro}$ does not process the capsid protein precursor P1 efficiently. Rather, the $3C^{pro}$ precursor, $3CD^{pro}$, is required for processing of P1. By regulating the quantity of $3CD^{pro}$ produced, the extent of capsid protein processing can be controlled. Because $3CD^{pro}$ and $3C^{pro}$ process Gln-Gly pairs in the remainder of the polyprotein with the same efficiency, an interesting question is why $3CD^{pro}$, which also contains $3D^{pol}$ protein, is further processed to produce $3C^{pro}$ (Fig. 11.15A). The answer is that $3CD^{pro}$ protein, while active as a protease, does not possess RNA polymerase activity and consequently some molecules must be cleaved to allow RNA replication.

Some viral precursor proteins are processed by cellular proteases. The genome of flaviviruses contains an open

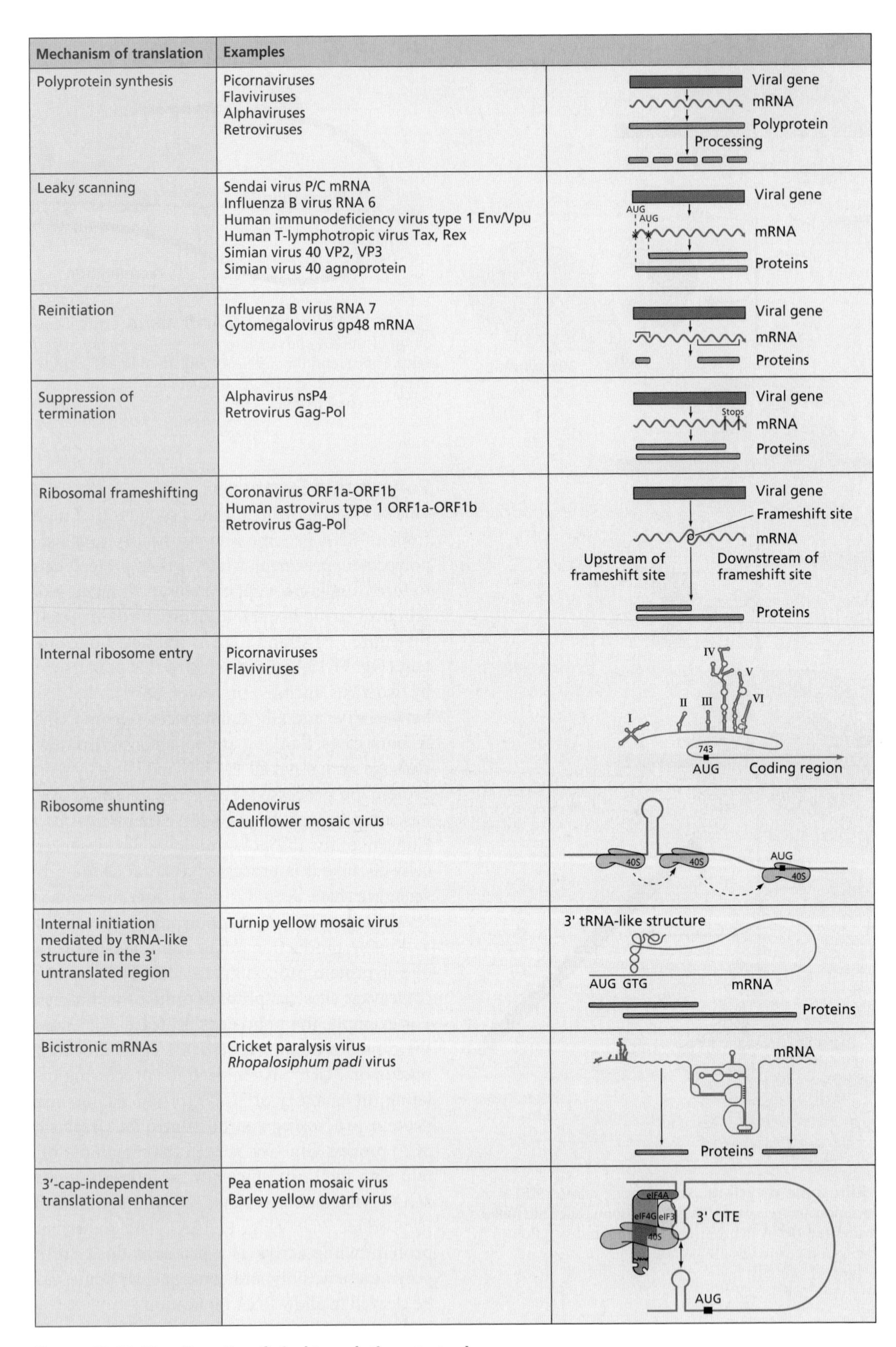

Mechanism of translation	Examples
Polyprotein synthesis	Picornaviruses Flaviviruses Alphaviruses Retroviruses
Leaky scanning	Sendai virus P/C mRNA Influenza B virus RNA 6 Human immunodeficiency virus type 1 Env/Vpu Human T-lymphotropic virus Tax, Rex Simian virus 40 VP2, VP3 Simian virus 40 agnoprotein
Reinitiation	Influenza B virus RNA 7 Cytomegalovirus gp48 mRNA
Suppression of termination	Alphavirus nsP4 Retrovirus Gag-Pol
Ribosomal frameshifting	Coronavirus ORF1a-ORF1b Human astrovirus type 1 ORF1a-ORF1b Retrovirus Gag-Pol
Internal ribosome entry	Picornaviruses Flaviviruses
Ribosome shunting	Adenovirus Cauliflower mosaic virus
Internal initiation mediated by tRNA-like structure in the 3' untranslated region	Turnip yellow mosaic virus
Bicistronic mRNAs	Cricket paralysis virus *Rhopalosiphum padi* virus
3'-cap-independent translational enhancer	Pea enation mosaic virus Barley yellow dwarf virus

Figure 11.14 The diversity of viral translation strategies.

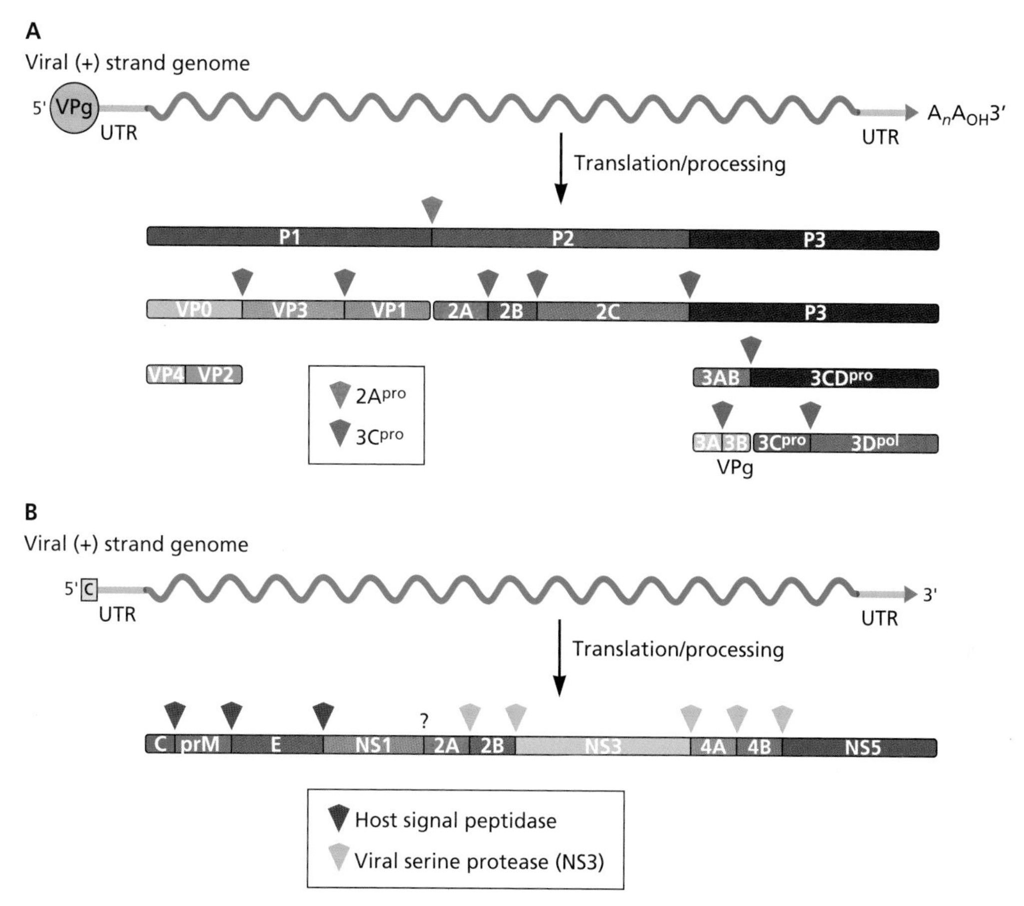

Figure 11.15 Polyprotein processing of picornaviruses and flaviviruses. (A) Processing map of protein encoded by the poliovirus genome. The viral RNA is translated into a long precursor polyprotein that is processed by two viral proteases, 2A^{pro} and 3C^{pro}, to form viral proteins. Cleavage sites for each protease are shown. **(B)** Cleavage map of protein encoded in the flavivirus genome. Processing of the flavivirus precursor polyprotein is carried out either by the host signal peptidase or by the viral protease NS3. UTR, untranslated region.

reading frame of >10,000 bases (Fig. 11.15B). This mRNA is translated into a polyprotein precursor that is processed by a viral serine protease and by host signal peptidase. The latter enzyme is located in the endoplasmic reticulum (ER), where it removes the signal sequence from proteins translocated into the lumen (Chapter 12). The viral proteins processed by the cellular signal peptidase must therefore be inserted into the ER.

Leaky Scanning

Although the vast majority of eukaryotic mRNAs are monocistronic (Fig. 11.1), leaky scanning allows some viral mRNAs to be functionally polycistronic, i.e., to encode more than one protein. In the scanning model of mRNA translation, 40S ribosomal subunits bind close to the mRNA 5′ end and initiate translation at the first AUG. In a mechanism called **leaky scanning**, some ribosomes bypass the first AUG codon and continue scanning to an alternative downstream AUG. Leaky scanning can allow the synthesis of multiple isoforms of a protein with common C termini, or distinct proteins, by translation of overlapping or nonoverlapping open reading frames, respectively. Translation of overlapping reading frames also occurs in many other viral mRNAs, and is the most frequent mechanism for translation of polycistronic mRNAs of RNA viruses.

The P/C gene of Sendai virus is the model for genes that encode mRNAs with such translational flexibility (Fig. 11.16). P protein is translated from an open reading frame beginning with an AUG codon at nucleotide 104. C proteins are produced from a different reading frame, which begins at nucleotide 81, and are completely different from P proteins. No less than four C proteins (called C′, C, Y1, and Y2) are produced by translation beginning at four in-frame initiation codons. The first start site is an unusual ACG codon, and the third, fourth, and fifth are AUG codons; the result is a nested set of proteins with a common C terminus.

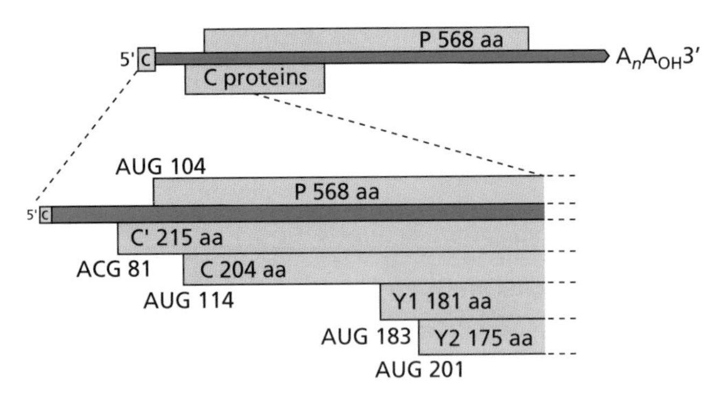

Figure 11.16 Leaky scanning and mRNA editing in the Sendai virus P/C gene. P and C protein open reading frames are shown as brown and blue boxes, respectively. An enlargement of the 5′ end of the mRNA is shown below, indicating the different start sites for four of the C proteins. aa, amino acids. Adapted from J. Curran et al., *Semin Virol* **8:**351–357, 1997, with permission.

The first three initiation sites on P/C mRNA are likely to be arranged to permit translation by leaky scanning. The first start site, $ACG^{81/C'}$, is surrounded by a good initiation context but is inefficient because of the unusual start codon. Some ribosomes bypass this initiator codon and initiate at the second, $AUG^{104/P}$ (CGC<u>AUG</u>G). Although the second is an AUG codon, the context is poor, and some ribosomes find their way to the third initiation codon, $AUG^{114/C}$, which has a better context (AAG<u>AUG</u>C). Consistent with this hypothesis, mutagenesis of $ACG^{81/C'}$ to AUG abolishes initiation at $AUG^{104/P}$ and $AUG^{114/C}$. When successive initiation codons are used in leaky scanning, they are increasingly efficient as start sites.

The last two C protein initiation codons, $AUG^{183/Y1}$ and $AUG^{201/Y2}$, are not likely to be translated by leaky scanning because they are in the poorest contexts of the five. Furthermore, mutagenesis of $ACG^{81/C'}$ to AUG has no effect on synthesis of Y1 and Y2 proteins. Translation of Y1 and Y2 proteins is initiated by ribosome shunting. An interesting question is how the different mechanisms for translation of P/C mRNA are coordinated such that, for example, shunting does not dominate at the expense of translation of upstream AUG codons. The answer to this question is not known, but Y protein synthesis relative to that of the other C proteins varies in different cell lines. This result suggests that cellular proteins might regulate ribosome shunting on P/C mRNA, although no such protein has been identified.

Leaky scanning may be promoted by mechanisms other than a suboptimal sequence surrounding the first AUG codon. Proximity of an AUG codon to the mRNA 5′ end (<30 nucleotides) or to a downstream AUG codon (within ~10 nucleotides) decreases efficiency of initiation.

Reinitiation

Upon termination of translation, the ribosome dissociates into 40S and 60S subunits and falls away from the mRNA. Ribosomes that translate a short open reading frame may remain associated with the mRNA and can reinitiate on a downstream AUG, resulting in two proteins from a single mRNA (Fig. 11.17). When the ribosome completes translation of the short open reading frame, it cannot reinitiate until it reacquires initiation proteins, including the ternary complex, as it moves downstream.

Reinitiation after translation of a long open reading frame is rare, and requires specialized signals in the mRNA or *trans*-acting proteins. Reinitiation of translation of longer, overlapping reading frames occurs on mRNA of influenza B virus RNA 7, which encodes two proteins, M1 protein and BM2 protein (Fig. 11.17). M1 protein is translated from the 5′-proximal AUG codon, while the BM2 protein AUG initiation codon is part of the termination codon for M1 protein (UAAUG).

Suppression of Termination

Although translational suppression in eukaryotic mRNAs is extremely rare, suppression of termination occurs during translation of many viral mRNAs as a means of producing a second protein with an extended C terminus. The Gag and Pol genes of Moloney murine leukemia virus are encoded in a single mRNA and separated by an amber termination codon, UAG (Fig. 11.18). The efficiency of suppression is about 4 to 10%. The Gag-Pol precursor is subsequently processed proteolytically to liberate the Gag and Pol proteins. Without this suppression mechanism, the viral enzymes reverse transcriptase and integrase could not be produced. In a similar way, translational suppression of a different termination codon, UGA, is required for the synthesis of nsP4 of alphaviruses (Fig. 11.18). In this example, the efficiency of synthesis is about 10% of that of the normally terminated nsP3 protein. Because nsP4 encodes the RNA-dependent RNA polymerase, suppression is essential for viral RNA replication.

Most translational suppression takes place when normal tRNAs misread termination codons. The misreading of the amber codon in Moloney murine leukemia virus Gag protein for a Gln codon is an example. Suppressor tRNAs that can recognize termination codons and insert a specific amino acid are rare. One example is a suppressor tRNA that inserts selenocysteine, the 21st amino acid, in place of a UGA codon.

The nucleotide sequence 5′ or 3′ of the termination codon can influence the efficiency of translational suppression. Two adenosines just 5′ to the stop codon stimulate read-through of many plant virus mRNAs. Downstream stimulators of suppression comprise either nucleotides adjacent to the codon or RNA secondary structures that begin ~8 nucleotides from the termination codon. In Sindbis virus, efficient suppression of the UGA codon requires only a single C residue 3′ of the termination codon. In contrast, read-through of the UAG codon in Moloney murine leukemia virus mRNA requires a purine-rich sequence 3′ to the termination codon, as well as a

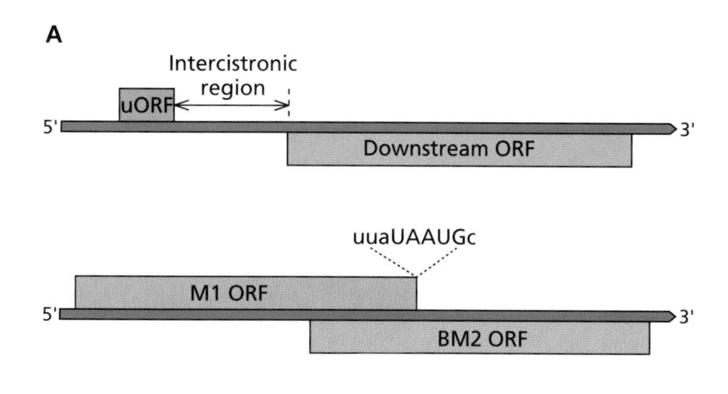

Figure 11.17 Reinitiation of translation. (A) (Top) Some mRNAs contain one or more short, upstream open reading frames (uORFs) that may be translated. Expression of the longer, downstream ORF depends on reinitiation. (Bottom) mRNA produced from influenza B virus RNA segment 7 encodes two proteins, M1 and BM2. The initiation AUG codon for BM2 overlaps the termination codon of M1. **(B)** Synthesis of BM2 occurs by reinitiation, which is dependent on a UGGGA RNA sequence, called the termination upstream ribosome-binding site, located 45 nucleotides upstream of the termination codon. This sequence is complementary to an RNA loop in 18S rRNA. The interaction between these two sequences prevents dissociation of the 40S subunit, allowing time for ternary complex recruitment and initiation of the downstream ORF.

pseudoknot structure farther downstream (see Chapter 6 for a description of pseudoknots).

The effect of bases at the 3′ side of the stop codon may influence suppression by regulating competition between release factor and near-cognate tRNAs that bind the stop codon. Secondary RNA structures may govern suppression by modulating mRNA-protein or mRNA-rRNA interactions, by sterically interfering with release factor function, or by blocking unwinding by ribosome-associated helicases. It has been suggested that the pseudoknot of Moloney murine leukemia

Figure 11.18 Suppression of termination codons of alphaviruses and retroviruses. (A) Structure of the termination site between Gag and Pol of Moloney murine leukemia virus. The stop codon that terminates synthesis of Gag is underlined; it is followed by a pseudoknot that is important for suppression of termination. Adapted from J. H. Strauss and E. G. Strauss, *Microbiol Rev* **58:**491–562, 1994. **(B)** Suppression of termination during the synthesis of alphavirus P123 to produce nsP4, the RNA-dependent RNA polymerase. The termination codon is shown on the RNA as a box.

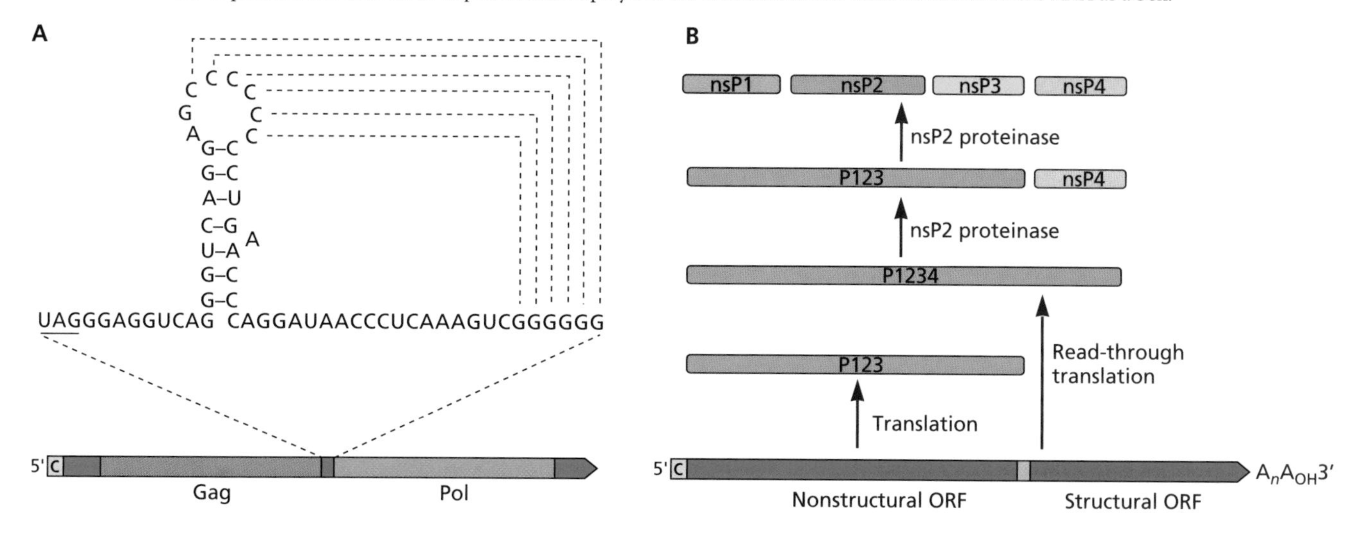

virus RNA causes the ribosome to pause, allowing the suppressor tRNA to compete with eRF1 at the suppression site. Maximal read-through efficiency also requires the interaction of viral reverse transcriptase with eRF1.

Suppression of termination is far more prevalent during translation of RNAs of RNA viruses than mRNAs of DNA viruses or cells. The RNA sequences and structures required for suppression are not found in most cellular mRNAs. For example, there is a strong bias against cytidine residues at the 3′ end of UGA termination codons in cellular mRNAs. Furthermore, suppression by tRNAs charged with selenocysteine has been found in <50 eukaryotic mRNAs.

Ribosomal Frameshifting

During ribosomal frameshifting, in response to signals in mRNA, ribosomes move into a different reading frame and continue translation. This mechanism was discovered in cells infected with Rous sarcoma virus and has since been described for many other viruses, including human immunodeficiency virus type 1, severe acute respiratory syndrome coronavirus, and herpes simplex virus. Frameshifting also occurs during translation of archaeal, bacterial, and eukaryotic mRNAs. This process may occur by shifting the reading frame 1 base toward the 5′ end (−1 frameshifting) or the 3′ end (+1 frameshifting) of the mRNA.

Frameshifting not only enables production of two proteins from one mRNA, but also can regulate their ratio. In the genome of retroviruses, the *gag* and *pol* genes may be separated by a stop codon (Fig. 11.18), or they may be in different reading frames, with *pol* overlapping *gag* in the −1 direction (Fig. 11.19). During synthesis of Rous sarcoma virus Gag, ribosomes frameshift before reaching the Gag stop codon and continue translating Pol, such that a Gag-Pol fusion is produced at about 10% of the frequency of Gag. Alteration of the frameshifting ratio by mutagenesis can be deleterious to viral replication.

Figure 11.19 Frameshifting on a retroviral mRNA. The structure of open reading frames is illustrated. Rous sarcoma virus mRNA encodes Gag and Pol proteins in reading frames that overlap by −1. Normal translation and termination produce the Gag protein; ribosomal frameshifting to the −1 frame results in the synthesis of a Gag-Pol fusion protein.

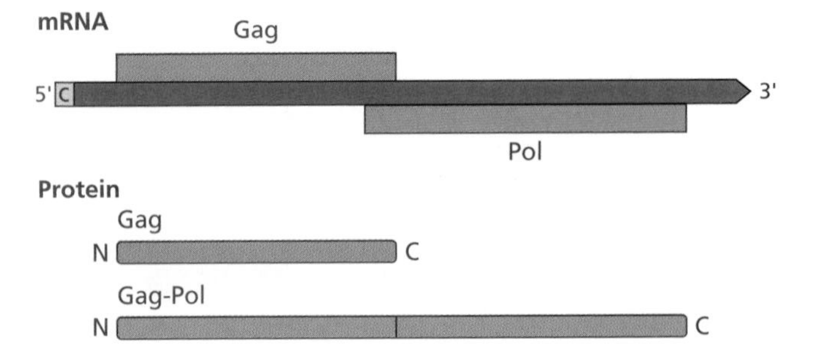

Studies on the requirements for frameshifting in retroviruses and coronaviruses have identified two essential components: a "slippery" homopolymeric sequence, which is a heptanucleotide stretch with two homopolymeric triplets of the form X-XXY-YYZ (e.g., in Rous sarcoma virus A-AAU-UUA); and an RNA secondary structure, usually a pseudoknot, 5 to 8 nucleotides downstream. The pseudoknot is thought to impede forward movement of the ribosome over the slippery sequence, creating tension in the mRNA that is relieved by disengagement of tRNAs followed by slippage and realignment to the −1 reading frame.

The tandem shift model for frameshifting has received substantial experimental support. In this model, two tRNAs in the zero reading frame (X-XXY-YYZ) slip back by 1 nucleotide during the frameshift to the −1 phase (XXX-YYY). Each tRNA base pairs with the mRNA in the first 2 nucleotides of each codon (Fig. 11.20). The peptidyl-tRNA is transferred to the P site, the −1 frame codon is decoded, and translation continues to produce the fusion protein. In this model, slippage occurs before peptide transfer, with the peptidyl- and aminoacyl-tRNAs bound to the P and A sites. However, it is possible that the shift occurs after peptide transfer but before translocation of the tRNAs, or when the aminoacyl-tRNA occupies the A site. These models cannot be distinguished by mutagenesis or by the sequence of the protein products.

Bicistronic mRNAs

Some viral mRNAs are bicistronic: they have two nonoverlapping open reading frames, and translation of each occurs by internal initiation. Examples include the mRNAs of members of the *Dicistroviridae*, including cricket paralysis virus and *Rhopalosiphum padi* (aphid) virus (Fig. 11.14). The upstream open reading frame begins with an AUG codon and is preceded by an IRES similar to those of picornaviruses. The downstream open reading frame, which encodes the viral capsid proteins, is translated independently from a completely different IRES. The 40S ribosomal subunit binds directly to the intergenic region that is partially folded to mimic a tRNA (Fig. 11.5A). The tRNA-like structure occupies the P site of the ribosome, and initiation occurs from the A site at a nonmethionine codon. The genome of canine picodicistrovirus is also bicistronic, but both open reading frames are translated from picornavirus-like IRESs.

Regulation of Translation during Viral Infection

Alterations in the cellular translation apparatus are commonplace in virus-infected cells. As part of the antiviral defense, or in response to stress caused by virus infection, the cell initiates measures designed to inhibit protein synthesis and limit virus production. Many viral genomes encode proteins or nucleic acids that neutralize this response, restore trantslation,

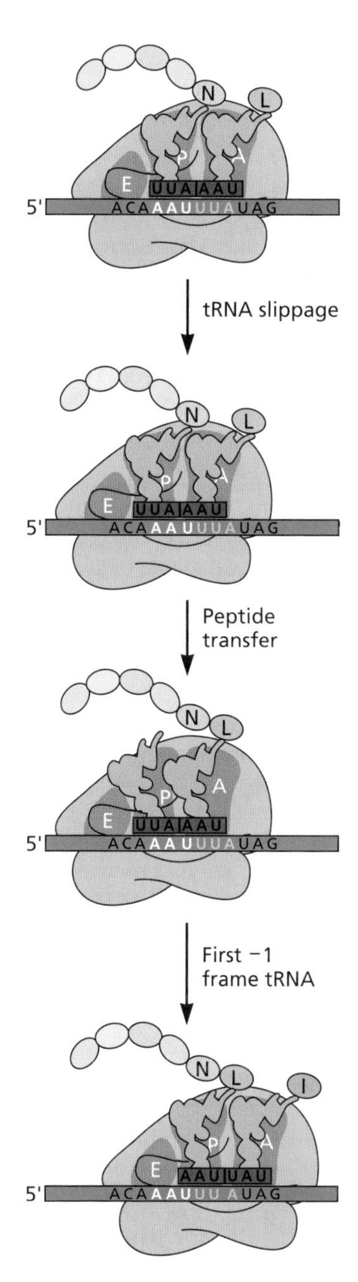

Figure 11.20 A model for −1 frameshifting. Slippage of the two tRNAs occurs after aminoacyl-tRNA enters the A site but before peptide transfer. Slippage allows the tRNAs to form only two base pairs with the mRNA. The site shown is that of Rous sarcoma virus. One-letter amino acid codes are used. Adapted from P. J. Farabaugh, *Microbiol Rev* **60:** 103–134, 1996.

and maximize virus reproduction. In addition, many viral gene products modify the host translation machinery to favor synthesis of viral proteins over those of the cell. As a result, not only can the entire synthetic capability of the cell be turned to the production of new virus particles, but translation of cellular antiviral proteins is restricted. These cellular and viral modifications of the translation apparatus may affect initiation, elongation, or termination. Some viral proteins inactivate eIF5B, eEF1A, or eEF2 to regulate 60S ribosome subunit recruitment and elongation; modulation of termination was discussed above.

Inhibition of Translation Initiation after Viral Infection

Phosphorylation of eIF2α

Translation initiation can be regulated by phosphorylation of the α subunit of the translation initiation protein eIF2 by four different cellular protein kinases that respond to virus infection or metabolic stress. One of these is the double-stranded-RNA-dependent protein kinase Pkr, which is induced by interferons (IFNs) produced as part of the rapid innate immune response of vertebrates to viral infection (discussed in Volume II, Chapter 3). IFNs diffuse to neighboring cells, bind to cell surface receptors, and activate signal transduction pathways that result in transcription of hundreds of cellular genes and the establishment of an **antiviral state**. IFN production by infected cells induces antiviral proteins in neighboring cells, thereby preventing viral reproduction and spread.

Pkr is a serine/threonine protein kinase composed of an N-terminal regulatory domain and a C-terminal catalytic domain (Fig. 11.21; see also Volume II, Chapter 3). Small quantities of an inactive form of Pkr are present in most uninfected mammalian tissues. Transcription of its gene is induced 5- to 10-fold by interferon. Pkr is activated by the binding of dsRNA to two dsRNA-binding motifs at the N terminus of the protein (Fig. 11.22). Such dsRNA is produced in cells infected by either DNA or RNA viruses. Binding to dsRNA leads to formation of Pkr dimers and autophosphorylation. This modification is thought to stabilize the dimer, which can then phosphorylate eIF2α in the absence of dsRNA. A cell protein, Pkr activator (Pact), may activate this protein kinase independently of dsRNA (Fig. 11.22).

Figure 11.21 Schematic structures of three eIF2α kinases. Ψ-kinase, pseudokinase domain; HisRS domain, histidyl-tRNA synthetase-like domain. Adapted from C. G. Proud, *Semin Cell Dev Biol* **16:** 3–12, 2005, with permission.

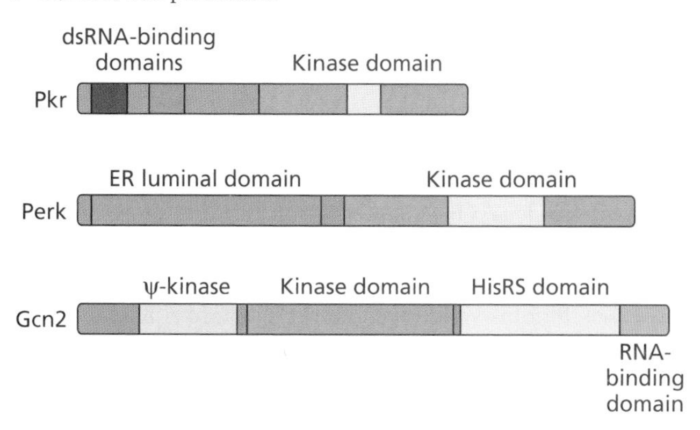

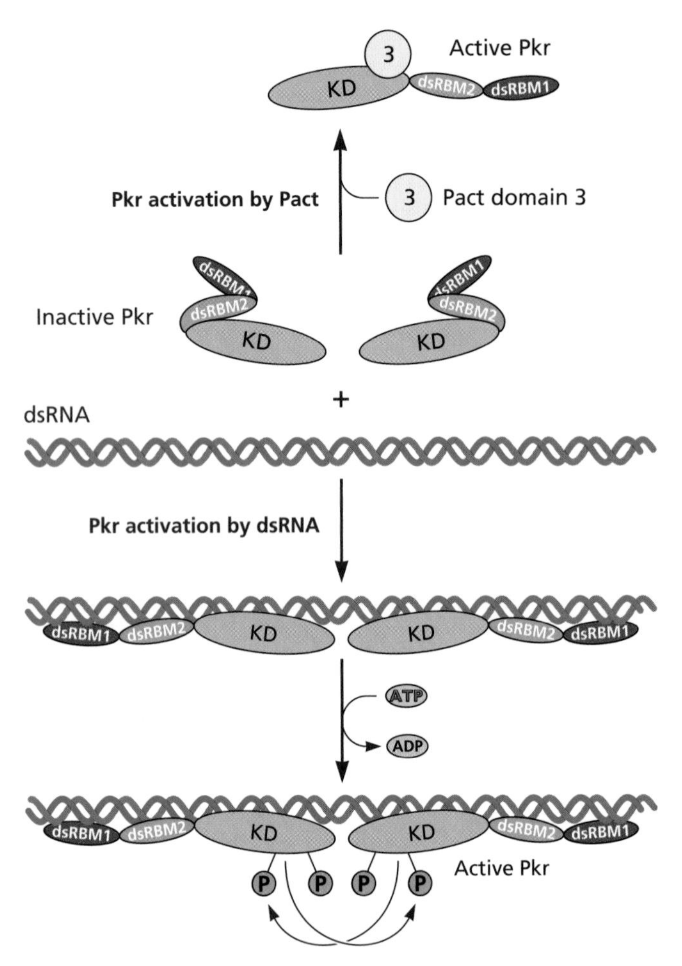

Figure 11.22 Model of activation of Pkr. Pkr is maintained in an inactive monomer by the interaction between a Pact domain 3-binding sequence in Pkr and dsRBM2. Pkr is activated when it binds Pact or dsRNA. When two or more molecules of inactive Pkr bind to one dsRNA molecule, cross-phosphorylation occurs because of the physical proximity of the molecules. Phosphorylation is thought to cause a conformational change in the kinase domain (KD) to allow phosphorylation of other substrates, including eIF2α. dsRBM, double-stranded-RNA-binding motif. Adapted from J. W. B. Hershey et al. (ed.), *Translational Control* (Cold Spring Harbor Laboratory Press, Cold Spring Harbor, NY, 1996), with permission.

Two other eIF2α protein kinases regulate translation during virus infection. In mammalian cells, general control nonderepressible 2 protein (Gcn2p) is activated during amino acid starvation when uncharged tRNA binds a histidyl-tRNA synthetase-like domain in the protein (Fig. 11.21). During infection with Sindbis virus, vesicular stomatitis virus, or adenovirus, Gcn2p is activated upon binding of viral RNA, leading to phosphorylation of eIF2α and restriction of virus reproduction. Consistent with a role in mediating antiviral responses, Sindbis virus reproduction is more efficient in cells lacking Gcn2p. Pkr-like ER kinase (Perk), a transmembrane protein of the ER, is a component of the unfolded protein response. Its luminal domain senses the equilibrium between unfolded and misfolded proteins and chaperone proteins. Under conditions of intracellular stress, such as occurs during infection of cells with enveloped viruses, Perk oligomerizes within the membrane, is activated, and phosphorylates eIF2α in the cytoplasm.

The initiation protein eIF2α is part of the ternary complex that also contains GTP and Met-$tRNA_i$ (Fig. 11.3). After GTP hydrolysis, the bound GDP must be exchanged for GTP to permit the binding of another molecule of Met-$tRNA_i$. This exchange is carried out by eIF2B (Fig. 11.23). When the α subunit of eIF2 is phosphorylated, eIF2-GDP binds eIF2B with such high affinity that it is effectively trapped; recycling of eIF2 stops, and ternary complexes are depleted. eIF2B is less abundant than eIF2, and phosphorylation of about 10 to 40% of eIF2 (depending on the cell type and the relative concentrations of eIF2 and eIF2B) results in the complete sequestration of eIF2B, leading to a block in protein synthesis. As viral translation is also impaired, the production of new virus particles is diminished.

Viral Regulation of Pkr

Most viral infections induce activation of eIF2α kinases and consequent phosphorylation of eIF2α. As global inhibition of translation would be a threat to successful viral reproduction, viral genomes encode one or more proteins that prevent eIF2α phosphorylation in different ways (Fig. 11.24).

RNA antagonists of Pkr. The 166-nucleotide adenovirus VA-RNA I, which accumulates to massive concentrations (up to 10^9 copies per cell) late in infection following transcription of the viral gene by RNA polymerase III, is a potent inhibitor of Pkr. An adenovirus mutant that cannot express the VA-RNA I gene grows poorly. In cells infected with this mutant virus, eIF2α becomes extensively phosphorylated, causing global translational inhibition. VA-RNA I binds the dsRNA-binding region of Pkr and blocks activation. It has been suggested that binding of VA-RNA I to Pkr prevents the interaction with authentic dsRNA and hence prevents activation of the kinase. The Epstein-Barr virus genome also encodes small RNAs that inhibit Pkr activation.

dsRNA-binding proteins. The vaccinia virus genome encodes a protein (E3L) that sequesters dsRNA. This protein contains the same dsRNA-binding motif as Pkr; it binds dsRNA and prevents it from activating the kinase. Deletion of the gene encoding the E3L protein renders the virus more sensitive to IFN and causes production of larger quantities of active Pkr in infected cells. The influenza virus NS1 protein and the reovirus σ3 protein also sequester dsRNA.

Inhibition of kinase function. The genomes of several viruses encode proteins that directly inhibit the kinase activity of Pkr or Perk, and some do so by acting as pseudosubstrates.

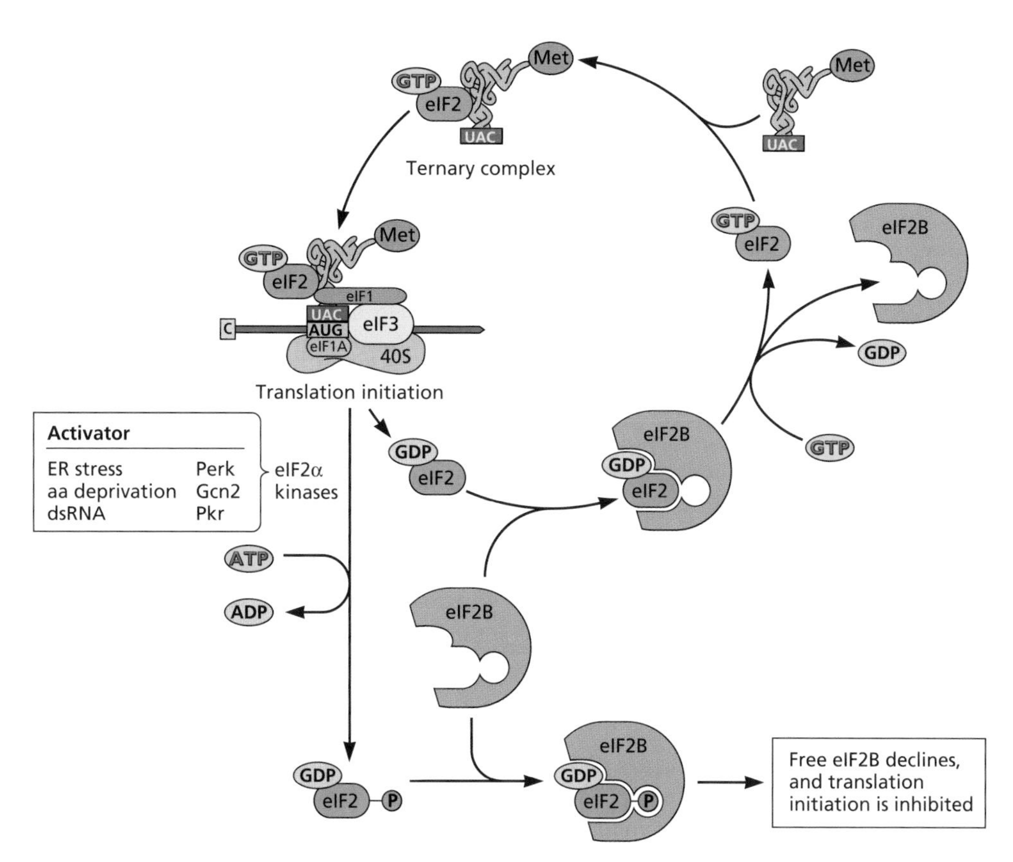

Figure 11.23 Effect of eIF2α phosphorylation on catalytic recycling. eIF2-GTP and tRNA-Met_i form the ternary complex required for translation initiation. During initiation, GTP is hydrolyzed to GDP, and in order for initiation to continue, eIF2 must be recharged with GTP. Such recycling is accomplished by eIF2B, which exchanges GTP for GDP on eIF2. When eIF2 is phosphorylated on the α subunit, it binds irreversibly to eIF2B, preventing the latter from carrying out its role in recycling active eIF2. As a result, the concentration of eIF2-GTP declines and translation initiation is inhibited. aa, amino acid.

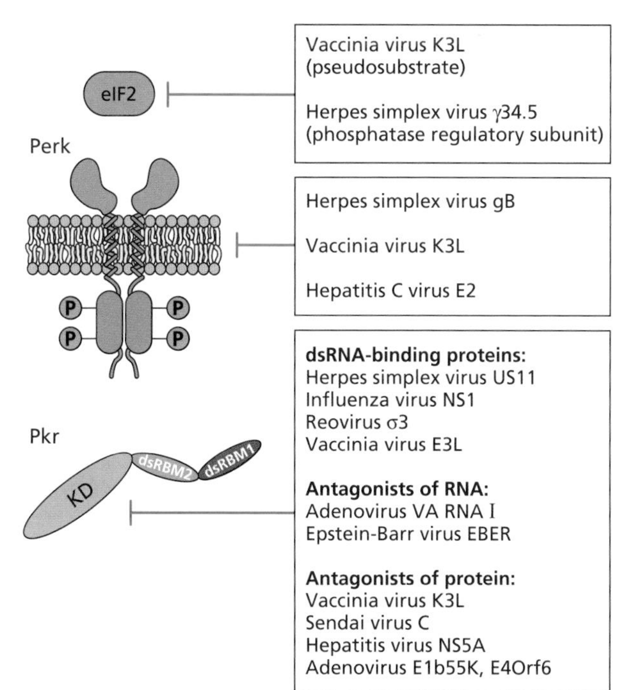

Figure 11.24 Some viral proteins and RNAs that counter inactivation of eIF2. Vaccinia virus K3L and herpes simplex virus γ34.5 interfere with phosphorylation of eIF2α by acting as a pseudosubstrate or by removing the phosphate from the protein, respectively. Other viral proteins that directly inhibit eIF2α kinases Perk and Pkr are shown. Viral proteins that prevent double-stranded activation of Pkr are also listed. dsRBM, double-stranded-RNA-binding motif; KD, kinase domain.

For example, vaccinia virus K3L protein has amino acid homology to the N terminus of eIF2α. The protein binds tightly to Pkr within the catalytic cleft and blocks autophosphorylation. The growth of vaccinia virus mutants lacking the K3L gene is severely impaired by IFN. The herpes simplex virus type 1 genome encodes proteins that bind to Pkr and Perk and directly inhibit kinase activity. Us11 binds to Pkr and blocks its activation, while the viral glycoprotein gB associates with the luminal domain of Perk and prevents its activation and subsequent phosphorylation of eIF2α.

Cellular proteins can also function as inhibitors of eIF2α kinase function. Influenza virus infection activates a cellular protein, $p58^{IPK}$, that binds Pkr and prevents autophosphorylation. In cells lacking this protein, eIF2α phosphorylation is increased and viral mRNA translation is reduced.

Dephosphorylation of eIF2α. Another mechanism for reversing the consequences of Pkr activation is dephosphorylation of its target. In herpes simplex virus-infected cells, Pkr is activated but eIF2α is not phosphorylated. During infection with viruses lacking the viral ICP34.5 gene, Pkr is activated and eIF2α becomes phosphorylated, causing global inhibition of protein synthesis. This viral protein associates with a type 1a protein phosphatase and acts as a regulatory subunit, redirecting the enzyme to dephosphorylate eIF2α (Volume II, Fig. 5.15). The effects of activated Pkr are reversed, ensuring continued protein synthesis. In a similar fashion, the E6 protein of human papillomavirus activates a phosphatase, leading to dephosphorylation of eIF2α.

Host and virus evolution. The inhibitory effect of eIF2α phosphorylation on viral reproduction has led to the acquisition of viral genes that antagonize this function. In turn, the *pkr* gene has been selected to evade the effects of viral inhibitors. Phylogenetic analysis of the *pkr* gene in primates indicates that it has undergone bursts of positive selection. Some of the observed amino acid substitutions prevent binding of Pkr by the vaccinia virus antagonist, the K3L protein. How such mutations become fixed in the viral genome can be illuminated by experiments in cell culture. The vaccinia virus K3L protein does not efficiently antagonize Pkr of human cells. Serial propagation of the virus in cell culture leads to amplification of the *k3l* gene, causing a 7 to 10% increase in genome size. These amplifications are transient; when amino acid changes are selected that increase the antagonism of K3L for Pkr, genome reduction takes place. The expanding and contracting viral genes that antagonize host defenses have been characterized as "genomic accordions."

Beneficial Effects of eIF2α Phosphorylation on Viral Reproduction

Inhibition of host translation by phosphorylation of eIF2α can be beneficial for virus reproduction because viral mRNAs can be selectively translated, and the host IFN response may be repressed. Consistent with this reasoning, eIF2α phosphorylation is not blocked in cells infected with some viruses. Translation of some viral mRNAs, such as those of the picornavirus-like viruses of insects discussed previously in this chapter, does not require eIF2α, because the secondary structure of the IRES of these viruses mimics an uncharged tRNA. Translation of other viral mRNAs does require $Met\text{-}tRNA_i$, yet can proceed when eIF2α is phosphorylated. Translation of classical swine fever virus mRNA is not inhibited by eIF2α phosphorylation, because eIF5B, independent of eIF2, can promote $Met\text{-}tRNA_i$ binding to the ribosome. A different mechanism is exemplified by Sindbis virus: translation of subgenomic mRNAs is not inhibited by eIF2α phosphorylation. Delivery of $Met\text{-}tRNA_i$ to the ribosome is accomplished by other cellular proteins, including ligatin, a protein that normally participates in cellular localization of phosphoglycoproteins. This unusual mechanism depends on placement of the AUG codon in the P site of the ribosome, an activity mediated by a stem-loop structure ~25 nucleotides downstream of the initiation codon.

Regulation of eIF4F

The eIF4F protein plays several important roles during 5′-end-dependent initiation, including recognition of the cap, recruitment of the 40S ribosomal subunit, and unwinding of RNA secondary structure. It is not surprising, therefore, that several viral proteins modify the activity of this protein. The cap-binding subunit eIF4E is frequently a target, probably because its activity can be modulated in at least two ways and because it is present in limiting quantities in cells. The cap-binding complex can also be inactivated by cleavage of eIF4G.

Cleavage of eIF4G

Poliovirus infection of mammalian cells in culture results in dramatic inhibition of cellular protein synthesis. By 2 h after infection, polyribosomes are disrupted and translation of nearly all cellular mRNAs declines (Fig. 11.25). Translationally competent extracts from infected cells can readily translate poliovirus mRNA but not capped mRNAs. Studies of these extracts demonstrated that they lack functional eIF4F: eIF4G is cleaved proteolytically. As the N-terminal domain of eIF4G binds eIF4E, which in turn binds the 5′ cap of cellular mRNAs, such cleavage prevents eIF4F from recruiting 40S ribosomal subunits (Fig. 11.26). Poliovirus mRNA is uncapped and is translated by internal ribosome binding, a process that does not require intact eIF4G. In fact, IRES-mediated initiation function appears to require the C-terminal fragment of eIF4G, which, as discussed above, is necessary to recruit 40S ribosomal subunits to the IRES. Consequently, cleavage of eIF4G not only inhibits translation of cellular mRNAs but also is a strategy for stimulating IRES-dependent translation. Cleavage of eIF4G is carried out by viral proteases

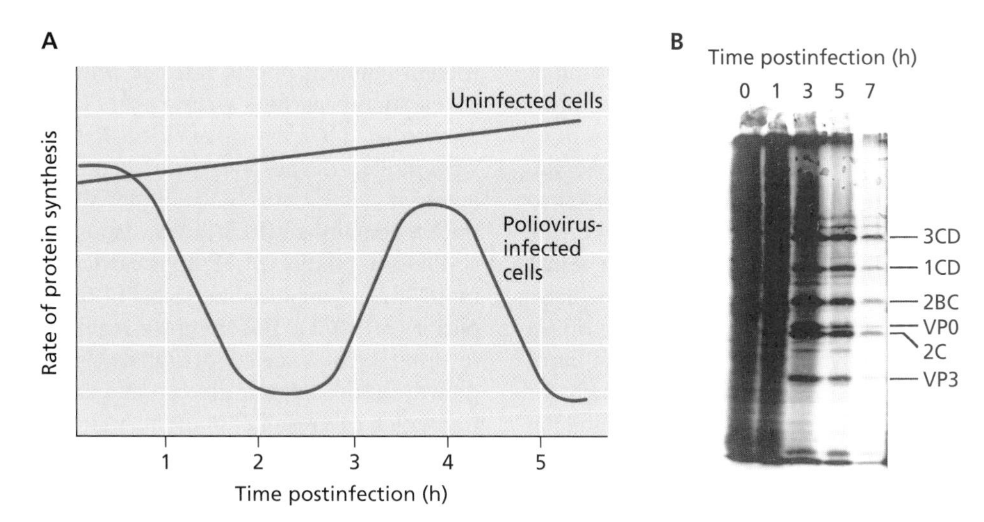

Figure 11.25 Inhibition of cellular translation in poliovirus-infected cells. (A) Rate of protein synthesis in poliovirus-infected and uninfected cells. During poliovirus infection, host cell translation is inhibited by 2 h after infection and is replaced by translation of viral proteins. Adapted from H. Fraenkel-Conrat and R. R. Wagner (ed.), *Comprehensive Virology* (Plenum Press, New York, NY, 1984), with permission. **(B)** Sodium dodecyl sulfate-polyacrylamide gel electrophoresis of [^{35}S]methionine-labeled proteins at different times after poliovirus infection. In this experiment, host translation was shut off by 5 h postinfection and was replaced by the synthesis of viral proteins, some of which are labeled at the right.

such as 2A^{pro} of poliovirus, rhinovirus, and Coxsackievirus and the L protease of foot-and-mouth disease virus.

Modulation of eIF4E Activity by Phosphorylation

Two protein kinases that are associated with eIF4G, mitogen-activated protein kinase interacting serine/threonine kinase 1 and 2 (Mnk1 and Mnk2), phosphorylate Ser209 of eIF4E. Inhibition of cellular translation during mitosis and heat shock correlates with reduced phosphorylation of eIF4E. It has been suggested that phosphorylation of eIF4E allows tighter binding to the 5′-terminal cap. However, the effect of phosphorylation on the function of eIF4E is unclear.

A decrease in eIF4E phosphorylation may be responsible for the inhibition of mRNA translation in cells infected

Figure 11.26 Regulation of eIF4F activity. The illustration shows regulation of eIF4F activity, and inhibition of translation, by dephosphorylation of eIF4E, interaction with two eIF4E-binding proteins, and proteolytic cleavage of eIF4G.

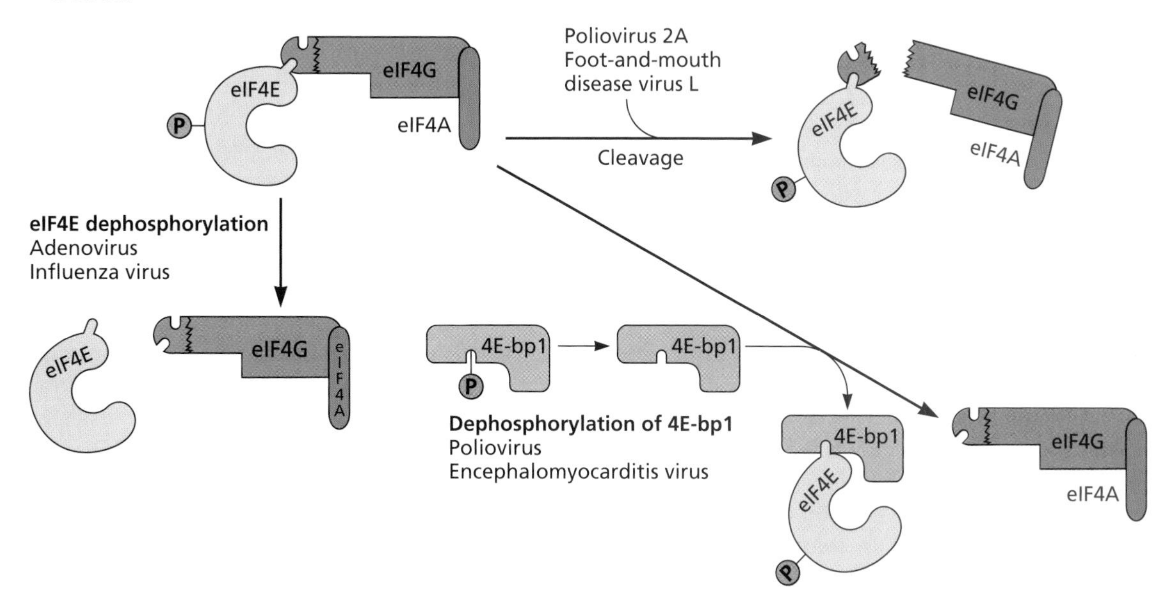

with some viruses. For example, cellular protein synthesis is inhibited at late times in adenovirus-infected cells, a result of virus-induced underphosphorylation of eIF4E. The viral L4 100-kDa protein binds to the C terminus of eIF4G, preventing binding of Mnk1, and hence presumably blocks phosphorylation of eIF4E. Adenoviral late mRNAs continue to be translated because they possess a reduced requirement for eIF4E. The majority of these viral mRNAs contain the tripartite leader (Fig. 10.12), a common 5′ noncoding region that mediates translation by ribosome shunting. Initiation by this mechanism is less dependent on eIF4F, presumably because the shunting of part of the 5′ untranslated region reduces the requirement for RNA-unwinding (helicase) activity associated with initiation by cap binding and scanning. Furthermore, adenovirus late mRNAs efficiently recruit the small quantities of phosphorylated eIF4E present late in infection, a feature of mRNAs with little RNA secondary structure near the 5′ cap. The tripartite leader therefore confers selective translation of viral over cellular mRNAs under conditions in which eIF4E is underphosphorylated. Adenovirus-induced translation inhibition not only boosts viral late mRNA translation, but also enhances cytopathic effects and consequently release of virus from cells.

Phosphorylation of eIF4E also regulates the innate immune response to infection. Mice that produce only a form of eIF4E that cannot be phosphorylated are less susceptible to infection with a number of RNA and DNA viruses. The animals produce more IFN, because nonphosphorylated eIF4E leads to reduced production of the inhibitor Ikbα, a regulator of Nf-κb. The genomes of herpesviruses and poxviruses encode proteins that promote phosphorylation of eIF4E, presumably to antagonize Nf-κb activation and reduce IFN production.

Modulation of eIF4E Activity by Binding Proteins

Three related low-molecular-weight cellular proteins, 4E-bp1, 4E-bp2, and 4E-bp3, bind to eIF4E and inhibit translation following 5′-end-dependent scanning, but not by internal ribosome entry (Fig. 11.26). The first was found to be identical to a previously described protein, called phosphorylated heat- and acid-stable protein regulated by insulin (Phas-I). This protein was known to be an important substrate for phosphorylation in cells treated with insulin and growth factors. Phosphorylation of 4E-bp *in vitro* blocks its association with eIF4E. When bound to 4E-bp, eIF4E cannot bind to eIF4G, and active eIF4F is not formed. eIF4G and 4E-bp proteins compete for binding to eIF4E. Treatment of cells with hormones and growth factors leads, via signal transduction pathways, to the phosphorylation of 4E-bp and its release from eIF4E. Translation of mRNAs with extensive secondary structure in the 5′ untranslated region is preferentially sensitive to the phosphorylation state of 4E-bp.

Some viral infections lead to alteration of the phosphorylation state of 4E-bp (Fig. 11.26). In contrast to the shutoff that occurs in poliovirus-infected cells, inhibition of cellular protein synthesis in cells infected with another picornavirus, encephalomyocarditis virus, occurs late in infection and is not mediated by cleavage of eIF4G. Rather, infection with this virus induces dephosphorylation of 4E-bp1. As a result, translation of cellular mRNAs is inhibited, but, because the viral mRNA contains an IRES, its translation is unaffected.

Phosphorylation of 4E-bp is carried out by a serine/threonine kinase, mammalian target of rapamycin kinase complex 1 (mTorC1). This complex regulates protein synthesis in response to a variety of signals (Fig. 11.27). Presence of growth factors, oxygen, glucose, and energy lead to increased translation as a result of phosphorylation of 4E-bp1 and ribosomal protein S6.

Many viral mRNAs are capped and therefore depend upon eIF4F for translation. As would be expected, mTorC1 is activated during infection, leading to increased protein synthesis under conditions (e.g., virus-induced stress) that would otherwise limit translation. Examples include inhibition of the tuberous sclerosis complex (Tsc) by the human papillomavirus E6 protein and stimulation of phosphatidylinositol 3-kinase (Pi3k) by the adenovirus E4 Orf1 protein.

Members of the *Herpesviridae* stimulate protein synthesis via multiple mechanisms. Both herpes simplex virus type 1 and human cytomegalovirus stimulate mTorC1 via inhibition of Tsc. The Us3 protein of herpes simplex virus type 1, a serine/threonine kinase, functions as an Akt mimic and phosphorylates Tsc2. In contrast, the human cytomegalovirus UL38 protein inactivates Tsc2 through direct binding. The vGpcr (viral G-protein coupled receptor) protein of Kaposi's sarcoma-associated herpesvirus and the LMP-2A protein of Epstein-Barr virus inactivate signaling pathways upstream of mTorC1. In contrast to many other viruses, human cytomegalovirus infection does not inhibit cellular protein synthesis, and furthermore, the abundance of eIF4F increases. These effects are a consequence in part of the viral UL38 protein, which activates mTorC1.

The protein 4E-bp1 is degraded by the proteasome in cells infected with herpes simplex virus type 1, but this action is not sufficient to promote assembly of eIF4F. Binding of eIF4E to eIF4G is stimulated by the viral ICP6 protein, which shares a domain with cellular chaperone protein hsp27 that controls eIF4F formation.

Modulation of eIF4E by miRNA

In response to enterovirus infection, synthesis of miR-141, a micro-RNA (miRNA) that targets mRNA encoding eIF4E, is induced as a result of the synthesis of a cellular transcription protein. Consequently, translation by 5′-end-dependent initiation is impaired. Translation of viral mRNAs is unaffected, because they are initiated by an IRES-dependent mechanism that does not require eIF4E. As expected, silencing of miR-141 reduces the production of infectious virus particles.

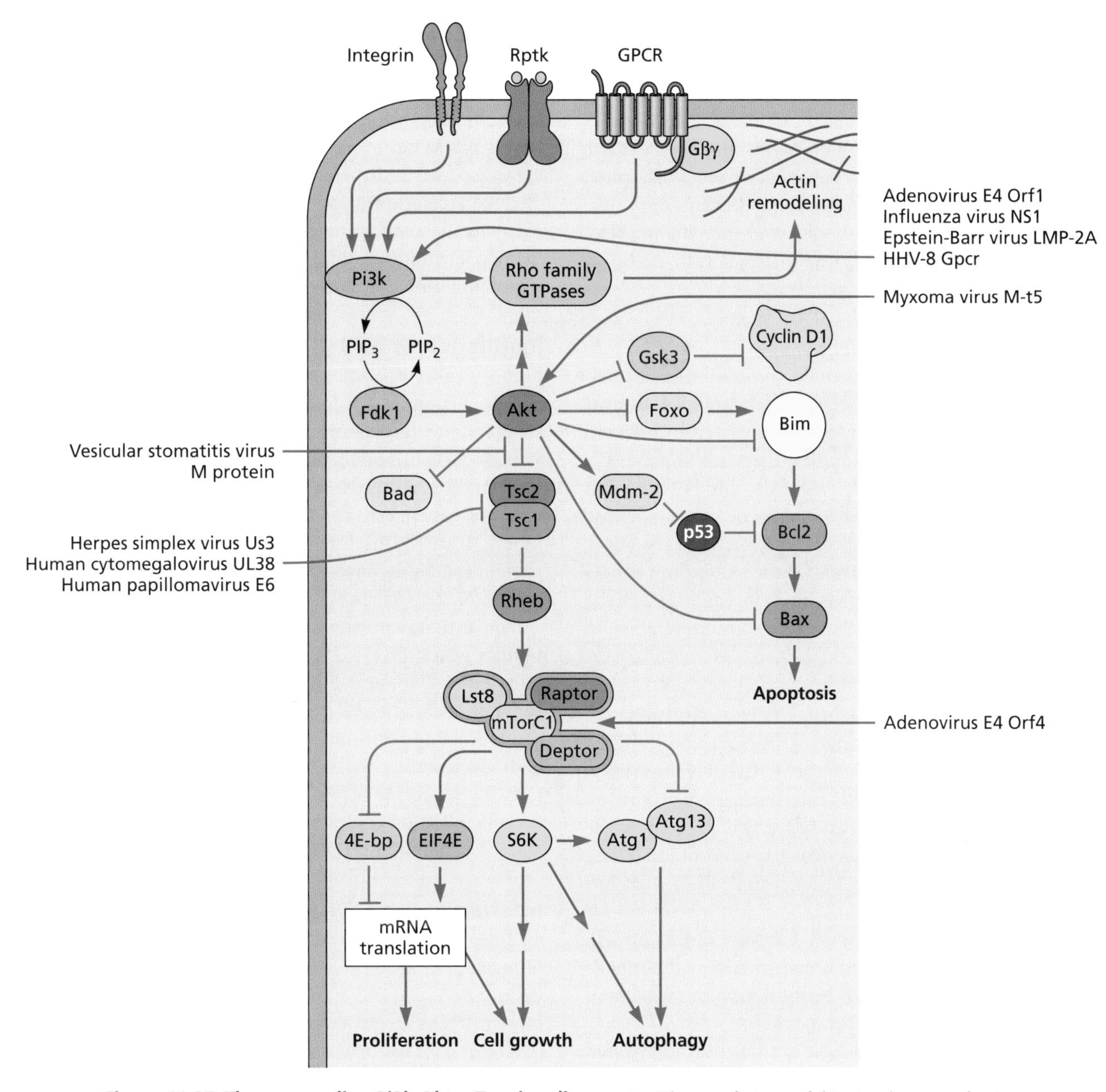

Figure 11.27 The mammalian Pi3k-Akt-mTor signaling route. The core features of this signaling transduction system are illustrated. Binding of ligand to any one of several types of plasma membrane receptors initiates signaling to Pi3k associated with the inner surface of the plasma membrane and phosphorylation and activation of this kinase. Once activated, phosphoinositol 3-kinases phosphorylate phosphoinositol present on membrane lipids to produce phosphoinositol 3,4,5-triphosphate (PIP_3). These modified lipids are bound by particular domains of other proteins, such as phophoinositide-dependent kinase 1 (Pdk1), which then transmit the signal to Akt. Synthesis of PIP_3 also leads to activation of small G proteins of the Rho (Ras homology) family that control actin polymerization and depolymerization, such as Rac (Ras-related C3 botulism toxin substrate 1) and Cdc42 (cell division control protein 42 homolog). Shown are consequences of Akt activation that promote cell growth and proliferation via activation of the mTor kinase present in mTorC1. Activated mTor facilitates translation by multiple mechanisms and also induces autophagy. Viral proteins that activate (green) or inhibit (red) are shown. Atg, autophagy-related protein; Bad, Bcl2-associated death protein; Bax, apoptosis-regulator Bcl2-associated protein; Bcl2, apoptosis-regulator Bcl2 (B cell CLL/lymphoma 2); Bim, Bcl2-interacting mediator of cell death; Deptor, Dep domain-containing mTor-interacting protein; 4E-bp, eukaryotic initiation factor 4E-binding protein; Foxo, forkhead box protein 1; Gpcr, G-protein coupled receptor; Gsk3, glycogen synthase kinase 3; HHV, human herpesvirus; Lst8, target of rapamycin complex subunit Lst8 homology; Mdm2, E3 ubiquitin ligase Mdm2 (double minute protein 2); Raptor, regulatory-associated protein of mTorC1; Rheb, Ras-homology enriched in protein; Rptk, receptor protein tyrosine kinase; S6k, ribosomal protein S6 kinase; Tsc, tuberous sclerosis protein.

A Viral Protein That Replaces eIF4F

The nucleocapsid (N) protein of hantaviruses can replace all components of eIF4F. N protein substitutes for eIF4E by binding the mRNA cap, and can bind directly to the 43S preinitiation complex, replacing eIF4G. N also replaces the helicase activity of eIF4A. A heptanucleotide sequence in the 5′ untranslated region of viral mRNAs is sufficient for preferential N-dependent translation of viral over nonviral mRNAs. These activities presumably ensure efficient translation of viral mRNAs.

A Viral Cap-Binding Protein

Influenza virus infection leads to the inhibition of cellular mRNA translation, in part via dephosphorylation of eIF4E. How capped viral mRNAs are translated in infected cells was revealed by the finding that the viral polymerase, consisting of PB1, PB2, and PA subunits, binds with high affinity to cap structures. The PB2 subunit of the polymerase binds eIF4G, which is required for viral mRNA translation. The viral NS1 participates in translation by binding eIF4G and poly(A)-binding protein, possibly bringing together the 5′ and 3′ ends of the viral mRNAs to ensure more efficient translation.

Regulation of Poly (A)-Binding Protein Activity

The poly (A)-binding protein plays a crucial role in mRNA translation, bringing together the ends of the mRNA (Fig. 11.4). In cells infected with enteroviruses, lentiviruses, and caliciviruses, viral proteases cleave this protein, while it is sequestered by the rubella virus capsid protein. These events are believed to contribute to inhibition of host cell translation. However, poly(A)-binding protein is required for IRES-mediated translation initiation. Cleavage of this protein in enterovirus-infected cells (and sequestration by the rubella virus capsid) may contribute to the inhibition of translation needed for the switch to viral RNA synthesis.

The 3′ ends of rotaviral mRNAs are not polyadenylated and therefore cannot interact with poly(A)-binding protein. Instead, these 3′ untranslated regions contain a conserved sequence that binds the viral protein nsP3. This protein also occupies the poly(A)-binding protein binding site of eIF4G, bringing together the 5′ and 3′ ends of mRNA. Host translation is inhibited because nsP3 displaces poly(A)-binding protein from eIF4G. These interactions are believed to favor translation of viral mRNAs. However, nsP3 is not required for translation of viral mRNAs or for virus reproduction.

Redistribution of poly(A)-binding protein is another mechanism for selective translation of viral mRNAs in infected cells. The herpes simplex virus type 1 ICP27 and UL47 proteins and the Kaposi's sarcoma-associated herpesvirus proteins SOX and K8.1 cause redistribution of this poly(A)-binding protein to the nucleus. This effect likely contributes to shutoff of host cell translation: in cells infected with viruses lacking the gene encoding the SOX protein, poly(A)-binding protein is not routed to the nucleus, and host translation is unimpaired. In contrast, poly(A)-binding protein is not found in the nucleus of cells infected with human cytomegalovirus, in which host cell translation is unaffected. The redistribution of poly(A)-binding protein and eIF4F to cytoplasmic replication factories in poxvirus-infected cells likely contributes to inhibition of host translation and favoring of viral mRNA translation.

Regulation of eIF3

Some viruses encode proteins that bind eIF3 and impair 5′-end-dependent translation. The spike glycoprotein of severe acute respiratory syndrome coronavirus, the rabies virus M protein, and the measles virus N proteins all bind subunits of eIF3. The eIF3α and eIF3β subunits are cleaved by the viral protease in cells infected with the picornavirus foot-and-mouth disease virus, further contributing to inhibition of host protein synthesis caused by cleavage of eIF4G. It is not known how viral mRNAs are translated under these conditions.

An antiviral mechanism comprises three IFN-induced human genes, *ISG54*, *ISG56*, and *ISG60*, which encode proteins (P54, P56, and P60) that bind subunits of eIF3 and prevent translation. The P56 protein binds the e subunit of eIF3, while P54 binds to the c and e subunits. Both P54 and P56 interfere with stabilization of the ternary complex (Met-tRNA$_i$–eIF2–GTP), and P54 also inhibits formation of the 48S initiation complex (Fig. 11.3). Both 5′-end-dependent and internal initiation are inhibited by P56.

Interfering with RNA

Cellular protein synthesis may also be interrupted by virus-induced alteration of cellular mRNAs. Among RNA viruses, influenza viral and hantaviral endonucleases cleave cellular mRNAs to provide primers for viral RNA synthesis (Chapter 6). This process leads to destabilization of cellular mRNAs and inhibition of translation. The nsp1 protein of severe acute respiratory syndrome coronavirus has a similar effect by binding 40S ribosomes and degrading cellular mRNAs (Fig. 10.19). In cells infected with vesicular stomatitis virus, nuclear export of cellular mRNAs is suppressed.

DNA viruses such as poxviruses encode decapping enzymes that destabilize cellular mRNAs, while the herpes simplex virus type 1 virion shutoff protein is an endonuclease that binds eIF4A and eIF4B, leading to increased mRNA turnover (Fig. 10.19). The SOX protein of Kaposi's sarcoma-associated herpesvirus also induces degradation of cellular mRNA, but by a different mechanism: it recruits the cellular Xrn1 exonuclease to polysomes. The SOX protein bypasses the regulatory steps of deadenylation and decapping typically required for activation of Xrn1. Instead, SOX first internally cleaves mRNAs, which are then degraded by Xrn1. Some cellular mRNAs are protected from SOX cleavage by a sequence within the 3′ untranslated region.

In response to the production of viral dsRNAs, the cellular antiviral response includes production of RNase L, which is activated by the products of 2′-5′-oligoadenylate synthetase and degrades both rRNA and mRNA (Volume II, Chapter 3). Viral genomes encode a variety of proteins that bind dsRNAs and inhibit the RNase L pathway, preventing degradation of mRNAs. The murine hepatitis virus ns2 gene encodes a protein that cleaves 2′,5′-oligoadenylate chains to limit activation of RNase L. AU-rich binding proteins bind to sequences in the 3′ noncoding region of mRNAs to modulate their stability. These proteins also bind to the 5′ noncoding region of enteroviruses, but viral mRNA degradation is blocked because they are degraded by the viral 3CD protease.

Stress-Associated RNA Granules

Another mechanism by which mRNA translation can be impaired in virus-infected cells is by sequestering of mRNA from the translation apparatus in processing (P) bodies and stress granules. P bodies and stress granules are two nonmembranous cytoplasmic aggregates composed of mRNA, cellular miRNAs, mRNA-binding proteins, 40S ribosomal subunits, and many proteins that participate in mRNA translation. These granules are believed to form when translation is inhibited by intracellular and extracellular stresses such as nutrient deprivation or viral infection. A critical trigger for their formation is phosphorylation of eIF2α. When stress conditions are alleviated, the mRNAs found in these aggregates may be deadenylated and degraded, or returned to the pool of translated RNAs. Stress granules and P bodies may interact and exchange proteins and mRNAs with each other and with the cytoplasm.

Stress granules contain hundreds of RNA-binding proteins, and >100 cellular genes encode proteins that participate in their assembly. Two components of stress granules include T cell-restricted intracellular antigen-1 (Tia-1) and the Rasgap SH3 domain-binding protein 1 (G3bp1) (Fig. 11.28). Reduction in concentrations of either protein impairs formation of stress granules, and overproduction of either component stimulates formation of these aggregates. Stress granule formation occurs during infection with different viruses, and in most cases, they are suppressed at some point in the infectious cycle. Stress granules form early in cells infected with poliovirus. Late in viral infection, the viral proteinase 3C^pro cleaves G3bp1, disassembling stress granules, an event required for efficient viral reproduction. The presence of a noncleavable form of G3bp1 prevents the disassembly of stress granules and impairs viral reproduction. The NS1 protein of influenza A viruses prevents formation of stress granules by antagonizing Pkr. Stress granule components may also be redirected to other cellular sites in virus-infected cells. For example, the nsP3 protein of Semliki Forest virus sequesters G3bp1 into viral replication complexes. Removal of the nsP3 sequences that are important for interaction with this cellular protein impair viral production, suggesting a role for the protein in viral reproduction.

P bodies are a second type of non-membrane-bound aggregate in the cytoplasm that is enriched for components of the RNA decay machinery. These aggregates, which are the sites of RNA deadenylation and mRNA repression, are composed of proteins such as the decapping enzymes and proteins that mediate mRNA deadenylation. Virus replication may also lead to alteration of P bodies and redirection of their components. The enterovirus 3C^pro proteinase cleaves several P-body components including Xrn1, Dcp1a, and Pan3, disrupting P-body formation. Influenza virus infection leads to dispersal of P bodies via the viral NS1 protein, which binds a cellular protein required for P-body assembly. Infection with some viruses leads to co-opting of P-body components. In cells infected with the flavivirus West Nile virus, a number of P-body proteins are sequestered in viral replication complexes as the number of P bodies diminish. Some of these cellular proteins may be required for viral RNA synthesis. Subgenomic flavivirus RNA (sfRNA) is a fragment of the 3′ untranslated region produced when exonucleolytic decay by Xrn1 stalls at a pseudoknot. The sfRNA enters P bodies, where it inhibits Xrn1 activity.

Perspectives

From the smallest to the largest, all viral genomes encode proteins that recruit the host cell translational machinery for production of proteins needed for viral reproduction. These viral proteins control or modify cellular translation proteins, ribosomes, and the signaling pathways that regulate their activities. The result is not only production of viral proteins, but also suppression of intrinsic immune defenses. Among all the viruses studied, every step of the translation process appears to be modified. The study of such modifications has revealed a great deal about how proteins are made and how this process is regulated.

Very early in infection, intrinsic defense responses are mounted, and protein synthesis is inhibited in an attempt to limit viral reproduction. Should infection proceed, cellular stress responses, which cause further reduction in translation, are activated. As viral proteins and RNAs are produced, modifications to the cellular translation apparatus take place to favor the production of viral proteins. The interplay of cellular and viral modifications is an important determinant of the outcome of infection. Studies of ancient viral and cellular proteins that participate in translation reveal an evolutionary arms race as viral proteins change to overcome host defenses, and cellular proteins change in response. The results reveal the remarkable plasticity of protein function, and how genes and genomes have been shaped by challenges from viruses.

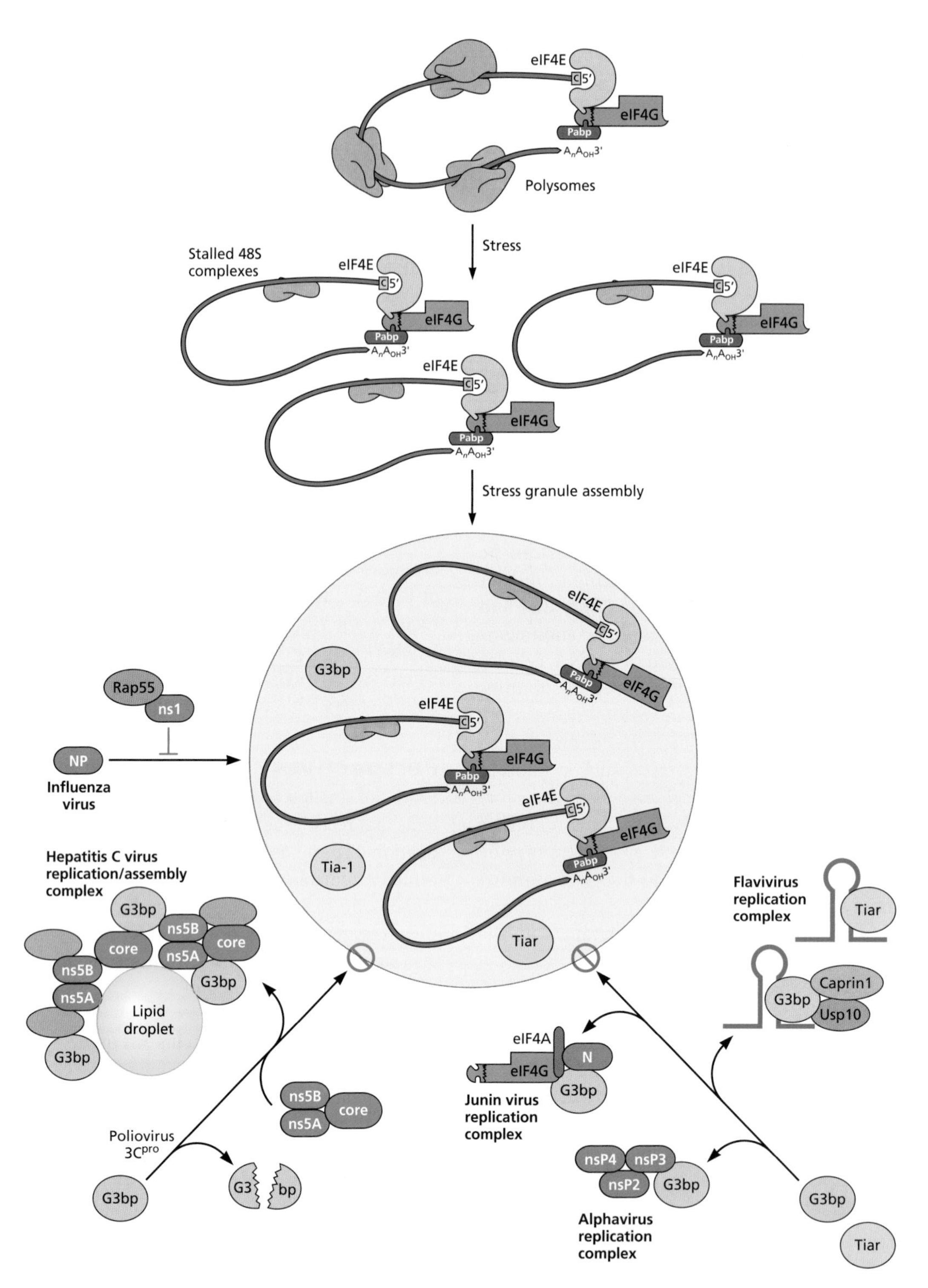

Figure 11.28 Inhibition of stress granule assembly by viral proteins. When protein synthesis is inhibited as a response to stress, stalled translational complexes are routed to stress granules. Three marker proteins for stress granules, T cell intracellular antigen-1 (Tia-1), Tia-1-related protein (Tiar), and G3bp, are shown. Infection by some viruses, such as West Nile virus, dengue virus, and poliovirus, may antagonize the formation of stress granules by interfering with the function of Tiar, Tia-1, or G3bp. RNA-associated protein 55 (Rap55) is a component of stress granules that is bound by influenza virus NS1 protein. G3bp may be cleaved by poliovirus $3C^{pro}$, or sequestered into the replication complexes that form in cells infected with hepatitis C virus, Junin virus, alphaviruses, or flaviviruses, blocking stress granule assembly. Caprin1 and Usp10 bind G3bp and may also have a role in stress granule formation.

Viral reproduction cycles often include inhibition of translation of cellular mRNAs. The vast majority of virus-induced modifications affect the initiation step of protein synthesis. Indeed, our detailed understanding of this step of translation has been a consequence of unraveling the effects of viral infection. Although elongation and termination require far fewer cellular proteins, there are nonetheless examples of viral modulation of these steps as well. We also have a growing appreciation of how viral infection affects the stability of cellular mRNAs, through cellular mRNA decay pathways. A description of these controls can be found in Chapter 10.

An intriguing recent addition to our knowledge of translational control in virus-infected cells concerns RNA-containing granules. They regulate the mRNA cycle, metabolism, and gene expression and are an important point of control during virus infection. The field of virus-RNA granule interactions is young, and many questions remain. The impact of RNA granule formation on virus reproduction, the effect of virus proteins on granules, and the roles of granule proteins in viral reproduction have barely been explored. As it has not been possible to purify RNA granules from cells, other approaches to understanding how they are built and how they function must be developed. An intriguing hypothesis is that the formation of stress granules is part of an integrated response that includes intrinsic antiviral mechanisms. Emerging evidence indicates that intrinsic immunity and stress responses are linked at many levels. An example is Pkr, which is an IFN response protein but might also sense the formation of stress granules. Stress granule proteins are localized with proteins such as Rig-I-like receptors that activate IFN responses. How stress responses and intrinsic immunity interact at multiple levels may well be a major goal of future research in this field.

References

Review Articles

Beier H, Grimm M. 2001. Misreading of termination codons in eukaryotes by natural nonsense suppressor tRNAs. *Nucleic Acids Res* **29:**4767–4782.

Dinman JD. 2012. Mechanisms and implications of programmed translational frameshifting. *Wiley Interdiscip Rev RNA* **5:**661–673.

Donnelly N, Gorman AM, Gupta S, Samali A. 2013. The eIF2α kinases: their structures and functions. *Cell Mol Life Sci* **70:**3493–3511.

Firth AE, Brierley I. 2012. Non-canonical translation in RNA viruses. *J Gen Virol* **93:**1385–1409.

Fonseca BD, Smith EM, Yelle N, Alain T, Bushell M, Pause A. 2014. The ever-evolving role of mTOR in translation. *Semin Cell Dev Biol* **36:**102–112.

Hinnebusch AG. 2014. The scanning mechanism of eukaryotic translation initiation. *Annu Rev Biochem* **83:**779–812.

Jackson RJ. 2013. The current status of vertebrate cellular mRNA IRESs. *Cold Spring Harb Perspect Biol* **5:**a011569. doi:10.1101/cshperspect.a011569.

Lloyd RE. 2013. Regulation of stress granules and P-bodies during RNA virus infection. *Wiley Interdiscip Rev RNA* **4:**317–331.

Mohr I, Sonenberg N. 2012. Host translation at the nexus of infection and immunity. *Cell Host Microbe* **12:**470–483.

Nicholson BL, White KA. 2014. Functional long-range RNA-RNA interactions in positive-strand RNA viruses. *Nat Rev Microbiol* **12:**493–504.

Walsh D, Mathews MB, Mohr I. 2013. Tinkering with translation: protein synthesis in virus-infected cells. *Cold Spring Harb Perspect Biol* **5:**a012351. doi:10.1101/cshperspect.a012351.

Walsh D, Mohr I. 2011. Viral subversion of the host protein synthesis machinery. *Nat Rev Microbiol* **9:**860–875.

Xie J, Kozlov G, Gehring K. 2014. The "tale" of poly(A) binding protein: the MLLE domain and PAM2-containing proteins. *Biochim Biophys Acta* **1839:** 1062–1068.

Papers of Special Interest

Burgui I, Yángüez E, Sonenberg N, Nieto A. 2007. Influenza virus mRNA translation revisited: is the eIF4E cap-binding factor required for viral mRNA translation? *J Virol* **81:**12427–12438.

Clippinger AJ, Maguire TG, Alwine JC. 2011. The changing role of mTOR kinase in the maintenance of protein synthesis during human cytomegalovirus infection. *J Virol* **85:**3930–3939.

Desmet EA, Anguish LJ, Parker JS. 2014. Virus-mediated compartmentalization of the host translational machinery. *mBio* **5:**e01463-14. doi:10.1128/mBio.01463-14.

Fitzgerald KD, Chase AJ, Cathcart AL, Tran GP, Semler BL. 2013. Viral proteinase requirements for the nucleocytoplasmic relocalization of cellular splicing factor SRp20 during picornavirus infections. *J Virol* **87:**2390–2400.

Ho BC, Yu SL, Chen JJ, Chang SY, Yan BS, Hong QS, Singh S, Kao CL, Chen HY, Su KY, Li KC, Cheng CL, Cheng HW, Lee JY, Lee CN, Yang PC. 2011. Enterovirus-induced miR-141 contributes to shutoff of host protein translation by targeting the translation initiation factor eIF4E. *Cell Host Microbe* **9:**58–69.

Jiang J, Laliberté JF. 2011. The genome-linked protein VPg of plant viruses—a protein with many partners. *Curr Opin Virol* **1:**347–354.

Liu J, Castelli LM, Pizzinga M, Simpson CE, Hoyle NP, Bailey KL, Campbell SG, Ashe MP. 2014. Granules harboring translationally active mRNAs provide a platform for P-body formation following stress. *Cell Rep* **9:**944–954.

Matsuda D, Mauro VP. 2014. Base pairing between hepatitis C virus RNA and 18S rRNA is required for IRES-dependent translation initiation in vivo. *Proc Natl Acad Sci U S A* **111:**15385–15389.

McKinney C, Perez C, Mohr I. 2012. Poly(A) binding protein abundance regulates eukaryotic translation initiation factor 4F assembly in human cytomegalovirus-infected cells. *Proc Natl Acad Sci U S A* **109:**5627–5632.

McKinney C, Zavadil J, Bianco C, Shiflett L, Brown S, Mohr I. 2014. Global reprogramming of the cellular translational landscape facilitates cytomegalovirus replication. *Cell Rep* **6:**9–17.

Wang QS, Jan E. 2014. Switch from cap- to factorless IRES-dependent 0 and +1 frame translation during cellular stress and dicistrovirus infection. *PLoS One* **9:**e103601. doi:10.1371/journal.pone.0103601.

Woo PC, Lau SK, Choi GK, Huang Y, Teng JL, Tsoi HW, Tse H, Yeung ML, Chan KH, Jin DY, Yuen KY. 2011. Natural occurrence and characterization of two internal ribosome entry site elements in a novel virus, canine picodicistrovirus, in the picornavirus-like superfamily. *J Virol* **86:**2797–2808.

Yamamoto H, Unbehaun A, Loerke J, Behrmann E, Collier M, Bürger J, Mielke T, Spahn CM. 2014. Structure of the mammalian 80S initiation complex with initiation factor 5B on HCV-IRES RNA. *Nat Struct Mol Biol* **21:** 721–727.

Yángüez E, Rodriguez P, Goodfellow I, Nieto A. 2012. Influenza virus polymerase confers independence of the cellular cap-binding factor eIF4E for viral mRNA translation. *Virology* **422:**297–307.

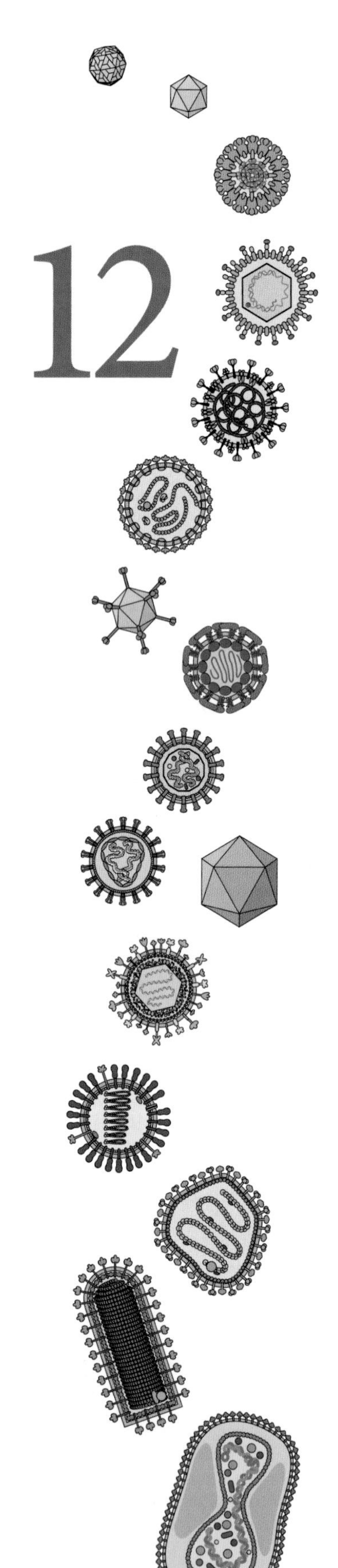

12 Intracellular Trafficking

Introduction

Assembly within the Nucleus

Import of Viral Proteins for Assembly

Assembly at the Plasma Membrane

Transport of Viral Membrane Proteins to the Plasma Membrane

Sorting of Viral Proteins in Polarized Cells

Disruption of the Secretory Pathway in Virus-Infected Cells

Signal Sequence-Independent Transport of Viral Proteins to the Plasma Membrane

Interactions with Internal Cellular Membranes

Localization of Viral Proteins to Compartments of the Secretory Pathway

Localization of Viral Proteins to the Nuclear Membrane

Transport of Viral Genomes to Assembly Sites

Transport of Genomic and Pregenomic RNA from the Nucleus to the Cytoplasm

Transport of Genomes from the Cytoplasm to the Plasma Membrane

Perspectives

References

LINKS FOR CHAPTER 12

⏭ *Video: Interview with Dr. Ari Helenius*
http://bit.ly/Virology_Helenius

⏭ *Movie 12.1: Mouse salivary glands infected with a virulent strain of dually fluorescent derivative of pseudorabies virus*
http://bit.ly/Virology_V1_Movie12-1

He who is everywhere is nowhere.

SENECA

Introduction

Successful viral reproduction requires the intracellular assembly of progeny virions from their protein, nucleic acid, and, in many cases, membrane components. In preceding chapters, we have considered molecular mechanisms that ensure the synthesis of the macromolecules from which virus particles are constructed in the host cell. Because of the structural and functional compartmentalization of eukaryotic cells, components of these particles are generally produced at multiple intracellular locations, and must be brought together for assembly. Intracellular trafficking and sorting of viral nucleic acids, proteins, and glycoproteins to the appropriate sites is therefore an essential prelude to the assembly of all animal viruses.

From our point of view, animal cells are very small, with typical diameters of 10 to 30 μm. However, in the microscopic world inhabited by viruses, an animal cell is large: the distances over which virion components must be transported within a cell are roughly equivalent to a mile on the macroscopic, human scale. The properties of the intracellular milieu prevent viral particles, genomes, or subassemblies from reaching the appropriate intracellular destinations during entry or egress within reasonable periods simply by diffusion (Box 12.1). Their movement therefore requires transport systems and a considerable expenditure of energy, supplied by the host cell. The cellular highways most commonly used for movement of viral components for assembly are those formed by microtubules (as is also true during entry). These filaments are polarized and highly organized within the cell, with (−) ends at the microtubule-organizing center (near the nucleus) and (+) ends at the cell periphery. They are traveled by cellular (−) end- and (+) end-directed motor proteins that carry cargo and convert the chemical energy of ATP into kinetic energy.

The intracellular trafficking of viral macromolecules must be appropriately directed so that the building blocks of virus particles are delivered to the correct assembly site. Assembly of viral particles can occur at any one of several intracellular addresses, depending on whether the particles are enveloped or naked and on the site and mechanism of genome replication (e.g., Fig. 12.1 and 12.2). All viral envelopes are derived from one of the host cell's membranes, which are modified by insertion of viral proteins. Many virus particles assemble at the plasma membrane, but some envelopes are derived from membranes of internal compartments. Consequently, assembly of enveloped viruses requires delivery of some viral proteins to the appropriate membrane, as well as transport of other proteins and the nucleic acid genome to that membrane. Other assembly sites are the cell nucleus and within the cytoplasm. These strategies impose less complex trafficking problems than does assembly of enveloped viruses at membrane sites, but additional mechanisms may be required for egress of progeny particles from the cell. In some cases, genome-containing nucleocapsids are formed in infected cell nuclei but assembly is completed at a cellular membrane. Such spatial and temporal separation of assembly reactions depends on appropriate coordination among multiple transport processes.

The need for movement of proteins and nucleic acids from one cellular compartment to another, or for insertion of proteins into specific membranes, is not unique to viruses. The majority of cellular RNA species are exported from

PRINCIPLES *Intracellular trafficking*

- Progeny genomes, structural proteins, and enzymes of virus particles must be concentrated at the intracellular site at which assembly takes place.
- The movement of viral components requires transport systems and a considerable expenditure of energy, supplied by the host cell.
- When viruses with DNA genomes are assembled in the nucleus, cytoplasmic proteins must be actively imported.
- The particles of many viruses that reproduce in animal cells include a lipid envelope derived from a host cell membrane host cell, and therefore assembly takes place at a cell membrane.
- All proteins destined for insertion into the plasma membrane enter the ER as they are translated, and signal sequences within the amino termini of such proteins guide this process.
- The ER lumen is the site of many essential protein modification and folding reactions.
- Elaborate quality control mechanisms in the ER ensure that proteins that are misfolded are transported to the cytoplasm and degraded.
- Viral glycoproteins may be proteolytically processed in the Golgi network, a reaction essential for the formation of infectious particles.
- Viral components are sorted to specialized surfaces in polarized cells, including epithelial cells and neurons.
- The matrix or tegument proteins of enveloped viruses, which lie between the inner surface of the membrane and the capsid, do not enter the secretory pathway, but are synthesized in the cytoplasm and directed to membrane assembly sites by specific signals.
- The envelopes of a variety of viruses are acquired from internal membranes of the infected cell, rather than from the plasma membrane.

BOX 12.1

DISCUSSION

Getting from point A to point B in heavy traffic

Within a cell, directional movement and coordination of such movements in space and time are very complicated processes. Distributions of high-molecular-weight reactants and products are rarely controlled by concentration gradients and diffusion, as they are *in vitro*. Indeed, the inside of a cell is so tightly packed with organelles and cytoskeletal structures (panel A in the figure) that it is simply inappropriate to think of the contents of the cytoplasm, the nucleus, or organelle lumens as "gels" or "suspensions."

Directional movement in cells is achieved by two general processes (panel B). Short-distance movement across membranes or in and out of capsids is measured in angstroms to nanometers and is accomplished primarily via protein channels. Movement through such channels (transporters, translocons, pores, and portals) generally requires energy supplied by hydrolysis of ATP (or other nucleoside triphosphates).

Long-distance movement of proteins, viral particles or their components, and organelles inside cells is measured in micrometers to meters. Such movement invariably requires energy and is mediated by molecular motors moving on cytoskeletal tracks; myosins move cargo on actin fibers, while dynein and kinesin move cargo on microtubules.

(A) A three-dimensional model of a section through a rat synaptic bouton (the site of neurotransmitter release at the end of an axon terminal) showing 60 proteins, and the plasma membrane (light beige). This model was constructed by combining the results of several complementary approaches: quantitative immunoblotting and quantitative mass spectrometry to measure the number of molecules of each protein; electron microscopy to determine the number, size, and positions of organelles; and super-resolution fluorescence microscopy to localize the proteins. Previously determined molecular structures of the proteins and their interactions were also used. Adapted from B. G. Wilhelm et al., *Science* **344:**1023–1028, 2014, with permission. Courtesy of S. Rizzoli, European Neuroscience Institute, Germany. **(B)** Summary of properties of short- and long-range transport of viral components in infected cells.

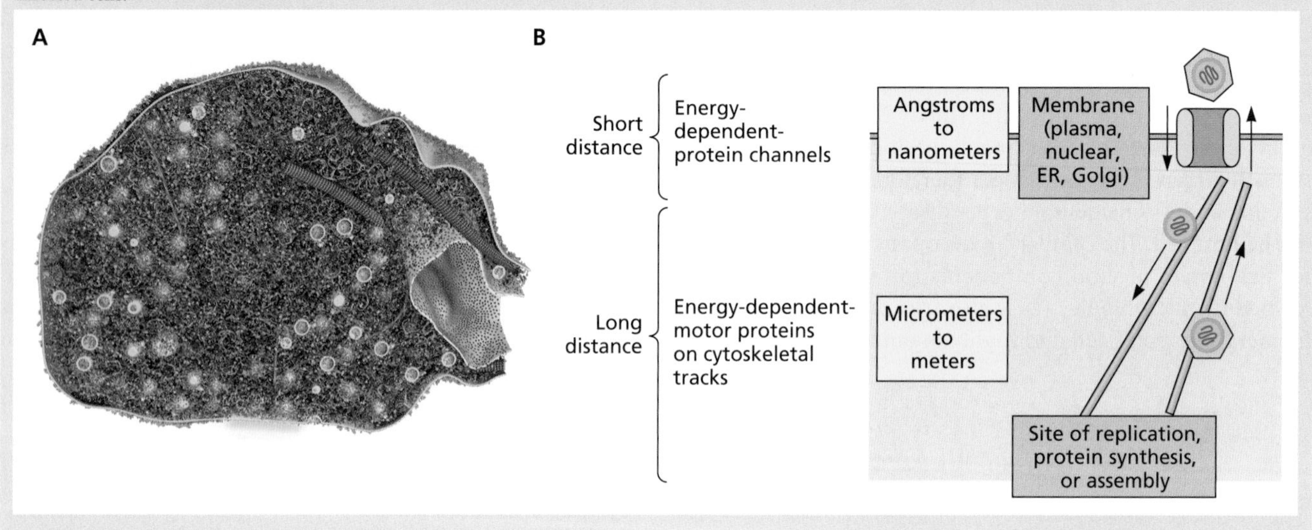

the nucleus, in which transcription takes place. Similarly, cellular proteins are made by translation of messenger RNAs (mRNAs) in the cytoplasm and must then be transported to their sites of operation. Eukaryotic cells are therefore constantly engaged in transport of macromolecules among their compartments via intracellular trafficking systems. The cellular systems that sort macromolecules to particular intracellular sites are just as indispensable for viral reproduction as the cellular biosynthetic machineries responsible for transcription, DNA synthesis, or translation. Indeed, the advances in our understanding of cellular trafficking mechanisms can be traced to initial studies of viral membrane or nuclear proteins. In the following sections, the cellular transport pathways required during viral reproduction are described in the context of the site at which virion assembly takes place.

Assembly within the Nucleus

Assembly of the majority of viruses with DNA genomes, including adenoviruses, papillomaviruses, and polyomaviruses, takes place within infected cell nuclei, the site of viral DNA synthesis. All structural proteins of these nonenveloped viruses are imported into the nucleus following synthesis in the infected cell cytoplasm (Fig. 12.1), allowing complete assembly within this organelle. In contrast, assembly of the structurally more elaborate herpesviruses, which harbor a DNA-containing nucleocapsid assembled within the nucleus, is completed at extranuclear sites. So too is that of some enveloped RNA viruses with genomes that are replicated in nuclei, such as orthomyxoviruses. In these cases, only a subset of viral structural proteins must be imported into the nucleus.

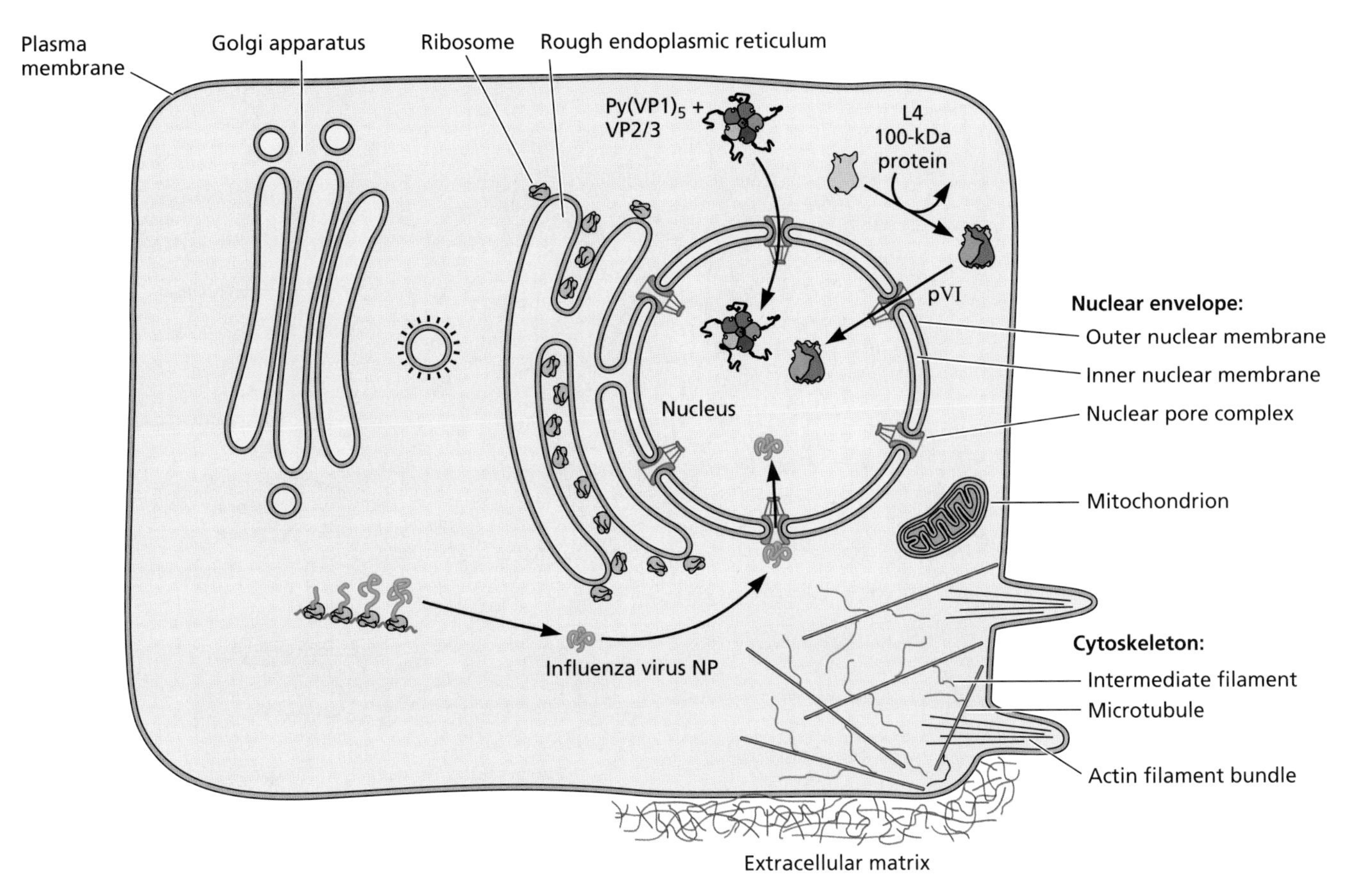

Figure 12.1 Localization of viral proteins to the nucleus. The nucleus and major membrane-bound compartments of the cytoplasm, as well as components of the cytoskeleton, are illustrated schematically and not to scale. Viral proteins destined for the nucleus are synthesized by cytoplasmic polyribosomes, as illustrated for the influenza virus NP protein. They engage with the cytoplasmic face of the nuclear pore complex and are translocated into the nucleus by the protein import machinery of the host cell. Some viral structural proteins enter the nucleus as preassembled structural units, as shown for polyomavirus [Py] VP1 pentamers associated with one molecule of either VP2 or VP3 and adenovirus hexon trimers formed with the assistance of the viral L4 100-kDa protein chaperone. Interaction of hexon trimers with the import receptors importin α/β is mediated by a second structural protein, protein pre-VI (pVI).

As far as we know, all viral structural proteins that enter the nucleus do so via the normal cellular pathways of nuclear protein import. These same pathways are responsible for import of both viral genomes (or nucleoproteins) and viral nonstructural proteins that function in the nucleus early in the infectious cycle (Chapter 5). Proteins destined for the nucleus carry nuclear localization signals (see Fig. 5.22), which are recognized by components of the cellular nuclear import machinery for subsequent transport into the nucleus.

Import of Viral Proteins for Assembly

The primary sequences of many such viral proteins destined for nuclear import contain putative nuclear localization sequences, which are characterized by clusters of basic amino acids. The majority of these sequences that have been verified experimentally conform to the simple or bipartite nuclear localization signals described in Chapter 5 (Fig. 5.22). However, some have noncanonical sequences, for example, the herpes simplex virus type 1 structural protein VP19C, or form only once the folded protein has assembled into structural units, as in porcine parvovirus VP2: this three-dimensional nuclear localization motif is present only in VP2 trimers, the major structural unit of the capsid.

A typical mammalian cell contains on the order of 3,000 to 4,000 nuclear pore complexes, each with a very high translocation capacity, with 10^3 translocation events/s. However, nuclear import also depends on the limited supply of soluble transport proteins. As large quantities of viral structural proteins must enter the nucleus prior to assembly, there is potential for competition among viral and cellular proteins for access to receptors or the nuclear pore complex proteins that mediate transport. Such competition is minimized in

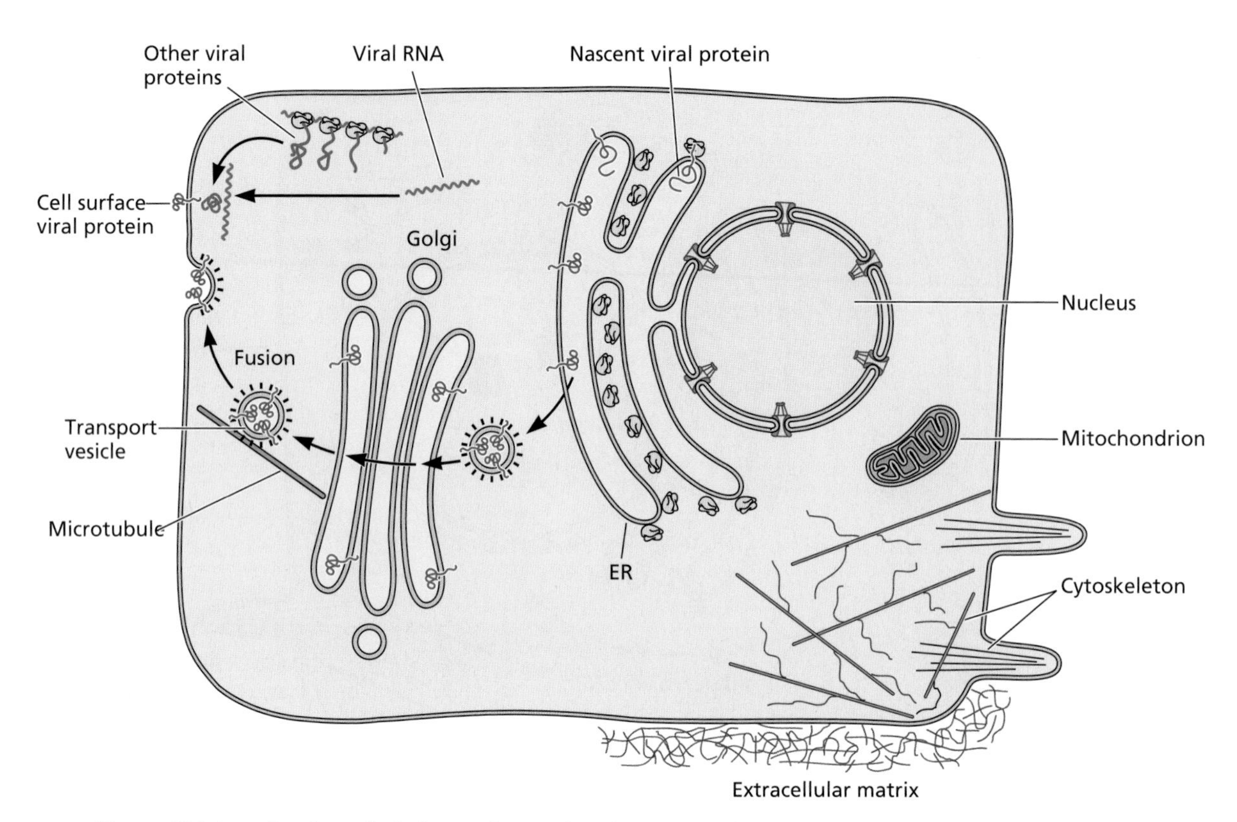

Figure 12.2 Localization of viral proteins to the plasma membrane. Viral envelope glycoproteins (red) are cotranslationally translocated into the ER lumen and folded and assembled within that compartment. They travel via transport vesicles to and through the Golgi apparatus and then to the plasma membrane. The internal proteins of the particle (blue) and the genome (green) are also directed to plasma membrane sites of assembly.

cells infected by the larger DNA viruses, such as adenoviruses and herpesviruses: by the time structural proteins are made during the late phase of infection, cellular protein synthesis is severely inhibited. The proteins of viruses that do not induce inhibition of cellular protein synthesis, such as those of the polyomaviruses, must enter the nucleus despite continual transport of cellular proteins. Whether import of viral proteins is favored in such circumstances, for example, by the presence of high-affinity nuclear localization signals, is not known.

Many viral structural proteins that enter infected cell nuclei form multimeric capsid components. Import of structural units of virus particles can depend on prior assembly in the cytoplasm to form the nuclear localization signal (see above) or to ensure efficient import. Pentamers of the major capsid protein (VP1) of simian virus 40 and polyomavirus specifically bind a common C-terminal sequence of either VP2 or VP3, the minor capsid proteins (Appendix, Fig. 23B). Such heteromeric assemblies are the substrates for import into the nucleus. Indeed, efficient nuclear localization of polyomavirus VP2 and VP3 proteins can occur only in cells in which VP1 is also made. Assembly of the heteromeric complex facilitates import of the minor structural proteins, even though each contains a nuclear localization signal. The increased density of these signals may allow more effective competition for essential components of the import pathway, or the nuclear localization signals may be more accessible in the complex.

Despite such potential advantages as increased efficiency of import of viral proteins and transport of the structural proteins in the appropriate stoichiometry, import of preassembled capsid components is not universal. For example, adenoviral hexons, trimers of viral protein II, are found only in the nucleus of the infected cell. Assembly of trimers requires a viral chaperone, the L4 100-kDa protein. However, when these two proteins are made in the absence of other adenoviral gene products, hexons do not enter the nucleus. This process requires a second structural protein, pre-VI (pVI), which interacts with hexons and cellular import receptors (Fig. 12.1).

Assembly at the Plasma Membrane

The particles of many viruses that reproduce in animal cells include a lipid envelope derived from a membrane of the host cell, although this structural feature is rare among

BOX 12.2

DISCUSSION

Does host cell architecture shape virus structure?

Many viruses that are important human pathogens, including hepatitis B and C viruses, human immunodeficiency virus type 1, and influenza A virus, are enveloped. In fact, the particles of >50% of the virus families that reproduce in animal cells include a lipid membrane, regardless of the nature of the viral genome. Furthermore, acquisition of the envelope and release of these viruses from the host cell are frequently accomplished in a single step. In contrast, the particles of only some 10% of plant virus families are enveloped (3 of 29 listed in the *Ninth Report of the International Committee on Taxonomy of Viruses* [2012]). Two of these families, *Bunyaviridae* and *Rhabdoviridae*, also include viruses that replicate in animal cells, but with significant differences in assembly and release.

In mammalian cells, rhabdoviruses, such as vesicular stomatitis virus, acquire their envelope, and are concomitantly released, by budding through the plasma membrane. However, plant rhabdoviruses form upon budding of internal components either into the endoplasmic reticulum (lettuce necrotic yellow virus) or through the inner nuclear membrane (potato yellow dwarf virus), and in both cases accumulate at these intracellular sites. In similar fashion, bunyavirus particles are released from infected animal cells via the secretory pathway following formation within Golgi compartments, but are not released from plant cells. For example, tomato spotted wilt virus particles accumulate in vesicles derived from Golgi and endoplasmic reticulum membranes until the cells are ingested by insect vectors (thrips) during feeding. In infected salivary gland cells of the insect host, tomato spotted wilt virus particles are formed and secreted from the plasma membrane like bunyavirus particles in mammalian cells.

Formation of an envelope provides an effective means of direct or indirect release from animal cells of progeny virus particles, which can then infect other cells in the organism via their accessible plasma membranes. In contrast, plant cells are surrounded by a structure that imposes formidable barriers to exit and entry by these mechanisms, the cell wall. This thick and rigid structure is built from microfibrils of cellulose organized into a network with the polysaccharides pectin and cross-linking glycans (see the figure). Neighboring cell walls are penetrated by the numerous microchannels (plasmodesmata) by which a plant cell is connected to its neighbors. Consequently, the acquisition of an envelope is of little benefit to viruses that reproduce in plant cells. Rather, the genomes of all plant viruses encode movement proteins that induce alterations of plasmodesmata to allow direct passage of virus particles (or genomes) from one cell to another (Box 13.14). Furthermore, the great majority of plant viruses are transmitted among host plants not by release into the environment but by vectors, most commonly insects.

Kormelink R, Garcia ML, Goodin M, Sasaya T, Haenni AL. 2011. Negative-strand RNA viruses: the plant-infecting counterparts. *Virus Res* **162:**184–202.

Two adjacent plant cells showing the plasma membrane components of the cell wall and a plasmodesma through the plasma membrane and its internal tube-like structure, the dermotubule derived from the endoplasmic reticulum. Plasmodesmata directly connect one plant cell to its neighbors. Adapted from Molecular Expressions (http://micro.magnet.fsu.edu/cells/plants/plasmodesmata.html), with permission.

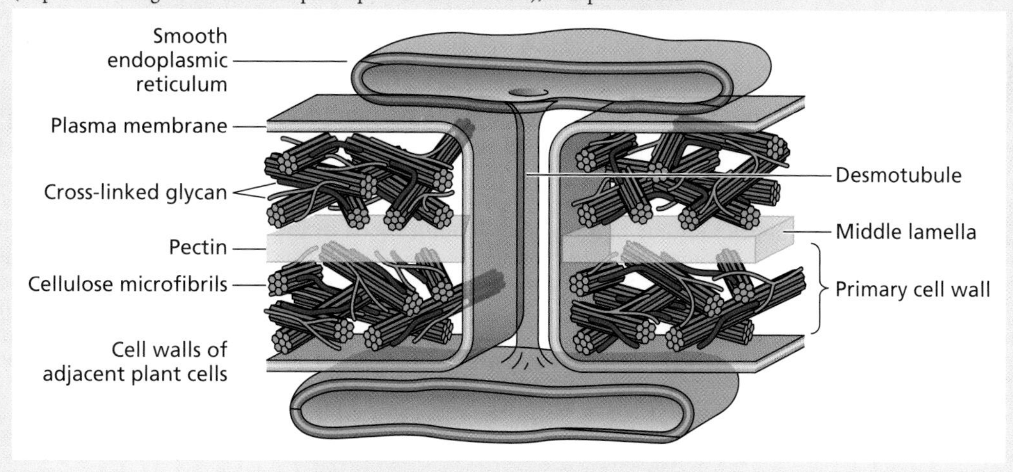

plant viruses (Box 12.2) . Assembly of the majority of such enveloped viruses takes place at the plasma membrane. Before such virus particles can form, viral integral membrane proteins must be transported to this cellular membrane. The first stages of the pathway by which viral and cellular proteins are delivered to the plasma membrane were identified more than 35 years ago, and the process is now understood quite well. Viruses with envelopes derived from the plasma membrane also contain internal proteins, which may be membrane associated, and, of course, nucleic acid genomes. These internal components must also be sorted to appropriate plasma membrane sites for assembly (Fig. 12.2).

Transport of Viral Membrane Proteins to the Plasma Membrane

Viral membrane proteins reach their destinations by the highly conserved, cellular **secretory pathway**. Many of the steps in the pathway have been studied by using viral membrane glycoproteins, such as the vesicular stomatitis virus G and influenza virus hemagglutinin (HA) proteins. These viral proteins offer several experimental advantages: they are synthesized in large quantities; their synthesis is initiated in a controlled fashion following infection; and their transport can be studied readily by genetic, biochemical, and imaging methods.

Entry into the first staging post of the secretory pathway, the endoplasmic reticulum (ER), is accompanied by membrane insertion of integral membrane proteins. Viral envelope proteins generally span the cellular membrane into which they are inserted only once, and therefore contain a single transmembrane domain. In viral proteins, transmembrane segments (described in Chapter 5) usually separate large extracellular from smaller cytoplasmic domains (Fig. 12.3). The former include the binding sites for cellular receptors, crucial for initiation of the infectious cycle, whereas the latter are important in virus assembly. Viral membrane proteins are usually oligomers (Chapter 4). Most interactions among the subunits of viral membrane proteins are noncovalent, but some examples of association via covalent interchain disulfide bonds are known. Oligomer assembly takes place during transit from the cytoplasm to the cell surface, as does the proteolytic processing necessary to produce some mature (functional) envelope glycoproteins from the precursors that enter the secretory pathway. For example, the human immunodeficiency virus type 1 Env protein and influenza virus HA0 precursor of pathogenic strains of avian influenza virus are cleaved within Golgi compartments. Viral (and cellular) proteins that travel the secretory pathway also possess distinctive structural features, including disulfide bonds and covalently linked oligosaccharide chains (Fig. 12.3). These characteristic covalent modifications (as well as oligomerization) take place as proteins travel through a series of specialized compartments that provide the chemical environments and enzymatic machinery necessary for their maturation, as illustrated in Fig. 12.4 for the influenza A virus HA0 protein. The first such compartment, the ER, is encountered by viral membrane proteins as they are synthesized.

Translocation of Viral Membrane Proteins into the Endoplasmic Reticulum

All proteins destined for insertion into the plasma membrane, or the membranes of such intracellular organelles as the Golgi apparatus, enter the ER as they are translated (Fig. 12.2). This membranous structure appears as a basketwork of tubules and sacs extending throughout the cytoplasm (Fig. 12.5A). The ER membrane demarcates a geometrically convoluted but continuous internal space, the **ER lumen**, from the remainder of the cytoplasm. The ER lumen is characterized by a chemically distinctive environment and is topologically equivalent to the outside of the cell. Proteins that enter the ER during their synthesis are therefore sequestered from the cytoplasmic environment as they are made.

Polyribosomes engaged in synthesis of proteins that will enter the secretory pathway become associated with the cytoplasmic face of the ER membrane soon after translation begins. Areas of the ER to which polyribosomes are bound form the **rough ER** (Fig. 12.5B). The association of polyribosomes with the ER membrane is directed by a short sequence in the nascent protein, termed the **signal peptide**. It is now taken for granted that the primary sequences of proteins

Figure 12.3 Primary sequence features and covalent modifications of the influenza virus HA protein. The primary sequence of the HA0 protein is depicted by the red line in the center, with the orange arrowhead indicating the site of the proteolytic cleavages that produce the HA1 and HA2 subunits from HA0 of pathogenic avian strains. The fusion peptide, the N-terminal signal sequence that is removed by signal peptidase in the ER, and the C-terminal transmembrane domain are hydrophobic. Disulfide bonds, one of which maintains covalent linkage between the HA1 and HA2 proteins following HA0 cleavage, are indicated, as are sites of N-linked glycosylation (oligosaccharides) and palmitoylation (Ac).

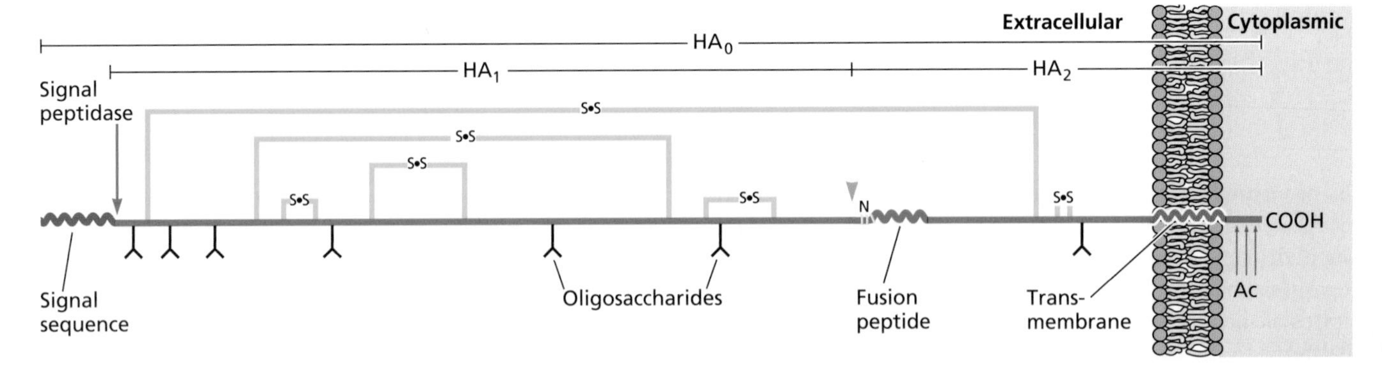

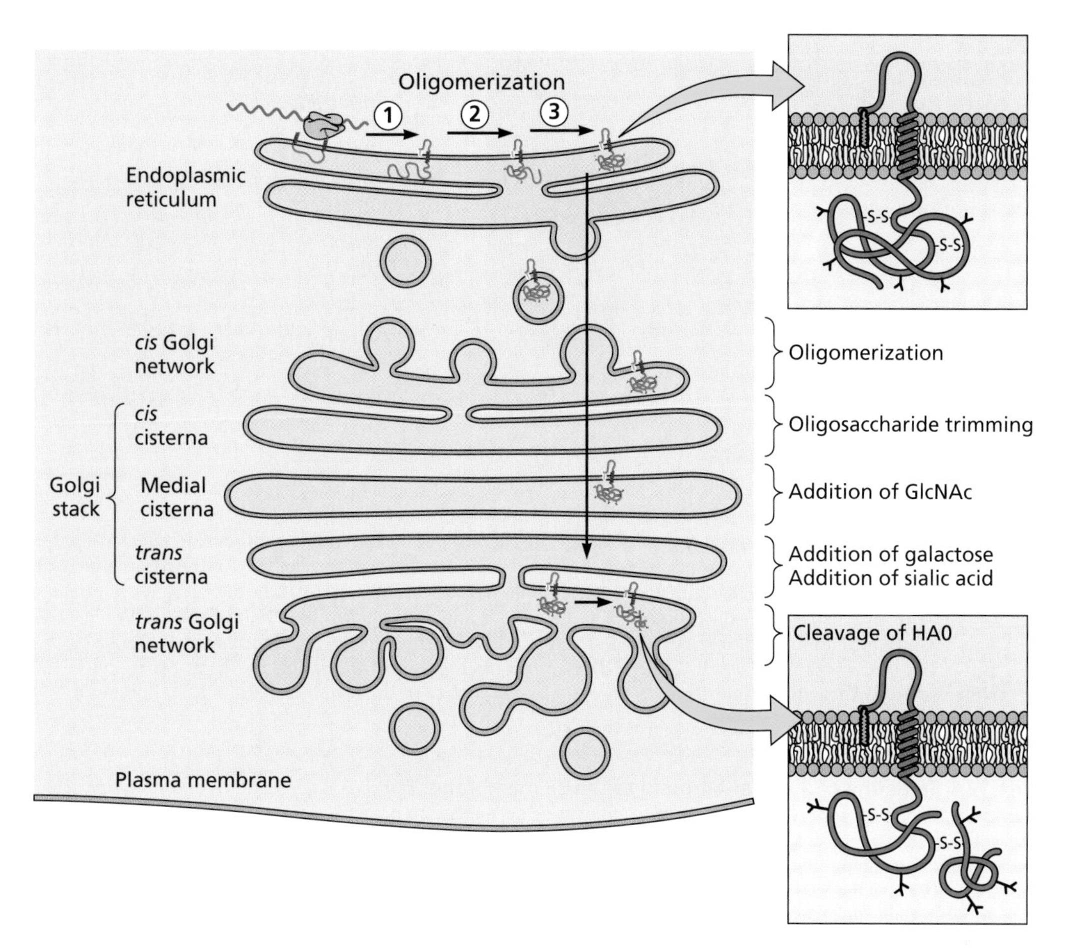

Figure 12.4 Maturation of influenza virus HA0 protein during transit along the secretory pathway. The modifications that occur during transit of the influenza virus HA0 protein through the various compartments of the secretory pathway are illustrated. In the ER, these are translocation and signal peptide cleavage (1), disulfide bond formation, and addition of N-linked core oligosaccharides (2), as the protein folds (3). The cytoplasmic domain acquires palmitate (orange) while the protein travels to the plasma membrane, but it has not been established when this modification takes place. For simplicity, the protein is depicted as a monomer, although oligomerization also takes place in the ER lumen. Note that the protein domain initially introduced into the ER lumen, in this case the N-terminal portion of the protein (type I orientation), corresponds to the extracellular domain of the cell surface protein.

include "zip codes" specifying the cellular addresses at which the proteins must reside to fulfill their functions, such as the nuclear localization signals discussed in the previous section. The signal peptides of proteins that enter the ER lumen were the first such zip codes to be identified, and established this paradigm some 35 years ago. Signal peptides are commonly found at the N termini of proteins destined for the secretory pathway. They are usually about 20 amino acids in length and contain a core of 15 hydrophobic residues. Signal peptides are often removed enzymatically during protein translocation into the ER by a protease located in the lumen, signal peptidase.

Translation of a protein that will enter the ER begins in the normal fashion and continues until the signal peptide emerges from the ribosome (Fig. 12.6). This signal then directs binding of the translation machinery to the ER membrane by means of two components: the signal peptide is recognized by the **signal recognition particle (SRP)**, which in turn binds to the cytoplasmic domain of an integral ER membrane protein termed the **SRP receptor**. Binding of the signal recognition particle to the ribosome temporarily halts translation, to allow the stalled translation complex to bind to the ER membrane. Following the initial docking of the complex at the membrane, the ribosome becomes tightly bound to the membrane and engaged with a protein translocation channel, which forms a gated, aqueous pore through the ER membrane. This interaction is coordinated with release of the signal recognition particle, association of the signal peptide with the

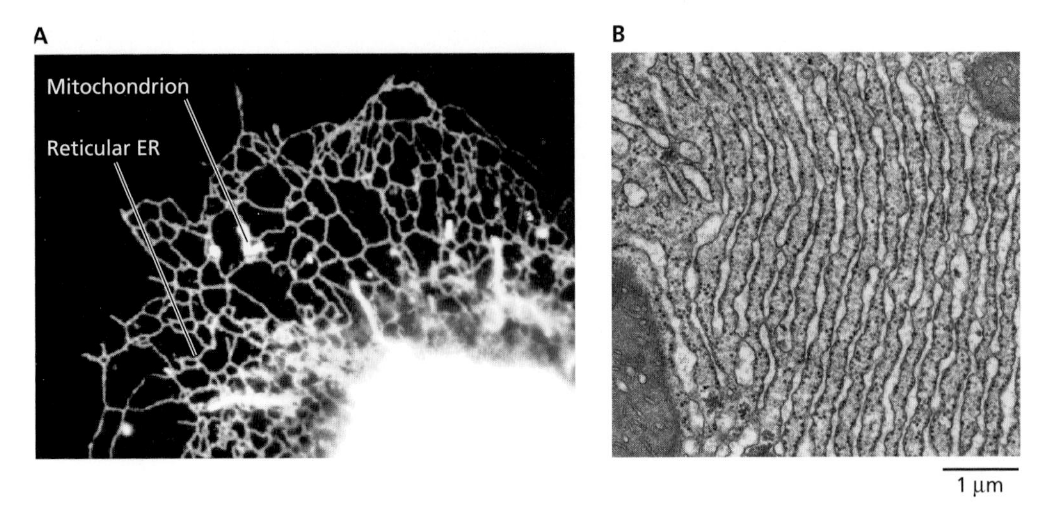

Figure 12.5 The endoplasmic reticulum. (A) The ER of a mammalian cell in culture. The reticular ER, which extends throughout the cytoplasm, was visualized by fluorescence microscopy of fixed African green monkey kidney epithelial cells stained with the lipophilic fluorescent dye 3,3′-dihexyloxacarbocyanine iodide. This dye also stains mitochondria. The ER membrane accounts for over half of the total membrane of a typical animal cell and possesses a characteristic lipid composition. Courtesy of M. Terasaki, University of Connecticut Health Center. **(B)** Electron micrograph of the rough ER in rat hepatocytes. Note the many ribosomes associated with the cytoplasmic surface of the membrane. From R. A. Rodewald, Biological Photo Service.

Figure 12.6 Targeting of a nascent protein to the ER membrane. Translation of an mRNA encoding a protein that will enter the ER lumen proceeds until the signal peptide (purple) emerges from the ribosome. The signal recognition particle (SRP), which contains a small RNA molecule and several proteins, binds to both the signal peptide and the ribosome to halt or pause translation, upon binding of GTP to one of the protein subunits (step 1). The nascent polypeptide-SRP-ribosome complex then binds to the SRP receptor in the ER membrane (step 2). This interaction triggers hydrolysis of GTP bound to SRP and to its receptor; release of SRP (step 3); and close association of the ribosome with, and binding of the hydrophobic signal peptide to, the heterotrimeric protein translocation channel (step 4). These interactions trigger opening of the cytosolic end of the channel. The luminal end of the translocation channel is also initially closed. Translation is then resumed, and the seal maintained at the luminal end of the channel early in translocation is reversed by binding of the chaperone Grp78. The growing polypeptide chain is transferred through the membrane as its translation continues (step 5). In some cases, signal peptidase removes the signal peptide cotranslationally (step 6). A lateral gate in the channel opens within the membrane for transfer of the transmembrane domain(s) of translocated proteins into the ER membrane.

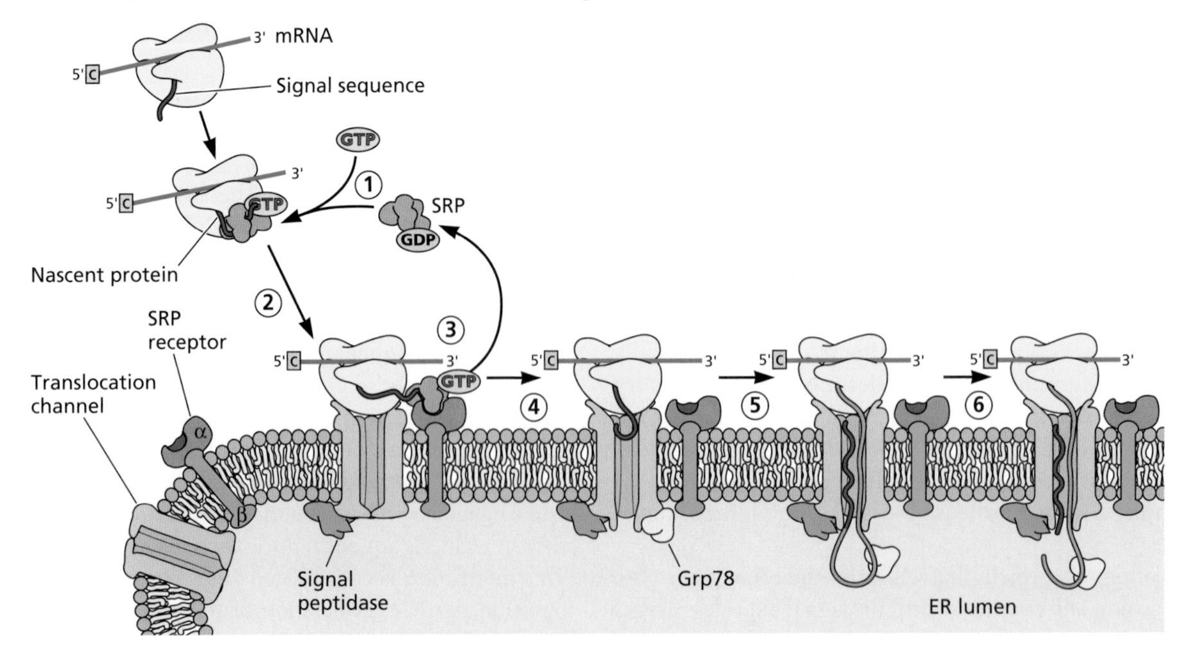

translocation channel, and resumption of translation. Because the ribosome remains bound to the membrane upon release of the signal recognition particle, continued translation facilitates movement of the growing polypeptide chain through the membrane. Such coupling of translation and translocation ensures that the protein crosses the membrane as an unfolded chain that can be accommodated within the translocation channel. Movement of the growing polypeptide through the membrane channel is facilitated by binding of the luminal chaperone Grp78 (Bip) to the nascent protein.

When a protein entering the ER is destined for secretion from the cell, translocation continues until the entire polypeptide chain enters the lumen. During translocation, the signal peptide is proteolytically removed by signal peptidase, releasing the soluble protein into the ER. In contrast, translocation of integral membrane proteins with a single transmembrane domain, such as viral envelope proteins, halts when a hydrophobic **stop transfer signal** is encountered in the nascent protein. This sequence may be the signal peptide itself or a second, internal hydrophobic sequence. In proteins that span the membrane multiple times, the number, location, and orientation of stop and start transfer signals within a protein determine the topology with which it is organized in the ER membrane. The programming of insertion of proteins into the ER membrane by signals built into their primary sequences ensures that every molecule of a particular protein adopts the identical topology in the membrane. As this topology is maintained during the several membrane budding and fusion reactions by which proteins reach the cell surface, the way in which a protein is inserted into the ER membrane determines its orientation in the plasma membrane.

Reactions within the ER

The folding and initial posttranslational modification of proteins that enter the secretory pathway take place within the ER. The lumen contains many enzymes that catalyze chemical modifications, such as disulfide bond formation and glycosylation, or that promote folding and oligomerization.

Glycosylation. Viral envelope proteins that travel the secretory pathway, like their cellular counterparts, are generally modified by the addition of oligosaccharides to either asparagine (N-linked glycosylation) or serine or threonine (O-linked glycosylation). Initial assembly of a typical oligosaccharide, its transfer to a protein, and its subsequent maturation by removal and addition of sugar residues require a large suite of enzymes. Consequently, the great majority of viral glycoproteins are glycosylated by host cell components, but the structural proteins of some large DNA viruses are modified in this way by viral enzymes (Box 12.3) .

The presence of oligosaccharides on a protein can be detected as changes in the protein's electrophoretic mobility, following exposure of cells to inhibitors of glycosylation, or of cell extracts to enzymes that cleave the oligosaccharide (Fig. 12.7A). The first steps in N-linked glycosylation take place as a polypeptide chain emerges into the ER lumen. Oligosaccharides rich in mannose preassembled on a lipid carrier are added to asparagine residues by an oligosaccharyltransferase (Fig. 12.7B). Subsequently, several sugar residues are trimmed from N-linked core oligosaccharides in preparation for additional modifications that take place as the protein travels from the ER to the plasma membrane.

Sites of N-linked glycosylation are characterized by the sequence NXS/T (where X is any amino acid except proline), but not every potential glycosylation site is modified. Even a single specific site within a protein is not necessarily modified with 100% efficiency. Each glycoprotein population therefore comprises a heterogeneous mixture of **glycoforms**, varying in whether a particular site is glycosylated, as well as in the composition and structure of the oligosaccharide present at each site. As many viral and cellular proteins contain a large number of potential N-linked glycosylation sites, particular proteins can exist in an extremely large number of glycoforms. This property complicates investigation of the physiological functions of oligosaccharide chains present on glycoproteins. Nevertheless, glycosylation has been assigned a wide variety of functions.

As essential components of receptors and ligands, oligosaccharides participate in many molecular recognition reactions. These processes include binding of certain hormones to their cell surface receptors; interactions of cells with one another; binding of virus particles, such as those of influenza A virus and herpesviruses, to their host cells; and later steps in virus entry. Some sugar units serve as signals, targeting proteins to specific locations, in particular to lysosomes. Glycosylation has also been suggested to fulfill more general functions, such as protecting proteins (and virus particles) that circulate in body fluids from degradation and host immune defenses. Many proteins contain such a large number of glycosylation sites that carbohydrate can contribute >50% of the mass of the mature protein, for example, the poliovirus receptor and the respiratory syncytial virus G protein. The hydrophilic oligosaccharides are present on the surface of such proteins, where they can form a sugar "shell," masking much of the proteins' surfaces, including epitopes recognized by antiviral antibodies (Box 12.4).

Studies of viral glycoproteins have established that glycosylation can be absolutely required for proper folding. For example, elimination (by mutagenesis) of all sites at which the vesicular stomatitis virus G or influenza virus HA0 proteins are glycosylated blocks the folding of these proteins and their exit from the ER (see "Protein folding and quality control" below). Before a protein folds, its hydrophobic amino acids, which are ultimately buried in the interior, are exposed. Such exposed hydrophobic patches on individual unfolded polypeptide chains tend to interact with one another nonspecifically, leading to aggregation. The hydrophilic oligosaccharide chains are thought to counter this tendency.

BOX 12.3

EXPERIMENTS

Self-glycosylation: virus-encoded enzymes for formation of glycoproteins

The paradigm for production of viral envelope glycoproteins is addition of N- (and O-) linked oligosaccharides as the proteins travel the secretory pathway. Such modification is the result of sequential action of several host cell glycosidases and glycolsyltransferases located in the ER and Golgi compartments (Fig. 12.4). Consequently, the genomes of enveloped viruses (some of which are quite small) typically do not encode such enzymes. One striking exception is provided by phycodnaviruses.

These viruses, such as *Paramecium bursaria* chlorella virus 1 (PBCV-1), have large, double-stranded DNA genomes encoding >350 proteins and share evolutionary history with other large DNA viruses, including poxviruses. The major capsid protein, Vp54, and two minor structural proteins are glycosylated. Sequencing of the PBCV-1 genome identified five potential glycosyltransferases, and several observations indicate that these enzymes, rather than the host cell machinery, modify the viral proteins:

- Sugars typically present in N- (and O-) linked oligosaccharides synthesized by cellular enzymes (e.g., *N*-acetylglucosamine) could not be detected in Vp54 glycans.
- Polyclonal antibodies against virus particles do not react with cellular glycoproteins and recognize Vp54 before, but not after, chemical removal of its oligosaccharides.
- Vp54 proteins specified by viral mutants resistant to the inhibitory effects of such antibodies exhibit differences in electrophoretic migration when glycosylated, but not following removal of glycans.
- The Vp54 genes of all such mutants are identical in sequence to the wild-type gene, but a subset carry substitutions or deletions in a gene encoding a candidate glycosyltransferase, α64r.
- The N-terminal domain of the α64r protein is structurally similar to glycosyltransferases that transfer sugars from a UDP carrier.

These and other observations establish that viral enzymes glycosylate the PBCV-1 Vp54 protein (at six Asn and Ser residues). The mechanisms of glycosylation, and when this process occurs during the infectious cycle, remain to be established.

Graves MV, Bernadt CT, Cerny R, Van Etten JL. 2001. Molecular and genetic evidence for a virus-encoded glycosyltransferase involved in protein glycosylation. *Virology* **285:**332–345.

Van Etten JL, Gurnon JR, Yanai-Balser GM, Dunigan DD, Graves MV. 2010. Chlorella viruses encode most, if not all, of the machinery to glycosylate their glycoproteins independent of the endoplasmic reticulum and Golgi. *Biochim Biophys Acta* **1800:**152–159.

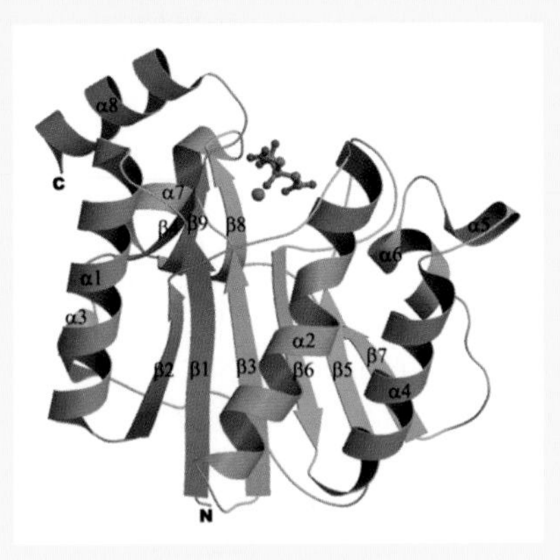

Crystal structure of the glycosyltransferase domain of the viral A64R protein with Mn2+ and citrate ions bound. These ions are shown in ball-and-stick representation with N, C, O, and Mn atoms colored blue, gray, red, and purple, respectively. This structure, which comprises a central β-sheet flanked by α-helices, is very similar to that of the catalytic domains of one of two groups of cellular glycosyltransferases. Such structural conservation is striking, as the sequence conservation is very low (<14%) and no relationship of the viral to cellular enzymes could be established by sequence analysis. Adapted from Y. Zhang et al., *Structure* **15:**1031–1039, 2007, with permission. Courtesy of M. Rossmann, Purdue University.

Disulfide bond formation. A second chemical modification that generally is restricted to proteins entering the secretory pathway, and essential for the correct folding of many, is the formation of intramolecular disulfide bonds between pairs of cysteine residues (Fig. 12.3). These bonds can make important contributions to the stability of a folded protein. However, they rarely form in the reducing environment of the cytoplasm. The more oxidizing ER lumen provides an appropriate chemical environment for disulfide bond formation. This compartment contains high concentrations of protein disulfide isomerase and other enzymes that catalyze the formation, reshuffling, or even breakage of disulfide bonds under appropriate redox conditions. As formation of the full and correct complement of disulfide bonds in a protein is often the rate-limiting step in its folding, these enzymes are important catalysts of this process.

The cellular enzymes that promote formation of disulfide bonds are present in the ER lumen. Consequently, this modification typically is limited to proteins that enter, or protein domains exposed to, this compartment. Remarkably, however, several viral membrane proteins present in mature virus particles of the poxvirus vaccinia virus and other viruses with large DNA genomes have stable disulfide bonds in their **cytoplasmic** domains: the genomes of these viruses encode all the enzymes necessary to catalyze the formation of disulfide bonds in the cytoplasm (Box 12.5).

Protein folding and quality control. A number of other cellular proteins assist the folding of the extracellular domains of viral membrane glycoproteins as they enter the lumen of the ER. In contrast to the enzymes described above, these proteins do not alter covalent structures. Rather, their primary function is to facilitate folding, largely by preventing improper associations among unfolded, or incompletely folded, polypeptide chains, such as the nonspecific, hydrophobic interactions described above. Such **molecular chaperones** play essential roles in the folding of individual polypeptides and in the oligomerization of proteins.

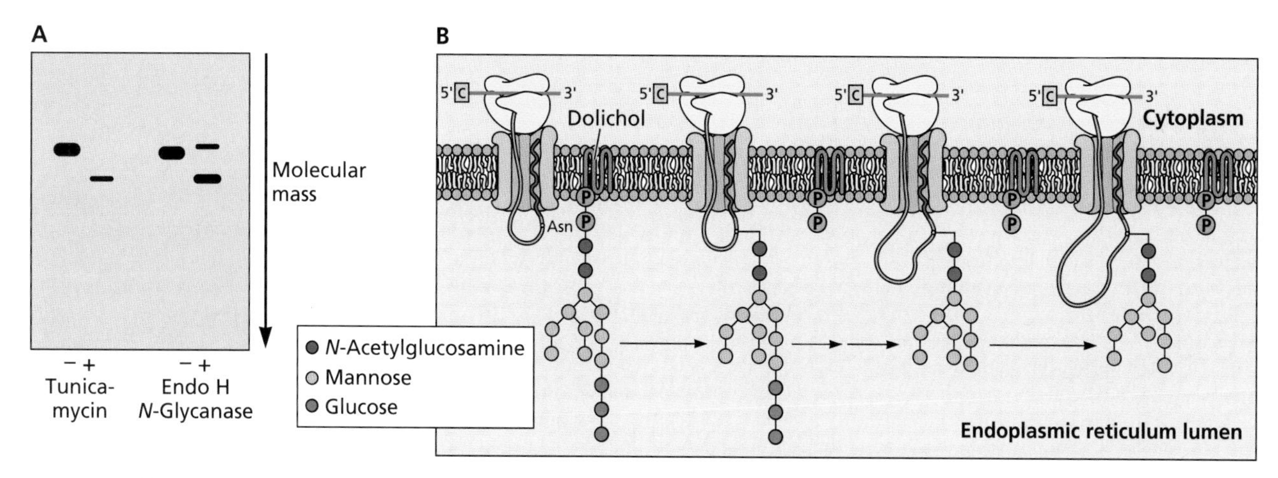

Figure 12.7 Detection and structure of N-linked oligosaccharides. (A) Detection of N-linked oligosaccharides using inhibitors or specific enzymes. Addition to cells of tunicamycin, an inhibitor of the first step in synthesis of the oligosaccharide precursor, prevents N-linked glycosylation, so that the mobility of glycoproteins is altered (left). *In vitro* treatment of glycoproteins with enzymes that cleave within the oligosaccharide, such as endoglycosidase H (Endo H) or *N*-glycanase, can also alter glycoprotein mobility (right). Glycosylation of a protein can also be assayed by incorporation of radioactively labeled monosaccharides. **(B)** The branched, mannose-rich oligosaccharide added via an N-glycosidic bond to asparagine residues of proteins is initially assembled on the lipid carrier dolichol phosphate (left). This common precursor is transferred to N-linked glycosylation sites as proteins are translocated into the ER. While within the ER, three glucose residues and one mannose residue are trimmed from the core oligosaccharide.

BOX 12.4

DISCUSSION

The evolving sugar "shield" of human immunodeficiency virus type 1

Mutational studies have implicated N-linked glycosylation at specific sites in the envelope proteins of several viruses in protection against host neutralizing antibodies. The Env protein of human immunodeficiency virus type 1 (HIV-1) provides a dramatic example of this phenomenon.

The SU (gp120) subunit of the HIV-1 Env protein carries a large number of oligosaccharide chains, which form a dense shell that masks much of the protein's surface (see the figure). These oligosaccharides govern several properties of HIV-1. For example, the tropism of the virus for CCr5 or CXCr4 coreceptors correlates with specific patterns of glycosylation in the variable loops of the SU subunit. However, a major function of such modification is to block access of host anti-HIV-1 antibodies to SU protein epitopes: high-resolution structural studies of the SU protein core have confirmed that N-linked oligosaccharides cover much of the protein's surface. Furthermore, the sugar chains are highly ordered, forming the outer surface of the Env spike. As predicted from this arrangement, N-linked glycosylation at specific sites blocks binding of monoclonal antibodies that recognize nearby sequences in the protein.

Several observations have led to the hypothesis that HIV-1 carries an evolving carbohydrate "shield" that enhances immune evasion. For example, the number of N-linked oligosaccharides added to SU tends to increase during the course of an HIV-1 infection, and the sites of N-linked glycosylation also change. Furthermore, broadly neutralizing antibodies that inhibit reproduction of multiple strains and clades of the virus recognize mannose-containing glycans, in some cases in conjunction with a protein epitope.

Chen B, Vogan EM, Gong H, Skehel JJ, Wiley DC, Harrison SC. 2005. Structure of an unliganded simian immunodeficiency virus gp120 core. *Nature* **433:**834–841.

Pejchal R, Doores KJ, Walker LM, Khayat R, Huang PS, Wang SK, Stanfield RL, Julien JP, Ramos A, Crispin M, Depetris R, Katpally U, Marozsan A, Cupo A, Maloveste S, Liu Y, McBride R, Ito Y, Sanders RW, Ogohara C, Paulson JC, Feizi T, Scanlan CN, Wong CH, Moore JP, Olson WC, Ward AB, Poignard P, Schief WR, Burton DR, Wilson IA. 2011. A potent and broad neutralizing antibody recognizes and penetrates the HIV glycan shield. *Science* **334:**1097–1103.

Scanlan CN, Offer J, Zitzmann N, Dwek RA. 2007. Exploiting the defensive sugars of HIV-1 for drug and vaccine design. *Nature* **446:**1038–1045.

Electron micrograph of HIV-1 particles, showing carbohydrates stained with ruthenium red (dark). Courtesy of Edwin P. Ewing, Jr., Centers for Disease Control and Prevention (CDC), Atlanta, GA (CDC Public Health Image Library).

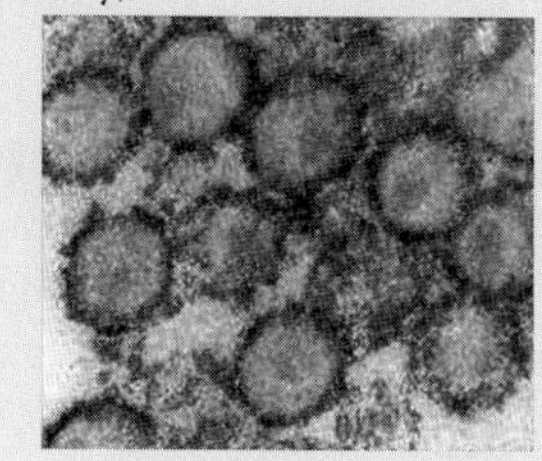

BOX 12.5

DISCUSSION

A viral thiol oxidoreductase system that operates in the cytoplasm

The intracellular mature virus particle of the poxvirus vaccinia virus is the first of two infectious particles assembled in infected cells. This particle carries an envelope containing viral membrane proteins surrounding an internal core in which the DNA genome is packaged. In 1999, it was reported that some viral core proteins synthesized in the cytoplasm, as well as the cytoplasmic domains of some membrane proteins, contain stable disulfide bonds. This property explained the previously reported sensitivity of vaccinia particles to disruption by reducing agents. In addition, it raised the intriguing question of how disulfide bonds could be introduced into viral proteins or domains that are **never** exposed to the major cellular site of thiol oxidation, the ER lumen. Within a few years, viral genes were shown to encode all the components necessary to catalyze formation of disulfide bonds. This viral thiol oxidoreductase system comprises three components, and the final substrates, which include the L1R and F9L proteins that are present in mature virus particles. The sequence in which the three viral enzymes act, summarized in the figure, was deduced from a variety of experimental observations.

The vaccinia virus E10R protein is a sulfhydryl oxidase that contains the motif CXXC common to proteins that participate in exchange of pairs of thiol groups for disulfide bonds. Such proteins include protein disulfide isomerase and other ER oxidoreductases that promote protein folding. The vaccinia virus enzyme belongs to a second class, which includes mitochondrial proteins that operate in the mitochondrial intermembrane space. As summarized in the figure, the viral enzyme system, like that of mitochondria, transfers electrons from substrates via intermediate oxidoreductases to an electron acceptor, typically O_2.

The proteins that comprise the viral thiol oxidoreductase pathway are conserved among all poxviruses. Sulfhydryl oxidases with low but readily discernible sequence identity were shown to be encoded in the genomes of other large DNA viruses, including African swine fever virus (an iridovirus) and mimivirus. An open reading frame exhibiting homology to the viral/mitochondrial family of sulfhydryl oxidases is also present in the genome of pandoraviruses. In all cases that have been examined, the viral enzymes are necessary for assembly of virus particles.

The abundance in the cytoplasm of compounds that reduce disulfide bonds, such as glutathione, indicates that viral proteins containing these bonds must be sequestered, for example, within the viral factories in which viral gene expression and protein synthesis take place.

Hakim M, Fass D. 2010. Cytosolic disulfide bond formation in cells infected with large nucleocytoplasmic DNA viruses. *Antioxid Redox Signal* **13:**1261–1271.

Locker JK, Griffiths G. 1999. An unconventional role for cytoplasmic disulfide bonds in vaccinia virus proteins. *J Cell Biol* **144:**267–279.

Senkevich TG, White CL, Koonin EV, Moss B. 2002. Complete pathway for disulfide bond formation encoded by poxviruses. *Proc Natl Acad Sci U S A* **99:**6667–6672.

The coupled oxidation-reduction (thiol-exchange) reactions among the proteins of the vaccinia virus disulfide bond formation are depicted in order (left to right). The transfer of electrons to oxygen via flavin adenine dinucleotide (FAD) (left) is based on homology of E10R with members of a family of FAD-containing sulfhydryl oxidases, and has not been demonstrated experimentally. These reactions are analogous to those catalyzed by enzymes present in the mitochondrial intermembrane space. Adapted from T. G. Senkevich et al., *Proc Natl Acad Sci U S A* **99:**6667–6672, 2002, with permission.

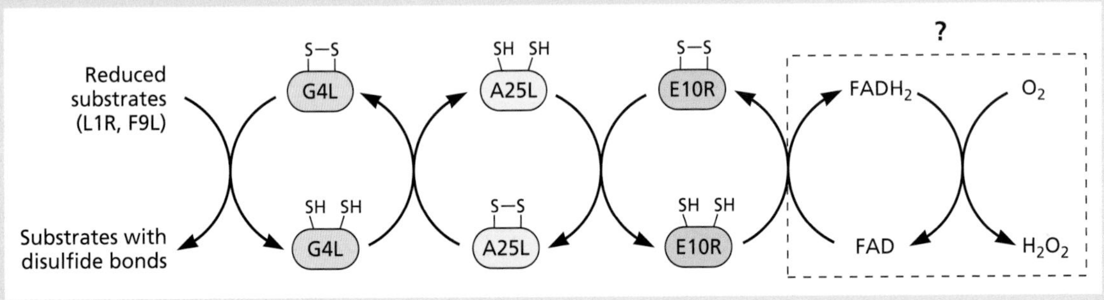

The ER chaperones, which include Grp78 and calnexin, are also crucial for the quality control processes that determine the fate of newly synthesized proteins translocated into the ER.

Grp78 is a member of the Hsp70 family of stress response proteins. It associates transiently with incompletely folded viral and cellular proteins. Binding of this chaperone, generally at multiple sites in a single nascent protein molecule, is thought to protect against misfolding and aggregation by sequestering sequences prone to nonspecific interaction, such as hydrophobic patches. The release of unfolded proteins from Grp78 is controlled by the hydrolysis of ATP bound to the chaperone. Multiple cycles of association with, and dissociation from, Grp78 probably take place as a protein folds. Once the sequences to which the chaperone binds are buried in the interior of the protein, such interactions cease. For example, molecules of vesicular stomatitis virus G, Semliki Forest virus E1, or influenza virus HA0 proteins that have acquired the full complement of correct disulfide bonds can no longer associate with Grp78. The ER contains many other folding catalysts and chaperones, some specific for particular proteins. Relatively little is known about the parameters that determine the chaperone(s) to which a newly synthesized protein binds, and the order in which chaperones operate. However, studies of specific viral glycoproteins in living cells indicate that the positions of oligosaccharides within the protein chain are one important determinant of chaperone selection (Box 12.6).

Calnexin is an integral membrane protein of the ER that also binds transiently to immature proteins. In contrast to Grp78, which recognizes protein sequences directly, calnexin distinguishes newly synthesized glycoproteins by binding to immature oligosaccharide chains. For example, the vesicular stomatitis virus G and influenza virus HA0 proteins bind to calnexin only when their oligosaccharide chains retain terminal glucose residues (Fig. 12.7B). In fact, formation of the mature oligosaccharide is intimately coupled with folding of glycoproteins and their retention within the ER (Fig. 12.8A). Proteins with sugars that include a single glucose residue are recognized by calnexin, but are released upon removal of the glucose by the enzyme glucosidase II. An enzyme that re-adds terminal glucose appears to be the "sensor" of the folded state of the glycoprotein: it recognizes incompletely folded proteins by virtue of exposed hydrophobic amino acids and specifically reglucosylates such proteins, controlling cycles of substrate binding and release from calnexin (Fig. 12.8A). This specificity ensures that only fully folded proteins can escape these chaperones and travel along the secretory pathway.

Proteins that are misfolded or not modified correctly cannot escape covalent or noncovalent associations with ER enzymes or molecular chaperones. For example, a temperature-sensitive vesicular stomatitis virus G protein remains bound to calnexin, and hence to the ER membrane, at a restrictive temperature. Consequently, egress of nonfunctional proteins from the ER to subsequent compartments in the secretory pathway is prevented. These interactions also target misfolded proteins for degradation. The mechanisms responsible for specific recognition of misfolded proteins, and induction of transport from the ER to the cytoplasm, are not fully understood. However, removal of multiple mannose residues, as well as

BOX 12.6

EXPERIMENTS

Selectivity of chaperones for viral glycoproteins entering the ER

The parameters that determine which of the many ER chaperones operate on individual proteins are not fully understood. However, analysis of the folding and chaperone association of viral glycoproteins suggests that the position of glycosylation sites can determine chaperone selection.

The Semliki Forest virus E1 and p62 (pre-E2) glycoproteins enter the ER cotranslationally and are cleaved from a precursor by signal peptidase. However, E1, which folds via three intermediates differing in their disulfide bonding, initially associates with Grp78, whereas nascent p62 molecules bind to calnexin. One major difference between E1 and p62 is that the latter contains glycosylation sites close to the N terminus: addition of oligosaccharides at such sites could allow recognition of nascent p62 by calnexin and preclude association with Grp78.

This hypothesis was tested by elimination of N-linked glycosylation sites of the influenza virus HA0 protein, either close to the N terminus or in more C-terminal positions. The wild-type protein does not bind to Grp78, but prevention of glycosylation at N-terminal sites (positions 8, 22, and 38) led to association with this chaperone.

As summarized in the figure, these observations suggest that nascent proteins that carry N-linked glycosylation sites close to the N terminus (p62, HA0) enter the calnexin folding pathway directly (**A**). Other proteins, such as the Semliki Forest virus E1 and the vesicular stomatitis virus G, associate initially with Grp78 and protein disulfide isomerase and are transferred to the calnexin pathway as they mature (**B**).

Molinari M, Helenius A. 2000. Chaperone selection during glycoprotein translocation into the endoplasmic reticulum. *Science* **288:**331–333.

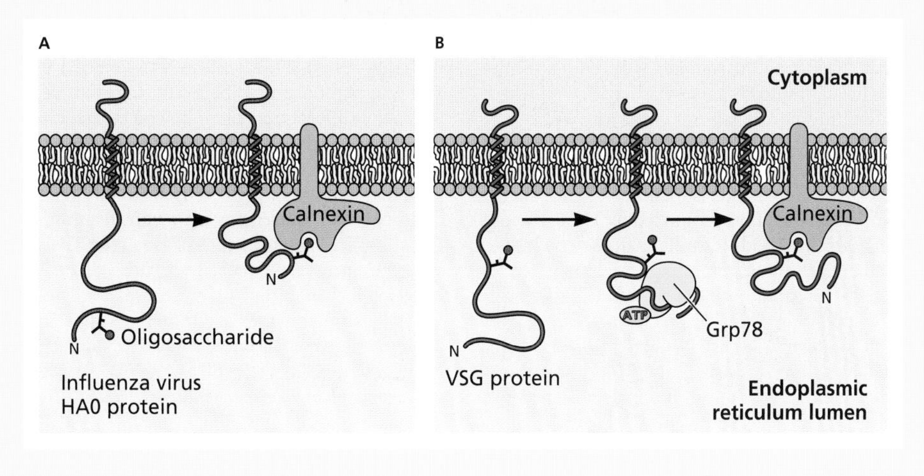

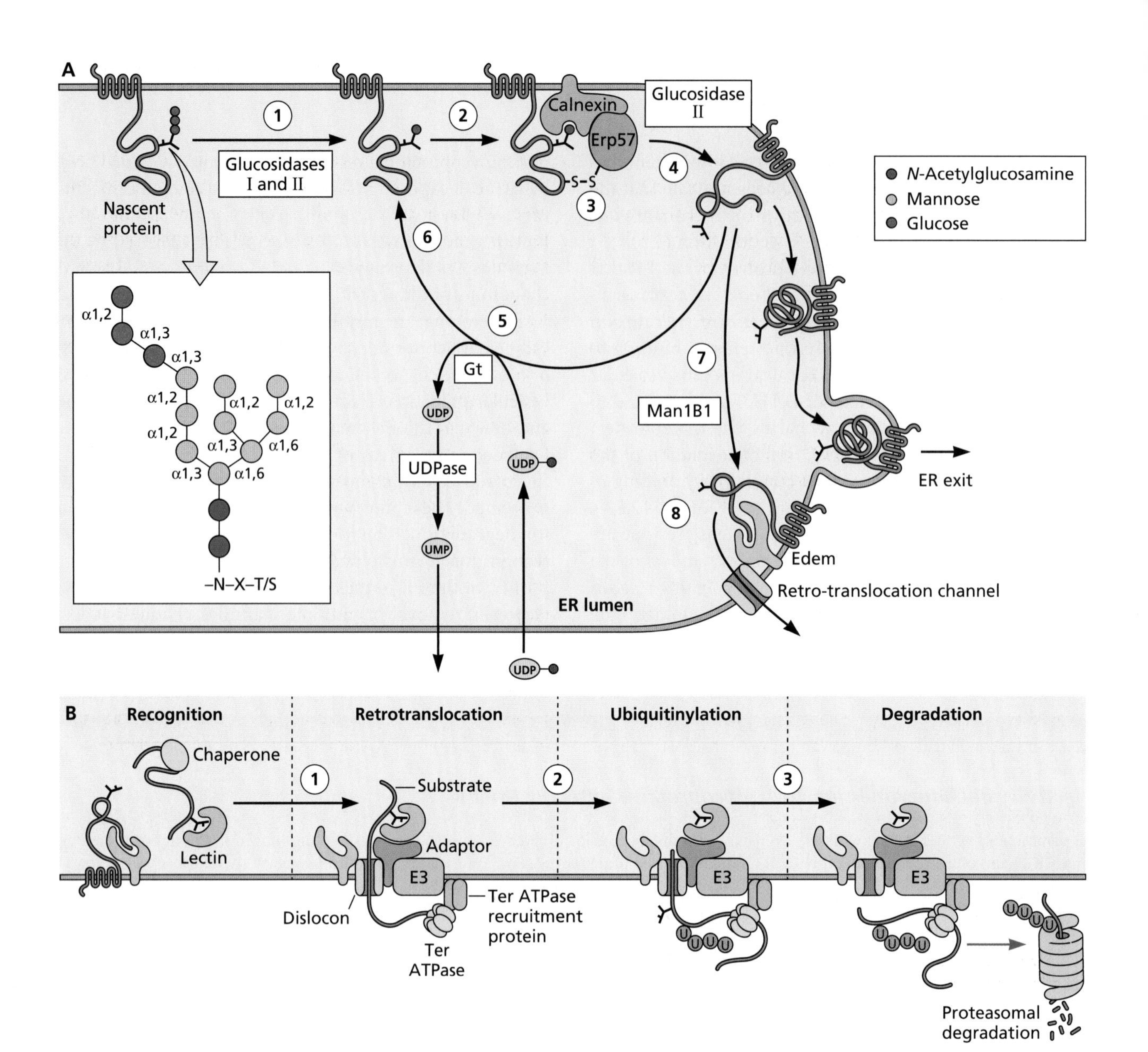

Figure 12.8 Integration of folding and glycosylation in the ER. The model illustrates the coordination of ER retention by calnexin with glycosylation and folding of a newly synthesized glycoprotein (red) containing an N-linked oligosaccharide, depicted as in Fig. 12.3. Trimming of terminal glucose residues by glucosidases I and II (1) yields a monoglucosylated chain, to which calnexin (or calreticulin) binds (2). Because the thiol oxidoreductase Erp57 associates with calnexin, the newly synthesized protein is brought into contact with Erp57, with which transient intermolecular disulfide bonds (-S-S-) can form. When the remaining glucose is removed by glucosidase II (3), the protein dissociates from the calnexin-Erp57 complex. If it has attained its native structure, the protein can leave the ER (4). However, if it is incompletely (or incorrectly) folded (5), the protein is specifically recognized by UDP-glucose glycoprotein transferase (Gt), which re-adds terminal glucose residues to the oligosaccharide (6) and therefore allows rebinding to calnexin. Cycles of binding and modification are repeated until the protein is either folded properly or targeted for degradation. Proteins that cannot escape this cycle by folding to the native conformation are subjected to progressive trimming of mannose residues by enzymes such as ER mannosidase 1 (Man1B1) and probably Edem1 and -2 (7). Removal of mannose residues prevents further reglucosylation and provides signals for recognition by one of several mannose-specific lectins, and direction to the ER membrane machinery for retrotranslocation and ubiquitinylation (the E3 complex) (8). Subsequently, nonglycosylated proteins that do not fold in the ER were found to be marked for retrotranslocation by addition of an O-linked mannose residue. Adapted from M. H. Smith et al., *Science* **334:**1086–1090, 2011, with permission. **(B)** Binding of a misfolded protein by a mannose-specific lectin (for example, amplified in osteosarcoma 9 [Os9] or ER lectin 1) is followed by association with adapter proteins, commonly Sel1L (suppressor/enhancer of lin-12-like proteins) (1). The adapter protein nucleates assembly of a large, membrane-associated complex that contains the protein destined for return to the cytoplasm and components required for retrotranslocation, such as derlin-1 or -2 and transitional ER-associated ATPase (Ter ATPase), which is thought to provide the necessary energy by hydrolysis of ATP (2). This complex also contains E3 ubiquitin ligases, such as synoviolin-1 (Sysn1, also known as Hrd1). These enzymes are thought to ubiquitinylate retrotranslocating protein chains upon entry into the cytoplasm (3) to target the proteins for degradation by the proteasome (4). Adapted from J. A. Olzmann et al., *Cold Spring Harb Perspect Biol* **5:**a013185, 2013, with permission.

the participation of several ER proteins, has been implicated in diversion of misfolded glycoproteins for translocation to the cytoplasm. These proteins, such as the stress-induced ER-degradation-enhancing mannosidase-like proteins Edem1 and -2, promote association of misfolded proteins with ER membrane components for ubiquitinylation and transport to the cytoplasm via retrotranslocation (Fig. 12.8B). Initial identification of proteins required for the latter step came from studies of herpesviral proteins that induce translocation of major histocompatibility complex (MHC) class I molecules from the ER to the cytoplasm (Box 12.7). Once the proteins reenter the cytoplasm, they are degraded by the proteasome. The quality control functions of resident ER chaperones and other proteins therefore ensure that nonfunctional proteins are cleared from the secretory pathway at an early step.

Oligomerization. Most viral membrane proteins are oligomers that must assemble as their constituent protein chains are folded and covalently modified. Such assembly generally begins in the ER, as the surfaces that mediate interactions among protein subunits adopt the correct conformation. For many proteins, these reactions are also completed within the ER. For instance, influenza virus HA0 protein monomers are restricted to the ER lumen, whereas trimers are found in this and all subsequent compartments of the secretory pathway. Indeed, several viral and some cellular heteromeric membrane proteins must oligomerize to exit the ER, because folding of one subunit depends on association with the other(s). This requirement has been characterized in some detail for the glycoproteins of alphaviruses, such as Sindbis virus: the association of the two envelope proteins within the ER is essential for the productive folding, and exit, of both (Fig. 12.9). Similarly, the herpes simplex virus type 1 envelope glycoproteins gH and gL must interact with one another for the transport of either from the ER, and in the absence of gL, gH cannot fold correctly.

Assembly of other viral membrane proteins is completed following exit from the ER: disulfide-linked dimers of the hepatitis B virus surface antigen form higher-order complexes in the next compartment in the pathway, and oligomers of the human immunodeficiency virus type 1 Env protein can be detected only in the Golgi apparatus. At present, we can discern no simple rules describing the relationship of oligomer assembly and transport of membrane proteins from the ER. Nevertheless, oligomerization begins, and in some cases must be completed, within the ER, where it can be facilitated by the folding catalysts and chaperones characteristic of this compartment.

BOX 12.7

TRAILBLAZER

How a herpesviral glycoprotein led to identification of proteins required for ER retrotranslocation

Two human cytomegalovirus (a betaherpesvirus) membrane glycoproteins, US2 and US11, were known to insert into the ER membrane and induce rapid transfer of MHC class I heavy chains from the ER to the cytosol (retrotranslocation). These cellular proteins become polyubiquitinylated and degraded by the cytosolic proteasome. In the case of the viral US11 protein, a glutamine residue (Glu192) in the transmembrane domain is essential for retrotranslocation of MHC class I proteins.

This property was exploited to purify human proteins that bound specifically to wild-type US11, but not to the viral protein carrying a Glu192 → Leu substitution. Several ER proteins bound to both US11 proteins, but only one associated specifically with wild-type US11. This protein, identified by mass spectrometry, showed some similarity to the yeast Der1p protein that is known to participate in degradation of misfolded ER proteins, and was named derlin-1. Overproduction of a dominant-negative derivative of derlin-1 inhibited US11-mediated retrotranslocation of MHC class I proteins.

In an alternative approach, components of a canine ER retrotranslocation channel were identified by virtue of their interaction with a cytoplasmic ATPase (ATPase p97) that was known to be essential for degradation of both misfolded ER proteins in yeast and MHC class I molecules in US11-producing human cells. The protein assembly identified in this way contained derlin-1 and a second ER membrane protein. These ER proteins were shown by immunoprecipitation to interact with both US11 and MHC class I proteins.

Subsequently, derlin-1 was shown to promote transport of other misfolded proteins from the ER to the cytoplasm, and many other proteins that participate in this process were identified (Fig. 12.8B). Nevertheless, the proteins that actually form the mammalian retrotranslocation channel have not been identified. The US11 protein acts as a virus-specific adapter, as shown in the figure.

Lilley BN, Ploegh HL. 2004. A membrane protein required for dislocation of misfolded proteins from the ER. *Nature* **429:**834–840.

Ye Y, Shibata Y, Yun C, Ron D, Rapoport TA. 2004. A membrane protein complex mediates retro-translocation from the ER lumen into the cytosol. *Nature* **429:**841–847.

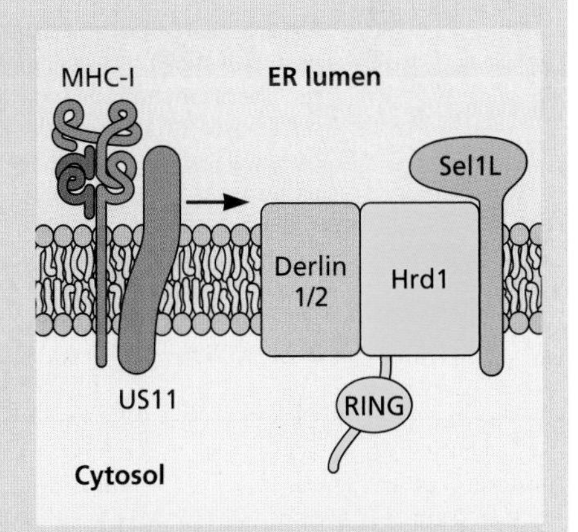

Model for the function of the human cytomegalovirus US11 protein as a virus-specific adapter for ER-associated degradation. This viral proteins binds to MCH class I proteins to direct them to the E3 complex that contains the E3 ubiquitin ligase Hrd1 (HMG-CoA reductase degradation 1) and derlin-1 for return to the cytoplasm, ubiquitinylation, and degradation.

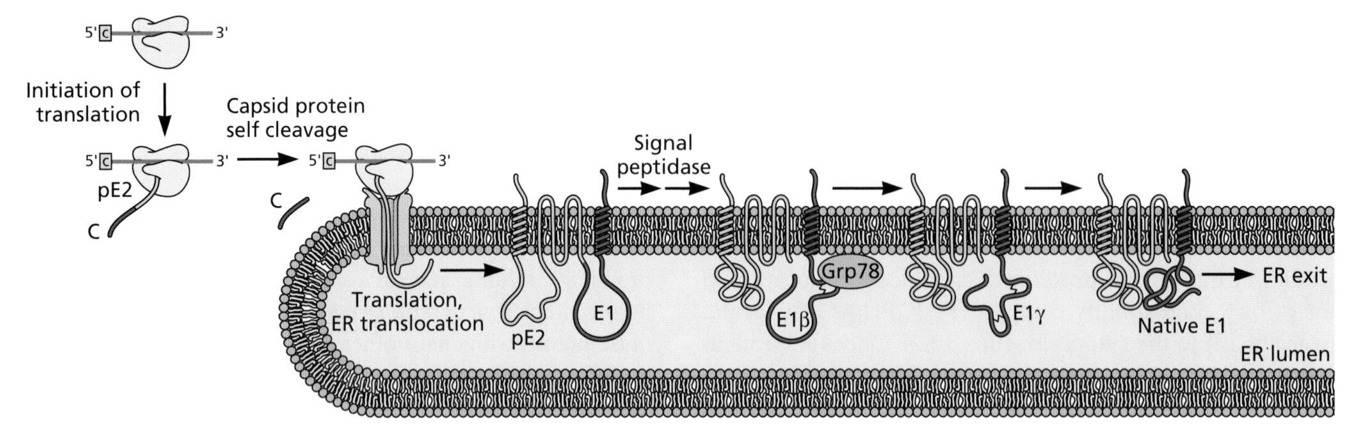

Figure 12.9 Folding of the two Sindbis virus envelope proteins depends on formation of heterodimers. The viral subgenomic mRNA encodes the precursor for the capsid (C) and envelope (pE2 and E1) proteins. Once the capsid protein emerges during translation by free cytoplasmic ribosomes, it is liberated by autoproteolysis. A hydrophobic sequence of pE2 exposed in this way directs association of the translating complex with the ER membrane, such that the envelope protein precursor is translocated across the ER membrane as its synthesis is completed. An unusual cleavage by signal peptidases releases the pE2 and E1 proteins. The nascent E1 protein becomes associated with Grp78, as well as thiol oxidoreductases, and folds via three intermediates (E1α [not shown], -β, and -γ) with different disulfide bonds. As shown, folding beyond E1β via E1γ to the native state depends on replacement of Grp78 by pE2, and folding of pE2 depends on interaction with E1: the pE2 protein misfolds when synthesized in the absence of E1 protein.

Vesicular Transport to the Cell Surface

The mechanism of vesicular transport. Viral membrane proteins, like their cellular counterparts, travel to the cell surface through a series of membrane-bound compartments and vesicles. The first step in this pathway, illustrated schematically in Fig. 12.10, is transport of the folded protein from the ER to the Golgi apparatus. Within the Golgi apparatus, proteins are sorted according to the addresses specified in their primary sequences or by their covalent modifications. **Transport vesicles**, and larger vesicular structures, which bud from one compartment and move to the next, carry cargo proteins between compartments of the secretory pathway (Box 12.8).

Figure 12.10 Compartments in the secretory pathway. The lumen of each membrane-bound compartment shown is topologically equivalent to the exterior of the cell. Proteins destined for secretion or for the plasma membrane travel from the ER to the cell surface via the Golgi apparatus. However, proteins can be diverted from this pathway to lysosomes or to secretory granules that carry proteins to the cell surface for regulated release. The return of proteins from the Golgi apparatus to the ER is indicated. The endocytic pathway discussed in Chapter 5 and the secretory pathway intersect in endosomes and the Golgi apparatus.

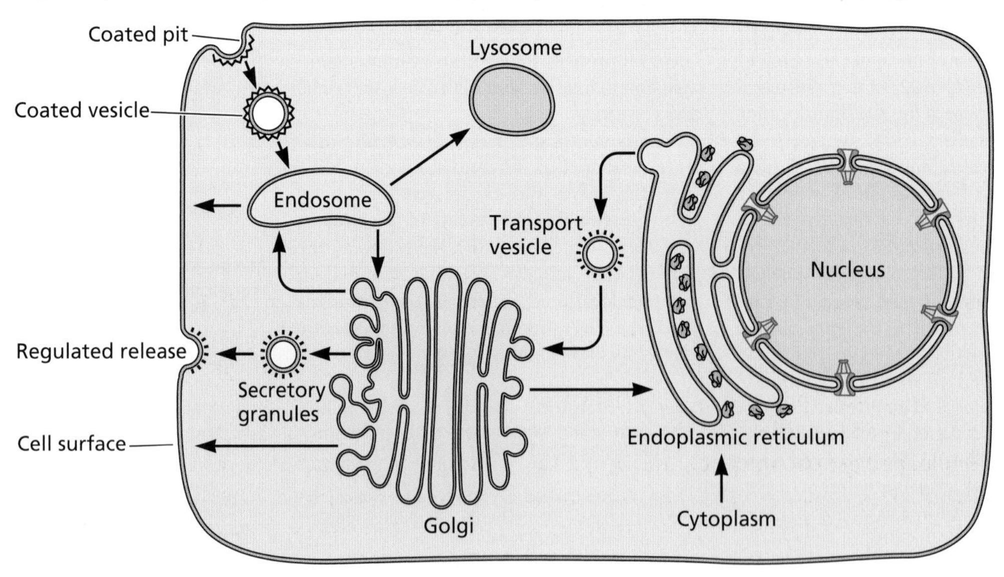

BOX 12.8

EXPERIMENTS

ER-to-Golgi transport in living cells

The vesicular stomatitis virus G protein made in cells infected with the mutant virus *ts*045 misfolds and is retained in the ER at high temperature (40°C). It refolds and is transported to the Golgi apparatus when the temperature is reduced to 32°C. This temperature-sensitive protein has therefore been used extensively to study transport through the secretory pathway. In initial studies of this process in living cells, the green fluorescent protein was attached to the cytoplasmic tail of the viral G protein. Control experiments established that this modification did not alter the temperature-sensitive folding or transport of the G protein. Time-lapse fluorescence microscopy of cells shifted from high to low temperature demonstrated that the chimeric G protein rapidly left the ER at multiple peripheral sites. The protein appeared in membranous structures, which were often larger than typical transport vesicles. These structures moved rapidly toward the Golgi in a stop-start manner, with maximal velocities of 1.4 μm/s. Such transport, but not formation of vesicles derived from the ER, was blocked when microtubules were depolymerized by treatment with nocodazole, or when the (−) end-directed microtubule motor dynein was inhibited. It was therefore concluded that vesicles and other membrane-bound structures that emerge from the ER at peripheral sites are actively transported along microtubules to the Golgi complex.

Extension of this approach to use fluorescent vesicular stomatitis virus as a model cargo in living cells that produce components of the secretory pathway fused to distinguishable fluorescent proteins has provided further insight into the mechanism of transport. For example, the viral G protein cargo was observed to leave the ER in vesicles with CopII coats, but the association between the cargo and this coat was reversed a short distance from the ER (panel B of the figure).

Presley JF, Cole NB, Schroer TA, Hirschberg K, Zaal KJ, Lippincott-Schwartz J. 1997. ER-to-Golgi transport visualized in living cells. *Nature* **389:**81–85.

Stephens DJ, Lin-Marq N, Pagano A, Pepperkok R, Paccaud JP. 2000. COPI-coated ER-to-Golgi transport complexes segregate from COPII in close proximity to ER exit sites. *J Cell Sci* **113:**2177–2185.

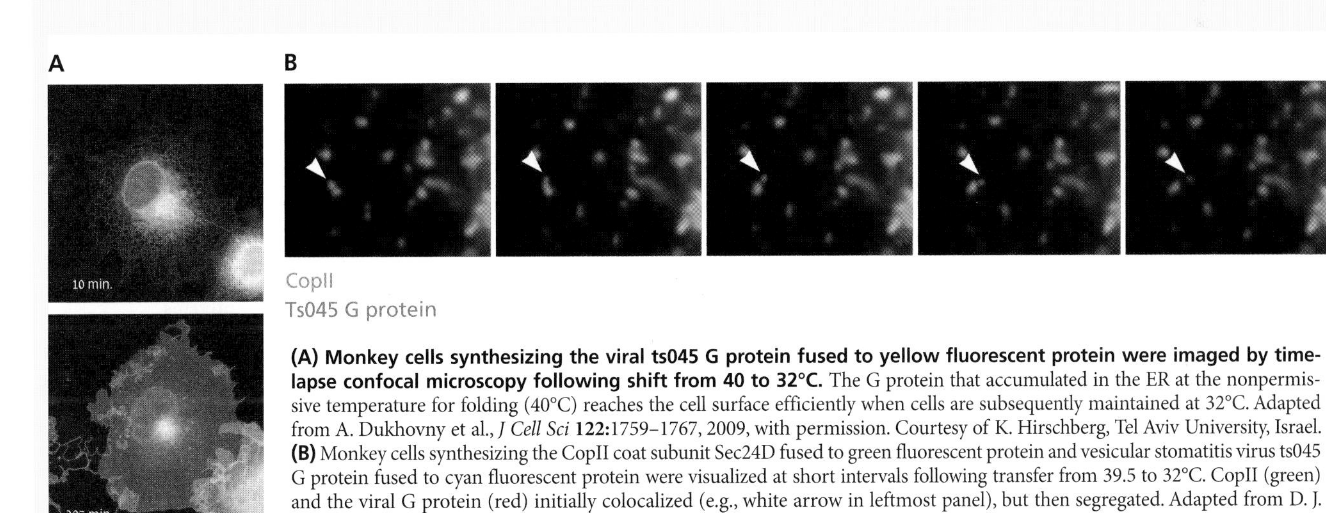

(A) Monkey cells synthesizing the viral ts045 G protein fused to yellow fluorescent protein were imaged by time-lapse confocal microscopy following shift from 40 to 32°C. The G protein that accumulated in the ER at the nonpermissive temperature for folding (40°C) reaches the cell surface efficiently when cells are subsequently maintained at 32°C. Adapted from A. Dukhovny et al., *J Cell Sci* **122:**1759–1767, 2009, with permission. Courtesy of K. Hirschberg, Tel Aviv University, Israel. **(B)** Monkey cells synthesizing the CopII coat subunit Sec24D fused to green fluorescent protein and vesicular stomatitis virus ts045 G protein fused to cyan fluorescent protein were visualized at short intervals following transfer from 39.5 to 32°C. CopII (green) and the viral G protein (red) initially colocalized (e.g., white arrow in leftmost panel), but then segregated. Adapted from D. J. Stephens et al., *J Cell Sci* **113:**2177–2185, 2000, with permission. Courtesy of D. J. Stephens, University of Bristol, United Kingdom.

Fusion of the vesicle membrane with that of the target compartment releases the cargo into the lumen of that compartment. Consequently, proteins that enter the secretory pathway upon translocation into the ER (and are correctly folded) are never again exposed to the cytoplasm of the cell. This strategy effectively sequesters proteins that might be detrimental, such as secreted or lysosomal proteases, and avoids exposure of disulfide-bonded proteins to a reducing environment.

Many soluble and membrane proteins that participate in vesicular transport have been identified and characterized by biochemical, molecular, and genetic methods. The properties of these proteins suggest that similar mechanisms control the budding and fusion of different types of transport vesicles. The general mechanism of vesicular transport is quite well understood. Budding of transport vesicles from the membranes of compartments of the secretory pathway requires proteins that form external coats of the vesicles, such as the protein complex called CopII, which initiates ER-to-Golgi transport, and small GTPases (Fig. 12.11A). The coat proteins induce membrane curvature and vesicle budding, and are subsequently removed by various mechanisms. The vesicle then moves to the next compartment, by either passive diffusion or active transport via microtubule-associated motor proteins over longer distances. When a transport vesicle encounters its target membrane, it docks as a result of specific interactions among Snare proteins present in the vesicle and target membranes. The coupled folding and assembly of Snare proteins on the two membranes lead to membrane fusion (Fig. 12.11B).

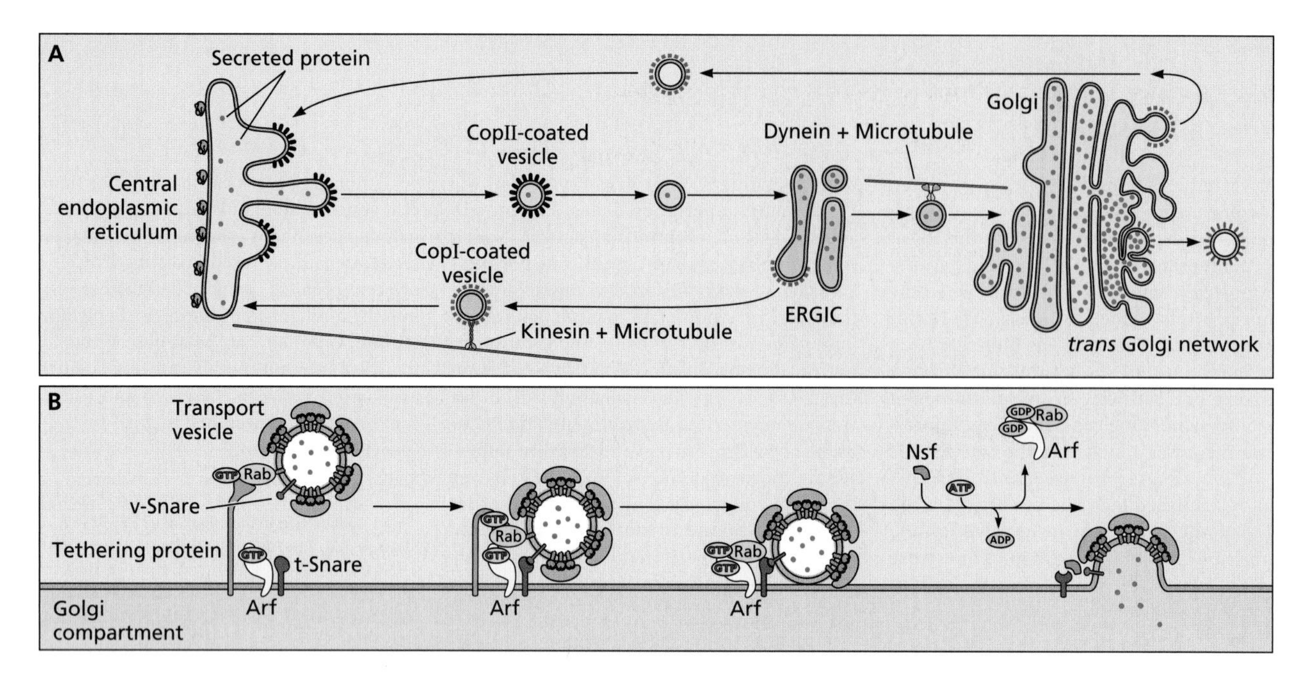

Figure 12.11 Protein transport from the ER to the Golgi apparatus. (A) Proteins leave the ER in transport vesicles at specialized ER exit sites, free of ribosomes. Vesicle formation is initiated by binding of cytoplasmic coat protein complex II (CopII), which contains a small GTPase (Sar1) and several other proteins, to the membrane. The vesicle membranes also carry proteins that direct them to appropriate destinations, such as particular v-Snares. Cargo is loaded by interactions between proteins of the CopII coat and either cytoplasmically exposed tails of cargo proteins or export receptors. The CopII coat induces budding and pinching off of vesicles, which move to the ER-Golgi intermediate compartment (ERGIC). Within this compartment, signals present in cargo proteins direct sorting for transport back to the ER, via CopI-coated vesicles, or for continued transport to the plasma membrane. The ERGIC matures into and/or fuses with the *cis*-Golgi. **(B)** Fusion of transport vesicle and target compartment membranes. Both vesicle (v-Snare) and target compartment (t-Snare) proteins govern the specificity of membrane fusion. The first step is thought to be tethering of a vesicle by interaction of a tethering protein with the GTP-bound Rab and/or components of the coat. Tethering proteins (e.g., Dsl1 and Cog proteins) contain multiple subunits, each built from several α-helical bundles to form extended structures. They interact with Snares, and may function as chaperones for the assembly of membrane-bridging complexes between the v- and t-Snares, a process known as docking. Membrane fusion takes places as v- and t-SNAREs finish zippering into highly stable helical bundles. Fusion is accompanied by ATP hydrolysis and disassembly of the fusion complex by the SNARE disassembly ATPase Nsf. The specificity of the v-Snare–t-Snare interaction contributes to the specificity of fusion, as do multisubunit tethering proteins. In addition, small GTP-binding proteins of the Rab family, each of which is associated with a specific organelle, provide, in the GTP-bound form, an "identity signal" recognized by proteins that participate in vesicle budding or fusion. Specific phosphoinositides (lipids) are also important determinants of the identity of some organelles, for example, of Golgi compartments.

A vesicle Snare and an appropriate Snare in the target membrane represent a minimal machinery for membrane fusion *in vitro*, but additional proteins perform essential regulatory functions in cells.

The high density of intracellular protein traffic requires considerable specificity during vesicle formation and fusion. For example, vesicles that transport proteins from the ER to the first compartment of the Golgi apparatus must take up only the appropriate proteins when budding from the ER, and must fuse only with the membrane of the *cis*-Golgi network. Specificity of cargo loading during formation of these (and other) transport vesicles is achieved by both direct association of cargo proteins with proteins of the vesicular coat and indirect interactions via transmembrane export receptors. Several types of protein establish the specificity of vesicular transport, including the Snare proteins resident in the vesicle and target compartment membranes.

Reactions within the Golgi apparatus. One of the most important staging posts in the secretory pathway is the Golgi apparatus, which is composed of a series of membrane-bound compartments. Proteins enter the Golgi apparatus from the ER via the *cis*-Golgi network, which is composed of connected tubules and sacs (Fig. 12.10). A similar structure, the *trans*-Golgi network, forms the exit face of this organelle. The *cis*- and *trans*-Golgi networks are separated by a variable number of cisternae termed the *cis*, medial, and *trans* compartments. Each of these compartments, which can comprise multiple cisternae, is the site of specific reactions, including those that form mature N-linked oligosaccharides (Fig. 12.12).

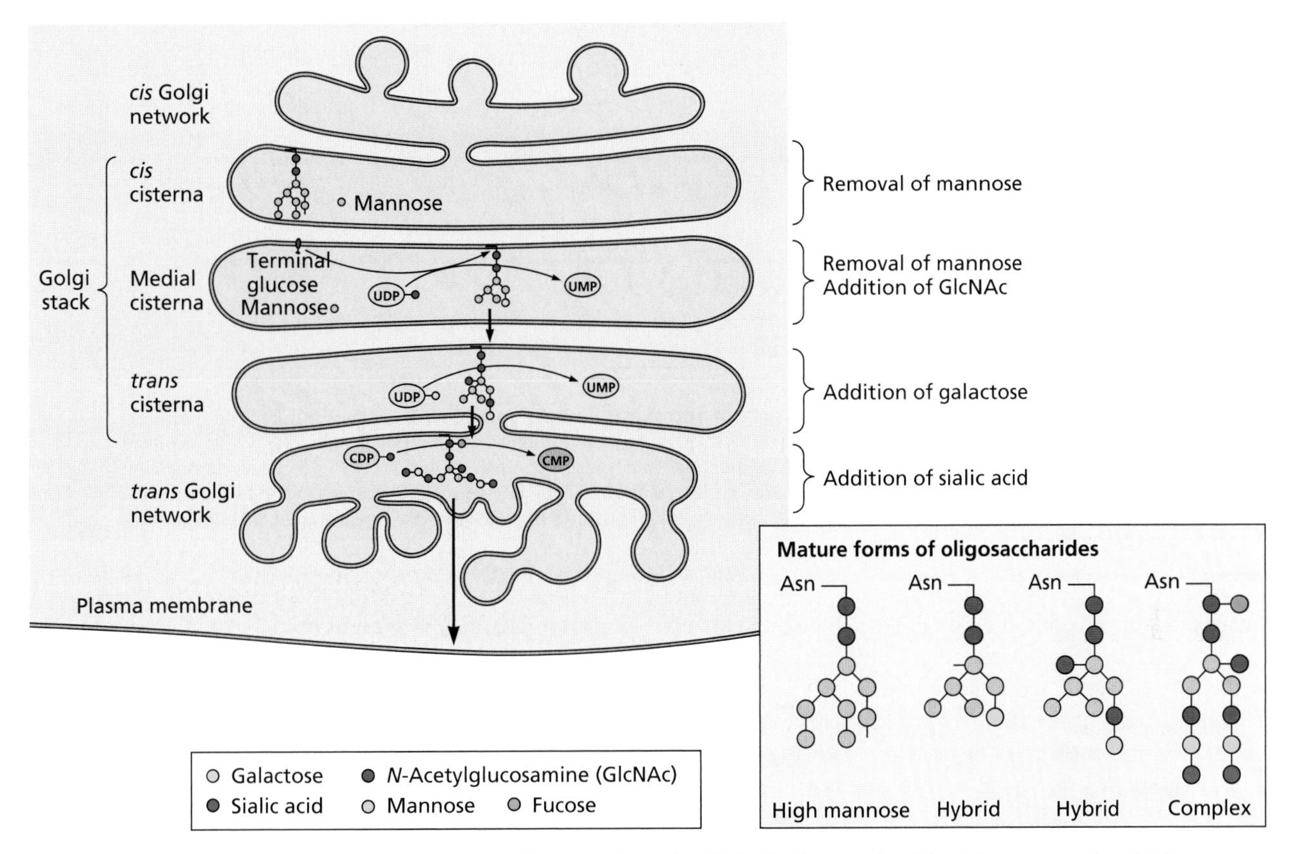

Figure 12.12 Compartmentalization of processing of N-linked oligosaccharides. The reactions by which mature N-linked oligosaccharide chains are produced from the high-mannose core precursor added in the ER (Fig. 12.7B) are shown in the Golgi compartment in which they take place. Trimming of terminal glucose and mannose residues of the common core precedes stepwise addition of the sugars found in the mature chains. The enzymes responsible for early reactions in maturation of oligosaccharides are located in *cis* cisternae, whereas those that carry out later reactions are present in the medial and *trans* compartments. Such spatial separation ensures that oligosaccharide processing follows a strict sequence as proteins pass through the compartments of the Golgi apparatus. Synthesis of O-linked oligosaccharides by glycosyltransferases, which add one sugar unit at a time to certain serine or threonine residues, also takes place in the Golgi apparatus.

A number of viral envelope glycoproteins are also processed proteolytically by cellular enzymes resident in late Golgi compartments. Retroviral Env glycoproteins are cleaved in the *trans*-Golgi network to produce the TM (transmembrane) and SU (surface unit) subunits from the Env polyprotein precursor (Fig. 12.13). Similarly, the HA0 protein of certain avian influenza A viruses is cleaved into the HA1 and HA2 chains (Fig. 12.4) in the same compartment. These and other viral membrane proteins (Table 12.1) are processed by members of a family of resident Golgi proteases that cleave after pairs of basic amino acids. The members of this family, which in mammalian cells include furins found in the *trans*-Golgi network, are serine proteases related to the bacterial enzyme subtilisin. Various furin family members have been shown by genetic and molecular methods to process viral glycoproteins; their normal function is to process cellular polyproteins, such as certain hormone precursors.

These proteolytic cleavages are not necessary for assembly but are essential for production of infectious particles. For example, proteolytic processing of envelope proteins of retroviruses and alphaviruses is necessary for infectivity, probably because sequences important for fusion and entry become accessible. Virulent strains of avian influenza A virus encode HA0 proteins that can be processed by the ubiquitous furin family proteases, such that virus particles carrying fusion-active HA protein are released (Volume II, Chapter 5). It seems likely that the common dependence on furin family proteases (Table 12.1), which act on proteins relatively late in the secretory pathway, helps minimize complications that would arise if viral glycoproteins were initially synthesized with their fusion peptides in an active conformation. Furthermore, exposure to the low-pH environment of *trans*-Golgi network compartments (pH ~6.0) can be a prerequisite for processing of viral envelope proteins.

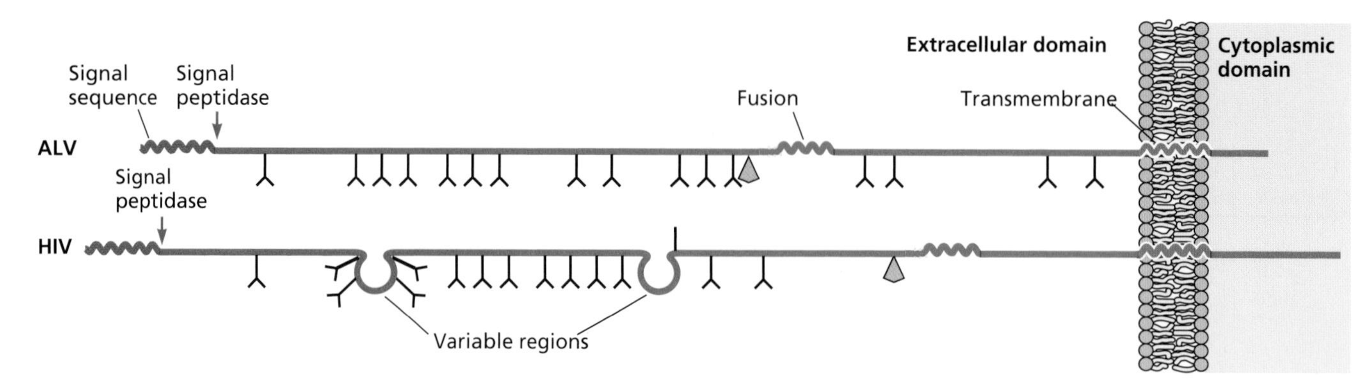

Figure 12.13 Modification and processing of retroviral Env polyproteins. Sequence features and modifications of the Env proteins of avian leukosis virus (ALV) and human immunodeficiency virus type 1 (HIV) are depicted as in Fig. 12.3. The variable regions of human immunodeficiency virus type 1 Env differ greatly in sequence among viral isolates. The translocation products shown here are cleaved by the ER signal peptidase (red arrows) and by furin family proteases in the *trans*-Golgi network (orange arrowheads). The latter liberates the transmembrane (TM) and surface unit (SU) subunits from the Env precursor.

This requirement is exemplified by the envelope proteins of flaviviruses, such as dengue virus (Fig. 12.14).

In the case of highly pathogenic influenza A viruses, the ion channel activity of the viral M2 protein helps to maintain HA in a fusion-incompetent conformation following cleavage. This HA protein switches to the fusion-competent conformation at a pH higher than that required by HA proteins of human influenza A viruses. The M2 protein, which forms a proton channel, is present at quite high concentrations in the membranes of secretory pathway compartments. By providing an exit channel for protons, and hence increasing the pH of normally acidic compartments, such as those of the *trans*-Golgi network, this protein prevents premature switching of proteolytically processed HA to the fusion-active conformation described in Chapter 5.

Table 12.1 Some viral envelope glycoprotein precursors processed by secretory pathway proteases

Virus family	Precursor glycoprotein	Membrane-associated cleavage products
Signal peptidase		
Alphavirus	Envelope polyprotein precursor	E1, pE2
Bunyavirus	Translation product of M mRNA	Gn Gc
Flavivirus	Polyprotein	prM, E
Furin family proteases		
Alphavirus	pE2	E2
Flavivirus	PrM	M
Hepadnavirus	preC	C antigen[a,b]
Herpesvirus[b,c]	pre-gB	gB
Orthomyxovirus[d]	HA0	HA1, HA2
Paramyxovirus	F0	F1, F2
Retrovirus	Env	TM, SU

[a]This cleavage product is largely secreted into the extracellular medium, but is also associated with plasma membrane of infected cells.

[b]Cleavage is not necessary for production of infectious virus particles in cells in tissue culture.

[c]Some alphaherpesviruses (e.g., varicella-zoster virus), and all known betaherpesviruses.

[d]Virulent strains of avian influenza A virus.

Although all the envelope proteins of viruses that assemble at the plasma membrane travel the cellular secretory pathway, there is considerable variation in the rate and efficiency of their transport. A champion is the influenza virus HA protein, which folds and assembles with a half time of only 7 min, with >90% of the newly synthesized molecules reaching the cell surface. Many other viral proteins do considerably less well. Parameters determining the rate and efficiency of transport may include the complexity of the protein and the inherent asynchrony of protein folding. With some exceptions (see "Inhibition of Transport of Cellular Proteins" below), cellular proteins continue to enter and traverse the secretory pathway as enveloped viruses assemble at the plasma membrane. Consequently, competition among viral and cellular proteins, which may vary with the nature and physiological state of the host cell, is also likely to affect the transport of viral proteins to the cell surface.

We have focused our discussion of viral envelope proteins on the well-understood maturation of their extracellular domains. However, the cytoplasmic portions of these proteins are also frequently modified. Many, including the influenza HA and human immunodeficiency virus Env proteins, are **acylated** by the covalent linkage of the fatty acid palmitate to cysteine residues in their cytoplasmic domains (Table 12.2). This modification can be necessary for optimal production of progeny virions. For example, inhibition of palmitoylation of the Sindbis virus E2 glycoprotein or of the human

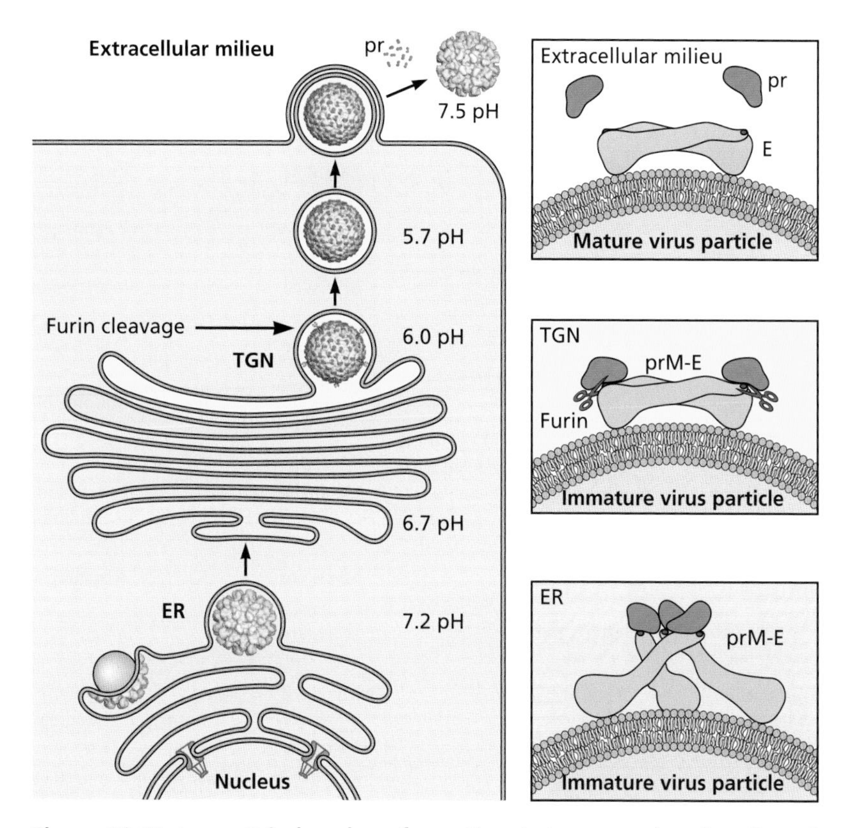

Figure 12.14 Low pH-induced conformational change and maturation of dengue virus particles. The envelope of mature particles of dengue virus (and other arthropod-borne flaviviruses) contains dimers of the envelope (E) protein that lie flat along the surface (see Fig. 4.23). However, this viral protein is initially inserted into membranes in association with the viral precursor membrane protein (prM) and forms heterotrimeric E-prM spikes on the surface of immature virus particles that bud into the ER lumen (step 1). The particles travel the secretory pathway, passing through compartments with decreasing internal pH, reaching pH 6.0 in the *trans*-Golgi network (TGN). The reduced pH induces a major reorganization of the surface proteins to form flat dimers (step 2) and conformational change that exposes a furin cleavage site in prM. Following cleavage, which is important for the infectivity of virus particles, a portion of prM (termed pr) remains associated with E proteins until its dissociation is triggered upon the release of particles into the neutral pH of the extracellular milieu (step 3). This model is based on comparison of the structures of mature virus particles and the immature particles containing uncleaved prM that are released when infected cells are exposed to compounds that increase the pH of the *trans*-Golgi network. Subsequent studies established that particles that carry uncleaved prM are quite prevalent in the populations released from animal cells, and in particular from mosquito cells. The impact of such heterogeneity on infectivity, antibody recognition, and pathogenesis is not yet clear. Adapted from I.-M. Yu et al., *Science* **319:**1834–1837, 2008, with permission. Reconstruction of virus particles courtesy of J. Chen, Purdue University.

immunodeficiency virus type 1 Env protein impairs virus assembly and budding. The bulky fatty acid chains attached to the short cytoplasmic tails may regulate envelope protein conformation or association with specific membrane domains.

Sorting of Viral Proteins in Polarized Cells

Proteins that are not specifically targeted to an intracellular address travel from the Golgi apparatus to the plasma membrane (Fig. 12.10). However, the plasma membrane is not uniform in all animal cells: differentiated cells often devote different parts of their surfaces to specialized functions, and the plasma membranes of such **polarized cells** are divided into correspondingly distinct regions. During infection by many enveloped viruses, the asymmetric surfaces of such cells are distinguished during entry and when components of virus particles are sorted to a specific plasma membrane region. In this section, we describe the final steps in the transport of proteins to specialized plasma membrane regions in two types

Table 12.2 Examples of acylated or isoprenylated viral proteins

Virus	Protein	Lipid	Probable function
Envelope proteins			
Alphavirus			
Sindbis virus	E2	Palmitate	Efficient budding of virus particles
Coronavirus			
Severe acute respiratory syndrome virus	S	Palmitate	Fusion
Hepadnavirus			
Hepatitis B virus	L (pre-S1)	Myristate	Initiation of infection
Orthomyxovirus			
Influenza A virus	HA	Palmitate	Fusion and infectivity
Retrovirus			
Human immunodeficiency virus type 1, Moloney murine leukemia virus	Env (TM)	Palmitate	Budding of virus particles
Other viral proteins			
Hepatitis delta satellite virus	Large delta antigen	Geranylgeranol	Interaction with HBV L protein; assembly; inhibition of HDV RNA replication
Papovavirus			
Simian virus 40	VP2	Myristate	Assembly
Picornavirus			
Poliovirus	VP0, VP4	Myristate	Assembly; uncoating
Retrovirus			
Human immunodeficiency virus type 1, murine leukemia virus	Gag, MA	Myristate	Membrane association, assembly, and budding
Rous sarcoma virus	pp60^{src}	Myristate	Membrane association, transformation

Figure 12.15 Polarized epithelial cells and neurons. (A) Tight junctions block the intercellular space between epithelial cells and delineate the apical and basolateral domains. Proteins destined for vesicular transport to these distinct domains are sorted, on the basis of the specific signals they carry, within the *trans*-Golgi network (TGN). The vesicles that carry cargo to both membrane domains can also arise from recycling endosomes (REs). AP, adapter protein. **(B)** The membrane of the axon of a neuron is equivalent to the apical domain of an epithelial cell, as indicated. It is demarcated from the remainder of the neuronal plasma membrane by the axon initial segment (gray membrane), which is not myelinated and contains bundled microtubules. The formation of axonal vesicular carriers is illustrated. Adapted from J. S. Bonifacino, *J Cell Biol* **204:**7–17, 2014, with permission.

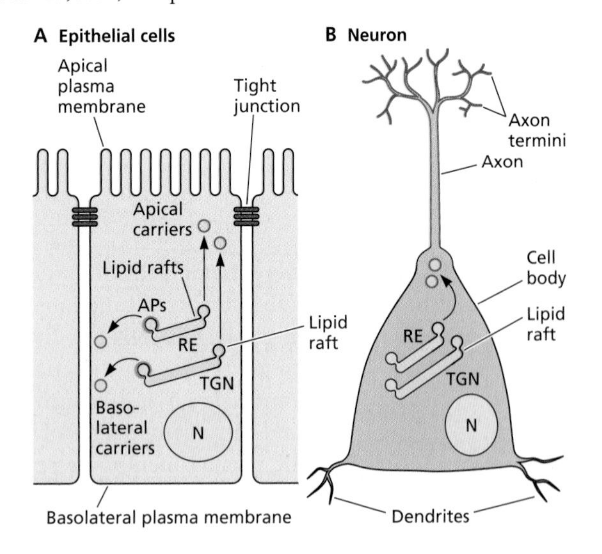

of polarized cells in which animal viruses often reproduce, epithelial cells and neurons (Fig. 12.15).

Epithelial Cells

Epithelial cells, which cover the external surfaces of vertebrates and line all their internal cavities (such as the respiratory and gastrointestinal tracts), are primary targets of virus infection. The cells of an epithelium are organized into close-knit sheets, by both the tight contacts they make with one another and their interactions with the underlying basal lamina, a thin layer of extracellular matrix (Fig. 2.3). Within the best-characterized epithelia, such as those that line the intestine, each cell is divided into a highly folded **apical domain** exposed to the outside world and a **basolateral domain** (Fig. 12.15). The former performs more-specialized functions, whereas the latter is associated with cellular housekeeping. These two domains differ in their protein and lipid content, in part because they are separated by specialized cell-cell junctions (tight junctions), which prevent free diffusion and mixing of components in the outer leaflet of the lipid bilayer. However, such physical separation does not explain how the polarized distribution of plasma membrane proteins is established and maintained.

Viruses have been important tools in efforts to elucidate the molecular mechanisms responsible for the polarity of typical epithelial cells, because certain enveloped viruses bud asymmetrically. For example, in all epithelial cells studied, influenza A virus buds apically and vesicular stomatitis virus buds basolaterally. Such polarized assembly and release of virus particles can facilitate virus spread within or among host organisms (Volume II, Chapter 2). The polarity of virus budding is generally the result of accumulation of envelope proteins at the specific membrane regions, such as HA and G in the apical and basolateral domains, respectively. The most common mechanism for selective localization appears to be signal-dependent sorting of proteins in the *trans*-Golgi network, for packaging into appropriately targeted transport vesicles. Signals necessary for basolateral targeting comprise short amino acid sequences located in the cytoplasmic domains of membrane proteins (for example, YXXØ, where X is any and Ø is a bulky hydrophobic amino acid). These signals are recognized by components of coated vesicles that confer specificity during cargo selection and vesicle fusion, such as adapter protein 1, which interacts with clathrin or other scaffolding proteins. Indeed, many basolateral targeting signals overlap with those that direct proteins for clathrin-dependent endocytosis, and certain proteins are transferred through endosomes from one membrane domain to the other, a process termed transcytosis (see Volume II, Fig. 4.23). The sorting of viral glycoproteins to basolateral membrane domains can also be governed by additional viral proteins. When made in the absence of other viral proteins, the two envelope proteins of measles virus (F and H) are transported to the basolateral membrane. However, when the viral matrix protein binds to the cytoplasmic tails of F and H, these proteins are redirected from the default basolateral sorting pathway, and accumulate at the apical surface of epithelial cells.

A rather diverse set of determinants direct proteins to the apical membrane. They include certain sequences present in transmembrane domains (as are found in the influenza A virus HA and NA proteins), N- or O-linked oligosaccharides present in external or cytoplasmic domains, and a lipid anchor (glycosylphosphatidylinositol) that is added to some proteins made in the cytoplasm. It is thought that such signals confer affinity for specialized microdomains, termed **lipid rafts**. Such rafts, which are dynamic assemblies that can incorporate particular proteins selectively, were initially shown to mediate apical transport of glycosylphosphatidylinositol-anchored proteins. The influenza virus HA and NA proteins associate specifically with lipid rafts via their transmembrane domains, which determine apical sorting. Cellular proteins known to participate in apical trafficking of viral glycoproteins, such as caveolin-1 and myelin, are also associated with these membrane microdomains. Inhibition of the activity or synthesis of these proteins disrupts transport of the influenza virus HA protein (and other proteins) from the Golgi complex to the apical membrane. Lipid rafts seem likely to be more generally important in targeting of viral membrane proteins and assembly in nonpolarized cells: measles virus glycoproteins are selectively enriched in lipid rafts in nonpolarized cells, and association of the human immunodeficiency virus type 1 Gag polyprotein with these membrane domains promotes production of virus particles.

Neurons

Neurons are probably the most dramatically specialized of the many polarized cells of vertebrates. The axon is typically long and unbranched, whereas the dendrites form an extensive branched network of projections (Fig. 12.15). Axons are specialized for the transmission of electrical and chemical impulses, ultimately via the formation of synaptic vesicles and release of their contents. In contrast, dendrites provide a large surface area for the receipt of signals from other neurons. The nucleus, the rough ER, and the Golgi are also located in the dendritic region and the cell body of a neuron. Although axonal and dendritic surfaces are not separated by tight junctions, proteins must be distributed asymmetrically in neurons. Several mechanisms contribute to the establishment and maintenance of neuronal polarity, including transport of vesicles in specific directions along the highly organized microtubules of the axon (**axonal transport**), transport of particular mRNAs to specific regions of the neuron, and sorting and targeting of membrane proteins for delivery to axonal or dendritic surfaces.

The directional movement of vesicles and many cellular organelles in neurons is dependent on polarized microtubules and motor molecules that travel toward either their (−) or (+) ends. Such motors therefore mediate transport both toward the cell body from axons and dendrites and away from the cell body (Fig. 12.16). Infection, assembly, and egress of viruses that infect neurons depend on these mechanisms. An important example is provided by the neurotropic alphaherpesviruses, a group that includes the human pathogens herpes simplex virus type 1 and varicella-zoster virus. Following entry into sensory neurons, herpesvirus nucleocapsids and some tightly associated tegument proteins are transported along axons to the nucleus by microtubule-based transport, mediated by (−) end-directed motors such as dynein (Fig. 12.16). Later in the infectious cycle, virion components must be moved in the opposite direction (toward the synapse) upon association with proteins of the kinesin family. The spread of herpesviruses from neuron to neuron occurs at or near sites of synaptic contact, indicating that virus particles must be targeted to specific areas within neurons for egress. This attribute can be exploited to define neuronal connections in a living animal by using the virus as a tracer (Volume II, Chapter 1). Whether assembly is completed within the cell body of infected neurons or following transport of components of virus particles to sites of egress has been a subject of considerable debate (Box 12.9), but in the case of pseudorabies virus, there is compelling evidence for the former mechanism (Fig. 12.16).

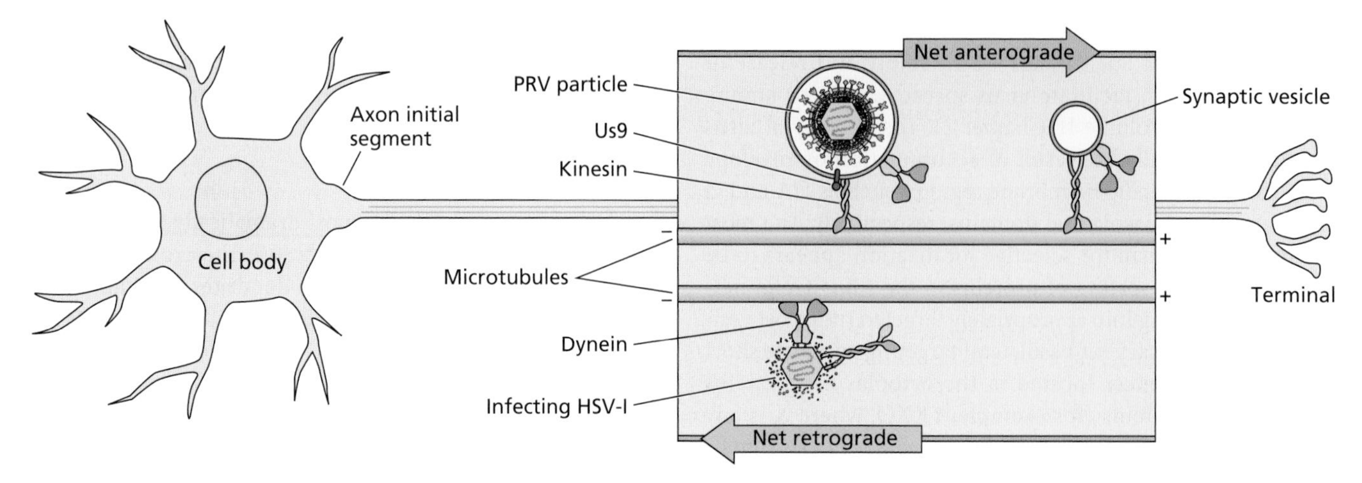

Figure 12.16 Axonal transport of herpesviral particles in neurons. At the beginning of an infectious cycle, nucleocapsids of alphaherpesviruses enter axon termini from epithelial cells (or other neurons) and are transported rapidly along microtubules in the retrograde direction upon association with the (−) end-directed motor dynein. Nucleocapsids associated with inner tegument proteins are transported efficiently, but which protein(s) carry dynein-binding sites is not yet clear. Later in the infectious cycle, the direction of transport must be reversed. There has been a long-standing debate about whether newly synthesized nucleocapsids are transported prior to or following secondary envelopment. In the case of pseuorabies virus (PRV), shown here, there is compelling evidence for the second mechanism. Efficient anterograde transport requires the glycoproteins gE and gI and the membrane protein Us9, which recruits the kinesin-3 motor Kif1A. It remains to be seen whether this model applies to human simplex virus type 1 (HSV-1).

Disruption of the Secretory Pathway in Virus-Infected Cells

Inhibition of Transport of Cellular Proteins

Some viral proteins interfere with transport to the plasma membrane of specific cellular proteins, notably MHC class I molecules. These proteins carry peptides derived from viruses (and other invaders) to the plasma membrane, where they alert cells of the adaptive immune response to infection. Prevention of transport of MHC class I proteins to the cell surface therefore helps prevent or delay the detection and destruction of infected cells (Volume II, Chapter 4). In the ER, the adenovirus E3 glycoprotein gp19 (Appendix, Fig. 1B) forms intramolecular disulfide bonds with these important components of the adaptive immune system and sequesters them within this organelle. Several herpesviral proteins also block transport of MHC class I molecules to the cell surface, including the human cytomegalovirus US11 and US2 gene products, discussed previously, which induce transport of cellular proteins from the ER to the cytosol for degradation by the cellular proteasome. The human immunodeficiency virus type 1 Vpu protein (Appendix, Fig. 29), a transmembrane phosphoprotein, also induces selective degradation of newly synthesized MHC class I proteins, and of CD4, by a similar mechanism. Such degradation of CD4, the major receptor for this virus, is important for assembly and release: tight binding of this cellular protein to the viral Env glycoprotein in the ER prevents transit of both proteins to the cell surface. Vpu also reduces the cell surface concentration of a third cellular protein, tetherin (also known as bone marrow stromal antigen 2 [Bst2]), which restricts release of human immunodeficiency virus from infected cells (Chapter 13). However, Vpu acts on tetherin not in the ER, but rather to induce its displacement from sites of virus budding in the plasma membrane, and to target the cellular protein for lysosomal degradation, thereby reducing recycling of tetherin from endosomes to the plasma membrane (Volume II, Chapter 7).

Drastic Effects on Compartments of the Secretory Pathway

Proteins encoded in the genomes of certain other viruses exert more-drastic effects on the cellular secretory pathway. For example, rotaviruses, which lack a permanent envelope but are transiently membrane enclosed during assembly, encode a protein that disrupts the ER membrane. This protein is thought to allow removal of the temporary envelope formed during assembly of virus particles. The replication of most (+) strand RNA viruses takes place in association with membranous structures derived from various cytoplasmic membranes of the host cell (Chapters 6 and 14). Such remodeling of cellular membranes can lead to dramatic reorganization of cytoplasmic compartments and inhibition of trafficking via the secretory pathway, effects well characterized in cells infected by poliovirus and other enteroviruses.

The Golgi complex is disrupted in poliovirus-infected cells, although such disassembly is not required for viral reproduction. Rather, it appears to be a consequence of diversion

BOX 12.9

DISCUSSION

Which herpesviral components are targeted to axons for anterograde transport?

Alphaherpesviruses (e.g., herpes simplex virus and pseudorabies virus) reproduce within polarized neurons. Infection begins with entry at mucosal surfaces and spread of virus particles between cells of the mucosal epithelium. The peripheral nervous system is infected via axon termini innervating this region, and subsequent trafficking of nucleocapsids to the cell body. It is here that a reactivatable, latent infection that persists for the life of the host can be established. A well-known but poorly understood phenomenon is that, upon reactivation from latent infection, alphaherpesviruses rarely enter the central nervous system, despite having what seems to be two rather similar choices: cross one synapse and infect the central nervous system (rare) or traffic back down the axon and cross to the initial site of infection, mucosal epithelial cells (very common). Inherent in this choice is the fact that progeny viral particles or their components must be targeted to axons.

The mechanisms by which newly synthesized components of virus particles, such as nucleocapsids and envelope proteins, are sorted to axons for anterograde transport have been the subjects of considerable controversy. In fact, different processes have been proposed for herpes simplex virus and pseudorabies virus, separate transport of the nucleocapsid plus tegument and viral glycoprotein, and transport of enveloped virus particles, respectively. Similar methods have been used to examine anterograde transport of the two viruses in several laboratories. These methods include confocal microscopy, imaging of nucleocapsids and glycoproteins that carry distinguishable fluorescent labels in live cells, and immunoelectron microscopy. Nevertheless, the controversy has not been resolved. Although counterintuitive, it is possible that different processes operate in cells infected by these two viruses, which are the most distantly related among the alphaherpesviruses. This hypothesis implies that envelopment of naked herpes simplex virus nucleocapsids takes place at the membrane of axonal growth cones, whereas nonenveloped pseudorabies virus nucleocapsids can travel only in the retrograde direction.

Feierbach B, Bisher M, Goodhouse J, Enquist LW. 2007. In vitro analysis of transneuronal spread of an alphaherpesvirus infection in peripheral nervous system neurons. *J Virol* **81:**6846–6857.

Granstedt AE, Brunton BW, Enquist LW. 2013. Imaging the tranport dynamics of single alphaherpesvirus particles in intact peripheral nervous system explants from infected mice. *mBio* **4:**e00358-13. doi:10.1128/mBio:00358-13.

Smith G. 2012. Herpesvirus transport to the nervous system and back again. *Annu Rev Microbiol* **66:**153–176.

Snyder A, Bruun B, Browne HM, Johnson DC. 2007. A herpes simplex virus gD-YFP fusion glycoprotein is transported separately from viral capsids in neuronal axons. *J Virol* **81:**8337–8340.

The salivary glands of mice were infected with a virulent strain of dually fluorescent derivative of pseudorabies virus: the minor capsid protein VP26 was fused to mRFP (red) and the envelope protein Us9 was fused to GFP (green). Tissues including the salivary gland and submandibular ganglia were removed 24 h after infection, and the ganglia were exposed for time-lapse epifluorescence microscopy. Movie 12.1 (http://bit.ly/Virology_V1_Movie12-1) shows that all anterograde-moving red particles (nucleocapsids) also contain the envelope protein Us9 (green), consistent with sorting for axonal transport after formation of complete virus particles. Adapted from A. E. Granstedt et al., *mBio* **4:**e00358-13, 2013, with permission.

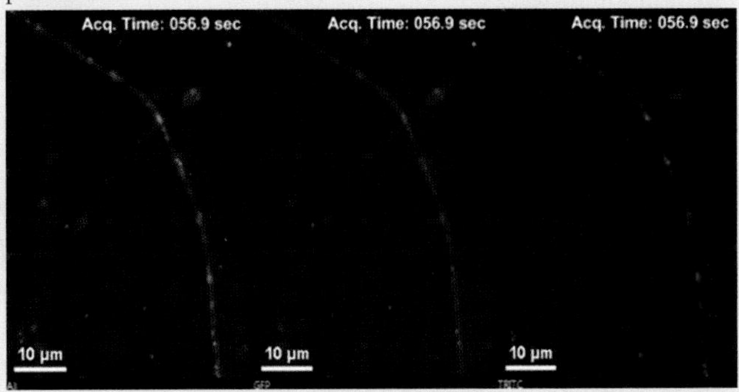

of membranes from earlier compartments in the secretory pathway. Poliovirus infection induces a transient increase in, but subsequent inhibition of, budding of CopII-coated vesicles from ER exit sites, where the viral 2B and 2BC proteins colocalize with cellular proteins that form this coat. The temporary acceleration in vesicular traffic from the ER may increase the supply of membranes and other components to the ER-Golgi intermediate compartment (ERGIC), the origin of the distinctive vesicles that serve as platforms for the replication of the viral RNA genome. These replication compartments are characterized by the presence of specific vesicle-associated proteins of the cell and the viral 3A protein, which recruits an enzyme that catalyzes synthesis of phosphoinositol 4-phosphate (PI4P)-containing lipids. The viral $3D^{pol}$ RNA polymerase localizes to the membranes of such replication compartments by virtue of its preferential binding to PI4P-containing lipids. The diversion of membranes from the ERGIC and disruption of Golgi compartments result in inhibition of protein traffic to the surface of infected cells, and may dampen antiviral responses mediated by MHC class I molecules and cytokines to facilitate survival of infected cells.

Later in infection, double-membrane vesicles 200 to 400 nm in diameter accumulate in the cytoplasm. These vesicles resemble autophagosomes, and can also be associated with replicating viral RNA genomes (Chapter 14). They may also play a role in nonlytic release of virus from cells (Chapters 13 and 14). The vesicles that serve as scaffolds for formation of replication complexes in cells infected by coronaviruses also exhibit properties of autophagosomes. The mechanisms by which infection by these viruses override the cellular circuits that normally prevent autophagy are not yet known. Nevertheless, it is clear that formation of autophagosomes facilitates virus reproduction: virus yield is reduced when synthesis of cellular proteins required for autophagy is prevented.

Induction or Inhibition of the Unfolded Protein Response

The quality control functions of the ER ensure that improperly folded proteins are retained in that organelle for degradation upon retrotranslocation to the cytoplasm. When the capacity of the ER protein folding and removal machinery is exceeded, three signaling pathways are activated by transmembrane receptors, because Grp78 is sequestered from their luminal domains by association with the large quantities of misfolded or unfolded proteins. These signal transduction cascades, collectively known as the unfolded protein response, lead to inhibition of translation and enhanced production of ER membranes and resident folding chaperones and catalysts, as well as ER proteins that clear improperly folded proteins (Fig. 12.17). However, when these measures fail to restore homeostasis, and the unfolded protein response is prolonged, apoptosis is induced via the Pkr-like ER kinase (Perk) and activating transcription factor 4 (Atf4).

Not surprisingly, the demands placed on the biosynthetic capacity of virus-infected cells can induce the unfolded protein response. For example, synthesis of the ER chaperone Grp78 is stimulated in cells infected by a variety of enveloped viruses, including bunyaviruses, flaviviruses, herpesviruses, influenza A virus, and paramyxoviruses. In most cases, accumulation of improperly folded viral proteins is likely to trigger activation of Atf6 (Fig. 12.17). However, during the early stages of infection, the major immediate early protein of the betaherpesvirus human cytomegalovirus directly activates transcription from the promoter of the Grp78 gene, and translation of Grp78 mRNA is also stimulated. A second viral protein, UL50, binds to inositol-requiring enzyme 1 (Ire1) to block signaling from this protein and increase synthesis of others that mediate ER-associated degradation late in infection.

From the point of view of successful virus reproduction, activation of the unfolded protein response is a mixed blessing: increased ER capacity would facilitate the production of large quantities of viral proteins, but attenuation of translation, induction of apoptosis, and increased ER-associated degradation (Fig. 12.17) could limit virus reproduction. This property may account for the differential impact of certain viruses on the various arms of the unfolded protein response. For example, in cells infected by the flaviviruses West Nile virus and dengue virus, signaling via Atf6 and Ire1 is increased, but Perk-mediated inhibition of translation and induction of apoptosis is blocked. The mechanisms that allow such discrimination among the three signaling pathways are not yet known, but clearly facilitate virus reproduction: the yield of infectious dengue virus particles is reduced by an order of magnitude in cells that lack Atf6, but increased to the same or a greater degree in *Perk*$^{-/-}$ cells.

Signal Sequence-Independent Transport of Viral Proteins to the Plasma Membrane

Many enveloped viruses contain matrix or tegument proteins lying between, and making contact with, the inner surface of the membrane of the particle and the capsid or nucleocapsid (Chapter 4). In contrast to the integral membrane proteins of enveloped viruses, such internal proteins of virus particles do not enter the secretory pathway, but are synthesized in the cytoplasm of an infected cell and directed to membrane assembly sites by specific signals.

Lipid-plus-Protein Signals

It has been known for many years that cytoplasmic proteins can be modified by the covalent addition of lipid chains (Table 12.2). Best characterized are the addition of the 14-carbon saturated fatty acid myristate to N-terminal glycine residues, and of unsaturated polyisoprenes, such as farnesol (C_{15}) or geranylgeranol (C_{20}), to a specific C-terminal sequence (Fig. 12.18). Palmitate is also added to some viral proteins that do not enter the secretory pathway. The discovery that transforming proteins of oncogenic retroviruses, the Src and Ras proteins, are myristoylated and isoprenylated, respectively, led to a resurgence of interest in these modifications. In this section, we focus on myristoylation and isoprenylation of viral structural proteins.

Myristoylation of the cytoplasmic Gag proteins of retroviruses and its consequences have been examined in detail. The internal structural proteins of these viruses, MA (matrix), CA (capsid), and NC (nucleocapsid), are produced by proteolytic cleavage of the Gag polyprotein following virus assembly. The Gag proteins of the majority of retroviruses are myristoylated at their N-terminal glycines. Mutations that prevent such acylation of murine leukemia virus or human immunodeficiency virus type 1 Gag proteins block interaction of the protein with the plasma membrane, induce cytoplasmic accumulation of Gag, and inhibit virus assembly and budding. In the case of the human immunodeficiency virus type 1 Gag protein, the myristoylated N-terminal segment and a highly basic sequence located a short distance downstream form a bipartite signal, which allows membrane binding *in vitro* and virus assembly and budding *in vivo* (Fig. 12.19). The MA domain of the Gag protein of this protein binds to phosphatidylinositol (4,5)-bisphosphate, and the acyl chains of other lipids enriched in the inner leaflet of the plasma membrane. This interaction accounts for the preferential association of

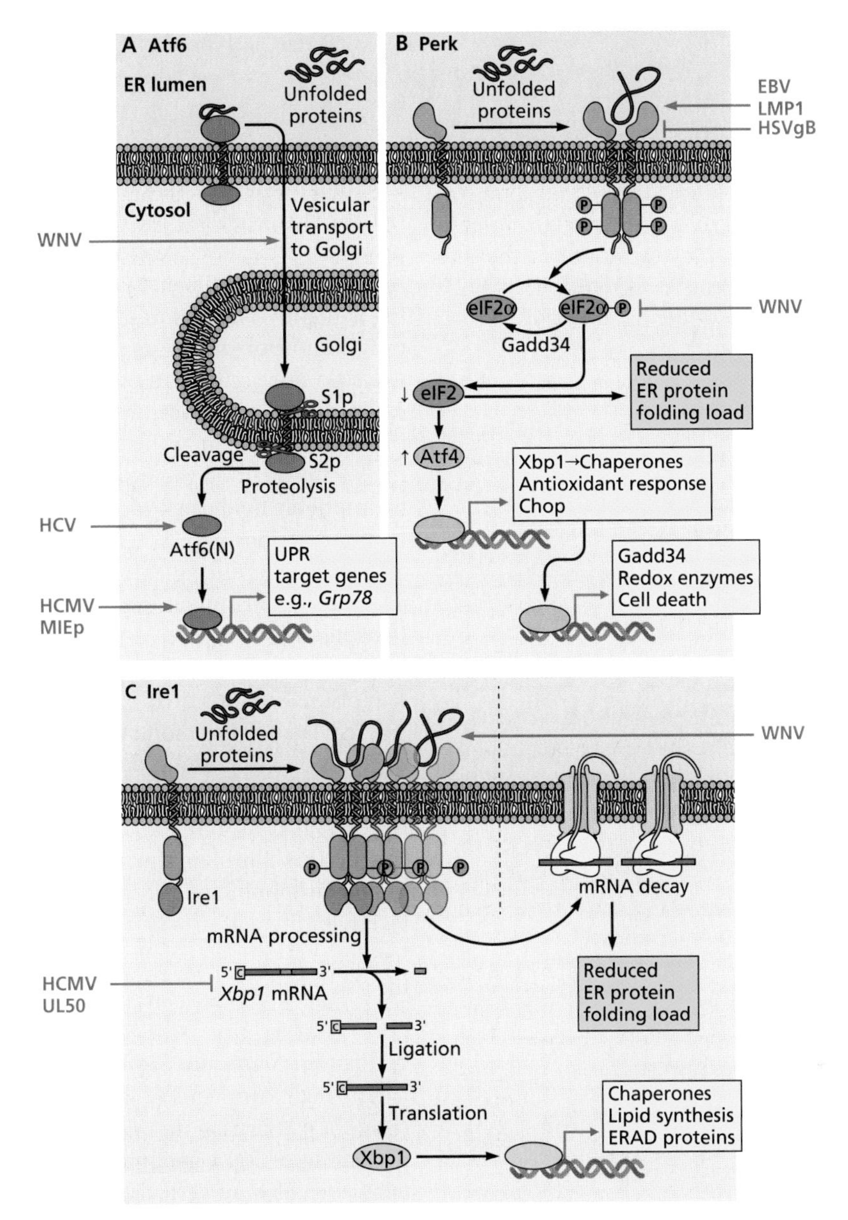

Figure 12.17 Modulation of the unfolded protein response in cells infected by flaviviruses and herpesviruses. Within the ER, Grp78 associates with the luminal domains of members of three families of signal transducers, Atf6 (activating transcription factor 6), Perk (double-stranded RNA activated protein kinase [Pkr]-like ER kinase), and Ire1 (inositol-requiring enzyme 1). Sequestration of the chaperone when concentrations of incompletely or improperly folded proteins in the ER lumen are high leads to activation of these signaling molecules, by different mechanisms. **(A)** Atf6 is released for transport to Golgi compartments, where it is cleaved by the site 1 (S1p) and site 2 (S2p) proteases. The N-terminal segment liberated into the cytoplasm enters the nucleus, where it activates transcription of specific genes. The majority of these genes encode ER proteins, including chaperones such as Grp78 and protein folding catalysts like Pdi, to increase the capacity of the ER to handle protein folding. **(B)** Signaling from Perk is initiated by dimerization and autophosphorylation of its cytoplasmic domain. The active kinase then phosphorylates the α subunit of the translation initiation protein eIF2 to inhibit translation (see Chapter 11). This response decreases the flow of newly synthesized proteins into the ER. However, some mRNAs, including Atf4 mRNA, are translated preferentially when eIF2 concentrations are limiting. Genes increased in expression in response to this regulator include those for X box-binding protein 1 (Xbp1) and transcriptional activator c/Ebp homologous protein (Chop). The latter protein in turn stimulates transcription of genes that encode proapoptotic proteins such as growth arrest and DNA damage-inducible 34 (Gadd34). Gadd34 also functions as a regulatory subunit of protein phosphatase to reverse phosphorylation and inhibition of eIF2. Consequently, prolonged signaling from Perk can induce cell death. **(C)** Ire1, the only unfolded protein signal transducer present in yeast, contains cytoplasmic kinase and RNase domains. Binding of unfolded proteins to the luminal domain is thought to induce oligomerization, autophosphorylation, and activation of the RNase. This enzyme initiates a very unusual splicing reaction by excision of the intron of Xbp1 mRNA, for subsequent ligation of its exons (by tRNA ligase in yeast and an as yet unidentified enzyme in mammalian cells). Subsequent synthesis of active Xbp1 leads to induction of transcription of genes that code for enzymes that catalyze lipid synthesis and proteins that facilitate removal of misfolded proteins from the ER. In addition to the exquisitely specific cleavages of Xbp1 mRNA, Ire1 initiates degradation of mRNAs associated with the ER by low-specificity endonucleolytic cleavage. As indicated, virus infection can lead to activation or inhibition of the three arms of the unfolded protein response, although in most cases the mechanisms of such modulation remain to be established. HCMV, human cytomegalovirus; EBV, Epstein Barr virus, HCV, hepatitis C virus; HSV, herpes simplex virus type 1; WNV, West Nile virus; MIEp, major immediate early protein; LMP1, latent membrane protein 1; ERAD, ER-associated degradation; UPR, unfolded protein response. Adapted from P. Walter and D. Ron, *Science* **334:**1081–1086, 2011, with permission.

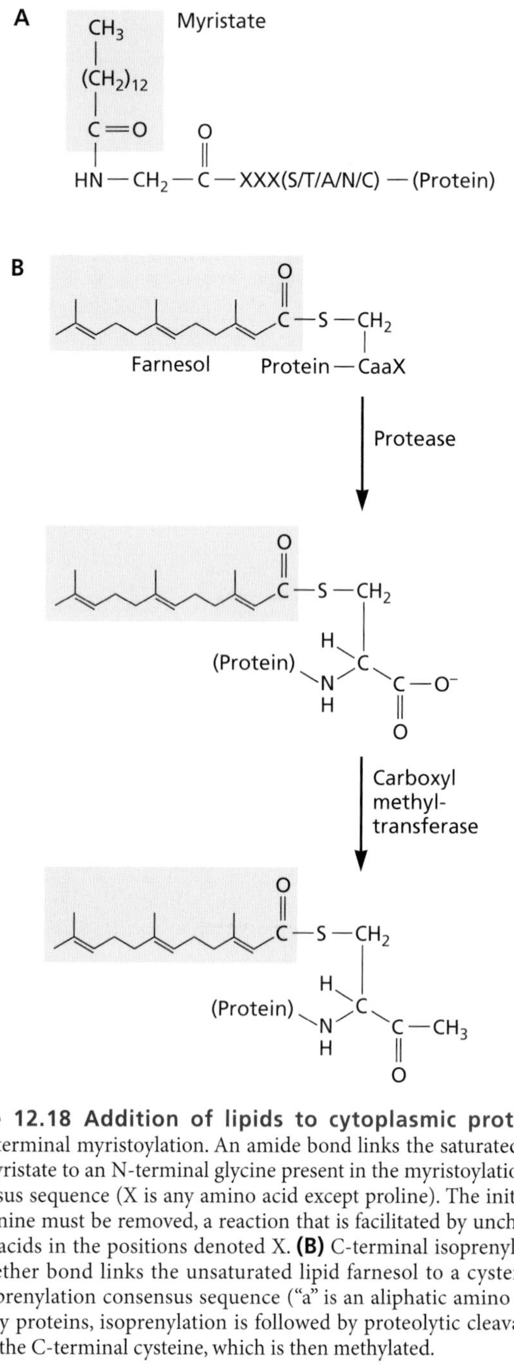

Figure 12.18 Addition of lipids to cytoplasmic proteins. (A) N-terminal myristoylation. An amide bond links the saturated fatty acid myristate to an N-terminal glycine present in the myristoylation site consensus sequence (X is any amino acid except proline). The initiating methionine must be removed, a reaction that is facilitated by uncharged amino acids in the positions denoted X. **(B)** C-terminal isoprenylation. A thioether bond links the unsaturated lipid farnesol to a cysteine in the isoprenylation consensus sequence ("a" is an aliphatic amino acid). In many proteins, isoprenylation is followed by proteolytic cleavage to expose the C-terminal cysteine, which is then methylated.

Gag with the plasma membrane. It also induces a conformational change that leads to exposure of the N-terminal myristate, and presumably tighter association of Gag with the membrane.

The hepatitis B virus large surface (L) protein is also myristoylated at its N terminus. However, in contrast to retroviral Gag, the L protein is present in the envelope of virus particles. Modification of its N terminus must therefore occur while it traverses the secretory pathway. In this case, myristoylation is not necessary for assembly or release of virus particles, but is required for infection of primary hepatocytes, presumably because it contributes to the initial interaction of the virus with, or its entry into, the host cell. More surprising is the myristoylation of structural proteins of poliovirus (VP4) and polyomavirus (VP2): although these virus particles do not contain an envelope, this modification is necessary for efficient assembly. In mature poliovirus particles, the myristate chain at the N terminus of VP4 interacts with the VP3 protein on the inside of the capsid (Fig. 4.12B). The hydrophobic lipid chain must therefore facilitate protein-protein interactions necessary for the assembly. The fatty acid is also important during entry into cells of poliovirus particles and their uncoating at the beginning of an infectious cycle (Chapter 5).

Among viral structural proteins, only the large delta protein of the hepatitis delta satellite virus has been found to be isoprenylated. Formation of the particles of this satellite virus depends on structural proteins provided by the helper virus, hepatitis B virus. The isoprenylation of large delta protein is necessary, but not sufficient, for its binding to the hepatitis B virus S protein during assembly of the satellite virus. This hydrophobic tail of large delta protein seems likely to facilitate interaction with the plasma membrane adjacent to regions that contain helper virus S protein in cells infected by the two viruses.

Protein Sequence Signals

The matrix proteins of members of several families of (−) strand RNA viruses are essential for correct localization and packaging of RNA genomes. During assembly, matrix proteins, such as M of vesicular stomatitis virus and M1 of influenza A virus, must bind to the inner surface of the plasma membrane of infected cells. These proteins are produced in the cytoplasm, but receive no lipid after translation. When the influenza virus M1 protein is synthesized in host cells in the absence of other viral proteins, it associates tightly with cellular membranes. Both this protein and the vesicular stomatitis virus M protein contain specific sequences that are necessary for their interaction with the plasma membrane *in vivo* or with lipid vesicles *in vitro*. This region of the influenza A virus M1 protein contains two hydrophobic sequences (Fig. 12.20A), which might form a hydrophobic surface in the folded protein. In addition to hydrophobic segments, membrane association of the vesicular stomatitis virus M protein requires a basic N-terminal sequence (Fig. 12.20B). This latter segment might participate directly in membrane binding, like the basic sequence of the human immunodeficiency virus type 1 Gag membrane-targeting signal, or it might stabilize a conformation of the internal sequence favorable for interaction with the membrane.

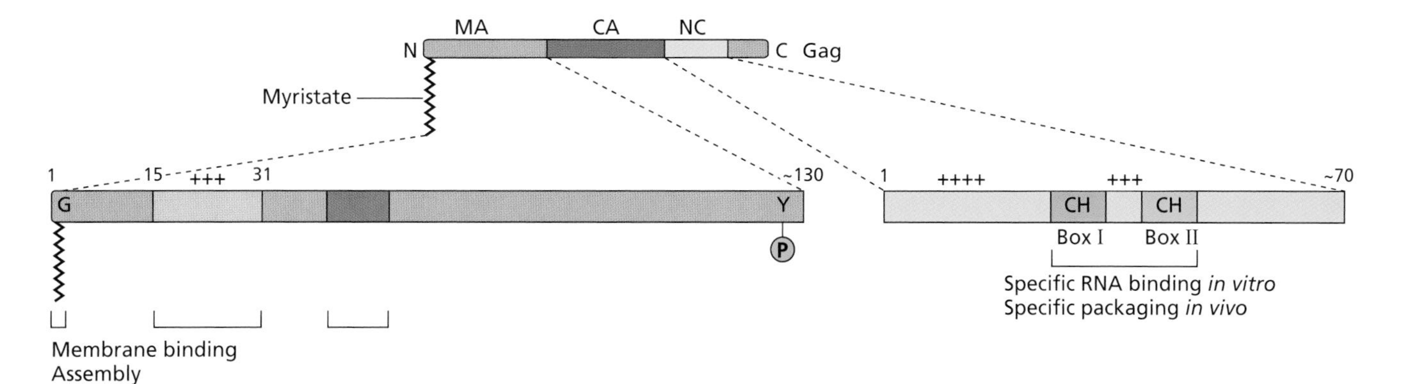

Figure 12.19 Human immunodeficiency virus type 1 Gag proteins and their targeting signals. The locations of the internal structural proteins MA (matrix), CA (capsid), and NC (nucleocapsid) in the Gag polyprotein are shown at the top. Sequence features, localization signals (MA), and the RNA-binding domain (NC) are shown below. The lengths of the MA and NC proteins are listed as approximate because of the variation among virus isolates. Specific amino acids are given in the single-letter code in the boxes, and a plus sign indicates a basic amino acid. The basic region of MA of retroviruses with simple genomes, such as avian sarcoma virus, is not required for membrane binding. The CH boxes of NC contain three cysteines and one histidine, and each coordinates one Zn^{2+} ion. CH box I is conserved among retroviruses, but CH box II is not.

In some cases, such as the vesicular stomatitis virus M protein, specificity for the plasma membrane is an intrinsic property, suggesting that these proteins might recognize phospholipids enriched in the inner leaflet of the plasma membrane, such as phosphatidylserine and phosphatidylinositol. However, binding of matrix proteins to the cytoplasmic tails of viral envelope glycoproteins can also be an important determinant of membrane association. The cytoplasmic domains of both the NA and HA proteins of influenza A virus stimulate membrane binding by M1 protein. Similarly, membrane binding by the matrix protein of Sendai virus (a paramyxovirus) is independently stimulated by the presence of either of the two viral glycoproteins (F or HN) in the membrane.

Interactions with Internal Cellular Membranes

The envelopes of a variety of viruses are acquired from internal membranes of the infected cell, rather than from the plasma membrane. The majority of these viruses assemble at the cytoplasmic faces of compartments of the secretory pathway (Table 12.3). Although a single budding reaction is typical, the more complex herpesviruses and poxviruses interact with multiple internal membranes during assembly and exocytosis (Chapter 13).

The diversity of the internal membranes with which these viruses associate during envelope acquisition and exocytosis is the result of variations on a single mechanistic theme: the site

Figure 12.20 Targeting signals of matrix proteins of influenza virus (A) and vesicular stomatitis virus (B). Sequence features of specific segments of the proteins and the boundaries of targeting and RNP-binding domains are shown. Amino acids are written in the one-letter code, and a plus sign indicates a basic amino acid. NES, nuclear export signal; NLS, nuclear localization signal; VSV, vesicular stomatitis virus.

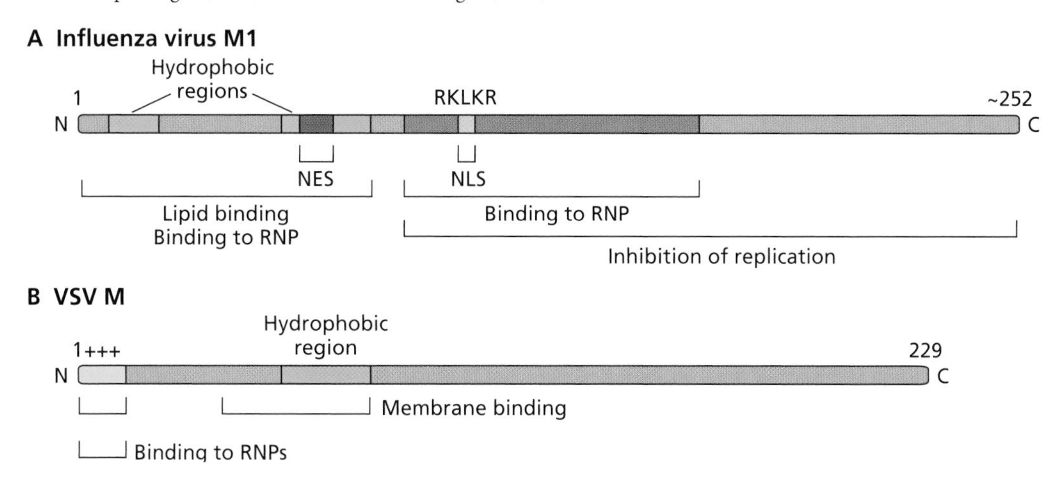

Table 12.3 Interactions of viruses with internal cellular membranes

Virus family	Example	Integral membrane protein(s)	Intracellular membrane(s)	Mechanism of envelopment
Bunyaviruses	Uukuniemi virus Hantaan virus	Gn Gc	*cis*-medial Golgi cisternae	Budding into Golgi cisternae
Coronavirus	Mouse hepatitis virus Severe acute respiratory syndrome coronavirus	M, S	ERGIC and *cis*-Golgi network	Budding into ERGIC and *cis*-Golgi network
Flavivirus	Dengue virus West Nile virus	E, prM	ER	Budding into ER
Hepadnavirus	Hepatitis B virus	L, M, S	ER and other compartments	Budding into multivesicular body
Herpesvirus	Herpes simplex virus type 1	gB, gH, UL34 gE-gI	Nuclear membrane *trans*-Golgi cisternae	Primary envelopment; budding of capsids from inner nuclear membrane Budding at *trans*-Golgi membrane
Poxvirus	Vaccinia virus, immature Vaccinia virus, intracellular mature virus	A14L, A13L, A17L A56R (HA), F13L, B5R	ER Late *trans*-Golgi cisternae and post-Golgi vesicles	Formation of mature virion Wrapping of mature virion
Rotavirus	Simian rotavirus	VP7, NS28	ER	Budding into ER

of assembly is determined by the intracellular location of viral envelope proteins (Fig. 12.21), just as assembly at the plasma membrane is the result of transport of such proteins to that site. Assembly of viruses at internal membranes therefore requires transport of envelope proteins to, and their retention within, appropriate intracellular compartments.

Localization of Viral Proteins to Compartments of the Secretory Pathway

The bunyaviruses, a family that includes Uukuniemi and Hantaan viruses, are among the best-studied viruses that assemble by budding into compartments of the secretory pathway. Bunyavirus particles contain two integral membrane glycoproteins, called Gn and Gc, which are encoded within a single open reading frame of the M genomic RNA segment. Like alphaviral envelope proteins, the bunyaviral polyprotein containing Gn and Gc is processed cotranslationally by signal peptidase as the precursor enters the lumen of the ER (Table 12.1). However, association of the glycoproteins with one another is required for transport of Gc to Golgi compartments: when synthesized alone, Gn accumulates in the Golgi complex as it does in infected cells, but Gc fails to leave the ER. The signals necessary for Golgi residence of Gn and associated Gc lie in the transmembrane domain of Gn.

Golgi cisternae are by no means the only compartments of the secretory pathway at which virus budding can occur. For example, rotaviruses transiently acquire an envelope by budding into and out of the ER, whereas coronaviruses bud into the ERGIC and the Golgi apparatus. In these and other cases, it is the presence of viral glycoproteins in specific cellular membranes (Table 12.3) that determines the site of assembly and budding.

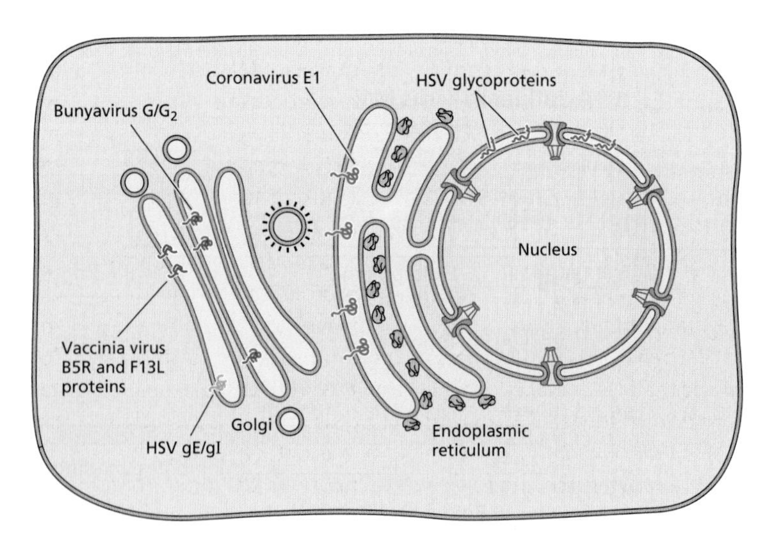

Figure 12.21 Sorting of viral glycoproteins to internal cell membranes. The destinations of membrane glycoproteins of viruses that bud into compartments of the secretory pathway (bunyaviruses and coronaviruses) or from the inner nuclear membrane and compartments of the *trans*-Golgi network (herpesviruses [HSV]) or are wrapped by cellular membranes during assembly (poxviruses) are indicated.

Localization of Viral Proteins to the Nuclear Membrane

Herpesviruses such as herpes simplex virus type 1 are the only enveloped viruses that are known to assemble initially within, and bud from, the nucleus. The first association of an assembling herpesvirus with a cellular membrane is therefore budding of the nucleocapsid through the inner membrane of the nuclear envelope. This process, which is described in Chapter 13, depends on association of particular viral proteins with the inner nuclear membrane (Fig. 12.21; Table 12.3).

Transport of Viral Genomes to Assembly Sites

Like the structural proteins and enzymes of virus particles, progeny genomes must be available, or concentrated, at the intracellular site at which assembly takes place. In several cases, this requirement is met by genome replication within the same cellular organelle or structure as assembly of new virus particles. The genomes of DNA viruses that are synthesized in infected cell nuclei are encapsidated within that organelle. Similarly, the replication of the genomes of many (+) strand viral RNAs (including those of picornaviruses and flaviviruses) and of large DNA viruses such as poxviruses, and assembly of virus particles, are restricted to specialized cytoplasmic structures derived from host cell membranes (Chapters 13 and 14). In contrast, the genomes of many (−) strand RNA viruses must be transported to the cytoplasmic faces of the appropriate membrane, most commonly the plasma membrane. Yet other RNA genomes must travel even farther: both influenza virus and retroviral genomic RNAs are synthesized within the infected cell nucleus, but progeny virus particles bud from the plasma membrane.

Transport of Genomic and Pregenomic RNA from the Nucleus to the Cytoplasm

Retroviral genomes are unspliced RNA transcripts synthesized in infected cell nuclei by host cell RNA polymerase II, as is hepadnaviral pregenomic RNA. These RNAs must be exported to the cytoplasm for assembly, a process that requires that the inefficient export of unspliced mRNAs characteristic of host cells be circumvented. Viral RNA-binding proteins promote export of unspliced RNA of retroviruses with complex genomes, such as human immunodeficiency virus type 1, whereas specific sequences that are recognized by cellular proteins direct export of genomic RNAs of other retroviruses and hepadnaviral pregenomic RNA (Chapter 10).

Perhaps the most elaborate requirements for transport of viral RNA species between nucleus and cytoplasm are found in influenza A virus-infected cells: both the direction of transport of genomic RNA and the nature of the viral RNA exported from the nucleus change as the infectious cycle progresses. When the cycle is initiated, viral genomic ribonucleoproteins (RNPs) enter the nucleus under the direction of the nuclear localization signal of the NP protein (Chapter 5). The mechanisms that ensure export of viral (+) strand mRNAs for translation are not well understood. With the switch to replication, genomic (−) strand RNA segments are synthesized in infected cell nuclei, where they accumulate as viral RNPs containing the NP protein and the three P proteins. These RNPs must be exported to allow virus assembly and completion of the infectious cycle, a reaction that requires the viral M1 and NEP proteins (Fig. 12.22). The M1 protein binds to both viral RNPs and NEP. Efficient transport of this complex to the cytoplasm requires two leucine-rich nuclear export signals present in NEP and one in M1. This requirement may restrict export of genome segments to the late phase of infection, when NEP accumulates, and ensure association of RNPs with the protein (M1) necessary for guiding them to the plasma membrane.

Transport of Genomes from the Cytoplasm to the Plasma Membrane

Accumulation of the RNA genomes of enveloped viruses at the appropriate cellular membrane depends on signals present in viral proteins bound to the RNA. The membrane-binding domain of influenza virus M1, described previously, allows association of the genomic RNPs with the plasma membrane following transport from the juxtanuclear region, at which they accumulate upon export from the nucleus. This region contains the microtubule organizing center, and the RNPs become associated with recycling endosomes and are transported along microtubules to the plasma membrane (Fig. 12.22). The concentration of RNPs on the membranes of recycling endosomes may also facilitate association and packaging of the full complement of genome segments during assembly (Chapter 13).

Although more limited in its RNA transport functions, the M protein of vesicular stomatitis virus shares several properties with the influenza virus M1 protein. Newly synthesized genomic RNA molecules assemble with the N, L, and NS proteins to form helical RNPs. Genomic RNA molecules within RNPs can serve as templates for additional cycles of replication or for mRNA synthesis. However, these RNPs eventually must travel to the plasma membrane for association with the G protein and incorporation into virus particles. Entry into the latter pathway is determined by the viral M protein, which associates with RNPs containing genomic RNA to induce formation of a tightly coiled RNA-protein "skeleton" (Fig. 12.23). Formation of this structure precludes replication and mRNA synthesis and allows transport of nucleocapsids to the plasma membrane via microtubules (Box 12.10). The membrane-binding domains of the M protein described above then direct association with the plasma membrane.

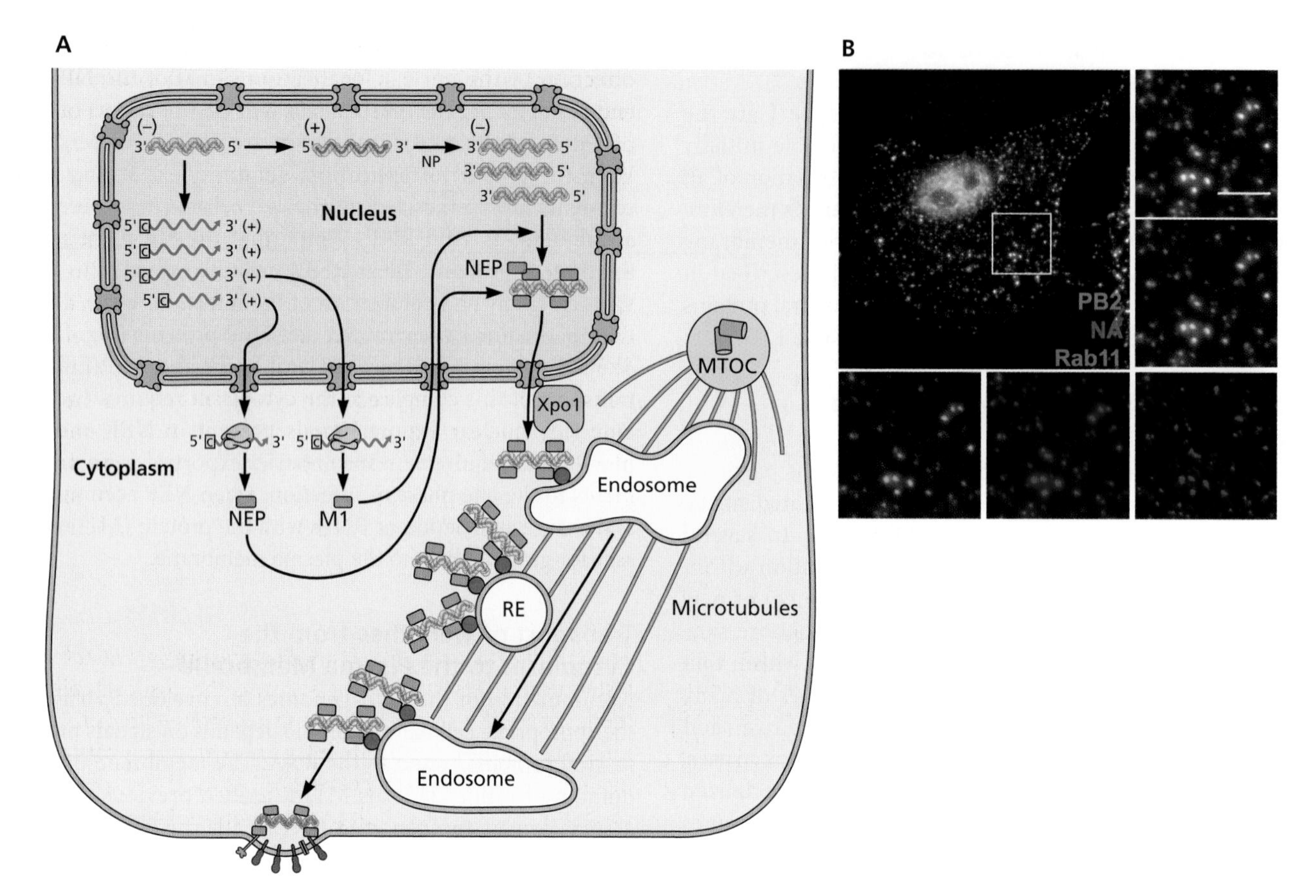

Figure 12.22 Transport of influenza A virus genomic RNA segments from the nucleus to the plasma membrane. (A) Genomic RNA segments are bound by the NP protein as they are synthesized (see Chapter 6) and subsequently by the M1 protein. M1 is the most abundant protein of the virus particle and enters the nucleus by means of a typical nuclear localization signal (Fig. 5.22). Binding of M1 to genomic RNPs (Fig. 12.20A) both inhibits RNA synthesis and promotes genomic RNP export. M1-containing RNPs are directed to the cellular Xpo1 export pathway upon binding of NEP, which contains two nuclear export signals. NEP possesses no intrinsic RNA-binding activity, but includes a C-terminal M1-binding domain. This domain is thought to allow recognition of RNPs to which the M1 protein is bound. Following export from the nucleus, RNPs accumulate in the region that contains the microtubule organizing center (MTOC) and, via interaction of the viral polymerase with the GTP-bound form of Rab11, become associated with recycling endosomes (REs), which are transported to the plasma membrane along microtubules. **(B)** It has been proposed that Rab11-containing organelles, such as REs, serve as platforms for association of the individual vRNA segments that comprise a genome with one another, based in part on colocalization. In the experiment shown here, human cells were infected with influenza virus for 10 h (later in the infectious cycle), and subjected to two-color, high-sensitivity fluorescence *in situ* hybridization (FISH) with oligonucleotides complementary to the PB2 and NA vRNA segments labeled with different fluorophores, Cy3 (green) and Cy5 (red), respectively, and then to immunostaining for Rab11. A high degree of colocalization of these vRNAs with one another (yellow puncta in the two-color FISH image, top, lower panel on right) and with Rab11 (white puncta in the merged FISH and immunostaining image, top, upper panel on right) was observed. Quantification of the degree of colocalization of the Rab11-associated vRNAs and the vRNAs that were not so associated suggested that interaction with Rab11 promoted vRNA colocalization. Consistent with a model in which Rab11-bound organelles provide a niche for assembly of individual vRNA segments in transit to the plasma membrane for assembly, synthesis of a dominant-negative derivative of Rab11 in infected cells reduced the efficiency of colocalization of the PB2 and NA vRNAs. Adapted from Y.-Y. Chou et al., *PloS Pathog* **9:**e1003358, 2013, with permission.

The retroviral proteins that mediate membrane association of genomic RNA are similar to the matrix proteins of these (−) strand RNA viruses in several respects. Once within the cytoplasm, unspliced retroviral RNA is translated on polyribosomes into the Gag and, at low frequency, Gag-Pol polyproteins. The functions of Gag include transport of unspliced RNA molecules to membrane assembly sites and packaging of the RNA into assembling capsids. Consequently, whether unspliced genomic RNA molecules continue to be translated or are redirected for assembly is controlled by the cytoplasmic concentration of Gag.

The NC segment of Gag contains an RNA-binding domain required for specific recognition of the RNA-packaging sequences described in Chapter 13. NC functions as an

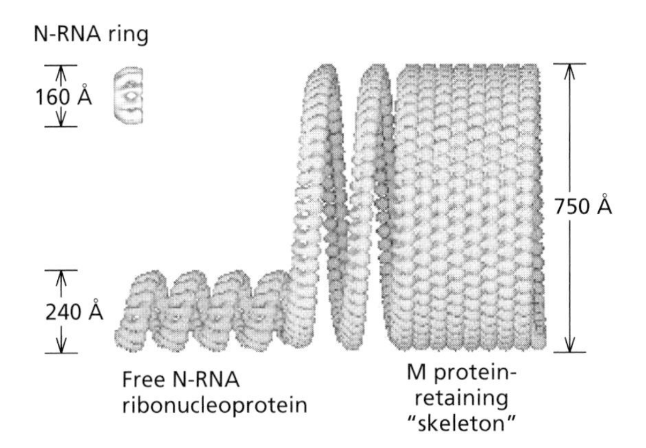

Figure 12.23 Models of the rabies virus nucleocapsid, showing the free nucleocapsid and the nucleocapsid present in virus particles. The models are based on cryo-electron microscopy and image reconstruction of the two forms of the nucleocapsid, as well as of rings of 9 or 10 molecules of the viral N protein and RNA assembled when the protein is produced in insect cells (160-Å N-RNA ring). The free nucleocapsid, which is the template for viral RNA synthesis, is a loosely coiled helix with a variable pitch and diameter of 240 Å. In contrast, the nucleocapsid helix incorporated into virus particles is tightly wound, with a small pitch and a much larger diameter (750-Å M protein-retaining "skeleton"). These structural transitions are induced by binding of the M protein to the free nucleocapsid. Adapted from G. Schoehn et al., *J Virol* **75:**490–498, 2001, with permission.

independent protein in mature virus particles, is very basic, and contains at least one copy of a zinc-binding motif (Fig. 12.19). This domain makes a major contribution to the specificity with which Gag or NC proteins bind to unspliced retroviral RNA and, in conjunction with basic amino acids located nearby, is responsible for the RNA-packaging activity of Gag. The N-terminal MA portion contains the signals described above that target the polyprotein to the plasma membrane. Binding of Gag to unspliced retroviral RNA therefore allows delivery of the genome to assembly sites at the plasma membrane. The interaction probably takes place at the juxtanuclear sites where newly synthesized Gag accumulates. Movement of Gag (and presumably associated viral RNA genomes) to the plasma membrane requires the kinesin Kif4, a motor that moves toward the (+) (plasma membrane) end of microtubules and binds to the viral protein. A variety of other cellular proteins, including the RNA-binding protein Staufen1 and proteins that interact with it, have been implicated in transport of retroviral genomes to the plasma membrane. However, whether transport is direct, or via association with membrane-bound structures such as multivesicular bodies, or whether these different mechanisms dominate in different cell types is not yet clear.

The influenza virus M1, vesicular stomatitis virus M, and retroviral Gag proteins each possess the ability to bind directly or indirectly to RNPs containing genomic RNA and to membranes. Such interactions commit genomic RNA to the assembly pathway, direct the RNA to the plasma membrane, and promote interactions among internal and envelope components of virus particles. These properties are essential at the end of an infectious cycle, when the primary task is assembly of progeny. On the other hand, they would be disastrous if the interactions could not be reversed before or at the beginning of a new cycle, when the infecting genome must reach nuclear (influenza virus) or cytoplasmic (vesicular stomatitis virus and retroviruses) sites distant from the plasma membrane. In the case of (−) strand viruses, matrix proteins are removed during virus entry. The retroviral mechanism is more elegant: following virus assembly and budding, Gag (and Gag-Pol) polyproteins are processed by the viral protease to the individual structural proteins shown in Fig. 12.19. Such cleavages place the RNA-binding domain of NC in protein molecules separate from membrane-binding signals of MA, so that matrix-free core RNPs can be released into the cell to initiate a new infectious cycle.

Perspectives

The cellular trafficking systems described in this chapter are just as crucial for virus reproduction as the host cell's biosynthetic capabilities. The trafficking requirements during the infectious cycle can be quite intricate, with transport of viral macromolecules (or structures built from them) over large distances, or in opposite directions during different periods of the infectious cycle. Assembly of progeny particles of all viruses depends on the prior sorting of their components by at least one cellular trafficking system.

The intracellular sorting of viral proteins or nucleic acids synthesized in large quantities in infected cells has provided important tools with which to study these processes, which are also essential to cellular physiology. Indeed, the fundamental principle of protein sorting, that a protein's final destination is dictated by specific signals within its amino acid sequence and/or covalently attached sugars or lipids, was established by analyses of viral proteins. Furthermore, the study of viral proteins that enter the secretory pathway has provided much of what we know of the reactions by which proteins are folded and processed within the ER, as well as those that clear misfolded proteins from the pathway. It therefore seems certain that viral systems will provide equally important insights into signals and sorting mechanisms that are presently less well characterized, such as those responsible for the direction of proteins to the specialized membrane regions of polarized cells.

One of the greatest current challenges in this field remains the elucidation of the mechanics of the movement of proteins, nucleic acids, nucleoproteins, or transport vesicles from one cellular compartment or site to another. The development and application of techniques that exploit fluorescent proteins to visualize transport in living cells is providing important new insights into these processes. Transport of components of virus particles to sites of assembly results in formation of

BOX 12.10

EXPERIMENTS

Movement of vesicular stomatitis virus nucleocapsids within the cytoplasm requires microtubules

Vesicular stomatitis virus (VSV) nucleocapsids that must be transported to the plasma membrane for assembly and budding of virus particles contain the (−) strand RNA genome and several viral proteins, including the P protein. To examine intracellular trafficking of nucleocapsids, a sequence coding for a green fluorescent protein (eGFP) was inserted into that for the hinge region of the P protein. Control experiments established that the P-eGFP fusion protein catalyzed both viral mRNA synthesis and genome replication, although it exhibited somewhat reduced activity. Furthermore, mutant particles containing P-eGFP in place of the P protein were infectious.

In infected cells, P-eGFP colocalized with newly synthesized viral RNA, as well as with the N and L proteins, in cytoplasmic structures of the size predicted for nucleocapsids. Time-lapse imaging of these P-eGFP-containing structures indicated that nucleocapsids move toward the cell periphery. Indeed, nucleocapsids were observed to be distributed throughout the cytoplasm in close association with microtubules (panel A of the figure). Treatment of infected cells with drugs that disrupt microtubules, such as nocodazole, dramatically altered this pattern: nucleocapsids became clustered in the absence of microtubules in large aggregates and did not reach the plasma membrane (panel B). Such drugs also reduced virus yield significantly, confirming the importance of microtubules in the transport of vesicular stomatitis virus nucleocapsids to sites of assembly.

Das SC, Nayak D, Zhou Y, Pattnaik AK. 2006. Visualization of intracellular transport of vesicular stomatitis virus nucleocapsids in living cells. *J Virol* **80:** 6368–6377.

Localization of P-eGFP-containing nucleocapsids (green) and microtubules (red) in cells infected by the mutant virus VSV-PeGFP and untreated (A) or treated with nocodazole prior to infection (B). Nuclei are in blue. Courtesy of Asit Pattnaik, University of Nebraska-Lincoln.

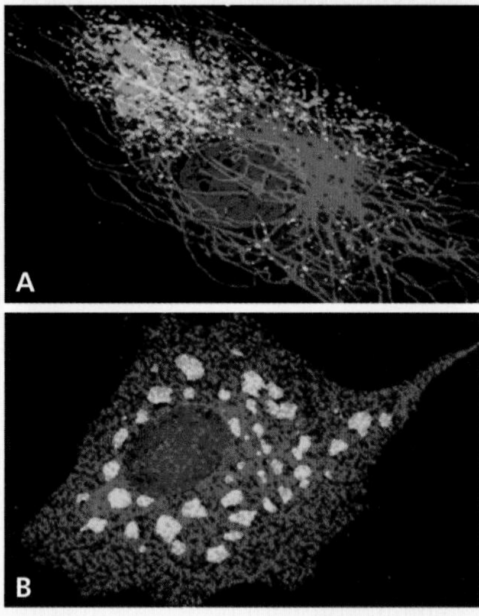

microenvironments containing high concentrations of viral structural proteins and the nucleic acid genome. Such microenvironments are ideal niches for the assembly of progeny particles from their multiple parts.

References

Reviews

Behnia R, Munro S. 2005. Organelle identity and the signposts for membrane traffic. *Nature* **438:**597–604.

Bonifacino JS. 2014. Adaptor proteins involved in polarized sorting. *J Cell Biol* **204:**7–17.

Brandenburg B, Zhuang X. 2007. Virus-trafficking—learning from single-virus tracking. *Nat Rev Microbiol* **5:**197–208.

Ellgaard L, Helenius A. 2003. Quality control in the endoplasmic reticulum. *Nat Rev Mol Cell Biol* **4:**181–191.

Gahmberg CG, Tolvanen M. 1996. Why mammalian cell surface proteins are glycoproteins. *Trends Biochem Sci* **21:**308–311.

Hebert DN, Molinari M. 2012. Flagging and docking: dual roles for *N*-glycans in protein quality control and cellular proteostasis. *Trends Biochem Sci* **37:**404–410.

Hutchinson EC, Fodor E. 2013. Transport of the influenza virus genome from nucleus to nucleus. *Viruses* **5:**2424–2446.

Jackson WT, Giddings TH, Jr, Taylor MP, Mulinyawe S, Rabinovitch M, Kopito RR, Kirkegaard K. 2005. Subversion of cellular autophagosomal machinery by RNA viruses. *PLoS Biol* **3:**e156. doi:10.1371/journal.pbio.0030156.

Jahn R, Scheller RH. 2006. SNARES—engines for membrane fusion. *Nat Rev Mol Cell Biol* **7:**631–643.

Kratchmarov R, Taylor MP, Enquist LW. 2012. Making the case: married versus separate models of alphaherpes virus anterograde transport in axons. *Rev Med Virol* **22:**378–391.

Lord C, Ferro-Novick S, Miller EA. 2013. The highly conserved COPII coat complex sorts cargo from the endoplasmic reticulum and targets it to the Golgi. *Cold Spring Harb Perspect Biol* **5:**a013367. doi:10.1101/cshperspect.a013367.

Pelham HR. 2001. Traffic through the Golgi apparatus. *J Cell Biol* **155:**1099–1101.

Ploubidou A, Way M. 2001. Viral transport and the cytoskeleton. *Curr Opin Cell Biol* **13:**97–105.

Rapoport TA. 2007. Protein translocation across the eukaryotic endoplasmic reticulum and bacterial plasma membranes. *Nature* **450:**663–669.

Suomalainen M. 2002. Lipid rafts and the assembly of enveloped viruses. *Traffic* **3:**705–709.

Smith MH, Ploegh HL, Weissmann JS. 2011. Road to ruin: targeting proteins for degradation in the endoplasmic reticulum. *Science* **334:**1086–1090.

Swanson CM, Malim MH. 2006. Retrovirus RNA trafficking: from chromatin to invasive genomes. *Traffic* **7:**1440–1450.

Tsai B, Ye Y, Rapoport TA. 2002. Retro-translocation of proteins from the endoplasmic reticulum into the cytosol. *Nat Rev Mol Cell Biol* **3:**246–255.

Walter P, Ron D. 2011. The unfolded protein response: from stress pathway to homeostatic regulation. *Science* **334:**1081–1086.

Yarbrough ML, Mata MA, Sakthivel R, Fontoura BM. 2014. Viral subversion of nucleocytoplasmic trafficking. *Traffic* **15:**127–140.

Papers of Special Interest

Import of Viral Proteins into the Nucleus

Forstová J, Krauzewicz N, Wallace S, Street AJ, Dilworth SM, Beard S, Griffin BE. 1993. Cooperation of structural proteins during late events in the life cycle of polyomavirus. *J Virol* **67:**1405–1413.

Zhao LJ, Padmanabhan R. 1988. Nuclear transport of adenovirus DNA polymerase is facilitated by interaction with preterminal protein. *Cell* **55:** 1005–1015.

Transport of Viral and Cellular Proteins via the Secretory Pathway

Balch WE, Dunphy WG, Braell WA, Rothman JE. 1984. Reconstitution of the transport of protein between successive compartments of the Golgi measured by the coupled incorporation of N-acetylglucosamine. *Cell* **39:**405–416.

Hammond C, Helenius A. 1994. Folding of VSV G protein: sequential interaction with BiP and calnexin. *Science* **266:**456–458.

Horimoto T, Nakayama K, Smeekens SP, Kawaoka Y. 1994. Proprotein-processing endoproteases PC6 and furin both activate hemagglutinin of virulent avian influenza viruses. *J Virol* **68:**6074–6078.

Kramer T, Greco TM, Taylor MP, Ambrosini AE, Cristea IM, Enquist LW. 2012. Kinesin-3 mediates axonal sorting and directional transport of alphaherpesvirus particles in neurons. *Host Cell Microbe* **12:**806–814.

Rothman JE, Lodish HF. 1977. Synchronised transmembrane insertion and glycosylation of a nascent membrane protein. *Nature* **269:**775–780.

Takeda M, Leser GP, Russell CJ, Lamb RA. 2003 Influenza virus hemagglutinin concentrates in lipid raft microdomains for efficient viral fusion. *Proc Natl Acad Sci U S A* **100:**14610–14617.

Takeuchi K, Lamb RA. 1994. Influenza virus M2 protein ion channel activity stabilizes the native form of fowl plague virus hemagglutinin during intracellular transport. *J Virol* **68:**911–919.

Transport of Viral Proteins to Intracellular Membranes

Alconada A, Bauer U, Sodeik B, Hoflack B. 1999. Intracellular traffic of herpes simplex virus glycoprotein gE: characterization of the sorting signals required for its *trans*-Golgi network localization. *J Virol* **73:**377–387.

Hobman TC, Lemon HF, Jewell K. 1997. Characterization of an endoplasmic reticulum retention signal in the rubella virus E1 glycoprotein. *J Virol* **71:** 7670–7680.

Melin L, Persson R, Andersson A, Bergström A, Rönnholm R, Pettersson RF. 1995. The membrane glycoprotein G1 of Uukuniemi virus contains a signal for localization to the Golgi complex. *Virus Res* **36:**49–66.

Överby AK, Popov VL, Pettersson RF, Neve EP. 2007. The cytoplasmic tails of Uukuniemi virus (*Bunyaviridae*) G_N and G_C glycoproteins are important for intracellular targeting and the budding of virus-like particles. *J Virol* **81:**11381–11391.

Saad JS, Loeliger E, Luncsford P, Liriano M, Tai J, Kim A, Miller J, Joshi A, Freed EO, Summers MF. 2007. Point mutations in the HIV-1 matrix protein turn off the myristyl switch. *J Mol Biol* **366:**574–585.

Impact of Virus Infection on the Secretory Pathway

Beske O, Reichelt M, Taylor MP, Kirkegaard K, Andino R. 2007. Poliovirus infection blocks ERGIC-to-Golgi trafficking and induces microtubule-dependent disruption of the Golgi complex. *J Cell Sci* **120:**3207–3218.

Burgert HG, Kvist S. 1985. An adenovirus type 2 glycoprotein blocks cell surface expression of human histocompatibility class I antigens. *Cell* **41:** 987–997.

Peña J, Harris E. 2011. Dengue virus modulates the unfolded protein response in a time-dependent manner. *J Biol Chem* **286:**14226–14236.

Schubert U, Antón LC, Bacík I, Cox JH, Bour S, Bennink JR, Orlowski M, Strebel K, Yewdell JW. 1998. CD4 glycoprotein degradation induced by human immunodeficiency virus type 1 Vpu protein requires the function of proteasomes and the ubiquitin-conjugating pathway. *J Virol* **72:**2280–2288.

Stahl S, Burkhart JM, Hinte F, Tirosh B, Mohr H, Zahedi RP, Sickmann A, Ruzsics Z, Budt M, Brune W. 2013. Cytomegalovirus downregulates IRE1 to repress the unfolded protein response. *PLoS Pathog* **9:**e1003544. doi:10.1371/journal.ppat.1003544.

Wiertz EJ, Jones TR, Sun L, Bogyo M, Geuze HJ, Ploegh HL. 1996. The human cytomegalovirus US11 gene product dislocates MHC class I heavy chains from the endoplasmic reticulum to the cytosol. *Cell* **84:**769–779.

Transport of Viral RNA Genomes to Sites of Assembly

Amorin MJ, Bruce EA, Read EK, Foeglein A, Mahen R, Stuart AD, Digard P. 2011. A Rab11- and microtubule-dependent mechanism for cytoplasmic transport of influenza A virus RNA. *J Virol* **85:**4143–4156.

Chong LD, Rose JK. 1994. Interactions of normal and mutant vesicular stomatitis virus matrix proteins with the plasma membrane and nucleocapsids. *J Virol* **68:**441–447.

Martin K, Helenius A. 1991. Nuclear transport of influenza virus ribonucleoproteins: the viral matrix protein (M1) promotes export and inhibits import. *Cell* **67:**117–130.

O'Neill RE, Talon J, Palese P. 1998. The influenza virus NEP (NS2 protein) mediates the nuclear export of viral ribonucleoproteins. *EMBO J* **17:** 288–296.

Tang Y, Winkler U, Freed EO, Torrey TA, Kim W, Li H, Goff SP, Morse HC, III. 1999. Cellular motor protein KIF-4 associates with retroviral Gag. *J Virol* **73:** 10508–10513.

Zhou W, Parent LJ, Wills JW, Resh MD. 1994. Identification of a membrane-binding domain within the amino-terminal region of human immunodeficiency virus type 1 Gag protein which interacts with acidic phospholipids. *J Virol* **68:**2556–2569.

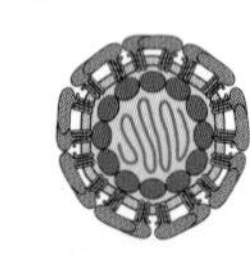

13 Assembly, Exit, and Maturation

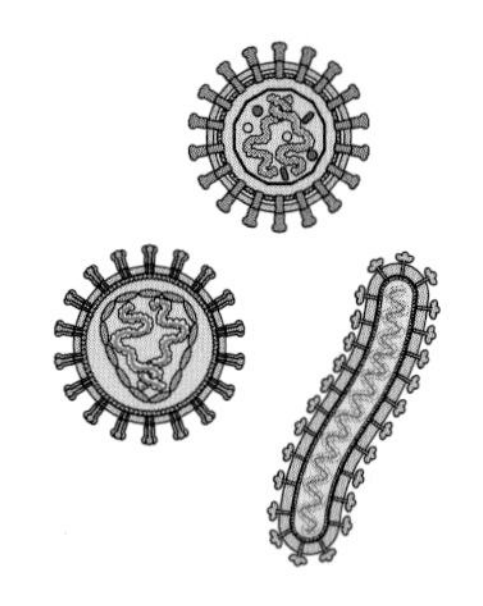

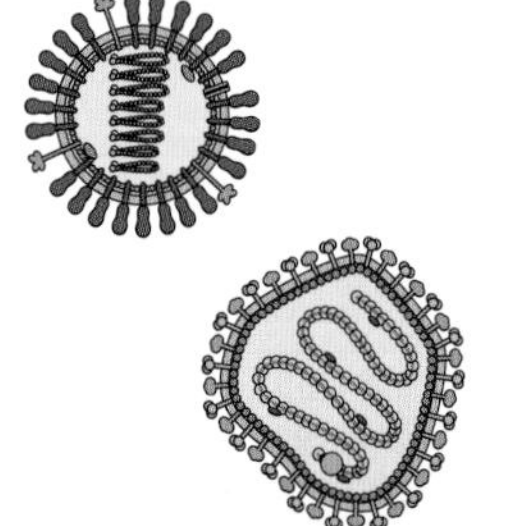

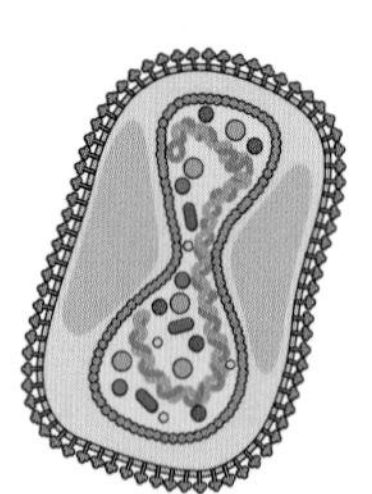

Introduction

Methods of Studying Virus Assembly and Egress
- Structural Studies of Virus Particles
- Visualization of Assembly and Exit by Microscopy
- Biochemical and Genetic Analyses of Assembly Intermediates
- Methods Based on Recombinant DNA Technology

Assembly of Protein Shells
- Formation of Structural Units
- Capsid and Nucleocapsid Assembly
- Self-Assembly and Assisted Assembly Reactions

Selective Packaging of the Viral Genome and Other Components of Virus Particles
- Concerted or Sequential Assembly
- Recognition and Packaging of the Nucleic Acid Genome
- Incorporation of Enzymes and Other Nonstructural Proteins

Acquisition of an Envelope
- Sequential Assembly of Internal Components and Budding from a Cellular Membrane
- Coordination of the Assembly of Internal Structures with Acquisition of the Envelope

Release of Virus Particles
- Assembly and Budding at the Plasma Membrane
- Assembly at Internal Membranes: the Problem of Exocytosis
- Release of Nonenveloped Viruses

Maturation of Progeny Virus Particles
- Proteolytic Processing of Structural Proteins
- Other Maturation Reactions

Cell-to-Cell Spread

Perspectives

References

LINKS FOR CHAPTER 13

- *Video: Interview with Dr. Wesley Sundquist*
  http://bit.ly/Virology_Sundquist
- *Movie 13.1: Active repulsion of vaccinia virus particles from infected cells*
  http://bit.ly/Virology_V1_Movie13-1
- *Covering up a naked virus*
  http://bit.ly/Virology_3-19-15
- *Cutting through mucus with the influenza virus neuraminidase*
  http://bit.ly/Virology_1-8-14

The probability of formation of a highly complex structure from its elements is increased, or the number of possible ways of doing it diminished, if the structure in question can be broken down in a finite series of successively smaller substrates!

J. D. BERNAL
in A. I. Oparin (ed.), *The Origins of Life on Earth*
(Pergamon, Oxford, United Kingdom, 1959)

Introduction

Virus particles exhibit considerable diversity in size, composition, and structural sophistication, ranging from those comprising a single nucleic acid molecule and one structural protein to complex structures built from many different proteins and other components. Nevertheless, successful reproduction of all viruses requires execution of a common set of *de novo* assembly reactions. These processes include formation of the structural units of the protective protein coat from individual protein molecules, assembly of the coat by interaction among the structural units, and incorporation of the nucleic acid genome (Fig. 13.1). In many cases, formation of internal virion structures must be coordinated with acquisition of a cellular membrane into which viral proteins have been inserted, or additional maturation steps must be completed to produce infectious particles. Assembly of even the simplest viruses is therefore a remarkable process that requires considerable specificity in, and coordination among, each of multiple reactions. In the extreme case of giant viruses, such as mimivirus and Pandoravirus, hundreds of proteins must interact appropriately with one another, with host cell membranes, and with the viral genome. Furthermore, virus reproduction is successful only if each of the assembly reactions proceeds with reasonable efficiency and if the overall pathway is irreversible. The diverse mechanisms by which viruses assemble represent powerful solutions to these problems associated with *de novo* assembly. Indeed, infectious virus particles are produced in prodigious numbers with great specificity and efficiency.

The architecture of a virus particle determines the nature of the reactions by which it is formed (Fig. 13.1). Despite variations in structure and biological properties, all virus particles must be well suited for protection of the nucleic acid genome in extracellular environments. They must also be metastable structures, that is, built in a way that allows their ready disassembly during entry into a new host cell. A number of elegant mechanisms resolve the apparently paradoxical requirements for very stable associations among virion components during assembly and transmission but the ready reversal of these interactions when appropriate signals are encountered upon infection of a host cell.

Like synthesis of viral nucleic acids and proteins, assembly of virus particles depends on host cell components, such as the cellular proteins that catalyze or assist the folding of individual protein molecules. Furthermore, the building blocks of virus particles are transported to the appropriate assembly site by cellular pathways (Chapter 12). Concentration of components of virus particles to a specific intracellular compartment or region undoubtedly facilitates virus production by increasing the rates of intermolecular assembly reactions. It is also likely to restrict the number of interactions in which particular components can engage, thereby increasing the specificities of these reactions.

The survival and propagation of a virus in a host population generally require dissemination of the virus beyond the cells initially infected. Progeny virus particles must therefore escape from the infected cell for transmission to new cells within the same host or to new hosts. The majority of viruses leave an infected cell by one of two general mechanisms: they are released into the external environment in various ways, or they are transferred directly from one cell to another.

PRINCIPLES *Assembly, exit, and maturation*

- Some structural units are formed from individual protein subunits; an alternative mechanism is assembly while individual protein coding sequences are covalently linked in a polyprotein precursor.
- The structural units of some protein shells assemble from individual units.
- An assembly line mechanism is well-suited for orderly formation of some virus particles.
- Accurate assembly of some large icosahedral protein shells requires scaffolding or chaperone proteins.
- Structural proteins contain the information necessary to specify assembly, but this process may be improved by the participation of cellular or viral chaperones.
- During encapsidation, viral genomes must be distinguished from cellular RNA or DNA and therefore often contain specific packaging signals.
- All viral genomes are packaged by one of two mechanisms, in conjunction with or following assembly of a protein shell (concerted or sequential encapsidation).
- Acquisition of an envelope by budding from the plasma membrane or internal membranes may be coordinated with, or follow, assembly of internal structures.
- Viral protein L domain sequences promote budding of enveloped viruses by recruitment of cellular proteins that participate in vesicular trafficking.
- When viral particles are assembled from polyproteins or precursor proteins, proteolytic cleavage by viral proteases is essential to produce infectious viral particles.
- Viruses may be released as free particles or spread from cell to cell without exposure to the extracellular milieu.

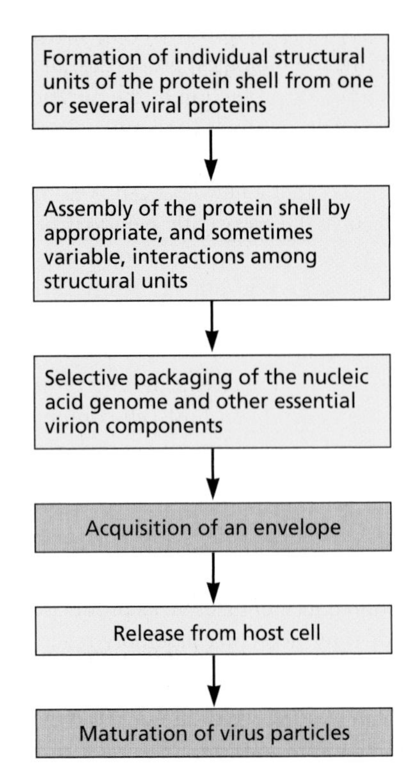

Figure 13.1 Hypothetical pathway of virus particle assembly and release. Reactions common to all viruses are shown in yellow, and those common to many viruses are shown in blue. The structural units that are often the first assembly intermediates are the homo- or hetero-oligomers of viral structural proteins from which virus particles are built (see Table 4.1). The arrows indicate a general sequence that applies to only some viruses. Packaging of the genome can be coordinated with assembly of the capsid or nucleocapsid, and for enveloped viruses, the assembly of internal components can be coordinated with acquisition of the envelope.

Methods of Studying Virus Assembly and Egress

Mechanisms of virus assembly and release can be understood only with the integration of information obtained by structural, biochemical, genetic, and imaging approaches. These methods are introduced briefly in this section.

Structural Studies of Virus Particles

The mechanisms by which virus particles form within, and leave, their host cells are intimately related to their structural properties. Our understanding of these processes therefore improves dramatically whenever the structure of a virus particle is determined. An atomic-level description of the contacts among the structural units that maintain the integrity of the particle identifies the interactions that mediate assembly and the ways in which these interactions must be regulated. For example, the X-ray crystal structure of the polyomavirus simian virus 40 described in Chapter 4 solved the enigma of how VP1 pentamers could be packed in hexameric arrays and identified three distinct modes of interpentamer contact. Assembly of the simian virus 40 capsid therefore must require specific variations in the ways in which pentamers associate, depending on their position in the capsid shell. Such subtle, yet sophisticated, regulation of the association of structural units was not anticipated and could be revealed only by high-resolution structural information.

Visualization of Assembly and Exit by Microscopy

While high-resolution structural studies of purified virus particles or individual proteins provide a molecular foundation for describing virus assembly, they offer no clues about how assembly (or exit) actually proceeds in an infected cell. Electron microscopy can be applied to investigation of these processes. Examination of thin sections of cells infected by a wide variety of viruses has provided important information about intracellular sites of assembly, the nature of assembly intermediates, and mechanisms of envelope acquisition and release of particles. This approach can be particularly useful when combined with immunocytochemical methods for identification of individual viral proteins or of the structures that they form via binding of specific antibodies attached to electron-dense particles of gold (Fig. 13.2A). More recently, intracellular viral structures and sites of assembly have been visualized by scanning electron and cryo-electron tomography (Chapter 4), which can capture three-dimensional information (Fig. 13.2B and C).

The labeling of viral proteins by fusion with green fluorescent protein or its derivatives (Chapter 2) (or of membranes with fluorescent lipophilic dyes) allows direct visualization of assembly and egress, an approach inconceivable even a few years ago. Such chimeric proteins and virus particles containing them can be observed in living cells, and their associations and movements can be recorded by video microscopy (see, for example, Chapter 12, Boxes 8 and 9). Consequently, these techniques overcome the limitations associated with traditional methods of microscopy, which provide only static views of populations of proteins or virus particles. On the other hand, the resolution that can be achieved by conventional fluorescence microscopy (<200 nm) is very low. Superresolution methods of fluorescence microscopy offer sufficiently improved resolution to visualize structural features of virus particle or assembly intermediates but have not yet been developed for routine live-cell imaging.

Biochemical and Genetic Analyses of Assembly Intermediates

Although of great value, the information provided by X-ray crystallography or microscopy is not sufficient to describe the dynamic processes of virus assembly and release; such understanding requires identification of the intermediates

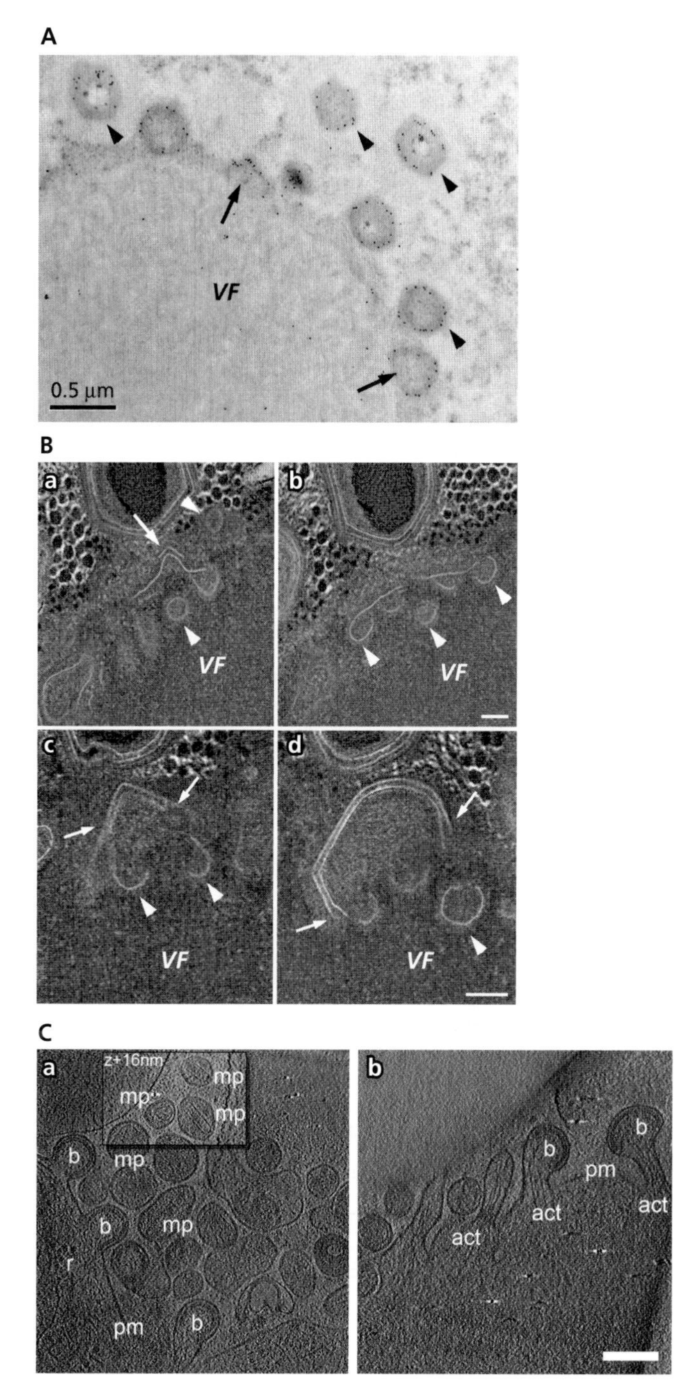

Figure 13.2 Examination of virus assembly by electron microscopy. Increasingly powerful methods of electron microscopy have been developed and applied to virus-infected cells. These approaches are illustrated using cells infected by mimivirus and human immunodeficiency virus type 1. **(A)** In immunoelectron microscopy, samples of infected cells suspended in a solid (but transparent) resin are sliced into sections, prior to reaction with gold-labeled antibodies against proteins of interest. This example shows 100- to 120-nm-thick sections of mimivirus-infected acanthamoeba and structures detected with antibodies that recognize the scaffolding protein of mimivirus (L425). This protein can be seen both in assembling capsids that form at the periphery of viral factories (VF), indicated by the arrows, and in closed capsids that are more distant from factors (arrowheads). **(B)** In scanning electron tomography, thick sections (250 to 400 nm in this example) of embedded samples are examined by electron microscopy at multiple tilt angles, and three-dimensional reconstructions are then computed from the images collected. Shown are 10-nm digital slices 40 nm apart derived from a tomogram of an amoeba infected with mimivirus for 8 h (a, b), in which can be seen viral factories and angular structures (arrows) forming on top of an open membrane sheet and surrounded by vesicles (arrowheads). By a later stage in assembly (c, d), angular structures (arrows) with truncated icosahedral symmetry (arrowheads) can be observed. These structures correspond to the scaffold protein-containing open structure shown in panel A. Adapted from Y. Mutsafi et al., *PLoS Pathog.* **9**(5):e1003367, 2013, doi:10.1371/journal.Ppat.1003367, with permission. Courtesy of A. Minsky, The Weitzmann Institute of Science, Israel. **(C)** In cryo-electron tomography, vitrified samples of infected cells are examined at a series of tilt angles and three-dimensional images are reconstructed (see Chapter 4). In this example, human cells with thin peripheral areas amenable to visualization by this method (≤500 nm) were infected with an adenovirus vector for expression of the human immunodeficiency virus type 1 Gag-Pol coding sequence. Shown are computational slices 1.6 nm in thickness through a cryo-electron tomogram with budding particles (b), mature particles (mp), the plasma membrane (pm), actin filaments (act), and ribosomes (r) indicated. The inset in the left panel is offset by 16 nm perpendicular to the image plane to show the morphology of mature particles with a discrete cone-shaped internal core. Adapted from L.-A. Carlson et al., *PLoS Pathog.* **6**(11):e1001173, 2010, doi:10.1371/journal.ppat.1001173, with permission. Courtesy of K. Grünewald, University of Oxford, United Kingdom.

and reactions in the pathway by which individual viral proteins and other components of virus particles are converted to mature infectious virus particles.

When extracts are prepared from the appropriate compartment of infected cells under conditions that preserve protein-protein interactions, a variety of viral assemblies can often be detected by techniques that separate them on the basis of mass and conformation (velocity sedimentation in sucrose gradients or gel filtration) or of density (equilibrium centrifugation). These assemblies range from the simplest structural units (see Table 4.1 for the definitions) to empty capsids and mature virus particles. Similar methods have identified various complexes formed by viral structural proteins in *in vitro* reactions. Furthermore, such structures can be organized into a sequence logical for assembly, from the least to the most complete. On the other hand, it is often quite difficult to **prove** that structures identified by these approaches, such as empty capsids, are true intermediates in the pathway.

By definition, the intermediates in any pathway do not accumulate unless the next reaction is rate limiting. For this reason, assembly intermediates are generally present within infected cells at low concentrations against a high background of the starting material (mono- or oligomeric structural proteins) and the final product (virus particles). This property makes it difficult to establish precursor-product relationships by pulse-chase experiments; the large pools of structural proteins initially labeled are converted only slowly and inefficiently into subsequent intermediates in the pathway. Genetic methods of analysis provide one powerful solution to this problem. Mutations that confer temperature sensitivity or other phenotypes that block a specific reaction have been invaluable in the elucidation of assembly pathways. A specific intermediate may accumulate in mutant-virus-infected cells and can often be purified and characterized more readily. Temperature-sensitive mutants can allow the reactions in a pathway to be ordered, and second-site suppressors of such mutations can identify viral proteins that interact with one another. Of even greater value is the combination of genetics with biochemistry, an elegant approach pioneered more than 35 years ago with the development of *in vitro* complementation for studies of the assembly of bacteriophage T4 (Box 13.1).

The difficulties inherent in kinetic analyses are compounded by the potential for formation of dead-end products and the unstable nature of some assembly intermediates. Dead-end assembly products are those that form by off-pathway (side) reactions. Because they are not true intermediates, they may accumulate in infected cells and be identified incorrectly as components in the pathway. Authentic intermediates by definition exist only transiently, and some may be fragile structures because they lack the complete set of intermolecular interactions that stabilize the virus particle. Less obvious is the conformational instability of some intermediates; such assemblies do not fall apart during isolation and purification but, rather, undergo irreversible conformational changes so that the structures studied experimentally do not correspond to **any** present in the infected cell. Such conformational change may well escape notice, as was initially the case for poliovirus empty capsids.

BOX 13.1

BACKGROUND

Late steps in T4 assembly

As illustrated, the head, tail, and tail fibers of this morphologically elaborate bacteriophage first form separately and then assemble with one another. The many genes encoding products that participate in building the T4 particle are listed by the reaction for which they are required. These gene products, and the order in which they act, were identified by genetic methods that included mapping of second-site suppressors of specific mutations (Chapter 3). The development of *in vitro* systems in which specific reactions were reconstituted was also of the greatest importance, allowing biochemical complementation. For example, noninfectious T4 particles lacking tail fibers accumulate in infected cells when the tail fiber pathway (right part of figure) is blocked by mutation. These incomplete particles can be converted to infectious phage when mixed *in vitro* with extracts prepared from cells infected with T4 mutated in the gene encoding the major head protein. The fact that the bacteriophages formed in this way were infectious established that assembly was accurate. This type of system was used to identify the genes encoding proteins that are required for assembly of heads or tails, as well as scaffolding proteins that are essential for assembly of the head, but not are present in the virus particle. Adapted from W. B. Wood, *Harvey Lect.* **73**:203–223, 1978, and W. B. Wood et al., *Fed. Proc.* **27**:1160–1166, 1968, with permission.

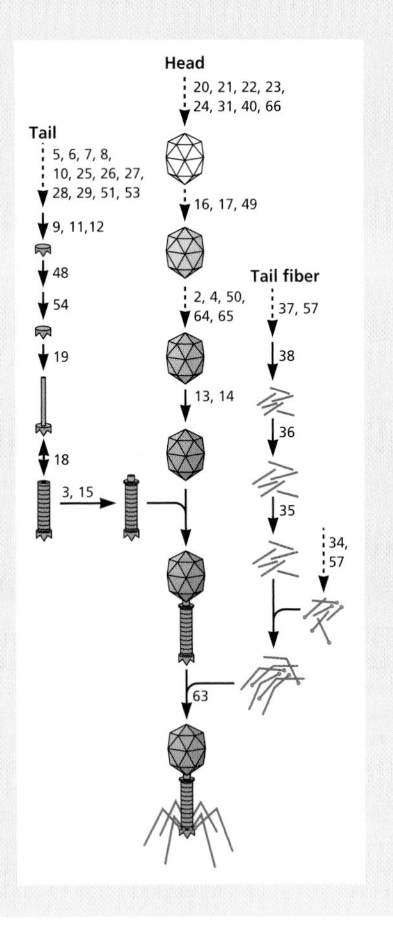

Methods Based on Recombinant DNA Technology

Modern methods of molecular biology and the application of recombinant DNA technology have greatly facilitated the study of virus assembly. Especially valuable is the simplification of this complex process, that can be achieved by the synthesis of an individual viral protein or small sets of proteins in the absence of other viral components (Box 13.2).

Assembly of Protein Shells

Although virus particles are far simpler in structure than any cell, they are built from multiple components, such as a capsid, a nucleoprotein core containing the genome, and a lipid envelope carrying viral glycoproteins. The first steps in assembly are therefore the formation of the various components of virus particles from their parts. To complete the construction of the virus particle, these intermediates must then associate in ordered fashion, in some cases after transport to the appropriate intracellular site. Application of the techniques described in the previous section has allowed us to delineate the pathways by which many viruses are assembled and to describe some specific reactions in exquisite detail. In this section, we draw on this large body of information to illustrate mechanisms for the efficient assembly of protective protein coats for genomes, the first reaction listed in Fig. 13.1.

Formation of Structural Units

In some cases, notably assembly of (−) strand RNA viruses, structures built entirely from proteins do not accumulate because fabrication of a protein shell is coordinated with binding of structural proteins to the viral genome. In other cases, the first assembly reaction is the formation of the structural units from which the capsid is constructed (Fig. 13.1). This process is relatively simple: individual structural units contain a small number of protein molecules, typically two to six, that must associate appropriately following (or during) their synthesis. Nevertheless, structural units are formed by several different mechanisms, and in some cases additional proteins are required to assist the reactions (Fig. 13.3).

Assembly from Individual Proteins

The structural units of some protein shells, including the VP1 pentamers of simian virus 40, assemble from their individual protein components (Fig. 13.3A). This straightforward mechanism is analogous to the formation of cellular structures containing multiple proteins, such as nucleosomes. In this kind of reaction, the surfaces of individual protein molecules that contact other molecules of either the same protein or a different protein are formed prior to assembly of the structural unit. This mechanism facilitates specific binding when appropriate protein molecules encounter one another; no energetically costly conformational change is required, and subunits that come into contact can simply interlock. Production of these structural units generally can be reconstituted *in vitro* or in cells that synthesize the component proteins. Such experiments confirm that all information necessary for accurate assembly is contained within the primary sequence and, hence, the folded structure of the protein subunits. On the other hand, the individual protein subunits must find one another in a dense intracellular environment in which the concentration of irrelevant (cellular) proteins is very high (20 to 40 mg/ml). Such a milieu offers opportunities for nonspecific binding of viral proteins to unrelated cellular proteins. This problem can

BOX 13.2

METHODS

Assembly of herpes simplex virus 1 nucleocapsids in a simplified system

The assembly and egress of herpesviruses from infected cells are complicated processes that comprise multiple steps (Fig. 13.8 and 13.23). To facilitate analysis of the initial reactions that lead to assembly of the protein shell, the viral genes that encode the proteins of the nucleocapsid were introduced into baculovirus vectors. Formation of the nucleocapsid was examined by electron microscopy of insect cells infected with various combinations of the recombinant baculoviruses. Empty capsids indistinguishable from those formed in herpes simplex virus 1-infected cells were observed when six viral genes were expressed together. Four of these encode the structural proteins VP5 (hexons and pentons), VP19C and VP23 (triplexes that link VP5 structural units), and VP26 (which caps hexons of VP5). By omission of individual recombinant baculoviruses, it was shown that VP26 is not necessary for nucleocapsid assembly. Furthermore, only partial or deformed structures assemble in the absence of VP24, VP21, and VP22a, the protease and scaffolding proteins (see "Viral Scaffolding Proteins: Chaperones for Assembly").

Tatman JD, Preston VG, Nicholson P, Elliot RM, Rixon FJ. 1994. Assembly of herpes simplex virus 1 capsids using a panel of recombinant baculoviruses. *J Gen Virol* **75**:1101–1113.

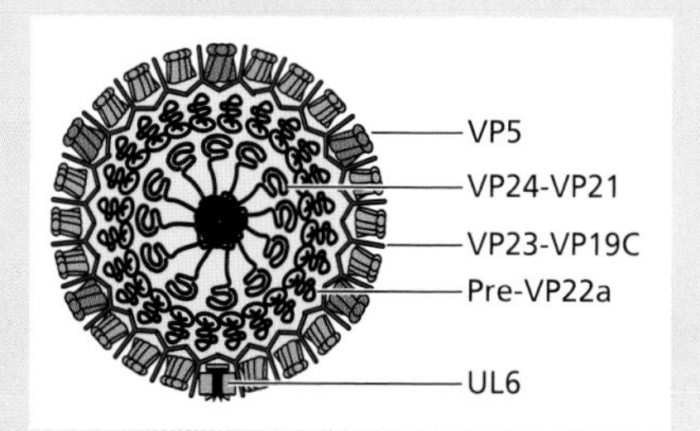

Herpes simplex type 1 procapsid showing the capsid proteins (VP5, VP23, VP19C), the portal (UL216), and the proteins that facilitate assembly and subsequent maturation (VP24, VP21 and VP22a).

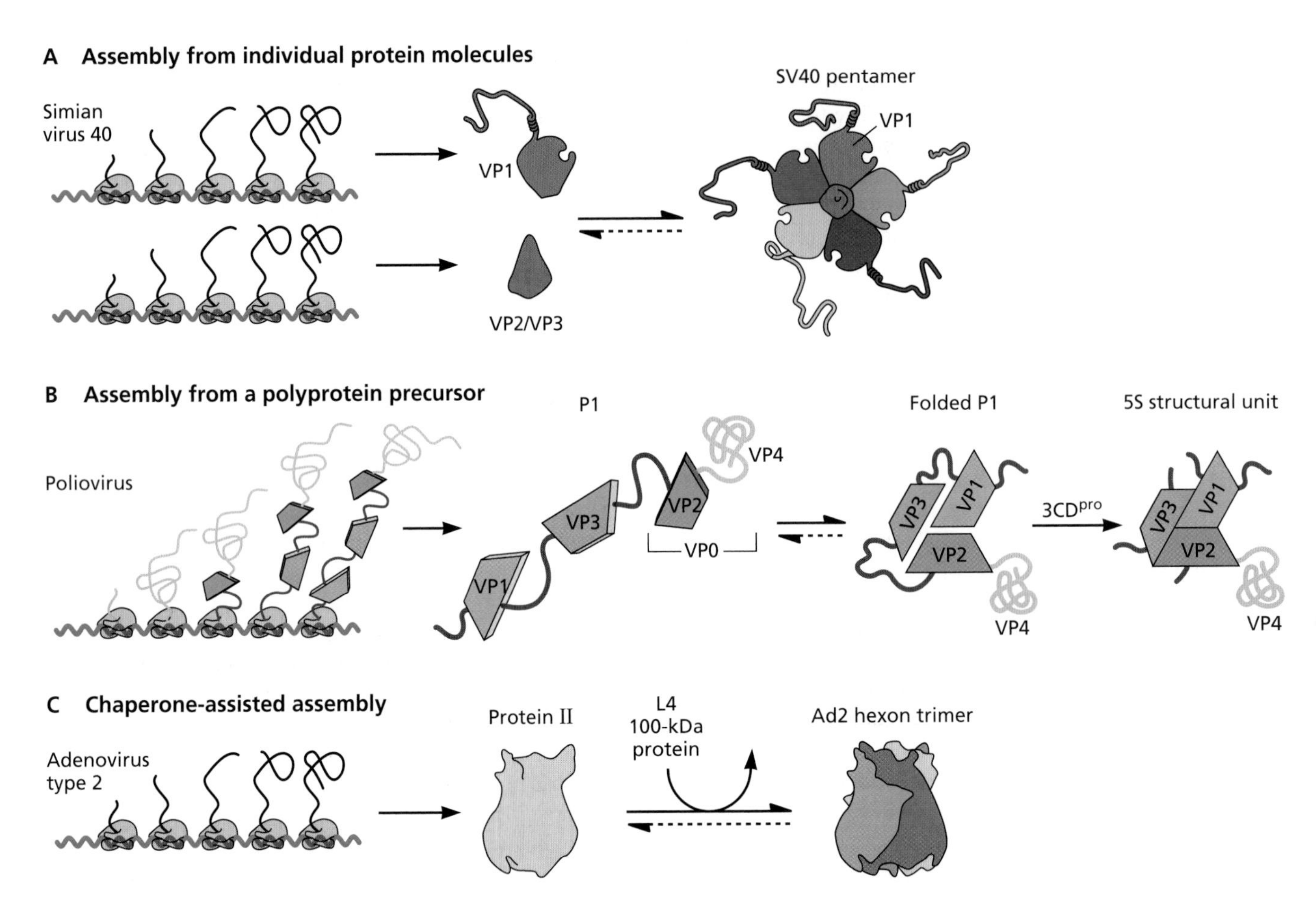

Figure 13.3 Mechanisms of assembly of viral structural units. **(A)** Assembly from folded protein monomers, illustrated with simian virus 40 (SV40) VP1 pentamers. The assembly reaction is the result of specific interactions among the proteins that form structural units. In many cases, the interactions have been described at atomic resolution (Chapter 4). These assembly reactions are driven in a forward direction by the high concentrations of protein subunits synthesized in infected cells, as indicated by the solid arrows. Other structural units that assemble in this way are the adenoviral fiber (trimer of protein IV) and penton (pentamer of protein III), the hepatitis B virus capsid **(C)** protein dimer, and the s3-μ1c hetero-oligomers of the outer capsid shell of reovirus. **(B)** Assembly from a polyprotein precursor, illustrated with the poliovirus polyprotein that contains the four proteins that form the heteromeric structural unit. The latter proteins are synthesized as part of the single polyprotein precursor from which all viral proteins are produced by proteolytic processing. For simplicity, only the P1 capsid protein precursor and its cleavage by the viral 3CD protease following the folding and assembly of the immature structural unit (VP0, VP3, and VP1) are shown. The flexible covalent connections between VP1, VP3, and VP0 in the P1 precursor, which are exaggerated for clarity, are severed by the protease to form the 5S structural unit. However, VP4 remains covalently linked to VP2 in VP0 until assembly is completed (see the text). **(C)** Assisted assembly. Some structural units are assembled only with the assistance of viral chaperones, such as the adenoviral L4 100-kDa protein, which is required for formation of the hexon trimer from the protein II monomer. Similarly, the herpes simplex virus 1 VP22a protein assists in the assembly of VP5 pentamers and hexamers.

be circumvented by both the synthesis of viral structural proteins in quantities far in excess of those incorporated in virus particles and their accumulation at specialized assembly sites, common features of virus-infected cells. Such high concentrations not only increase the probability that viral proteins will encounter one another by random diffusion but also provide a sufficient reservoir to compensate for any loss by nonspecific binding to cellular components. Another benefit of high protein concentration is that the formation of structural units proceeds efficiently (Fig. 13.3A), driving the assembly pathway in the productive direction.

Assembly from Polyproteins

An alternative mechanism for the production of structural units is assembly while covalently linked in a polyprotein precursor. This mechanism, exemplified by formation of picornaviral capsids, circumvents the need for protein subunits to meet by random diffusion and avoids competition from nonspecific binding reactions. The first poliovirus intermediate, which sediments as a 5S particle, is the immature structural unit that contains one copy each of VP0, VP3, and VP1 (Fig. 13.3B). It is thought that folding of their central β-barrel domains (Fig. 4.11) takes place during synthesis of their precursor (P1). The poliovirus structural unit can then form by intramolecular interactions among the surfaces of these β-barrel domains, before the covalent connections that link the proteins are severed by the viral $3CD^{pro}$ protease.

Retrovirus assembly illustrates an elegant and effective variation on the polyprotein theme. Mature retrovirus particles contain three protein layers. An inner coat of NC protein, which packages the dimeric RNA genome, is enclosed within the capsid built from the CA protein. The capsid is in turn surrounded by the MA protein, which lies beneath the inner surface of the viral envelope (see Appendix Fig. 30A). These three structural proteins are synthesized as the Gag polyprotein precursor, which contains their sequences in the order of the protein layers that they form in virus particles, with MA at the N terminus (Fig. 13.4). Retrovirus particles assemble from such Gag polyprotein molecules by a unique mechanism that allows orderly construction of the three protein layers and, as we shall see, coordination of this reaction with encapsidation of the genome and acquisition of the envelope.

Figure 13.4 Radial organization of the Gag polyprotein in immature human immunodeficiency virus type 1 particles. The model for the arrangement of the Gag polyprotein shown to the right of the cryo-electron micrograph of a virus-like particle assembled from Gag was deduced from radial density measurements of digitized images of the particles. The plot indicates density as a function of distance from the particle center, in angstroms. Courtesy of T. Wilk, European Molecular Biology Laboratory.

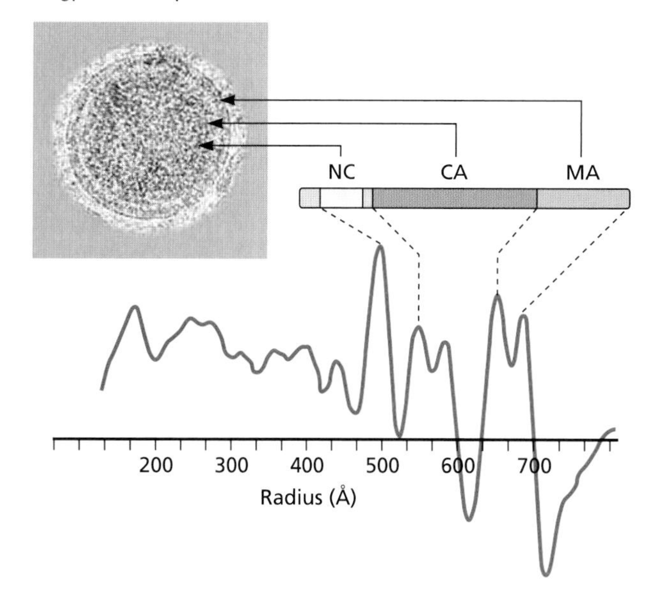

Participation of Cellular and Viral Chaperones

Chaperones are specialized proteins that facilitate the folding of other proteins by preventing improper, nonspecific associations among sticky patches exposed on nascent and newly synthesized proteins. The first chaperone to be identified, the product of the *Escherichia coli groEL* gene, was discovered because it is essential for reproduction of bacteriophages T4 and lambda (Fig. 13.5A). The participation of chaperones resident in the lumen of the endoplasmic reticulum (ER) in folding and assembly of oligomeric viral glycoproteins is well established (Chapter 12). Cytoplasmic and nuclear chaperones are probably equally important for the formation of structural units or later reactions in virus assembly. A number of viral structural proteins have been shown to interact with one or more cellular chaperones, but in most cases, a role for these proteins in viral assembly is based on "guilt by association." However, some cellular chaperones have been directly implicated in assembly reactions (Fig. 13.5). For example, association of molecules of the Gag protein of the betaretrovirus Mason-Pfizer monkey virus with one another and accumulation of capsids depend upon interaction of Gag with the cytoplasmic chaperone TriC, which facilitates proper folding of the polyprotein.

Chaperones are abundant in all cells, and some accumulate to concentrations even greater than those of the very numerous ribosomes. Nevertheless, the genomes of several viruses encode proteins with chaperone activity, some with sequences and functions homologous to those of cellular proteins (Table 13.1). Some viral chaperones are essential participants in the reactions by which structural units are formed. For example, assembly of adenoviral hexon trimers, which form the faces of the icosahedral capsid, depends on such an accessory protein, the viral L4 100-kDa protein (Fig. 13.3C).

Capsid and Nucleocapsid Assembly

The accumulation of viral structural units within the appropriate compartment of an infected cell sets the stage for the assembly of more-elaborate capsids or nucleocapsids (see Table 4.1 for nomenclature). For reasons discussed previously, the reactions by which these structures are formed are often not understood in detail. Nevertheless, several different mechanisms for their assembly can be distinguished.

Intermediates in Assembly

A striking feature of well-characterized pathways of bacteriophage assembly (Box 13.1) is the sequential formation of progressively more elaborate structures; heads, tails, and tail fibers are each assembled in stepwise fashion via defined intermediates. Such an assembly line process appears ideally suited for orderly formation of virus particles, which can be large and architecturally intricate. Discrete intermediates also form during the assembly of some icosahedral animal viruses.

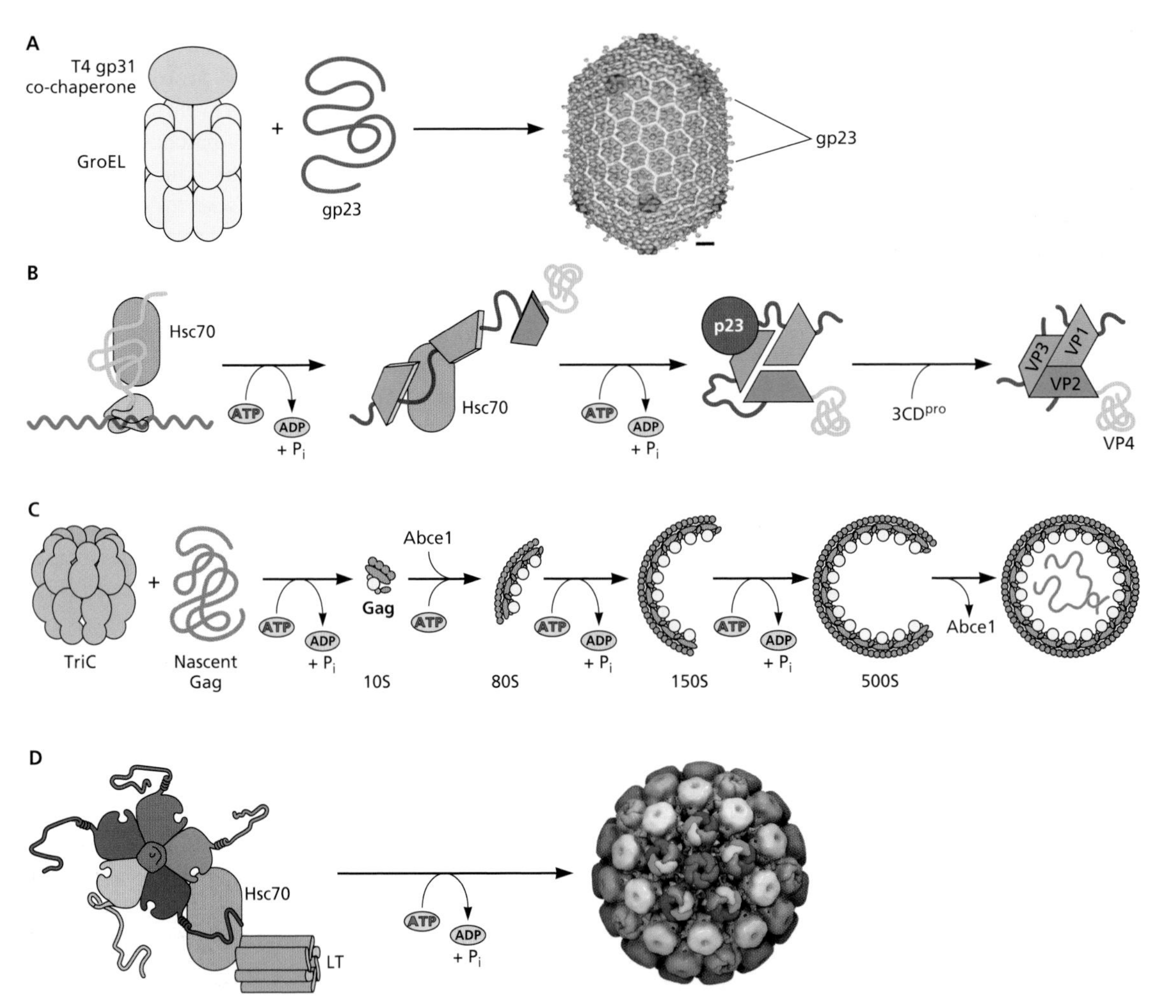

Figure 13.5 Some assembly reactions assisted by cellular chaperones. (A) The *E. coli* chaperone GroEL, which comprises two rings of eight identical subunits, promotes folding of the bacteriophage T4 major coat protein (gp23) and its assembly to form the prohead (Box 13.1). GroEL normally functions in concert with the co-chaperone GroES, but the bacteriophage protein gp31 replaces GroES for the folding of gp23. GroEL is also necessary for reproduction of bacteriophage λ. Adapted from A. Fokine et al., *Proc. Natl. Acad. Sci. U. S. A.* **101:**6003–6008, 2004, with permission. Courtesy of M. Rossmann, Purdue University. **(B)** The P1 polyprotein precursor of poliovirus (and other picornaviruses) associates with the host cell cytoplasmic chaperone Hsc70. This chaperone promotes productive folding in cycles of binding to, and release from, hydrophobic patches in the nascent protein, governed by ATP binding and hydrolysis. The interaction of P1 with a second chaperone, Hsp90, and its co-chaperone p23 is required for cleavage of P1 by the viral $3C^{pro}$ protease. It is thought that Hsp90 promotes the folding of P1 to a conformation that allows recognition of the cleavage sites for the viral enzyme. **(C)** The contribution to the folding of Gag of the cytoplasmic chaperone TriC, a ring-like structure built from eight different subunits, is based on studies of Mason-Pfizer monkey virus (see the text). Whether TriC also facilitates folding of Gag proteins of other retroviruses is not known. However, Gag proteins of several primate retroviruses, including human immunodeficiency virus type 1, associate with the cellular protein ATP-binding cassette sub-family E member 1 (Abce1) *in vitro* and in cells in which the viral protein is produced. Abce1 interacts with several intermediates (80S, 150S, and 500S) in the assembly of virus-like particles but not with this final product of assembly. The Abce1 protein contains ATP-binding domains, and depletion of ATP from Gag-producing cells leads to the accumulation of the 80S and 150S assembly intermediates. These observations suggest that Abce1 serves as a chaperone or scaffolding protein for assembly of retrovirus particles. **(D)** The VP1 and VP3 proteins of the polyomavirus simian virus 40 form pentamers efficiently *in vitro* or when made in *E. coli*. Such pentamers assemble into capsid-like particles but only when incubated with ATP, the cellular chaperone Hsc70, and the viral large T antigen (LT) (or with the bacterial chaperones GroEL and -ES). The viral protein contains a J domain, which is similar in sequence to a specific domain in cellular chaperones of the DnaJ family, and like these cellular proteins, LT stimulates the ATPase activity of Hsp70 chaperones. This N-terminal domain of LT is also present in the small T antigen (sT). LT and sT are associated with VP1 and the cellular chaperone during the late phase of infection, but the folding or assembly reactions that they assist have not been identified. The image of the simian virus particle was created by Jason Roberts, Doherty Institute, Melbourne, Australia.

Table 13.1 Some viral chaperones and scaffolding proteins

Protein	Properties and/or function(s)
Chaperones	
Adenovirus type 2 L4 100-kDa protein	Formation of hexon trimmers
African swine fever virus CAP80	Productive folding of the major capsid protein p73
Herpes simplex virus 1 VP22a	Formation of VP5 pentamers
Simian virus 40 LT antigen	N-terminal J domain necessary for assembly of virus particles; binds to and stimulates the activities of cellular Hsc70 proteins
Scaffolding proteins	
Adenovirus type 5 L1 52/55-kDa proteins	Necessary for formation of capsids; present in immature, but not mature, particles; may be required for encapsidation
Adenovirus-associated virus type 2 assembly-activating protein (AAP)	Interacts with common C terminus VP1, VP2, and VP3; promotes capsid assembly; forms high-molecular-weight oligomers
Herpes simplex virus 1 VP22a	Forms a scaffold-like structure that organizes assembly of the empty nucleocapsid

A stepwise assembly mechanism has been well characterized for poliovirus; the 5S structural unit described in the previous section is the immediate precursor of a 14S pentamer, which in turn is incorporated into virus particles in a two-step process (Fig. 13.6). The pentamer is stabilized by extensive protein-protein contacts and by interactions mediated by the myristate chains present on the five VP0 N termini (Fig. 4.12C). The contribution of the lipids to pentamer stability is so great that this structure does not form at all when myristoylation of VP0 is prevented. Formation of the very stable 14S assembly intermediate is irreversible under normal conditions, a property that imposes the appropriate directionality on the entire assembly pathway (Fig. 13.6).

Discrete assembly intermediates like the poliovirus pentamer have been difficult to identify in cells infected by many viruses. In some cases, the absence of intermediates can be attributed to coordination of assembly of protein shells with binding of the structural proteins to the nucleic acid genome. This mode of assembly is exemplified by the ribonucleoproteins of (−) strand RNA viruses, which assemble as genomic RNA is synthesized. Nucleocapsid formation depends on interactions of the protein components with both the nascent RNA and other protein molecules previously bound to the RNA.

Methods that permit the synthesis of subsets of structural proteins have begun to provide insights into how such ribonucleoproteins assemble. The vesicular stomatitis virus N protein, which is a dimer in the helical nucleocapsid, aggregates when synthesized alone in *E. coli*. However, when the viral P protein is also made, aggregation does not occur, and discrete, disk-like oligomers assemble. The assembly contains 10 molecules of the N protein, 5 molecules of the P protein, and an RNA molecule (of bacterial origin) of some 90 nucleotides (Fig. 4.6). The disk-like oligomer is equivalent to one turn of the ribonucleoprotein helix formed in vesicular stomatitis virus-infected cells. No further assembly takes place in bacterial cells, perhaps because the viral proteins cannot be posttranslationally modified in the appropriate fashion. However, N-RNA complexes purified from virus particles or insect cells synthesizing N are competent to form bullet-shaped structures with the morphology of the virion ribonucleoprotein (Fig. 13.7; compare to Fig. 4.6). This property indicates that the interactions of N protein molecules with one another and with RNA are sufficient to specify assembly of the helical ribonucleoprotein.

Assembly of the protein shells of many enveloped viruses, including retroviruses, is coordinated with binding of structural proteins to a cellular membrane. This property makes isolation of intermediates a technically demanding task. Nevertheless, new methods for separation of intermediates make it possible to examine assembly reactions of these viruses. Some assembly reactions can also be studied by using simplified experimental systems. When synthesized in a cell-free transcription-translation system, the human immunodeficiency virus type 1 Gag protein multimerizes through a series of discrete intermediates to form 750S particles (Fig. 13.5C) that resemble virus-like particles released when Gag is the only viral gene expressed in mammalian cells. These observations illustrate the power of simplified approaches to the study of virus assembly. An important caveat is that such experimental systems must faithfully reproduce reactions that take place within infected cells. There is good reason to conclude that the *in vitro* assembly of Gag particles meets this crucial criterion: the assembly phenotypes exhibited by altered Gag proteins *in vitro* correspond closely to those observed in infected cells, and binding of Gag to the cellular chaperone Abcel is required for the assembly of later intermediates in both cases.

Self-Assembly and Assisted Assembly Reactions

The primary sequences of viral structural proteins contain all the information necessary to specify assembly, including intricate reactions like the alternative 5- and 6-fold packing of VP1 pentamers in the simian virus 40 capsid; when synthesized in *E. coli*, VP1 is isolated as pentamers that assemble into capsid-like structures *in vitro*. Such self-assembly of structural proteins is the primary mechanism for formation of protein shells, but other viral components or cellular proteins can assist the process.

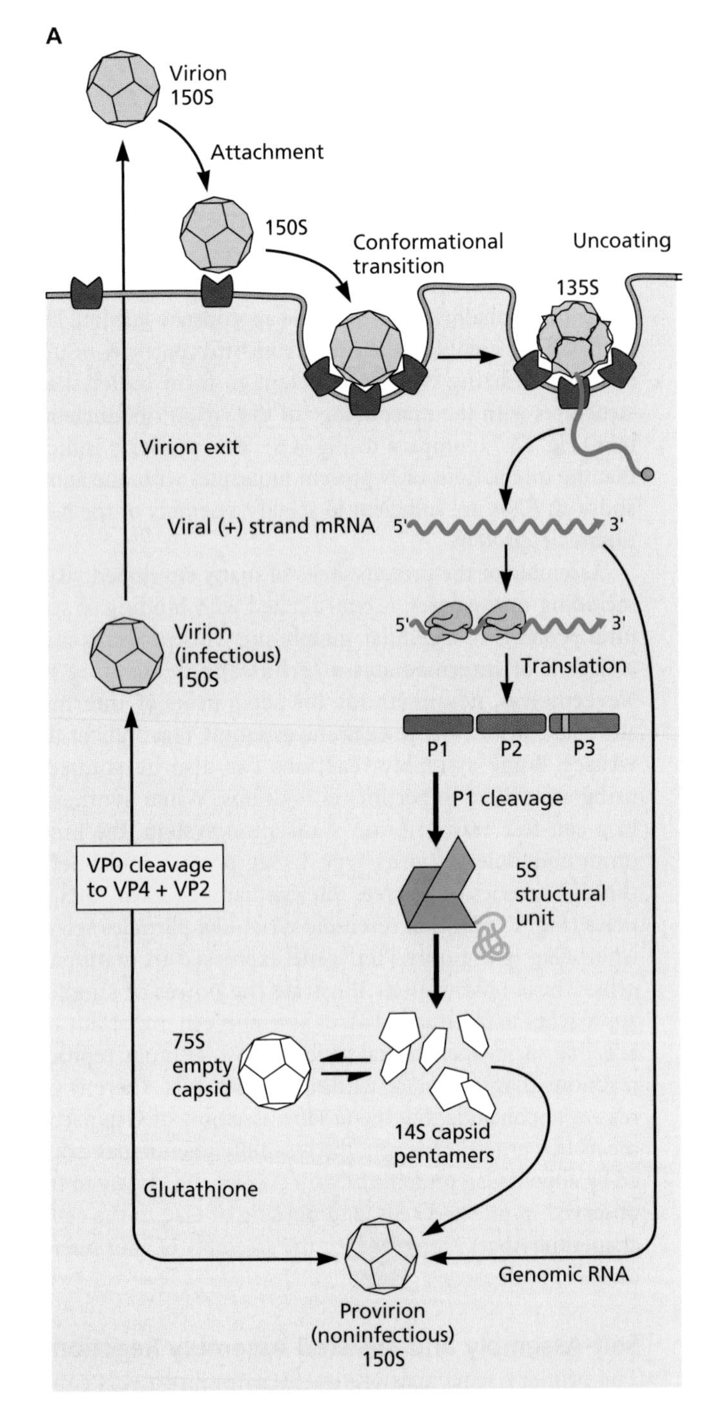

B

Figure 13.6 Assembly of poliovirus in the cytoplasm of an infected cell. (A) Most of the assembly reactions are essentially irreversible, because of proteolytic cleavage (formation of 5S structural units and mature virus particles) or extensive stabilizing interactions in the assembled structure (formation of 14S pentamers and of provirions). Stable, empty capsids, originally considered the precursors of provirions, do not possess the same conformation as the mature virus particle, as symbolized by the white color, and are dead-end products. Formation of the capsid shell from 14S pentamers is coordinated with genome encapsidation and requires replication of genomic RNA. The conformational transition upon attachment to the poliovirus receptor, for which the virus particle is primed by cleavage of VP0 to VP2 and VP4, is also illustrated. Some evidence for this mechanism is summarized in the text. In addition, in a cell-free system for the synthesis of infectious poliovirus particles, exogenously added 14S pentamers assemble with newly synthesized viral (+) strand RNA to form virus particles with antigenic sites characteristic of those produced in infected cells. In contrast, exogenously added empty capsids undergo no further assembly, even when genomic RNA is synthesized, confirming that they are dead-end products. **(B)** The sites of substitutions that render the reproduction of poliovirus resistant to depletion of glutathione are shown as colored spheres on one structural unit of a pentamer, in which VP1, VP2, and VP3 are colored blue, green, and red, respectively. The other structural units are shown in gray. Most of these substitutions lie at the interface between adjacent structural units. Adapted from H.-C. Ma et al., *PLoS Pathog.* **10:**e1004052, doi:10.1371/journal.ppat.1004052, 2014, with permission. Courtesy of P. Jiang and E. Wimmer, State University of New York, Stonybrook, MY.

Viral and Cellular Components That Regulate Self-Assembly

Interactions among viral structural proteins may be the mortar for the construction of virus particles, but other components of the particle often provide an essential foundation or the blueprint for correct assembly. As we have seen, assembly of the nucleocapsids of (−) strand RNA viruses is both coordinated with and dependent on synthesis of genomic RNA. The RNA serves as a template for productive and repetitive binding of nucleocapsid proteins to one another. Interactions of retroviral Gag proteins with RNA mediated by the NC RNA-binding domain also appear to be essential to initiate assembly of the Gag protein shell (Box 13.3). In other cases, the viral genome plays a more subtle yet equally important

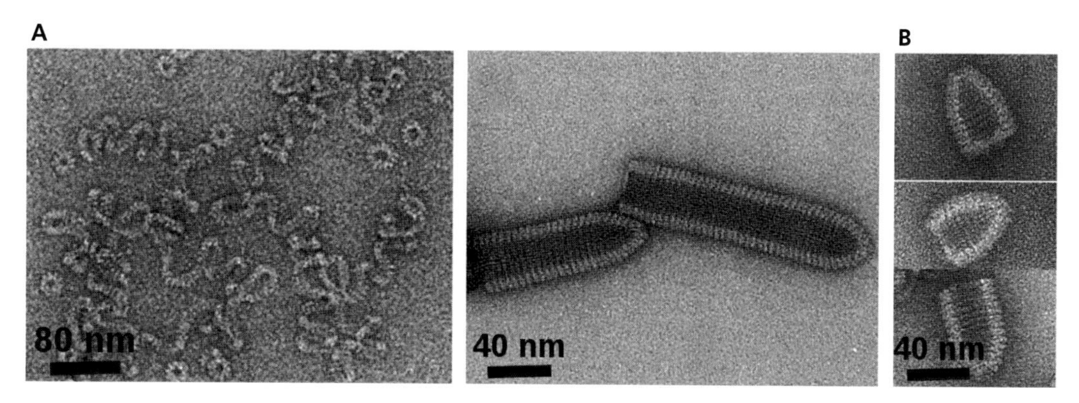

Figure 13.7 Formation of bullet-shaped particles by the vesicular stomatitis virus N protein. (A) At neutral pH and physiological ionic strength, N-RNA complexes purified from vesicular stomatitis virus particles form loosely coiled ribbons (left), but at lower pH and ionic strength, they assemble as bullet-shaped particles (right). These particles exhibit the morphology of the ribonucleoprotein in virus particles but vary in diameter. **(B)** The formation of bullet-shaped particles by the N protein synthesized in insect cells in the absence of other viral proteins and bound to cellular RNA indicates that this viral protein contains the information to specify assembly of these nonuniform structures. Adapted from A. Derfosses et al., *Nat. Commun.* **4**:1429, doi:10.1038/ncomms2435, 2013, with permission. Courtesy of I. Gutsche, UJF-EMBL-CNRS, Grenoble, France.

BOX 13.3

DISCUSSION

A scaffolding function for RNA

When synthesized in the absence of any other viral component, retroviral Gag polyproteins direct assembly and release of the virus-like particles shown in the figure. It was therefore assumed for many years that this protein contains all the information necessary and sufficient for assembly of particles. However, the results of subsequent experiments indicate that RNA acts as a scaffold during Gag assembly.

In vitro studies of the ability of truncated Gag proteins to multimerize with the full-length protein initially underscored the importance of the nucleocapsid (NC) RNA-binding domain for efficient assembly. The association of Gag with RNA is also required for multimerization in this system. The apparent contradiction between these findings and efficient assembly of Gag in mammalian cells in the absence of genomic RNA was subsequently resolved; virus-like particles contain cellular RNAs when they form in cells infected by a Moloney murine leukemia virus mutant with a deletion in the signal that directs packaging of the RNA genome. Furthermore, RNase digestion of cores assembled from Gag in wild-type Moloney murine leukemia virus-infected cells was shown to dissociate these structures. These observations indicate that interactions of Gag molecules with RNA, as well as with one another, are required for assembly and to maintain particle stability.

Campbell S, Vogt VM. 1995. Self assembly in vitro of purified CA-NC proteins from Rous sarcoma virus and human immunodeficiency virus type 1. *J Virol* **69:**6487–6497.

Muriaux D, Mirro J, Harvin D, Rein A. 2001. RNA is a structural element in retrovirus particles. *Proc Natl Acad Sci U S A* **98:**5246–5251.

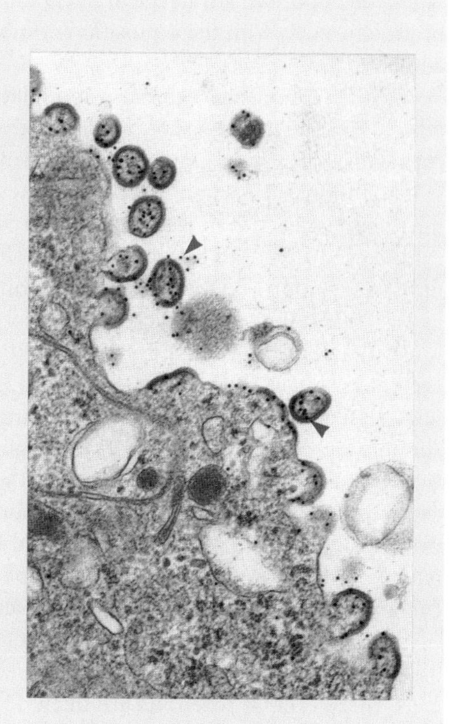

Electron micrograph showing a thin section (fixed and stained) of a human T cell synthesizing the viral Gag polyproteins. Prior to electron microscopy, viral particles (red arrowheads) were labeled with polyclonal antibodies (attached to gold beads) recognizing the CA protein. Bar, 1.0 μm. N, nucleus; M, mitochondrion. Courtesy of J. J. Wang, Institute of Biomedical Sciences, Academica Sinica, Taipei, Taiwan, and B. Horton and L. Ratner, Washington University School of Medicine, St. Louis, MO.

role, ensuring that the interactions among structural units are those necessary for infectivity. For example, poliovirus empty capsids lack internal structural features characteristic of the mature virus particle, because VP0 is not cleaved to form VP4 and VP2. The RNA genome is thought to participate in the autocatalytic cleavage of this precursor, which is essential for the production of infectious particles. Association of structural proteins with a cellular membrane is essential for the assembly of some virus particles, a situation exemplified by many retroviruses; the sequences of MA that specify Gag myristoylation and binding to the cytoplasmic surface of the plasma membrane (described in Chapter 12) are also required for assembly.

Binding of structural proteins to the genome or to a cellular membrane might simply raise their local concentrations sufficiently to drive self-assembly, might organize the proteins in such a way that their interactions become cooperative, or might induce conformational changes necessary for the productive association of structural units. These mechanisms, which are not mutually exclusive, have not been distinguished experimentally, but there is evidence for induction of conformational transitions in specific cases. We do not understand adequately the molecular mechanisms by which binding of structural proteins to other components directs or regulates particle assembly. However, such a requirement offers the important advantage of integration of the formation of protein shells with the acquisition of other essential virion components.

Cellular components can also modulate the fidelity with which viral structural proteins bind to one another. The capsid-like structures assembled when simian virus 40 VP1 is made in insect or mammalian cells are much more regular in appearance than those formed *in vitro* by bacterially synthesized VP1. Modification of VP1 (by acetylation and phosphorylation) or the participation of chaperones, such as Hsc70 and the J domain of the viral large T antigen (LT), must therefore improve the accuracy with which VP1 pentamers associate to form capsids (Fig. 13.5D). Similarly, *in vitro* self-assembly of poliovirus structural proteins is very slow, proceeding at least 2 orders of magnitude more slowly than that observed in infected cells. Furthermore, the empty capsids that form have the altered conformation described previously, unless the reaction is seeded by 14S pentamers isolated from infected cells. This property indicates that the appropriate folding, modification, and/or interactions of the viral structural proteins are critical for subsequent assembly reactions to proceed productively. Within infected cells, these crucial reactions are likely to be modulated by cellular chaperones, such as Hsc70, which is associated with the polyprotein during its folding to form 5S structural units. Remarkably, a small host cell molecule, glutathione, also facilities assembly of poliovirus (Fig. 13.6); depletion of glutathione reduces both virus yield and the accumulation of pentamers. This molecule binds to VP1 and VP3, and mutations that confer resistance to glutathione depletion carry amino acid substitutions at the interfaces between adjacent structural units in the pentamer, consistent with stabilization of this intermediate upon binding of glutathione. It is clear from these examples that host cells provide a hospitable environment for productive virus assembly, one that is not necessarily reproduced when viral structural proteins are assembled *in vitro*.

Viral Scaffolding Proteins: Chaperones for Assembly

Accurate assembly of some large icosahedral protein shells, such as those of adenoviruses and herpesviruses, is mediated by proteins that are not components of mature virus particles (Table 13.1). Because these proteins participate in reactions by which the capsid or nucleocapsid is constructed but are then removed, they are termed **scaffolding proteins**. Among the best characterized is the precursor of the herpes simplex virus 1 VP22a protein.

This protein is the major component of an interior core present in assembling nucleocapsids (Fig. 13.8A). In the absence of other viral proteins, it forms specific scaffold-like structures and appears as an ordered sphere in immature nucleocapsids isolated from infected cells. Self-association of pre-VP22a stimulates binding to VP5, the protein that forms the hexameric and pentameric structural units of the nucleocapsid. The interactions of VP5 with the scaffolding protein guide and regulate the intrinsic capacity of VP5 hexamers (and other nucleocapsid proteins) for self-assembly; omission of the scaffolding protein from a simplified assembly system (Box 13.2) leads to the production of partial and deformed nucleocapsid shells.

One of the 12 vertices of the herpesviral nucleocapsid comprises not a VP5 pentamer but rather the portal through which the DNA enters (Fig. 13.8A; see also Fig. 4.29). This unique structural unit, a dodecamer of the UL6 protein, must be incorporated at just one vertex during assembly. The reaction requires interaction of the portal with the major scaffolding protein; a small molecule that blocks the interaction prevents assembly of portal-containing nucleocapsids in infected cells. Although the portal is dispensable for the formation of **procapsids** or nucleocapsids, the results of *in vitro* studies indicate that it can be incorporated only during the initial stages of assembly. The mechanism that ensures that each nucleocapsid contains only one portal remains an enigma.

Once nucleocapsids have assembled, scaffolding proteins must be discarded, so that viral genomes can be accommodated (Fig. 13.8A). The virion protease VP24 is essential for such DNA encapsidation. This protein is incorporated into the assembling nucleocapsid as a precursor (Fig. 13.8B).

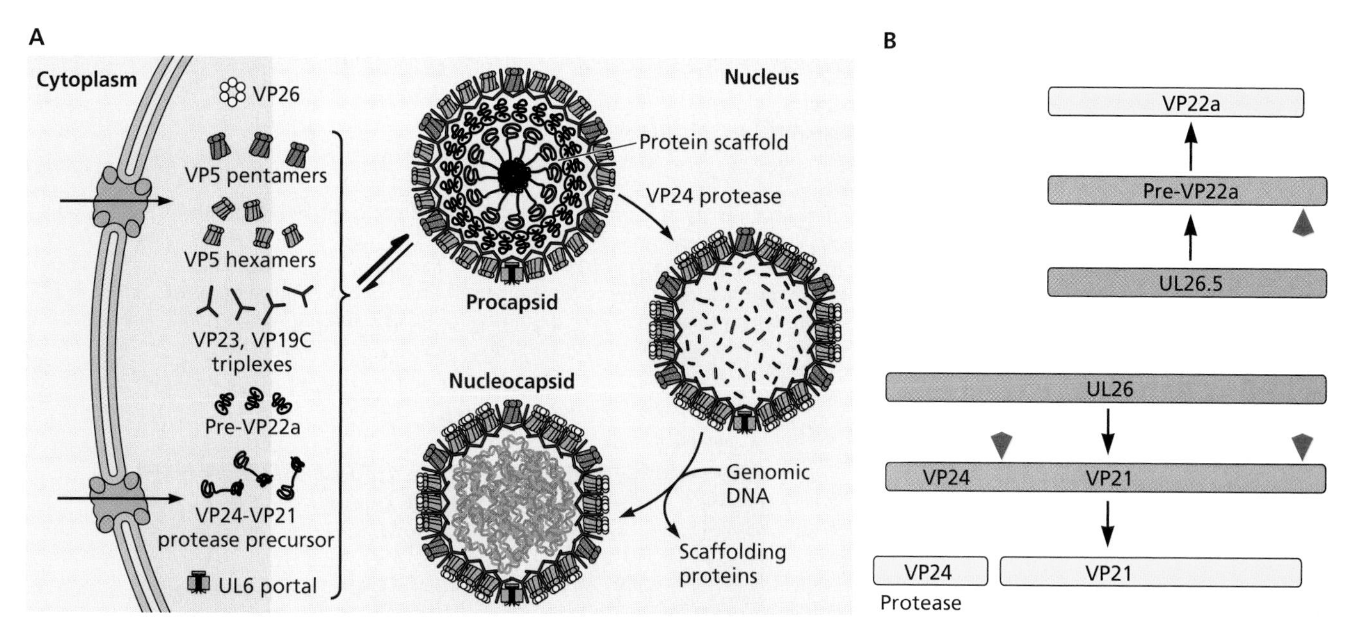

Figure 13.8 Assembly of herpes simplex virus 1 nucleocapsids. (A) Assembly begins as soon as nucleocapsid proteins accumulate to sufficient concentrations in the infected cell nucleus. Intermediates include pentamers and hexamers of the major capsid protein VP5, which form pentons and hexons in the capsid, and triplexes of the minor proteins VP23 and VP19C. Whether structural units assemble prior to transport into the nucleus is not clear. Viral proteins essential for assembly of the nucleocapsid but not present in mature virus particles, namely, the scaffolding protein (pre-VP22a) and the viral protease precursor (VP24-VP21), must also enter the nucleus. Assembly of nucleocapsids depends on the formation of an internal scaffold around which the protein shell assembles. The VP5 and pre-VP22a proteins form a core via hydrophobic interactions, to which additional VP5 hexamers and the triplexes of VP19 and VP23 are added. These structures are required for further assembly, which occurs by sequential formation of partial dome-like structures and the spherical immature nucleocapsid. Subsequent reactions require the viral protease to remove the scaffolding protein, allowing entry of the DNA genome and morphological transitions. As discussed in the text, encapsidation is concurrent with cleavage of the concatemeric products of herpesviral DNA replication. **(B)** Overlapping sequences of scaffolding proteins. The UL26 and UL26.5 reading frames are shown in purple, and their primary translation products are shown in light brown. The initiating methionine of the VP22a protein is within the larger reading frame that encodes the VP24-VP21 polyprotein. Consequently, VP21 and VP22a are identical in sequence, except that the former contains a unique N-terminal segment. All proteolytic cleavages at sites indicated by the red arrowheads, including those that liberate the protease itself from the VP24-VP21 precursor, are carried out by the VP24 protease. The cleavage at the C-terminal site in VP22a disengages the scaffolding from the capsid proteins.

The protease precursor possesses some activity and initiates cleavage to produce VP24, which then cleaves the scaffolding protein to remove a short C-terminal sequence that is required for binding to VP5. Such processing presumably disengages scaffolding from structural proteins, once assembly of the nucleocapsid is complete. The protease also degrades the scaffolding protein so that encapsidation of the genome can begin.

The proteolytic cleavages that liberate the VP5 structural units from their association with the scaffold also induce major changes in the organization and stability of the nucleocapsid shell. Studies of nucleocapsid assembly *in vitro* and in cells infected by a mutant virus encoding a temperature-dependent viral protease suggest that the uncleaved spherical precursor is analogous to the well-characterized **procapsids** that are formed during the assembly of certain DNA-containing bacteriophages (Box 13.4).

Assembly of simpler protein shells can also depend critically on a viral protein. In addition to its many other functions, simian virus 40 LT participates in assembly of virus particles (Fig. 13.5D). This protein does not form a scaffold, but an N-terminal domain of LT appears to be essential for organization of the capsid; alterations within this domain block production of particles and induce the accumulation of an incomplete structure that contains the viral chromatin and VP1. The N-terminal segment of LT possesses chaperone activity, which may ensure the productive binding of VP1 pentamers to one another and to other components of the particle during assembly.

BOX 13.4

EXPERIMENTS

Visualization of structural transitions during assembly of DNA viruses

The assembly of viruses that package double-stranded DNA genomes into a preformed protein shell exhibits several common features, regardless of the host organism. These include the presence of a portal for DNA entry in the capsid or nucleocapsid precursor and probably the mechanism of DNA packaging (see the text). In addition, as illustrated for bacteriophage λ and herpes simplex virus 1, formation of DNA-containing structures is accompanied by major reorganizations of the protein shell. **(A) Cryo-electron micrographs of the bacteriophage λ prohead and the DNA-containing mature capsid.** The former comprises hexamers and pentamers of the capsid protein gpE organized with $T = 7$ icosahedral symmetry and is assembled prior to encapsidation of the DNA genome. It is smaller than the mature capsid (270 and 315 Å in diameter, respectively), but its protein shell is considerably thicker. Packaging of the DNA genome leads to an expansion of the capsid as a result of reorganization of gpE hexamers. This change is accompanied by binding of the gpD protein, which contributes to capsid stabilization. Adapted from T. Dokland and H. Murialdo, *J. Mol. Biol.* **233:**682–694, 1993, with permission. **(B) Cryo-electron micrographs of herpes simplex virus 1 nucleocapsid precursor and mature nucleocapsid, viewed along a 2-fold axis of icosahedral symmetry.** Some copies of the proteins that form the particles' surfaces are colored as follows: VP5 hexons, red; VP5 pentons, yellow; and triplexes containing one molecule of VP19C and two of VP23, green. The precursor nucleocapsid is spherical (rather than icosahedral), and its protein shell is thicker. Furthermore, the VP5 hexamers are not organized in a highly regular, symmetric manner in the precursor, resulting in a more open protein shell. The precursor nucleocapsid also lacks the VP26 protein, which binds to the external surfaces of VP5 hexamers, but not pentamers, in the mature nucleocapsid. Adapted from A. C. Steven et al., *FASEB J.* **10:**733–742, 1997, with permission.

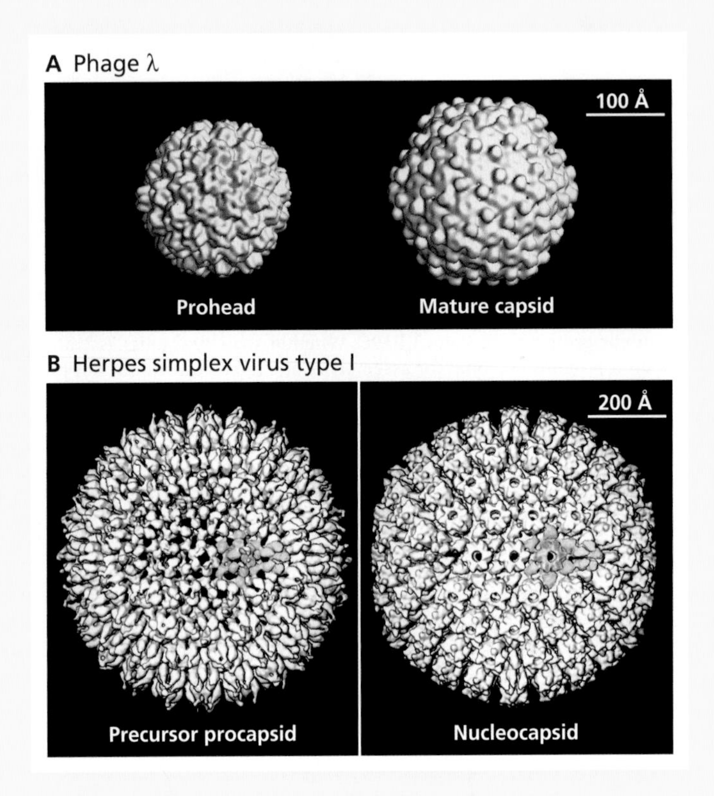

Selective Packaging of the Viral Genome and Other Components of Virus Particles

Concerted or Sequential Assembly

Incorporation of the viral genome into assembling particles, often called packaging, requires specific recognition of genomic RNA or DNA molecules. All viral genomes are packaged by one of two mechanisms, concerted or sequential assembly.

In concerted assembly, the structural units of the protective protein shell assemble productively only in association with the genomic nucleic acid. The nucleocapsids of (−) strand RNA viruses form by a concerted mechanism (Fig. 13.9), as do retrovirus particles (Fig. 13.10) and those of other (+) strand RNA viruses. In many cases, these assembly reactions are coordinated with synthesis of the viral genome. In the alternative mechanism, sequential assembly, the genome is inserted into a preformed protein shell. The formation of herpesviral nucleocapsids provides a clear example of this packaging mechanism (Fig. 13.8). Mutations that inhibit viral DNA synthesis or that prevent DNA packaging do not block assembly of capsid-like structures that lack DNA. These phenotypes establish unequivocally that the DNA genome must enter preformed nucleocapsids. In contrast to concerted assembly, encapsidation of the genome in a preformed structure requires specialized mechanisms to maintain or open a portal for entry of the nucleic acid to pull or push the genome

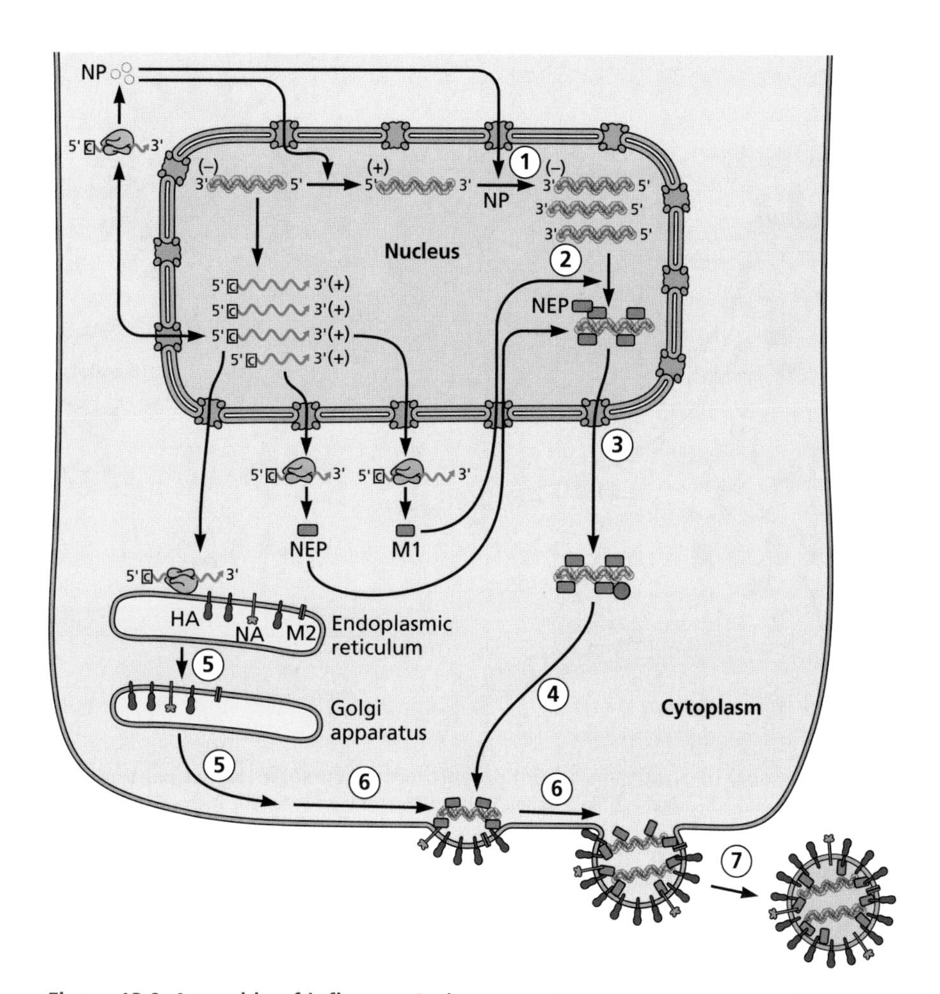

Figure 13.9 Assembly of influenza A virus. Assembly proceeds in stepwise fashion within different compartments of an infected cell. As (−) strand genomic RNA is synthesized in the nucleus, it is packaged by the NP RNA-binding protein (step 1). These ribonucleoproteins may serve as templates for mRNA synthesis, participate in further cycles of replication, or bind the M1 protein (step 2). The latter interaction prevents further RNA synthesis and allows binding of the viral nuclear export protein (NEP) and export of the nucleocapsid to the cytoplasm (step 3). The M1 protein also binds to the cytoplasmic face of the plasma membrane via specific sequences and directs the nucleocapsid to the plasma membrane (step 4). The plasma membrane carries the viral HA, NA, and M2 proteins, which reach this site via the cellular secretory pathway (step 5). The M1 protein probably controls budding (step 6) via recruitment of cellular components (see the text). Fusion of the membrane bud releases the enveloped particle (step 7). Only two of the eight genome segments are illustrated for clarity.

into the capsid (see the next section). The herpesviral portal UL6, which is present at only 1 of the 12 vertices of the nucleocapsid (Fig. 13.8; see also Fig. 4.29), fulfills the latter function. The DNA genomes of adenovirus (Fig. 13.11) and adenovirus-associated virus also appear to be packaged into preformed capsids.

Recognition and Packaging of the Nucleic Acid Genome

During encapsidation, viral nucleic acid genomes must be distinguished from the cellular DNA or RNA molecules present in the compartment in which assembly takes place. This process requires a high degree of discrimination among similar nucleic acid molecules. For example, retroviral genomic RNA constitutes much less than 1% of an infected cell's cytoplasmic mRNA population and bears all the hallmarks of cellular messenger RNAs (mRNAs), yet it is **the** RNA packaged in the great majority of retrovirus particles. Such discrimination is the result of specific recognition of sequences or structures unique to the viral genome, termed **packaging signals**. These can be defined by genetic analyses as the sequences that are necessary for incorporation of the nucleic acid into the assembling virus particle or sufficient to direct incorporation of foreign nucleic acid. The organization

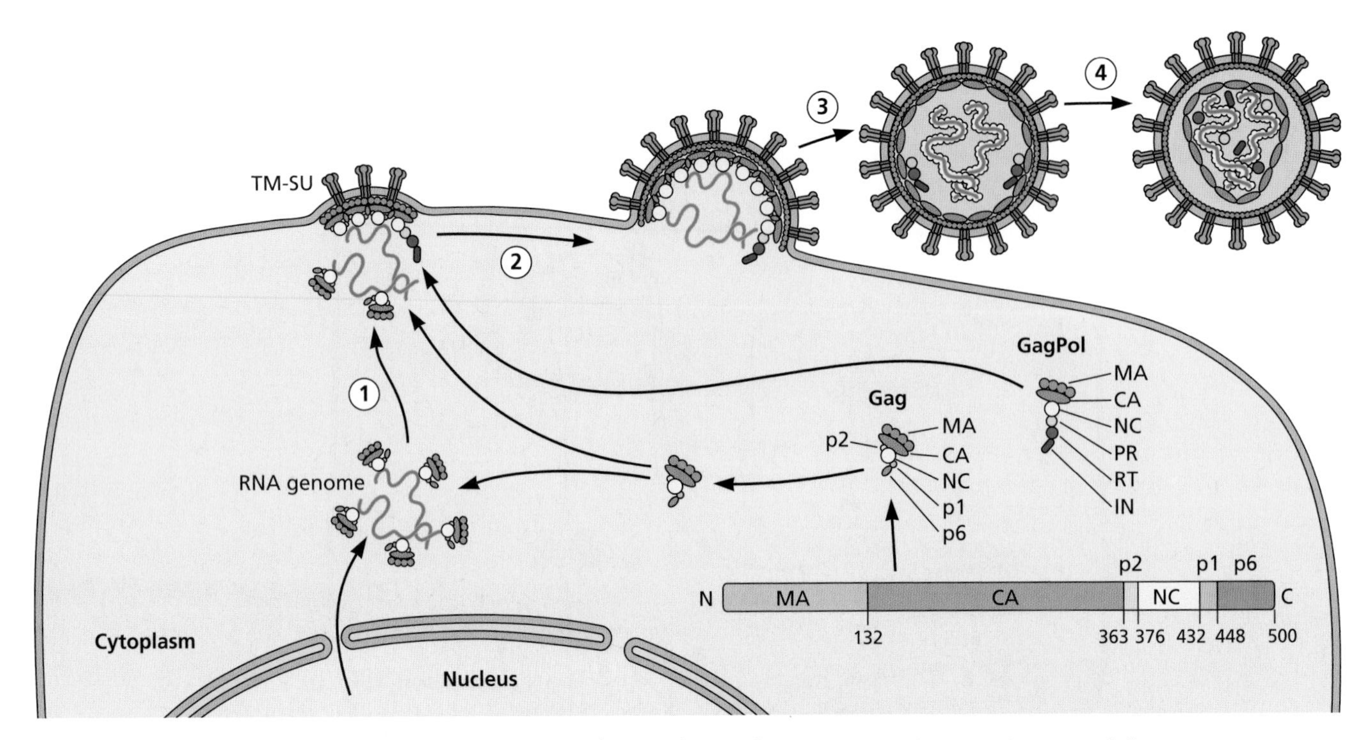

Figure 13.10 Assembly of a retrovirus from polyprotein precursors. The Gag polyprotein of all retroviruses contains the MA, CA, and NC proteins linked by spacer peptides that vary in length and position. The proteins are in the order (from the N to the C terminus) of the protein shells of the virus particle, from the outer to the inner. The organization of human immunodeficiency virus type 1 Gag is illustrated on the right. A minor fraction, about 1 in 10, of Gag translation products carry the retroviral enzymes, denoted by PR, RT, and IN, at their C termini. The association of Gag molecules with the plasma membrane, with one another, and with the RNA genome via binding of NC segments initiates assembly at the inner surface of the plasma membrane (step 1). In some cases, such as human immunodeficiency virus type 1, the MA segment also binds specifically to the internal cytoplasmic domain of the TM-SU glycoprotein. Assembly of the particle continues by incorporation of additional molecules of Gag (step 2). This pathway is typical of many retroviruses, but some (e.g., betaretroviruses) complete assembly of the core in the interior of the cell prior to its association with the plasma membrane. The dimensions of the assembling particle are determined by interactions among Gag polyproteins. Eventually, fusion of the membrane around the budding particle (step 3) releases the immature noninfectious particle. Cleavage of Gag and Gag-Pol polyproteins by the viral protease (PR) produces infectious particles (step 4) with a morphologically distinct core (see Fig. 13.25).

of the packaging signals of several viruses is therefore quite well understood.

Nucleic Acid Packaging Signals

DNA signals. The products of adenoviral or polyomaviral DNA synthesis are genomic DNA molecules that can be incorporated into assembling virus particles without further modification. These DNA genomes contain discrete packaging signals with several common properties (Fig. 13.12). The signals comprise repeats of short sequences, some of which are also part of viral promoters or enhancers; they are positioned close to an origin of replication, and their ability to direct DNA encapsidation depends on this location. They differ in whether they are recognized directly or indirectly by viral proteins.

The encapsidation signal of the adenoviral genome, which is located close to the left inverted repeat sequence and origin, comprises a set of repeated sequences. The sequences are recognized by the viral late proteins IVa_2 and L4 22-kDa, while the L1 52/55-kDa protein is recruited by interaction with the IVa_2 protein. Cooperative binding of these proteins to the repeated sequence is thought to form a higher-order nucleoprotein structure that promotes packaging of the genome. The results of genetic experiments have established the importance of these proteins in assembly: mutations that prevent production of the proteins block the formation of mature virus particles generally with accumulation of empty, immature capsids, consistent with a sequential encapsidation mechanism (Fig. 13.11)

The simian virus 40 DNA-packaging signal is located in the regulatory region of the genome that contains the origin of replication, the enhancer, and early and late promoters. Several sequences within this region contribute to the encapsidation signal, which includes multiple binding sites for the

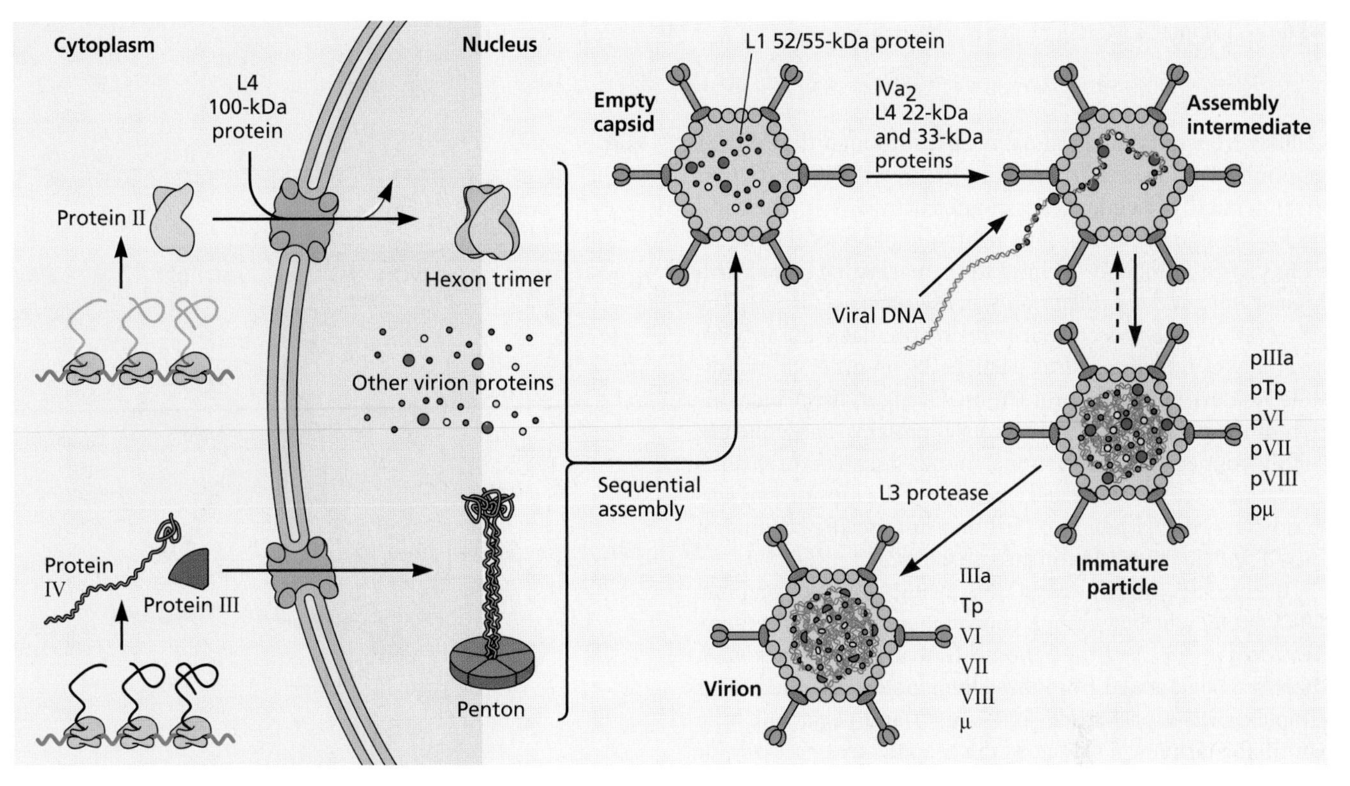

Figure 13.11 Adenovirus assembly. Synthesis and assembly of hexons and pentons and their transport into the nucleus set the stage for assembly. The L4 100-kDa protein is required for formation of hexons, but its molecular function is not known. These structural units together with the proteins that stabilize the capsid assemble into empty capsids. The L1 52-/ 55-kDa proteins are necessary for the formation of structures that can complete assembly and decrease in concentration as assembly proceeds. The DNA is then inserted into this structure via the packaging signal located near the left end of the genome. The viral IVa_2 and L4 22-kDa proteins bind specifically to this sequence *in vitro* and are required for assembly in infected cells. Premature breakage of DNA in the process of insertion would yield the structure designated "Assembly intermediate," in which an immature capsid is associated with a DNA fragment derived from the left end of the viral genome. Core proteins are encapsidated with the viral genome to yield noninfectious young virus particles. Mature particles are produced upon cleavage of the precursor proteins listed to the right of the young particle.

Figure 13.12 Viral DNA-packaging signals. (A) Human adenovirus type 5 (Ad5). The locations of the repeated sequences (blue arrows) of the packaging signal relative to the left inverted terminal repeat (ITR), the origin of replication (Ori), and the E1A transcription unit are indicated. The repeated sequences are AT rich and functionally redundant, and several overlap enhancers that stimulate transcription of viral genes. The viral IVa_2 protein binds directly to the 3′ portion of the sequence that is conserved in each of the repeats. Once the IVa_2 protein is associated, the L4 22-kDa protein interacts with the 5′ segment of the conserved sequences. The positions of transcriptional enhancers within this region are also shown. Enhancer 1 stimulates transcription of the immediate early E1A gene, whereas enhancer II increases the efficiency of transcription of all viral genes. **(B)** Simian virus 40 (SV40). The region of the genome containing the enhancer, origin of replication (Ori), and packaging signal is shown, with positions (base pairs) in the circular genome indicated below. The Sp1-binding sites within the packaging sequence are required for genome packaging.

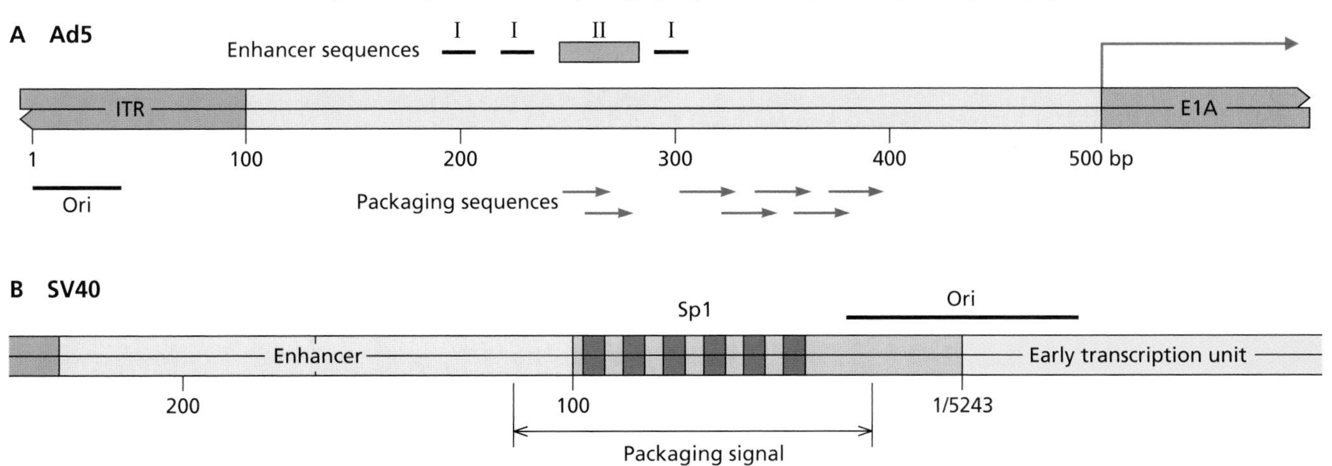

cellular transcriptional regulator Sp1. Although the cellular genome contains many such binding sites, the particular arrangement of sequences recognized by Sp1 in the viral packaging signal is unique. The internal proteins of the simian virus capsid (VP2 and -3) bind to the packaging signal with high affinity and specificity only in the presence of Sp1. The observation that this cellular protein stimulates the *in vitro* assembly of infectious virus particles by an order of magnitude is consistent with a role in mediating indirect recognition of the packaging signal by capsid proteins. Subsequently, highly cooperative interactions among the structural units appear to drive the concerted assembly of the capsid, concomitantly with displacement of Sp1 and nonspecific binding of capsid proteins to viral minichromosomes.

The products of herpesviral DNA synthesis are not genomic DNA molecules but, rather, concatemers containing many head-to-tail copies of the genome. Individual genomes must therefore be liberated from these long molecules. The herpes simplex virus 1 packaging signals *pac1* and *pac2*, which lie within the terminal *a* repeats of the genome, are necessary for both recognition of the viral DNA and its cleavage within the adjacent direct repeats (Fig. 13.13A). It is generally thought that cleavage is concomitant with genome encapsidation. One model (Fig. 13.13B) proposes that a protein complex formed on the unique short *pac* sequence interacts with the portal in the nucleocapsid. Following the first DNA cleavage, a unit-length genome is reeled into the nucleocapsid from the concatemer prior to the second DNA cleavage. This mechanism is analogous to that by which concatemeric DNA products of bacteriophage T4 replication are cleaved and packaged by a multisubunit terminase, which hydrolyzes ATP and associates transiently with the portal protein of the preformed capsid. The products of at least seven herpes simplex virus 1 genes are dedicated to encapsidation of the viral genome. The UL15, UL28, and UL33 proteins, which interact with one another and with the portal protein, exhibit the properties predicted for the terminase. For example, the UL15 protein contains a sequence motif characteristic of ATPases that is essential for encapsidation of viral genomes, while the UL28 gene product binds to *pac* sequences required for DNA cleavage.

RNA signals. Because it is also an mRNA, the retroviral genome must be distinguished during encapsidation from both cellular and subgenomic viral mRNAs. In addition, two genomic RNA molecules must interact with one another, for the retroviral genome is packaged as a dimer. This unusual property is thought to help retroviruses survive extensive damage to their genomes (Chapter 7). In virus particles, the dimeric genome is in the form of a 70S complex held together by many noncovalent interactions between the RNA molecules. However, most attention has focused on sequences that allow the formation of stable dimers, termed the **dimer**

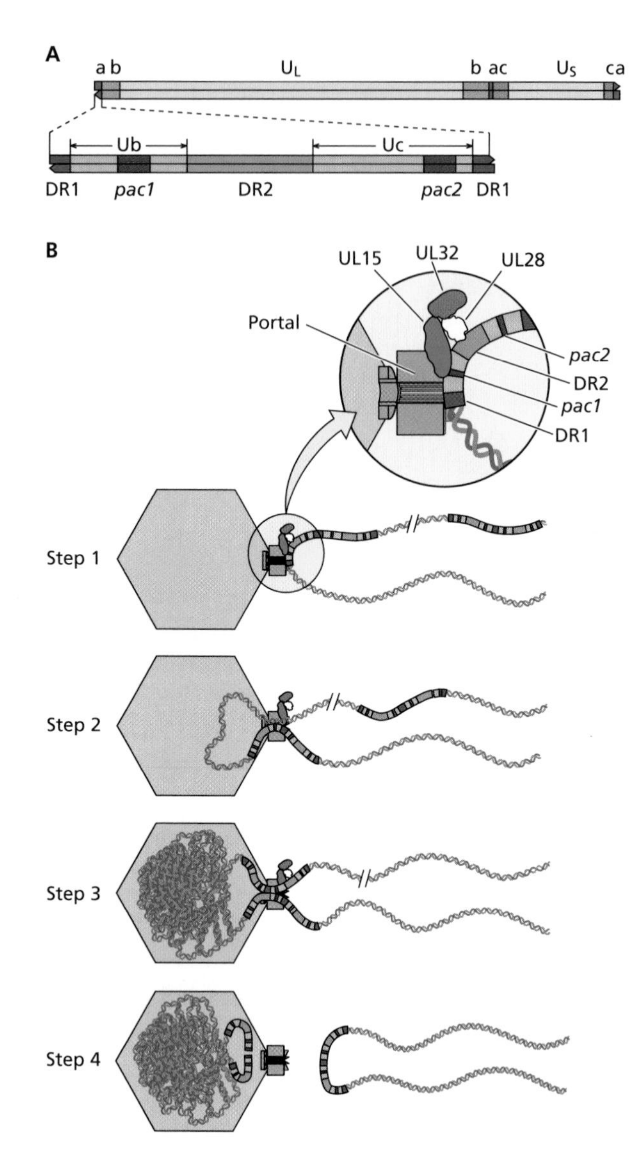

Figure 13.13 Packaging of herpes simplex virus 1 DNA. (A) Organization of the *a* repeats of the viral genome, showing the locations of the *pac1* and *pac2* sequences within the nonrepeated sequences Ub and Uc and relative to the flanking direct repeats 1 and 2 (DR1 and DR2). One to several copies of the *a* sequence are present at the end of the unique long (U_L) segment and at the internal L-S junction, but only one copy lies at the end of the unique short (U_S) region. **(B)** Model of herpes simplex virus 1 DNA packaging, in which encapsidation is initiated by formation of a terminase, which includes the proteins indicated, bound to the packaging sequence. This protein-DNA complex is oriented to interact with the portal of the nucleocapsid (step 1). The DNA is then reeled into the capsid (steps 2 and 3) until a headful threshold is reached and an *a* sequence in the same orientation (i.e., 1 genome equivalent) is encountered (step 3), when cleavage in DR1 sequences takes place (step 4). In addition to the proteins shown, the UL17 protein is essential for DNA cleavage, but its function in this process is not known. It is also necessary for recruitment of the UL15 protein subunit of the terminase. The UL25 protein is also necessary for efficient packaging of the genome. In its absence, DNA cleavage does occur, but fewer nucleocapsids are formed. It has therefore been suggested that one function of this protein is to stabilize the protein shell so that it can withstand the pressure exerted by the encapsidated genome.

linkage sequence. *In vitro* experiments with human immunodeficiency virus type 1 RNA have provided evidence for base pairing between loop sequences of a specific hairpin (SL1) within the dimer linkage sequence (Fig. 13.14A) and the formation of an intermolecular, four-stranded helical structure (known as a G tetrad or G quartet). The effects of mutations in or duplication of this sequence indicate that it nucleates formation of genome RNA dimers *in vivo* and that dimerization is required for efficient genome packaging. Indeed, the dimer linkage sequence lies within the series of hairpin loops that comprise the RNA-packaging signal (Fig. 13.14A).

Sequences necessary for packaging of retroviral genomes, termed psi (ψ), vary considerably in complexity and location. In some cases, exemplified by Moloney murine leukemia virus, a ψ sequence of about 350 contiguous nucleotides (Fig. 13.14B) is both necessary and sufficient for RNA encapsidation. As this sequence lies downstream of the 5′ splice site, only unspliced genomic RNA molecules are recognized for packaging. The human immunodeficiency virus type 1 genome also contains a primary packaging sequence (Fig. 13.14A) that distinguishes the full-length genome from spliced viral RNA molecules. However, this sequence fails to direct packaging when it is incorporated into heterologous RNA species, indicating that it is not sufficient. Additional sequences required for genomic RNA encapsidation lie within tar and adjacent sequences and at more-distant locations. One function of sequences upstream of ψ is to participate in a structural switch that governs the accessibility of the dimer initiation sequence and the initiation codon of the Gag coding sequence and hence determine whether full-length (+) strand RNA is dimerized and packaged or translated (Box 13.5). Sequestration of the complete ψ sequence present in the subgenomic mRNA of avian retroviruses (Fig. 13.14B) in a folded structure may also account for the inefficient encapsidation of the mRNA.

The NC domain of Gag mediates selective and efficient encapsidation of genomic RNA during retroviral assembly and can facilitate annealing of RNA molecules. The central region of NC containing the zinc-binding motif(s) and adjacent basic sequences binds specifically to RNAs that carry ψ sequences *in vitro* and is necessary for selective packaging of the genome in infected cells. Structural studies of NC proteins bound to RNA-packaging signals indicate that NC binds specifically to short RNA sequences. The zinc-binding motifs form a compact structure that makes specific contacts with bases and is complementary in charge and shape to the bound RNA. The Moloney murine leukemia virus ψ signal region contains 13 copies of the sequence recognized specifically by its NC protein, a higher frequency than elsewhere in the genome. However, this high-affinity NC-binding signal is exposed only after dimerization of the RNA, a property that promotes selective packaging of genome dimers.

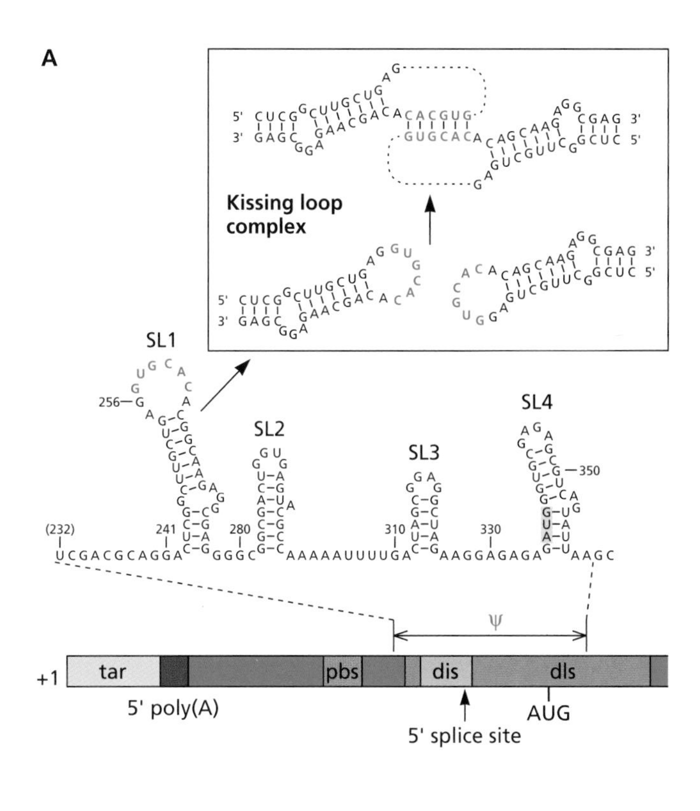

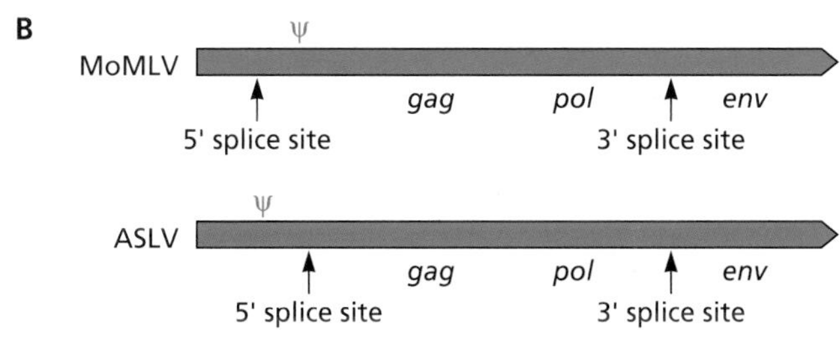

Figure 13.14 Sequences important for the packaging of retroviral genomes. (A) The 5′ end of the human immunodeficiency virus type 1 genome is shown to scale at the bottom, indicating the positions of tar, the 5′ polyadenylation signal [5′ poly (A)], the tRNA primer-binding site (pbs), the 5′ splice site, a packaging signal designated ψ, the sequence that forms the dimer linkage structure (dls), and the dimerization initiation site (dis), which can initiate dimerization *in vitro*. The four hairpins (SL1 to SL4) formed by the ψ sequence are shown above. The SL1 hairpin is the dimer initiation sequence. The loop-loop "kissing" complex proposed to form when two genomic RNA molecules dimerize via the self-complementary sequence shown in red is depicted at the top. The ψ sequence, which includes intronic sequences and therefore is present only in unspliced RNA, appears to be necessary but not sufficient for encapsidation of genomic RNA. **(B)** Locations within the RNA genomes of sequences necessary for the encapsidation of Moloney murine leukemia virus (MoMLV) and avian sarcoma/leukosis virus (ASLV) RNAs, designated ψ. The ψ signal resides only upstream of the 5′ splice site. Even though both genomic and subgenomic RNAs contain this sequence, spliced mRNA molecules are not encapsidated efficiently.

BOX 13.5

EXPERIMENTS

Dimerization-induced conformational change and encapsidation of the human immunodeficiency virus type 1 genome

The figure shows models of the secondary structure of the 5′ leader RNA determined using nuclear magnetic resonance (NMR) methods for analysis of RNAs of >50 nucleotides. At low ionic strength, the RNA was predominantly monomeric, and the NMR signals indicated that the AUG codon for initiation of translation of Gag was present in a hairpin (left). With increasing ionic strength, RNA dimers were formed and the NMR signals of the AUG-containing region were consistent with base pairing to the U5 sequence, displacing the dimerization initiation site (dis; right). These models were confirmed by various observations, including the results of site-directed mutagenesis. For example, mutations in the AUG region designed to disrupt base pairing promoted dimer formation. Furthermore, the AUG-U5 base-pairing interaction increased the affinity with which NC bound to RNA, and mutations that precluded such base pairing severely impaired RNA packaging. It has therefore been proposed that U5:AUG base pairing serves as a regulatory switch that governs the accessibility of both dis and the Gag AUG and hence dimerization and packaging versus translation of full-length (+) strand viral RNA. How this switch might be induced in infected cells is not known.

Lu K, Heng X, Garyu L, Monti S, Garcia EL, Karytonchy S, Dorjseuren B, Kulandaivel G, Jones S, Hirimath A, Divakaruni SS, LaColti C, Barton S, Tummillo D, Hosic A, Edme K, Albrecht S, Telesnitsky A, Summers MF. 2011. NMR detection of structures in the 5′-leader RNA that regulate genome packaging. *Science* **334:**242–245.

Shown are models of the secondary structures of an RNA comprising residues 1 to 356 of human immunodeficiency virus type 1 RNA determined by NMR methods. Those on the left and right predominate when the RNA is monomeric and dimeric, respectively. Adapted from K. Lu et al., *Science* **334:**242–245, 2011, with permission.

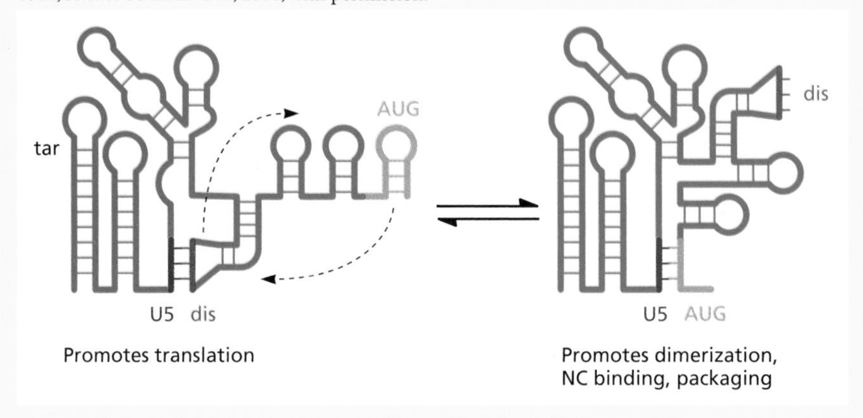

Other parameters that govern genome encapsidation. Specific signals may be required to mark a viral genome for encapsidation, but their presence does not guarantee packaging. The fixed dimensions of the closed icosahedral capsids or nucleocapsids of many viruses impose an upper limit on the size of viral nucleic acid that can be accommodated. Consequently, even when they contain appropriate packaging signals, nucleic acids that are more than 5 to 10% larger than the wild-type genome cannot be encapsidated. This property has important implications for the development of viral vectors. In some cases, the length of the DNA that can be accommodated in the particle (a "headful") is a critical parameter. This size limitation is exemplified by the coupled cleavage and encapsidation of genomic herpesviral DNA molecules from the concatemeric products of replication, as both specific sequences and a headful of DNA are recognized. Indeed, the packaging of some viral DNA genomes, such as T4 DNA, depends solely on the latter parameter (Box 13.6).

The specificity with which the viral genome is incorporated into assembling structures may also be the result of the coupling of encapsidation with its synthesis. As mentioned previously, such coordination is typical of the assembly of (−) strand RNA viruses (see, e.g., Fig. 13.9). Coordination of replication and encapsidation may contribute to the great specificity with which picornaviral genomes are packaged; not only abundant cytoplasmic cellular RNA species, but also (−) strand viral RNA and viral mRNA lacking VPg are excluded from virus particles. However, no packaging signal has yet been identified in the poliovirus genome. Encapsidation of the (+) strand, Vpg-linked RNA genome, which initiates assembly of virus particles (Fig. 13.6), is coordinated with genome replication, and both processes take place in association with cytoplasmic vesicles unique to poliovirus-infected cells (Chapter 14). Such sequestration of genomic RNA molecules with the viral proteins that must bind to them could promote specific packaging by reducing competition from cytoplasmic cellular RNAs. However, selective encapsidation of newly synthesized genomes is also facilitated by interaction of an essential component of the viral replication machinery (the viral 2C protein ATPase) with the capsid protein VP3. As packaging of flavivirus RNA also depends on replication of genomic RNA, coincident genome synthesis and assembly may be a general feature of (+) strand RNA viruses.

BOX 13.6

BACKGROUND

Packaging a headful of viral DNA

During assembly of herpesvirus and several bacteriophages with large, double-stranded DNA genomes, including bacteriophage T4, the linear genome is cleaved from concatemeric products of viral genome replication as it is inserted into a preformed protein shell. While encapsidation of T4 DNA is coordinated with cleavage of concatemers, the T4 genome exhibits several unusual features:

- the linear genomes do not have unique terminal sequences;
- the genetic map is circular, even though the genome is linear;
- the terminal sequences, which are different in each DNA molecule, are repeated at each end of DNA; and
- each virus particle contains more than a genome's length of DNA.

It was deduced from these properties that the T4 genome is circularly permuted and terminally redundant. These properties can be accounted for by essentially random cleavage of head-to-tail concatemers (the preferred substrate for DNA packaging) resulting in encapsidation of DNA molecules that are **longer** than the unique sequence in the genome (see the figure). No specific DNA sequence dictates the cleavages that liberate linear DNA during encapsidation. Rather, the first cleavage occurs randomly, and the second takes place once the phage T4 head has been filled with DNA. As predicted by this "headful" packaging mechanism, when head size is increased or decreased by mutation in specific genes (or other manipulations), longer and shorter DNA molecules, respectively, are encapsidated. Furthermore, when sequences are deleted from, or inserted into, the genome, the length of the terminal repeats increases or decreases to the corresponding degree. These properties demonstrate directly that a fixed length of DNA, a headful, is incorporated during assembly.

A headful of DNA is packaged during assembly of other bacteriophage and animal viruses with double-stranded DNA genomes, including herpesviruses. Structural studies of bacteriophage P22 virus particles revealed that tight spooling of DNA in the nucleocapsid induces major conformational change in the portal, through which DNA enters. It has therefore been proposed that the change in portal structure provides the signal that the nucleocapsid is full to activate termination of DNA encapsidation.

Lander GC, Tang L, Casjens SR, Gilcrease EB, Prevelige P, Poliakov A, Potter CS, Carragher B, Johnson JE. 2006. The structure of infectious P22 virion shows the signal for headful DNA packaging. *Science* **312:**1791–1795.

A head-to-tail concatemer, in which the unique genome sequence is represented by ABCDEFGH, is shown at the top. Initial cleavage between H and A is followed by packaging of a headful length that is longer than the length of the unique genome sequence, and the second cleavage. Repetition of this process yields a population of particles with encapsidated DNA molecules of the same length but that are circularly permuted and terminally redundant.

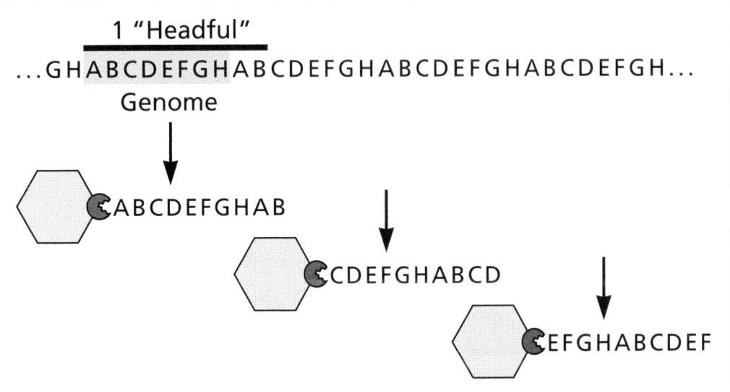

Packaging of Segmented Genomes

Segmented genomes pose an intriguing packaging problem. The best-studied example among animal viruses is the influenza A virus genome, which comprises eight molecules of RNA. It has been appreciated for many years that formation of an infectious virus particle requires incorporation of at least one copy of each of the eight genomic segments. However, it proved difficult to distinguish random packaging from a selective mechanism for inclusion of a full complement of genomic RNAs.

Packaging of the bacteriophage ϕ6 genome provides clear precedent for a selective mechanism. The genome of this bacteriophage comprises one copy of each of three double-stranded RNA segments designated S, M, and L. The (+) strand of each segment is packaged prior to synthesis of the double-stranded RNA segments, as with the packaging and synthesis of the reovirus genome (Chapter 6). The particle-to-PFU ratio of ϕ6 is close to 1, indicating that essentially all particles contain a complete complement of genome segments. Such precise packaging appears to be the result of the serial dependence of packaging of the (+) strand RNA segments. In *in vitro* reactions, the S segment packages alone, but entry of M RNA requires the presence of S RNA within particles, and packaging of the L segments is dependent on the prior entry of both S and M RNAs.

A random-packaging mechanism in which any eight RNA segments of the influenza virus genome were incorporated into virus particles would yield a maximum of 1 infectious particle for every 400 or so assembled ($8!/8^8$). This ratio might seem impossibly low, but it is within the range of ratios of noninfectious to infectious particles found in virus preparations. Furthermore, if packaging of more than eight RNA segments were possible, the proportion of infectious particles would increase significantly. For example, with 12 RNA molecules per particle, 10% would contain the complete viral genome. Particles containing more than eight RNA segments have been isolated, consistent with random packaging. Nevertheless, it has become clear in the last decade that the packaging of influenza virus genome segments **is** selective. For example, eight RNA segments were observed in all particles by electron tomography (Fig. 13.15). Furthermore, estimation of the copy number of each viral RNA segment by single-molecule fluorescent *in situ* hybridization established that each particle contains one copy of each of the eight segments of the genome (Box 13.7).

A selective mechanism implies that each of the eight (−) strand genome RNAs (vRNAs) carries a unique signal that ensures its packaging. These sequences comprise the short 5′ and 3′ noncoding regions of each segment but extend short distances into adjacent coding regions. The extreme 5′- and 3′-terminal sequences are highly conserved among vRNA segments of influenza A virus isolates and might distinguish vRNAs from viral and cellular mRNAs. Although the mechanisms by which the full complement of the eight vRNA segments that comprise the viral genome is selected for packaging is not fully understood, the available evidence implicates direct interactions among RNA segments. This mechanism was first suggested by the observation that deletion of packaging signals from the vRNA segments for the polymerase proteins impaired the encapsidation not only of these vRNAs but also of others. The greater importance of some vRNA packaging signals (those of the PA, PB2, M, and NP RNA segments) is consistent with a hierarchical mechanism of packaging. Each vRNA segment associates with at least one other *in vitro*, and specific base pairing between two segments (vRNA2 and vRNA8) has been shown to be necessary for the packaging of both into virus particles in infected cells and for optimal virus reproduction. It therefore appears that base-pairing interactions drive formation of an organized assembly containing one molecule of each vRNA segment prior to encapsidation. Such interactions may occur during transport of vRNAs to the plasma membrane (Chapter 12) and are consistent with the nonrandom organization of vRNAs observed in budding virus particles (Fig. 13.15).

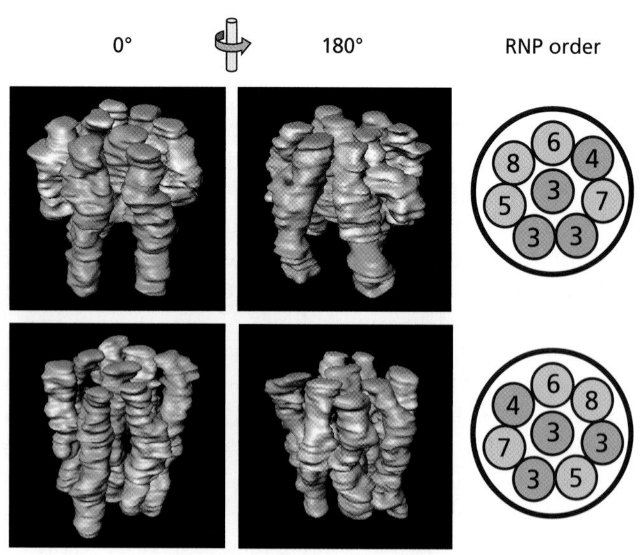

Figure 13.15 Organization of ribonucleoproteins in influenza A virus particles. Purified influenza A virus particles were examined by scanning transmission electron tomography. Shown are three-dimensional models of the viral ribonucleoproteins (vRNPs) observed in the particles. All particles examined contained 8 vRNPs, most of which could be distinguished by their lengths and are designated 3, 3, 3, 4, 5, 6, 7, and 8 in terms of decreasing length. Longer (3, 4) and shorter RNPs are shown in pink and gray, respectively. The RNPs are packaged in the 7+1 arrangements illustrated at the right and associated with one another in all particles examined. Adapted from T. Noda et al., *Nat. Commun.*, 2012, doi:10.1038/ncomms1647. Courtesy of Y. Kawaoka, University of Tokyo, Tokyo, Japan.

Incorporation of Enzymes and Other Nonstructural Proteins

In many cases, the production of infectious particles requires essential viral enzymes or other proteins that are important in establishing an efficient infectious cycle. Some of these proteins are also structural proteins. For example, the herpes simplex virus 1 VP16 protein is both a major component of the tegument and the activator of transcription of viral immediate early genes.

A simple, yet elegant, mechanism ensures entry of retroviral enzymes (protease [PR], reverse transcriptase [RT], and integrase [IN]) into the assembling core. In most cases, the precursors to these enzymes are synthesized as C-terminal extensions of the Gag polyprotein. The organizations and complements of these translation products, designated Gag-Pol, vary among retroviruses, but the important point is that they contain not only Pol but also the sequences specifying Gag-Gag interactions, which can direct incorporation of Gag-Pol molecules into assembling particles (Fig. 13.10). The low efficiency with which Gag-Pol polyproteins are translated determines their concentrations relative to Gag in the cell and in virus particles (1:9). The enzymes present in other virus particles, such as the RNA-dependent RNA polymerases of (−) strand RNA viruses (see Table 4.3), are synthesized as individual molecules and therefore must enter assembling particles by noncovalent binding to the genome or to structural proteins.

All retroviral capsids also contain the cellular tRNA primer for reverse transcription, brought into particles by its base pairing with a specific sequence in the RNA genome and by specific binding to RT. In some cases, including human immunodeficiency virus type 1, the host amino acyl tRNA

BOX 13.7

EXPERIMENTS

Counting the number of unique viral RNA segments in influenza virus particles

To provide single-molecule sensitivity, "counting" experiments exploited fluorescent *in situ* hybridization (FISH), with individual viral RNA segments detected by hybridization to multiple oligonucleotides complementary to different sequences of the RNA.

Purified influenza A virus particles were immobilized at low density on glass slides by binding to surface-bound antibodies against the viral HA protein. Particles were fixed and permeabilized prior to hybridization. The specificity of the method was first established by showing that fluorophore (Cy3 or Cy5)-labeled probes for two viral RNA segments each detected >50 spots, but a probe against a cellular RNA yielded few, if any.

Hybridization with a mixture of Cy3- and Cy5-labeled probes against a single segment established that the efficiency of colocalization was high (see the figure). When Cy3-labeled oligonucleotides complementary to the PB2 RNA segment and Cy5-labeled probes for each of the other 7 segments were applied, efficient colocalization of PB2 with each of the other RNAs was observed. Calculations based on the colocalization efficiencies indicated that over 50% of the virus particles examined contained all 8 genomic RNA segments.

The copy number of each segment in viral particles was examined by using photobleaching. This approach relied on the stepwise loss of fluorescent intensity as a function of time when multiple probes were hybridized to an RNA segment. Control experiments with a recombinant virus containing 2 copies of the HA RNA segment established that this approach can distinguish clearly 1 from 2 copies of an RNA segment per particle. The results obtained when the other RNA segments were "counted" in this way indicted that a single copy of each of the eight genomic segments was packaged in >90% of the hundreds of virus particles examined. These experiments provided compelling evidence for selective and efficient packaging of all the parts of this segmented genome in each virus particle.

Chou YY, Vufabakhsh R, Doganay S, Gao Q, Ha T, Palese P. 2012. One influenza virus particle packages eight unique viral RNAs as shown by FISH analysis. *Proc Natl Acad Sci U S A* **109:**9101–9106.

Images of influenza A virus particles labeled by FISH with 23 Cy3-labeled and 25 Cy5-labeled oligonucleotides complementary to the NA RNA segment. As illustrated in the merged image, the great majority (90%) of particles are labeled with both probes (right). Scale bar = 5 μm. Adapted from Y.-Y. Chou et al., *Proc. Natl. Acad. Sci. U. S. A.* **109:**9101–9106, 2012, with permission. Courtesy of P. Palese, Mount Sinai School of Medicine, New York, NY.

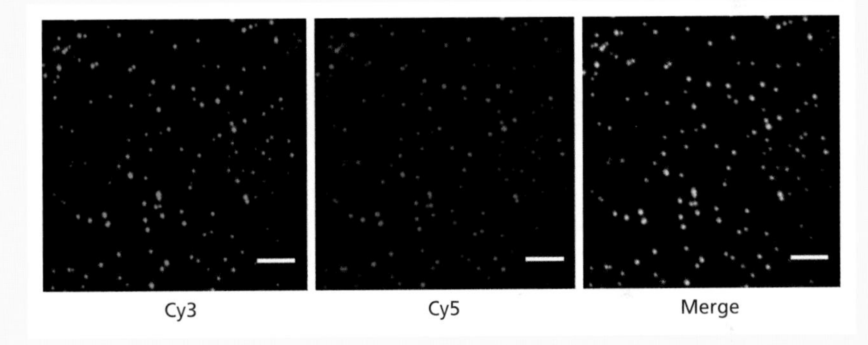

synthetase that aminoacylates the particular tRNA used as primer is also encapsidated (Chapter 7). The absence from virus particles of other amino acid tRNA synthetases and the similar concentrations of the enzyme and its tRNA substrate in human immunodeficiency virus type 1 particles suggest that the synthetase may be recognized by viral components during packaging (Box 7.3).

Acquisition of an Envelope

The formation of many types of virus particle requires envelopment of capsids or nucleocapsids by a lipid membrane carrying viral proteins. Most such enveloped viruses assemble by virtue of specific interactions among their components at a cellular membrane before budding and pinching off of a new virus particle. There is considerable variety in the interactions of viral proteins with membranes (and with one another) that induce membrane curvature (bud formation) (Fig. 13.16). Whether particles assemble at the plasma or an internal membrane is determined by the destination of viral proteins that enter the cellular secretory pathway (Chapter 12). Enveloped viruses assemble by one of two mechanisms, distinguished by whether acquisition of the envelope follows assembly of internal structures or whether these processes take place simultaneously.

Sequential Assembly of Internal Components and Budding from a Cellular Membrane

The assembly of the internal structures of most enveloped virus particles and their interaction with a cellular membrane modified by insertion of viral proteins are spatially and temporally separated. This class of assembly pathways is exemplified by (−) strand RNA viruses, such as influenza A virus (Fig. 13.9) and vesicular stomatitis virus. Influenza A virus ribonucleoproteins containing individual genomic RNA segments, NP protein, and the polymerase proteins are assembled in the infected cell nucleus as genomic RNA segments are synthesized and are then transported to the cytoplasm (Chapter 12). The viral glycoproteins HA and NA and the M2 membrane protein travel separately to specialized regions in

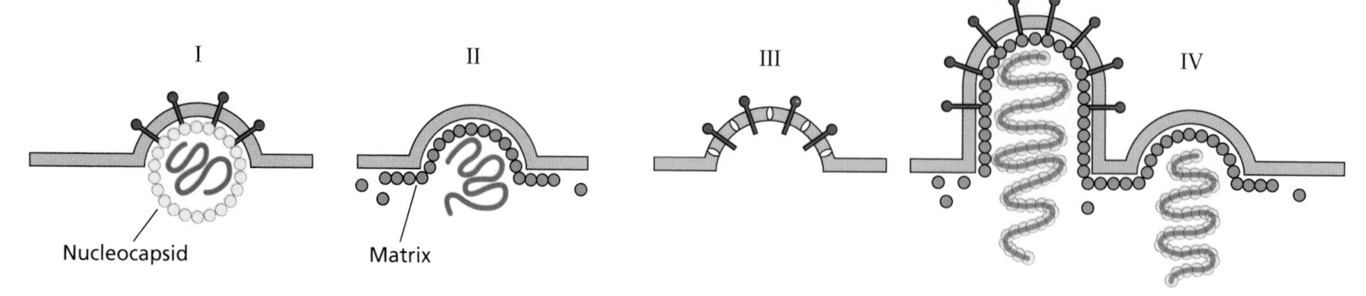

Figure 13.16 Interaction of viral proteins responsible for budding at the plasma membrane. Four distinct budding strategies have been identified. In type I budding, exemplified by alphaviruses, such as Sindbis virus, both the envelope glycoproteins and the internal capsid are essential. Quite detailed structural pictures of alphaviruses are now available (Chapter 4). Certain altered or chimeric envelope proteins that reach the membrane normally do not support budding. These observations indicate that lateral interactions among the envelope heterodimers, as well as those of the heterodimers with the capsid, cooperate to drive budding. Type II budding, such as Gag-dependent budding of many retroviruses, requires only the internal Gag polyprotein. For other viruses, type II budding requires only capsid proteins. Conversely, budding can be driven solely by envelope proteins (type III), a mechanism exemplified by the envelope proteins of the coronavirus mouse hepatitis virus. Type IV budding is driven by matrix proteins, but its proper functioning depends on additional components. For example, in the case of rhabdoviruses and orthomyxoviruses, internal matrix proteins alone can drive budding. However, this process is inefficient or results in deformed or incomplete particles in the absence of envelope glycoproteins or the internal ribonucleoprotein. Adapted from H. Garoff et al., *Mol. Microbiol. Rev.* **62:**1171–1190, 1998, with permission.

the plasma membrane (lipid rafts) via the cellular secretory pathway (Fig. 13.9; Chapter 12). The M1 protein interacts with both viral nucleocapsids and the inner surface of the plasma membrane to direct the assembly of progeny particles at that membrane. Vesicular stomatitis virus assembles in a similar fashion. The matrix proteins of these (−) strand RNA viruses therefore provide the links between ribonucleoproteins and the modified cellular membrane necessary for assembly and budding.

The cellular membranes destined to form the envelopes of virus particles contain viral integral membrane proteins that play essential roles in the attachment of virus particles to, and their entry into, host cells. In simple enveloped alphaviruses, direct binding of the cytoplasmic portions of the viral glycoproteins to the single nucleocapsid protein (see Fig. 4.24) is necessary for acquisition of the envelope during budding from the plasma membrane. The crucial role and specificity of these interactions in the final steps in assembly are illustrated by the failure of a chimeric Sindbis virus containing the coding sequence for the E1 glycoprotein of a different togavirus to bud efficiently. The heterodimeric glycoproteins (E1 plus E2) are formed and transported to the plasma membrane. However, these chimeras exhibit an altered conformation and fail to bind to nucleocapsids at the plasma membrane. Binding of viral glycoproteins to internal components also appears to be important for the production of structurally more complicated enveloped viruses. For example, interactions between the influenza virus M1 protein and the cytoplasmic tails of the HA and NA glycoproteins are necessary for formation of virus particles with normal size and morphology.

Coordination of the Assembly of Internal Structures with Acquisition of the Envelope

The alternative pathway of acquiring an envelope, in which assembly of internal structures and budding from a cellular membrane are largely coincident in space and time, is exemplified by many retroviruses. Assembling cores of the majority first appear as crescent-shaped patches at the inner surface of the plasma membrane. These structures extend to form a closed sphere as the plasma membrane wraps around and the assembling particle is eventually pinched off (Fig. 13.10). Formation of the assembling particles depends on the interaction of Gag polyprotein molecules with one another to form the protein core, with the RNA genome via the NC portion, and with the plasma membrane via the MA segment.

Specific segments of Gag mediate the orderly association of polyprotein molecules with one another and are required for proper assembly. These sequences include an essential C-terminal multimerization domain of the CA segment; substitutions that disrupt the CA dimer interface block assembly of the CA protein *in vitro* and severely inhibit Gag assembly and formation of virus particles in infected cells. The capsids of retroviruses can be spherical, conical, or cylindrical (Fig. 4.17), and specific CA sequences that determine the morphology of mature particles have also been identified. Certain sequences present only in the Gag polyprotein also govern morphology, for their removal results in the assembly of misshapen particles.

As discussed previously, Gag multimerization during particle assembly of human immunodeficiency virus type 1 (and many other retroviruses) is regulated by binding of the NC domain to the RNA genome. This process is also promoted by interaction of Gag with the plasma membrane via the MA membrane-binding signals (Fig. 12.18). Elimination of the signal for myristoylation prevents assembly, as does alteration of the sequence predicted to lie at the interfaces of the MA trimers formed in crystals (Fig. 4.26). It has been suggested that MA trimerization increases the accessibility of the myristate chain. Conversely, efficient membrane binding of Gag depends on sequences other than the membrane-binding region of MA, such as a sequence in the N-terminal portion of NC. Because this sequence is not required for production of stable Gag or its transport to the plasma membrane, it may promote Gag-Gag or Gag-RNA interactions that lead to cooperative and stable binding of Gag molecules to the membrane. In this context, it is noteworthy that Gag-Gag interactions have been observed in living cells only at the plasma membrane.

In some cases, the MA segment of Gag also binds to the cytoplasmic tail of the viral envelope glycoprotein. For example, association of the assembling human immunodeficiency virus type 1 core with the TM-SU glycoprotein requires the N-terminal 100 amino acids of MA. Such Gag-Env interactions ensure specific incorporation of viral glycoproteins into virus particles. Nonetheless, they do not appear to be universal; glycoprotein-containing virus particles are produced even when the C-terminal tails of TM of other retroviruses (e.g., avian sarcoma virus) are deleted. Nor can a model based solely on Gag-Env interactions account for the ease with which "foreign" viral and cellular glycoproteins are included in the envelopes of all retroviruses. The final reaction, fusion of membrane regions juxtaposed as the particle assembles (Fig. 13.10), is shared with other viruses that assemble at the plasma membrane. This process is considered in the next section.

Release of Virus Particles

Many enveloped viruses assemble at, and bud from, the plasma membrane. Consequently, the final assembly reaction, fusion of the membrane around the internal viral components, releases the newly formed virus particle into the extracellular environment. When the envelope is derived from an intracellular membrane, the final step in assembly, budding, is also the first step in egress, which must be followed by transport of the particles to the cell surface. The assembly of enveloped viruses is therefore both mechanistically coupled and coincident with (or at least shortly followed by) their exit from the host cell. In some cases, nondestructive budding permits a long-lasting relationship with the host cell. The progeny of many retroviruses are released throughout the lifetime of an infected cell, which is not harmed (but may be permanently altered [see Volume II, Chapter 6]). The egress of some viruses without envelopes from certain cell types also occurs by specific mechanisms. However, reproduction of such viruses more commonly results in destruction (lysis) of the host cell. Large quantities of assembled virus particles may accumulate within infected cells for hours, or even days, prior to their release.

Assembly and Budding at the Plasma Membrane

The release of enveloped virus particles from the plasma membrane is an intricate process that comprises induction of membrane curvature by viral components (bud formation), bud growth, and fusion of the membrane (scission) to liberate virus particles. As discussed in the previous section, interactions among internal viral proteins and the membrane (and/or viral glycoproteins within it) induce membrane curvature and bud formation. However, with some exceptions (see the next section), viral proteins are not sufficient for membrane scission. In fact, it is now clear that the cellular endosomal complex required for transport (Escrt) mediates the release of many viruses.

Escrt-Dependent Budding

Common sequence motifs are required for budding. A major breakthrough in our understanding of how particles of some viruses bud from the plasma membrane came with the identification of mutants of human immunodeficiency virus type 1 with an unusual assembly phenotype; amino acid substitutions in the p6 region unique to the Gag polyprotein did not impair assembly of immature particles, but the particles remained attached to the host cell by a thin membrane stalk (Fig. 13.17A). It was therefore concluded that these Gag sequences are required for the fusion reaction that separates the viral envelope from the plasma membrane. Subsequently, functionally analogous sequences, termed late-assembly (L) domains, were identified in Gag proteins of several other retroviruses. These L domains are not conserved in their location within Gag or in amino acid sequence but nevertheless can substitute for one another to promote budding.

Retroviral L domains contain a small number of short, core sequence motifs, such as PTAP and PPXY. The recognition of such motifs, and their ability to function independently of position or sequence context, led to identification of L domains in the proteins required for the budding of viruses of several different families (Table 13.2). These **L domain sequences** promote budding by recruitment of cellular proteins that participate in specific steps in vesicular trafficking.

The activity of viral L domains depends on vesicular sorting proteins. The autonomous activity (and in some cases the sequence) of L domains suggested that these sequences mediate protein-protein interactions. Cellular proteins that bind to each of the prototype sequences have now been identified (Fig. 13.17B). The PTAP motif was first shown to

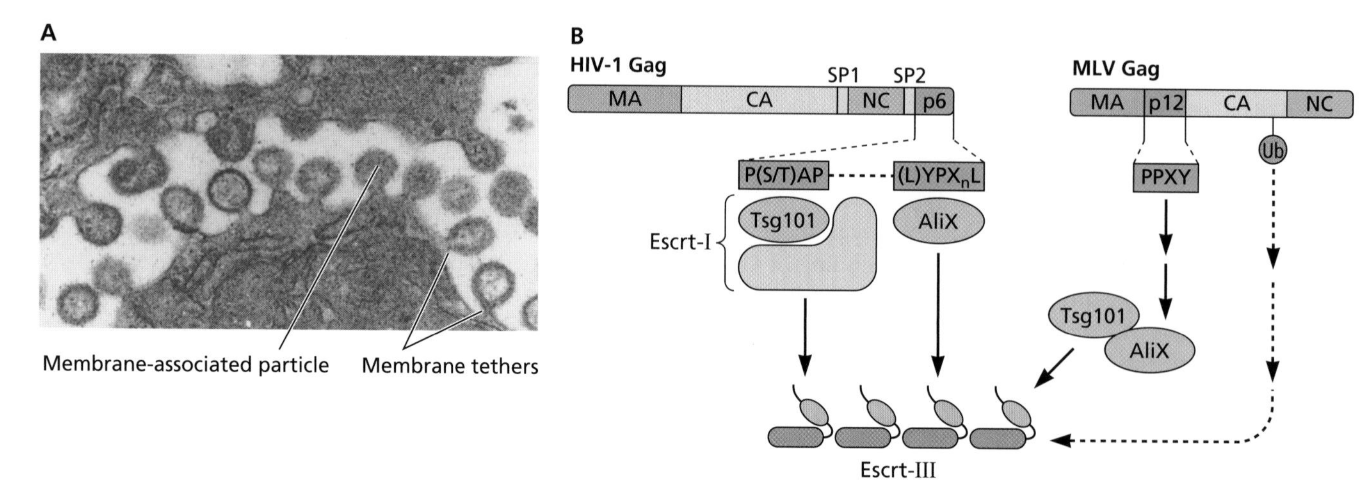

Figure 13.17 L domains and release of retroviral particles. (A) Electron micrograph of monkey Cos-7 cells containing a human immunodeficiency virus type 1 mutant provirus from which Gag p6 cannot be expressed. The plasma membrane-associated particles exhibit normal morphology but remain tethered to the membrane. Adapted from H. G. Göttlinger et al., *Proc. Natl. Acad. Sci. U. S. A.* **88:**3195–3199, 1991, with permission. Courtesy of H. Göttlinger, University of Massachusetts Medical Center. **(B)** Summary of the association of cellular trafficking proteins with core sequence motifs of L domains present in retroviral Gag proteins (and other proteins) required for release of viral particles. Interactions are shown by direct contact between motifs and proteins and by double-headed arrows. The various adapter proteins, such as Escrt-I, Alix, and likely Nedd4 family ubiquitin ligases, recruit Escrt-III to sites of budding. For example, interaction of Alix with the Escrt-III protein Chmp4 (charged multivesicular body protein 4) is required for budding of human immunodeficiency virus type 1. Mammalian cells contain 12 different Escrt-III-like proteins, of which 2 (Chmp2 and Chmp4) are essential for release of human immunodeficiency virus type 1. These proteins are auto-inhibited by interaction of C-terminal segments with a long α-helical core domain. They form homo- and heteromeric filaments upon relief of auto-inhibition, an activity that is thought to drive membrane constriction.

recruit the product of tumor susceptibility gene 101 (Tsg101), an interaction that is essential for budding of human immunodeficiency virus type 1. Mammalian Tsg101 participates in sorting and trafficking of cellular proteins from late endosomes to structures called multivesicular bodies, which fuse with lysosomes. As their name implies, multivesicular bodies contain vesicles within vesicles. The formation of these structures and budding of virus particles are topologically equivalent processes; in both cases, membranes invaginate away from the cytoplasm and fusion releases vesicles with cytoplasmic contents into a lumen or the extracellular space. Recruitment of Tsg101 by the PTAP L domain therefore suggested that the cellular machinery that mediates sorting and trafficking of endocytic vesicles is diverted to promote budding and release of virus particles. In fact, Tsg101 proved to be the human homolog of one subunit of the heteromeric protein Escrt-I, first identified because it is required for sorting of yeast proteins to the vacuole/lysosome. The other subunits of human Escrt-I are also required for release of human immunodeficiency virus type 1. Escrt-I is but one of several multiprotein assemblies that participate in trafficking by way of multivesicular bodies and are necessary for release of retroviruses and other enveloped viruses, including arenaviruses, filoviruses, paramyxoviruses, and rhabdoviruses. Of these, the filamentous protein Escrt-III and the ATPases that associate with it (Vsp4A or -B) act late in budding to drive membrane constriction and fission. The subunits of Escrt-III form a filamentous spiral in the bud neck that is thought to constrict this structure and juxtapose the membranes to promote scission (Fig. 13.18). It therefore appears that formation and release of virus particles with very different structures, genomes, and composition are driven by the same cellular components and mechanism.

Table 13.2 Common sequence motifs required for budding of enveloped virus particles

L domain motif	Escrt component	Viral protein
P (T/S)AP	Tsg101	Human immunodeficiency virus type 1 Gag Murine leukemia virus Gag3 Ebola virus GP40 Bluetongue virus NS
YPXnL	Alix	Human immunodeficiency virus type 1 Gag Rous sarcoma virus Gag Sendai virus M Yellow fever virus NS3
PPXY	Nedd4	Rous sarcoma virus Gag Ebolavirus GP40 Vesicular stomatis virus M

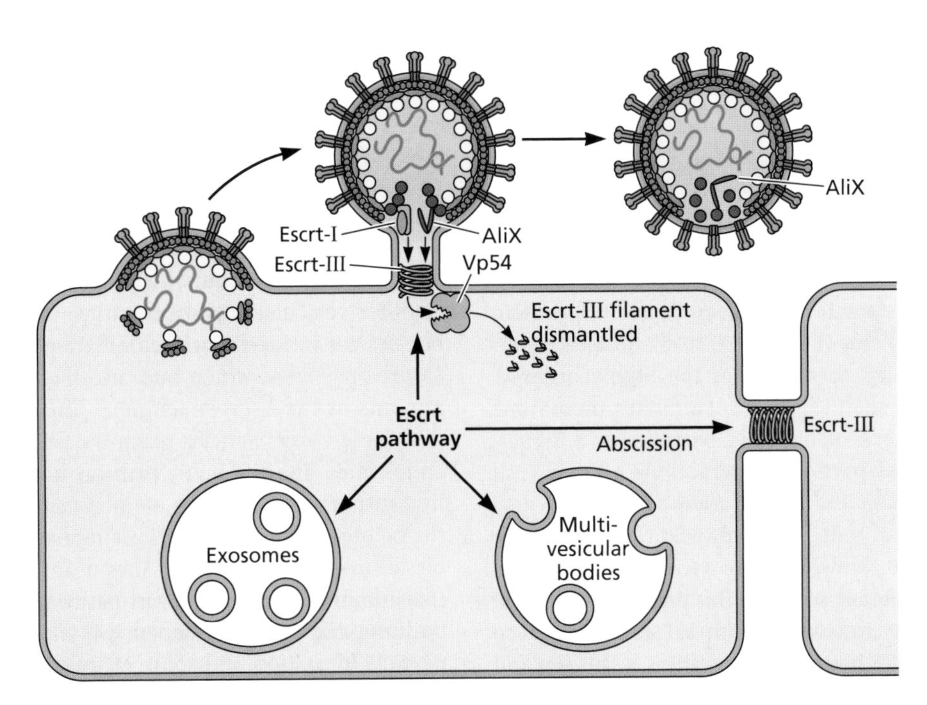

Figure 13.18 Functions of the Escrt pathway in uninfected and virus-infected cells. The Escrt machinery (>30 proteins) catalyzes the membrane fission reactions required for several important cellular processes, including formation of multivesicular bodies and exosomes, as well as the scission reaction that severs the connection between daughter cells in the final step of cell division (abscission). The Escrt pathway is also necessary for the budding and release of retroviruses and a variety of other enveloped viruses; structural proteins of these viruses recruit the core Escrt-III proteins via different adapter proteins (Table 13.2). A model for how cooperation among the various Escrt pathway proteins that interact with L domains present in the human immunodeficiency virus type 1 Gag protein might promote the release of virus particles is illustrated. This model is based on the observations that retrovirus particles contain ubiquitin, which, when attached to Gag, can function as an L domain and that the adapter proteins Alix and Escrt-I can bind to both Gag and ubiquitin and to each other. They may therefore cooperate to recruit Escrt-III for membrane scission at the bud neck. This protein is shown as a filament at this location (middle) and as depolymerized and auto-inhibited individual subunits after release of the virus particle (right). The Vsp4A and -B ATPases can depolymerize Esct-III, presumably to allow its participation in multiple cycles of membrane budding. Vsp4 is required for the release of retrovirus particles, but its molecular role in this process is not yet clear. Adapted from J. Votteler and W. I. Sundquist, *Cell Host Microbe* **14:**2320241, 2013, with permission.

A second L domain (the YPXL domain) that is present in Gag proteins of several retroviruses and facilitates the release of multiple viruses (Table 13.2) also recruits components of the Escrt pathway, in this case via the adapter protein Alix (Fig. 13.17B). A small fraction of retroviral Gag is ubiquitinylated, and a third type of L domain (PPXY) recruits specific ubiquitin ligases (Fig. 13.17B). A catalytically active ubiquitin ligase is necessary for release of retroviruses with this Gag L domain sequence. Furthermore, ubiquitinylation of human immunodeficiency virus type 1 Gag at sites C terminal to the CA domain is necessary for efficient release; substitutions that prevent modification at these sites lower the rate of release and induce the accumulation of virus particles tethered to the plasma membrane. As ubiquitin is recognized by several of the endocytic trafficking proteins, this modification might promote the assembly of the machine that mediates budding and release of retroviruses (Fig. 13.18).

After the identification of its role in release of enveloped virus particles, the Escrt pathway was shown to mediate an analogous reaction in uninfected cells, scission of the thin intercellular bridges between daughter cells during the final step in division (Fig. 13.18) (See the interview with Dr. Wesley Sundquist: http://bit.ly/Virology_Sundquist). Escrt proteins were first identified in budding yeast, and Escrt-III components are also required for cell division of a subset of archaea. These ancient and conserved proteins are thus available in many different species and types of cell in which viruses reproduce.

Escrt-Independent Budding

Although Escrt-dependent budding is a common mechanism for release of enveloped particles from the plasma membrane, it is not the only one; the structural proteins of other enveloped viruses, including influenza viruses and togaviruses, contain no L domains, and budding of these particles is not inhibited by dominant negative derivatives of Escrt pathway proteins.

Budding of togaviruses, such as Ross River and Sindbis viruses, is driven by interactions between capsid (C) protein and envelope glycoproteins (E1 and E2) in the plasma membrane. It is thought that formation of the highly ordered, external glycoprotein shell (Fig. 4.24) facilitates membrane constriction at the neck of budding particles and, hence, scission and release of particles. Interactions among viral glycoproteins (HA, NA) and the internal protein M1 also mediate the assembly of influenza virus particles and induce membrane curvature. However, the viral M2 protein is required for final membrane scission. This protein is recruited to budding particles by interaction with M1 and localizes to the bud neck. It can alter membrane curvature at this site and may do so by membrane insertion of an amphipathic α-helix.

Nonstructural Proteins Can Facilitate Release

Release from the plasma membrane can also depend on viral proteins other than major structural proteins. For example, in some cell types, efficient release of human immunodeficiency virus type 1 requires the viral Vpu protein. In the absence of Vpu, particles accumulate in intracellular vacuoles or are attached to the infected cell surface. This viral protein was shown to counteract the action of an antiviral protein that tethers virus particles to the cell surface and is produced when cells are exposed to interferon α. The organization of this protein, termed tetherin (or bone marrow stromal antigen 2 [Bst2]), suggests that interactions between tetherin molecules inserted in the plasma membrane and the viral envelope are responsible for retaining virus particles at the cell surface. Vpu associates with tetherin in the *trans*-Golgi network to reduce transport of the cellular protein to the plasma membrane. In some cell types, this association leads to ubiquitinylation of tetherin, which is then sorted in an Escrt-dependent manner for lysosomal degradation.

Tetherin is now known to limit the release of many other retroviruses, as well as filoviruses, rhabdoviruses, and herpesviruses, even though the latter buds at an internal membrane (see below). This interferon-inducible protein is an important component of antiviral defense and may also serve as a sensor of viral infection (Volume II, Chapter 3).

Assembly at Internal Membranes: the Problem of Exocytosis

Cytoplasmic Compartments of the Secretory Pathway

Several enveloped viruses are assembled at the cytoplasmic surfaces of compartments of the secretory pathway under the direction of specifically located viral glycoproteins and form by budding of the particle into the lumen of one of these compartments (Table 12.3). These particles therefore lie within membrane-bound organelles. It is generally assumed that such virus particles must be packaged within cellular transport vesicles for travel along the secretory pathway to the cell surface, but few details have been reported. On the other hand, there is accumulating evidence for the participation of other vesicular transport pathways. For example, release but not the intracellular accumulation of infectious hepatitis C virus particles, which bud into the ER, depends on components of the Escrt machinery. This observation suggests that these virus particles reach the cell surface via recycling endosomes. The endocytic pathway has also been implicated in transport to the plasma membrane of immature capsids of the betaretrovirus Mason-Pfizer monkey virus, which assembles at internal cytoplasmic sites near the centrioles. Proteins that function late in the Escrt pathway are also required for budding and release of hepatitis B virus, although the site of particle formation and route of travel to the cell surface are not yet clear.

The budding of virus particles into internal compartments of the secretory pathway is initiated by interactions among the cytoplasmic domains of viral membrane proteins and internal components of the particle. Consequently, this process generally begins as soon as the integral membrane and cytoplasmic viral proteins attain sufficient concentrations in the infected cell. For example, the concentration of viral membrane proteins (surface proteins) determines the fate of hepadnaviral cores, which contain the capsid (C) protein, a DNA copy of the pregenomic RNA, and the viral polymerase (Appendix; Fig. 11). Early in infection, the concentration of the large surface protein (L) in membranes is too low for efficient envelopment of cores, and these structures enter the nucleus, where they contribute to the pool of viral DNA templates for transcription (Fig. 13.19). As the concentration of the L protein increases, it interacts with cores, and enveloped particles form. The ability of hepadnaviral cores to bind to this viral glycoprotein is also regulated by the nature of the nucleic acid that they contain; the synthesis of DNA from the pregenomic RNA induces significant conformational changes in the exterior surfaces of the C protein, notably a more open geometry of a hydrophobic pocket that is lined with residues required for envelopment and thought to make contact with viral envelope proteins.

Although budding into internal compartments imposes the need for subsequent transport and release of virus particles, this mechanism may confer some advantages. Intracellular budding may reduce the concentration of viral glycoproteins exposed on the surface of the infected cell. This property would decrease the likelihood that an infected cell would be recognized by components of the immune system

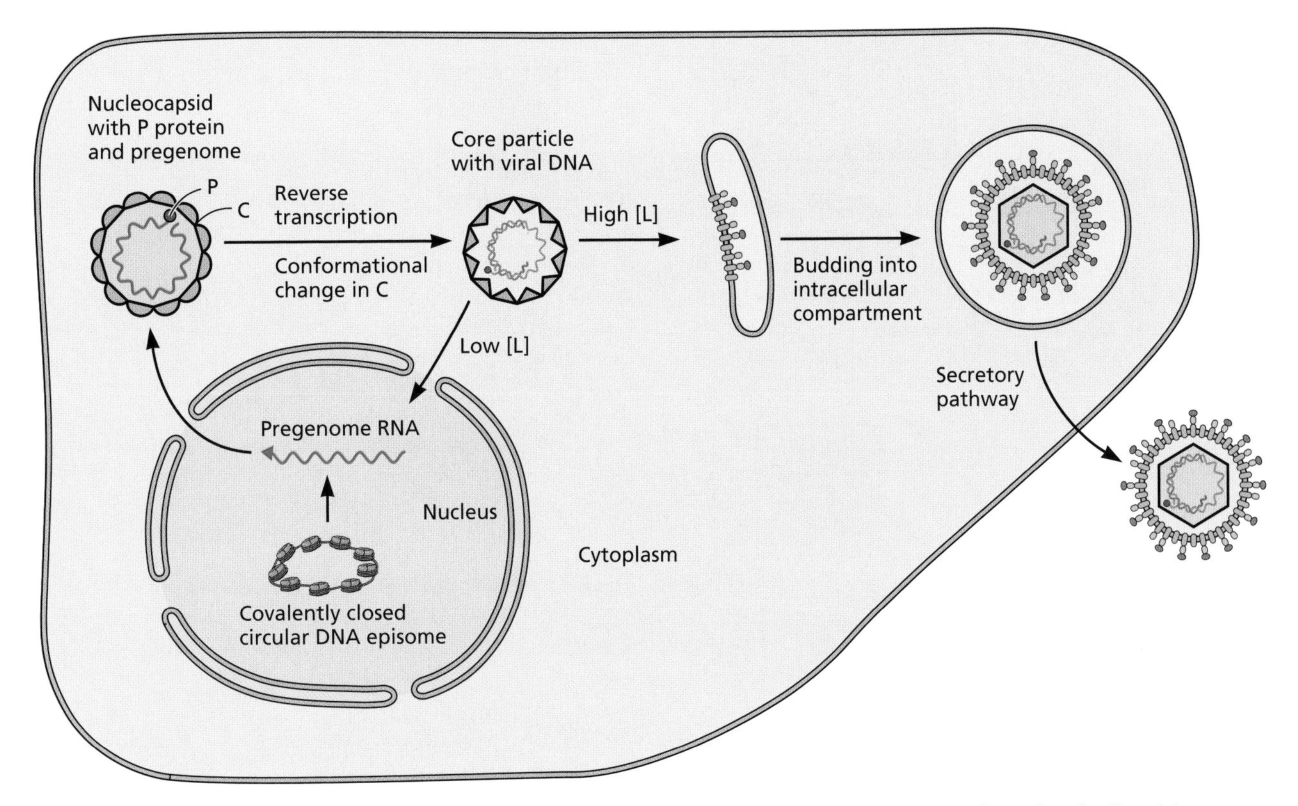

Figure 13.19 Model of hepatitis B virus envelopment. The pregenome RNA synthesized in infected cell nuclei (see Chapter 7) is exported to the cytoplasm, where it is incorporated into particles built from the capsid (C) protein. Reverse transcription to produce the DNA genome induces a conformational change in the C protein that allows interaction of capsid with the large surface protein (L) inserted into internal membranes. Whether core particles containing DNA enter the nucleus or become enveloped by budding into compartments of the secretory pathway is determined by the concentration of the L protein. The L, middle (M), and small (S) envelope glycoproteins accumulate in membranes of the ER-Golgi intermediate compartment, into which subviral particles that contain only lipid and envelope proteins (primarily S) also appear to bud. As the S, as well as the L, protein is required for envelopment, it is generally accepted that virus particles are also formed by budding into this same compartment of the secretory pathway. However, the results of recent experiments indicate that cellular proteins that participate in endocytic trafficking (see the text) participate in hepatitis B virus budding and release.

before the maximal number of progeny particles was assembled and released. Alternatively, the simpler cytoplasmic surfaces of internal membranes, which are not burdened with cytoskeletal structures and the proteins that attach them to the extracellular matrix, may make for more-facile assembly or budding reactions, or the distinctive lipid composition of internal membranes may confer some (as-yet-unknown) special property that is advantageous to these viruses.

Envelopment by a Virus-Specific Mechanism

The interaction of components of the poxvirus vaccinia virus with internal cellular membranes during assembly is most unusual. One remarkable feature is the assembly of two **different** infectious particles, which have been termed the intracellular mature and the extracellular enveloped virions, that differ in the number and origin of their lipid membranes. Furthermore, the initial acquisition of a membrane early in assembly occurs by a virus-specific mechanism that appears to be shared with other large DNA viruses that assemble in the cytoplasm, such as mimivirus. Finally, infectious particles leave the host cell by at least three distinct routes.

Vaccinia virus assembly includes the formation of several intermediates, such as crescents (see below) and immature particles, and major morphological rearrangements as infectious particles are formed (Fig. 13.20). The assembly pathway was elucidated initially by electron microscopy in some of the earliest studies of vaccinia virus. Numerous viral proteins that participate in the various assembly reactions have been identified by genetic experiments (Table 13.3). Synthesis of viral DNA genomes and structural proteins takes place in discrete cytoplasmic domains termed viral factories. The first morphological sign of assembly is the appearance within viral factories of rigid, curved structures 10 to 15 nm

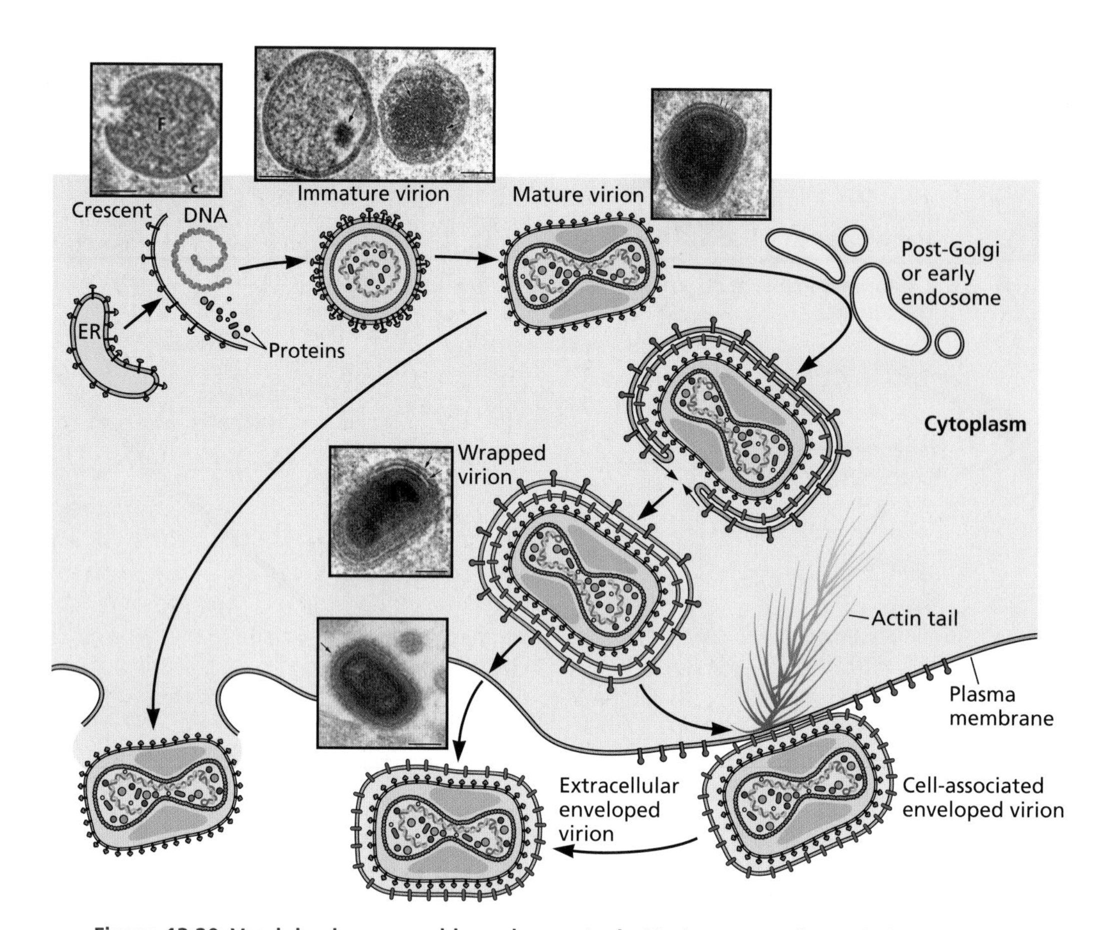

Figure 13.20 Vaccinia virus assembly and exocytosis. Viral structures observed when HeLa cells infected with vaccinia virus for 10 or 24 h were prepared for electron microscopy by quick freezing and negative staining while frozen are shown in a schematic model of assembly and exocytosis. Assembly begins with the formation of crescents by diversion of membrane from the ER. That shown in the electron micrograph (c) is present in a viral factory (F) and encircles a dense focus of viral material (viroplasm). The viral D13 protein, which is associated with the outer leaflets of crescents, maintains the curvature and rigidity of the crescent membrane as it enlarges and eventually closes with the incorporation of viral DNA and proteins from viral factories to form immature virus particles, two examples of which are shown. As the D13 protein is lost during the morphological transitions that form the brick-shaped mature virus particle, it is considered a scaffolding protein. The mature particle is released from infected cells only upon lysis. However, a significant proportion of these structures acquire additional membranes by wrapping in membranes derived from a late or post-Golgi compartment to form the intracellular enveloped virus particle. The additional double membrane is indicated by the arrows in the electron micrograph. This particle is transported to the plasma membrane, where fusion with this membrane forms the cell-associated enveloped virus, which has lost one outer membrane layer. This particle induces formation of actin tails. Adapted from B. Sodeik and J. Krijnse-Locker, *Trends Microbiol.* **10:**15–24, 2002, with permission. Electron micrographs are adapted from C. Risco et al., *J. Virol.* **76:**1839–1855, 2000, with permission. Bars, 100 nm.

thick (Fig. 13.20). There has never been any doubt that these structures, termed crescents, contain at least one lipid bilayer. In contrast, the origin of this membrane has been a subject of much debate. The current consensus is that crescents contain a single lipid membrane that is derived from the ER membrane but by a virus-specific mechanism (Box 13.8). As the crescents enlarge, they retain their original curvature and therefore eventually form spheres surrounding viral macromolecules present in viral factories, including the DNA genome (Fig. 13.2). Such immature particles then undergo major morphological transitions to form brick-shaped mature virions (Fig. 13.20). This maturation process requires several distinct reactions, including proteolytic cleavage of several structural proteins by a viral protease(s) (Table 13.3), the action of the viral redox system (Chapter 12), and removal of at least one crescent-associated protein. These changes resemble those that occur during assembly of herpesviruses and adenoviruses.

Table 13.3 Some proteins implicated in vaccinia virus assembly[a]

Assembly reaction	Protein(s)	Function(s)/properties
Crescent formation	AG	Localization of major viral membrane proteins to viral factories
	A11	Colocalizes with ER at viral membranes in viral factories
	A14, A17	Essential for this step; integral membrane proteins; phosphorylated by F10; form a disulfide bond-stabilized lattice
	D13	Imparts crescent curvature and rigidity; interacts with A17
	F10	Essential for ER membrane remodeling and appearance of crescents; dual-specificity protein kinase
Assembly of IV	Complex of 7 proteins, including F10	Association of viroplasm with crescent membranes
	A32	Genome encapsidation; required for packaging DNA and I6
	I6	Genome encapsidation; binds specifically to terminal hairpins in DNA
Formation of MV	A4	Core assembly during morphogenesis; present in outer palisade layer of core wall
	A3	Formation of morphologically normal and transcriptionally active cores; proteolytically processed during morphogenesis
	A19	Efficient processing of core proteins
	G1	IV-to-MV transition; metalloprotease
	I7	IV-to-MV transition; cysteine protease, required for processing A3 and other proteins
	E10, A2.5, G4	IV-to-MV transition; thiol redox proteins (see Chapter 12)
Formation of wrapped virus particles	A27	Essential for this step; disulfide-bonded trimer bound to MV membrane
	B5, F13	Required for efficient wrapping; transmembrane proteins sorted to intracellular wrapping site(s)

[a]IV, immature virus particles; MV, mature virus particles.

The mature virus is released only upon lysis of the infected cell. However, some of these particles become engulfed by the membranes of a second intracellular compartment, probably a *trans*-Golgi or early endocytic compartment, to form the wrapped virus particle (Fig. 13.20). The mature particle is transported to the site(s) of wrapping via microtubules. The remodeling of organelle membranes to form the wrapped particle depends on a number of viral proteins that are present only in this type of particle (Table 13.3) and that appear to be sorted to wrapping sites via the secretory pathway. The wrapped particle can be released from the cell as the two-membrane-containing extracellular enveloped virus particle, following transport to the cell surface and fusion of its outer membrane with the plasma membrane (Fig. 13.20). As the mature and the extracellular enveloped virions bind to different cell surface receptors, the release of two types of infectious particle may increase the range of cell types that can be infected. A significant proportion of enveloped virus particles are not released following membrane fusion but, rather, remain attached to the host cell surface as cell-associated enveloped virions. The mechanisms of transport and egress that produce these cell-associated particles are amazing processes that depend on major reorganization of components of the host cell cytoskeleton.

Wrapped virus particles initially travel from sites of assembly to the plasma membrane on microtubules, carried by a cellular motor protein of the kinesin family. The interaction of these particles with the motor depends on the viral A36 protein present in their outer membrane, which binds to the light chain of the kinesin motor. Such active transport allows movement of the large wrapped virus particles to the cell periphery in less than 1 min (compared to an estimated 10 h that would be required by passive diffusion!). Remodeling of the dense layer of cortical actin that lies beneath the plasma membrane (Fig. 2.4) is also required to deliver these particles to the cell surface. This phenomenon is induced by a viral protein that modulates the cellular signaling pathway that regulates the dynamics of cortical actin.

The particles formed by fusion of wrapped virus particles with the plasma membrane remain cell associated because of a remarkable activity: they induce a further dramatic reorganization of the actin cytoskeleton just below the site of fusion. The number of typical actin stress fibers is significantly decreased, as the virus induces the formation of

BOX 13.8

DISCUSSION

The enigma of how the vaccinia virus crescent membrane formed

It is simple to visualize how reorganization and fusion of internal cellular membranes can "wrap" structures in a double membrane, as during formation of wrapped particles of vaccinia virus (Fig. 13.20). In contrast, it is not at all obvious how viral structures containing a **single** lipid bilayer, namely, the crescent that is the first structure built during vaccinia virus assembly (see the figure), can arise by a non-budding mechanism. This conundrum led to the early proposal that the crescent membrane is synthesized *de novo* from cellular lipids. No mechanism for such *de novo* assembly has been identified, and it is generally agreed that crescents are derived from preexisting cellular membranes. There is accumulating evidence that the ER is the source of the crescent membrane.

- Several of the major viral membrane proteins are inserted into the ER membrane in infected cells, and one (A9) is present, near sites of assembly, in tubular structures that contain the ER luminal enzyme protein disulfide isomerase. Furthermore, when a heterologous signal sequence was added to the N terminus of A9, the signal sequence was cleaved off and only the truncated protein was detected in immature and mature virus particles. As signal peptidase, which removes signal sequences, resides in the ER (Chapter 12), this observation provides compelling support for the view that viral membrane proteins travel from the ER to the crescent membrane.
- Inhibition of transport from the ER to the Golgi compartments via the secretory pathway did not impair assembly of immature or mature virus particles (although the subsequent wrapping step was blocked).
- Repression of synthesis in infected cells of the major components of the crescent membrane, namely, the transmembrane proteins A14 and A17 and the scaffold protein D13, led to formation of only irregular membranes or small vesicles and tubules and the accumulation of dense, virus-specific inclusions.
- Several other viral proteins are required for assembly of the crescent membrane and immature virus particles. Of these, the L2 protein is synthesized in infected cells before viral factories appear and is associated with the ER at sites at which crescent membranes will form. Repression of L2 synthesis resulted in complete inhibition of vaccinia virus reproduction and assembly of mature virus particles. Some immature virus-like particles did assemble but contained greatly reduced quantities of several viral membrane proteins. Very short crescents associated with the major crescent proteins and the ER chaperone calnexin were also observed.

These observations indicate that the crescent membrane is constructed from the ER under the direction of viral proteins, but the mechanism remains an enigma.

Hussain M, Weisberg AS, Moss B. 2006. Existence of an operative pathway from the endoplasmic reticulum to the immature poxvirus membrane. *Proc Natl Acad Sci U S A* **103:**19506–19511.

Maruri-Avidal L, Weisberg AS, Bisht H, Moss B. 2013. Analysis of viral membranes formed in cells infected by a vaccinia virus L2 deletion mutant suggests their origin from the endoplasmic reticulum. *J Virol* **87:**1861–1871

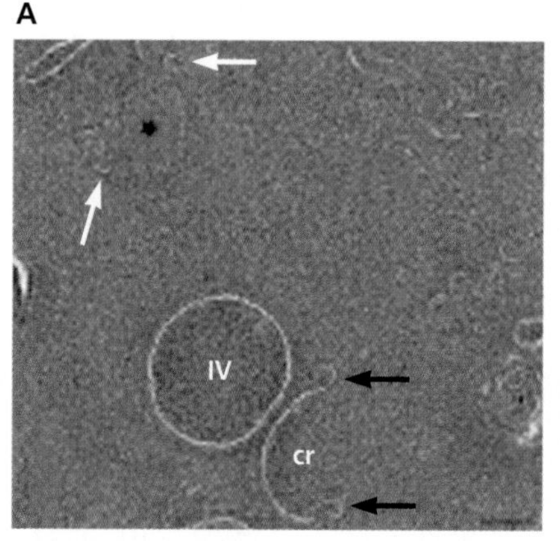

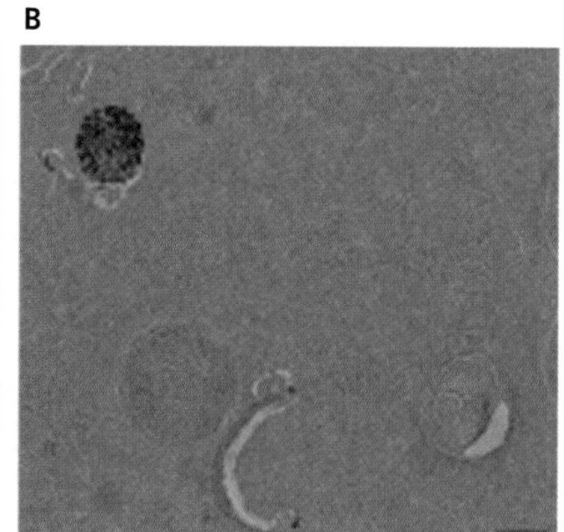

Cryo-electron tomography of vaccine virus-infected HeLa cells showing a 0.9-nm slice taken from a 200-nm-thick cryo-section, with an immature virus particle (IV), a cresent (cr), and a small patch of scaffold protein (star) indicated. The black arrows indicate the ends of the crescent membrane that typically curl away from the coated scaffold region, and the white arrows the small membrane curls that are seen close to patches of the scaffold protein. In the right panel, the crescent membrane and scaffold have been rendered in green and red, respectively. Adapted from C. Suarez et al., *Cell. Microbiol.* **15:**1883–1895, 2010. Courtesy of J. Krijnse Locker, Heidelberg University, Heidelberg, Germany.

new, filamentous, actin-containing structures. Each of these, which are termed actin tails, is in contact with a single virus particle (Fig. 13.21A and B). Viral particles attached to the tips of actin tails are propelled by the polymerization of actin at the front end of the tail and its depolymerization at the back end. As the infection progresses, they can be seen on large microvilli induced by the actin tails (Fig. 13.21B). Formation of actin tails in vaccinia virus-infected cells requires the same viral protein (A36) that allows transport of wrapped virions along microtubules. This protein is phosphorylated at specific positions by the cellular tyrosine kinase Src, which plays an important role in the regulation of actin dynamics in uninfected cells. Phosphorylation of A36 triggers its dissociation from kinesin and allows binding of cellular proteins that promote actin polymerization (Fig. 13.22).

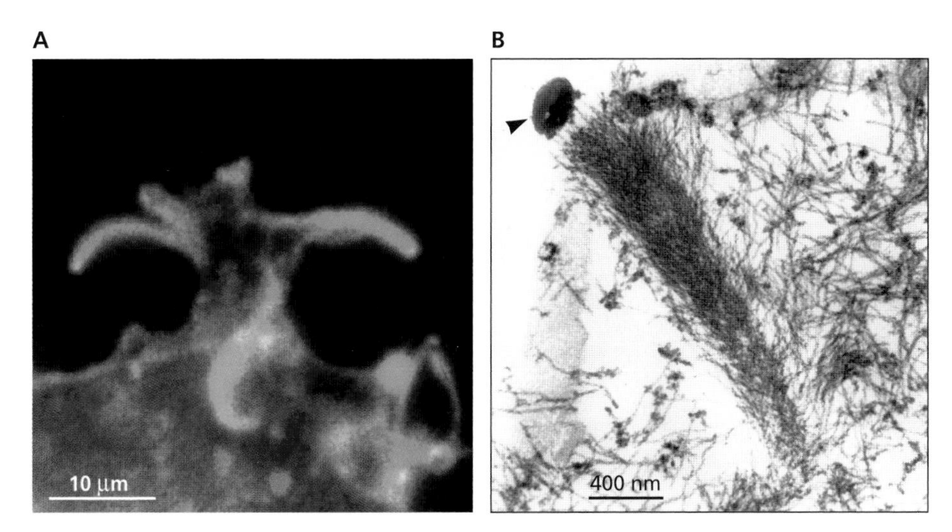

Figure 13.21 Movement of vaccinia virus on actin tails. (A) Immunofluorescence micrograph of virus particles (red) at the ends of the cell surface projections containing actin tails (green). The coincidence of the tips of the projecting actin and viral particles gives yellow-orange signals, indicating that the particles are projected from the cell surfaces on the tips of actin tails. When infected cells are plated with uninfected cells, such actin-containing structures to which virus particles are attached can be seen extending from the former into the latter. **(B)** Electron micrograph of a virus particle attached (arrowhead) to an actin tail. From S. Cudmore et al., *Nature* **378:**636–638, 1995, with permission. Courtesy of S. Cudmore and M. Way, European Molecular Biology Laboratory.

The formation of vaccinia-actin tails is necessary for efficient spread of the virus; mutants that cannot induce these structures form only small plaques on cells in culture. Cellular projections containing actin tails with virus particles at their tips can extend from infected cells toward neighboring uninfected cells, suggesting that they may facilitate direct cell-to-cell spread of infectious particles. More importantly for rapid spread of vaccinia virus, they mediate a remarkable mechanism of repulsion of virus particles from infected cells (Box 13.9).

Intranuclear Assembly

The problem of egress is especially acute for the enveloped herpesviruses, because the nucleocapsids assemble in the nucleus. The pathway by which the virus leaves the cell has been a topic of fierce controversy, centered on where and when the viral envelope is acquired. A large body of evidence now favors the less intuitive double-envelopment model summarized in Fig. 13.23.

The first step in egress is exit of nucleocapsids from the nucleus, which is achieved not by transport through nuclear pore complexes but, rather, by an unusual budding mechanism (Fig. 13.23), which was shown subsequently to export large ribonucleoproteins containing certain cellular mRNAs (Box 13.10). In the case of herpes simplex virus 1, a subset of the tegument proteins, including VP16, associates with the nucleocapsid prior to budding. Late in infection, two viral proteins act in concert with the cellular protein kinase C to induce disruption of the nuclear lamina (Fig. 13.24) and subsequently drive budding of the nucleocapsid through the inner nuclear membrane. The de-envelopment reaction that subsequently releases nucleocapsids into the cytoplasm (Fig. 13.23) requires either the gB or the gH glycoprotein and the cellular ATPase torsinA.

The second envelopment, in which particles acquire their envelopes, takes place at the cytoplasmic surfaces of compartments of the *trans*-Golgi network. Viral membrane proteins, including those necessary for secondary envelopment (e.g., gD, gE/gI, gM, and the UL20 protein), are sorted to these cellular compartments via the secretory pathway (Chapter 12). Some tegument proteins accumulate at the sites of secondary envelopment and are required for this step. Others associate with the nucleocapsid in the cytoplasm. The latter proteins include the US3 required for nuclear exit and the UL36 and UL37 proteins, which are required for transport of nucleocapsids through the cytoplasm. Once the nucleocapsid reaches the *trans*-Golgi network, interactions between these two classes of tegument protein must take place prior to envelopment. Nevertheless, the proteins that mediate such final assembly of the tegument have not yet been identified, nor have those that induce membrane budding and fusion. Some recent observations hint that such viral proteins may function via some components of the cellular Escrt machinery that mediates the release of simpler enveloped viruses.

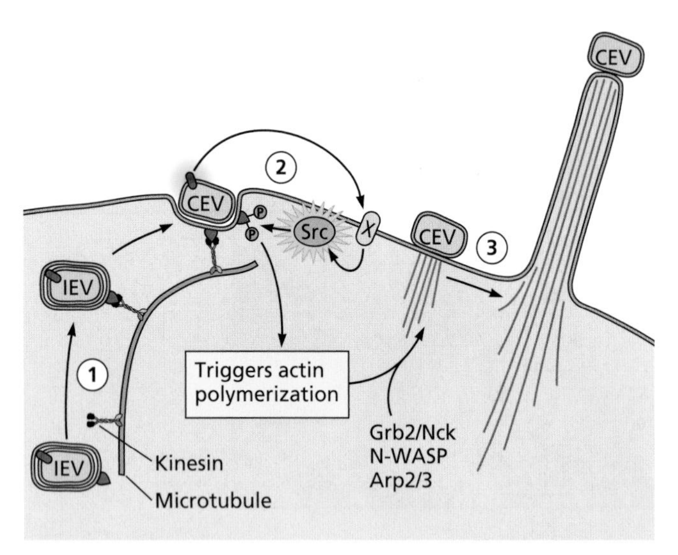

Figure 13.22 Model for the switch from microtubule- to actin-dependent transport of vaccinia virus particles. The A36R protein (red) present in the outer membranes of wrapped virus particles binds to the light chain of the kinesin motor, which then transports the particles to the cell periphery. Remodeling of cortical actin by viral proteins allows close approach of the particles to the plasma membrane. Fusion of the outer membrane of wrapped virus particles with the plasma membrane releases cell-associated virus particles, which carry the B5R glycoprotein (blue) in their new outer membrane. This viral protein activates the cellular Src tyrosine kinase, presumably via interaction with one or more cellular membrane proteins (X). Src then phosphorylates the membrane-associated A36R protein, a modification shown by genetic experiments to be essential for formation of actin tails. Furthermore, A36 remains bound to kinesin in vaccinia virus-infected cells that lack Src or that are treated with inhibitors of this kinase. Phosphorylated A36R binds via adapter (Grb and Nck) and scaffolding (N-Wasp) proteins to proteins that induce actin polymerization. Such polymerization drives the formation of actin tail-containing protrusions that project cell-associated virus particles away from the host cell. The viral F11 protein-induced inhibition of signaling via the small G protein RhoA leads to increased microtubule dynamics and facilitates transport of progeny virus particles to the plasma membrane. Such inhibition also stimulates the migration of vaccinia virus-infected cells, a property that promotes spread of progeny virus particles. CEV, Cell-associated enveloped virions; IEV, Intracelleular enveloped virion. Adapted from A. Hall, *Science* **306:**65–67, 2006, with permission.

Release of Nonenveloped Viruses

The most usual fate of host cells permissive for reproduction of nonenveloped viruses is death (but see Volume II, Chapter 5). In natural infections, the host defenses are an important cause of infected-cell destruction. However, infection by these viruses destroys host cells more directly; they are cytopathic to cells in culture. Although the mechanisms by which reproduction of nonenveloped viruses induces death and lysis of host cells are not well understood, some viral proteins that induce rupture of particular membranes and/or cell lysis have been identified.

The VP4 protein of the polyomavirus simian virus 40, which accumulates late in infection, perforates membranes *in vitro* by forming pores in them. It is considered a **viroporin**, a class of small, hydrophobic proteins that are encoded in the genomes of a variety of viruses; examples include the influenza A virus M2 protein and Vpu of human immunodeficiency virus type I (see Volume II, Chapter 7). Simian virus 40 VP4 associates with the nuclear envelope, where it induces release of nuclear contents into the cytoplasm. Such activity assuredly contributes to the escape of newly assembled virus particles from the nucleus and might contribute to lysis of the host cell. However, the viral agnoprotein is also likely to be important; the analogous protein of the human polyomavirus JC virus has been reported to form pores in the plasma membrane to facilitate release of progeny virus particles. A small viral protein is also necessary for efficient nuclear disruption and lysis of cells infected by human adenovirus. This adenovirus death protein (ADP) accumulates in the nuclear envelope late in infection and stimulates release of virus particles, but its mechanism of action is not clear. The severe inhibition of cellular protein synthesis toward the end of the infectious cycle and disruption of cytoplasmic intermediate filaments upon cleavage of their components by the viral L3 protease are likely to facilitate release of adenovirus particles by compromising the structural integrity of the infected cell.

While cell lysis is the most common means of escape of naked viruses, there is evidence that some are released in the absence of any cytopathic effect. When poliovirus replicates in polarized epithelial cells resembling those lining the gastrointestinal tract (a natural site of infection), progeny virus particles are released exclusively from the apical surface by a nondestructive mechanism. The viral 2BC and 3A proteins induce the formation of infected-cell-specific vesicles that closely resemble autophagosomes (see Chapter 14). Coxsackie B virus particles are also released in autophagosomes. It has been proposed that these vesicles, which contain two membranes and virus particles late in infection provide a route for nonlytic release of particles assembled to the cytoplasm (Fig. 13.25A). Another pathogenic picornavirus, hepatitis A virus, leaves liver cells in culture in enveloped particles that resemble exosomes (Fig. 13.25B). This "wolf-in-sheep's-clothing" strategy for release prevents recognition of virus particles by neutralizing antibodies *in vitro* and presumably aids spread of the virus in the liver.

Maturation of Progeny Virus Particles

Proteolytic Processing of Structural Proteins

The products of assembly of several viruses are noninfectious particles. In all cases, proteolytic processing of specific proteins with which the particles are initially built converts them to infectious virions. The maturation reactions are carried out

BOX 13.9

EXPERIMENTS

Repulsion of virus particles from infected cells accelerates vaccinia virus spread

Vaccinia virus particles are spread by mechanisms that include increased migration of cells induced by infection and propulsion toward neighboring cells on actin projections (Fig. 13.21). Measurement of the rate of increase in the size of vaccinia virus plaques in various cell lines indicated that the virus crossed one cell every 1.2 h. This rate of spread is considerably higher than can be explained by either the assembly of progeny virus particles or the induction of infected-cell motility, both of which require 5 to 6 h after the initial infection. Mutant viruses defective for formation of actin tails infected new cells only every 5 to 6 h, consistent with the kinetics of the infectious cycle. This finding implicated actin tail formation in the rapid spread of vaccinia virus.

Virus particles containing a structural protein fused to enhanced green fluorescent protein (eGFP) and cells producing actin fused to Mcherry fluorescent protein were used to investigate the mechanism of rapid spread. Green virus particles were detected on red actin tails in cells that contained no viral factories or progeny virus particles. These structures appeared before viral factories, and particles on a red actin tail induced the formation of a new actin tail upon recontact with the same cells (see the figure and Movie 13.1: http://bit.ly/Virology_V1_Movie13-1), suggesting a mechanism of active repulsion of virus particles from infected cells.

The viral A33 and A36 proteins, which are required for formation of actin tails, are made early in the infectious cycle and accumulate in the plasma membrane at the edges of plaques. Mutant viruses that direct the synthesis of these proteins late rather than early in infection produce only small plaques. Furthermore, the synthesis of just these two proteins in uninfected cells allowed the formation of actin tails within 15 to 30 min after exposure to extracellular enveloped virus particles.

These observations identified a previously unrecognized mechanism of spread of vaccinia virus particles, repulsion from infected cells on actin tails toward neighboring cells. This process prevents superinfection and hence accelerates the rate of spread of the virus.

Doceul V, Hollinshead M, van der Linden L, Smith GL. 2010. Repulsion of superinfecting virions: a mechanism for rapid virus spread. *Science* **327:**873–876.

Simian BSC-1 cells synthesizing actin fused to cherry fluorescent protein (red) were infected at a low multiplicity of infection with vaccinia virus with a structural protein (A5L) fused to eGFP (green). Shown are actin tails formed at the times indicated (min) after infection, and before the appearance of large green viral factories at 55 min. Scale bar = 5 μm. The time-lapse movie shows such a cell and induction of a new actin tail when a virus particle at the tip of a red actin tail recontacts the cell surface. Adapted from C. Doceul et al., *Science* **327:**878–867, with permission. Courtesy of G. L. Smith, Imperial College, London, United Kingdom.

by virus-encoded enzymes and take place late in assembly of particles or following their release from the host cell. Proteolytic cleavage of structural proteins introduces an irreversible reaction into the assembly pathway, driving it in a forward direction. This modification can also make an important contribution to resolving the contradictory requirements of assembly and virus entry. One consequence of proteolytic processing is the exchange of covalent linkages between specific protein sequences for much weaker noncovalent interactions, which can be disrupted in a subsequent infection. A second is the liberation of new N and C termini at each cleavage site and, hence, opportunities for additional protein-protein contacts. Such changes in chemical bonding among structural proteins clearly facilitate virus entry, for the proteolytic cleavages that introduce them are necessary for infectivity. Accordingly, viral proteases and the structural consequences of their actions are of considerable interest. Moreover, these enzymes are excellent targets for antiviral drugs, as exemplified by the success of therapeutic agents that inhibit the human immunodeficiency virus type 1 protease.

Cleavage of Polyproteins

The alterations in the structure of the virus particle following proteolytic processing and their functional correlates are best understood for small RNA viruses, such as the picornavirus poliovirus. A single cleavage to liberate VP4 and VP2 from VP0 converts noninfectious provirions to mature virus particles (Fig. 13.5). As the viral proteases are not incorporated into particles, VP0 cleavage may be catalyzed by a specific feature of the capsid itself, with internal genomic RNA participating in the reaction. The structural changes induced by such maturation cleavage can be described in great detail, for the structures of mature and empty particles in which VP0 has not been cleaved have been determined at high resolution. Cleavage of VP0 allows the extensive internal structures of the particle (Fig. 4.12C) to be established and consequently is important for the stability of the virion.

Cleavage of VP0 to VP4 and VP2 is also necessary for the release of the RNA genome into a new host cell. The conformational transitions that mediate entry of the genome

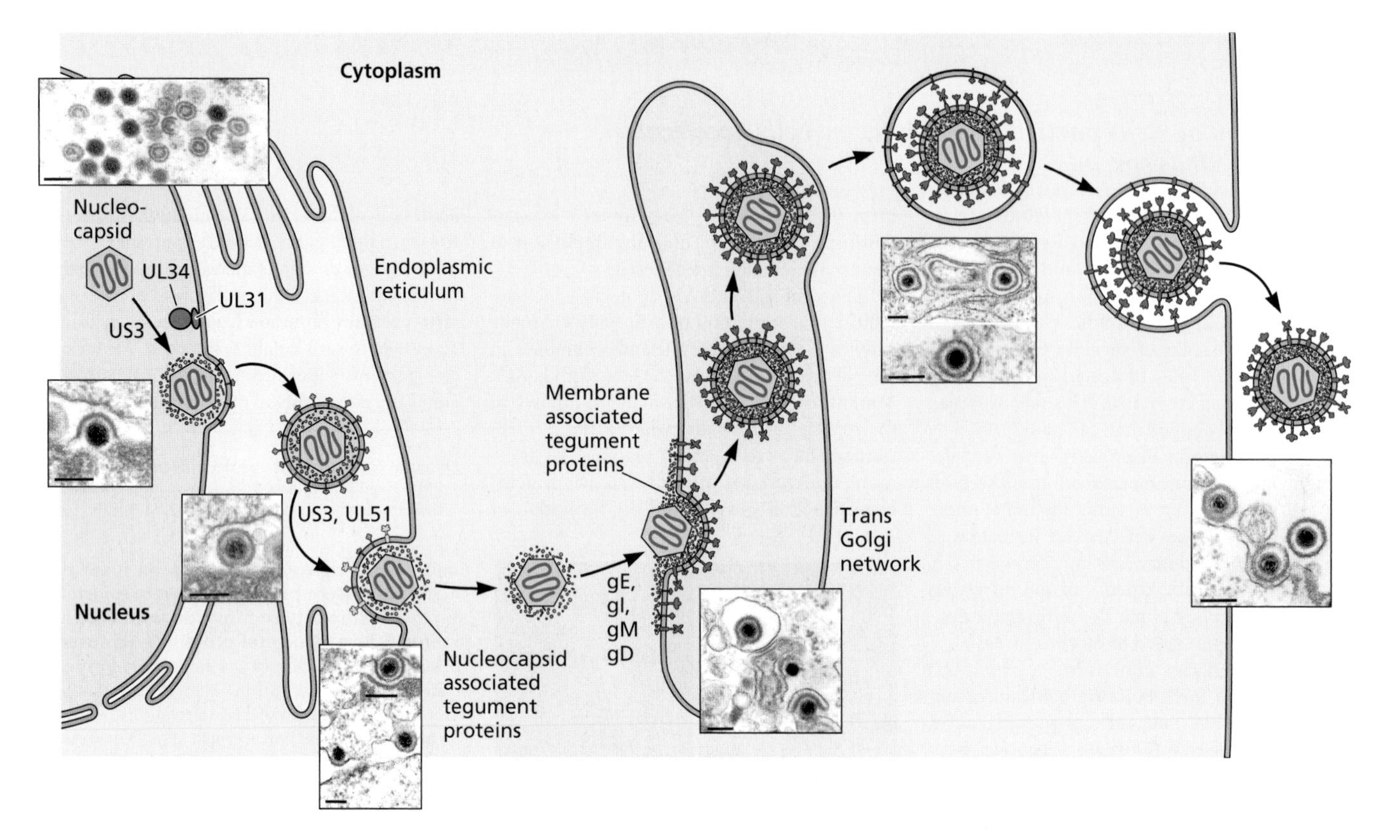

Figure 13.23 Pathway of herpesvirus egress. The mature nucleocapsid assembled within the nucleus (Fig. 13.8) initially acquires an envelope by budding through the inner nuclear membrane. The dense meshwork of protein filaments that abuts the inner nuclear membrane (the nuclear lamina) is dramatically reorganized and perforated (Fig. 13.24), presumably to allow juxtaposition of the nucleocapsid and membrane. Such disruption of the nuclear lamina requires the viral UL31 phosphoprotein and the UL34 transmembrane protein. These proteins, often called the nuclear export complex, associate with one another at the inner surface of the inner nuclear membrane and bind the proteins that form the lamina (lamins A/C and B) and cellular protein kinase C. This enzyme phosphorylates the lamins, while the viral US3 kinase phosphorylates the nuclear membrane protein emerin, which binds to lamins and has been implicated in the maintenance of nuclear integrity. These modifications are thought to disrupt the interactions that form the nuclear lamina. The UL31 protein interacts with nucleocapsids, and the UL31-UL34 assembly is sufficient to induce deformation and scission of membranes *in vitro*. These activities are inhibited by substitutions in the UL34 protein that block nuclear exit of nucleocapsids but not their assembly with UL31 at the inner nuclear membrane, suggesting that the two viral proteins drive budding through this nuclear membrane. Upon fusion with the outer nuclear membrane, this membrane is lost as unenveloped nucleocapsids are released into the cytoplasm. Some tegument proteins interact with the nucleocapsid in the cytoplasm, whereas others, including the UL11, UL46, and UL49 proteins, concentrate at sites of secondary envelopment. The latter are presumably localized at membranes of *trans*-Golgi compartments by interactions with the cytoplasmic domains of viral glycoproteins, such as the binding of the UL11 and UL49 proteins to the cytoplasmic domains of gE and gD. The myristoylated UL11 protein accumulates at the membranes of *trans*-Golgi compartments and directs other tegument proteins to sites of secondary envelopment. The viral envelope is acquired upon budding of tegument-containing structures into compartments of the *trans*-Golgi network. Virus particles formed in this way are thought to be transported to the plasma membrane in secretory transport vesicles and released upon membrane fusion, as illustrated. Viral gene products implicated in specific reactions are indicated. The reactions are illustrated in the electron micrographs of cells infected by the alphaherpesvirus pseudorabies virus. Bar, 150 nm. Adapted from T. C. Mettenleiter, *J. Virol.* **76:**1537–1547, 2002, with permission. Courtesy of T. C. Mettenleiter, Federal Research Center for Virus Diseases of Animals, Insel Riems, Germany.

BOX 13.10

TRAILBLAZER

Budding as a mechanism for export of cellular mRNAs from the nucleus

Until quite recently, nuclear pore complexes were thought to provide the only routes for transport of molecules and macromolecules between the nucleus and cytoplasm. These large and architecturally elaborate structures (Fig. 5.23) allow free passage of molecules of less than 20 kDa and bidirectional transport up concentration gradients of larger proteins and RNAs (almost always packaged as ribonucleoproteins [RNPs]). Many viral genomes reach the nucleus by way of nuclear pore complexes (Chapter 5). Toward the end of infectious cycles, progeny genomes of several viruses, including influenza viruses and retroviruses, are exported to the cytoplasm via these structures, whereas particles that complete assembly in the nucleus (for example, adenoviruses and polyomaviruses) escape that organelle upon destruction and lysis of the host cell.

When initially discovered, nuclear budding of newly assembled nucleocapsids of alphaherpesviruses was a virus-specific mechanism with no counterpart in uninfected cells. Subsequently, however, budding of cellular mRNAs was discovered in studies of the development of neuromuscular junction synapses in the body wall muscles of *Drosophila* larvae. This process depends on the secreted signaling protein wingless (Wnt). Binding of Wnt to its receptor, *Drosophila* frizzled-2 (Dfz2), on a postsynaptic cell induces endocytosis of the receptor, its cleavage, and import of the C-terminal segment (Dfz2C) into the nucleus. Application of various imaging techniques established that the latter protein is associated with large, RNA-containing granules in the space between the inner and outer nuclear membranes. These granules could be seen leaving the nucleus. They contained multiple mRNAs coding for postsynaptic proteins, and mutations that impaired the formation of the granules prevented proper synaptic differentiation. Like budding of herpesviral nucleocapsids, formation of Dfz2C-containing granules and invaginations of the inner nuclear membrane require protein kinase C and phosphorylation of lamins. Furthermore, exit of both the mRNA-containing granules in *Drosophila* larvae and herpesviral nucleocapsids in mammalian cells to the cytoplasm depends on torsinA.

A budding mechanism might be necessary for export of very large structures, be they ribonucleoprotein granules or viral assemblies. Such a mechanism might also ensure simultaneous export of multiple mRNAs coding for proteins that operate together for subsequent cotransport and cotranslation. Despite previous reports of perinuclear granules and inner nuclear membrane invaginations in both animal and plant cells, budding from the nucleus of other cellular mRNAs has not yet been demonstrated.

Speese SD, Ashly J, Joki V, Nunnari J, Barniw R, Alaman B, Koon A, Chang Y-T, Li Q, Moore MJ, Budnik V. 2012. Nuclear envelope budding enables large ribonucleoprotein particle export during synaptic Wnt signaling. *Cell* **149:**832–846.

Nuclei from Drosophila larval body wall muscle were examined by live-cell imaging after incubation with an RNA-specific dye (E46) (green; left panel) and then visualized after being fixed and stained with antibodies to lamin C (red; middle panel). These images are shown superimposed on the right. The arrow indicates an RNA-containing granule. Such granules were observed moving away from the nucleus during time-lapse imaging. Adapted from S. Speese et al., *Cell* **149:**832–836, 2012, with permission. Courtesy of V. Budnik, University of Massachusetts Medical School.

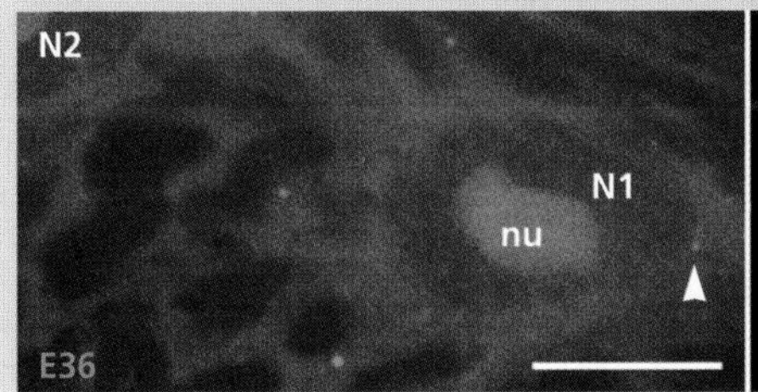

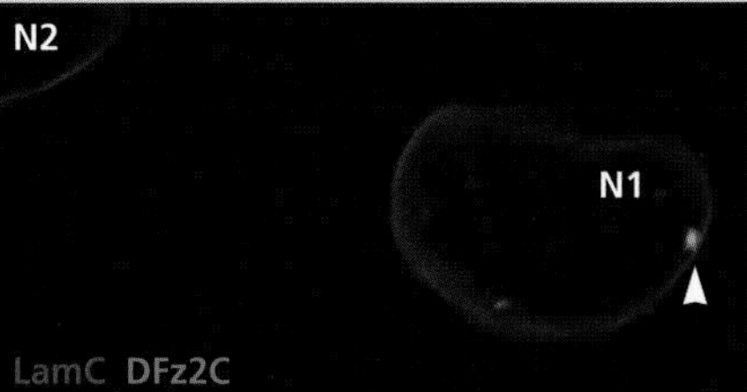

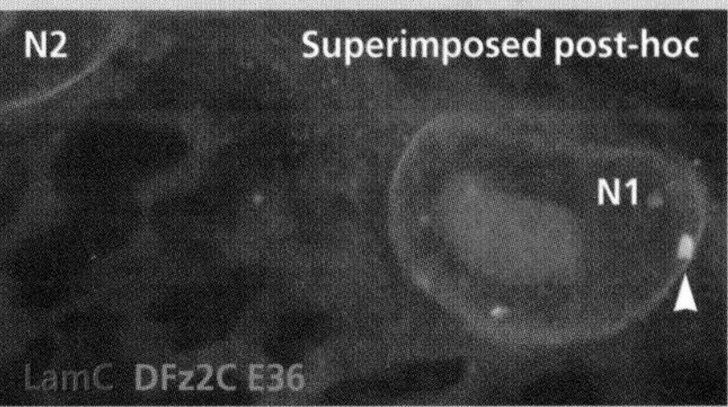

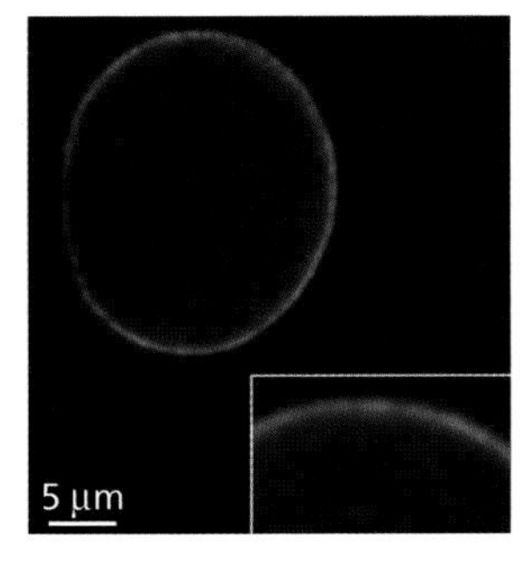

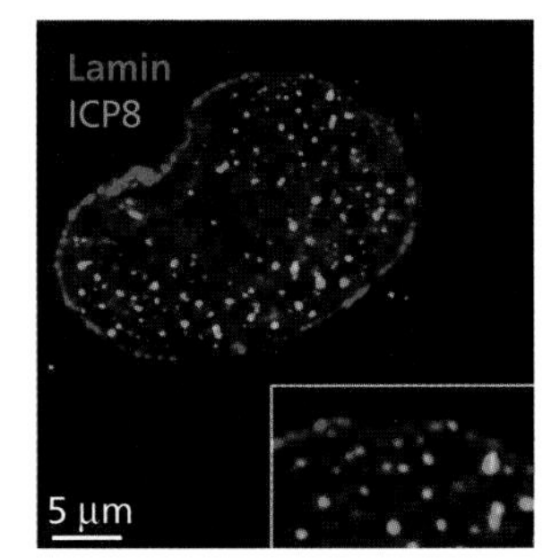

Figure 13.24 Disruption of the nuclear lamina in herpes simplex virus 1-infected cells. Human cells mock infected or infected with herpes simplex virus 1 for 16 h were examined by indirect immunofluorescence. The cellular lamin A/C and viral 1CP8 proteins are in red and green, respectively. The insets show magnified regions of equal sizes. Adapted from M. Simpson-Holley et al., *J. Virol.* **79:**12840–12851, with permission. Courtesy of D. Knipe, Harvard University Medical School.

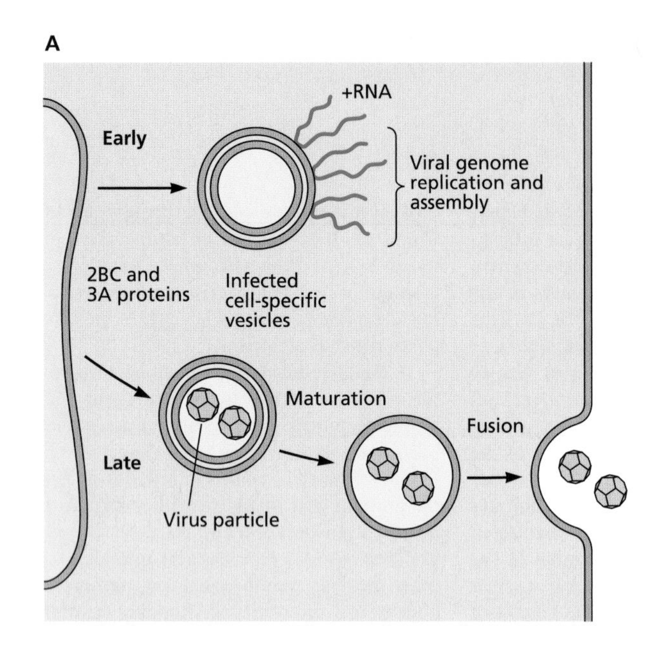

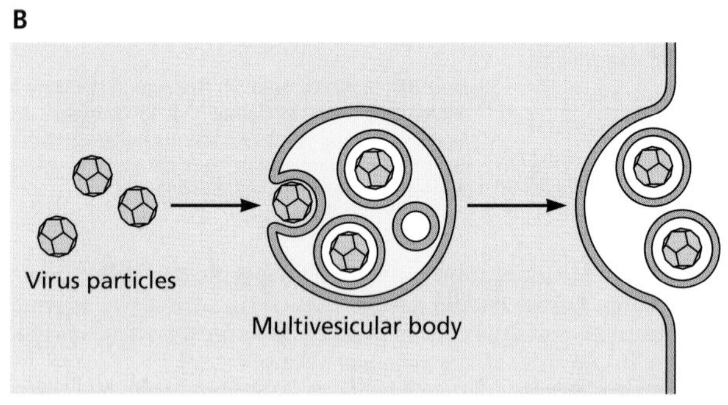

Figure 13.25 Models for nonlytic release of picornavirus particles. (A) Synthesis of the poliovirus 2BC and 3A proteins leads to formation of infected-cell-specific vesicles that resemble autophagosomes. The surfaces of these vesicles are sites of genome replication and assembly (top). It has been proposed that as autophagosome-like vesicles are formed from these membranes later in infection, they enclose virus particles. Maturation of such particle-containing vesicles in a manner analogous to the maturation of autophagosomes would result in complete or partial degradation of the inner membrane. Subsequent fusion of the mature vesicle with the plasma membrane would release virus particles. This model is based on the observation that RNA interference-mediated knockdown of proteins required for the formation of autophagosome-like vesicles reduced the yield of extracellular virus particles to a greater degree than the yield of intracellular particles. **(B)** Hepatitis A virus, also a member of the family *Picornaviridae*, is a common cause of hepatitis and is transmitted by an enteric route. Virus particles released from hepatocytes infected in culture were found to be enclosed within membrane vesicles that carried 1 to 4 particles. Such membrane-enclosed particles were also observed in the blood of humans suffering from hepatitis A virus infection. These particles are infectious and resistant to inhibition by neutralizing antibodies. The enveloped virus particles resemble exosomes in size, and their formation requires cellular proteins that participate in the formation of multivesicular bodies and exosomes, such as Alix and Escrt-III (Fig. 13.17). It has therefore been proposed that hepatitis A virus particles bud into multivesicular bodies upon interaction of the capsid with such proteins. Fusion of the multivesicular body with the plasma membrane would result in release of virus particle enclosed within cellular membrane that is not modified by insertion of viral proteins.

following attachment of the virus to its receptor are not fully understood. However, many alterations that impair receptor binding and entry map to those regions of the capsid proteins that adopt their final organization only upon VP0 cleavage. Cleavage of VP0 therefore not only stabilizes the virus particle but also "spring-loads" it for the conformational transitions that take place during the entry and release of the genome.

Following or during release of most retrovirus particles, the Gag polyprotein is processed by the viral protease, concomitantly with substantial morphological and conformational rearrangements (Fig. 13.26; Box 13.11). Such processing plays an essential part in the mechanisms by which most infectious retroviruses are assembled and released. As we have seen, interactions among Gag polyproteins and the viral RNA and between their NC and MA domains and the plasma membrane build and organize an assembling retrovirus particle. Efficient and orderly assembly also depends on "spacer" peptides that are removed during proteolysis. Furthermore, the membrane-binding signal of MA of human immunodeficiency virus type 1 is exposed when MA is part of Gag but is blocked by a C-terminal α-helix of MA in the mature protein. It is therefore very unlikely that retrovirus particles could be constructed correctly from mature Gag proteins. Indeed, alterations that increase the catalytic activity of the viral protease inhibit budding and production of infectious particles, indicating that premature processing of the polyproteins is detrimental to assembly. On the other hand, the covalent connection of the structural proteins that is so necessary during assembly is incompatible with the release of the internal core following fusion of the viral envelope with the membrane of a new host cell. Such covalent linkage also precludes efficient activity of virion enzymes, which are incorporated as Gag-Pol polyproteins. In some particles, including those of

Figure 13.26 Morphological rearrangement of retrovirus particles upon proteolytic processing of the Gag polyprotein. These two cryo-electron micrographs show the maturation of human immunodeficiency virus type 1 virus particles. **(Left)** The immature particles contain a Gag polyprotein layer below the viral membrane and its external spikes. **(Right)** Processing of Gag converts such particles to mature virus particles with elongated cone-shaped internal capsids. Courtesy of G. Jensen and W. Sundquist, University of Utah School of Medicine.

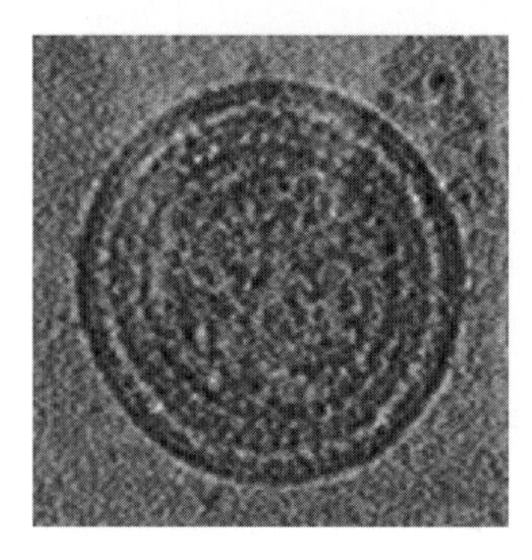

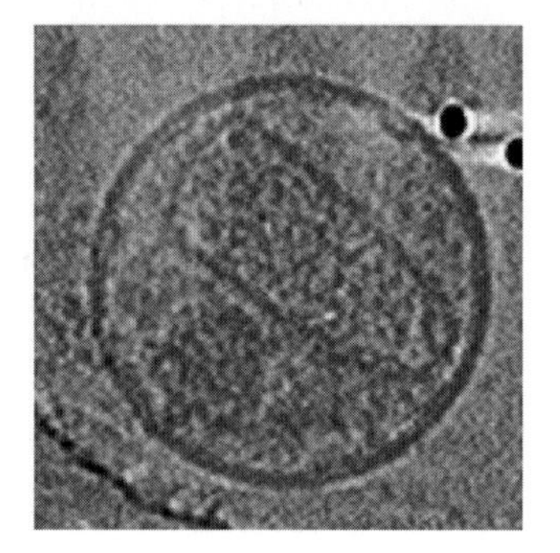

BOX 13.11

DISCUSSION

Model for refolding of the human immunodeficiency virus type 1 CA protein on proteolytic processing of Gag

The model for the radial organization of the human immunodeficiency virus type 1 Gag polyprotein (left), which contains the spacer peptides p1, p2 and p6, is based on cryo-electron micrographs like those shown in Fig. 13.4. The three-dimensional structures of the processed proteins, the MA trimer (red), the CA dimer (blue), and monomeric NC (violet) bound to the SL3 packaging signal (green), shown on the right, are derived from high-resolution structures discussed in this and preceding chapters.

In the X-ray crystal structure of the N-terminal portion of mature CA (right), the charged N terminus is folded back into the protein by a β hairpin formed by amino acids 1 to 13 and forms a buried salt bridge with the carboxylate of Asp51. The lack of a charged N terminus prior to cleavage of CA from MA and the steric difficulties of burying the N terminus of CA attached to an MA extension (left) indicate that the β hairpin and buried salt bridge can form only after proteolytic cleavage. Furthermore, the viral protease recognizes the cleavage site between MA and CA in an extended conformation. As the N-terminal β hairpin of mature CA forms a CA-CA interface, it has been proposed that proteolytic cleavage and the consequent refolding of the N terminus of CA facilitate the rearrangements to form the conical core during maturation of virus particles. Alteration of amino acids in this interface inhibits core assembly and formation of infectious viral particles in infected cells, consistent with this model. Courtesy of T. L. Stemmler and W. Sundquist, University of Utah.

von Schwedler UK, Stemmler TL, Klishko VY, Li S, Albertine KH, Davis DR, Sundquist WI. 1998. Proteolytic refolding of the HIV-1 capsid protein amino-terminus to facilitate viral core assembly. *EMBO J* **17:**1555–1568.

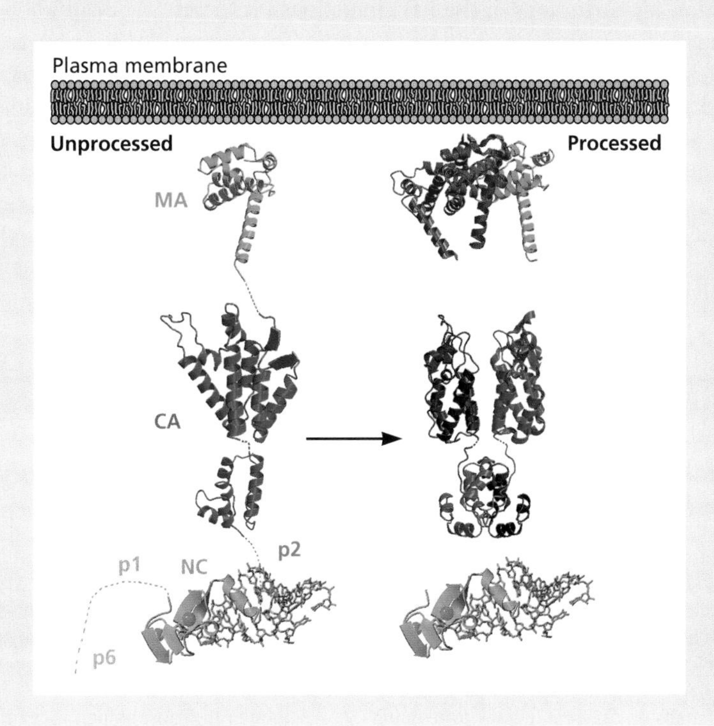

Moloney murine leukemia virus, the protease also removes a short C-terminal segment of the cytoplasmic tail of the TM envelope protein to activate the fusionogenic activity of TM. The retroviral proteases that sever such connections therefore are absolutely necessary for the production of virions, even though they are dispensable for assembly.

The retroviral proteases belong to a large family of enzymes with two aspartic acid residues at the active site (aspartic proteases). The viral and cellular members of this family are similar in sequence, particularly around the active site, and are also similar in three-dimensional structure. All aspartic proteases contain an active site formed between two lobes of the protein, each of which contributes a catalytic aspartic acid. The retroviral proteases are homodimers in which each monomer corresponds to a single lobe of their cellular cousins. Consequently, the active site is formed only upon dimerization of two identical subunits. This property undoubtedly helps avoid premature activity of the protease within infected cells, in which the low concentration of the polyprotein precursors mitigates against dimerization. Indeed, dimerization of the protease appears to be rate limiting for maturation of virus particles. Fusion of the protease to the NC domain of Gag also inhibits dimerization. Consequently, synthesis of the protease as part of a polyprotein precursor not only allows incorporation of the enzyme into assembling particles but also contributes to regulation of its activity. These properties raise the question of how the protease is activated, a step that requires its cleavage from the polyprotein. Polyproteins containing the protease (e.g., made in bacteria) possess some activity, sufficient to liberate fully active enzyme at a very low rate *in vitro*. It is therefore thought that such activity of the polyproteins initially releases protease molecules within the particle. Furthermore, it has been shown, using Gag-Pol proteins yielding distinguishable cleavage products, that the initial proteolytic cleavages are intramolecular. The high local concentrations of protease molecules within the assembling particle would facilitate subsequent dimerization of protease molecules to form the fully active enzyme.

Cleavage of Precursor Proteins

Like its retroviral counterpart, the adenoviral protease converts noninfectious particles to infectious particles, in this case by cleavage at multiple sites within six structural proteins (Fig. 13.10). Although the adenoviral enzyme does not process polyprotein precursors, the cleavage of so many proteins alters protein-protein interactions necessary for assembly in preparation for early steps in the next infectious cycle; particles that lack the protease are not infectious. This enzyme is a cysteine protease containing an active-site cysteine and two additional cysteines, all highly conserved. One mechanism by which its activity is regulated is by interaction with a small peptide, a product of cleavage of the structural protein pVI, or with pVI itself. The pVI peptide binds covalently via a disulfide bond to the proteases both *in vitro* and in virus particles to increase the catalytic efficiency of the enzyme over 1,000-fold. A second cofactor is the viral DNA genome, along which the protease-pVI assembly moves rapidly by one-dimensional diffusion. Several lines of evidence indicate that this movement facilitates the association of the activated protease with its far more numerous substrates and their cleavage within virus particles.

Other Maturation Reactions

Newly assembled virus particles appear to undergo few maturation reactions other than proteolytic processing. However, the trimming of certain oligosaccharides, or formation of disulfide bonds, is known to be required for the infectivity in some cases. Moreover, a surprising extracellular assembly process has been identified recently (Box 13.12).

Terminal sialic acid residues are removed from the complex oligosaccharides added to the envelope HA and NA glycoproteins of influenza A virus during their transit to the plasma membrane. The influenza A virus receptor is sialic acid, which is specifically recognized by the HA protein. Consequently, newly synthesized virus particles have the potential to aggregate with one another and with the surface

BOX 13.12

EXPERIMENTS

A notable example of virus maturation: extracellular assembly of specific structures

Acidianus two-tailed virus was discovered in an acidic hot spring (pH 1.5, 85 to 93°C) at Pozzuoli, Italy, where it reproduces in the thermophilic archaeon *Acidianus convivator*. (A) The virus particles isolated from this source have a lemon-shaped body with filamentous tails of different lengths protruding from each end. (B) However, when the virus was propagated in host cells grown in culture at 75°C, the released particles lacked such tails. Remarkably, tails formed over 1 week when such particles were incubated at 75°C in the **absence** of host cells (left to right). Moreover, this extracellular assembly reaction was complete in less than 1 h when particles were incubated at the temperatures optimal for host cell growth, 85 to 90°C.

Although the morphological changes that accompany maturation of virus particles are well documented (see the text), *Acidianus* two-tailed virus represents the first example of extracellular assembly. This capacity implies that the tailless particles released from host cells contain all the components and information necessary for tail assembly. The tails are presumed to facilitate attachment of virus particles to host cells.

Häring M, Vestergaard G, Rachel R, Chen L, Garrett RA, Prangishvili D. 2005. Independent virus development outside a host. *Nature* **436:**1101–1102.

Electron micrographs of *Acidianus* two-tailed virus particles isolated from a hot spring (A) or released from host cells infected in culture at 75°C and maintained in cell-free medium at 75°C for 0, 2, 5, 6, and 7 days (B, left to right). Scale bars, 0.5 μm **(A)** and 0.1 μm **(B)**. From M. Häring et al., *Nature* **436:**1101–1102, 2005, with permission. Courtesy of David Prangishvili, Institut Pasteur, Paris, France.

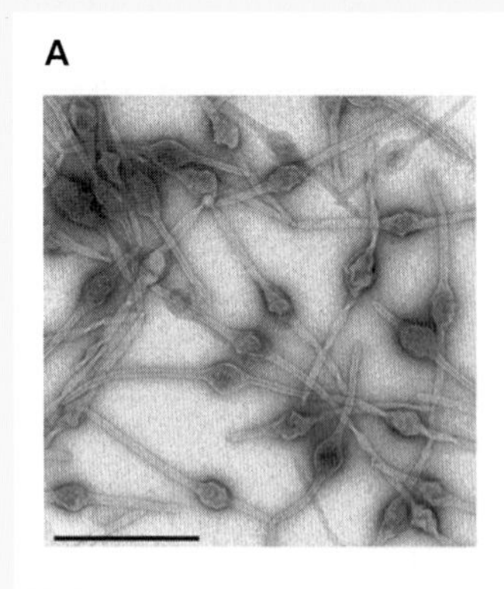

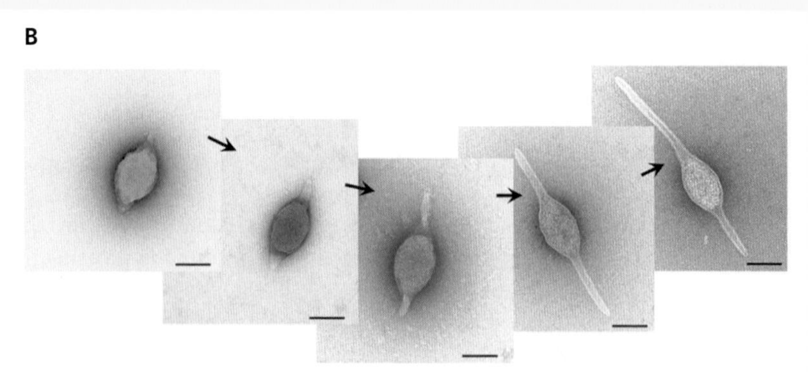

of the host cell by binding of an HA molecule on one particle to a sialic acid present in an envelope protein of another particle or in cell surface proteins. Such aggregation is observed when the viral neuraminidase is inactivated. The neuraminidase eliminates such binding of newly synthesized virus particles to one another and to cell surface proteins. The activity of this enzyme, which removes terminal sialic acid residues from oligosaccharide chains, is essential for effective release of progeny virus particles from the surface of a host cell. This requirement has been exploited to develop new drugs (e.g., oseltamivir phosphate [Tamiflu]) designed specifically to inhibit the viral neuraminidase.

The capsids of nonenveloped papillomaviruses, which are built from 72 pentamers of the major structural protein L1, are stabilized by intermolecular disulfide bonds between specific L1 cysteine residues. This protein does not travel the secretory pathway, raising the question of how such cysteines become oxidized. When the human papillomavirus type 16 (or 18) L1 and the minor capsid (L2) proteins are made in mammalian cells, they assemble to form particles that lack disulfide bonds and are less stable and less infectious than mature capsids. Disulfide bonds form spontaneously at a low rate when immature particles are incubated 37°C and more quickly in the presence of oxidizing agents. This process is accompanied by increased stability and infectivity and the appearance of more regularly structured particles. Papillomaviruses are thought to be released slowly during natural infections as the outer layers of the epithelia in which they replicate are shed. It is therefore likely that newly assembled capsids are exposed to an oxidizing environment for a considerable period (several days) prior to release.

Cell-to-Cell Spread

All progeny virions must infect a new host cell in which the infectious cycle can be repeated. Many viruses are released as free particles by the mechanisms described in preceding sections and must travel within the host until they encounter a susceptible cell. The new host cell may be an immediate neighbor of that originally infected or a distant cell reached via the circulatory or nervous systems of the host. Virus particles are designed to withstand such intercellular passage, but they are susceptible to several host defense mechanisms that can destroy them (Volume II, Chapters 2 to 4). Localized release of virus particles only at points of contact between an infected cell and its uninfected neighbor(s) can minimize exposure to these host defense mechanisms. Furthermore, some viruses can spread from one cell to another by mechanisms that circumvent the need for release of progeny virus particles into the extracellular environment.

In some cases, virus particles can be transferred directly from an infected cell to its neighbors (Box 13.13), a strategy that avoids exposure to host defense mechanisms targeted against extracellular virus. Such cell-to-cell spread, which depends on the viral fusion machinery, is defined operationally as infection that still occurs when released virus particles are neutralized by addition of antibodies. In the case of herpes simplex virus 1, the glycoproteins that promote fusion during entry (gB, gH, and gL) and another glycoprotein (gD) are required. The latter protein binds to the cell surface protein nectin-1, which is localized to cell-cell junctions. Two additional glycoproteins (as well as other proteins, including UL34) are also necessary for efficient cell-to-cell spread but have no known role in entry of extracellular particles. Mutant viruses that lack the gE or gI gene form only small plaques when transfer of free virus particles from one cell to another is prevented. They are also defective for both lateral spread of infection in polarized epithelial cells and the spread of infection from an axon terminal to an uninfected neuron in animals. Such cell-to-cell spread of herpesviruses

BOX 13.13

BACKGROUND

Extracellular and cell-to-cell spread

(A) Many viruses spread from one host cell to another as extracellular virus particles released from an infected cell. Such extracellular dissemination is necessary to infect another naive host. Some viruses, notably alphaherpesviruses, paramyxoviruses, and some retroviruses, can also spread from cell to cell without passage through the extracellular environment **(B)** and can therefore be disseminated by both mechanisms **(C)**.

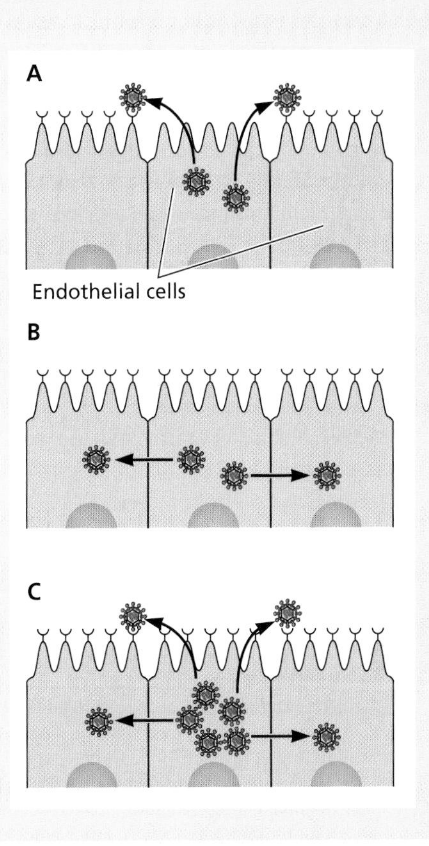

is thought to occur at specialized sites, such as tight junctions of epithelial cells, and by synaptic contacts between individual neurons, but the molecular mechanism is not clear.

Direct cell-to-cell spread is the predominant mechanism for transmission of human immunodeficiency virus type 1 and other retroviruses. Specialized sites of close intercellular contact called **virological synapses** assemble when an infected cell contacts an uninfected neighbor. Virological synapses form at lipid raft regions of the plasma membrane that are enriched in cholesterol and sphingomyelin (see Chapter 12) and also the sites of release of viral particles by budding into the extracellular space. In virological synapses, the viral Gag and Env proteins accumulate in the donor cell membrane, and the CD4 and CxCr4 coreceptors in those of acceptor cells. Env-CD4 interactions are required for intercellular transfer of human immunodeficiency virus type 1 (Volume II, Fig. 7.14). This mode of transmission is some 2 to 3 orders of magnitude more efficient than infection via entry of extracellular virions. Formation of stable filopodia or nanotube contacts between uninfected and infected cells as a result of strong association of the viral Env protein of released virus particles with its receptor on uninfected cells has also been observed. Virus particles then travel toward the uninfected cell along the outer surfaces of such bridges.

Persistent measles virus (a paramyxovirus) infection of the brain is associated with subacute sclerosing panencephalitis (Volume II, Appendix, Fig. 15). Little infectious virus can be recovered from brain tissue of patients with this disease, although the genomic RNA and viral proteins are present. Indeed, budding of virus particles does not take place from the surfaces of infected mouse or human neurons in culture, which contain nucleocapsids accumulating at presynaptic membranes, nor does spread of measles virus between cultured neurons require the viral receptors (cell-cell contact and the fusion protein are necessary). In neurons, measles virus therefore spreads without the release and attachment to infected cells of free virus particles; rather, it spreads from cell to cell, most likely through synapses. Measles virus may also spread directly between other cell types as virological synapses form between infected dendritic cells and uninfected T cells.

There are other examples of more-radical mechanisms of transfer. In astrocytes (supporting cells of the central nervous system), measles virus spreads by inducing the formation of syncytia, sheets of neighboring cells fused to one another (Fig. 13.27A). Certain cell types infected by human immunodeficiency virus type 1 also form syncytia when they would not normally do so (Fig. 13.27B). Even more remarkable are the mechanism by which formation of actin tails repels vaccinia virus particles from infected cells to accelerate their spread to uninfected neighbors (Box 13.9) and the actions of the movement proteins encoded in the genomes of all plant viruses (Box 13.14).

Figure 13.27 Formation of syncytia. (A) Cell-to-cell spread by measles virus. Human astrocytoma cells were infected at a low multiplicity of infection with a recombinant measles virus encoding a green fluorescent protein. The autofluorescence of this protein identifies infected cells (a) and allows the spread of the virus to be monitored in living cells. With increasing time, the virus spreads to cells neighboring those initially infected and can be clearly seen in the processes connecting the cells that become infected (b to f). The arrows point to an extended astrocyte process of a newly infected cell (b), the weak autofluorescence of the nucleus of a cell in a very early phase of infection (c), the nucleus from the same cell 5 and 7 h later (d and e, respectively), and an extended astrocytic process issuing from the cell shown in panels d and e (f). From W. P. Duprex et al., *J. Virol.* **73**:9568–9575, 1999. Courtesy of W. D. Duprex, Queen's University, Belfast, United Kingdom. **(B)** Syncytia formed by human immunodeficiency virus type 1 in a T cell line. The photograph shows a large syncytium of SupT1 cells (a $CD4^+$ T-lymphotropic cell line) that had been infected with a viral vector that expresses the *env* gene. Large quantities of the Env protein accumulate at the cell surfaces, mediating fusion. A single cell is indicated for comparison. Courtesy of Matthias Schnell, Philip McKenna, and Joseph Kulkosky, Thomas Jefferson University School of Medicine, Philadelphia, PA.

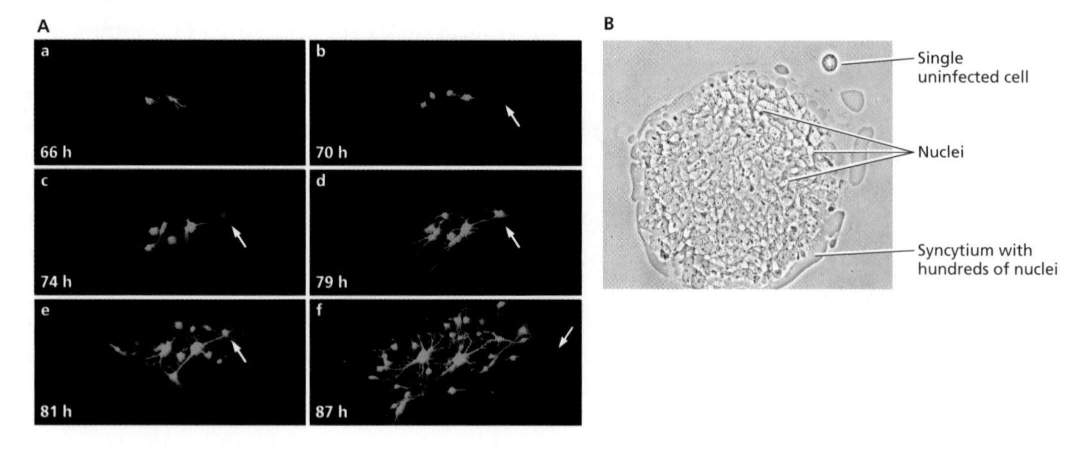

BOX 13.14

DISCUSSION

Intercellular transport by plant virus movement proteins

Plant cells are encased in a thick and rigid cell wall. The possible impact of this feature on the architecture of plant viruses is discussed in Box 12.2. The cell wall hampers release of progeny virus particles and cell-to-cell spread by the mechanisms described in the text. In fact, the spread of viruses within an infected plant relies on a second characteristic feature of plant cells, the specialized structures (plasmodesmata) that connect neighboring cells to one another. Each cell has many (up to 10^5) plasmodesmata, and these structures can be formed by all plant cells. Consequently, a plant is, at least in principle, a monstrous syncytium ideal for the local and systemic spread of a virus.

Plasmodesmata comprise the central desmotubule derived from the endoplasmic reticulum within a second membrane originating from the plasma membrane. The intervening space, termed the cytoplasmic sleeve, which contains many proteins, including actin and myosin, is the major conduit for intercellular transport. The inherent size exclusion limits of plasmodesmata for passive diffusion vary with cell type and physiological state but are very low (some 1 to 7 kDa). However, plasmodesmata are dynamic, and proteins of up to 50 kDa can travel through them under appropriate conditions. This "normal" expanded capacity is still insufficient for the passage of even the smallest plant virus particles (members of the *Nanoviridae*, such as fava bean necrotic yellow virus, with a molecular mass of $\sim 1.6 \times 10^3$ kDa) or of viral genomes in the form of ribonucleoproteins. This potential barrier to virus spread and propagation is circumvented by the movement proteins encoded in the genomes of all plant viruses.

The first such protein to be identified was the 30-kDa movement protein of tobacco mosaic virus; progeny virus particles of temperature-sensitive strains of the virus with substitutions in the 30-kDa protein coding sequence are unable to move from the infected cell. The plant virus movement proteins can be compared and grouped on the basis of various properties, including sequence and interactions with other viral components. However, they fall into two broad functional classes (see the figure). A large group, including the movement protein of tobacco mosaic virus, induces transient increases in the size exclusion limit of plasmodesmata indirectly by a variety of mechanisms. Depending on the virus, such proteins allow transport from cell to cell of either progeny virus particles or ribonucleoproteins containing the viral genome. Members of the second class of movement proteins actually form tubules within plasmodesmata, displacing desmotubules. These structures provide tracks for cell-to-cell spread, but what viral components are transported and how are not yet clear.

Benitez-Alfonso V, Faulkner C, Ritzenthuler C, Maule AJ. 2010. Plasmodesmata: gateways for local and systemic virus infection. *Mol Plant Microbe Interact* **23:**1403–1412.

Niehl A, Heinlein M. 2011. Cellular pathways for viral transport through plasmodesmata. *Protoplasma* **248:**75–99.

The organization of a plasmodesma in an uninfected cell (left) and the impact of plant virus movement proteins that increase the size exclusion limit indirectly (middle) or form transport tubules (right) are illustrated schematically. Pdlp, plasmodesmata receptor-like proteins; PD proteins, unidentified proteins that localize to plasmodesmata; TMV, tobacco mosaic virus; GFLV, grapevine fan leaf virus.

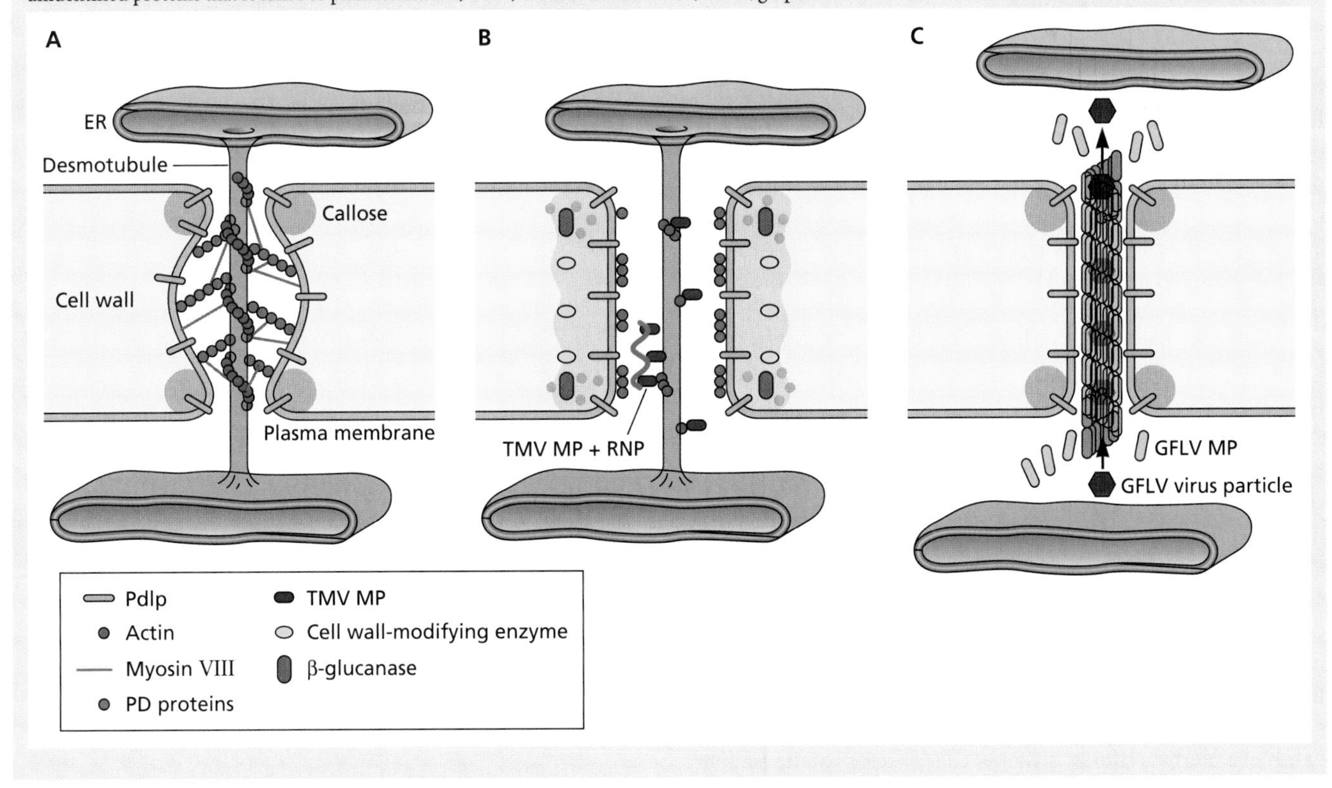

The production of "decoys," noninfectious particles released in large quantities, is one alternative strategy to avoid host defense mechanisms during transmission. The vast majority of particles detected in hepatitis B virus-infected humans are empty. Another strategy would be to disguise virus particles with normal products of a host cell, as during releases of hepatitis A virus in exosomes. Some viral envelopes retain cellular proteins, such as major histocompatibility complex class II proteins and the adhesion receptor Icam-1, in their membranes. The latter protein substantially increases the infectivity of human immunodeficiency virus type 1 particles. However, the importance of such a masking strategy for the spread of a virus from one cell to another in the host has yet to be documented.

Perspectives

The assembly of even the simplest virus is a complicated process in which multiple reactions must be completed in the correct sequence and coordinated in such a way that the overall pathway is irreversible. These requirements for efficient production and release of stable structures must be balanced with the fabrication of virus particles primed for ready disassembly at the start of a new infectious cycle. The integration of information collected by the application of structural, imaging, biochemical, and genetic methods of analysis has allowed an outline of the dynamic processes of assembly, release, and maturation for many viruses. Despite the considerable structural diversity of virions, the repertoire of mechanisms for successful completion of the individual reactions is limited. Furthermore, we can identify common mechanisms that ensure that assembly proceeds efficiently and irreversibly or that resolve the apparent paradox of great particle stability during assembly and release but facile disassembly at the start of the next infectious cycle. These mechanisms include high concentrations of virion components at specific sites within the infected cell and proteolytic cleavage of viral proteins at one or more steps in the production of infectious particles. Indeed, for some smaller viruses, the structural changes that accompany the production of infectious virions from noninfectious precursor particles can be described in atomic detail. Such information has revealed unanticipated relationships between structures that stabilize virus particles and interactions that prime them for conformational rearrangements during entry.

On the other hand, the pathways for assembly, production, and release of even the simplest virus particles cannot be described fully. These reactions are difficult to study in infected cells, and even the simplest proved more difficult to reconstitute *in vitro* than originally anticipated. The latter observation emphasizes the crucial contributions to virus assembly that can be made by cellular proteins that assist protein folding and oligomerization (chaperones) or that covalently modify virion proteins. Historically, assembly reactions have received less attention than mechanisms of viral gene expression or replication of viral genomes. However, the development of new structural and imaging methods, coupled with the experimental power and flexibility provided by modern molecular biology, has revitalized investigation of the essential processes of assembly, release, and maturation of virus particles. This renaissance has been further stimulated by the success of therapeutic agents designed to inhibit virus-specific reactions crucial for the production of infectious particles.

References

Chapters in Books

Wood WB, King J. 1979. Genetic control of complex bacteriophage assembly, p 581–633. *In* Fraenkel-Conrat H, Wagner RR (ed), *Comprehensive Virology*, vol 13. Plenum Press, New York, NY.

Reviews

Bieniasz PD. 2005. Late budding domains and host proteins in enveloped virus release. *Virology* **344:**55–63.

Brandenburg B, Zhuang X. 2007. Virus trafficking—learning from single-virus tracking. *Nat Rev Microbiol* **5:**197–208.

Bruss V. 2006. Envelopment of the hepatitis B virus nucleocapsid. *Virus Res* **106:**199–209.

Condit R, Moussatche N, Traktman P. 2006. In a nutshell: structure and assembly of the vaccinia virion. *Adv Virus Res* **66:**31–124.

Gerber M, Isel C, Moules V, Marquest R. 2014. Selective packaging of the influenza A virus genome and consequences for genetic reassortment. *Trends Microbiol* **22:**446–455.

Jiang P, Liu Y, Ma H-C, Paul AV, Wimmer E. 2014 Picornavirus morphogenesis. *Microbiol Mol Biol Rev* **78:** 418-437

Johnson DC, Huber MT. 2002. Directed egress of animal viruses promotes cell-to-cell spread. *J Virol* **76:**1–8.

Kirkegaard K, Jackson WT. 2005. Topology of double-membraned vesicles and the opportunity for non-lytic release of cytoplasm. *Autophagy* **1:**182–184.

Lyles DS. 2013. Assembly and budding of negative strand RNA viruses. *Adv Virus Res* **85:**57–90.

Martin-Serrano J. 2007. The role of ubiquitin in retroviral egress. *Traffic* **8:**1297–1303.

Meng B, Lever AML. 2013. Wrapping up the bad news—HIV assembly and release. *Retrovirology* **10:**5.

Mettenleiter TC, Müller F, Granzow H, Klupp BG. 2013. The way out: what we know and do not know about herpesvirus nuclear egress. *Cell Microbiol* **15:**170–178.

Moss B, Ward BM. 2001. High-speed mass transit for poxviruses on microtubules. *Nat Cell Biol* **3:**E245–E246.

Nieva JL, Madan V, Carrasco L. 2012. Viroporins: structure and biological functions. *Nat Rev Microbiol* **10:**563–574.

Rossman JS, Lamb RA. 2011. Influenza virus assembly and budding. *Virology* **411:**229–236.

Rossman JS, Lamb RA. 2013. Viral membrane scission. *Annu Rev Cell Dev Biol* **29:**551–569.

Steven AC, Heymann JB, Cheng N, Trus BL, Conway JF. 2005. Virus maturation: dynamics and mechanism of a stabilizing structural transition that leads to infectivity. *Curr Opin Struct Biol* **15:**227–236.

Strauss JH, Strauss EG, Kuhn RJ. 1995. Budding of alphaviruses. *Trends Microbiol* **3:**346–350.

Sullivan CS, Pipas JM. 2001. The virus-chaperone connection. *Virology* **287**:1–8.

Sundquist WI, Kräusslich H-G. 2012. HIV-1 assembly, budding and maturation. *Cold Spring Harb Perspect Med* **2**:a006924.

Votteiler J, Sundquist WI. 2013. Virus budding and the ESCRT pathway. *Cell Host Microbe* **14**:232–241.

Weber J. 1995. The adenovirus endopeptidase and its role in virus infection. *Curr Top Microbiol Immunol* **199**:227–235.

Weissenhorn W, Poudevigne E, Effastin G, Bassereau P. 2013. How to get out: ssRNA enveloped viruses and membrane fission. *Curr Opin Virol* **3**:159–167.

Wileman T. 2007. Aggresomes and pericentriolar sites of virus assembly: cellular defense or viral design? *Annu Rev Microbiol* **61**:149–167.

Papers of Special Interest

Assembly of Protein Shells

Basavappa R, Syed R, Flore O, Icenogle JP, Filman DJ, Hogle JM. 1994. Role and mechanism of the maturation cleavage of VP0 in poliovirus assembly: structure of the empty capsid assembly intermediates at 2.9Å resolution. *Protein Sci* **3**:1651–1669.

Desai P, DeLuca NA, Glorioso JC, Person S. 1993. Mutations in herpes simplex virus 1 genes encoding VP5 and VP23 abrogate capsid formation and cleavage of replicated DNA. *J Virol* **67**:1357–1364.

Desfosses A, Ribeiro EA, Jr, Schoehn G, Blondel D, Gulligay D, Jamin M, Ruigrok RWH, Gutsche I. 2013. Self-organization of the vesicular somatic virus nucleocapsid into a bullet shape. *Nat Commun* **4**:1429.

Edvardsson B, Everitt E, Jörnvall E, Prage L, Philipson L. 1976. Intermediates in adenovirus assembly. *J Virol* **19**:533–547.

Khromykh AA, Varnavski AN, Sedlak PL, Westaway EG. 2001. Coupling between replication and packaging of flavivirus RNA: evidence derived from the use of DNA-based full-length cDNA clones of Kunjin virus. *J Virol* **75**:4633–4640.

Li H, Dou J, Ding L, Spearman P. 2007. Myristoylation is required for human immunodeficiency virus type 1 Gag-Gag multimerization in mammalian cells. *J Virol* **81**:12899–12910.

Ng SC, Bina M. 1984. Temperature-sensitive BC mutants of SV40: block in virion assembly and accumulation of capsid-chromatin complexes. *J Virol* **50**:471–477.

Reicin ES, Ohagen A, Yin L, Hoglund S, Goff SP. 1996. The role of Gag in human immunodeficiency virus type 1 virion morphogenesis and early steps of the viral life cycle. *J Virol* **70**:8645–8652.

Verlinden Y, Cuconati A, Wimmer E, Rombaut B. 2000. Cell-free synthesis of poliovirus: 14S subunits are the key intermediates in the encapsidation of poliovirus RNA. *J Gen Virol* **81**:2751–2754.

Yuen LKC, Consigli RA. 1985. Identification and protein analysis of polyomavirus assembly intermediates from infected primary mouse embryo cells. *Virology* **144**:127–136.

Assembly Chaperones and Scaffolds

Cepko CL, Sharp PA. 1982. Assembly of adenovirus major capsid protein is mediated by a non-virion protein. *Cell* **31**:407–415.

Chromy LR, Pipas JM, Garcia RL. 2003. Chaperone-mediated *in vitro* assembly of polyomavirus capsids. *Proc Natl Acad Sci U S A* **100**:10477–10482.

Desai P, Watkins SC, Person S. 1994. The size and symmetry of B capsids of herpes simplex virus type 1 are determined by the gene products of the UL26 open reading frame. *J Virol* **68**:5365–5374.

Dokland T, McKenna R, Hag LL, Bowman BR, Incardona NL, Fane BA, Rossmann MG. 1997. Structure of a viral procapsid with molecular scaffolding. *Nature* **389**:308–313.

Gao M, Matusick-Kumar L, Hurlburt W, DiTusa SF, Newcomb WW, Brown JC, McCann PC, III, Deckma I, Colonno RJ. 1994. The protease of herpes simplex virus type 1 is essential for functional capsid formation and viral growth. *J Virol* **68**:3702–3712.

Hasson TB, Soloway PD, Ornelles DA, Doerfler W, Shenk T. 1989. Adenovirus L1 52- and 55-kilodalton proteins are required for assembly of virions. *J Virol* **63**:3612–3621.

Thibaut HJ, van der Linden L, Jiang P, Thys B, Canela M-D, Aguado L, Rombaut B, Wimmer E, Paul A, Perez-Perez M-J, van Kuppervald FJM, Neyts J. 2014. Binding of glutathione to enterovirus capsids is essential for virion morphogenesis. *PLoS Pathog* **10**(4):e1004039. doi:10.1371/journal.ppat.1004039.

Packaging the Viral Genome

Ansardi DC, Marrow CD. 1993. Poliovirus capsid proteins derived from P1 precursors with glutamine-valine sites have defects in assembly and RNA encapsidation. *J Virol* **67**:7284–7297.

Beard PM, Taus NS, Baines JD. 2002. DNA cleavage and packaging proteins encoded by genes U_L28, U_L15, and U_L33 of herpes simplex virus type 1 form a complex in infected cells. *J Virol* **76**:4785–4791.

Chou YY, Keaton NS, Gao O, Palese P, Singer R, Lionnet T. 2013. Colocalization of different influenza viral RNA segments in the cytoplasm before viral budding as shown by single-molecule sensitivity FISH analysis. *PLoS Pathog* **9**(5):e1003358. doi:10.1371/journal.ppat.1003358.

Frilander M, Bamford DH. 1995. In vitro packaging of the single-stranded RNA genomic precursors of the segmented double-stranded RNA bacteriophage ϕ6: the three segments modulate each other's packaging efficiency. *J Mol Biol* **246**:418–428.

Gordon-Shaag A, Ben-Nun-Shaul O, Roitman V, Yosef Y, Oppenheim A. 2002. Cellular transcription factor Sp1 recruits simian virus 40 capsid proteins to the viral packaging signal, *ses. J Virol* **76**:5915–5924.

Liang Y, Huang T, Ly H, Parslow TG, Liang Y. 2008. Mutational analysis of packaging signals of influenza virus PA, PB1, and PB2 genome RNA segments. *J Virol* **82**:229–236.

Lu K, Hung X, Garyu L, Monti S, Garcia EL, Kharytonchyk S, Dorjsuren B, Kulandaivel G, Jones S, Hiremath A, Divakaruni SS, LaCotti C, Barton S, Tummillo D, Hosic A, Edme K, Albrecht S, Telesrutsky A, Summers MF. 2011. NMR detection of structures in the HIV-1 5′-leader RNA that regulate genome packaging. *Science* **334**:242–245.

Nugent CI, Johnson KL, Sarnow P, Kirkegaard K. 1999. Functional coupling between replication and packaging of poliovirus replicon RNA. *J Virol* **73**:427–435.

Schmid SI, Hearing P. 1997. Bipartite structure and functional independence of adenovirus type 5 packaging elements. *J Virol* **71**:3375–3384.

Zhang W, Imperiale MJ. 2000. Interaction of the adenovirus IVa2 protein with viral packaging sequences. *J Virol* **74**:2687–2693.

Acquisition of an Envelope

Feng Z, Hensley L, KcKNight KL, Hu F, Madden V, Ping L, Jeong S-H, Walker C, Lanford RE, Lemon SM. 2013. A pathogenic picornavirus acquires an envelope by hijacking cellular membranes. *Nature* **496**:367–371.

Finzi A, Orthwein A, Mercier J, Cohen EA. 2007. Productive human immunodeficiency virus type 1 assembly takes place at the plasma membrane. *J Virol* **81**:7476–7490.

Freed EO, Martin MA. 1996. Domains of the human immunodeficiency virus type 1 matrix and gp41 cytoplasmic tail required for envelope incorporation into virions. *J Virol* **70**:341–351.

Jose J, Przybyla L, Edwards TJ, Perera R, Burgner JW, II, Kuhn RJ. 2012. Interactions of the cytoplasmic domain of Sindbis virus E2 with nucleocapsid cores promote alphavirus budding. *J Virol* **86**:2585–2599.

Justice PA, Sun W, Li Y, Ye Z, Grigera PR, Wagner RR. 1995. Membrane vesiculation function and exocytosis of wild-type and mutant matrix proteins of vesicular stomatitis virus. *J Virol* **69**:3156–3160.

Roseman AM, Beriman JA, Wyane SA, Bulter JG, Crowther RA. 2005. A structural model for maturation of the hepatitis B virus core. *Proc Natl Acad Sci U S A* **102**:15821–15826.

Suárez C, Welsch S, Chlanda P, Hagen W, Hoppe S, Kolovoa A, Pagnier I, Rault D, Krijnse Locker J. 2013. Open membranes are the precursors for assembly of large DNA viruses. *Cell Microbiol* **15:**1883–1895.

Watanabe T, Sorensen EM, Naito A, Scholt M, Kim S, Ahlquist P. 2007. Involvement of host cellular multivesicular body functions in hepatitis B virus budding. *Proc Natl Acad Sci U S A* **104:**10205–10210.

Maturation of Virus Particles

Bernstein H, Bizub D, Skalka AM. 1991. Assembly and processing of avian retroviral Gag polyproteins containing linked protease dimers. *J Virol* **65:**6165–6172.

Buck CB, Thompson CD, Pang Y-YS, Lowry DR, Schiller JT. 2005. Maturation of papillomavirus capsids. *J Virol* **79:**2839–2846.

Webster A, Hay RT, Kemp G. 1993. The adenovirus protease is activated by a virus-coded disulfide-linked peptide. *Cell* **72:**97–104.

Release and Spread of Virus Particles

Bigalke JM, Heuser T, Nicastro D, Heldwein EE. 2014. Membrane deformation and scission by the HSV-1 nuclear egress complex. *Nat Commun* **5:**4131. doi:10.1038/ncomms5131.

Chen P, Hübner W, Spinelli MA, Chen BK. 2007. Predominant mode of human immunodeficiency virus transfer between T cells is mediated by sustained Env-dependent neutralization-resistant virological synapses. *J Virol* **81:**12582–12595.

Farnsworth A, Wisner TW, Webb M, Roller R, Cohen G, Eisenberg R, Johnson DC. 2007. Herpes simplex glycoproteins gB and gH function in fusion between the virion envelope and the outer nuclear membrane. *Proc Natl Acad Sci U S A* **104:**10187–10192.

Johnson DC, Webb M, Wisner TW, Brunetti C. 2001. Herpes simplex virus gE/gI sorts nascent virions to epithelial cell junctions, promoting virus spread. *J Virol* **75:**821–833.

Neil SJD, Zang T, Beiniasz PD. 2008. Tetherin inhibits retrovirus release and is antagonized by HIV-1 Vpu. *Nature* **451:**425–431.

Raghava S, Giorda KM, Romano FB, Heuck AP, Hebert DN. 2011. The SV40 late protein VP4 is a viroporin that forms pores to disrupt membranes for viral release. *PLoS Pathog* 7(6)**:**e1002116. doi:10.1371/journal.ppat.1002116.

Roper RL, Wolfe EJ, Weisberg A, Moss B. 1998. The envelope protein encoded by the A33R gene is required for formation of actin-containing microvilli and efficient cell-to-cell spread of vaccinia virus. *J Virol* **72:** 4192–4204.

Sherer NM, Lehmann MJ, Siminez-Soto LF, Horensavitz C, Pypaert M, Mothes W. 2007. Retroviruses can establish filopodial bridges for efficient cell-to-cell transmission. *Nat Cell Biol* **9:**310–316.

Tollefson AE, Scaria A, Hermiston TW, Ryerse JS, Wold LH, Wold WS. 1996. The adenovirus death protein (E3–11.6K) is required at very late stages of infection for efficient cell lysis and release of adenovirus from infected cells. *J Virol* **70:**2296–2306.

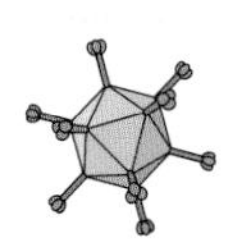

14 The Infected Cell

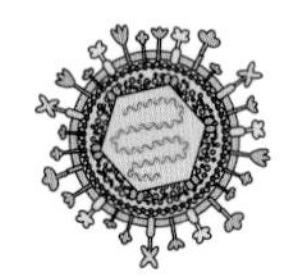

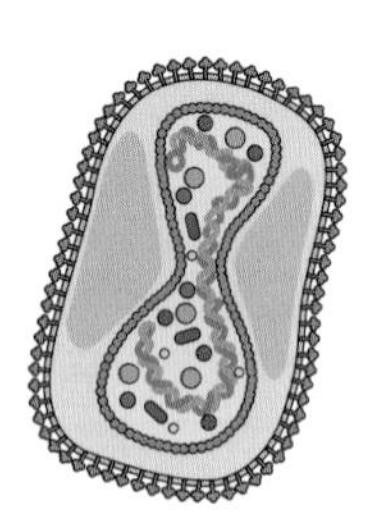

Introduction

Signal Transduction
- Signaling Pathways
- Signaling in Virus-Infected Cells

Gene Expression
- Inhibition of Cellular Gene Expression
- Differential Regulation of Cellular Gene Expression

Metabolism
- Methods To Study Metabolism
- Glucose Metabolism
- The Citric Acid Cycle
- Electron Transport and Oxidative Phosphorylation
- Lipid Metabolism

Remodeling of Cellular Organelles
- The Nucleus
- The Cytoplasm

Perspectives

References

LINKS FOR CHAPTER 14

⏭ *Video: Interview with Dr. Thomas Shenk*
http://bit.ly/Virology_Shenk

⏭ *Herpes and the sashimi plot*
http://bit.ly/Virology_Twiv339

> *He hath eaten me out of house and home.*
>
> WILLIAM SHAKESPEARE
> King Henry IV, Part II

Introduction

In previous chapters, we have described the reactions that comprise viral infectious cycles, from initial attachment to a receptor on the surface of a susceptible cell to assembly and release of progeny particles. The focus has been on the mechanisms that ensure successful viral gene expression, replication of viral genomes, and production of virus particles. These processes depend to a greater or lesser degree on the host cell's metabolic and biosynthetic capabilities, signal transduction pathways, and trafficking systems. Consequently, productive virus infection inevitably redirects, and frequently compromises, normal cellular physiology, and indeed can result in lysis and death of the infected cell within a matter of hours to days. Some of the mechanisms by which viral gene products fashion cellular systems to virus-specific ends have been touched on in previous chapters. Here, we present an integrated description of cellular responses to illustrate the marked, and generally irreversible, impact of virus infection on the host cell.

The initial responses of a host cell to virus infection are rapid, initiated upon contact of a virus particle with a receptor or immediately following entry of virus particles (or components thereof) into the cell. A major consequence of entry is the recognition of viral components by cellular proteins specialized for detection of microbial invaders (pattern recognition receptors). Such recognition initiates signal transduction cascades that mobilize host defenses, such as those mediated by interferons. These defensive responses, which can include alterations in expression of large sets of cellular genes, for example, of up to 1,000 or so interferon-inducible genes, are described in Volume II, Chapter 3. Virus infection also elicits alterations in host cell processes that facilitate production and release of progeny virus particles. Infection may modify expression of cellular genes, redirect metabolic pathways, disrupt trafficking of cellular macromolecules, or remodel cellular components and organelles to promote specific reactions in an infectious cycle. The extent and magnitude of such alterations depend on properties of the host cell, such as whether it is normal or transformed, quiescent or proliferating, as well as whether an infection is productive: when an infection is latent or persistent, only a subset of viral genes is expressed and their products promote survival of infected cells, rather than the widespread reprogramming of cellular gene expression observed in cells productively infected by many viruses.

Our understanding of the cellular response to viral infection has deepened enormously since the development of the techniques of systems biology as well as improved imaging methods. Indeed, application of these approaches has revealed just how different an infected cell that is supporting virus reproduction can be from its uninfected cell counterpart.

Signal Transduction

Signaling Pathways

All cells, be they individual organisms (e.g., bacteria, archaea, and protozoa) or but one of millions in a multicellular animal or plant, must be capable of sensing their environment and responding in an appropriate manner. They must also possess mechanisms to perceive internal cues that provide information about the need for particular metabolites, the integrity of the genome, or the presence of microbes. In multicellular organisms, the coordination of the properties and behaviors of individual cells with those of local neighbors, or more distant cells, is critical for successful differentiation and development, and for maintaining homeostasis among functionally specialized organs and tissues. Cells therefore possess elaborate sensing mechanisms that monitor, and when appropriate, initiate a response to, information about the external and internal milieus. These **signal transduction pathways** govern and integrate every aspect of cell physiology and conduct, from the rate of metabolic reactions to the decisions to move in a particular direction, to divide, or to differentiate.

PRINCIPLES ***The infected cell***

- A single signal transduction pathway can be modified in cells infected by many different viruses.
- Infection of cells with a single virus can result in modification of multiple signaling pathways.
- Inhibition of cellular gene expression is a common outcome of viral infection.
- Cellular gene expression can be inhibited in virus infected cells by blocking cellular mRNA production, inhibiting translation, or inducing increased degradation of cellular mRNAs.
- A second common pattern is differential regulation, in which expression of some genes is increased and that of others decreased.
- The rates of glucose uptake and metabolism are increased in cells infected by a wide variety of viruses, but the products of this pathway have virus-specific uses.
- Virus-induced changes in lipid metabolism are required for energy production, formation of replication centers, generation of viral envelopes with unusual characteristics, or maturation of virus particles.
- Remodeling of the nucleus or cytoplasm during virus infection can facilitate genome replication, assembly of progeny virus particles, or both.

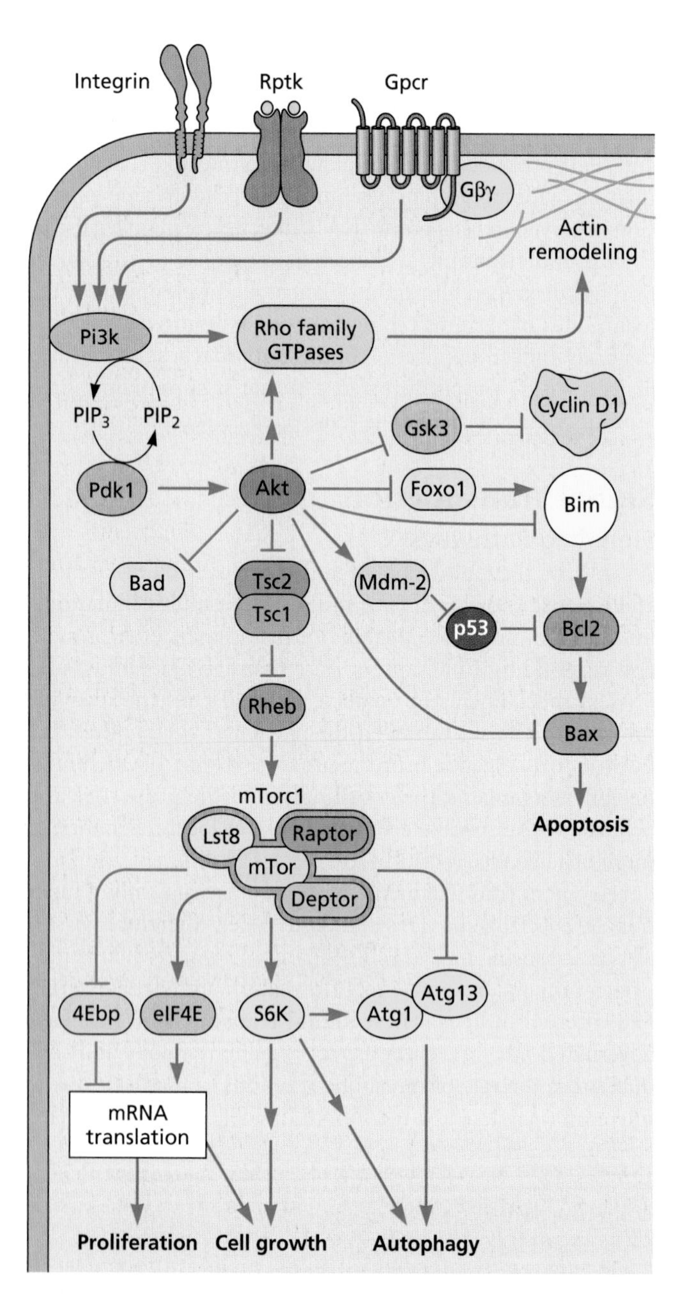

Figure 14.1 The mammalian Pi3k-Akt-mTor signaling route. The core features of this signaling transduction system are illustrated. Binding of ligand to any one of several types of plasma membrane receptors initiates signaling to Pi3k, which is associated with the inner surface of the plasma membrane, and activation of this kinase via phosphorylation. Mammalian cells contain three classes of Pi3ks, distinguished by their intracellular distributions, mechanisms of activation, and substrate specificity. Shown is the most common, class I Pi3ks, which comprise a regulatory (p85) and a catalytic (p110) subunit. Once activated, these kinases phosphorylate phosphoinositol present on membrane lipids to produce phosphoinositol 3,4,5-triphosphate (PIP_3). These modified lipids are bound by particular domains of other proteins, such as phophoinositide-dependent kinase 1 (Pdk1), which then transmit the signal to Akt. Synthesis of PIP_3 also leads to activation of small G proteins of the Rho (Ras homology) family that control actin polymerization and depolymerization, such as Rac (Ras-related C3 botulism toxin substrate 1) and Cdc42 (cell division control protein 42 homolog). Activation of Akt modulates numerous substrates and several processes. Shown are consequences that promote cell growth and proliferation via activation of the mTor kinase present in mTorC1. Activated mTor facilitates translation by multiple mechanisms and also induces **autophagy**, a process that helps cells survive extreme forms of stress, such as amino acid starvation. The signaling hubs Pi3k, Akt, and mTor are connected to, and regulated by, other signaling systems and to one another by various feedback circuits. Atg, autophagy-related protein; Bad, Bcl2-associated death protein; Bax, apoptosis-regulator Bcl2-associated protein; Bcl2, apoptosis-regulator Bcl2 (B cell CLL/lymphoma 2); Bim, Bcl2-interacting mediator of cell death; Deptor, Dep domain-containing mTor-interacting protein; 4Ebp, eukaryotic initiation factor 4E-binding protein; eIF4G, eukaryotic translation initiation factor 4G; Foxo1, forkhead box protein 1; Gsk3, glycogen synthase kinase 3; Lst8, target of rapamycin complex subunit Lst8 homology; Mdm2, E3 ubiquitin ligase Mdm2 (double minute protein 2); Raptor, regulatory-associated protein of mTor complex 1; Rheb, Ras-homology enriched in protein; S6k, ribosomal protein S6 kinase; Tsc, tuberous sclerosis protein.

Consequently, a considerable fraction of a cell's coding capacity is devoted to genes that encode signaling proteins: of the some 30,000 human genes, 918 (~1.7%) specify protein kinases, just one of the several classes of signal transduction protein. Extensive alteration in cellular signaling is an inevitable consequence of virus infection.

In signal transduction pathways, detection of an informational molecule, such as a metabolite (e.g., glucose), hormone, or growth factor, by a receptor initiates amplification of the signal as it is transmitted to effectors. Amplification is achieved by the actions of protein kinases that catalyze sequential phosphorylation and activation of additional kinases or other substrates, and the synthesis of small, diffusible molecules that act as messengers, for example, cyclic AMP and phosphoinositol 3-phosphate (PI3P). Proteins that operate in any cellular process may be effectors, but those that regulate gene expression are common targets. Many of the numerous signaling pathways of mammalian cells respond to more than a single input, regulate multiple molecular processes, and communicate with one another. The phosphoinositol 3-kinase (Pi3k)-Akt pathway exemplifies these properties: it receives input from multiple membrane receptors and regulates many aspects of cell metabolism, proliferation, and survival both directly and via connections to other pathways, such as that centered on the serine/threonine protein kinase mammalian target of rapamycin (mTor) (Fig. 14.1). The kinases Pi3k and Akt are focal points or hubs in the signaling network, with multiple inputs and outputs.

Signaling in Virus-Infected Cells

Much of our understanding of the impact of virus infection on host cell signal transduction cascades comes from investigation of the functions of viral gene products in cells in culture. In such infected cells, alterations in signaling are both rapid and substantial. For example, quantitative analysis of protein phosphorylation by mass spectrometry revealed changes in the frequency of phosphorylation at specific sites on 175 cellular proteins within a minute of exposure of host $CD4^+$ T cells to human immunodeficiency virus type 1. Furthermore, it is clear that viral infection can effect changes in signaling that promote every reaction in the infectious cycle. Many viral gene products intervene to block defensive responses of the host that would inhibit virus reproduction. In fact, every virus that has been examined has been found to direct synthesis of at least one viral gene product that impairs detection of infection or blocks the initial antiviral responses (Volume II, Chapter 3). Although most alterations are transient (because infected cells generally do not survive), some viral proteins can induce permanent changes in cellular signaling systems that allow cells to proliferate indefinitely. This process, termed transformation, is essential for **oncogenesis** and is described in Volume II, Chapter 6.

In this section, we focus on modulations of signaling pathways that facilitate virus reproduction, and use specific examples to illustrate two general principles: the same signal transduction pathway can be modified in cells infected by many different viruses, and individual viruses can modulate multiple signaling pathways.

Activation of Common Signaling Pathways

A core set of processes, including entry into a host cell, translation of viral mRNAs, and synthesis of viral nucleic acids, are common to all viral infectious cycles. Consequently, it is not unexpected that the same signal transduction pathway can be modulated in cells infected by viruses belonging to different families. One example of this phenomenon is activation of the transcriptional regulator Nf-κb in cells infected by several viruses with DNA genomes and some retroviruses to facilitate transcription of viral DNA templates (Fig. 8.11). However, viral gene products also frequently block this pathway, because Nf-κb is critical for activation of innate immune defenses (Volume II, Chapter 3). Signaling via Pi3k and Akt regulates a broad range of cellular processes (Fig. 14.1) and is modulated following infection by a large number of viruses. We therefore illustrate the varied impact of infection on one signal transduction cascade using this pathway.

Among many other aspects of cell physiology, this signaling pathway regulates remodeling of the cytoskeleton by polymerization and depolymerization of actin fibers (Fig. 14.1). Such resculpting of these structural components of the cell is essential for movement of cells; formation of extensions, such as lamellipodia; and other processes that require reorganization of the external surface of the cell, including virus entry. Attachment of viruses belonging to numerous families to their cognate cell surface receptors induces rapid activation (phosphorylation) of Pi3k. This response is required for efficient virus entry, as inhibition of Pi3k or of downstream effectors (see Fig. 14.2) impairs this process. Although Pi3k is activated in all cases, the downstream pathways are virus specific, because the mechanisms of entry differ from virus to virus. Attachment of human adenovirus to its integrin receptor leads to signaling from Pi3k via small G proteins to induce actin reorganization and facilitate endocytosis of virus particles. In other cases, it is signaling from Pi3k to Akt that has been implicated in promoting entry of virus particles by endocytosis (Fig. 14.2). Attachment of influenza A virus particles, which leads to clustering of lipid rafts and associated receptor protein tyrosine kinases and subsequent activation of Pi3k, stimulates not only actin remodeling, but also the acidification of endosomes necessary for disassembly of virus particles. In contrast, activation of the Pi3k-Akt pathway by binding of Zaire ebolavirus to its receptor does not promote virus entry directly, but rather prevents the diversion of endosomal virus particles to cytoplasmic vesicles in which fusion of viral and cellular membranes cannot occur. Presumably, these distinct outputs of Pi3k and Akt signaling are determined by the virus-specific mechanisms of activation of the kinases. Entry of all viruses that reproduce in mammalian cells depends on some degree of refashioning of the plasma membrane and associated cytoskeleton. It therefore seems likely that subversion of the normal function of Pi3k, Akt, or both in regulating membrane transactions will prove to be a more general response to the encounter of host cells with virus particles.

Signaling initiated by activation of Pi3k also facilitates later steps in virus reproduction. This kinase signals to not only Akt but also a second kinase, mTor, present in mTor complex 1 (mTorC1). Outputs from these downstream hubs in the cascade increase the rate of translation (and hence support cell growth and proliferation), modulate metabolic pathways, and promote cell survival (Fig. 14.1). All these responses would be expected to be beneficial for completion of viral infectious cycles. In fact, in every case that has been examined, virus infection has been observed to activate signaling via Pi3k to Akt, and in many cases, mTorC1. Modulation of such kinases can also contribute to pathogenesis (Box 14.1).

The genomes of a number of DNA viruses and retroviruses include oncogenes. The products of such genes can induce permanent activation of cell proliferation, a process termed transformation, and sometimes acquisition of the ability to form tumors in animals. These viral proteins stimulate cell proliferation by a variety of mechanisms, and typically also activate the Pi3k-Akt-mTorC1 signaling cascade to support cell growth and promote cell survival (Volume II, Chapter 6). Infection by many other viruses with both RNA and DNA genomes also circumvents the normal mechanisms of regulation of Pi3k (or downstream signaling molecules) to facilitate translation of viral mRNAs and/or to block apoptosis, a defense to virus infection of last resort. The genomes of several viruses, including human adenovirus type 5, hepatitis C virus, and rotavirus, encode the proteins that bind directly to the regulatory subunit of Pi3k to activate the kinase (Fig. 14.3A). Infection by human enterovirus 71, a pathogenic picornavirus, also leads to activation of Pi3k via its regulatory subunit, but in this case the interaction is with a cellular protein (Sam68) that is induced to relocalize from the nucleus to the cytoplasm. Other viral proteins act upstream of Pi3k or by mechanisms that are not yet known (Fig. 14.3B).

While various responses have been ascribed to activation of the Pi3k pathway by specific viral proteins, or in cells infected by different viruses (Fig. 14.3), only certain outputs were examined in each case, and it is therefore possible that the consequences of the increased activity of this cascade are more far-reaching. Furthermore, the genomes of a variety of viruses encode proteins that intervene downstream of Pi3k to maintain mTor activity and consequently efficient translation (Chapter 11).

Infection with a Particular Virus Modulates Multiple Signal Transduction Pathways

Virus reproduction is invariably accompanied by alterations in more than a single signaling relay, typically with one or more pathways blocked and others stimulated. Prominent among those that are inhibited in infected cells are pathways that detect microbes and mediate cellular defenses (Volume II, Chapter 3). Concurrently, signaling cascades that govern other processes are modulated to support the reactions necessary for expression and replication of viral genomes and assembly of progeny virus particles.

Infection by viruses with even relatively simple genomes and mechanisms of reproduction that depend on a minimal set of cellular systems and components leads to modification of several signaling pathways. For example, infection with Coxsackie B

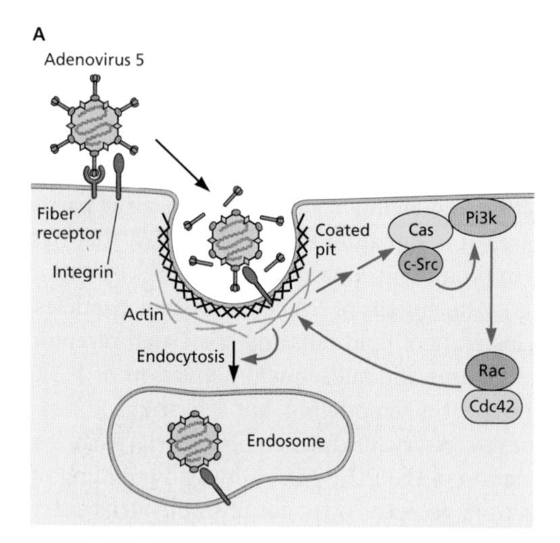

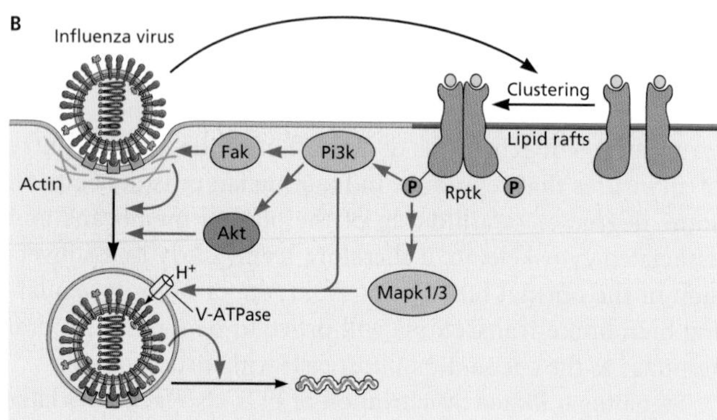

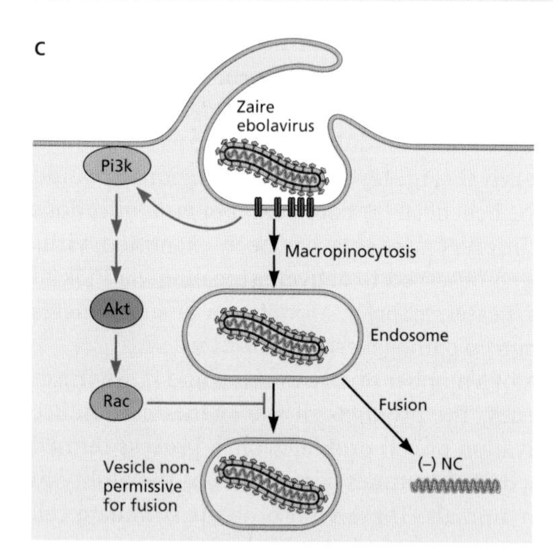

Figure 14.2 Signaling via Pi3k facilitates virus entry by a variety of mechanisms. Pi3k is activated following attachment to cellular receptors of adenoviruses, filoviruses, flaviviruses, influenza viruses, herpesvirus, and poxviruses, among others. Shown are three examples of the consequences of activated signaling from Pi3k. **(A)** Binding of a nonenveloped adenovirus type 5 particle to the αv integrin receptor leads to activation of Pi3k upon association of its p85 regulatory subunit with phosphorylated Crk-associated subunit (Cas), which is a substrate of the c-Src tyrosine kinase. Signaling initiated by the action of Pi3k results in actin remodeling via activation of the small G proteins Rac and Cdc42. Inhibition of phosphorylation of Cas or production of a dominant-negative derivative of Rac inhibits adenovirus entry, emphasizing the importance of this cellular signaling pathway for efficient internalization. **(B)** Attachment of an influenza A virus particle to its sialic acid receptor also induces activation of Pi3k to promote actin remodeling and endocytosis. In this case, these processes depend on signaling via Akt and focal adhesion kinase (Fak), and Pi3k is activated following clustering of lipid rafts and their associated receptor protein tyrosine kinases (Rptks) in the plasma membrane. Such concentration of receptors facilitates their activation by cross-phosphorylation, and also activates Mapk1 and -3. These kinases, in conjunction with signals transmitted from Pi3k, increase the activity of the vacuolar ATPase (V-ATPase) present in the membrane of endosomes, which pumps protons into the vesicles. The increased flux of protons reduces pH in the endosomal lumen and facilitates disassembly of virus particles for release of genome RNA segments into the cytoplasm. **(C)** Binding of the filovirus Zaire ebolavirus to its receptor induces entry via macropinocytosis and also activates signaling via Pi3k, Akt, and Rac by a mechanism that is not yet known. Such signal transduction indirectly facilitates release of the viral genomes into the cytoplasm by blocking diversion of endosomes containing virus particles to a nonproductive pathway.

BOX 14.1

EXPERIMENTS

A viral pathogenic RNA blocks Akt activation to induce apoptosis

In cells infected by arthropod-transmitted flaviviruses such as West Nile and dengue viruses, degradation of the viral (+) strand RNA genome by the cellular 5′ ⊠ 3′ exonuclease Xrn1 produces small, noncoding RNAs derived from the 3′ untranslated region of the genome (Box 10.9). Initial characterization of this subgenomic (sg) RNA of West Nile virus established that it is necessary for formation of plaques and killing of infected cells in culture, and for pathogenicity in mice.

Production of dengue virus sgRNA (of some 400 nucleotides) is prevented by a deletion that removes a segment predicted to form two hairpin stem-loop structures near the end of the 3′ untranslated region. This mutation does not have much effect on synthesis of viral proteins, replication of the genome, or the yield of infectious virus particles following infection of mammalian cells. However, the ability of the mutant virus to form plaques on these cells was impaired (see the figure), as was induction of apoptosis when assessed by three different assays. At the time of peak genome RNA concentration in cells infected by the wild-type virus, Akt was inactivated (loss of phosphorylation at a specific residue) and the concentration of the antiapoptotic protein Bcl-2 was reduced. Neither of these alterations was observed in cells infected by the mutant virus. Introduction of a plasmid that directed synthesis of sgRNA into cells infected by the mutant led to a partial restoration of plaque formation and increases in cell death and the number of apoptotic cells.

These observations indicate that the dengue virus sgRNA inhibits signaling from Akt late in infection, and are consistent with a model in which this function of the sgRNA promotes apoptosis as a result of reduced concentrations of Bcl-2 (see Fig. 14.1). How a small RNA blocks activation of Akt remains to be established.

Liu Y, Liu H, Zou J, Zhang B, Yuan Z. 2014. Dengue virus subgenomic RNA induces apoptosis through the Bcl-2-mediated PI3k/Akt signaling pathway. *Virology* **448:**15–25.

Pijlman GP, Funk A, Kondratieva N, Leung J, Torres S, van der Aa L, Liu WJ, Palmenberg AC, Shi PY, Hall RA, Khromykh AA. 2008. A highly structured, nuclease-resistant, noncoding RNA produced by flaviviruses is required for pathogenicity. *Cell Host Microbe* **4:**579–591.

Rubyh JA, Pijlman GP, Wilusz J, Khromykh AA. 2014. Noncoding subgenomic flavivirus RNA: multiple functions in West Nile virus pathogenesis and modulation of host responses. *Viruses* **6:**404–427.

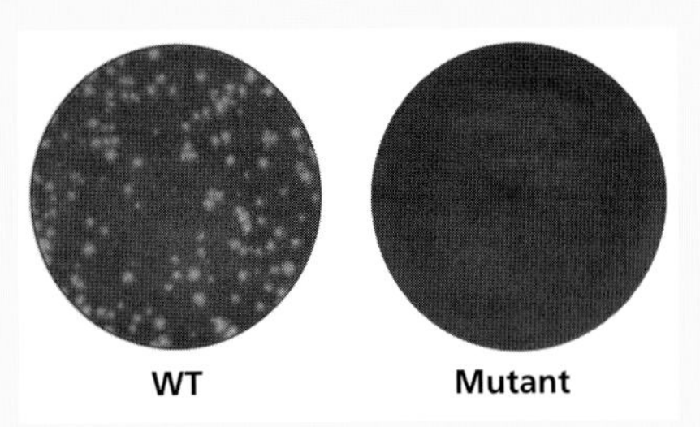

Plaques formed on BHK-21 cells by wild-type dengue virus (WT) and a mutant with a deletion near the 3′ end of the 3′ untranslated RNA that prevents production of sgRNA. Cells were fixed and stained with crystal violet. Adapted from Y. Liu et al., *Virology* **448:**15–25, 2014, with permission. Courtesy of Z. Yuan, Chinese Academy of Sciences, Wuhan, China.

Figure 14.3 Common activation of the Pi3k-Akt-mTor relay in virus-infected cells. (A) Direct association (double-headed arrows) of several viral proteins with the p85 regulatory subunit activates Pi3k and hence Akt to promote cell survival and block apoptosis, and in the case of adenovirus 5 (Ad5) E4 Orf1 protein, to activate mTor and stimulate translation. **(B)** The hepatitis B virus (HBV) X protein activates Pi3k in the same way, but subsequent signal transduction induces increased production of cyclin D1, which promotes cell proliferation and autophagy. The human herpesvirus 8 (HHV8)-encoded G protein-coupled receptor (GPCR) and formation of replication complexes of the alphavirus Sindbis virus also stimulate Pi3k-dependent signaling, by unknown mechanisms.

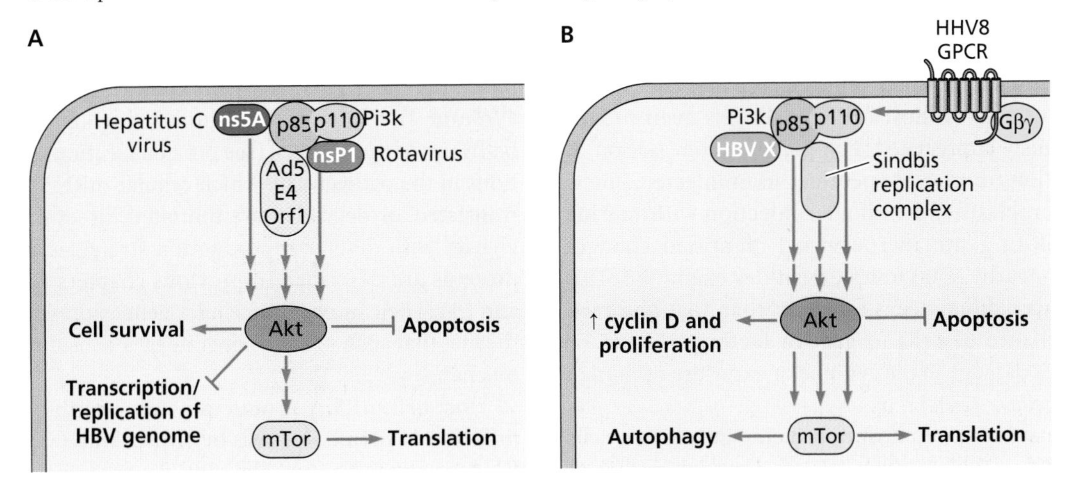

virus, a pathogenic picornavirus with an RNA genome of some 7.5 kb that is replicated by viral proteins in the cytoplasm, leads to signaling via not only Pi3k and mTor, but also the mitogen-activated protein kinase (Mapk) and the Nf-κb pathways and the tyrosine kinase Src. As might be anticipated from the additional participation of nuclear components such as the primers for synthesis of viral mRNA from the (−) strand genomic RNA segments and splicing proteins in influenza A virus reproduction, infection by this virus leads to activation of a larger number of signaling pathways than does picornavirus infection.

In general, it appears that the more elaborate the strategy for viral reproduction, the greater the impact of infection on signaling pathways. However, direct comparisons of the responses of signal transduction cascades in a particular cell type to infection with different viruses have not been reported. Furthermore, how radically cellular signaling systems are altered will also be determined by the origin and proliferation state of the host cell. Many human cells in routine use in the laboratory are derived from tumors (Chapter 2), and consequently are abnormal in many respects, including unrestrained proliferation and permanent activation of signaling circuits that promote cell growth and progression through the cell cycle. In contrast, in natural infections, many host cells proliferate only slowly or are quiescent (withdrawn from the cell cycle). Successful virus reproduction in such cells is therefore likely to depend to a greater degree on activation of signaling pathways that control these processes than does reproduction in tumor-derived cell lines.

The most common bit of information that is transmitted during biological signaling is the presence (or absence) of a phosphate group on specific amino acids in a protein. Application of methods of mass spectrometry that allow detection and very accurate quantification of differences in concentration among samples of thousands of phosphopeptides can therefore provide global, unbiased views of changes in signal transduction pathways under specific conditions. The results of recent applications of these methods to comparison of uninfected and virus-infected cells suggest that the impact of particular viruses on host cell signaling is even broader than previously appreciated. For example, comparison of the concentrations of phosphopeptides in uninfected, quiescent mouse fibroblasts and 18 h after infection with murine herpesvirus 68 (a gammaherpesvirus) identified changes in 86% of the nearly 2,500 unique peptides examined. This infection-induced difference is far larger than that observed following exposure of cells to growth factors ($<13\%$) or assaults, such as damage to the genome or exposure of human cells to *Salmonella* ($\sim24\%$).

Large-scale analyses of phosphoproteins in infected cells can also identify cellular substrates of signaling pathways that are important for virus reproduction. The abundance of phosphorylation sites in 175 host proteins was observed to increase or decrease within 1 min of exposure of unstimulated $CD4^+$ T cells to human immunodeficiency virus type 1. Bioinformatics analyses of the amino acid sequences spanning these phosphorylation sites indicated activation of signaling via Mapk and calmodulin-dependent kinase II (CamkII), in agreement with previous studies, but inhibition of signaling from protein kinase A. Subsequent functional analysis of production of proteins with increased phosphorylation in infected cells established the important contribution of host proteins not previously implicated in reproduction of human immunodeficiency virus type 1, notably a specific set of splicing proteins.

Gene Expression

Altered host cell gene expression is a universal consequence of virus infection. The altered patterns range from inhibition of the synthesis or translation of the majority if not all cellular mRNAs, to differential increases or decreases in expression of particular sets of cellular genes as an infection proceeds. These changes may be the result of modulation of any of the several reactions by which mammalian pre-mRNAs are produced, used as template for protein synthesis, and turned over.

The impact of virus infection on cellular gene expression and the mechanism(s) by which this process is altered vary with the strategies by which viral genes are expressed. For example, transcription of viral genes from DNA templates by the cellular transcriptional machinery and processing of the transcripts in the same manner as cellular pre-mRNAs precludes inhibition of these reactions (although their selectivity may be redirected). Furthermore, the genomes of such viruses often encode powerful activators of transcription that promote viral gene expression, but can also exert broad effects on cellular mRNA synthesis.

Inhibition of Cellular Gene Expression

Viral genomes typically encode one or more proteins (or RNAs) that inhibit cellular gene expression indirectly, by blocking mechanisms that expedite antiviral responses (Volume II, Chapter 3) or modulating signal transduction pathways of the host cell (see previous section). However, reactions in the pathways by which cellular mRNAs are produced, translated, or degraded are inhibited directly by proteins of viruses with diverse reproduction strategies. Many of these proteins are described in previous chapters (Chapters 8, 10, and 11). Their impact on cellular gene expression emphasizes the fact that such proteins operate by a considerable variety of mechanisms (Fig. 14.4).

Host cell mRNA production can be inhibited following infection of permissive cells by viruses with (+) or (−) strand RNA genomes because the viral genomes are expressed and replicated with minimal dependence on cellular systems. Such inhibition facilitates selective and efficient synthesis of

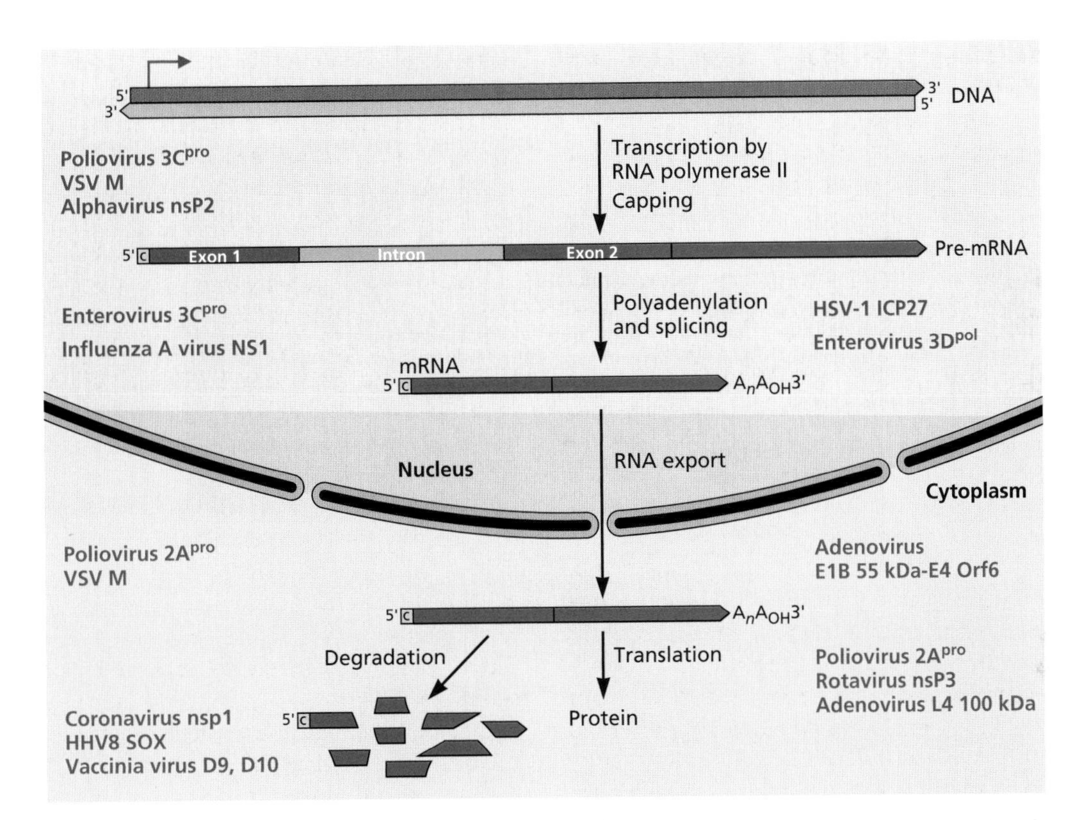

Figure 14.4 Inhibition of cellular gene expression by viral proteins. Viral proteins can inhibit the reactions by which mammalian mRNAs are made in the nucleus or exported to the cytoplasm and translated, or can induce accelerated mRNA degradation. (Transcription) The poliovirus $3C^{pro}$, nsP2 of Old World alphavirus, and the vesicular stomatitis virus (VSV) M protein all target components of the basal transcriptional machinery, Tbp, RNA polymerase II, and transcription initiation protein IID (TfIId), respectively. The first two are degraded in infected cells, but the mechanism by which the VSV M protein inactivates TfIId has not been established. Cleavage of Tbp by poliovirus $3C^{pro}$ also inhibits transcription by RNA polymerases I and III. (Polyadenylation) Both $3C^{pro}$ of enterovirus 71 (a picornavirus) and influenza A virus NS1 protein block polyadenylation of cellular pre-mRNAs, by inducing degradation of a subunit of cleavage stimulatory protein (Cstf) and sequestering cleavage and polyadenylation specificity protein (Cpsf), respectively. In both cases, the 3′ poly(A) sequences of viral mRNAs are synthesized by the viral RNA polymerase (Chapter 6). (Splicing) Splicing, essential for the production of the majority of cellular mRNAs, is perturbed by herpes simplex virus 1 (HSV-1) ICP27, which inhibits an early reaction in splicing and disrupts the nuclear foci in which splicing proteins are concentrated. This process is blocked in cells infected by enterovirus 71 as a result of interaction of the viral $3D^{pol}$ protein with a core component of the cellular splicing machinery, Prp8 (pre-mRNa processing protein 8). (Export) Export of RNAs from the nucleus is disrupted by a second poliovirus protease, $2A^{pro}$, and the VSV M protein, while export of cellular mRNAs is selectively blocked by the virus-specific E3 ubiquitin ligase containing the adenoviral E1B 55 kDa and E4 Orf6 proteins and several proteins co-opted from cellular enzymes of this type. (Translation) Translation of cellular mRNAs is also selectively inhibited in cells infected by a variety of viruses. The viral proteins shown block the function of proteins critical for initiation of translation of cellular mRNAs, because the viral mRNAs carry distinctive features that reduce or eliminate the dependence of their translation on these proteins (Chapter 11). (Degradation) The genomes of larger DNA and RNA viruses encode proteins that initiate mRNA degradation by removal of the 5′ cap (vaccinia virus D9 and D10 proteins) or endonucleolytic cleavage (coronavirus nsp1 and human herpesvirus 8 SOX protein). In some cases, cellular but not viral RNAs are degraded (see text).

viral proteins by reducing competition of cellular with viral mRNAs for components of the translation machinery, and can also mitigate host responses that impair virus reproduction. Although the specific reactions that are disrupted by viral proteins vary, the targets are cellular proteins necessary for production of cellular mRNAs in the nucleus or their export to the cytoplasm. Such proteins include essential components of the transcriptional initiation machinery, such as TATA-binding protein (Tbp), which is cleaved and inactivated by the poliovirus protease $3C^{pro}$; and a catalytic subunit of RNA polymerase II targeted for degradation by alphavirus nsP2. Proteins necessary for splicing or export of viral mRNAs to the cytoplasm can also be destroyed or inhibited by viral proteins (Fig. 14.4). The impact of such inhibition can be both substantial and widespread. For instance, only some 7% of the >5,000 cellular cytoplasmic poly(A)-containing mRNAs examined by microarray hybridization could be detected by 18 h after infection with the alphavirus Sindbis virus (Fig. 14.5A).

In some cases, viral proteins also block translation. This property is illustrated by poliovirus proteins that inhibit not only production of cellular mRNAs but also their translation.

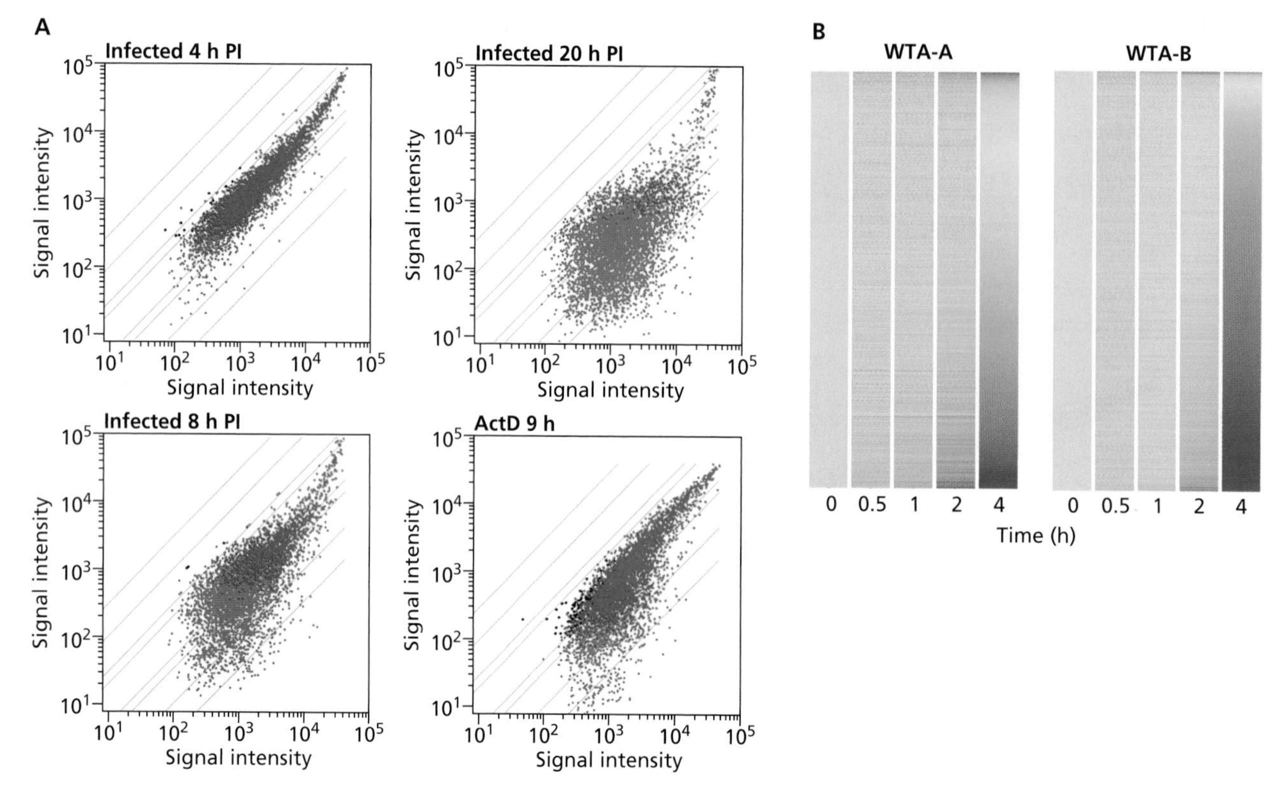

Figure 14.5 Decreases in cellular mRNA concentration in virus-infected cells. (A) Murine NIH 353 cells were infected with the alphavirus Sindbis virus (20 PFU/cell) (Infected). Total cell RNA was isolated after the periods of infection (PI) indicated, and from uninfected cells exposed to the transcriptional inhibitor actinomycin D (ActD) for 9 h. The concentrations of mRNAs were examined by hybridization to microarrays, using bacterial RNA added to each sample to provide an internal control. The mRNA signals are shown with those from infected or actinomycin D-treated cells plotted against the signals from uninfected cells. The mRNAs that changed little in concentration and decreased are shown in red and blue, respectively. Note that Sindbis virus infection leads to reductions in the concentration of far more cellular mRNAs than does exposure to actinomycin D. Adapted from R. Gorchakov et al., *J Virol* **79:**9397–9409, 2005. Courtesy of I. Frolov, University of Texas Medical Branch, with permission. **(B)** Human HeLa cells were infected with purified vaccinia virus under conditions that ensure infection of all cells, and total polyadenylated RNA was isolated at the times indicated. cDNAs were prepared and subjected to high-throughput sequencing. The number of read counts for each cellular mRNA (individual horizontal lines) is shown as the fold change from time zero, after normalization to the total number of reads for two independent infections (A and B). Colors from yellow to blue and to red indicate decreases and increases, respectively. Adapted from Z. Yang et al., *Proc Natl Acad Sci U S A* **107:**11513–11518, 2010, with permission. Courtesy of B. Moss, National Institute of Allergy and Infectious Diseases, National Institutes of Health.

Such seeming redundancy ensures efficient synthesis of viral proteins: inhibition of transcription and export of RNAs from the nucleus blocks the flow of newly synthesized cellular mRNAs into the cytoplasm, but those mRNAs made before infection, many of which are quite stable, are still present and potential templates for translation. The unusual mechanism of initiation of translation of viral mRNAs and concomitant cleavage of the initiation protein eIF4G by the viral protease $2A^{pro}$ eliminates such potential competition from cellular mRNAs (Chapter 11). The very short infectious cycle of poliovirus (and other picornaviruses; some 8 h) may necessitate particularly effective measures to prevent synthesis of cellular proteins.

The genomes of poxviruses and other large DNA viruses that reproduce in the cytoplasm are expressed by virally encoded transcription and RNA-processing systems that synthesize viral mRNAs with 5′ caps and 3′ poly(A) tails (Chapters 8 and 10). As might therefore be anticipated, loss of cellular mRNAs has been observed using high-throughput methods following infection by such viruses. Only ~10% of the RNA sequences in amoebae infected by mimivirus for 6 h were cellular in origin. Similarly, the concentrations of the majority of cellular mRNAs decreased by 4 h after vaccinia virus infection (Fig. 14.5B). The effect of infection on host cell transcription is not clear. However, the viral enzymes that remove 5′ caps from mRNA to initiate exonucleolytic degradation by the cellular nuclease Xrn1 are thought to make a major contribution to decreasing cellular mRNA concentrations.

Cellular gene expression can also be impaired in several different ways when viral mRNA synthesis depends on host cell components. Such selective inhibition targets reactions that are less critical for production of viral mRNAs than for those of the host cell, for example, pre-mRNA splicing in cells infected by herpesviruses or influenza viruses: in both cases, the majority of viral mRNAs comprise a single exon and cannot be spliced (Chapter 10). Similarly, the adenovirus L4 100-kDa protein induces selective translation of viral major late mRNAs. These mRNAs share a common 5′ untranslated region that bypasses the requirement for specific translation initiation proteins reduced in activity in infected cells (Chapter 11). Infection by gammaherpesviruses, such as Epstein-Barr virus and human herpesvirus 8, and severe acute respiratory syndrome coronavirus induces selective degradation of cellular mRNAs, a very effective mechanism for favoring synthesis of viral proteins. In the case of human herpesvirus 8, such degradation is initiated by a viral endonuclease (the SOX protein) that acts in conjunction with Xrn1. How cellular and viral mRNAs are distinguished is not known. However, destruction of cellular mRNAs triggers changes in cellular proteins that participate in production of host cell mRNAs to further restrict expression of cellular genes (Box 14.2).

Despite favoring translation of viral mRNAs, decreases in the production or acceleration of turnover of cellular mRNAs in virus-infected cells would not be expected to change the population of host cell proteins radically: many cellular mRNAs are quite stable, with half-lives longer than the time required for production of progeny particles of some viruses, such as picornaviruses or poxviruses (6 to 8 h). Indeed, when the concentrations of cellular proteins were assessed early and late after vaccinia virus infection of human cells, $<$10% were observed to change significantly.

Virus reproduction depends on many stable cellular proteins, including ribosomal proteins and structural proteins of the cytoskeleton. However, it may also be necessary to maintain the production of much less stable host proteins for optimal reproduction of particular viruses, even in the face of widespread inhibition of cellular gene expression. The mechanisms that allow such selective expression of specific sets of cellular genes in virus-infected cells are not well understood, but synthesis of particular proteins can be maintained

BOX 14.2

DISCUSSION

Domino effects of the human herpesvirus 8 SOX protein on cellular and viral gene expression

Synthesis of an early human herpesvirus 8 gene product, the SOX protein, in infected cells leads to degradation of the majority of cellular mRNAs. One consequence of such degradation is release of the poly(A)-binding protein PabpC1, which is normally bound to the 3′ poly(A) tails of cytoplasmic mRNAs to regulate their stability. This cellular protein then becomes relocalized to the nucleus (see the figure), where it leads to accumulation of poly(A), hyperadenylation of mRNAs, and impaired export of mRNAs from the nucleus.

In the nucleus, PabpC1 associates with a viral noncoding RNA, the polyadenylated nuclear (PAN) RNA (see the figure). This viral RNA accumulates to very high concentrations in infected cell nuclei (~500,000 copies per nucleus) and is not transported to the cytoplasm. The interaction of PabpC1 with PAN RNA may contribute to such accumulation by stabilizing PAN RNA: the RNA-degradation function of SOX and nuclear PabpC1 increase the nuclear concentrations of PAN RNA. This viral RNA does not contribute to inhibition of cellular gene expression. Rather, PAN RNA promotes synthesis of viral late proteins and release of DNA-containing viral particles: both were impaired when PAN RNA concentrations were reduced by introduction of complementary oligonucleotides and RNAse H.

The direct action of SOX and subsequent indirect consequences of release of PabpC1 therefore ensure efficient synthesis of viral late proteins.

Borah S, Darricarrère N, Darnell A, Myoung J, Steitz JA. 2011. A viral nuclear noncoding RNA binds re-localized poly(A) binding protein and is required for late KSHV gene expression. *PLoS Pathog* 7:e1002300. doi:10/1371/journal.ppat.10023000.

Kumar GR, Glaunsinger BA. 2010. Nuclear import of cytoplasmic poly(A) binding protein restricts gene expression via hyperadenylation and nuclear retention of mRNA. *Mol Cell Biol* **30:**4996–5008.

The localization of PabpC1 (red) and PAN RNA (green) determined by immunofluorescence and *in situ* hybridization, respectively, in human cells latently infected with human herpesvirus 8 (top) or following entry into the lytic replication cycle by synthesis of the viral regulator (RTA) that is necessary and sufficient to induce this switch (bottom). Adapted from S. Borah et al., *PLoS Pathog* 7:e1002300, 2011, with permission. Courtesy of J. Steitz, Yale University.

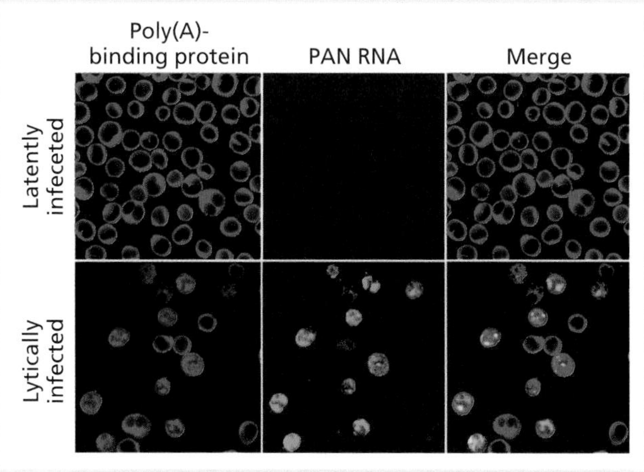

in various ways. For example, a small number of cellular mRNAs enriched in those for proteins that participate in signal transduction or regulation of apoptosis initially escape destruction in vaccinia virus-infected cells (Fig. 14.5B), and translation of specific host cell mRNAs is stimulated following hepatitis C virus infection.

Differential Regulation of Cellular Gene Expression

Although widespread inhibition of cellular gene expression is a common outcome of infection, more-subtle alterations also occur, particularly when viral gene expression depends on the host cell transcription and RNA-processing machineries. The scale and complexity of such modulation of host cell gene expression have become apparent only since the development of genome-wide methods for measurement of mRNA concentration, initially microarray hybridization and more recently high-throughput RNA sequencing. The latter method does not require hybridization to DNA and hence selection of DNA probes. It can therefore also provide information about noncoding RNAs. Alterations in the concentrations of micro-RNAs, long intergenic noncoding RNAs, small nucleolar RNAs, antisense RNAs, and transcripts of pseudogenes have been observed in cells infected by several viruses, but the significance of such broad impacts on the host cell RNA population is not yet clear.

These methods of RNA profiling are typically applied to total cell RNA populations or those enriched in mRNA by selection for the presence of a 3′ poly(A) tail. Consequently, they measure steady-state concentrations of mRNA (or other RNAs). Changes in this parameter are generally interpreted in terms of increases or decreases in transcription of individual genes, although they could be the result of alterations in any of the reactions by which an mRNA is produced, or in its rate of turnover (Box 14.3). More-precise information about the mechanisms that result in modulation of RNA accumulation in virus-infected cells can be collected by isolation of specific populations of RNA, for example, mRNAs that are serving as translational templates, prior to application of high-throughput quantification methods (Box 14.4).

The impact of infection on patterns of cellular gene expression varies with virus. Expression of some 800 host genes is altered in activated $CD4^+$ T cells infected by human immunodeficiency virus type 1. In contrast, the concentrations of ~10,000 cellular mRNAs increased or decreased by at least a factor of 2 in primary mouse fibroblasts infected by the herpesvirus murine cytomegalovirus, which possesses a large genome that encodes several transcriptional or posttranscriptional regulators. Furthermore, cellular mRNAs can even accumulate when viral gene products block production or increase turnover of the majority. This phenomenon is illustrated by the detection of increased concentrations of 400 or so cellular mRNAs by 7 h after herpes simplex virus 1 infection of normal human cells, despite the action of the viral Vhs endonuclease that leads to mRNA degradation (Chapter 10). When combined with various types of bioinformatics analysis, such as classification of differentially expressed genes by their functional annotations (gene ontology analysis), the results of these descriptive studies can help identify cellular gene products and pathways that promote or counter virus reproduction, or that correlate with virus pathogenicity or the responses of individual hosts to infection (Box 14.5).

In most cases, indirect effects of infection on cellular RNA populations, for example, as a result of modulation of signal transduction pathways, have not been distinguished from the direct actions of viral gene products. One exception is provided by cells infected by adenovirus. Infection of quiescent, normal human fibroblasts by human adenoviruses is followed by increases or decreases of at least 2-fold in expression of 10% of cellular genes. Many of these changes are associated with reentry of quiescent cells into the cell cycle, or support genome replication and expression. Transcription of a subset of genes repressed in infected cells, particularly interferon-sensitive and other genes associated with antiviral defenses, is inhibited by the viral E1B 55-kDa protein. How-

BOX 14.3

BACKGROUND

Multiple parameters govern the steady-state concentration of a cellular mRNA

The steady-state concentration of a cellular mRNA is determined by the balance between the overall rate of production of the mRNA and the rate at which it is degraded. Consequently, changes in concentration measured by microarray hybridization or high-throughput sequencing can be the result of alterations in the rate of either synthesis of the mRNA or its degradation, or changes in both parameters.

As summarized in the figure, the appearance of a functional mRNA available for translation in the cytoplasm is the end result of several processes. The rate of turnover of the mRNA can also be influenced by various parameters. Consequently, it is not correct to ascribe alterations in mRNA concentration measured by these techniques to increased or decreased transcription.

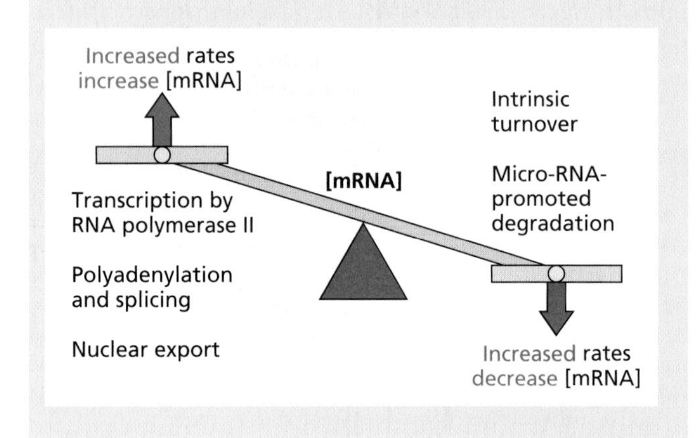

BOX 14.4

METHODS

Increasing the specificity of RNA profiling

Microarray hybridization or high-throughput sequencing of total cell RNA or mRNA populations does not yield direct information about the mechanism(s) responsible for changes in RNA concentration or their consequences for the protein repertoire of a cell. The first of these limitations can be circumvented by isolation of specific classes of RNA prior to analysis.

The impact of a particular condition, such as virus infection, on translation can be assessed by purification of mRNAs present in polyribosomes and therefore serving as templates for protein synthesis. Methods that fractionate macromolecules on the basis of size and shape, such as density gradient centrifugation, readily separate larger polyribosomes from free mRNAs and ribosomal subunits (panel A of the figure). When combined with parallel analysis of total mRNA populations, this approach can establish whether cellular gene expression is subject to translational regulation, as has been observed in permissive cells infected by hepatitis C virus or human cytomegalovirus.

To distinguish transcriptional from post-transcriptional regulation, it is necessary to examine newly synthesized RNA (a small fraction of the total population in a cell). Selection of this population can be achieved by incorporation of modified derivatives of uridine, such as 4-thiouridine or 5-ethynyluridine (EU). RNA containing these modified bases, at the relatively low frequency of 1 per 50 to 100 nucleotides, is biotinylated after purification and then separated from the bulk RNA pool by binding to streptavidin attached to beads (panel B). When the uridine derivative is supplied to uninfected and infected cells for short periods (≤1 h), newly transcribed RNAs can be isolated and compared to identify changes in transcription. Application of this approach to normal murine cells infected by murine cytomegalovirus has established an almost perfect concordance between changes in the total RNA and the newly synthesized RNA pools during the early phase of infection, indicating that the rates of turnover of cellular mRNAs were not modulated at this time. Particles of murine cytomegalovirus and other herpesviruses contain substantial quantities of viral RNA species that are incorporated nonspecifically and transferred into new host cells. Tagging and isolation of newly synthesized RNA therefore facilitates elucidation of the pattern of viral gene expression early after infection, because the viral RNAs introduced from virus particles are not labeled.

This approach can also be used to measure the rates of turnover of mRNAs, by using a pulse-chase method in which the uridine derivative is supplied for a period sufficient for incorporation into mature mRNA, and the cells are then exposed to medium containing a vast excess of unmodified uridine. Measurement of the concentrations of the tagged mRNAs as a function of time after this switch (the "chase") allows rates of decay to be determined.

Colman H, Le Berre-Scoul C, Hernandez C, Pierredon S, Bihouée A, Houlgatte R, Vagner S, Rosenberg AR, Féray C. 2013. Genome-wide analysis of host mRNA translation during hepatitis C virus infection. *J Virol* **87:**6668–6677.

Marcinowski L, Lidschreiber M, Windhager L, Rieder M, Bosse JB, Rädle B, Bonfert T, Györy I, de Graaf M, Prazeres da Costa O, Rosenstiel P, Friedel CC, Zimmer R, Ruzsics Z, Dölken L. 2012. Real-time transcriptional profiling of cellular and viral gene expression during lytic cytomegalovirus infection. *PLoS Pathog* **8:**e1002908. doi:10.1371/journal.ppat.1002908.

McKinney C, Zavadil J, Bianco C, Shiflett L, Brown S, Mohr I. 2014. Global reprogramming of the cellular translational landscape facilitates cytomegalovirus replication. *Cell Rep* **16:**9–17.

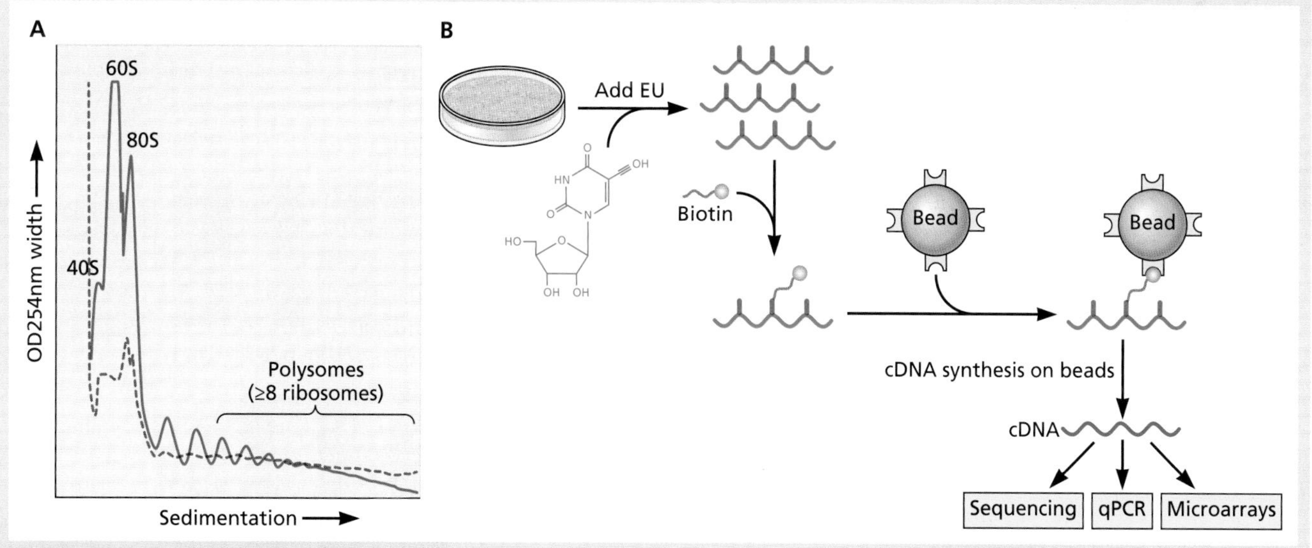

(A) Comparison of the polyribosome profiles of normal human fibroblasts infected by human cytomegalovirus for 48 h (red) and uninfected cells (gray). Cell extracts were prepared under conditions that do not disrupt polyribosomes and sedimented through 15 to 50% linear sucrose gradients. The absorbance at 254 nm (shown) was monitored during collection of gradient fractions. Adapted from C. McKinney et al., *Cell Rep* **16:**9–17, 2014, with permission. **(B)** Isolation of newly synthesized RNA is achieved by incorporation of 5-ethynyluridine (EU; structure shown above the first step) for a short period followed by *in vitro* biotinylation, and subsequent separation from the bulk RNA population by binding to bead-bound streptavidin. The bead-bound RNA is then converted to cDNA for analysis by microarray hybridization, high-throughput sequencing, or quantitative PCR (qPCR).

BOX 14.5

DISCUSSION

Insights into virus-host interactions from RNA profiling studies

It is well established that gene expression profiling can make quite fine distinctions among multiple subtypes of particular cancers, for example, breast cancers, and provide important information about prognosis. When combined with genome-wide analysis of mutations and changes in copy number, this approach has the potential to inform rationally based personalized treatment. As the two examples below illustrate, RNA profiling is also beginning to yield new insights into the interactions of viruses with their hosts.

In a very rare subset of individuals infected with human immunodeficiency virus type 1, reproduction of the virus (that is, synthesis of genomes) is undetectable by all but extremely sensitive assays. Studies of such "elite controllers" have identified various features of the immune response that correlate with restriction of virus reproduction (Volume II, Chapter 7). Unexpectedly, two groups of elite controllers were distinguished by the results of microarray hybridization of RNA isolated from $CD4^+$ T cells of uninfected individuals, human immunodeficiency virus type 1-infected patients undergoing highly active antiretroviral therapy (HAART), and elite controllers. Members of one group of elite controllers exhibited gene expression patterns closely resembling those of virus-negative individuals, while the expression patterns of the second group clustered with those of patients receiving HAART (panel A of the figure). Subsequently, the elite controllers in the first group were shown to have higher $CD4^+$ T cell counts and reduced $CD8^+$ T cell responses compared with the members of the second group. The number of elite controllers with gene expression profiles like those of uninfected persons was small (4 of the 12 elite controllers examined), but further studies of the properties of their T cells may provide clues in the search for a cure for human immunodeficiency virus type 1.

Infection of humans by highly pathogenic avian influenza virus is occurring more frequently since the first case of H5N1 avian virus in humans was reported in 1997 (Volume II, Chapter 11). So far, such outbreaks, which are associated with high case-fatality ratios, have been sporadic, because the avian viruses are not transmitted from human to human. The pathogenicity of these avian viruses in mammals cannot be predicted a priori from their genome sequences, and is traditionally assessed in such model animals as mice and ferrets. In one complementary approach to such traditional methods, the gene expression profiles of lung tissue of mice infected with influenza virus strains that differ in pathogenicity were compared. These viruses were

- an H5N1 avian influenza virus strain that is highly pathogenic in laboratory animals but poorly transmissible
- an H7N9 avian influenza virus that is somewhat less pathogenic but more transmissible
- an H7N7 avian influenza virus strain with lethality in animals similar to that of the H7N9 virus
- a strain from the 2009 H1N1 pandemic in humans, which was observed to be much less pathogenic in mice than any of the avian virus isolates

Gene expression differences in lung tissue in mice infected by these four influenza virus strains were observed. However, the

(A) Clustering of genes differentially expressed in uninfected individuals, human immunodeficiency virus type 1 (HIV-1)-infected patients receiving HAART, and elite controllers, indicated by the green, blue, and red squares, respectively, at the left. These differences are based on the results of microarray hybridization of duplicate samples of RNA isolated from $CD4^+$ T cells. Note the clear separation of the profiles from uninfected persons and infected patients, and the dispersal of those from elite controllers in both clusters. Adapted from F. Vigneault et al., *J Virol* **85:**3015–3019, 2011, with permission. Courtesy of M. Lichterfeld, Massachusetts General Hospital. **(B)** Mice were infected with the types of influenza virus strain indicated at the top, and RNA was isolated from lung tissue after 1 or 3 days of infection and from mock-infected animals. Differences between the infected and mock-infected samples identified by microarray hybridization are shown for cytokine response, lipid metabolism, and coagulation genes. Adapted from J. Morrison et al., *J Virol* **88:**10556–10568, 2014, with permission. Courtesy of M. Katze, University of Washington.

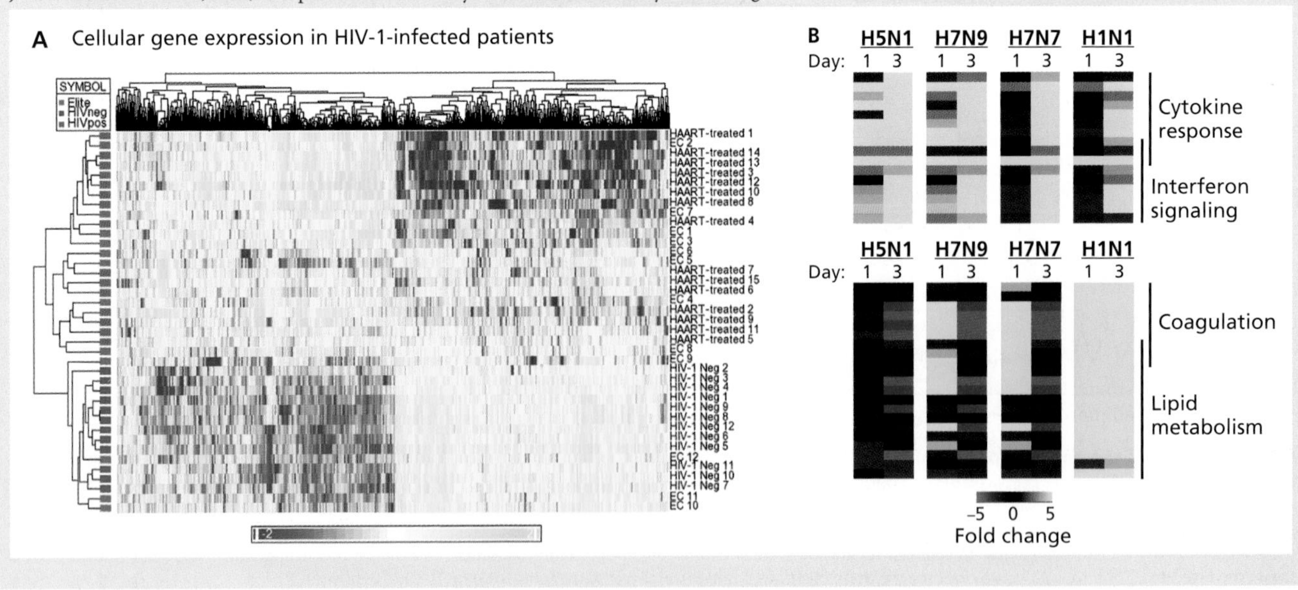

most striking was the correlation of high pathogenicity with the increased expression of genes that encode cytokines, interferons, and proteins that mediate the interferon response and the decreased expression of genes associated with lipid metabolism and coagulation (panel B). The latter changes were also observed in animals infected by the highly pathogenic 1918 H1N1 influenza virus, although their significance is not yet known. It has been suggested that such a gene expression signature of highly pathogenic influenza viruses could help identify new therapeutics that target specific host proteins or pathways.

Morrison J, Josset L, Tchitchek N, Chang J, Belser JA, Swayne DE, Pantin-Jackwood MJ, Tumpey TM, Katze MG. 2014. H7N9 and other pathogenic avian influenza viruses elicit a three-pronged transcriptomic signature that is reminiscent of 1918 influenza virus and is associated with a lethal outcome in mice. *J Virol* **88:**10556–10568.

Vigneault F, Woods M, Buzon MJ, Li C, Pereyra F, Crosby SD, Rychert J, Church G, Martinez-Picado J, Rosenberg ES, Telenti A, Yu XG, Lichterfeld M. 2011. Transcriptional profiling of CD4 T cells identifies distinct subgroups of HIV-1 elite controllers. *J Virol* **85:**3015–3019.

ever, the majority of the changes can be attributed to the viral 243R E1A protein (Fig. 8.18), as they also occur when cells are infected by a mutant virus that directs efficient synthesis of only this protein. The E1A protein associates with the promoters of host cell genes altered in expression in infected cells, where it modulates the recruitment to promoters of the cellular repressor Rb and acetylation of histones at specific residues (Fig. 14.6).

Metabolism

Host cells supply not only the molecular machinery needed for synthesis of viral nucleic acids and proteins (at a minimum the translational machinery), but also the essential building blocks, nucleotides and amino acids. Assembly of enveloped viruses also requires cellular membranes and the lipids from which these structures are constructed. The production of large quantities of viral macromolecules and virus particles, often within a short period (a day or less), imposes heavy demands on the host cell's biosynthetic systems that manufacture nucleotides, amino acids, and, in many cases, fatty acids. Synthesis of these molecules consumes energy, typically supplied by the hydrolysis of ATP, as does production of viral macromolecules: synthesis of a single peptide bond, for example, consumes the equivalent of 4 molecules of ATP, and energy is also expended during the folding of viral proteins and intracellular transport of viral nucleic acids and proteins during the infectious cycle. Consequently, virus infection can lead to alterations in the pathways by which cells generate energy from molecular fuels (**catabolism**), as well as those that make the precursors of nucleic acids, proteins, and membranes (**anabolism**). Perhaps not surprisingly, the impact of infection on host metabolism is virus specific, ranging from relatively simple alterations in the rates of particular reactions to extensive redirection of multiple pathways to virus-specific ends. Indeed, infection by some viruses has been associated with development of metabolic diseases.

Methods To Study Metabolism

Some of the earliest studies of host cell responses to virus infection examined rates of catabolism by measuring the uptake of molecular oxygen or release of lactic acid (Fig. 14.7), the end product of anaerobic glycolysis. It has therefore been appreciated for decades that virus infection modulates cellular energy metabolism, often increasing the rate of glycolysis. However, this aspect of virus-host cell interactions was difficult to study in detail until the development of methods for simultaneous and comprehensive measurement of the concentrations of large numbers of metabolites and of changes in flux through individual pathways and reactions, so-called metabolomics. In the past decade or so, application of these methods to virus-infected cells (and their mock-infected counterparts) has revealed just how extensive the modulation of catabolism or anabolism can be, and unexpected ways in which virus infection can redirect metabolic networks.

Substrates and intermediates of metabolic pathways turn over as they are converted to other compounds, and many do so at high rates. Accurate measurement of the concentrations of metabolites under a particular condition therefore requires that metabolic pathways be halted quickly as samples are collected. This imperative is typically met by rapidly transferring cells to ice-cold organic solvents, a process that also contributes to extraction of metabolites. These compounds are then separated and identified by a variety of analytic techniques, most commonly liquid or gas chromatography followed by one- or two-dimensional mass spectrometry. The ions separated in this way are identified by comparison of their properties to the contents of reference libraries of metabolites, and can be quantified. Accurate comparison among samples is facilitated by examination of multiple experimental replicates and the addition of internal standards.

This approach can identify changes in the concentration of many metabolites, and hence indicates that the rates of particular pathways may be altered following virus infection. More-precise information can be obtained by supplying infected cells with a metabolic precursor labeled with a heavy atom (such as ^{13}C or ^{15}N). Because mass spectrometry separates compounds on the basis of mass and charge, the flow of the heavy atom to other metabolites can be then traced as a function of time. This method allows measurement of the

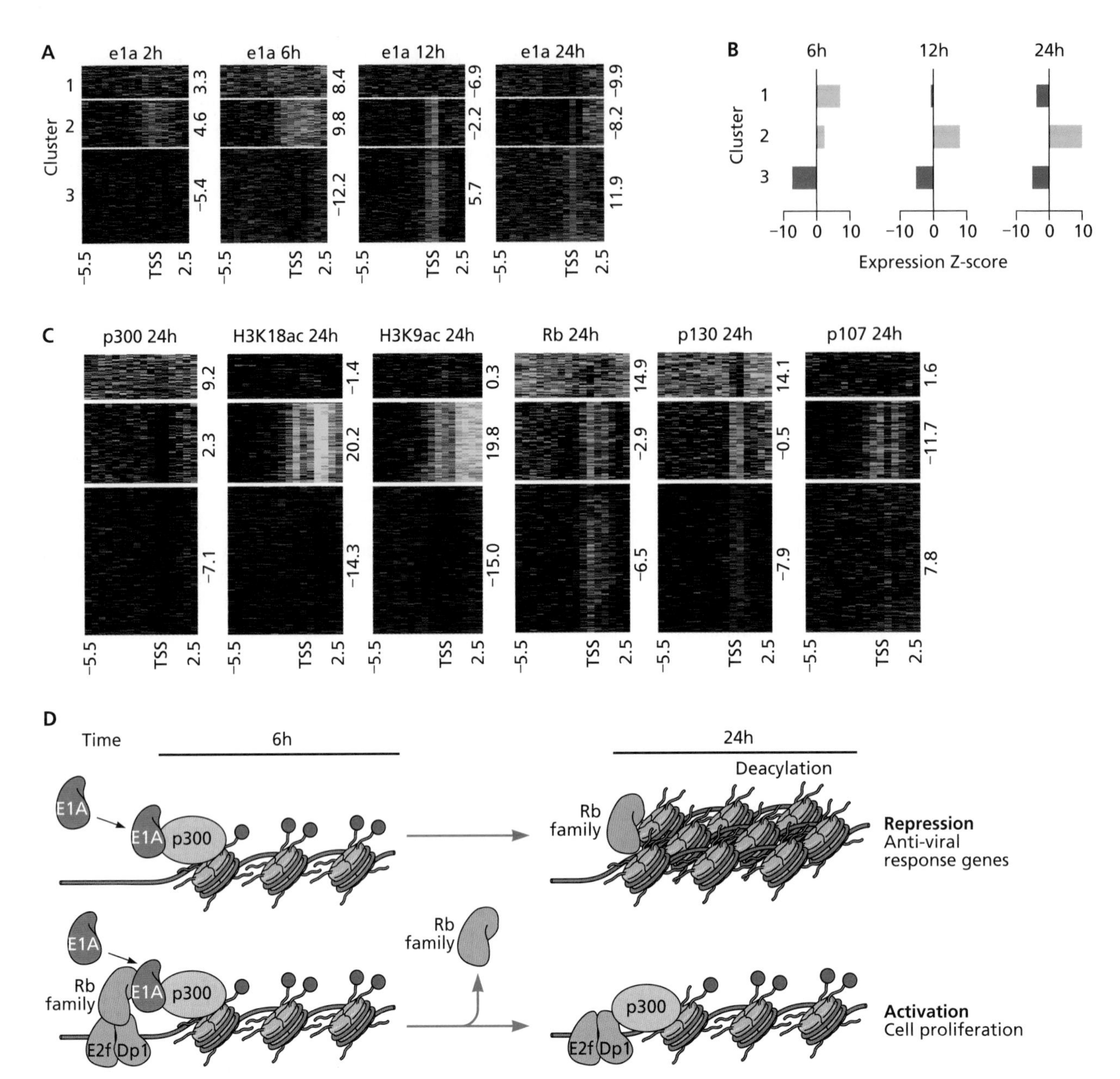

Figure 14.6 Reprogramming of promoter-associated transcriptional regulators by an adenovirus protein. (A) Contact-inhibited (quiescent) normal human fibroblasts were infected by a mutant of human adenovirus type 2 that directs synthesis in infected cells of only the smaller (243R) E1A protein. At the times after infection indicated, DNA bound to the E1A protein was isolated by chromatin immunoprecipitation (Box 8.1) and hybridized to microarrays containing probes that span from −5.5 kb to +2.5 kb (relative to the transcriptional start site [TSS]) of some 17,000 human promoters (so-called tiling arrays). The patterns of enrichment or loss of E1A across the promoters for 70% of the promoters to which E1A bound defined three clusters as indicated. **(B)** Comparison of the RNA profiles of infected and mock-infected cells showed that the expression of the genes present in each of the three clusters defined by kinetic patterns of E1A protein association exhibited different responses to synthesis of the E1A protein (note the change of scale at 24 h postinfection). **(C)** Chromatin immunoprecipitation was performed with antibodies that recognize cellular proteins that bind to the E1A protein (p300/Cbp, Rb, and the related proteins p130 and p107) or histone H3-bearing acetyl groups at specific lysine residues. The recovered DNA was hybridized to the same arrays. **(D)** These data indicated mechanisms by which the E1A protein alters expression of cellular genes. For example, the large increase in expression of genes in cluster 2 by 24 h postinfection correlated with loss of the transcriptional repressor Rb (and its relatives) and concomitant large increases in the association of histone H3 acetylated at Lys9 or Lys18, posttranslational modifications associated with activation of transcription. This cluster was enriched in genes associated with cell proliferation (growth and progression through the cell cycle) and DNA synthesis, and with promoters that contain binding sites for transcriptional activators of the E2F family, the primary targets of repression of gene expression by Rb proteins (Chapter 8). These E1A-induced alterations in proteins associated with the promoters of cluster 2 genes are consistent with ability of the E1A protein to overcome Rb-mediated inhibition of transcription of E2F-dependent genes observed in simplified experimental systems. Adapted from R. Ferrari et al., *Science* **321:**1086–1088, 2008, with permission. Panels A to C courtesy of S. Kurdistani, University of California, Los Angeles.

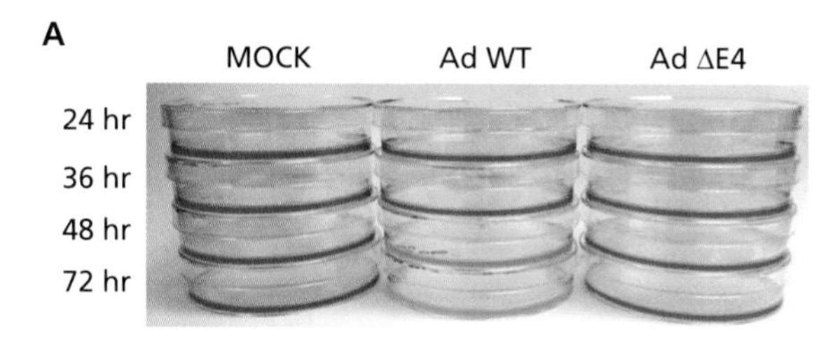

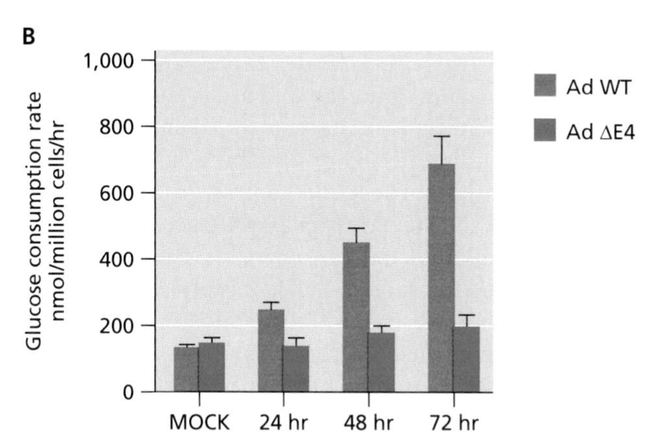

Figure 14.7 Increased glycolysis in virus-infected cells. **(A)** Infection by a variety of viruses (including adenoviruses, hepdnaviruses, herpesviruses, orthomyxoviruses, papillomaviruses, polyomavirus, and retroviruses) increases the rate of glycolysis, as illustrated for adenovirus. Plates of human breast epithelial cells infected with human adenovirus type 5 (Ad5) or a mutant that lacks the E4 gene (Ad ΔE4) or mock-infected (mock) were incubated for the periods indicated at the left. The pH of the medium is indicated by the color of the indicator it contains, where red and yellow indicate neutral and acidic pH, respectively. The increasing acidity of the medium of Ad5-infected cells is the result of the increased production of lactic acid, the product of glycolysis under anaerobic (or hypoxic) conditions. **(B)** This change was accompanied by increased rates of glucose consumption (but decreased O_2 uptake). Adapted from M. Thai et al., *Cell Metab* **19:**694–701, 2014, with permission. Courtesy of H. Christofk, University of California, Los Angeles.

rates of metabolic reactions of interest. It has been particularly valuable in tracing some unusual fates of common metabolites in virus-infected cells.

Comparison of the concentrations of metabolic enzymes (or of the mRNAs that encode them) in uninfected and infected cells can also indicate virus-induced perturbations of particular metabolic pathways, or mechanisms by which the production or consumption of metabolites is modulated. Application of these approaches, in conjunction with examination of the effects of inhibition of individual enzymes on viral reproduction, has illuminated various ways in which virus infection deranges the metabolic homeostasis of the host cell.

Glucose Metabolism

During glycolysis, the 6-carbon sugar glucose, the major product of breakdown of dietary carbohydrate, is converted to 2 molecules of the 3-carbon compound pyruvate (Fig. 14.8). From 1 molecule of glucose, this series of 10 reactions generates energy in the form of 2 molecules of ATP, which can be used directly in numerous reactions and processes, and 2 molecules of NADH (reduced nicotinamide adenine dinucleotide). The latter compound can be used for production of

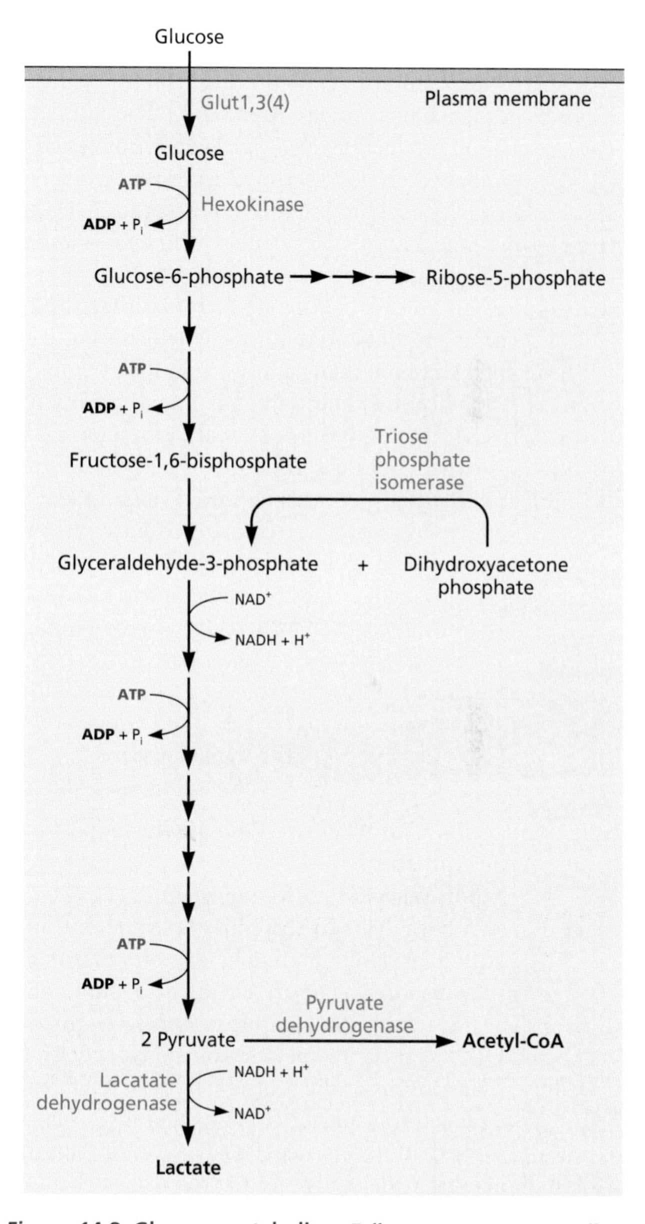

Figure 14.8 Glucose metabolism. Following transport into cells via glucose transporters (Gluts) and phosphorylation by hexokinase, glucose can enter glycolysis or, in rapidly growing cells in which nucleotide biosynthesis is required, the pentose phosphate pathway. The numbers of reactions in these pathways are indicated by the arrows, and the products, some intermediates, and some enzymes are listed. The 6-carbon molecule fructose-1,6-bisphosphate is converted to two 3-carbon compounds, both of which can be converted to pyruvate. Consequently, glycolysis produces 2 molecules of pyruvate from 1 of glucose, with a net energy yield of 2 ATP and 2 NADH.

additional ATP by the electron transport chain and oxidative phosphorylation in mitochondria, or be consumed in other metabolic reactions. One important function of glycolysis is to allow synthesis of ATP, and in some cells glucose is the only (red blood cells) or preferred (neurons) source of energy. However, this pathway also yields intermediates and products that allow synthesis of much larger quantities of ATP, or provide carbon skeletons for biosynthetic reactions. For example, glucose-6-phosphate is not only the substrate of the second glycolytic reaction, but also the precursor for synthesis of ribose, and hence nucleotides, RNA, and DNA. Similarly, pyruvate, the final product of glycolysis, is often converted to acetyl coenzyme A (acetyl-CoA), which serves as a precursor for synthesis of fatty acids and sterols or enters the citric acid cycle. This cycle generates energy and precursors to amino acids and bases (purines and pyrimidines).

Our understanding of the impact of virus infection on glycolysis, particularly the mechanisms by which viral gene products regulate this important pathway, is far from complete. Nevertheless, it is clear that infection of mammalian cells by a variety of viruses increases the rate of glycolysis by altering the concentration, activity, or other properties of cellular enzymes and other proteins that execute the initial metabolism of glucose. An increased rate of flux from glucose to pyruvate (Fig. 14.8) is a common response to virus infection, but is not inevitable, a perhaps counterintuitive fact emphasized by the opposite effects on this pathway of infection by two human herpesviruses (Box 14.6). In this section, we use some specific examples to illustrate the ways in which glucose metabolism can be perturbed in virus-infected cells and the association of such responses to virus infection with human disease.

Virus Infection Can Alter the Rate of Glycolysis by Several Mechanisms

Regardless of its final fate, glucose must enter cells before it can be metabolized. This hydrophilic molecule is transported across the plasma membrane by any one of a number of glucose transporters (12 are encoded in the human genome). These include the ubiquitous protein glucose transporter 1 (Glut1); tissue-specific transporters such as Glut3, present in neurons of the central nervous system; and the insulin-regulated transporter Glut4, which is present in skeletal muscle and adipose tissue. Localization of Glut4 to the plasma membrane depends on the presence of the hormone insulin, which is synthesized in and released from specialized cells (β cells) in the pancreas in response to high concentrations of blood glucose. Glucose is phosphorylated to glucose-6-phosphate upon entry into cells, a reaction that is irreversible under normal physiological conditions. This modification ensures retention of glucose within cells, and activates it for subsequent glycolysis or entry into the pentose phosphate pathway (Fig. 14.8). One parameter contributing to the increased rate of glycolysis observed in cells infected by hepatitis C virus, herpesviruses, and human immunodeficiency virus type 1 is an increased rate of glucose uptake.

In several cases, such accelerated transport of glucose into virus-infected cells can be attributed to elevated concentrations of Glut1 or Glut3, as a result of alterations in signal transduction pathways that regulate transcription. For example, in cells infected by the gammaherpesvirus Epstein-Barr virus or human papillomaviruses, the concentration of hypoxia-inducible factor 1α (Hif-1α) is increased. This transcriptional regulator activates expression of the genes encoding Glut1 and Glut3 (as well as several that encode glycolytic enzymes). This mechanism resembles the normal responses of uninfected cells to low availability of oxygen (hypoxia) and nutrients. In contrast to modulation of Glut1 or Glut3, the increased uptake of glucose into human cytomegalovirus-infected fibroblasts is mediated by the insulin-regulated transporter Glut4. This protein is not made in uninfected fibroblasts, but transcription of the gene that encodes it is turned on in infected cells as a result of increased production of the transcriptional regulator Chrebp (carbohydrate response element-binding protein): human cytomegalovirus infection therefore overrides the mechanisms that normally control production of Glut4. The restriction of the transporter to intracellular vesicles unless insulin stimulates signaling via Akt is also circumvented, because infection activates this signal transduction pathway. Inhibition of synthesis of Chrebp by RNA interference reduced glucose uptake by infected cells, prevented production of Glut4 mRNA, and reduced the yield of progeny virus particles. These observations emphasize the importance of the switch from Glut1 to Glut4 for efficient human cytomegalovirus reproduction, a necessity that would not have been predicted from the relatively modest increase (some 3-fold) in the affinity of Glut4 for glucose.

In principle (the law of mass action), increased intracellular concentrations of glucose as a result of more efficient transport across the plasma membrane could account for increased rates of glycolysis in virus-infected cells. Nevertheless, in several cases, the rate of flux through this pathway is also accelerated by increases in the intracellular concentration or activity of one or more glycolytic enzymes. For example, the activity of phosphofructokinase 1, which catalyzes the committed reaction in this pathway (Fig. 14.8), is increased in cells infected by human cytomegalovirus. The concentrations of several glycolytic enzymes (or their mRNAs) are also elevated following infection by this virus, and early during acute infection of hepatocytes in culture with hepatitis C virus.

Virus Infection Can Redirect the Utilization of Glycolytic Intermediates and Products

Acceleration of glycolysis in virus-infected cells can help provide the additional energy required for production and transport of viral macromolecules. However, this response to

BOX 14.6

EXPERIMENTS

Members of the same virus family can exert different effects on metabolism: glycolysis in cells infected by two human herpesviruses

The impact of infection with the alpha- and betaherpesvirus herpes simplex virus 1 and human cytomegalovirus, respectively, on carbon metabolism in normal human fibroblasts or epithelial cells was compared initially by measuring the concentration of >80 metabolites as a function of time after infection. Analysis of these data identified some host-cell-type-specific responses and some changes common to infection by the two viruses, such as increased concentrations of dTTP in infected cells. However, major differences were also detected, notably increased accumulation of glycolytic intermediates in cells infected by herpes simplex virus 1, but decreased concentrations in human cytomegalovirus-infected cells (panel A of the figure).

The reasons for this difference were investigated further, for example, by supplying infected cells with ^{13}C-labeled glucose and monitoring its incorporation into downstream metabolites as a function of time thereafter. In human cytomegalovirus-infected cells, the uptake of glucose and the labeling of glycolytic intermediates such as fructose-1,6-bisphosphate were increased, indicating stimulation of glycolytic flux (panel B). However, these parameters were decreased in herpes simplex virus-infected cells, accounting for the buildup of glycolytic intermediates (panel A). This response was accompanied by increased concentrations of intermediates in the pentose phosphate pathway and of its product, ribose-5-phosphate (panel B). Synthesis of pyrimidines is also increased in herpes simplex virus 1-infected cells (see text). The increased production of pyruvate in human cytomegalovirus-infected cells because of the acceleration of glycolytic flux supports increased production of fatty acids, following synthesis of the precursor, acetyl-CoA (see the text).

It has been proposed that the quite different fates of carbon from glucose in cells infected by these two human herpesviruses is the result of the much shorter replication cycle of herpes simplex virus 1 than of human cytomegalovirus, some 24 and 96 h, respectively, to attain the maximal yield of progeny virus particles. The relatively rapid reproduction of herpes simplex virus and some 10-fold-higher yield of virus particles require synthesis of a large number of viral DNA genomes (and large quantities of viral RNAs) in a short period, and hence impose a greater demand for nucleotide precursors from the host cell.

Vastag L, Koyuncu E, Grady SL, Shenk TE, Rabinowitz JD. 2011. Divergent effects of human cytomegalovirus and herpes simplex virus-1 on cellular metabolism. *PLoS Pathog* **7:**e1002124. doi:10.1371/journal.ppat.1002124.

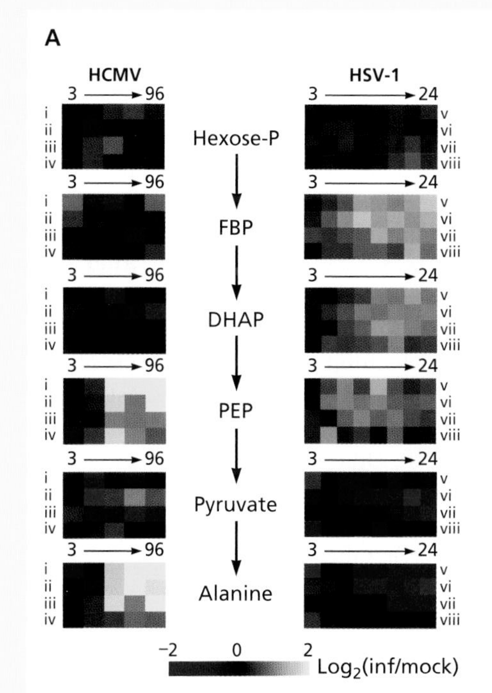

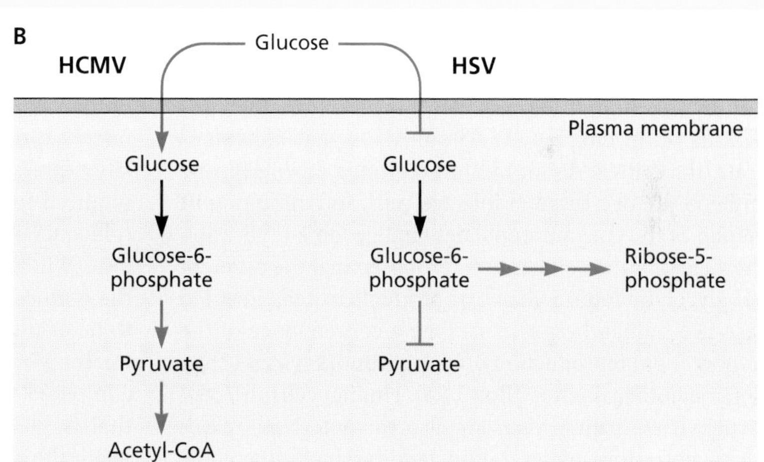

(A) Concentrations of glycolytic intermediates at the times indicated after infection with human cytomegalovirus (HCMV) or herpes simplex virus 1 (HSV-1) are shown relative to the concentrations measured in mock-infected cells. i, ii, v, and vi, human foreskin fibroblasts infected by two strains of HCMV (i, ii) or HSV-1 (v, vi); iii, iv, vii, and viii, human embryonic lung fibroblasts infected by two different strains of HCMV (iii, iv) or HSV-1 (vii, viii). From L. Vastag et al., *PLoS Pathog* **7:**e1002124, 2011, with permission. Courtesy of L. Vastag, Castleton State College, Vermont. **(B)** Summary of the effects of the two viruses on glycolytic flux and the concentration of specific metabolites, with increases and decreases shown in green and red, respectively.

infection can also foster virus reproduction by increasing the supply of precursors for biosynthesis of lipids and nucleotides.

Under aerobic conditions in mammalian cells, pyruvate (the final product of glycolysis) enters mitochondria and is converted to acetyl-CoA. This compound is an activated carrier of units of 2 carbon atoms that is a major source of energy following its entry into the citric acid cycle (see below), and is also the precursor for synthesis of fatty acids and sterols. In cells infected by human cytomegalovirus, most of the acetyl-CoA produced from pyruvate is rerouted

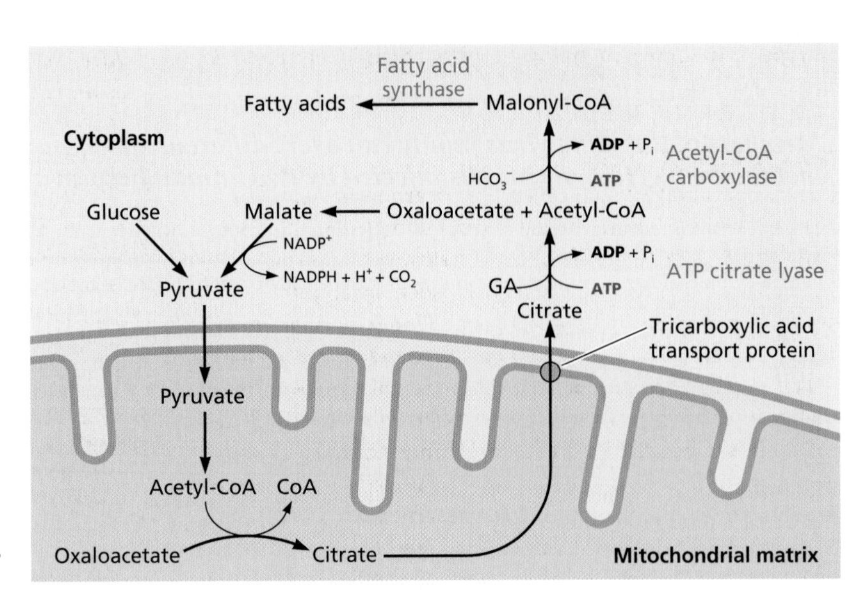

Figure 14.9 Diversion of acetyl-CoA for fatty acid synthesis in human cytomegalovirus-infected cells. Acetyl-CoA is the precursor for synthesis of fatty acids. However, this process takes place in the cytoplasm, whereas acetyl-CoA is produced within the mitochondrial matrix, as indicated. As this metabolite cannot be transported across the mitochondrial membrane, it is first converted to citrate via the first reaction in the citric acid cycle. Citrate then enters the cytoplasm via the tricarboxylate transport protein, where it is reconverted to acetyl-CoA and oxaloacetate by ATP citrate lyase. Formation of malonyl-CoA by acetyl-CoA carboxylase initiates fatty acid synthesis, which is catalyzed by the multi-active-site enzyme fatty acid synthase. As shown, pyruvate can be regenerated from oxaloacetate, with production of NAPDH, which is consumed during fatty acid synthesis.

from mitochondria to the cytoplasm by the shuttle shown in Fig. 14.9 to promote biosynthesis of fatty acids. In contrast, an important function of glycolysis during the early phase of herpes simplex virus 1 infection is to allow increased synthesis of ribose and deoxyribose (and hence nucleotides and nucleic acids): the concentrations of intermediates and enzymes of the pentose phosphate pathway are elevated in fibroblasts infected by this human herpesvirus. Furthermore, the concentrations of an enzyme (and its mRNA) that consumes aspartate, the critical donor of amino groups during purine biosynthesis, are decreased in infected cells, and inhibition of production of the enzyme stimulates viral genome replication and the yield of virions. It has been proposed that the different fates of glycolytic intermediates or products in cells infected by these two human herpesviruses are a consequence of the much more rapid reproduction of herpes simplex virus 1 than of human cytomegalovirus (Box 14.6). Human cells infected by the poxvirus vaccinia virus are also characterized by elevated concentrations of several nucleotides, including TMP, dATP, and dGTP, but whether flux through the pentose phosphate pathway is increased is not clear.

Human Disease Associated with Virus-Induced Alterations in Glucose Metabolism

In humans (and other mammals), the liver makes a critical contribution to glucose homeostasis: it is a major site of both the synthesis of glycogen, the polymer in which excess glucose is stored, and *de novo* synthesis of the sugar from 2-carbon compounds. The latter activity is essential to maintain blood glucose concentrations when food intake is low or during prolonged exercise or stress. Among the detrimental effects of infection by some hepatotropic viruses are major perturbations of glucose homeostasis that can lead to development of disease.

Infection by the hepadnavirus hepatitis B virus is strongly associated with the development of type 2 diabetes (insulin-independent) in certain populations (for example, Asians) and with increased blood glucose levels (hyperglycemia) in patients with acute infection. Such a systemic perturbation in glucose concentrations arises because a high proportion of liver cells can be infected by this virus. In both human hepatocytes in culture and the livers of transgenic mice, expression of the viral X gene is sufficient to stimulate expression of several cellular genes that encode enzymes required for synthesis of glucose from 2-carbon precursors (gluconeogenesis), notably phosphoenolpyruvate carboxykinase, which catalyzes the first (and committed) reaction in the synthesis of glucose from pyruvate. This response is likely to be of physiological significance, because mice transgenic for the viral X gene exhibit increased synthesis of glucose compared to control mice. They are also glucose intolerant; that is, glucose is not removed effectively from the blood when the animals are either fed or starved. Both activation of glucogenesis and glucose intolerance are symptoms of type 2 diabetes.

Insulin resistance, another diagnostic marker for development of type 2 diabetes, is exhibited by some 25% of patients with chronic hepatitis C virus infection. Insulin-dependent signaling is disrupted by the viral C (capsid) protein, which induces altered phosphorylation and increased degradation of a critical signal transducer in this pathway. However, hepatitis C virus infection of human hepatocytes in culture also results in increased concentrations of phosphoenolpyruvate carboxykinase (and its mRNA) and reduced quantities of Glut4 and cell-surface Glut2. These observations suggest that increased glucose synthesis and reduced uptake of glucose by infected hepatocytes may also promote development of type 2 diabetes in patients with chronic hepatitis C virus infection.

The Citric Acid Cycle

The citric acid cycle (also called the tricarboxylic acid [TCA] or Krebs cycle) is an important central hub of carbon metabolism, serving as the final common pathway for oxidation of carbon from glucose and other fuels (such as fatty acids and the carbon skeletons of amino acids). Under normal conditions, 1 acetyl group, which enters the cycle as acetyl-CoA, is oxidized to CO_2 in 1 turn of the cycle with generation of energy in the form of the reduced electron carriers NADH and $FADH_2$ (reduced flavin adenine dinucleotide), and GTP. However, the 8 reactions that comprise the citric acid cycle are amphibiotic; that is, they also yield precursors for biosynthesis of a great variety of compounds (e.g., Fig. 14.10). Although such biosynthetic reactions remove intermediates from the cycle, their concentrations are normally almost constant, because the citric acid cycle is replenished by several reactions. Despite its importance in both catabolism and anabolism, the citric acid cycle in virus-infected cells has received relatively little attention: only the impacts of infection by some of the larger DNA viruses have been examined in detail.

Figure 14.10 The citric acid cycle and some alterations induced in virus-infected cells. Acetyl-CoA produced by oxidation of pyruvate enters the citric acid cycle when its 2 carbon atoms are transferred to oxaloacetate to form citrate. The subsequent 8 reactions of the cycle accomplish complete oxidation of the acetate group to 2 molecules of CO_2, with production of energy in the form of GTP and the reduced electron carriers NADH and $FADH_2$, as shown. Reactions consuming intermediates in the cycle or replenishing them that are stimulated in virus-infected cells are indicated. HSV, herpes simplex virus 1; HCMV, human cytomegalovirus; VACV, vaccinia virus.

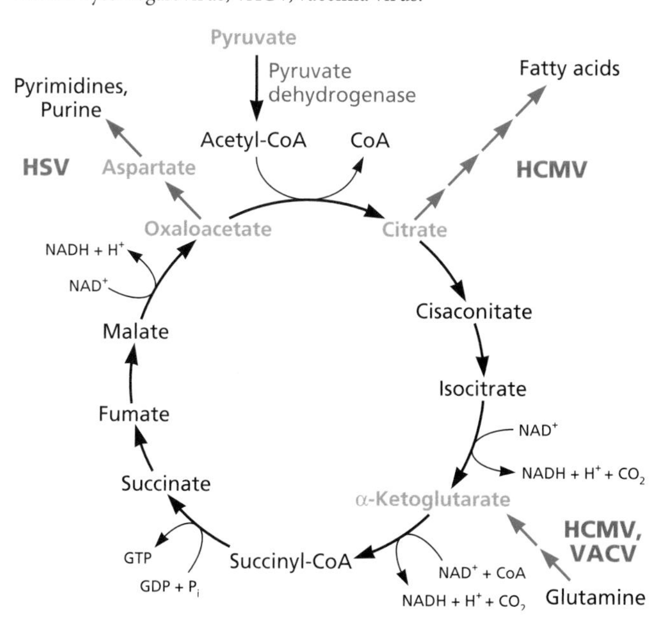

Alternative Mechanisms of Entry of Carbon Atoms from Glucose into the Citric Acid Cycle in Cells Infected by Alpha- and Betaherpesviruses

When infected cells are supplied with ^{13}C-labeled glucose, it is possible to trace the flow of the heavy carbon atoms from this metabolite into intermediates in the citric acid cycle (or in other pathways). Such studies established that 2 carbon atoms are incorporated into citrate (6 carbon atoms) from glucose in cells infected by herpes simplex virus 1, but that 3 carbon atoms of citrate become labeled in human cytomegalovirus-infected cells. This difference indicates that pyruvate (the product of glycolysis) is first converted to acetyl-CoA or oxaloacetate, respectively (Fig. 14.11). The choice between these fates of pyruvate is strongly regulated by the concentration of acetyl-CoA. This important intermediate in metabolism is produced by catabolism of a variety of fuels, and its concentration is therefore an indicator of how well a cell is supplied with a source of energy. It is a positive regulator of pyruvate carboxylase, but an allosteric inhibitor of pyruvate dehydrogenase (Fig. 14.11). In human cytomegalovirus-infected cells, much acetyl-CoA is consumed for synthesis of fatty acids (see "Lipid Metabolism" below). Consequently, the mitochondrial concentration of this compound would be low, favoring activation of pyruvate dehydrogenase. However, the mechanisms responsible for the differential regulation of these enzymes in cells infected by these two herpesviruses have not been examined.

Even though synthesis of oxaloacetate catalyzed by pyruvate carboxylase is a reaction that replenishes the citric acid cycle, the concentration of intermediates in this cycle decreases as the herpes simplex virus infectious cycle proceeds. This reduction is the result of diversion of aspartate from participation in the cycle to synthesis of pyrimidine nucleotides, such as UTP (Fig. 14.10). Inhibition of pyruvate carboxylase or the first enzyme in the pathway by which pyrimidines are made from aspartate impaired production of infectious particles of herpes simplex virus, but not of human cytomegalovirus, emphasizing the very different needs for cellular metabolic pathways for efficient reproduction of these human herpesviruses.

Enhanced Replenishment of the Citric Acid Cycle by Metabolism of Glutamine

Analysis of the flux of labeled carbon atoms from glucose in cells infected by human cytomegalovirus indicated that most of the citrate produced from acetyl-CoA and pyruvate leaves the mitochondria for conversion to oxaloacetate and acetyl-CoA in the cytoplasm (Fig. 14.9). Halting of the citric acid cycle as a result of such withdrawal of citrate is prevented by the enhanced uptake of glutamine and its conversion to the citric acid cycle intermediate α-ketoglutarate, because of increased production and activity of the enzymes that catalyze this process in infected cells.

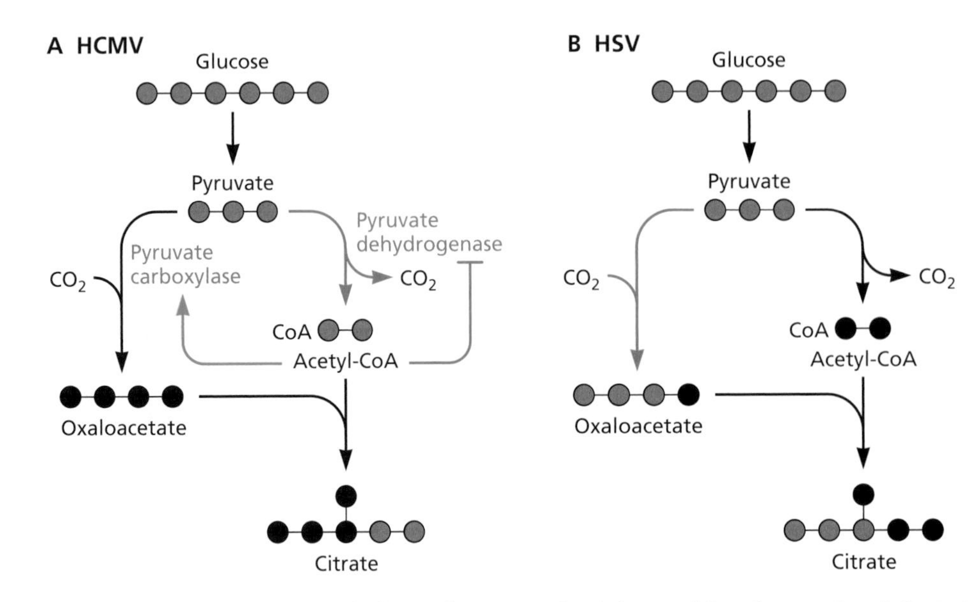

Figure 14.11 Alternative metabolism of pyruvate in alpha- and betaherpesvirus-infected cells. When all carbon atoms of glucose are labeled with ^{13}C (red), so too are all 3 carbon atoms of pyruvate. This metabolite can be converted to acetyl-CoA by pyruvate dehydrogenase or to oxaloacetate by pyruvate carboxylase to produce citrate with 2 or 3 ^{13}C-labeled carbon atoms, respectively. These compounds can be distinguished by mass spectrometry. Measurement of the pools of citrate with different numbers of ^{13}C-labeled carbon atoms indicated that citrate is produced by pyruvate dehydrogenase in cells infected by human cytomegalovirus (HCMV) **(A)**, but by pyruvate carboxylase in herpes simplex virus 1 (HSV)-infected cells **(B)**. Adapted from L. Vastag et al., *PLoS Pathog* **7**:e1002124, 2011, with permission.

Glutamine is also necessary for maximally efficient reproduction of the poxvirus vaccinia virus: its removal from the medium of infected cells reduces the yield of infectious virus particles by >3 orders of magnitude. However, the absence of glucose had no effect, indicating that vaccinia virus reproduction does not depend on catabolism of glucose. The result of experiments in which infected cells were supplied with other metabolites indicated that one fate of glutamine is replenishment of the citric acid cycle via synthesis of α-ketoglutarate. However, continual operation of the cycle to produce energy requires a source of acetyl-CoA (Fig. 14.10). The increased reduction in synthesis of acetyl-CoA from pyruvate observed in vaccinia virus-infected cells indicates that glycolysis is not the source of acetyl-CoA. Rather, it has been proposed that in cells infected by this poxvirus, this crucial compound enters the citric acid cycle by a baroque mechanism in which the fatty acid palmitate is first synthesized in the cytoplasm and then degraded to acetyl-CoA by oxidation in mitochondria (Box 14.7). Such oxidation is a process that itself generates considerable energy.

Electron Transport and Oxidative Phosphorylation

Most of the ATP consumed in a cell is produced as a result of transfer of electrons from the reduced electron carriers NADH and $FADH_2$ through a series of acceptor and donor groups to the final acceptor, molecular oxygen, which is reduced to water (Fig. 14.12). This electron transport system comprises four extremely large, multiprotein assemblies (usually named complexes I, II, III, and IV) and is located in the inner mitochondrial membrane, as is ATP synthase. Transfer of electrons through complexes I, III, and IV results in translocation of protons across the inner mitochondrial membrane, and hence generation of proton and pH gradients across this membrane. The rotary machine ATP synthase harnesses the electrochemical energy of such gradients to synthesize ATP from ADP and phosphate, as protons flow through the enzyme back into the mitochondrial matrix. Normally, there is tight coupling among the electron transfer reactions. However, when cells are hypoxic or experience some other forms of stress, such coupling is compromised, leading to increased formation of damaging reactive oxygen species, such as superoxide and hydroxyl free radical, and decreased synthesis of ATP.

Infection by several viruses has been reported to modulate one or more of these processes. For example, the rates of electron transport and ATP synthesis increase early after infection with the alphavirus Sindbis virus, but the ATP concentration then decreases as the infectious cycle progresses, and oxygen consumption, a surrogate for the rate of ATP production, is decreased following human herpesvirus 8 infection. A more common response to infection is increased production of

BOX 14.7

EXPERIMENTS

Vaccinia virus infection stimulates both synthesis and degradation of long-chain fatty acids

Studies of fatty acid metabolism in human cells infected by the poxvirus vaccinia virus revealed an unusual, if not unique, combination of metabolic pathways to produce ATP.

Inhibition of either of two enzymes needed for synthesis of fatty acids (acetyl-CoA carboxylase or fatty acid synthase) with small-molecule inhibitors reduced substantially the yield of infectious particles from vaccinia virus-infected cells, indicating that *de novo* synthesis of long-chain fatty acids is necessary for efficient reproduction of this poxvirus. Vaccinia virus particles are enveloped, and their assembly is initiated with formation of the virus-specific crescent membrane in infected cells (Chapter 13). Consequently, the final product of fatty acid synthesis, palmitate (C_{16}), might be expected to facilitate production of membrane components such as phospholipids. However, an inhibitor of the enzyme that catalyzes the first reaction in phospholipid synthesis did not impair vaccinia virus reproduction. Rather, it was demonstrated that

- the absence of glucose from the medium had no effect on the production of infectious vaccinia virus particles, indicating that glycolysis does not provide (directly or indirectly) energy to support the infectious cycle
- inhibition of entry of palmitate into mitochondria, the site of fatty acid oxidation, **did** reduce the yield of virions, particularly when glucose was absent, indicating that fatty acid oxidation might provide acetyl-CoA for energy generation via the citric acid cycle, electron transport, and oxidative phosphorylation
- consistent with this possibility, an inhibitor of a critical enzyme in the fatty acid oxidation pathway reduced virus yield in a dose-dependent manner
- O_2 consumption, a surrogate for the rate of ATP production, increased within a short period following infection, but this increase was blocked when entry of palmitate into mitochondria was prevented

It was therefore concluded that oxidation of palmitate in mitochondria is the primary means of energy generation in vaccinia virus-infected cells, and that this palmitate is first produced in the cytoplasm (see the figure). Other experiments indicated that the citric acid cycle is maintained by uptake of glutamine and its conversion to α-ketoglutarate, a process that might provide acetyl-CoA for palmitate synthesis following shuttling to the cytoplasm in the form of citrate. Inhibition of this mechanism of energy generation specifically impaired late reactions in the infectious cycle and assembly and morphogenesis of virus particles.

The synthesis of palmitate from acetyl-CoA so that this fatty acid can then be degraded by oxidation to produce acetyl-CoA for entry into the citric acid cycle might seem to represent a nonproductive, futile process. However, complete oxidation of 1 molecule of palmitate,

$$\textit{Palmitoyl-CoA} + 23\ O_2 + 108\ P_i + 108\ \textit{ADP} \rightarrow \textit{CoA} + 108\ \textit{ATP} + 16\ CO_2 + 23\ H_2O$$

yields 108 ATP (calculated assuming 1 NADH and 1 $FADH_2$ generate 2.5 and 1.5 molecules, respectively, of ATP via the electron transport chain and oxidative phosphorylation).

Synthesis of 1 molecule of palmitate,

$$8\ \textit{Acetyl-CoA} + 7\ \textit{ATP} + 14\ \textit{NADPH} + 14\ H^+ \rightarrow \textit{Palmitate} + 8\ \textit{CoA} + 7\ \textit{ADP} + 7\ P_i + 14\ \textit{NADP}^+ + 6\ H_2O$$

consumes 7 ATP directly and 49 indirectly (because 1 NADPH is equivalent to 1 NADH + 1 ATP), a total of 56 molecules of ATP.

Consequently, nearly twice as much ATP is produced as is consumed. Furthermore, the net yield from 1 molecule of palmitate, 52 molecules of ATP, is considerably greater than that generated by complete oxidation of 1 molecule of glucose, 30 to 32 molecules of ATP.

Greseth MD, Traktman P. 2014. *De novo* fatty acid biosynthesis contributes significantly to establishment of a bioenergetically favorable environment for vaccinia virus infection. *PLoS Pathog* **10:**e1004021. doi:10.1371/journal.ppat.1004021.

A model for production and utilization of palmitate in vaccinia virus-infected cells based on the observations summarized above. The red bars indicate inhibitors of reaction or pathways that reduced the yield of infectious virus particles. Acc, acetyl-CoA carboxylase; Cpt1, carnitine palmitoyltransferase 1; Fas, fatty acid synthase. Adapted from M. D. Greseth and P. Traktman, *PLoS Pathog* **10:**e1004021, 2014, with permission.

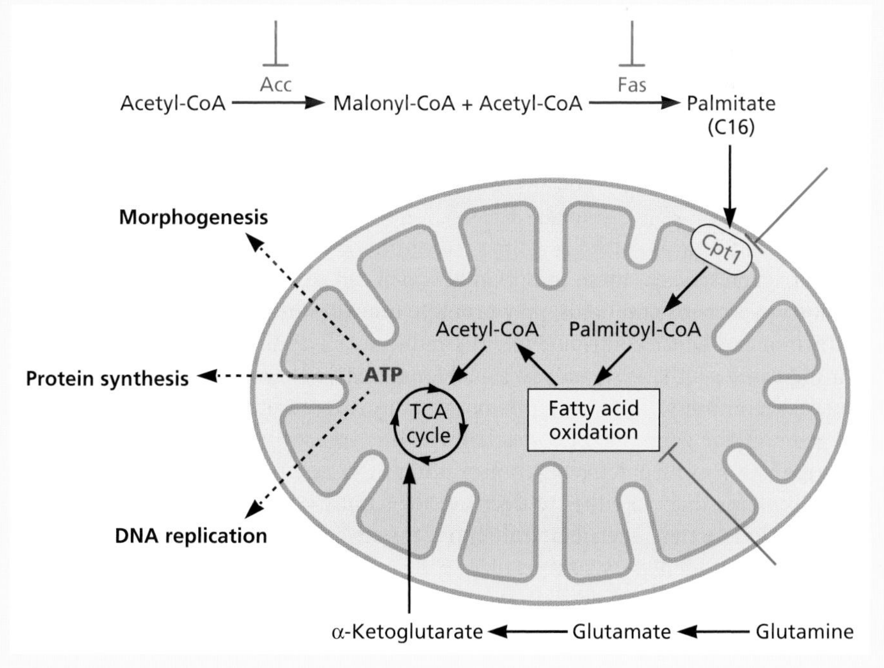

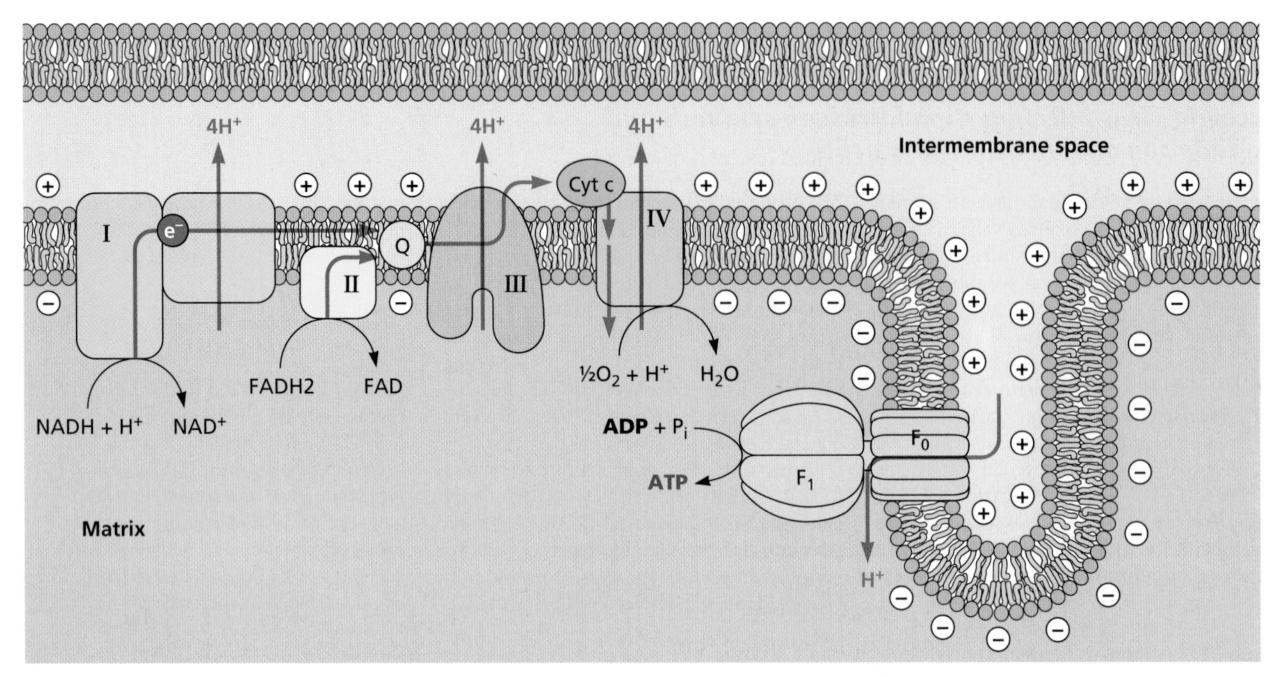

Figure 14.12 The electron transport chain and oxidative phosphorylation. The electron transport chain resides in the inner mitochondrial membrane and comprises four extremely large, multiprotein assemblies called complexes I to IV. Mammalian complex I, for example, is some 1,000 kDa. Furthermore, there is considerable evidence indicating that the complexes associate with one another to form supramolecular assemblies, an organization that would facilitate transfer of electrons among them. Electrons that enter the chain from NADH at complex I are transferred sequentially to multiple electron acceptors in each of complexes I, III, and IV to the ultimate acceptor, molecular oxygen, which is reduced to water. Such transfer of electrons to carriers of increasing reduction potential is accompanied by transfer of protons across the impermeable inner mitochondrial membrane. Electrons from reduced FAD (e.g., bound to the citric acid cycle enzyme succinate dehydrogenase) are transferred to carriers in complex II. As a result of the smaller number of electron transfer reactions between complex II and oxygen, entry of electrons at this site generates less energy for production of ATP than does entry via complex I. Shuttling of electrons from complex I or II is mediated by ubiquinone (complex Q or Q), a lipid-soluble quinone that carries an isoprenoid side chain. The accumulation of protons in the intermembrane space produces both a chemical gradient and an electrical gradient that favor flow of proteins back into the mitochondrial matrix. Because the inner mitochondrial membrane is impermeable to protons, these ions can enter only by flow through the hydrophilic, proton-specific channels present in the F_0 domain of ATP synthase. The flow of protons drives rotation of a cylinder of α-helical subunits and hence of proteins of the F_1 domain that are connected (via a shaft) to the rotary unit of F_0. The F_1 domain subunits include those with the active sites for synthesis of ATP. This reaction is driven by conformational change in the enzymatic subunits as their rotation brings them into contact with a stationary arm of the F_0 domain. ATP synthase is therefore a rotary machine that harnesses the proton motive force generated by the electron transport chain for synthesis of ATP.

reactive oxygen species. These compounds, which can oxidize and damage proteins and lipids, may promote mitochondrial dysfunction and hence contribute to virus-induced cell death. Their increased synthesis serves as a signal for oxidative stress to trigger compensating mechanisms, notably increased expression of the gene that encodes Hif-1α. This transcriptional regulator in turn activates transcription of genes that encode proteins that act either to decrease the supply of the initial electron carriers, such as an inhibitor of pyruvate dehydrogenase, or to increase synthesis of ATP by glycolysis, for example, Glut1, Glut3, and hexokinase (Fig. 14.8). As we have seen, this transcriptional regulator has been implicated in the acceleration of glycolysis observed in cells infected by a variety of viruses. Reactive oxygen species are also important in signaling pathways that activate innate immune defenses (Volume II, Chapter 3), so their increased concentrations in cells infected by these viruses may facilitate recruitment of such antiviral defenses.

Release of mitochondrial proteins, such as cytochrome *c*, into the cytoplasm initiates the protease (caspase) cascade that executes the apoptotic program. Because this process could terminate viral infectious cycles prematurely, it is targeted by proteins encoded in the genomes of many, if not all, viruses. Mechanisms by which viral proteins block apoptosis, an important antiviral defense, are described in Volume II, Chapter 3.

Lipid Metabolism

The oxidation of fatty acids, carboxylic acids with hydrocarbon chains of 4 to 35 carbon atoms, is an important source of energy, and lipids in the form of triacylglycerols are the primary energy store in most organisms. Lipids also serve as detergents, transporters, hormones, and intracellular signaling molecules, while phospholipids and cholesterol (a sterol) are major components of cell membranes. Consequently, most cells synthesize fatty acids and other lipids. Membranes derived from those of the host cell are the foundations of the envelopes present in many virus particles (Chapters 4, 12, and 13). Furthermore, infection by enveloped and some nonenveloped viruses leads to quite dramatic reorganization and expansion of membrane-bounded structures (see "Remodeling of Cellular Organelles" below). It is now clear that lipid metabolism is modulated following infection of mammalian cells by a number of these viruses, although to virus-specific ends.

Regulation of Fatty Acid Oxidation in Virus-Infected Cells

Lipids are stored in the form of triacylglycerols, in which 1 molecule of glycerol is esterified to 3 fatty acid chains. When an organism requires energy, these stores are mobilized with release of fatty acids for transport in the blood bound to serum albumin (Fig. 14.13). Once they enter cells (for example, of cardiac or skeletal muscle), fatty acids are linked to acetyl-CoA to form acyl-CoAs and transported into mitochondria, where they undergo repeated cycles of oxidative removal of 2 carbon units (as acetyl-CoA) and production of energy, in the form of 1 molecule each of the reduced electron carriers NADH and $FADH_2$ per cycle. Because fatty acids are highly reduced, their complete oxidation yields more than twice the energy than can be extracted from the same mass of carbohydrate.

Degradation of fatty acids is important for reproduction of vaccinia virus (Box 14.7): inhibition of the enzymes responsible for import of acyl-CoAs into mitochondria reduced the yield of virions by >10-fold. The rate of oxidation of the 16-carbon fatty acid palmitate increases during infection with human immunodeficiency virus type 1, and inhibition of this pathway impairs production of both viral genomic RNA and infectious virus particles. This process is also necessary for efficient reproduction of the flavivirus dengue virus. In this case, palmitate is obtained by an unusual mechanism of processing of intracellular triacylglycerols, which are stored in lipid droplets (Box 14.8).

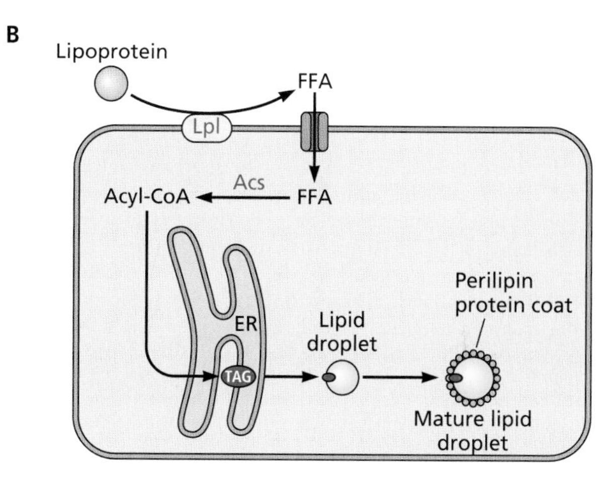

Figure 14.13 Storage and mobilization of fatty acids. (A) Fatty acids are transported and stored as triacylglycerols, in which the carbon atoms of glycerol (green) are linked to 3 fatty acid chains. As illustrated, these chains vary in length and degree of saturation. **(B)** Fatty acids in the form of triacylglycerols are transported in the blood as lipoproteins, phospholipid-bonded particles that contain lipid-binding apolipoproteins and also other lipids, notably cholesterol. At cell surfaces, plasma membrane-associated lipoprotein lipases (Lpl) hydrolyze triacylglycerols to release free fatty acids (FFA) for entry into the cell via dedicated channels. Fatty acids can also be transported bound to serum albumin. Within a cell, fatty acids are thioesterified to acetyl-CoA by acyl-CoA synthases (Acs). Acyl-CoA molecules may be transported into mitochondria for oxidation and production of energy. They can also be packaged for storage within the cell. In this process, triacylglycerols (TAG) are formed following entry of acyl-CoAs into the ER, and released from the ER associated with one of several proteins (red oval). Such immature lipid droplets coalesce and become coated by the protein perilipin to form mature lipid droplets.

Infection by Several Enveloped Viruses Stimulates Fatty Acid Synthesis

Comparison of the concentrations of enzymes of fatty acid synthesis, notably the multiple-active-site enzyme fatty acid synthase, or their mRNAs, and direct measurement of intermediates such as malonyl-CoA, have established that biogenesis of these lipids is accelerated in response to infection by several enveloped viruses, including the flaviviruses dengue and hepatitis C viruses, some herpesviruses, and human immunodeficiency virus type 1. Furthermore, the major perturbations of lipid metabolism in the livers of patients infected with hepatitis B or C virus contribute to the development of such symptoms as steatosis (accumulation of fat), obesity, and hepatocellular carcinoma. We illustrate the mechanisms by which lipid synthesis is increased in virus-infected cells and the consequences, using two well-characterized examples.

Human cytomegalovirus infection induces synthesis of very-long-chain fatty acids for assembly of infectious virus particles. As discussed previously, infection of human cells with human cytomegalovirus increases the flux of carbon from glucose to acetyl-CoA, the product of the oxidation of

BOX 14.8

DISCUSSION

Dengue virus infection induces autophagy to mobilize fatty acids for energy generation

Efficient reproduction of the flaviviruses dengue virus and hepatitis C virus depends on autophagy (literally "self-eating") in infected cells. This process is normally induced in response to starvation and is characterized by the formation of autophagosomes, which are bounded by two membranes. These structures engulf cellular components and can deliver them to lysosomes for degradation and recycling of essential materials, such as amino acids. The unanticipated function of autophagy in dengue virus-infected cells became clear when the impact of infection on lipid metabolism was examined.

While autophagosomes do not appear to be co-opted to serve as viral replication centers, they are associated with lipid droplets in human hepatocytes in dengue virus-infected cells, and the area occupied by these storage depots for triacylglycerides is reduced. This decrease is the result of delivery of lipids via autophagosomes to lysosomes (see the figure), with concomitant reduction in the concentration of triglycerides (but not other lipids) in infected cells, as they are hydrolyzed by lysosomal lipases to liberate fatty acids (and glycerol). The rate of fatty acid oxidation is also increased, and inhibition of transport of fatty acids into mitochondria (the site of fatty acid oxidation) impaired replication of the viral RNA genome and production of infectious virus particles. Furthermore, addition of exogenous free fatty acids rescued the defects in virus reproduction caused by inhibition of autophagy, or of both autophagy and transport of fatty acids into mitochondria.

Infection by dengue virus therefore appears to evoke a response normally restricted to extreme conditions (e.g., starvation) to mobilize fatty acids stored as triglycerides in lipid droplets for oxidation and energy generation.

Heaton NS, Randall G. 2010. Dengue virus-induced autophagy regulates lipid metabolism. *Cell Host Microbe* **8**:422–432.

A plasmid encoding the autophagosomal protein Lc3 fused to green fluorescent protein (GFP) (green) was introduced into established human hepatocyte cells, which were then infected with dengue virus for 24 h. At that time, the cells were stained with LysoTracker (red), a dye that detects lysosomes in living cells **(A)**, or with Oil Red O, which stains neutral lipids **(B)**. White arrows indicate acidified autophagosomes (A) and localization of neutral lipids to these vesicles **(B)**. Adapted from N. S. Heaton and G. Randall, *Cell Host Microbe* **8**:422–432, 2010, with permission. Courtesy of G. Randall, University of Chicago.

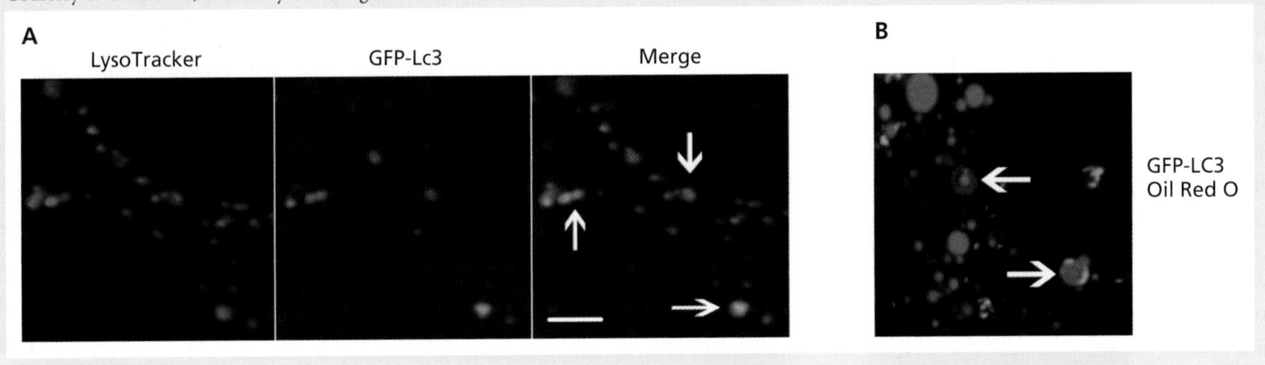

pyruvate by pyruvate dehydrogenase (Fig. 14.8). However, much of this acetyl-CoA does not enter the citric acid cycle, but rather is shuttled to the cytoplasm in the form of citrate, where it is converted to malonyl-CoA, the committed precursor for synthesis of fatty acids (Fig. 14.9). Flux through this pathway is accelerated by a factor of 20. These changes are crucial for efficient virus reproduction: inhibition of either the enzymes that catalyze synthesis of malonyl-CoA (acetyl-CoA carboxylase) or fatty acid synthase reduced the yield of infectious virus particles by several orders of magnitude.

The formation of viral envelopes, and of the cytoplasmic membrane-bound compartments at which particles acquire their final envelope, imposes an increased demand for lipid synthesis. However, human cytomegalovirus infection does not simply increase production of fatty acids in infected cells, but also alters their nature: very-long-chain fatty acids (with carbon chains of $\geq$26) are increased nearly 10-fold in concentration in infected cells with no change in the abundance of the C_{14}-C_{24} fatty acids. Such a skewed distribution of fatty acids was even more pronounced in the envelope of virus particles. Inhibition of the enzymes that make long-chain fatty acids from those with shorter hydrocarbon chains (elongases) led to production of virus particles with a reduced content of long-chain fatty acids and poor infectivity. It therefore appears that the final budding of human cytomegalovirus particles is at a membrane enriched in lipids with long-chain fatty acids, but how the presence of such lipids promotes assembly and initiation of a new infectious cycle is not yet clear.

These changes in lipid metabolism can be traced to the increased availability of activators necessary for the transcription of many genes of lipid synthesis, Chrebp (described previously) and Srebps (sterol regulatory element-binding proteins). The latter regulators are synthesized as inactive precursors that remain associated with the endoplasmic reticulum (ER) membrane until needed, when they are transported to Golgi compartments and cleaved to release active Srebps (Fig. 14.14). Human cytomegalovirus infection leads to increased cleavage and release of Srebps, as a result of increased synthesis of the Pkr-like ER-associated kinase (Perk) described in Chapter 12. In addition, the activation of signaling via mTor inactivates a negative regulator of Srebps, lipin-1, which retains these proteins in the cytoplasm.

The net result of the many perturbations of host cell carbon metabolism characteristic of human cytomegalovirus-infected cells is to channel fuels like glucose to increased synthesis of lipids, a redirection that depends on intervention in many metabolic reactions and the coordinated modulation of multiple signal transduction pathways. Such coordination appears to be facilitated by altered properties of the antiviral protein viperin in infected cells (Box 14.9).

Hepatitis C virus infection stimulates fatty acid synthesis and increases lipid retention to induce steatosis in hepatocytes. Infection of hepatocytes with hepatitis C virus results in several perturbations in lipid metabolism. Export of fatty acids and cholesterol in the form of lipoproteins is inhibited, a response likely to be related to the reduced concentrations of serum cholesterol seen in hepatitis C virus-infected patients: normally, cholesterol and fatty acids from the diet are packaged into lipoproteins in the liver, for subsequent transport to tissues and organs where needed. Concomitantly, the synthesis of fatty acids and cholesterol is stimulated, because of increased expression of genes that encode such enzymes as fatty acid synthase and those needed for cholesterol production. The expression of the gene encoding SrebpC1, the Srebp family member required for transcription of genes for enzymes of both fatty acid and cholesterol synthesis, is also increased. These changes in lipid metabolism are important for viral reproduction. For example, inhibition of Srebp release from the ER by incubation of infected cells with a cholesterol derivative severely inhibited synthesis of viral RNA.

The viral C protein leads to elevated concentrations of active Srebp1 via the Akt signaling pathway and the transcriptional regulator forkhead box protein 1 (Foxo1) (Fig. 14.15A), as well as decreased concentration of transcriptional activators needed for transcription of genes that encode enzymes of fatty acid oxidation. The viral nonstructural proteins NS4B and NS5A have also been reported to increase the concentration of mature (active) Srebp1, but how they might do so is not known. Furthermore, the reactive oxygen species made in infected cells are thought to contribute to inhibition of lipoprotein release. *In toto*, these alterations in lipid

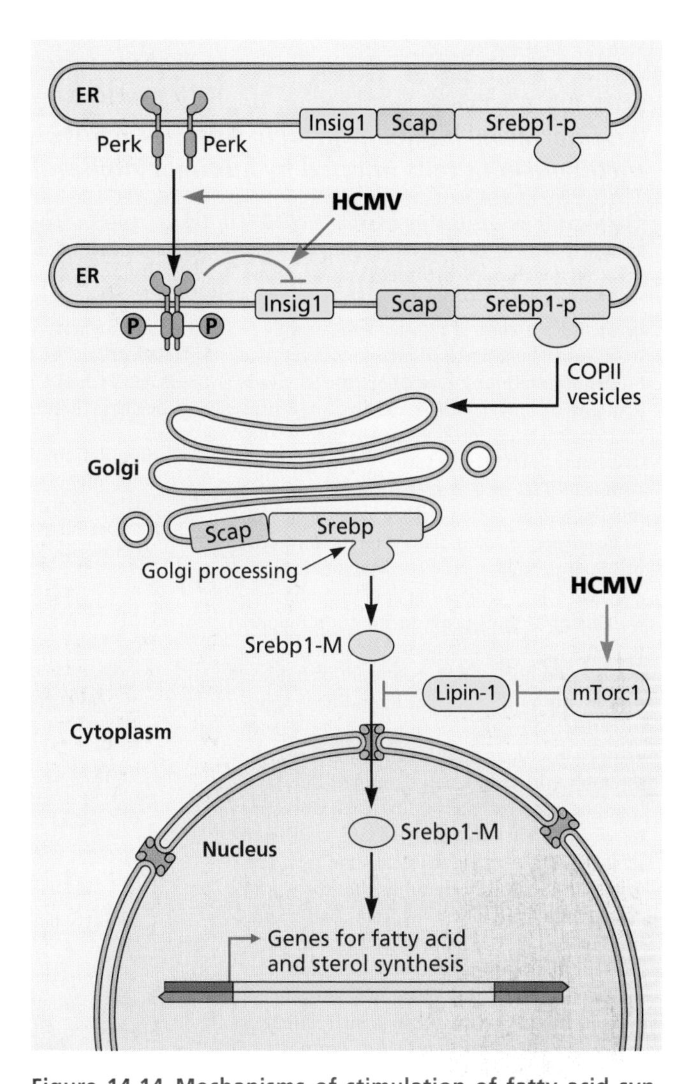

Figure 14.14 Mechanisms of stimulation of fatty acid synthesis in human cytomegalovirus (HCMV)-infected cells. Transcriptional activators of the Srebp family are required for expression of many genes that encode enzymes and other proteins required for synthesis of fatty acids. The Srebps are synthesized as inactive precursors (Srebp-ps) that are sequestered in the ER membrane by association with Srebp cleavage activation protein (Scap), which in turn binds to the protein insulin-inducible gene 1 (Insig1) when cholesterol is present. These interactions are disrupted in uninfected cells when cholesterol concentrations are low, allowing exit of Srebp-p from the ER and transport to Golgi compartments. At this location, proteolytic cleavages release active, mature Srebps (Srebp-M) into the cytoplasm for translocation into the nucleus and stimulation of transcription of Srebp-responsive genes, such as that encoding fatty acid synthase. This mechanism is overridden in human cytomegalovirus-infected cells, apparently in large part because of increased synthesis of the ER membrane enzyme Perk and its activation (see Chapter 12). The data available indicate that one consequence is reduced concentration of Insig1, and hence stimulation of production of active Srebp. Increased signaling from mTorC1 in infected cells also results in inactivation of the negative regulator of Srebps, lipin 1.

BOX 14.9

DISCUSSION

An interferon-inducible protein initiates redirection of lipid metabolism in cells infected by human cytomegalovirus

The expression of the interferon-inducible gene viperin is increased in human cells infected by human cytomegalovirus, independently of production of interferon. Viperin includes a central region with homology to a motif present in enzymes that use S-adenosylmethionine as a cofactor (the radical SAM family) and binds to iron-sulfur clusters in other proteins (hence the alternative name for viperin: radical SAM domain containing 2, or Rsad2). Although viperin impairs human cytomegalovirus reproduction when its gene is artificially expressed in cells prior to infection, more-recent studies indicate a critical role in redirecting lipid metabolism.

Human cytomegalovirus infection induces not only synthesis of viperin, but also its localization to mitochondria as a result of binding to the viral mitochondrial inhibitor of apoptosis (vMIA) protein. In this organelle, viperin associates with and inhibits the multienzyme assembly (the trifunctional protein) that catalyzes the last 3 reactions in the 4-reaction cycle by which 2 carbon units are removed from long-chain (C ≥ 12) fatty acids during oxidation. Such inhibition requires the iron-sulfur cluster-binding motif of viperin. As summarized in the figure, inhibition of fatty acid oxidation and the resulting decrease in intracellular ATP concentrations activate AMP-dependent protein kinase and a subsequent regulatory cascade that leads to increased synthesis of fatty acids to facilitate assembly of infectious virus particles. The inhibitory and stimulatory responses shown are blocked when viperin production is prevented in infected cells using RNA interference, and reproduced when viperin is targeted to mitochondria in uninfected cells.

As noted in the text, formation of active Srebps, which, like Chrebp, stimulate transcription from lipogenic genes, is increased in human cytomegalovirus-infected cells. Although this response does not depend on viperin, expression of lipogenic genes is not stimulated when viperin cannot be made in infected cells. The transcriptional regulators Srebp1 and Chrebp may therefore operate synergistically to increase fatty acid synthesis.

Seo JY, Cresswell P. 2013. Viperin regulates cellular lipid metabolism during human cytomegelavirus infection. *PLoS Pathog* **9:**e1003497. doi:10.1371/journal.ppat.1003497.

Shenk T, Alwine JC. 2014. Human cytomegalovirus: coordinating cellular stress, signaling, and metabolic pathways. *Annu Rev Virol* **1:**355–374.

Regulation of lipid metabolism in human cytomegalovirus (HCMV)-infected cells, with inhibition and stimulation indicated by red bars and green arrows, respectively, and metabolites or proteins increased and decreased in concentration shown in green and red parentheses, respectively. AmpK, AMP-dependent protein kinase; ACC, acetyl-CoA carboxylase; FA, fatty acid.

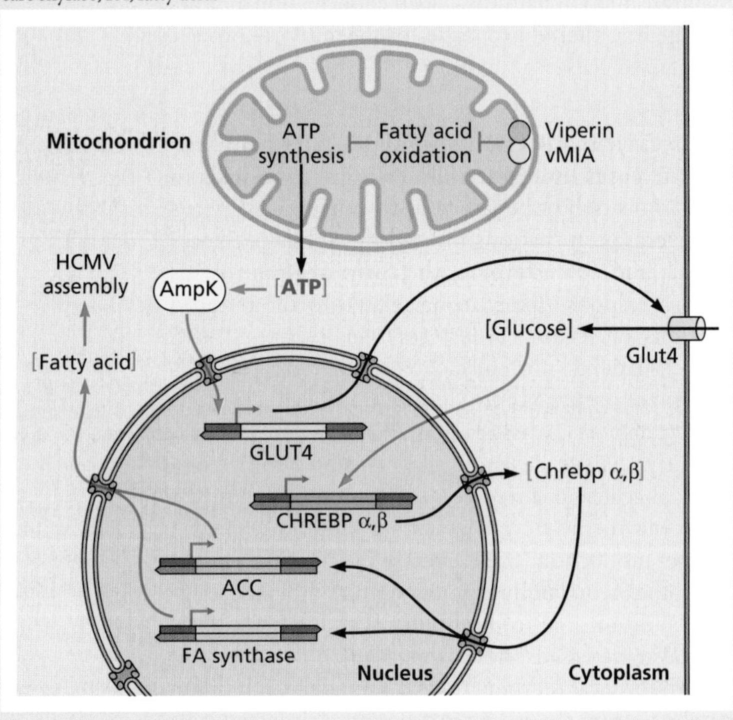

metabolism greatly increase the accumulation of lipid droplets (Fig. 14.15B), which serve as platforms for assembly of virus particles (see "Remodeling of Cellular Organelles" below).

Infection by Nonenveloped Viruses Can Also Reprogram Lipid Metabolism

Even when virus particles lack an envelope, lipid metabolism can be perturbed in the infected cell, and one such nonenveloped virus has been associated with development of obesity in humans (Box 14.10). A characteristic feature of cells infected by various viruses with (+) strand RNA genomes is the formation of cytoplasmic membranous structures that are the sites of viral genome replication and/or assembly of virus particles. This process can require reshaping of the repertoire of lipids in the infected cell, a phenomenon illustrated using the picornavirus poliovirus.

Synthesis of phospholipids (major constituents of cellular membranes), particularly phosphatidylcholine, is stimulated strongly within a short period after infection with poliovirus (or other picornaviruses). This response is the result of a greatly increased rate of import of fatty acids triggered by viral protein 2A and their utilization for

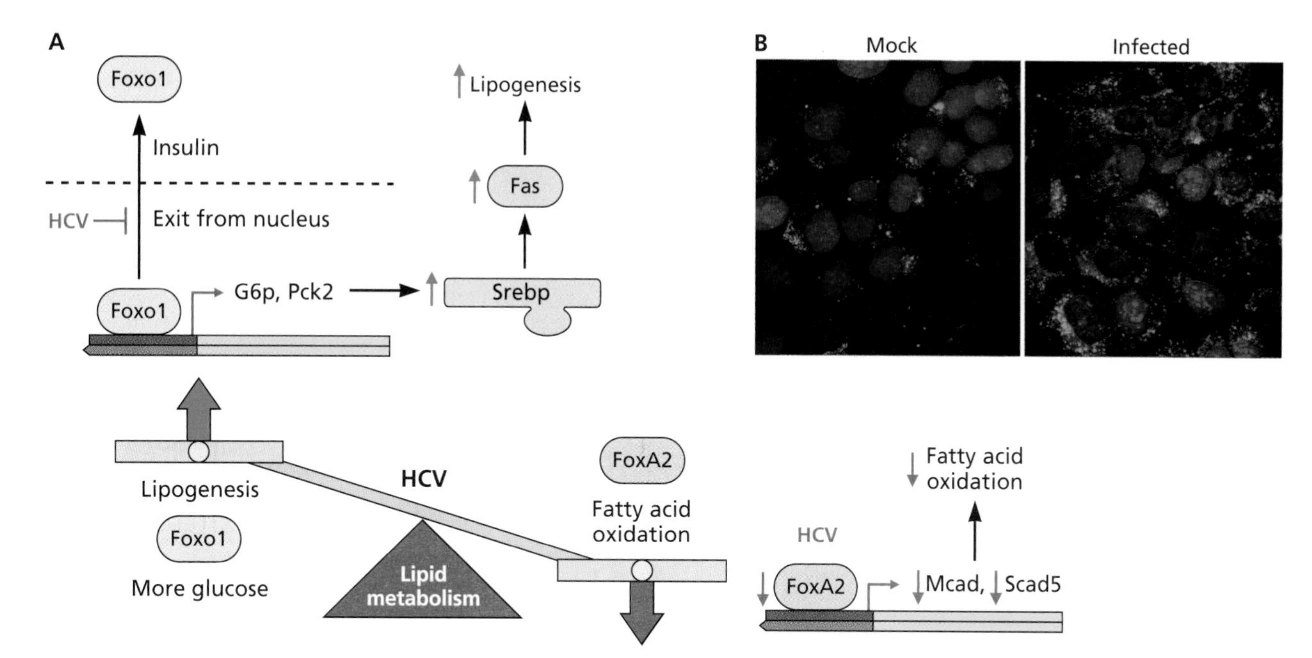

Figure 14.15 Increased synthesis and accumulation of fatty acids in hepatitis C virus-infected cells. (A) Model for the opposite regulation of fatty acid synthesis and breakdown in hepatitis C virus (HCV)-infected hepatocytes. Synthesis of fatty acids is stimulated, because of increased production of the enzyme fatty acid synthase (Fas). As a result of the increased availability of nuclear Srebp1 in hepatitis C virus-infected cells, translocation of the transcriptional activator Foxo1 from the nucleus to the cytoplasm (normally stimulated by insulin) is blocked. Consequently, expression of genes that encode gluconeogenic enzymes and production of glucose are increased, a response that induces increased accumulation of active Srebp1. At the same time, production of enzymes that catalyze fatty acid oxidation, such as medium- and short-chain acyl-CoA dehydrogenases (Mcad, Scad), is reduced because of decreased concentrations of FoxA2, and hence of expression of the genes that encode these enzymes. The net result is that fatty acid synthesis greatly outpaces degradation, and neutral lipids accumulate in lipid droplets. The accumulation of the protein perilipin, a component of lipid droplets that protects against removal of lipids by the action of lipases, is also increased in infected cells, while activation of lipases is prevented. As the model predicts, viral RNA genome replication is impaired by small interfering RNA-mediated knockdown of Foxo1 or overproduction of FoxA2. Adapted from S. K. Bose et al., *J Virol* **88:**4195–4203, 2014, with permission. **(B)** Human hepatocytes from a hepatocellular carcinoma were infected with hepatitis C virus or were mock-infected, and neutral lipids were examined by staining with the lipophilic fluorescent dye Bodipy 493/503 (green). Nuclei are stained blue. Courtesy of R. Ray, Washington University, St. Louis.

synthesis of phosphatidylcholine (and presumably other phospholipids). The complement of fatty acids incorporated into phospholipids is also shifted in favor of those with longer acyl chains (C_{16} or C_{18}), because the activity of a long-chain acyl-CoA synthase (Acsl3) is increased. Newly imported fatty acids are seen associated with a viral protein in structures resembling replication centers (Fig. 14.16), and inhibition of production of Ascl3 by RNA interference impairs replication of a poliovirus replicon. These observations indicate that poliovirus replication factories possess a unique lipid composition, but how this property favors their formation or function is not yet clear.

Remodeling of Cellular Organelles

Infection of cells in culture by many viruses causes changes in morphology that are obvious even when cells are observed by low-power light microscopy (see Fig. 2.8). The more dramatic changes, such as rounding up of cells and their detachment from the surfaces of tissue culture dishes, are the result of severe perturbations of cellular physiology and metabolism, and induction of cell death. They are typically seen late in infection, but even at earlier times, virus infection can induce large-scale reorganization of cellular organelles or their components. Such remodeling of host cell architecture supports fabrication of infected cell-specific structures in which replication of viral genomes and/or assembly of virus particles take place. Cytoplasmic organelles or the machinery needed for their formation may also be co-opted to facilitate release of progeny virus particles from infected cells. In this section, such alterations of cellular morphology are described in the context of individual organelles of the host cell.

The Nucleus

Altered nuclear morphology is a common feature of cells infected by viruses with DNA genomes that are replicated within nuclei. These organelles become enlarged as the infectious cycle progresses and, in many cases, filled with large arrays of mature and assembling virus particles

BOX 14.10

DISCUSSION

Does infection by human adenovirus type 36 contribute to obesity in humans?

Human adenovirus type 36 (Ad36), first isolated in 1979, received relatively little attention until several reports of increased body weight following infection of experimental animals. It was initially observed that Ad36-infected chickens and mice showed large increases in body weight and fat accumulation, in contrast to animals infected in parallel with an avian adenovirus. The Ad36-infected animals also exhibited decreased concentrations of serum triglycerides and improved glycemic control. Weight gain and obesity were also reported following infection of rats, hamsters, and nonhuman primates.

This response to Ad36 infections is likely to be the result of direct effects on adipocytes (fat storage cells). Infection of preadipocytes in culture induces differentiation to adipocytes that produce increased concentrations of enzymes of fatty acid synthesis, accumulate triglycerides, and exhibit increased rates of glucose uptake. Inhibition of synthesis of the viral E4 Orf1 protein in Ad36-infected cells by RNA interference blocks these changes. Furthermore, this viral protein is sufficient to stimulate glucose uptake by activation of signaling via Pi3k and Akt to increase the availability of Glut4.

Ad36 is found worldwide, with a prevalence of some 15% in the United States. Despite the consistent observations made in experimental animals, a clear connection between infection of humans with this virus and obesity has not been established. In some studies, the presence of antibodies against the virus was more common in obese than in normal individuals (e.g., 64 vs. 32% in a study of 203 adults in Italy), but no such differences were detected in several other studies (e.g., of 509 Dutch and Belgian adults). It has been suggested that these conflicting observations might reflect the multifactorial nature of obesity, and hence differences in parameters such as genetic makeup, the microbiome, diet, lifestyle, and race among the individuals participating in the various studies.

Esposito S, Preti V, Consolo S, Nazzari E, Principi N. 2012. Adenovirus 36 infection and obesity. *J Clin Virol* 55:95–100.

(Fig. 14.17), while cellular chromatin may become condensed or dispersed to the nuclear periphery and silenced by epigenetic mechanisms. Prior to appearance of these particles, nuclear constituents are reorganized and often relocated as viral replication compartments (also called replication centers) form. These sites of viral genome replication have been observed in cells infected by all nuclear replicating DNA viruses that have been examined.

Nuclear replication centers contain incoming and replicating viral DNA genomes, proteins of both viral and cellular

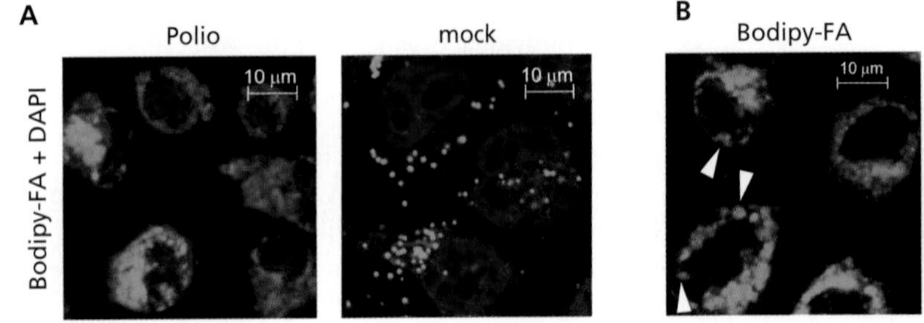

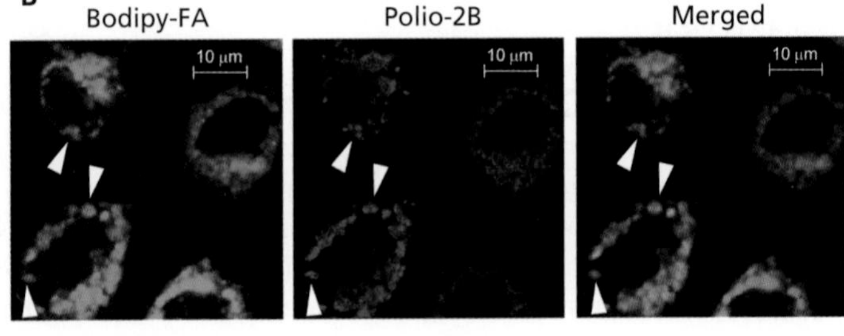

Figure 14.16 Increased import of fatty acids into poliovirus-infected cells. (A) HeLa cells infected with poliovirus for 4 h or mock-infected were incubated with the fluorescent fatty acid Bodipy-FA (green), which is thought to mimic fatty acids with 18 carbon atoms, for 30 min, and nuclei were stained in blue. The greatly increased accumulation of exogenous fatty acids in infected cells is clearly evident. **(B)** Infected cells were treated as described for panel A, and the poliovirus 2B protein was then visualized by immunofluorescence (blue). This viral protein localizes to viral replication centers, the discrete domains indicated by the white arrowheads, where the fluorescent fatty acid also accumulates. Adapted from J. A. Nchoutmboube et al., *PLoS Pathog* **9**:e1003401, 2013, with permission. Courtesy of G. A. Belov, University of Maryland.

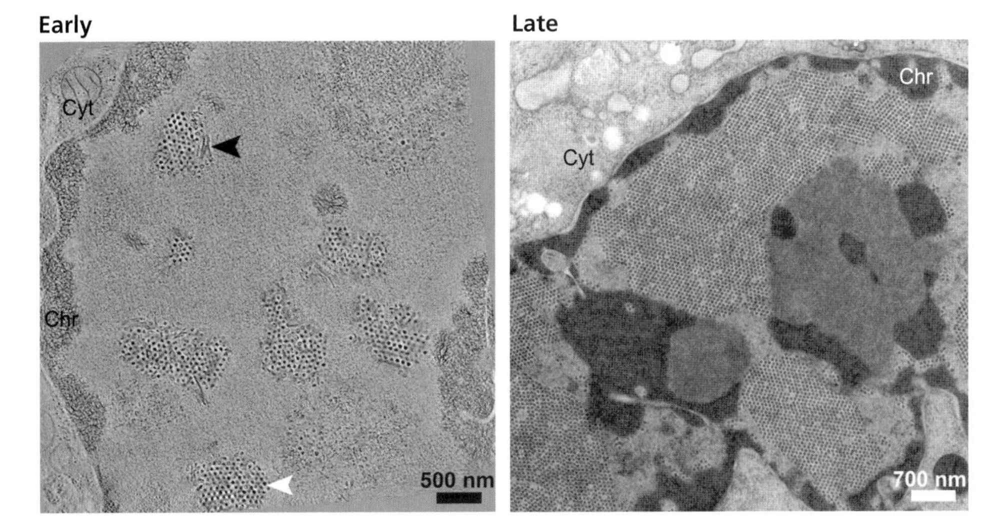

Figure 14.17 Reorganization of nuclei in polyomavirus-infected cells. Murine 3T3 cells infected with polyomavirus for 32 h were frozen under high pressure, stained at low temperature, and embedded in plastic, and sections were examined by electron tomography. Shown are 1-nm sections extracted from a 2 × 2 montage of six serial sections (1.8 μm thick) of individual infected cell nuclei. These nuclei represent earlier and later stages in the infectious cycle, as defined operationally by the sizes and numbers of clusters of virus particles present: infection of individual cells proceeds asynchronously, allowing multiple stages in the cycle to be observed in a single sample. Once the major structural protein VP1 is made (left), clusters of virus particles (white arrowhead) partially fill the interchromatin space. Each cluster is associated with tubular structures (black arrowhead) that have been shown also to be built from viral structural proteins and viral genomes. Cellular chromatin is condensed. As infection proceeds, virus particles form dense arrays that fill the interchromatin space (right). Chr, condensed chromatin, Cyt, cytoplasm. Adapted from K. D. Erickson et al., *PLoS Pathog* **8:**e1002630, 2012. Courtesy of R. L. Garcea, University of Colorado, Boulder.

origin needed for viral DNA synthesis, and a virus-specific constellation of other proteins (Chapter 9). They are also commonly associated with newly synthesized viral transcripts and the viral proteins required for efficient expression of viral genes. Although establishment of replication compartments can facilitate synthesis and transcription of viral DNA in several ways (Chapters 8 and 9), we know relatively little about how these structures are assembled in infected cells.

It has been established that one herpesviral genome is sufficient to initiate fabrication of a replication compartment. It seems likely that this is also the case for other DNA viruses with genomes replicated in the nucleus. Entering viral genomes associate with nuclear foci formed by Pml proteins (Pml bodies; Chapter 9). In many cases, viral proteins then induce degradation, dispersion, or inactivation of Pml body proteins as viral replication centers form. The latter structures enlarge as viral DNA synthesis takes place and may eventually coalesce into large reticular networks that occupy much of the nucleus. In cells infected by some herpesviruses, nuclear cages that encase assembling viral nucleocapsids are fashioned from Pml proteins late in infection (Fig. 14.18). Cellular proteins that participate in such normal processes as DNA synthesis, recombination, repair, and transcription and

Figure 14.18 Example of a Pml-containing nuclear structure in DNA virus-infected cells. Cages of Pml (green) surrounding herpesviral nucleocapsids (yellow [immature] and orange [mature]) in human melanoma cells overproducing a specific Pml isoform (Pml-IV). Cells were infected with varicella-zoster virus, the causative agent of chickenpox in humans, for 48 h, and serial sections were then examined by scanning electron microscopy. The three-dimensional reconstruction shown was produced from tracings of 18 individual sections. A larger view of the upper cage in the top panel is shown at the right, with the Pml cage in transparent green to illustrate the encasing of immature (im) and mature (m) capsids. Cellular chromatin is shown in blue. Adapted from M. Reichelt et al., *PLoS Pathog* **8:**e1002740, 2012, with permission. Courtesy of M. Reichelt, Stanford University School of Medicine.

Varicella zoster virus-infected cells

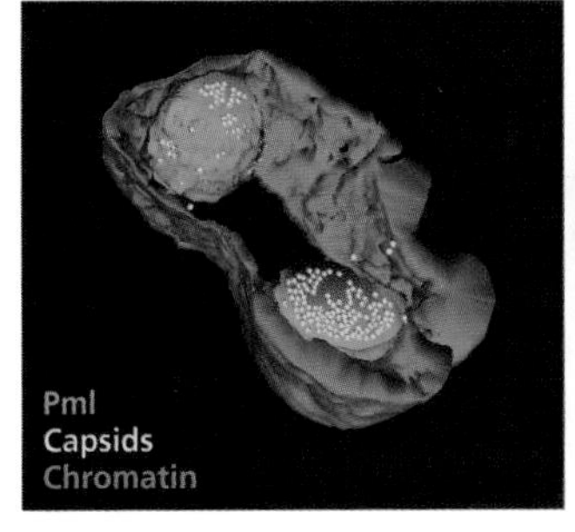

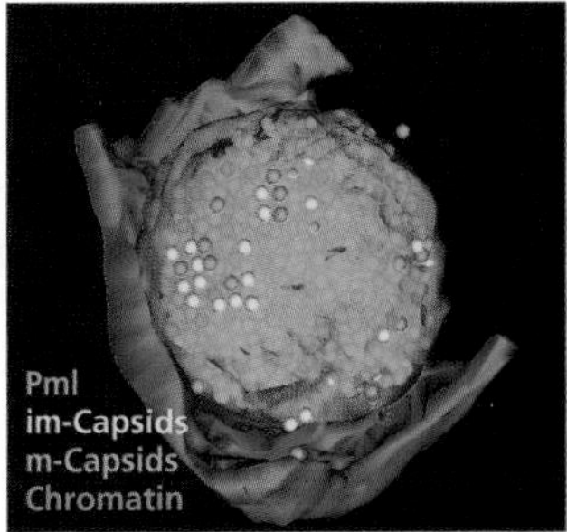

pre-mRNA processing are recruited to replication centers in virus-specific fashion (Chapters 8 and 9). The sequence of reactions that initiate formation of nuclear replication (or assembly) compartments and lead to recruitment of the cellular (and viral) proteins necessary for viral genome replication and expression has not been elucidated for any nuclear DNA viruses. Nevertheless, their formation can be essential for efficient viral DNA synthesis and reproduction. Interferon treatment blocks the appearance of replication centers in cells infected by certain adenovirus mutants, and the efficiency of viral genome replication is reduced considerably.

Infection can also result in the sculpting of other virus-specific nuclear domains or structures by reorganization of host cell components. For instance, nucleoli (the sites of ribosomal RNA synthesis) are disrupted in adenovirus-infected cells as several viral proteins, including the core proteins, accumulate in them. Core protein V associates with the abundant nucleolar protein nucleophosmin and induces its redistribution to the nucleoplasm. This activity of protein V is necessary for efficient assembly of virus particles in normal human cells, but how dispersal of nucleolar protein facilitates adenovirus reproduction is not clear. During the late phase of adenovirus infection, cellular small nuclear ribonucleoproteins that participate in splicing initially associate with the peripheral zones of viral replication centers, where transcription of viral DNA takes place, but then appear in distinct foci. These enlarged interchromatin granules contain spliced viral late mRNA (Fig. 14.19A), and their formation correlates with export of these viral mRNA to the cytoplasm. A very different type of infected cell-specific nuclear edifice has been observed following infection by herpes simplex virus 1, virus-induced chaperone-enriched (VICE) domains. These dynamic domains are defined by the presence of cellular chaperones, such as Hsc70, and are first seen adjacent to assembling viral replication compartments as viral early genes are expressed (Fig. 14.19B). They also contain proteasomes and ubiquitin. The viral ICP22 protein is sufficient for recruitment of Hsc70 into VICE domains and is present in these foci in virus-infected cells. These domains may serve as safe depots for storage and disposal of misfolded proteins in herpes simplex virus-infected cells, or they may be sites of storage of the cellular chaperones needed during assembly of virus particles, when large quantities of structural units must be built from individual protein subunits (Chapter 13). It seems likely that continued application of increasingly powerful methods of microscopy and proteomics will reveal new, virus-specific structures in infected cell nuclei.

The nucleus is also the site of replication of the (−) strand RNA genome of influenza A virus, as well as synthesis of viral mRNA. A characteristic feature of influenza A virus-infected cells is disruption of the architecture of the nucleolus from early in infection. The viral proteins NS1 and NP localize to nucleoli via specific targeting signals, and conversely, nucleolar proteins become associated with viral genome-containing ribonucleoproteins. Nucleolar localization of NP has been reported to be important for efficient viral genome replication and mRNA synthesis, but the molecular

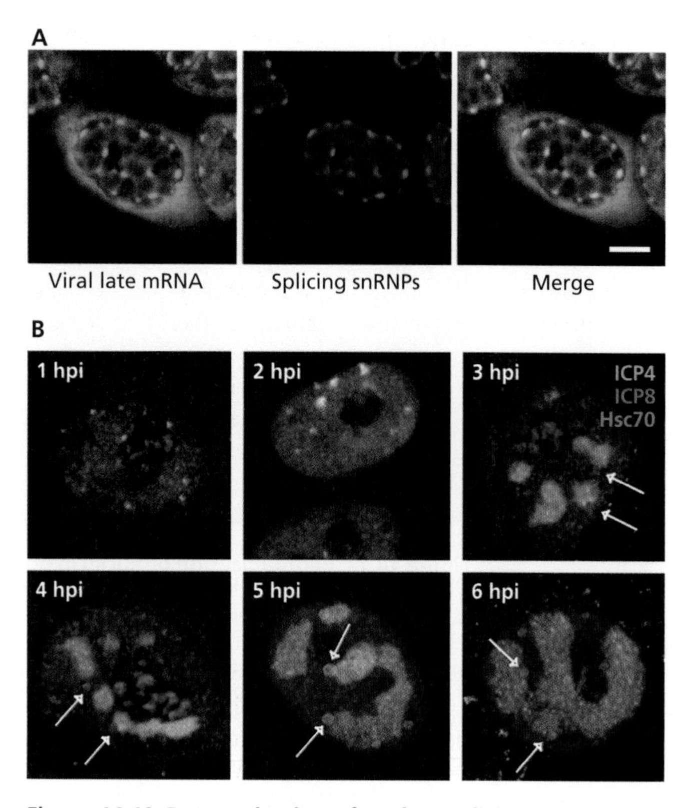

Figure 14.19 Reorganization of nuclear splicing components and chaperones in DNA virus-infected cells. (A) Formation of enlarged chromatin granules is characteristic of adenovirus-infected cell nuclei during the late phase of infection, as shown for HeLa cells infected with human adenovirus type 5 for 22 h. Infected cells were fixed and viral late mRNAs visualized by fluorescence *in situ* hybridization (green), with an oligonucleotide complementary to the sequence spanning exons 1 and 2 of the tripartite leader sequence common to all major late mRNAs (see Fig. 10.11). Small nuclear proteins (snRNPs) that participate in splicing were visualized by immunofluorescence using an antibody that recognizes a protein common to all of them (red). Bar, 10 μm. Adapted from E. Bridge et al., *Virology* **311**:40–50, 2003, with permission. Courtesy of E. Bridge, Miami University, Ohio. **(B)** Virus-induced chaperone-enriched (VICE) domains assemble as the herpes simplex virus 1 infectious cycle progresses, in this example after infection of established monkey cells with 10 PFU/cell. At the hours postinfection (hpi) indicated, cells were fixed and the viral immediate ICP4 (green) and early ICP8 (red) proteins and the cellular chaperone Hsc70 (blue) were visualized by indirect immunofluorescence. The VICE domains that form adjacent to developing replication centers, which contain the viral proteins, are indicated by white arrows. Adapted from C. M. Livingston et al., *PLoS Pathog* **5**:e1000619, 2009, with permission. Courtesy of S. Weller, University of Connecticut Health Center.

consequences of the association of nucleolar and viral components are not known.

The Cytoplasm

Reproduction of a considerable variety of viruses is completed in the cytoplasm, and is typically accompanied by remodeling of one or more cytoplasmic components. The result may be construction of infected cell-specific platforms for replication of viral genomes, or remodeling of membrane-bound organelles for envelopment of virus particles, or their release. Such reorganization of cytoplasmic membranes also occurs in cells infected by enveloped viruses with genomes that are synthesized in the nucleus, such as herpes simplex virus 1 and human immunodeficiency virus type 1. Furthermore, cytoplasmic components can be altered even when most steps in the reproduction of nonenveloped viruses take place in the nucleus, a phenomenon illustrated by the cleavage of cytoskeletal filaments by the adenoviral protease late in infection (Chapter 13).

Cytoplasmic Viral Factories

A definitive feature of cells infected by poxvirus and other large DNA viruses that are reproduced in the host cell cytoplasm, such as mimivirus, is the establishment of sizable, viral DNA-containing foci, termed viral factories. Such viral factories contain not only the viral genome and all components of the viral DNA synthesis, transcription, and mRNA-processing machines (Chapter 8 and 9), but also cellular translation proteins, such as the initiation proteins eIF4E and eIF4G. It appears that all reactions necessary for production of progeny viral genomes and expression of viral genes take place in viral factories. A single viral genome is sufficient to seed formation of such a structure (Fig. 9.21A), but, as with nuclear viral replication compartments, it is not known how viral factories are assembled and remodeled as infection proceeds. Early in infection, vaccinia virus DNA factories are often, but not invariably, bound by rough ER. This membrane is dispersed later in infection, when assembly of virus particles is initiated by viral proteins that induce formation of crescent membranes derived from the ER membrane. This process is described in detail in Chapter 13.

Replication and Assembly Platforms

Replication of a number of viral RNA genomes (and often assembly reactions) takes place on or in infected cell-specific frameworks constructed from internal membranes or lipids of the host cell. Such structures, often called replication complexes, contain viral genomes, viral RNA polymerases, and other nonstructural proteins, and may be fashioned from the membranes of the ER, Golgi, or other cytoplasmic organelles or from lipid droplets. For instance, bunyavirus infection induces the formation of tubular-like sheets from Golgi compartment membranes, whereas cells infected by the flavivirus dengue virus are characterized by the presence of an elaborate collection of vesicles with single and double membranes and more-convoluted membranous sheets derived from the ER membrane (Fig. 14.20). These membranous elements illustrate just how great an impact virus infection can have on the morphology of the host cell cytoplasm. Their properties have been examined in some detail in mammalian cells infected by flaviviruses.

The dengue virus nonstructural proteins required for replication of the (+) strand RNA genome (the RNA polymerase and the helicase) and double-stranded RNA replication intermediates accumulate within vesicular invaginations into the ER membrane called vesicular packets (Fig. 14.20A). These properties indicate that such vesicles are sites of viral genome replication. More-convoluted membranous sheets are associated with the viral protease and have been proposed to be the sites of synthesis and processing of the viral polyprotein (Appendix, Fig. 10). The mechanisms by which membranes of the host cell ER are refashioned during dengue virus infection are not well understood. However, the important contribution of this process to viral reproduction is illustrated by the finding that replication of a viral replicon is blocked by inhibition of the lipid kinase phosphatidylinositol 1,4-phosphate, which is bound by the viral NS5A protein. The sites of budding of dengue virus particles into the ER are located close to the vesicular packets and the pores that connect them to the ER (Fig. 14.20B). This spatial arrangement may facilitate selective encapsidation of the viral RNA genome.

Although it is also a member of the *Flaviviridae*, hepatitis C virus infection leads to the appearance of a rather different membranous framework (Fig. 14.21A), larger vesicles (150 nm compared to some 90 nm in dengue virus-infected cells) with double membranes, many of which can be seen as protrusions connected to the outer membrane of the ER by a thin stalk. These vesicles contain active viral replication complexes. Their formation is thought to require the concerted action of several viral nonstructural proteins and also depends on the cellular chaperone cyclophilin A. This cellular protein interacts with the viral NS5A protein and is necessary for viral genome replication. Inhibitors of the chaperone strongly impair viral replication and are in late phases of testing as antivirals. The mechanism by which cyclophilin A promotes formation of the infected cell-specific vesicles and hence viral genome replication is not yet clear. The membranous platforms of hepatitis C virus genome replication are closely associated with lipid droplets (Fig. 4.21B) and the ER sites containing the viral C protein, at which budding of virus particles initiates.

In these examples, both viral genome replication and assembly take place in the infected cell cytoplasm. However, infected cell-specific membranous structures can also be induced when the viral genome is replicated in the nucleus, a phenomenon exemplified by formation of cytoplasmic assembly compartments in herpesvirus-infected cells. These structures are unusually large vesicles that contain cellular proteins normally

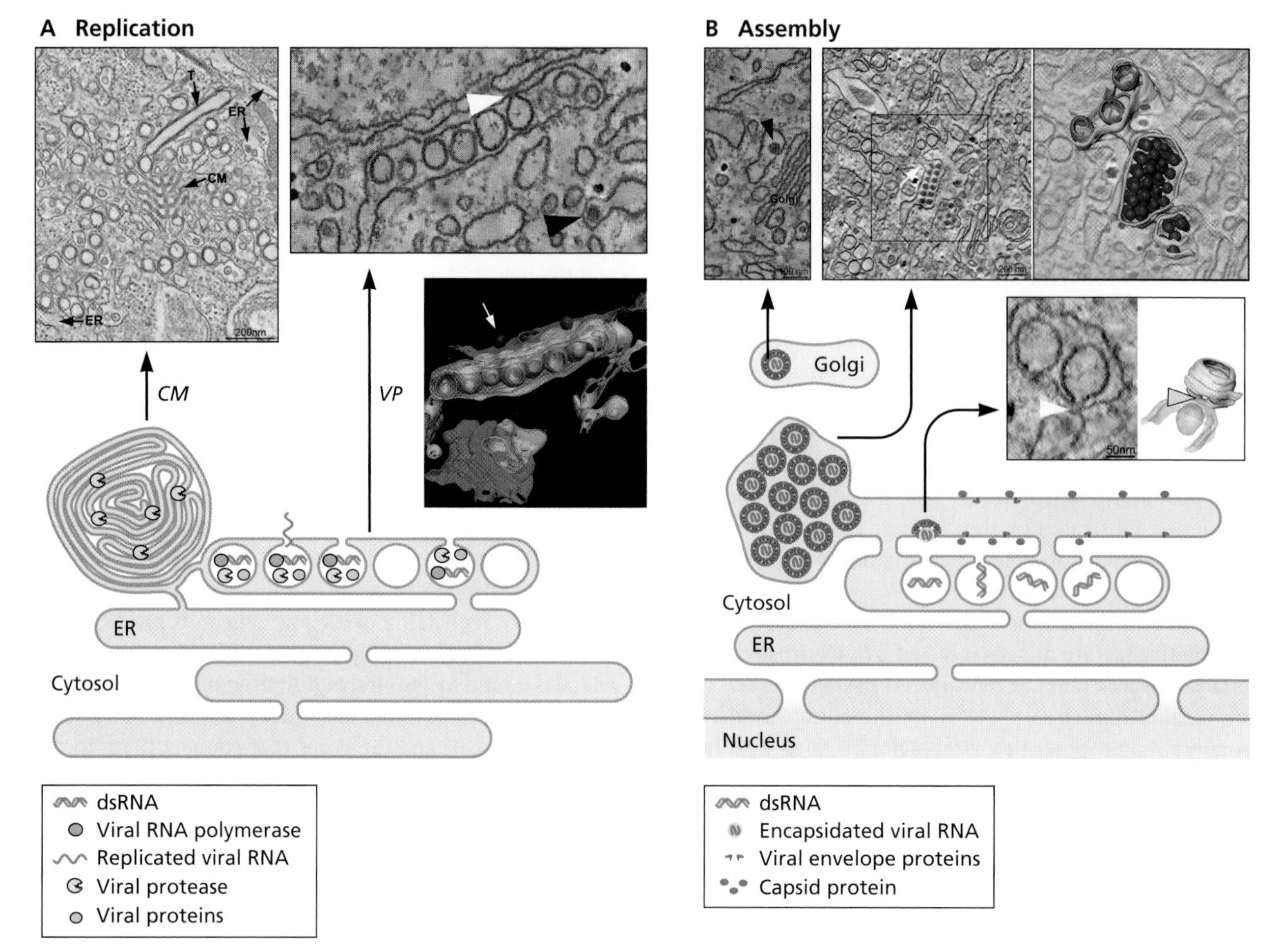

Figure 14.20 Dengue virus cytoplasmic replication and assembly compartments. Human hepatoma cells infected with dengue virus 2 for 24 h (A) or 26.5 h (B) were examined by transmission electron microscopy or by electron tomography and three-dimensional reconstruction. The latter reconstructions are shown in color. **(A)** The model for replication compartments is based on electron microscopic images like those above. The white arrowhead in the top right panel indicates a putative virus budding site shown in the tomogram in the panel below (red). CM, convoluted membranes; VP, vesicle packet; dsRNA, double-stranded RNA. **(B)** The model of assembly compartments is based on electron microscopic images like those shown above the schematic. The tomogram (far right, top) shows the reconstruction of the area boxed in the image shown to its left. The lower right panel shows a single section with a virus-induced vesicle invaginated into the ER and budding of a virus particle into the ER lumen opposite the neck of the invaginated vesicle. The three-dimensional reconstruction (right) shows the continuity between the membranes of the ER and a virus-induced vesicle (yellow) and what is probably a budding virus particle (pink). Electron micrographs and three-dimensional reconstructions reproduced from S. Welsch et al., *Cell Host Microbe* **5**:365–376, 2009, with permission. Courtesy of R. Bartenschlager, University of Heidelberg, Germany. Adapted from L. Chatel-Chaix and R. Bartenschlager, *J Virol* **88**:5907–5911, 2014, with permission.

present in Golgi compartments or endosomes, and viral late proteins. In cells infected by human cytomegalovirus, *de novo* synthesis of fatty acids is necessary for the formation of assembly compartments, and efficient envelopment of nucleocapsids to produce progeny virus particles.

The replication of the genomes of RNA viruses that do not acquire an envelope can also occur in association with host cell membranes. Infection by poliovirus (and other picornaviruses) induces inhibition of the secretory pathway (Chapter 12) and a transient increase in budding of coatomer protein II (CopII)-coated vesicles. The viral 2BC and 3A proteins co-opt membranes of the ER-Golgi intermediate compartment to establish infected cell-specific vesicles enriched in lipids that contain phosphatidylinositol 1,4-phosphate, which is bound specifically by the viral RNA polymerase. Viral genome replication and initial assembly of virus particles take place on the surfaces of these membranous replication complexes (Chapter 6). However, later in infection, double-walled autophagosomes that are associated with viral replication proteins accumulate. Inhibition of the formation of autophagosomes

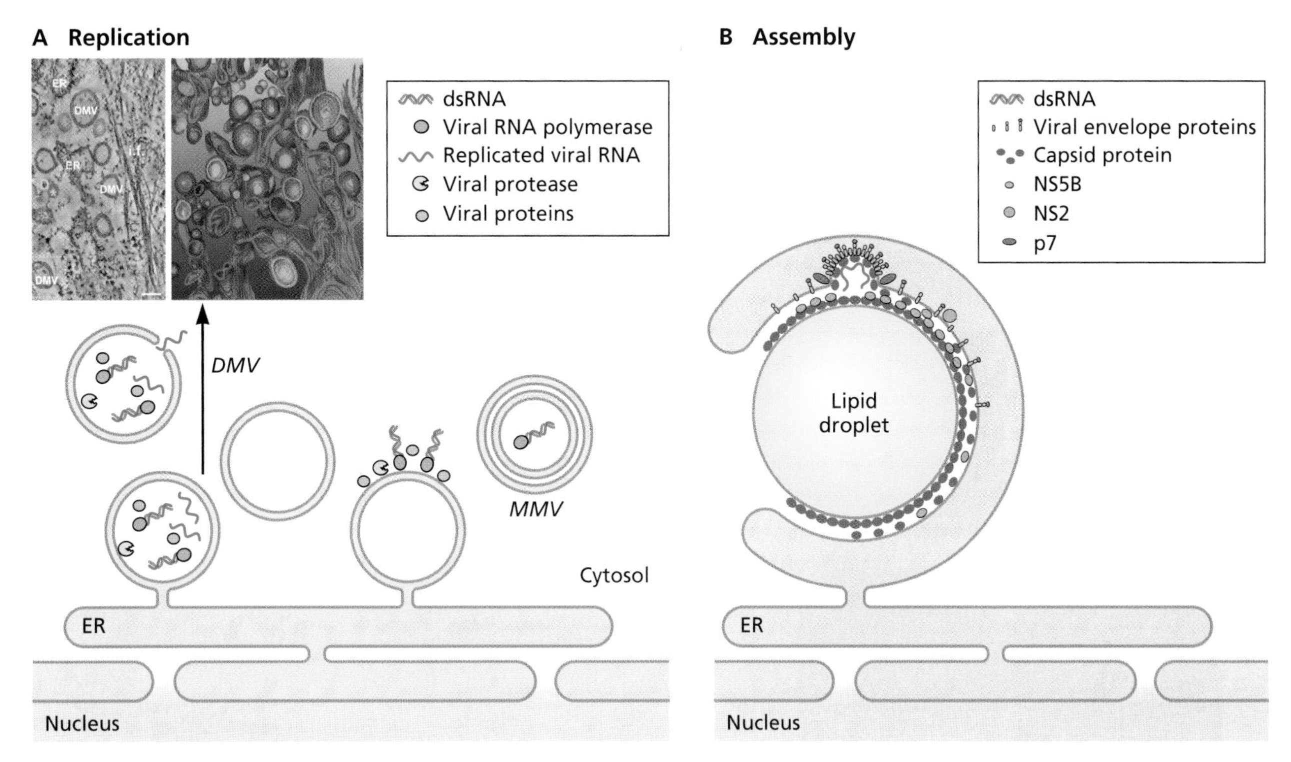

Figure 14.21 Hepatitis C virus replication and assembly compartments. (A) The model of replication compartments is based on electron tomography of human hepatoma cells infected with hepatitis C virus for 16 h. The left part of the image at the top shows one slice of a tomogram, with a three-dimensional reconstruction at the right showing a double membrane vesicle (DMV). i.f., intermediate filament. Bar, 100 nm. dsRNA, double-stranded RNA; MMV, multimembrane vesicle. Electron microscopic image reproduced from I. Romero-Brey et al., *PLoS Pathog* **8:**e1003056, 2012. Courtesy of R. Bartenschlager, University of Heidelberg, Germany. **(B)** The model of assembly compartments is based on transmission electron microscopy and indirect immunofluorescence of human hepatoma cells containing a hepatitis virus replicon that produces infectious virus particles.Adapted from L. Chatel-Chaix and R. Bartenschlager, *J Virol* **88:**5907–5911, 2014, with permission.

and their subsequent acidification reduce production of mature (infectious) virus particles, which contain VP2 and VP4 cleaved from VP0. It has therefore been proposed that the vesicles bound by a single membrane are precursors to autophagosomes, which, upon acidification, provide an environment conducive to maturation of virus particles and promote their subsequent nonlytic release (Fig. 14.22B).

Once synthesis of viral proteins begins, virus-specific inclusion bodies, termed viroplasms, are also observed in cells infected by the nonenveloped rotaviruses, which possess segmented, double-stranded RNA genomes. Viroplasms contain at least 7 (of the 12) viral proteins, and viral genomic RNA segments and mRNAs. In contrast to the virus-specific vesicular structures described above, these rotavirus-induced platforms are built of cellular lipids and proteins derived from lipid droplets. They are dynamic assemblies usually seen near the infected cell nucleus (Fig. 14.23A), and closely associated with cellular microtubules. Viroplasms first appear as small foci, but enlarge as infection progresses, because of fusion and synthesis of additional viral proteins. Their formation requires the viral NSP2 and NSP5 proteins, which are sufficient to induce assembly of viroplasm-like structures in the absence of other viral components. These proteins and VP2 recruit the other viral proteins and also cellular proteins present in lipid droplets, such as perilipin (Fig. 14.23B). Viroplasms are the sites of the initial reactions in viral genome replication and assembly of virus particles. Partially assembled, double-layered particles accumulate within these inclusions, and are then released to enter the ER for formation of the complete, three-layered particles (Chapters 4 and 6). When assembly of viroplasms is prevented, for example, by mutations in the viral genome or exposure of infected cells to inhibitors of lipid droplet formation, the yield of infectious virus particles is reduced, as is virus-induced cell death.

Cytoplasmic Vesicles and the Release of Virus Particles

In addition to viral genome replication and assembly, release of many types of virus particles depends on usurpation of cytoplasmic vesicles, or the cytoplasmic machines by which

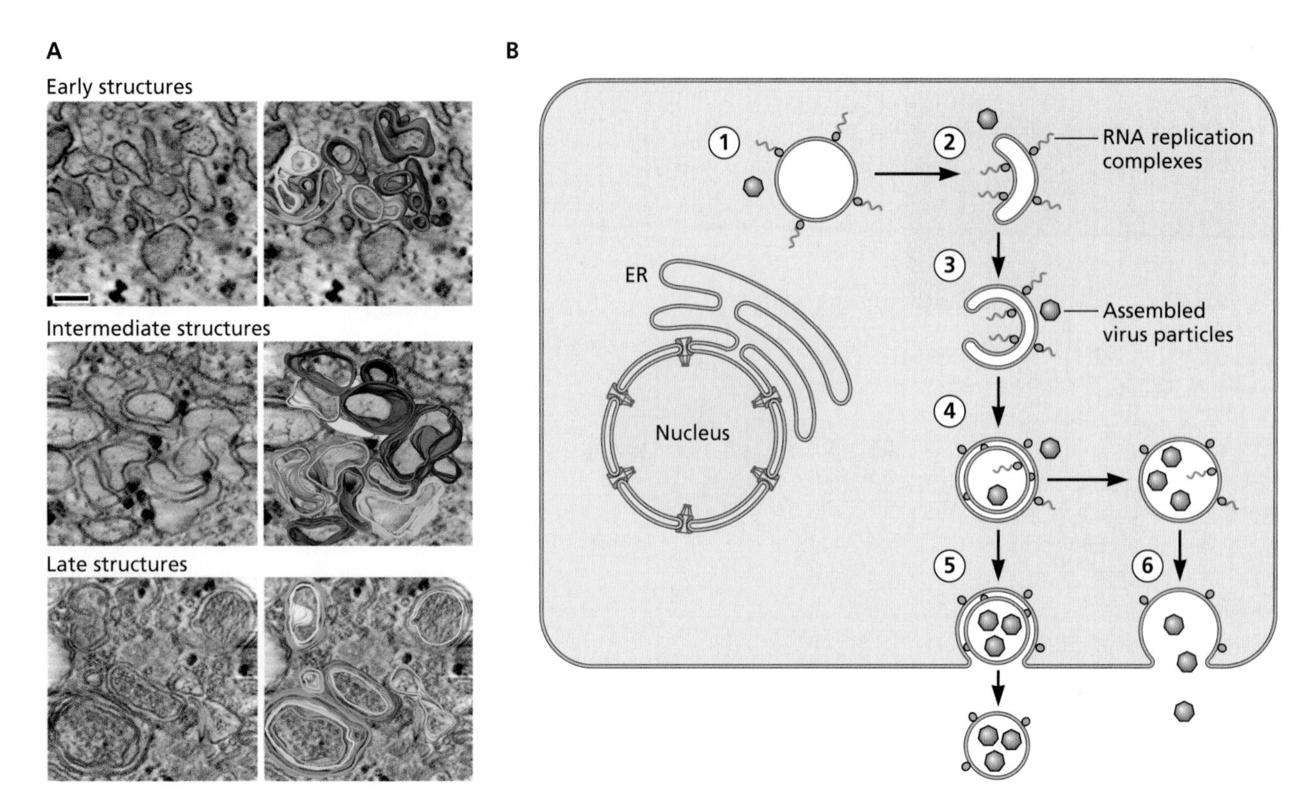

Figure 14.22 Co-option of cytoplasmic membranes in poliovirus-infected cell. (A) HeLa cells were infected with poliovirus and 260-nm-thick sections examined by electron tomography at early (3 h), intermediate (4 h), and late (7 h) times after infection. Shown are central slices in tomograms (left-hand panels) and sections with three-dimensional reconstructions overlaid (right-hand panels). In the reconstructions, single membranes are shown in blue and the inner and outer membranes of double-membrane vesicles in yellow and green. As shown, single- and double-membrane structures predominate early and late, respectively, in the infectious cycle, but occasional double-membrane vesicles can be seen from intermediate times. When sites of viral RNA synthesis were visualized using 5-bromouridine triphosphate (BrUTP) incorporation into newly synthesized viral RNA and immunoelectron microscopy with anti-BrUTP antibodies, single-membrane replication centers were seen to be most active. Adapted from G. A. Belov et al., *J Virol* **86:**302–312, 2012, with permission. Courtesy of E. Ehrenfeld, National Institutes of Health. **(B)** Model for the transition from single-membrane structures that support very active viral genome replication to double-membrane autophagic vesicles. The latter are proposed to develop upon membrane invagination into single-membrane vesicles (steps 1 to 4), so that viral genome replication and assembly can occur on and within vesicles, as has been reported. Autophagic vesicles would fuse with the plasma membrane to release vesicle-enclosed virus particles (step 5). They may also mature into autolysomes, which possess an acidic, degradative environment, with loss of one of the two autophagosomal membranes. Subsequent fusion with the plasma membrane would allow nonlytic release of mature poliovirus particles (step 6). Adapted from A. L. Richards and W. T. Jackson, *PLoS Pathog* **9:**e1003262, 2013, with permission.

they are formed. In particular, components of the multivesicular body system are co-opted to allow release of the particles of a wide variety of enveloped viruses (Chapter 13). Autophagy, a survival mechanism normally invoked under extreme conditions, may also allow nonlytic release of poliovirus and other picornaviruses.

Perspectives

Since the earliest virological experiments with host cells in culture, the considerable impact of virus infection has been documented and exploited, for example, by using cytopathic effects to search for previously unrecognized viruses. Elucidation of the molecular details of the reproduction of individual viruses established that progression through the infectious cycle is often accompanied by inhibition of fundamental cellular processes and reorganization of cellular architecture. However, a more complete appreciation of the magnitude and diversity of host cell responses has come only relatively recently, with the increasing application of the methods of systems biology.

These approaches allow the identification and quantification of very large numbers of RNAs, proteins, protein modifications, or metabolites in a single sample. Consequently, they provide extraordinarily detailed comparisons between cells infected by a particular virus and their uninfected counterparts. These methods have been applied to cells infected by a limited repertoire of viruses for only a relatively short time. Nevertheless, they have not only established the very large scale of alterations in cellular processes induced by infection, from modulation of multiple signaling pathways to stimulation or inhibition of expression of thousands of cellular genes, but also revealed unanticipated

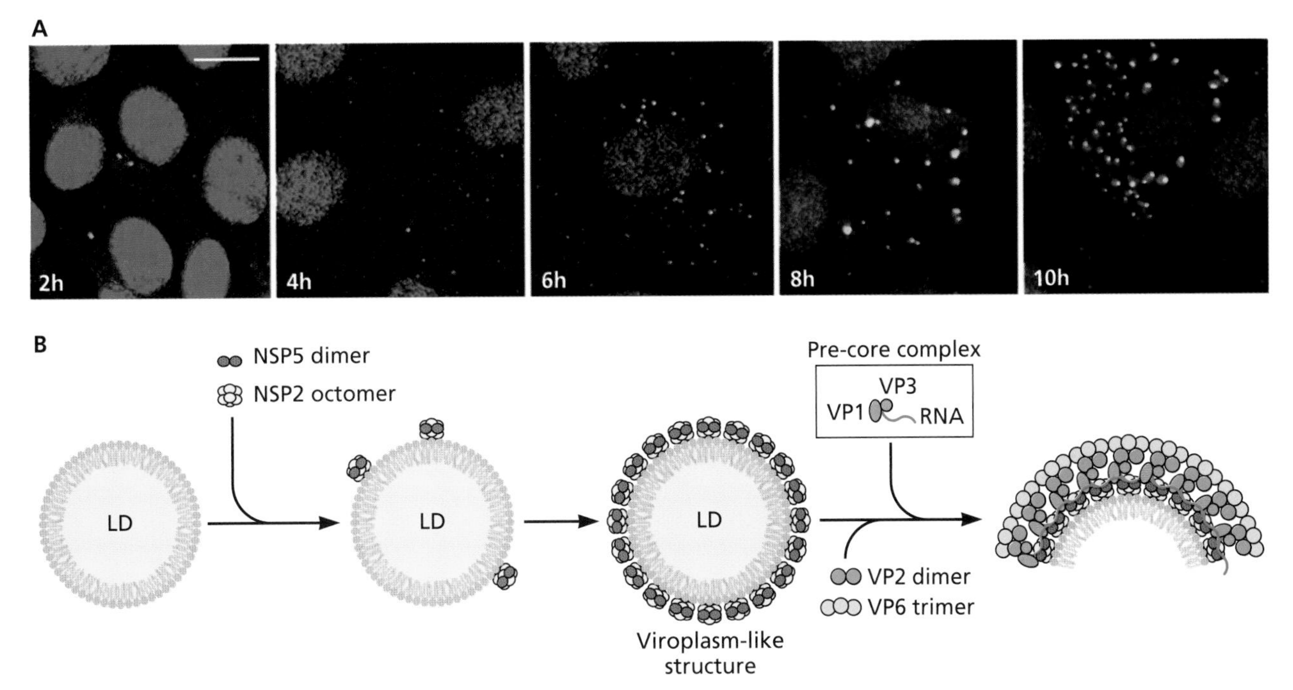

Figure 14.23 Initial rotavirus assembly on lipid droplets. (A) The kinetics of viroplasm development in established monkey cells infected with a bovine rotavirus were examined by indirect immunofluorescence using antibodies that recognize the viral protein NSP5 (green) and the cellular protein perilipin (red), a component of lipid droplets. Structures in which these proteins are both localized increase in size and number as the infectious cycle progresses. They have been termed viroplasms, and contain other viral proteins, cellular lipids, and a second cellular protein found in lipid droplets. Bar, 10 μm. Courtesy of U. Desselberger, University of Cambridge, United Kingdom. **(B)** Model for the initial assembly of rotavirus particles on lipid droplets, in which it is proposed that small, viroplasm-like structures that carry the viral NSP5 and NSP2 proteins serve as platforms for recruitment of viral structural proteins and viral single-stranded (+) RNA segments associated with VP1 and VP3 (pre-core complexes). By a poorly understood process that may include lipid degradation, such structures mature into viroplasms that contain double-layered particles. Adapted from W. Cheung et al., *J Virol* **84:**6782–6798, 2010, with permission.

redirection of cellular pathways. The latter phenomenon is epitomized by effects on cellular energy metabolism.

It might seem obvious that virus-infected cells consume large quantities of energy in the form of ATP for synthesis of viral macromolecules and intracellular transport of components of virus particles for assembly, and consequently, that infection would induce changes in cellular gene expression and metabolic pathways to promote catabolism. Cells infected by a variety of RNA and DNA viruses do indeed take up and metabolize glucose at increased rates, but this compound is not necessarily used for energy metabolism: in cells infected by human herpesviruses, it serves primarily as a source of precursors for synthesis of nucleotides or fatty acids. Furthermore, energy can also be supplied by apparently unique, virus-specific mechanisms, such as the synthesis of the fatty acid palmitate for its subsequent oxidation in poxvirus-infected cells, or the mobilization of lipid stores by induction of autophagy in cells infected by dengue virus. Additional surprises, as well as a better understanding of the mechanisms by which viral gene products directly or indirectly regulate or redirect particular cellular processes and pathways, can be anticipated.

We can now describe in considerable detail some of the striking ways in which virus reproduction and redirection is accompanied by remodeling of architectural features of the host cell. These advances are the result of improvements in the methods by which infected cells can be visualized, notably those of electron tomography and three-dimensional reconstruction. A considerable variety of infected cell-specific structures fashioned from either nuclear or cytoplasmic components have been implicated in facilitating viral genome replication and gene expression, or assembly of progeny virus particles. In some cases, viral proteins necessary for formation of such infected cell-specific platforms have been identified, but much remains to be learned about how such proteins induce reorganization of host cell components.

In this chapter, we have focused on the impact of virus infection on fundamental processes that all cells must carry out to survive and prosper, such as gene expression and generation of energy. However, the cells of multicellular organisms are specialized for particular tasks, and therefore also exhibit cell-type-specific properties and molecular functions. Virus infection can result in major perturbations, even loss, of such specialized functions. Such changes can result in far-reaching consequences for the host and contribute to viral pathogenesis, and are considered in Volume II.

References

Review Articles

Alirezaei M, Flynn CT, Whitton JL. 2012. Interactions between enteroviruses and autophagy in vivo. *Autophagy* **8:**973–975.

Anand SK, Tikoo SK. 2013. Viruses as modulators of mitochondrial functions. *Adv Virol* **2013:**738794. doi:10.1155/2013/738794.

Buchkovich NJ, Yu Y, Zampieri CA, Alwine JC. 2008. The TORrid affairs of viruses: effects of mammalian DNA viruses on the PI3K-Akt-mTOR signalling pathway. *Nat Rev Microbiol* **6:**266–275.

Diehl N, Schaal H. 2013. Make yourself at home: viral hijacking of the PI3K/Akt signaling pathway. *Viruses* **5:**3192–3212.

Fernández de Castro I, Volonté L, Risco C. 2013. Virus factories: biogenesis and structural design. *Cell Microbiol* **15:**24–34.

Heaton NS, Randall G. 2011. Multifaceted roles for lipids in viral infection. *Trends Microbiol* **19:**368–373.

Hsu NY, Ilnytska O, Belov G, Santiana M, Chen YH, Takvorian PM, Pau C, van der Schaar H, Kaushik-Basu N, Balla T, Cameron CE, Ehrenfeld E, van Kuppeveld FJ, Altan-Bonnet N. 2010. Viral reorganization of the secretory pathway generates distinct organelles for RNA replication. *Cell* **141:**799–811.

Parvaiz F, Manzoor S, Tariq H, Javed F, Fatima K, Qadri I. 2011. Hepatitis C virus infection: molecular pathways to insulin resistance. *Virol J* **8:**474. doi:10.1186/1743-422X-8-474.

Reaves ML, Rabinowitz JD. 2011. Metabolomics in systems microbiology. *Curr Opin Biotechnol* **22:**17–25.

Richards AL, Jackson WT. 2013. How positive-strand RNA viruses benefit from autophagosome maturation. *J Virol* **87:**9966–9972.

Schmid M, Speiseder T, Dobner T, Gonzalez RA. 2014. DNA virus replication compartments. *J Virol* **88:**1404–1420.

Shenk T, Alwine JC. 2014. Human cytomegalovirus: coordinating cellular stress, signaling, and metabolic pathways. *Annu Rev Virol* **1:**355–374.

Syed GH, Amako Y, Siddiqui A. 2010. Hepatitis C virus hijacks host lipid metabolism. *Trends Endocrinol Metab* **21:**33–40.

Papers of Special Interest

Signal Transduction

Izmailyan R, Hsao JC, Chung CS, Chen CH, Hsu PW, Liao CL, Chang W. 2012. Integrin β1 mediates vaccinia virus entry through activation of PI3K/Akt signaling. *J Virol* **86:**6677–6687.

Li E, Stupack DG, Brown SL, Klemke R, Schlaepfer DD, Nemerow GR. 2000. Association of $p130^{CAS}$ with phosphatidylinositol-3-OH kinase mediates adenovirus cell entry. *J Biol Chem* **275:**14729–14735.

Liu X, Cohen JI. 2013. Varicella-zoster virus ORF12 protein activates the phosphatidylinositol 3-kinase/Akt pathway to regulate cell cycle progression. *J Virol* **87:**1842–1848.

Saeed MF, Kolokoltsov AA, Freiberg AN, Holbrook MR, Davey RA. 2008. Phosphoinositide-3 kinase-Akt pathway controls cellular entry of Ebola virus. *PLoS Pathog* **4:**e1000141. doi:10.1371/journal.ppat.1000141.

Tiwari V, Shukla D. 2010. Phosphoinositide 3 kinase signaling may affect multiple steps during herpes simplex virus type-1 entry. *J Gen Virol* **91**(Pt 12)**:**3002–3009.

Gene Expression

Imbeault M, Giguère K, Ouellet M, Tremblay MJ. 2012. Exon level transcriptomic profiling of HIV-1-infected $CD4^+$ T cells reveals virus-induced genes and host environment favorable for viral replication. *PLoS Pathog* **8:**e1002861. doi:10.1371/journal.ppat.1002861.

Marcinowski L, Lidschreiber M, Windhager L, Rieder M, Bosse JB, Rädle B, Bonfert T, Györy I, de Graaf M, Prazeres da Costa O, Rosenstiel P, Friedel CC, Zimmer R, Ruzsics Z, Dölken L. 2012. Real-time transcriptional profiling of cellular and viral gene expression during lytic cytomegalovirus infection. *PLoS Pathog* **8:**e1002908. doi:10.1371/journal.ppat.1002908.

Wojcechowskyj JA, Didigu CA, Lee JY, Parrish NF, Sinha R, Hahn BH, Bushman FD, Jensen ST, Seeholzer SH, Doms R. 2013. Quantitative phosphoproteomics reveals extensive cellular reprograming during HIV-1 entry. *Cell Host Microbe* **13:**613–623.

Metabolism

Bose SK, Kim H, Meyer K, Wolins N, Davidson NO, Ray R. 2014. Forkhead box transcription factor regulation and lipid accumulation by hepatitis C virus. *J Virol* **88:**4195–4203.

Cui L, Lee YH, Kumar Y, Xu F, Lu K, Ooi EE, Tannenbaum SR, Ong CN. 2013. Serum metabolome and lipidome changes in adult patients with primary dengue infection. *PLoS Negl Trop Dis* **7:**e2373. doi:10.1371/journal.pntd.0002373.

Fieldsteel AH, Preston WS. 1953. The effect of meningopneumonitis virus infection on glycolysis by mouse and rat brain. *J Infect Dis* **93:**236–242.

Mazzon M, Castro C, Roberts LD, Griffin JL, Smith GL. 2015. A role for vaccinia virus protein C16 in reprogramming cellular energy metabolism. *J Gen Virol* **96**(Pt 2)**:**395–407.

Shin HJ, Park YH, Kim SU, Moon HB, Park do S, Han YH, Lee CH, Lee DS, Song IS, Lee DH, Kim M, Kim NS, Kim DG, Kim JM, Kim SK, Kim YN, Kim SS, Choi CS, Kim YB, Yu DY. 2011. Hepatitis B virus X protein regulates hepatic glucose homeostasis via activation of inducible nitric oxide synthase. *J Biol Chem* **286:**29872–29881.

Yu Y, Maguire TG, Alwine JC. 2014. ChREBP, a glucose-responsive transcriptional factor, enhances glucose metabolism to support biosynthesis in human cytomegalovirus-infected cells. *Proc Natl Acad Sci U S A* **111:** 1951–1956.

Remodeling of Cellular Organelles

Cheung W, Gill M, Esposito A, Kaminski CF, Courousse N, Chwetzoff S, Trugnan G, Keshavan N, Lever A, Desselberger U. 2010. Rotaviruses associate with cellular lipid droplet components to replicate in viroplasms, and compounds disrupting or blocking lipid droplets inhibit viroplasm formation and viral replication. *J Virol* **84:**6782–6798.

Katsafanas GC, Moss B. 2007. Colocalization of transcription and translation within cytoplasmic poxvirus factories coordinates viral expression and subjugates host functions. *Cell Host Microbe* **2:**221–228.

Li J, Liu Y, Wang Z, Liu K, Wang Y, Liu J, Ding H, Yuan Z. 2011. Subversion of cellular autophagy machinery by hepatitis B virus for viral envelopment. *J Virol* **85:**6319–6333.

Miyanari Y, Atsuzawa K, Usuda N, Watashi K, Hishiki T, Zayas M, Bartenschlager R, Wakita T, Hijikata M, Shimotohno K. 2007. The lipid droplet is an important organelle for hepatitis C virus production. *Nat Cell Biol* **9:** 1089–1097.

APPENDIX

Structure, Genome Organization, and Infectious Cycles

Adenoviruses

Family *Adenoviridae*

Selected Genera	Examples
Mastadenovirus	Human adenovirus type 5
Aviadenovirus	Fowl adenovirus 1

Human serotypes are very widespread in the population. Infection by these viruses is often asymptomatic but can result in respiratory disease in children (members of species B and C), conjunctivitis (members of species B and D), and gastroenteritis (species F serotypes 40 and 41). Human adenoviruses 40 and 41 are the second leading cause (after rotaviruses) of infantile viral diarrhea. Adenoviruses share capsid morphology and linear double-stranded DNA genomes, but the members of the genera differ in size, organization, and coding sequences. The *Mastadenovirinae* comprise over 65 adenoviruses of humans and other mammals, including mice, sheep, and dogs, and some are oncogenic in rodents. Study of human adenovirus transformation of cultured cells has provided fundamental information about mechanisms that control progression through the cell cycle and oncogenesis. Characteristic features of the replication of these viruses include stereotyped temporal control of viral gene expression and an unusual mechanism of initiation of viral DNA synthesis (protein priming). Mastadenoviral genomes also include genes transcribed by cellular RNA polymerase III.

Figure 1 Structure and genome organization of human adenovirus type 5. (A) Virion structure. The electron micrograph shows a negatively stained human adenovirus type 5 particle (courtesy of M. Bisher, Princeton University, Princeton, NJ). Bar = 50 nm. **(B) Genome organization.** The DNA genome length is 36 to 38 kbp. Green and tan arrows represent primary products of RNA polymerase II and III transcription, respectively, and are labeled in bold type. Coding sequences for viral proteins or families of major late mRNAs are also indicated. Hatched lines show splicing of the major late (ML) tripartite leader. ITR, inverted terminal repetition; Ori, origin of replication.

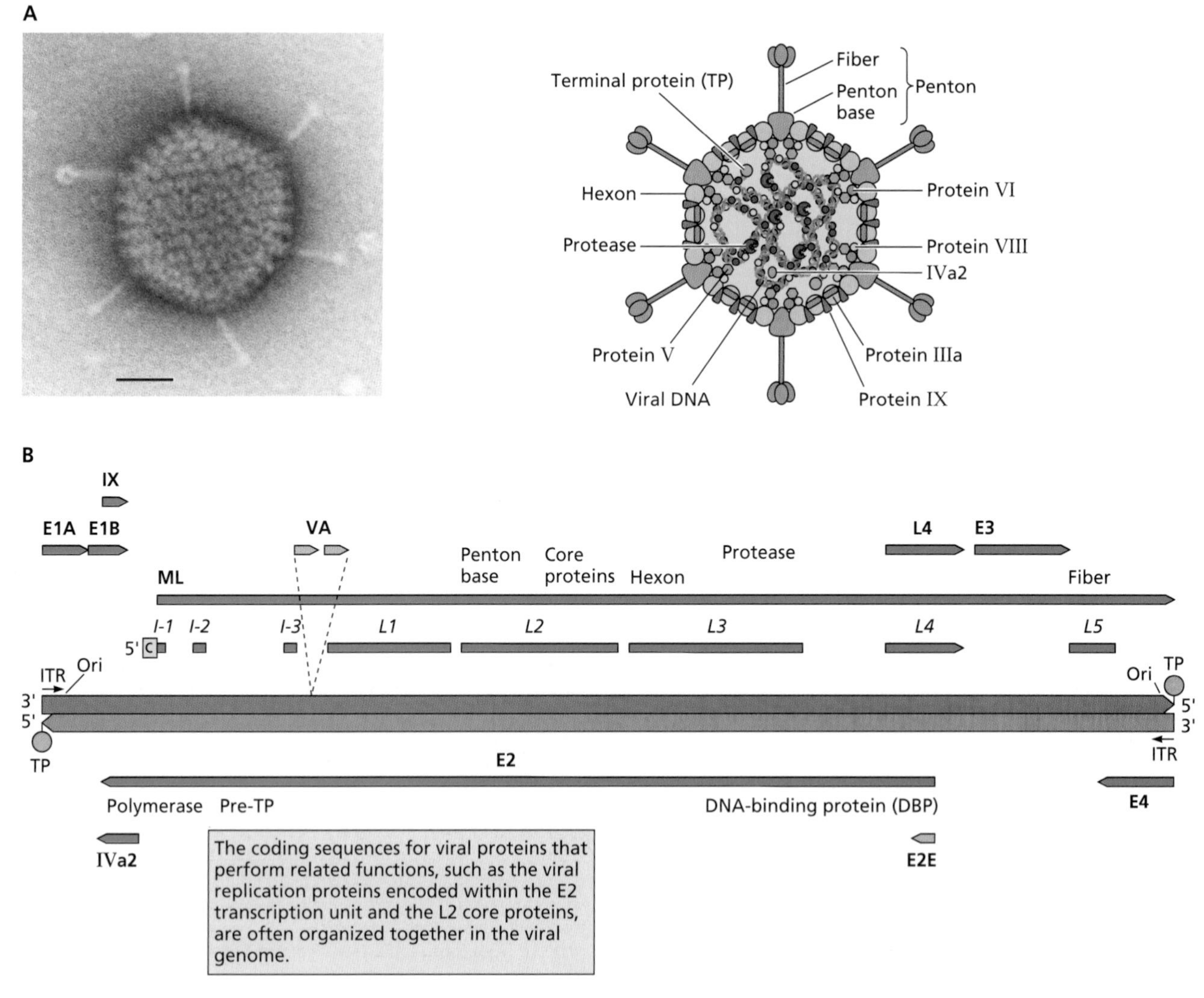

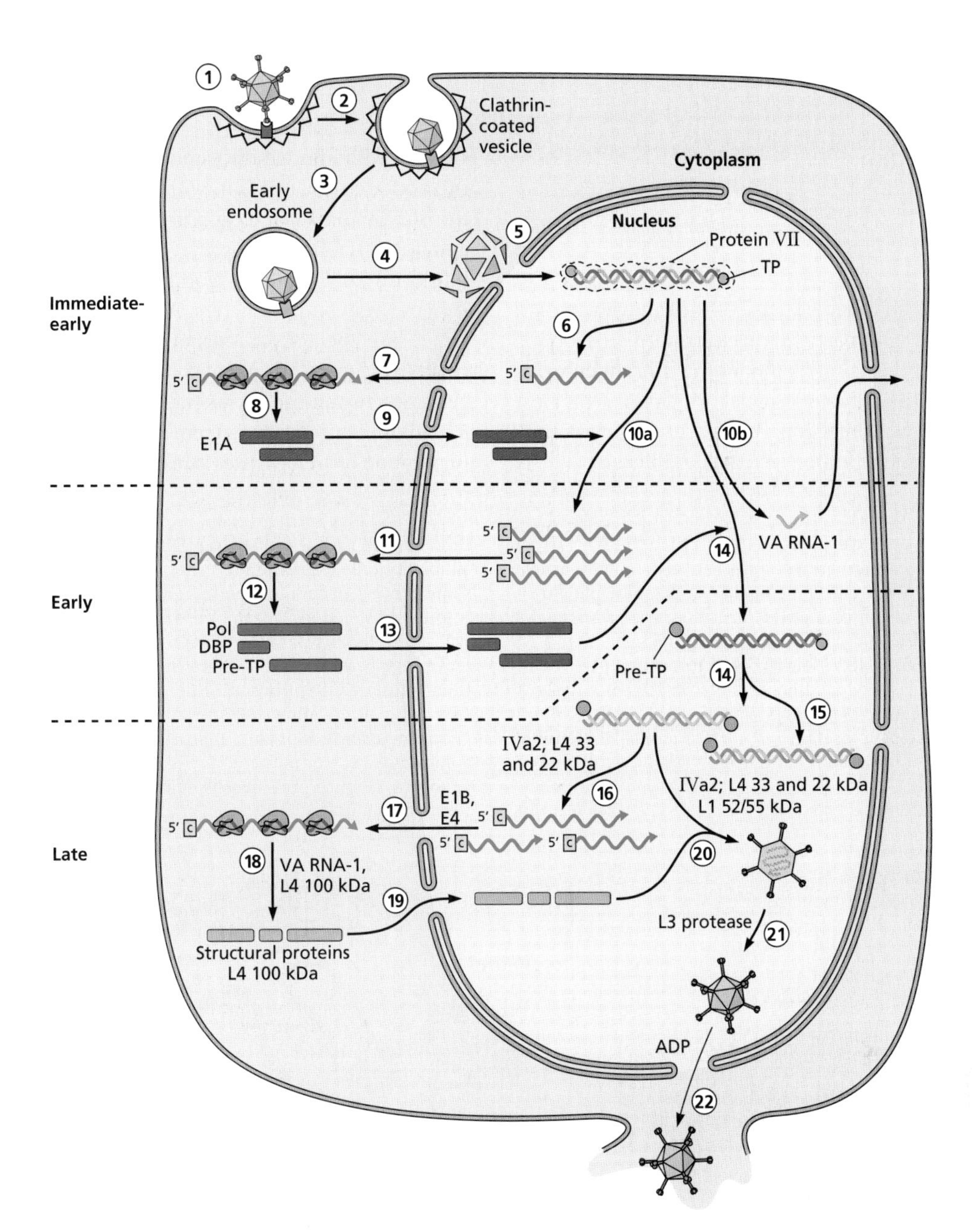

Figure 2 Single-cell reproductive cycle of human adenovirus type 5. **(1)** The virus attaches to a permissive human cell via interaction between the fiber and (with most serotypes) the coxsackie-adenovirus receptor on the cell surface (purple cylinder). **(2)** The particle then enters the cell via receptor-mediated endocytosis. **(3)** A second protein of the virus particle, penton, then interacts with a cell integrin (blue cylinder). **(4)** Partial disassembly takes place prior to entry of particles into the cytoplasm, a step that requires a membrane-lytic region of protein VI. **(5)** Following further uncoating, the viral genome associated with core protein VII is imported into the nucleus. **(6)** The host cell RNA polymerase II system transcribes the immediate early E1A gene. **(7, 8)** E1A proteins are synthesized by the cellular translation machinery, following alternative splicing and export of E1A mRNAs to the cytoplasm. **(9)** These proteins are imported into the nucleus, where they regulate transcription of both cellular and viral genes. **(10a)** The larger E1A protein stimulates transcription of the viral early genes by cellular RNA polymerase II. **(10b)** Transcription of the VA genes by host cell RNA polymerase III also begins during the early phase of infection. **(11, 12)** The early pre-mRNA species are processed, exported to the cytoplasm, and translated. **(13)** These early proteins are imported into the nucleus. **(14)** The viral replication proteins cooperate with a limited number of cellular proteins in viral DNA synthesis. **(15, 16)** Replicated viral DNA molecules can serve as templates for further rounds of replication or for transcription of late genes. Some late promoters are activated simply by viral DNA replication, but maximally efficient transcription of the major late transcription unit (Fig. 1, ML) requires the late IVa2 and L4 proteins. **(17)** Processed late mRNA species are selectively exported from the nucleus as a result of the action of the E1B 55-kDa and E4 Orf6 proteins. **(18)** Their efficient translation requires the major VA RNA, VA RNA-I, which counteracts a cellular defense mechanism, and the late L4 100-kDa protein. **(19)** The latter protein also serves as a chaperone for assembly of trimeric hexons as they and the other structural proteins are imported into the nucleus. **(20)** Within the nucleus, capsids are assembled from these proteins and the progeny viral genomes to form noninfectious immature virus particles. Assembly requires a packaging signal located near the left end of the genome, as well as the IVa2, L1 52/55-kDa, and L4 22/33-kDa proteins. Immature particles contain the precursors of the mature forms of several proteins. **(21)** Mature virions are formed when these precursor proteins are cleaved by the viral L3 protease, which is assembled into the core. **(22)** Progeny virus particles are released, usually upon destruction of the host cell via mechanisms that are not well understood, although the E3 adenovirus death protein (ADP) facilitates exit of particles from the nucleus.

Arenaviruses

Family *Arenaviridae*

Genus	Example
Arenavirus	Lymphocytic choriomeningitis virus

The *Arenaviridae* are so called because of the sandy (*arenosus*; Latin) appearance of viral particles by electron microscopy. A prototype member of this family, lymphocytic choriomeningitis virus, has been used to elucidate essential principles of the host immune response to viral infection. Arenaviruses cause chronic, usually asymptomatic, infections in rodents, the natural host. Contact with infected mice and rats (typically by bite) can result in zoonotic transmission, with outcomes in humans ranging from asymptomatic infection to febrile illness, aseptic meningitis, and often fatal hemorrhagic fevers. Arenaviruses are categorized into Old World and New World serogroups, based on geographical and genetic parameters. Arenaviruses are enveloped and have a bisegmented RNA genome consisting of a large (L) and a small (S) segment. Only four proteins are encoded in these viral genomes, two in each segment. Two genes, one encoded in the 5′ end of each segment, are present in an ambisense orientation. Replication of arenaviruses is restricted to the cytoplasm.

Figure 3 Structure and genome organization. (A) Virion structure. Cryo-electron micrograph of a negatively stained particle of the arenavirus lymphocytic choriomeningitis virus. These pleiomorphic particles are ~120 nm in diameter. (Image courtesy of Michael J. Buchmeier and Benjamin W. Neuman, School of Biological Sciences, University of Reading, United Kingdom.) **(B) Genome organization.** The viral genome comprises two segments: large (L; 7.2 kb) and short (S; 3.5 kb). The L segment encodes the RNA-dependent RNA polymerase (L) and an accessory protein (Z) that functions in genome packaging, particle assembly, and budding. The S segment encodes a surface glycoprotein (GP), which binds to the viral receptor and mediates target cell recognition and entry, and a histone-like nucleocapsid protein (NP) that, with the viral RNAs, forms the ribonucleocapsid. For simplicity, only expression of genes on the S segment is shown, but the same process occurs for the L segment. Upon entry of the viral RNA into the host cell cytoplasm, the viral L protein, which enters the cell with the infecting particle, binds to the 3′ end of the RNA (shown as an orange ball) and synthesizes the (+) strand NP mRNA, which is then translated. Replication of the genomic RNA into a complementary antigenome allows synthesis of GP mRNA. This mechanism of gene expression results in temporal control of viral gene expression, a common feature of the reproductive cycles of many viruses with DNA genomes. (Figure provided by Juan Carlos de la Torre, The Scripps Research Institute.)

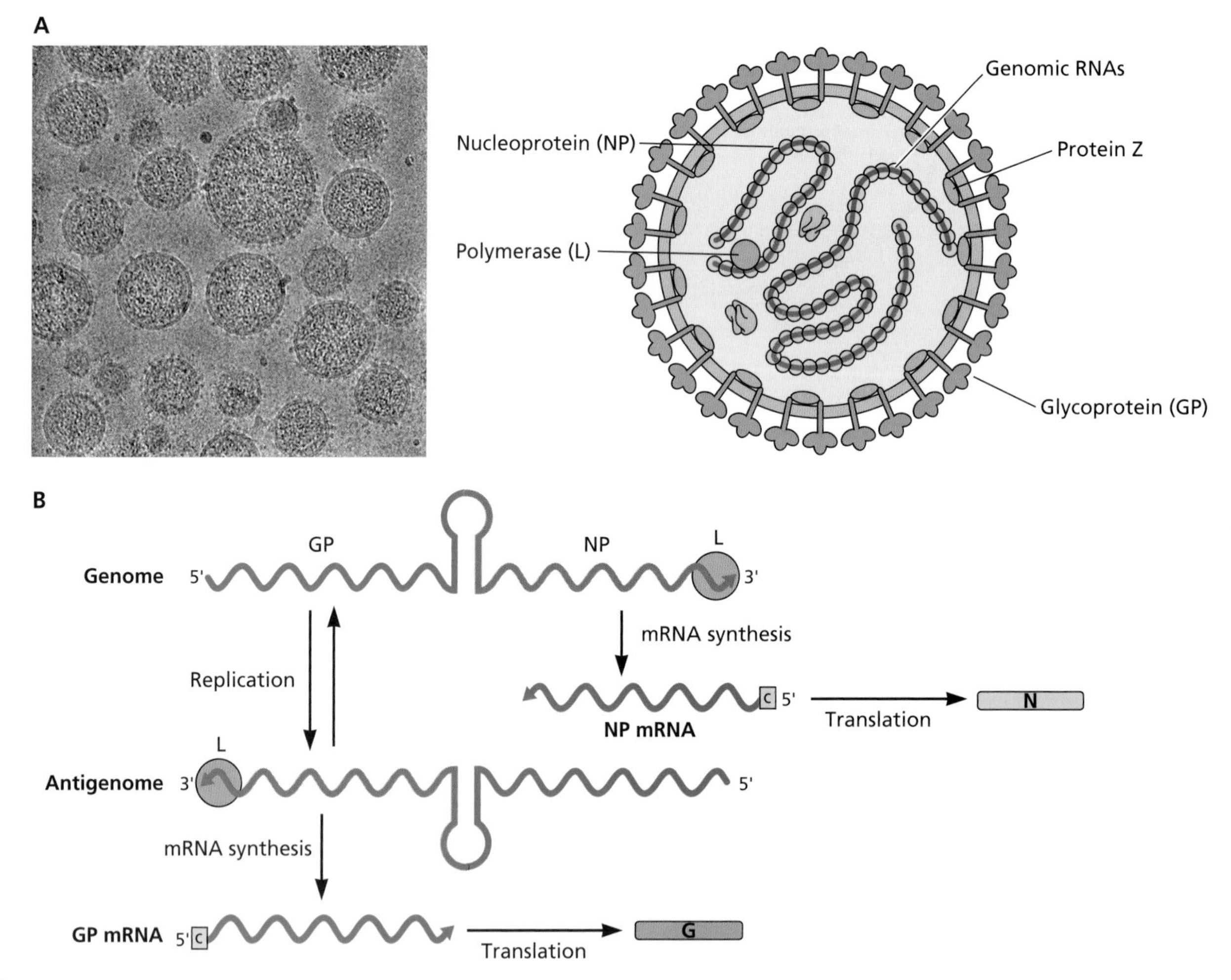

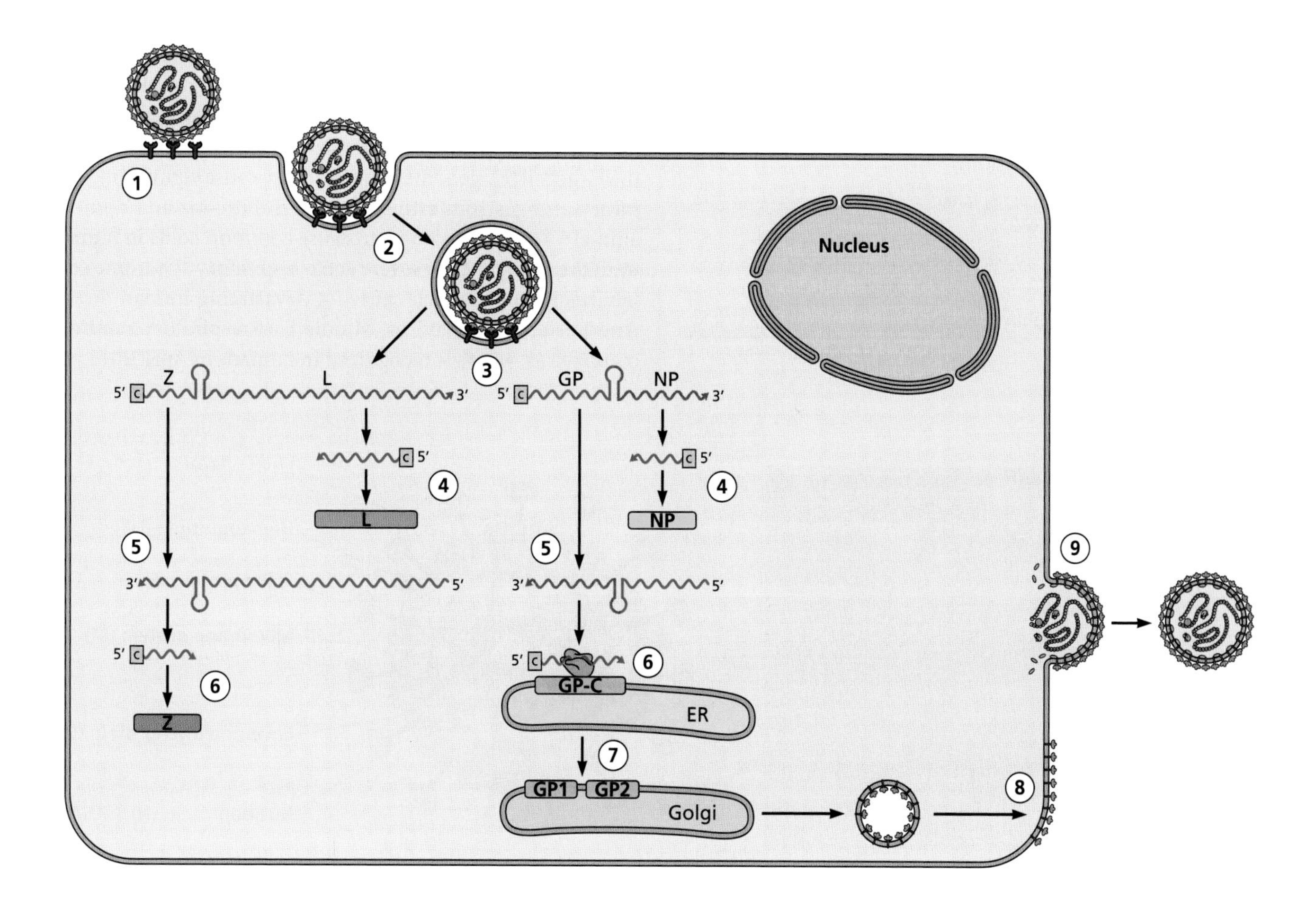

Figure 4 Single-cell reproductive cycle. **(1, 2)** The virion binds to a cellular receptor (alpha dystroglycan), which induces receptor-mediated endocytosis. **(3)** Low-pH-triggered membrane fusion between the viral and cellular membranes releases the viral genome segments into the cytoplasm. **(4)** NP and L mRNAs are synthesized using the viral genome as a template. **(5)** In addition, the ambisense viral genome (Fig. 3B) is the template for synthesis of a complementary antigenome by the viral RNA-dependent RNA polymerase (L). **(6)** The antigenome serves both as a template for production of progeny viral genomes, as well as the template for the synthesis of the other two viral mRNAs, Z and GP-C. In all cases, the intergenic region (IGR) that separates the two coding regions of each segment serves to terminate mRNA synthesis. **(7)** GP-C, which is translated by ER-bound ribosomes, is cleaved into GP-1 and GP-2 by a cellular protease (MBTPS1) as it traverses the secretory pathway. **(8)** GP-1 and GP-2 then associate to form the spikes that form the outer surface of the viral particles. **(9)** The small RING finger protein Z facilitates budding through interaction with cellular proteins, enabling release of extracellular viral particles.

Coronaviruses

Family *Coronaviridae*

Genera	Examples
Alphacoronavirus	Human coronavirus 229E
Betacoronavirus	Murine coronavirus
	Severe acute respiratory syndrome-related coronavirus (SARS-CoV)
	Middle East respiratory syndrome coronavirus (MERS-CoV)
Gammacoronavirus	Avian coronavirus

Coronaviruses are enveloped RNA viruses that infect mammals and birds. The name derives from the fringe of club-shaped spikes observed in electron micrographs that give the virus particles the appearance of a solar corona. These viruses have the largest RNA genomes known. They cause significant respiratory and gastrointestinal disease in humans and domestic animals. They were known to cause common colds in humans, until the emergence of severe acute respiratory syndrome coronavirus in 2002, which caused a devastating human disease. Another new coronavirus, Middle East respiratory syndrome coronavirus, was first recognized in humans in April 2012.

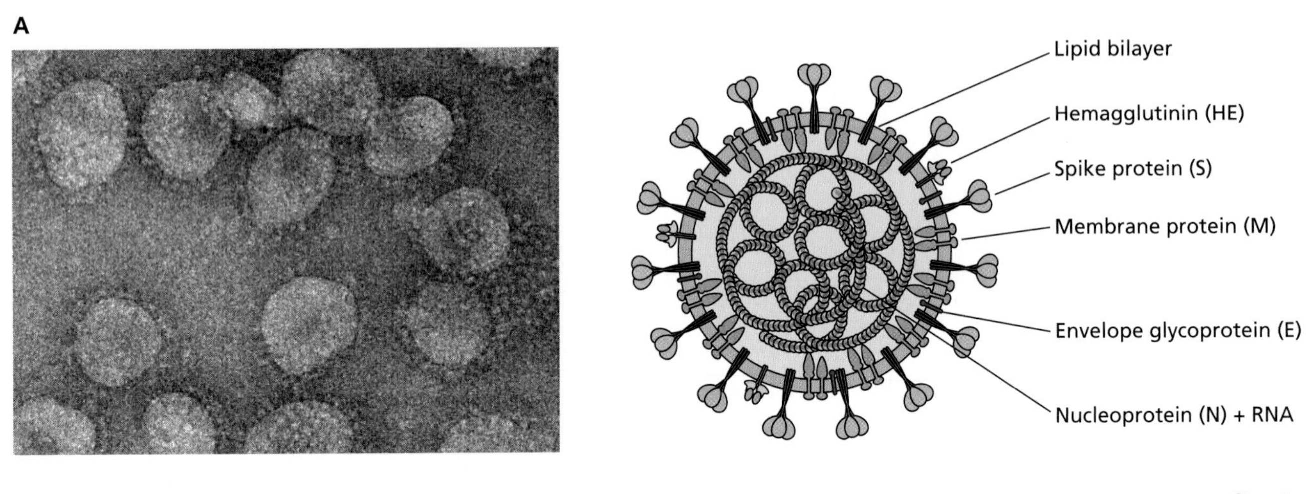

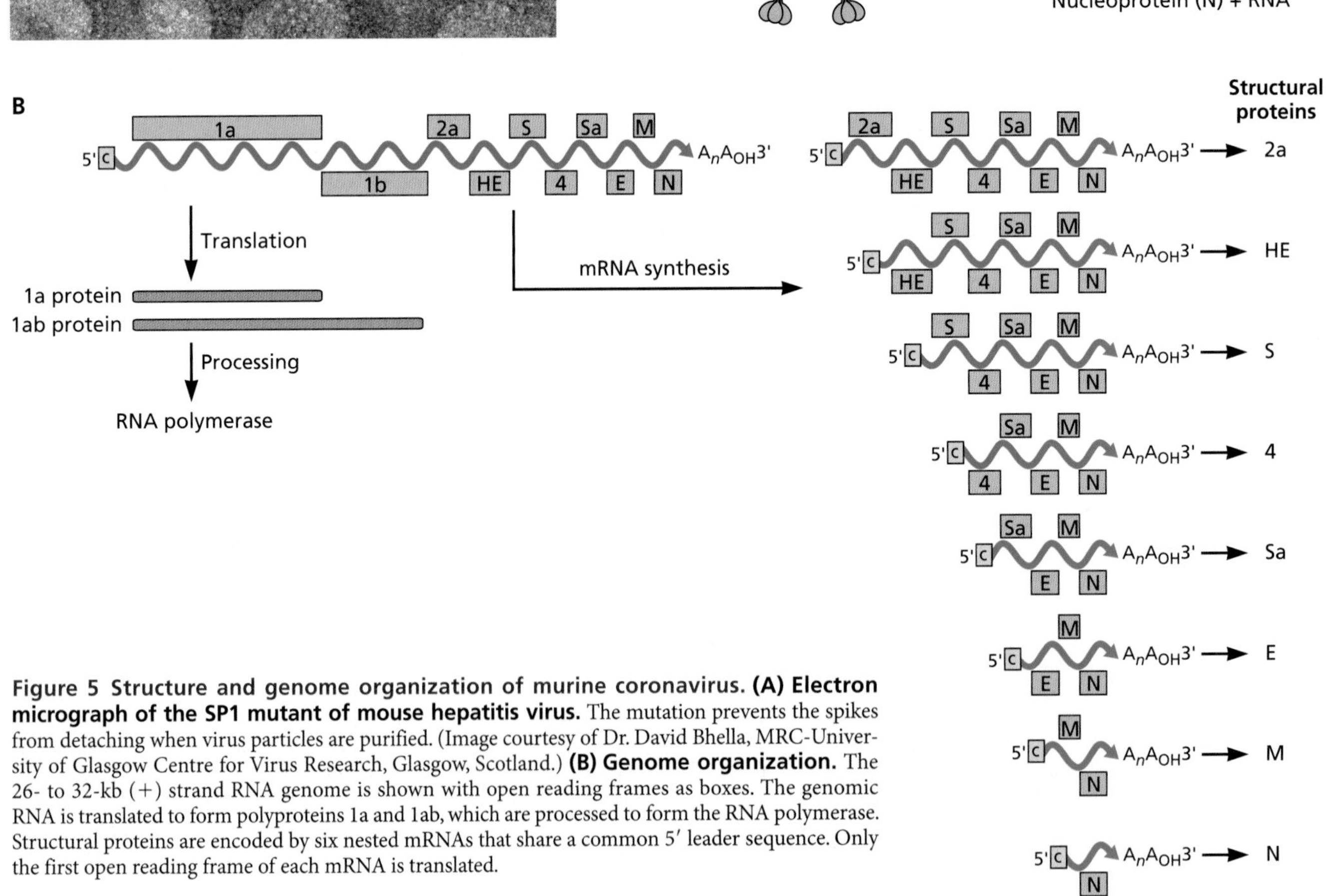

Figure 5 Structure and genome organization of murine coronavirus. (A) Electron micrograph of the SP1 mutant of mouse hepatitis virus. The mutation prevents the spikes from detaching when virus particles are purified. (Image courtesy of Dr. David Bhella, MRC-University of Glasgow Centre for Virus Research, Glasgow, Scotland.) **(B) Genome organization.** The 26- to 32-kb (+) strand RNA genome is shown with open reading frames as boxes. The genomic RNA is translated to form polyproteins 1a and 1ab, which are processed to form the RNA polymerase. Structural proteins are encoded by six nested mRNAs that share a common 5′ leader sequence. Only the first open reading frame of each mRNA is translated.

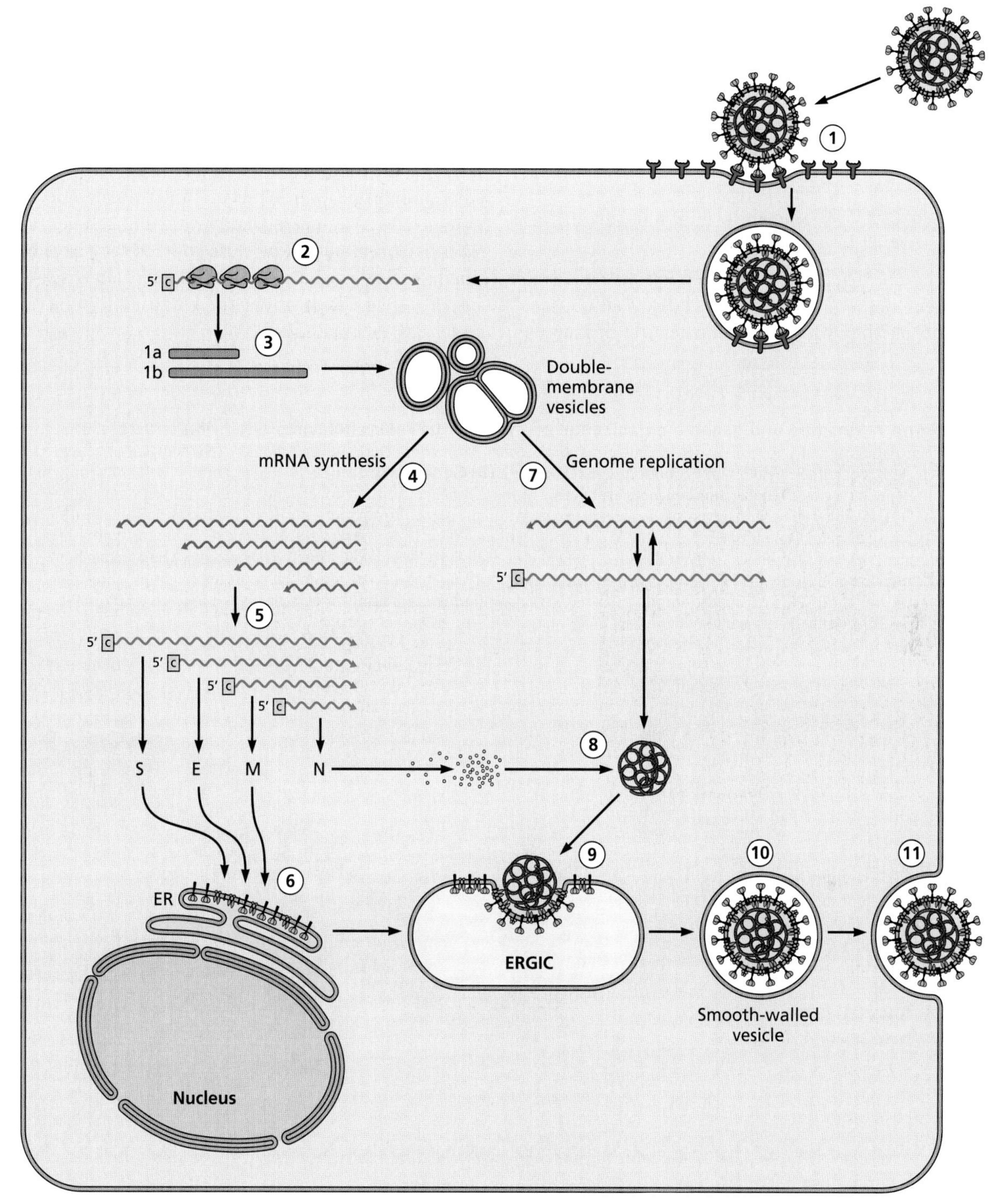

Figure 6 Single-cell reproductive cycle. **(1)** The virion binds to a cell surface receptor, and fusion of the viral and cell membrane occurs either at the cell surface or from within endosomes, depending on the virus. **(2)** Fusion is triggered by low pH, leading to delivery of the nucleocapsid into the cytoplasm. **(3)** The viral genome is translated to produce the 1a and 1ab proteins (the latter by ribosomal frameshifting). These are autoproteolytically processed by viral proteases to produce a variety of viral proteins, including the RNA-dependent RNA polymerase, proteins that remodel cellular membranes to form structures that are sites of viral RNA synthesis, enzymes that catalyze multiple steps in the synthesis of the 5′-terminal cap structure of mRNA, and an exonuclease that provides a proofreading function. **(4)** The other viral proteins are encoded by a nested set of mRNAs that share a common 5′ leader sequence. Discontinuous RNA synthesis occurs during (−) strand RNA synthesis. Most of the (+) strand template is not copied, probably because it loops out as the polymerase completes synthesis of the leader RNA. **(5)** The resulting (−) strand RNAs, with leader sequences at the 3′ ends, are then copied to form mRNAs. These mRNAs serve as templates for structural and nonstructural proteins. **(6)** The membrane-bound proteins M, S, and E are inserted into the ER, and then move to the site of viral assembly, the ER-Golgi intermediate compartment. **(7)** Full length (−) strand RNAs are produced, and these are templates for the synthesis of full length (+) strands, **(8)** which are encapsidated by N protein. **(9)** The nucleocapsid buds into the ER-Golgi intermediate compartment (ERGIC), acquiring a membrane that contains S, E, and M proteins. **(10)** Virus particles are transported to the plasma membrane in smooth-walled vesicles and **(11)** released from the cells by exocytosis as the transport vesicle fuses with the plasma membrane.

Filoviruses

Family *Filoviridae*

Genera	Examples
Marburgvirus	Lake Victoria marburgvirus
Ebolavirus	Zaire ebolavirus
	Sudan ebolavirus

Members of the *Filoviridae* are enveloped viruses with (−) strand RNA genomes. The virus particles possess an unusual filamentous morphology, which led to the name of this family (*Filum* is the Latin word for thread). They are agents of serious hemorrhagic fever in humans and primates. Because these viruses have a high case fatality ratio and can be transmitted from person to person by close contact, they have been classified as select agents by the U.S. Centers for Disease Control and Prevention. Research on these viruses must be carried out under BSL-4 containment.

Figure 7 Structure and genome organization of the filovirus Zaire ebolavirus. (A) Virion structure. The electron micrograph shows an image of ebolavirus particles inside of an infected, cultured monkey cell (courtesy of Elizabeth R. Fischer, Rocky Mountain Laboratories, NIAID, NIH). **(B) Genome organization.** The genome is ~19 kb long and contains 7 genes. Conserved sequences are present at the 3′ (leader) and 5′ (trailer) ends of the viral genome. Each gene is flanked by short, conserved sequences that specify initiation and termination of mRNA synthesis. In some cases the termination and initiation sequences of neighboring genes overlap. The (−) strand RNA is the template for synthesis of leader RNA and 7 monocistronic mRNAs (capped and polyadenylated) encoding the 7 viral proteins. The fourth gene of ebolaviruses encodes an mRNA that is translated to form a secreted, nonstructural glycoprotein (orange). Editing of the gene 4 mRNA is required to produce an mRNA encoding the membrane-associated GP. The secreted GP is not encoded by marburgvirus genomes.

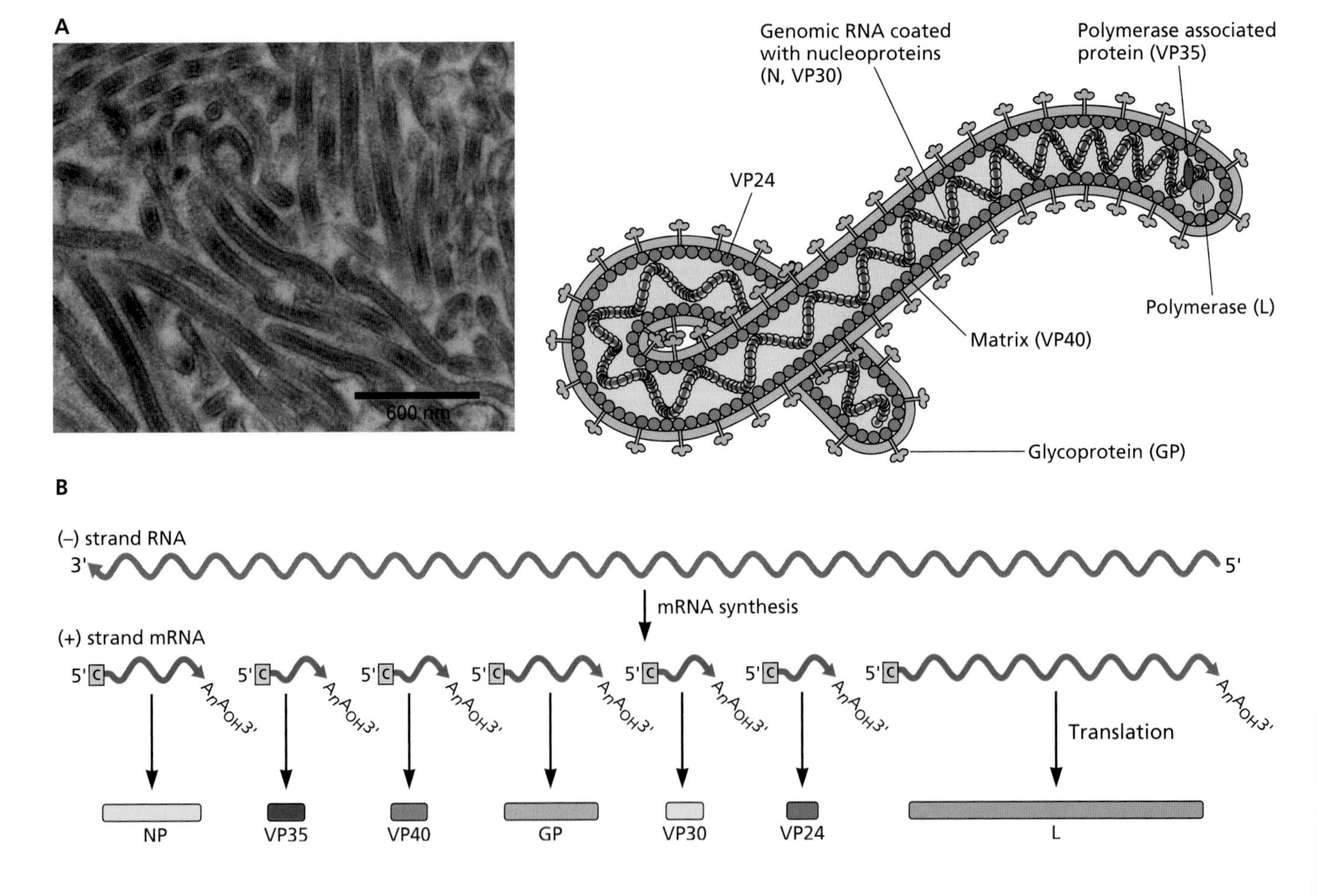

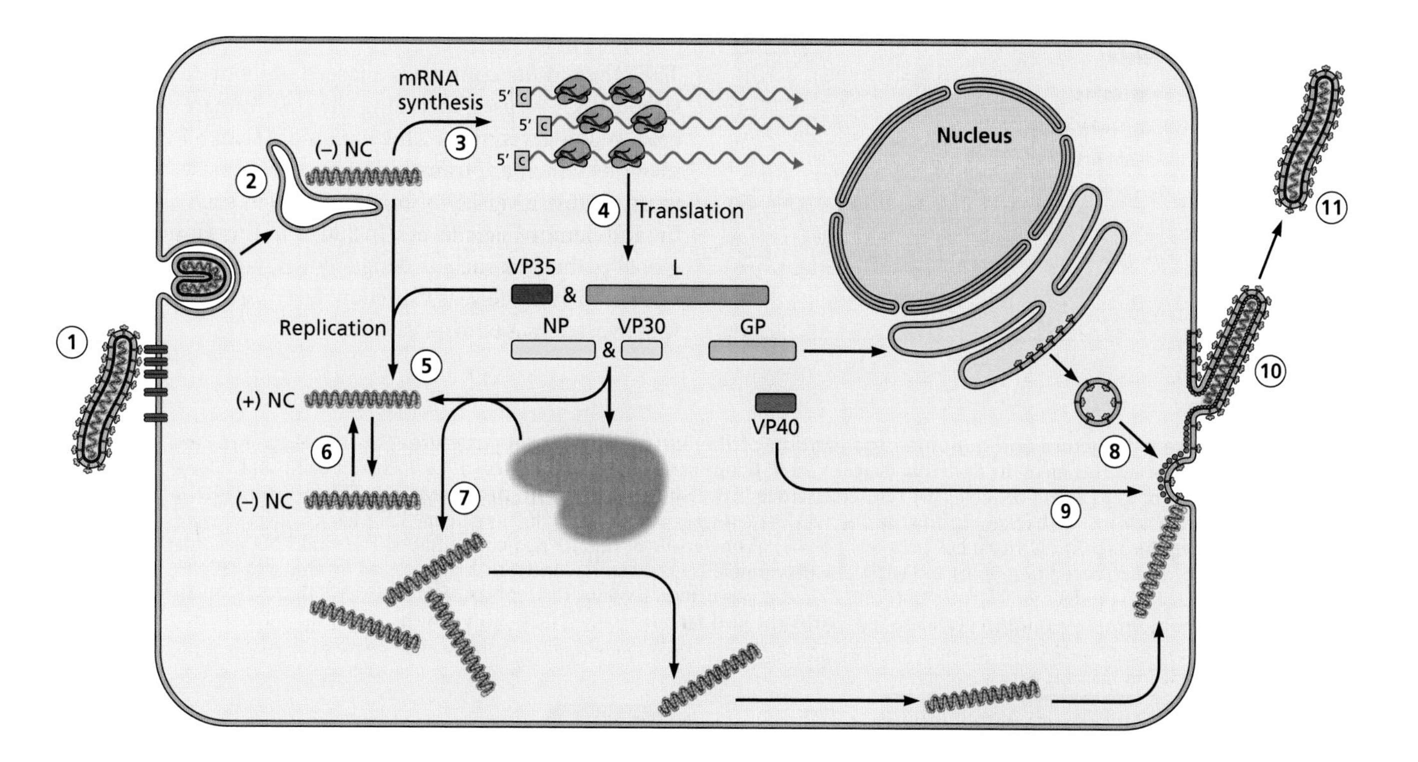

Figure 8 Single-cell reproductive cycle of ebolavirus. **(1)** Virus particles bind to a cell surface receptor, followed by uptake via macropinocytosis and trafficking to late endosomes. **(2)** Viral GP is cleaved by endosomal cysteine proteases (cathepsins) and then binds to the Niemann-Pick disease type C integral membrane protein, NPC1, which is exposed in the endosomal lumen. This interaction leads to fusion of the viral and endosomal membranes for delivery of the viral nucleocapsid into the cytoplasm. **(3)** The nucleocapsid, which is composed of the viral RNA and L (the RNA-dependent RNA polymerase), NP, and VP30 proteins, is the template for the synthesis of seven viral mRNAs in the cytoplasm. The capped and polyadenylated mRNAs are synthesized in a 3′ to 5′ direction from the (−) strand RNA template, by a process of initiation and termination as the polymerase complex recognizes conserved start and stop sequences on the template. **(4)** These mRNAs are translated. **(5)** The concentrations of viral proteins, especially NP, regulate the switch from mRNA synthesis to genome replication, which begins with synthesis of full-length (+) strand copies of the viral RNA. **(6)** These (+) strands are encapsidated by NP and, in turn, serve as templates for the synthesis of full length (−) strand RNAs. **(7)** Inclusion bodies are the sites of RNA synthesis and nucleocapsid assembly. **(8)** Assembly begins with the synthesis of GP in the ER and transport to lipid raft microdomains at the plasma membrane. **(9)** Octamers of VP40 are produced and transported to GP-containing lipid rafts by the endocytic multivesicular bodies. VP40 interacts with the C terminus of NP and serves to direct viral (−) strand nucleocapsids to sites of virus budding. **(10)** Nucleocapsids form parallel to the plasma membrane, and **(11)** virus particles are released by budding.

Flaviviruses

Family *Flaviviridae*

Genera	Examples
Flavivirus	Yellow fever virus
	Dengue virus
	West Nile virus
Hepacivirus	Hepatitis C virus
	GB virus B
Pestivirus	Bovine viral diarrhea virus
Pegivirus	GB virus A, C, D

The *Flaviviridae* comprises a large family of enveloped, (+) strand RNA viruses, including the first human virus discovered, yellow fever virus. There are more than 50 viral species, many of which are transmitted by arthropod vectors. Flaviviruses cause a variety of human diseases, such as encephalitis, and hemorrhagic fevers. Included in this family are major global pathogens such as dengue virus, Japanese encephalitis virus, and West Nile virus. Yellow fever virus vaccine was the first live, attenuated viral vaccine.

Figure 9 Structure and genome organization of flaviviruses. (A) Virion structure. The cryo-electron micrograph reconstruction of the flavivirus, dengue virus (50-nm) particles. (Image courtesy of Dr. Richard J. Kuhn and Valorie Bowman, Department of Biological Sciences, Purdue University.) **(B) Genome organization.** The (+) strand RNA genome is from 9.6 kb (*Hepacivirus* genus) to 12.3 kb (*Pestivirus* genus) in length. The genome RNA has a 5′ cap structure (except for hepatitis C virus) but lacks a 3′ poly(A) characteristic of cellular and viral mRNAs. The viral RNA genome has 5′ and 3′ nocoding regions and encodes a polyprotein (~3,400 amino acids) that is processed by viral and cellular proteases to produce viral structural (C, M, E) and nonstructural proteins (NS1, NS2A, NS2B, NS3, NS4A, NS4B, NS5). Cleavage sites for host signal peptidase and a virus-encoded serine protease, NS2B-3, are shown.

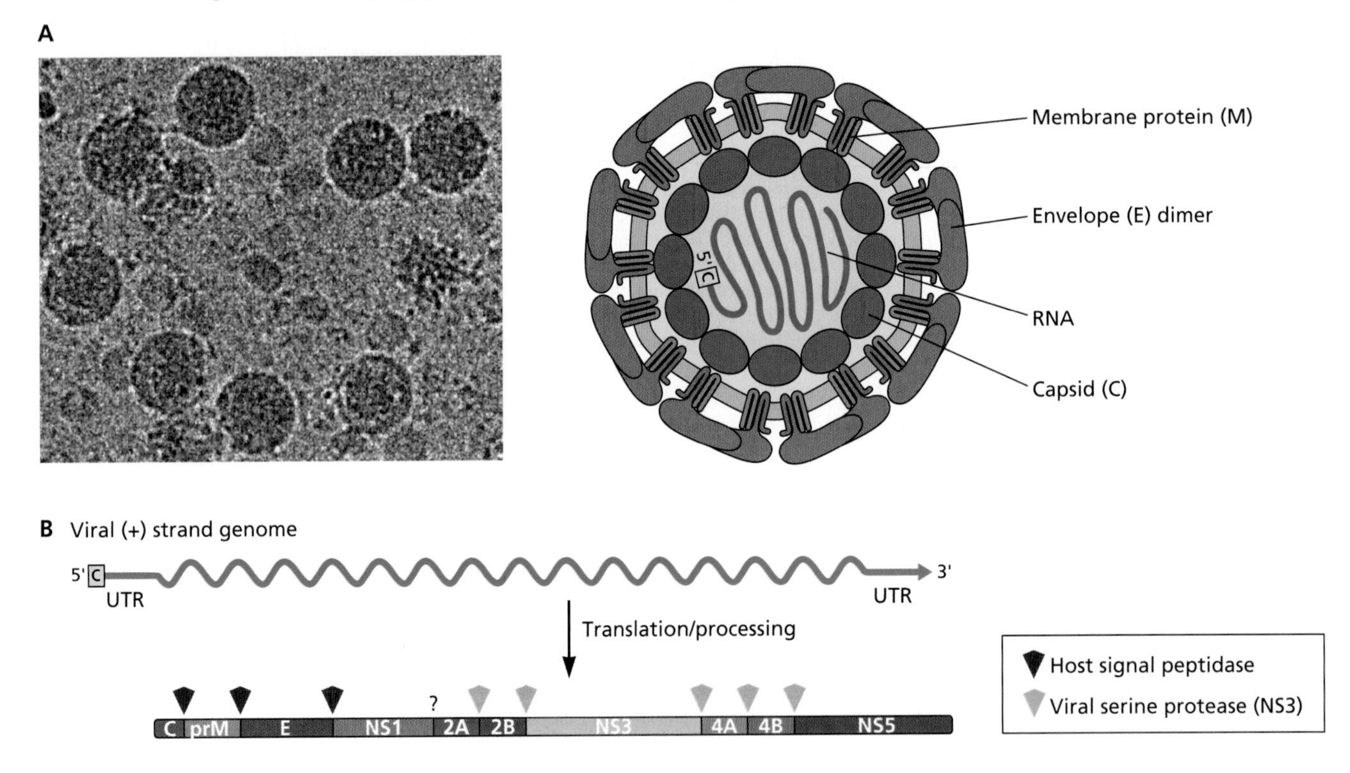

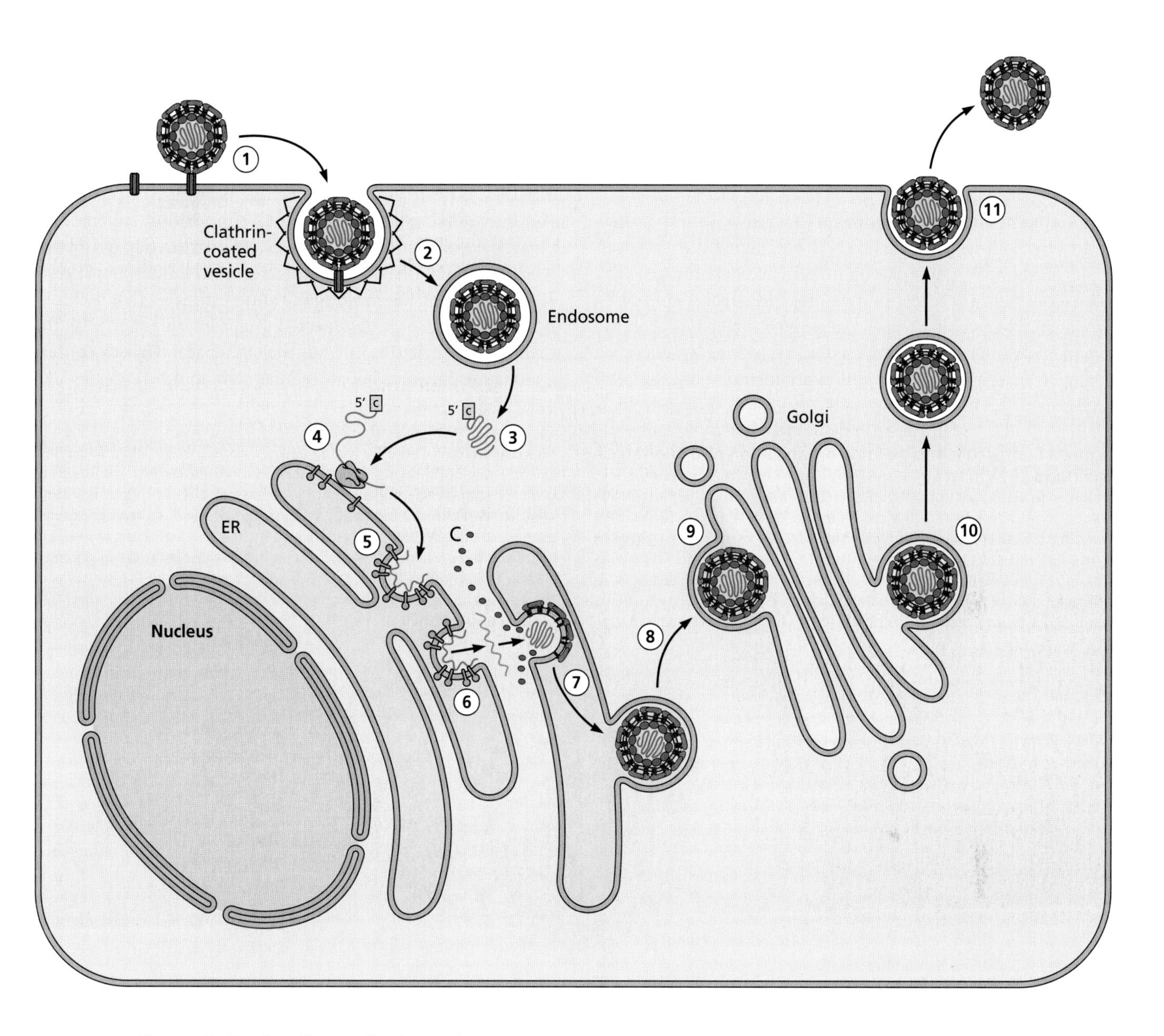

Figure 10 Single-cell reproductive cycle. **(1)** The virion binds to a cell surface receptor and **(2)** is taken into the cell by receptor-mediated endocytosis. **(3)** Fusion of the viral and cell membranes is triggered by low pH in the late endosome and the viral (+) strand RNA is released into the cytoplasm, **(4)** where it is associated with cellular membranes and translated into a polyprotein that is co- and posttranslationally cleaved into at least 10 proteins. **(5)** The viral NS proteins recruit the viral genome to a replication complex, which consists of ER-derived membrane vesicles that are invaginations of the ER that are open to the cytoplasm. **(6)** Replication begins with the synthesis of a genome-length (−) strand RNA, which is then copied to produce new (+) strand RNA genomes. These sites of RNA replication are near the sites of virus assembly; together both form a continuous network, possibly to couple replication and assembly. **(7)** The assembly process begins when C protein dimers associate with viral (+) strand RNA. This complex then buds into ER membranes containing the E-prM proteins. **(8)** The newly formed immature virus particles are transported to the cell surface by the secretory pathway. **(9)** During transport through the secretory pathway, particles undergo a series of maturation steps including glycosylation of prM and E, low-pH-induced rearrangement of E-prM, and prM cleavage. **(10)** Mature virus particles are transported to the cell surface in vesicles. **(11)** Particles are released from the cell surface by exocytosis.

Hepadnaviruses

Family *Hepadnaviridae*

Genera	Examples
Orthohepadnavirus	Human hepatitis B virus
Avihepadnavirus	Duck hepatitis B virus

The hepadnaviruses all show very narrow host specificity and marked tropism for liver tissue. Hepadnaviruses can replicate following inoculation of primary hepatocytes with virus-containing serum, but most hepadnaviruses cannot be propagated in established cell lines. Hepadnaviruses replicate via an RNA intermediate and, like the retroviruses, encode a reverse transcriptase. Both families are included in the group called **retroid viruses**. Natural infections may be acute or persistent, depending on host age, inoculum dose, and other (undefined) parameters that influence the host immune response. Sera of infected individuals typically carry numerous small round particles and some rodlike particles, both of which include viral surface antigens but lack a capsid and genome. Relatively few mature 42-nm virions, called Dane particles, are found in these sera. Approximately 5% of the world's population has been infected with human hepatitis B virus; the World Health Organization estimates that 400 million are now chronically infected. Persistent infection with the orthohepadnaviruses but not the avihepadnaviruses confers an increased risk for hepatocellular carcinoma.

Figure 11 Structure and genome organization of orthohepadnaviruses. (A) Virion structure. The electron micrograph shows negatively stained woodchuck hepatitis virus, a mammalian hepadnavirus related to human hepatitis B virus (courtesy of W. Mason and T. Gales, Fox Chase Cancer Center, Philadelphia, PA). **(B) Genome organization.** The relaxed circular DNA genome of human hepatitis B virus is shown at the center. It comprises a complete (−) strand of 3,227 nucleotides and an incomplete (+) strand, which is only about two-thirds genome length and can have variable 3′ ends. The viral reverse transcriptase protein (indicated by a blue ball) is covalently attached to a short (8- to 9-nucleotide) single-strand terminal redundancy at the 5′ end of the (−) strand. The approximate locations of direct repeats, DR1 and DR2, in the DNA are indicted. The locations of the 5′ ends of the pregenome and mRNAs that are synthesized by host cell Pol II are shown surrounding the genome, all of which end at the same location (marked by a "t" in the genome). The outermost, colored rings show the locations of the open reading frames, which are all organized in the same direction (clockwise in the figure).

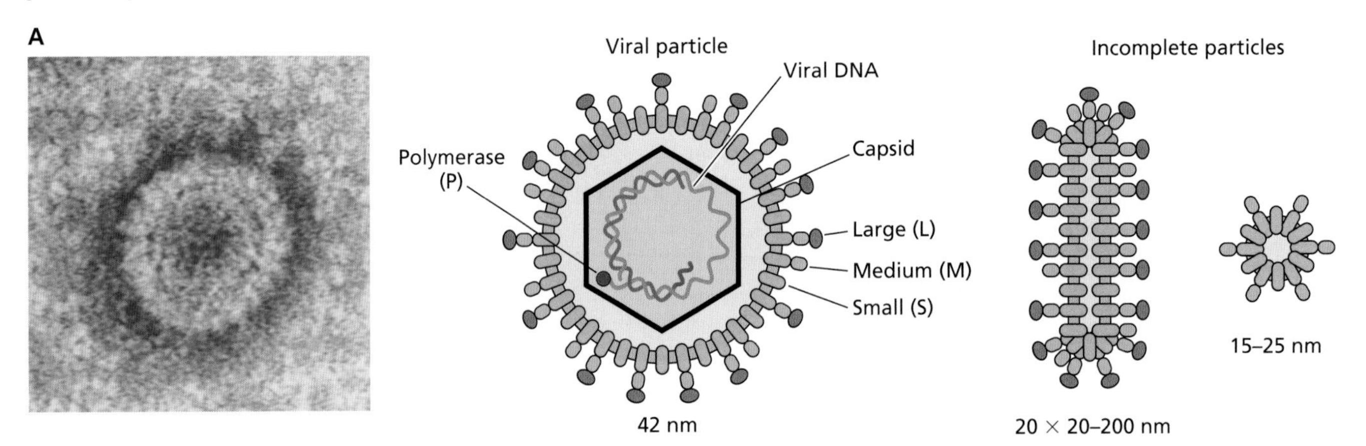

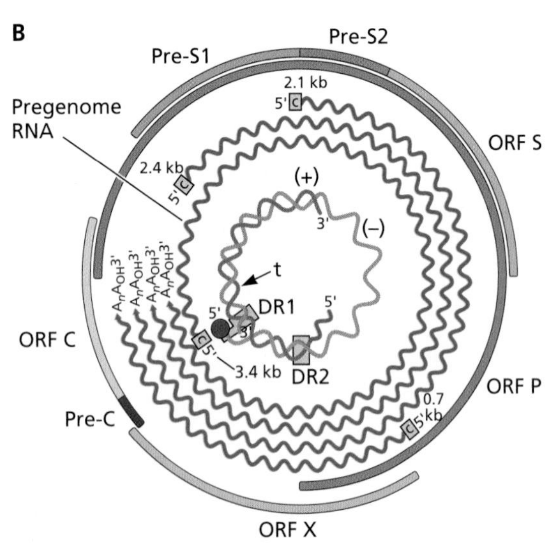

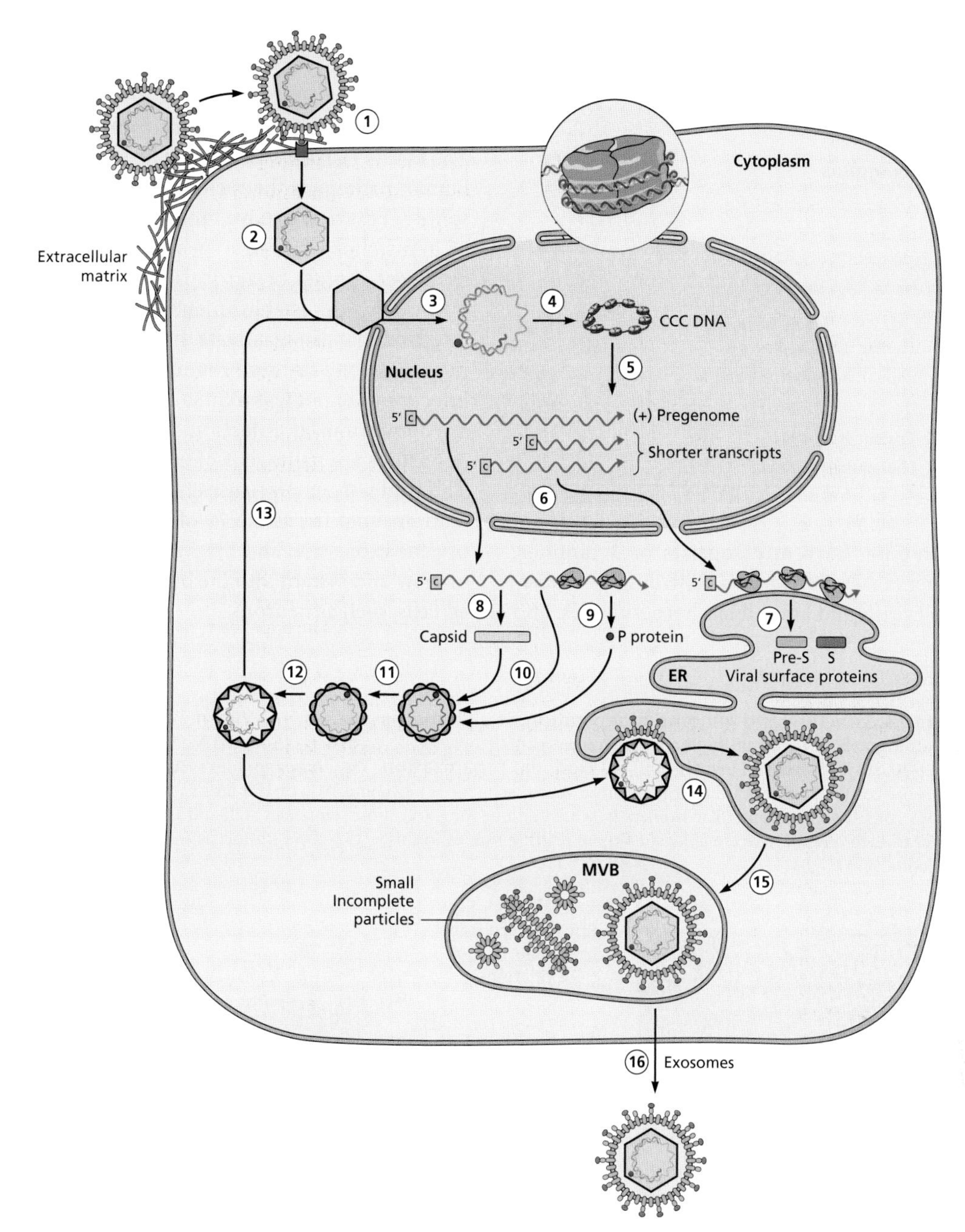

Figure 12 Single-cell reproductive cycle of hepatitis B virus. **(1)** The virion attaches to a susceptible hepatocyte, most likely via weak interaction with cell-associated heparin sulfate proteoglycans and then through recognition of a specific cell surface receptor, the sodium taurocholate-cotransporting polypeptide in human cells. **(2)** Details of entry and transport of the core (capsid encased genome) to the nucleus are unknown, although host cell calveolin-1 has been implicated in entry and the viral core is presumed to be transported along microtubules. **(3)** The core binds to proteins in the nuclear pore and the viral genome is released into the nucleus. **(4)** Repair of the gapped (+) DNA strand is likely accomplished by cellular enzymes. The product is a covalently closed circular form called CCC DNA, which associates with histones to form a minichromosome. **(5)** The (−) strand of CCC DNA is the template for transcription by cellular RNA polymerase II of a longer-than-genome-length RNA called the pregenome and shorter, subgenomic transcripts, all of which serve as mRNAs. **(6)** Viral mRNAs are transported from the nucleus. **(7)** Subgenomic viral mRNAs, which encode the viral envelope protein, are translated by ribosomes bound to the endoplasmic reticulum (ER). Proteins destined to become anchored in the viral envelope, as well as in incomplete particles, enter the secretory pathway. **(8)** The pregenome RNA is translated to produce capsid protein. **(9)** The P protein, the viral reverse transcriptase, is also produced from pregenome RNA but at low efficiency; the ratio of capsid to P protein translation is 200 to 300 to 1. Following its synthesis, P binds to the packaging signal at the 5′ end of its own transcript, where viral DNA synthesis is eventually initiated. **(10)** Concurrently with capsid formation, and aided by the host heat shock protein chaperones Hsp90/70, the RNA-P protein complex is packaged and DNA replication is primed from a tyrosine residue in the polymerase. **(11)** Reverse transcription of the pregenome occurs within the capsid. **(12)** After completion of DNA synthesis, the newly assembled "cores" acquire the ability to interact with envelope proteins. **(13)** However, at early times after infection, core particles are transported to the nucleus, where the viral genomes are deposited and give rise to additional copies of CCC DNA. Eventually, 10 to 30 molecules of CCC DNA accumulate, leading to a concomitant increase in viral mRNA concentrations. **(14)** At later times, and possibly as a consequence of the accumulation of sufficient envelope proteins, the core particles acquire envelopes as they bud into the ER, where viral surface proteins have been synthesized. **(15)** Viral assembly is believed to be completed in multivesicular bodies (MVB). **(16)** Progeny enveloped virus particles, and numerous small genome-lacking incomplete particles, are released from the cell by exocytosis.

Herpesviruses

Family *Herpesviridae*

Subfamilies and Selected Genera	Examples
Alphaherpesviruses	
Simplexvirus	Human herpes simplex virus type 1 and 2
Varicellovirus	Varicella-zoster virus
Betaherpesviruses	
Cytomegalovirus	Human cytomegalovirus
Roseolovirus	Human herpesvirus 6 and 7
Gammaherpesviruses	
Lymphocryptovirus	Epstein-Barr virus
Rhadinovirus	Human herpesvirus 8

The order *Herpesvirales* currently consists of 3 families, 3 subfamilies, 17 genera, and 90 species. The family *Alloherpesviridae* comprises fish and amphibian herpesviruses, and the family *Malacoherpesviridae* comprises viruses of oysters. The family *Herpesviridae*, listed here, includes the well-known human pathogens that belong to all three subfamilies. While some herpesviruses have broad host ranges, most are restricted to infection of a single species and spread in the population by direct contact or aerosols. The hallmark of herpesvirus infections is the establishment of a lifelong, latent or quiescent infection that can reactivate to spread to other hosts and often may cause one or more rounds of disease. Many herpesvirus infections are not apparent, but if the host's immune defenses are compromised, infections can be devastating. Some herpesviruses are pathogens of economically important animals. The study of herpesviruses has provided fundamental information about the assembly of complex virions, the regulation of gene expression and mechanisms of immune system modulation, and insight into the biology of terminally differentiated cells, such as neurons.

Figure 13 Structure and genome organization of alphaherpesviruses. (A) Virion structure. Cryo-electron tomograph of a slice through a single herpes simplex virus type 1 particle. (Adapted from E. Grunwald et al., *Science* **302:**1396–1398, with permission.) **(B) Genome organization.** The herpes simplex virus type 1 genome can "isomerize" or recombine via the large inverted repeat sequences (TRL and IRL, or IRS and TRS) such that all populations consist of four equimolar isomers in which unique long and short sequences (UL and US) are inverted with respect to each other. There are at least 84 open reading frames in this ~152-kbp genome, as well as three origins of replication (Ori).

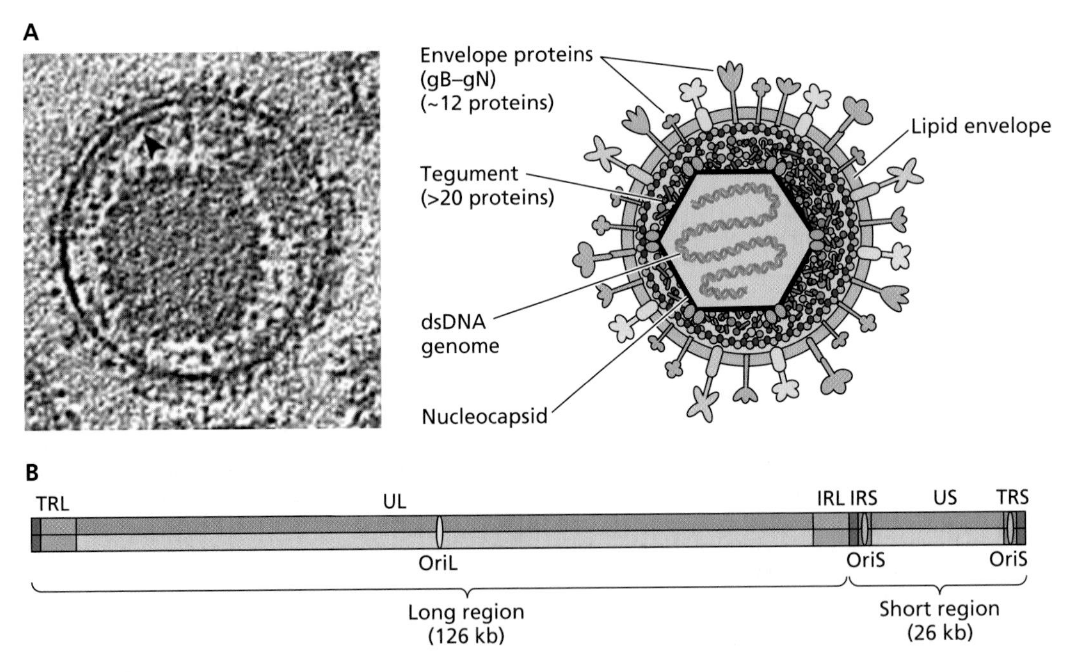

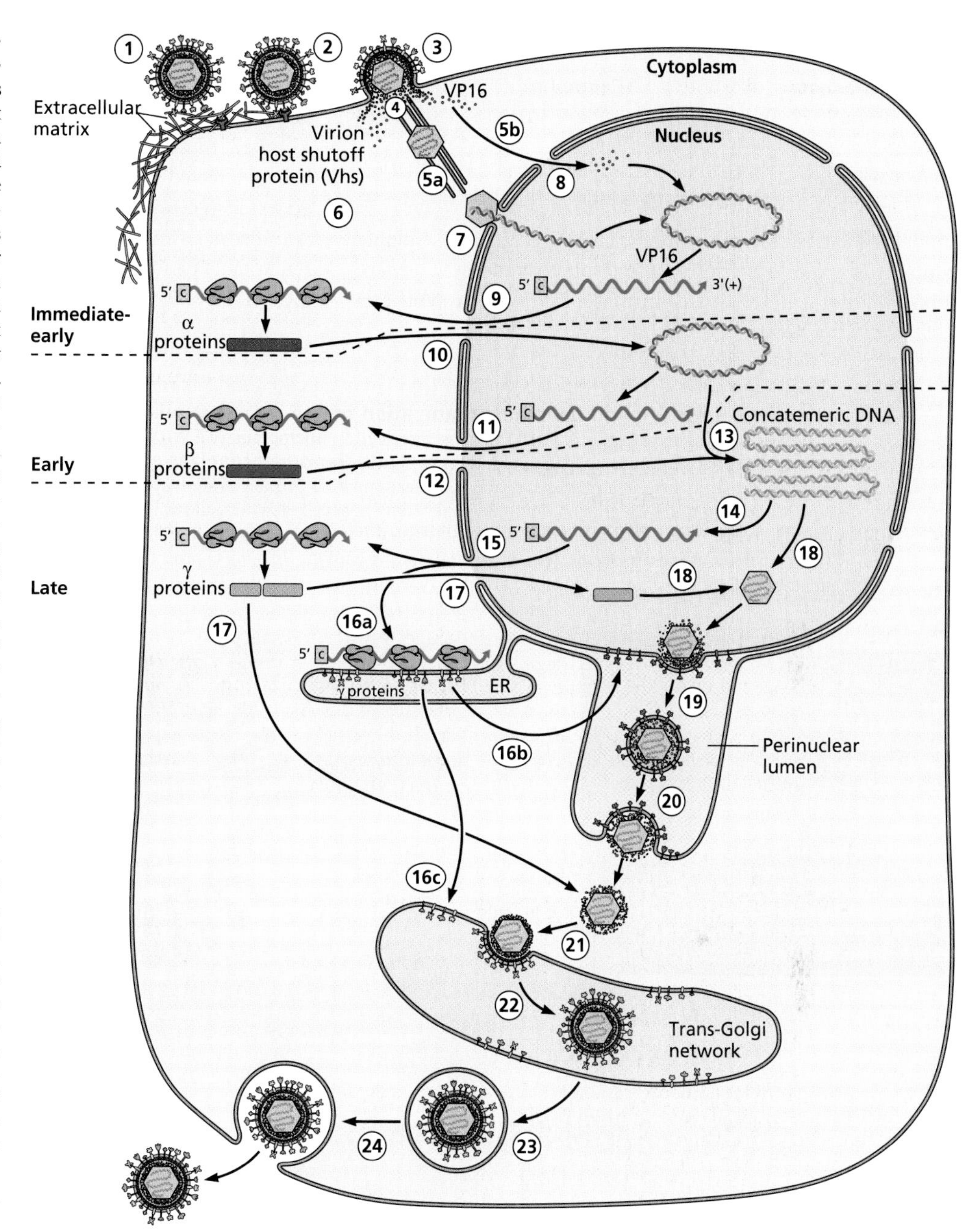

Figure 14 Single-cell reproductive cycle of herpes simplex virus type 1. **(1)** Virions bind to the extracellular matrix (heparan sulfate or chondroitin sulfate proteoglycans) via gB and gC. **(2)** Another viral membrane protein (gD) interacts with a second cellular receptor (such as nectin-1). **(3)** Particles can enter cells via a pH-independent fusion of viral envelope with the plasma membrane or alternatively (not shown) via an endocytic pathway that is similar to phagocytosis. Viral and plasma membrane fusion is mediated by viral membrane glycoproteins (gD, gB, gH, and gL). **(4)** After membrane fusion, some tegument proteins and the nucleocapsid are released into the cytoplasm. **(5a)** Viral nucleocapsids with associated inner tegument proteins attach to microtubules and are transported to the nucleus. **(5b)** Certain tegument proteins are transported to the nucleus independently of the nucleocapsids. **(6)** Other proteins, such as Vhs, remain in the cytoplasm. **(7)** Viral nucleocapsids dock at the nuclear pore, releasing DNA into the nucleus, where it is rapidly circularized. **(8)** VP16 interacts with host transcription proteins to stimulate transcription of immediate early genes by host cell RNA polymerase II. **(9)** Some immediate early mRNAs are spliced and all are transported to the cytoplasm, where they are translated. **(10)** The immediate early proteins (α proteins) are transported to the nucleus, where they activate transcription of early genes and regulate transcription of immediate early genes. **(11)** Early gene transcripts, which are rarely spliced, are transported to the cytoplasm, where they are translated. The early proteins (β proteins) function primarily in DNA replication and production of substrates for DNA synthesis. **(12)** Some early proteins function in the cytoplasm, and some are transported to the nucleus. **(13)** Viral DNA synthesis is initiated from viral origins of replication. **(14)** DNA replication and recombination produce long, concatemeric DNA, the template for late gene expression. **(15)** Most late mRNAs are not spliced but nevertheless are transported to the cytoplasm, where they are translated. Late proteins (γ proteins) are primarily structural proteins and additional proteins needed for virus assembly and particle egress. **(16a)** Some late proteins are made on, and inserted into, membranes of the rough endoplasmic reticulum. **(16b)** Many of these membrane proteins are modified by glycosylation. Some precursor viral membrane proteins are localized both to the outer and inner nuclear membranes, as well as membranes of the endoplasmic reticulum. **(16c)** The precursor glycoproteins are also transported to the Golgi apparatus for further modification and processing. **(17)** Some late proteins are transported to the nucleus for assembly of the nucleocapsid and DNA cleavage to release genomes concomitant with packaging, while some remain in the cytoplasm. **(18)** Newly replicated viral DNA is packaged into nucleocapsids. **(19)** DNA-containing nucleocapsids, together with some tegument proteins, bud from the inner nuclear membrane into the perinuclear lumen, acquiring an envelope thought to contain precursors to viral membrane proteins. **(20)** Immature enveloped particles fuse with the outer nuclear membrane from within, releasing the nucleocapsid into the cytoplasm. **(21)** This structure is transported to the *trans*-Golgi network or an endosome that contains mature viral membrane proteins. Tegument proteins added in the nucleus remain with the nucleocapsid, and others are added in the cytoplasm. **(22)** As nucleocapsids bud into the Golgi or endosome compartment, they acquire an envelope containing mature viral envelope proteins and the complete tegument layer (secondary envelopment). **(23)** The enveloped virus particle then buds into a vesicle that is transported to the plasma membrane for **(24)** release by exocytosis.

Orthomyxoviruses

Family *Orthomyxoviridae*

Selected Genera	Examples
Influenzavirus A	A/PR/8/34(H1N1)
Influenzavirus B	B/Lee/40
Influenzavirus C	C/California/78

Influenza viruses of humans are the causative agents of a highly contagious and often serious acute respiratory illness. They are unusual among RNA viruses in that all viral RNA synthesis occurs in the cell nucleus. Initiation of viral mRNA synthesis with a capped primer derived from host cell mRNA was first observed in cells infected with influenza viruses. The viral genomes undergo extensive reassortment when a host cell is infected with two distinct strains, and coding sequences are expressed via a remarkable panoply of unusual strategies, including RNA splicing, overlapping reading frames, and leaky scanning.

Figure 15 Structure and genomic organization of the orthomyxovirus influenza A virus. (A) Virion structure. Colorized negative stained transmission electron micrograph (TEM) of the influenza virus A/CA/4/09 (courtesy of CDC Public Health Image Library, ID#11214). **(B) Genome organization.** The (−) strand RNA genome comprises eight segments, each of which encodes at least one viral protein as shown. Some of the (+) strand mRNA of the smallest genomic RNA segments, 7 and 8, is spliced by host cell enzymes, allowing the production of two proteins from each. The NS (nonstructural) proteins were so named because they were thought initially not to be incorporated into virus particles. An accessory protein with proapoptotic activity, PB1-F2, is produced from the PB1 RNA by translation of an overlapping open reading frame. The PB1 N40 protein is translated from a third open reading frame in PB1 RNA.

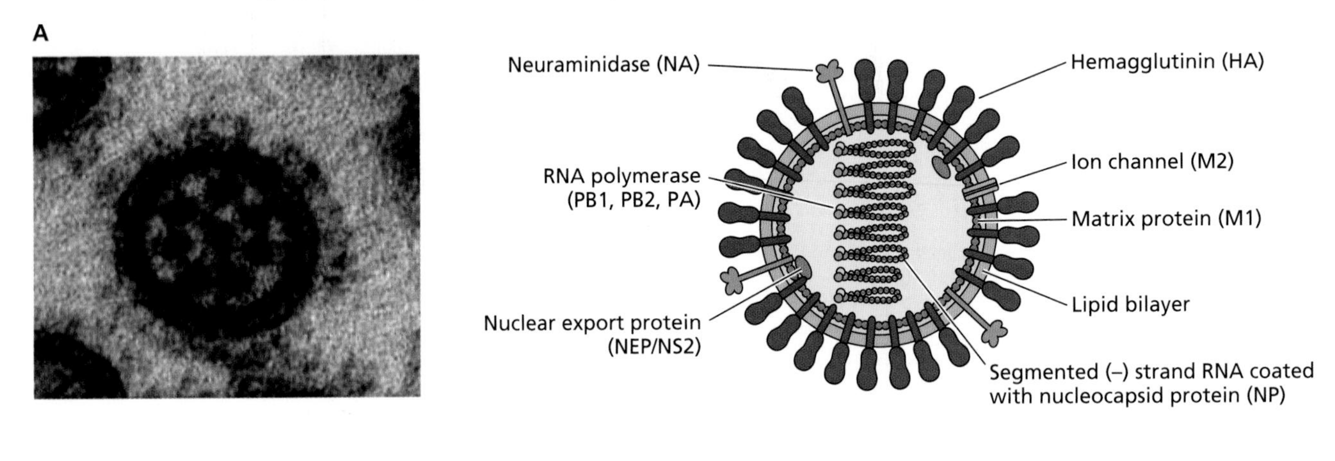

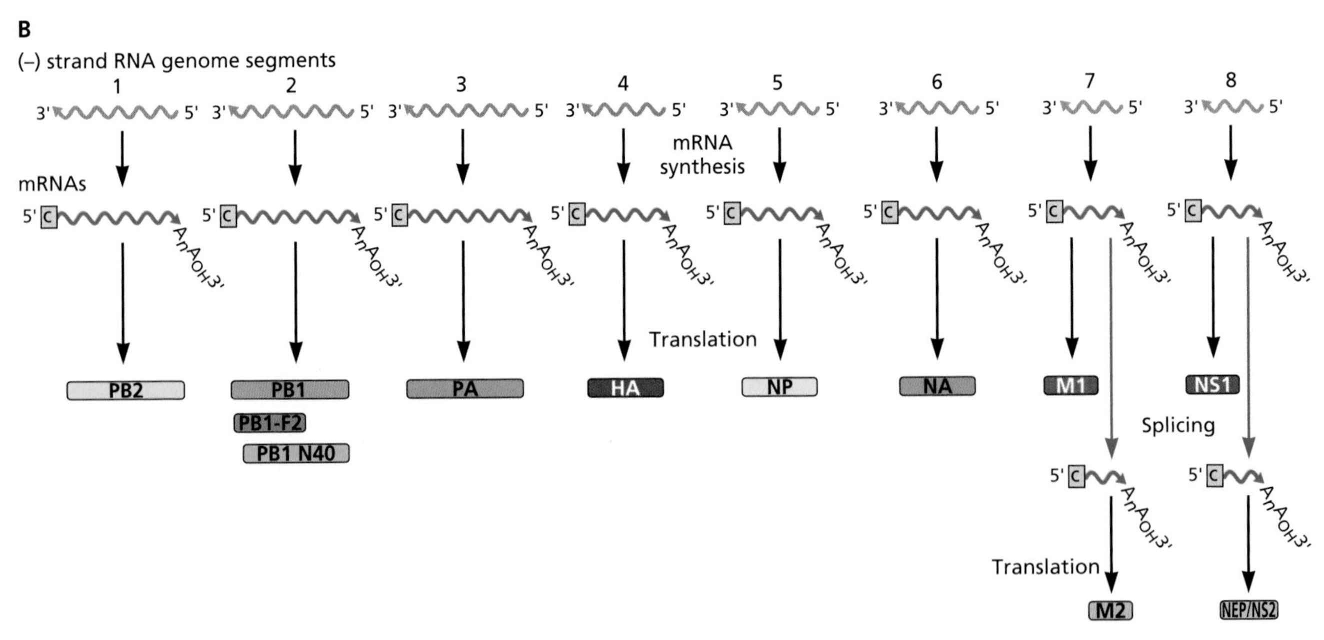

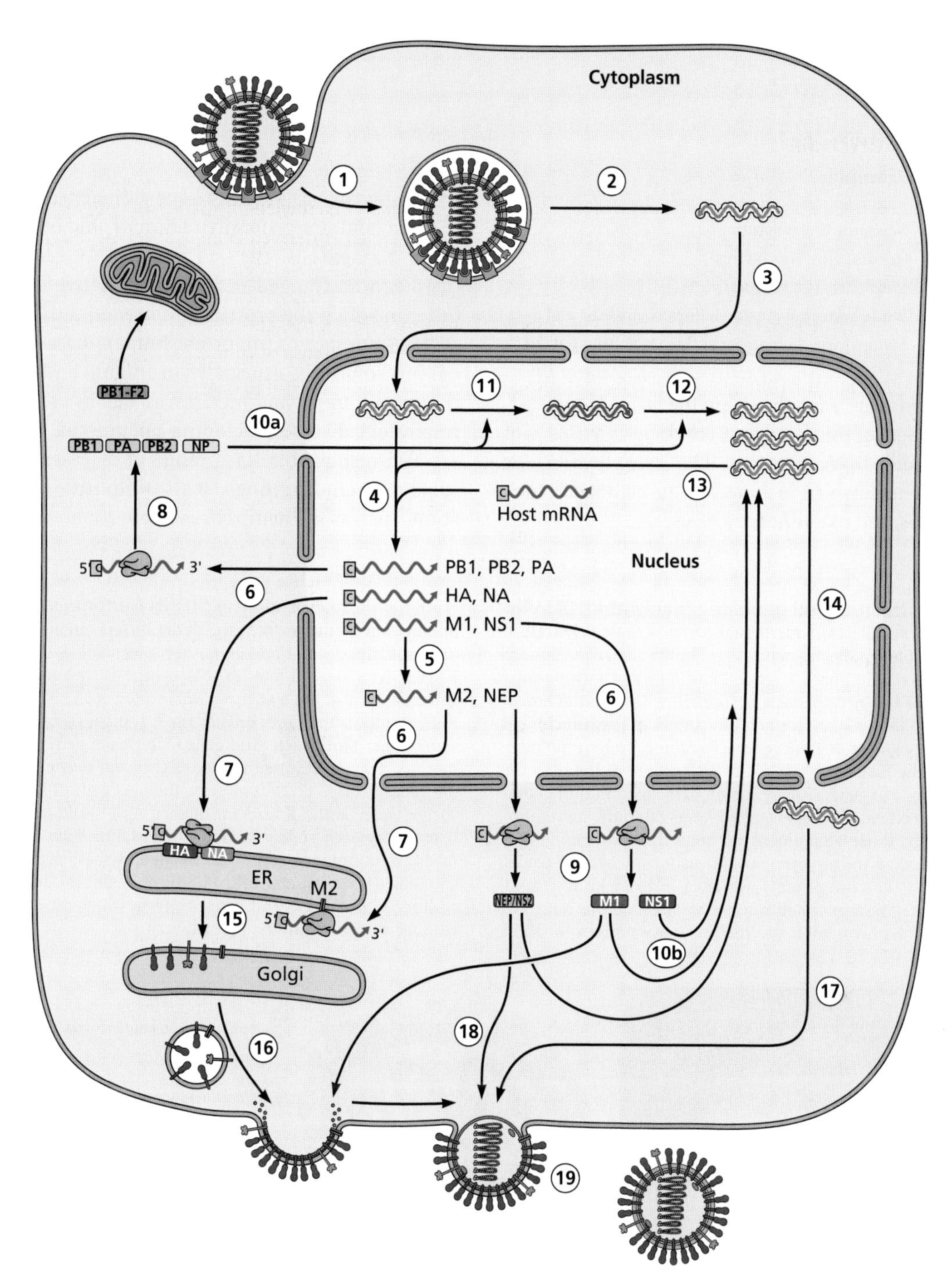

Figure 16 Single-cell reproductive cycle of influenza A virus. **(1)** The virion binds to a sialic acid-containing cellular surface protein or lipid and enters the cell via receptor-mediated endocytosis. **(2)** Upon acidification of the vesicle, the viral membrane fuses with the membrane of the vesicle, releasing the eight viral nucleocapsids into the cytoplasm (for simplicity, only one is shown). **(3)** The viral nucleocapsids containing (−) strand genomic RNA, multiple copies of the NP protein, and the P proteins are transported into the nucleus. **(4)** The (−) strand RNAs are copied by RNA polymerase entering with the virus particles into mRNAs, using the capped 5′ ends of host pre-mRNAs (or mRNAs) as primers to initiate synthesis. **(5)** Some of the mRNA encoding NS2/NEP and M2 is spliced, and **(6)** the mRNAs are transported to the cytoplasm. **(7)** The mRNAs specifying the viral membrane proteins (HA, NA, and M2) are translated by ribosomes bound to the endoplasmic reticulum (ER). These proteins enter the host cell's secretory pathway, where HA and NA are glycosylated. **(8, 9)** All other mRNAs are translated by ribosomes in the cytoplasm. **(10a)** The PA, PB1, PB2, and NP proteins are imported into the nucleus, where they participate in the synthesis of **(11)** full-length (+) strand RNAs and then of **(12)** (−) strand genomic RNAs, both of which are synthesized in the form of nucleocapsids. **(13)** Some of the newly synthesized (−) strand RNAs enter the pathway for mRNA synthesis. **(10b)** The M1 protein and the NS1 protein are transported into the nucleus. **(14)** Binding of the M1 protein to newly synthesized (−) strand RNAs shuts down viral mRNA synthesis and, in conjunction with the NS2/NEP protein, induces export of progeny nucleocapsids to the cytoplasm. **(15)** The HA, NA, and M2 proteins are transported to the cell surface and **(16)** become incorporated into the plasma membrane. **(17, 18)** The nucleocapsids associated with the M1 protein and the NS2/NEP protein are transported to the cell surface and interact with regions of the plasma membrane that contain the HA, NA, M1, and M2 proteins. **(19)** Assembly of virus particles is completed at this location by budding from the plasma membrane.

Paramyxoviruses

Family *Paramyxoviridae*

Selected Genera	Examples
Avulavirus	Newcastle disease virus
Henipavirus	Hendra virus, Nipah virus
Morbillivirus	Measles virus, rinderpest virus
Respirovirus	Sendai virus, human parainfluenza viruses 1 and 3
Rubulavirus	Mumps virus, human parainfluenza viruses 2 and 4

The *Paramyxoviridae* are a family within the order *Mononegavirales*. The members of this family are enveloped viruses with (−) single-stranded RNA genomes. The *Paramyxoviridae* comprise eight genera, which include human pathogens such as measles, mumps, and parainfluenza viruses.

Paramyxovirus particles have a diameter of approximately 150 nm, and are generally pleiomorphic but can be spherical or filamentous. The (−) strand RNA genome contains 6 to 10 genes arranged in the same relative order, and in the order in which the proteins are needed during the infectious cycle. A number of important human diseases are caused by paramyxoviruses, especially in infants and children. These comprise mumps and measles, as well as others that result in respiratory diseases, including pneumonia. Paramyxoviruses are also responsible for a range of diseases in other animal species, including dogs, seals, dolphins, birds, and cattle. Some, such as the henipaviruses, are zoonotic pathogens.

Figure 17 Virion structure and genome organization. (A) Virion structure. The cryo-electron micrograph shows a negatively stained paramyxovirus. The pleiomorphic particles are ~120 to 150 nm in diameter. (Courtesy of Linda Stannard/Science Photo Library, with permission.) The surface of the virus particle is studded with the attachment protein that binds to the cellular receptor. The nature of this attachment protein differs somewhat among the genera. For some, such as the morbilliviruses, this is the hemagglutinin (H). Others, such as those in the *Rubula*- and *Respirovirus* genera, possess both hemagglutination activity and the ability to cleave sialic acid (called hemagglutinin-neuraminidase [HN]). Finally, those attachment proteins that possess neither activity are simply called the glycoprotein (G), as for the henipaviruses. The RNP consists of the viral genome, which is 15 to 19 kb in length, wrapped around the virus-encoded nucleoprotein (N), the large (L) RNA-dependent RNA polymerase, and the accessory protein for RNA synthesis, the phosphoprotein (P). **(B) Genome structure.** While the number and names of the viral genes differs among the paramyxovirus genera, the order of these genes is constant. The viral RNA-dependent RNA polymerase initiates mRNA synthesis by binding to the encapsidated genome at the leader region, located at the 3′ end of the genome. RNA synthesis then proceeds as the L protein recognizes start and stop signals that flank each viral gene. After each gene is copied, the polymerase pauses to release the new mRNA and may either dissociate from the genome or go on to transcribe the next gene. If L dissociates, it must "begin again" at the 3′ leader sequence. As a result, sequentially less RNA is made for each gene as a factor of distance from the 3′ end. All viral mRNAs are capped and polyadenylated by the L protein during synthesis. Leaky scanning and mRNA editing result in the translation of two additional proteins, C and V, respectively, which are encoded in alternate reading frames within the P gene.

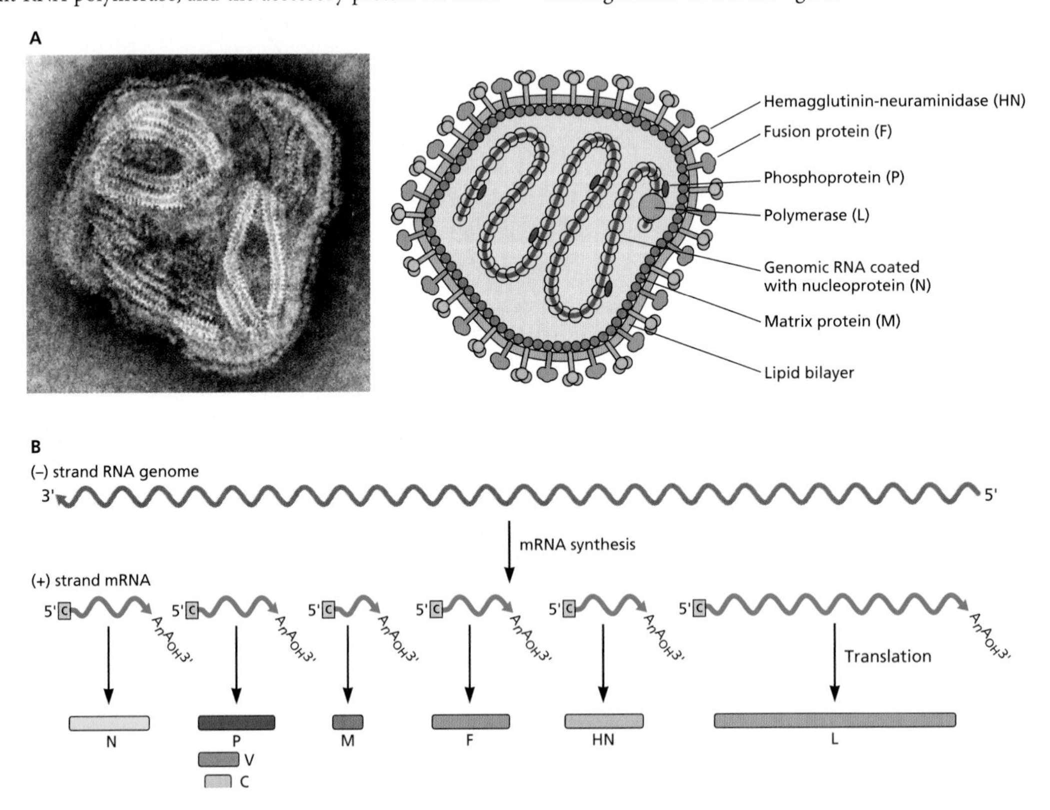

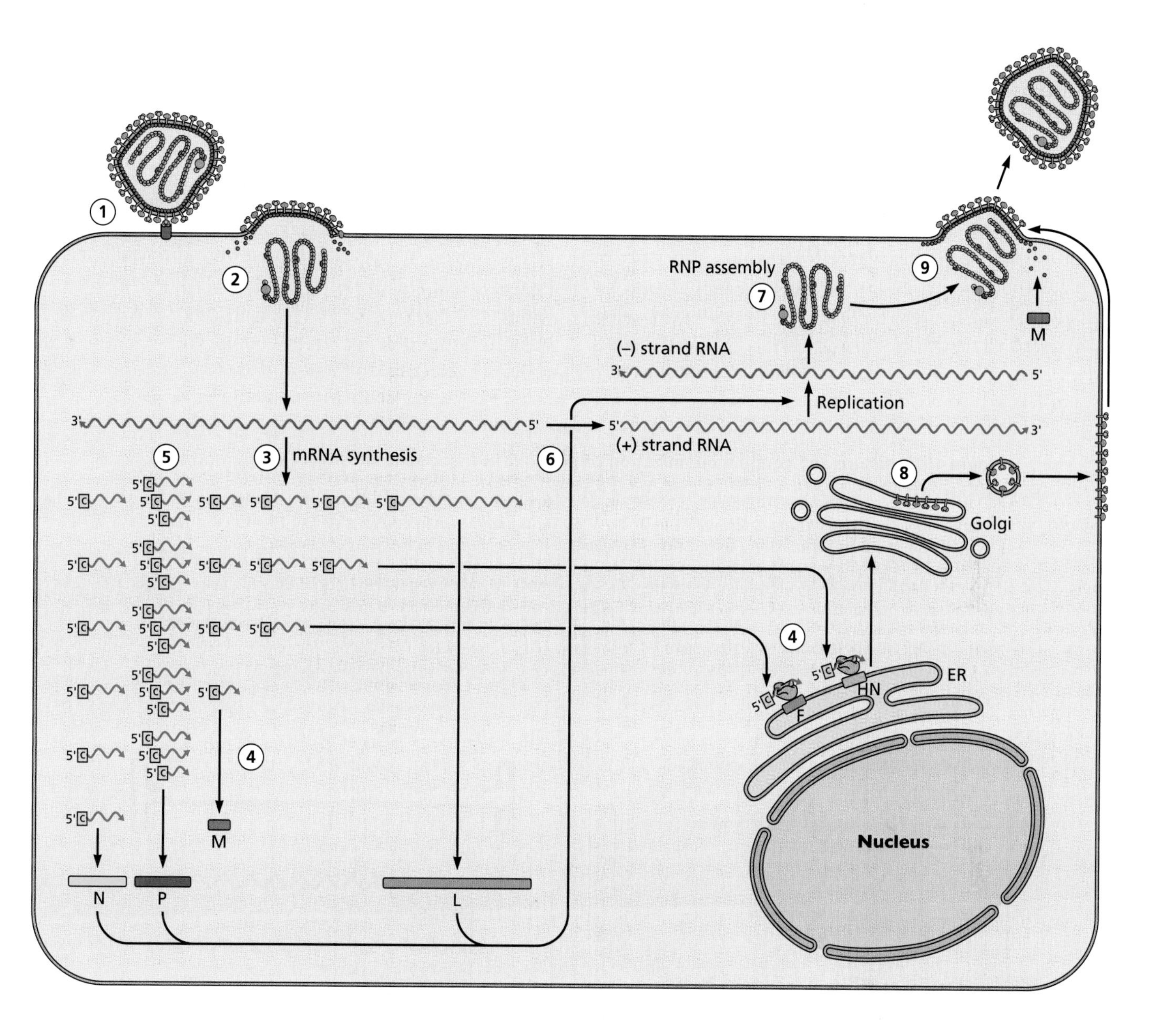

Figure 18 Single-cell reproductive cycle. **(1)** The virion attaches by binding to specific receptors on the surface of the cell. The identities of these receptors are known, and they vary among different paramyxoviruses; multiple receptors may be used by individual paramyxoviruses. **(2)** Upon binding of the virus to the receptor, the viral membrane fuses with the cellular membrane, releasing the single-stranded, negative-sense genome into the cytoplasm. **(3)** mRNAs are synthesized from this (−) strand RNA template. Separating each gene are transcription termination signals; consequently, the viral RNA-dependent RNA polymerase must re-engage with genomic RNA to continue transcription of the next downstream gene. As a result, a gradient of mRNA abundance is established, decreasing with each successive gene. **(4)** The capped mRNAs are translated. **(5)** Multiple proteins are made from the coding sequence of the P gene (Fig. 17B). **(6)** The N, P, and L proteins drive the replication of the incoming genome to produce a full-length (+) strand RNA, which then is the template for production of progeny viral genomes. **(7)** The ribonucleoprotein assembles in the cytoplasm when free N subunits associate with the genome to form a helical structure. **(8)** Viral glycoproteins are modified posttranslationally as they are transported through the endoplasmic reticulum and the Golgi network to the surface of the surface of the infected cell. **(9)** The RNP acquires its envelope at the cell surface as it buds through the plasma membrane; the viral M protein is thought to mediate association of the RNP with the viral glycoproteins.

Parvoviruses

Family *Parvoviridae*

Selected Genera	Examples
Parvovirus	Minute virus of mice
Erythrovirus	Human B19 virus
Dependovirus	Human adenovirus-associated viruses

Members of the family *Parvoviridae* are among the smallest of the animal viruses with DNA genomes. They are of particular interest because of the unique structure of their genomic DNA and its mechanism of replication. Most parvoviruses, such as the well-studied minute virus of mice, can reproduce autonomously, although they require the host cell to go through S phase in order to do so. Reproduction of dependoviruses requires a helper adenovirus or herpesvirus to induce S phase and to provide components that promote dependovirus gene expression and replication. These viruses can establish a latent infection during which their DNA is integrated into the host cell genome in an inactive state, to be activated upon subsequent infection with a helper. Because of their ability to persist and lack of pathogenicity, human adenovirus-associated viruses have been developed as vectors for gene therapy.

Figure 19 Structure and genome organization of adenovirus-associated virus (AAV). (A) Virion structure. The electron micrograph shows AAV4 (courtesy of Mavis Agbandje-McKenna, University of Florida, Gainesville, FL). Shown schematically to the right, the capsid comprises 60 protein subunits, primarily (~90%) VP3, which contains the same sequences as the C termini of VP1 and VP2. Virus particles contain either (+) or (−) single-stranded DNA. **(B) Genome organization.** The best-characterized DNA genome, that of AAV2, comprises ca. 4,600 nucleotides and includes terminal repeats (TR) of 145 nucleotides, the first 125 of which contain palindromic sequences. The TR is required in *cis* for genome replication, transcription, and encapsidation, and plays a role in integration into the host DNA during establishment of a latent infection. Use of multiple initiation codons and alternative splicing results in synthesis of multiple Rep (tan bars) and structural proteins (purple bars), respectively. ORF, open reading frame. (Adapted from R. M. Linden and K. Berns, p. 68–84, *in* S. Faisst and J. Rommelaere [ed.], *Contributions to Microbiology*, vol. 4, *Parvoviruses: from Molecular Biology to Pathology and Therapeutic Uses* [S. Karger, Basel, Switzerland, 2000] with permission.)

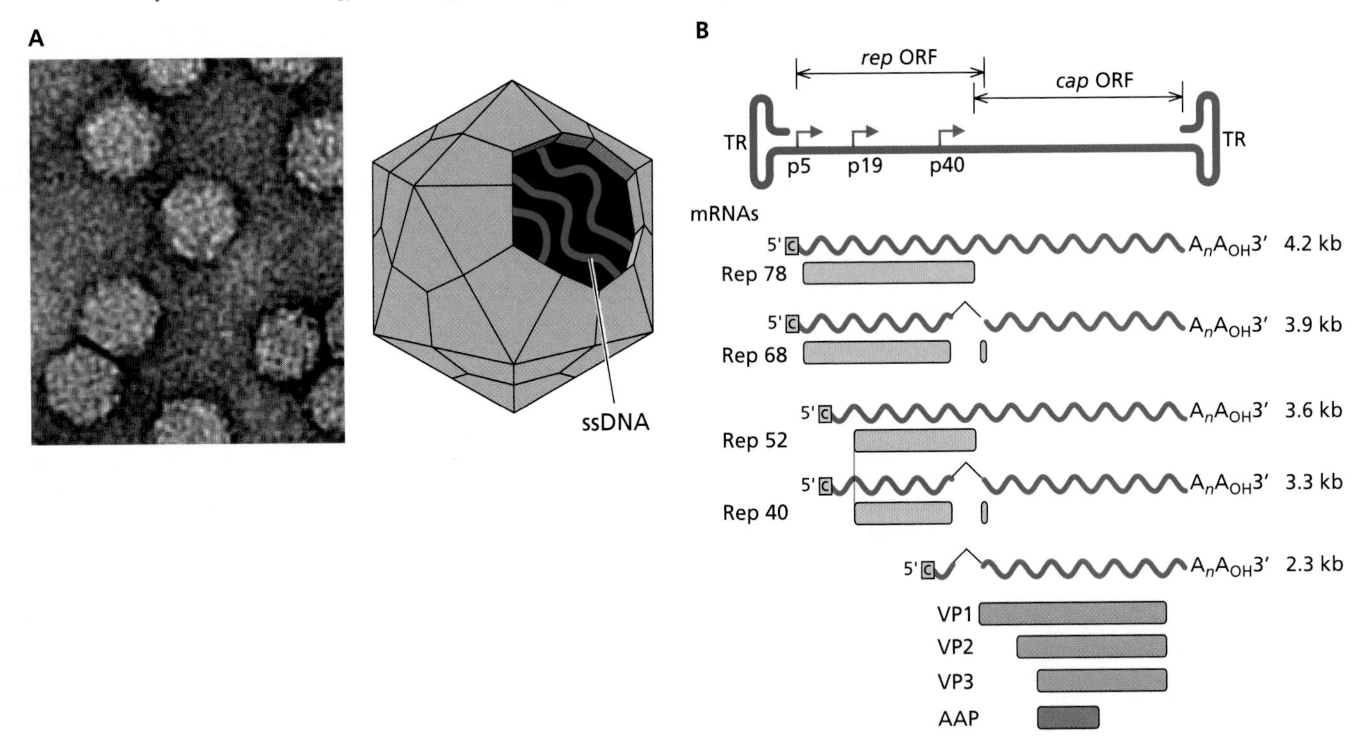

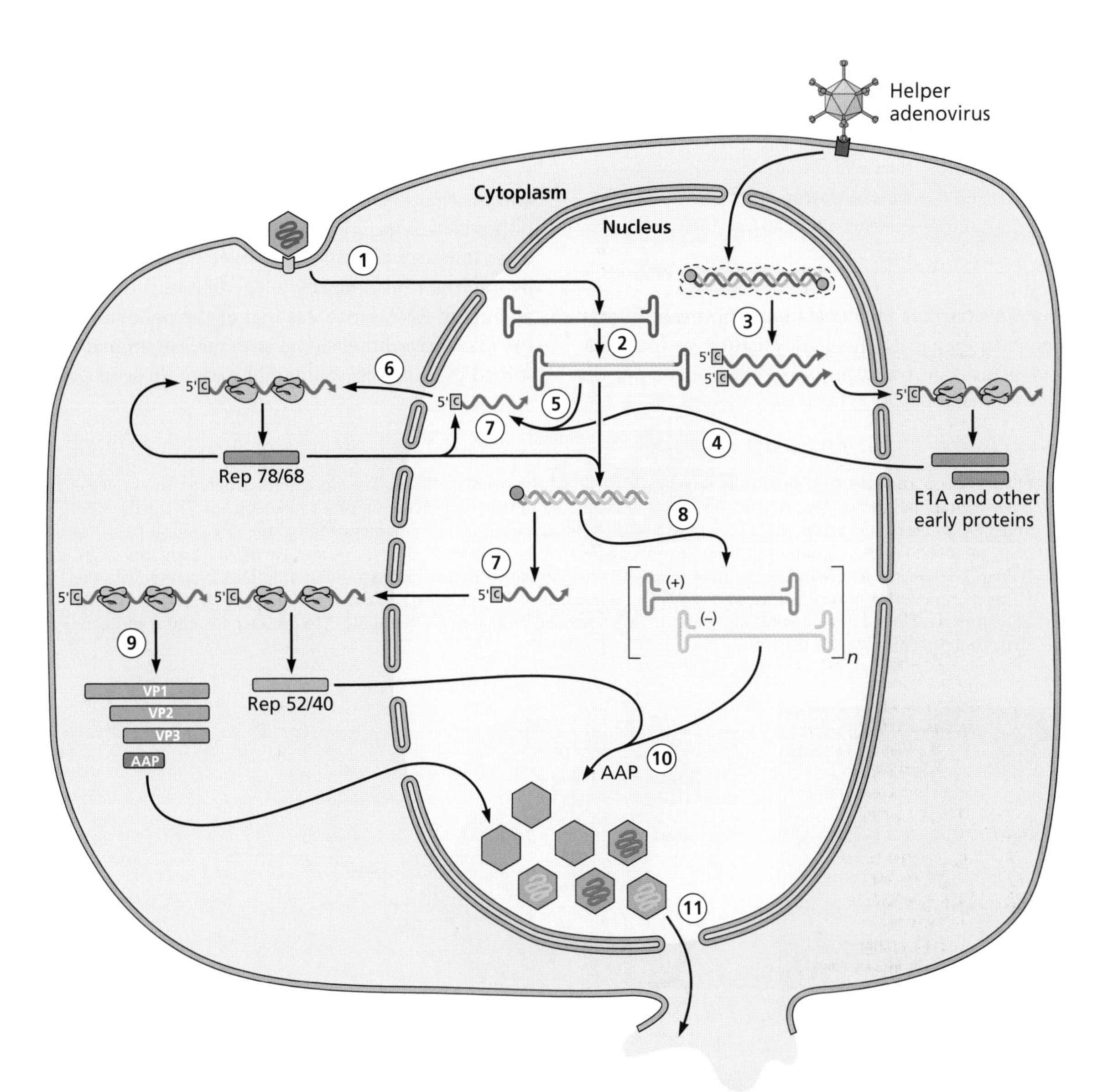

Figure 20 Single-cell reproductive cycle. Heparan sulfate proteoglycans are the primary cell surface receptors for AAV2. **(1)** However, the processes of adsorption, uncoating, and entry of the DNA into the nucleus are poorly understood for all *Parvoviridae*. **(2)** Cellular replication proteins convert the single-stranded viral DNA genome into a largely double-stranded molecule. **(3)** Upon coinfection with a helper virus, AAV undergoes a productive infection. With an adenovirus helper, this response is dependent on the expression of early genes E1A, E1B, E4, and E2A, which induce S phase and the concomitant production of cellular DNA replication proteins needed for viral DNA synthesis. **(4)** The adenovirus EIA transcriptional activator also induces transcription from the p5 promoter, **(5, 6)** leading to the production of Rep78/68 mRNA and proteins. **(7)** These proteins then function as powerful transcriptional activators (rather than repressors as in latency) and induce transcription from both the p5 and p19 promoters. **(8)** Viral DNA is replicated by a single-strand displacement mechanism that is initiated by recognition of the terminal resolution site (*trs*) by the Rep78/68 proteins, which remain linked covalently to the DNA through subsequent steps of DNA synthesis. A very large number of replicating forms (ca. 106 double-stranded genomes/cell) can be produced within a short time. **(9)** The capsid proteins produced in the cytoplasm self-associate in the nucleus during assembly of progeny particles. **(10)** Newly synthesized viral genomes are then encapsidated. The (+) or (−) strand genomes are encapsidated in equal numbers in progeny virus particles. **(11)** As with the adenovirus helper, progeny virus particles are released, usually upon destruction of the cell.

Picornaviruses

Family *Picornaviridae*

Selected Genera	Examples
Enterovirus	Poliovirus
Rhinovirus	Human rhinovirus A
Cardiovirus	Encephalomyocarditis virus
Aphthovirus	Foot-and-mouth disease virus
Hepatovirus	Hepatitis A virus

The family *Picornaviridae* includes many important human and animal pathogens. Because they cause serious disease, poliovirus and foot-and-mouth disease virus are the best-studied picornaviruses. These two viruses have had important roles in the development of virology. The first animal virus discovered, in 1898, was foot-and-mouth disease virus. The plaque assay was developed using poliovirus, and the first RNA-dependent RNA polymerase identified was poliovirus 3Dpol. Polyprotein synthesis was discovered in experiments with poliovirus-infected cells, as was translation by internal ribosome entry. The first infectious DNA clone of an animal RNA virus was that of the poliovirus genome, and the first three-dimensional structures of animal viruses determined by X-ray crystallography were those of poliovirus and rhinovirus.

Figure 21 Structure and genomic organization. (A) Virion structure. The electron micrograph shows negatively stained poliovirus. The capsid consists of 60 structural units (each made up of a single copy of VP1, VP2, VP3, and VP4, colored blue, green, red, and yellow, respectively) arranged in 12 pentamers. One of the icosahedral faces has been removed in the diagram to illustrate the locations of VP4 and the viral RNA. (Courtesy of N. Cheng and D. M. Belnap, National Institutes of Health, Bethesda, MD.) **(B) Genome organization.** Polioviral RNA is shown with the VPg protein covalently attached to the 5′ end. The genome is of (+) polarity and encodes a polyprotein precursor. The polyprotein is cleaved during translation by two virus-encoded proteases, 2Apro and 3Cpro, to produce structural and nonstructural proteins, as indicated.

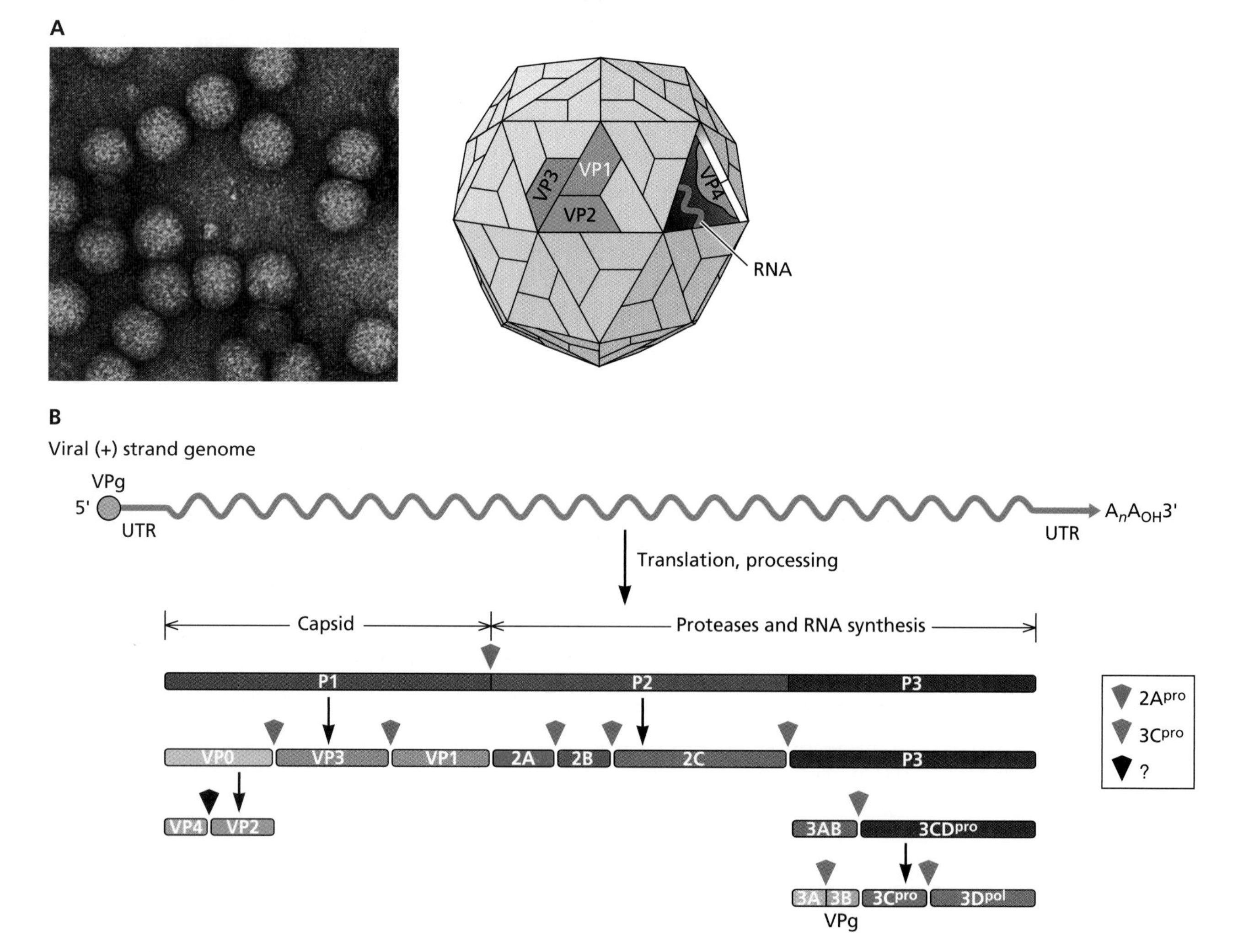

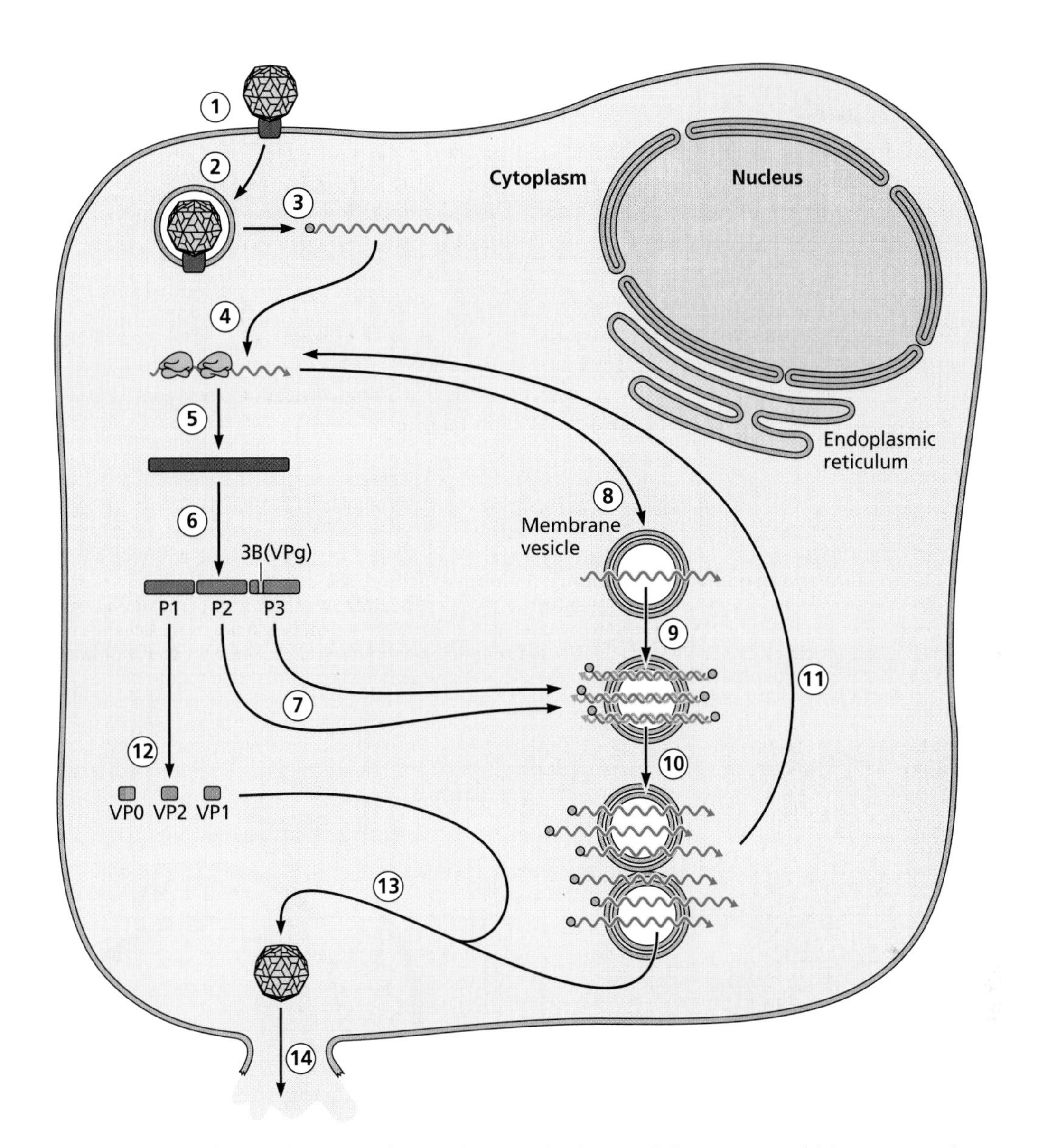

Figure 22 Single-cell reproductive cycle. **(1)** The virion binds to a cellular receptor and **(2)** enters an endosome. **(3)** Release of the poliovirus genome occurs from within early endosomes located close to the plasma membrane (within 100 to 200 nm). **(4)** The VPg protein, depicted as a small orange circle at the 5′ end of the virion RNA, is removed, and the RNA associates with ribosomes. **(5)** Translation is initiated at an internal site 741 nucleotides from the 5′ end of the viral mRNA, and a polyprotein precursor is synthesized. **(6)** The polyprotein is cleaved during and after its synthesis to yield the individual viral proteins. Only the initial cleavages are shown here. **(7)** The proteins that participate in viral RNA synthesis are transported to membrane vesicles. RNA synthesis occurs on the surfaces of these infected-cell-specific membrane vesicles. **(8)** The (+) strand RNA is transported to these membrane vesicles, **(9)** where it is copied into double-stranded RNAs. **(10)** Newly synthesized (−) strands serve as templates for the synthesis of (+) strand genomic RNAs. **(11)** Some of the newly synthesized (+) strand RNA molecules are translated after the removal of VPg. **(12)** Structural proteins are formed by partial cleavage of the P1 precursor and **(13)** associate with (+) strand RNA molecules that retain VPg to form progeny virus particles, **(14)** which are released from the cell upon lysis.

Polyomaviruses

Family *Polyomaviridae*

Genera	Examples
Orthopolyomavirus	Simian virus 40
Wukipolyomavirus	KI virus

The family *Polyomaviridae* includes mouse polyomaviruses, simian virus 40, and the human JC and BK viruses, which are orthopolyomaviruses isolated from a patient with progressive multifocal leukoencephalopathy and an immunosuppressed recipient of a kidney transplant, respectively. The genus *Wukipolyomavirus* includes the human Karolinska Institute (KI) polyomavirus and human polyomaviruses 6 and 7, among others.

Under some conditions, mouse polyomavirus infection of the natural host results in formation of a wide variety of tumors (hence the name). A characteristic property of the members of this family is an ability to transform cultured cells or to induce tumors in animals. Investigation of such transforming activity has provided much information about mechanisms of oncogenesis, including the discovery of the cellular tumor suppressor protein p53. These viruses, particularly simian virus 40, have also been important in elucidation of cellular mechanisms of transcription and its regulation and characterization of the mammalian DNA synthesis machinery.

Figure 23 Structure and genome organization. (A) Virion structure. The electron micrograph shows negatively stained simian virus 40 particles (from F. A. Andered et al., *Virology* **32:**511–523, 1967, with permission). As shown on the right, the double-stranded DNA genome is organized into approximately 25 nucleosomes by the cellular core histones. One molecule of either VP2 or VP3, which possess a common C-terminal sequence, is associated with each VP1 pentamer. **(B) Genome organization.** The 5,243-bp simian virus 40 genome is shown, with locations of the origin of viral DNA synthesis (Ori) and of the early and late mRNAs indicated. The late mRNA species generally contain additional open reading frames in their 5′-terminal exons, such as that encoding the agnoprotein (LP1). The structural proteins VP2, VP3, and VP4 are encoded within the same open reading frame.

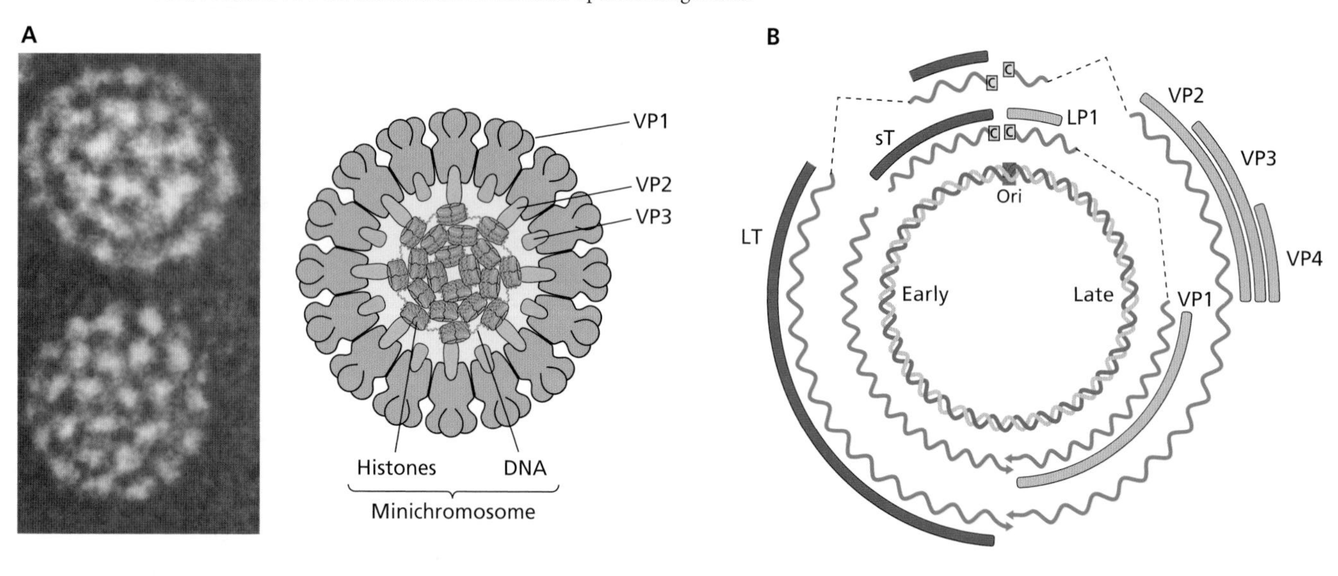

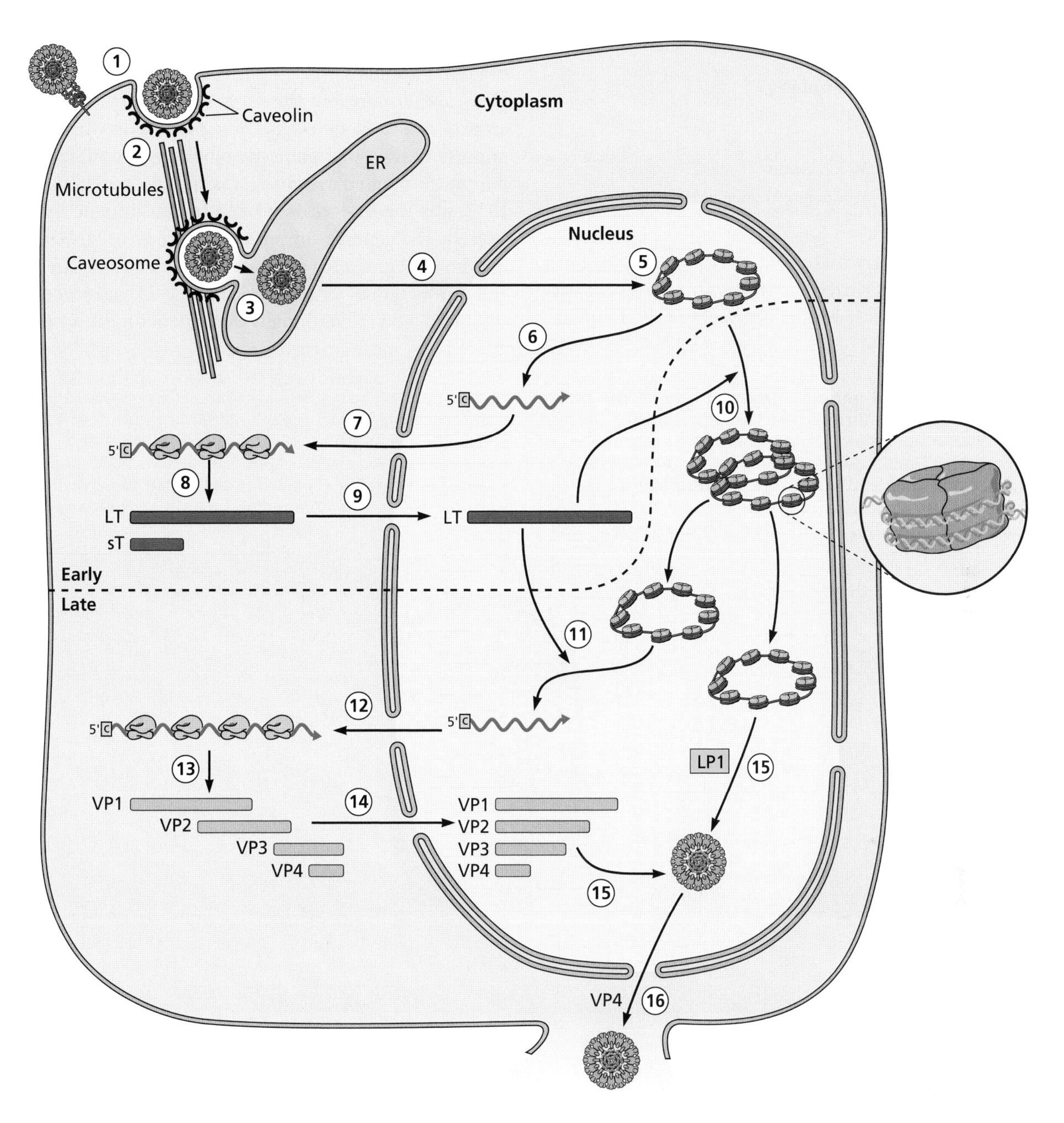

Figure 24 Single-cell reproductive cycle of simian virus 40. **(1)** The virus particle attaches to susceptible monkey cells upon binding of VP1 to the ganglioside Gm1 (a glycolipid) on the surface. **(2)** The particle is then endocytosed in caveolae, transported to the endoplasmic reticulum, and **(3)** enters that organelle. **(4)** Subsequently, it is transported to the nucleus and uncoated by unknown mechanisms. **(5)** The viral genome packaged by cellular nucleosomes is found within the nucleus. **(6)** The early transcription unit is transcribed by host cell RNA polymerase II. **(7)** After alternative splicing and export to the cytoplasm, **(8)** the early mRNAs are translated to produce the early proteins LT and sT. **(9)** The former is imported into the nucleus, **(10)** where it binds to the origin of replication to initiate DNA synthesis. Apart from LT, all components needed for viral DNA replication are provided by the host cell. As they are synthesized, daughter viral DNA molecules associate with cellular nucleosomes to form the viral nucleoproteins often called minichromosomes. **(11)** LT also stimulates transcription of the late gene from replicated viral DNA templates. **(12)** Processed late mRNAs are exported to the cytoplasm and **(13)** translated to produce the structural proteins VP1, VP2, and VP3, as well as VP4. **(14)** The structural proteins are imported into the nucleus and **(15)** assemble around viral minichromosomes to form virus particles. **(16)** Release of progeny virus particles is facilitated by VP4.

Poxviruses

Family *Poxviridae*

Selected Genera	Examples
Orthopoxvirus	Vaccinia virus
Avipoxvirus	Fowlpox virus
Leporipoxvirus	Myxoma virus
Yabapoxvirus	Yaba monkey tumor virus

Poxviruses infect most vertebrates and invertebrates, causing a variety of diseases of veterinary and medical importance. The best-known poxviral disease is smallpox, a devastating human disease that has been eradicated by vaccination. The origins of modern vaccinia virus, the virus used in smallpox virus vaccines, are obscure, but this virus is widely studied in the laboratory as a model poxvirus. Myxoma virus, which causes an important disease of domestic rabbits, was described in 1896. Rabbit fibroma virus, which was first described by Shope in 1932, was the first virus proven to cause tissue hyperplasia (warts). The genomes of poxviruses are large DNA molecules that include genes for all proteins needed for DNA synthesis and production of viral mRNAs. These viruses replicate in the cytoplasm and are minimally dependent on the host cell.

Figure 25 Structure and genome organization of the poxvirus vaccinia virus. (A) Virion structure. The electron micrograph shows the mature virion in cross section (courtesy of David J. Vaux, Sir William Dunn School of Pathology, Oxford University, Oxford, United Kingdom). **(B) Genome organization.** Shown are details for the 191-kb genome of the Copenhagen strain of vaccinia virus, with open reading frames identified in a small section of the genome. The two strands of the DNA genome are covalently connected by terminal, unpaired loops at the ends of inverted terminal repeated sequences (ITR). The genome includes ~185 unique protein-coding sequences. Those that encode structural proteins and essential enzymes are clustered in the center; those that affect virulence, host range, or immunomodulation are predominantly near the ends.

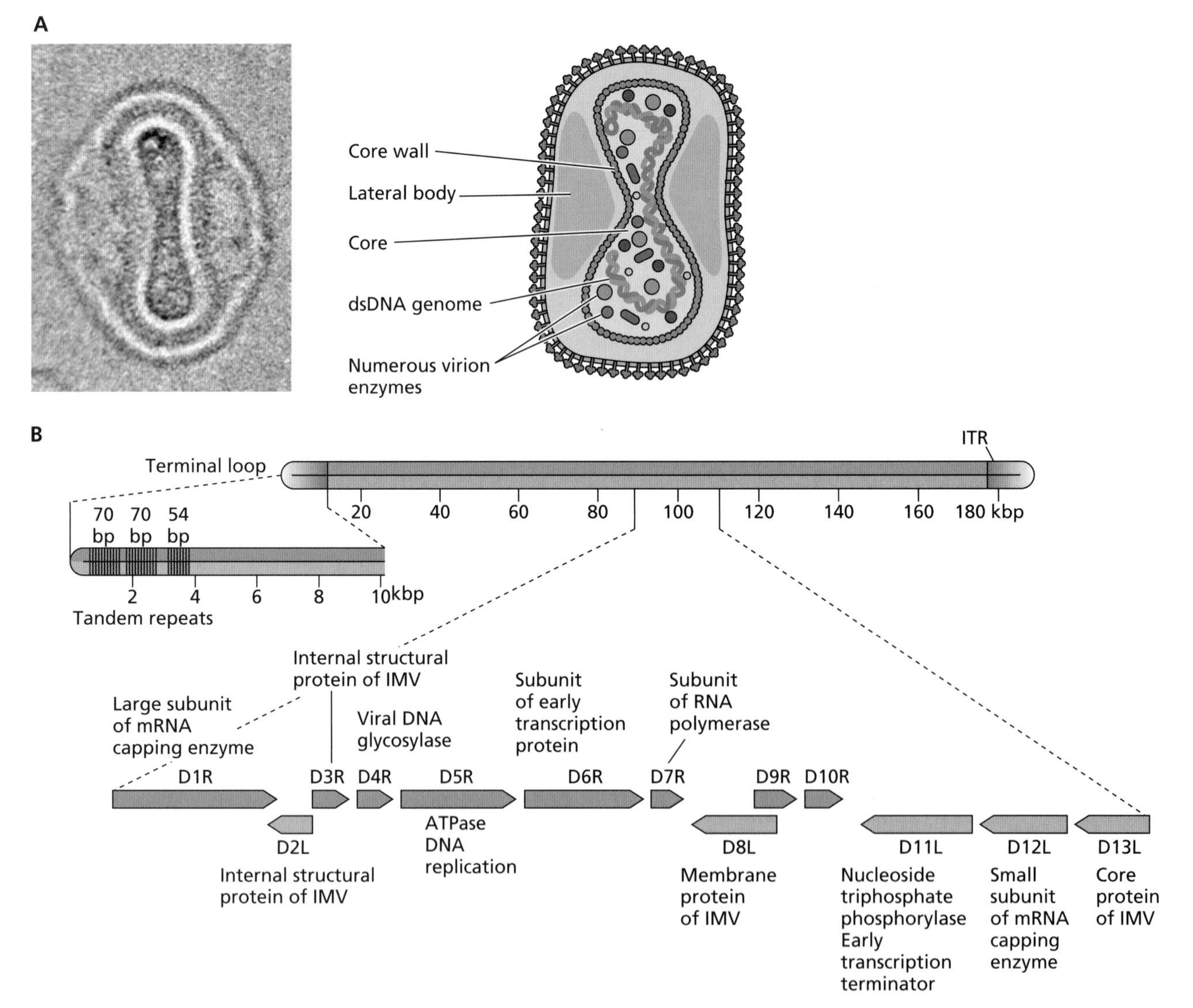

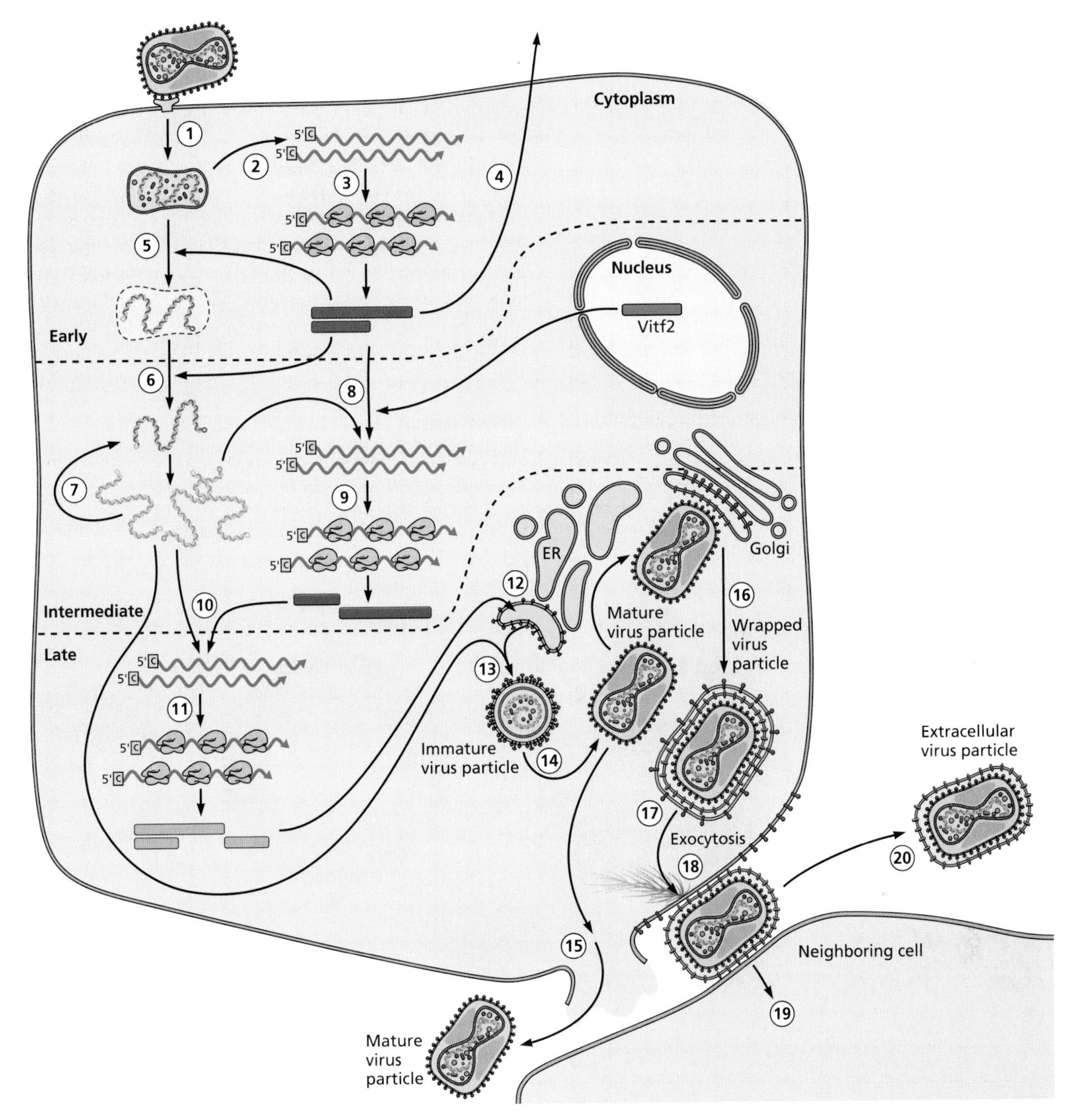

Figure 26 Single-cell reproductive cycle of vaccinia virus. (**1**) After receptor binding and fusion of viral and plasma membranes, or fusion following endocytosis, the viral core is released into the cytoplasm. (**2**) Early mRNAs are synthesized by the viral RNA polymerase aided by initiation proteins that enter the cell with virus particles. (**3**) These mRNAs are translated by the cellular protein-synthesizing machinery. (**4**) Some early proteins, which have sequence similarity to cellular growth factors and can induce proliferation of neighboring host cells, are secreted. Other early proteins counteract host immune defense mechanisms. (**5**) Some early viral proteins induce a second uncoating reaction in which the viral genome is released from the core in a nucleoprotein complex, and (**6**) others mediate replication of the genome. Newly synthesized viral DNA molecules can serve as templates (**7**) for additional cycles of genome replication and (**8**) for transcription of viral intermediate genes. Transcription of intermediate genes requires viral initiation proteins, which are products of early genes, and a cellular protein (Vitf2), which relocates from the infected cell nucleus to the cytoplasm. (**9**) Proteins made upon translation of intermediate mRNAs include those necessary for (**10**) transcription of late genes. (**11**) Late mRNAs are translated to produce viral structural proteins, enzymes, and other essential proteins that are needed early in subsequent infections and must be incorporated into virus particles during assembly. (**12**) Assembly of progeny particles begins in specialized sites, termed viral factories, that form upon viral DNA synthesis. These sites contain cellular membranes, probably derived from the endoplasmic reticulum, which are initially reorganized by specific viral protein to form crescents, (**13**) the precursor to spherical DNA-containing particles, called immature viral particles. (**14**) These particles then mature into brick-shaped intracellular mature virus particles upon proteolysis and release from viral factories. (**15**) Such particles are released only upon cell lysis. (**16**) However, they can acquire a second, double membrane from a *trans*-Golgi or early endosomal compartment to form intracellular wrapped virus particles. (**17**) The latter particles move to the cell surface on microtubules where (**18**) fusion with the plasma membrane forms cell-associated particles (**19**) that induce actin polymerization for direct transfer to surrounding cells or (**20**) that dissociate from the membrane as the extracellular virus particle. Association of the extracellular virus particle with a host cell in the next cycle of infection is thought to result in rupture of the outer membrane, giving rise to the mature virus particle.

Reoviruses

Family *Reoviridae*

Selected Genera	Examples
Subfamily *Spinareovirinae*	
Orthoreovirus	Mammalian orthoreovirus
Coltivirus	Colorado tick fever virus
Subfamily *Sedoreovirinae*	
Orbivirus	Bluetongue virus
Rotavirus	Rotavirus A

Reoviridae is one of nine families of viruses with double-stranded RNA genomes. Included in this family are the human pathogens rotaviruses and Colorado tick fever virus. Reoviruses are the best studied of all the double-stranded RNA viruses. Some of the first *in vitro* research on RNA synthesis was done using reoviruses, and the 5′-terminal cap structure of mRNA was discovered in studies of reovirus mRNAs.

Figure 27 Structure and genomic organization of an orthoreovirus. (A) Virion structure. Electron micrograph of negatively stained reovirus particles (courtesy of S. McNulty, Queen's University, Belfast, United Kingdom). The locations of six virion proteins are indicated on the illustration to the right. **(B) Genome organization.** The double-stranded genome comprises 10 segments, named according to size: large (L), medium (M), and small (S). The S1 RNA encodes two proteins: σ1s protein is translated from a second initiation codon in a different reading frame from σ1. Two proteins are also produced from the M3 RNA: protein μNSC is produced by translation at a second initiation codon in the same reading frame as μNS.

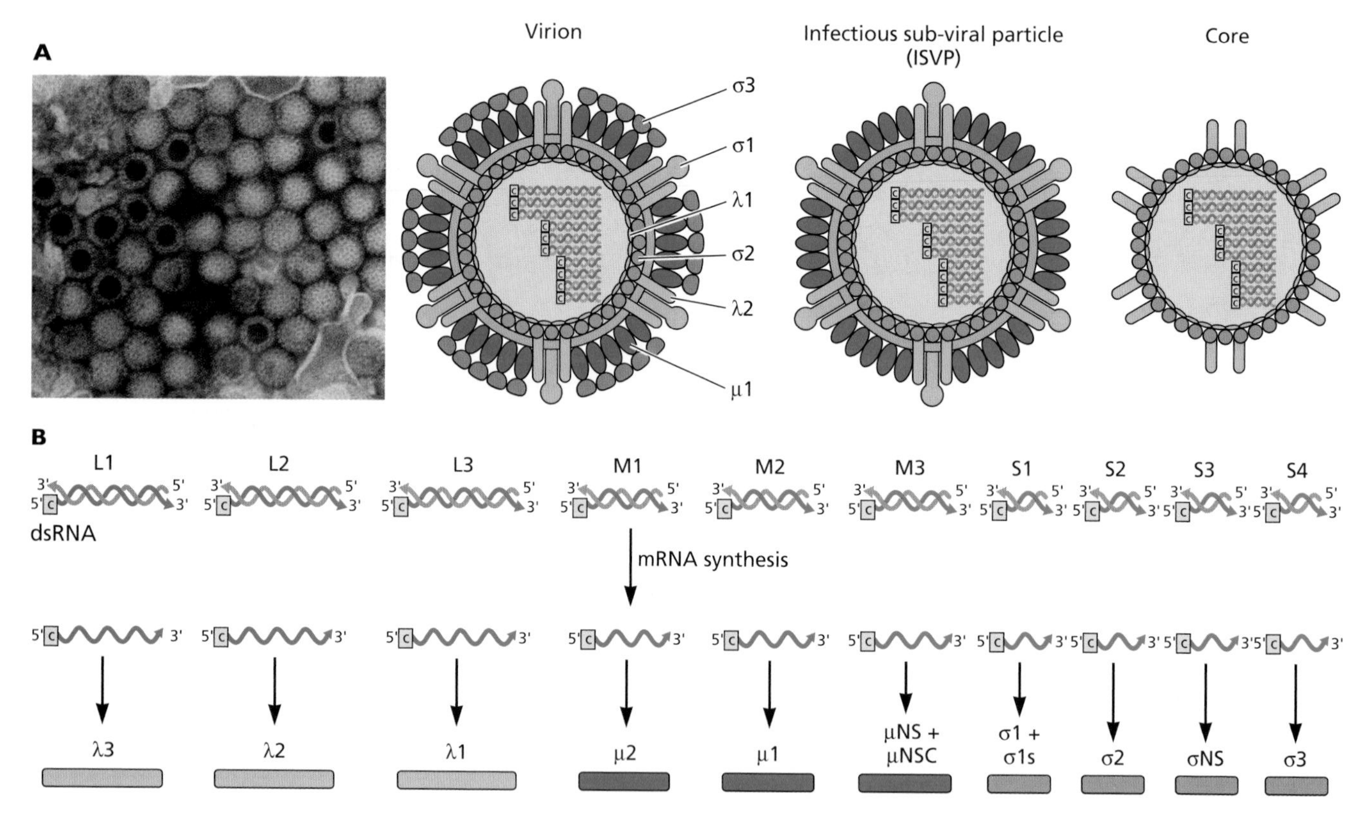

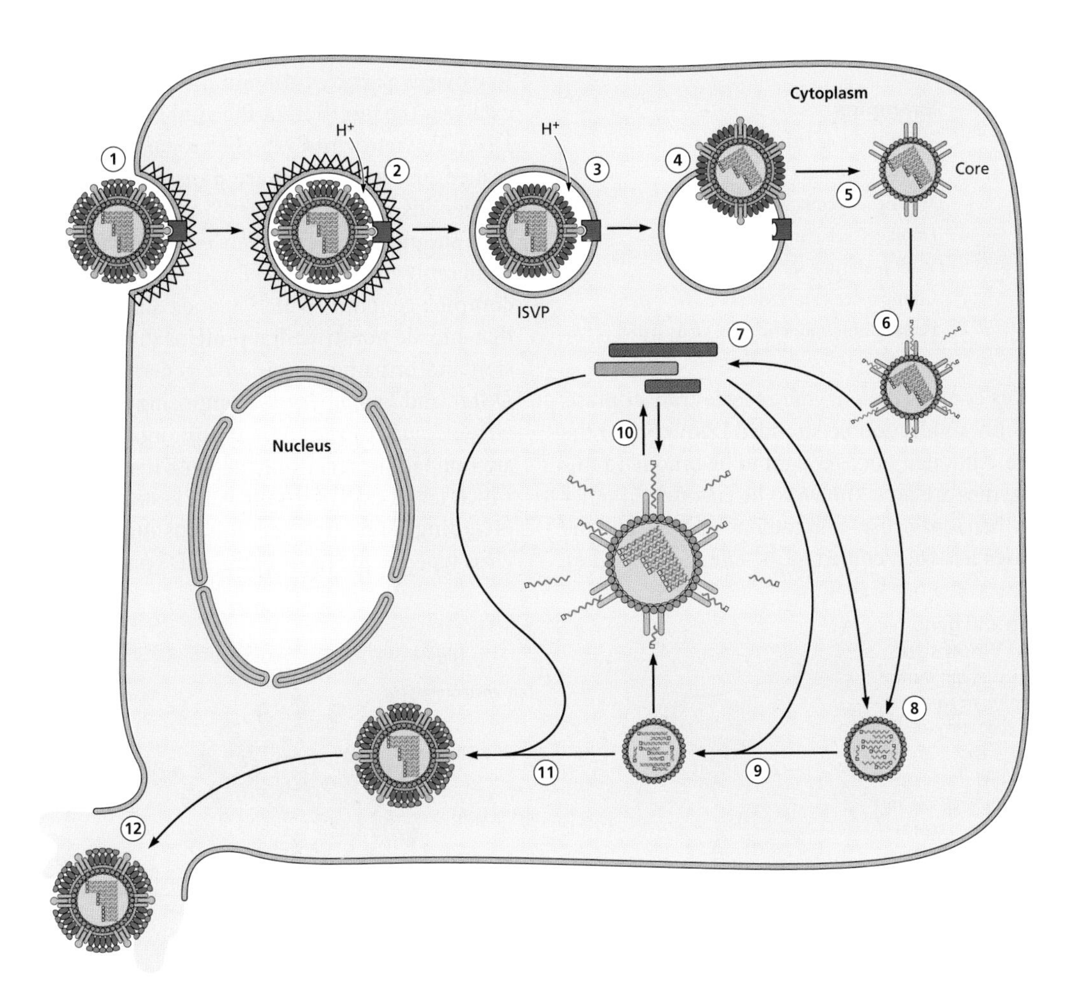

Figure 28 Single-cell reproductive cycle of orthoreovirus. **(1)** The virion binds to the cellular receptor and **(2)** enters the cell via receptor-mediated endocytosis. **(3)** In endosomes and lysosomes, the virion undergoes acid-dependent proteolytic cleavage to form the ISVP, **(4)** which penetrates the endosomal membrane, **(5)** releasing the core into the cytoplasm. **(6)** Synthesis of 10 capped viral mRNAs begins within the core particle. **(7)** These mRNAs are translated and associate with newly synthesized viral proteins **(8)** to form RNase-sensitive subviral particles in which reassortment may occur. **(9)** Each of the 10 mRNAs is a template for (−) strand RNA synthesis, leading to the production of an RNase-resistant subviral particle that contains 10 double-stranded RNAs. **(10)** Viral mRNAs produced within subviral particles are used for the synthesis of viral proteins and the assembly of additional virus particles. **(11)** In the final steps of capsid assembly, preformed complexes of outer capsid proteins are added to subviral particles. **(12)** Mature virus particles are released from the cell by lysis.

Retroviruses

Family *Retroviridae*

Genera	Examples
Alpharetrovirus	Avian leukosis virus
Betaretrovirus	Mouse mammary tumor virus
Gammaretrovirus	Murine leukemia virus
Deltaretrovirus	Human T cell lymphotropic virus
Epsilonretrovirus	Walleye dermal sarcoma virus
Spumavirus	Chimpanzee foamy virus
Lentivirus	Human immunodeficiency virus type 1

Retrovirus particles contain the enzyme reverse transcriptase, which mediates synthesis of a double-stranded DNA copy of the viral RNA genome. Although once thought to be unique to this family, similar enzymes are now known to be encoded in other viral genomes (i.e., hepadnaviruses and caulimoviruses), and the term **retroid viruses** has been coined to include these families.

Retrovirus particles contain a second enzyme, integrase, that catalyzes the insertion of the viral DNA into many sites in host DNA. The retroviruses can be propagated as integrated elements (called proviruses) that are transmitted in the germ line or as exogenous infectious agents. Alpha- and gammaretroviruses have **simple** genomes that encode only the three genes common to all retroviruses—*gag*, *pol*, and *env*. All of the others have more **complex** genomes, which include auxiliary or accessory genes that encode nonstructural proteins that affect viral gene expression and/or pathogenesis. Five genera, *Alpha-*, *Beta-*, *Gamma-*, *Delta-*, and *Epsilonviruses*, comprising the subfamily *Orthoretroviruses*, cause cancer in their host organisms. The spumaviruses are nonpathogenic, while the lentiviruses are serious pathogens which target cells of the immune system in a number of species, including humans. The lentivirus human immunodeficiency virus type 1 is the cause of the AIDS pandemic.

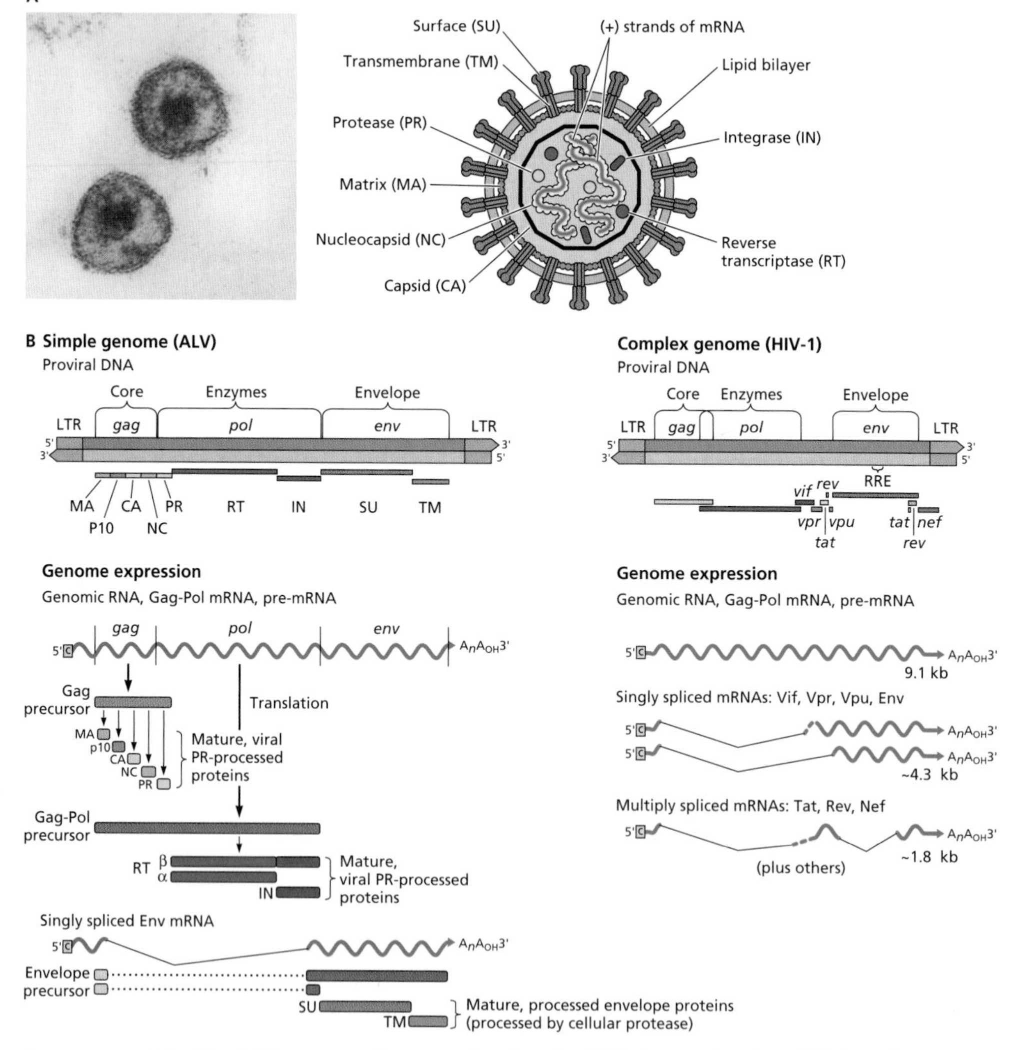

Figure 29 Structure and genomic organization. (A) Virion structure. The electron micrograph shows a negatively stained alpharetrovirus, Rous sarcoma virus (courtesy of R. Katz and T. Gales, Fox Chase Cancer Center, Philadelphia, PA). Envelope protein projections are not visible in this image. **(B) Genome organization.** (Left) A retrovirus with a simple genome (avian leukosis virus [ALV]). Proviral genes are located in different reading frames (indicated by different horizontal positions below the DNA) and are also overlapping. Colored boxes delineate open reading frames. LTR, long terminal repeats that include transcription signals. Origins of RNA and protein products are shown below. (Right) A retrovirus with a more complex genome illustrated with the lentivirus human immunodeficiency virus type 1 (HIV-1). Proviral genes are located in all three reading frames, as indicated by the overlaps. Human immunodeficiency virus type 1 mRNAs fall into one of three classes. The first type is an unspliced transcript of 9.1 kb, identical in function to that synthesized from the simple retrovirus genome shown at the left. The second type comprises singly spliced mRNAs (average length, 4.3 kb) that result from splicing from a 5′ splice site upstream of *gag* to any one of a number of 3′ splice sites near the center of the genome. One of these mRNAs specifies the Env polyprotein precursor, as illustrated for the singly spliced mRNA of the retrovirus with a simple genome. The others specify the human immunodeficiency virus type 1 accessory proteins. The third type comprises a complex class of mRNAs (average length, 1.8 kb) derived by multiple splicing from 5′ and 3′ splice sites throughout the genome. They include mRNAs that specify the regulatory proteins Tat and Rev and are the first to accumulate after infection.

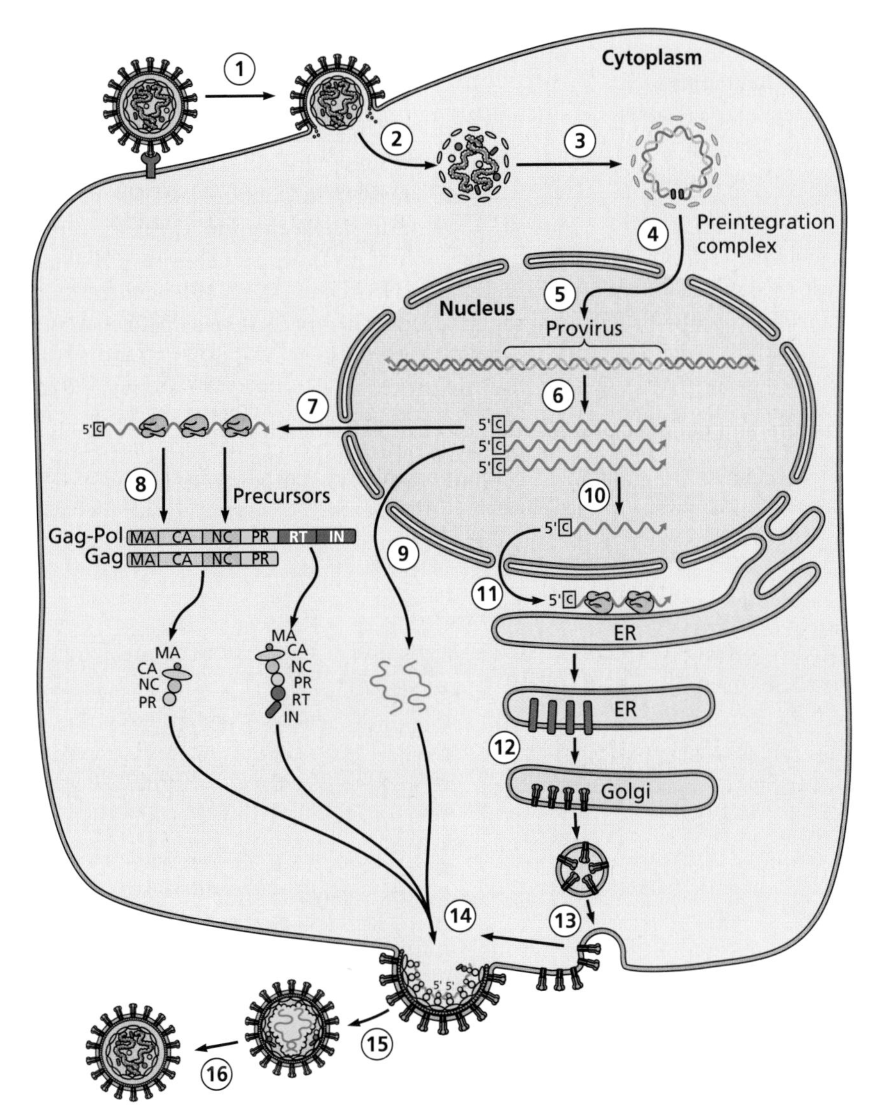

Figure 30 Single-cell reproductive cycle of a retrovirus with a simple genome. **(1)** The virus attaches by binding of the viral envelope protein to specific receptors on the surface of the cell. The identities of receptors are known for many retroviruses. **(2)** The viral core is deposited into the cytoplasm following fusion of the virion and cell membranes. Entry of some beta- and gammaretroviruses may occur via the endocytic pathways. **(3)** The viral RNA genome is reverse transcribed by the virion reverse transcriptase (RT) within a subviral particle. The product is a linear double-stranded viral DNA with ends that are shown juxtaposed in preparation for integration. **(4)** Viral DNA and integrase (IN) protein gain access to the nucleus with the help of intracellular trafficking machinery or, in some cases, by exploiting nuclear disassembly during mitosis. **(5)** Integrative recombination, catalyzed by IN, results in site-specific insertion of the viral DNA ends, which can take place at many locations in the host genome, with distinct, characteristic general preferences for different viral IN proteins. **(6)** Transcription of integrated viral DNA (the **provirus**) by the host cell RNA polymerase II system produces full-length RNA transcripts. **(7)** Some full-length RNA molecules are exported from the nucleus to the cytoplasm and serve as mRNAs. **(8)** These mRNAs are translated by cytoplasmic ribosomes to form the viral Gag and Gag-Pol polyprotein precursors at a ratio of approximately 10:1. **(9)** Some full-length RNA molecules are destined to become encapsidated as progeny viral genomes. The mechanism for sequestering RNAs for this purpose is unknown, but there is evidence that a fraction of the ASV Gag protein traffics through the nucleus, where it could perform this function. **(10)** Other full-length RNA molecules are spliced within the nucleus to form mRNA for the Env polyprotein. **(11)** Env mRNA is translated by ribosomes bound to the endoplasmic reticulum (ER). **(12)** The Env proteins are transported through the Golgi apparatus, where they are glycosylated and cleaved by cellular enzymes to form the mature SU-TM complex. **(13)** Mature envelope proteins are delivered to the surface of the infected cell. **(14)** Virion components (two copies of the viral RNA, Gag and Gag-Pol precursors, and SU-TM) assemble at budding sites with the help of *cis*-acting signals encoded in each. Type C retroviruses (e.g., alpharetroviruses and lentiviruses) assemble at the inner face of the plasma membrane, as illustrated. Other types (A, B, and D) assemble on internal cellular membranes. **(15)** The nascent particles bud from the surface of the cell. **(16)** Maturation (and infectivity) requires the action of the virus-encoded protease (PR), which is itself a component of the core precursor polyprotein. During or shortly after budding, PR cleaves at specific sites within the Gag and Gag-Pol precursors to produce the mature viral proteins. This process causes a characteristic condensation of the virus cores.

Rhabdoviruses

Family *Rhabdoviridae*

Genera	Examples
Vesiculovirus	Vesicular stomatitis virus Indiana
Lyssavirus	Rabies virus
Cytorhabdovirus	Lettuce necrotic yellows virus

Among the 175 known rhabdoviruses are the causative agents of rabies, one of the oldest recognized infectious diseases, and economically important diseases of fish. The host range of these viruses is very broad: they infect many vertebrates, invertebrates, and plants. The genome of vesicular stomatitis virus has been a model for the replication and expression of viral genomes that consist of a single molecule of (−) strand RNA. The first RNA-dependent RNA polymerase discovered in a virus particle was that of vesicular stomatitis virus.

Figure 31 Structure and genomic organization of vesicular stomatitis virus. (A) Virion structure. The electron micrograph shows negatively stained vesicular stomatitis virus (courtesy of J. Rose, Yale University School of Medicine, New Haven, CT). **(B) Genome organization.** The (−) strand RNA is the template for synthesis of leader RNA and five monocistronic mRNAs (capped and polyadenylated). Two proteins are produced from the P/C mRNA from upstream and downstream translation initiation codons.

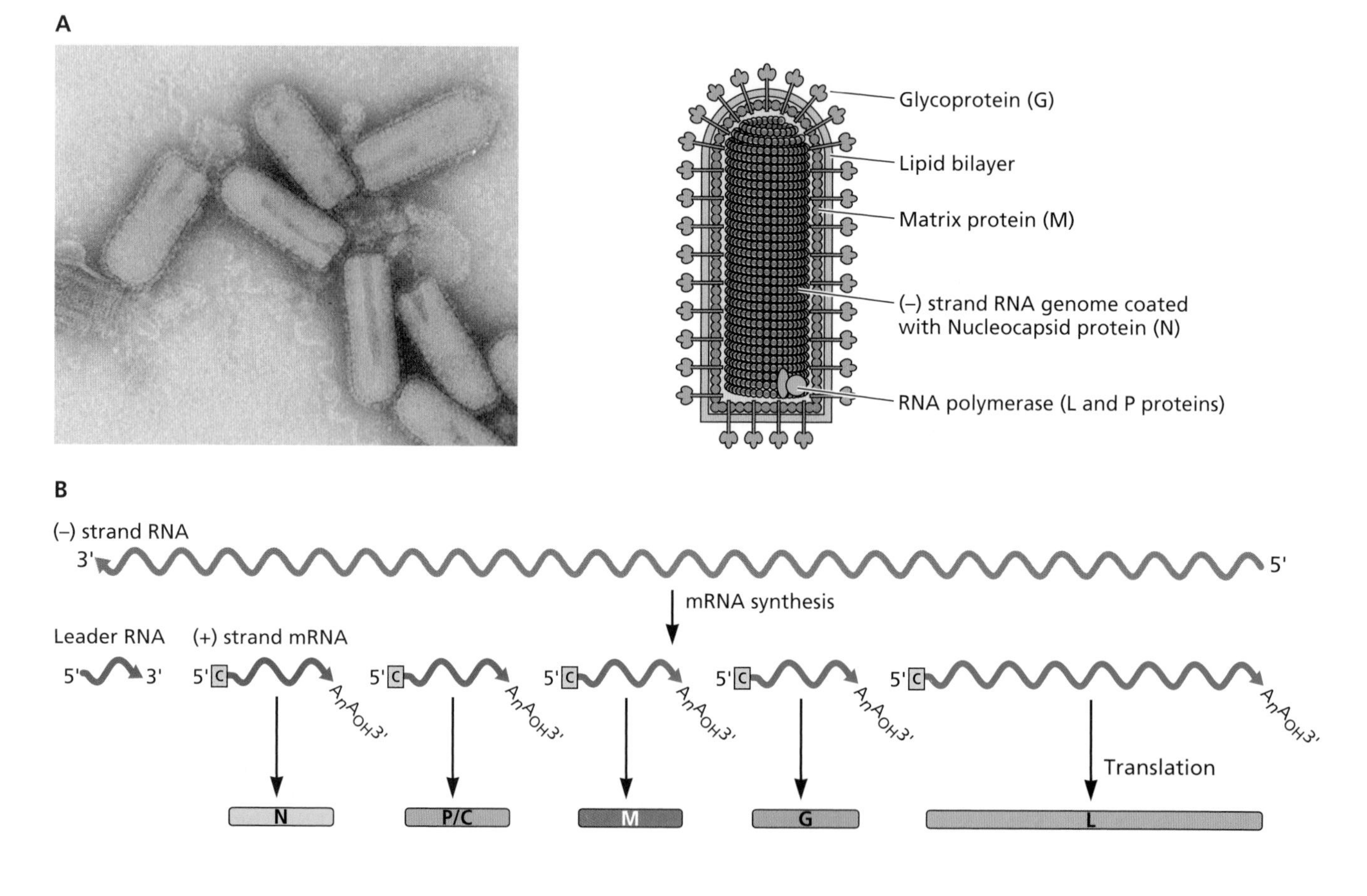

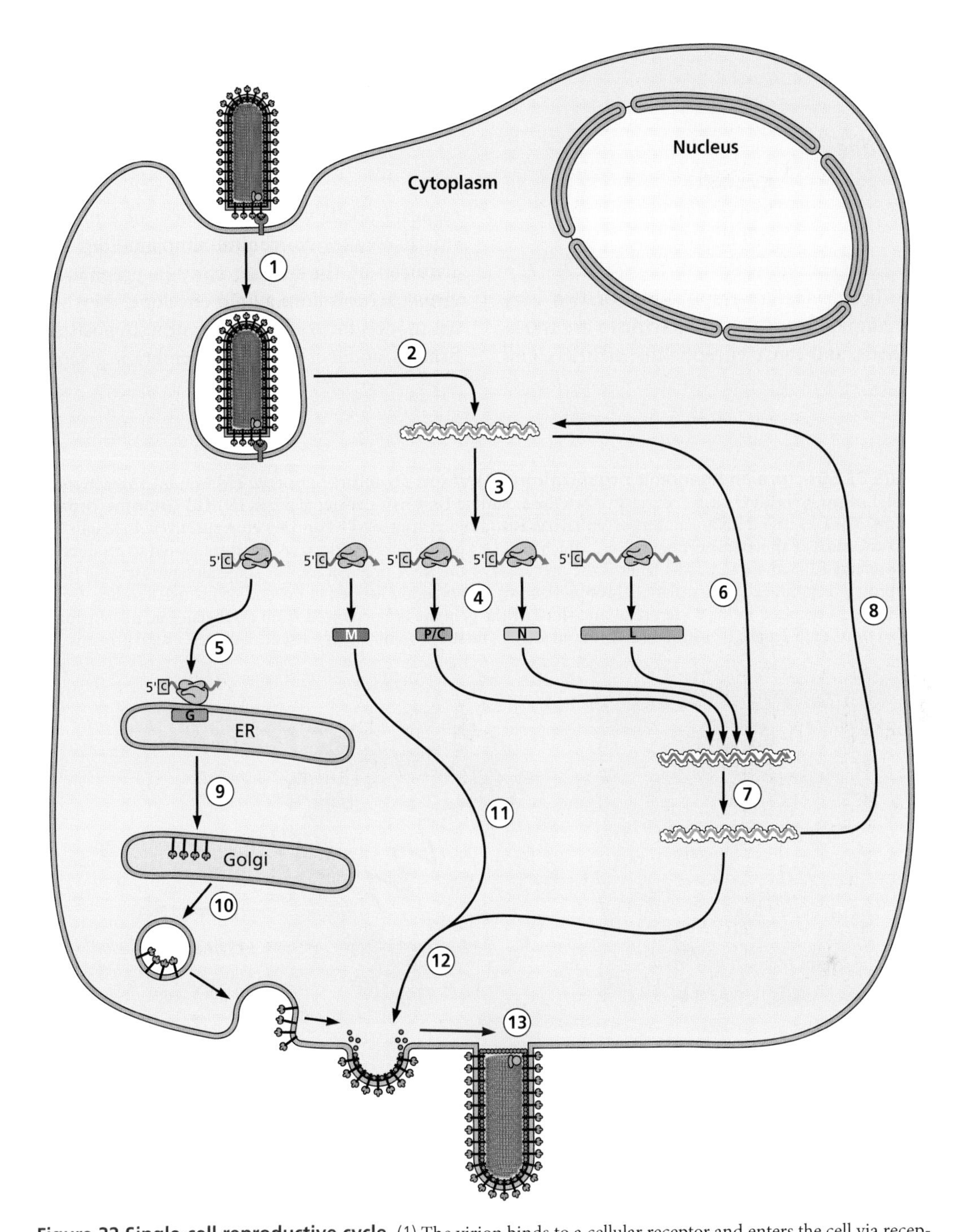

Figure 32 Single-cell reproductive cycle. (1) The virion binds to a cellular receptor and enters the cell via receptor-mediated endocytosis. **(2)** The viral membrane fuses with the membrane of the endosome, releasing the helical viral nucleocapsid. This structure comprises (−) strand RNA coated with nucleocapsid protein molecules and a small number of L and P protein molecules, which catalyze viral RNA synthesis. **(3)** The (−) strand RNA is copied into five subgenomic mRNAs by the L and P proteins. **(4)** The N, P/C, M, and L mRNAs are translated by free cytoplasmic ribosomes, **(5)** while G mRNA is translated by ribosomes bound to the endoplasmic reticulum. Newly synthesized N, P, and L proteins participate in viral RNA replication. **(6)** This process begins with synthesis of a full-length (+) strand copy of genomic RNA, which is also in the form of a ribonucleoprotein containing the N, L, and P proteins. **(7)** This RNA, in turn, serves as a template for the synthesis of progeny (−) strand RNA in the form of nucleocapsids. **(8)** Some of these newly synthesized (−) strand RNA molecules enter the pathway for viral mRNA synthesis. **(9)** Upon translation of G mRNA, the G protein enters the secretory pathway, **(10)** in which it becomes glycosylated and travels to the plasma membrane. **(11 and 12)** Progeny nucleocapsids and the M protein are transported to lipid rafts in the plasma membrane, **(13)** where association with regions containing the G protein is followed by budding to release virus particles.

Togaviruses

Family *Togaviridae*

Genera	Examples
Alphavirus	Sindbis virus
Rubivirus	Rubella virus

Members of the *Togaviridae* are responsible for two very different kinds of human disease. All alphaviruses are transmitted by arthropods, and cause encephalitis, arthritis, and rashes. Rubella virus is the agent of a mild rash disease but can also cause congenital abnormalities in the fetus when acquired by the mother early in pregnancy. Because these virus particles have a lipid envelope, they have been important models for studying the synthesis, posttranslational modification, and localization of membrane glycoproteins.

Figure 33 Structure and genomic organization. (A) Virion structure. The cryo-electron micrograph shows the alphavirus Ross River virus (courtesy of N. Olson, Purdue University, West Lafayette, IN). **(B) Genome organization.** The (+) strand RNA genomes of alphaviruses and rubiviruses are 11.7 and 9.8 kb, respectively, in length. The first two-thirds of alphavirus genomic RNA, which is of (+) polarity and carries a 5′ cap, is translated to produce the polyproteins P123 and P1234. The latter is the precursor of the RNA polymerase. For some alphaviruses, the P1234 polyprotein is produced by translational suppression of a stop codon located at the end of the nsP3 coding region. The proteins encoded in the 3′-terminal one-third of the genome are produced from a subgenomic mRNA that is copied from a full-length (−) strand RNA intermediate. The subgenomic mRNA encodes the structural proteins.

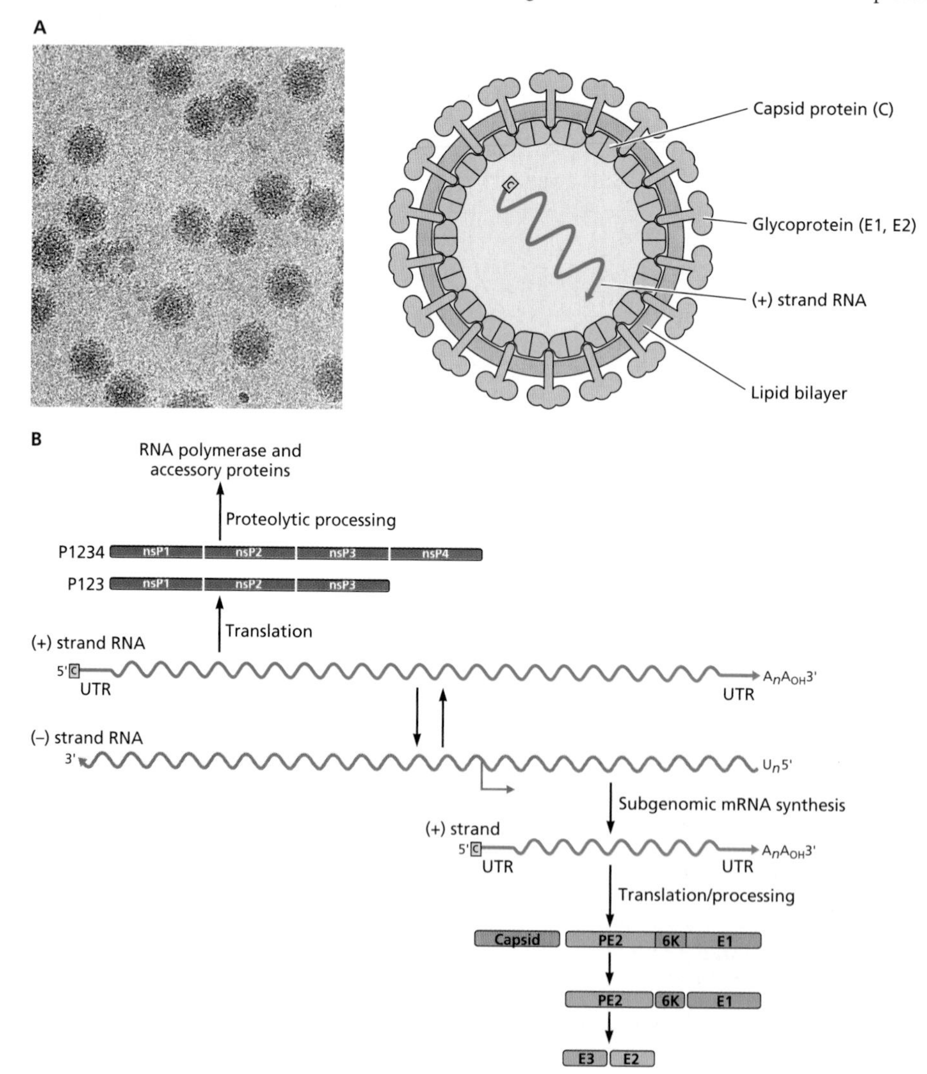

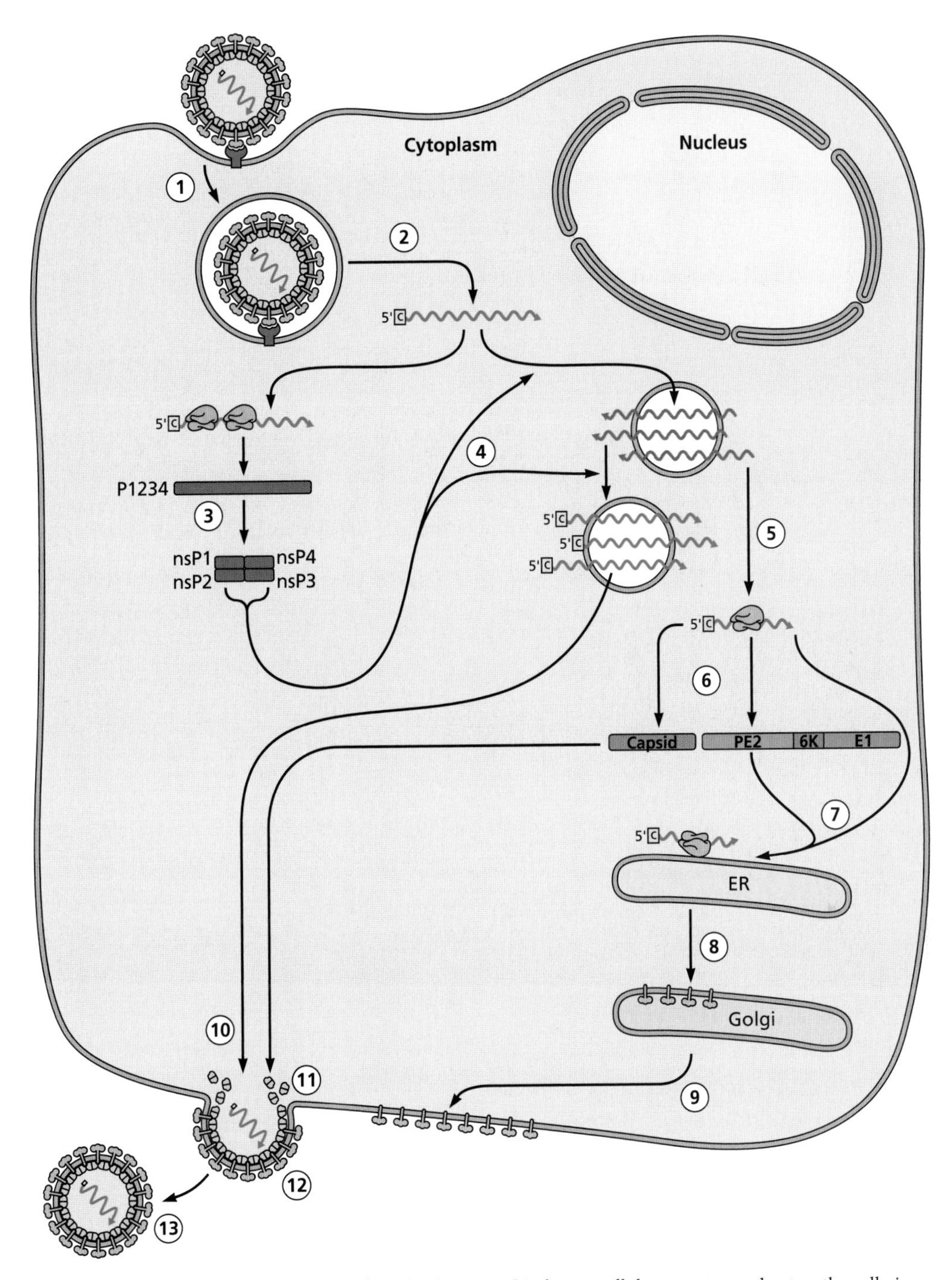

Figure 34 Single-cell reproductive cycle. (1) The virion binds to a cellular receptor and enters the cell via receptor-mediated endocytosis. (2) Upon acidification of the vesicle, the viral nucleocapsid is released into the cytoplasm and subsequently disassembled to release the (+) strand viral RNA, (3) which is translated to form the polyprotein P1234. (4) Sequential cleavage of this polyprotein at different sites by a viral proteinase produces RNA polymerases with different specificities. (5) These viral enzymes then copy (+) strands into full-length (−) and (+) strands and catalyze synthesis of the subgenomic mRNA. Viral RNA synthesis takes place on membranous structures that first accumulate at the plasma membrane and later move to the cell interior. (6) The subgenomic mRNA is translated by free cytoplasmic ribosomes to produce the capsid protein. (7) Proteolytic cleavage to liberate the capsid protein exposes a hydrophobic sequence of PE2 that induces the ribosomes to associate with the endoplasmic reticulum (ER). (8) As a result, the PE2-6K-E1 polyprotein enters the secretory pathway. (9) The glycoproteins are transported to the cell surface. (10) The capsid protein and (+) strand genomic RNA assemble to form nucleocapsids that migrate to the plasma membrane and (11) associate with viral glycoproteins. (12) The nucleocapsid acquires an envelope by budding at this site, and (13) virus particles are released.

Glossary

3′ untranslated region The region of an mRNA downstream of the translation termination codon. *(Chapter 11)*

5′ untranslated region The region of an mRNA upstream of the translation initiation codon. *(Chapter 11)*

Acylation Posttranslational addition of saturated or unsaturated fatty acids to a protein. *(Chapter 12)*

Affinity The measure of the strength with which one molecule associates with another noncovalently. *(Chapter 5)*

Allele specific Complementing only a specific change; refers to suppressor mutations. *(Chapter 3)*

Alternative splicing Splicing of different combinations of exons in a pre-mRNA, generally leading to synthesis of mRNAs with different protein-coding sequences. *(Chapter 10)*

Ambisense Producing mRNAs from both (−) strand genomic RNA and the complementary (+) strand; refers to viral genomes. *(Chapter 6)*

Amphipathic Having both hydrophilic and hydrophobic portions. *(Chapter 2)*

Anabolism The metabolic reactions by which larger molecules are built from simpler ones, with the consumption of energy. *(Chapter 14)*

Aneuploid Abnormal in chromosome morphology and number. *(Chapter 2)*

Apical domain The specialized surface of an epithelial cell exposed to the environment. Also called apical surface. *(Chapters 2, 9, and 12)*

Asymmetric unit The unit from which capsids or nucleocapsids of a virus particle are built. Also called protomer or structural unit. *(Chapter 4)*

Attenuated An infection in which normally severe symptoms or pathology are mild or inconsequential; a state of reduced virulence. *(Chapter 1)*

Autophagy The controlled degradation, in response to stress, of proteins and other cellular components taken into double-membrane vesicles (autophagosomes) that fuse with lysosomes, literally self-eating. *(Chapter 14)*

Avidity The sum of the affinities (strengths) of multiple noncovalent interactions. *(Chapter 5)*

Bacteriophages Viruses that infect bacteria; derived from the Greek word *phagein,* meaning "to eat." *(Chapter 1)*

Basal lamina A thin layer of extracellular matrix bound tightly to the basolateral surface of cells; the basal lamina is linked to the basolateral membrane by integrins. *(Chapter 2)*

Basolateral domain The nonspecialized surface of an epithelial cell that contacts an internal basal lamina or adjacent or underlying cells in the tissue. Also called basolateral surface. *(Chapters 2, 9, and 12)*

Capping The addition of m^7G via a 5′–5′ phosphodiester bond to the 5′ ends of cellular and viral transcripts made in eukaryotic cells. *(Chapter 10)*

Cap structure The m^7G linked via a 5′–5′ phosphodiester bond to the 5′ ends of the majority of viral and cellular mRNAs made in eukaryotic cells. *(Chapter 11)*

Capsid The outer shell of viral proteins that surrounds the genome in a virus particle. *(Chapters 1 and 4)*

Cap snatching Cleavage of cellular RNA polymerase II transcripts by a viral endonuclease to produce capped primers for viral mRNA synthesis. *(Chapters 6 and 10)*

Catabolism The reactions that break down complex molecules into simpler ones to generate energy directly or indirectly. *(Chapter 14)*

Caveolae Flask-shaped invaginations of the plasma membrane of many types of cells that contain the protein caveolin and are rich in lipid rafts; caveolae internalize membrane components, extracellular ligands, bacterial toxins, and some animal viruses. *(Chapter 5)*

Centrosome An organelle that is the main microtubule-organizing center. *(Chapter 5)*

Chaperone A protein that facilitates the folding of other polypeptide chains, the assembly of multimeric proteins, or the formation of macromolecular assemblies (e.g., chromatin). Also called molecular chaperone. *(Chapters 4, 12, and 13)*

Chemokines Small proteins that attract and stimulate cells of the immune defense; produced by many cells in response to infection. Also called chemotactic cytokines. *(Chapter 5)*

Coactivator A protein that stimulates transcription by RNA polymerase II without binding to a specific DNA sequence; generally interacts with sequence-specific transcriptional activators. *(Chapter 8)*

Codon Three contiguous bases in an mRNA template that specify the amino acids incorporated into protein. *(Chapter 11)*

Complementation The ability of gene products of two different, individually nonreproducing mutants to interact functionally in the same cell to permit virus reproduction. *(Chapter 3)*

Concatemer A DNA molecule comprising multiple, tandem copies of a viral genome (or other DNA sequence) joined end to end. *(Chapter 9)*

Constitutive transport elements Sequences in certain unspliced viral mRNAs that direct export from the nucleus by host cell proteins. *(Chapter 10)*

Continuous cell lines Cultures of a single cell type that can be propagated indefinitely in culture. *(Chapter 2)*

Copy choice A mechanism of recombination in which an RNA or DNA polymerase first copies the 3′ end of one parental strand and then exchanges one template for another at the corresponding position on a second parental strand. *(Chapters 6 and 7)*

Coreceptor A cell surface molecule that is required, in addition to the receptor, for entry of virus particles into cells. *(Chapter 5)*

Core promoter The minimal set of DNA sequences required for accurate initiation of transcription by RNA polymerase II. *(Chapter 8)*

Culling Removing and destroying diseased or potentially exposed animals to prevent further spread of infection. *(Chapter 1)*

Cytopathic effects The morphological changes induced in cells by viral infection. *(Chapter 2)*

Cytoskeleton The intracellular structural network composed of actin filaments, microtubules, and intermediate filaments. *(Chapter 2)*

Defective interfering RNAs Subgenomic RNAs that replicate more rapidly than full-length RNA and therefore compete for the components of the RNA synthesis machinery and interfere with the replication of full-length RNAs. *(Chapter 6)*

Deletion mutation Loss of one or more bases in a nucleic acid. *(Chapter 3)*

Diploid cell strains Cell cultures that consist of a homogeneous population of a single type and that can divide up to 100 times before dying. *(Chapter 2)*

Eclipse period The phase of viral infection during which the viral nucleic acid is uncoated from its protective shell and no infectious virus can be detected inside cells. *(Chapter 2)*

Efficiency of plating The plaque titer divided by the number of virus particles in the sample. *(Chapter 2)*

Elongation Stepwise incorporation of ribonucleoside monophosphates or deoxyribonucleoside monophosphates into the 3′-OH end of the growing RNA or DNA chain in the 5′ → 3′ direction. *(Chapter 6)*

Endemic Having a disease pattern typical of a particular geographic area; persisting in a population for a long period without reintroduction of the causative virus from outside sources. *(Chapter 1)*

Endogenous proviruses Proviruses that enter the germ line at some point in the history of an organism and are thereafter inherited in normal Mendelian fashion by every cell in that organism and by its progeny. *(Chapter 7)*

Endosome A vesicle that transports molecules from the plasma membrane to the cell interior. *(Chapter 5)*

Enhancer A DNA sequence containing multiple elements that can stimulate RNA polymerase II transcription over long distances, independently of orientation or location relative to the site of transcriptional initiation. *(Chapter 8)*

Envelope The host cell-derived lipid bilayer carrying viral glycoproteins that forms the outer layer of many virus particles. *(Chapter 4)*

Epidemic A pattern of disease characterized by rapid and sudden appearance of cases spreading over a wide area. *(Chapter 1)*

Epitope A short contiguous sequence or unique conformation of a macromolecule that can be recognized by the immune system; also called an antigenic determinant; a T cell epitope is a short peptide recognized by a particular T cell receptor, while a B cell epitope is recognized by the antigen-binding domain of antibody and is part of an intact protein. *(Chapter 2)*

Exons Blocks of noncontiguous coding sequences (generally short) present in many cellular and viral pre-mRNAs. *(Chapter 10)*

Foci (plural) Clusters of cells that are derived from a single progenitor and share properties, such as unregulated growth, that cause them to pile up on one another. One such cluster is called a **focus**. *(Chapter 2)*

Fusion peptide A short hydrophobic amino acid sequence (20 to 30 amino acids) that is thought to insert into target membranes to initiate fusion. *(Chapter 5)*

Fusion pore An opening between two lipid bilayers formed by the action of fusion proteins; it allows exchange of material across membranes. *(Chapter 5)*

Glycoforms The total set of forms of a protein that differ in the number, location, and nature of oligosaccharide chains. *(Chapter 12)*

Glycoprotein A protein carrying covalently linked sugar chains (oligosaccharides). *(Chapter 4)*

G_0 state A state in which the cell has ceased to grow and divide and has withdrawn from the cell cycle. Also called resting state. *(Chapter 9)*

Half-life The time required for decay of a molecule or macromolecule to half of the original concentration. *(Chapter 10)*

Helical symmetry The symmetry of regularly wound structures defined by the relationship $P = \mu \times \rho$, where P = pitch of the helix, μ = the number of structural units per turn, and ρ = the axial rise per unit. *(Chapter 4)*

Helper virus A virus that provides viral proteins needed for the reproduction of a coinfecting defective virus. *(Chapter 6)*

Hemagglutination The linking of multiple red blood cells by virus particles, resulting in a lattice; basis of a method to measure virus concentration. *(Chapter 2)*

Heterogeneous nuclear RNAs Nuclear precursors to mRNAs that are larger than mRNAs and heterogeneous in size. *(Chapter 10)*

Homologous recombination The exchange of genetic information between any pair of related DNAs at sites with identical sequences. *(Chapter 9)*

Host range A listing of species and cells (hosts) that are susceptible to and permissive for infection. *(Chapter 5)*

Icosahedral symmetry The symmetry of the icosahedron, the solid with 20 faces and 12 vertices related by axes of two-, three-, and five-fold rotational symmetry. *(Chapter 4)*

Indirectly anchored proteins Proteins that are indirectly bound to the plasma membrane by interacting with either integral membrane proteins or the charged sugars of membrane glycolipids. *(Chapter 2)*

Infectious DNA clone A double-stranded DNA copy of the viral genome carried on a bacterial plasmid or other vector. *(Chapter 3)*

Initiation codon The codon at which translation of an mRNA begins, most commonly AUG. *(Chapter 11)*

Initiator A short DNA sequence that is sufficient to specify the site at which RNA polymerase II initiates transcription. *(Chapter 8)*

Insertion mutation Addition of one or more nucleotides to a nucleic acid sequence. *(Chapter 3)*

Integral membrane proteins Proteins that are embedded in a lipid bilayer, with external and internal domains connected by one or more membrane-spanning domains. *(Chapters 2 and 4)*

Internal ribosome entry site (IRES) An internal binding site for 40S ribosomal subunits and initiation of translation present in some viral and few cellular mRNAs. *(Chapter 11)*

Introns Noncoding sequences that separate coding sequences (exons) in many cellular and viral pre-mRNAs. *(Chapter 10)*

Inverted terminal repetitions Sequences that are present in the opposite orientation at the ends of certain linear viral DNA genomes. *(Chapter 9)*

Koch's postulates Criteria developed by the German physician Robert Koch in the late 1800s to determine if a given agent is the cause of a specific disease. *(Chapter 1)*

Lariat An intermediate in pre-mRNA splicing containing the intron and 3′ exon, with the branch point A residue of the intron linked via a 2′–5′ phosphodiester bond to the nucleotide at the 5′ end of the intron. *(Chapter 10)*

Latent infection Long term infection in which the viral genome is maintained with limited expression of viral genes and without loss of host cell viability. *(Chapters 8 and 9)*

Latent period The phase of viral infection during which no extracellular virus can be detected. *(Chapter 2)*

L domain sequences Short amino acid sequences required for membrane fusion during budding of enveloped viruses. *(Chapter 13)*

Lectin A protein that binds to a specific sugar. *(Chapter 5)*

Lipid raft A microdomain of the plasma membrane that is enriched in cholesterol and saturated fatty acids and is more densely packed and less fluid than other regions of the membrane. *(Chapters 2 and 12)*

Long terminal repeat A direct repeat of genetic information that is present in the proviral DNA of retroviruses; it is formed by reverse transcription of the RNA template and includes *cis*-acting elements required for viral DNA integration and its subsequent transcription. *(Chapter 7)*

Lysogenic Pertaining to a bacterium that carries the genetic information of a quiescent bacteriophage, which can be induced to reproduce, and subsequently lyse, the bacterium. *(Chapter 1)*

Lysogeny The phenomenon by which the lysogenic state is established and maintained in bacteria. *(Chapter 1)*

Lysosome A vesicle in the cell that contains enzymes that degrade sugars, proteins, nucleic acids, and lipids. *(Chapter 5)*

Marker rescue Replacement of all local nucleic acids that include a mutation with wild-type nucleic acid. *(Chapter 3)*

Marker transfer Introduction of a mutation by replacement of a segment of viral nucleic acid with one containing the mutation. *(Chapter 3)*

Membrane-spanning domain A segment of an integral membrane protein that spans the lipid bilayer; often α-helical. *(Chapters 2 and 4)*

Metagenomic analysis Sequencing of samples recovered directly from the environment, and containing many genomes. *(Chapter 1)*

Metastable structure A structure that has not attained the lowest free energy state. *(Chapter 4)*

Microdomains Regions of the plasma membrane with distinct lipid and protein composition. *(Chapter 2)*

Missense mutation A change in a single nucleotide or codon that results in the production of a protein with a single amino acid substitution. *(Chapter 3)*

Molecular chaperone *See* Chaperone.

Monocistronic Encoding one polypeptide; refers to mRNA. *(Chapter 11)*

Monoclonal antibody An antibody of a single specificity made by a clone of antibody-producing cells. *(Chapter 2)*

Monolayer A layer of cultured cells growing in a cell culture dish. *(Chapter 2)*

Multiplicity of infection The number of infectious virus particles added per cell. *(Chapter 2)*

Negative [(−)] strand The strand of DNA or RNA that is complementary in sequence to the (+) (coding) strand. *(Chapter 1)*

Neutralize To block (by antibodies) the infectivity of virus particles. *(Chapter 2)*

Nonsense mutation A substitution mutation that produces a translation termination codon. *(Chapter 3)*

Nuclear localization signal Amino acid sequence that is necessary and sufficient for import of a protein into the nucleus. *(Chapter 5)*

Nucleocapsid A nucleic acid-protein assembly packaged within the virus particle; the term is used when this complex is a discrete substructure of a complex particle. *(Chapter 4)*

Obligate parasites Organisms that are absolutely dependent on another living organism for reproduction. *(Chapter 1)*

Okazaki fragments Short (100–200 nucleotides) DNA segments elongated from RNA primers during discontinuous synthesis of the lagging strand at a replication fork. *(Chapter 9)*

Oligomerization Association of polypeptide chains, which may be the same or different, to form a protein with multiple subunits. *(Chapter 8)*

Oligosaccharide A short linear or branched chain of sugar residues (monosaccharides); also called a glycan. *(Chapter 4)*

Oncogenesis The processes leading to cancer. *(Chapter 14)*

One-hit kinetics A linear relationship between plaque count and virus concentration that indicates that one infectious particle is sufficient to initiate infection. *(Chapter 2)*

One-step growth curve A single reproduction cycle that occurs synchronously in every infected cell. *(Chapter 2)*

Origins (of replication) Specific sites at which replication of DNA begins. *(Chapter 9)*

Packaging Incorporation of the viral genome during assembly of virus particles. *(Chapter 13)*

Packaging signal Nucleic acid sequence or structural feature directing incorporation of a viral genome into a virus particle. *(Chapter 13)*

Particle–to–plaque-forming-unit (PFU) ratio The inverse value of the absolute efficiency of plating: the ratio of the total number of particles to the number that are infectious. *(Chapter 2)*

Pathogen Disease-causing virus or microorganism. *(Chapter 1)*

Permissive Able to support virus reproduction when viral nucleic acid is introduced; refers to cells. *(Chapter 2)*

Plaque A circular zone of infected cells that can be distinguished from the surrounding monolayer. *(Chapter 2)*

Plaque-forming units per milliliter A measure of virus infectivity. *(Chapter 2)*

Plaque purified Prepared from a single plaque (refers to virus stock); when one infectious virus particle initiates a plaque, the viral progeny within the plaque are clones. *(Chapter 2)*

Polarized cells Differentiated cells with surfaces divided into functionally specialized regions. *(Chapter 12)*

Polyadenylation The addition of ~200 A residues to the 3′ ends of cellular and viral transcripts made in eukaryotic cells. *(Chapter 10)*

Poly(A) tail The segment of ~200 A residues present at the 3′ ends of most cellular and many viral mRNAs. *(Chapters 10 and 11)*

Polycistronic Encoding several polypeptides; refers to mRNA. *(Chapter 11)*

Polyclonal antibodies The antibody repertoire against the many epitopes of an antigen produced in an animal. *(Chapter 2)*

Polysome An mRNA bound to multiple ribosomes that are synthesizing proteins from the mRNA template. *(Chapter 11)*

Portal A specialized structure for entry and/or exit of a viral genome into a preassembled protein shell. *(Chapter 4)*

Positive [(+)] strand The strand of DNA or RNA that corresponds in sequence to that of the messenger RNA. Also known as the sense strand. *(Chapter 1)*

Pregenomic mRNA The hepadnaviral mRNA that is reverse transcribed to produce the DNA genome. *(Chapter 7)*

Preinitiation complex (transcription) A promoter-bound assembly of an RNA polymerase and initiator proteins competent to initiate transcription. *(Chapter 8)*

Preinitiation complex (translation) The 40S ribosomal subunit bound to translation initiation proteins and initiator tRNA. *(Chapter 11)*

Primary cell cultures Cell cultures prepared from animal tissues; these cultures include several cell types and have a limited life span, usually no more than 5 to 20 cell divisions. *(Chapter 2)*

Primary cells Cells that have been freshly derived from an organ or tissue. *(Chapter 1)*

Primase An enzyme that synthesizes RNA primers for DNA synthesis. *(Chapter 9)*

Primer A free 3′-OH group required for initiation of synthesis of DNA from DNA or RNA templates and initiation of synthesis of some viral RNA genomes. *(Chapters 6 and 9)*

Prions Infectious agents comprising an abnormal isoform of a normal cellular protein but no nucleic acid; implicated as the causative agents of transmissible spongiform encephalopathies. *(Chapter 1)*

Procapsid A closed, protein-only structure into which viral genomes are inserted; precursor to a capsid or nucleocapsid. *(Chapter 13)*

Processivity The ability of an enzyme to copy a nucleic acid template over long distances from a single site of initiation. *(Chapters 7, 8, and 9)*

Promoter A set of DNA sequences necessary for initiation of transcription by a DNA-dependent RNA polymerase. *(Chapter 8)*

Promoter occlusion The mechanism by which access to a promoter is blocked by passage of a transcribing RNA polymerase. *(Chapter 8)*

Proofreading Correction of mistakes made during chain elongation by exonuclease activities of DNA-dependent DNA polymerases. *(Chapter 6)*

Prophage The genome of the quiescent bacteriophage in a lysogenic bacterium. *(Chapter 1)*

Proteasome A complex containing multiple proteases with different specificities that is responsible for degradation of polyubiquitin-tagged proteins to amino acids and small peptides. *(Chapter 8)*

Proteoglycans Proteins linked to glycosaminoglycans, which are unbranched polysaccharides made of repeating disaccharides. *(Chapter 2)*

Proviral DNA *See* Provirus.

Provirus Retroviral DNA that is integrated into its host cell genome and is the template for formation of retroviral mRNAs and genomic RNA. Also called proviral DNA. *(Chapter 7)*

Pseudodiploid Having two RNA genomes per virus particle that give rise to only one DNA copy, as is the case for retroviruses. *(Chapter 7)*

Pseudoreversion Phenotypic reversion caused by second-site mutation; also known as suppression. *(Chapter 3)*

Quasiequivalence The arrangement of structural units in a virus particle such that similar interactions among them are allowed. *(Chapter 4)*

Quasispecies Virus populations that exist as dynamic distributions of nonidentical but related replicons. *(Chapter 6)*

Reactivation A switch from a latent to a productive infection; usually applied to herpesviruses. *(Chapter 8)*

Reassortants Viral genomes that have exchanged segments after coinfection of cells with viruses with segmented genomes. *(Chapter 3)*

Reassortment The exchange of entire RNA molecules between genetically related viruses with segmented genomes. *(Chapters 3 and 6)*

Receptor The cellular molecule to which a virus attaches to initiate infection. *(Chapter 5)*

Replication centers Specialized nuclear structures in which viral DNA genomes are replicated. Also called replication compartments. *(Chapter 9)*

Replication forks The sites of synthesis of nascent DNA chains that move away from origin as replication proceeds. *(Chapter 9)*

Replication intermediate An incompletely replicated DNA molecule containing newly synthesized DNA. *(Chapter 9)*

Replication licensing Mechanisms that ensure that replication of cellular DNA is initiated at each origin once, and only once, per cell cycle. *(Chapter 9)*

Replicon A unit of replication in large genomes, defined by discrete origin and termini. *(Chapter 9)*

Resolution The minimal size of an object that can be distinguished by microscopy or other methods of structural analysis. *(Chapter 4)*

Resting state *See* G_0 state.

Retroelement A nucleic acid sequence that has been copied into DNA from an intermediate by reverse transcription. *(Chapter 7)*

Retroid viruses Viruses that replicate their genomes via reverse transcription. *(Chapter 7)*

Revert To change to the parental, or wild-type, genotype or phenotype. *(Chapter 3)*

Ribosome A molecular machine composed of RNA and protein that is the site of protein synthesis. *(Chapter 11)*

Ribozyme An RNA molecule with catalytic activity. *(Chapters 6 and 10)*

RNA-dependent RNA polymerase The protein assembly required to carry out RNA synthesis. *(Chapter 6)*

RNA editing The introduction into an RNA molecule of nucleotides that are not specified by a cellular or viral gene. *(Chapter 10)*

RNA interference A mechanism of posttranscriptional regulation of gene expression by small RNA molecules that induce mRNA degradation or inhibition of translation. *(Chapters 3 and 10)*

RNA processing The series of co- or posttranscriptional covalent modifications that produce mature mRNAs from primary transcripts. *(Chapter 10)*

RNA pseudoknot An RNA secondary structure formed when a single-stranded loop region base pairs with a complementary sequence outside the loop. *(Chapter 6)*

Satellites Small, single-stranded RNA molecules that lack genes required for their reproduction but do reproduce in the presence of another virus, which provides essential components (the **helper virus**). *(Chapter 1)*

Satellite virus A satellite with a genome that encodes one or two proteins. *(Chapter 1)*

Scaffolding protein A viral protein that is required for assembly of an icosahedral protein shell but is absent from mature virions. *(Chapter 13)*

Secretory pathway The series of membrane-demarcated compartments (e.g., the endoplasmic reticulum and Golgi apparatus), tubules, and vesicles through which secreted and membrane proteins travel to the cell surface. *(Chapter 12)*

Self-priming A mechanism by which some viral DNA genomes serves as primers, as well as templates, for DNA synthesis. *(Chapter 9)*

Semiconservative replication Production of two daughter DNA molecules, each containing one strand of the parental template and a newly synthesized complementary strand. *(Chapter 9)*

Serotype A virus type as defined on the basis of neutralizing antibodies. *(Chapter 2)*

Signal peptide A short sequence (generally hydrophobic) that directs nascent proteins to the endoplasmic reticulum. The signal may be removed, or retained as a transmembrane domain. *(Chapter 12)*

Signal transduction cascade or pathway A chain of sequential physical interactions among, and biochemical modification of, membrane-bound, cytoplasmic, and nuclear proteins. *(Chapter 14)*

Single-exon mRNAs mRNAs produced without splicing, because their precursors lack introns and splice sites. *(Chapter 10)*

siRNAs *See* Small interfering RNAs.

Site-specific recombination Exchange of DNA sequences at short DNA sequences that are specifically recognized by proteins that catalyze recombination. *(Chapter 9)*

Small interfering RNAs Small RNA molecules that base-pair with mRNAs to induce mRNA cleavage or inhibition of translation. Abbreviated siRNAs. *(Chapters 3 and 10)*

Small nuclear ribonucleoproteins Structures that contain small nuclear RNAs and several proteins; several participate in pre-mRNA splicing. *(Chapter 10)*

S phase The phase of the cell cycle in which the DNA genome is replicated. *(Chapter 9)*

Spliceosome The large complex that assembles on an intron-containing pre-mRNA before splicing; in mammalian cells, it comprises the small nuclear ribonucleoproteins containing U1, U2, U4, U5, and U6 small nuclear RNAs and ~150 proteins. *(Chapter 10)*

Splice sites Sites at which pre-mRNA sequences are cleaved and ligated during splicing; defined by short consensus sequences. *(Chapter 10)*

Splicing The precise ligation of blocks of noncontiguous coding sequences (exons) in cellular or viral pre-mRNAs with excision of the intervening noncoding sequences (introns). *(Chapter 10)*

Stop transfer signal A hydrophobic sequence that halts translocation of a nascent protein across the endoplasmic reticulum membrane; serves as a transmembrane domain. *(Chapter 12)*

Structural unit *See* Asymmetric unit.

Substitution mutation Replacement of one or more nucleotides in a nucleic acid. *(Chapter 3)*

Subunit A single folded protein of a multimeric protein. *(Chapter 4)*

Supercoiling The winding of one duplex DNA strand around another. *(Chapter 9)*

Suppression *See* Pseudoreversion.

Susceptible Producing the receptor(s) required for virus entry; refers to cells. *(Chapter 2)*

Tegument The layer interposed between the nucleocapsid and the envelope of herpesvirus particles. *(Chapter 4)*

Telomere A region of repeated sequences at the ends of linear, cellular chromosomes that is maintained (copied) by a specialized enzyme and that protects against loss of genetic information. *(Chapter 9)*

Termination codon Codons at which translation of an mRNA ceases, with release of both the nascent protein and ribosomes. *(Chapter 11)*

Termini Sites at which DNA replication stops. *(Chapter 9)*

Tight junctions The areas of contact between adjacent epithelial cells, circumscribing the cells at the apical edges of their lateral membranes. *(Chapter 2)*

Topology The geometric arrangement of, and connections among, secondary-structure units in a protein. *(Chapter 4)*

Transcription Copying of DNA carrying genetic information into a complementary RNA. *(Chapters 6 and 8)*

Transcriptional control region Local and distant DNA sequences necessary for initiation and regulation of transcription. *(Chapter 8)*

Transcytosis A mechanism of transport in which material in the intestinal lumen is endocytosed by M cells, transported to the basolateral surface, and released to the underlying tissues. *(Chapter 2)*

Transfection Introduction of viral nucleic acid into cells by transformation, resulting in the infection of cells. *(Chapter 3)*

Transfer RNAs Adapter molecules that align each amino acid with its corresponding codon on the mRNA. Abbreviated tRNAs. *(Chapter 11)*

Transport vesicles Membrane-bound structures with external protein coats that bud from compartments of the secretory pathway and carry cargo in anterograde or retrograde directions. *(Chapter 12)*

tRNAs *See* Transfer RNAs.

Tropism The predilection of a virus to invade, and reproduce, in a particular cell type. *(Chapter 5)*

Tumor suppressor gene A cellular gene encoding a protein that negatively regulates cell proliferation; mutational inactivation of both copies of the genes is associated with tumor development. *(Chapter 9)*

Two-hit kinetics A parabolic relationship between plaque count and virus concentration which indicates that two different types of virus particle must infect a cell to ensure reproduction. *(Chapter 2)*

Type-specific antigens Epitopes, defined by neutralizing antibodies, that distinguish and define viral serotypes (e.g., poliovirus types 1, 2, and 3). *(Chapter 2)*

Uncoating The release of viral nucleic acid from its protective protein coat or lipid envelope; in some cases, the liberated nucleic acid is still associated with viral proteins. *(Chapter 5)*

Vaccination Inoculation of healthy individuals with attenuated or related microorganisms, or their antigenic products, to elicit an immune response that will protect against later infection by the corresponding pathogen. *(Chapter 1)*

Variolation Inoculation of healthy individuals with material from a smallpox pustule, or in modern times from a related or attenuated cowpox (vaccinia) virus preparation, through a scratch on the skin (called scarification). *(Chapter 1)*

Viral pathogenesis The processes by which viral infections cause disease. *(Chapter 2)*

Virion An infectious virus particle. *(Chapters 1 and 4)*

Viroids Unencapsidated, small, circular, single-stranded RNAs that replicate autonomously when inoculated into plant cells. *(Chapter 1)*

Viroporin Hydrophobic viral protein that forms pores in cellular membranes; many facilitate release of progeny virus particles. *(Chapter 13)*

Viruses Submicroscopic, obligate parasitic pathogens comprising genetic material (DNA or RNA) surrounded by a protective protein coat. *(Chapter 1)*

Virus reproduction The sum total of all events that occur during the infectious cycle. *(Chapter 2)*

Virus titer The concentration of a virus in a sample. *(Chapter 2)*

Wild type The original (often laboratory-adapted) virus from which mutants are selected and which is used as the basis for comparison. *(Chapter 3)*

Zoonotic Transmitted among humans and other vertebrates; refers to infections and diseases. *(Chapter 1)*

Index

A

A (aminoacyl or acceptor) site, 350, 352, 354–355, 357, 360–362, 368–369
A9 protein, vaccinia virus, 448
A13L protein, vaccinia virus, 410
A14 protein, vaccinia virus, 448
A14L protein, vaccinia virus, 410
A17 protein, vaccinia virus, 448
A17L protein, vaccinia virus, 410
A33 protein, vaccinia virus, 451
A36 protein, vaccinia virus, 448, 450–451
A50R protein, vaccinia virus, 279
A56R protein, vaccinia virus, 410
AAP (assembly-activating protein), adeno-associated virus type 2, 425
AAV. *See* Adeno-associated virus
Abce1 protein, 424
Abundance, of viruses, 4
Acanthamoeba castellani, 116
Acanthamoeba polyphaga, 114
Accommodation, ribosome, 360–361
Acetyl coenzyme A (acetyl-CoA), 480–485
Acid-catalyzed membrane fusion, 139–142
Acidianus bottle virus, 82
Acidianus convivator, 456
Acidianus two-tailed virus, 456
Acidic activation domain of VP16 protein, herpes simplex virus, 252
Acquired immunodeficiency syndrome (AIDS). *See* Human immunodeficiency virus; Human immunodeficiency virus type 1
Acsl3, 491
Actin, 118, 133, 137, 145, 147–148, 467
Actin-dependent transport of vaccinia virus particles, 447–450
Actinomycin D, 214
Activator, 225, 232–233, 235–237, 245, 247–251, 253–254, 257–258
  coactivator, 235, 248, 251
  transactivator of transcription (Tat), 237, 239–245, 248, 262
Activator protein 1 (Ap1), 232, 233, 249
Acylation of proteins, 400, 402
Adapters, export, 327, 329–331
Adar1 (adenosine deaminase acting on RNA 1), 326
Adeno-associated virus
  DNA synthesis, 268, 276, 296
  genome, 268
    conversion for transcription, 226–227
    organization, 520
  integration, 296, 306
  inverted terminal repetition (ITR), 276–278, 296
  origin recognition proteins, 277
  origin replication protein (AAV5), 285
  overview, 520–521
  recombination, 306
  Rep 78/68 protein, 268, 276–277, 283, 296–297, 306
  reproductive cycle, single-cell, 521
  structure, 92, 94, 96, 520
  viral vectors, 74–78, 296
  virus particle appearance (AAV4), 82
Adeno-associated virus type 2
  assembly-activating protein (AAP), 425
  DNA synthesis, 268, 276, 296
  genome, 268, 520
  inverted terminal repetition (ITR), 276–278, 296
  Rep 78/68 protein, 276, 306
  reproductive cycle, single-cell, 521
  structure, 92, 94, 96
Adeno-associated virus type 4, virus particle appearance, 82
Adeno-associated virus type 5, origin replication protein, 285
Adenosine deaminase acting on RNA (Adar1), 326
Adenovirus
  alternative splicing, 322, 325, 326, 331–332, 334
  assembly, 382, 384, 422, 423, 425, 428
    genome packaging, 432–433
  attachment to host cells, 127, 128–129, 467
  avian adenovirus, growth in embryonated eggs, 36
  blockage of RNA interference, 345
  capsid structure, 128–129
  cell cycle arrest, 292
  cellular gene expression and, 473, 474, 477, 478
  cleavage of precursor proteins, 456
  cytopathic effect, 34, 35
  cytoskeletal filament cleavage, 495
  diseases, 502
  DNA polymerase, 277–278, 280, 281, 286–287, 303
  DNA synthesis, 268, 277–278, 282, 286–288, 292, 294, 307
  E1A protein, 235, 246–248, 250–253, 289, 477, 478
  E2 protein, 232, 246, 251, 280, 290
  E2 transcription factor (E2f), 232, 247, 251, 253
  E3 glycoprotein gp19, 404
  E4 Orf1 protein, 374
  eIF4E activity modulation, 374
  entry into cells, 128, 144
  genome organization, 56, 502
  genome packaging, 432–433
  genome structures critical for function, 62
  hemagglutination, 41
  host cell signaling and, 467
  import of viral genome into nucleus, 149, 151, 153
  infectious DNA clones, 68
  information retrieval from genome, 63
  inhibition of cellular DNA synthesis, 291
  inhibition of cellular mRNA export, 334
  IVa_2 protein, 231, 247, 253, 432–433
  L1 protein, 432–433
  L3 protease, 450
  L4 protein, 383, 384, 423, 425, 432, 473
  major late (ML) transcription unit, 333, 334
  movement within cells, 148
  origin recognition protein, 277
  overview, 502–503
  packaging genome, 105
  particle-to-PFU ratio, 40
  polyadenylation, 315, 330
  Pre-TP, 277–278, 281, 286, 290
  pre-VI (pVI) protein, 384, 456
  promoter, 232
  protease, 450, 456
  protein priming, 165, 502
  proteins in virion, 117
  Rb protein, 251–253
  release of virus particles, 450
  replication, 216
  replication centers, 292
  replication system, 286–287
  reproductive cycle, single-cell, 503
  reprogramming promoter-associated transcriptional regulators, 478

Adenovirus *(continued)*
  RNA polymerase II transcription units, 257–259
  RNA polymerase III, transcription by, 502
  single-stranded-DNA-binding protein, 280, 286–287, 292
  splicing, 317–318
  structure, 85–86, 99–101, 502
  transcription, 226, 228, 235, 246–248, 250–253
  transcriptional regulators, 247, 254
  uncoating, 144, 153
  VA-RNA I, 257–259, 370
  viral mRNAs in virus particle, 117
  viral oncotherapy, 75
  viral vectors, 74–76
  virion enzymes, 116
  virus-associated (VA) RNAs, 345
Adenovirus death protein (ADP), 450
Adenovirus type 2, human
  assembly, 422
  late promoter, 228
  mapping initiation site and transcription *in vitro,* 229
  virion enzymes, 116
Adenovirus type 5, human
  DNA synthesis, 268, 277, 282, 287, 294
  E1B protein, 295
  E4 Orf3 protein, 294
  E4 Orf6 protein, 295
  genome organization, 502
  host cell signaling and, 467
  L1 protein, 425
  one-step growth curve, 49
  origin of replication, 282
  origin recognition protein, 277
  overview, 502–503
  packaging signal, 433
  reproductive cycle, single-cell, 503
  scaffolding proteins, 425
  single-stranded-DNA-binding protein, 287
  structure, 502
  viral vectors, 74
  virus-associated RNA I (VA-RNA I), 257–259
Adenovirus type 36, human
  E4 Orf1, 492
  obesity, 492
Adenovirus type 40, 502
Adenovirus type 41, 502
ADP (adenovirus death protein), 450
ADP-ribosylation factor 1 (Arf1), 117
Aequorea victoria, 40
Affinity, receptor, 124
Affinity isolation, 126
African swine fever virus
  CAP80 protein, 425
  sulfhydryl oxidases, 392
Ago2, 339, 341
Aichi virus, 357
AIDS. *See* Human immunodeficiency virus; Human immunodeficiency virus type 1
Alix protein, 442
Alkaline nuclease, 291, 306
Alkylating agents, 66
Allele-specific suppressors, 71
Alloherpesvirus, 514
α-amanitin, 166, 168, 181
Alphacoronavirus, 506
Alphaherpesvirus
  attachment to host cells, 131–132
  axonal transport, 148, 403–404, 405
  citric acid cycle and, 483, 484
  genome organization, 514
  latency, 256
  latency-associated miRNAs, 343
  structure, 514
Alpharetrovirus, 530
  reverse transcriptase domain and subunit relationships, 199
Alphavirus
  budding at plasma membrane, 440
  cell-to-cell spread, 457
  cellular gene expression inhibition, 472
  cytopathic effect, 34
  diseases, 534
  E2 protein, 402
  entry into cells, 142
  envelope glycoprotein proteolytic processing in Golgi, 399–400
  furin family protease, 400
  fusion protein, 140
  genome organization, 534
  genome structure, 157
  information retrieval from genome, 63
  membrane fusion, 138
  mRNA cap, 313, 314
  mRNA synthesis, 159
  oligomerization of proteins, 395
  particle-to-PFU ratio, 40
  replication, 159
  RNA synthesis, 171–172
    cellular sites for, 179
  RNA-dependent RNA polymerase, 161
    template specificity, 169, 171–172
  structure, 109–111, 534
  suppression of termination, 367
  unbalanced production of (+) and (−) strand RNA, 178
Alternative splicing, 317, 322–325
  adenovirus, 322, 325, 326, 331–332, 334
  bovine papillomavirus type 1, 322–323, 324
  regulation by viral proteins, 331–332, 334
  retrovirus, 323–325
Amantadine, 109
Ambisense RNA, 56, 59, 60
Ambisense RNA genome, 157, 159, 174
Aminoacyl-tRNA, 360–361, 368, 369
Aminoacyl-tRNA synthase, 350–351
  encapsidation of, 438–439
Amphipathic molecules, 29
Anabolism, 477
Anaerobic glycolysis, 477
Anellovirus
  genome organization, 57
  genome size, 65
Aneuploid, 33
Animal virus, 4
  classification of, 17–21
  landmarks in the study of, 22
  latent infection, 254, 256–258
  one-step growth analysis, 49–50
  particle-to-PFU ratio, 39, 40
  plaque assay, 36–37
  viral genomes, 19
Anthropomorphic characterization of viruses, 18
Antibody
  epitope, 42
  monoclonal, 42, 43
  polyclonal, 42, 43
  serological methods, 42–45
Anticodon loop, tRNA, 351
Antigens
  direct and indirect methods for detection, 43
  neutralization antigenic sites, 42
  serological methods, 42–45
  type-specific, 42
Antiquity, viral infections in, 6–7
Antiviral state, 369
Ap1 (activator protein 1), 232, 233, 249
Aphthovirus, 128, 522
Apical domain, epithelial cell plasma membrane, 402–403
Apical surface, 27
Apobec3, 117, 327
Apoptosis, regulation of, 466–467, 469, 474, 486, 490
Archaea
  polycistronic messenger RNA (mRNA), 349–350
  viruses, 19
Arenavirus
  cytopathic effect, 34
  diseases, 504
  genome organization, 59, 60, 504
  overview, 504–505
  release of virus particles, 442
  reproductive cycle, single-cell, 505
  RNA polymerase, 174
  RNA synthesis, 174, 177
  structure, 504
  zoonoses, 504
Arf1 (ADP-ribosylation factor 1), 117
Argonaute (Ago) family, 339, 341
Arterivirus
  genome organization, 58
  information retrieval from genome, 63
  ribonucleoproteins, 160
  synthesis of nested subgenomic mRNAs, 172
Arthropod vectors
  alphavirus, 534
  flavivirus, 510
Asfarvirus, genome organization, 56
ASLV. *See* Avian sarcoma/leukosis virus
Assay of viruses, 36–46
  measurements of infectious units, 36–39
    efficiency of plating, 39
    end-point dilution assay, 38–39, 40
    fluorescent-focus assay, 37
    infectious-centers assay, 37–38
    plaque assay, 36–38
    transformation assay, 38, 39
  measurements of viral particles/components, 39–46
    hemagglutination, 41
    imaging particles, 40–41
    serological methods, 42–45
    viral enzyme activity measurement, 41–42
    viral nucleic acid detection, 45–46
  nucleic acid detection, 45–46
    DNA microarrays, 45
    high-throughput sequencing, 45–46
    polymerase chain reaction, 45
  serological methods, 42–45
    enzyme immunoassay, 44–45
    hemagglutination inhibition, 42–43
    immunostaining, 43–44
    virus neutralization, 42
Assembly of virus, 380–414
  bacteriophage, 423–424
  bacteriophage T4, 420
  chaperone participation, 423, 424, 428–429

concerted, 430, 431, 432
coordination with envelope acquisition, 440–441
in cytoplasm
poliovirus, 426
replication and assembly platforms, 495–497
dead-end products, 420
envelope acquisition, 439–441
extracellular, 456
Gag protein function in, 412–413
herpes simplex virus type 1, 421
interactions with internal cellular membranes, 409–411
localization of viral proteins to compartments of the secretory pathway, 410
localization of viral proteins to nuclear membrane, 411
intermediates, 423, 425
methods of studying, 418–421
biochemical and genetic analyses of intermediates, 418–420
microscopy, 418–419
recombinant DNA technology-based, 421
structural studies, 418
nucleocapsids, hepadnavirus, 218
within nucleus, 382–384
import of viral proteins for assembly, 383–384
overview, 417–418
at plasma membrane, 384–409
disruption of secretory pathway, 404–406
overview, 384–385
release of virus particles, 441–444
signal-sequence-independent transport of viral proteins, 406, 408–409
sorting of viral proteins in polarized cells, 401–404
transport of viral membrane proteins to plasma membrane, 386–401
protein incorporation into particles, 438–439
protein shells
assisted assembly reactions, 428–429
capsid and nucleocapsid assembly, 423–425
self-assembly, 425–428
structural unit formation, 421–423
visualization of structural transitions, 430
secretory pathway disruption, 404–406
effects on compartments, 404–406
inhibition of cellular protein transport, 404
unfolded protein response, 406, 407
sequential, 430–431
signal-sequence-independent transport of viral proteins, 406, 408–409
lipid-plus-protein signals, 406, 408–409
protein sequence signals, 408–409
sorting of viral proteins in polarized cells, 401–404
epithelial cells, 402–403
neurons, 402, 403–404
transport of viral genomes to assembly sites, 411–414
influenza virus, 411–412
microtubules and, 414
transport from cytoplasm to plasma membrane, 411–413
transport from nucleus to cytoplasm, 411
Assembly-activating protein (AAP), adeno-associated virus type 2, 425
Astrocytes, 458
Astroviridae
genome organization, 58
information retrieval from genome, 63
Asymmetric units, 90
Atf family, 248
Atf-2, 251, 253
Atf-6, 406–407
Atk, 466–470, 489, 492
Atm (ataxia telangiectasia mutated) kinase, 301
ATP hydrolysis, 271, 275, 286
ATP synthase, 484, 486
ATPase, 133
att sites, 190
Attachment to host cells, 123–132. *See also* Receptor(s)
avidity, 124
enveloped viruses, 129–132
alphaherpesviruses, 131–132
human immunodeficiency virus type 1, 131
influenza virus, 129–131
general principles, 123–124
nonenveloped viruses, 126–129
glycolipids, 129
via protruding fibers, 128–129
via surface features, 126–128
rate of attachment, 124
receptor identification, 124–126, 127
Attenuated vaccine, 9
AUG codon, 350–357, 365–366, 372, 436
bicistronic mRNAs, 368
choice of initiation codon, 354
leaky scanning and, 365–366
Autocatalytic reaction, 95
Autointegration, 210
Autonomously replicating sequences, 270
Autophagosomes, 405–406, 488, 496–497
Autophagy, 488
nonlytic release of virus particles, 498
Autoregulation, transcription, 235, 237–240, 248–249
Avery, Oswald, 14
Aviadenovirus, 502
Avian adenovirus, growth in embryonated eggs, 36
Avian coronavirus, 506
Avian influenza virus
H5N1, 4
H7N9, 43
host range, 131
RNA profiling, 476
Avian leukosis virus. *See also* Avian sarcoma/leukosis virus
entry into cells, 142
genome organization, 530
integration site, 341
splicing, 323
Avian sarcoma/leukosis virus (ASLV)
enhancer, 237
epigenetic silencing of integrated proviral DNA, 238
genome packaging, 435
integrase, 204
structure, 211, 213
integration sites, 208
reverse transcriptase, 193
domain and subunit relationships, 199
transcription, 235–237
viral mRNAs in virus particle, 117
Avidity, of virus binding, 124
Avihepadnavirus, 512
Avipoxvirus, 526
Avulavirus, 518
Axon, 403–405
Axonal transport, 148, 403–404, 405

B

B5R protein, vaccinia virus, 410
B19 virus, human, 520
Bacteria
polycistronic messenger RNA (mRNA), 349–350
RNA polymerase, 225
Bacteriophage
assembly, 423–424
biomass on earth, 4
burst concept, 14, 46
CRISPR/Cas system, 340
discovery of, 12
DNA injection, 150
genome sequence, 60
Hershey-Chase experiment, 15
historical aspects, 14–16
landmarks in molecular biology, 6
lessons from, 14–16
lysogeny, 15, 16
one-step growth cycle, 14, 17, 46, 48–49
particle-to-PFU ratio, 39
protein priming, 165, 278
T phages, 15
transduction of host genes, 16
Bacteriophage ϕ6
de novo priming of RNA synthesis, 165
RNA polymerase, 169, 170, 176
semiconservative replication, 176
Bacteriophage ϕ29, 216
model for transition between initiation and elongation during DNA synthesis, 281
protein priming of DNA synthesis, 278, 281
Bacteriophage ϕX174, rolling-circle replication of, 278
Bacteriophage HK97, 95
Bacteriophage λ, 15, 423
genetic map, 190
gpD protein, 430
gpE protein, 430
integration, 190
lysogeny, 254–256
lytic cycle, 254–256
origin of replication, 272
recombinase, 304, 306
replication and recombination, 304–306
repressors, 255–256
rolling-circle replication, 305
visualization of structural intermediates during assembly, 430
Bacteriophage M13, 87
Bacteriophage MS2, 60
Bacteriophage Mu, 16, 205
Bacteriophage P1, 16
Bacteriophage P22
capsid, 94
packaging viral genome, 437
Bacteriophage PRD1, 97
protein priming of DNA synthesis, 278
Bacteriophage Qβ, 170
Bacteriophage T3, transcription, 226
Bacteriophage T4
assembly, 420, 423–424
burst concept, 46

Bacteriophage T4 *(continued)*
coordination of late genes with viral DNA synthesis, 251
entry into cells, 150
genome packaging, 434, 437
origin-independent, recombination-dependent replication, 304
packaging genome, 105
structure, 13, 111–112, 114
transcription, 226
Bacteriophage T7
coordination of late genes with viral DNA synthesis, 251
RNA polymerase, 68–69, 163, 358
transcription, 226
Bacteriophage T7 RNA polymerase, 68–69, 163, 358
Baculovirus, genome size, 65
Baculovirus vector, 421
Baf (barrier-to-autointegration factor), 210
Bafilomycin A1, 147
Baltimore, David, 189
Baltimore classification system, 3, 20–21, 55
Bang, Olaf, 11
Barrier-to-autointegration factor (Baf), 210
Basal lamina, 28
Base analogs, as mutagens, 66
Base substitutions, mutational intermediates for, 199
Basolateral domain, epithelial cell plasma membrane, 402–403
Basolateral surfaces, 27
B-cell lymphoma, 343
Bcl-2, 469
Beijerinck, Martinus, 11
Benefits, of viruses, 4
Bernal, J. D., 417
β-barrel jelly roll, 94–96, 98–100, 107, 114
Betacoronavirus, 506
Betaherpesvirus
citric acid cycle and, 483, 484
micro-RNAs
inhibition of antiviral defenses, 345
latency-associated miRNAs, 343
overview, 514
Betaretrovirus, 102, 530
Biochemical analyses of assembly intermediates, 418–420
Bioinformatics, 45–46
Biomass, of viruses, 4
Birnavirus
RNA polymerase, 176
semiconservative replication, 176
BK virus, 524
attachment to host cells, 129
large T antigen, 344
micro-RNA (miRNA), 342, 344
Blackcurrent reversion virus, 360
Bluetongue virus, 528
capping enzyme, 313–314
packaging genome, 104
structure, 101–102
VP4 protein, 313–314
BM2 protein, influenza virus, 366, 367
Borna disease virus, import of viral genome into nucleus, 123
Bornavirus, genome organization, 59
Bovine diarrhea virus
pathogenic viruses produced by recombination, 184, 185
RNA polymerase, 169
Bovine herpesvirus, attachment to host cells, 131
Bovine papillomavirus type 1
alternative polyadenylation and splicing control, 322–323, 324
transcriptional regulators, 247
Bovine papillomavirus type 5
origin recognition proteins, 277
Bovine viral diarrhea virus, 510
Brd proteins, 209
Brd4 protein, 301
Brg-1, 244, 252
BSL-4 containment, for filovirus, 508
Bst2 (bone marrow stromal antigen 2), 444
Budding of cellular mRNAs from nucleus, 453
Budding of virus particles, 402, 406
from cellular membrane, 439–440
exocytosis, 444–450
from nucleus, 411
at plasma membrane
Escrt-dependent budding, 441–443
Escrt-independent budding, 444
sequence motifs required for budding, 441–442
from plasma membrane, 385, 402–403, 406, 414
polarity, 403
into secretory pathway compartments, 410
into secretory pathway cytoplasmic compartment, 444–445
strategies, 440
Bunyavirus
budding from plasma membrane, 403
cap snatching, 315
envelope acquisition, 385
envelope glycoprotein proteolytic processing in Golgi, 400
Gc protein, 410
genome organization, 59, 60
Gn protein, 410
interaction with internal cellular membranes, 410
plant viruses, 385
priming by capped RNA fragments, 166–167
protein priming, 166
RNA synthesis, 166–168
RNA-dependent RNA polymerase, 161, 163, 166, 168
Burst concept, 14, 46

C

C protein
alphavirus, 109
hepadnavirus, 214, 216, 221, 444–445
hepatitis B virus, 444–445
hepatitis C virus, 495
Semliki Forest virus, 109
togavirus, 444
CA protein
human immunodeficiency virus type 1, 117, 409, 455
retrovirus, 406, 423, 440–441
Calicivirus
genome organization, 58
protein priming, 165
RNA synthesis, 165, 171
VPg-dependent ribosomal recruitment, 353
whales, 4
Calmodulin-dependent kinase II (CamkII), 470
Calnexin, 392–394
Campbell, Allan, 190, 225
Cancer
cervical, papillomavirus-associated, 338
hepatocellular, 342
human herpesvirus 8-associated, 343
oncogene, 467
oncogenic retroviruses, 189
signal transduction and, 467, 470
tumor suppressor gene, 288
viral oncogenesis promotion by cellular miRNA-155, 341
viral oncotherapy, 73, 74, 75, 76
viruses and, 3, 5
Canyons, capsid, 97, 98, 127–128
Cap, primer, 161
Cap analogs, 353–354
Cap snatching, 315
Cap structure, 349
3′-cap-independent translational enhancer (3′ CITE), 360
CAP80 protein, African swine fever virus, 425
Cap-binding complex (Cbc), 331
Cap-binding site, 176
Cap-dependent endonuclease in virus particles, 116
Cap-independent translational enhancer (CITE), 360
Capping, 311, 312–315
acquisition of cap structures from cellular RNAs, 315
assembly line for, 313, 314
cap snatching, 315
decapping enzymes, 336, 338, 341
identification of cap structures on viral mRNAs, 313
messenger RNA, viral, 168, 262
overview, 311–312
synthesis of cap structures by cellular enzymes, 312–313, 314
synthesis of cap structures by viral enzymes, 313–315
Capping enzyme, 312–316
Capsid, 86–104
assembly, 423–425
asymmetric, 102–104
canyons, 97, 98, 127–128
closed structure, 86
covalent joining of subunits, 91
defined, 83
helical, 86, 88–89
icosohedral, 86, 89–102
adeno-associated virus type 2, 92, 94, 96
adenovirus, 99–101
poliovirus, 94–98
reovirus, 101–102
simian virus 40, 98–99
open structure, 86
packaging genome within, 104–106
retrovirus, 102–104
size, 65
viral chain mail, 91, 95
Car receptor, 127, 128–129
Carbohydrate chains, membrane protein, 31
Cardiovirus, 522
capsid structure, 128
inhibition of cellular mRNA export, 334
Cas1, 340
Cascade, 340
Caspar, Donald, 90, 93
Caspase, 486
Catabolism, 477
Catenated DNA, 274

Cauliflower mosaic virus, replication cycle of, 222
Caulimovirus, transcription, 226
Caveolae, 124, 133, 137, 148
Caveolin, 30, 145, 403
Caveolin-mediated endocytosis, 133, 136–137, 137
Caveosome, 137
Cbc (cap-binding complex), 331
CCR5, 127, 135, 136, 391
CD4, 124, 127, 131, 134, 135, 136, 239, 240, 404, 458, 467
gene expression in infected, 474
CD55, 132, 148
CD155, 126, 128, 131, 145, 148
Cell(s)
host cell architecture shaping viral structure, 385
as host for infectious cycle, 25, 26
host response to infection, 464–500
gene expression, 470–478
metabolism, 477, 479–492
remodeling of cellular organelles, 491–499
signal transduction, 465–470
movement of viral and subviral particles within cells, 147–148
one-step growth curve, 25
plant cell structure, 385
RNA synthesis in, 58
types of epithelia, 27
viral entry into, 31
Cell cultures, 16, 32–35
evidence of viral growth in cultured cells, 33–35
types of, 32–33
Cell cycle, 289–292, 297–300
arrest, 292
Epstein-Barr virus replication, 298–299
G_1 phase, 289, 299
halting for DNA repair, 303
once-per-cell-cycle human papillomavirus replication, 302
Cell membrane proteins, 30–31
Cell metabolism, regulation of viral enhancers by, 233
Cell surface
architecture of, 27–31
vesicular transport to, 396–401
mechanism, 396–398
reactions with Golgi, 398–401
Cell wall, 385
Cell-to-cell spread, 457–460
masking strategy, 459
vaccinia virus, 449, 451
Cellular membrane, budding of virus from, 439–440
Cellular tethers, in retroviral integration, 208–209
Cellular transport pathways, 123
Central dogma, 21, 189
Centrosome, 133, 148
Cervical cancer, papillomavirus-associated, 338
Chain mail, viral, 91, 95
Chaperone, 117
adenovirus L4 protein, 383, 384
assembly assisted by, 422, 423–424
endoplasmic reticulum, 388–390, 392–395, 406–407
GroEL, *Escherichia coli,* 423–424
Grp78, 388–389, 392–393, 396, 406–407
selectivity of chaperones for viral glycoproteins entering endoplasmic reticulum, 393
TriC, 423–424
virus-induced chaperone-enriched (VICE) domains, 494
Chargaff, Erwin, 55
Chase, Martha, 14–15
Chemokine, 135
Chemokine genes, 51
Chemotactic cytokines, 135
Chimpanzee foamy virus, 530
ChIP (chromatin immunoprecipitation assay), 227, 252
Chk2, 299, 301, 304
Chlorella, 350
Chrebp (carbohydrate response element-binding protein), 480, 488–489, 490
Chromatin
cellular tethers in retroviral integration, 208–209
euchromatin, 236
heterochromatin, 226
replication of chromatin templates, simian virus 40 DNA synthesis, 274–275
structure, 226, 228
transcription, 254
viral, 228
Chromatin assembly factor 1, 275
Chromatin immunoprecipitation assay (ChIP), 227, 252
Chronic fatigue syndrome, 197
cI repressor, bacteriophage λ, 255–256
Circovirus
genome organization, 57
genome size, 65
mutation rate, 302
cis-acting replication elements (cre), 169
CITE (cap-independent translational enhancer), 360
Citric acid cycle, 483–484
Classical swine fever virus, 372
Classification, viruses, 17–21
animal viruses, 17–21
Baltimore, 3, 20–21, 55
classical system, 17–20
complexities of, 18
by genome types, 18–19
by neutralization tests, 42
Clathrin, 124, 145
Clathrin-mediated endocytosis, 133, 135–136, 403
Cleavage and polyadenylation specificity protein (Cpsf), 315–317, 324–325, 329, 331, 333, 335
Clinical isolates, 66
Closterovirus, genome organization, 64
Clustered regularly interspersed short palindromic repeat (CRISPR) system, 340
Coactivator, 235, 248, 251
Coated pits, 124, 133, 135–136
Coatomer protein II (CopII)-coated vesicles, 496
Codon, 349
initiation codon, 349–355, 357, 359, 365–366, 372
stop, 361, 366–368
termination, 349, 350, 362, 366–368
Collagen, 28
Colorado tick fever virus, 528
Coltivirus, 528
Common cold, 506
Complementation, 67
Concatemer, 278–279, 288–289, 296, 301, 303, 305–307, 437
Concerted processing and joining step, in retroviral DNA integration, 206
Confocal microscopy, 44
Conservative replication, 176, 179
Constitutive transport elements (CTEs), 328–329, 331
Continuous cell lines, 33
Convertase, 138
CopII, 397–398, 405, 496
Copy choice, 184, 198
Core promoters, 228
Coreceptor, 123, 124, 127
Corepressors, 235
Coronavirus
autophagosome-like vesicles, 405–406
budding at plasma membrane, 440
classification, 18
cytopathic effect, 34
diseases, 506
genome organization, 58, 59, 506
genome size, 65
information retrieval from genome, 63
interaction with internal cellular membranes, 410
M protein, 410
MERS, 52
overview, 506–507
reproductive cycle, single-cell, 507
ribonucleoproteins, 160
RNA synthesis, 171
membranous sites for, 179, 182
S protein, 402, 410
structure, 506
synthesis of nested subgenomic mRNAs, 172
Covalently closed circular DNA, hepadnavirus, 214
Cowpea mosaic virus, 87
Cowpox, 8–9
Coxsackievirus
attachment to host cells, 126, 128, 132
cleavage of eIF4G, 373
host cell signaling and, 468, 470
internal ribosome entry site (IRES), 359
release of virus particles, 450
signaling via cell receptors, 148
Cpsf (cleavage and polyadenylation specificity protein), 315–317, 324–325, 329, 331, 333, 335
Cre (*cis*-acting replication elements), 169
Creb (cyclic AMP response element (CRE)-binding protein), 232
Creb-binding protein (Crebbp), 251, 252
Crescents, vaccinia virus, 445–448
Crick, Francis, 86, 89
viral classification by, 20–21, 55
Cricket paralysis virus
bicistronic mRNA, 368
internal ribosome entry site (IRES), 357
CRISPR system, 340
CRISPR-Cas9 (clustered regularly interspersed short palindromic repeat-CRISPR-associated nuclease 9), 72–73
Crm-1, 327
Cro, bacteriophage λ, 255–256
CrRNA, 340
Cryo-electron microscopy, 82, 83–86, 430
Cryo-electron tomography, 448
Cstf (cleavage stimulatory protein), 316, 317, 326, 330–331
CTEs (constitutive transport elements), 328–329, 331
Cul5 protein, 295
Culling, 4
Cultivation of viruses, 32–36
cell culture, 32–35
embryonated eggs, 35–36
laboratory animals, 36

CXCR4, 124, 127, 135, 391, 458
Cyclic AMP, 466
Cyclic AMP response element (CRE)-binding protein (Creb), 232
Cyclin, 242
Cyclin A, 283
Cyclin T, 242, 245
Cyclin-dependent kinase 2 (Cdk2), 283
Cyclin-dependent kinase 9 (Cdk9), 242, 245, 253
Cyclophilin A, 117, 118, 495
Cystic fibrosis, 76
Cystic fibrosis transmembrane conductance regulator gene, 76
Cytokine genes, 51
Cytomegalovirus, 514
  assembly compartments, 496
  cell cycle arrest, 292
  citric acid cycle and, 483
  cytopathic effect, 35
  DNA polymerase, 303
  glucose metabolism and, 480–482
  host cell gene expression and, 475
  lipid metabolism, 487–489
  protein synthesis stimulation, 374
  UL38 protein, 374
  UL50 protein, 406–407
  unfolded protein response and, 406–407
  US2 protein, 395, 404
  US11 protein, 395, 404
  viral mRNAs in virus particle, 117
Cytopathic effects, 33–35
  end-point dilution assays, 40
Cytopathic viruses, burst concept and, 46
Cytoplasm
  assembly of virus in, 426, 444–449
  release of virus particles, 496–497
  remodeling, 495–498
  replication and assembly platforms, 495–497
  retrotranslocation from ER, 394–395, 406
  secretory pathway compartments, 444–445
  transport of viral genomes to assembly sites
    transport from cytoplasm to plasma membrane, 411–413
    transport from nucleus to cytoplasm, 411
  vesicular stomatitis virus nucleocapsid movement via microtubules, 414
  viral factories, 495
  viral thio oxidoreductase system operating in, 392
Cytoplasmic domain, of viral membrane proteins, 390
Cytoplasmic vesicles, 497–498
Cytorhabdovirus, 532
Cytoskeleton, 28, 147–148, 382–384. *See also* Microtubules
  remodeling, 467

D

D4R protein, vaccinia virus, 291
D5R protein, vaccinia virus, 279, 287
D9 protein, poxvirus, 336, 338
D10 protein, poxvirus, 336, 338
D13 protein, vaccinia virus, 448
Dane particles, 512
Darwin, Charles, 9
Dcp1a, 336
Dc-sign, 127, 132
Deadenylation-dependent degradation, 336
Death domain-associated protein (Daxx), 238–239
Decapping, 336, 338, 341
Decay-accelerating protein, 132
Defective interfering RNAs, 185
Delbrück, Max, 14, 46
Deletion
  mutations, 70
  in viral vectors, 74–76
Deltaretrovirus, 530
Dendrites, 403
Dengue virus
  apoptosis blockage, 469
  autophagy and, 499
  cytoplasmic replication and assembly compartments, 495–496
  envelope glycoprotein proteolytic processing in Golgi, 400–401
  fatty acid metabolism, 487–488
  interaction with internal cellular membranes, 410
  overview, 510
  RNA helicase, 170
  RNA synthesis, 165
    membranous sites for, 182
  structure, 109, 112, 510
  subgenomic flavivirus RNAs (sfRNAs), 337
  subgenomic RNA (sgRNA), 469
  unfolded protein response and, 406
Deoxynucleotide monophosphates (dNMPs), 267, 281
Deoxynucleotide triphosphates (dNTPs), 267, 274, 289–292, 307
Dependovirus, 520
Deproteinized viral DNA molecules, 68
Der1p protein, 395
Desmotubules, 459
Destructive replication, 197
D'Hérelle, Félix, 12
Diabetes
  hepatitis B virus and, 482
  hepatitis C virus and, 482–483
Dicer, 339, 341, 342, 345
Dicistrovirus, bicistronic mRNA of, 349, 368
Difference mapping, 86
Dimer linkage sequence, 434–435
Dimerization, packaging viral genome and, 434–436
Diploid cell strains, 33
Discontinuous DNA synthesis, 269–270, 273–275, 278–279, 307
Disease, 3
Disintegration step, in retroviral DNA integration, 206
Disulfide bond formation
  in cytoplasm, 392
  in endoplasmic reticulum, 390
  maturation of progeny virus particles, 457
DNA
  catenated, 274
  as genetic material, 12
  genome (*see* DNA genome)
  packaging signals, 432–434
  priming via DNA, 275–277, 279, 304
  relaxation of, 274
  size of molecules, 65
  (+) strand, 57, 214, 216, 219–221
  (−) strand, 57, 214, 216–221
  (+) strong-stop, 195
  (−) strong-stop, 193
  supercoiling, 274
  synthesis (*see* DNA synthesis)
DNA damage response proteins, 293–294
DNA genome, 18–19, 56–58. *See also* DNA virus
  circular, 57
  conversion of genomes to templates for transcription, 226–228
  devolution, 61, 64
  double-stranded, 19, 56–57, 97
  gapped DNA, 57, 65, 214
  import into nucleus, 151
  replication, 266–307 (*see also* DNA synthesis, viral)
  single-stranded, 19, 57–58, 302
  size, 65, 267
DNA integration. *See* Integration
DNA ligase, in vaccinia virus replication, 279, 287
DNA ligase I, 279
DNA ligase IV, 288, 303
DNA microarrays, 45
DNA polymerase, 58
  cellular, 267, 270, 292, 297, 301–302
    eukaryotic, 270
    fidelity, 301–302
    lagging strand synthesis, 272–274
    leading strand synthesis, 272–274
    mammalian, 270
    mechanism of action, 272–273
    postintegration repair, 207
    proofreading, 301–302
  polymerase chain reaction, 45
  primer requirement, 267
  processivity, 270
  proofreading, 65, 182, 301–303
  reverse transcriptase compared, 198
  RNA-dependent RNA polymerase compared, 161
  viral, 267, 277–278, 280–281, 286–287, 289–292, 299, 301–303, 307
    adenovirus, 277–278, 280, 281, 286–287
    fidelity, 302–303
    hepadnavirus P protein, 214–220
    proofreading, 303
DNA polymerase α-primase, 270, 272–273, 275, 277, 282, 286, 298
DNA polymerase I, 272
  Escherichia coli DNA polymerase I Klenow fragment, 163, 200
DNA polymerase III
  cellular tethers in retroviral integration, 208
  E. coli, 272
DNA polymerase-α, 270, 272–273, 275, 280, 282
DNA polymerase-β, 270
DNA polymerase-δ, 270
DNA polymerase-ϵ, 270
DNA polymerization
  slow by retroviral reverse transcriptase, 198
DNA primer, 275–277, 279, 304
DNA repair
  double-stranded breaks, 303–304
  inhibition of, 303
  nonhomologous end-joining pathway, 207
  replication centers and, 293–294
  retroviral DNA integration, 206–207
DNA replication. *See* DNA synthesis
DNA synthesis
  bidirectional, 269–270
  cellular, 269–271
    DNA polymerases, 270
    eukaryotic replicons, 269–270
    general features, 269–270
    inhibition of, 291–292

origins of replication, 270
replication proteins, 270–271
discontinuous, 269–270, 273–275, 278–279, 307
error frequency, 183
lagging strand synthesis, 272–274
leading strand synthesis, 272–274
Okazaki fragments, 269–270, 272–273, 279
priming, 267–270, 272–273, 275–279, 281, 286, 307
DNA as primer, 275–277, 279, 304
Pre-TP, adenovirus, 277–278, 281, 286, 290
proofreading and, 302–303
protein, 275, 278, 281
RNA, 268–270, 272–273, 275–276, 279, 306
proteins involved in, 267–268
provirus, 276, 296
replication bubble, 270, 284, 285, 306
replication fork, 267–270, 272–275, 278, 280, 283, 294, 303–305
in S phase of cell cycle, 289
semidiscontinuous, 270
sliding clamp, 272
termination and resolution, 274
viral (*see* DNA synthesis, viral)
DNA synthesis, viral, 271–307
concatemers, 278–279, 288–289, 296, 301, 303, 305–307
coordination of late genes with viral DNA synthesis, 251–254
delayed synthesis of structural proteins preventing premature packaging of DNA templates, 291
exponential accumulation of genomes, 288–296
herpes simplex virus type 1, 268, 276, 278, 280–283, 286–292, 294, 303–304, 306–307
independent of cellular proteins, 291
limited replication of genomes, 296–301
location of synthesis, 292–296
model for transition between initiation and elongation, 281
origin of replication, 268, 271–286, 294, 296–300
common features, 280
Epstein-Barr virus, 278, 280, 297–299, 300
number of origins, 278, 280
papillomavirus, 276, 278, 280, 286, 299
properties, 278–280
recognition of, 280–286
simian virus 40, 268, 271–275, 282, 284–285, 307
origin-independent replication, 279
origins of genetic diversity in DNA viruses, 301–307
fidelity of replication by viral DNA polymerases, 301–303
inhibition of DNA repair, 303–304
recombination of viral genomes, 304–307
papillomavirus, 268, 276–278, 280, 283–286, 288–290, 294, 299–302, 307
preinitiation complex, 278, 280–282, 298
priming, 267–270, 272–273, 275–279, 281, 286, 307
DNA as primer, 275–277, 279, 304
Pre-TP, adenovirus, 277–278, 281, 286, 290
proofreading and, 302–303
protein, 275, 278, 281
RNA, 268–270, 272–273, 275–276, 279, 306
self-priming, 276–277, 279
provirus, 276, 296
replication center, 292–294, 296, 303–304
replication intermediate, 276–277, 306
resolution and processing of viral replication products, 287–288
rolling-circle replication, 277–278, 288
simian virus 40 (SV40), 268, 271–275
helicase, 271–272, 274, 277, 280, 282–284
lagging-strand synthesis, 272–274
leading-strand synthesis, 272–273
mechanism, 271–275
mechanism of unwinding and translocating along DNA, 284
origin of replication, 268, 271–275, 282, 284–285, 307
origin recognition, 271–272
replication of chromatin templates, 274–275
termination and resolution, 274, 275
unwinding, 271–272, 274, 280, 284
single-stranded-DNA-binding protein, 272, 280, 284, 286–287, 292, 294, 306
strand displacement, 268, 277–278, 284
synthesis machines, 286–287, 290–291
template availability and structure, 254
terminal resolution site, 276–277, 283, 296
vaccinia virus, 268, 277, 279, 287, 290–293
DNA target sites, for retroviral DNA integration, 207–208
DNA template
availability and structure, 254
delayed synthesis of structural proteins preventing premature packaging, 291
RNA synthesis from, 224–263
unwinding of, 228, 230
viral proteins regulating viral DNA template transcription, 237–257
DNA virus
assembly
location, 411
sequential, 429, 430
visualization of structural intermediates during assembly, 430
cytoplasmic membrane reorganization, 495
DNA synthesis, 271–307
evolution, 64
genome packaging, 432–434, 436–437
import of viral genome into nucleus, 123
infectious DNA clones, 68
mutation rate, 66
nucleus remodeling, 491–494
one-step growth curve, 49
origins of genetic diversity in DNA viruses, 301–307
packaging signals, 432–434
transcription, 224–263
activators, 225, 232–233, 235–237, 245, 247–251, 253–254, 257–258
adenovirus, 226, 228, 235, 246–248, 250–253
advantages of temporal gene expression, 253
autoregulation, 235, 237–240, 248–249
by cellular machinery alone, 235–237
cellular RNA polymerase, 225–226
common strategies, 245–247
conversion of genomes to templates for transcription, 226–228
coordination of late genes with viral DNA synthesis, 251–254
DNA-dependent RNA polymerase, viral, 260–262
DNA-looping model, 232, 234
early phase, 245, 247, 250–252, 254, 259
enhancers, 228–229, 232–234, 236–237, 239–240, 242–243, 245–247, 249, 254, 262
entry into transcriptional programs, 254–257
herpes simplex virus type 1, 246, 247, 248–250, 251, 252, 254, 256–259
histone associations, 226–228
inhibition of cellular transcriptional machinery, 259–260
initiation, 226, 228–234
late phase, 245–247, 251, 253–254, 257, 259
mapping human adenovirus type 2 initiation site and transcription *in vitro*, 229
mapping vaccinia virus transcripts, 261
preinitiation complexes, 228, 230, 245, 248, 250–251
regulation, 228–235
repressors, 235–236, 247–249, 252–256
by RNA polymerase II, 228–235
by RNA polymerase III, 226, 257–259
strategies, 226
transcriptional cascade, 237, 245–254
transcriptional control region, 228, 232, 236, 240–242, 262
unusual functions of cellular transcription components, 260
vaccinia virus, 247, 260–262
viral proteins governing transcription, 237–257
viral proteins regulating viral DNA template transcription, 237–257
viral factory, 495
DNA virus vectors, 76–77, 78
DNA-binding proteins
common properties of proteins that regulate transcription, 234–235
single-stranded-DNA-binding protein, 272, 280, 284, 286–287, 292, 294, 306
VP16 protein, herpes simplex virus, 246–252, 256–258
Zta protein, Epstein-Barr virus, 247–249, 256
DNA-dependent DNA polymerase, 267, 270, 307. *See also* DNA polymerase
Escherichia coli DNA polymerase I Klenow fragment, 163
mammalian, 270
primer requirement, 267
processivity, 270
proofreading capability, 182
reverse transcriptase compared, 198
RNA-dependent RNA polymerase compared, 161
DNA-dependent protein kinase, 207
DNA-dependent RNA polymerase, 181, 260, 262. *See also specific polymerases*
bacteriophage T7 RNA polymerase, 163, 358
properties, 225–226
RNA-dependent RNA polymerase compared, 161
vaccinia virus, 116
DNA-looping model, 232, 234
DNA-mediated transformation, 68, 125, 127
DNA-mediated transformation assay, 287
DNase, vaccinia virus, 291
dNMPs (deoxynucleotide monophosphates), 267, 281
Dnmts, 238–239
dNTPs (deoxynucleotide triphosphates), 267, 274, 289–292, 307

Dose-response curve of the plaque assay, 37, 38
Double-stranded DNA genome, 19, 56–57, 97
 architectural relationships among viruses, 97
 evolution, 97
Double-stranded RNA genome, 19, 58, 70, 157, 159, 175–178
Doyle, Sir Arthur Conan, 25
Drosophila
 budding as mechanism of mRNA export from nucleus, 453
 gypsy retrotransposon, 203
dsRNA-binding proteins, Pkr regulation by, 370
Duck hepatitis B virus, 214, 512
 reverse transcriptase, 217
 reverse transcription process, 218
Dulbecco, Renato, 36, 65
dUTPase, 291, 303
Dynamin, 133, 145
Dynein, 133, 148, 382, 397, 398, 403, 404

E

E protein
 dengue virus, 112
 flavivirus, 109, 410
 tick-borne encephalitis virus, 106, 108
E (exit) site, 350, 352, 355, 361
E1 protein
 alphavirus, 109
 bovine papillomavirus type 5, 277
 papillomavirus, 284, 286, 300–302
 Semliki Forest virus, 109, 392–393
 Sindbis virus, 110, 440
 togavirus, 444
 ubiquitin-activating enzyme, 295
E1A protein, adenovirus, 235, 246–248, 250–253, 289, 477, 478
E1B protein, adenovirus, 295, 334
E2 protein
 adenovirus, 232, 246, 251, 280, 290
 alphavirus, 109, 402
 bovine papillomavirus type 1, 247
 bovine papillomavirus type 5, 277
 papillomavirus, 286, 300–301
 Sindbis virus, 110, 400, 402
 togavirus, 444
 ubiquitin-conjugating enzyme, 295
E2 transcription factor (E2f), 232, 247, 251, 253, 289–290
E3 glycoprotein gp19, adenovirus, 404
E3 ubiquitin ligase, 295, 334
E3L protein, vaccinia virus, 370
E4 Orf1 protein, adenovirus, 374, 492
E4 Orf6 protein, adenovirus type 5, 295
E4 protein, adenovirus, 334, 336
E6 protein, human papillomavirus, 295–296, 338, 374
E7 protein, papillomavirus, 289, 301, 338
E10R protein, vaccinia virus, 392
Early phase transcription, 245, 247, 250–252, 254, 259
EBER, 226, 258–259
EBNA-1, 248–249, 277, 298–300
Ebola virus
 attachment, 467
 cell entry at lipid rafts, 30
 entry into cells, 137
 fusion protein, 141
 genome organization, 508
 host cell signaling and, 467
 structure, 89, 508
 vaccines, 77
 Zaire, 508
Echovirus
 attachment to host cells, 126, 128
 exit from caveosome, 137
Eclipse period, 48, 49
Ecological homeostasis, 3
Efficiency of plating, 39
EF-Ts, 170
EF-Tu, 170
Eggs, embryonated, 35–36
eIF1, 351–352, 361, 363
eIF1A, 351–352, 361, 363
eIF2, 352, 355, 369–372, 376
eIF2α
 dephosphorylation, 372
 host and viral evolution, 372
 phosphorylation, 369–370, 371, 377
eIF2B, 370–371
eIF3, 351–353, 355, 359, 361–363
 regulation, 376
eIF4E, 352–355, 360, 372–376, 495
 activity modulation, 373–374
  by binding proteins, 374
  by miRNA, 374
  by phosphorylation, 373–374
eIF4F, 336, 352–353, 355, 359–360
 regulation, 372–376
 viral protein replacing, 376
eIF4G, 352–353, 355, 359–360, 362, 372–374, 376, 472, 495
 cleavage, 372–373
eIF4H, 336
eIF6, 351–352
Elastin, 28
Electron microscopy, 12, 13, 40, 82–86
 cryo-electron microscopy, 430
 resolution, 83
 viral assembly, studies of, 418
Electron tomography, 438
Electron transport, 484, 486
ELISA (enzyme-linked immunosorbent assay), 44
Ellerman, Vilhelm, 11
Ellis, Emory, 14, 46
EloB/c proteins, 295
Elongation
 RNA synthesis, 161
 translation, cellular, 360–361
Embryonated eggs, 35–36
Encapsidation. *See* Packaging viral genome
Encephalomyocarditis virus, 522
 eIF4E activity modulation, 374
 inhibition of cellular mRNA export, 334
 internal ribosome entry site (IRES), 357, 359
 L protein, 334
End point, 39
Endemic virus, 6
Enders, John, 16, 32
Endocytic pathways, 123
Endocytosis
 clatrin-dependent, 403
 uncoating during, 135–137
  caveolin-mediated endocytosis, 136–137
  clathrin-mediated endocytosis, 133, 135–136
  lipid raft-mediated endocytosis, 136–137
  macropinocytosis, 137
Endogenous provirus, 202, 203
Endogenous reactions, 191
Endoplasmic reticulum, 385
 chaperones, 388–390, 392–395, 406–407
 disruption, 404–405
 ER-Golgi intermediate compartment (ERGIC), 405
 ER-to-Golgi transport, 396–398
  vesicular stomatitis virus G protein, 397
 integration of folding and glycosylation in, 394
 reactions within, 389–396
  disulfide bond formation, 390
  glycosylation, 389–390, 394
  oligomerization, 395–396
  protein folding and quality control, 390, 392–395, 396
 retrotranslocation, 394–395, 406
 rotavirus budding into and out of ER, 410
 rough, 386, 388, 403
 selectivity of chaperones for viral glycoproteins entering, 393
 signal peptide, 386–389
 structure, 386, 388
 translocation channel, 387–389, 395
 translocation of viral membrane proteins to, 386–389
 unfolded protein response, 406–407
 viral RNA synthesis on, 179
 virus budding, 410, 444
 virus replication and assembly, 33–35
Endoplasmic reticulum lumen, 386–387, 391–397, 401, 410
Endornavirus, genome organization, 58
Endosomal fusion receptor, 142
Endosome, 133
 alphavirus RNA synthesis on surface, 179
 caveosome compared, 137
 disrupting the endosomal membrane, 144
 early, 135
 fusion with lysosomes, 136
 late, 135–136, 320
 pore formation in endosomal membrane, 145
End-point dilution assay, 38–39, 40
Enhancer, 75, 228, 229, 232–234, 236, 246–247, 249, 254, 262, 322
 adenovirus, 251
 avian leukosis virus, 237
 defined, 229
 human immunodeficiency virus type 1, 237, 239–240, 242–243, 245
 mechanism of action, 234
 regulation by host cell metabolism, 233
 simian virus 40, 229, 232–233
 3′-cap-independent translational enhancer (3′ CITE), 360
Enterovirus, 522
 attachment to host cells, 126
 cytopathic effect, 35
 protease $3C^{pro}$, 377
 secretory pathway disruption, 404
 structure, 94
Entry into cells, 31, 132–148
 architecture of cell surface, 27–31
 bacteriophage, 150
 cytoskeleton remodeling, 467
 host cell signaling and, 467
 membrane fusion, 137–147
  acid-catalyzed, 139–142
  disrupting the endosomal membrane, 144
  endosomal fusion receptor, 142

hemifusion intermediate, 139
pore formation in endosomal membrane, 145
receptor priming for low-pH fusion, 142
release of viral ribonucleoprotein, 142–143
uncoating in cytoplasm by ribosomes, 143–144
uncoating in the lysosome, 145–147
movement of viral and subviral particles within cells, 147–148
uncoating at plasma membrane, 132–135
uncoating during endocytosis, 135–137
caveolin-mediated endocytosis, 136–137
clathrin-mediated endocytosis, 135–136
lipid raft-mediated endocytosis, 136–137
macropinocytosis, 137
Env protein
acylation of, 400
foamy virus, 215
human immunodeficiency virus type 1, 386, 391, 395, 400–402, 458
modification and processing of retroviral Env polyproteins, 399–400
Moloney murine leukenia virus, 402
retrovirus, 323
Envelope, viral, 18, 106–114
acquisition, 439–441 (*see also* Release of virus particles)
budding from cellular membrane, 439–440
coordination of assembly of internal structures, 440–441
sequential assembly from internal components, 439–440
acquisition of, 384–385
defined, 83
glycoproteins, 106–109
overview, 81, 106
viral glycoprotein proteolytic processing in Golgi, 399–401
Enveloped virus
assembly, 425
attachment to host cells, 129–132
entry into cells
fusion and plasma membrane, 132–135
membrane fusion, 137–147
envelope acquisition, 439–441
fatty acid synthesis/metabolism, 487–490
Environmental metagenomic analyses, 61
Enzyme(s)
cellular
incorporation into particles, 438
viral, 116
incorporation into particles, 438
nucleic acid metabolism, 291
Enzyme activity, measurement of viral, 41–42
Enzyme immunoassay, 44–45
Enzyme-linked immunosorbent assay (ELISA), 44
Epidemics, 10
Epigenetic silencing of integrated proviral DNA, 236, 238–239
Episome, 17, 270
Epstein-Barr virus, 297, 299–300
hepadnavirus, 215
papillomavirus, 301–302
Epithelial, 27–28
Epithelial cell
plasma membrane, 402–403
apical domain, 402–403
basolateral domain, 402–403
sorting of viral proteins in, 402–403
Epitope, 42
Epitope mapping, 42
Epsilon, hepadnavirus, 216
Epsilonretrovirus, 530
Epstein-Barr virus, 514
cell cycle arrest, 292
cellular gene expression inhibition, 473
EBER, 226, 258–259
EBNA-1, 277, 298–300
episome, 297, 299–300
latency, 256
licensing of replication, 298–299
LMP-2A protein, 374
micro-RNAs
inhibition of antiviral defenses, 345
viral oncogenesis promotion by cellular miRNA-155, 341
origin of replications, 278, 280, 297–299, 300
origin for plasmid maintenance (OriP), 278, 297–300
OriLyt, 278, 280, 299
origin recognition protein, 277
transcription, 247–249, 256, 258, 259
visualization of duplication and partitioning of plasmid replicons in living cells, 300
Zta protein, 247–249, 256, 292
Equilibrium centrifugation, 419
ER. *See* Endoplasmic reticulum
ER-degradation-enhancing mannosidase-like proteins (Edem1/Edem2), 394–395
eRF1, 361–363, 368
eRF3, 361–363
ER-Golgi intermediate compartment (ERGIC), 405, 410
Escherichia coli
chaperone GroEL, 423–424
clustered regularly interspersed short palindromic repeat (CRISPR) system, 340
DNA polymerase III, 272
origin of replication (OriC), 272
reverse transcription in, 202
RNase P, 321
Escherichia coli DNA polymerase I Klenow fragment, 163, 200
Escrt (endosomal complex required for transport)
Escrt-dependent budding, 441–443
Escrt-independent budding, 444
functions of pathway, 443
Euchromatin, 236
Eukaryotic DNA polymerases, 270
Eukaryotic replicons, 269–270
Evolution
double-stranded DNA viruses, 97
eIF2α and host and viral evolution, 372
origin of diversity in RNA virus genomes, 182–185
misincorporation of nucleotides, 182–183
reassortment, 183
recombination, 183–185
RNA editing, 185
origins of genetic diversity in DNA viruses, 301–307
RNA-dependent RNA polymerase II, 181
virus DNA in host genomes, 4–5
Exocytosis, 444–450
Exon, 317–323, 325–326, 329–332. *See also* Splicing
export from nucleus of single-exon viral mRNAs, 329–330, 332
overview, 311–312
ExoN protein, nidovirus, 183
Exon shuffling, 317
Exon skipping, 322, 323
Exonuclease, 336–337
proofreading, 302–303
Exoribonuclease, 335, 336
Export adapters, 327, 329–331
Export receptor, 327, 331, 341
Exportin-1, 327, 330
Expression vectors, internal ribosome entry site (IRES) in, 358
Extracellular fluid, uptake of macromolecules from, 137
Extracellular matrix, 27–28, 402

F

F factor, 17
F protein
measles virus, 403
paramyxovirus, 132, 134–135
Sendai virus, 409
F13L protein, vaccinia virus, 410
$FADH_2$ (reduced flavin adenine dinucleotide), 483–487
Farnesol X receptor (Fxr), 233
Fatty acid synthase, 487
Fatty acid synthesis/metabolism, 477, 480–492, 496, 499
oxidation, regulation of, 487
stimulation of synthesis by enveloped virus, 487–490
storage and mobilization of fatty acid, 487
Fava bean necrotic yellow virus, 459
Feline leukemia virus, 209
Fibrin, 28
Fibronectin, 28, 31
Field isolates, 66
Filopodia, 148
Filovirus
BSL-4 containment, 508
diseases, 508
genome organization, 59, 508
information retrieval from genome, 63
NP protein, 89
overview, 508–509
proteins in virion, 117
release of virus particles, 442, 444
reproductive cycle, single-cell, 509
RNA editing, 185
structure, 89, 508
Filter systems, 12
Fingers domain, RNA-dependent RNA polymerase, 161, 163–165, 169
Fire, Andrew, 338
Flavivirus
apoptosis blockage, 469
cytoplasmic replication and assembly compartments, 495–497
diseases, 510
E protein, 410
envelope glycoprotein proteolytic processing in Golgi, 400–401
fatty acid metabolism, 487–488
furin family protease, 400
fusion protein, 140
genome organization, 58, 59, 64, 510
information retrieval from genome, 63
interaction with internal cellular membranes, 410

Flavivirus *(continued)*
membrane fusion, 138, 141
mRNA cap, 314
mRNA degradation, 336
mRNA synthesis, 159
overview, 510–511
packaging of genome, 436
pathogenic viruses produced by recombination, 184
polyprotein processing, 363, 365
prM protein, 410
replication, 159
reproductive cycle, single-cell, 511
RNA degradation, 337
RNA helicase, 170
RNA synthesis, 164, 165, 171
cellular sites for, 179
RNA-dependent RNA polymerase, 164
structure, 109, 111, 510
subgenomic flavivirus RNAs (sfRNAs), 337, 377
uncapped mRNAs, 312
unfolded protein response and, 406–407
Flock house virus, 345
Flotillin, 145
Fluorescence microscopy, 41, 44
superresolution methods, 418
viral assembly, studies of, 418
Fluorescent proteins, 40–41
Fluorescent-focus assay, 37, 43
Foamy virus, 215, 222
prototype foamy virus, 215
integrase structure, 210, 211–213
reverse transcriptase domain and subunit relationships, 199
Foci, 38, 39
Foot-and-mouth disease virus
attachment to host cells, 127, 128
cleavage of eIF4G, 373
discovery, 11
fragmentation of genome, 64
internal ribosome entry site (IRES), 358–359
L protease, 373
outbreaks, 4, 5
overview, 522
RNA polymerase bound to ribavirin, 183
RNA-RNA interactions, 358–359
Forkhead box protein 1 (Foxo1), 489
Fowl adenovirus 1, 502
Fowlpox virus, 526
Fraenkel-Conrat, Heinz, 32
Frameshift mutations, mutational intermediates for, 199
Frameshifting, ribosomal, 368, 369
Franklin, Benjamin, 123
Free energy conformation, 81
Frosch, Paul, 11
Fullerene cone model, 103
Furin family proteases, 399–400
Fusion peptide, 132, 134, 136, 139, 140
Fusion pore, 138, 139
Fusion protein, 138, 140–142
paramyxovirus, 132, 134–135

G

G protein
respiratory syncytial virus, 389
vesicular stomatitis virus, 386, 389, 392–393, 397, 411
G_0 state, 288
G_1 phase, 289, 299
G3bp1 protein, 377
Gag protein
assembly and, 423–428, 432, 435–436, 438, 440–441
cell-to-cell viral spread and, 458
foamy virus, 215
functions, 412–413
Gag-dependent budding, 440–441
hepadnavirus, 218
human immunodeficiency virus type 1, 402, 403, 406, 408–409, 458
multimers, 425, 430–441
radial organization in immature virus particles, 423
Mason-Pfizer monkey virus, 423–424
Moloney murine leukemia virus, 366
murine leukemia virus, 402
myristoylation, 406, 408, 428
proteolytic processing, 454–455
radial organization in immature virus particles, 423
retrovirus, 323, 406, 408, 412–413, 423
L domains, 441–443
Rous sarcoma virus, 368
scaffolding function of RNA and, 427
Gag-Pol polyprotein
incorporation into viral particle, 438
Moloney murine leukemia virus, 366
proteolytic processing, 454–455
retrovirus, 204, 412
ribosomal frameshifting, 368
Gammacoronavirus, 506
Gammaherpesvirus
cellular gene expression inhibition, 473
glucose metabolism and, 480
micro-RNAs
inhibition of antiviral defenses, 345
latency-associated miRNAs, 343
overview, 514
Gammaretrovirus, 197, 530
cellular tethers in retroviral integration, 209
koala retrovirus, 203
reverse transcriptase domain and subunit relationships, 199
structure, 102
viral vectors, 74, 76
Ganciclovir, 75
Gangliosides, as receptors for viruses, 129
Gapped DNA, 57, 65, 214
Gapped intermediate, in retrovirus DNA integration, 206
gB protein, herpes simplex virus type 1, 410, 449, 457
GB virus A, 510
GB virus B, 510
GB virus C, 510
GB virus D, 510
Gc protein, bunyavirus, 410
Gcn2p (general control nonderepressible 2 protein), 369–370
gD protein, herpes simplex virus type 1, 457
gE protein, herpes simplex virus type 1, 457
gE-gI proteins, herpes simplex virus type 1, 410
Geminivirus
rolling-circle replication of, 277
transcription, 226
Gene expression
advantages of temporal regulation, 253
cellular, 470–478
differential regulation of, 474–478
inhibition of, 470–474
DNA virus, 225
mapping vaccinia virus transcripts, 261
posttranscriptional regulation
inhibition of cellular mRNA production, 333–335
to block antiviral responses, 334
to facilitate viral mRNA production, 334
inhibition of cellular mRNA export, 334–335
inhibition of polyadenylation and splicing, 333–335
of viral gene expression, 330–333
regulation mRNA export, 332–333
regulation of alternative polyadenylation by viral proteins, 330–331
regulation of alternative splicing, 331–332, 334
small RNAs that inhibit, 338–345
transcriptional regulation, 225
Gene order, mapping by UV irradiation, 175
Gene repression, 16
Gene therapy, 6, 75–78
General control nonderepressible 2 protein (Gcn2p), 369–370
Genetic analysis, 65–78
of assembly intermediates, 418–420
classical methods, 66–67
engineering mutations, 67–73
engineering viral genomes, 73–78, 79
Genetic engineering
viral vectors, 73–78
Genome, viral, 31–32, 55–79. *See also* DNA genome; RNA genome
classification by, 18–19, 18–21, 55
complexity, 55–59
diversity, RNA virus
misincorporation of nucleotides, 182–183
reassortment, 183
recombination, 183–185
RNA editing, 185
engineering, 73–78, 79
engineering mutations into, 67–73
exponential accumulation of genomes, 288–296
genetic analysis, 65–78
classical methods, 66–67
engineering mutations, 67–73
engineering viral genomes, 73–78, 79
information encoded in, 56, 60
information not contained in, 56
information retrieval from, 60–61, 63
limited replication of genomes, 296–301
membrane directly surrounding, 107
origin, 61, 64
packaging (*see* Packaging viral genome)
principles, 55
recombination, 306–307
segmented, 64, 67, 437–438
sequencing, 60
size limit, 65
structure, 55–60, 61, 62
critical structures for viral function, 62
secondary and tertiary, 60, 61, 72
Genome isomers, herpes simplex virus type 1, 288, 306–307
Genome sequences, 46

Genome-wide transcriptional profiling, 51
Genomic RNA, transport from nucleus to cytoplasm, 411
Gey, George, 34
gH protein, herpes simplex virus type 1, 395, 410, 449, 457
gI protein, herpes simplex virus type 1, 457
Giant virus, 20
  assembly, 417
  structure, 114, 116
Gibbon ape leukemia virus, 203
gL protein, herpes simplex virus type 1, 395, 457
Glucocorticoid receptor, 232
Gluconeogenesis, 482
Glucose metabolism, 479–483
  rate of glycolysis, 480
  redirection of glycolytic intermediates, 480–482
  virus-induced alterations in, 482–483
Glucose transporter 1 (Glut1), 480, 486
Glucose transporter 2 (Glut2), 482
Glucose transporter 3 (Glut3), 480, 486
Glucose transporter 4 (Glut4), 480, 482, 492
Glucosidase II, 393–394
Glutamine, 483–484
Glutathione, 428
Gly-Asp-Asp sequence of RNA polymerase motif C, 163
Glycoforms, 389
Glycolipids, 29
  as receptors for viruses, 129
Glycolsyltransferase, 390
Glycolysis, 479–482
Glycolytic intermediates, 480–482
Glycoproteins, 31
  cellular
    binding of enveloped viruses by transmembrane, 129–132
    envelope, 106–109
    glycoslyation (*see* Glycosylation)
    in virus particles, 117
  viral
    envelope glycoprotein proteolytic processing in Golgi, 399–401
    glycoslyation (*see* Glycosylation)
    selectivity of chaperones for viral glycoproteins entering endoplasmic reticulum, 393
Glycosaminoglycans, 27
Glycosidase, 390
Glycosylation
  in endoplasmic reticulum, 389–390, 394
  functions of, 389
  in Golgi apparatus, 390
  N-linked, 386–387, 389, 391, 393–394, 398–399, 403
  O-linked, 389, 394, 399, 403
  self-glycosylation, 390
GM1 ganglioside, 127
Gn protein, bunyavirus, 410
Gold particles, 87
Golgi apparatus, 384–387, 395
  budding of viruses, 410
  cis-Golgi network, 398–399, 410
  disruption of, 404–405
  ER-Golgi intermediate compartment (ERGIC), 405
  ER-to-Golgi transport, 396–398
    vesicular stomatitis virus G protein, 397
  glycosylation in, 390
  reactions within, 398–401
  trans-Golgi network, 398–400, 402–403, 410, 444, 447, 449, 452
  vesicular transport to cell surface, 396–401
  virus budding, 410
Gp5 protein, bacteriophage HK97, 95
gp23 protein, bacteriophage T4, 112, 424
gp120, 136
gpD protein, bacteriophage λ, 430
gpE protein, bacteriophage λ, 430
Granules, stress-associated RNA, 377, 378
Grass, Günter, 349
Green fluorescent protein, 40–41
GroEL, *Escherichia coli,* 423–424
Gropius, Walter, 81
Grp78 chaperone, 388–389, 392–393, 396, 406–407
GTP hydrolysis, 352, 360–361, 370
GTPase, 133
  vesicle transport and, 397–398
Guanine, methylation during capping, 313
Guide RNA, 342
Gypsy retrotransposon, 203

H

H protein, measles virus, 403
HA protein, influenza virus, 41, 106, 108, 129–132, 139–140, 386, 399–400, 402, 403, 409, 439–440, 444, 456–457
HA0 protein, influenza virus, 386–387, 389, 392–393, 395, 399
Hairpin structure, 158, 160
  clustered regularly interspersed short palindromic repeat (CRISPR) system, 340
  in pre-miRNAs, 339
  U6 snRNA, 321
  in vaccinia virus DNA synthesis, 279
Hairpinning, 138
Hammerhead RNAs, 181
Hantaan virus, 410
Hantavirus
  endonuclease, 376
  interference with cellular RNA, 376
  N protein, 376
  replacement of eIF4F, 376
Hcf (host cell factor), 232, 248–250
HECT domain, 295
HeLa [Henrietta Lacks] cells, 33, 34
Helical capsid, 86, 88–89
Helical symmetry, 86, 88
Helicase, 169–170
  adenovirus Rep 78/68, 277
  in doubled-strand DNA break repair, 304
  eIF4A, 353
  herpes simplex virus type 1, 277, 283, 287
  inhibition of, 170
  Mcm (minichromosome maintenance complex), 277, 284, 298–299, 302
  papillomavirus E1, 284, 286
  RNA-dependent, 321
  role in splicing, 321
  simian virus 40, 271–272, 274, 277, 280, 282–284
  (+) strand RNA virus, 159
  structure, 170
  vaccinia virus D5R protein, 279, 287
Helper virus, 185
Hemagglutination, 41
Hemagglutination assay, 41
Hemagglutination inhibition, 42–43
Hemagglutinin (HA) protein
  acylation of, 400, 402
  influenza virus, 41, 106, 108, 129–132, 139–140, 386, 399–400, 402, 403, 409, 439–440, 444, 456–457
Hemifusion intermediate, 139
Hendra virus, 518
Henipavirus, 518
Hepacivirus, 510. *See also* Hepatitis C virus
Hepadnavirus
  assembly of nucleocapsids, 218
  C protein, 444–445
  classification, 19
  diseases, 512
  furin family protease, 400
  genome conversion for transcription, 226–227
  genome organization, 57, 214–215, 512
  genome size, 65
  import into nucleus, 149
  information retrieval from genome, 63
  interaction with internal cellular membranes, 410
  L protein, 402, 444–445
  life cycle, 214–215, 216
  overview, 512–513
  P protein, 214–220
  posttranscriptional regulatory element (PRE), 330, 332
  pregenomic RNA synthesis by RNA polymerase II, 411
  release of virus particles, 444–445
  replication cycle, 222
  reproductive cycle, single-cell, 513
  reverse transcription
    cis-acting signals, 216, 218, 221
    critical steps in, 218–221
    elongation, 218–220
    essential components, 216–218
    first template exchange, 218, 220
    function in life cycle, 214–215
    host protein involvement, 218
    initiation, 218, 220
    P protein, 214–220
    pregenomic mRNA, 214, 216–218, 220
    primers, 216–221
    retroviral compared, 217
    RNase H degradation of RNA template, 219, 220
    second template exchange, 220, 221
    translocation of primer for (+) strand synthesis, 219–221
  structure, 512
Heparan sulfate, 28, 127, 131, 132
Hepatitis, seronegative, 47
Hepatitis A virus, 522
  release of virus particles, 450, 454, 460
Hepatitis B virus
  C protein, 444–445
  decoy particles, 460
  diseases, 512
  enhancers, 233
  envelopment, 445
  experimental studies in animals, 36
  genome organization, 512
  genome size, 65
  genome structures critical for function, 62
  glucose metabolism and, 482
  interaction with internal cellular membranes, 410
  L protein, 402, 408, 410, 444–445
  lipid metabolism, 487
  M protein, 410

Hepatitis B virus *(continued)*
micro-RNA inhibition of reproduction, 341
oligomerization of proteins, 395
overview, 512–513
posttranscriptional regulatory element (PRE), 332
protein priming, 165
release, 444–445
reproductive cycle, single-cell, 513
RNA editing, 327
S protein, 408, 410
structure, 512
Hepatitis C virus
budding into ER, 444
C protein, 495
classification, 18
cytoplasmic replication and assembly compartments, 495, 497
fatty acid metabolism, 487–491
genome organization, 510
glucose metabolism and, 480, 482–483
host cell gene expression and, 474, 475
host cell signaling and, 467–468
internal ribosome entry site (IRES), 355, 357
micro-RNAs
inhibition of reproduction, 341
liver-specific reproduction of hepatitis C virus promoted by cellular miRNA-122, 342
protection of genome from degradation and promotion of replication, 343
overview, 510
RNA helicase, 170
RNA polymerase, 165
RNA synthesis, 169
cellular sites for, 182
structure, 97
uncapped mRNAs, 312
Hepatitis delta satellite virus
genome organization, 58
information retrieval from genome, 63
large delta antigen, 402, 408
ribozymes, 176–177, 181
RNA editing, 185
RNA synthesis, 176–177, 180, 260
Hepatitis G virus, 4
Hepatocellular carcinoma, 512
Hepatocyte nuclear factor (Hnf), 233
Hepatovirus, 522
Heraclitus, 311
Herpes simplex virus. *See also* Herpes simplex virus type 1
attachment to host cells, 127
citric acid cycle and, 483
cytopathic effect, 35
eIF2a dephosphorylation, 372
entry into cells, 137
frameshifting, 368
genome size, 65
growth in embryonated eggs, 36
inhibition of cellular mRNA export, 334
inhibition of cellular mRNA splicing, 333
particle-to-PFU ratio, 40
protein synthesis stimulation, 374
thymidine kinase gene, 75
viral vectors, 76
Herpes simplex virus type 1
alkaline nuclease, 291, 306
assembly, 421, 425, 428–429
genome packaging, 434
procapsids, 428–429
scaffolding proteins, 425, 428–429
UL6 protein, 428
visualization of structural transitions during, 430
attachment to host cells, 131–132
budding of virus particles from nucleus, 411
cell-to-cell spread, 457
cell-type-specific production of stable viral intron, 319
cellular gene expression and, 474
cellular proteins in virus particle, 118
cytoplasmic membrane reorganization, 495
DNA synthesis, 268, 276, 278, 280–283, 286–292, 294, 303–304, 306–307
enzymes of nucleic acid metabolism, 291
export of single-exon mRNAs, 330
gB protein, 410, 449, 457
gD protein, 457
gE protein, 457
gE-gI proteins, 410
genome
circularized, 289
concatemerized, 289
features, 281
isomers, 288, 306–307
organization, 514
packaging, 434
recombination, 306–307
gH membrane glycoprotein, 395
gH protein, 410, 449, 457
gI protein, 457
gL protein, 395, 457
glucose metabolism and, 481, 482
helicase, 277, 283, 287
histone association, 227
ICP0 protein, 291, 295
ICP4 protein, 291, 294
ICP6 protein, 374
ICP27, 330, 332–334
ICP27 protein, 291, 294, 376
ICP36 protein, 291
import of viral genome into nucleus, 149
inhibition of cellular mRNA production, 333–334
interaction with internal cellular membranes, 410
interference with cellular RNA, 376
kinase inactivation, 372
latency-associated transcripts (LATs), 256–257, 319
latent infection, 256–258
lytic cycle, 257
nuclear lamina disruption by, 449, 452
nuclear localization sequences, 383
oligomerization of proteins, 395
origin recognition protein, 277
OriL, 280–282
OriS, 280–282, 287
pac1/pac2, 434
packaging signals, 434
primase, 276, 277, 287
procapsid, 428–429
reactivation, 257–258
recombination-dependent replication, 304, 306
redistribution of poly(a)-binding protein, 376
release of virus particles, 449, 452
replication centers, 292–294
replication system, 286–287
reproductive cycle, single-cell, 515
splicing, 318
structure, 94, 112–113, 115, 514
transcription, 246, 247, 248–250, 251, 252, 254, 256–259
UL6 protein, 421, 431
UL8 protein, 286
UL9 protein, 268, 277, 283, 287
UL12 protein, 291
UL15 protein, 434
UL23 protein, 291
UL28 protein, 434
UL33 protein, 434
UL34 protein, 410, 457
UL39 protein, 291
UL40 protein, 291
UL47 protein, 376
Us3 protein, 374
Vhs protein, 336, 338, 376, 474
viral vectors, 74
virion enzymes, 116
virion host shut-off protein (Vhs), 336, 338, 376, 474
virus-induced chaperone-enriched (VICE) domains and, 494
VP5 protein, 421, 428–430
VP16 protein, 438, 449
VP19C protein, 383, 421, 430
VP21 protein, 421
VP22a protein, 421, 425, 428–429
VP23 protein, 421, 430
VP24 protein, 421, 428–429
VP26 protein, 421, 430
Herpes simplex virus type 2, attachment to host cells, 131
Herpesvirus
ADP-ribosylation factor 1 (Arf1), 117
assembly, 384, 428–429
axonal transport, 403–404, 405
binding of virus particle, 389, 411
cell-to-cell spread, 457–458
cellular gene expression inhibition, 473
counting number of genomes expressed and replicated, 293
cytopathic effect, 34
diseases, 514
DNA polymerase, 303
DNA synthesis, 268, 276–278, 287–288, 291–293, 303, 306–307
concatemers, 288
eIF4E phosphorylation, 374
entry into cells, 137
envelope, 108
enzymes of nucleic acid metabolism, 291
fatty acid metabolism, 487
furin family protease, 400
genome organization, 56, 58, 514
genome sequence, 60
glucose metabolism and, 480–481
headful packaging, 436–437
histone associations, 227, 228
import of genome into nucleus, 151
infectious DNA clones, 68
information retrieval from genome, 63
inhibition of cellular DNA synthesis, 291–292
integration into host cell telomere, 297
interaction with internal cellular membranes, 410
latency-associated miRNAs, 343, 345
latent infection, 254, 256–258, 297, 514
localization of viral proteins to nuclear membrane, 411

micro-RNAs
inhibition of antiviral defenses, 345
inhibition of reproduction, 341
origin recognition protein, 277
overview, 514–515
packaging genome, 105
particle-to-PFU ratio, 40
polyadenylation, 330
protein synthesis stimulation, 374
proteins in virion, 117
reactivation, 254, 257–258, 514
release of virus particles, 444, 449, 452–453
reproductive cycle, single-cell, 515
RNA polymerase II transcription, 258
scaffolding proteins, 425, 428–429
single-exon mRNAs, 330
size, 13
structure, 112–113, 115, 514
transcription, 226
transcriptional regulators, 247, 254
UL20 protein, 449
UL36 protein, 449
UL37 protein, 449
unfolded protein response and, 406–407
US3 protein, 449
viral mRNAs in virus particle, 117
viral oncotherapy, 75
virion enzymes, 116
Herpesvirus saimiri, 258
Hershey, Alfred, 14–15
Hershey-Chase experiment, 15, 18, 68
Heterochromatin, 226
Heterogeneous nuclear ribonucleoprotein I (hnRNPI), 359
Heterogeneous nuclear RNAs (hnRNAs), 317
Hexameric array, 98
Hexokinase, 479, 486
Hexon, 85–86, 99–101, 115, 318, 326, 333
High-throughput sequencing, 45–46
Histone(s), 59
hepadnavirus and, 214
herpesvirus association with, 228
packaging viral genome and, 105
posttranslational modification, 226, 236
in virus particles, 117
Histone code, 236
Histone deacetylases, 238, 253
HIV. *See* Human immunodeficiency virus
HIV-1. *See* Human immunodeficiency virus type 1
HN protein, Sendai virus, 409
hnRNAs (heterogeneous nuclear RNAs), 317
hnRNPI (heterogeneous nuclear ribonucleoprotein I), 359
Holoenzyme, 228
Homeostasis
glucose, 482
metabolic, 479
Homer, 6–7
Homologous recombination, 77
Horne, Robert, 17
Host cell factor (Hcf), 232, 248–250
Host cell response to infection, 464–500
gene expression, 470–478
differential regulation of cellular, 474–478
inhibition of cellular, 470–474
metabolism, 477, 479–492
citric acid cycle, 483–484
electron transport, 484, 486
glucose, 479–483
lipid, 485, 486–492
methods to study, 477, 479
oxidative phosphorylation, 484, 486
remodeling of cellular organelles, 491–499
cytoplasm, 495–499
nucleus, 491–495
signal transduction, 465–470
signaling in virus-infected cells, 466–470
signaling pathways, 465–466
Host defenses, 25, 32
Host range, receptors and, 123–124
Hosts
animal cells as, 16–17
organisms as, 14
House hepatitis virus, 410
Hsc70 protein, 424, 428, 494
Hsp70 protein, 424
Hsp90 protein, 424
Human adenovirus. *See* Adenovirus
Human B19 virus, 520
Human body
number of virus particles in, 4
viral DNA in, 4–5
Human coronavirus 229E, 506
Human cytomegalovirus. *See* Cytomegalovirus
Human enterovirus 71, host cell signaling and, 468
Human genomes, viral DNA in, 4–5
Human herpesvirus 6, 514
integration, 296–297
U94 protein, 297
Human herpesvirus 7, 514
Human herpesvirus 8, 60, 514
ATP synthesis in host cells, 484
attachment to host cells, 127
cellular gene expression inhibition, 473
K3 protein, 295
K5 protein, 295
micro-RNAs
inhibition of antiviral defenses, 345
latency-associated miRNAs, 343, 345
RNA editing, 326
SOX protein, 336, 473
Human immunodeficiency virus. *See also* Human immunodeficiency virus type 1
intasomes, 213
integrase inhibitors, 206, 212
number of genomes, 4
preintegration complex, 210
receptors and coreceptors, 124, 127
Human immunodeficiency virus type 1
amino acyl tRNA synthase encapsidation, 438–439
attachment to host cells, 127, 131
axonal transport, 403–404
blocking infection with soluble cell receptors, 136
CA protein, 117, 409, 455
capsid, 102–104
cell entry at lipid rafts, 30
cell-to-cell spread, 458
cellular gene expression and, 474, 476
cellular tethers in retroviral integration, 208–209
cyclophilin A chaperone, cellular, 117
cytoplasmic membrane reorganization, 495
electron microscopy, 419
enhancer, 237, 239–240, 242–243, 245
entry into cells, 134, 135
Env protein, 386, 391, 395, 400–402, 458
envelope acquisition, 440–441
envelope glycoprotein proteolytic processing in Golgi, 400–401
epigenetic silencing of integrated proviral DNA, 238–239
export of RNAs from nucleus, 327–328
frameshifting, 368
fusion protein, 141
Gag protein, 402, 403, 406, 408–409, 458
multimers, 425, 440–441
proteolytic processing of polyprotein, 454–455
radial organization in immature virus particles, 423
genome
dimerization, 435–436
organization, 530
glucose metabolism and, 480
host cell signaling and, 467, 470
integrase structure, 211, 213
integrase-interacting protein 1 (INi-1), 210
integration sites, 208
lipid metabolism, 487
LTR, 237, 239–242
MA protein, 109–110, 114, 402, 409, 455
maturation, 454
micro-RNA inhibition of reproduction, 341
NC protein, 409, 455
Nef protein, 117
oligomerization of proteins, 395
primer-binding site (pbs), 194
protease, 451
proteins in virion, 117
provirus, 238–239
receptors and coreceptors, 124, 127, 131, 135
refolding of CA protein, 455
release of virus particles, 442, 444
Rev protein, 327–330, 332
reverse transcriptase, 217
domain and subunit relationships, 199
error rate, 199
flipping, 201
structure, 163, 200–201, 204
template primer interaction, 200–201
RNA editing, 326–327
RNA profiling, 476
SU (gp120), 131, 134, 135, 391
sugar shield, 391
TAR (transactivation response) element, 241, 243–245
Tat protein, 237, 239–245, 248, 262
TM-Su protein, 441
transcription regulation, 237–246
transcriptional control region, 240
transport of RNA from nucleus to cytoplasm, 411
transport through M cells, 27
viral vectors, 78
virion enzymes, 116
Vpr protein, 117
Vpu protein, 404, 444, 450
Human immunodeficiency virus type 2, 136
Human papillomavirus, 295, 338. *See also* Papillomavirus
cervical cancer, 338
E6 protein, 295–296, 338, 374
E7 protein, 338
glucose metabolism and, 480
integration, 338, 339
maturation of progeny virus particles, 457
mRNA stability, 338, 339
once-per-cell-cycle replication, 302
random replication, 302
replication, 302

Human parainfluenza virus 1, 518
Human parainfluenza virus 2, 518
Human parainfluenza virus 3, 518
Human parainfluenza virus 4, 518
Human polyomavirus 6, 524
Human polyomavirus 7, 524
Human rhinovirus 14, 94
Human rhinovirus A, 522
Human T-lymphotrophic virus, 530
  Tax protein, 237, 247, 251
  transcriptional regulators, 247
Hydroxylamine, 66
Hypoxia-inducible factor 1α (Hif-1α), 480

I

Icam-1 (intercellular adhesion molecule 1), 127, 132, 460
Icosahedral symmetry, 86, 89, 91
Icosohedral capsid, 86, 89–102
  packing in simple structures, 91
  quasiequivalence, 90–91, 97
  triangulation number, 90–94
ICP0 protein, herpes simplex virus type 1, 246–247, 254, 256–258, 291, 295
ICP4 protein, herpes simplex virus type 1, 246–247, 254, 256–257, 291, 294
ICP6 protein, herpes simplex virus type 1, 374
ICP22, herpes simplex virus, 257, 259
ICP27 protein, herpes simplex virus type 1, 291, 294, 330, 332–334, 376
ICP36 protein, herpes simplex virus type 1, 291
ICTV (International Committee on Taxonomy of Viruses), 20
ID_{50} (infectious dose 50%), 39, 40
Ikbα, 374
Imaging particles, 40–41
Immunofluorescence, 43–44
Immunohistochemistry, 44
Immunostaining, 43–44
Importin-α, 150
Importin-β, 150
In vitro, meaning of term, 34
In vitro assembly of tobacco mosaic virus, 32
In vivo, meaning of term, 34
Indirectly anchored proteins, 30
Induction, 16
Infection, host cell response to, 464–500
  gene expression, 470–478
    differential regulation of cellular, 474–478
    inhibition of cellular, 470–474
  metabolism, 477, 479–492
    citric acid cycle, 483–484
    electron transport, 484, 486
    glucose, 479–483
    lipid, 485, 486–492
    methods to study, 477, 479
    oxidative phosphorylation, 484, 486
  remodeling of cellular organelles, 491–499
    cytoplasm, 495–499
    nucleus, 491–495
  signal transduction, 465–470
    signaling in virus-infected cells, 466–470
    signaling pathways, 465–466
Infectious centers assay, 37–38
Infectious cycle, 25
Infectious DNA clone
  DNA viruses, 68
  RNA viruses, 68–70
Infectious dose 50% (ID_{50}), 39, 40
Infectious sub-viral particle (ISVP), reovirus, 528–529
Influenza A virus
  genome organization, 516
  overview, 516–517
  reproductive cycle, single-cell, 517
  structure, 516
Influenza B virus, 516
Influenza C virus, 516
Influenza virus
  1918 strain, 51, 131
  assembly, 431, 439–440
  attachment to host cells, 127, 129–131, 467
  binding of virus particle, 389
  BM2 protein, 366.21
  budding from plasma membrane, 403
  cellular gene expression inhibition, 473
  commercial rapid antigen detection assays, 45
  diseases, 516
  endonuclease, 376
  entry into cells, 136, 137, 138, 139–140
  envelope, 109
    acquisition, 439–440
    glycoprotein proteolytic processing in Golgi, 399–400
  export of mRNAs from nucleus, 330
  fusion protein, 141
  genome organization, 516
  growth in embryonated eggs, 36
  H1N1, 476–477
  H5N1, 476
  H7N9, 476
  HA0 protein, 386–387, 389, 392–393, 395, 399
  hemagglutination assay, 41
  hemagglutination inhibition assays, 42–43
  hemagglutinin (HA) protein, 106, 108, 386, 399–400, 402, 403, 409, 439–440, 444, 456–457
  hemifusion intermediate, 139
  historical aspects, 7
  host cell signaling and, 467
  host range, 131
  immunofluorescence assay, 44
  import of viral genome into nucleus, 149, 151
  inhibition of cellular mRNA production, 333
  interference with cellular RNA, 376
  M1 protein, 366, 367, 408–409, 411–412, 413, 440, 444
  M2 protein, 109, 143, 318, 400, 439, 444, 450
  maturation, 456–457
  micro-RNA inhibition of reproduction, 341
  moving-template model for mRNA synthesis, 173, 176
  NA protein, 403, 409, 439–440, 444, 456–457
  NEP protein, 411–412
  NP protein, 411–412, 494
  NS1 protein, 330, 333, 370, 376, 377
  nucleus remodeling, 494–495
  oligomerization of proteins, 395
  overview, 516–517
  P body dispersal, 377
  packaging of genome, 437–439
  particle-to-PFU ratio, 40
  poly(A) addition, 173
  priming by capped RNA fragments, 166–168
  propagation of, 35
  reassortment of RNA segments, 67
  receptors, 124
  recovery from cloned DNA, 69, 70
  reinitiation of translation, 366, 367
  release of viral ribonucleoprotein, 143
  release of virus particles, 444
  removal of terminal sialic acid residues, 456–457
  reproductive cycle, single-cell, 517
  ribonucleoprotein, 89–90, 411–412
    organization in particles, 438
    structure, 160
  RNA profiling, 476–477
  RNA synthesis, 164, 165, 166–168, 169, 173, 176
  RNA-dependent RNA polymerase, 161, 168, 169
  segment reassortment, 183
  sorting in epithelial cells, 403
  splicing, 318
  stress granule prevention, 377
  structure, 89, 109, 516
  transcriptional profiling, 51
  translation of capped viral mRNAs, 376
  transport of viral genomes to assembly sites, 411–412
    transport from cytoplasm to plasma membrane, 411, 412
    transport from nucleus to cytoplasm, 411, 412
  vaccine, 9
  virion enzymes, 116
Initiation, transcription, 226, 228–234
Initiation, translation, 351–360
  5′-end-dependent initiation, 352–355
  5′-end-independent initiation, 355–360
  inhibition after viral infection, 369–376
  methionine-independent initiation, 354
Initiation codon, 349–355, 357, 359, 365–366, 372
Inositol-requiring enzyme 1 (Ire1), 406–407
Insertion mutations, 70
Insertional mutagenesis, 16
Intasome
  assembly, 213
  characterization, 211–212, 214
Integral membrane proteins, 30, 106, 108
  viral interaction with internal cellular membranes, 410
Integrase, 16
  cellular tethers, 208–209
  host DNA target sites, 207–209
  host proteins that regulate integration, 209–210
  incorporation into viral particle, 438
  prototype foamy virus (PFV), 210, 211–213
  retrovirus, 189, 204–214
    integrase-catalyzed steps, 205–210
    structure and mechanism, 210–214
  steps catalyzed by, 205–210
    concerted cleavage and ligation, 206
    disintegration step, 206
    joining step, 206–207
    processing step, 205–207
  structure and mechanism, 210–214
    catalytic core domain, 210, 211, 212
    C-terminal domain, 210, 211, 212
    intasome assembly, 213
    intasome characterization, 211–212, 214
    multimeric form, 210–211
    N-terminal domain, 210, 211, 212
  turnover rate, 211
  in virus particles, 116
  in vitro assay, 206

Integrase inhibitors, 206, 212
Integration
adeno-associated virus, 296
bacteriophage λ, 190
human herpesvirus 6, 296–297
human papillomavirus, 338, 339
retroviral, 204–214
site-specific, 190
Integration sites, retroviral, 207–208
Integrins, 28, 31
viral attachment and, 127, 128, 129, 132, 144
Intercalating agents, as mutagens, 66
Intercellular adhesion molecule 1 (Icam-1), 127, 132, 460
Interferon(s)
antiviral state and, 369
inhibition of cellular mRNA export by vesicular stomatitis virus M protein, 334
mechanism of action, 313
Pkr induction, 326, 369
single-exon mRNAs, 330
Interferon inducible protein, 444
Intermediate filaments, 147
Internal membranes, assembly of viruses at, 444–450
Internal ribosome entry site (IRES), 60, 63, 352, 355–358
discovery, 356
host cell proteins in IRES function, 358–359
mechanism of internal initiation, 355, 358
types, 357
use in expression vectors, 358
International Committee on Taxonomy of Viruses (ICTV), 20
Intracellular trafficking, 380–414
assembly at the plasma membrane, 384–409
disruption of secretory pathway, 404–406
overview, 384–385
signal-sequence-independent transport of viral proteins, 406, 408–409
sorting of viral proteins in polarized cells, 401–404
transport of viral membrane proteins to plasma membrane, 386–401
assembly within the nucleus, 382–384
import of viral proteins for assembly, 383–384
interactions with internal cellular membranes, 409–411
localization of viral proteins to compartments of the secretory pathway, 410
localization of viral proteins to nuclear membrane, 411
movement in heavy traffic, 382
overview, 381–382
transport of viral genomes to assembly sites, 411–414
influenza virus, 411–412
microtubules and, 414
transport from cytoplasm to plasma membrane, 411–413
transport from nucleus to cytoplasm, 411
Intron, 317–323, 326–329, 331–333, 335, 339. *See also* Splicing
cell-type-specific production of stable viral intron, 319
export from nucleus of intron-containing mRNAs, 327–329
group I, 181
group II, 181
latency-associated transcript (LAT), 319
miRNAs from, 339
overview, 311–312
self-splicing, 181, 321
Tetrahymena, 181, 321
Inverted terminal repetition (ITR), adeno-associated virus type 2, 276–278, 296
Ion channel, 143
Ire1 (inositol-requiring enzyme 1), 406–407
IRES. *See* Internal ribosome entry site
Iridovirus, genome organization, 57
Isoprenylated viral proteins, 402, 406, 408
ISVP (infectious sub-viral particle), reovirus, 528–529
ITR (inverted terminal repetition), adeno-associated virus type 2, 276–278, 296
IVa_2 protein, adenovirus, 231, 247, 253, 432–433
Ivanovsky, Dimitrii, 10, 11

J

Jacob, François, 15, 16, 17
Japanese encephalitis virus, 337, 510
JC virus, 524
attachment to host cells, 129
micro-RNA (miRNA), 342
release of virus particles, 450
Jelly roll, 94–98–22, 107, 114
Jenner, Edward, 8
Jingmen tick virus (JMTV), 64
Joining step, in retroviral DNA integration, 206–207
Just in time inventory control, 253

K

K3 protein, human herpesvirus 8, 295
K3L protein, vaccinia virus, 372
K5 protein, human herpesvirus 8, 295
Kaposi's sarcoma, 343
Kaposi's sarcoma-associated herpesvirus, 374, 376
interference with cellular RNA, 376
redistribution of poly(a)-binding protein, 376
SOX protein, 376–377
Karolinska Institute (KI) polyomavirus, 524
Kif4, 413
Kinase inhibition, 370, 372
Kinesin, 133, 148, 151, 382, 398, 403, 404, 413, 447–448, 450
Klenow fragment, of *Escherichia coli* DNA polymerase I, 163, 200
Klug, Aaron, 90, 93
Koala retrovirus, 203
Koch, Robert, 10
Koch's postulates, 3, 10

L

L domains, retrovirus, 441–443
L protease, foot-and-mouth disease virus, 373
L protein
arenavirus, 174
bunyavirus, 168
encephalomyocarditis virus, 334
hepadnavirus, 402, 444–445
hepatitis B virus, 402, 408, 410, 444–445
vesicular stomatitis virus, 163, 168, 169, 173, 313, 411, 414
L1 protein, adenovirus, 331–334, 425, 432–433
L3 protease, adenovirus, 450
L4 protein, adenovirus, 383, 384, 423, 425, 432–433
Laboratory animals, experimental infection of, 36
Lacks, Henrietta, 34
Lagging strand synthesis, 272–274
Lake Victoria marburgvirus, 508
λ-2 protein, reovirus, 313
Laminin, 28
Large delta antigen, hepatitis delta satellite virus, 402, 408
Large T antigen, 246, 247, 254
BK virus, 344
simian virus 40 (SV40), 271–275, 277, 280–285, 289–290, 303, 307, 342–343, 345, 424–425, 428–429
functions and organization, 280–283
mechanism of unwinding and translocation along DNA, 284–285
properties, 280–283
regulation of activity, 283
Lassa virus, 160
Late phase transcription, 245–247, 251, 253–254, 257, 259
Latency-associated miRNAs of herpesviruses, 343, 345
Latency-associated transcripts (LATs), herpes simplex virus, 256–257, 319
Latent infection
herpes simplex virus type 1, 256–258
herpesvirus, 254, 256–258, 514
integration into host cell telomere, 297
Latent period, 48
Lateral flow immunochromatographic assay, 44–45
Law of mass action, 480
LD_{50} (lethal dose 50%), 39, 40
Leading strand synthesis, 272–274
Leaky scanning, 63, 365–366
Leber's congenital amaurosis, 77
Lectins, 132
Ledgf (lens epithelium-derived growth factor), 208–209
Lens epithelium-derived growth factor (Ledgf), 208–209
Lentivirus, 530
reverse transcriptase
domain and subunit relationships, 199
structure, 102
viral vectors, 74, 76, 78
Leporipoxvirus, 526
Lethal dose 50% (LD_{50}), 39, 40
Lettuce necrotic yellows virus, 385, 532
Leucine zipper, 236
Licensing of replication, 298, 299
Ligase IV, 207
LINEs (long interspersed nuclear elements), 202–204
Linkage map, 190
Lipid metabolism, 485, 486–492
fatty acid storage and mobilization, 487
fatty acid synthesis/metabolism, 477, 480–492, 496, 499
Lipid raft-mediated endocytosis, 136–137
Lipid rafts, 30, 124, 129, 136–137
sorting in epithelial cells, 403
Lipids
lipid-plus-protein signals, 406, 408
phosphoinositol 4-phosphate (PI4P)-containing, 405

Lister, Joseph, 10
Live-cell imaging of single fluorescent virions, 40–41
Liver-specific reproduction of hepatitis C virus promoted by cellular miRNA-122, 342
Living entity, virus as, 18
LMP-2A protein, Epstein-Barr virus, 374
Loeffler, Friedrich, 3, 11
Long terminal repeats (LTRs), 195–197
  avian sarcoma and leukosis virus, 235–237
  human immunodeficiency virus type 1, 237, 239–242
  proviral DNA flanked by, 205
  retrotransposon, 202–203
Loops, virus attachment and, 128
LTRs. *See* Long terminal repeats
Luria, Salvador, 3, 14
Luria's Credo, 3
Lwoff, André, 15, 16, 17
Lymphocryptovirus, 514
Lymphocytic choriomeningitis virus
  diseases, 504
  genome organization, 504
  reproductive cycle, single-cell, 505
  structure, 504
Lysogenic, 15, 16
Lysogeny, 15, 16
  bacteriophage λ, 254–256
Lysosomal lipase, 488
Lysosome, 136
  alphavirus RNA synthesis on surface, 179
  endosome fusion with, 136
  uncoating in, 145–147
Lyssavirus, 532
Lysyl-tRNA synthetase, 192
Lytic cycle
  bacteriophage λ, 254–256
  herpes simplex virus type 1, 257
Lytic response to infection, 16

M

M cells, 27
M protein
  coronavirus, 410
  hepatitis B virus, 410
  rabies virus, 376
  vesicular stomatitis virus, 89, 334, 408–409, 411, 413
M1 protein
  influenza virus, 366, 367, 408–409, 411–413, 440, 444
  myxovirus, 323
M2 protein, influenza virus, 109, 143, 318, 400, 439, 444, 450
MA protein
  human immunodeficiency virus type 1, 109–110, 114, 402, 409, 455
  murine leukemia virus, 402
  retrovirus, 109, 406, 423, 428, 440–441, 454
MacLeod, Colin, 14
Macropinocytosis, 137
Major histocompatibility complex (MHC) class I proteins
  inhibition of transport to cell surface, 404–405
  translocation from ER to cytoplasm, 395
Major histocompatibility complex (MHC) class II proteins, in virus envelopes, 460
Malacoherpesvirus, 514
Mamavirus, 20, 65
Mammalian cell, 26
Mammals
  DNA double-strand break repair, 303
  DNA replication in
    DNA polymerases, 270
    origin of replication, 270
    origin recognition complex, 299
    proteins required for DNA synthesis, 270–271
  DNA-dependent DNA polymerases, 270
  nuclear pore complexes, 383
  Pi3k-Akt-mTor signaling route, 375, 466–470
  ribosomes, 350
Mapk (mitogen-activated protein kinase), 470
Mapping
  gene order by UV irradiation, 175
  mutations, 66–67
  vaccinia virus transcripts, 261
Marburgvirus, 508
Marek's disease virus, 296
Mason-Pfizer monkey virus
  assembly, 423–424, 444
  constitutive transport elements (CTEs), 328
  Gag protein, 423–424
  structure, 102
Mass spectrometry, 477, 479
Mastadenovirus, 502
Maturation of progeny virus particles, 450–457
  disulfide bond formation, 457
  extracellular assembly, 456
  proteolytic processing of structural proteins, 450–451, 454–456
    cleavage of polyproteins, 451, 454–456
    cleavage of precursor proteins, 456
  removal of terminal sialic acid residues, 456–457
McCarty, Maclyn, 14
Mcm (minichromosome maintenance complex), 277, 284, 298–299, 302
Measles virus, 518
  cell-to-cell spread, 458
  eIF3 regulation, 376
  F protein, 403
  H protein, 403
  immunofluorescence assay, 44
  N protein, 376
  sorting in epithelial cells, 403
  syncytia formation, 458
  viral vectors, 74
Med23, 251, 253
Megavirus chilensis, 65
Mello, Craig, 338
Membrane(s), cellular
  budding of virus from, 439–440
  release of virus particles, 444–450, 496–497
  transport vesicles, 384, 396–398, 403, 413
  viral RNA synthesis on, 179, 182
  virus interactions with internal cellular membranes, 409–411
    localization of viral proteins to compartments of the secretory pathway, 410
    localization of viral proteins to nuclear membrane, 411
  virus replication and assembly platforms, 495–497
Membrane fusion, 137–147
  acid-catalyzed, 139–142
  disrupting the endosomal membrane, 144
  endosomal fusion receptor, 142
  hemifusion intermediate, 139
  pore formation in endosomal membrane, 145
  receptor priming for low-pH fusion, 142
  release of viral ribonucleoprotein, 142–143
  uncoating in cytoplasm by ribosomes, 143–144
  uncoating in the lysosome, 145–147
Membrane proteins, 30–31
Membrane proteins, viral
  acylation of, 400, 402
  envelope glycoprotein proteolytic processing in Golgi, 399–401
  oligomerization, 395–396
  release of virus particles and, 444
  transport to plasma membrane, 386–401
Membrane-spanning domain, 30–31, 106, 108–109
Mengovirus RNA-dependent RNA polymerase, 159–160
MERS. *See* Middle East respiratory syndrome coronavirus (MERS-CoV)
Messenger RNA (mRNA). *See also* Transcription; Translation
  CRISPR-Cas9 and, 73
  nested mRNAs, 63
  as positive (+) strand, 56
  pre-miRNA, 339, 341–343, 345
  small interfering RNAs and, 72
  splicing, 63
Messenger RNA (mRNA), cellular
  budding mechanism of export from nucleus, 453
  cap, 349
  5′ untranslated region (5′ UTR), 349
  half-life, 335–336
  initiation codon, 349
  internal ribosome entry site (IRES), 352
  juxtaposition of mRNA ends, 363
  polycistronic, 349–350
  poly(A) tail, 349
  posttranscriptional regulation by viral proteins, 333–335
    to block antiviral responses, 334
    to facilitate viral mRNA production, 334
    inhibition of cellular mRNA export, 334–335
    inhibition of polyadenylation and splicing, 333–335
  ribosome assembly on mRNA, 352–355
  RNA profiling, 474–477
  secondary structure, 353–354
  stability regulation by viral proteins, 336–338
  steady-state concentration, 474
  structure of, 349–350
  termination codon, 349
  3′ untranslated region (3′ UTR), 349
  translation (*see* Translation, cellular)
  turnover in cytoplasm, 335–336
  viral inhibition of translation, 471–474
  viral interference with, 376–377
Messenger RNA (mRNA), viral
  bicistronic, 368
  capping, 168, 262, 311, 312–315
    acquisition of cap structures from cellular RNAs, 315
    assembly line for, 313, 314
    cap snatching, 315
    identification of cap structures on viral mRNAs, 313
    overview, 311–312
    synthesis of cap structures by cellular enzymes, 312–313, 314
    synthesis of cap structures by viral enzymes, 313–315

export from nucleus, 311, 327–331
balance between export and splicing, 329
cellular export machinery, 327
constitutive transport elements (CTEs), 328–329, 331
export adapters, 327, 329–331
export receptors, 327, 331, 341
intron-containing mRNAs, 327–329
regulation by viral proteins, 332–333
Rev protein and, 327–330
RNA signals, 328–329
single-exon viral mRNAs, 329–330, 332
half-life, 335–336
internal ribosome entry site (IRES), 352, 355–359
plant virus mRNAs, 353
polyadenylation, 315–317
bovine papillomavirus type 1, alternative polyadenylation and splicing control in, 322–323, 324
formation of end by termination of transcription, 316, 317
identification of poly(A) sequences on viral mRNAs, 315
overview, 311–312, 315
regulation of alternative polyadenylation by viral proteins, 330–331
during viral mRNA synthesis, 316, 317
of viral pre-mRNA by cellular enzymes, 315–317
of viral pre-mRNA by viral enzymes, 316, 317
pregenomic mRNA, hepadnavirus, 214, 216–218, 220
pre-mRNA (*see* Pre-mRNA, viral)
RNA editing, 325–327
single-exon mRNAs, 329–330, 332
splicing
adenovirus, 322, 325, 326
alternative, 317, 322–325, 331–332, 334
balance between export and splicing, 329
bovine papillomavirus type 1, 322–323, 324
constitutive, 322
discovery, 317–318
mechanism, 318–322
overview, 311–312
regulation of alternative splicing by viral proteins, 331–332, 334
retrovirus, 323–325
RNA-dependent helicase participation in, 321
RNA-RNA interactions, 320
small nuclear RNA (snRNA) participation in, 320–321
splice sites, 318–319, 321–323, 326–327, 329, 331, 334
synthesis (*see* Messenger RNA synthesis, viral)
translation (*see* Translation, viral)
turnover in cytoplasm, 335–339
half-life, 335–336
regulation of stability by viral proteins, 336, 338
stabilization facilitating transformation, 338, 339
viral classification by, 20–21, 55
in virus particle, 117
Messenger RNA synthesis, viral
ambisense RNA virus, 174
double-stranded RNA virus, 175–178
hepatitis delta satellite virus, 176–177, 180
influenza virus, 174, 176
mapping gene order by UV irradiation, 175
mRNA release from rotavirus particles, 175, 178
nested subgenomic mRNAs, 172–173
(+) strand RNA virus, 171
(−) strand RNA virus, 173–174
strategies of representative virus families, 159
vesicular stomatitis virus, 173–176
Metabolism, 477, 479–492
anabolism, 477
catabolism, 477
citric acid cycle, 483–484
electron transport, 484, 486
glucose, 479–483
lipid, 485, 486–492
methods to study, 477, 479
oxidative phosphorylation, 484, 486
Metabolomics, 477
Metagenomic analyses, 20
Metastable structures, virus particles as, 81
Metavirus, 203
Methionine-independent initiation of translation, 354
Methyltransferase, 313, 316
Met-tRNA$_i$, 351–352, 354–355, 360–361, 370, 372
Micrococcal nuclease, 360
Microdomains, 30
receptor, 124
Microfilaments, 147
Microorganisms as pathogenic agents, 9
Micro-RNA (miRNA), 75, 338–346
cellular, 341–342
liver-specific reproduction of hepatitis C virus promoted by cellular miRNA-122, 342
protecting hepatitis C virus genome, 343
viral oncogenesis promotion by cellular miRNA-155, 341
databases, 342
eIF4E modulation by, 374
pre-miRNAs, 339, 341–343, 345
pri-miRNA, 341, 345
production and function of, 339–345
synthesis, 225–226
viral, 342–345
inhibition of antiviral defenses by, 345
latency-associated miRNAs of herpesviruses, 343, 345
polyomavirus, 342–343, 344, 345
RNA inference blocked by, 345
simian virus 40 (SV40), 342–343, 345
Microtubule organizing center (MTOC), 381, 412
Microtubules, 133, 147–148, 381–382, 397, 412, 413
axonal transport, 403–404
rotavirus and, 497
transport of genomes from cytoplasm to plasma membrane, 411
vaccinia virus movement, 447–450
vesicular stomatitis virus movement within cytoplasm, 414
Middle East respiratory syndrome coronavirus (MERS-CoV), 52, 506
receptor identification, 125, 126
Mimivirus, 19, 20
assembly, 417
contributions to translational machinery, 350
electron microscopy, 419
structure, 114
sulfhydryl oxidases, 392
viral factory, 495
virus particle appearance, 82
Minichromosome, 105, 226, 254, 275
hepadnavirus, 214
polyomavirus, 524
Minichromosome maintenance complex (Mcm), 277, 284, 298–299, 302
Minichromosome maintenance element, 301
Minus (−) strand, 21, 56
Minute virus of mice, 520
miRBase database, 342
miRNA. *See* Micro-RNA
Misfolding of proteins, 392–397, 406–407, 413
Misincorporation of nucleotides, 182–183
Mismatch repair, 302–303
Missense mutations, 70
Mitogen-activated protein kinase (Mapk), 470
Mnk1/Mnk2, 373–374
Molecular chaperones, 390. *See also* Chaperone
Molecular cloning, identification of receptor genes by, 125, 127
Molecular mimicry, 194
Molecular parasites, viruses as, 17
Molecular replacement, 85
Moloney murine leukemia virus
Env protein, 402
fusion protein, 141
Gag, 366
genome packaging, 435
import of viral genome into nucleus, 153
Pol, 366
proteolytic processing of polyprotein, 455
pseudoknot, 367–368
RNA polymerase II transcription, 258
structure, 102
suppression of termination, 366–368
Monocistronic messenger RNA (mRNA), cellular, 349–350
Monoclonal antibodies, 42, 43
identification of receptors for virus particles, 124–125, 127
Monoclonal antibody-resistant variants, 42
Monod, Jacques, 15, 16
Monolayer, 33
Mononegavirus, 105
Montagu, Lady Mary Wortley, 8
Morbillivirus, 518
Mouse hepatitis virus, budding at plasma membrane, 440
Mouse mammary tumor virus, 530
Mouse polyomavirus, attachment to host cells, 129
Movement of viral and subviral particles within cells, 147–148
MRN complex, 301, 303–304
mRNA. *See* Messenger RNA (mRNA)
MTOC (microtubule organizing center), 381, 412
mTor (mammalian target of rapamycin), 466–470, 489
mTor complex 1 (mTorC1), 374, 467
Multimeric structural unit, 94
Multiple isomorphous replacement, 85
Multiplex PCR, 45
Multiplicity of infection (MOI), 25, 50, 124, 293
Mumps virus, 518
growth in embryonated eggs, 36
historical aspects, 7
immunofluorescence assay, 43
Murine coronavirus
genome organization, 506
structure, 506

Murine gammaherpesvirus 68, 258
Murine hepatitis virus, 377
Murine leukemia virus, 530
  cellular tethers in retroviral integration, 209
  Gag protein, 402
  integrase, 204, 210, 211
  integration sites, 208
  MA protein, 402
  preintegration complex, 210
  reverse transcriptase domain and subunit relationships, 199
  viral vectors, 74
Murine retrovirus vectors, 78
Murray Valley encephalitis virus, 337
Mutagen, 66
Mutagenesis, 70–71
  site-directed, 131
Mutation. *See also specific types*
  engineering into viral genomes, 67–73
  functional analysis, 67
  induced, 66
  mapping, 66–67
  reassortment in segmented genomes, 67
  relationship to observed phenotype, 71
  in retroviruses, 198–199
  reverse transcriptase and, 198–199
  reversion analysis, 71–72
  second-site, 71–72
  spontaneous, 66
  types, 70
  in viral vector genomes, 74
Mutation rate, RNA *vs.* DNA viruses, 66
Myelin, 403
Myristoylation of proteins, 406, 408, 428
Myxobacteria, reverse transcription in, 202
Myxoma virus, 73, 526
Myxovirus, 323

N

N linked, 31
N protein
  hantavirus, 376
  measles virus, 376
  paramyxovirus, 174
  respiratory syncytial virus, 106
  vesicular stomatitis virus, 89, 105–106, 158, 160, 168, 169, 173–175, 411, 414, 425, 427
NA protein, influenza virus, 403, 409, 439–440, 444, 456–457
NADH (reduced nicotinamide adenine dinucleotide), 479, 483–487
Nanochemistry, 87
Nanomachines, 82
Nanovirus, 459
NC protein
  human immunodeficiency virus type 1, 409, 455
  retrovirus, 406, 412–413, 435, 454–455
NCLDVs (nucleocytoplasmic large DNA viruses), 19–20
Nectins, 131–132
Nedd4 protein, 442
Nedd8 protein, 295
Nef protein, human immunodeficiency virus type 1, 117
Negative (−) strand, 21, 56
NEP protein, influenza, 411–412
Nested mRNAs, 63
  synthesis of subgenomic, 172–173
Neuraminidase, influenza, 129
Neuron(s)
  axonal transport, 403–404, 405
  cell-type-specific production of stable viral intron, 319
  retrograde transport, 405
  sorting of viral proteins in, 402, 403–404
Neutralization, 42, 105
Neutralization antigenic sites, 42
Newcastle disease virus, 36, 518
Newton, Isaac, 157
Nf-κb (nuclear factor κb), 232, 239, 240, 245, 251, 374, 467, 470
Nidovirus
  ExoN protein, 183
  genome organization and expression, 172
  replication fidelity, 183
  synthesis of nested subgenomic mRNAs, 172
Niemann-Pick type C1 disease, 142
NIH-Chongquing virus, 47
Nipah virus, 518
Nitrous acid, 66
N-linked glycosylation, 386–387, 389, 391, 393–394, 398–399, 403
Nodamura virus, 95
Nodavirus
  genome organization, 58
  packaging genome, 104
Nomenclature
  complexities of viral, 18
  International Committee on Taxonomy of Viruses (ICTV), 20
  virus architecture, 82–83
Nonenveloped virus
  attachment to host cells, 126–129
  fatty acid synthesis/metabolism, 490–491
  release of virus particles, 450, 454
Nonhomologous end-joining, 207, 303
Nonsense mutations, 70
Norwalk virus, 36
NP protein
  arenavirus, 174
  filovirus, 89
  influenza virus, 89, 90, 174, 411–412, 494
NS protein, vesicular stomatitis virus, 411
NS1 protein
  influenza virus, 330, 333, 370, 376, 377
  myxovirus, 323
NS3 protein, bovine diarrhea virus, 184
NS28 protein, simian rotavirus, 410
Nsb1, 304
Nsp1 protein, severe acute respiratory syndrome coronavirus, 336, 376
NSP2 protein, rotavirus, 497, 499
nsP3 protein
  rotavirus, 376
  Semliki Forest virus, 377
NSP5, rotavirus, 497, 499
Nuclear export factor 1 (Nxf1), 328–331, 334
Nuclear factor κb (Nf-κb), 232, 239, 240, 245, 251, 374, 467, 470
Nuclear lamina disruption by herpes simplex virus type 1, 449, 452
Nuclear localization signal, 149, 151, 383–384, 387, 409, 411, 412
Nuclear magnetic resonance, 83, 436
Nuclear membrane, localization of viral proteins to, 411
Nuclear pore complex, 149–150, 152, 153, 383
Nucleic acid detection, 45–46
  DNA microarrays, 45
  high-throughput sequencing, 45–46
  polymerase chain reaction, 45
Nucleic acid metabolism, viral enzymes of, 291
Nucleic acid purification columns, contaminated, 47
Nucleocapsid, 158–159
  assembly, 423–425, 428–429
    hepadnavirus, 218
  defined, 83
  helical, 86, 88–89
  movement via microtubules, 414
  rabies virus, 413
Nucleocapsid protein, retrovirus, 191
Nucleocytoplasmic large DNA viruses (NCLDVs), 19–20
Nucleosome
  histone code, 236
  packaging viral genome and, 105
  remodeling and HIV-1 Tat protein, 245
  simian virus 40 genome associated with, 274–275
  structure, 226
  virus association with, 226–228
Nucleotides
  misincorporation of, 182–183
  RNA editing, 325–327
Nucleus
  budding as mechanism of mRNA export from nucleus, 453
  budding of viruses from, 411
  export of RNAs from, 311, 327–331
    balance between export and splicing, 329
    cellular export machinery, 327
    constitutive transport elements (CTEs), 328–329, 331
    export adapters, 327, 329–331
    export receptors, 327, 331, 341
    intron-containing mRNAs, 327–329
    regulation by viral proteins, 332–333
    Rev protein and, 327–330
    RNA signals, 328–329
    single-exon viral mRNAs, 329–330, 332
  import of viral genomes into, 123, 148–153
    DNA genomes, 151
    influenza virus ribonucleoprotein, 151
    nuclear import pathway, 150–151
    nuclear localization signals, 149, 151
    nuclear pore complex, 149–150, 152, 153
    retroviral genomes, 151, 153
    strategies, 149
  localization of viral proteins to, 383, 411
  release of virus particles assembled at, 449, 452–453
  remodeling, 491–495
  transport of genomic and pregenomic RNA from nucleus to cytoplasm, 411
  virus assembly within, 382–384
    import of viral proteins for assembly, 383–384
Nup53, 334
Nup98, 334
Nxf1 (nuclear export factor 1), 328–331, 334

O

O linked, 31
Obesity, human adenovirus type 36 and, 492
Oceans, viruses in, 4

Octamer-binding protein 1 (Oct-1), 232, 248–249, 252
Octamer-binding protein 2 (Oct-2), 249
Okazaki fragments, 269–270, 272–273, 279
Oligomerization
in endoplasmic reticulum, 395–396
viral protein, 395–396
Oligomerization state, 235
Oligo(U) primer, 161
Oligosaccharide, 106, 386–387, 389–391, 393–394, 398–399, 403
compartmentalization of processing of N-linked, 398–399
detection and structure of N-linked, 391
glycoslyation (*see* Glycosylation)
self-glycosylation, 390
as sugar shell, 389, 391
Oligosaccharyltransferase, 389
O-linked glycosylation, 389, 394, 399, 403
Oncogene, 467
Oncogenesis, promotion by cellular miRNA-155, 341
Oncolytic viruses, 74, 75
Oncotherapy, viral, 73, 74, 75, 76
One-hit kinetics, 37, 39
One-step growth curve (cycle), 25, 46–50, 51
animal viruses, 49–50
bacteriophages, 14, 17, 46, 48–49
initial concept, 46, 48
Operator, bacteriophage λ, 255–256
Orbivirus, 101
Orc (origin recognition complex), 270, 298, 299
Organelles, remodeling of cellular, 491–499
cytoplasm, 495–499
nucleus, 491–495
Origin for plasmid maintenance (OriP), 278, 297–300
Origin of replication, 267–287, 294, 296–300
bacteriophage λ, 272
bidirectional, 269–270
Escherichia coli (OriC), 272
eukaryotic, 269–270, 299
mammalian, 270
mutations, 72
origin for plasmid maintenance (OriP), 278, 297–300
origin-independent, recombination-dependent replication, 304, 306
replication licensing, 298, 299
simian virus 40, 268, 271–275, 282, 284–285, 307
viral, 268, 271–286, 294, 296–300
common features, 280
Epstein-Barr virus, 278, 280, 297–299, 300
number of origins, 278, 280
papillomavirus, 276, 278, 280, 286, 299
properties, 278–280
recognition of, 280–286
simian virus 40, 268, 271–275, 282, 284–285, 307
Origin recognition complex (Orc), 270, 298, 299
Origin recognition proteins, 271, 277–278, 282–283, 285, 287, 299, 307
Origin-independent, recombination-dependent replication, 304, 306
OriL, herpes simplex virus type 1, 280–282
OriLyt, Epstein-Barr virus, 278, 280, 299
OriP (origin for plasmid maintenance), 278, 297–300
OriS, herpes simplex virus type 1, 280–282, 287
Orthohepadnavirus, 512
Orthomyxovirus. *See also* Influenza virus
assembly, 382
attachment to host cells, 129
budding at plasma membrane, 440
diseases, 516
envelope, 108
furin family protease, 400
genome organization, 59, 60, 64, 516
hemagglutination, 41
information retrieval from genome, 63
membrane fusion, 138
overview, 516–517
particle-to-PFU ratio, 40
reproductive cycle, single-cell, 517
structure, 89, 516
virion enzymes, 116
Orthopolyomavirus, 524
Orthopoxvirus, 526
Orthoreovirus
genome organization, 528
reproductive cycle, single-cell, 529
structure, 101, 528
uncoating, 145
Oxidative phosphorylation, 484, 486
Oxidoreductases, 392

P

P (processing) bodies, 335, 377
P protein
hepadnavirus, 214–220
influenza virus, 169
vesicular stomatitis virus, 168, 169, 173–175, 414, 425
P (peptidyl) site, 350, 352, 354–355, 357, 360–361, 363, 368, 372
P1 polyprotein, poliovirus, 423–424
p21 protein, 343
p53 protein, 524
P54 protein, 376
P56 protein, 376
P60 protein, 376
p300, 251
P1234 polyprotein, alphavirus, 314–315
pac1/pac2, herpes simplex virus type 1, 434
Packaging signals, 431–435
dimer linkage sequence, 434–435
DNA signals, 432–434
psi (ψ) sequence, 435
RNA signals, 434–435
Packaging viral genome, 60, 104–106, 430–439
bacteriophage, 434, 437
by cellular proteins, 105–106
concerted assembly, 430, 431, 432
delayed synthesis of structural proteins preventing premature packaging of DNA templates, 291
dimerization, 434–436
direct contact with a protein shell, 104–105
headful packaging, 436–437
influenza virus, 437–439
packaging signals
DNA signals, 432–434
RNA signals, 434–435
recognition of nucleic acid genome, 431–438
segmented genome, 437–438
sequential assembly, 430–431
size limitation, 436
by specialized viral proteins, 105
synthesis coupled with encapsidation, 436
Pact (Pkr activator), 369–370
Palindromes, as target sites for retroviral DNA integration, 207–208
Palm domain, RNA-dependent RNA polymerase, 161–162, 164, 165
Palmitate, 484–485, 487, 499
PAMs (protospacer adjacent motifs), 340
PAN (polyadenylated nuclear RNA), 473
Pan3, 336
Pandoravirus, 19, 20
assembly, 417
contributions to translational machinery, 350
genome size, 65, 97
structure, 114
sulfhydryl oxidases, 392
Pandoravirus salinus, 60, 65
Papillomavirus. *See also* Human papillomavirus
assembly, 382
cervical cancer, 338
chromatin thethering, 209
cytopathic effect, 34
disulfide bond formation, 457
DNA synthesis, 268, 276–278, 280, 283–286, 288–290, 294, 299–302, 307
E1 protein, 284, 286, 300–302
E2 protein, 286, 300–301
E7 protein, 289, 301
episome, 301–302
genome organization, 56
infectious DNA clones, 68
information retrieval from genome, 63
integration, 338, 339
limited and amplifying replication from a single origin, 299–301
maturation of progeny virus particles, 457
micro-RNA inhibition of reproduction, 341
minichromosome, 226
mRNA stabilization, 338
nucleosome association, 226, 228
origin of replication, 276, 278, 280, 286, 299
packaging genome, 105
particle-to-PFU ratio, 40
polyadenylation, 330
release of virus particles, 457
splicing, 317
transcription, 226
transcriptional regulators, 247
Papovavirus
origin recognition proteins, 277
VP2 protein, 402
Paralytic dose 50% (PD_{50}), 39
Paramecium bursaria chlorella virus 1 (PBCV-1), 390
Paramyxovirus
cell-to-cell spread, 457–458
cytopathic effect, 34
diseases, 518
entry into host cells, 132, 134–135
F protein, 132, 134–135
furin family protease, 400
genome organization, 59, 60, 518
hemagglutination, 41
information retrieval from genome, 63
membrane fusion, 138
overview, 518–519
packaging genome, 106

Paramyxovirus *(continued)*
release of virus particles, 442
reproductive cycle, single-cell, 519
RNA editing, 185
RNA synthesis, 174
RNA-dependent RNA polymerase, 161
rule of six, 174
structure, 89, 518
uncoating at plasma membrane, 132, 134–135
viral oncotherapy, 75
zoonoses, 518
Particle-to-PFU ratio, 39, 40
Parvovirus
DNA synthesis, 267–268, 276–277, 280, 302, 307
origin recognition proteins, 277
priming via DNA, 276–277
genome conversion for transcription, 226–227
genome organization, 57, 520
genome structure, 60, 62
import into nucleus, 149, 151
mutation rate, 302
nomenclature, 20
overview, 520–521
reproductive cycle, single-cell, 521
splicing, 317
structure, 92, 94, 520
transcription, 226
viral oncotherapy, 75
viral vectors, 74
virus particle appearance, 82
Parvovirus-like hybrid virus (PHV-1), 47
Pasteur, Louis, 8, 9–10
Pathogenesis, 4, 10, 32
PBCV-1 (*Paramecium bursaria* chlorella virus 1), 390
pbs. *See* Primer-binding site
Pcna (proliferating-cell nuclear antigen), 272–273, 277
PCR (polymerase chain reaction), viral nucleic acid detection, 45
Pegivirus, 510
Pentamers, VP1 simian virus 40, 98–99
Pentose phosphate pathway, 482
Peptidyltransferase, 350
Perilipin, 497
Perk (Pkr-like ER kinase), 369–370, 489
Permissive cell, 31
Permissive conditions, 66
Peroxisome proliferator-activated receptor (Ppar-α), 233
Pestivirus, 510
internal ribosome entry site (IRES), 355
Peyer's patches, 27
Phage. *See* Bacteriophage
Phenotype
mutant, 71
reversion, 71
Phosphatidylcholine, 490–491
Phosphatidylinositol 1,4-phosphate, 495, 496
Phosphoenolpyruvate carboxykinase, 482
Phosphofructokinase, 480
Phosphoinositol 3-kinase (Pi3K), 374, 466–470, 492
Phosphoinositol 3-phosphate (PI3P), 466
Phosphoinositol 4-phosphate (PI4P)-containing lipids, 405
Phospholipid bilayer, 29–30
Phospholipid synthesis, 485, 490
Phosphopeptides, 470
Phosphoproteins, 470
Phosphorylation, oxidative, 484, 486
PHV-1 (parvovirus-like hybrid virus), 47
Phylogenetic trees, 46, 48
Pi3k (phosphoinositol 3-kinase), 374, 466–470, 492
Pi3k-Akt-mTor signaling route, 375
Picornavirus
assembly, 423
attachment to host cells, 126–128
cellular gene expression inhibition, 473–474
cytopathic effect, 34
cytoplasmic replication and assembly compartments, 496–498
diseases, 522
eIF3 regulation, 376
eIF4E activity modulation, 374
genome organization, 58, 59, 522
genome structure, 60, 61, 157
pseudoknot, 169
host cell signaling and, 470
infectivity reduction, environmental conditions and, 81
information retrieval from genome, 63
inhibition of cellular mRNA export, 334
internal ribosome entry site (IRES), 352, 355–357
lipid metabolism, 490–491
long-range RNA-RNA interactions, 360
messenger RNAs, 355
mRNA synthesis, 159
overview, 522–523
packaging genome, 104
particle-to-PFU ratio, 40
polyadenylation, 317
polyprotein processing, 363, 365
protein priming, 116–117, 165
release of virus particles, 450, 454
replication, 159
coupled with encapsidation, 436
reproductive cycle, single-cell, 523
resistance to detergents, 81
RNA polymerase, fidelity checkpoint for, 183
RNA synthesis, 164, 165, 171
structure, 94–95, 97, 522
uncapped mRNAs, 312
viral oncotherapy, 75
VP0 protein, 402
VP4 protein, 402
Pithovirus, 114, 116
Pithovirus sibericum, 116
Pkr, 326–327, 369–372, 377, 379
activation of, 369–370
viral regulation of, 370, 372
dsRNA-binding proteins, 370
inhibition of kinase function, 370, 372
RNA antagonists, 370
Pkr activator (Pact), 369–370
Pkr-like ER kinase (Perk), 369–370, 489
Plant cell structure, 385
Plant virus, 4, 7, 14
classification, 19
dose-response curve of the plaque assay, 37
enveloped, 385
long-distance RNA-RNA interactions, 61
messenger RNAs, 353
movement proteins, 459
reverse transcription by, 202
viral genomes, 19
VPg-dependent ribosomal recruitment, 353
Plaque, 36–37
Plaque assay, 25, 36–38, 522
calculating virus titer from, 38
dose-response curve, 37
fluorescent-focus assay, 37
infectious-centers assay, 37–38
Plaque-forming units (PFU) per milliliter, 37, 38
particle-to-PFU ratio, 39, 40
Plaque-purified virus stock, 37
Plasma membrane
assembly of virus at, 384–409
disruption of secretory pathway, 404–406
localization of viral proteins to, 384
overview, 384–385
signal-sequence-independent transport of viral proteins, 406, 408–409
sorting of viral proteins in polarized cells, 401–404
transport from cytoplasm to plasma membrane, 411–413
transport of viral membrane proteins to plasma membrane, 386–401
epithelial cell, 402–403
membrane proteins, 30–31
microdomains, 30
properties, 29–30
release of virus particles at, 441–444
Escrt-dependent budding, 441–443
Escrt-independent budding, 444
nonstructural proteins facilitating release, 444
sequence motifs required for budding, 441–442
uncoating at, 132–135
Plasmid
infectious DNA clones, 67–70
origin for plasmid maintenance (OriP), 278, 297–300
visualization of duplication and partitioning of Epstein-Barr virus plasmid replicons in living cells, 300
Plasmodesmata, 385, 459
Plus (+) strand, 21, 56
PML (progressive multifocal leukoencephalopathy), 524
Pml bodies, 294–296
Pmls (promyelocytic leukemia proteins), 294–296, 493
Pol protein
Moloney murine leukemia virus, 366
retrovirus, 204, 323
Rous sarcoma virus, 368
Polarized cells, sorting of viral proteins in, 401–404
epithelial cells, 402–403
neurons, 402, 403–404
Poliovirus
2A protease, 334
2B protein, 405
2BC protein, 405, 450, 454, 496
2C protein, 436
3A protein, 405, 450, 454, 496
assembly, 420, 422–426
attachment to host cells, 126–128
attenuated vaccine strains, 66
autophagosome-like vesicles, 405
autophagy and nonlytic release of virus particles, 498
capsid, 94–98
cell culture, 32–35

cellular gene expression inhibited by, 471
cis-acting replication elements (cre), 169
cleavage of eIF4G, 372–373
co-option of cytoplasmic membranes, 498
cytopathic effect, 35
cytoplasmic replication and assembly compartments, 496–498
disease, 522
eIF4E activity modulation, 374
end-point dilution assay, 40
entry into cells, 145, 146
experimental studies in animals, 36
genome organization, 522
genome structure, 72
pseudoknot, 169
genome structures critical for function, 62
Golgi disruption, 404–405
historical aspects, 7
host range, 123–124
infectious cycle of, 26
infectious DNA clones, 68–69
infectivity reduction, environmental conditions and, 81
inhibition of cellular transcriptional machinery, 260
inhibition of translation, 471–472
internal ribosome entry site (IRES), 357, 358
lipid metabolism, 490–492
movement within cells, 148
myristoylation of proteins, 408
neutralization antigenic sites on capsid, 42
neutralization tests, 42
overview, 522–523
P1 polyprotein, 423–424
packaging of genome, 96, 97, 436
particle-to-PFU ratio, 40
plaques, 37
polyprotein, 422–424
cleavage of, 451, 454
processing, 363
protease 2A^pro^, 472
protease 3CD^pro^, 423–424
protease 3C^pro^, 363, 365, 377, 471
protein priming, 165–167
proteins in virion, 117
receptor, 123–124
receptor glycosylation, 389
release of virus particles, 450, 454
replication inhibition by siRNA, 73
reproductive cycle, single-cell, 523
ribosome-RNA polymerase collisions, 179
RNA polymerase, 405
RNA recombination, 184
RNA synthesis, 165–171
cellular sites for, 179, 182
host cellular proteins required for, 170
(−) strand synthesis, 167
RNA-dependent RNA polymerase, 159, 171
fidelity, 183
processivity, 168
structure, 161–164
template specificity, 169
Sabin vaccine, 184
secretory pathway inhibition, 496
serotypes, 42
size, 13
stress-associated RNA granules, 377
structure, 522
transport through M cells, 27
vaccine strains, 183
vaccines, 34
viral vectors, 74
VP0 protein, 402, 423, 425–426, 451, 454, 497
VP1 protein, 423, 426
VP2 protein, 408, 426, 451, 497
VP3 protein, 408, 423, 426, 436
VP4 protein, 402, 408, 426, 451, 497
VPg protein, 94, 165–167, 278, 436
Poly(A) addition. *See* Polyadenylation
Polyadenylated nuclear RNA (PAN), 473
Polyadenylation, 315–317
bovine papillomavirus type 1, alternative polyadenylation and splicing control in, 322–323, 324
formation of end by termination of transcription, 316, 317
hepatitis delta satellite virus, 177
identification of poly(A) sequences on viral mRNAs, 315
influenza virus, 173
inhibition of cellular mRNA production by viral proteins, 333–335
overview, 311–312, 315
poliovirus, 171
regulation of alternative polyadenylation by viral proteins, 330–331
vesicular stomatitis virus, 173, 176
during viral mRNA synthesis, 316, 317
of viral pre-mRNA by cellular enzymes, 315–317
of viral pre-mRNA by viral enzymes, 316, 317
Poly(A)-binding protein, 170, 316, 333, 353, 362, 376, 473
Poly(rC)-binding protein, 170, 179, 359
Polycistronic messenger RNA, 349–350
Polyclonal antibodies, 42, 43
Polymerase. *See also* DNA polymerase; RNA polymerase
activity assay, 41
misincorporation of nucleotides, 182–183
proofreading capability, 182
two-metal mechanism of catalysis, 162
Polymerase chain reaction (PCR), viral nucleic acid detection, 45
Polyomavirus
alternative splicing, 322
assembly, 382, 384
attachment to host cells, 129
cytopathic effect, 34
diseases, 524
DNA synthesis, 267–269, 272, 280–281, 288–289, 294, 307
entry into cells, 133
genome organization, 56, 524
genome size, 97
infectious DNA clones, 68
information retrieval from genome, 63
micro-RNA (miRNA), 342–343, 344, 345
minichromosome, 226
movement within cells, 148
nucleosome association, 226, 228
nucleus reorganization, 493
overview, 524–525
packaging genome, 105
particle-to-PFU ratio, 40
release of virus particles, 450
reproductive cycle, single-cell, 525
splicing, 317
structure, 524
transcription, 226, 246
transcriptional regulators, 247
VP1 protein, 384
VP2 protein, 384
VP3 protein, 384
VP4 protein, 450
Poly(A) polymerase, 315–317, 324–325, 345
Poly(U) polymerase, 161
Polyprotein
assembly of structural unit from, 422–423
cleavage of, 451, 454–456
Gag-Pol, 438
incorporation into viral particle, 438
Moloney murine leukemia virus, 366
proteolytic processing, 454–455
retrovirus, 204, 412
ribosomal frameshifting, 368
poliovirus, 422–424
synthesis, 63, 363, 365
Polypurine tract (ppt), 195
Polypyrimidine-tract-binding protein (Ptb), 359
Polyribosome, 386, 475
Polysome, 361, 376
Poly(A) tail. *See also* Polyadenylation
RNA profiling, 474
role in 5′-end-dependent initiation of translation, 352–353
Porcine parvovirus
nuclear localization sequences, 383–384
VP2, 383
Portal, 112–113, 115
Positive autoregulatory loop, 237
Positive (+) strand, 21, 56
Posttranscriptional regulation
inhibition of cellular mRNA production, 333–335
to block antiviral responses, 334
to facilitate viral mRNA production, 334
inhibition of cellular mRNA export, 334–335
inhibition of polyadenylation and splicing, 333–335
of viral gene expression, 330–333
regulation mRNA export, 332–333
regulation of alternative polyadenylation by viral proteins, 330–331
regulation of alternative splicing, 331–332, 334
Posttranslational modification
of histones, 226, 236
ubiquitinylation of proteins, 295
Posttranstricptional regulatory element (PRE), 330, 332
Potato yellow dwarf virus, 385
Potyvirus, 353
Poxvirus
cellular gene expression inhibition, 471–473
classification, 20
cytopathic effect, 34
diseases, 526
DNA polymerase, 303
DNA synthesis, 267–268, 277, 290–291, 303, 307
DNA-dependent RNA polymerase, 260, 262
eIF4E phosphorylation, 374
envelope, 108
enzymes of nucleic acid metabolism, 291
genome organization, 56–57, 526
genome sequence, 60
genome structures critical for function, 62

Poxvirus (continued)
glutamine metabolism and, 484
growth in embryonated eggs, 36
infectious DNA clones, 68
information retrieval from genome, 63
inhibition of cellular DNA synthesis, 291
interaction with internal cellular membranes, 410
mRNA degradation, 336, 338
overview, 526–527
particle-to-PFU ratio, 40
reproductive cycle, single-cell, 527
structure, 113, 115, 526
transcription, 225, 226
transcriptional regulators, 247
viral factory, 495
viral oncotherapy, 75
viral thio oxidoreductase system operating in cytoplasm, 392
viral vectors, 77
virion enzymes, 116
pp60 protein, Rous sarcoma virus, 402
Ppar-α (peroxisome proliferator-activated receptor), 233
ppt (polypurine tract), 195
PPXY motif, 441–443
PRD1, 106
PRE (posttranscriptional regulatory element), 330, 332
Pregenomic mRNA, hepadnavirus, 214, 216–218, 220
Pregenomic RNA, transport from nucleus to cytoplasm, 411
Preinitiation complex, 228, 230, 245, 248, 250–251, 278, 280–282, 298, 352–353, 359, 376
Preintegration complex, 205–207, 210
Pre-miRNA, 339, 341–343, 345
Pre-mRNA, viral, 310–346
capping, 311–315, 312–315
acquisition of cap structures from cellular RNAs, 315
assembly line for, 313, 314
cap snatching, 315
identification of cap structures on viral mRNAs, 313
overview, 311–312
synthesis of cap structures by cellular enzymes, 312–313, 314
synthesis of cap structures by viral enzymes, 313–315
export from nucleus, 311, 327–331
regulation by viral proteins, 332–333
polyadenylation, 311–312, 315–317
bovine papillomavirus type 1, alternative polyadenylation and splicing control in, 322–323, 324
formation of end by termination of transcription, 316, 317
identification of poly(A) sequences on viral mRNAs, 315
overview, 311–312, 315
regulation of alternative polyadenylation by viral proteins, 330–331
during viral mRNA synthesis, 316, 317
of viral pre-mRNA by cellular enzymes, 315–317
of viral pre-mRNA by viral enzymes, 316, 317
splicing, 311–312
adenovirus, 322, 325, 326
alternative, 317, 322–325
balance between export and splicing, 329
bovine papillomavirus type 1, 322–323, 324
constitutive, 322
discovery, 317–318
mechanism, 318–322
overview, 311–312
regulation of alternative splicing by viral proteins, 331–332, 334
retrovirus, 323–325
RNA-dependent helicase participation in, 321
RNA-RNA interactions, 320
small nuclear RNA (snRNA) participation in, 320–321
splice sites, 318–319, 321–323, 326–327, 329, 331, 334
Pre-TP, adenovirus, 277–278, 281, 286, 290
Pre-VI (pVI) protein, adenovirus, 384, 456
Primary cell cultures, 33
Primase, 270, 272–273, 275–277, 279, 282, 286–287, 298
DNA polymerase α-primase, 270, 272–273, 275, 277, 282, 286, 298
lagging strand synthesis, 272–273
leading strand synthesis, 272–273
viral, 276, 277, 279, 287
Primer
cap, 161
DNA synthesis, 267–270, 272–273, 275–279, 281, 286, 307
DNA as primer, 275–277, 279, 304
Pre-TP, adenovirus, 277–278, 281, 286, 290
proofreading and, 302–303
protein, 275, 278, 281
RNA, 268–270, 272–273, 275–276, 279, 306
self-priming, 276–277, 279
hepadnavirus reverse transcription, 216–221
initiation of RNA synthesis, 164, 165–168
oligo(U), 161
RNA-dependent RNA polymerase, 157, 161, 164
tRNA primer encapsidation, 438
Primer tRNA, retrovirus, 191–192, 193
Primer-binding site (pbs)
human immunodeficiency virus type 1, 194
retrovirus, 192
Priming, protein, 165–167, 275, 278, 281, 502
pri-miRNA, 341, 345
Prions, 20
prM protein, flavivirus, 410
Procapsid, 428–429
Processed pseudogenes, 202
Processing, RNA, 310–346
Processing step, in retroviral DNA integration, 205–207
Processivity
DNA-dependent DNA polymerase, 270
of retrovirus reverse transcriptase, 198
RNA-dependent RNA polymerase, 168
Progressive multifocal leukoencephalopathy (PML), 524
Proliferating-cell nuclear antigen (Pcna), 272–273, 277
Promoter(s). See also Transcription
bacteriophage T7 RNA polymerase, 68–69
cellular regulation of activity, 252–254
core, 228, 231
DNA-dependent RNA polymerase, 260, 262
herpes simplex virus type 1, 248, 250, 251
herpesvirus, 246
RNA polymerase II, 228, 231, 322
RNA polymerase III, 257, 259
simian virus 40 (SV40), 231, 252–253, 254
TATA sequence, 228, 231–232, 234, 237, 249–250, 259–260
use of term, 158
vaccinia virus, 260, 262
in viral vectors, 73–74, 75
virus-associated RNA I (VA-RNA I), 257–259
Zta gene, Epstein-Barr virus, 248–249
Promoter occlusion, 236
Promyelocytic leukemia proteins (Pmls), 294–296
Proofreading, 65, 182, 183
DNA polymerase, cellular, 301–303
DNA polymerase, viral, 303
of single-stranded DNA virus genomes, 302
translation, 351
Propagation, 21, 25, 32–36
Prophage, 15, 16, 190
Prostate cancer, 197
Protease, 123
adenovirus, 456
envelope glycoprotein proteolytic processing in Golgi, 399–401
furin family, 399–400
human immunodeficiency virus type 1, 451
incorporation into viral particle, 438
membrane fusion, 138
poliovirus 3CD^{pro}, 423–424
retrovirus, 454–455
in virus particles, 116
Protein coat. See Capsid
Protein disulfide isomerase, 392
Protein folding, 389–390, 392–397, 400, 406–407
misfolding, 392–397, 406–407, 413
unfolded protein response, 406, 407
Protein IX, adenovirus, 100–101
Protein kinase, in signal transduction pathways, 466–470
Protein kinase A, 470
Protein kinase C, 449
Protein priming, 165–167, 275, 278, 281, 502
Protein shells, assembly, 421–430
assisted assembly reactions, 428–429
capsid and nucleocapsid assembly, 423–425
self-assembly, 425–428
structural unit
chaperone participation, 423, 424
formation, 421–423
from individual proteins, 421–422
from polyproteins, 422–423
visualization of structural transitions, 430
Protein synthesis, 348–379
cellular (see Translation, cellular)
mechanisms of eukaryotic, 349–363
overview, 349
proofreading, 351
viral (see Translation, viral)
Protein V, 494
Protein VII, adenovirus, 105
Proteins, cellular
common properties of proteins that regulate transcription, 234–235
genome packaging by, 105–106
glycoproteins, 31
hepadnavirus P protein folding, 218
incorporation into particles, 438
inhibition of transport, 404
in IRES function, 358–359

membrane proteins, 30–31
replication proteins, 270–271
retrovirus DNA integration and, 209–210
RNA synthesis, viral, 170
role in viral replication, 51
translation proteins, 351
ubiquitinylation, 295
vesicular sorting proteins, 441–443
viral induction of replication proteins, 288–290
in viral particle, 117–118
Proteins, viral, 31
acylated, 400, 402
assembly of protein shell
chaperone participation, 423
from individual proteins, 421–422
from polyproteins, 422–423
cellular gene expression inhibited by, 470–474
chaperones, 423, 425
common properties of proteins that regulate transcription, 234–235
envelope, 106–109
enzymes of nucleic acid metabolism, 291
folding, 389–390, 392–397, 400, 406–407
genome packaging by, 105
glycoproteins, 106–109
inactivation of Rb protein, 288–290
incorporation into particles, 438
induced synthesis of cellular replication proteins, 288–290
inhibition of cellular protein transport, 404
interaction with internal cellular membranes, 410–411
isoprenylated, 402, 406, 408
lipid-plus-protein signals, 406, 408
localization to compartments of the secretory pathway, 410
localization to nuclear membrane, 411
myristoylation of, 406, 408
origin recognition proteins, 271, 277–278, 282–283, 285, 287, 299, 307
proteolytic processing of structural proteins, 450–451, 454–456
cleavage of polyproteins, 451, 454–456
cleavage of precursor proteins, 456
regulating transcription of viral DNA templates, 237–257
scaffolding proteins, 425, 428–429
selectivity of chaperones for viral glycoproteins entering endoplasmic reticulum, 393
signal-sequence-independent transport of viral proteins, 406–409
in viral particle, 116–117
Proteoglycans, 28
Proteolytic processing of structural proteins
cleavage of polyproteins, 451, 454–456
cleavage of precursor proteins, 456
Protospacer adjacent motifs (PAMs), 340
Prototype foamy virus, 215
integrase structure, 210, 211–213
reverse transcriptase domain and subunit relationships, 199
Provirus, 78
adeno-associated virus, 296
avian sarcoma and leukosis virus, 236–237
endogenous, 202, 203
epigenetic silencing, 236, 238–239
human immunodeficiency virus type 1, 238–239
replication, 276, 296
retrovirus, 197, 204–205, 276, 530–531
chromatin, 226
splicing, 317
transcription, 225
Pseudodiploid, 191
Pseudogenes, 474
processed, 202
Pseudoknot
Moloney murine leukemia virus, 367–368
RNA, 158, 160, 169
Pseudorabies virus
attachment to host cells, 131
axonal transport, 403–404, 405
counting number of genomes expressed and replicated, 293
plaques, 37
Pseudoreversion, 71–72
Psi (ψ) sequence, 435
PTAP motif, 441–442
Ptb (polypyrimidine-tract-binding protein), 359
p-Tefb, 242, 245
Pulse-chase experiments, 420
pVI (pre-VI) protein, adenovirus, 384, 456
Pyruvate, metabolism of, 483, 484
Pyruvate carboxylase, 483, 484
Pyruvate dehydrogenase, 483, 484, 486, 488

Q

Quantitative PCR, 45
Quasiequivalence, 90–91, 97
Quasispecies, 183

R

Rabbit fibroma virus, 526
Rabies virus, 532
cytopathic effect, 34
eIF3 regulation, 376
M protein, 376
nucleocapsid, 413
Rae1, 334
Ran, 150–151, 152, 327, 329–330, 334
RAP94, 262
Rapamycin, 374
Ras protein, retrovirus, 406
Rb protein, 277, 282, 288–290, 292
adenovirus, 251–253
functional inactivation of, 288–290
Rbx1EloB protein, 295
Rcc-1, 151, 152
Reactivation, 254, 257–258, 514
Reactive oxygen species, 484, 486, 489
Real-time PCR, 45
Reassortment, 67, 183
Receptor(s), 123
affinity, 124
alternate, 132
attachment to host cells
enveloped viruses, 129–132
nonenveloped viruses, 126–129
via protruding fibers, 128–129
via surface features, 126–128
blocking HIV-1 infection with soluble cell receptors, 136
coreceptor, 123, 124, 127
definition, 123, 124
endosomal fusion receptor, 142
export, 327, 331, 341
host range and, 123–124
microdomains, 124
multiple, 132
tropism and, 123
virus-induced signaling via cell receptors, 148
Receptor genes, cellular, 125–126
Receptor-mediated endocytosis, adenovirus, 128
Recombinant DNA technology, 5–6
assembly, studies of, 421
for mutation introduction, 70
Recombinase, 189, 210. *See also* Integrase
bacteriophage λ, 304, 306
Recombination
bacteriophage λ integration, 190
base-pair-dependent, 184
base-pair-independent, 184
copy choice, 184, 198
general mechanisms, 304, 306
homologous, 77
origin-independent, recombination-dependent replication, 304, 306
origins of genetic diversity in DNA viruses, 304–307
pathogenic viruses produced by, 184
retroviral, 197–198
during reverse transcription, 197–198
RNA, 183–185
site-specific, 304, 306
viral genome, 306–307
Recombination frequency, 67
Recombination mapping, 66–67
Regulation
gene expression, advantages of temporal regulation, 253
gene expression, cellular, 470–478
differential regulation of, 474–478
inhibition of, 470–474
posttranscriptional
inhibition of cellular mRNA production, 333–335
to block antiviral responses, 334
to facilitate viral mRNA production, 334
inhibition of cellular mRNA export, 334–335
inhibition of polyadenylation and splicing, 333–335
of viral gene expression, 330–333
regulation mRNA export, 332–333
regulation of alternative polyadenylation by viral proteins, 330–331
regulation of alternative splicing, 331–332, 334
transcription, 228–235
translation during viral infection, 368–377
Reinitiation of viral translation, 366, 367
Relaxation of DNA, 274
Release of virus particles, 441–450. *See also* Budding of virus particles
assembled at internal membranes, 444–450
assembled at plasma membrane, 441–444
Escrt-dependent budding, 441–443
Escrt-independent budding, 441–443
nonstructural proteins facilitating release, 444
sequence motifs required for budding, 441–442
assembled in nucleus, 449, 452–453
by cell lysis, 450
double-envelopment model, 449, 452
envelopment by virus-specific mechanism, 445–450

Release of virus particles *(continued)*
herpesvirus, 444, 449, 452–453
nonenveloped virus, 450, 454
nonlytic, 450, 454
Remodeling of cellular organelles, 491–499
cytoplasm, 495–499
nucleus, 491–495
Reovirus
capped mRNAs, 175
conservative replication, 176, 179
diseases, 528
entry into cells, 136, 145, 147
genome organization, 58, 64, 528
λ-2 protein, 313
mRNA cap, 312–313
one-step growth curve, 49
overview, 528–529
particle-to-PFU ratio, 40
reproductive cycle, single-cell, 529
RNA polymerase, 169, 175
RNA synthesis, 175, 177
σ3 protein, 370
structure, 101–102, 528
uncoating, 145
viral oncotherapy, 75
virion enzymes, 116
Reovirus type 1, virion enzymes, 116
Rep 78/68 protein, adeno-associated virus, 268, 276–277, 283, 296–297, 306
Replicase, 158. *See also* RNA-dependent RNA polymerase
Replication
error frequency in DNA replication, 183
origin-independent, recombination-dependent replication, 304, 306
rolling-circle, 277–278, 288
semi-conservative, 267, 278
strategies of representative RNA virus families, 159
viral, 271–307
determining a role for cellular proteins, 51
DNA genomes, 56–58
errors, 65
RNA genomes, 58–59
in vitro, 307
Replication bubble, 270, 284, 285, 306
Replication center, 292–294, 296, 303–304
Replication compartments, 292, 293, 306
Replication complexes, 495–497
Replication factor C (Rf-C), 272–273, 277
Replication fork, 267–270, 272–275, 278, 280, 283, 294, 303–305
Replication intermediate, 276–277, 306
Replication licensing, 298, 299
Replication protein A (Rp-A), 271–273, 284–286, 298
Replicons
Epstein-Barr virus plasmid, 300
eukaryotic
general features, 269–270
origins of replication, 270
parvoviral integration into, 296–297
visualization of duplication and partitioning of Epstein-Barr virus plasmid replicons in living cells, 300
Repressor, 235–236, 247–249, 252–256
bacteriophage λ, 255–256
corepressors, 235
titration of cellular, 252–254
Reproduction of viruses, burst concept and, 46
Resolution, 83
Respiratory syncytial virus
commercial rapid antigen detection assays, 45
G protein, 389
N protein, 106
packaging genome, 106
Respirovirus, 518
Restrictive conditions, 66
Reticulocyte lysate, 360
Retinoblastoma *(rb)* gene, 288. *See also* Rb protein
Retroelements, 202–204
Retrograde axon transport, 405
Retroid viruses, 202, 512, 530
Retrotranslocation, 394–395, 406
Retrotransposons, 189, 202–204
Retrovirus
alternative splicing, 323–325
assembly, 423, 425
Gag protein and, 423–428, 432, 435–436, 438, 440–441
from polyprotein precursors, 432
recognition of nucleic acid genome, 431, 434–436
attachment to host cells, 131
budding at plasma membrane, 440
CA protein, 406, 423, 440–441
capsid, 102–104
cell-to-cell spread, 457, 458
classification, 19
constitutive transport elements (CTEs), 328–329, 331
dimerization, genome, 434–436
diploid genome, 191, 192
diseases, 530
DNA genome, 215
DNA integration, 204–214
cellular tethers, 208–209
characteristic features of, 205
gapped intermediate, 206
host proteins that regulate, 209–210
integrase structure and mechanism, 210–214
integrase-catalyzed steps, 205–210
preintegration complex, 205–207, 210
repair of integration intermediate, 206–207
selection of host DNA target sites, 207–209
in vitro assays, 206
endogenous, 197, 202, 203
enhancer, 237
entry into cells, 142
Env protein, 399
envelope, 109
envelope glycoprotein proteolytic processing in Golgi, 399–400
enzyme activity assay, 41
export of RNAs from nucleus, 328, 331
frameshifting, 368
furin family protease, 400
Gag protein, 406, 408, 412–413, 423
assembly and, 423–428, 432, 435–436, 438, 440–441
cell-to-cell viral spread and, 458
Gag-dependent budding, 440–441
L domains, 441–443
multimers, 425, 440–441
proteolytic processing, 454–455
radial organization in immature virus particles, 423
Gag-Pol polyprotein, 412
proteolytic cleavage of, 198
genome conversion for transcription, 226–227
genome organization, 59, 191, 192, 530
complex genome, 530
simple genome, 530
genome structures critical for function, 62
human endogenous, 4
import of viral genome into nucleus, 151, 153
information retrieval from genome, 63
integrase, 189, 204–214
koala, 203
L domains, 441–443
MA protein, 406, 423, 428, 440–441, 454
membrane fusion, 138
modification and processing of retroviral Env polyproteins, 400
mutation rate, 198–199
NC protein, 406, 412–413, 435, 454–455
one-step growth curve, 49
origin of name, 189
overview, 530–531
packaging of genome, 434–436
dimer linkage sequence, 433–434
psi (ψ) sequence, 435
pathogenic viruses produced by recombination, 184
primer-binding site, 192
protease, 204, 454–455
proteins in virion, 117
provirus, 197, 204–205, 276, 530–531
chromatin, 226
splicing, 317
transcription, 225
pseudodiploid, 191
Ras protein, 406
recognition of nucleic acid genome, 431, 434–436
recombination, 197–198
release of virus particles, 441–443
replication cycle, 222
reproductive cycle, single-cell, 531
reverse transcriptase, 198–202
catalytic properties, 198–200
domain structure, 198, 199
fidelity, 198–199
flipping, 201
number of copies, 192
processivity, 198
structure, 198–202
variable subunit organization, 198, 199
reverse transcription, 68, 189–202
discovery, 189
essential components, 191–193
first template exchange, 193, 195
genome conversion for transcription, 226–227
hepadnavirus compared, 217
impact, 189
initiation of (+) strand DNA synthesis, 195, 196
initiation of (−) strand DNA synthesis, 193, 195
intermediates, 191
recombination during, 197–198
second template exchange, 195–197
steps, 193–198
ribonucleoproteins, 160
RNA polymerase II transcription, 258
RNA synthesis by RNA polymerase II, 411
RNA-dependent DNA polymerase, 163
Src protein, 406

structure, 530
suppression of termination, 367
transcription, 226
regulation, 237–246
transcriptional control region, 228
transcriptional regulators, 247
transfer RNA in virion, 117
transformation assay, 38
tRNA primer encapsidation, 438
Vif proteins, 295
viral mRNAs in virus particle, 117
viral oncotherapy, 75
viral vectors, 75, 78, 79
virion enzymes, 116
Rev protein, human immunodeficiency virus type 1, 327–330, 332
Reverse transcriptase
assay, 41
avian sarcoma/leukosis virus (ASLV), 193
catalytic properties, 198–200
low fidelity, 198–199
RNase H, 200
slow DNA polymerization, 198
direction reversal, 201
DNA genome evolution, 64
duck hepatitis B virus, 217
errors by, 198–199
evolutionary origin, 202
gapped DNA genomes and, 57
human immunodeficiency virus type 1, 217
domain and subunit relationships, 199
error rate, 199
flipping, 201
structure, 163, 200–201, 204
template primer interaction, 200–201
incorporation into viral particle, 438
pauses during synthesis, 198
primer tRNAs for, 192
retrovirus, 68, 192–193, 198–202
catalytic properties, 198–200
discovery, 189
domain structure, 198, 199
fidelity, 198–199
flipping, 201
number of copies, 192
processivity, 198
structure, 198–202
variable subunit organization, 198, 199
RNase H activity, 193, 200, 201, 210
sequence motifs, 191
uses of, 189
Reverse transcriptase inhibitors, 197, 199
Reverse transcription
errors, 189
essential components, 191–193
genomic RNA, 191
primer tRNA, 191–192, 193
reverse transcriptase, 192–193
foamy virus, 215
hepadnavirus, 214–222
cis-acting signals, 216, 218, 221
critical steps in, 218–221
elongation, 218–220
essential components, 216–218
first template exchange, 218, 220
function in life cycle, 214–215
host protein involvement, 218
initiation, 218, 220
P protein, 214–220
pregenomic mRNA, 214, 216–218, 220
primers, 216–221
retroviral compared, 217
RNase H degradation of RNA template, 219, 220
second template exchange, 220, 221
translocation of primer for (+) strand synthesis, 219–221
LINEs (long interspersed nuclear elements), 202–204
process, 189–198
retroelements, 202–204
retrotransposons, 202–204
retrovirus, 189–202
discovery, 189
essential components, 191–193
first template exchange, 193, 195
genome conversion for transcription, 226–227
hepadnavirus compared, 217
impact, 189
initiation of (+) strand DNA synthesis, 195, 196
initiation of (−) strand DNA synthesis, 193, 195
intermediates, 191
recombination during, 197–198
second template exchange, 195–197
steps, 193–198
SINEs (short interspersed nuclear elements), 202, 204
steps, 193–198
Reversion, 71
Reversion analysis, 71–72
Rev-responsive element (RRE), 326–330
Rf-C (replication factor C), 272–273, 277
Rhabdovirus
budding at plasma membrane, 440
cytopathic effect, 34
diseases, 532
envelope acquisition, 385
genome organization, 59, 60, 532
information retrieval from genome, 63
inhibition of cellular mRNA export, 334
inhibition of cellular transcriptional machinery, 260
overview, 532–533
packaging genome, 106
proteins in virion, 117
release of virus particles, 442, 444
reproductive cycle, single-cell, 533
RNA-dependent RNA polymerase, 164
structure, 89, 532
viral oncotherapy, 75
viral vectors, 76
virion enzymes, 116
Rhadinovirus, 514
Rhinovirus, 522
attachment to host cells, 126–127
cleavage of eIF4G, 373
internal ribosome entry site (IRES), 359
RNA synthesis, 169
structure, 94
Rhopalosiphum padi virus, bicistronic mRNA of, 368
Ribavirin, 65
mode of action, 183
Ribonucleoprotein, 158–159
influenza virus, 89–90, 411–412, 438
intracellular transport, 411–412
release of viral, 142–143
structure, 160
vesicular stomatitis virus, 411
Ribonucleotide reductase, 290–291
Ribosomal frameshifting, 63, 361, 368, 369
Ribosomal RNA (rRNA) synthesis, 225–226
Ribosome
accommodation, 360–361
assembly on mRNA, 352–355
E (exit) site, 350, 352, 355, 361
leaky scanning, 365–366
overview, 349
P (peptidyl) site, 350, 352, 354–355, 357, 360–361, 363, 368, 372
polyribosome, 386, 475
polysome, 361, 376
recycling, 363
RNA polymerase collision, 179, 182
scanning, 352, 354–355, 374
A (aminoacyl or acceptor) site, 350, 352, 354–355, 357, 360–362, 368–369
structure, 350–351
uncoating in cytoplasm by, 143–144
VPg-dependent ribosomal recruitment, 353
Ribosome shunting, 354–355
Ribozyme
examples, 181
hepatitis delta satellite virus, 176–177, 181
spliceosome, 321
Rinderpest virus, 518
RING domain, 295
RNA
ambisense, 56, 59, 60
polyadenylated nuclear RNA (PAN), 473
scaffolding function, 427
size of molecules, 65
strand terminology, 21, 56
synthesis (*see* RNA synthesis)
viral, 31
export from nucleus, 311, 327–331
export of RNAs from nucleus, 311, 327–331
packaging signals, 434–435
secondary structure, 157–158, 160, 169
RNA editing, 63, 325–327
as antiviral defense mechanism, 326–327
genome diversity from, 185
overview, 311
RNA genome, 18–19, 58–60
ambisense, 157, 159, 174
circular, 57, 58
double stranded, 19, 58, 70, 157, 159, 175–178
error frequency, 183
evolution, 61
mutation rate, 66
negative (−) strand RNA, 19, 21, 59, 64, 69–70
origin of diversity in, 182–185
misincorporation of nucleotides, 182–183
reassortment, 183
recombination, 183–185
RNA editing, 185
positive (+) strand RNA, 19, 21, 58–59, 64, 68–69
positive (+) strand RNA with DNA intermediate, 59
quasispecies, 183
segmented, 64, 67, 157, 158–159
single-stranded RNA, 19
size, 65
(+) strand, 157–159, 171
(−) strand, 157–159, 173–174
RNA granules, stress-associated, 377, 378

RNA helicase, 169–170
  eIF4A, 353
  inhibition of, 170
  structure, 170
RNA interference (RNAi), 72, 238
  blockage of human adenovirus type 36 E4 Orf1, 492
  as host antiviral defense mechanism, 336
  overview, 311, 338
  RNA-mediated induction of mRNA degradation, 336
  viral gene products that block, 345
RNA packaging, Gag protein function in, 412–413
RNA polymerase, 31, 57, 58–59
  alphavirus, 171–172
  bacterial, 225
  bacteriophage T7 RNA polymerase, 163, 358
  DNA-dependent, 181, 260, 262
    bacteriophage T7 RNA polymerase, 163, 358
    properties, 225–226
    RNA-dependent RNA polymerase compared, 161
    vaccinia virus, 116
  errors, 65
  infectious DNA clones and, 68–70
  poliovirus, 405
  release of influenza virus ribonucleoprotein, 143
  ribosome and RNA polymerase collision, 179, 182
  vaccinia virus, 116
RNA polymerase I, 69, 225–226
RNA polymerase II, 31, 58, 69
  closed initiation complex, 230, 231
  genome conversion for transcription by, 226–227
  hepatitis delta satellite virus RNA synthesis, 260
  holoenzyme, 228
  inhibition of, 259–260
  initiation of transcription, 226, 228–234
  mapping human adenovirus type 2 initiation site and transcription *in vitro*, 229
  nomenclature for site-specific DNA-binding proteins, 232
  nucleosome organization and, 226, 228
  phosphorylation of paused as signal for capping, 312–313
  preinitiation complexes, 228, 230, 245, 248, 250–251
  promoters, 228, 231, 322
  properties, 225–226
  recruitment to viral replication centers, 294
  regulation, 228–235
  retrovirus RNA synthesis, 411
  splicing and, 322
  transcription by, 228–235
  transcriptional control elements, 228
  yeast, 225
RNA polymerase II, cellular, 197
  hepadnavirus transcription, 214
  hepatitis delta satellite virus RNA synthesis, 176–177, 180, 181
  influenza virus mRNA synthesis and, 166
RNA polymerase III
  adenovirus gene transcription by, 502
  miRNA synthesis, 339
  promoter(s), 257, 259
  properties, 225–226
  transcription of viral genes, 226, 257–259
  transcription units, 258
  virus-associated RNA I (VA-RNA I), 257–259, 345, 370
RNA primer
  DNA synthesis, 268–270, 272–273, 275–276, 279, 306
  synthesis of, 275–276
RNA processing, 310–346
RNA profiling, 474–477
RNA recombination, 183–185
RNA replication, error frequency in, 183
RNA silencing. *See also* RNA interference (RNAi)
  overview, 311, 338
  viral gene products that block, 345
RNA sponge, 341
RNA synthesis, 13
  from DNA templates, 224–263
  error frequency, 183
  RNA-dependent RNA synthesis by RNA polymerase II, 260
  viroids, 260
RNA synthesis, viral
  ambisense RNA, 174
  capping, 168
  cellular proteins required for, 170
  cellular sites of, 179, 182
  double-stranded RNA, 175–178
  elongation, 168–169
  hepatitis delta satellite virus, 176–177, 180
  initiation, 164–168
    capped RNA fragments as primers, 166–168
    de novo, 164–165
    primer-dependent, 165–168
    protein priming, 165–167
    swinging-gate model of, 165
  machinery, 159–164
  mechanisms, 164–170
  nested subgenomic mRNAs, 172–173
  origin of genome diversity and, 182–185
  paradigms for, 170–179
  ribosome and RNA polymerase collision, 179, 182
  RNA template, 156–185
    naked or nucleocapsid RNA, 158–159, 161
    nature of, 157–159
    secondary structure, 157–158, 160, 169
    specificity, 169, 171–172
    unwinding, 169–170
  (+) strand RNA, 171
  (−) strand RNA, 173–174
  strategies of representative virus families, 159
  surface catalysis, 182
  template specificity, 169, 171–172
  unbalanced production of (+) and (−) strand RNA, 177–178
  unwinding RNA template, 169–170
RNA template
  nature of, 157–159
    naked or nucleocapsid RNA, 158–159, 161
    secondary structure, 157–158, 160, 169
  reverse transcription (*see also* Reverse transcription)
    template-primer interaction, 200–201
  RNA synthesis from, 156–185
  RNA-dependent RNA polymerase interaction, 162
  RNase H degradation of in hepadnavirus reverse transcription, 218–220
  specificity in viral RNA synthesis, 169, 171–172
  unwinding, 169–170
RNA virus
  assembly location, 411
  cytoplasmic membrane reorganization, 496–498
  genome diversity
    misincorporation of nucleotides, 182–183
    reassortment, 183
    recombination, 183–185
    RNA editing, 185
  infectious DNA clones, 68–70
  mutation rate, 66
  one-step growth curve, 49
  packaging signals, 434–435
RNA virus vectors, 77–78, 79
RNA world, 61, 181
RNA-dependent DNA polymerase, 189, 193. *See also* Reverse transcriptase
RNA-dependent RNA polymerase, 157
  activity assays, 161
  cap-binding site, 176
  capping enzyme cooperation with, 313
  cis-acting elements, 169
  DNA-dependent DNA polymerase compared, 161
  DNA-dependent RNA polymerase compared, 161
  error frequency, 183
  Gly-Asp-Asp sequence of motif C, 163
  identification of, 159–161
  incorporation into viral particle, 438
  influenza virus, 161, 168, 169
  mengovirus, 159–160
  misincorporation of nucleotides, 182–183
  names for, 158
  nucleotide binding, 183
  poliovirus, 159, 161–164, 169, 171, 183
    fidelity, 183
  primer, 157, 161, 164
  processivity, 168
  ribonuceloproteins and, 158–159
  sequence relationships among, 161, 163
  structure, 161–164
  template specificity, 169, 171–172
  two-metal mechanism of catalysis, 162
  vesicular stomatitis virus, 163, 168, 173–176, 532
RNA-dependent RNA polymerase II, 181
RNA-induced silencing complex (Risc), 339, 341–343
RNase, resistance of ribonucleoprotein complexes to, 158
RNase A, 315
RNase H, 473
  hepadnavirus, 218–220
  reverse transcriptase, 193, 200, 201, 210
RNase L, 197, 377
RNase P, 181, 321
RNP. *See* Ribonucleoprotein
Robbins, Frederick, 32
Roberts, Richard, 317
Rolling-circle replication, 277–278, 288
  hepatitis delta satellite virus, 176–177
Roseola infantum, 297
Roseolovirus, 514
Ross River virus, 534
  budding, 444
  structure, 109
Rotavirus, 528
  assembly, 497, 499
  budding into and out of ER, 410
  commercial rapid antigen detection assays, 45
  ER membrane disruption, 404
  host cell signaling and, 468
  infectivity reduction, environmental conditions and, 81

interaction with internal cellular membranes, 410
mRNA release from particles, 175, 178
NSP2 protein, 497, 499
nsP3 protein, 376
NSP5 protein, 497, 499
packaging genome, 105
RNA synthesis, 175
structure, 13, 101
VP1 protein, 499
VP2 protein, 497, 499
VP3 protein, 499
Rotavirus A, 528
Rough endoplasmic reticulum, 386, 388, 403
Rous, Peyton, 11
Rous sarcoma virus, 11–12
frameshifting, 368
Gag, 368
growth in embryonated eggs, 36
Pol, 368
pp60 protein, 402
splicing, 323
structure, 530
transformation assay, 38, 39
Rp-A (replication protein A), 271–273, 284–286, 298
RPE65 gene, 77
RPO30 gene product, vaccinia virus, 262
RRE (Rev-responsive element), 326–330
Rubella virus, 35, 534
Rubivirus, 534
Rubulavirus, 518
Rule of six, 174
Runoff assay, 229

S

S phase, 289, 299
S protein
coronavirus, 402, 410
hepatitis B virus, 408, 410
severe acute respiratory syndrome coronavirus, 402
Sabin, Albert, 66
Saccharomyces cerevisiae
origin of replication, 270
poly(A)-binding protein (Pabp1), 353
S-adenosylmethionine (SAM), 490
Sam68, 468
SARS. *See* Severe acute respiratory syndrome coronavirus (SARS-CoV)
Satellite RNAs, 181
Satellite virus, 20
Satellites, 20
Scaffolding function of RNA, 427
Scaffolding proteins, 403, 425, 428–429
Scanning, 352, 354–355, 374
leaky, 365–366
Second messengers, 148
Secondary structure
messenger RNA, cellular, 353–354
RNA, 60, 61, 72
RNA, viral, 157–158, 160, 169
Second-site mutations, 71–72
Secretory pathway, 386–401, 413. *See also* Endoplasmic reticulum; Golgi apparatus
assembly at internal membranes, 444–445
compartments, 396–397, 409
localization of viral proteins to, 410
cytoplasmic compartments, 444–445
disruption, 404–406
effects on compartments, 404–406
inhibition of cellular protein transport, 404
unfolded protein response, 406, 407
localization of viral proteins to compartments of the, 410
poliovirus inhibition of, 496
signal peptides, 386–389
vesicular transport to cell surface, 396–401
Segment reassortment, 183
Segmented genomes, 64, 67, 157, 158–159
packaging viral genome, 437–438
Self-assembly of protein shells, 425–428
Self-priming of viral DNA synthesis, 276–277, 279
Semiconservative replication, 176, 179, 267, 278
Semidiscontinuous DNA synthesis, 270
Semliki Forest virus
E1 protein, 392–393
entry into cells, 143
mRNA cap, 314
nsP3 protein, 377
particle-to-PFU ratio, 40
structure, 109
Sendai virus, 518
F protein, 409
HN protein, 409
leaky scanning, 365–366
mRNA editing, 366
P/C mRNA, 365–366
structure, 89
Seneca, 381
Seneca Valley virus, 73, 104
Sequence databases, 46
Sequencing
high-throughput, 45–46
viral genomes, 60
Serine/arginine-rich splicing factor 3 (SRp20), 359
Serological methods, 42–45
enzyme immunoassay, 44–45
hemagglutination inhibition, 42–43
immunostaining, 43–44
virus neutralization, 42
Serotypes, 42
Serum response factor (srf), 232
Severe acute respiratory syndrome coronavirus (SARS-CoV), 506
eIF3 regulation, 376
frameshifting, 368
interaction with internal cellular membranes, 410
nsp1 protein, 336, 376
RNA synthesis, membranous sites for, 182
S protein, 402
Sf2 protein, adenovirus, 331
sgRNAs (single-stranded guide RNAs), 72
Shakespeare, William, 465
Sharp, Phillip, 317
Sialic acid, 124, 127, 129–131
removal of terminal sialic acid residues, 456–457
σ3 protein, reovirus, 370
Signal peptidase, 410
Signal peptide, 386–389
Signal recognition particle (SRP), 387–389
Signal transduction, 465–470
genes encoding signaling proteins, 466
Pi3k-Akt-mTor signaling route, 466–470
signaling in virus-infected cells, 466–470
signaling pathways, 465–466
modulation of multiple pathways, 468, 470
virus-induced signaling via cell receptors, 148
Signals
lipid-plus-protein, 406, 408
nuclear localization signal, 383–384, 387, 409, 411, 412
protein sequence, 408–409
signal-sequence-independent transport of viral proteins, 406, 408–409
Silencing
epigenetic silencing of integrated proviral DNA, 236, 238–239
overview, 311
Simian immunodeficiency virus (SIV), 136
Simian rotavirus
interaction with internal cellular membranes, 410
NS28 protein, 410
VP7 protein, 410
Simian virus 5, 141
Simian virus 40 (SV40)
assembly, 384, 421–422, 424–425, 428–429, 432, 434
attachment to host cells, 127, 129
capsid, 98–99
DNA synthesis, 268, 271–275
lagging-strand synthesis, 272–274
leading-strand synthesis, 272–273
mechanism, 271–275
mechanism of unwinding and translocating along DNA, 284
origin of replication, 268, 271–275, 282, 284–285, 307
origin recognition, 271–272
origin recognition proteins, 277
replication of chromatin templates, 274–275
termination and resolution, 274, 275
unwinding, 271–272, 274, 280, 284
enhancers, 229, 232–233
entry into cells, 133
genome organization, 524
helicase, 271–272, 274, 277, 280, 282–284
large T antigen, 271–275, 277, 280–285, 289–290, 303, 307, 342–343, 345, 424–425, 428–429
functions and organization, 280–283
mechanism of unwinding and translocation along DNA, 284–285
properties, 280–283
regulation of activity, 283
micro-RNA (miRNA), 342–343, 345
mutations in Ori, 72
nuclear localization signal, 149
origin of replication, 268, 271–275, 282, 284–285, 307
mapping of, 271
structure, 272
overview, 524–525
packaging genome, 105
packaging signal, 432–434
particle-to-PFU ratio, 40
polyadenylation, 315
promoters, 231, 252–253, 254
reproductive cycle, single-cell, 525
signaling via cell receptors, 148
structure, 418, 524
transcription, 246, 247, 254
viral DNA replication, 254
VP1 protein, 384, 418, 421–422, 424–425, 428–429
VP2 protein, 384, 402, 434
VP3 protein, 384, 424, 428, 434
VP4 protein, 450

Simplexvirus, 514. *See also* Herpes simplex virus type 1; Herpes simplex virus type 2
Sindbis virus, 534
ATP synthesis in host cells, 484
benefits of eIF2α phosphorylation, 372
budding, 440, 444
cellular gene expression inhibition, 472
E1 protein, 440
E2 protein, 400, 402
envelope glycoprotein proteolytic processing in Golgi, 400
genome structure and expression, 171
heparan sulfate binding, 132
inhibition of cellular transcriptional machinery, 260
mRNA cap, 314
oligomerization of proteins, 395–396
RNA synthesis, 171
structure, 109–110
SINEs (short interspersed nuclear elements), 202, 204
Singer-Nicholson fluid mosaic model of membrane structure, 30
Single-exon viral mRNAs, 329–330, 332
Singleplex PCR, 45
Single-stranded DNA genome, 19, 57–58, 302
Single-stranded guide RNAs (sgRNAs), 72
Single-stranded-DNA-binding protein, 272, 280, 284, 286–287, 292, 294, 306
siRNA. *See* Small interfering RNA
Site-directed mutagenesis, 131
Site-specific integration of DNA, 190
Site-specific recombination, 304, 306
Skin, cross section of, 28
Sliding clamp proteins, 272
Small interfering RNA (siRNA), 72, 73, 245
epigenetic silencing of integrated proviral DNA, 238
guide strand identification, 342
production and function, 338–339, 345
VP16 knockdown by, 118
Small nuclear ribonucleoproteins (snRNPs), 320–325
Small nuclear RNAs (snRNAs)
export from nucleus, 327
splicing participation, 320–321
Small-angle X-ray scattering, 86
Smallpox, 526
historical aspects, 7–8
lesions, 8
vaccine, 8–9
Snare proteins, 397–398
Snf, 210, 235, 245
snRNAs. *See* Small nuclear RNAs
snRNPs (small nuclear ribonucleoproteins), 320–325
Sorting of viral proteins in polarized cells, 401–404
epithelial cells, 402–403
neurons, 402, 403–404
SOX protein
human herpesvirus 8, 336, 473
Kaposi's sarcoma-associated herpesvirus, 376–377
Sp1 (stimulatory protein 1), 232, 248, 251, 253, 434
Species C human adenovirus
virus-associated RNA I (VA-RNA I), 257–259
Splice sites, 318–319, 321–323, 326–327, 329, 331, 334
Spliceosome, 320–321, 327, 329, 333
Splicing
alternative, 317, 322–325
adenovirus, 322, 325, 326
bovine papillomavirus type 1, 322–323, 324
regulation by viral proteins, 331–332, 334
retrovirus, 323–325
balance between export and splicing, 329
bovine papillomavirus type 1, 322–323, 324
constitutive, 322
discovery, 317–318
mechanism, 318–322
overview, 311–312
regulation of alternative splicing by viral proteins, 331–332, 334
RNA-dependent helicase participation in, 321
RNA-RNA interactions, 320
small nuclear RNA (snRNA) participation in, 320–321
splice sites, 318–319, 321–323, 326–327, 329, 331, 334
Spontaneous-generation hypothesis, 9
Spumavirus, 199, 215, 530
Sputnik, 20
SR proteins, adenovirus, 331–332, 334
Src tyrosine kinase, 406, 448
SrebpC1, 489
Srebps, 490
SRP (signal recognition particle), 387–389
SRP receptor, 387–388
SRp20 (serine/arginine-rich splicing factor 3), 359
Stanley, Wendell, 12
Staufen1, 413
Steatosis, 487
Stimulatory protein 1 (Sp1), 232, 248, 251, 253, 434
Stop codon, 361, 366–368
Stop transfer signal, 389
Strand displacement, 268, 277–278, 284
Strand displacement synthesis, 198
(+) strand RNA virus
assembly, 430
assembly location, 411
envelope, 109
mRNA caps, 313–315
packaging of genome, 436, 437
structure, 95
uncapped mRNAs, 312
(−) strand RNA virus
assembly, 421, 425, 426, 430, 431, 432, 439–440
assembly location, 411
envelope, 109
envelope acquisition, 439–440
packaging genome, 105
replication coupled with encapsidation, 436
RNA-dependent RNA polymerase, 438
Streptococcus thermophilus, 340
Stress-associated RNA granules, 377, 378
Structural asymmetry, in HIV-1 reverse transcriptase, 200
Structural unit
assembly, 421–423
chaperone participation, 423, 424
from individual proteins, 421–422
from polyproteins, 422–423
defined, 83
icosohedral capsid, 90, 91
multimeric, 94
Structure, 80–119
capsid, 86–104
cellular macromolecules, 117–118
different in different hosts, 112
envelopes, 106–114
glycoproteins, 106–109
enzymes, 116
genome packaging, 104–106
by cellular proteins, 105–106
direct contact with a protein shell, 104–105
by specialized viral proteins, 105
host cell architecture shaping viral structure, 385
large structurally complex viruses, 111–116
bacteriophage T4, 111–112, 114
giant viruses, 114, 116
herpesviruses, 112–113, 115
poxvirus, 113, 115
metastable structures, virus particles as, 81
methods of study, 83–86
difference mapping, 86
electron microscopy, 82–86
X-ray crystallography, 83, 85, 88–89, 92, 95–96, 99–102, 104, 108, 111, 114, 119
nomenclature, 82–83
nongenomic viral nucleic acid, 117
proteins in virus particle, 116–117
size and shape variation, 82
virion functions, 81–82
SU (gp120), human immunodeficiency virus type 1, 131, 134, 135, 391
Substitution mutations, 70
Subunit, capsid protein
covalent joining, 91
defined, 83
helical capsids, 86, 88–89
icosohedral capsids, 90–91
Sulfhydryl oxidases, 392
Sulfobolus turreted icosohedral virus, 107
Sumo proteins, 295
Supercoiling, 274
Super-resolution microscopy, 44
Suppression, 71
Suppression of termination, 63, 366–368
Suppressor mutations, 71–72
Suppressor tRNAs, 366, 368
Susceptible cell, 31
Suspension cultures, 33
SV40. *See* Simian virus 40
Swi, 210, 235, 245
Synapse, virological, 458
Synaptic button, 382
Synchronous infection, 48, 124
Syncytia formation, 458
Syndecan, 28
Systems biology, 25, 50–51

T

T antigen, 246, 247, 254
T cells, human immunodeficiency virus in, 238–240
T number, 90–94
Tandem shift model for frameshifting, 368, 369
Tap, 328
TAR (transactivation response) element, 241, 243–245
TAR RNA-binding protein (Trbp), 342
Target size, in mapping gene order by UV irradiation, 175
Targeted gene editing with CRISPR-Cas9, 72–73

Tat protein, HIV, 237, 239–245, 248, 262
nucleosome remodeling and, 245
RNA structure recognition, 241
transcriptional elongation stimulation by, 241–245
TATA sequence, 228, 231–232, 234, 237, 249–250, 259–260
TATA-binding protein (Tbp), 228, 231, 234, 250, 252–253, 260, 471
Tax protein, human T-lymphotrophic virus type 1, 237, 251
T-cell-restricted intracellular antigen-1 (Tia-1), 377
Tegument proteins, 449
Telomerase, 270
Telomere, 270, 297
Telomere repeat-binding protein 2 (Trf2), 299
Temin, Howard, 189
Temperature-sensitive mutants, 66, 420
Template exchange
copy choice, 184
retrovirus reverse transcription
first template exchange, 193, 195
second template exchange, 195–197
Terminal resolution site, 276–277, 283, 296
Termination
suppression of, 366–368
translation, cellular, 361–362
Termination codon, 70, 349, 350, 362, 366–368
Termini, DNA replication, 267
Teschovirus, internal ribosome entry site (IRES), 355
Tetherin, 444
Tethers, in retroviral integration, 208–209
Tetrahymena thermophila, 181
TfIIb (transcription factor IIb), 231, 250, 259–260
TfIIc (transcription factor IIc), 259–260
TfIId (transcription factor IId), 228, 231, 240, 243–244, 248, 250, 253, 260, 262
TfIIf (transcription factor IIf), 231, 250
TfIIh (transcription factor IIh), 231, 234
TfIIIa (transcription factor IIIa), 328
TfIIIb (transcription factor IIIb), 260
TfIIIc (transcription factor IIIc), 260
Theiler, Max, 14
Thio oxidoreductase system, 392
3A protein, poliovirus, 405, 450, 454, 496
3′-cap-independent translational enhancer (3′ CITE), 360
Thumb domain, RNA-dependent RNA polymerase, 161–163, 165, 169
Thymidine kinase, 290–291
Thymidylate kinase, 291
Tia-1 (T-cell-restricted intracellular antigen-1), 377
Tick-borne encephalitis virus, E protein, 106, 108
Tight junctions, 27, 402, 458
Titer, virus, 36–39
calculating from plaque assay, 38
definition, 36
efficiency of plating, 39
end-point dilution assay, 38–39
transformation assay, 38
TM protein, human immunodeficiency virus type 1, 131, 135
TM-Su protein, human immunodeficiency virus type 1, 441
Tn5, 205
Tobacco mosaic disease, 11
Tobacco mosaic virus, 12, 459
lesions, 14
structure, 13, 83, 86, 88–89
in vitro assembly of, 32
Togavirus
budding, 444
C protein, 444
diseases, 534
E1 protein, 444
E2 protein, 444
genome organization, 58, 59, 534
one-step growth curve, 49
overview, 534–535
reproductive cycle, single-cell, 535
structure, 534
Tomato bushy stunt virus, 83, 95, 345
Tomato spotted wilt virus, 385
Topoisomerase I, 272, 274–275
Topoisomerase II, 274–275, 287–288
Topology, 95–96, 99, 109, 111, 114
Torque teno (TT) virus, 65
TorsinA, 449
Totivirus, genome organization, 58
Tournier, Paul, 17
Transcriptase, 158. *See also* RNA-dependent RNA polymerase
Transcription, 224–263
activators, 225, 232–233, 235–237, 245, 247–251, 253–254, 257–258
cellular, viral inhibition of, 470–471
closed initiation complex, 230, 231
DNA virus, 224–263
activators, 225, 232–233, 235–237, 245, 247–251, 253–254, 257–258
adenovirus, 226, 228, 235, 246–248, 250–253
advantages of temporal gene expression, 253
autoregulation, 235, 237–240, 248–249
by cellular machinery alone, 235–237
cellular RNA polymerase, 225–226
common strategies, 245–247
conversion of genomes to templates for transcription, 226–228
coordination of late genes with viral DNA synthesis, 251–254
DNA-dependent RNA polymerase, viral, 260–262
DNA-looping model, 232, 234
early phase, 245, 247, 250–252, 254, 259
enhancers, 228–229, 232–234, 236–237, 239–240, 242–243, 245–247, 249, 254, 262
entry into transcriptional programs, 254–257
herpes simplex virus type 1, 246, 247, 248–250, 251, 252, 254, 256–259
histone associations, 226–228
inhibition of cellular transcriptional machinery, 259–260
initiation, 226, 228–234
late phase, 245–247, 251, 253–254, 257, 259
mapping human adenovirus type 2 initiation site and transcription *in vitro,* 229
mapping vaccinia virus transcripts, 261
preinitiation complexes, 228, 230, 245, 248, 250–251
regulation, 228–235
repressors, 235–236, 247–249, 252–256
by RNA polymerase II, 228–235
by RNA polymerase III, 226, 257–259
strategies, 226
transcriptional cascade, 237, 245–254
transcriptional control region, 228, 232, 236, 240–242, 262
unusual functions of cellular transcription components, 260
vaccinia virus, 247, 260–262
viral proteins governing transcription, 237–257
viral proteins regulating viral DNA template transcription, 237–257
enhancers, 228, 229, 232–234, 236, 246–247, 249, 254, 262
adenovirus, 251
avian leukosis virus, 237
defined, 229
human immunodeficiency virus type 1, 237, 239–240, 242–243, 245
mechanism of action, 234
regulation by host cell metabolism, 233
simian virus 40, 229, 232–233
entry into transcriptional programs, 254–257
Epstein-Barr virus, 247–249, 256, 258, 259
eukaryotic systems, 225–226
initiation, 226, 228–234
preinitiation complexes, 228
read-through, 261
regulation, 228–235
activators, 225, 232–233, 235–237, 245, 247–251, 253–254, 257–258
advantages of temporal gene expression, 253
autoregulation, 235, 237–240, 248–249
common properties of proteins that regulate transcription, 234–235
enhancers, 228–229, 232–234, 236–237, 239–240, 242–243, 245–247, 249, 254, 262
by host cell metabolism, 233
inhibition of cellular transcriptional machinery, 259–260
nomenclature for site-specific DNA-binding proteins, 232
patterns, 237
positive autoregulatory loop, 237
recognition of local and distant regulatory elements, 229, 232
repressors, 235–236, 247–249, 252–256
transcriptional cascade, 237, 245–254
viral proteins regulating viral DNA template transcription, 237–257
repressors, 235–236, 247–249, 252–256
reverse (*see* Reverse transcription)
TATA sequence, 228, 231–232, 234, 237, 249–250, 259–260
termination, 262
transcriptional elongation stimulation by Tat protein of HIV-1, 241–245
transient-expression assays, 241
unwinding of DNA template, 228, 230
use of term, 158
Transcriptional cascade, 237, 245–254
Transcriptional control region, 228, 232, 236, 240–242, 262
Transcytosis, 27
Transduction of host genes, 16
Transfection, 67–68, 69, 70–71
Transfer RNA (tRNA), 349
aminoacyl-tRNA, 360–361, 368, 369
in elongation phase of translation, 360–361
mimicry, 194
retrovirus primer tRNA, 191–192, 193
structure, 350–351

Transfer RNA (tRNA) *(continued)*
suppressor tRNAs, 366, 368
synthesis, 225–226
viral encoded, 350
in viral particle, 117
Transformation
cancerous, signal transduction and, 467, 470
DNA-mediated, 68, 125, 127
mRNA stabilization facilitating, 338
oncogenic, 189
Transformation assay, 38, 39
Transient-expression assays, 241
Translation, cellular
elongation, 360–361
5′-end-dependent initiation, 352–355
initiation codon, 354
methionine-independent initiation, 354
mRNA secondary structure, 353–354
poly(A) tail, role of, 352–353
ribosome assembly on mRNA, 352–355
ribosome shunting, 354–355
VPg-dependent ribosomal recruitment, 353
5′-end-independent initiation, 355–360
host cell proteins in IRES function, 358–359
internal ribosome entry site, 355–358
mechanism of internal initiation, 355, 358
3′-cap-independent translational enhancer (3′ CITE), 360
inhibition of initiation, 369–376
juxtaposition of mRNA ends, 363
machinery, 350–351
ribosomes, 350–351
transfer RNAs (tRNAs), 350–351
translation proteins, 351
mechanisms, 349–363
proofreading, 351
regulation during viral infection, 368–377
cap-binding protein, 376
eIF2α dephosphorylation, 372
eIF2α phosphorylation, 369–370, 371
eIF3 regulation, 376
eIF4E activity modulation, 373–374
eIF4F regulation, 372–376
eIF4G cleavage, 372–373
inhibition of initiation, 369–376
interfering with RNA, 376–377
by miRNA, 374
Pkr regulation, 370, 372
poly (A)-binding activity, 376
stress-associated RNA granules, 377, 378
ribosome accommodation, 360
ribosome recycling, 363
termination, 361–362
viral inhibition, 471–474
in vitro, 360
Translation, viral
bicistronic mRNAs, 368
diversity of strategies, 362–368
leaky scanning, 365–366
polyprotein synthesis, 363, 365
reinitiation, 366, 367
ribosomal frameshifting, 368, 369
suppression of termination, 366–368
Translocation
of primer for (+) strand synthesis hepadnavirus reverse transcription, 219–221
of template-primer during reverse transcription, 200
Translocation channel, 387–389, 395
Transmembrane glycoproteins, binding of enveloped viruses by, 129–132
Transmembrane signaling, 124
Transport vesicle, 384, 396–398, 403, 413, 444
Transposase, 16
Transposons
retrotransposons, 202–204
Trex1 (transcription and export complex 1), 329–331, 335
Trf2 (telomere repeat-binding protein 2), 299
Triangulation number, 90–94
TriC chaperone, 423–424
tRNA. *See* Transfer RNA
Tropism
definition, 124
receptors and, 123
Tsc (tuberous sclerosis complex), 374
Tsg101 protein, 442
Tuberous sclerosis complex (Tsc), 374
Tubulin, 133
Tulip mosaic virus, 7, 8
Tumor suppressor gene, 288
Turnip mosaic virus, 353
Turnip yellow mosaic virus, 354
2B protein, poliovirus, 405
2BC protein, poliovirus, 405, 450, 454, 496
2C protein, poliovirus, 436
Two-hit kinetics, 37
Two-metal mechanism of polymerase catalysis, 162
Twort, Frederick, 12
Ty3 retrotransposon, 204, 208, 260
Ty5 retrotransposon, 208
Tymovirus, 354
Type-specific antigens, 42
Tyrosine kinases, 145

U

U1 snRNA, 320–321, 323
U1 snRNP, 322–325
U2 snRNA, 320–321
U2 snRNP, 323
U2-associated protein (U2af), 324
U4 snRNA, 320–322
U5 snRNA, 321
U5 snRNP, 320
U6 snRNA, 320–321
U94 protein, human herpesvirus 6, 297
Ubiquitin, 184, 334
Ubiquitin ligase, 295
Ubiquitin-activating enzyme, 295
Ubiquitination, 295, 444
Ubiquitin-conjugating enzyme, 295
Ubiquity of viruses, 3–4
UL6 protein, herpes simplex virus type 1, 113, 115, 421, 428, 431
UL8 protein, herpes simplex virus type 1, 286
UL9, herpes simplex virus type 1, 268, 277, 283, 287
UL12 protein, herpes simplex virus type 1, 291
UL15 protein, herpes simplex virus type 1, 434
UL20 protein, herpesvirus, 449
UL23 protein, herpes simplex virus type 1, 291
UL28 protein, herpes simplex virus type 1, 434
UL33 protein, herpes simplex virus type 1, 434
UL34 protein, herpes simplex virus type 1, 410, 457
UL36 protein, herpesvirus, 449
UL37 protein, herpesvirus, 449
UL38 protein, human cytomegalovirus, 374
UL39 protein, herpes simplex virus type 1, 291
UL40 protein, herpes simplex virus type 1, 291
UL47 protein, herpes simplex virus type 1, 376
UL50 protein, human cytomegalovirus, 406–407
Ultraviolet (UV) irradiation
mapping gene order by, 175
as mutagen, 66
Uncoating, 123
in cytoplasm by ribosomes, 143–144
during endocytosis, 135–137
caveolin-mediated endocytosis, 136–137
clathrin-mediated endocytosis, 135–136
lipid raft-mediated endocytosis, 136–137
macropinocytosis, 137
in the lysosome, 145–147
at plasma membrane, 132–135
Unfolded protein response, 406, 407
Unwinding of DNA. *See also* Helicase
simian virus 40 helicase, 271–272, 274, 277, 280, 284
transcription, 228, 230
Unwinding RNA template, 169–170
Upstream stimulatory factor (Usf), 232
Uracil DNA glycolase, 291, 303
Uridylylation of VPg, 165–166
US2 protein, human cytomegalovirus, 395, 404
US3 protein
herpes simplex virus type 1, 374
herpesvirus, 449
US11 protein, human cytomegalovirus, 395, 404
Usf (upstream stimulatory factor), 232
Uukuniemi virus, 410

V

Vaccination, origin of term, 8
Vaccine
attenuated, 9
historical aspects, 7–8
polio, 16
smallpox, 8–9
yellow fever virus, 510
Vaccinia virus, 8–9
A9 protein, 448
A13L protein, 410
A14 protein, 448
A14L protein, 410
A17 protein, 448
A17L protein, 410
A33 protein, 451
A36 protein, 448, 450–451
A56R protein, 410
actin-dependent transport of, 447–450
assembly, 445–450
proteins implicated in, 447
B5R protein, 410
blockage of RNA interference, 345
cell-to-cell spread, 449, 451
cellular gene expression inhibition, 471–474
crescents, 445–448
D4R protein, 291
D5R protein, 279, 287
D13 protein, 448
DNA synthesis, 268, 277, 279, 287, 290–293
DNA-dependent RNA polymerase, 116, 260, 262
E3L protein, 370

E10R protein, 392
entry into cells, 137
enzymes of nucleic acid metabolism, 291
F13L protein, 410
fatty acid metabolism and, 485, 487
genome organization, 526
glutamine metabolism and, 484
interaction with internal cellular membranes, 410
K3L protein, 372
mRNA cap, 312–313, 316
mRNA degradation, 336, 338
nucleotide concentration and, 482
origin of, 526
overview, 526–527
polyadenylation, 317
primase, 279, 287
promoter, 260, 262
recombinant, 69
release of virus particles, 445–451
replication system, 286–287
reproductive cycle, single-cell, 527
structure, 113, 115, 526
transcription, 260–262
transcriptional regulators, 247
VETF protein, 247, 262
viral thio oxidoreductase system operating in cytoplasm, 392
viral vectors, 74, 76, 77
virion enzymes, 116
virus factory, 292–293, 445, 495
Van Leeuwenhock, Antony, 9
Varicella-zoster virus, 514
axonal transport, 403
Varicellovirus, 514
Variolation, 8
VA-RNA I, adenovirus, 257–259, 370
Vectors, viral. *See* Viral vectors
Velocity sedimentation, 419
Vesicular packets, 495
Vesicular sorting proteins, 441–443
Vesicular stomatitis virus
assembly, 425, 427, 440
budding from plasma membrane, 403
entry into cells, 136
envelope acquisition, 385
ER-to-Golgi transport, 397
G protein, 78, 386, 389, 392–393, 397, 411
genome organization, 532
inhibition of cellular mRNA export, 334
inhibition of cellular transcriptional machinery, 260
L protein, 163, 168, 169, 173, 313, 411, 414
M protein, 89, 334, 408–409, 411, 413
mapping gene order by UV irradiation, 175
mRNA cap, 313
mRNA map, 175
N protein, 89, 105–106, 158, 160, 168, 169, 173–175, 411, 414, 425, 427
NS protein, 411
nucleocapsid movement via microtubules, 414
P protein, 168, 169, 173–175, 414, 425
packaging genome, 105–106
poly(A) addition, 173, 176
ribonucleoproteins (RNPs), 411
RNA synthesis, 173–176
RNA-dependent RNA polymerase, 163, 168, 173–176, 532
start-stop model of mRNA synthesis, 173–174
structure, 13, 88–89, 532
transport of genome to plasma membrane, 411
viral vectors, 74, 77–78
virion enzymes, 116
Vesiculovirus, 532
VETF protein, vaccinia virus, 247, 262
Vhs protein, herpes simplex virus type 1, 336, 338, 376, 474
Vibrio cholerae, 340
VICE (virus-induced chaperone-enriched) domains, 494
Vif proteins, retrovirus, 295
Viperin, 490
Viral factory, 292–293, 495
Viral genome. *See* Genome, viral
Viral mitochondrial inhibitor of apoptosis (vMIA) protein, 490
Viral oncogenesis, promotion by cellular miRNA-155, 341
Viral vectors, 5–6, 73–78
adeno-associated virus, 296
clinical uses, 73, 74, 75–76, 77
design, 73–75
DNA virus vectors, 76–77, 78
RNA virus vectors, 77–78, 79
table of, 76
-viridae suffix, 20
Virion, 17
defined, 81, 83
enzymes, 116
functions, 81–82
structure, 80–119
Virion host shutoff protein (Vhs), 336, 338, 376, 474
VIRmiRNA database, 342
Viroid, 17, 20, 181, 260
Virological synapse, 458
Virome, 45
Viroplasms, rotavirus, 497, 499
Viroporin, 450
Virus
anthropomorphic characterization, 18
cataloging, 17–21
defined, 17
definitive properties, 12
discovery, 10–12, 47
intracellular parasitism, 14–17
as living entity, 18
origin of term, 11
prehistory, 6–10
propagation strategy, 21
reasons to study, 3–6
size, 12, 13
structural simplicity, 12–14
structure, 80–119 (*see also* Structure)
ubiquity, 3–4
Virus factory, 292–293, 445, 495
Virus neutralization assay, 42
Virus particles
forming progeny, 32
measurement of, 39–41
Virus reproduction, 25, 26
Virus stock, plaque-purified, 37
Virus titer, 36–39
Virus-induced chaperone-enriched (VICE) domains, 494
Vision, restoring with viral gene therapy, 77
Vitf2, 262
VP0 protein
picornavirus, 402
poliovirus, 423, 425–426, 451, 454, 497
VP1 protein
bunyavirus, 166
poliovirus, 94–98, 423, 426
polyomavirus, 384
rhinovirus, 169
rotavirus, 499
simian virus 40, 98–99, 384, 418, 421–422, 424–425, 428–429
VP2 protein
papovavirus, 402
poliovirus, 94–98, 426, 451, 497
polyomavirus, 384
porcine parvovirus, 383
rotavirus, 497, 499
simian virus 40 (SV40), 384, 402, 434
VP3 protein
bluetongue virus, 101–102
poliovirus, 94–98, 408, 423, 426, 436
polyomavirus, 384
rotavirus, 499
simian virus 40 (SV40), 384, 418, 424, 428, 434
VP4 protein
bluetongue virus, 313–314
picornavirus, 402
poliovirus, 94, 96, 98, 408, 426, 451, 497
polyomavirus, 450
simian virus 40 (SV40), 450
VP5 protein, herpes simplex virus type 1, 112–113, 421, 428–430
VP7 protein, simian rotavirus, 410
VP16 protein, herpes simplex virus type 1, 118, 246–252, 256–258, 438, 449
acidic activation domain, 252
reactivation from latency and, 256–257
VP19C protein, herpes simplex virus type 1, 113, 383, 421, 430
VP21 protein, herpes simplex virus type 1, 421
VP22a protein, herpes simplex virus type 1, 421, 425, 428–429
VP23 protein, herpes simplex virus type 1, 113, 421, 430
VP24 protein, herpes simplex virus type 1, 421, 428–429
VP26 protein, herpes simplex virus type 1, 421, 430
VP54 protein, *Paramecium bursaria* chlorella virus 1 (PBCV-1), 390
VPg protein
calicivirus, 353
picornavirus, 159
poliovirus, 94, 165–167, 216, 278, 436
potyvirus, 353
turnip mosaic virus, 353
VPg-dependent ribosomal recruitment, 353
Vpr protein, human immunodeficiency virus type 1, 117
Vpu protein, human immunodeficiency virus type 1, 404, 444, 450

W

Walleye dermal sarcoma virus, 530
Washington, George, 8
Watson, James, 86, 89
Weller, Thomas, 32

West Nile virus, 510
  apoptosis blockage, 469
  interaction with internal cellular membranes, 410
  mRNA cap, 314
  mRNA degradation, 336
  P bodies and, 377
  RNA synthesis, 165
  structure, 109
  subgenomic flavivirus RNAs (sfRNAs), 337
  unfolded protein response and, 406–407
Western equine encephalititis virus, one-step growth curve, 49
Whales, 4
Wheat germ extract, 360
Wild type virus, 66
Williams, Robley, 32
WIN compounds, 145
Wollman, Elie, 17
Woodchuck hepatitis virus, 512
Woodchuck hepatitis virus posttranscriptional regulatory element, 332
Wukipolyomavirus, 524

X

XMRV (xenotropic murine leukemia virus-related virus), 197, 204
Xpo1, 327–328, 330
Xpo5, 345
X-ray crystallography, 83, 85, 88–89, 92, 95–96, 99–102, 104, 108, 111, 114, 119, 418
X-ray diffraction, 85, 88
Xrcc4, 207
Xrn1, 335–338, 343, 376–378, 469, 473

Y

Yaba monkey tumor virus, 526
Yabapoxvirus, 526
Yeast
  autonomously replicating sequence, 270
  origin of replication, 270
  RNA polymerase II, 225
Yellow fever virus
  historical aspects, 7, 14
  overview, 510
  subgenomic flavivirus RNAs (sfRNAs), 337
  vaccine, 9, 51, 510
YPXnL motif, 442–443

Z

Zoonosis, 3, 4
  arenavirus, 504
  paramyxovirus, 518
Zta protein, Epstein-Barr virus, 247–249, 256, 292